Brief Contents

D0806956

Contents

5 Chemical Accounting   140

6 Gases, Liquids, Solids . . . and Intermolecular Forces   169

7 Acids and Bases   196

8 Oxidation and Reduction   224

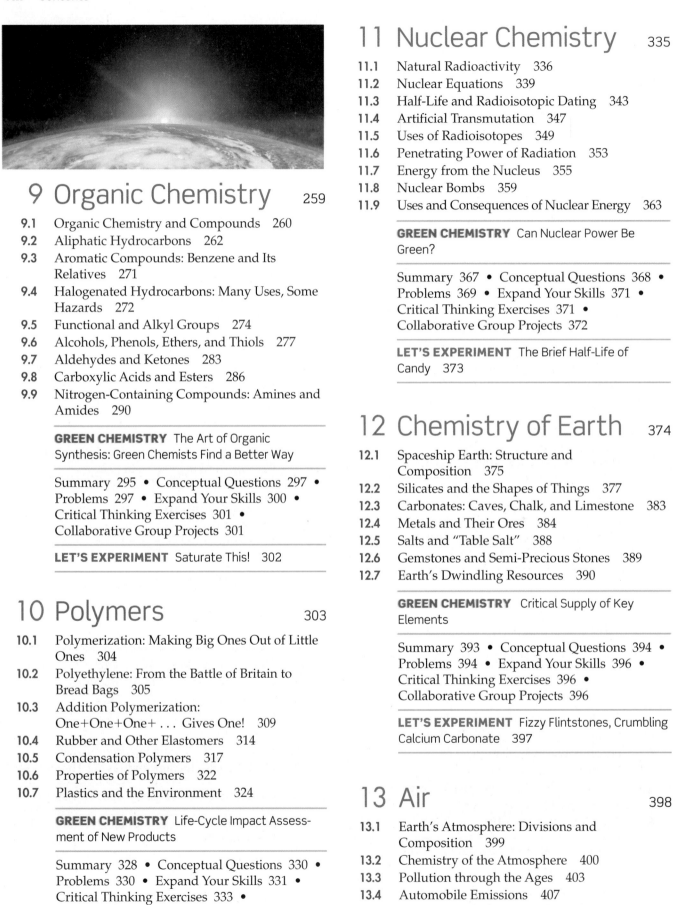

14 Water 431

15 Energy 459

16 Biochemistry 501

17 Nutrition, Fitness, and Health 548

Green Chemistry

The fifteenth edition of *Chemistry for Changing Times* is pleased to present the green chemistry essays listed below. The topics have been carefully chosen to introduce students to the concepts of green chemistry—a new approach to designing chemicals and chemical transformations that are beneficial for human health and the environment. The green chemistry essays in this edition highlight cutting-edge research by chemists, molecular scientists, and engineers to explore the fundamental science and practical applications of chemistry that is "benign by design." These examples emphasize the responsibility of chemists for the consequences of the new materials they create and the importance of building a sustainable chemical enterprise.

Preface

Chemistry for Changing Times is now in its fifteenth edition. Times have changed immensely since the first edition appeared in 1972 and continue to change more rapidly than ever—especially in the vital areas of biochemistry (neurochemistry, molecular genetics), the environment (sustainable practices, climate change), energy, materials, drugs, and health and nutrition. This book has changed accordingly. We have updated the text and further integrated green chemistry throughout. Green Chemistry essays throughout the text have been updated for relevancy. Learning objectives and end-of-chapter problems are correlated to each essay. In preparing this new edition, we have responded to suggestions from users and reviewers of the fourteenth edition, as well as used our own writing, teaching, and life experiences. The text has been fully revised and updated to reflect the latest scientific developments in a fast-changing world.

New to This Edition

- The *Let's Experiment!* activities (formerly *Chemistry@ Home*) have been revised to improve clarity, to maximize success of the experiment, and to increase relevance to everyday life.
- In Chapter 4, a new, clearer approach to drawing Lewis structures is presented.
- Determination of oxidation number in Chapter 8 has been greatly simplified.
- Chapter 9 now includes an introductory section that clearly differentiates between the general properties of organic compounds and the inorganic compounds that were covered in Chapters 1–8. Coverage of thiols, sulfur-containing organic compounds that are important in biochemistry, is also included.
- Chapter 12 now includes discussions of gems and related minerals, salt, and precious metals.
- Chapter 16 (Biochemistry) has had three new sections added. The use of carbohydrates, fats, and protein as foodstuffs is discussed directly after the coverage of structures of these biochemical molecules. Students no longer need to refer back to a previous chapter to find structures of the molecules involved.
- Chapters 17 and 19 have been logically and cohesively combined into a single chapter, "Nutrition, Fitness, and Health."
- A number of the new end-of-chapter problems are multiple-choice premise-and-conclusion problems requiring critical thinking (e.g., "The premise is correct but the conclusion is wrong.").

Revisions

- Almost every worked Example is now accompanied by *two* exercises that are closely related to the material covered in the Example. The B exercise is usually somewhat more challenging than the A exercise.
- More than 25% of the end-of-chapter problems have been revised or replaced in their entirety. Where practical, the revised/replacement problems highlight current events or modern issues that are chemistry-related.
- Brief answers to the odd-numbered end-of-chapter problems are provided in an Answer Appendix. In addition to being vetted by accuracy checkers, those answers have been carefully reviewed by one or more authors.
- Review Questions are now called *Conceptual Questions*, as they deal largely with chapter concepts. Routine end-of-chapter problems are now followed by more challenging problems in a section called *Expand Your Skills*.
- Chapter 14 includes expanded descriptions of some of the unique properties of water, and better organization of water pollutants and ways of purifying water.
- The global perspective has been added or enhanced in many chapters, broadening students' views of some of the challenges facing humanity.

To the Instructor

Our knowledge base has expanded enormously since this book's first edition, never more so than in the last few years. We have faced tough choices in deciding what to include and what to leave out. We now live in what has been called the Information Age. Unfortunately, information is not knowledge; the information may or may not be valid. Our focus, more than ever, is on helping students evaluate information. May we all someday gain the gift of wisdom.

A major premise of this book is that a chemistry course for students who are not majoring in science should be quite different from a course offered to science majors. It must present basic chemical concepts with intellectual honesty, but it need not—probably should not—focus on esoteric theories or rigorous mathematics. It should include lots of modern everyday applications. The textbook should be appealing to look at, easy to understand, and interesting to read.

A large proportion of the legislation considered by the U.S. Congress involves questions having to do with science or technology, yet only rarely does a scientist or engineer enter politics. Most of the people who make important decisions regarding our health and our environment are not trained in science, but it is critical that these decision makers be scientifically literate. In the judicial system, decisions often depend on scientific evidence, but judges and jurors frequently have little education in the sciences. A chemistry course for students who are not science majors should emphasize practical applications of chemistry to problems involving, most notably, environmental pollution, radioactivity, energy sources, and human health. The students who take liberal arts chemistry courses include future teachers, business leaders, lawyers, legislators, accountants, artists, journalists, jurors, and judges.

Objectives

Our main objectives for a chemistry course for students who are not majoring in science are as follows:

- To attract lots of students from a variety of disciplines. If students do not enroll in the course, we can't teach them.
- To help students become literate in science. We want our students to develop a comfortable knowledge of science so that they may become productive, creative, ethical, and engaged citizens.
- To use topics of current interest to illustrate chemical principles. We want students to appreciate the importance of chemistry in the real world.
- To relate chemical problems to the everyday lives of our students. Chemical problems become more significant to students when they can see a personal connection.
- To acquaint students with scientific methods. We want students to be able to read about science and technology with some degree of critical judgment. This is especially important because many scientific problems are complex and controversial.
- To show students, by addressing the concepts of sustainability and green chemistry, that chemists seek better, safer, and more environmentally friendly processes and products.
- To instill an appreciation for chemistry as an open-ended learning experience. We hope that our students will develop a curiosity about science and will want to continue learning throughout their lives.

Accuracy Reviewers

David F. Maynard, *California State University, San Bernardino*

Green Chemistry Contributors

We are enormously grateful to Thomas Goodwin, Hendrix College, who reviewed and revised the green chemistry essays for the fifteenth edition. We thank him for his dedication to this project. We also thank the team of green

Questions and Problems

Worked-out Examples and accompanying exercises are given within most chapters.

Each Example carefully guides students through the process for solving a particular type of problem. It is then followed by one or more exercises that allow students to check their comprehension right away. Many Examples are followed by two exercises, labeled A and B. The goal in an A exercise is to apply to a similar situation the method outlined in the Example. In a B exercise, students must often combine that method with other ideas previously learned. Many of the B exercises provide a context closer to that in which chemical knowledge is applied, and they thus serve as a bridge between the Worked Examples and the more challenging problems at the end of the chapter. The A and B exercises provide a simple way for the instructor to assign homework that is closely related to the Examples. Answers to all the in-chapter exercises are given in the Answers section at the back of the book.

Answers to all odd-numbered end-of-chapter problems, identified by blue numbers, are given in the Answers section at the back of the book. The end-of chapter problems include the following:

- Conceptual Questions for the most part simply ask for a recall of material in the chapter.
- A set of matched-pair problems is arranged according to subject matter in each chapter.
- Expand Your Skills Problems are not grouped by type. Some of these are more challenging than the matched-pair problems and often require a synthesis of ideas from more than one chapter. Others pursue an idea further than is done in the text or introduce new ideas.

Acknowledgments

For more than four decades, we have greatly benefited from hundreds of helpful reviews. It would take far too many pages to list all of those reviewers here, but they should know that their contributions are deeply appreciated. For the fifteenth edition, we are especially grateful to the following reviewers:

Christine Seppanen, *Riverland Community College*

chemists listed below who contributed the green essays and helped to integrate each essay's content into the chapter with learning objectives, end-of-chapter problems, summaries, and section references.

Eric Beckman, *University of Pittsburgh*
Amy S. Cannon, *Beyond Benign*
David Constable, *ACS Green Chemistry Institute*
Scott Cummings, *Kenyon College*
Joseph Fortunak, *Howard University*
Tom Goodwin, *Hendrix College*
Michael Heben, *University of Toledo*
Phil Jessop, *Queen's University*
Margaret Kerr, *Worcester State University*
Karen Larson, *Clarke University*
Irv Levy, *Gordon College*
Doris Lewis, *Suffolk University*

Jennifer MacKellar, *ACS Green Chemistry Institute*
Alex Mayer, *Michigan Technological University*
Lallie C. McKenzie, *Chem11 LLC*
Martin Mulvihill, *University of California–Berkeley*
Katie Privett, *Green Chemistry Centre of Excellence York, UK*
Douglas Raynie, *South Dakota State University*
Robert Sheldon, *Delft University of Technology*
Galen Suppes, *University of Missouri*
David Vosburg, *Harvey Mudd College*
John Warner, *Warner Babcock Institute*
Rich Williams, *Environmental Science & Green Chemistry Consulting, LLC*

Reviewers of This Edition

Amy Albrecht, *Charleston Southern University*
Joseph Cradlebaugh, *Jacksonville University*
Jeannie Eddleton, *Virginia Tech*

Katherine Leigh, *Dixie State University*
David Perry, *Charleston Southern University*

We also appreciate the many people who have called, written, or e-mailed with corrections and other helpful suggestions. Cynthia S. Hill prepared much of the original material on biochemistry, food, and health and fitness.

We owe a special debt of gratitude to Doris K. Kolb (1927–2005), who was an esteemed coauthor from the seventh through the eleventh editions. Doris and her husband, Ken, were friends and helpful supporters long before Doris joined the author team. She provided much of the spirit and flavor of the book. Doris's contributions to *Chemistry for Changing Times*—and indeed to all of chemistry and chemical education—will live on for many years to come, not only in her publications, but in the hearts and minds of her many students, colleagues, and friends.

Throughout her career as a teacher, scientist, community leader, poet, and much more, Doris was blessed with a wonderful spouse, colleague, and companion, Kenneth E. Kolb. Over the years, Ken did chapter reviews, made suggestions, and gave invaluable help for many editions. All who knew Doris miss her greatly. Those of us who had the privilege of working closely with her miss her wisdom and wit most profoundly. Let us all dedicate our lives, as Doris did hers, to making this world a better place.

We also want to thank our colleagues at the University of Wisconsin–River Falls, Murray State University, Winona State University, and Bradley University for all their help and support through the years. Thank you to Amy Cannon and Kate Anderson who coordinated the *Let's Experiment!* material. The *Let's Experiment!* demonstrations help bring the subject matter to life for students.

We also owe a debt of gratitude to the many creative people at Pearson who have contributed their talents to this edition. Jessica Moro, Senior Courseware Portfolio Analyst, has been a delight to work with, providing valuable guidance throughout the project. She showed extraordinary skill and diplomacy in coordinating all the many facets of this project. Courseware Director Barbara Yien and Development Editor Ed Dodd contributed greatly to this project,

especially in challenging us to be better authors in every way. We treasure their many helpful suggestions of new material and better presentation of all the subject matter. We are grateful to Project Managers Erin Hernandez and Norine Strang and Content Producer Cynthia Abbott for their diligence and patience in bringing all the parts together to yield a finished work. We are indebted to our copyeditor, Mike Gordon, whose expertise helped improve the consistency of the text; and to the proofreader Clare Romeo and accuracy checkers whose sharp eyes caught many of our errors and typos. We also salute our art specialist, Andrew Troutt, for providing outstanding illustrations, and our photo researcher, Jason Acibes, who vetted hundreds of images in the search for quality photos.

Terry W. McCreary would like to thank his wife, Geniece, and children, Corinne and Yvette, for their unflagging support, understanding, and love. Rill Ann Reuter is very thankful to her husband, Larry, and her daughter, Vicki, for their patience and support, especially during this project. Marilyn D. Duerst would like to thank her husband, Richard, for his patience and encouragement, and all six of their daughters, Karin, Sue, Linda, Rebecca, Christine and Sarah, for their enthusiasm and support.

Finally, we also thank all those many students whose enthusiasm has made teaching such a joy. It is gratifying to have students learn what you are trying to teach them, but it is a supreme pleasure to find that they want to learn even more. And, of course, we are grateful to all of you who have made so many helpful suggestions. We welcome and appreciate all your comments, corrections, and criticisms.

Terry W. McCreary
tmccreary@murraystate.edu

Marilyn D. Duerst
marilyn.d.duerst@uwrf.edu

Rill Ann Reuter
rreuter@winona.edu

To the Student

Tell me, what is it you plan to do
with your one wild and precious life?
—American poet Mary Oliver (b. 1935)
"The Summer Day," from *New and Selected Poems*
(Boston, MA: Beacon Press, 1992)

Welcome to Our Chemical World!

Learning chemistry will enrich your life—now and long after this course is over—through a better understanding of the natural world, the scientific and technological questions now confronting us, and the choices you will face as citizens in a scientific and technological society.

Skills gained in this course can be exceptionally useful in many aspects of your life. Learning chemistry involves thinking logically, critically, and creatively. You will learn how to use the language of chemistry: its symbols, formulas, and equations. More importantly, you will learn how to obtain meaning from information. The most important thing you will learn is how to learn. Memorized material quickly fades into oblivion unless it is arranged on a framework of understanding.

Chemistry Directly Affects Our Lives

How does the human body work? How does aspirin cure headaches, reduce fevers, and lessen the chance of a heart attack or stroke? How does penicillin kill bacteria without harming our healthy body cells? Is ozone a good thing or a threat to our health? Do we really face climate change, and if so, how severe will it be? Do humans contribute to climate change, and if so, to what degree? Why do most weight-loss diets seem to work in the short run but fail in the long run? Why do our moods swing from happy to sad? Chemists have found answers to questions such as these and continue to seek the knowledge that will unlock other secrets of our universe. As these mysteries are resolved, the direction of our lives often changes—sometimes dramatically. We live in a chemical world—a world of drugs, biocides, food additives, fertilizers, fuels, detergents, cosmetics, and plastics. We live in a world with toxic wastes, polluted air and water, and dwindling petroleum reserves. Knowledge of chemistry will help you better understand the benefits and hazards of this world and will enable you to make intelligent decisions in the future.

We Are All Chemically Dependent

Even in the womb we are chemically dependent. We need a constant supply of oxygen, water, glucose, amino acids, triglycerides, and a multitude of other chemical substances.

Chemistry is everywhere. Our world is a chemical system—and so are we. Our bodies are durable but delicate systems with innumerable chemical reactions occurring constantly within us that allow our bodies to function properly. Learning, exercising, feeling, gaining or losing weight, and virtually all life processes are made possible by these chemical reactions. Everything that we ingest is part of a complex process that determines whether our bodies work effectively. The consumption of some substances can initiate chemical reactions that will stop body functions. Other substances, if consumed, can cause permanent handicaps, and still others can make living less comfortable. A proper balance of the right foods provides the chemicals that fuel the reactions we need in order to function at our best. Learning chemistry will help you better understand how your body works so that you will be able to take proper care of it.

Changing Times

We live in a world of increasingly rapid change. Isaac Asimov (1920–1992), Russian-born American biochemist and famous author of popular science and science fiction books, once said that "The only constant is change, continuing change, inevitable change, that is the dominant factor in society today. No sensible decision can be made any longer without taking into account not only the world as it is, but the world as it will be." We now face some of the greatest problems that humans have ever encountered, and these dilemmas seem to have no perfect solutions. We are sometimes forced to make a best choice among only bad alternatives, and our decisions often provide only temporary solutions. Nevertheless, if we are to choose properly, we must understand what our choices are. Mistakes can be costly, and they cannot always be rectified. It is easy to pollute, but cleaning up pollution is enormously expensive. We can best avoid mistakes by collecting as much information as possible and evaluating it carefully before making critical decisions. Science is a means of gathering and evaluating information, and chemistry is central to all the sciences.

Chemistry and the Human Condition

Above all else, our hope is that you will learn that the study of chemistry need not be dull and difficult. Rather, it can enrich your life in so many ways—through a better understanding of your body, your mind, your environment, and the world in which you live. After all, the search to understand the universe is an essential part of what it means to be human. We offer you a challenge first issued by American educator Horace Mann (1796–1859) in his 1859 address at Antioch College: "Be ashamed to die until you have won some victory for humanity."

In Memoriam

The fifteenth edition of *Chemistry for Changing Times* is dedicated to the memory of John W. Hill, who died of lymphoma on August 7, 2017. The reader may have noticed that the title of the book has been changed to *Hill's Chemistry for Changing Times*. This is a tribute to the professor, gentleman, and our friend, who was the leading edge of liberal arts chemistry for over four decades.

I met John Hill when I was a yet-untenured assistant professor. He had taken a sabbatical to teach here at Murray State University, selecting our consumer-chemistry course as his assignment. John was one of the very few instructors I've known who reveled in teaching what some disparagingly call "chemistry for poets." John enjoyed bringing chemistry to the ordinary student, the one who would most likely take a single science course in her curriculum. And he was very, very good at it.

Not long after he began teaching at University of Wisconsin–River Falls, he was assigned to their liberal arts chemistry course. He had no difficulty preparing notes, but he wasn't satisfied with the textbooks he was able to find. His notes, along with uncounted hours of literature searching and writing, eventually became the first edition of *Chemistry for Changing Times*, in 1972.

The amount of work John put into the earlier editions was staggering. Hand-writing or manually typing the entire manuscript; sending the work to the publisher by snail mail; preparing sketches for figures; reviews, proof pages, figures, and photos obtained and delivered by the same slow process; hand-marking hundreds of proof pages; and crossing his fingers, hoping that he'd not missed anything critical. It's difficult to appreciate that level of effort when we consider the tools we have at our disposal today.

Personally John was a quiet, modest man who enjoyed writing of all sorts, including a few children's books. He loved silly jokes, especially the sort of pun that would elicit a terrible groan from anyone within earshot. I doubt that he ever realized how much of a difference his professional works made to millions (literally) of students. It was a privilege to know him and work with him. John will be greatly missed by all who knew him.

Terry W. McCreary

I first met John Hill in August of 1981, when I applied for a one-year teaching position that suddenly had opened up at the University of Wisconsin–River Falls. In the interview, John quickly observed that I, too, had a passion and the personality for teaching non-science students. I eagerly accepted the position, and one year eventually turned into thirty-four years at UW-RF. During that span of time, I taught the course for non-science majors for more than sixty academic terms, using updated editions of this textbook, and never tired of it.

John and I engaged in numerous discussions over the years about ways to improve and deepen student learning, and how chemical demonstrations could enhance student engagement in the classroom, as that was my forte. He jokingly called me "Mrs. Wizard." We wrote a children's book together nearly twenty years ago that included experiments for the readers to perform at home, which was great fun. John was a soft-spoken man, with infinite patience and a closet full of T-shirts with silly science-related sayings, which he unashamedly wore to class. It was truly a pleasure and honor to be a colleague of John W. Hill.

Marilyn D. Duerst

My work with John Hill initially began with a review I did for an earlier edition of *Chemistry for Changing Times*. Indeed, I did not actually meet him in person until after I had worked on several editions of the book, but we had many informative exchanges first via snail mail and then over the phone and e-mail. I always enjoyed those discussions, and they were often very thought-provoking.

John worked hard not only to present students with correct information, but also to present it in a clear and unambiguous way. Rather than just presenting the bald facts, as so many books do, *Chemistry for Changing Times* also includes considerable historical information about how those facts were determined, helping students to understand why we know what we know.

Chemistry was not a static subject for John. He constantly looked for information about new developments and how they affect our everyday lives. Understanding the role and relevance of chemistry is important for all of us, including non-science students. We are all citizens of this world, and our actions will affect future generations.

It was my privilege to have the opportunity to work with John W. Hill.

Rill Ann Reuter

About the Authors

John W. Hill

John Hill received his Ph.D. from the University of Arkansas. As an organic chemist, he published more than 50 papers, most of which have an educational bent. In addition to *Chemistry for Changing Times*, he authored or coauthored several introductory-level chemistry textbooks, all of which have been published in multiple editions. He presented over 60 papers at national conferences, many relating to chemical education. He received several awards for outstanding teaching and was active in the American Chemical Society, both locally and nationally.

Terry W. McCreary

Terry McCreary received his Ph.D. in analytical chemistry from Virginia Tech. He has taught general and analytical chemistry at Murray State University since 1988 and was presented with the Regents Excellence in Teaching Award in 2008. He is a member of the Kentucky Academy of Science and has served as technical editor for the *Journal of Pyrotechnics*. McCreary is the author of several laboratory manuals for general chemistry and analytical chemistry, as well as *General Chemistry* with John Hill, Ralph Petrucci, and Scott Perry, and *Experimental Composite Propellant*, a fundamental monograph on the preparation and properties of solid rocket propellant. In his spare time, he designs, builds, and flies rockets with the Tripoli Rocketry Association, of which he was elected president in 2010. He also enjoys gardening, machining, woodworking, and astronomy.

Marilyn D. Duerst

Marilyn D. Duerst majored in chemistry, math, and German at St. Olaf College, graduating in 1963, and earned an M.S. from the University of California–Berkeley in 1966. For over five decades, her talents in teaching have flourished in every venue imaginable, with students aged four to 84, but were focused on non-science majors, preservice and inservice teachers. She taught at the University of Wisconsin–River Falls from 1981 to 2015; in 2006 she was presented with the Outstanding Teaching Award. Now a Distinguished Lecturer in Chemistry, emerita, from UW–RF, she is a Fellow of the American Chemical Society, an organization in which she has long been active both locally and nationally, particularly in outreach activities to the public. In 1999, she co-authored a book for children with John W. Hill entitled *The Crimecracker Kids and the Bake-shop Break-in*. Marilyn is a birder, rockhound, and nature photographer; she collects sand, minerals and elements, has traveled four continents, and studied a dozen languages.

Rill Ann Reuter

Rill Ann Reuter earned her B.A. in Chemistry from Connecticut College and her M.S. in Biochemistry from Yale University. She worked in academic research laboratories at Yale University, Princeton University, and the University of Massachusetts Medical School for 12 years, with a primary emphasis on nucleic acid research. After moving to Minnesota in 1980, she taught at Saint Mary's University of Minnesota, the College of Saint Teresa, and Winona State University and did research on photosynthesis. She retired from Winona State in 2015 as Professor Emerita of Chemistry. Over the years, she has taught large numbers of general chemistry, non-science, and pre-nursing students. She was active in local and regional science fairs for 35 years and is a member of the American Chemical Society. She has a keen interest in history, politics, and classical music.

About Our Sustainability Initiatives

Pearson recognizes the environmental challenges facing this planet, as well as acknowledges our responsibility in making a difference. Along with developing and exploring digital solutions to our market's needs, Pearson has a strong commitment to achieving carbon-neutrality. As of 2009, Pearson became the first carbon- and climate-neutral publishing company. Since then, Pearson remains strongly committed to measuring, reducing, and offsetting our carbon footprint.

The future holds great promise for reducing our impact on Earth's environment, and Pearson is proud to be leading the way. We strive to publish the best books with the most up-to-date and accurate content, and to do so in ways that minimize our impact on Earth. To learn more about our initiatives, please visit https://www.pearson.com/social-impact.html.

Engage students with contemporary and relevant applications of chemistry

Hill's Chemistry for Changing Times has defined the liberal arts chemistry course and remains the most visually appealing and readable introduction to the subject. For the **15th Edition,** new co-authors Marilyn D. Duerst and Rill Ann Reuter join author Terry W. McCreary to introduce new problem types that engage and challenge students to develop skills they will use in their everyday lives, including developing scientific literacy, analyzing graphs and data, and recognizing fake vs. real news. New up-to-date applications focus on health, wellness, and the environment, helping non-science and allied-health majors to see the connections between the course materials and their everyday lives. Enhanced digital tools and additional practice problems in **Mastering Chemistry** and the **Pearson eText** ensure students master the basic content needed to succeed in this course.

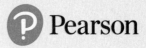

Connect chemistry

GREEN CHEMISTRY Atom Economy
Margaret Kerr, Worcester State University

Principles 1, 2

Learning Objectives • Explain how the concept of atom economy can be applied to pollution prevention and environmental protection. • Calculate the atom economy for chemical reactions.

Imagine yourself in the future. Your job is related to environmental protection, which requires that you provide information to practicing chemists as they design new processes and reactions. Waste management is one of your top concerns. Although waste typically has been addressed only after production of desired commodities, you realize that a greener approach can mean minimizing the waste from the start. What methods would you use in this job? What topics are important? How can chemists provide new products while protecting the environment?

Intrinsic in greener approaches to waste management is the concept of *atom economy*, a calculation of the number of atoms conserved in the desired product rather than gone in waste. In 1998, Barry Trost of Stanford University won a Presidential Green Chemistry Challenge Award for his work in developing this concept. By calculating the number of atoms that will not become part of the desired product and, therefore, will enter the waste stream, chemists can precisely determine the minimum amount of waste that will be produced by chemicals used in a reaction before even running the reaction.

You have learned how to write and balance chemical equations (Section 5.1), and you also can calculate molar mass, convert from mass to moles, and determine the amount of product formed from given amounts of reactants (Section 5.3 and Section 5.4). Other

product. Reactant atoms that do not appear in the product are considered waste. The % A.E. is given by the following relationship:

$$\% \text{ A.E.} = \frac{\text{molar mass of desired product}}{\text{molar masses of all reactants}} \times 100\%$$

A reaction can have a poor atom economy even when the percent yield is near 100%.

Consider the following two ways to make butene (C_4H_8), a compound that is an important chemical feedstock in the plastics industry. First, butene can be made using butylbromide (C_4H_9Br) and sodium hydroxide (NaOH).

$$C_4H_9Br + NaOH \longrightarrow C_4H_8 + H_2O + NaBr$$

In this reaction, a Br atom and a H atom are eliminated from the butylbromide to form the final product. Generally, in reactions like this one, only one product is desired and all other products are not used. We can calculate the atom economy for this reaction.

$$\% \text{ A.E.} = \frac{\text{molar mass } C_4H_8}{\text{molar mass } C_4H_9Br + \text{molar mass NaOH}} \times 100\%$$

$$\% \text{ A.E.} = \frac{56.11 \text{ g/mol}}{137.02 \text{ g/mol} + 40.00 \text{ g/mol}} \times 100\% = 31.7\%$$

pg.162

UPDATED! Green Chemistry Essays reflect current events and recent scientific findings that provide students with a way to interpret environmental issues through a chemical perspective. The essays emphasize recycling as a theme throughout the book and include discussions on problems of atmospheric pollution and preservation of the benign greenhouse effect. Auto-graded assessments tied to the essays are now available in the Mastering™ Chemistry end-of-chapter materials.

Let's Experiment!, located at the end of each chapter, provide students with safe and interesting activities they can do on their own to observe how chemistry is relevant to their day-to-day lives. Videos of the experiments are available in the Pearson eText and assignable in Mastering Chemistry.

LET'S EXPERIMENT! Polymer Bouncing Ball

Materials Needed:

• 2 small plastic cups (4 oz)
• Measuring spoons
• Warm water
• Borax
• 2 wooden craft sticks

• White craft glue
• Cornstarch
• Food coloring (if desired)
• Plastic bag with zip lock (for storage)

Did you know that the earliest balls were made of wood and stone? What are most bouncy balls made of today?

Many bouncing balls are made out of rubber, but they can also be made out of leather or plastic and can be hollow or solid. This experiment will use common, inexpensive ingredients to make a ball that bounces.

Polymers are molecules made up of repeating chemical units. Glue is made up of the polymer polyvinyl acetate (PVA). In this experiment, borax (boric acid) is responsible for hooking the molecules together and cross-linking the molecules, providing the ball with its putty-like and bouncy properties.

To start, label the two cups *Borax Solution* and *Ball Mixture*.
• For the borax solution, pour 2 tablespoons of warm water and $\frac{1}{2}$ teaspoon of borax powder into the cup. Use a craft stick to stir the mixture to dissolve the borax. Add food coloring, if desired.
• For the ball mixture, pour 1 tablespoon of glue into the cup. Add $\frac{1}{2}$ teaspoon of the borax solution you just made and 1 tablespoon of cornstarch. Do not stir. Allow the ingredients to interact on their own for 10–15 seconds. Then use the other craft stick to stir them together to fully mix them. Once the mixture becomes too thick to stir, take it out of the cup and start molding the ball with your hands.

The ball will start out sticky and messy but will solidify as you knead it. Once the ball is less sticky, go ahead and bounce it! To keep your ball from drying out, store your ball in a plastic zip lock bag.

Questions

1. Does your ball bounce? How high?
2. Does making a polymer ball cause a chemical or physical reaction? Explain.
3. Describe how changing the amounts of each ingredient would affect the ball mixture.
4. Is this ball biodegradable? Why or why not?

to the real world

WHY IT MATTERS

The incandescent light bulb is very inefficient with respect to the energy it uses; as much as 95% of the electric energy it uses is changed to heat, not light. Though compact fluorescent bulbs, containing mercury, were popular for about a decade, now LED bulbs are taking over the light bulb market. They are more expensive, but supposedly will last 10 to 20 years, and use a lot less energy for the same brightness effect. Use of such bulbs will be a "greener" way to light your surroundings.

WHY IT MATTERS

An isotonic, or "normal," intravenous solution must have the proper concentration of solute to avoid damage to blood cells. High concentrations cause blood cells to shrivel (*crenation*) as water is drawn out of them by osmosis. Low concentrations cause the cells to swell (*hemolysis*) and even burst.

pgs. 79, 160

REVISED AND UPDATED! Why It Matters
presents contemporary, relevant, and up-to-date applications with a concentration on health, wellness, and the environment to resonate with non-science and allied-health majors taking the course.

Nuclear Chemistry

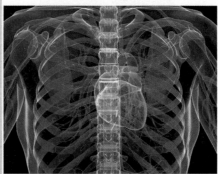

11.1 Natural Radioactivity
11.2 Nuclear Equations
11.3 Half-Life and Radioisotopic Dating
11.4 Artificial Transmutation
11.5 Uses of Radioisotopes
11.6 Penetrating Power of Radiation
11.7 Energy from the Nucleus
11.8 Nuclear Bombs
11.9 Uses and Consequences of Nuclear Energy

Have You Ever Wondered?

1 Is radiation entirely a human-made problem? **2** How do we measure radioactivity and its effect on people? **3** Do irradiated foods contain radioactive material? **4** Are we exposed to dangerous radiation during X-rays and other medical procedures? **5** What causes radiation sickness, and how serious is it? **6** Can we minimize or recycle radioactive wastes to make nuclear power generation a more sustainable process?

The photos above show images from PET (positron emission tomography). At the upper right of this page is an image from a CAT (computerized axial tomography). The picture to the side shows an example of an X-ray, which is used for examining bones and internal organs. All of these diagnostic techniques use radiation or a radioactive isotope. With these techniques, we can now examine tissues and diagnose diseases that

THE HEART OF MATTER Many people associate the term *nuclear energy* with fearsome images of a mighty force: giant mushroom clouds from nuclear explosions that devastated cities and nuclear power plant accidents at Three Mile Island, Pennsylvania, in 1979; Chernobyl, Ukraine, in 1986; and Fukushima, Japan,

343

REVISED AND UPDATED! Chapter Openers concentrate on wellness applications such as diet, exercise, supplements, natural remedies, and medications to help students connect chemistry with their everyday lives.

pg. 335

Build students' critical thinking and

🌐 Critical Thinking Exercises

Apply knowledge that you have gained in this chapter and one or more of the FLaReS principles (Chapter 1) to evaluate the following statements or claims.

12.1. An economist has said that we need not worry about running out of copper, because it can be made from other metals.

12.2. A citizen testifies against establishing a landfill near his home, claiming that the landfill will leak substances into the groundwater and contaminate his well water.

12.3. A citizen lobbies against establishing an incinerator near her home, claiming that plastics burned in the incinerator will release hydrogen chloride into the air.

12.4. An environmental activist claims that we could recycle all goods, leaving no need for the use of raw materials to make new ones.

12.5. A salesperson tells you that *ceramics* is just a fancy word for glass.

pg. 396

Critical Thinking Exercises encourage students to think critically about the scientific process and evaluate whether specific statements they might see in their daily lives meet the rational and objective standards of scientific rigor as outlined by the FLaReS method (Falsifiability, Logic, Replicability, Sufficiency).

These items are also assignable in Mastering Chemistry with answer-specific feedback designed to help students understand the scientific process.

problem-solving skills

EXAMPLE 6.1   Determining Intermolecular Forces

What is the major kind of force that exists between (a) NH_2Cl molecules; (b) CF_4 molecules; and (c) an H_2O molecule and an H_2S molecule?

Solution

a. NH_2Cl molecules are similar to NH_3 molecules, in that they are both trigonal pyramidal and polar. They also have the requirements for hydrogen bonding: H covalently bonded to N in one molecule, and N in a polar bond (NH bond) in a neighboring molecule. The major force is therefore hydrogen bonding.

b. Despite the fact that CF_4 molecules contain highly electronegative fluorine atoms, they are nonpolar because the fluorine atoms are symmetrically arranged around the carbon atom, similar to CH_4. Therefore, the only forces that exist between CF_4 molecules are dispersion forces.

c. Both H_2O and H_2S are bent molecules with polar bonds, so both are polar. Water molecules have hydrogen atoms covalently bonded to oxygen. However, H_2S does not contain N, O, or F atoms in a polar bond. Therefore, the major force here is a dipole–dipole force.

⟩ Exercise 6.1A

What is the major kind of force between (a) SiH_4 molecules and (b) SF_2 molecules?

⟩ Exercise 6.1B

What is the major kind of force between a H_3CCHO molecule (H and O bonded to C) and a water molecule?

pg. 175

NEW! Examples throughout the book

guide students through the process for solving a particular type of problem. Every Example in the book follows a consistent model with two follow-up exercises—the first requires the student to apply a similar situation to the method outlined in the Example, and the second asks the student to combine that method with ideas previously learned.

REVISED! End-of-Chapter problems

expand their application of chemistry and its relevance to students. Additional Problems immediately follow the End-of-Chapter Problems, giving instructors one set of "traditional" problems and a follow-up set of more applied, contemporary problems.

Expand Your Skills

73. Evaluate each of the following as possible scientific hypotheses.
   a. If the temperature of a cup of tea is increased, then the quantity of sugar that can be dissolved in it will be increased.
   b. If the rate of photosynthesis, as measured by the quantity of oxygen produced, is related to the wavelength (color) of light, then light of different colors will cause a plant to make different quantities of oxygen.
   c. If the rate of metabolism in animals is related to the temperature, then raising the surrounding temperature will cause an increase in animal metabolism.
   d. If I meditate hard enough, I will pass this chemistry exam.

74. The nucleus of a hydrogen atom is 1.75 fm in diameter. The atom is larger than the nucleus by a factor of about 145,000. (a) Use exponential notation to express each measurement in terms of an SI base unit. (b) What is the volume of a hydrogen nucleus in fm^3? Of a hydrogen atom in nm^3? The volume of a sphere of radius $r = 4/3 \, \pi r^3$.

75. A certain chemistry class is 1.00 microcenturies (μcen) long. What is its length in minutes?

76. A unit of beauty, a *helen*, thought to have been invented by British mathematician W.A.H. Rushton, is based on Helen of Troy (from Christopher Marlowe's play *Doctor Faustus*, which described her as having "the face that launched a thousand ships"). How many ships could be launched by a face with 1.00 millihelens of beauty?

gram of ice, and (b) swallowing the ice and allowing it to melt in his stomach "uses up" that same amount of food energy. Although both (a) and (b) are correct, the diet does not work. Explain. (*Hint*: See page 30 for a discussion of energy units, then calculate the amount of energy in food calories needed to melt a kilogram of ice.)

79. A particular brand of epoxy glue is used by mixing two volumes of liquid epoxy resin (density 2.25 g/mL) with one volume of liquid hardener (density 0.94 g/mL) before application. If the epoxy glue is to be prepared by mass rather than volume, what mass in grams of hardener must be mixed with 10.0 g of resin?

For Problems 80 and 81, classify each numbered statement as (a) an experiment, (b) a hypothesis, (c) a scientific law, (d) an observation, or (e) a theory. (It is not necessary to understand the science involved to do these problems.)

80. In the early 1800s, many scientific advances came from the study of gases. (1) For example, Joseph Gay-Lussac reacted hydrogen and oxygen to produce water vapor, and he reacted nitrogen and oxygen to form either dinitrogen oxide (N_2O) or nitrogen monoxide (NO). Gay-Lussac found that hydrogen and oxygen react in a 2:1 volume ratio and that nitrogen and oxygen can react in 2:1 or 1:1 volume ratios depending on the product. (2) In 1808, Gay-Lussac published a paper in which he stated that the relative volumes of gases in a chemical reaction are present in the ratio of small integers provided that all gases are measured at the same temperature and pressure. (3) In 1811, Amedeo

Give students anytime, anywhere access with Pearson eText

Pearson eText is a simple-to-use, mobile-optimized, personalized reading experience available within Mastering. It allows students to easily highlight, take notes, and review key vocabulary all in one place—even when offline. Seamlessly integrated videos, rich media, and interactive self-assessment questions engage students and give them access to the help they need, when they need it. Pearson eText is available within Mastering when packaged with a new book; students can also purchase Mastering with Pearson eText online. For instructors not using Mastering, Pearson eText can also be adopted on its own as the main course material.

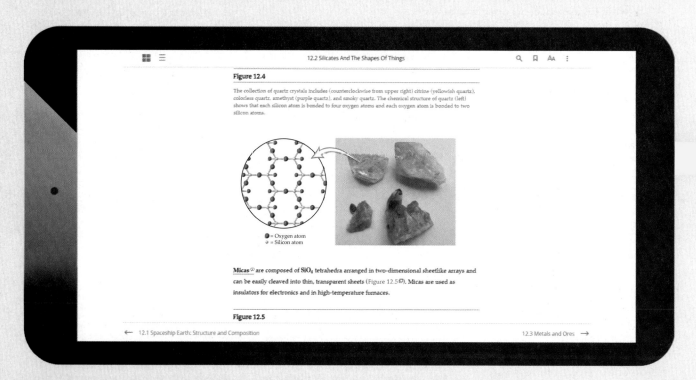

Reach every student with Mastering Chemistry

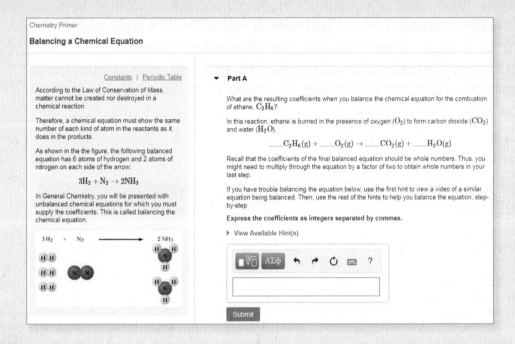

The Chemistry Primer in Mastering Chemistry helps students remediate their chemistry math skills and prepare for their first college chemistry course. Scaled to students' needs, remediation is only suggested to students that perform poorly on an initial assessment. Remediation includes tutorials, wrong-answer specific feedback, video instruction, and stepwise scaffolding to build students' abilities.

With Learning Catalytics, you'll hear from every student when it matters most. You pose a variety of questions that help students recall ideas, apply concepts, and develop critical-thinking skills. Your students respond using their own smartphones, tablets, or laptops. You can monitor responses with real-time analytics and find out what your students do — and don't — understand, to help students stay motivated and engaged.

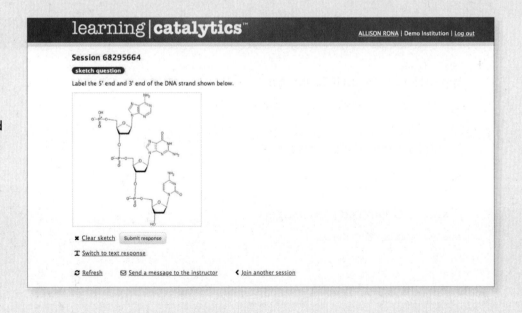

Instructor support you can rely on

Hill's Chemistry for Changing Times includes a full suite of instructor support materials in the Instructor Resources area in Mastering Chemistry. Resources include lecture presentations, images, reading quizzes, clicker questions, and worked examples in PowerPoint; all images from the text; videos and activities; virtual lectures; and a test bank.

Download instructor resources from the links below.

PowerPoint Presentation Tools

Chapter 7 Image PowerPoint	pptx, 25.9 MB
Chapter 7 Lecture Outline PowerPoint	zip, 16.4 MB
Chapter 7 Reading Quiz Clicker PowerPoint	pptx, 3.4 MB
Chapter 7 Review Clicker PowerPoint	pptx, 3.3 MB

JPEG Images

Appendices Labeled JPEG Images
Labeled images from appendices in the text.
zip, 11.5 MB

Chapter 7 Labeled JPEG Images
Labeled images from the text.
zip, 24.7 MB

Chapter 7 Unlabeled JPEG Images
Unlabeled images from the text.
zip, 6.2 MB

Chemistry

Chemistry is everywhere, not just in laboratories. Did you know that a kitchen is a laboratory where you eat the product? Cornstarch thickens a stir-fry dish, bread dough rises, a delicious brown crust forms on meat, all because of chemistry! Every natural and manufactured product you can think of, from solar cells to quartz crystals, is a result of chemistry.

Have You Ever Wondered?

1 Why should I study chemistry? **2** Is it true that chemicals are bad for us? **3** Why do scientists so often say "more study is needed"? **4** Why do scientists bother with studies that have no immediate application? **5** Can we change lead into gold?

You will find an answer to each of these questions at the appropriate point within this chapter. Look for the answers in the margins.

A SCIENCE FOR ALL SEASONS Join us on a journey toward a horizon of infinite possibilities. We will explore chemistry, a field of endeavor that pervades every aspect of our lives. Look around you. Everything you see is made of chemicals: the food we eat, the air we breathe, the clothes we wear, the buildings that shelter us, the vehicles we ride in, and the medicines that help keep us healthy.

Everything we *do* also involves chemistry. Whenever we eat a sandwich, bathe, listen to music, drive a car, or ride a bicycle, we use chemistry. Even when we are asleep, chemical reactions go on constantly throughout our bodies.

Chemistry also affects society as a whole. Developments in health and medicine involve a lot of chemistry. The astounding advances in biotechnology—such as

▲ Organic foods are not chemical-free. In fact, they are made entirely of chemicals!

genetic engineering, new medicines, improvements to nutrition, and much more—have a huge chemical component. Understanding and solving environmental problems require knowledge and application of chemistry. The worldwide issues of ozone depletion and climate change involve chemistry.

So, what exactly is chemistry anyway? We explore that question in some detail in Section 1.5. And just what is a chemical? The word *chemical* may sound ominous, but it is simply a name for any material. Gold, water, salt, sugar, air, coffee, ice cream, a computer, a pencil—all are chemicals or are made entirely of chemicals.

Material things undergo changes. Sometimes these changes occur naturally—maple leaves turn yellow and red in autumn. Often, we change material things intentionally, to make them more useful, as when we light a candle or cook an egg. Most of these changes are accompanied by changes in energy. For example, when we burn gasoline, the process releases energy that we can use to propel an automobile. Chemistry helps to define life. How do we differentiate a living collection of chemicals from the same assembly of chemicals in a dead organism or sample of inanimate matter that was never alive? A living set of molecules can replicate itself and has a way to harvest energy from its surroundings.

Our bodies are marvelous chemical factories. They take the food we eat and turn it into skin, bones, blood, and muscle, while also generating energy for all of our activities. This amazing chemical factory operates continuously, 24 hours a day, for as long as you live. Chemistry affects your own life in every moment, and it also transforms society as a whole. Chemistry shapes our civilization.

1.1 Science and Technology: The Roots of Knowledge

Learning Objectives • Define *science*, *chemistry*, *technology*, and *alchemy*.
• Describe the importance of green chemistry and sustainable chemistry.

① Why should I study chemistry?

Chemistry is a part of many areas of study and affects everything you do. Knowledge of chemistry helps you to understand many facets of modern life.

Chemistry is a *science*, but what is science? **Science** is essentially a process, a search for understanding of and explanations for natural phenomena through careful observation and experimentation. It is the primary means by which we obtain new knowledge. Science accumulates knowledge about nature and our physical world, and it generates theories that we use to explain that knowledge. **Chemistry** is that area of knowledge that deals with the behavior of matter and how it interacts with other matter and with some forms of energy.

Science and technology often are confused with one another. **Technology** is the application of knowledge for practical purposes. Technology arose in prehistoric times, long before science. The discovery of fire was quickly followed by cooking foods, baking pottery, and smelting ores to produce metals such as copper. The discovery of fermentation led to beer and winemaking. Such tasks were accomplished without an understanding of the scientific principles involved.

About 2500 years ago, Greek philosophers attempted to formulate *theories* of chemistry—rational explanations of the behavior of matter. These philosophers generally did not test their theories by experimentation. Nevertheless, their view of nature—attributed mainly to Aristotle—dominated Western thinking about the workings of the material world for the next 20 centuries.

The experimental roots of chemistry lie in **alchemy**, a primitive form of chemistry that originated in the Arab world around 700 C.E. and spread to Europe in the Middle Ages. Alchemists searched for a "philosopher's stone" that would turn cheaper metals into gold and for an elixir that would bring eternal life. Although they never achieved these goals, alchemists discovered many new chemical substances and perfected techniques, such as distillation and extraction, that are still used today.

Toward the end of the Middle Ages, a real science of chemistry began to see light. The behavior of matter began to be examined through experimentation. Theories that arose from that experimentation gradually pushed aside the authority of early philosophers. The 1800s saw a virtual explosion of knowledge as more scientists studied the behavior of matter in breadth and depth. Through the 1950s and early 1960s, science in general and chemistry in particular saw increasing relevance in our lives. Laboratory-developed fertilizers, alloys, drugs, and plastics were incorporated into everyday living. DuPont, one of the largest chemical companies in the world, used its slogan "Better Living Through Chemistry" with great effect through the 1970s.

For most of human history, people exploited Earth's resources, unfortunately giving little thought to the consequences. Rachel Carson (1907–1964), a biologist, was an early proponent of environmental awareness. The main theme of her book *Silent Spring* (1962) was that our use of chemicals to control insects was threatening to destroy all life—including ourselves. People in the pesticide industries and their allies strongly denounced Carson as a propagandist, though some scientists rallied to support her. By the late 1960s, however, the threatened extinction of several species of birds and the disappearance of fish from many rivers, lakes, and areas of the ocean caused many scientists to move into Carson's camp. Popular support for Carson's views became overwhelming.

In response to growing public concern, chemists have in recent years developed the concept of **green chemistry**, which uses materials and processes that are intended to prevent or reduce pollution at its source. This approach was further extended in the first decade of the twenty-first century to include the idea of **sustainable chemistry**—chemistry designed to meet the needs of the present generation without compromising the needs of future generations. Sustainability preserves resources and aspires to produce environmentally friendly products from renewable resources.

Chemicals themselves are neither good nor bad. Their misuse can indeed cause problems, but properly used, chemicals have saved countless millions of lives and have improved the quality of life for the entire planet. Chlorine and ozone kill bacteria that cause dreadful diseases. Drugs and vaccinations relieve pain and suffering. Fertilizers such as ammonia increase food production, and petroleum provides fuel for heating, cooling, lighting, and transportation. In short, chemistry has provided ordinary people with necessities and luxuries that were not available even to the mightiest rulers in ages past. Chemicals are essential to our lives—life itself would be impossible without chemicals.

▲ Rachel Carson's *Silent Spring* was one of the first publications to point out a number of serious environmental issues.

2 **Is it true that chemicals are bad for us?**

Everything you can see, smell, taste, or touch is either a chemical or is made of chemicals. Chemicals are neither good nor bad, objectively. They can be put to good use, bad use, or anything in between.

▲ A century ago, contaminated drinking water was often the cause of outbreaks of cholera and other diseases. Modern water treatment uses chemicals to remove solid matter and kill disease-causing bacteria, making water safe to drink.

SELF-ASSESSMENT Questions

Select the best answer or response.

1. Which of the following would *not* be a technological advancement made possible by understanding chemistry?
   a. Cooking pans coated with a nonstick surface like Teflon®
   b. The ability to change lead or other metals into gold
   c. Lengthening the life span of human beings using medicines
   d. Alternate fuel sources to lessen our dependence on petroleum

2. Alchemy is
   a. philosophical speculation about nature
   b. chemistry that is concerned with environmental issues
   c. the forerunner of modern chemistry
   d. the application of knowledge for practical purposes

3. The main theme of Rachel Carson's *Silent Spring* was that life on Earth could be destroyed by
   a. botulism   **b.** nuclear war   **c.** overpopulation   **d.** pesticides

4. A goal of green chemistry is to
   a. produce cheap green dyes
   c. reduce pollution
   b. provide great wealth for corporations
   d. turn deserts into forests and grasslands

5. Which of the following chemicals are bad?
   a. Trinitrotoluene (TNT)
   c. Botulism toxin
   b. Hydrogen cyanide
   d. None of these

Answers: 1, b; 2, c; 3, d; 4, c; 5, d

1.2 Science: Reproducible, Testable, Tentative, Predictive, and Explanatory

Learning Objective • Define *hypothesis, scientific law, scientific theory,* and *scientific model,* and explain their relationships in science.

We have defined science, but science has certain characteristics that distinguish it from other studies.

Scientists often disagree about what is and what will be, but does that make science merely a guessing game in which one guess is as good as another? Not at all. Science is based on *evidence,* on observations and experimental tests of our assumptions. However, it is not a collection of unalterable facts. We cannot force nature to fit our preconceived ideas. Science is good at correcting errors; establishing truths is somewhat more challenging, for science is an unfinished work. The things we have learned from science fill millions of books and scholarly journals, but what we know pales in comparison with what we do not yet know.

Scientific Data Must Be Reproducible

Scientists collect data by making careful observations. Data must be *reproducible*—the data reported by a scientist must also be observable by other scientists. Careful measurements are required, and conditions are thoroughly controlled and described. Scientific work is not fully accepted until it has been verified by other scientists, a process called *peer review.*

Observations, though, are just the beginning of the intellectual processes of science. There are many different paths to scientific discovery, one of which is shown in Figure 1.1. However, there is no general set of rules. Science is not a straightforward process for cranking out discoveries.

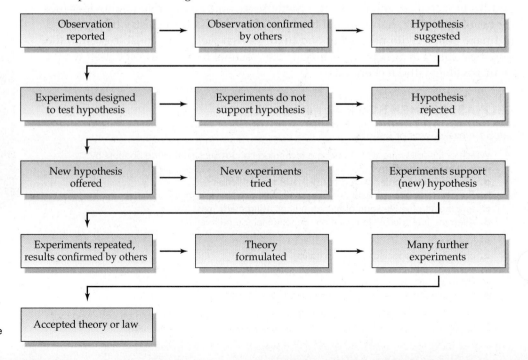

► **Figure 1.1** A possible scientific process. It may be many years from "Observation reported" to "Accepted theory or law." Obtaining new objective knowledge often takes much time and effort.

Scientific Hypotheses Are Testable

Scientists do not merely state what they feel may be true. They develop *testable* **hypotheses** (educated guesses; *hypothesis,* in the singular) as tentative explanations of observed data. They test these hypotheses by designing and performing experiments. Experimentation distinguishes science from the arts and humanities. In the humanities, people still argue about some of the same questions that were debated thousands of years ago: What is truth? What is beauty? These arguments persist because the proposed answers cannot be tested and confirmed objectively.

Like artists and poets, scientists are often imaginative and creative. The tenets of science, however, are *testable.* Experiments can be devised to answer most scientific questions. Ideas can be tested and thereby either verified or rejected. Some ideas may be accepted for a while, but rejected when further studies are performed. For example, it was long thought that exercise caused muscles to tire and become sore from a buildup of lactic acid. Recent findings suggest instead that lactic acid *delays* muscle tiredness and that the cause of tired, sore muscles may be related to other factors, including leakage of calcium ions inside muscle cells, which weakens contractions. Through many experiments, scientists have established a firm foundation of knowledge, allowing each new generation to build on the past.

Large amounts of scientific data are often summarized in brief verbal or mathematical statements called **scientific laws**. For example, Robert Boyle (1627–1691), an Irishman, conducted many experiments on gases. From these experiments, he established *Boyle's law*, which said that the volume of the gas decreased when the pressure applied to the gas was increased. Mathematically, Boyle's law can be written as $PV = k$, where P is the pressure on a gas, V is its volume, and k is a constant. If P is doubled, V will be cut in half. Scientific laws are *universal.* Under the specified conditions, they hold everywhere in the observable universe.

3 **Why do scientists so often say, "More study is needed"?**

More data help scientists refine a hypothesis so that it is better defined, clearer, or more applicable.

Scientific Theories Are Tentative and Predictive

Scientists organize the knowledge they accumulate on a framework of detailed explanations called theories. A **scientific theory** represents the best current explanation for a phenomenon. In essence, a law says, "this is what happens," while a theory says, "this is *why* it happens."

Some people think that science is absolute, but nothing could be further from the truth. A theory is always *tentative.* Theories may have to be modified or even discarded as a result of new observations. For example, the atomic theory proposed in the early 1800s was extensively modified as we learned that atoms are made up of even smaller particles. The body of knowledge that is a large part of science is rapidly growing and always changing.

Theories organize scientific knowledge and are also useful for their *predictive* value. Predictions based on theories are tested by further experiments, both by the original investigators and by other scientists. Theories that make successful predictions are usually widely accepted by the scientific community. A theory developed in one area is often found to apply in others.

Scientific Models Are Explanatory

Scientists often use models to help *explain* complicated phenomena. A **scientific model** uses tangible items or pictures to represent invisible processes. For example, the invisible particles of a gas can be visualized as marbles or pool balls, or as dots or circles on paper.

We know that when a glass of water is left standing for a period of time, the water disappears through the process of evaporation (Figure 1.2). Scientists explain evaporation with the *kinetic-molecular theory,* which proposes that a liquid is composed of tiny particles called molecules that are in constant motion and are held together by forces of attraction. The molecules collide with one another like pool balls on a pool table. Sometimes, a "hard break" in pool causes one ball to fly off the

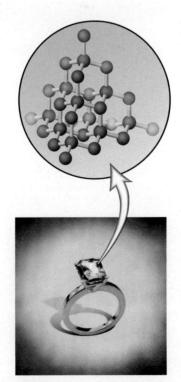

▲ A molecular model of diamond shows the tightly linked, rigid structure that explains why diamonds are so hard.

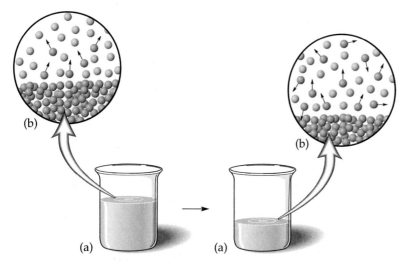

● = Water molecule

● = Air (nitrogen or oxygen) molecule

▲ **Figure 1.2** The evaporation of water. (a) When a container of water is left standing open to the air, the water slowly disappears. (b) Scientists explain evaporation with a model that shows the motion of molecules.

table. Likewise, some of the molecules of a liquid gain enough energy through collisions to overcome the attraction to their neighbors, escape from the liquid, and disperse among the widely spaced molecules in air. The water in the glass gradually disappears. This model gives us more than a name for evaporation; it gives us an understanding of the phenomenon.

When performing experiments, developing theories, and constructing models, it is important to note that an apparent connection—a *correlation*—between two items is not necessarily evidence that one *causes* the other. For example, many people suffer from allergies in the fall, when goldenrod is in bloom. However, research has shown that the main cause of these allergies is ragweed pollen. There is a correlation between the blooming of goldenrod and autumnal allergies, but goldenrod pollen is not the cause. Ragweed happens to bloom at the same time.

The Limitations of Science

Some people say that we could solve many of our problems if we would only attack them using the methods of science. Why can't the procedures of the scientist be applied to social, political, ethical, and economic problems? And why do scientists disagree over environmental, social, and political issues?

What Science Is—and Is Not

Responsible news media generally try to be fair, presenting both sides of an issue regardless of where the prevailing evidence lies. In science, the evidence often indicates that one side is simply wrong. Scientists strive for accuracy, not balance. The idea of a flat Earth is not given equal credence to that of a (roughly) spherical Earth. Only ideas that have survived experimental testing and peer review are considered valid. Ideas that are beautiful, elegant, or even sacrosanct can be invalidated by experimental data. For example, until 1543, the idea that the Sun revolved around the Earth was considered sacrosanct.

Science is not a democratic process. Majority rule does not determine what constitutes sound science. Science does not accept notions that are proven false or remain untested by experiment.

Disagreement often results from the inability to control *variables*. A **variable** is something that can change over the course of an experiment. If, for example, we wanted to study in the laboratory how the volume of a gas varies with changes in pressure, we could hold constant factors such as temperature and the amount and kind of gas. If, on the other hand, we wanted to determine the effect of low levels of a particular pollutant on the health of a human population, we would find it almost impossible to control such variables as individuals' diets, habits, and exposure to other substances, all of which affect health. Although we could make observations, formulate hypotheses, and conduct experiments on the health effect of the pollutant, interpretation of the results would be difficult and subject to disagreement.

SELF-ASSESSMENT Questions

Select the best answer or response.

1. To gather information to support or discredit a hypothesis, a scientist
   a. conducts experiments
   b. consults an authority
   c. establishes a scientific law
   d. formulates a scientific theory

2. The statement "mass is always conserved when chemical changes occur" is an example of a scientific
   a. experiment
   b. hypothesis
   c. law
   d. theory

3. A successful theory
   a. can be used to make predictions
   b. eventually becomes a scientific law
   c. is not subject to further testing
   d. is permanently accepted as true

4. Which of the following is *not* a hypothesis?
   a. A quarter is heavier than a nickel.
   b. Ice floats on water because of the air bubbles that get trapped during the freezing process.
   c. Oxygen reacts with silver to form tarnish.
   d. Synthetic hormones have the same effect in an organism as the naturally occurring ones.

5. Which of the following is a requirement of scientific research?
   a. It must be approved by a committee of scientists and politicians.
   b. It must benefit the Earth and improve human life.
   c. It must be experimentally tested and peer reviewed for validity.
   d. It must be balanced and weigh the pros and cons of the results.

6. Social problems are difficult to solve because it is difficult to
   a. control variables
   b. discount paranormal events
   c. form hypotheses
   d. formulate theories

Answers: 1, a; 2, c; 3, a; 4, a; 5, c; 6, a

1.3 Science and Technology: Risks and Benefits

Learning Objectives • Define *risk* and *benefit*, and give an example of each.
• Estimate a desirability quotient from benefit and risk data.

Most people recognize that society has benefited from science and technology, but many seem not to realize that there are risks associated with every technological advance. How can we determine when the benefits outweigh the risks? One approach, called **risk–benefit analysis**, involves the estimation of a *desirability quotient* (DQ).

$$DQ = \frac{Benefits}{Risks}$$

A **benefit** is anything that promotes well-being or has a positive effect. Benefits may be economic, social, or psychological. A **risk** is any hazard that can lead to loss or injury. Some of the risks associated with modern technology have led to disease, death, economic loss, and environmental deterioration. Risks and benefits may involve one individual, a group, or society as a whole.

Every technological advance has both benefits and risks. For example, a car provides the benefit of rapid, convenient transportation. But driving a car involves risk—individual risks of injury or death in a traffic accident and societal risks such as pollution and climate change. When one considers the number of people who drive cars, it is clear that most people consider the benefits of driving a car to outweigh the risks.

Weighing the benefits and risks connected with a product is more difficult when considering a group of people. For example, pasteurized low-fat milk is a safe, nutritious beverage for many people of northern European descent. Some people in this group can't tolerate lactose, the sugar in milk. And some are allergic to milk proteins. But since these problems are relatively uncommon among people of northern European descent, the benefits of milk are large and the risks are small, resulting in a large DQ for this group. However, adults of other ethnic backgrounds often are lactose-intolerant, and for them, milk has a small DQ.

Other technologies provide large benefits and present large risks. For these technologies the DQ is uncertain. An example is the conversion of coal to liquid fuels. Most people find liquid fuels to be very beneficial in transportation, home heating, and industry. There are great risks associated with coal conversion, however, including risks

WHY IT MATTERS

For most people of northern European ancestry, pasteurized low-fat milk is a wholesome food. Milk's benefits far outweigh its risks. Other ethnic groups have high rates of lactose intolerance among adults, and the desirability quotient for milk is much smaller.

to coal-mine workers, air and water pollution, and exposure of conversion plant workers to toxic chemicals. The result, again, is an uncertain DQ and political controversy.

There are yet other problems in risk–benefit analysis. Some technologies benefit one group of people while presenting a risk to another. For example, gold plating and gold wires in computers and other consumer electronics benefit the consumer, providing greater reliability and longer life. But when the devices are scrapped, small-scale attempts to recover the gold often produce serious pollution in the area of recovery. Difficult political decisions are needed in such cases.

Other technologies provide current benefits but present future risks. For example, although nuclear power now provides useful electricity, improperly stored wastes from nuclear power plants might present hazards for centuries. Thus, the use of nuclear power is controversial.

Science and technology obviously involve *both* risks and benefits. The determination of benefits is almost entirely a social judgment. Although risk assessment also involves social and personal decisions, it can often be greatly aided by scientific investigation. Understanding the chemistry behind many technological advancements will help you make a more accurate risk–benefit analysis for yourself, your family, your community, and the world.

 CONCEPTUAL EXAMPLE 1.1   Risk–Benefit Analysis

The drug ketorolac is a prescription NSAID (non-steroidal anti-inflammatory drug) that is said to be as effective as some opioids for treating moderate to severe pain. However, because of the side effects and potential for stroke and heart attack, the FDA recommends it for short-term use only. Do risk–benefit analyses of the use of ketorolac in treating the pain of **(a)** a 24-year-old male following an appendectomy and **(b)** a 52-year-old female who suffers from high blood pressure and the chronic pain of arthritis.

Solution

a. Pain from an appendectomy or similar procedure is generally short-term, and the likelihood of stroke or a heart attack in the short term for a young, healthy person is probably low. The DQ is probably moderate to high.

b. Treating the pain of a chronic condition, such as arthritis, would require long-term use of the drug. Also, the patient's age and high blood pressure probably make a stroke or heart attack much more likely. Also, there are other drugs that may not be as effective but are much safer to use long-term. The DQ is low in this case.

❭ Exercise 1.1A

Chloramphenicol is a powerful antibacterial drug that often destroys bacteria unaffected by other drugs. It is highly dangerous to some individuals, however, causing fatal aplastic anemia in about 1 in 30,000 people. Do risk–benefit analyses of administering chloramphenicol to **(a)** sick farm animals, resulting in milk and meat which might contain residues of the drug, and **(b)** a person with Rocky Mountain spotted fever facing a high probability of death or permanent disability.

❭ Exercise 1.1B

The drug thalidomide was introduced in Europe in the 1950s as a sleeping aid. It was found to be a *teratogen*, a substance that causes birth defects, and it was removed from the market after children whose mothers took it during pregnancy were born with deformed limbs. Recently, thalidomide has been investigated as an effective treatment for the lesions caused by leprosy and for Kaposi's sarcoma (a form of cancer often diagnosed in patients with AIDS). Do risk–benefit analyses of prescribing thalidomide for treatment of leprosy in **(a)** women aged 25–40 and **(b)** women aged 55–70.

Risks of Death

Our perception of risk often differs from the actual risk we face. Some people fear flying but readily assume the risk of an automobile trip. The odds of dying from various causes are listed in Table 1.1.

TABLE 1.1 Approximate Lifetime Risks of Death in the United States

Action	Lifetime Risk[a]			Details/Assumptions
All causes	1	or	1 in 1	We all die of something.
Cigarettes	0.25	or	1 in 4	Cigarette smoking, 1 pack/day
Heart disease	0.20	or	1 in 5	Heart attacks, congestive heart failure
All cancers	0.14	or	1 in 7	All cancers
Motor vehicles	0.01	or	1 in 100	Death in motor vehicle accident
Home accidents	0.01	or	1 in 100	Home accident death
Natural forces	0.0003	or	1 in 3360	Heat, cold, storm, earthquakes, etc.
Peanut butter (aflatoxin)	0.00060	or	1 in 1700	4 tablespoons peanut butter a day
Airplane accidents	0.00005	or	1 in 20,000	Death in aircraft crashes
Terrorist attack	0.00077	or	1 in 1300	One 9/11-level attack per year [b]
Terrorist attack	0.000077	or	1 in 13,000	One 9/11-level attack every 10 years

[a]The odds of dying of a particular cause in a given year are calculated by dividing the population by the number of deaths by that cause in that year.

[b]Unlikely scenario

Science is a unified whole. Common scientific laws apply everywhere and on all levels of organization. The various areas of science interact and support one another. Accordingly, chemistry is not only useful in itself but is also fundamental to other scientific disciplines. The application of chemical principles has revolutionized biology and medicine, has provided materials for powerful computers used in mathematics, and has profoundly influenced other fields, such as the production of new materials. The social goals of better health, nutrition, and housing are dependent to a large extent on the knowledge and techniques of chemists. Recycling of basic materials—paper, glass, and metals—involves chemical processes.

Chemistry is indeed a central science (Figure 1.3). There is no area of our daily lives that is not affected by chemistry. Many modern materials have been developed by chemists, and even more amazing materials are in the works.

Chemistry is also important to the *economies* of industrial nations. In the United States, the chemical industry makes thousands of consumer products, including personal-care products, agricultural products, plastics, coatings, soaps, and detergents. It produces 80% of the materials used to make medicines. The U.S. chemical industry is one of the country's largest industries, with sales of more than $800 billion in 2016, accounting for about 10% of all U.S. exports. It employs more than 826,000 workers,

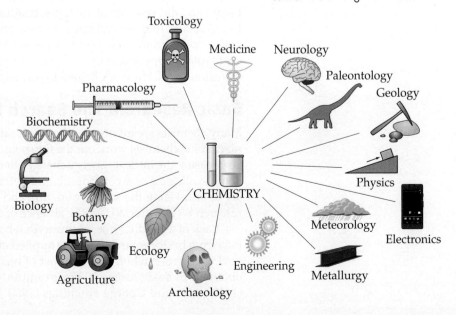

▼ **Figure 1.3** Chemistry has a central role among the sciences.

including scientists, engineers, and technicians, and generates nearly 11% of all U.S. patents. Contrary to the popular belief that chemicals are highly dangerous, workers in the chemical industry are five times safer than the average worker in the U.S. manufacturing sector.

SELF-ASSESSMENT Questions

Select the best answer or response.

1. Our perception of risk is
   a. always based on sound science
   b. always higher than actual risk
   c. always lower than actual risk
   d. often different from actual risk

2. Among the following, the highest risk of death (see Table 1.1) is associated with
   a. heart disease   **b.** peanut butter
   c. cancer   **d.** automobile accidents

3. Chemistry is called "the central science" because
   a. the chemical industry is based mainly in the U.S. Midwest
   b. it is the core course in all college curricula
   c. it is fundamental to other scientific disciplines
   d. it is the source of many environmental problems

4. The U.S. chemical industry is
   a. patently unsafe
   b. the source of nearly all pollution
   c. the source of many consumer products
   d. unimportant to the U.S. economy

Answers: 1, d; 2, a; 3, c; 4, c

▲ George Washington Carver (1864–1943), research scientist, in his laboratory at Tuskegee Institute.

1.4 Solving Society's Problems: Scientific Research

Learning Objective • Distinguish basic research from applied research.

Chemistry is a powerful force in shaping society today. Chemical research not only plays a pivotal role in other sciences, but it also has a profound influence on society as a whole. There are two categories of research: *applied research* or *basic research*. The two often overlap, and it isn't always possible to label a particular project as one or the other.

Applied Research

Some chemists test polluted soil, air, and water. Others analyze foods, fuels, cosmetics, detergents, and drugs. Still others synthesize new substances for use as drugs or pesticides or formulate plastics for new applications. These activities are examples of **applied research**—work oriented toward the solution of a particular problem in an industry or the environment.

Among the most monumental accomplishments in applied research were those of George Washington Carver. Born in slavery, Carver attended Simpson College and later graduated from Iowa State University. A botanist and agricultural chemist, Carver taught and did research at Tuskegee Institute. He developed more than 300 products from peanuts, from peanut butter to hand cleaner to insulating board. He created other new products from sweet potatoes, pecans, and clay. Carver also taught Southern farmers to rotate crops and to use legumes to replenish the nitrogen removed from the soil by cotton crops. His work helped to revitalize the economy of the South.

Basic Research: The Search for Knowledge

Many chemists are involved in **basic research**, the search for knowledge for its own sake. Some chemists work out the fine points of atomic and molecular structure. Others measure the intricate energy changes that accompany complex chemical reactions. Some synthesize new compounds and determine their properties. Done for the sheer joy of unraveling the secrets of nature and discovering order in our universe, basic research is characterized by the absence of any predictable, marketable product.

Lack of a product doesn't mean that basic research is useless. Far from it! Findings from basic research often *are* applied at a later time, though this is not the main goal of the researcher. In fact, most of our modern technology is based on results obtained in basic research. For example, in the 1940s and 1950s Gertrude Elion (1918–1999) and George Hitchings (1905–1998) examined the role of compounds called *purines* in cell chemistry. This led to the discovery of new drugs that facilitate organ transplants and treat gout, malaria, herpes, cancer, and AIDS. Basic research in the 1960s involving a device called the *maser* led to the development of today's global positioning system (GPS). In the 1920s, basic research led to discovery of *nuclear spin*, a property of certain atomic nuclei. This work led to the development of *magnetic resonance imaging* (MRI), used by virtually every hospital today to view internal body structure. Without the base of factual information obtained from basic research, technological innovation would be haphazard and slow.

Applied research is carried out mainly by industries seeking immediate gain by developing a novel, better, or more salable product. The ultimate aim of such research is usually profit for the stockholders. Basic research is conducted mainly at universities and research institutes. Most support for this research comes from federal and state governments and foundations, although some larger industries also support it.

SELF-ASSESSMENT Questions

Classify each of the following as (a) applied research or (b) basic research.

1. A chemist develops a faster-acting and longer-lasting asthma remedy.
2. A researcher investigates the effects of different acids on the breakdown of starches.
3. A scientist seeks to extract biologically active compounds from sea sponges.
4. Scientists create rBST to improve milk production in cows.

Answers: 1, a; 2, b; 3, b; 4, a

1.5 Chemistry: A Study of Matter and Its Changes

Learning Objective • Differentiate: mass and weight; physical and chemical changes; and physical and chemical properties.

As we have noted, chemistry is often defined as the study of matter and the changes it undergoes. Because the entire physical universe is made up of matter and energy, the field of chemistry extends from atoms to stars, from rocks to living organisms. Now let's look at matter a little more closely.

Matter is the stuff that makes up all material things. It is anything that occupies space and has mass. Scientifically, **mass** is a measure of the *inertia* of an object; the greater the mass of an object, the greater is its inertia, meaning that it is more difficult to change the object's velocity. You can easily deflect a fist-sized tennis ball coming toward you at 30 meters per second, but you would have difficulty stopping an iron cannonball of that size moving at the same speed. A cannonball has more mass than a tennis ball of equal size. Wood, sand, water, air, and people all occupy space and have mass and are, therefore, matter.

The mass of an object does not vary with location. An astronaut has the same mass on the moon, or in "weightless" orbit, as on Earth. In contrast, **weight** measures a force. On Earth, it measures the force of attraction between our planet and the mass in question. On the moon, where gravity is one-sixth that on Earth, an astronaut weighs only one-sixth as much as on Earth (Figure 1.4). Weight varies with gravity; mass does not. Nonetheless, in this book we often will use the term "weight" to mean "mass" for two reasons. First, it can be a bit awkward or confusing to say, "The student massed about 15 grams of sugar," rather than "The student weighed about 15 grams of sugar." Second, most chemical reactions occur under constant gravity on the surface of the Earth, so weight remains virtually the same anywhere on Earth. Ordinarily the American units of ounces and pounds are used when expressing weight in this book; grams, kilograms, etc., refer to mass.

4 Why do scientists bother with studies that have no immediate application?

The results of basic research may not have an immediate practical use. However, basic research extends our understanding of the world and can be considered an investment in the future. And basic research often *finds* a practical application. For example, *zone refining* is a method developed in the early twentieth century for producing extremely pure solids. It had little practical application until the advent of integrated circuits. Zone refining is now used to produce the ultrapure silicon needed for every electronic device on Earth.

WHY IT MATTERS

Gertrude Elion won a Nobel Prize for her basic research on purines. Her work has led to many applications in pharmaceuticals and medicine. Many important modern applications started out with basic research. In 1991, Elion became the first woman to be inducted into the National Inventors Hall of Fame.

▲ **Figure 1.4** Although Astronaut Harrison Schmitt's moon suit weighs as much on Earth as he does (180 pounds each), he finds it fairly easy to move, because lunar gravity pulls him and his suit only one-sixth as much as gravity does on Earth.

Q *Based on the information in the caption, how much do Schmitt and his moon suit weigh, in pounds, on the moon? What would be the mass in kilograms of him in his suit on Earth?*

 CONCEPTUAL EXAMPLE 1.2 Mass and Weight

Gravity on the planet Mercury is three-eighths that on Earth. **(a)** What would be the mass on Mercury of a person who had a mass of 80 kilograms (kg) on Earth? **(b)** What would be the weight in pounds on Mercury of a person who weighed 160 pounds (lb.) on Earth?

Solution

a. The person's mass would be the same as on Earth (80 kg). The quantity of matter has not changed.
b. The force of attraction between Mercury and the person is only three-eighths, or 0.375 times, that between Earth and the person, so the person would weigh only 0.375×160 lb. = 60 lb.

〉Exercise 1.2A

At the surface of Venus, the force of gravity is 0.903 times that at Earth's surface. **(a)** What would be the mass of a 1.00 kg object on Venus? **(b)** How much (in pounds) would a man who weighed 198 lb. on Earth weigh on the surface of Venus?

〉Exercise 1.2B

A man who weighs 198 pounds on Earth would weigh 475 pounds on the surface of Jupiter. How many times greater than Earth's gravity is Jupiter's gravity?

Physical and Chemical Properties

We can use our knowledge of chemistry to change matter to make it more useful. Chemists can change crude oil into gasoline, plastics, pesticides, drugs, detergents, and thousands of other products. Changes in matter are accompanied by changes in energy. Often, we change matter to extract part of its energy. For example, we burn gasoline to get energy to propel our automobiles.

To distinguish between samples of matter, we can compare their properties (Figure 1.5). A **physical property** of a substance is a characteristic or behavior that can be observed or measured without forming new types of matter. Color, odor,

▶ **Figure 1.5** A comparison of the physical properties of two elements. Copper (left) can be hammered into thin foil or drawn into wire. Iodine (right) consists of brittle gray crystals that crumble into a powder when struck.

Q *What additional physical properties of copper and iodine are apparent from the photographs?*

TABLE 1.2 **Some Examples of Physical Properties**

Property	Examples
Temperature	Water freezes at 0 °C and boils at 100 °C.
Mass	A nickel has a mass of 5 g. A penny has a mass of 2.5 g.
Color	Sulfur is yellow. Bromine is reddish-brown.
Taste	Acids are sour. Bases are bitter.
Odor	Benzyl acetate smells like jasmine. Hydrogen sulfide smells like rotten eggs.
Boiling point	Water boils at 100 °C. Ethyl alcohol boils at 78.5 °C.
Hardness	Diamond is exceptionally hard. Sodium metal is soft.
Density	1.00 g/mL for water, 19.3 g/cm^3 for gold.

TABLE 1.3 **Some Examples of Chemical Properties**

Substance	Typical Chemical Property
Iron	Rusts (combines with oxygen to form iron oxide)
Carbon	Burns (combines with oxygen to form carbon dioxide)
Silver	Tarnishes (combines with sulfur to form silver sulfide)
Nitroglycerin	Explodes (decomposes to produce a mixture of gases)
Carbon monoxide	Is toxic (combines with hemoglobin, causing anoxia)
Neon	Is inert (does not react with anything)

and hardness are physical properties (Table 1.2). When a **chemical property** is observed, new types of matter with different compositions are formed. Burning wood is a chemical change because new substances such as carbon dioxide, water vapor, and ash (mostly potassium oxide) are formed (Table 1.3).

When a **physical change** occurs, there is a change in the physical appearance of matter without changing its chemical identity or composition. An ice cube can melt to form a liquid, but it is still water. Melting is a physical change, and the temperature at which it occurs—the melting point—is a physical property.

A **chemical change** or chemical reaction involves a change in the chemical identity of matter into other substances that are chemically different. In exhibiting a chemical property, matter undergoes a chemical change. At least one substance in the original matter is replaced by one or more new substances. Iron metal reacts with oxygen from the air to form rust (iron oxide). When sulfur burns in air, sulfur—a yellow solid made up of one type of atom—and odorless oxygen gas (from air), which is made up of another type of atom, combine to form sulfur dioxide, a smelly, choking gas made up of molecules, each containing one sulfur atom and two oxygen atoms. A *molecule* is a group of atoms bound together as a single unit. (You'll learn more about atoms and molecules in the next section.)

It is difficult at times to determine whether a change is physical or chemical. We can decide, though, on the basis of what happens to the composition or structure of the matter involved as well as the reversibility of the change. Physical changes are often reversible while chemical changes are not. *Composition* refers to the types of atoms that are present and their relative proportions, and *structure* refers to the arrangement of those atoms with respect to one another or in space. A chemical change results in a change in composition or structure, whereas a physical change does not. To answer the question "Is this a chemical change?" simply consider whether new substances with new properties have been formed. If new substances are formed, the change is a chemical change.

CONCEPTUAL EXAMPLE 1.3   Chemical Change and Physical Change

Identify each of the following as a physical change or chemical change.
   a. A piece of wood is sanded to make it smooth, forming sawdust.
   b. Lemon juice is added to warmed milk to make cottage cheese.
   c. Molten aluminum is poured into a mold, where it solidifies.
   d. Sodium chloride (table salt) is broken down into sodium metal and chlorine gas.

Solution
We examine each change and determine whether there has been a change in composition or structure. In other words, we ask, "Have new substances that are chemically different been created?" If so, the change is chemical; if not, it is physical.
   a. Physical change: Both the smoothed wood and the sawdust are still wood, chemically the same as the wood before sanding.
   b. Chemical change: Cottage cheese and milk have very different compositions.
   c. Physical change: Whether it is solid or liquid, aluminum is the same substance with the same composition.
   d. Chemical change: New substances, sodium and chlorine, are formed.

› Exercise 1.3A
Identify each of the following as a physical change or chemical change.
   a. A steel wrench left out in the rain becomes rusty.
   b. A stick of butter melts.
   c. Charcoal briquettes are burned.

› Exercise 1.3B
Identify each of the following as a physical change or chemical change.
   a. A food processor changes a beef roast into ground beef.
   b. Pancake batter is cooked on a griddle.
   c. The plastic filament in a 3D printer is used to make a small figure.

SELF-ASSESSMENT Questions

Select the best answer or response.

1. Which of the following is *not* an example of matter?
   a. methane gas   **b.** steam
   c. silver   **d.** a rainbow

2. Which of the following is an example of matter?
   a. sunlight
   b. air pollution
   c. magnetic field
   d. ultraviolet light from a tanning lamp

3. Two identical items are taken from Earth to Mars, where they have
   a. both the same mass and the same weight as on Earth
   b. neither the same mass nor the same weight as on Earth
   c. the same mass but not the same weight as on Earth
   d. the same weight but not the same mass as on Earth

4. What kind of change alters the identity of a material?
   a. lowering the temperature   **b.** physical
   c. hammering the material   **d.** chemical

5. Bending glass tubing in a hot flame involves
   a. applied research   **b.** a chemical change
   c. an experiment   **d.** a physical change

6. Creating solid ice from liquid water involves
   a. adding energy to the water
   b. a physical change
   c. a chemical change
   d. creating a new substance

7. Which of the following is a chemical property?
   a. Iodine vapor is purple.
   b. Copper tarnishes.
   c. Salt dissolves in water.
   d. Balsa wood is easily carved.

8. Which of the following is an example of a physical change?
   a. A cake is baked from flour, baking powder, sugar, eggs, shortening, and milk.
   b. Cream left outside a refrigerator overnight turns sour.
   c. Sheep are sheared, and the wool is spun into yarn.
   d. Spiders eat flies and make spider silk.

9. Which of the following is an example of a chemical change?
   a. An egg is broken and poured into an eggnog mix.
   b. A tree is pruned, shortening some branches.
   c. A tree grows larger from being watered and fertilized.
   d. Frost forms on a cold windowpane.

Answers: 1, d; 2, b; 3, c; 4, d; 5, d; 6, b; 7, b; 8, c; 9, c

1.6 Classification of Matter

Learning Objective • Classify matter according to state and as mixture, substance, compound, and/or element.

In this section, we examine three of the many ways of classifying matter. First, we look at the physical forms or *states of matter*.

The States of Matter

There are three familiar states of matter: solid, liquid, and gas (Figure 1.6). They can be classified by bulk properties (a *macro* view) or by arrangement of the particles that compose them (a *molecular*, or *micro*, view). A **solid** object maintains its shape and volume regardless of its location. A **liquid** occupies a definite volume but assumes the shape of the portion of a container that it occupies. If you have 355 milliliters (mL) of a soft drink, you have 355 mL regardless of whether the soft drink is in a can, in a bottle, or, through a mishap, on the floor—which demonstrates another property of liquids. Unlike solids, liquids flow readily. A **gas** maintains neither shape nor volume. It expands to fill completely whatever container it occupies. Like liquids, gases flow; unlike liquids and solids, gases are easily compressed. For example, enough air for many minutes of breathing can be compressed into a steel tank for SCUBA diving.

Bulk properties of solids, liquids, and gases are explained using the kinetic-molecular theory (Chapter 5). In solids, the particles are close together and in fixed positions. In liquids, the particles are close together, but they are free to move about. In gases, the particles are far apart and are in rapid random motion.

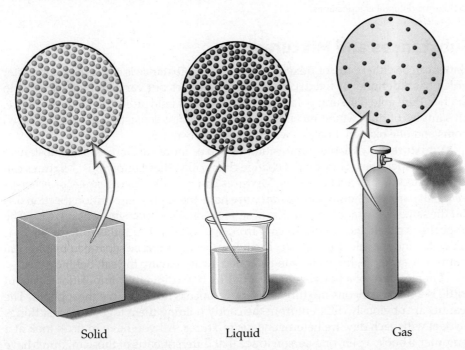

Solid   Liquid   Gas

▲ **Figure 1.6** The kinetic-molecular theory can be used to interpret (or explain) the bulk properties of *solids, liquids,* and *gases*. In solids, the particles are close together and in fixed positions. In liquids, the particles are close together, but they are free to move about. In gases, the particles are far apart and are in rapid random motion.

Q *Based on this figure, why does a quart of water vapor weigh so much less than a quart of liquid water?*

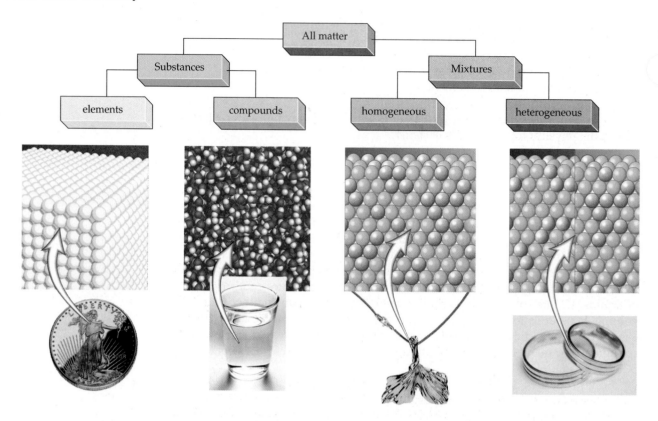

▲ **Figure 1.7** A scheme for classifying matter according to its composition. From left to right: The coin is pure gold, an *element*. Water is a *compound*. The pendant is made of rose gold, a *homogeneous mixture* of gold and copper; all parts of the pendant have the same composition. The ring at the far right is a *heterogeneous mixture* of rose gold and white gold (which contains gold and silver). Rose gold and white gold have different compositions.

Substances and Mixtures

Matter can be either pure or mixed (Figure 1.7). A **substance** is a pure form of matter which has a definite, or fixed, composition that does not vary from one sample to another. Pure gold (24-karat gold) consists entirely of gold atoms, so it is a substance. All samples of pure water are composed of molecules consisting of two hydrogen atoms and one oxygen atom, so water is also a substance.

A **mixture** is a variable composition of two or more substances. The substances retain their identities. They do not change chemically; they simply mix. Mixtures can be separated by physical changes. Mixtures can be either *homogeneous* or *heterogeneous*. All parts of a **homogeneous mixture** have the same composition (Section 6.4) and the same appearance. A solution of salt in water is a homogeneous mixture. The proportions of salt and water can vary from one solution to another, but the water is still water and the salt remains salt. The two substances can be separated by physical means. For example, the water can be boiled away, leaving the salt behind.

Different parts of a **heterogeneous mixture** have different compositions. Peanut brittle is a heterogeneous mixture. The appearance is not the same throughout. The peanuts are obviously different from the candy holding them together. Most things we deal with each day are heterogeneous mixtures. All you have to do is look at a computer, a book, a pen, or a person to see that different parts of those mixtures have different compositions. (Your hair is quite different from your skin, for example.)

Elements and Compounds

A *substance* is either an element or a compound. An **element** is one of the fundamental substances from which all material things are constructed. Elements cannot be broken down into simpler substances by any chemical process. There are 118 known elements, of which this book deals with only about a third. A list of all elements appears near the front of this book. Oxygen, carbon, sulfur, aluminum, and iron are familiar elements.

5 Can we change lead into gold?

One element cannot be changed into another element by chemical reactions. An element cannot be created or destroyed, although an element can be extracted from a mixture or compound containing the element.

A **compound** is a substance made up of two or more elements chemically combined in a fixed ratio. For example, water is a compound because it has fixed proportions of hydrogen (H) and oxygen (O). Aluminum oxide (the "sand" on sandpaper), carbon dioxide, and iron disulfide (FeS_2, "fool's gold") are other compounds. Just as a whole dictionary of words can be constructed from 26 letters, tens of millions of compounds have been made from the 118 elements. A distinguishing property of a compound is that it can only be separated into its elements by a chemical change.

Because elements are so fundamental to our study of chemistry, we find it useful to refer to them in a shorthand form. Each element can be represented by a **chemical symbol** made up of one or two letters derived from the name of the element. Symbols for the elements are listed in the Table of Atomic Masses near the front cover. The first letter of a symbol is always capitalized. The second (if there are two) is always lowercase. It makes a difference. For example, Co is the symbol for cobalt, a metallic element, but CO is the formula for carbon monoxide, a poisonous compound.

A chemical symbol in a formula stands for one atom of the element. If more than one atom is included in a formula, a subscript number is used after the symbol. For example, the formula H_2 represents two atoms of hydrogen, and the formula CH_4 represents a compound with molecules each of which contains one atom of carbon and four atoms of hydrogen. Mixtures do not have such formulas, because a mixture does not have a fixed composition.

The elements' symbols are the alphabet of chemistry. Most are based on the English names of the elements, but a few are based on Latin and Greek names. For example, the symbol for gold is Au, from the Latin word *aurum*, and lead's symbol is Pb, from the Latin word *plumbum*. (The latter is also the origin of the word *plumbing*, because lead was used for plumbing pipes and fittings until the 1960s.)

▲ Lead plumbing pipes such as this one were used in home and commercial construction up to the 1960s, and lead-based solder was used to join copper pipes until the Safe Drinking Water Act of 1986. Lead is soft and easy to work, but even tiny traces of lead in drinking water can lead to impaired cognition, as was seen in Flint, Michigan, starting in 2015.

CONCEPTUAL EXAMPLE 1.4   Elements and Compounds

Which of the following represent elements and which represent compounds?

C   Cu   HI   BN   In   HBr

Solution
The periodic table shows that C, Cu, and In represent elements. (Each is a single symbol.) HI, BN, and HBr are each composed of symbols of two elements, and represent compounds.

> **Exercise 1.4A**
Which of the following represent elements, and which represent compounds?

Hf   No   CuO   NO   HF   Fm

> **Exercise 1.4B**
How many *different* elements are represented in the list in Exercise 1.4A?

Atoms and Molecules

An **atom** is the smallest characteristic part of an element. Each element is composed of atoms of a particular kind. For example, the element copper is made up of copper atoms, and gold is made up of gold atoms. All copper atoms are alike in a fundamental way and are different from gold atoms.

The smallest characteristic part of most compounds is a molecule. A **molecule** is a group of atoms bound together as a unit. All the molecules of a given compound have the same atoms in the same proportions. For example, all water molecules have two hydrogen atoms and one oxygen atom, as indicated by the formula, H_2O. We will discuss atoms in some detail in Chapter 2, and much of the focus of many later chapters is on molecules.

SELF-ASSESSMENT Questions

1. Which state of matter has a definite volume but not a definite shape?
   a. gas   **b.** compound   **c.** liquid   **d.** solid

2. Which of the following is a mixture?
   a. carbon   **b.** copper   **c.** silver   **d.** soda water

3. Which of the following is *not* an element?
   a. aluminum   **b.** brass   **c.** lead   **d.** sulfur

4. A compound can be separated into its elements
   a. by chemical means   **b.** by mechanical means
   c. by physical means   **d.** using a magnet

5. The smallest part of an element is a(n)
   a. atom   **b.** corpuscle   **c.** mass   **d.** molecule

6. The symbol for sodium is
   a. S   **b.** Na   **c.** Sd   **d.** So

7. Ar is the symbol for
   a. air   **b.** argon   **c.** aluminum   **d.** arsenic

Answers: 1, c; 2, d; 3, b; 4, a; 5, a; 6, b; 7, b

1.7 The Measurement of Matter

Learning Objective • Assign proper units of measurement to observations, and manipulate units in conversions.

Accurate measurements of such properties as mass, volume, time, and temperature are essential to the compilation of dependable scientific data. Such data are of critical importance in all science-related fields. Measurements of temperature and blood pressure are routinely made in medicine, and modern medical diagnosis depends on a whole battery of other measurements, including careful chemical analyses of blood and urine.

Some form of standardization of units is critical. For example, if we did not agree on the meaning of "one pound," trade would be impossible. The measurement system agreed upon by scientists since 1960 is the *International System of Units*, or **SI units** (from the French *Système International*), a modernized version of the metric system established in France in 1791. Most countries use metric measures in everyday life. In the United States, these units are used mainly in science laboratories, although they are increasingly being used in commerce, especially in businesses with an international component. One-liter and two-liter bottled beverages are found everywhere, and sporting events often use metric measurements, such as the 100-meter dash. Figure 1.8 compares some metric and customary units.

Because SI units are based on the decimal system, it is easy to convert from one unit to another. All measured quantities can be expressed in terms of the seven base units listed in Table 1.4. We use the first six in this text.

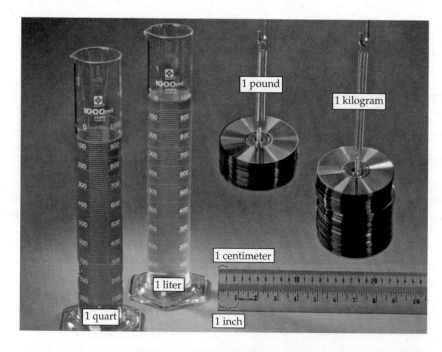

▶ **Figure 1.8** Comparisons of metric and customary units of measure. The meter stick is 100 centimeters long and is slightly longer than the yardstick below it (1 m = 1.09 yd.).

Q *From this figure, approximately how many pounds are in a kilogram? Which two units are closer in size, the inch and centimeter or the quart and the liter?*

TABLE 1.4 The Seven SI Base Units

Physical Quantity	Name of Unit	Symbol of Unit
Length	meter [a]	m
Mass	kilogram	kg
Time	second	s
Temperature	kelvin	K
Amount of substance	mole	mol
Electric current	ampere	A
Luminous intensity	candela	Cd

[a]Spelled *metre* in most countries.

Exponential Numbers: Powers of 10

Scientists deal with objects smaller than atoms and as large as the universe. We usually use exponential notation to describe the sizes of such objects. A number is in **exponential notation** when it is written as the product of a coefficient and a power of 10, such as 1.6×10^{-19} or 35×10^{6}. A number is said to be expressed in **scientific notation** when the coefficient has a value between 1 and 10 or −1 and −10.

An electron has a diameter of about 10^{-15} meter (m) and a mass of about 10^{-30} kilogram (kg). At the other extreme, a galaxy typically measures about 10^{23} m across and has a mass of about 10^{41} kg. It is difficult even to imagine numbers so small or so large. The accompanying figure offers some perspectives on size. Appendix A.2 provides a more detailed discussion of scientific notation.

Because measurements using the basic SI units can be of awkward magnitude, we often use exponential numbers. However, it is sometimes more convenient to use prefixes (Table 1.5) to indicate units larger or smaller than the base unit. The following examples show how prefixes and powers of 10 are interconverted.

EXAMPLE 1.5 Prefixes and Powers of 10

Convert each of the following measurements to a unit that replaces the power of 10 by a prefix.
a. 2.89×10^{-6} g   **b.** 4.30×10^{3} m

Solution
Our goal is to replace each power of 10 with the appropriate prefix from Table 1.5. For example, $10^{-3} = 0.001$, which corresponds to milli (unit). (It doesn't matter what the unit is; here we are dealing only with the prefixes.)
a. 10^{-6} corresponds to the prefix *micro-*; that is, $10^{-6} \times$ (unit) = micro (unit). So we have 2.89 micrograms (μg).
b. 10^{3} corresponds to the prefix *kilo-*; that is, $10^{3} \times$ (unit) = kilo (unit). So we have 4.30 km.

⟩ Exercise 1.5A
Convert each of the following measurements to a unit that replaces the power of 10 by a prefix.
a. 7.24×10^{3} g   **b.** 4.29×10^{-6} m   **c.** 7.91×10^{-3} s

⟩ Exercise 1.5B
Convert each of the following measurements to a measurement that uses the base unit and a power of 10.
a. 3.8 ns   **b.** 7.54 mm   **c.** 2.9 kA

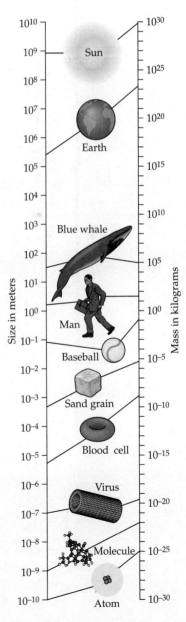

▲ Comparison of very large and very small objects is much easier with exponential notation.

TABLE 1.5 Approved Numerical Prefixes[a]

Exponential Expression	Decimal Equivalent	Prefix	Pronounced	Symbol
10^{12}	1,000,000,000,000	*tera-*	TER-uh	T
10^{9}	1,000,000,000	*giga-*	GIG-uh	G
10^{6}	1,000,000	*mega-*	MEG-uh	M
10^{3}	1,000	*kilo-*	KIL-oh	k
10^{2}	100	*hecto-*	HEK-toe	h
10^{1}	10	*deka-*	DEK-uh	da
10^{-1}	0.1	*deci-*	DES-ee	d
10^{-2}	0.01	*centi-*	SEN-tee	c
10^{-3}	0.001	*milli-*	MIL-ee	m
10^{-6}	0.000001	*micro-*	MY-kro	μ
10^{-9}	0.000000001	*nano-*	NAN-oh	n
10^{-12}	0.000000000001	*pico-*	PEE-koh	p
10^{-15}	0.000000000000001	*femto-*	FEM-toe	F

[a]The most commonly used prefixes are shown in color.

 EXAMPLE 1.6 Prefixes and Powers of 10

Express each of the following measurements in terms of an SI base unit in both decimal and exponential form.

   a. 4.12 cm

   b. 947 ms

   c. 3.17 nm

Solution

   a. Our goal is to find the power of 10 that relates the given unit to the SI base unit. That is, centi (base unit) = 10^{-2} × (base unit):

$$4.12 \text{ cm} = 4.12 \text{ centimeter} = 4.12 \times 10^{-2} \text{ m} = 0.0412 \text{ m}$$

   b. To change millisecond (ms) to the base unit second, we replace the prefix *milli-* with 10^{-3}; as a shortcut, you can move the decimal place to the left for each negative power or to the right for each positive power:

$$947 \text{ ms} = 0.947 \text{ s}$$

   c. To change nanometer (nm) to the base unit meter, we replace the prefix *nano-* with 10^{-9}. The answer in exponential form is 3.17×10^{-9} m, and in decimal notation, it is 0.00000000317.

> **Exercise 1.6A**

Convert the following measurements to an SI base unit.

   a. 7.45 nm   **b.** 5.25 μs   **c.** 1.415 km

   d. 2.06 mm   **e.** 6.19×10^{6} mm

> **Exercise 1.6B**

Convert the following measurements to an SI base unit.

   a. 57 km

   b. 11 mA

Mass

The SI base unit for mass is the **kilogram (kg)**, about 2.2 pounds (lb.), or about the mass of a 1 L soft drink. This base unit is unusual in that it already has a prefix. A more convenient mass unit for most laboratory work is the gram (g), about the mass of a large paperclip.

$$1 \text{ kg} = 1000 \text{ g} = 10^3 \text{ g}$$

The milligram (mg) is a suitable unit for small quantities of materials, such as some drug dosages or spices added while cooking food. Other drugs or vitamins that we take require a very small amount, such as a microgram (μg), which is commonly abbreviated as mcg on the label.

$$1 \text{ mg} = 10^{-3} \text{ g} = 0.001 \text{ g} \qquad 1 \, \mu\text{g} = 10^{-6} \text{ g} = 0.000001 \text{ g}$$

Chemists can now detect masses in the nanogram (ng), picogram (pg), and even femtogram (fg) ranges.

Length, Area, and Volume

The SI base unit of length is the **meter (m)**, a unit slightly longer than 1 yard (yd.). The kilometer (km) is used to measure distances along highways.

$$1 \text{ km} = 1000 \text{ m}$$

In the laboratory, we usually find lengths smaller than the meter to be more convenient. For example, we use the centimeter (cm), which is about the width of a typical calculator button, or the millimeter (mm), which is about the thickness of the cardboard backing in a notepad.

$$1 \text{ cm} = 0.01 \text{ m} \quad 1 \text{ mm} = 0.001 \text{ m}$$

For measurements at the atomic and molecular levels, we use the micrometer (μm), the nanometer (nm), and the picometer (pm). For example, a hemoglobin molecule, which is nearly spherical, has a diameter of 5.5 nm or 5500 pm, and the diameter of a sodium atom is 372 pm.

The units for area and volume are derived from the base unit of length. The SI unit of area is the square meter (m^2), but we often find square centimeters (cm^2) or square millimeters (mm^2) to be more convenient for laboratory work. Notice that although 1 cm $= 10^{-2}$ m, 1 cm^2 is *not* equal to 10^{-2} m^2.

$$1 \text{ cm}^2 = (10^{-2} \text{ m})^2 = 10^{-4} \text{ m}^2 \qquad 1 \text{ mm}^2 = (10^{-3} \text{ m})^2 = 10^{-6} \text{ m}^2$$

Similarly, the SI unit of volume is the cubic meter (m^3), but two units more likely to be used in the laboratory are the cubic centimeter (cm^3 or cc) and the cubic decimeter (dm^3). A cubic centimeter is about the volume of a sugar cube, and a cubic decimeter is slightly larger than 1 quart (qt). As with area, 1 cm^3 is *not* equal to 10^{-2} m^3.

$$1 \text{ cm}^3 = (10^{-2} \text{ m})^3 = 10^{-6} \text{ m}^3 \qquad 1 \text{ dm}^3 = (10^{-1} \text{ m})^3 = 10^{-3} \text{ m}^3$$

The cubic decimeter is commonly called a liter. A **liter (L)** is 1 cubic decimeter, or 1000 cubic centimeters.

$$1 \text{ L} = 1 \text{ dm}^3 = 1000 \text{ cm}^3$$

The milliliter (mL), or cubic centimeter, is frequently used in laboratories. A milliliter is about one "squirt" from a medicine dropper.

$$1 \text{ mL} = 1 \text{ cm}^3$$

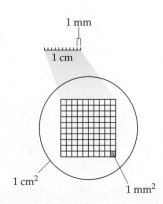

▲ Units of area such as mm^2 and cm^2 are derived from units of length.

Q *How many mm are in 1 cm? How many mm^2 are in 1 mm^2?*

Time

The SI base unit for measuring intervals of time is the **second (s)**. Extremely short time periods are expressed using SI prefixes: millisecond (ms), microsecond (μs), nanosecond (ns), and picosecond (ps).

$$1 \text{ ms} = 10^{-3} \text{ s} \qquad 1 \text{ }\mu\text{s} = 10^{-6} \text{ s} \qquad 1 \text{ ns} = 10^{-9} \text{ s} \qquad 1 \text{ ps} = 10^{-12} \text{ s}$$

Long time intervals, in contrast, are usually expressed in traditional, non-SI units: minute (min), hour (h), day (d), and year (y).

$$1 \text{ min} = 60 \text{ s} \qquad 1 \text{ h} = 60 \text{ min} \qquad 1 \text{ d} = 24 \text{ h} \qquad 1 \text{ y} = 365 \text{ d}$$

Problem Solving: Estimation

Many chemistry problems require calculations that yield numerical answers. When you use a calculator, it will always give you an answer—but the answer may not be correct. You may have punched a wrong number or used the wrong function. You can learn to estimate answers so you can tell whether your calculated answers are reasonable. At times an estimated answer is good enough, as the following example and exercises illustrate. This ability to estimate answers can be important in everyday life as well as in a chemistry course. Note that estimation does *not* require a detailed calculation. Only a rough calculation—or none at all—is required.

 EXAMPLE 1.7   Mass, Length, Area, Volume

Without doing detailed calculations, determine which of the following are a reasonable **(a)** mass (weight) and **(b)** height for a typical two-year-old child.

$$\text{Mass:} \quad 10 \text{ mg} \quad 10 \text{ g} \quad 10 \text{ kg} \quad 100 \text{ g}$$
$$\text{Height:} \quad 85 \text{ mm} \quad 85 \text{ cm} \quad 850 \text{ cm} \quad 8.5 \text{ m}$$

Solution

a. In customary units, a two-year-old child should weigh about 20–25 lb. Since 1 kg is a little more than 2 lb., the only reasonable answer is 10 kg.

b. In customary units, a two-year-old child should be about 30–36 in. tall. Since 1 cm is a little less than 0.5 in., the answer must be a little more than twice that range, or a bit more than 60–72 cm. Thus the only reasonable answer is 85 cm.

〉 Exercise 1.7A

Without doing a detailed calculation, determine which of the following is a reasonable area for the front cover of your textbook.

$$500 \text{ mm}^2 \quad 50 \text{ cm}^2 \quad 500 \text{ cm}^2 \quad 50 \text{ m}^2$$

〉 Exercise 1.7B

Without doing a detailed calculation, determine which of the following is a reasonable volume for your textbook.

$$1600 \text{ mm}^3 \quad 16 \text{ cm}^3 \quad 1600 \text{ cm}^3 \quad 1.6 \text{ m}^3$$

Problem Solving: Unit Conversions

It is easy to convert from one metric unit to another using the *unit conversion* method of problem solving. If you are not familiar with this method, you should study it now in Appendix A.3, which also discusses the conversion of common units to metric units.

You can find a discussion of **significant figures**, a way of indicating the precision of measurements, in Appendix A.4. In the following problems and throughout the text, we will simply carry three digits in most calculations.

EXAMPLE 1.8 **Unit Conversions**

Convert

(a) 1.83 kg to grams and   **(b)** 729 microliters to milliliters.

Solution

a. We start with the given quantity, 1.83 kg. The prefix *kilo-* means 1000, which gives us the equivalence 1 kg = 1000 g. From this we can form a conversion factor (see Appendix A.3) that allows us to cancel the unit kg and end with the unit g.

$$1.83 \text{ kg} \times \frac{1000 \text{ g}}{1 \text{ kg}} = 1830 \text{ g}$$

b. Here we start with the given quantity, 729 μL, which we want to be in mL. We do not have a single factor to change from μL to mL, so one way to approach the problem is to use two conversion factors. According to Table 1.5, *micro-* is 10^{-6} and *milli-* is 10^{-3}. That is, 1 μL = 10^{-6} L and 1 mL = 10^{-3} L. Set up the first conversion factor with μL (starting unit) on the bottom and L on top. The second factor will have L on the bottom and mL (desired unit) on top. Both μL and L cancel, leaving us with an answer in the correct unit.

$$729 \text{ } \mu\text{L} \times \frac{10^{-6} \text{ L}}{1 \text{ } \mu\text{L}} \times \frac{1 \text{ mL}}{10^{-3} \text{ L}} = 0.729 \text{ mL}$$

❯ Exercise 1.8A

Convert

   a. 0.755 m to millimeters   **b.** 205.6 mL to liters

   c. 0.206 g to micrograms   **d.** 7.38 microamperes to amperes

❯ Exercise 1.8B

Convert

   a. 0.409 kg to mg   **b.** 245 ms to nanoseconds

SELF-ASSESSMENT Questions

1. The SI unit of length is the
  a. foot   **b.** kilometer
  c. pascal   **d.** meter

2. The SI unit of mass is the
  a. dram   **b.** gram
  c. grain   **d.** kilogram

3. The prefix that means 10^{-2} is
  a. centi-   **b.** deci-
  c. micro-   **d.** milli-

4. An inch is about 250% longer than a
  a. meter   **b.** centimeter
  c. millimeter   **d.** decimeter

5. One cubic centimeter is equal to one
  a. deciliter   **b.** dram
  c. liter   **d.** milliliter

6. One quart is slightly less than one
  a. deciliter   **b.** kiloliter
  c. liter   **d.** milliliter

7. A rope is 5.775 cm long. What is its length in millimeters?
  a. 0.05775 mm   **b.** 0.5775 mm
  c. 5.775 mm   **d.** 57.75 mm

8. Which of the following has a mass of roughly 1 g?
  a. an ant   **b.** an orange
  c. a peanut   **d.** a watermelon

9. Your doctor recommends that you take 1000 μg of vitamin B_{12}. That is the same as
  a. 1 centigram   **b.** 1 decigram
  c. 1 gram   **d.** 1 milligram

10. Which of the following has a thickness of roughly 1 cm?
  a. a brick   **b.** a cell phone
  c. a knife blade   **d.** this textbook

11. Which of the following has a volume of roughly 250 mL (0.250 L)?
  a. 1 cup   **b.** 1 gallon
  c. 1 pint   **d.** 1 tablespoon

Answers: 1. d; 2. d; 3. a; 4. b; 5. d; 6. c; 7. d; 8. c; 9. d; 10. b; 11. a

Nanoworld

For over two centuries, chemists have been able to rearrange atoms to make molecules, which have dimensions in the *picometer* (10^{-12} m) range. This ability has led to revolutions in the design of drugs, plastics, and many other materials. Over the last several decades, scientists have made vast strides in handling materials with dimensions in the *micrometer* (10^{-6} m) range. The revolution in electronic devices—computers, cell phones, and so on—was spurred by the ability of scientists to produce computer chips by photolithography on a micrometer scale.

Now many scientists focus on *nanotechnology*. The prefix *nano-* means one-billionth (10^{-9}). Nanotechnology bridges the gap between picometer-sized molecules and micrometer-sized electronics. A nanometer-sized object contains just a few hundred to a few thousand atoms or molecules, and such objects often have different properties than large objects of the same substance. For example, bulk gold is yellow, but a ring made up of nanometer-sized gold particles appears red. Carbon in the form of graphite (pencil lead) is quite weak and

soft, but carbon nanotubes can be 100 times stronger than steel. Nanotubes, so called because of their size (only one 10,000th the thickness of a human hair) and their shape (hollow tube), have thousands of potential uses but are extremely expensive to make. Scientists can now manipulate matter on every scale, from nanometers to meters, greatly increasing the scope of materials design. We will examine some of the many practical applications of nanotechnology in subsequent chapters.

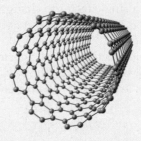

▲ Molecular view of a carbon nanotube—stronger than steel and more expensive than gold.

1.8 Density

Learning Objective • Calculate the density, mass, or volume of an object given the other two quantities.

In everyday life, we might speak of lead as "heavy" or aluminum as "light," but such descriptions are imprecise at best. What we really mean is that lead is heavy (or aluminum is light) for its "size" or for the space it takes up. Scientists use the term *density* to describe this important property. The **density**, d, of a substance is the quantity of mass, m, per unit of volume, V.

$$d = \frac{m}{V}$$

For substances that don't mix, such as oil and water, the concept of density allows us to predict which will float on the other. It isn't the mass of each material but the *mass per unit volume*—the density—that determines the result. Density is the property that explains why oil (lower density) floats on water (higher density) in a bottle of Italian salad dressing. Figure 1.9 shows other examples.

We can rearrange the equation for density to give

$$m = d \times V \quad \text{and} \quad V = \frac{m}{d}$$

These equations are useful for calculations. Densities of some common substances are listed in Table 1.6. They are customarily reported in grams per milliliter (g/mL) for liquids and grams per cubic centimeter (g/cm³) for solids. Values listed in the table are used in some of the following examples and exercises and in some of the end-of-chapter problems.

▲ One cubic centimeter of copper weighs 8.94 g, so the density of copper is 8.94 g/cm³.

Q *What is the mass of 2 cm³ of copper? What is the density of 2 cm³ of copper?*

TABLE 1.6 Densities of Some Common Substances at Specified Temperatures

Substance*	Density	Temperature
Solids		
Copper (Cu)	8.94 g/cm³	25 °C
Gold (Au)	19.3 g/cm³	25 °C
Magnesium (Mg)	1.738 g/cm³	20 °C
Water (ice) (H_2O)	0.917 g/cm³	0 °C
Liquids		
Ethanol (CH_3CH_2OH)	0.789 g/mL	20 °C
Hexane ($CH_3CH_2CH_2CH_2CH_2CH_3$)	0.660 g/mL	20 °C
Mercury (Hg)	13.534 g/mL	25 °C
Urine (a mixture)	1.003–1.030 g/mL	25 °C
Water (H_2O)	0.9998 g/mL	0 °C
Water	1.000 g/mL	4 °C
Water	0.998 g/mL	20 °C

*Formulas are provided for possible future reference.

▲ **Figure 1.9** At room temperature, the density of water is approximately 1.00 g/mL. A coin sinks in water, but a cork floats on water.

Q *Is the density of these coins less than, equal to, or greater than 1.00 g/mL? Is the density of the corks less than, equal to, or greater than 1.00 g/mL?*

As with problems involving mass, length, area, and volume (Example 1.7 on page 23), it is often sufficient to estimate answers to problems involving densities—for example, in determining relative volumes of materials or whether a material will float or sink in water. Such estimation is illustrated in the following example and exercises. Note again that estimation does *not* require a detailed calculation.

EXAMPLE 1.9 Mass, Volume, and Density

Density can provide a wealth of information without even doing a detailed calculation. This example and the accompanying exercises can be worked without using a calculator.

During the gold rush, prospectors would often find *fool's gold* (iron pyrite), which looks almost exactly like gold. However, iron pyrite has a density of approximately 3.3 g/cm³. Using Table 1.6, find the density of gold. If a miner found a nugget of fool's gold that fit in the palm of his hand, **(a)** would there be a significant difference in mass compared with a similar-sized nugget of pure gold? **(b)** If the fool's gold weighed the same as the gold nugget, would you see a difference in size?

Solution

a. From Table 1.6, we can see that the density of gold is 19.3 g/ml. That means gold is nearly 6 times heavier than a similarly sized piece of iron pyrite. This allowed prospectors to easily identify which ore they had found.

b. As seen above, the density of gold is roughly 6 times that of iron pyrite. Because gold is denser, it can pack in tighter and take up less space. Comparing two ore samples of equal mass, we find that the gold would be one-sixth as large as the pyrite.

> **Exercise 1.9A**
Lead has a density of 11.34 g/cm³. Will lead float or sink in hexane? In mercury?

> **Exercise 1.9B**
Based on what you know about crude oil, which is most likely to be its density: 0.88 g/mL, 1.25 g/mL, or 1.83 g/mL? (Hint: See Table 1.6 for useful information.)

EXAMPLE 1.10   Density, Mass, or Volume

Mass and volume give density. What is the density of iron if 156 g of iron occupies a volume of 20.0 cm³?

Solution

The given quantities are

$$m = 156\ g \quad \text{and} \quad V = 20.0\ cm^3$$

We use the equation that defines density.

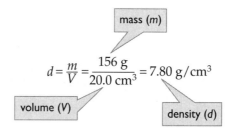

$$d = \frac{m}{V} = \frac{156\ g}{20.0\ cm^3} = 7.80\ g/cm^3$$

Likewise, density and mass give you volume; volume and density give you mass, according to the following rearrangements of our original density equation.

$$V = \frac{m}{d} \quad \text{or} \quad m = dV$$

> **Exercise 1.10A**

Find the density of a salt solution if 210 mL of solution has a mass of 234 g.

> **Exercise 1.10B**

What metal do you most likely have if a 10 cm³ piece weighs 17.38 g? (See Table 1.6.)

You can do your own demonstration of a density difference. Make a classic French vinaigrette salad dressing by adding 10 mL of red wine vinegar to 30–40 mL of extra virgin olive oil. Add seasonings such as salt, pepper, oregano, thyme, Dijon mustard, and garlic. Which material—oil or vinegar—forms the top layer? Enjoy!

SELF-ASSESSMENT Questions

1. The density of a wood plank floating on a lake is
 a. less than that of air
 b. less than that of water
 c. more than that of water
 d. the same as that of water

2. A pebble and a lead sinker both sink when dropped into a pond. This shows that the two objects
 a. are both less dense than water
 b. are both denser than water
 c. have different densities
 d. have the same density

3. Ice floats on liquid water. A reasonable value for the density of ice is
 a. 0.92 g/cm³
 b. 1.08 g/cm³
 c. 1.98 g/cm³
 d. 4.90 g/cm³

4. A prospector panning for gold swirls a mixture of mud, gravel, and water in a pan. He looks for gold at the bottom because gold
 a. dissolves in water and sinks
 b. has a high density and sinks
 c. is less dense than rocks
 d. is repelled by the other minerals

5. Which of the following metals will sink in a pool of mercury ($d = 13.534\ g/cm^3$)?
 a. aluminum ($d = 2.6\ g/cm^3$)
 b. iron ($d = 7.9\ g/cm^3$)
 c. lead ($d = 11.34\ g/cm^3$)
 d. uranium ($d = 19.5\ g/cm^3$)

6. If the volume of a rock is 80.0 cm³ and its mass is 160.0 g, its density is
 a. 0.05 g/cm³   **b.** 0.50 g/cm³
 c. 2.0 g/cm³   **d.** 20 g/cm³

1.9 Energy: Heat and Temperature

Learning Objectives • Distinguish between heat and temperature. • Explain how the temperature scales are related.

The physical and chemical changes that matter undergoes are almost always accompanied by changes in energy. **Energy** is required to make something happen that wouldn't happen by itself. Energy is the ability to change matter, either physically or chemically.

Two important concepts in science that are related, and sometimes confused, are *temperature* and *heat*. When two objects at different temperatures are brought together, heat flows from the warmer to the cooler object until both are at the same temperature. **Heat** is energy on the move, the energy that flows from a warmer object to a cooler one. **Temperature** is a measure of how hot or cold an object is. Temperature tells us the direction in which heat will flow; heat automatically flows from a higher-temperature region to a lower-temperature region. For example, if you touch a hot test tube, heat will flow from the tube to your hand. If the tube is hot enough, your hand will be burned. Figure 1.10 illustrates the difference between heat and temperature.

▲ **Figure 1.10** Temperature and heat are different phenomena. Both the baby's bathtub and the hot tub are at roughly the same temperature—about 40 °C—but it took much more heat to warm the hot tub to that temperature than it did the baby's bathtub.

The SI base unit of temperature is the **kelvin (K)**. For laboratory work, we often use the more familiar **Celsius (°C)** scale. On this temperature scale, the freezing point of water is 0 degrees Celsius and the boiling point is 100 °C. The interval between these two reference points is divided into 100 equal parts, each a *degree Celsius*. The Kelvin scale is called an *absolute scale* because its zero point is the coldest temperature possible, or absolute zero. (This fact was determined by theoretical considerations and has been confirmed by experiment, as we will see in Chapter 6.)

The zero point on the Kelvin scale, 0 K, is equal to −273.15 °C, which we often round to −273 °C. A kelvin is the same size as a degree Celsius, so the freezing point of water on the Kelvin scale is 273 K. The Kelvin scale has no negative temperatures, and we don't use a degree sign with the K. To convert from degrees Celsius to kelvins, simply add 273.15 to the Celsius temperature.

$$K = °C + 273.15$$

EXAMPLE 1.11 Temperature Conversions

Diethyl ether boils at 36 °C. What is the boiling point of this liquid on the Kelvin scale?

Solution

$$K = °C + 273.15$$

$$K = 36 + 273.15 = 309 \text{ K}$$

⟩ **Exercise 1.11A**
What is the boiling point of ethanol (78 °C) expressed in kelvins?

⟩ **Exercise 1.11B**
At atmospheric pressure, liquid nitrogen boils at −196 °C. Express that temperature in kelvins.

The Fahrenheit temperature scale is widely used in the United States. Figure 1.11 compares the three temperature scales. Note that on the Fahrenheit scale, there are 180 Fahrenheit degrees between the freezing point and the boiling point of water. On the Celsius scale, there are 100 Celsius degrees between these two points. Thus, 1 Celsius degree equals 1.8 Fahrenheit degrees. Conversions between Fahrenheit and Celsius temperatures use this relationship, which is discussed in Appendix A.5.

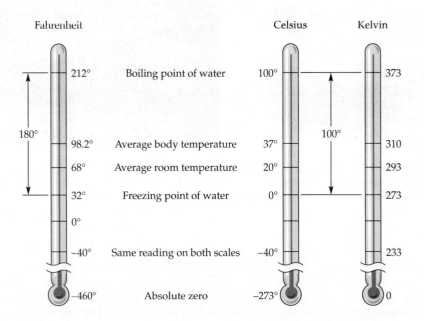

▶ **Figure 1.11** A comparison of the Fahrenheit, Celsius, and Kelvin temperature scales.

The SI-derived unit of energy is the **joule (J)**, but the **calorie (cal)** is the more familiar unit in everyday life.

$$1 \text{ cal} = 4.184 \text{ J}$$

A calorie is the amount of heat required to raise the temperature of 1 g of water by 1 °C.

The "calorie" used for expressing the energy content of foods is actually a **kilocalorie (kcal)**.

$$1000 \text{ cal} = 1 \text{ kcal} = 4184 \text{ J}$$

A dieter might be aware that a banana split contains 1500 "calories." If the dieter realized that this was really 1500 *kilo*calories, or 1,500,000 calories, giving up the banana split might be easier!

 EXAMPLE 1.12   Energy Conversions

When 1.00 g of gasoline burns, it yields about 10.3 kcal of energy. What is this quantity of energy in kilojoules (kJ)?

Solution

$$10.3 \text{ kcal} \times \frac{1000 \text{ cal}}{1 \text{ kcal}} \times \frac{4.184 \text{ J}}{1 \text{ cal}} \times \frac{1 \text{ kJ}}{1000 \text{ J}} = 43.1 \text{ kJ}$$

〉 Exercise 1.12A
Each day, the average European woman consumes food with an energy content of 7525 kJ. What is this daily intake in kilocalories (food calories)?

〉 Exercise 1.12B
It takes about 12 kJ of energy to melt a single ice cube. How much energy in kilocalories does it take to melt a tray of 12 ice cubes?

SELF-ASSESSMENT Questions

1. Heat is
   a. an element
   b. the energy flow from a hot object to a cold one
   c. a measure of energy intensity
   d. a temperature above room temperature

2. The SI scale for temperature is known as the
   a. Celsius scale   b. Fahrenheit scale
   c. Kelvin scale   d. Richter scale

3. Water boils at
   a. 0 °C
   b. 100 °F
   c. 273 °C
   d. 373 K

4. To convert a temperature on the Kelvin scale to degrees Celsius, we
   a. add 100
   b. add 273
   c. subtract 100
   d. subtract 273

5. The SI unit of energy is the
   a. calorie
   b. coulomb
   c. joule
   d. watt

Normal Body Temperature

Carl Wunderlich (1815–1877), a German physician, first recognized that fever is a symptom of disease. He recorded thousands of temperature measurements from healthy people and reported the average value as 37 °C. When this value was converted to the Fahrenheit scale, it somehow (improperly) acquired an extra significant figure, and 98.6 °F became widely (and incorrectly) known as the *normal body temperature*. In recent years, millions of measurements have revealed that *average* normal body temperature is actually 98.2 °F and that body temperatures in healthy people range from 97.7 °F to 99.5 °F.

1.10 Critical Thinking

Learning Objective • Use critical thinking to evaluate claims and statements.

One of the hallmarks of science is the ability to think critically—to evaluate statements and claims in a rational, objective fashion. For example, a diet plan advertisement shows people who have lost a lot of weight. Somewhere in the ad, in tiny print, you may find a statement such as "results may vary" or "results not typical." Such ads also often advise combining the plan with exercise.

Does an ad like this prove that a diet plan works? You can use critical thinking to evaluate the ad's claim. **Critical thinking** involves gathering facts, assessing them, using logic to reach a conclusion, and then evaluating the conclusion. The ability to think critically can be learned, and it will serve you well in everyday life as well as in science courses. Critical thinking is important for workers in all types of professions and employees in all kinds of industries. We will use an approach adapted from one developed by James Lett.[1] This approach is outlined below and followed by several examples.

You can use the acronym **FLaReS** (ignore the vowels) to remember four principles used to test a claim: *f*alsifiability, *l*ogic, *r*eplicability, and *s*ufficiency.

- **Falsifiability:** Can any conceivable evidence show the claim to be false? It must be possible to think of evidence that would prove the claim false. A hypothesis that cannot be falsified is of no value. Science cannot prove anything true in an absolute sense, although it can provide overwhelming evidence. Science *can* prove something false.

- **Logical:** Any argument offered as evidence in support of a claim must be sound. An argument is sound if its conclusion follows inevitably from its premises and if its premises are true. It is unsound if a premise is false, and it is unsound if there is a single exception in which the conclusion does not necessarily follow from the premises.

- **Replicability:** If the evidence for a claim is based on an experimental result, it is necessary that the evidence be replicable in subsequent experiments or trials. Scientific research is almost always reviewed by other qualified scientists. The peer-reviewed research is then published in a form that enables others to repeat the experiment. Sometimes, bad science slips through the peer-review process, but such results eventually fail the test of replicability. For example, in May 2005 the journal *Science* published research by Korean biomedical scientist Hwang Woo-suk in which he claimed to have tailored embryonic stem cells so that every patient could receive custom treatment. Others were unable to reproduce Hwang's findings, and evidence emerged that he had intentionally fabricated results. *Science* retracted the article in January 2006. Research results published in media that are not peer reviewed have little or no standing in the scientific community—nor should you rely on them.

- **Sufficiency:** The evidence offered in support of a claim must be adequate to establish the truth of that claim, with these stipulations: (1) The burden of evidence for any claim rests on the claimant, (2) extraordinary claims demand extraordinary evidence, and (3) evidence based on an authority figure or on testimony is never adequate.

If a claim passes all four FLaReS tests, then it *might* be true. On the other hand, it could still be proven false. However, if a claim fails even one of the FLaReS tests, it is likely to be false. The FLaReS method is a good starting point for evaluating the myriad claims that you will find on the Internet and elsewhere. We apply one or more of the FLaReS principles to evaluate a claim in each of the following examples.

---

[1]Lett used the acronym FiLCHeRS (ignore the vowels) as a mnemonic for six rules: falsifiability, logic, comprehensiveness, honesty, replicability, and sufficiency. See Lett, James, "A Field Guide to Critical Thinking," *The Skeptical Inquirer*, Winter 1990, pp. 153–160.

GREEN CHEMISTRY Green Chemistry: Reimagining Chemistry for a Sustainable World

Jennifer MacKellar and David Constable, ACS Green Chemistry Institute®

Principles 1, 2, 3, 4, 5, 6, 7, 8, 9, 10, 11, 12

Learning Objectives • Define green chemistry. • Describe how green chemistry reduces risk and prevents environmental problems.

The famous naturalist, John Muir, once said, "When one tugs at a single thing in nature, he finds it attached to the rest of the world." In a world of shifting climate and unprecedented environmental disasters, we realize that nothing could be truer. But what can we do to change the current path? What would a sustainable planet look like?

Imagine a community where human society lives in harmony with nature. People enjoy good health without diminishing the resources that support them. Land, water, and air are managed in ways that retain the ecological value of the environment. Sustainable communities realize the importance of balancing economic, environmental, and social issues and thrive because of it.

The twentieth century saw enormous technical advances in medicine, manufacturing, and chemical processing that led to an increased quality of life in the United States. However, this standard of living has taken a significant toll on the environment and human health. Improper chemical use and disposal and general environmental pollution have significantly degraded the world's ecosystems, threatening the security of future generations. Since the publication of Rachel Carson's Silent Spring in 1962 (Section 1.1) and the birth of the environmental movement, an increasing number of governmental regulations focus on pollution prevention and environmental protection.

But what if we could prevent hazardous waste altogether? What if we could design chemicals that would meet our needs and desires without destroying our environment? Sustainable and green chemistry, in simple terms, is just a different way of thinking about how chemistry and chemical engineering can be done. Over the years, different principles have been proposed that can be used when thinking about the design, development, and implementation of chemical products and processes. These principles enable scientists and engineers to protect and benefit the economy, people, and the planet by finding creative and innovative ways to reduce waste, conserve energy, and replace hazardous substances. (See the inside front cover.)

It is important to note that green chemistry and engineering principles should be applied across the life of a product. Doing so means the use of more sustainable or renewable feedstocks all the way through to designing for end of life or the final disposition of the product. Green chemistry is applicable across all chemistry disciplines including organic, inorganic, analytical, and physical chemistry and biochemistry.

Governments and scientific communities throughout the world recognize that the practice of green chemistry leads not only to a cleaner and more sustainable Earth but also to economic benefits and positive social impacts. These benefits encourage businesses and government to support development of greener products and processes. The United States, specifically, has recognized green chemistry innovations since 1996 through an annual Presidential Green Chemistry Challenge Award program.

In 2016, Verdezyne, Inc. won a Presidential Green Chemistry Challenge Award for developing a yeast that can produce a chemical to make high performance nylon 6,12 for hair brushes, tooth brushes, coatings, fragrances and automotive and aviation oils. They started with a plant-based feedstock with low greenhouse gas emissions.

The use of green-chemistry practices makes a more sustainable future possible, but it is important to consider, too, that today's crises call for innovative thinking and a willingness to change. Albert Einstein said it best, "Problems cannot be solved at the same level of awareness that created them." Despite all of the research advances in green chemistry, mainstream business has yet to fully embrace greener and more sustainable technology. Green chemists are working very hard to take their innovations out of the lab and into the boardroom by creating products that can be embraced by today's industry leaders and consumers.

Sustainability and the health of our environment have an impact on all of us. The environment spans age, race, income levels, and cultural differences, producing an impact on all of life. Green chemistry is a solutions-based approach to addressing some of our pressing global challenges.

▲ In 2013, the Elevance biorefinery in Gresik, Indonesia, began producing specialty-chemical products commercially. The feedstocks for the biorefinery are renewable—mainly vegetable oils.

CRITICAL THINKING Examples

1.1. A practitioner claims outstanding success in curing cancer with a vaccine made from a patient's own urine. However, the vaccine works only if the patient's immune system is still sufficiently strong. Is the claim falsifiable?

Solution

> *No. The practitioner can always say that the patient's immune system was no longer sufficiently strong.*

1.2. A website claims, "The medical-industrial complex suppresses natural treatments because they cure the patient. A cured patient no longer needs a doctor, and the doctor loses income." Is the claim logical?

Solution

> *No. Doctors do not cure all patients. Many patients remain ill and some die.*

1.3. Many "psychics" claim to have precognition—the ability to predict the future. These predictions often are made near the end of one year and concern the next year. Apply all the FLaReS tests to evaluate such claims of premonitions of these psychics.

Solution

> 1. *Are the predictions falsifiable? Yes. You could keep the written prediction and then check it yourself at the end of the year.*
> 2. *Is precognition logical? No. If psychics could really predict the future, they should win lotteries, ward off all kinds of calamities by providing advance warning, and so on, far more often than the average person.*
> 3. *Are such predictions reproducible? No. Psychics often make broad, general predictions, such as "Tornadoes will hit Oklahoma in April and May" or "A prominent politician will lose his office in a stunning upset." (Tornadoes in Oklahoma are quite common in those months, and few years have gone by without an election upset.) The psychics later make claim-specific references to "hits" while ignoring "misses."*
> 4. *Is the claim sufficient? No. Precognition has never been demonstrated in proper testing. Any such claim is extraordinary and would require extraordinary evidence. The burden of proof for the claim rests on the claimant.*

 Summary

Section 1.1—**Technology** is the practical application of knowledge, while **science** is an accumulation of knowledge about nature and our physical world, along with our explanations of that knowledge. The roots of chemistry lie in **alchemy**, a mixture of chemistry and magic practiced from the Middle Ages to the 1700s. **Green chemistry** uses materials and processes that are intended to prevent or reduce pollution at its source. **Sustainable chemistry** is designed to meet the needs of the present generation without compromising the needs of future generations.

Section 1.2—Science is *reproducible, testable, predictive, explanatory,* and *tentative.* A set of confirmed observations about nature may lead to a **hypothesis**—a tentative explanation of the observations. A **scientific law** is a brief statement that summarizes large amounts of data and is universally true. There are very few laws and many theories. If a hypothesis stands up to testing and further experimentation, it may become a **scientific theory**—the best current explanation for a phenomenon. A theory is always tentative and can be rejected if it does not continue to stand up to further testing. A valid theory can be used to predict new scientific facts. Scientists disagree over social and political issues partly because of the inability to control **variables**, factors that can change during an experiment.

Section 1.3—Science and technology have provided many benefits for our world, but they have also introduced new

risks. A **risk–benefit analysis** can help us decide which outweighs the other.

Section 1.4—Most chemists are involved in some kind of research. The purpose of **applied research** is to make useful products or to solve a particular problem, while **basic research** is carried out simply to obtain new knowledge or to answer a fundamental question. Basic research often also leads to a useful product or process.

Section 1.5—Chemistry is the study of matter and the changes it undergoes. **Matter** is anything that has mass and takes up space. **Mass** is a measure of the amount of matter in an object, while **weight** represents the gravitational force of attraction for an object. **Physical properties** of matter can be observed without making new substances. When **chemical properties** are observed, new substances are formed. A **physical change** does not entail a change in chemical composition. A **chemical change** does involve such a change.

Section 1.6—Matter can be classified according to physical state. A **solid** has definite shape and volume. A **liquid** has definite volume but takes the shape of its container. A **gas** takes both the shape and the volume of its container. Matter also can be classified according to composition. A **substance** always has the same composition, no matter how it is made or where it is found. A **mixture** can have different compositions depending on how it is prepared. Substances are either elements or compounds. An **element** is composed of atoms of one type, an **atom** being the smallest characteristic particle of an element. There are more than a hundred elements, each of which is represented by a **chemical symbol** consisting of one or two letters derived from the element's name. A **compound** is made up of two or more elements, chemically combined in a fixed ratio. Most compounds exist as **molecules**, groups of atoms bound together as a unit.

Section 1.7—Scientific measurements are expressed with **SI units**, which comprise an agreed-on standard version of the metric system. Base units include the **kilogram (kg)** for mass and the **meter (m)** for length. Although it is not the SI standard unit of volume, the **liter (L)** is widely used. Prefixes make basic units larger or smaller by factors of 10.

Section 1.8—**Density** can be thought of as "how heavy something is for its size." It is the amount of mass per unit volume.

$$d = \frac{m}{V}$$

Density can be used as a conversion factor. The density of water is almost exactly 1 g/mL.

Section 1.9—When matter undergoes a physical or chemical change, a change in **energy** is involved. Energy is the *ability* to cause these changes. **Heat** is energy flow from a warmer object to a cooler one. **Temperature** is a measure of how hot or cold an object is. The **kelvin (K)** is the SI base unit of temperature, but the **Celsius (°C)** scale is more commonly used. On the Celsius scale, water boils at 100 °C and freezes at 0 °C. The SI unit of heat is the **joule (J)**. A more familiar unit, the **calorie (cal)**, is the amount of heat needed to raise the temperature of 1 g of water by 1 °C. The food "calorie" is actually a **kilocalorie (kcal)**, or 1000 calories.

Section 1.10—Claims may be tested by applying critical thinking. The acronym **FLaReS** represents four principles used to test a claim: *f*alsifiability, *l*ogical, *r*eplicability, and *s*ufficiency.

GREEN CHEMISTRY—Green chemistry means preventing waste, using nonhazardous chemicals and renewable resources, conserving energy, and making products that degrade after use. Green chemistry takes into account the sources and ultimate destinations of chemicals, and the impact they will have on humans and the planet.

Learning Objectives

Learning Objectives	Associated Problems
• Define *science, chemistry, technology,* and *alchemy*. (1.1)	1, 4
• Describe the importance of green chemistry and sustainable chemistry (1.1)	43, 44
• Define *hypothesis, scientific law, scientific theory,* and *scientific model,* and explain their relationships in science. (1.2)	2, 3, 71, 72, 74, 75
• Define *risk* and *benefit*, and give an example of each. (1.3)	5
• Estimate a desirability quotient from benefit and risk data. (1.3)	7, 13–18
• Distinguish basic research from applied research. (1.4)	8, 9
• Differentiate: mass and weight; physical and chemical change; and physical and chemical properties. (1.5)	29–32
• Classify matter in three ways. (1.6)	33–44
• Assign proper units of measurement to observations and manipulate units in conversions. (1.7)	10–12, 19–28, 45–52, 69, 70, 73, 78–80
• Calculate the density, mass, or volume of an object given the other two quantities. (1.8)	53–64, 76, 77, 81–88
• Distinguish between heat and temperature. (1.9)	67, 68
• Explain how the temperature scales are related. (1.9)	65, 66
• Use critical thinking to evaluate claims and statements. (1.10)	6
• Define green chemistry.	95
• Describe how green chemistry reduces risk and prevents environmental problems.	96–98

Conceptual Questions

1. State five distinguishing characteristics of science. Which characteristic best serves to distinguish science from other disciplines?

2. Why do experiments have to be done to support a hypothesis?

3. Why can't scientific methods always be used to solve social, political, ethical, and economic problems?

4. How does technology differ from science?

5. What is risk–benefit analysis?

6. What sorts of judgments go into **(a)** the evaluation of benefits and **(b)** the evaluation of risks?

7. What is a DQ? What does a large DQ mean? Why is it often difficult to estimate a DQ?

8. What derived units of **(a)** mass and **(b)** length are commonly used in the laboratory?

9. What is the SI-derived unit for volume? What volume units are most often used in the laboratory?

10. Following is an incomplete table of SI prefixes, their symbols, and their meanings. Fill in the blank cells. The first row is completed as an example.

Prefix	Symbol	Definition
tera-	T	10^{12}
_____	M	_____
centi-	_____	_____
_____	μ	_____
milli-	_____	_____
_____	_____	10^{-1}
_____	K	_____
nano-	_____	_____

11. Identify the following research projects as either applied or basic.
   a. A Virginia Tech chemist tests a method for analyzing coal powder in less than a minute, before the powder goes into a coal-fired energy plant.
   b. A Purdue engineer develops a method for causing aluminum to react with water to generate hydrogen for automobiles.
   c. A worker at University of Illinois Urbana-Champaign examines the behavior of atoms at high temperatures and low pressures.

12. Identify the following work as either applied research or basic research.
   a. An engineer determines the strength of a titanium alloy that will be used to construct notebook-computer cases.
   b. A biochemist runs experiments to determine how oxygen binds to the blood cells of lobsters.
   c. A biologist determines the number of eagles that nest annually in Kentucky's Land Between the Lakes area over a six-year period.

 Problems

A Word of Advice *You cannot learn to work problems by reading them or by watching your instructor work them, just as you cannot become a piano player solely by reading about piano-playing skills or by attending performances. Working problems will help you improve your understanding of the ideas presented in the chapter and your ability to synthesize concepts, as well as allow you to practice your estimation skills. Plan to work through the great majority of these problems.*

Risk–Benefit Analysis

13. Penicillin kills bacteria, thus saving the lives of thousands of people who otherwise might die of infectious diseases. Penicillin causes allergic reactions in some people, which can lead to death if the resulting condition is not treated. Do a risk–benefit analysis of the use of penicillin by society as a whole.

14. Sodium sulfite is a compound sometimes added to dried fruit, some fruit juices, and wines as a preservative. About 1% of the population is sensitive to sulfites. Consider the use of sulfites in **(a)** fruit juice and **(b)** wine. How might these uses differ in risk–benefit analysis?

15. Paints used on today's automobiles are often composed of two components and harden several hours after mixing. One component is an *isocyanate* that can sensitize a person if it is breathed (even in small amounts) over a long time period, causing an allergic reaction. Do a risk–benefit analysis of the use of these paints by

(a) an automotive factory worker and **(b)** a hobbyist who is restoring a 1965 Mustang. How might the DQ for both of these people be increased?

16. X-rays are widely used in medicine and dentistry. Do a risk–benefit analysis for **(a)** a dental patient who has one set of X-rays done per year and **(b)** a dentist who stays in the room with a patient while X-rays are taken, three times each day.

17. MRSA (methicillin-resistant *staphylococcus aureus* bacteria) infections are becoming more common as the prescribing of antibiotics increases. **(a)** Is the DQ for the routine use of antibiotics to treat sore throats high or low? Justify your answer briefly. **(b)** How does this DQ differ from the DQ for the use of antibiotics to treat influenza (higher or lower)? Justify your answer briefly.

18. The virus called HIV causes AIDS, a devastating and often deadly disease. Several drugs are available to treat HIV infection, but all are expensive. Used separately and in combination, these drugs have resulted in a huge drop in AIDS deaths. An expensive new drug shows promise for treating patients with HIV/AIDS, especially in preventing passage of the HIV virus from a pregnant woman to her fetus. What is the DQ for administering this drug to **(a)** a man who thinks he may be infected with HIV, **(b)** a pregnant woman who is HIV positive, and **(c)** an unborn child whose mother has AIDS?

Mass and Weight

19. Which are realistic masses for a cellular telephone and a laptop computer, respectively?

100 mg; 100 g   100 g; 100 mg   100 g; 2 kg   1 kg; 2 kg

20. In Europe, A2-sized paper measures 594 mm × 420 mm. Is this larger or smaller than the standard $8\frac{1}{2}$ in. × 11 in. paper used in the United States?

21. Which of the following are likely to be paired incorrectly? Add or remove zeroes to the incorrect masses to make them more reasonable: **(a)** large paper clip, 1000 mg; **(b)** pair of wire-frame glasses, 20 g; **(c)** carpet covering a large living room, 10 kg; **(d)** case of bottled water, 1 kg.

22. Sample X on the moon has exactly the same mass as Sample Y on Earth. Do the two samples weigh the same? Explain.

Length, Area, and Volume

23. Which of the following is a reasonable volume for a teacup?

25 mL   250 mL   2.5 L   25 L

24. Which of the following is a reasonable approximation for the volume of a peanut?

$0.5\ \text{cm}^3$ $50\ \text{cm}^3$ $500\ \text{cm}^3$ $5\ \text{m}^3$

25. Earth's oceans contain $3.50 \times 10^8\ \text{mi}^3$ of water and cover an area of $1.40 \times 10^8\ \text{mi}^2$. What is the volume of ocean water in cubic kilometers? (Hint: Refer to the discussion of area and volume on page 22 to convert from cubic miles to cubic kilometers.)

26. What is the area of Earth's oceans in square kilometers? See Problem 25.

27. Consider the two tubes shown at right. The aluminum tube has an outside diameter of 0.998 in. and an inside diameter of 0.782 in. Without doing a detailed calculation, determine whether the aluminum tube will fit inside another aluminum tube with an inside diameter of 26.3 mm.

28. Which of the following could be the inside diameter of the paper tube shown above (with Problem 27): 19.9 mm, 24.9 mm, and/or 18.7 mm? Justify your response.

Physical and Chemical Properties and Changes

29. Identify the following as physical or chemical properties.
   a. Unlike ordinary glass, tempered glass shatters into rounded bits.
   b. Bread grows mold when it is allowed to stand in the open.
   c. Hydrogen cyanide is toxic because it reacts with hemoglobin in the blood.
   d. ABS plastic filament melts easily in a 3D printer.

30. Identify the following as physical or chemical properties.
   a. Milk chocolate melts in the mouth.
   b. An iron nail rusts when it gets wet.
   c. Titanium chips burn with a brilliant white light.
   d. Yogurt is made by allowing a bacterial culture to ferment in warm milk.

31. Identify the following changes as physical or chemical.
   a. Bits of yellow plastic are melted and forced into a mold to make a yellow toy.
   b. Waste oil from restaurants is converted to biodiesel fuel that burns differently from cooking oil.
   c. Orange juice is prepared by adding three cans of water to one can of orange juice concentrate.

32. Identify the following changes as physical or chemical.
   a. Bacteria are killed by chlorine added to a swimming pool.
   b. Brown print is created in an inkjet printer by mixing cyan, magenta, and yellow inks.
   c. Sugar is dissolved in water to make syrup.

Substances and Mixtures

33. Identify each of the following as a substance or a mixture.
   a. adhesive tape   **b.** uranium in an atomic bomb
   c. distilled water   **d.** carbon dioxide gas

34. Identify each of the following as a substance or a mixture.
   a. oxygen   **b.** sodium sulfite
   c. smog   **d.** a carrot
   e. pure milk chocolate **f.** rabbit fur

35. Which of the following mixtures are homogeneous, and which are heterogeneous?
   a. cashews   **b.** window glass
   c. an envelope   **d.** Scotch whiskey

36. Which of the following mixtures are homogeneous, and which are heterogeneous?
   a. gasoline
   b. Italian salad dressing
   c. rice pudding
   d. an intravenous glucose solution

37. Every sample of the sugar glucose (no matter where on Earth it comes from) consists of 8 parts (by mass) oxygen, 6 parts carbon, and 1 part hydrogen. Is glucose a substance or a mixture? Explain.

38. An advertisement for shampoo says, "Pure shampoo, with nothing artificial added." Is this shampoo a substance or a mixture? Explain.

Elements and Compounds

39. Which of the following represent elements, and which represent compounds?
   a. KF   **b.** Fe   **c.** F   **d.** Fr

40. Which of the following represent elements, and which represent compounds?
   a. Li   **b.** CO   **c.** Cf   **d.** CF_4

41. Without consulting tables, name each of the following.
   a. C   **b.** Mg   **c.** He   **d.** N

42. Without consulting tables, write a symbol for each of the following.
   a. oxygen   **b.** phosphorus
   c. potassium   **d.** argon

43. In his 1789 textbook, *Traité élémentaire de Chimie*, Antoine Lavoisier (Chapter 2) listed 33 known elements, one of which was *baryte*. Which of the following observations best shows that baryte cannot be an element?
   a. Baryte is insoluble in water.
   b. Baryte melts at 1580 °C.

c. Baryte has a density of 4.48 g/cm³.

d. Baryte is formed in hydro-thermal veins and around hot springs.

e. Baryte is formed as a solid when sulfuric acid and barium hydroxide are mixed.

f. Baryte is formed as a sole product when a particular metal is burned in oxygen.

44. A yellow liquid is heated over a Bunsen burner. A brown gas is given off, and the yellow liquid becomes colorless. Which statement is correct?

a. The yellow liquid must be a compound.

b. The brown gas must be an element.

c. The yellow liquid could be either a mixture or a compound.

d. The brown gas must be a compound.

The Metric System: Measurement and Unit Conversion

45. Change the unit for each of the following measurements by replacing the power of 10 with an appropriate SI prefix.

a. 4.4×10^{-9} s   b. 8.5×10^{-2} g   c. 3.38×10^{6} m

46. Convert each of the following measurements to the SI base unit.

a. 45 mg   b. 125 ns
c. 10.7 μL   d. 12.5346 kg

47. Carry out the following conversions.

a. 5.52×10^{4} mL to L   b. 325 mg to g
c. 27 cm to m   d. 27 mm to cm
e. 78 μs to ms

48. Carry out the following conversions.

a. 546 mm to m   b. 65 ns to μs
c. 87.6 mg to kg   d. 46.3 dm³ to L
e. 181 pm to μm

49. Indicate which is the larger unit in each pair.

a. mm or cm   b. kg or g   c. dL or μL

50. There are about 7,000,000,000,000,000,000,000,000,000,000 (7 octillion) atoms in an adult human body. What is this quantity in scientific notation?

51. Express the mass of a 1.4 kg book in **(a)** mg, **(b)** g, and **(c)** ng.

52. What is the volume in liters of **(a)** a 352 mL soft drink and **(b)** a 26 kL tank of gasoline?

53. A typical four-year-old child is about _____ tall.

a. 10 centimeters   b. 10 millimeters
c. 100 meters   d. 1 meter

54. The wing of a housefly weighs about _____.

a. 10 μg   b. 200 mg   c. 0.5 kg   d. 50 kg

55. In 2015 the Lucara Diamond Company discovered a 1111-carat diamond (below left), the second-largest gem-quality diamond ever found, at its mine in Botswana, Africa. However, it is only about one-third the size of the Cullinan diamond, at 3106 ct uncut. Found in 1905 in the Premier Mine in Pretoria, South Africa, the Cullinan was cut into nine separate stones (one shown below right),

many of which are in the British Crown Jewels. What is the mass of each diamond **(a)** in grams and **(b)** in pounds? 1 carat (ct) = 200 mg.

56. A typical inkjet printer cartridge holds 1.20 fluid ounces (fl oz.) of ink. What is the volume of the ink **(a)** in milliliters and **(b)** in microliters? 1 fl oz. = 29.6 mL.

Density

(You may need data from Table 1.6 for some of these problems.)

57. It is found that 122 mL of canola oil weighs 112 g. What is the density in grams per milliliter of canola oil?

58. A nylon rod weighing 392 g has a volume of 341 cm³. What is the density of nylon?

59. The rubber in a pencil eraser has a density of 1.43 g/cm³. What is the mass of an eraser with a volume of 13.2 cm³?

60. A student measures 94.1 mL of concentrated nitric acid, which has a density of 1.51 g/mL. What is the mass of the acid?

61. What is the volume of **(a)** 227 g of hexane (in milliliters) and **(b)** a 454 g block of ice (in cubic centimeters)?

62. What is the volume of **(a)** a 475 g piece of copper (in cubic centimeters) and **(b)** a 253 g sample of mercury (in milliliters)?

63. A 40 mL quantity each of mercury and hexane, as well as 80 mL of water, is placed in the 250 mL beaker shown below. The three liquids do not mix with one another. Sketch the relative locations and levels of the three liquids in the container.

64. A piece of red maple wood (d = 0.49 g/cm³), a piece of balsa wood (d = 0.11 g/cm³), a copper coin, a gold coin, and a piece of ice are dropped into the mixture of Problem 63. Where will each solid end up in the beaker?

65. A metal stand specifies a maximum load of 450 lb. Will it support an aquarium that weighs 59.5 lb. and is filled with 37.9 L of seawater? $d = 1.03$ g/mL.

66. E85 fuel is a mixture of 85% ethanol and 15% gasoline by volume. What mass in kilograms of E85 ($d = 0.758$ g/mL) can be contained in a 14.0 gal tank?

67. A "crystal ball" is 148 mm in diameter and is made of glass that has a density of 3.18 g/cm³. What is the mass of this object? (The volume of a sphere is $4\pi r^3/3$ where r is the radius.)

68. The base of a paper carton of skim milk is 77 mm × 77 mm. Skim milk has a density of 1.031 g/mL. How tall must the carton be to hold 0.500 kg of skim milk?

Energy: Heat and Temperature

69. Liquid nitrogen, used for freezing sperm samples, has a temperature of 77 K. What is this temperature in Celsius?

70. Normal body temperature is about 37 °C. What is this temperature in kelvins?

71. The label on a 100 mL container of orange juice packaged in New Zealand reads in part, "energy . . . 161 kJ." What is that value in kilocalories ("food calories")?

72. To vaporize 1.00 g of sweat (water) from your skin, the water must absorb 584 calories. What is that value in kilojoules?

Expand Your Skills

73. Evaluate each of the following as possible scientific hypotheses.
    a. If the temperature of a cup of tea is increased, then the quantity of sugar that can be dissolved in it will be increased.
    b. If the rate of photosynthesis, as measured by the quantity of oxygen produced, is related to the wavelength (color) of light, then light of different colors will cause a plant to make different quantities of oxygen.
    c. If the rate of metabolism in animals is related to the temperature, then raising the surrounding temperature will cause an increase in animal metabolism.
    d. If I meditate hard enough, I will pass this chemistry exam.

74. The nucleus of a hydrogen atom is 1.75 fm in diameter. The atom is larger than the nucleus by a factor of about 145,000. **(a)** Use exponential notation to express each measurement in terms of an SI base unit. **(b)** What is the volume of a hydrogen nucleus in fm³? Of a hydrogen atom in nm³? The volume of a sphere of radius $r = 4/3\,\pi r^3$.

75. A certain chemistry class is 1.00 microcenturies (μcen) long. What is its length in minutes?

76. A unit of beauty, a *helen*, thought to have been invented by British mathematician W.A.H. Rushton, is based on Helen of Troy (from Christopher Marlowe's play *Doctor Faustus*, which described her as having "the face that launched a thousand ships"). How many ships could be launched by a face with 1.00 millihelens of beauty?

77. English chemist William Henry studied the amounts of gases absorbed by water at different temperatures and under different pressures. In 1803, he stated a formula, $p = k_Hc$, which related the concentration of the dissolved gas at a constant temperature to the partial pressure of that gas in equilibrium with that liquid. This relationship describes a scientific
    a. hypothesis   b. law   c. observation   d. theory

78. A classmate is planning to try a new diet. On this diet he can eat anything he wants, but he must swallow a spoonful of crushed ice (about 5 grams) with each bite of food. His reasoning is (a) it takes about 79 calories of heat to melt one gram of ice, and (b) swallowing the ice and allowing it to melt in his stomach "uses up" that same amount of food energy. Although both (a) and (b) are correct, the diet does not work. Explain. (*Hint*: See page 30 for a discussion of energy units, then calculate the amount of energy in food calories needed to melt a kilogram of ice.)

79. A particular brand of epoxy glue is used by mixing two volumes of liquid epoxy resin (density 2.25 g/mL) with one volume of liquid hardener (density 0.94 g/mL) before application. If the epoxy glue is to be prepared by mass rather than volume, what mass in grams of hardener must be mixed with 10.0 g of resin?

For Problems 80 and 81, classify each numbered statement as (a) an experiment, (b) a hypothesis, (c) a scientific law, (d) an observation, or (e) a theory. (It is not necessary to understand the science involved to do these problems.)

80. In the early 1800s, many scientific advances came from the study of gases. (1) For example, Joseph Gay-Lussac reacted hydrogen and oxygen to produce water vapor, and he reacted nitrogen and oxygen to form either dinitrogen oxide (N_2O) or nitrogen monoxide (NO). Gay-Lussac found that hydrogen and oxygen react in a 2:1 volume ratio and that nitrogen and oxygen can react in 2:1 or 1:1 volume ratios depending on the product. (2) In 1808, Gay-Lussac published a paper in which he stated that the relative volumes of gases in a chemical reaction are present in the ratio of small integers provided that all gases are measured at the same temperature and pressure. (3) In 1811, Amedeo Avogadro proposed that equal volumes of all gases measured at the same temperature and pressure contain the same number of molecules. (4) By midcentury, Rudolf Clausius, James Clerk Maxwell, and others had developed a detailed rationalization of the behavior of gases in terms of molecular motions.

81. Potassium bromate acts as a conditioner in dough made from wheat flour. Flour contains a protein called *gluten*. Proteins consist of amino-acid units linked in long chains. (1) Dough to which potassium bromate has been added rises better, producing a lighter, larger-volume loaf. (2) Potassium bromate acts by oxidizing the gluten and

cross-linking tyrosine (an amino acid) units of the protein molecules, enabling the dough to retain gas better. (3) A scientist wonders if sodium bromate might cross-link separate tyrosine molecules, forming a dimer (a molecule consisting of two subunits). (4) She adds sodium bromate to a solution of tyrosine. (5) She notes that a precipitate (solid falling out of solution) is formed.

82. A graduated cylinder is found to weigh 88.32 g. When about 50 mL of water is added, the cylinder and water weigh 137.29 g. Next, about 50 mL of antifreeze is added, and the cylinder and its contents now weigh 192.79 g. The cylinder is now found to contain 91.1 mL of liquid. What is the density in g/mL of the cylinder's contents? (Volumes of liquids mixed together are not always additive.)

83. A block of balsa wood ($d = 0.11$ g/cm^3) is 7.6 cm × 7.6 cm × 94 cm. What is the mass of the block in grams?

84. Arrange the following in order of increasing length (shortest first): (1) a 1.21 m chain, (2) a 75 in. board, (3) a 3 ft., 5 in. rattlesnake, and (4) a yardstick.

85. Arrange the following in order of increasing mass (lightest first): (1) a 5 lb. bag of potatoes, (2) a 1.65 kg cabbage, and (3) 2500 g of sugar.

86. One of the people in the photo has a mass of 47.2 kg and a height of 1.53 m. Which one is it likely to be?

87. Some metal chips having a volume of 3.29 cm^3 are placed on a piece of paper and weighed. The combined mass is found to be 18.43 g. The paper itself weighs 1.21 g. Calculate the density of the metal.

88. A 10.5 in. (26.7 cm) iron skillet has a mass of 7.00 lb. (3180 g). What is its volume in cubic centimeters? (See Table 1.6.)

89. A 5.79 mg piece of gold is hammered into gold leaf of uniform thickness with an area of 44.6 cm^2. What is the thickness of the gold leaf? (See Table 1.6.)

90. A bundle of cylindrical stainless steel ($d = 7.48$ g/cm^3) rods weighs 1850 kg. Each rod is 6.1 m long and 25.4 mm in diameter. How many rods are in the bundle? The volume of a cylinder $= \pi r^2 h$, where r is the radius and h is the height.

91. What is the mass, in metric tons, of a cube of gold that is 36.1 cm on each side? (1 metric ton = 1000 kg.) (See Table 1.6.)

92. A collection of gold-colored metal beads has a mass of 425 g. The volume of the beads is found to be 48.0 cm^3. Using the given densities and those in Table 1.6, identify the metal: bronze ($d = 9.87$ g/cm^3), copper, gold, or nickel silver ($d = 8.86$ g/cm^3).

93. The density of a planet can be approximated from its radius and estimated mass. Calculate the approximate average density, in grams per cubic centimeter, of **(a)** Jupiter, which has a radius of about 7.0×10^4 km and a mass of 1.9×10^{27} kg; **(b)** Earth, radius of 6.4×10^3 km and mass of 5.98×10^{24} kg; and **(c)** Saturn, radius 5.82×10^4 km and mass 5.68×10^{26} kg. The volume of a sphere is $4\pi r^3/3$ where $r = $ radius. How does the density of each planet compare with that of water?

94. An artist's conception of extrasolar planet HAT-P-1 is shown at right. This planet orbits a star 450 light years from Earth. The planet has about half the mass of Jupiter, and its radius is about 1.38 times that of Jupiter. (See Problem 93.) Calculate the approximate average density, in grams per cubic centimeter, of HAT-P-1. How does this density compare with that of water?

95. Define green chemistry.

96. List six of the Twelve Principles of Green Chemistry.

97. How does green chemistry reduce risk? (Choose one.)
    a. by reducing hazard and exposure
    b. by avoiding the use of chemicals
    c. by increasing exposures to higher levels
    d. by reducing human errors

98. For companies implementing green chemistry, how does green chemistry save money?
    a. by lowering costs
    b. by using less energy
    c. by creating less waste
    d. all of the above

Critical Thinking Exercises

Apply knowledge that you have gained in this chapter and one or more of the FLaReS principles to evaluate the following statements and claims.

1.1 An alternative health practitioner claims that a nuclear power plant releases radiation at a level so low that it cannot be measured, but that this radiation is harmful to the thyroid gland. He sells a thyroid extract that he claims can prevent the problem.

1.2 A doctor claims that she can cure a patient of arthritis by simply massaging the affected joints. If she has the patient's complete trust, she can cure the arthritis within a year. Several of her patients have testified that the doctor has cured their arthritis.

1.3 German physicist Jan Hendrik Schön of Bell Labs claimed in 2001 that he had produced a molecular-scale transistor. Other scientists noted that his data contained irregularities.

Attempts to perform experiments similar to Schön's did not give similar results.

1.4 A mercury-containing preservative called thiomersal was used to prevent bacterial and fungal contamination in vaccines from the 1930s until its use was phased out starting in 1999. In the United States, a widespread concern arose that these vaccines caused autism. There was no such concern in the United Kingdom, although the vaccine with the same preservative was used there. Is the concern valid?

1.5 A woman claims that she has memorized the New Testament. She offers to quote any chapter of any book entirely from memory.

1.6 The label on a bag of Morton solar salt crystals has the statement "No chemicals or additives."

Collaborative Group Projects

Prepare a PowerPoint, poster, or other presentation (as directed by your instructor) to share with the class.

1. Prepare a brief biographical report on one of the following.
 a. Frederick Sanger
 b. Rachel Carson
 c. Albert Einstein
 d. George Washington Carver

(The following problem is best done in a group of four students.)

2. Make copies of the following form. Student 1 should write a word from the list in the first column of the form and its definition in the second column. Then, she or he should fold the first column under to hide the word and pass the sheet to Student 2, who uses the definition to determine what word was defined and place that word in the third column. Student 2 then folds the second column under to hide it and passes the sheet to Student 3, who writes a

definition for the word in the third column and then folds the third column under and passes the form to Student

3. Finally, Student 4 writes the word corresponding to the definition given by Student 3.

Compare the word in the last column with that in the first column. Discuss any differences in the two definitions. If the word in the last column differs from that in the first column, determine what went wrong in the process.
 a. hypothesis
 b. theory
 c. mixture
 d. substance

Text Entry	Student 1	Student 2	Student 3	Student 4
Word	Definition	Word	Definition	Word

LET'S EXPERIMENT! Rainbow Density Column

Materials Needed

- $\frac{1}{4}$ cup dark corn syrup
- $\frac{1}{4}$ cup dishwashing liquid
- $\frac{1}{4}$ cup water
- $\frac{1}{4}$ cup vegetable oil

- $\frac{1}{4}$ cup rubbing alcohol
- A tall 12-ounce glass or clear plastic cup
- Food coloring
- Measuring cup

When you drop an ice cube into your glass of water, what happens? Everyone knows that ice floats . . . but why? The simple answer is density.

Water and ice, even though they are made of the same substance, have different *densities*. Density is a measurement of the ratio of the mass of something to its volume. Remember the density equation (density = mass/volume).

For this experiment you will need to locate different liquids that can be found in kitchen and bathroom cabinets. One way you could determine which liquids have higher or lower densities would be to weigh them, but this experiment involves layering the liquids to learn something about the densities of these common household items.

Before you start your experiment, prepare $\frac{1}{4}$ cup of rubbing alcohol with a few drops of food coloring (either blue or green) and prepare $\frac{1}{4}$ cup of water with food coloring (red or orange).

Take the 12-ounce clear glass. (A slim glass works best.) Being careful not to get syrup on the side of the glass, pour the syrup into the center of the bottom of the glass. Pour enough syrup to fill the glass one-sixth of the way.

After you have added the syrup, tip the glass slightly and pour an equal amount of the dishwashing liquid slowly down the side of the glass. Be careful to add the next liquids slowly. Tip the glass slightly and, pouring slowly down the side of the glass, add first the colored water, then the vegetable oil, and finally the colored rubbing alcohol.

Questions

1. Suggest a reason for the liquids staying separated.
2. Can you determine what the relationship is between the density of a liquid and its position in the glass?
3. Stir up the liquids in the glass and watch what happens to the layers. Have any of the layers mixed?
4. Did you notice a change after a few minutes? Why?
5. Why is it useful to know a substance's density?

Atoms

2

At the heart of every "high-tech" device, including cars, refrigerators, printers, smart phones, and tablet computers, lies a wafer of silicon in one or more integrated circuits (ICs, or "chips"). The silicon used for electronics must be so pure that these devices could not be manufactured until zone-refining purification was developed in the early 1950s. Silicon is an *element* made of a single type of atom. For making ICs, only about one atom of impurity can be present for every billion atoms of silicon.

But how do we even know that atoms exist?

Have You Ever Wondered?

1 What are atoms and why do we care what they are? **2** How do we know atoms are not figments of our imaginations? **3** Why is it often difficult to destroy hazardous wastes? **4** Is light made of atoms? **5** Why is the table of elements called the "periodic table"? **6** What's the difference between atoms and molecules?

41

The false-color image on the previous page left shows individual silicon atoms in a hexagonal arrangement, using a recently developed technique called *scanning tunneling microscopy* (STM). However, the behavior of matter led scientists to the atomic nature of matter over 200 years ago. In this chapter, we will see some of that history, and we will explore the properties of atoms.

ARE THEY REAL? We hear something about atoms almost every day. The twentieth century saw the start of the so-called Atomic Age. The terms *atomic power*, *atomic energy*, and *atomic clock* are a part of our ordinary vocabulary. But just what are atoms?

All material things in the world, including you the reader, are made up of atoms, and atoms determine how matter behaves. Because chemistry studies the behavior of matter, it is necessary for anyone studying chemistry to understand atoms to gain insights into how everything works.

So how do chemists define atoms? An atom is the smallest portion of an element with the properties of that element. For example, one carbon atom—if you could see it in action—would behave the same way as an entire chunk of diamond (which is made only of carbon atoms).

Atoms are not all alike. Each element has its own kinds of atoms. On Earth, about 90 elements occur in nature. Over two dozen more have been synthesized by scientists. As far as we know, the entire universe is made up of these same few elements.

So how small are atoms? The smallest speck of matter that can be detected by the human eye is made up of more atoms than you could count in a lifetime. For example, a one-ounce gold coin such as the Krugerrand or the Canadian Maple Leaf contains nearly 100,000,000,000,000,000,000,000 (10^{23}) gold atoms. Figure 2.1 tries to place this wondrously large number into context.

1 What are atoms and why do we care what they are?

Atoms are the basic building blocks of matter and control the behavior of all matter.

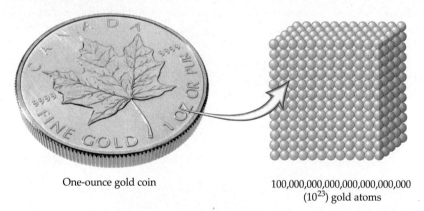

One-ounce gold coin

100,000,000,000,000,000,000,000 (10^{23}) gold atoms

▲ **Figure 2.1** If you had as many grains of sand as there are atoms of gold in a one-ounce gold coin, you could cover the entire Earth, including the seabeds, with sand—and have some left over!

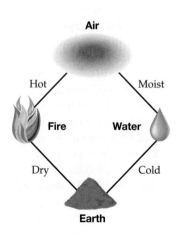

▲ **Figure 2.2** The Greek view of matter was that there were only four elements (in bold) connected by four "principles." The Chinese further broke down the earth component into wood and metal.

2.1 Atoms: Ideas from the Ancient Greeks

Learning Objective • Explain the ancient Greeks' ideas about the characteristics of matter.

Ancient Greek philosophers broke our planet—not literally, of course—into four main components or elements: air, water, earth, and fire (Figure 2.2). You can imagine that two of these philosophers—namely Leucippus, who lived in the fifth century B.C.E., and his student Democritus (ca. 460–ca. 370 B.C.E.)—might have been looking at a pool of water and thinking about the water being separated into drops, and then

▲ **Figure 2.3** A sandy beach.

Q *Sand looks continuous—infinitely divisible—when you look at a beach from a distance. Is it really continuous? Pavement appears continuous from a moving car, but is this perception correct? Water looks continuous, even when viewed up close. Is it really continuous? Is a cloud continuous? Is air?*

each drop being split into smaller and smaller drops, even after they became much too small to see. Is water infinitely divisible? Would you eventually get a drop, or particle, that—if divided—would no longer be water?

Leucippus reasoned that there must ultimately be tiny particles of water that could not be divided any further. After all, from a distance, the sand on the beach looked continuous, but closer inspection showed it to be made up of tiny grains (Figure 2.3).

Democritus expanded on Leucippus's idea. He called the particles *atomos* (meaning "cannot be cut"), from which we derive the modern name *atom*. Democritus thought that each kind of atom was distinct in shape and size (Figure 2.4). He thought that real substances were mixtures of various kinds of atoms.

Unfortunately for Democritus, the famous Greek philosopher Aristotle (ca. 384– ca. 322 B.C.E.) had declared that matter was *continuous* (infinitely divisible) rather than *discrete* (consisting of tiny indivisible particles). The people of that time had no way of determining which view was correct. To most of them, Aristotle's continuous view of matter seemed more logical and reasonable. Also, Aristotle was considered an authority in many areas, and ideas from authorities were accepted much more readily than were ideas from "just anyone." So this view prevailed for about 2000 years, even though it was wrong.

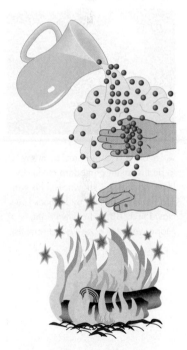

▲ **Figure 2.4** Democritus imagined that "atoms" of water might be smooth, round balls and that atoms of fire might have sharp edges.

SELF-ASSESSMENT Questions

1. Consider how the qualities *discrete* and *continuous* apply to materials at the macroscopic (visible to the unaided eye) level. Which of the following is best described as continuous?
   a. coffee (the beverage)
   b. a bag of coffee beans
   c. a pile of leaves
   d. a tiled floor

2. The view that matter is continuous rather than atomic prevailed for centuries because it was
   a. actually correct
   b. easy to understand
   c. tested by faulty experiments
   d. proposed by Democritus

Answers: 1, a; 2, b

2.2 Scientific Laws: Conservation of Mass and Definite Proportions

Learning Objectives • Describe the significance of the laws of conservation of mass and definite proportions. • Calculate the amounts of elements from the composition of a compound.

▲ Lavoisier is considered by many to be the father of modern chemistry. Here, Lavoisier is shown with his wife, Marie, in a painting by Jacques Louis David in 1788. In her own right, Marie was also a very good scientist.

2 How do we know atoms are not figments of our imaginations?

If matter had no "smallest particles," the law of conservation of mass, the law of definite proportions, and the law of multiple proportions would be almost impossible to explain. Atomic theory provides a simple explanation for these laws and for many others.

Much practical knowledge about chemical substances resulted from the work of alchemists over many centuries. However, there was little progress in understanding atoms and the atomic view until the Age of Enlightenment, beginning in the 1700s. During this period, a cascade of some colorful, committed, and—by and large—self-funded science "hobbyists" studied gases and developed quantitative methods. These methods led to the beginning of modern chemistry as we know it today.

Most notably, we need to go back to Robert Boyle (Ireland, 1627–1691), who published *The Sceptical Chymist* in 1661. Boyle worked on gases and proposed that elements were made up of various types and sizes of "corpuscles" (atoms), which could organize into groups to create different chemical substances. He also distinguished compounds from mixtures. Boyle stated that if a substance could be broken down into simpler substances, then it was not an element.

Boyle's work led to further study by Joseph Black (France and Scotland, 1728–1799), who demonstrated that air was not a single substance but a mixture. This idea was followed by the work of Joseph Priestley (England, 1733–1804) and Carl Scheele (Sweden, 1742–1786), both of whom discovered oxygen at about the same time. There is evidence that, while Scheele arrived at the discovery first, Priestley had formally published his discovery ahead of Scheele. In the days before email and the Internet, information flowed very slowly, making an exact chronology of discoveries difficult to establish. It should be noted that Priestley, while not particularly strong in theory, was a wonderful experimentalist and discovered nine additional new gases.

Henry Cavendish (England, 1731–1810), a contemporary of Priestley, was a detail-oriented person but was painfully shy and slow to publish, which resulted in a lack of recognition for his work. Yet it was Cavendish's concern for accuracy that led to the development of quantitative methods that demonstrated the 2:1 volume ratio of hydrogen to oxygen gases from water with better than 2% accuracy.

Building on the work of his predecessors, Antoine-Laurent Lavoisier (France, 1743–1794) was the first to weigh all the substances present before and after a reaction, and find that matter was *conserved*—the amount remained constant. He was also the first to interpret the meaning of these results correctly, in modern terms. Accordingly, some consider Lavoisier the father of modern chemistry. In a historical twist, while Lavoisier, a tax collector by day, was being executed during the French Revolution, Joseph Priestley was at sea, in exile, on his way to America. Already a friend of Benjamin Franklin, Priestley became a resident of Pennsylvania.

Chemistry was a collective—but not necessarily collaborative—achievement of a number of contemporaries. This era of chemistry was only possible through the development of ideas and written communication of those developments. Simply stated, scientists always have built knowledge on the knowledge of those who came before them.

The Law of Conservation of Mass

Lavoisier summarized his findings as the **law of conservation of mass** (or of matter), which states that matter is neither created nor destroyed during a chemical change (Figure 2.5). The total mass of the reaction products is always equal to the total mass of the reactants (starting materials). We will discuss chemical reactions in more detail beginning with Chapter 4. For now, some simple examples and discussion will illustrate the conservation of mass.

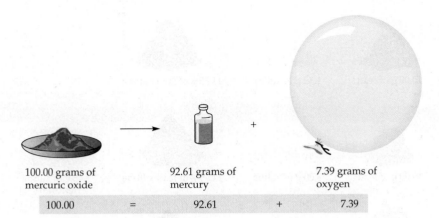

100.00 grams of
mercuric oxide

92.61 grams of
mercury

7.39 grams of
oxygen

| 100.00 | = | 92.61 | + | 7.39 |

◀ **Figure 2.5** Although mercuric oxide (a red solid) has none of the properties of mercury (a silver liquid) or oxygen (a colorless gas), when 100.00 g of mercuric oxide is decomposed by heating, the products are 92.61 g of mercury and 7.39 g of oxygen. Properties are completely changed in this reaction, but there is no change in mass.

Q *When 20.00 g of mercuric oxide decomposes, 1.478 g of oxygen forms. What mass of mercury forms?*

The law of conservation of mass is the basis for many chemical calculations. Because we cannot create materials from nothing, we can make new materials only by changing the way atoms are combined. For example, we can obtain iron metal from iron ore only because the ore contains iron atoms (as we will see in Chapter 12). Furthermore, we cannot get rid of wastes by the destruction of matter. We must put wastes somewhere, or change their forms to something more benign. Such transformations of matter from one form to another are what chemistry is all about.

The Law of Definite Proportions

By the end of the eighteenth century, Lavoisier and other scientists noted that many substances were compounds, composed of two or more elements. Each compound had the same elements in the same proportions, regardless of where it came from or who prepared it. The painstaking work of Joseph Louis Proust (France, 1754–1826) convinced most chemists of the general validity of these observations. In one set of experiments, for example, Proust found that the compound called basic copper carbonate was always composed of 57.48% copper, 5.43% carbon, 0.91% hydrogen, and 36.18% oxygen by mass. The composition was the same regardless of whether the compound was prepared in the laboratory or obtained from natural sources (Figure 2.6).

To summarize these and many other experiments, Proust formulated a new scientific law in 1799. The **law of definite proportions** (also known as Proust's law of constant combination, constant composition, or defined proportions) states that a compound always contains the same elements in certain definite proportions and in no others. This law explains Cavendish's 2:1 ratio of hydrogen to oxygen in water and is an example of theory catching up to experimental data. In other cases, experiments have been used to confirm theory, as will be seen later.

(a)　　　　　　　　(b)　　　　　　　　(c)

▲ **Figure 2.6** (a) The compound known as *basic copper carbonate* has the formula $Cu_2(OH)_2CO_3$ and occurs in nature as the mineral *malachite*. (b) It is formed as a patina on copper roofs. (c) It can also be synthesized in the laboratory. Regardless of its source, basic copper carbonate always has the same composition. Analysis of this compound led Proust to formulate the law of definite proportions.

▶ **Figure 2.7** An example showing how Berzelius's experiments illustrate the law of definite proportions.

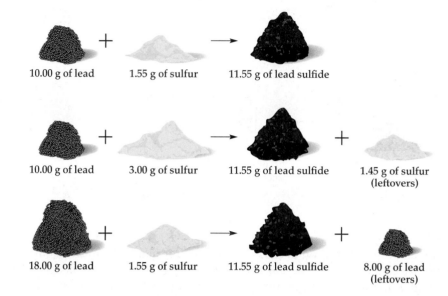

10.00 g of lead   1.55 g of sulfur   11.55 g of lead sulfide

10.00 g of lead   3.00 g of sulfur   11.55 g of lead sulfide   1.45 g of sulfur (leftovers)

18.00 g of lead   1.55 g of sulfur   11.55 g of lead sulfide   8.00 g of lead (leftovers)

Q *What mass of lead would completely react with 3.10 g of sulfur? What mass of lead sulfide forms?*

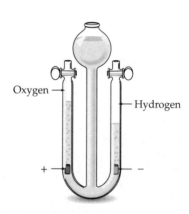

Oxygen
Hydrogen

▲ **Figure 2.8** Electrolysis of water. Hydrogen and oxygen are always produced in a volume ratio of 2:1.

Q *If 24 cubic feet of hydrogen gas is produced by electrolysis, how much oxygen gas will be produced?*

Jöns Jakob Berzelius (Sweden, 1779–1848) was a physician (and possibly a hypochondriac). In his spare time he carried out experiments much like the ones illustrated in Figure 2.7. In these experiments, a defined amount of lead (a grayish soft metal) was heated with various amounts of sulfur (a yellow solid) to form lead sulfide (a shiny black solid). Excess sulfur was washed away with carbon disulfide, a solvent that dissolves sulfur but not lead sulfide. As long as he used at least 1.55 g of sulfur with 10.00 g of lead, Berzelius always got exactly 11.55 g of lead sulfide. Any sulfur in excess of 1.55 g was left over; it did not react. If Berzelius used more than 10.00 g of lead with 1.55 g of sulfur, he got 11.55 g of lead sulfide, with some lead left over.

Although Cavendish's 2:1 ratio of hydrogen to oxygen for water was a volume ratio (using an apparatus similar to that shown in Figure 2.8) and Berzelius's experiment gave a mass ratio, both substantiate the law of definite proportions. This scientific law led to rapid developments in chemistry and dealt a deathblow to the ancient Greek idea of water as an element. It is also important to note that the large amount of carefully obtained data provided a depository to be used in testing of other theories, such as Dalton's atomic theory of matter.

The law of definite proportions is the basis for chemical formulas (Chapter 4), such as H_2O for water. Constant composition also means that substances have constant properties. Pure water always dissolves salt or sugar, and at normal pressure it always freezes at $0\,°C$ and boils at $100\,°C$.

SELF-ASSESSMENT Questions

1. If a chemical alteration takes place in a sealed container, the mass of the container and its contents
a. increases
b. decreases
c. remains unchanged
d. is unpredictable

2. When diamond—a pure form of carbon—is burned in an oxygen atmosphere, each carbon atom combines with two oxygen atoms. If a 2.000 g (10.00 carat) diamond is burned in a closed container with only 6.000 g of oxygen, how much will the container's contents weigh after the burning process?
a. 2.000 g
b. 6.000 g
c. 8.000 g
d. uncertain

3. The ancient Greeks thought that water was an element. In 1800, William Nicholson and Anthony Carlisle decomposed water into hydrogen and oxygen. Their experiment proved that
a. electricity causes decomposition
b. the Greeks were correct
c. hydrogen and oxygen are elements
d. water is not an element

4. Jim and Jane are a husband and wife who like clothes—so much so that each time Jim buys a new hat he also buys a pair of gloves, while every time Jane buys a new purse she also buys a pair of shoes. This 2:1 ratio of hats to gloves and purses to shoes is analogous to:
a. Dalton's atomic theory
b. Democritus's atomic theory
c. the law of definite proportions
d. the law of conservation of mass

5. The composition of water (solid [ice], liquid, or gas [water vapor]) always yields 88.81% oxygen and 11.19% hydrogen by mass, which illustrates:
   a. Boyle's gas laws
   b. Democritus's atomic theory
   c. the law of definite proportions
   d. the law of conservation of mass

6. In collaboration, Joseph Louis Gay-Lussac (France, 1778–1850) and his friend Alexander von Humboldt (Germany, 1769–1859) passed electric sparks through a mixture of oxygen and hydrogen gases to produce water vapor. They noticed that their experiments always required exactly twice as much volume of hydrogen as oxygen. This finding is best described by:
   a. Boyle's gas laws
   b. the law of diminishing returns
   c. the law of definite proportions
   d. the law of conservation of mass

7. When 60.0 g of carbon is burned in 160.0 g of oxygen, 220.0 g of carbon dioxide is formed. What mass of carbon dioxide is formed when 60.0 g of carbon is burned in 750.0 g of oxygen?
   a. 60.0 g   **b.** 160.0 g
   c. 220.0 g   **d.** 810.0 g

Answers: 1, c; 2, c; 3, d; 4, c; 5, c; 6, c; 7, c

2.3 John Dalton and the Atomic Theory of Matter

Learning Objectives • Explain why the idea that matter is made of atoms is a theory. • Describe how atomic theory explains the laws of multiple proportions and conservation of mass.

John Dalton (England, 1766–1844) was not the first to come up with the concept of atoms. This honor goes to Greek philosophers like Democritus. The Greek philosophers, however, were not encumbered by experimental data. Accordingly, their atomic models were wonderfully creative—and completely incorrect. For example, it was their belief that vinegar was composed of spiky atoms and olive oil was made of smooth atoms. While other predecessors of Dalton—like the already mentioned Boyle—wrote about the particulate nature of matter, Dalton's contribution was that he placed the concept of atoms in the framework of the law of definite proportions. He stated that the unvarying ratios come from matter's being made of atoms.

As Dalton continued his work, he discovered another law that his theory would have to explain. Proust had stated that a compound contains elements in certain proportions and only those proportions. Dalton's new law, called the **law of multiple proportions**, stated that elements might combine in *more* than one set of proportions, with each set corresponding to a different compound. For example, carbon combines with oxygen in a mass ratio of 1.00:2.66 (or 3.00:8.00) to form carbon dioxide, a gas that is a product of respiration and of the burning of coal or wood. But Dalton found that carbon also combines with oxygen in a mass ratio of 1.00:1.33 (or 3.00:4.00) to form carbon monoxide, a poisonous gas produced when a fuel is burned in the presence of a limited air supply. In short, when the same elements combine in different proportions, the compounds they form have very different properties. The law of multiple proportions also stated that for two (or more) different compounds made from the same two elements, the ratios of elements followed a definite fixed pattern.

Dalton then used his **atomic theory** to explain the various laws. Table 2.1 lists the important points of Dalton's atomic theory, with some modern modifications that we will consider later.

▲ John Dalton, in addition to developing the atomic theory, carried out important investigations of the behavior of gases. All of his contributions to science were made in spite of the fact that he was color-blind.

Explanations Using Atomic Theory

Dalton's theory clearly explains the difference between elements and compounds. *Elements* are composed of only one kind of atom. For example, a sample of the element phosphorus contains only phosphorus atoms. *Compounds* are made up of two or more kinds of atoms chemically combined in definite proportions. (We will see exactly what *kind* means in Section 3.5.) For example, water—a compound—contains both hydrogen atoms and oxygen atoms, in the fixed ratio of 2:1.

3 Why is it often difficult to destroy hazardous wastes?

Hazardous wastes that are compounds or mixtures can be converted to other compounds or mixtures—but the elements from those compounds or mixtures are still present and may be hazardous as well. For example, insecticide containing arsenic oxide can be broken down into the elements arsenic and oxygen—but the element arsenic is poisonous and can't be broken down into something else.

TABLE 2.1 Dalton's Atomic Theory and the Modifications that Have Been Made

Dalton's Atomic Theory	Modern Modifications
1. All matter is composed of extremely small and indestructible particles called atoms.	1. Dalton assumed atoms to be indivisible, but atoms are divisible. (See Chapter 3.)
2. Elements are made of just one type of atom that is unique to that element.	2. Dalton assumed that all the atoms of a given element were identical in all respects, including mass. We now know this to be incorrect, due to subtleties at the subatomic scale. (See page 50.)
3. Compounds are formed when atoms of different elements combine in fixed proportions.	3. Unmodified. Chemists still agree with Dalton. For example, carbon monoxide has a 1:1 ratio of carbon to oxygen, and carbon dioxide has a 2:1 ratio of carbon to oxygen.
4. A chemical reaction involves a *rearrangement* of atoms, but there is no breaking, destroying, or creating of atoms.	4. Unmodified for *chemical* reactions (Chapter 5), but it does not hold for *nuclear* reactions (Chapter 11).

At this moment, we should formally introduce the concept of a molecule. In the simplest possible terms, a molecule is a combination of two or more atoms in fixed ratios. Some elements exist in nature as molecules, such as oxygen (O_2) and hydrogen (H_2). Other molecular substances are compounds, resulting when two or more atoms of different elements combine. Water (H_2O) combines atoms of oxygen and hydrogen; table sugar (sucrose) combines atoms of carbon, hydrogen, and oxygen in the ratio of 12:22:11, for a formula of $C_{12}H_{22}O_{11}$. Chemical reactions are simply rearrangements of atoms through the breaking up and formation of molecules.

Figure 2.9 illustrates how atomic theory can be applied to the law of definite proportions. The figure illustrates that water is formed from two hydrogen atoms and one oxygen atom; that is, hydrogen and oxygen combine in the defined proportion of 2:1. The arrow is read as "forms" or "yields." In addition, note that any unreacted oxygen atoms are still present (left over). In other words, we have the same overall number of hydrogen and oxygen atoms on each side of the arrow. Thus, this figure illustrates that the atomic theory upholds the law of conservation of mass, as both the right side and left side (separated by the arrow) have the same mass.

The central part of Dalton's atomic theory and its application to the law of definite proportions was the concept that each element was made up of atoms with a single unique weight. In modern times, the concept of weights has been replaced by *relative atomic masses* or, simply, *atomic masses*. You will find a table of atomic masses on the front endpaper of this book. We will use these modern values in some examples showing how Dalton's atomic theory explains the laws of conservation of mass and of constant composition, as well as others.

Finally, atomic theory explains the law of multiple proportions. For example, 1.00 g of carbon combines with 1.33 g of oxygen to form carbon monoxide, or with 2.66 g of oxygen to form carbon dioxide. The law of multiple proportions says that in this example, the ratio of oxygen's mass in carbon dioxide (2.66 g) to oxygen's mass in carbon monoxide (1.33 g) is a small whole number; 2.66 g/1.33 g = 2. The reason this is true is that carbon dioxide has twice the mass

6 oxygen atoms
4 hydrogen atoms
Total mass: 100

6 oxygen atoms
4 hydrogen atoms
Total mass: 100

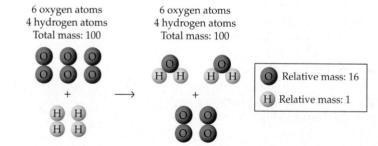

O Relative mass: 16
H Relative mass: 1

▲ **Figure 2.9** The law of definite proportions and the law of conservation of mass interpreted in terms of Dalton's atomic theory.

Q *If 10 molecules of diatomic oxygen (O_2) and 28 molecules of diatomic hydrogen (H_2) react, how many molecules of H_2O (water) can form? Which element is left over? How many molecules of the leftover element are there?*

TABLE 2.2   The Law of Multiple Proportions

Compound	Representation[a]	Mass of N per 1.000 g of O	Ratio of the Masses of N[b]
Dinitrogen oxide	●●●	1.750 g	$(1.750 \div 0.4375) = 4.000$
Nitrogen monoxide	●●	0.8750 g	$(0.8750 \div 0.4375) = 2.000$
Nitrogen dioxide	●●●	0.4375 g	$(0.4375 \div 0.4375) = 1.000$

[a] ● = nitrogen atom and ● = oxygen atom

[b] We obtain the ratio of the masses of N that combine with a given mass of O by dividing each quantity in the third column by the smallest (0.4375 g).

WHY IT MATTERS!

A molecule with one carbon (C) atom and two oxygen (O) atoms is carbon dioxide (CO_2), a gas that you exhale and that provides the "fizz" in soft drinks. Cooled to about $-80\,°C$, it becomes dry ice (shown above), used to keep items frozen during shipping. But a molecule with one fewer O atom is deadly carbon monoxide, CO. Just 0.2% CO in the air is enough to kill. Unfortunately, most fuels produce a little CO when they burn, which is why a car engine should never be left running in a closed garage.

of oxygen per gram of carbon that carbon monoxide does. This in turn is because one atom of carbon combines with *one* atom of oxygen to form carbon monoxide, whereas one atom of carbon combines with *two* atoms of oxygen to form carbon dioxide.

Using modern values, we assign an oxygen atom a relative mass of 16.0 and a carbon atom a relative mass of 12.0. One atom of carbon combined with one atom of oxygen (in carbon monoxide) means a mass ratio of 12.0 parts carbon to 16.0 parts oxygen, or 3.00:4.00. One atom of carbon combined with two atoms of oxygen (in carbon dioxide) gives a mass ratio of 12.0 parts carbon to $2 \times 16.0 = 32.0$ parts oxygen, or 3.00:8.00. Because all oxygen atoms have the same average mass, and all carbon atoms have the same average mass, the ratio of oxygen in carbon dioxide to oxygen in carbon monoxide is 8.00 to 4.00, or 2:1. The same law holds true for other compounds formed from the same two elements. Table 2.2 shows another example of the law of multiple proportions, involving nitrogen and oxygen.

Isotopes: Atoms of an Element with Different Masses

In Chapter 3, we will explain the reasons for using the word *average* in the discussion above, as this is where part of Dalton's atomic theory has hit a slight snag. As we have noted, Dalton's second assumption, that all atoms of an element are alike, has been modified.

Atoms of an element *can* have different masses, and such atoms are called *isotopes*. For example, most carbon atoms have a relative atomic mass of 12 (carbon-12), but 1.1% of carbon atoms have a relative atomic mass of 13 (carbon-13). Isotopes exist because atoms are composed of even smaller particles. (See Chapter 3.)

SELF-ASSESSMENT Questions

1. There are three different oxides of iron. For a fixed mass of iron in FeO, Fe_3O_4, and Fe_2O_3, the masses of oxygen in the three oxides give the ratio 6:8:9. This illustrates
   a. Avogadro's law
   b. the law of constant composition
   c. the law of multiple proportions
   d. the law of the west

2. Which part of Dalton's atomic theory had to be modified by modern discoveries?
   a. Atoms are rearranged in a chemical reaction.
   b. Atoms are neither created nor destroyed in a chemical reaction.
   c. Compounds are formed by atoms combining in different proportions.
   d. All atoms of an element are exactly alike.

3. When atoms of different elements combine in fixed proportions, they form a(n)
   a. molecule
   b. isotope
   c. ion
   d. polyatomic ion

4. Which pair could be used to demonstrate the law of multiple proportions?
   a. CoO and Co_2O_3
   b. MgS and CaS
   c. MgF_2 and CaF_2
   d. NaCl and KBr

2.4 The Mole and Molar Mass

Learning Objectives • Describe what a mole is and how it is used. • Convert between the masses and the moles of a substance.

The Mole and Avogadro's Number

At this point in our conversation, some house cleaning is required so that we can use the modern concept of the mole. We started off by asking just how small an atom was and stating that it was amazingly small. However, we need to talk about much larger masses, grams of elements. The reason is that, as a whole, we are more comfortable working in grams than atoms in the chemistry laboratory. This raises the question of just how many atoms there are in x grams of an element. To answer this question requires that we measure the same entity at both the incredibly small atomic scale and the more tangible macroscopic (gram) scale.

Modern scientific instrumentation allows us to make such measurements. In fact, modern instrumentation can measure the space occupied by a single carbon atom in a diamond. We can do this because in most solids, atoms are highly ordered. In a diamond, the carbon atoms are spaced equally apart from each other. This means that for a diamond of known dimensions we know how many carbon atoms it contains. By also knowing the mass of the diamond, we know how many atoms are needed to make up this mass.

Using this approach, we can determine that a diamond weighing 12.011 g (in a bit we'll see why this number is used) contains 6.022×10^{23} atoms of carbon. The quantity, 6.022×10^{23} atoms, is known as **Avogadro's number**, and 6.022×10^{23} atoms of carbon equal one *mole* of carbon. This concept applies to all elements, so that 6.022×10^{23} atoms of any element equal one **mole** of that element.

Avogadro's number is very large, and best imagined using analogies. For example, a mole—that is, Avogadro's number—of bowling balls would weigh more than Earth (1.314×10^{25} lb., or 5.972×10^{24} kg). A mole of pennies, stacked, would reach the moon . . . not once, but over a trillion times!

Chemists use the concept of a mole just as an egg farmer or baker uses the terms *dozen* and *gross*. Rather than 12 or 144, however, chemists use 6.022×10^{23}.

Molar Mass

So how much does a mole of 6.022×10^{23} atoms or molecules weigh? As stated previously, the answer will depend on the element (or elements, in the case of molecules). We also have to pick one element to be used as a point of reference. Here, the choice was easy—carbon. Carbon is very common, it is central to the existence of living organisms on Earth, and it has a very simple isotopic composition—nearly 99 out of 100 carbon atoms found in nature are carbon-12 isotopes. (Again, you will find out more about isotopes in Chapter 3.) One mole, or 6.022×10^{23} atoms, of pure carbon-12 isotope has a mass of exactly 12.0000 grams. Therefore, the **molar mass**— the mass of one mole—of carbon-12 is 12.0000 grams. The mass of a sample and the number of atoms or molecules in that sample can be related to one another using the concept of a mole (Figure 2.10).

Stated differently, the molar mass of an element is the mass of 6.022×10^{23} atoms of that element, while the molar mass of a compound is the mass of 6.022×10^{23} molecules of that compound. Because carbon-12 is used as a reference, molar masses of all other elements (and compounds) are related to it. Similarly, all elements' atomic masses are expressed relative to carbon-12, and typically, they are expressed in **atomic mass units (amu)**.

To find the molar mass of an element, simply refer to the periodic table. The number under the symbol for the element is its **atomic mass**, or the mass of one atom of that element in amu. The molar mass of the element is the *same* number,

4 **Is light made of atoms?**

Light is not matter and is not made of atoms. Light is a form of energy. Energy—including chemical energy, electrical energy, and nuclear energy—is defined as the ability to do work. We will learn more about energy in Chapter 15.

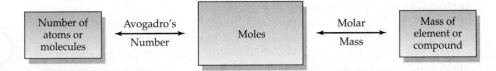

▲ **Figure 2.10** Connecting mass to mole to molecules (and other building blocks of matter) and back highlights the central role of the mole. The number of atoms or molecules is converted to moles through Avogadro's number. The mass of an element or a compound is converted to moles by molar mass. This means, as illustrated, that moles serve as a "hub" while going between the number of atoms (and molecules) and the mass of those atoms (and molecules), and vice versa.

Q *Using Figure 2.10, determine the mass of 1.807×10^{24} atoms of helium. With the aid of Figure 2.10, determine the number of carbon atoms in a 3 g diamond.*

but has units of grams instead of amu. Thus, the mass of one atom of boron (B) is 10.811 amu, while the mass of a mole of boron is 10.811 grams. One atom of mercury (Hg) weighs 200.59 amu, and one mole of mercury atoms weighs 200.59 grams.

It is worthwhile to warn you at this time not to confuse two words: molar and molecular, as in *molar* mass and *molecular* weight. Pay attention to the units. For instance, the *molar* mass of atomic hydrogen is 1.0079 grams per mole, so the molar mass of hydrogen gas (H_2) is $2 \times 1.0079 = 2.0158$ grams per mole. On the other hand, the *molecular* weight, or weight of a single molecule, of H_2 is 2.0158 amu.

Problem Solving: Mass, Atom Ratios, and Moles

We can use proportions, such as those determined by Dalton, or moles to calculate the amount of a substance needed to combine with (or form) a given quantity of another substance. To learn how to do this, let's look at some examples.

 CONCEPTUAL EXAMPLE 2.1 Mass Ratios, Atom Ratios and Moles

Hydrogen gas for fuel cells can be made by decomposing *methane* (CH_4), which is the main component of natural gas. Decomposition of methane is found to produce carbon (C) and hydrogen (H) in a ratio of 3.00 parts by mass of carbon to 1.00 part by mass of hydrogen. What mass in grams of hydrogen can be made from 90.0 g of methane?

Solution Using Mass Ratios

We can express the ratio of parts by mass in any unit we choose—pounds, grams, kilograms—as long as it is the same unit for both elements. Using grams as the unit, we see that 3.00 g of carbon and 1.00 g of hydrogen would be produced if 4.00 g of CH_4 were decomposed. To convert grams of CH_4 to grams of H, we use a conversion factor that has 1.00 g H in the numerator and 4.00 g CH_4 in the denominator.

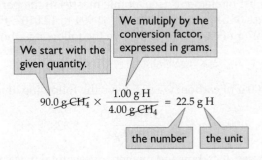

Solution Using Atom Ratios

If we look at the formula of methane, CH_4, we can see that for each carbon atom there are four hydrogen atoms. This means that, in methane, the carbon-to-hydrogen atom ratio by mass is always 1:4. From the periodic table, we can see that carbon is about 12 times more massive than hydrogen. This means that if we convert the 1:4 atom ratio to the mass ratio we will get 12:4, or 3:1. Thus, for every 4.0 g of methane decomposed, 1.0 g of hydrogen will be produced, giving the same conversion factor shown in the solution above. Note that we have arrived at the same conclusion as when pursuing the mass-ratio solution above.

Solution Using Moles

Recall that a mole is simply a fixed number (Avogadro's number) of atoms or molecules. This means that one mole of methane, CH_4, contains one mole of carbon atoms and four moles of hydrogen atoms. If we now look at the periodic table, we can see that a mole of carbon weighs 12.0107 g (molar mass of carbon) and that a mole of hydrogen weighs 1.00794 g. For methane, this means that for every 12.0107 grams (one mole) of carbon, we have 4.03176 g (four times 1.00794 g, or four moles) of hydrogen. Thus, once again, we have come very close to the carbon-to-hydrogen mass ratio of 3:1 and to an exact 1:4 atomic ratio for carbon to hydrogen. The mass ratio is not exact (2.99390:1.0000) because of the presence of isotopes, which have slightly different masses that affect the average mass of an element.

❭ Exercise 2.1A

Phosphine gas can be decomposed to give phosphorus and hydrogen in a mass ratio of 10.3:1.00. If the relative mass of phosphorus is 31.0 when the mass of hydrogen is taken to be 1.00, how many hydrogen atoms are combined with each phosphorus atom in the gas?

❭ Exercise 2.1B

Nitrous oxide (laughing gas) can be decomposed to give 7.00 parts by mass of nitrogen and 4.00 parts by mass of oxygen. If enough nitrous oxide is decomposed to yield 36.0 g of oxygen, what mass in grams of nitrogen is obtained?

 CONCEPTUAL EXAMPLE 2.2 Mass, Atom, or Mole Ratios

Diamonds (just like graphite) can be burned in pure oxygen, and if the temperature is high enough, this reaction produces only carbon dioxide (CO_2). At first, scientists thought that the diamond had simply disappeared. This reaction, though, was a critical step in providing evidence that diamonds were simply a pure form of carbon. If we were to completely burn a 10.0000 carat (2.00000 g) diamond with enough pure oxygen, how much CO_2 would be produced?

Solution Using Mass Ratios

Because diamonds are a pure form of carbon, a 2.00000 g diamond is, in reality, 2.00000 g of carbon. In CO_2, the product of the reaction, for every carbon atom there are two oxygen atoms. The list of elements (see atomic masses in the periodic table) tells us that the relative mass of oxygen to carbon is 15.9994 to 12.0107. For CO_2, this means that for every 12.0107 g of carbon we will have the following amount of oxygen:

$$2 \times 15.9994 \text{ g O} = 31.9988 \text{ g O}$$

Thus, for 2.00000 g of carbon, we will have the following amount of oxygen:

$$2.00000 \text{ g C} \times \frac{31.9988 \text{ g O}}{12.0107 \text{ g C}} = 5.32838 \text{ g O}$$

Adding the masses of carbon and oxygen gives a total mass of 7.32838 g of CO_2.

Solution Using Atom Ratios

In burning a diamond, two atoms of oxygen are consumed for every one atom of carbon. Therefore, we get the carbon-to-oxygen atom ratio of 1:2. If we now look at the periodic table, it can be seen that oxygen is 15.9994/12.0107 times more massive than carbon. This means that the 1:2 *atom* ratio gives a *mass* ratio of:

$$1 : 2(15.9994/12.0107) \quad \text{or} \quad 1 : 2.66419$$

Thus, 2.00000 g of carbon commands the use of 2 × 2.66419 or 5.32838 g of oxygen, and the total mass of the produced CO_2 is 7.32838 g.

Solution Using Moles

The burning of a diamond yields one mole of CO_2 for every one mole of carbon burned. This means that we have a simple 1:1 mole ratio between C and CO_2. The molar mass of carbon is 12.0107, while the molar mass for CO_2 is the molar mass of carbon plus two times the molar mass of oxygen, which is:

$$12.0107 + (2 \times 15.9994) = 44.0095 \text{ g/mole}$$

A 10.0000 carat diamond has the following moles of carbon:

$$2.00000 \text{ g}/12.0107 \text{ g/mole} = 0.66518 \text{ moles}$$

Therefore, 0.66518 moles of CO_2 (which is how much CO_2 is produced) weighs:

$$0.66518 \text{ moles} \times 44.0095 \text{ g/mole} = 7.32838 \text{ g}$$

This Example shows that more than one method exists when it comes to solving problems. Of course, correct conversions between ratios of atoms, masses, and moles must be applied. In the end, all three alternate solutions presented resulted in the same answer. Scientists often have personal preferences for the methods they use.

› Exercise 2.2A

Limestone ($CaCO_3$) is an important ingredient in the manufacture of iron, cement, and other commodities. When one mole of limestone is heated, one mole of quicklime (CaO) and one mole of carbon dioxide (CO_2) are formed. What mass of quicklime can be obtained from 200 grams of limestone?

› Exercise 2.2B

What mass of carbon dioxide can be obtained from 200 grams of limestone?

Dalton arrived at his atomic theory by basing his reasoning on experimental findings, and with modest modification, the theory has stood the test of time and of modern, highly sophisticated instrumentation. Formulation of so successful a theory was quite a triumph for a Quaker schoolteacher in 1803. With that said, it can be seen from the above examples that the concept of moles greatly aids us in our application of Dalton's atomic theory to common chemical calculations. The importance of the mole will be further highlighted in Chapter 5 (Chemical Accounting).

SELF-ASSESSMENT Questions

1. Consider carbon monoxide (CO) and carbon dioxide (CO_2). They both contain carbon and oxygen. If a sample of CO contains 36.0321 g of C and 47.9982 g of O, what mass in grams of O would there be in CO_2 if it contained 36.0321 g C?
   a. 15.9994
   b. 47.9982
   c. 36.0321
   d. 95.9964

2. According to Dalton, elements are distinguished from each other by
   a. their density in the solid state
   b. their nuclear charge
   c. the shapes of their atoms
   d. the weights of their atoms

3. Dalton postulated that atoms were indestructible to explain why
   a. the same two elements can form more than one compound
   b. no two elements have the same atomic mass
   c. mass is conserved in chemical reactions
   d. nuclear fission (splitting) is impossible

4. Dalton viewed chemical change as
   a. a change of atoms from one type into another
   b. creation and destruction of atoms
   c. a rearrangement of atoms
   d. a transfer of electrons

5. How many types of atoms are present in a given compound?
   a. at least two
   b. hundreds
   c. three or more
   d. It depends on the mass of the compound.

6. Which statement in regard to a mole of carbon is *incorrect*?
   a. It contains Avogadro's number of carbon atoms.
   b. It is a quantity of carbon.
   c. It weighs 12.0107 g.
   d. It contains 12 atoms of carbon.

7. Which statement in regard to molar mass is *incorrect*?
   a. The molar mass of carbon-12 is 12.0000 g/mole.
   b. It is the mass of 6.022×10^{23} atoms of an element.
   c. It is the mass of 6.022×10^{23} molecules of a compound.
   d. It can only be applied to atoms of an element.

Questions 8–10 refer to the following figures:

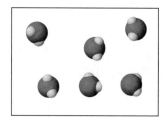
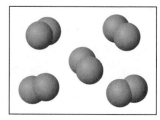

a.   b.

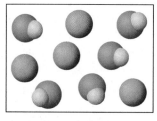

c.

8. Which figure represents an element?

9. Which figure represents a single compound?

10. Which figure represents a mixture?

Answers: 1, d; 2, d; 3, c; 4, c; 5, a; 6, d; 7, d; 8, b; 9, a; 10, c

2.5 Mendeleev and the Periodic Table

Learning Objective • Describe how the elements are arranged in the periodic table and why the arrangement is important.

During the eighteenth and nineteenth centuries, new elements were discovered with surprising frequency. By 1830, there were 55 known elements, all with different properties that demonstrated no apparent order. Dalton had set up a table of relative atomic masses in his 1808 book *A New System of Chemical Philosophy*. Dalton's rough values were improved in subsequent years, notably by Berzelius, who published a table of atomic weights containing 54 elements in 1828. Most of Berzelius's values agree well with modern values.

Emerging Patterns and Periodicity within the Known Elements

As early as 1816, Johann Wolfgang Döbereiner (Germany, 1780–1849) found a triadic relationship between elements based on their atomic masses. The average mass of the lightest and heaviest atoms of the triad equaled the mass of the middle element of the triad. Some examples of the triads he found are Li/Na/K; Ca/Sr/Ba; Cl/Br/I; and S/Se/Te. You can check to see how well these triads work with the modern periodic table the inside front cover of the textbook. Although useful at the time, these types of relationships have been replaced by the relationships present in the modern periodic table.

The next trend noticed in the available atomic masses was that those masses were separated by intervals of eight times the atomic mass of hydrogen. This pattern was discovered by both the English chemist John Newlands (1837–1898) and French mineralogist Alexandre-Emile Béguyer de Chancourtois (1820–1886). It is sometimes referred to as *Newlands's law of octaves*, even though Béguyer de Chancourtois was the first to publish his idea, in 1863, two years before Newlands did.

CONCEPTUAL EXAMPLE 2.3 Predicting Periodic Properties

Show how Döbereiner's triad of lithium, potassium, and sodium could be used to predict the approximate value of the atomic mass of sodium.

Solution

The atomic mass of lithium (Li) is 6.941 u and that of potassium (K) is 39.0983 u. The average of the two is (6.941 amu + 39.0983 amu) ÷ 2 = 23.020 amu, quite close to the modern value for sodium (Na) of 22.9898 amu.

⟩ Exercise 2.3A

Another of Döbereiner's triads was the *salt formers*: chlorine, bromine, and iodine. Show that the relative atomic mass of the middle element in this triad is close to the average of the relative atomic masses of the other two elements.

⟩ Exercise 2.3B

Which triad more accurately predicts the atomic mass of the middle element in that triad: Mg/Ca/Sr or He/Ne/Ar?

▲ Dmitri Mendeleev, the Russian chemist who arranged the 63 known elements into a periodic table similar to the one we use today, was considered one of the great teachers of his time. Unable to find a suitable chemistry textbook, he wrote his own, *The Principles of Chemistry*. Mendeleev also studied the nature and origin of petroleum and made many other contributions to science. Element 101 is named mendelevium (Md) in his honor.

The Periodic Table: Mendeleev and Meyer

The patterns discovered by these scientists were followed by the work of Julius Lothar Meyer (Germany, 1830–1895) and Dmitri Ivanovich Mendeleev (Russia, 1834–1907). They both linked chemical properties to atomic mass, and in doing so they developed the periodic law. In 1862, Meyer published his periodic table of 28 elements, while in 1869 Mendeleev published his *first* periodic table (Figure 2.11) for the known 63 elements (with corrected atomic masses). Mendeleev, however,

Tabelle II.

Reihen	Gruppe I. — R^2O	Gruppe II. — RO	Gruppe III. — R^2O^3	Gruppe IV. RH^4 RO^2	Gruppe V. RH^3 R^2O^5	Gruppe VI. RH^2 RO^3	Gruppe VII. RH R^2O^7	Gruppe VIII. — RO^4
1	H = 1							
2	Li = 7	Be = 9,4	B = 11	C = 12	N = 14	O = 16	F = 19	
3	Na = 23	Mg = 24	Al = 27,3	Si = 28	P = 31	S = 32	Cl = 35,5	
4	K = 39	Ca = 40	— = 44	Ti = 48	V = 51	Cr = 52	Mn = 55	Fe = 56, Co = 59, Ni = 59, Cu = 63.
5	(Cu = 63)	Zn = 65	— = 68	— = 72	As = 75	Se = 78	Br = 80	
6	Rb = 85	Sr = 87	?Yt = 88	Zr = 90	Nb = 94	Mo = 96	— = 100	Ru = 104, Rh = 104, Pd = 106, Ag = 108.
7	(Ag = 108)	Cd = 112	In = 113	Sn = 118	Sb = 122	Te = 125	J = 127	
8	Cs = 133	Ba = 137	?Di = 138	?Ce = 140	—	—	—	— — — —
9	(—)	—	—	—	—	—	—	
10	—	—	?Er = 178	?La = 180	Ta = 182	W = 184	—	Os = 195, Ir = 197, Pt = 198, Au = 199.
11	(Au = 199)	Hg = 200	Ti = 204	Pb = 207	Bi = 208	—	—	
12	—	—	—	Th = 231	—	U = 240	—	— — — —

der chemischen Elemente.

▲ **Figure 2.11** Mendeleev's periodic table. In this 1898 version, he wrongly "corrected" the atomic weight of tellurium to be less than that of iodine.

5 **Why is the table of elements called the "periodic table"?**

Periodic indicates that something occurs regularly. The physical and chemical properties of the elements are periodic—that is, many of those properties are similar for elements in a given column of the table.

daringly left spaces where he predicted yet-undiscovered elements should be! He even predicted the properties of these elements, based on the periodic law.

Three of the missing elements were soon discovered and named scandium, gallium, and germanium. As can be seen in Table 2.3, Mendeleev's predictions for germanium—called "eka-silikon" at the time—were amazingly successful. Mendeleev's great genius lay in his ability to connect the *knowns* as well as to *leave blanks* and predict the existence of the *unknowns*. The predictive value of his periodic table led to its wide acceptance. This is why Mendeleev tends to be given almost sole credit for developing the periodic table.

TABLE 2.3 **Properties of Germanium: Predicted and Observed**

Property	Predicted by Mendeleev for Eka-Silikon (1871)	Observed for Germanium (1886)
Atomic mass	72	72.6
Density (g/cm^3)	5.5	5.47
Color	Dirty gray	Grayish white
Density of oxide (g/cm^3)	EsO_2 : 4.7	GeO_2 : 4.703
Boiling point of chloride	$EsCl_4$: below 100 °C	$GeCl_4$: 86 °C

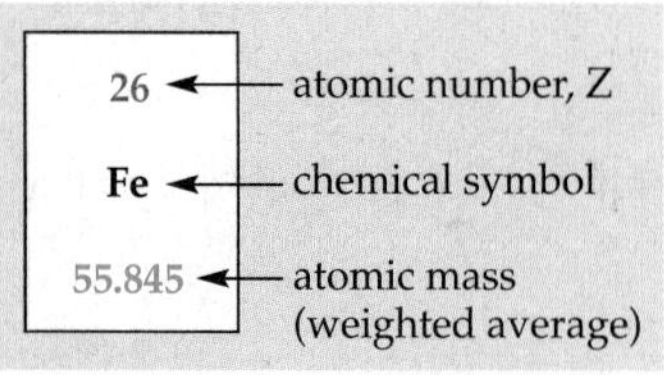

26 ◄——— atomic number, Z

Fe ◄——— chemical symbol

55.845 ◄——— atomic mass (weighted average)

▲ **Figure 2.12** Representation of an element on the modern periodic table.

Mendeleev continued to tinker with the periodic table for much of his life, eventually placing silver, gold, and several other metals slightly out of the previously proposed order. Doing so allowed sulfur, selenium, and tellurium, which have similar chemical properties, to appear in the same column. This rearrangement also put iodine in the same column as chlorine and bromine, which iodine resembles chemically.

The modern periodic table contains 118 elements. The latest one to be added was number 118, which was given the name *oganesson* late in 2016. Each element is represented by a box in the periodic table, as shown in Figure 2.12. We will discuss the periodic table and its theoretical basis in Chapter 3.

SELF-ASSESSMENT Questions

1. Which of the following is *not* true of Mendeleev's periodic table?
   a. It includes new elements that he had just discovered.
   b. Most elements are arranged generally in order of increasing atomic mass.
   c. He left gaps for predicted new elements.
   d. He placed some heavier elements before lighter ones.

2. Meyer and Mendeleev arranged elements in their periodic tables
   a. alphabetically
   b. according to their chemical properties
   c. in the order they were discovered
   d. according to density

3. Relative masses in the modern periodic table are based on the value for
   a. the carbon-12 isotope
   b. the hydrogen atom
   c. naturally occurring oxygen
   d. the oxygen-16 isotope

4. The number of elements known today is approximately
   a. 12
   b. 100
   c. 1000
   d. 30 million

2.6 Atoms and Molecules: Real and Relevant

Learning Objective • Distinguish atoms from molecules.

Are atoms real? Certainly they are real as a concept, and a highly useful one at that. And scientists today can observe computer-enhanced images of individual atoms. These portraits provide powerful (though still indirect) evidence that atoms exist.

Are atoms relevant? Much of modern science and technology—including the production of new materials and the technology of pollution control—is based on the concept of atoms. We have seen that atoms are conserved in chemical reactions. Thus, material things—things made of atoms—can be recycled, for the atoms are not destroyed no matter how we use them. The one way we might "lose" a material from a practical standpoint is to spread its atoms so thinly that it would take too much time and energy to put them back together again. The essay on recycling (see below) describes a real-world example of such loss.

Let's go back to Leucippus and Democritus and their musings on that Greek beach. We now know that if we keep dividing drops of water, we will ultimately obtain a small particle—called a *molecule*—that is still water. A **molecule** is a group of atoms chemically bonded, or connected, to one another. Molecules are represented by chemical formulas. The symbol H represents an *atom* of hydrogen. The formula H_2 represents a *molecule* of hydrogen, which is composed of two hydrogen atoms. The formula H_2O represents a molecule of water, which is composed of two hydrogen atoms and one oxygen atom. If we divide a water molecule, we will obtain *two atoms* of hydrogen and *one atom* of oxygen . . . and we no longer have water.

And if we divide those atoms . . . but that is a story for another time. (See Chapter 11.)

Dalton regarded the atom as indivisible, as did his successors up until the discovery of radioactivity in 1895. We examine the evolving concept of the atom in Chapter 3.

6 What's the difference between atoms and molecules?

A molecule is made of atoms that are combined in fixed proportions. A carbon atom is the smallest particle of the element carbon, and a carbon dioxide molecule is the smallest particle of the compound carbon dioxide. As you have probably figured out, molecules can be broken down into their atoms.

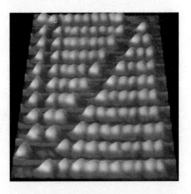

▲ A molecular abacus, created from C60 molecules on copper. Scanning tunneling microscopy (STM) makes it possible to manipulate individual molecules and atoms on a surface.

Recycling

Because it is an element, iron cannot be created or destroyed in a chemical reaction, but that does not mean it always exists in its elemental state. Let's consider two different pathways for the recycling of iron.

1. Hematite, an iron ore, is obtained from a mine and converted into pig iron, which is used to make steel cans. Once discarded, the cans rust. Iron ions (ions are explained in Chapter 4) slowly leach from the rust into groundwater and eventually wind up in the ocean. These dissolved ions can be absorbed by and incorporated into marine plants. Marine creatures that eat the plants will in turn absorb and incorporate iron ions. The original iron atoms are now widely separated in space. Some of them might even become part of the hemoglobin in your

blood. The iron has been recycled, but it will never again resemble the original pig iron. This means that for new products to be made of iron, more iron must be mined. Mining makes it a nonsustainable practice in the long run. It also is not a chemically green approach, due to the environmental damage caused by iron-mining practices.

2. Iron ore is taken from a mine and converted to pig iron and then into steel. The steel is used in making an automobile, which is driven for a decade and then sent to the junkyard. There, it is compacted and sent to a recycling plant, where the steel is recovered and ultimately used again in a new automobile. Once the iron was removed from its ore, it was conserved in its elemental metallic form. In

this form of recycling, the iron continues to be useful for a long time. This is a sustainable practice because it allows the mined iron to be reused over and over again. It is also a chemically green practice because recycling iron places much less stress on the environment than iron-mining practices do.

SELF-ASSESSMENT Questions

1. We can recycle materials because atoms
   a. always combine in the same way
   b. are conserved
   c. are indivisible
   d. combine in multiple portions

2. A molecule is
   a. a collection of like atoms
   b. conserved in chemical reactions
   c. a group of atoms chemically bonded to one another
   d. indivisible

3. Which statement is correct in regard to recycling?
   a. Compounds are never altered.
   b. New elements other than gold may be created.
   c. Elements remain unaltered.
   d. Only iron can be recycled.

Answers: 1, b; 2, c; 3, c

 Summary

Section 2.1—The concept of atoms was first suggested in ancient Greece by Leucippus and Democritus. However, this idea was rejected for almost 2000 years in favor of Aristotle's view that matter was continuous.

Section 2.2—The **law of conservation of mass** resulted from careful experiments by Cavendish, Lavoisier, and others, who found that no change in mass occurred in a chemical reaction. Boyle said that if a proposed element could be broken down into simpler substances, it was not an element. The **law of definite proportions** (or law of constant composition) states that a given compound always contains the same elements in exactly the same proportions by mass. It was based on work by Proust and Berzelius.

Section 2.3—In 1803, Dalton explained the laws of definite proportions and conservation of mass with his **atomic theory**, which had four main points: (1) Matter is made up of tiny particles called atoms; (2) atoms of the same element are alike; (3) compounds are formed when atoms of different elements combine in certain proportions; and (4) during chemical reactions, atoms are rearranged, not created or destroyed.

Dalton also discovered the **law of multiple proportions**, which states that different elements might combine in two or more different sets of proportions, each set corresponding to a different compound. His atomic theory also explained this new law. These laws can be used to perform calculations to find the amounts of elements that combine or are present in a compound or reaction. Dalton's atomic theory is very similar to its modern version. The most significant change has been the recognition that atoms of an element are not necessarily identical, but may have slightly different masses; such atoms are known as **isotopes**.

Section 2.4—A **mole** of a substance consists of **Avogadro's number**, or 6.022×10^{23}, particles (atoms, molecules, or compounds) of that substance. The mass of 6.022×10^{23} atoms or molecules is known as the **molar mass**. Both the atomic masses reported in the periodic table (in **atomic mass units [amu]**) and the molar masses we use in calculations (in grams per mole) are relative to the atomic mass and the molar mass, respectively, of carbon-12.

Section 2.5—Berzelius published a table of atomic weights in 1828, and his values agree well with modern ones. In 1869, Mendeleev published his version of the **periodic table**, a systematic arrangement of the elements that allowed him to predict the existence and properties of undiscovered elements. In the modern periodic table, each element is listed along with its symbol and the average mass of an atom of that element.

Section 2.6—Because atoms are conserved in chemical reactions, matter (which is made of atoms) can always be recycled. Atoms can be lost, in effect, if they are scattered in the environment. A **molecule** is a group of atoms chemically bonded together. Just as an atom is the smallest particle of an element, a molecule is the smallest particle of most compounds.

GREEN CHEMISTRY—When chemists design products and processes, potential impacts on human health and the environment of the incorporated elements should be considered. Identifying and replacing toxic and rare elements can lead to greener technologies and enhance sustainability.

GREEN CHEMISTRY It's Elemental
Lallie C. McKenzie, Chem11 LLC

Principles 3, 7, 10

Learning Objectives • Identify elements that could be classified as hazardous or rare. • Explain how green chemistry can change technologies that rely on hazardous or rare elements.

In this chapter, you learned that everything is made up of atoms and that an atom is the smallest particle that is characteristic of a given element. Although the universe is composed of an uncountable number of compounds and mixtures, the basic building blocks are elements. The periodic table (Section 2.5) organizes the elements by relative mass and, more importantly, by their predictable chemical properties.

What do elements have to do with green chemistry? When chemists design products and processes, they take advantage of the entire periodic table, including elements used rarely or not at all in nature or our bodies. It is important to think about impacts of the chosen elements on human health and the environment. Two important factors to evaluate are inherent hazard and natural abundance. Hazard can be different for the element itself or when it is in a compound, so the element's form can determine whether it causes harm. For example, elemental sodium is very reactive with water, but sodium chloride is a compound that we eat. Also, as you learned from the law of conservation of mass (Section 2.2), elements cannot be created or destroyed. Therefore, the amounts on the planet are limited and may be scarce.

Green chemistry principles #3, #7, and #10 apply to products and processes that involve hazardous and rare elements. Understanding the impacts of toxic elements and compounds on human health and the environment helps us to develop safer alternatives and reduce or eliminate the use of these materials. Also, designing new technologies that rely on relatively abundant elements can prevent depletion of resources. Finally, reclaiming materials after use is critical if they can cause harm or are scarce.

The use of three hazardous elements—lead, cadmium, and mercury—has been reduced through green chemistry. These elements are known to harm systems of the human body and have a negative impact on children's development. Although they are important in many products, the urgent need to limit exposure to these elements has led to developing safer alternatives. For example, lead is no longer in paint, compounds with lower toxicity now substitute for lead stabilizers and cadmium pigments in plastics, and new chemicals are being used instead of traditional solder in electronics. Substitution of metal hydrides for cadmium in batteries has diminished cadmium use.

In addition, light emitting diode (LED) bulbs produce light approximately 90% more efficiently than incandescent and have no mercury. Because coal-fired power plants emit mercury, advances in alternative energy sources also reduce our exposure to this element.

Many current technologies may be limited by their need for scarce elements, but green chemistry can promote new opportunities. Although almost all of the elements are found in the earth's crust, eight account for more than 98% of the total. In fact, almost three-quarters of the Earth's crust is comprised of two elements, silicon and oxygen. Currently, the permanent magnets in computer hard drives, wind turbines, and hybrid cars include the scarce elements neodymium and dysprosium. Also, many products depend on cerium, yttrium, neodymium, and lanthanum. These elements are drawn from a single source and can be hard to separate from other materials. Current supplies are much lower than needed to support projected growth. Approaches based on earth-abundant elements (such as silicon, iron, and aluminum) would support technology and protect resources.

Green chemistry supports the design of materials where the individual elemental components can be recycled easily. Atoms of a material can be reclaimed directly through recycling or spread through the environment. If released, hazardous materials can have a negative effect on health and the environment, so recycling of these elements should be a priority. Reclaiming raw materials reduces the demand for natural resources, including scarce elements.

Green chemistry approaches are especially important when products and processes present health or environmental risks. Identifying and replacing hazardous and rare elements can lead to greener technologies and a sustainable future for everyone.

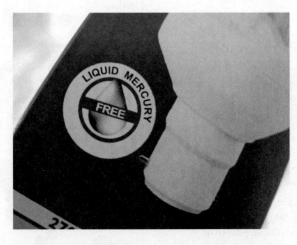

◄ New technologies have led to mercury-free efficient light bulbs like this one. These products save energy and do not contain toxic elements that were used for traditional fluorescent lighting.

Learning Objectives	Associated Problems
• Explain the ancient Greeks' ideas about the characteristics of matter. (2.1)	1–3
• Describe the significance of the laws of conservation of mass and definite proportions. (2.2)	4–8, 15–28, 47, 48, 56
• Calculate the amounts of elements from the composition of a compound. (2.2)	33, 34, 47-50
• Explain why the idea that matter is made of atoms is a theory. (2.3)	8
• Know how atomic theory explains the laws of multiple proportions and conservation of mass. (2.3)	9–11, 29–40, 52, 53
• Describe what a mole is and how it is used. (2.4)	12, 41–44, 54
• Convert between the masses and the moles of a substance. (2.4)	12, 41–43
• Describe how the elements are arranged in the periodic table and why the arrangement is important. (2.5)	13, 55
• Distinguish atoms from molecules. (2.6)	14, 45, 46
• Identify elements that can be classified as hazardous or rare.	57
• Explain how green chemistry can change technologies that rely on hazardous or rare elements.	58, 59

Conceptual Questions

1. Distinguish between **(a)** the atomic view and the continuous view of matter and **(b)** the ancient Greek definition of an element and the modern one.

2. What was Democritus's contribution to atomic theory? Why did the idea that matter was continuous (rather than discrete) prevail for so long? What discoveries finally refuted the idea?

3. Consider the following *macroscopic* (visible to the unaided eye) objects. Which are best classified as *discrete* (like Democritus's description of matter), and which are *continuous* (like Aristotle's description)?
   a. people
   b. cloth
   c. calculators
   d. milk chocolate
   e. M&M candies

4. What were Boyle's contributions to modern chemistry?

5. Cavendish found that water was composed of two hydrogen atoms for every one oxygen atom. Explain how this supports the law of definite proportions.

6. Describe Lavoisier's contribution to the development of modern chemistry.

7. Fructose (fruit sugar) is always composed of 40.0% carbon, 53.3% oxygen, and 6.7% hydrogen, regardless of the fruit that it comes from. What law does this illustrate?

8. Outline the main points of Dalton's atomic theory, and apply them to the following laws: **(a)** conservation of mass, **(b)** definite proportions, and **(c)** multiple proportions.

9. What law does the following list of compounds illustrate? N_2O, NO, NO_2, N_2O_4.

10. According to the law of multiple proportions, for a fixed mass of the first element in each of the following compounds, what is the relationship between (ratio of) the
    a. masses of oxygen in ClO_2 and in ClO?
    b. masses of fluorine in ClF_3 and ClF?
    c. masses of oxygen in P_4O_6 and P_4O_{10}?

11. In the accompanying figure, the blue spheres represent phosphorus atoms, and the red ones represent oxygen atoms. The box labeled "Initial" represents a mixture. Which one of the other three boxes (A, B, or C) could *not* represent that mixture after a chemical reaction occurred? Explain briefly.

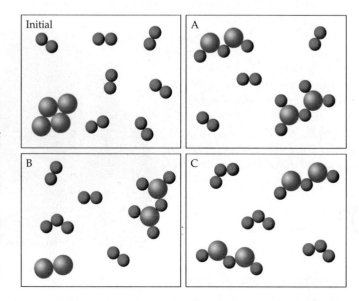

12. **a.** How is Avogadro's number linked with the concept of the mole? **b.** How many atoms are in a mole of oxygen atoms? **c.** How many molecules are in a mole of caffeine? **d.** How much does a mole of diatomic nitrogen (N_2) weigh?

13. Explain how the periodic law allowed Mendeleev to construct his periodic table and to predict the existence and properties of elements yet to be discovered.

14. Lavoisier considered *alumina* an element. In 1825, the Danish chemist Hans Christian Oersted (1777–1851) isolated aluminum metal by reacting aluminum chloride with potassium. Later experiments showed that alumina is formed by reacting aluminum metal with oxygen. What did these experiments prove?

 Problems

The Law of Conservation of Mass

15. **a.** If a closed container contains a mouse as well as enough food, water, and oxygen for the mouse to live for three weeks, how much will the container weigh one and two weeks later after the mouse has eaten, drunk, and exercised (respiration is CO_2 emission), and why? **b.** If the mouse were in a wire cage and only the weights of the mouse, food, and water were considered, would you come to the same answer as in **a**. Why?

16. An iron nail dissolves in a solution of hydrochloric acid. The nail disappears. Have the iron atoms been destroyed? If so, how? If not, where are they?

17. If a 4.00 g effervescent antacid tablet is dissolved in 100 mL (100 g) of water in a 100 g glass, how much will the glass and its contents weigh once the antacid pill has dissolved? Justify your answer.
    a. 104 g   **b.** 204 g
    c. a bit less than 204 g   **d.** a bit more than 204 g

18. Explain how the following two combustion (burning) reactions appear to disobey the law of conservation of mass:
    a. When a diamond is burned, the mass that remains is zero.
    b. When a piece of wood is burned, the leftover ashes weigh less than the wood.

    Upon further investigation, it is noted that CO_2 is vented off from both reactions. With this additional information, state how you can prove the diamond is made of pure carbon and that the wood is not.

19. The first noble gas compounds were prepared in 1962. Since then, a number of these compounds containing xenon and krypton have been made, including xenon trioxide, XeO_3, a high explosive. Xenon trioxide contains 7.32 g of xenon for every 2.68 g of oxygen. What mass of oxygen will be needed to convert 100.0 g of xenon to xenon trioxide?

20. Dinitrogen oxide (N_2O, or "laughing gas") contains 28.01 g of nitrogen in every 44.01 g of dinitrogen oxide. What mass of nitrous oxide can be formed from 48.7 g of nitrogen?

21. An overly enthusiastic chemistry student burns 88.20 g of propane (C_3H_8; 36.03 g C for each 8.07 g H) and states that 291 g of CO_2 (12.01 g C for each 32.00 g O) was produced. Explain to the student why her answer is impossible.

22. A student heats 2.796 g of zinc powder with 2.414 g of sulfur. He reports that he obtains 4.169 g of zinc sulfide and recovers 1.041 g of unreacted sulfur. Show by calculation whether his results obey the law of conservation of mass.

23. When 1.00 g zinc and 0.80 g sulfur are allowed to react, all the zinc is used up, 1.50 g of zinc sulfide (ZnS) is formed,

and some unreacted sulfur remains. What is the mass of *unreacted* sulfur?
    a. 0.20 g   **b.** 0.30 g   **c.** 0.50 g
    d. impossible to determine from this information alone

24. A city has to come up with a plan to dispose of its solid wastes. The solid wastes consist of many different kinds of materials, and the materials consist of many different elements, in both compounds and mixtures. The options for disposal include burying the wastes in a landfill, incinerating them, and dumping them at sea. Which method, if any, will get rid of the atoms that make up the wastes? Which method, if any, will immediately change the chemical form of the wastes?

The Law of Definite Proportions

25. When 18.02 g of water is decomposed by electrolysis, 16.00 g of oxygen and 2.02 g of hydrogen are formed. According to the law of definite proportions, what masses, in grams, of **(a)** hydrogen and **(b)** oxygen will be formed by the electrolysis of 489 g of water?

26. Hydrogen from the decomposition of water has been suggested as the fuel of the future (Chapter 15). What mass in kilograms of water would have to be electrolyzed to produce 907 kg (one ton) of hydrogen? (See Problem 25.)

27. Given a plentiful supply of air, 3.0 parts carbon will react with 8.0 parts oxygen by mass to produce carbon dioxide. Use this mass ratio to calculate the mass of carbon required to produce 14 kg of carbon dioxide.

28. If a diamond is burned in an oxygen-containing atmosphere, the carbon in the diamond is converted to CO_2. How much CO_2 will be produced if a 0.4 g (2 carat) diamond is burned? (See Problem 27.)

John Dalton and the Atomic Theory of Matter

29. All of the following are main points of Dalton's atomic theory except:
    a. All atoms of a specified element are identical.
    b. Atoms of different elements differ in size and mass.
    c. Compounds are formed through distinct whole-number combinations of atoms.
    d. Equal masses of two different compounds should have the same number of atoms.

30. The Greeks thought that atoms had different shapes, with water and oil atoms being smooth and atoms associated with fire and spiciness having sharp edges. Does this view agree with Dalton's atomic theory? Explain your answer.

31. Use Dalton's atomic theory to explain what happens to the carbon atoms when a diamond is burned, and what

happens to oxygen and hydrogen atoms when water is electrolyzed.

32. According to Dalton's atomic theory, when elements react, their atoms combine in (choose one)
   a. a simple whole-number ratio that is unique for each set of elements
   b. exactly a 1:1 ratio
   c. one or more simple whole-number ratios
   d. pairs
   e. random proportions

33. Hydrogen and oxygen combine in a mass ratio of about 1:8 to form water. If every water molecule consists of two atoms of hydrogen and one atom of oxygen, the mass of a hydrogen atom must be _____ the mass of an oxygen atom.
   a. $\frac{1}{16}$
   b. $\frac{1}{8}$
   c. 8 times
   d. 16 times

34. The elements fluorine and nitrogen combine in a mass ratio of 57:14 to form a compound. If every molecule of the compound consists of three atoms of fluorine and one atom of nitrogen, what fraction or multiple of the mass of a fluorine atom is the mass of a nitrogen atom?
   a. $\frac{19}{14}$
   b. 3 times
   c. $\frac{14}{19}$
   d. 14 times

Multiple Proportions

35. Which of the following pairs can be used to illustrate the law of multiple proportions?
   a. CH_4 and CO_2
   b. H_2O and HF
   c. NO and NO_2
   d. ZnO and ZnS

36. A compound containing only oxygen and rubidium has 0.187 g of O per 1.00 g of Rb. The relative atomic masses are 16.0 for O and 85.5 for Rb. Which is a possible O-to-Rb mass ratio for a different oxide of rubidium?
   a. 8.0:85.5
   b. 16.0:85.5
   c. 32.0:85.5
   d. 16.0:171

37. A sample of an oxide of tin with the formula SnO consists of 0.742 g of tin and 0.100 g of oxygen. A sample of another oxide of tin consists of 0.555 g of tin and 0.150 g of oxygen. What is the formula of the second oxide?

38. Consider three oxides of nitrogen, X, Y, and Z. Oxide X has an oxygen-to-nitrogen mass ratio of 2.28:1.00, oxide Y has an oxygen-to-nitrogen mass ratio of 1.14:1.00, and oxide Z has an oxygen-to-nitrogen mass ratio of 0.57:1.00. What is the whole-number ratio of masses of oxygen in these compounds, given a fixed mass of nitrogen?

39. Iron forms a number of oxides in nature. A sample of wüstite (FeO) contains 0.558 g of Fe and 0.160 g O, while samples of hematite and magnetite contain 0.558 g of Fe and 0.240 g O and 3.35 g Fe and 1.28 O, respectively. Given

what we know about wüstite, what are the iron-to-oxygen ratios for hematite and magnetite? Do your conclusions agree with the law of multiple proportions? Explain your answer.

40. Two compounds, V and W, are composed only of hydrogen and carbon. Compound V is 80.0% carbon by mass and 20.0% hydrogen by mass. Compound W is 83.3% carbon by mass and 16.7% hydrogen by mass. What is the ratio of masses of hydrogen in these compounds, given a fixed mass of carbon?

The Mole and Molar Mass

41. A pure diamond is made only of carbon atoms. How many moles of carbon are there in a 0.60 g (3.0 carat) diamond? From how many atoms of carbon is this diamond composed?

42. A water molecule is made up of one oxygen atom and two hydrogen atoms. What is the mass of one mole of water? How many atoms of oxygen and of hydrogen are there in one mole of water?

43. What is the mass of five moles of carbon-12? What is the mass of five moles of carbon?

44. What is the mass of hydrogen relative to carbon-12? What is the mass of hydrogen relative to carbon?

Mass and Atom Ratios

45. A blue solid called *azulene* is thought to be a pure compound. Analyses of three samples of the material yield the following results.

	Mass of Sample	Mass of Carbon	Mass of Hydrogen
Sample 1	1.000 g	0.937 g	0.0629 g
Sample 2	0.244 g	0.229 g	0.0153 g
Sample 3	0.100 g	0.094 g	0.0063 g

Could the material be a pure compound? Explain.

46. A colorless liquid is thought to be a pure compound. Analyses of three samples of the material yield the following results.

	Mass of Sample	Mass of Carbon	Mass of Hydrogen
Sample 1	1.000 g	0.862 g	0.138 g
Sample 2	1.549 g	1.295 g	0.254 g
Sample 3	0.988 g	0.826 g	0.162 g

Could the material be a pure compound? Explain.

Expand Your Skills

47. A compound of uranium and fluorine is used to generate uranium for nuclear power plants. The gas can be decomposed to yield 2.09 parts by mass of uranium for every 1 part by mass of fluorine. If the relative mass of a uranium atom is 238 and the relative mass of a fluorine atom is 19,

calculate the number of fluorine atoms that are combined with one uranium atom.

48. In one experiment, 3.06 g hydrogen was allowed to react with an excess of oxygen to form 27.35 g water. In a second experiment, electric current broke down a sample of water

into 1.45 g hydrogen and 11.51 g oxygen. Are these results consistent with the law of definite proportions? Show why or why not.

49. Two experiments were performed in which sulfur was burned completely in pure oxygen gas, producing sulfur dioxide and leaving some unreacted oxygen. In the first experiment, burning 0.312 g of sulfur produced 0.623 g of sulfur dioxide. In the second experiment, 1.305 g of sulfur was burned. What mass of sulfur dioxide was produced?

50. Use Figure 2.5 to calculate the mass of mercuric oxide that would be needed to produce 100.0 g of mercury metal.

51. See Table 2.1. Another compound of nitrogen and oxygen contains 0.5836 g of nitrogen per 1.000 g of oxygen. Calculate the ratio of mass of N in this compound to mass of N in nitrogen dioxide. What *whole-number* ratio does this value represent?

52. Gasoline can be approximated by the formula C_8H_{18}. An environmental advocate points out that burning one gallon (about 7 lb.) of gasoline produces about 19 lb. of carbon dioxide, a greenhouse gas that raises the temperature of the atmosphere. Explain this seeming contradiction of the law of conservation of mass.

53. In the experiment shown in the accompanying photos, about 15 mL of hydrochloric acid solution was placed in a flask and approximately 3 g of sodium carbonate was put into a balloon. The opening of the balloon was then carefully stretched over the top of the flask, taking care not to allow the sodium carbonate to fall into the acid in the flask. The flask was placed on an electronic balance, and the mass

of the flask and its contents was found to be 38.61 g. The sodium carbonate was then slowly shaken into the acid. The balloon began to fill with gas. When the reaction was complete, the mass of the flask and its contents, including the gas in the balloon, was found to be 38.61 g. What law does this experiment illustrate? Explain.

54. Fritz Haber was awarded a Nobel Prize for the processes he invented in which nitrogen and hydrogen gases are combined to make ammonia (NH_3), a valuable chemical and a vital nutrient in modern agriculture. It is known that ammonia is 82.2% nitrogen by mass.
   a. If 28.0 g of nitrogen and 6.05 g of hydrogen are combined in a closed system, what mass of ammonia will be produced?
   b. What mass of nitrogen is required to fully consume 47.5 g of hydrogen, and what mass of ammonia will be produced?
   c. If we start with 16.0. g of nitrogen and 2.02 g of hydrogen, what mass of nitrogen would remain if all the hydrogen were consumed?

55. Using Döbereiner's triadic relationship, calculate the mass of Se
   a. based on Mendeleev's values from his periodic table (Figure 2.11)
   b. based on values in the modern periodic table in the inside front cover of this book

56. Wastes from gold processing that contain potassium cyanide (KCN) can be rendered nonhazardous by heating at very high temperatures in the presence of oxygen. Wastes from battery production containing lead (II) sulfate, or $PbSO_4$, cannot be rendered safe by this treatment. Suggest a reason why this is so.

57. Identify whether the following elements are hazardous, rare, or neither.
   a. silicon   b. neodymium   c. mercury
   d. oxygen   e. lead

58. Give two examples of how green chemistry has helped to reduce the use of cadmium in consumer products.

59. Why is it important to recycle mercury-based fluorescent light bulbs instead of putting them in landfills?

 Critical Thinking Exercises

Apply knowledge that you have gained in this chapter and one or more of the FLaReS principles (Chapter 1) to evaluate the following statements or claims.

2.1. In a science fiction movie, a woman proceeds through nine months of pregnancy in minutes. She takes in no nutrients during this time. She dies during labor, and an emergency C-section is performed to save the child. The child lives for only a matter of hours, rapidly aging, and dies a withered old man. A classmate claims that the movie is based on a government-documented but secret alien encounter.

2.2. When water is electrolyzed, from each one molecule of water you always get one atom of oxygen and two atoms of hydrogen. However, when 1,000,000 U.S. dollars are exchanged for Canadian dollars, the amount of Canadian dollars may vary.

2.3. A health food store has a large display of bracelets made of copper metal. Some people claim that wearing such a bracelet will protect the wearer from arthritis or rheumatoid diseases.

2.4. An old cookbook claims that cooking acidic food, such as spaghetti sauce, in an iron pot provides more nutrients than cooking the same food in an aluminum pot.

2.5. A company markets a device it calls a "water ener-oxizer." It claims the device can supply so much energy to drinking water that the mass of oxygen in the water is increased, thereby providing more oxygen to the body.

2.6. According to a mystic, one can tell whether someone is guilty or innocent by using a diamond. His claim is that, in a guilty person's hand, the diamond appears dull, while it sparkles when handled by an innocent person, due to the person's goodness.

Collaborative Group Projects

Prepare a PowerPoint, poster, or other presentation (as directed by your instructor) to share with the class.

1. Prepare a brief report on early Greek contributions to and ideas in the field of science, focusing on the work of one of the following: Aristotle, Leucippus, Democritus, Thales, Anaximander, Anaximenes of Miletus, Heraclitus, Empedocles, or another Greek philosopher of the time before 300 B.C.E.

2. Prepare a brief report on the phlogiston theory. Describe what the theory was, how it explained changes in mass when something is burned, and why it was finally abandoned.

3. Write a brief essay on recycling of one of the following: metals, paper, plastics, glass, food wastes, or grass clippings. Contrast a recycling method that maintains the properties of an element with one that changes them.

 LET'S EXPERIMENT! Reaction in a Bag: Demonstrating the Law of Conservation of Matter

Materials Needed

- Alka-Seltzer tablets (8)
- $\frac{1}{4}$ cup of water
- 2 small plastic cups
- Sealable plastic bag (1)
- Kitchen scale

Can you observe the law of conservation of matter if a gas is produced in a chemical reaction?

The law of conservation of matter states that atoms are not created or destroyed during a chemical reaction. For your experiment, we are challenging you to demonstrate this law through the dissolution of Alka-Seltzer tablets in water.

Alka-Seltzer tablets function as an antacid and pain reliever, with aspirin as the analgesic. When the tablets are dissolved in water, citric acid and sodium bicarbonate are dissolved and react, forming water and carbon dioxide. The sodium citrate formed and the aspirin already in the product can be consumed to help with heartburn and pain relief. In this experiment, you can observe whether atoms are lost in the reaction, and, therefore, you are observing the law of conservation of matter.

First, measure $\frac{1}{8}$ cup of water into a small plastic cup. Record the weight of the water and the cup using a kitchen scale. (Hint: Set it to grams.) Drop four Alka-Seltzer tablets into the water and immediately record the total weight. Write down your observations as the tablets dissolve in the water. What observations can you make about the mixture?

After the reaction is completed, measure the final weight of the mixture. Did you see a difference between the weight measurements before and after the reaction?

For the next step, run the reaction again, but this time place four Alka-Seltzer tablets and $\frac{1}{8}$ cup of water (again in a small plastic cup) into a large plastic bag, with the tablets out of the water at first. (Hold the tablets in one corner of the bag.) Squeeze most of the air out of the bag, seal it, and weigh the bag and its contents. Then allow the tablets to fall into the cup of water. Weigh the bag and its contents both before and after the reaction.

Questions

1. What observations can be made about the mixture in the bag as compared with the mixture that was not in the bag?
2. Did you see a change in the weight of the mixture in the sealed bag? Why or why not?
3. What does this experiment tell you about the law of conservation of matter?

Atomic Structure

Neon lights—the signature of entertainment, restaurants, bars, and the big city. Glass tubes filled with neon have a characteristic red-orange colorwhen current is applied. Other colors are created by coating the inside of the tubes with phosphors or filling with another gas. Why red-orange for neon? The answer lies in the structure of its atoms—the subject of this chapter.

Have You Ever Wondered?

1 If we can't see atoms, how do we know they are made of even smaller particles? **2** Why are accidental discoveries so common in science? **3** What is radioactivity? **4** If atoms are mostly empty space, why are most solids hard? **5** Most light bulbs produce white light. Why does a neon light appear red-orange? **6** What is "periodic" about the periodic table?

IMAGES OF THE INVISIBLE   Atoms are exceedingly tiny particles, much too small to see even with an optical microscope. It is true that scientists can obtain images of individual atoms, but they use special instruments such as a scanning tunneling microscope (STM). Even so, we can see only outlines of atoms and their arrangements in a substance. If atoms are so small, how can we possibly know anything about their inner structures?

Although scientists have never examined the interior of an atom directly, they have been able to obtain a great deal of *indirect* information. By designing clever experiments and exercising their powers of deduction, scientists have been able to develop amazingly detailed information about what an atom's interior is like.

Why do we care about the structure of particles as tiny as atoms? The reason is that the arrangement of the various parts of atoms determines the properties of different kinds of atoms and the matter they make up. Only by understanding atomic structure can we learn how atoms combine to make millions of different substances. With such knowledge, we can modify and synthesize materials to meet our needs more precisely.

Perhaps of greater interest to you is the fact that your understanding of chemistry (as well as much of biology and other sciences) depends, at least in part, on your knowledge of atomic structure. Let's begin by going back to the time of John Dalton.

3.1 Electricity and the Atom

Learning Objectives • Explain the electrical properties of an atom. • Describe how the properties of electricity explain the structure of atoms.

▲ In addition to his work in electrochemistry, Michael Faraday also devised methods for liquefying gases such as chlorine and invented the electrical transformer.

The first decade of the nineteenth century marked significant changes in scientists' concept of the atom. The Greeks had believed that atoms were the smallest particles of anything, such as sand, water, rocks, or air. Dalton, who set forth his atomic theory in 1803, defined an atom as the smallest invisible particle *of an element*, which was a new concept, since it related atoms to elements, not to compounds or mixtures. Dalton envisioned atoms as extremely hard and indivisible "balls," a theory that intuitively made sense at the time, since solids are "hard." So how does electricity relate to atoms, and how did the discovery of electric current expand our understanding of atoms?

Static electricity had been observed since ancient times, but continuous electric current was born with the nineteenth century. In 1800, Alessandro Volta (Italy, 1745–1827) invented an electrochemical cell much like a modern battery. If the poles of such a cell are connected by a wire, current flows through the wire. The current is sustained by chemical reactions inside the cell. Volta's invention soon was applied in many areas of science and everyday life.

Electrolysis

Also in the year 1800, William Nicholson (England, 1753–1815) and Anthony Carlisle (England, 1768–1840) used electric current (electrolysis) to decompose water into hydrogen and oxygen gas—an astonishing discovery, as water had been considered an "element" since the time of the Greeks. Soon thereafter, Humphry Davy (1778–1829), a British chemist, built a powerful battery that he used to pass electricity through molten (melted) salts. Davy quickly discovered several new elements. In 1807, he was the first to use electrolysis to form highly reactive potassium metal from molten potassium hydroxide. Shortly thereafter, he produced sodium metal by passing electricity through molten sodium hydroxide. Within a year, Davy had also produced magnesium, strontium, barium, and calcium metals for the first time. The science of electrochemistry was born! (We will study electrochemistry in more detail in Chapter 8.)

Davy's protégé, Michael Faraday (England, 1791–1867), greatly extended this new science. Lacking in formal education, Faraday consulted others to define many of the terms we still use today. His physician helped coin the term **electrolysis** for chemical reactions caused by electricity (Figure 3.1) and the name **electrolyte** for a compound that conducts electricity when melted or dissolved in water. The carbon rods or metal strips inserted into a molten compound or a solution to carry the electric current are called **electrodes**. The electrode that bears a positive charge is the **anode** and the negatively charged electrode is the **cathode**. The entities that carry the electric current through a melted compound or a solution are called *ions* (which will be discussed in more detail in Chapter 4). An **ion** is an atom or a group of atoms bonded together that has an electric charge. An ion with a negative charge is an **anion**; it travels toward the anode. A positively charged ion is a **cation**; it moves toward the cathode.

Faraday studied what happens when the metal electrodes from a battery are inserted into an aqueous solution of various salts, not just the melted salt. It had been known since the time of the alchemists that copper metal dissolves in strong acids, forming a blue solution. When Faraday placed the electrodes from a battery into such a blue solution, pure copper metal appeared on the surface of the cathode. This observation and many other analogous studies led him to the very important idea that the particles that carry the "charge" of the battery (which we now call *electrons*) can *go into* or *be removed from* atoms. Therefore, those little particles must be *in* atoms. Because of Faraday's work, most scientists throughout the nineteenth century believed that atoms were small, hard particles, with those tiny charged particles stuck loosely on the outside of the atom (or possibly inside), like balls of unbaked chocolate chip cookies, or raisins in a pudding. It took several decades for further information about atomic structure to be discovered.

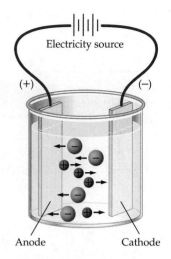

▲ **Figure 3.1** An electrolysis apparatus. The electricity source (for example, a battery) directs electrons through wires from the anode to the cathode. Cations (+) are attracted to the cathode (−), and anions (−) are attracted to the anode (+). This migration of ions is the flow of electricity through the solution.

Cathode-Ray Tubes

In 1875, William Crookes (1832–1919), an English chemist, was able to construct a low-pressure gas-discharge tube that would allow electricity to pass through it. Crookes's tube was the forerunner of neon tubes used for signs and of cathode-ray tubes used in early television sets and fluorescent light tubes. Metal electrodes are sealed in the tube, which is connected to a vacuum pump so that most of the air can be removed from it. Crookes observed a beam of green fluorescence when a strong electrical current struck a screen coated with zinc sulfide. This beam seemed to leave the cathode and travel to the anode, and was called a **cathode ray**.

Thomson's Experiment: Mass-to-Charge Ratio

Considerable speculation arose about the nature of cathode rays, and many experiments were undertaken. Were these rays beams of particles, or did they consist of a wavelike form of energy, much like visible light? The answer came (as scientific answers should) from an experiment, performed by the English physicist Joseph John Thomson (1856–1940) in 1897. His experiment is shown in Figure 3.2. Thomson showed that cathode rays were deflected in an electric field (Figure 3.2a). The beam was attracted to the positive plate and repelled by the negative plate. Thomson concluded that cathode rays consisted of negatively charged particles. His experiments also showed that the particles were the same no matter from what material the electrodes were made or what type of gas was in the tube. He concluded that all atoms have these negative particles in them. Thomson named these negatively charged units **electrons**. Cathode rays, then, are beams of electrons emanating from the cathode of a gas-discharge tube.

Cathode rays are deflected in magnetic fields as well as in electric fields (Figure 3.2b). The greater the charge on a particle, the more it is deflected in an electric (or magnetic) field. The heavier the particle, the less it is deflected by the field. By measuring the amount of deflection in fields of known strength, Thomson was able to calculate the *ratio* of the mass of the electron to its charge. (He could not measure either the mass or the charge separately.) For this research in the conduction of electricity by gases, Thomson was awarded the Nobel Prize in Physics in 1906.

1 If we can't see atoms, how do we know they are made of even smaller particles?

The existence of electrons and other subatomic particles was deduced from observations of the behavior of electricity and its interaction with matter. The existence and behavior of the subatomic particles make atoms behave the way they do.

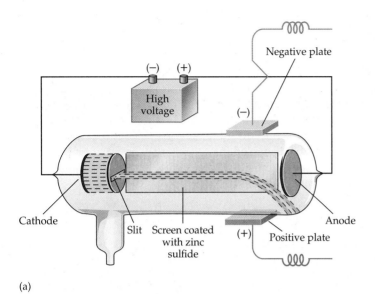

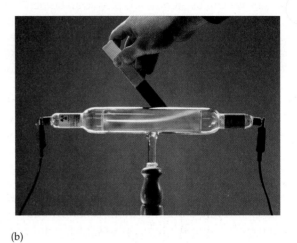

(a)   (b)

▲ **Figure 3.2** Thomson's apparatus, showing deflection of the cathode rays (a beam of electrons). Cathode rays are invisible but can be observed because they produce a green fluorescence when they strike a zinc sulfide-coated screen. The diagram (a) shows deflection of the beam in an electric field. Notice the direction that the beam deflects. The photograph (b) shows the deflection in a magnetic field. Cathode rays travel in straight lines unless some kind of external field is applied.

Goldstein's Experiment: Positive Particles

In 1886, German scientist Eugen Goldstein (1850–1930) performed experiments with gas-discharge tubes that had perforated cathodes (Figure 3.3). He found that although electrons were formed and sped off toward the anode as usual, positive particles were also formed and shot in the opposite direction, toward the cathode. Some of these positive particles went through the holes in the cathode. In 1907, a study of the deflection of these positive particles in a magnetic field indicated that the mass of these particles depended on the gas in the cathode. The lightest particles, formed when there was a little hydrogen gas in the tube, were later shown to have a mass 1837 times that of an electron.

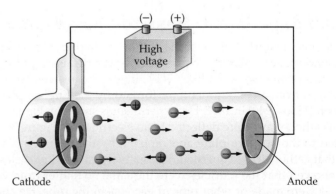

▲ **Figure 3.3** Goldstein's apparatus for the study of positive particles. This apparatus, with holes in the cathode, is a modification of the cathode-ray tube in Figure 3.2. Some positive ions, attracted toward the cathode, pass through the holes in the cathode. The deflection of these particles (not shown) in a magnetic field can be studied in the region to the left of the cathode.

Millikan's Oil-Drop Experiment: Electron Charge

The charge on the electron was determined in 1909 by Robert A. Millikan (America, 1868–1953), a physicist at the University of Chicago. Millikan observed electrically charged oil drops in an electric field. A diagram of his apparatus is shown in

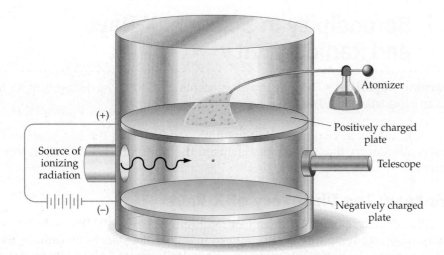

▲ **Figure 3.4** The Millikan oil-drop experiment. Oil drops irradiated with X-rays pick up electrons and become negatively charged. Their fall due to gravity can be balanced by adjusting the voltage of the electric field. The charge on the oil drop can be determined from the applied voltage and the mass of the oil drop. The charge on each drop is that of some whole number of electrons.

Figure 3.4. A spray bottle is used to form tiny droplets of oil. Some acquire negative charges, either from irradiation with X-rays or from static electricity. Either way, the droplets acquire one or more extra electrons.

Some of the oil droplets pass into a chamber, where they can be viewed through a telescope. The negative plate at the bottom of the chamber repels the negatively charged droplets, and the positive plate attracts them. By manipulating the charge on each plate and observing the behavior of the droplets, the charge on each droplet can be determined. Millikan took the smallest possible difference in charge between two droplets to be the charge of an individual electron. For his research, he received the Nobel Prize in Physics in 1923.

From Millikan's value for the charge and Thomson's value for the mass-to-charge ratio, the mass of the electron was readily calculated. That mass was found to be only 9.1×10^{-28} g. To have a mass of 1 g would require more than 1×10^{27} electrons—that's a 1 followed by 27 zeros, or a billion billion billion electrons. More important, the electron's mass is much smaller than that of the lightest atom.

In practice, we seldom use the actual mass and charge of the electron. For many purposes, the electron is considered the unit of electrical charge. The charge is shown as a superscript minus sign (meaning $^{1-}$). In indicating charges on ions, we use $^-$ to indicate a net charge due to one "extra" electron, $^{2-}$ to indicate a charge due to two "extra" electrons, and so on.

SELF-ASSESSMENT Questions

1. The use of electricity to cause a chemical reaction is called
   a. anodizing
   b. corrosion
   c. electrolysis
   d. hydrolysis

2. During electrolysis of molten sodium chloride, sodium cations move to the
   a. anode, which is negatively charged
   b. anode, which is positively charged
   c. cathode, which is negatively charged
   d. cathode, which is positively charged

3. Cathode rays are
   a. composed of positively charged particles
   b. composed of particles with a mass of 1 amu
   c. deflected by electric fields
   d. different for every element

4. J. J. Thomson measured the electron's
   a. atomic number
   b. charge
   c. mass
   d. mass-to-charge ratio

5. In his oil-drop experiment, Millikan determined the charge of an electron by
   a. measuring the diameters of the tiniest drops
   b. noting that drops had either a certain minimum charge or multiples of that charge
   c. using an atomizer that could spray out individual electrons
   d. viewing the smallest charged drops microscopically and noting that they had one extra electron

Answers: 1. c; 2. c; 3. c; 4. d; 5. b

2 Why are accidental discoveries so common in science?

Serendipity—a "happy accident" like drilling for water and striking oil—happens to everyone at times. Scientists, however, are trained observers. When something unexpected happens, they often wonder why and investigate. The same accident might happen to an untrained person and go unnoticed—though even scientists often miss important finds at first. Or, if something unusual is noticed, its significance might not be grasped.

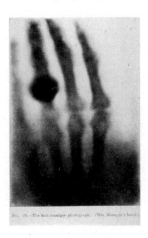

Fɪɢ. 29.—The first roentgen photograph. (Mrs. Röntgen's hand.)

▲ The use of X-rays in medical diagnosis began shortly after they were discovered by Wilhelm Röntgen in 1895. In fact, the first known X-ray photograph is the one seen here—of Mrs. Röntgen's hand!

▲ Marie Sklodowska Curie and Pierre Curie in their laboratory. They shared the 1903 Nobel Prize in Physics with Antoine Henri Becquerel.

3.2 Serendipity in Science: X-Rays and Radioactivity

Learning Objective • Describe the experiments that led to the discovery of X-rays and an explanation of radioactivity.

Let's take a look at two serendipitous discoveries during the last years of the nineteenth century that profoundly changed the world and are related to atomic structure.

Röntgen: The Discovery of X-Rays

In 1895, German scientist Wilhelm Conrad Röntgen (1845–1923) was working in a dark room, studying the glow produced in certain substances by cathode rays. To his surprise, he noted this glow on a chemically treated piece of paper some distance from the cathode-ray tube. The paper even glowed when a wall separated the tube from the paper. Clearly, this type of ray could travel through walls. When he waved his hand between the radiation source and the glowing paper, he was able to see an image of the bones of his hand in darker shades on the paper. He called these mysterious rays **X-rays**. He found, in general, that X-rays were absorbed more by bone or other hard and fairly dense materials than by soft, less dense tissues.

We now know that X-rays are a form of electromagnetic radiation—energy with electric and magnetic components. Electromagnetic radiation also includes visible light, radio waves, microwaves, infrared radiation, ultraviolet light, and gamma rays (Section 3.3). These forms differ in energy, with radio waves lowest in energy, followed by microwaves, infrared radiation, visible light, ultraviolet light, X-rays, and, the most energetic, gamma rays.

Today, X-rays are one of the most widely used tools in the world for medical diagnosis. Not only are they employed for examining decayed teeth, broken bones, and diseased lungs, but they are also the basis for such procedures as mammography and computerized tomography (Chapter 11). How ironic that Röntgen himself made no profit at all from his discovery. He considered X-rays a "gift to humanity" and refused to patent any part of the discovery. However, he did receive much popular acclaim and in 1901 was awarded the first Nobel Prize in Physics.

The Discovery of Radioactivity

Certain chemicals exhibit *fluorescence* after exposure to strong sunlight. They continue to glow even when taken into a dark room. In 1895, Antoine Henri Becquerel (1852–1908), a French physicist, was studying fluorescence by wrapping photographic film in black paper, placing a few crystals of the fluorescing chemical on top of the paper, and then placing the package in strong sunlight. If the glow was like ordinary light, it would not pass through the paper. On the other hand, if it were similar to X-rays, it would pass through the black paper and fog the film.

While working with a uranium compound, Becquerel made an important accidental discovery. When placed in sunlight, the compound fluoresced and fogged the film. On several cloudy days when exposure to sunlight was not possible, he prepared samples and placed them in a drawer. To his great surprise, the photographic film was fogged even though the uranium compound had not been exposed to sunlight. Further experiments showed that the radiation coming from the uranium compound was unrelated to fluorescence but was a characteristic of the element uranium.

Other scientists immediately began to study this new radiation. Becquerel's graduate student from Poland, Marie Sklodowska (1867–1934), gave the phenomenon a name: *radioactivity*. **Radioactivity** is the spontaneous emission of radiation or actual particles from certain unstable elements. Marie later married Pierre Curie (1859–1906), a French physicist. Together they discovered the radioactive elements polonium and radium.

After her husband's death in 1906, Marie Curie continued to work with radioactive substances, winning the Nobel Prize in Chemistry in 1911. For more than 50 years, she was the only person ever to have received two Nobel Prizes.

SELF-ASSESSMENT Questions

1. When Wilhelm Röntgen saw an X-ray image of his hand, the bones were darker than the glowing paper because the X-rays were
  a. absorbed more by bone than by flesh
  b. radiation that is less energetic than visible light
  c. made up of fast-moving atoms
  d. a type of radioactivity

2. Radioactivity arises from elements that
  a. absorb radio waves and release stronger radiation
  b. are unstable and emit radiation
  c. fluoresce
  d. undergo fission (splitting) of their atoms when they absorb energy

Answers: 1, a; 2, b

3.3 Three Types of Radioactivity

Learning Objective • Distinguish the three main kinds of radioactivity: alpha, beta, and gamma.

Scientists' experiments soon showed that three types of radiation emanated from various radioactive elements. Ernest Rutherford (1871–1937), a New Zealander who early in his career worked under J. J. Thomson (Section 3.1), chose the names *alpha*, *beta*, and *gamma* for the three types of radioactivity. When passed through a strong magnetic or electric field, the alpha type was deflected in a manner indicating that it consisted of a beam of positive particles (Figure 3.5). Later experiments showed that an **alpha (α) particle** has a mass four times that of a hydrogen atom and a charge twice the magnitude of, but opposite in sign to, that of an electron. An alpha particle is in fact identical to the nucleus (Section 3.4) of a helium atom, and is often symbolized by He^{2+}. These discoveries were quite important, since scientists had believed up until that time that the supposedly solid mass of the atom was indivisible. If particles the size of helium atoms could be emitted from larger atoms, atoms must consist of many sub-particles. In short, atoms were no longer believed to be indivisible.

The second type of radioactivity is called beta radiation and was shown to be made up of negatively charged particles identical to those of cathode rays. Therefore, a **beta (β) particle** is an electron, although it has much more energy than an electron

3 **What is radioactivity?**

Radioactivity is the emission of radiation or particles caused by the spontaneous disintegration of unstable atomic nuclei. Usually those atoms have either too many or too few neutrons (see Section 3.5). Exposure to the radiation—whether alpha, beta, or gamma rays—given off by radioactive material is usually hazardous unless appropriate safety precautions have been taken.

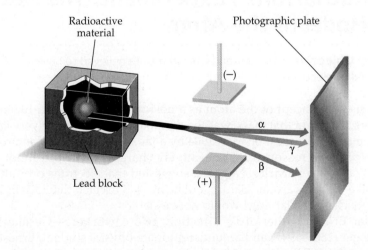

▲ **Figure 3.5** Behavior of a beam of radioactivity in an electric field.

Q *Why are alpha rays deflected upward in this figure? Why are beta rays deflected downward?*

TABLE 3.1 Types of Radioactivity

Name	Greek Letter	Mass (amu)	Charge
Alpha	α	4	2+
Beta	β	$\dfrac{1}{1837}$	1−
Gamma	γ	0	0

WHY IT MATTERS

Some people fear anything labeled "radiation"—a fear that is scientifically unfounded. It is important to distinguish among ordinary visible radiation (light), infrared radiation (heat), and the three types of radiation given off by radioactive atoms. As with light, small amounts of alpha, beta, or gamma rays generally can be absorbed by materials without causing problems thereafter. For example, alpha particles can be stopped by skin or a notebook. The main danger from radioactivity is from radioactive atoms that are ingested, or that have stuck to or been absorbed by materials or tissue, and that continuously give off damaging alpha, beta, or gamma rays. After the Fukushima Daiichi nuclear disaster in March 2011, a large amount of radioactive material was released from the damaged nuclear reactors. In the picture, workers from the Fukushima plant are being checked for radioactive particles on their clothing. Cleaning up the radioactive material from the accident will likely take decades and cost billions of dollars. A positive use of gamma rays or ultraviolet radiation is to kill harmful microorganisms in food, making the food safer. These topics are covered in more detail in Chapter 11.

in an atom. **Gamma (γ) rays** are not deflected by a magnetic field. They are a form of electromagnetic radiation, much like the X-rays used in medical work but even more energetic and more penetrating. The three types of radioactivity are summarized in Table 3.1. The masses are indicated in atomic mass units (amu). (See Section 2.4.) Biologists may use the units "daltons" for the atomic masses.

The discoveries of the late nineteenth century paved the way for an entirely new picture of the atom, which developed rapidly during the early years of the twentieth century.

SELF-ASSESSMENT Questions

1. The mass of an alpha particle is about
   a. the same as that of an electron
   b. four times that of a hydrogen atom
   c. the same as that of a hydrogen atom
   d. the same as that of a beta particle

2. Compared with the charge on an electron (designated 1−), the charge on an alpha particle is
   a. 4−   b. 1−   c. 1+   d. 2+   e. 4+

3. A beta particle is a(n)
   a. boron atom   b. electron
   c. helium nucleus   d. proton

4. The mass of a gamma ray is
   a. −1 amu   b. 0 amu
   c. 1 amu   d. 4 amu

Answers: 1, b; 2, d; 3, b; 4, b

3.4 Rutherford's Experiment: The Nuclear Model of the Atom

Learning Objective • Understand why atoms are believed to have a tiny nucleus surrounded by electrons.

J. J. Thomson's concept of the atom as a positively charged, ball-like mass, with electrons scattered in that mass, was the dominant theory into the very early 1900s. But that model was questioned in 1904 by a Japanese scientist, Hantaro Nagaoka (1865–1950), who reasoned that negatively charged particles cannot penetrate positively charged particles. Nagaoka suggested that electrons orbit the positive charge, like the rings around Saturn, but he himself rejected that idea in 1908, since electrically charged, flat rings would not be stable.

Under Ernest Rutherford's direction, two coworkers—German physicist Hans Geiger (1882–1945) and an undergraduate physics student, Ernest Marsden (England 1889–1970)—bombarded a very thin sheet of gold foil with alpha particles from a radioactive source (Figure 3.6). The vast majority of the alpha particles passed through the gold foil; some were deflected somewhat, and occasionally, one was sent right back in the direction from which it had come!

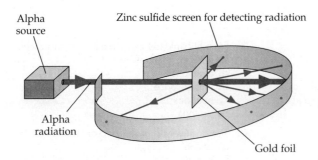

▲ **Figure 3.6** Rutherford's gold-foil experiment. Most alpha particles passed right through the gold foil, but now and then a particle was deflected.

4 If atoms are mostly empty space, why are most solids hard?

Even though most of the mass of atoms is concentrated in the tiny nucleus, the electrons in the outermost area of atoms are moving with tremendous speed in their orbitals; they are *not* simply located in some specific spot, or lazily floating about. This forms what science fiction could call a force field. Even when packed together tightly, as in a piece of metal, the atoms cannot penetrate each other.

To explain the results, Rutherford came to two important conclusions that shook the scientific community at that time (Figure 3.7):

1. Atoms are mostly empty space, with the electrons in the outermost region of atoms.
2. Atoms have a very tiny **nucleus** in the center that is positively charged and incorporates nearly all the mass of an atom.

Rutherford's new theory was consistent with his observations, since atoms that consisted of mostly empty space would have a low probability that positively charged alpha particles would strike and be repelled by a small, positively charged nucleus. Rutherford concluded that the electrons had so little mass that they were no match for the alpha particle "bullets." It would be analogous to a mouse trying to stop a charging hippopotamus.

Rutherford's nuclear theory of the atom, set forth in 1911, was revolutionary. To picture Rutherford's model, visualize a sphere as big as the largest covered football stadium. The nucleus at the middle of the sphere is as small as a pea but weighs several million tons. A few flies flitting here and there throughout the sphere represent the electrons.

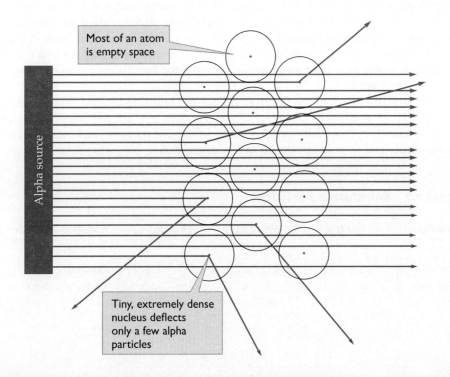

◀ **Figure 3.7** Model explaining the results of Rutherford's gold-foil experiment. Most of the alpha particles pass right through the foil, because the gold atoms are mainly empty space. But some alpha particles are deflected as they pass close to a dense, positively charged atomic nucleus. Once in a while, an alpha particle encounters an atomic nucleus head-on and is knocked back in the direction from which it came.

SELF-ASSESSMENT Questions

1. The most persuasive evidence from Rutherford's gold-foil experiment was that
  a. all the alpha particles were deflected a bit
  b. most of the alpha particles went right through the foil, but a tiny fraction were deflected right back toward the source
  c. some gold atoms were converted to lead atoms
  d. some gold atoms were split

2. Rutherford's gold-foil experiment showed that
  a. gold atoms consisted of electrons in a "pudding" of positive charge
  b. the nucleus was huge and positively charged
  c. the nucleus was tiny and positively charged
  d. some gold atoms were shattered by alpha particles

3. From his gold-foil experiment, Rutherford concluded that atoms consist mainly of
  a. electrons
  b. empty space
  c. neutrons
  d. protons

4. The positively charged part of the atom is its
  a. cation
  b. nucleus
  c. outer electrons
  d. widespread cloud of electrons

Answers: 1, b; 2, c; 3, b; 4, b

3.5 The Atomic Nucleus

Learning Objectives • List the particles that make up the nucleus of an atom, and give their relative masses and electric charges. • Identify elements and isotopes from their nuclear particles.

In 1914, Rutherford suggested that the smallest positive particle is the unit of positive charge in the nucleus, and in 1920 he proposed the name *proton* for such particles. The **proton** has a charge equal in magnitude to that of the electron and has nearly the same mass as a hydrogen atom. The nucleus of a hydrogen atom consists of one proton, and the nuclei of larger atoms contain greater numbers of protons.

But most atomic nuclei are heavier than is indicated by the number of positive charges (number of protons). For example, the helium nucleus has a charge of 2+ (and therefore two protons, according to Rutherford's theory), but its mass is *four* times that of hydrogen. This excess mass puzzled scientists at first. Rutherford had suggested in 1920 that some other neutral particles must be in atoms. Then, in 1932, English physicist James Chadwick (1891–1974) discovered a particle with about the same mass as a proton but with no electrical charge. It was called a **neutron**, and its existence explains the unexpectedly high mass of the helium nucleus. Whereas the hydrogen nucleus contains only one proton of mass 1 amu, the helium nucleus contains not only two protons (2 amu) but also two neutrons (2 amu), giving it a total mass of 4 amu.

With the discovery of the neutron, the list of "building blocks" that we use for "constructing" atoms was complete. The properties of these particles are summarized in Table 3.2. (There are dozens of other subatomic particles, but most exist only momentarily and are not important to our discussion.)

TABLE 3.2 Subatomic Particles

Particle	Symbol	Mass (u)	Charge	Location in Atom
Proton	p^+	1	1+	Nucleus
Neutron	n	1	0	Nucleus
Electron	e^-	$\dfrac{1}{1837}$	1−	Outside nucleus

Atomic Number

The number of protons in the nucleus of an atom of any element is the **atomic number (Z)** of that element. This number determines the kind of atom—that is, the identity of the element—and where it is found on any periodic table or list of the elements. Dalton thought that the mass of an atom determined the element. We now know it is not the mass but the number of protons that determines the identity of an element. For example, every atom with 26 protons is an atom of iron (Fe), whose atomic number $Z = 26$. Any atom with 50 protons is an atom of tin (Sn), which has $Z = 50$. In a neutral atom (one with no electrical charge), the positive charge of the protons is exactly neutralized by the negative charge of the electrons. The attractive forces between the unlike charges help hold the atom together.

A proton and a neutron have almost the same mass, 1.0073 amu and 1.0087 amu, respectively. This difference is so small that it usually can be ignored. Thus, for many purposes, we can assume that the masses of the proton and the neutron are the same, 1 amu. The proton has a charge equal in magnitude but opposite in sign to that of an electron. This charge on a proton is written as 1+. The electron has a charge of 1− and a mass of 0.00055 amu. The electrons in an atom contribute so little to its total mass that their mass is usually disregarded (treated as if it were zero). For neutral atoms, the number of electrons is equal to the number of protons in the nucleus.

Isotopes

We can now more precisely define *isotopes*, a term we mentioned in Section 2.3. Atoms of a given element can have different numbers of neutrons in their nuclei. For example, most hydrogen atoms have a nucleus consisting of a single proton and no neutrons. However, about 1 hydrogen atom in 6700 has a neutron as well as a proton in its nucleus. This heavier hydrogen atom is called a **deuterium** atom. A third, rare isotope of hydrogen is **tritium**, which has two neutrons and one proton in the nucleus.

Whether it has one neutron, two neutrons, or no neutrons, any atom with $Z = 1$ (with one proton) is a hydrogen atom. Atoms that have this sort of relationship—having the same number of protons but different numbers of neutrons—are called **isotopes** (Figure 3.8).

Most, but not all, elements exist in nature in isotopic forms. For example, tin is present in nature in 10 different isotopic forms. It also has 15 radioactive isotopes that do not occur in nature. Aluminum, on the other hand, has only one naturally occurring isotope.

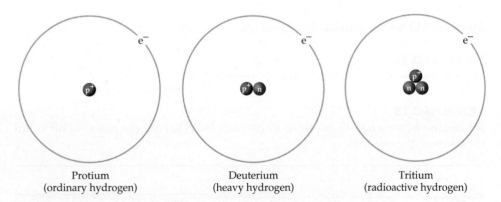

| Protium (ordinary hydrogen) | Deuterium (heavy hydrogen) | Tritium (radioactive hydrogen) |

▲ **Figure 3.8** The three isotopes of hydrogen. Each has one proton and one electron, but they differ in the number of neutrons in the nucleus.

Q *What is the atomic number and mass number of each of the isotopes? What is the mass of each, in atomic mass units, to the nearest whole number?*

Isotopes usually have little or no effect on ordinary chemical reactions. For example, all three hydrogen isotopes react with oxygen to form water. Because these isotopes differ in mass, compounds formed with them have different physical properties, but such differences are usually slight. Water in which both hydrogen atoms are deuterium is called *heavy water*, often represented as D_2O. Heavy water boils at 101.4 °C (instead of 100 °C) and freezes at 3.8 °C (instead of 0 °C). In nuclear reactions, however, isotopes are of the utmost importance, as we shall see in Chapter 11.

Symbols for Isotopes

Collectively, the two main nuclear particles, protons and neutrons, are called **nucleons**. Isotopes are represented by symbols with subscripts and superscripts.

$$^{A}_{Z}X$$

In this general symbol, Z is the nuclear charge (atomic number, or number of protons), and A is the **mass number**, or the **nucleon number**, because it is the number of particles in the nucleus. (*Nucleon number* is the term recommended by the International Union of Pure and Applied Chemistry, IUPAC.) We prefer "mass number" because it best represents the mass of the atom's nucleus, since amu means *atomic mass unit*. In other words, A = the total number of particles in the nucleus of a given atom, or the protons plus neutrons. As an example, the isotope with the symbol

$$^{35}_{17}Cl$$

has 17 protons and 35 nucleons. The number of neutrons is therefore $35 - 17 = 18$.

Isotopes often are named by adding the mass number as a suffix to the name of the element. The three hydrogen isotopes are represented as

$$^{1}_{1}H \quad ^{2}_{1}H \quad ^{3}_{1}H$$

and named hydrogen-1 (sometimes called protium), hydrogen-2 (deuterium), and hydrogen-3 (tritium), respectively.

 EXAMPLE 3.1   Number of Neutrons

How many neutrons are there in the $^{235}_{92}U$ nucleus?

Solution
We simply subtract the atomic number Z (number of protons) from the mass number A (number of protons plus number of neutrons).

$$A - Z = \text{number of neutrons}$$
$$235 - 92 = 143$$

There are 143 neutrons in the $^{235}_{92}U$ nucleus.

〉 Exercise 3.1A
How many neutrons are there in the $^{39}_{19}K$ nucleus?

〉 Exercise 3.1B
A bromine isotope has 45 neutrons in its nucleus. What are the mass number and the name of the isotope?

 CONCEPTUAL EXAMPLE 3.2   Isotopes

Refer to the following isotope symbols, in which the letter X replaces each element symbol. **(a)** Which represent isotopes of the same element? **(b)** Which have the same mass number? **(c)** Which have the same number of neutrons?

$$^{16}_{8}X \quad ^{16}_{7}X \quad ^{14}_{7}X \quad ^{14}_{6}X \quad ^{12}_{6}X$$

Solution

a. Isotopes of the same element have the same atomic number (subscript). Therefore, $^{16}_{7}X$ and $^{14}_{7}X$ are isotopes of nitrogen (N), and $^{14}_{6}X$ and $^{12}_{6}X$ are isotopes of carbon (C).

b. The mass number is the superscript, so $^{16}_{8}X$ and $^{16}_{7}X$ have the same mass number. The first is an isotope of oxygen, and the second an isotope of nitrogen. $^{14}_{7}X$ and $^{14}_{6}X$ also have the same mass number. The first is an isotope of nitrogen, and the second an isotope of carbon.

c. To determine the number of neutrons, we subtract the atomic number from the mass number. We find that $^{16}_{8}X$ and $^{14}_{6}X$ each has eight neutrons ($16 - 8 = 8$ and $14 - 6 = 8$, respectively).

⟩ Exercise 3.2A

Which of the following represent isotopes of the same element?

$$^{90}_{37}X \quad ^{90}_{35}X \quad ^{88}_{37}X \quad ^{88}_{38}X \quad ^{93}_{38}X$$

⟩ Exercise 3.2B

How many different elements are represented in Exercise 3.2A? What elements are they?

Did you wonder why some atoms are radioactive after reading Section 3.3? Why do some atoms emit beta particles and some emit alpha particles? Beta particles are emitted by atoms that have a neutron-to-proton ratio that is too high for stability. One of the excess neutrons breaks apart, releasing a high-speed electron (the beta particle), and a proton remains in the atom. Alpha emission usually happens with the heaviest elements, and involves the simultaneous emission of two protons and two neutrons, equivalent to a helium nucleus. (See Chapter 11.)

SELF-ASSESSMENT Questions

1. The three basic components of an atom are
 a. protium, deuterium, and tritium
 b. protons, neutrinos, and ions
 c. protons, neutrons, and electrons
 d. protons, neutrons, and ions

2. The identity of an element is determined by its
 a. atomic mass
 b. number of atoms
 c. nucleon number
 d. number of protons

3. How many electrons are required to equal the mass of one proton?
 a. 1
 b. 18
 c. about 1800
 d. 1×10^{27}

4. Which particles have approximately the same mass?
 a. electrons and neutrons
 b. electrons and protons
 c. electrons, neutrons, and protons
 d. neutrons and protons

5. A change in the number of protons in an atom produces a(n)
 a. compound
 b. atom of a different element
 c. ion
 d. different isotope of the same element

6. A change in the number of neutrons in an atom produces a(n)
 a. compound
 b. atom of a different element
 c. ion
 d. different isotope of the same element

7. The number of protons plus neutrons in an atom is its
 a. atomic mass   **b.** atomic number
 c. isotope number   **d.** mass number

8. How many protons (p) and how many neutrons (n) are there in a $^{15}_{7}N$ nucleus?
 a. 7 p, 7 n   **b.** 7 p, 8 n
 c. 7 p, 15 n   **d.** 8 p, 8 n

9. A neutral atom has 22 electrons and 48 neutrons. Its atomic number is
 a. 22   **b.** 48
 c. 26   **d.** 4

10. An atom has 14 neutrons. It is
 a. definitely silicon
 b. definitely nitrogen
 c. definitely aluminum
 d. not possible to identify the element from the number of neutrons

11. The symbol for the isotope potassium-40 is
 a. $^{21}_{9}K$ **b.** $^{39}_{19}K$
 c. $^{40}_{19}K$ **d.** $^{19}_{40}K$

5 Most light bulbs produce white light. Why does a neon light appear red-orange?

When a solid—like the tungsten filament in a soon-to-be-outmoded incandescent lamp—is heated, it quickly goes through a stage of red heat, then one of yellow. Finally, white light is emitted, which consists of all the visible wavelengths. When a gas such as neon is zapped with electricity, any of its atoms' 10 electrons jump to higher energy levels; when they fall to any one of many lower energy levels, photons are emitted with wavelengths primarily in the red, orange, and yellow parts of the visible spectrum.

3.6 Electron Arrangement: The Bohr Model (Orbits)

Learning Objectives • Understand how transitions of electrons in energy levels relate to absorption and emission. • Arrange the electrons in a given atom in general energy levels (shells).

Let's turn our attention once more to electrons. Rutherford demonstrated that atoms have a tiny, positively charged nucleus with electrons outside it. Evidence soon accumulated that the electrons were not randomly distributed but were arranged in an ordered fashion. We will examine that evidence soon. But first, let's take a side trip into some colorful chemistry and physics to provide a background for our study of the arrangement of electrons in atoms.

Fireworks and Flame Tests

Chemists of the eighteenth and nineteenth centuries developed flame tests that used the colors of flames to identify several elements (Figure 3.9). Sodium salts produce a persistent yellow flame, potassium salts a fleeting lavender flame, and lithium salts a brilliant red flame. It was not until the structure of atoms was better understood, however, that scientists could explain why each element emits a distinctive color of light.

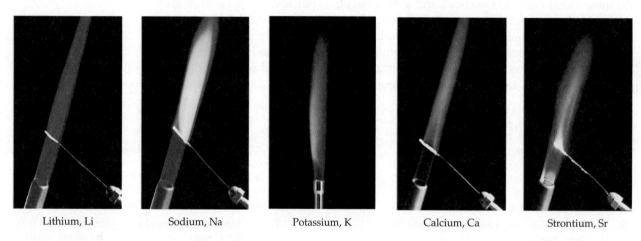

| Lithium, Li | Sodium, Na | Potassium, K | Calcium, Ca | Strontium, Sr |

▲ **Figure 3.9** Certain chemical elements can be identified by the characteristic colors their compounds impart to flames. Five examples are shown here.

Fireworks originated earlier than flame tests, in ancient China. The brilliant colors of aerial displays still mark celebrations of holidays throughout the world. The colors of fireworks are attributable to specific elements. A breathtaking photograph of such a pyrotechnic display was placed on the opening page of this chapter. Brilliant reds are produced by strontium compounds, whereas barium compounds are used to produce green, sodium compounds yield yellow, and copper salts produce blue or violet. However, the colors of fireworks and flame tests are not as simple as they seem to the unaided eye.

Continuous and Line Spectra

▲ **Figure 3.10** A glass prism separates white light into a continuous spectrum, or rainbow, of colors just as sunlight passing through raindrops causes a rainbow in the sky.

When white light from an incandescent lamp or an LED (light-emitting diode) lamp is passed through a prism, it produces a visible spectrum of colors (Figure 3.10) in the same order as seen in a rainbow (which is an atmospheric phenomenon). The different colors of light correspond to different wavelengths (the distance from crest to crest). Violet light has the shortest wavelengths in the visible spectrum, and red light has the longest, but there is no sharp transition in moving from one color to the next. All wavelengths are present in a continuous visible spectrum (Figure 3.11, top), ranging from about 350 nm to about 750 nm. White light is simply a combination of all the visible spectral colors.

If the light from a colored flame test is passed through a prism, only narrow colored lines are observed (Figure 3.11, bottom). A similar phenomenon was observed in the later 1800s, when a low-pressure gas-discharge tube containing a specific gaseous element was zapped with a high voltage and viewed through a prism or a diffraction grating. Each line corresponds to light of a particular wavelength, and the pattern of lines emitted by an element is called its **line spectrum**. The line spectrum of an element is characteristic of that element and can be used to identify it. Not all the lines in the spectrum of an atom are visible; some are infrared (wavelengths longer than about 750 nm) or ultraviolet radiation (wavelengths shorter than about 350 nm).

WHY IT MATTERS

The incandescent light bulb is very inefficient with respect to the energy it uses; as much as 95% of the electric energy it uses is changed to heat, not light. Though compact fluorescent bulbs, containing mercury, were popular for about a decade, now LED bulbs are taking over the light bulb market. They are more expensive, but supposedly will last 10 to 20 years, and use a lot less energy for the same brightness effect. Use of such bulbs will be a "greener" way to light your surroundings.

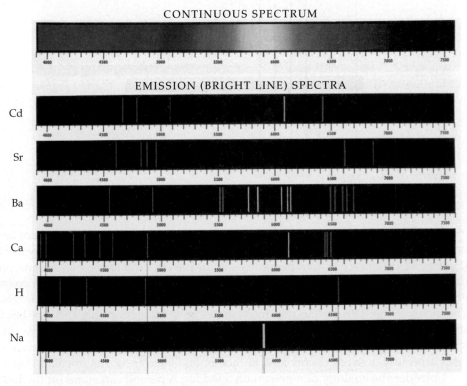

▲ **Figure 3.11** Line spectra of selected elements, in which the light emitted by excited atoms appears as colored lines. A continuous spectrum is shown at the top for comparison. The numbers are wavelengths of light given in Angstrom units (1 A = 10^{-8} cm), but more commonly are in nanometer units (1 nm = 10^{-9} m). Each element has its own characteristic line spectrum that differs from all others and can be used to identify the element.

The visible line spectrum of hydrogen is fairly simple, consisting of four lines in the visible spectrum (although many people can only see three lines, because they cannot see colors that close to the ultraviolet range). To explain the hydrogen spectrum, Danish physicist Niels Bohr (1885–1962) worked out a model for the arrangement of electrons in the hydrogen atom.

Bohr's Explanation of Line Spectra

Bohr presented his explanation of line spectra in 1913. He suggested that electrons cannot have just any amount of energy but can have only certain specified amounts, so electrons can exist only in specific **energy levels**. We say that the energy of an electron is *quantized*. A **quantum** (plural: *quanta*) is a tiny unit of energy whose value depends on the frequency of the radiation.

▶ **Figure 3.12** Possible electron transitions between energy levels in atoms to produce the lines found in spectra. The three colored arrows correspond to three of the colored lines in the hydrogen spectrum seen in Figure 3.11. Some other transitions result in emissions in the ultraviolet and infrared regions of the electromagnetic spectrum.

Q *Which transition involves the greater change in energy— a drop from energy level 4 to energy level 2, or a drop from energy level 5 to energy level 4? What color of light is absorbed when an electron in a hydrogen atom is boosted from energy level 2 to energy level 3?*

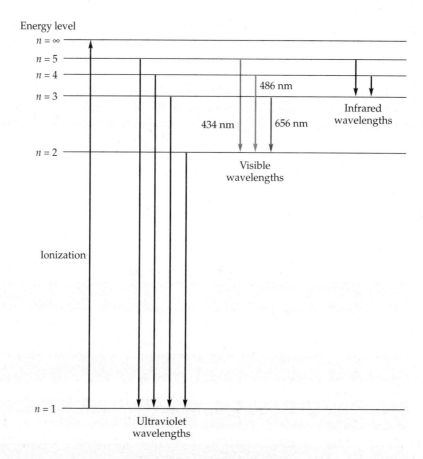

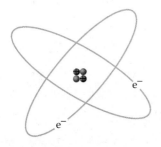

▲ **Figure 3.13** The nuclear atom, as envisioned by Bohr. Most of its mass is in an extremely small nucleus. Bohr thought that electrons orbited about the nucleus in the same way that planets orbit the sun. However, electron movements are in reality more complicated, as seen in Figure 3.14.

In absorbing a quantum of energy (for example, when atoms of the element are heated or zapped with a high voltage), an electron is elevated to a higher energy level (Figure 3.12). Because electrons prefer to be in the lowest, most stable energy state, an excited electron quickly falls to a lower energy level and, in doing so, gives up or emits energy equivalent to the difference between the higher and the lower energy levels. The energy released for various transitions shows up as a line spectrum. Each line has a specific wavelength corresponding to a quantum of energy equivalent to the energy difference between the higher and the lower energy levels. An electron moves practically instantaneously from one energy level to another, and there are no intermediate stages.

Consider the analogy of a person on a ladder. A person can stand on the first rung, the second rung, the third rung, and so on, but is unable to stand between rungs. As the person goes from one rung to another, the potential energy (energy due to position) changes by definite amounts. As an electron moves from one energy level to another, its total energy (both potential and kinetic) also changes by definite amounts, or quanta. The common phrase "a quantum leap" comes from the idea suggested by Figure 3.12, where an electron can "leap" from one energy level to another. A quantum of energy is too small to notice in daily life, but the fact that energy is quantized is central to the operation of lasers, integrated circuits, LEDs, and much of modern technology.

Bohr based his model of the atom on the laws of planetary motion that had been set down by the German astronomer Johannes Kepler (1571–1630) three centuries before. Bohr imagined the electrons to be orbiting about the nucleus much as planets orbit the sun (Figure 3.13). Different energy levels were pictured as different orbits. Bohr received the 1922 Nobel Prize in Physics for his planetary model of the atom with its quantized electron energy levels. Element 107 is named bohrium (Bh) in his honor.

As it turns out, the electrons in an atom don't really move around the nucleus in fixed, circular paths. The modern understanding of the structure of an atom is different, as we will see in the next section. But Bohr's model does help explain why elements produce line spectra.

Ground States and Excited States

The electron in a hydrogen atom is usually in the first, or lowest, energy level. An electron tends to stay in its lowest possible energy level (nearest the nucleus), and we say that the atom is in its **ground state**. When a flame or other source supplies energy to an atom (of hydrogen, for example) and an electron jumps from the lowest possible level to a higher level, the atom is said to be in an **excited state**. An atom in an excited state eventually emits a quantum of energy—often a **photon**, or "particle," of light—as the electron falls back down to one of the lower levels and ultimately reaches the ground state. See Figure 3.12 illustrating a variety of **transitions**; note that the three dominant colors in the visible part of the hydrogen spectrum are created by electrons falling from the third, fourth, or fifth energy level down to the second energy level. The larger the fall, the more energy released. So the electronic transition from the fifth to the second results in a violet-colored spectral line, with the shortest wavelength and highest energy.

Bohr's theory was spectacularly successful in explaining the line spectrum of hydrogen. It established the important idea of energy levels in atoms.

Atoms larger than hydrogen have more than one electron, and Bohr was also able to deduce that a given energy level of an atom could contain at most only a certain number of electrons. The maximum number of electrons that can be in a given level is indicated by the formula

$$\text{Maximum number of electrons} = 2n^2$$

where n is the energy level being considered. For the first energy level the maximum number of electrons is $2 \times 1^2 = 2 \times 1 = 2$. For the second energy level the maximum number is $2 \times 2^2 = 2 \times 4 = 8$. For the third level, the maximum number is $2 \times 3^2 = 2 \times 9 = 18$.

The various main energy levels are often called **shells**. The first energy level is the first shell, the second energy level is the second shell, and so on.

WHY IT MATTERS

Solar energy is becoming more important as fossil fuels become scarcer and more expensive. When light hits a photovoltaic "solar" cell, photons are absorbed, promoting electrons to excited states. Those excited-state electrons are captured by the photovoltaic cell to produce an electrical current. Scientists are working to create new materials that will absorb more of the sun's photons and more efficiently convert those photons into electrical energy, and at lower cost.

 EXAMPLE 3.3   Electron-Shell Capacity

What is the maximum number of electrons in the fifth shell (fifth energy level)?

Solution
The maximum number of electrons in a given shell is given by $2n^2$. For $n = 5$, we have

$$2 \times 5^2 = 2 \times 25 = 50 \text{ electrons}$$

〉Exercise 3.3A
What is the maximum number of electrons in the fourth shell (fourth energy level)?

〉Exercise 3.3B
What is the lowest energy level that can hold 18 electrons?

Building Atoms: Main Shells

Imagine building up atoms by adding one electron to the proper shell as *each* proton is added to the nucleus, keeping in mind that electrons will go to the lowest energy level (shell) available. For hydrogen, with a nucleus containing only one proton, the single electron goes into the first shell. For helium, with a nucleus having two protons, both electrons go into the first shell. According to Bohr, two electrons is the maximum population of the first shell, and that level is filled in the helium atom. (We can ignore the neutrons in the nucleus because they are not involved in this process.)

With lithium, two electrons go into the first shell, and the third must go into the second shell. This process of adding electrons is continued until the second shell is filled with eight electrons, as in a neon atom, which has two of its ten electrons in the first shell and the remaining eight in the second shell.

A sodium atom has eleven electrons. Two are in the first shell, the second shell is filled with eight electrons, and the remaining electron is in the third shell. We can represent the **electron configuration** (or arrangement) of electrons in atoms of the first eleven elements as follows (implying that the elements between lithium and sodium are simply adding one more electron in the second shell, one by one):

Element	1st shell	2nd shell	3rd shell
H	1		
He	2		
Li	2	1	
. . .	. . .	. . .	
Ne	2	8	
Na	2	8	1

Sometimes the main-shell electron configuration is abbreviated by writing the symbol for the element followed by the number of electrons in each shell, separated by commas, starting with the lowest energy level. Following the color scheme from the preceding table, the main-shell configuration for sodium is simply

Na 2, 8, 1

We can continue to add electrons to the third shell until we get to argon. The main-shell configuration for argon is

Ar 2, 8, 8

After argon, the notation for electron configurations is enhanced with more detail. This notation is discussed in greater detail in Section 3.8. Meanwhile, Example 3.4 shows how to determine a few main-shell configurations.

 EXAMPLE 3.4   Main-Shell Electron Configurations

What is the main-shell electron configuration for fluorine?

Solution

Fluorine has the symbol F. It has nine electrons. Two of these electrons go into the first shell, and the remaining seven go into the second shell.

F 2, 7

❭ **Exercise 3.4A**

Write the main-shell electron configurations for **(a)** boron and **(b)** aluminum. These elements are in the same column of the periodic table. What do you notice about the number of electrons in their outermost shells?

❭ **Exercise 3.4B**

Write the main-shell electron configurations for **(a)** nitrogen and **(b)** sulfur. These elements are not in the same column of the periodic table. Did you expect a similarity in the number of electrons in their outermost shells? Why or why not?

SELF-ASSESSMENT Questions

1. When electrified in a gas-discharge tube, different elements emit light that forms narrow lines that are
   a. characteristic of each particular element
   b. the same as those of the hydrogen spectrum
   c. dependent on the amount of gas present in the tube
   d. the same for all elements, but with differing intensities

2. When an electron in an atom goes from a higher energy level to a lower one, the electron
   a. absorbs energy
   b. changes its volume
   c. changes its charge
   d. releases energy

3. An atom in an excited state is one with an electron that has
   a. moved to a higher energy level
   b. been removed

c. bonded to another atom to make a molecule
d. combined with a proton to make a neutron

4. The maximum number of electrons in the first shell of an atom is
   a. 2   b. 6   c. 8   d. unlimited

5. The maximum number of electrons in the third shell of an atom is
   a. 6   b. 8   c. 18   d. 32

6. The main-shell electron configuration of magnesium is
   a. 2, 8, 2   b. 2, 8, 4
   c. 2, 8, 12   d. 2, 8, 8, 8, 2

7. The main-shell electron configuration of sulfur is
   a. 2, 8   b. 2, 8, 6
   c. 2, 8, 8, 8, 6 d. 2, 8, 8, 8

Answers: 1, a; 2, d; 3, a; 4, a; 5, c; 6, a; 7, b

3.7 Electron Arrangement: The Quantum Model (Orbitals/Subshells)

Learning Objectives • Relate the idea of a quantum of energy to an orbital.
• Write an electron configuration (in subshell notation) for a given atom.

Although Bohr's simple planetary model of the atom explained the spectrum of the hydrogen atom, it could never explain the spectra of any other elements. It was eventually replaced by more sophisticated models in which electrons are treated as *both* particles and waves. The hypothesis that the electron should have wavelike properties was first suggested in 1924 by Louis de Broglie (1892–1987), a young French physicist. Although it was hard to accept because of Thomson's evidence that electrons are particles, de Broglie's hypothesis was experimentally verified within a few years.

Erwin Schrödinger (1887–1961), an Austrian physicist, used highly mathematical *quantum mechanics* in the 1920s to develop equations that describe the properties of electrons in atoms. Fortunately, we can make use of some of Schrödinger's results without understanding his elaborate equations. The solutions to these equations express an electron's location in terms of the probability of finding that electron in a given volume of space. These variously shaped volumes of space, called **orbitals**, replaced the planetary orbits of the Bohr model.

Suppose you had a camera that could photograph electrons, and you left the shutter open while an electron zipped about the nucleus. The developed picture would give a record of where the electron had been. (Doing the same thing with an electric fan would give a blurred image of the rapidly moving blades, an image resembling a disk.) The electrons in the first shell would appear as a fuzzy ball (often referred to as a *charge cloud* or an *electron cloud* [Figure 3.14]).

Scientists realized that each electron orbital could contain a maximum of two electrons and that some shells could contain more than one kind of orbital. So, many more transitions are possible, both for absorption of energy (*excitation*) and release of energy (*emission*), than what is observed for the hydrogen atom. This fact partially explains why the flame tests for elements with many more electrons than hydrogen (Figure 3.9) produce more complex line spectra (Figure 3.11).

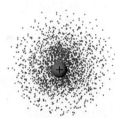

An *s* orbital

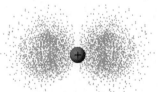

A *p* orbital

▲ **Figure 3.14** The modern charge-cloud representation of an atomic orbital is more accurate than the Bohr model (Figure 3.13). An electron cloud is the "fuzzy" region around an atomic nucleus where an electron is likely to be. The type of orbital determines the general shape of the cloud. The illustrations show the approximate "shape" of an *s* and a *p* orbital.

High-energy laser beams can also excite electrons, and the spectra obtained can be used to identify the elements present in a given sample. The modern instrumental method called *laser-induced breakdown spectroscopy* (LIBS) uses laser beams to identify elements in minerals or other solid samples. A *scanning electron microscope* (SEM) with an *electron-dispersive spectroscopy* (EDS) attachment uses a high-energy electron beam to identify the elemental composition of the tiniest specks in samples.

Building Atoms by Orbital Filling

Orbitals in the same shell that have the same letter designation make up a **subshell** (sublevel). The first shell of electrons in atoms contains a single spherical orbital called 1*s*, which can hold two paired electrons. The second shell has two subshells, the 2*s* subshell and the 2*p* subshell. The 2*s* subshell has just one spherical orbital and two electrons in it. The 2*p* subshell has three dumbbell-shaped orbitals, one in each of three dimensions (Figure 3.15). Each of those 2*p* orbitals can contain two electrons. The third shell contains nine orbitals distributed among three subshells: a spherical 3*s* orbital in the 3*s* subshell, three dumbbell-shaped 3*p* orbitals in the 3*p* subshell, and five 3*d* orbitals with more complicated shapes in the 3*d* subshell.

▲ A hand-held LIBS analyzer uses laser beams to excite atoms in solid samples and analyze the resultant emissions, and can be used to identify the elements present in those samples within seconds. Geologists use such instruments in the field for metal exploration.

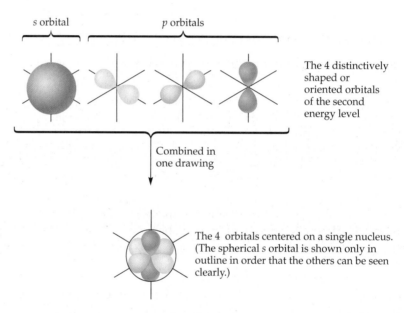

▲ **Figure 3.15** Electron orbitals of the second main shell. In these drawings, the nucleus of the atom is located at the intersection of the axes. The eight electrons that would be placed in the second shell of Bohr's model are distributed among these four orbitals in the current model of the atom, with two electrons per orbital.

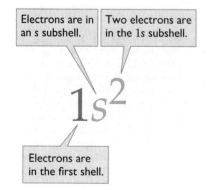

In building up the *electron configurations* of atoms of the various elements, the lower sublevels (subshells) are filled first.

Hydrogen has only one electron in the *s* orbital of the first shell. The electron configuration is

$$\text{H} \quad 1s^1$$

Helium has two electrons, and its electron configuration is

$$\text{He} \quad 1s^2$$

Lithium has three electrons—two in the first shell and one in the second shell. Because the *s* subshell is slightly lower in energy than the *p* subshell of the same shell, the third electron goes into the 2*s* orbital, and lithium's electron configuration is

$$\text{Li} \quad 1s^2 2s^1$$

Skipping to nitrogen, we fill the $1s$ and $2s$ subshells first. The remaining electrons go into the $2p$ subshell, and the seven electrons are arranged as follows:

$$\text{N} \quad 1s^2 2s^2 2p^3$$

Let's review what this notation means:

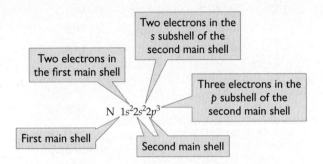

For argon, with its 18 electrons, the configuration is

$$\text{Ar} \quad 1s^2 2s^2 2p^6 3s^2 3p^6$$

Note that the highest occupied subshell, $3p$, is filled. However, the order of sub-shell filling for elements with Z greater than 18 is not intuitively obvious. In potassium (Z = 19), for example, the $4s$ subshell fills before the $3d$ subshell. The reason for this seeming oddity is that $3d$ orbitals, most of which are shaped like a three-dimensional four-leafed clover (a double dumbbell), have more energy than the spherical $4s$ orbitals. So the $4s$ subshell fills first. Remember that we fill orbit-als from the lower energy first, and then go on to the higher energy. The order in which the various subshells are filled is shown in Figure 3.16. Table 3.3 gives the electron configurations for the first 20 elements. The electrons in the outermost main shell, called *valence electrons* (Section 3.8), are shown in color.

So, which model should we use from now on? Which is most important? For some purposes in this text, we use only main-shell configurations of atoms to describe the distribution of electrons in atoms. At other times, the electron clouds of the quantum-mechanical model are more useful. Even Dalton's model sometimes proves to be the best way to describe certain phenomena (the behavior of gases, for example). The choice of model is always based on which one is most helpful in understanding a particular concept. This, after all, is the purpose of scientific models. In Chapter 4, we will see that the main-shell configurations of atoms aids in achieving a basic understanding of the two main types of chemical bonds.

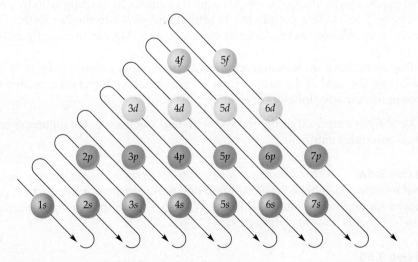

▲ **Figure 3.16** An order-of-filling chart for determining the electron configurations of atoms.

TABLE 3.3 **Electron Configurations for Atoms of the First 20 Elements**

Name	Atomic Number (Z)	Electron Configuration
Hydrogen	1	$1s^1$
Helium	2	$1s^2$
Lithium	3	$1s^2 2s^1$
Beryllium	4	$1s^2 2s^2$
Boron	5	$1s^2 2s^2 2p^1$
Carbon	6	$1s^2 2s^2 2p^2$
Nitrogen	7	$1s^2 2s^2 2p^3$
Oxygen	8	$1s^2 2s^2 2p^4$
Fluorine	9	$1s^2 2s^2 2p^5$
Neon	10	$1s^2 2s^2 2p^6$
Sodium	11	$1s^2 2s^2 2p^6 3s^1$
Magnesium	12	$1s^2 2s^2 2p^6 3s^2$
Aluminum	13	$1s^2 2s^2 2p^6 3s^2 3p^1$
Silicon	14	$1s^2 2s^2 2p^6 3s^2 3p^2$
Phosphorus	15	$1s^2 2s^2 2p^6 3s^2 3p^3$
Sulfur	16	$1s^2 2s^2 2p^6 3s^2 3p^4$
Chlorine	17	$1s^2 2s^2 2p^6 3s^2 3p^5$
Argon	18	$1s^2 2s^2 2p^6 3s^2 3p^6$
Potassium	19	$1s^2 2s^2 2p^6 3s^2 3p^6 4s^1$
Calcium	20	$1s^2 2s^2 2p^6 3s^2 3p^6 4s^2$

EXAMPLE 3.5 Subshell Notation

Without referring to Table 3.3, use subshell notation to write the electron configurations for **(a)** oxygen and **(b)** sulfur. What do the electron configurations of these two elements have in common?

Solution

a. Oxygen has eight electrons. We place them in subshells, starting with the lowest energy level. Two go into the 1s orbital and two into the 2s orbital. That leaves four electrons to be placed in the 2p subshell. The electron configuration is $1s^2 2s^2 2p^4$.

b. Sulfur atoms have 16 electrons each. The electron configuration is $1s^2 2s^2 2p^6 3s^2 3p^4$. Note that the total of the superscripts is 16 and that we have not exceeded the maximum capacity for any sublevel.

Both O and S have electron configurations with four electrons in the highest energy sublevel (outermost subshell).

⟩ Exercise 3.5A
Without referring to Table 3.3, use subshell notation to write out the electron configurations for **(a)** fluorine and **(b)** chlorine. What do the electron configurations for these two elements have in common?

⟩ Exercise 3.5B
Use Figure 3.16 to write the electron configurations for **(a)** titanium (Ti) and **(b)** tin (Sn).

1. The shape of a 1s orbital is
   a. circular
   b. a dumbbell
   c. spherical
   d. variable

2. The maximum number of electrons in an atomic orbital
   a. depends on the main shell
   b. depends on the subshell
   c. is always two
   d. is always eight

3. Which of the following subshells has the highest energy?
   a. 2p   **b.** 2s   **c.** 3p   **d.** 3s

4. What is the lowest-numbered main shell to have d orbitals?
   a. 1   **b.** 2   **c.** 3   **d.** 4

5. Which of the following subshells has the lowest energy?
   a. 4d   **b.** 5s   **c.** 5p   **d.** 5d

6. Which subshell has a total of three orbitals?
   a. s   **b.** p   **c.** d   **d.** f

7. In what main shell is a 3d subshell located?
   a. 1   **b.** 3   **c.** 5   **d.** none

8. The electron capacity of the 4p subshell is
   a. 3   **b.** 4   **c.** 6   **d.** 10

Answers: 1. c; 2. c; 3. c; 4. c; 5. b; 6. b; 7. b; 8. c

3.8 Electron Configurations and the Periodic Table

Learning Objective • Describe how an element's electron configuration relates to its location in the periodic table.

In general, the physical and chemical properties of elements can be correlated with their electron configurations. Because the number of protons equals the number of electrons in neutral atoms, the periodic table tells us about electron configuration as well as atomic number. The electron configuration is critical to understanding why and how bonding between atoms occurs, and we will explore bond formation using electron configurations in Chapter 4.

The modern periodic table (inside front cover) has horizontal rows and vertical columns.

- Each vertical column is a **group** or *family*. Elements in a group have similar chemical properties.
- A horizontal row of the periodic table is called a **period**. The properties of elements change in a recurring manner across a period.

In the United States, the groups are often indicated by a numeral followed by the letter A or B.

- An element in an A group is a **main-group element**.
- An element in a B group is a **transition element**.

IUPAC recommends numbering the groups from 1 to 18. Both systems are indicated on the periodic table on the inside front cover, but this book uses the traditional U.S. system. We prefer this system because then the total number of electrons in the outermost shell of a main group element equals its column number. This keeps the relationship between the electronic configuration and column number simple.

Family Features: Outer Electron Configurations

The outermost electrons are called **valence electrons**. They are the most important electrons in an atom; when atoms collide, it is obviously the outermost electrons of the atoms that come into closest contact with each other. Those electrons are the ones involved in chemical changes and in bonding between atoms to form new compounds (Chapter 4). The period in which an element appears in the periodic table tells us how many main shells an atom of that element has. Phosphorus, for example, is in the third period, so the phosphorus atom has three main shells. The U.S. group number for phosphorus is group 5A. Thus, we can deduce that it has five valence

6 What is periodic about the periodic table?

Periodic means recurring at regular intervals. Figure 3.17 shows why there are periodic, or recurring, trends in the periodic table. Elements in the same column have the same number of electrons in their outermost shells, and thus have similar chemical and physical characteristics.

electrons. Two of these are in an *s* orbital, and the other three are in *p* orbitals. We can indicate the *outer electron configuration* of the phosphorus atom (its valence electrons) as

$$P \quad 3s^2 3p^3$$

The valence electrons determine most of the chemistry of an atom. Because all the elements in the same group of the periodic table have the same number of valence electrons, they should have similar chemistry, and they do.

Figure 3.17 relates the subshell configurations to the groups in the periodic table, but does not include H and He. Those two elements could simply be called the 1s group. On the left of Figure 3.17 is the 1s block, consisting of the first two groups. For the atoms in these two groups, the last electron added goes into an *s* orbital, s^1 and s^2, respectively. On the far right is the *p* block. Its six groups correspond to the six electrons that can go into a *p* subshell. Between these blocks are the transition metals, whose ten groups correspond to the five *d* orbitals in a *d* subshell (the *d* block), and the inner transition metals, where the seven *f* orbitals in the *f* subshell get filled (the *f* block).

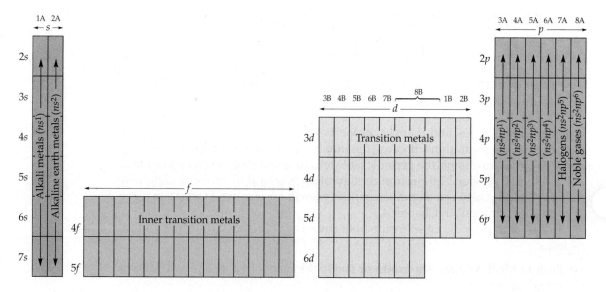

▲ **Figure 3.17** Valence-shell electron configurations and the periodic table.

Family Groups

Elements within a group or family have similar properties, due to similar outer electron configurations.

The metals in group 1A are called **alkali metals**, and have one valence electron ns^1, where *n* denotes the number of the outermost main shell. They react vigorously with water, producing hydrogen gas. There are noticeable trends in properties within this family. For example, lithium is the hardest metal in the group. Sodium is softer than lithium; potassium is softer still; and so on down the group. Lithium is also the least reactive toward water. Sodium, potassium, rubidium, and cesium are progressively more reactive. Francium is highly radioactive and extremely rare; few of its properties have been measured.

Hydrogen is the odd one in group 1A. It is not an alkali metal but rather a typical nonmetal. Based on its properties, hydrogen probably should be put in a group of its own.

Group 2A elements are known as **alkaline earth metals**. The metals in this group have the outer electron configuration ns^2. Most are fairly soft and moderately reactive with water. Beryllium is the odd member of the group in that it is rather hard and does not react with water. As in other families, there are trends in properties within the group. For example, magnesium, calcium, strontium, barium, and radium are progressively more reactive toward water.

Group 3A does not have a unique group name, and is referred to simply as the *boron family*. Similarly we have the *carbon family* (group 4A), the *nitrogen family* (group 5A), and the *oxygen family* (group 6A), which used to be called the *chalcogen family*, which means "ore former." (In Chapter 12 we will learn about a wide variety of minerals that are formed with oxygen and sulfur.) Group 7A elements are the **halogens**, and consist of reactive elements. All have seven valence electrons with the configuration ns^2np^5. Halogens react vigorously with alkali metals to form crystalline solids called **salts**. (This is discussed further in Chapter 4.) There are trends in the halogen family as you move down the group. Fluorine is most reactive toward alkali metals, chlorine is the next most reactive, and so on. Fluorine and chlorine are greenish gases at room temperature, bromine is a dark reddish liquid, and iodine is a grayish/violet solid. Astatine, like francium, is highly radioactive and extremely rare; few of its properties have been determined. An extremely small amount of tennessine (Ts, number 117) have been created in a laboratory in Russia.

Members of group 8A, to the far right on the periodic table, have a complete set of valence electrons and therefore undergo few, if any, chemical reactions. They are called *noble gases*, for their lack of chemical reactivity.

▲ (Upper) Lithium, an alkali metal, reacts with water to form hydrogen gas. (Lower) Potassium, another alkali metal, undergoes the same reaction but much more vigorously.

 EXAMPLE 3.6 **Valence-Shell Electron Configurations**

Use subshell notation to write the electron configurations for the outermost main shell of **(a)** strontium (Sr) and **(b)** arsenic (As).

Solution

  a. Strontium is in group 2A and thus has two valence electrons in an *s* subshell. Because strontium is in the fifth period of the periodic table, its outer main shell is $n = 5$. Its outer electron configuration is therefore $5s^2$.
  b. Arsenic is in group 5A and the fourth period. Its five outer (valence) electrons are in the $n = 4$ shell, and the configuration is $4s^24p^3$.

❯ Exercise 3.6A

Use subshell notation to write the configurations for the outermost main shell of **(a)** cesium (Cs), **(b)** antimony (Sb), and **(c)** silicon (Si).

❯ Exercise 3.6B

The valence electrons in aluminum have the configuration $3s^23p^1$. In the periodic table, gallium is directly below aluminum, and indium is directly below gallium. Use only this information to write the configurations for the valence electrons in **(a)** gallium and **(b)** indium.

Metals and Nonmetals

Elements in the periodic table are divided into two classes by a heavy, stepped line. (See inside front cover.) Those to the left of the line are *metals*. A **metal** has a characteristic luster (shininess) and generally is a good conductor of heat and electricity. Except for mercury, which is a liquid, all metals are solids at room temperature. Metals generally are **malleable**; that is, they can be hammered into thin sheets. Most also are **ductile**, which means that they can be drawn into wires.

Elements to the right of the stepped line are *nonmetals*. A **nonmetal** lacks metallic properties. Several nonmetals are gases (e.g., oxygen, nitrogen, fluorine, and chlorine). Others are solids (e.g., carbon, sulfur, phosphorus, and iodine). Bromine is the only nonmetal that is a liquid at room temperature.

Some of the elements bordering the stepped line are called *semimetals*, or **metalloids**, and these have intermediate properties that may resemble those of both metals and nonmetals. There is a lack of agreement on just which elements fit in this category.

▲ (Upper) Metals such as gold are easily shaped, and they conduct heat and electricity well. (Lower) Nonmetals such as sulfur are usually brittle, melt at low temperatures, and often are good insulators.

Learning Objectives • Distinguish the conversion of solar energy into electrical energy in a solar cell from the conversion of solar energy into the chemical bond energy of a solar fuel. • Explain why splitting water into the elements hydrogen and oxygen requires an energy input and why producing water by the reaction of hydrogen and oxygen releases energy.

Imagine a world powered by a clean fuel manufactured using sunlight and water that produces no carbon dioxide emissions when used. This has been the dream of chemists around the world, who have been working for many years to develop a "solar fuel" that might someday replace some of the fossil fuels (oil, coal, and natural gas) that are so important to modern society.

Sunlight is a free and abundant power source. The amount of solar energy reaching the Earth's surface in one hour is as much as all of the fossil fuel energy humans use in one year. The goal for chemists is to develop efficient methods to capture just a tiny part of this sunlight and convert it into a useful form. Of course, the most common approach is to convert solar energy into electricity using solar panels, devices constructed from photovoltaic cells that are usually made of silicon. But sunshine is intermittent, so this solar electricity is only available on sunny days.

A different approach is to use the energy of sunlight to promote a chemical change, converting radiant solar energy into chemical energy in the form of fuel that can be stored and used even when the sun has gone down. One idea for a solar fuel is hydrogen. Using sunlight to make hydrogen is one of the grand challenges of chemistry.

A full hydrogen energy cycle uses solar energy to split water (H_2O) into hydrogen (H_2) and oxygen (O_2) and then uses the hydrogen as a clean fuel to produce either heat (when burned) or electricity (using a fuel cell). Splitting water requires energy input to break the chemical bonds that hold together the O and H atoms in water molecules. One of the simplest ways to do this is by electrolysis (Section 3.1), which requires electrical energy. If the electricity is produced using a solar panel, then the hydrogen formed is a solar fuel. This approach is currently expensive, and much research in chemistry is aimed at discovering new and inexpensive photovoltaic materials—including plastics—from which to construct solar panels. Chemists also seek ways to make the electrodes and other components from abundant elements instead of from the precious metal platinum (Pt).

For a different approach to hydrogen production, chemists are turning to green plants for inspiration. Plants absorb energy from the sun and store it in the form of chemical bonds of compounds known as carbohydrates. This photosynthetic reaction, which is essential to sustain life on the planet, relies on the ability of molecules in the plant to split

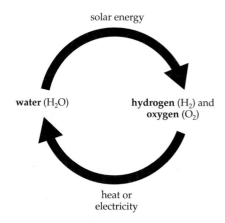

▲ **Clean energy cycle.** Solar energy is used to produce hydrogen and oxygen from water (top). Hydrogen fuel reacts with oxygen to produce water and releases energy as either heat or electricity.

water using sunlight. Chemists are hoping to unlock the secrets of the leaf to develop "artificial photosynthesis," which uses solar energy to produce hydrogen fuel from water. Scientists at Harvard devised a system to complete the process of making liquid fuel from sunlight, carbon dioxide, and water. Natural photosynthesis converts about one percent of solar energy into the carbohydrates used by plants, while the Harvard procedure turns ten percent of the energy in sunlight into fuel.

In natural photosynthesis, plants employ many different molecules to capture and convert solar energy. Chlorophyll molecules absorb part of the solar spectrum, giving a leaf its green color. To mimic this process, chemists are designing colorful synthetic dyes to capture photons from the sun. When a dye absorbs light, one of its electrons is excited, which generates an excited-state molecule. The challenge is to harness this excitation energy to split water before it is simply emitting as a photon (Section 3.6). Plants also employ a cluster of manganese, calcium, and oxygen atoms to crack apart water molecules and produce oxygen. Chemists have been active in trying to mimic this reaction as well, using compounds made in the laboratory.

Solar fuels such as hydrogen embody several green chemistry principles. Hydrogen fuel is only as green as the method used to make it. If produced using sunlight as the energy source and water as the chemical feedstock, though, the fuel can be clean and renewable (Principle 7). Unlike fossil fuels, hydrogen is a carbon-free fuel that, if produced using sunlight, can prevent carbon dioxide emissions (Principle 1). A solar-hydrogen system relies on compounds that increase efficiency and facilitate both the water-splitting chemistry and the use of hydrogen (Principle 7). If these materials are earth-abundant and do not cause harm to the environment, this process closes the clean-fuel cycle.

PHOTOSYNTHESIS
carbon dioxide + water + sunlight →
carbohydrate + oxygen

ARTIFICIAL PHOTOSYNTHESIS
water → hydrogen + oxygen

Two ways to make solar fuels. Green plants use photosynthesis to produce carbohydrate fuels. Chemists are trying to produce hydrogen fuel. Both reactions use water and solar energy.

SELF-ASSESSMENT Questions

1. Which pair of elements has the most similar chemical properties?
   a. Ca and Cd   **b.** Cl and Br
   c. P and S   **d.** Sb and Sc

2. Elements in the same column of the periodic table have the same
   a. number of protons in the nucleus
   b. total number of electrons
   c. number of electrons in the outermost shell
   d. number of neutrons in the nucleus

3. The valence-shell electron configuration of the halogens is
   a. ns^1 **b.** ns^2np^3
   c. ns^2np^5 **d.** ns^2np^6

4. The alkali metals are members of group
   a. 1A   **b.** 2A   **c.** 3A   **d.** 8A

5. What is the total number of valence electrons in an atom of bromine?
   a. 5   **b.** 7
   c. 17   **d.** 35

6. Which of the following sets of elements exhibits the most similar chemical properties?
   a. Ca, Zn, Kr   **b.** Pt, Au, Hg
   c. Li, Na, K   **d.** N, O, F

7. In the ground state, atoms of a fourth-period element must have
   a. a 3d sublevel
   b. electrons in the fourth shell
   c. four valence electrons
   d. properties similar to those of other elements in the fourth period

Answers: 1, b; 2, c; 3, c; 4, a; 5, b; 6, c; 7, b

Summary

Section 3.1—Davy, Faraday, and others showed that matter is electrical in nature. They were able to decompose compounds into elements by electrolysis or by passing electricity through molten salts. **Electrodes** are carbon rods or metal strips that carry electricity into the **electrolyte**—the solution or compound that conducts electricity. The electrolyte contains ions—charged atoms or groups of atoms. The **anode** is the positive electrode, and **anions** (negatively charged ions) move toward it. The **cathode** is the negative electrode, and **cations** (positively charged ions) move toward it. Experiments with cathode rays in gas-discharge tubes showed that matter contained negatively charged particles, which were called electrons. Thomson determined the mass-to-charge ratio for the electron. Goldstein's experiment showed that matter also contained positively charged particles. Millikan's oil-drop experiment measured the charge on the electron, so its mass could then be calculated.

Section 3.2—In his studies of cathode rays, Röntgen accidentally discovered **X-rays**, a highly penetrating form of radiation now used in medical diagnosis. Becquerel unexpectedly discovered another type of radiation that comes from certain unstable elements. Marie Curie named this new discovery *radioactivity* and studied it extensively.

Section 3.3—Radioactivity was soon classified as one of three different types. **Alpha (α) particles** have four times the mass of a hydrogen atom and a positive charge twice that of an electron. **Beta (β) particles** are energetic electrons. **Gamma (γ) rays** are a form of energy like X-rays, but are more penetrating.

Section 3.4—Rutherford's experiments with alpha particles and gold foil showed that most of the alpha particles emitted toward the foil passed through it. A few were deflected or, occasionally, bounced back almost directly toward the source. This indicated that all the positive charge and most of the mass of an atom must be in a tiny core, which Rutherford called the nucleus.

Section 3.5—Rutherford called the smallest unit of positive charge the proton; it has roughly the mass of a hydrogen atom, and a charge equal in size but opposite in sign to the electron. In 1932, Chadwick discovered the neutron, a nuclear particle as massive as a proton but with no charge. The number of

protons in an atom is the **atomic number (Z)**. Atoms of the same element have the same number of protons but may have different numbers of neutrons; such atoms are called isotopes. Different isotopes of an element are nearly identical chemically. Protons and neutrons collectively are called nucleons, and the number of nucleons is called the mass number or nucleon number. The difference between the mass number and the atomic number is the number of neutrons. The general symbol for an isotope of element X is written $^A_Z X$, where A is the mass number and Z is the atomic number.

Section 3.6—Light from the sun or from an incandescent lamp produces a continuous spectrum, containing all colors. Light from a gas-discharge tube produces a line spectrum, containing only certain colors. Bohr explained line spectra by proposing that an electron in an atom resides only at certain discrete energy levels, which differ from one another by a quantum, or discrete unit of energy. Electrons in atoms drop to a lower energy state, or ground state, after being in a higher energy state, or excited state. The energy emitted when electrons move to lower energy levels manifests as a line spectrum characteristic of the particular element, with the lines corresponding to specific wavelengths (colors) of light. Bohr also deduced that the energy levels of an atom could hold at most $2n^2$ electrons, where n is the number of the energy level, or shell. A description of the shells occupied by the electrons of an atom is one way of giving the atom's electron configuration, or arrangement of electrons.

Section 3.7—De Broglie hypothesized that electrons have wave properties. Schrödinger developed equations that described each electron's location in terms of an orbital—a volume of space that the electron usually occupies. Each orbital holds at most two electrons. Orbitals in the same shell and with the same energy make up a subshell, or sublevel, each of which is designated by a letter. An s orbital is spherical and a p orbital is dumbbell-shaped. The d and f orbitals have more complex shapes. The first main shell can hold only one s orbital; the second can hold one s and three p orbitals; and the third can hold one s, three p, and five d orbitals. In writing an electron configuration using subshell notation, we give the shell number and subshell letter, followed by a superscript indicating the number of electrons in that subshell. The order of the shells and subshells can be remembered with a chart or by looking at the periodic table.

Section 3.8—Elements in the periodic table are arranged vertically in groups and horizontally in periods. Electrons in the outermost shell of an atom, called valence electrons, determine the reactivity of that atom. Elements in a group usually have the same number of valence electrons and similarities in properties. We designate groups by a number and the letter A or B. The A-group elements are main-group elements. The B-group elements are transition elements. Group 1A elements are the alkali metals. Except for hydrogen, they are all soft, low-melting, highly reactive **metals**. Group 2A consists of the alkaline earth metals, which are fairly soft and reactive. Group 7A elements, the halogens, are reactive **nonmetals**. Group 8A elements, the noble gases, react

very little or not at all. A stepped line divides the periodic table into metals, which conduct electricity and heat and are shiny, malleable, and ductile, and nonmetals, which tend to lack the properties of metals. Some elements bordering the stepped line, called **metalloids**, have properties intermediate between those of metals and nonmetals.

GREEN CHEMISTRY—Hydrogen is a promising clean fuel because it can be manufactured using sunlight and water and because using it does not produce carbon dioxide as waste. Using a solar cell, energy from sunlight can be converted into electricity, which can then be used to split water into hydrogen and oxygen; this splitting of water can also be accomplished by a process that mimics photosynthesis. The hydrogen can be burned to produce heat or used in a fuel cell to produce electricity.

Learning Objectives

Objective	Associated Problems
Explain the electrical properties of an atom. (3.1)	1, 27
Describe how the properties of electricity explain the structure of atoms. (3.1)	3, 27, 37, 38
Describe the experiments that led to the discovery of X-rays and an explanation of radioactivity. (3.2)	2
Distinguish the three main kinds of radioactivity: alpha, beta, and gamma. (3.3)	2, 28
Understand why atoms are believed to have a tiny nucleus surrounded by electrons. (3.4)	3, 5, 6, 19, 20
List the particles that make up the nucleus of an atom, and give their relative masses and electric charges. (3.5)	13–18
Identify elements and isotopes from their nuclear particles. (3.5)	31–36
Understand how transitions of electrons in energy levels relate to absorption and emission. (3.6)	10–12, 43, 44
Arrange the electrons in a given atom in energy levels (shells). (3.6)	7, 40–44
Relate the idea of a quantum of energy to an orbital. (3.7)	8, 9, 30, 36
Write an electron configuration (in subshell notation) for a given atom. (3.7)	45–47, 59, 60
Describe how an element's electron configuration relates to its location in the periodic table. (3.8)	57, 58, 61–63
Explain why splitting water into the elements hydrogen and oxygen requires an energy input and producing water by the reaction of hydrogen and oxygen releases energy.	64
Distinguish the conversion of solar energy into electrical energy in a solar cell from the conversion of solar energy into the chemical-bond energy of a solar fuel.	65–67

Conceptual Questions

1. What type of subatomic particle makes up cathode rays?
2. What is radioactivity? How did the discovery of radioactivity contradict Dalton's atomic theory?
3. How was Goldstein's experiment different from Thomson's, and how did it reveal different information about the atom?
4. Compare Dalton's model of the atom with the nuclear model of the atom.
5. What were Rutherford's two surprising conclusions, when he analyzed the results of his "gold foil" experiments?
6. In Rutherford's model of the atom, where are the protons, the neutrons, and the electrons found?
7. How did Bohr's theory change the concept of the atom?
8. What theory uses wave properties to describe the motion of particles at the atomic and subatomic levels?
9. Explain what is meant by the term *quantum*.
10. Which atom absorbs more energy—one in which an electron moves from the second shell to the third shell, or an otherwise identical atom in which an electron moves from the first to the third shell? Why is this true?
11. Describe the two-step process of electron transitions that results in an emission of radiation, such as light, when an atom absorbs energy from heat, electricity, or a laser beam.
12. Explain why more than one line is observed in the emission spectrum of hydrogen.

 Problems

Components of Atoms

13. How many electrons are present in a neutral atom of potassium with 19 protons in its nucleus?

14. A neutral atom with 14 protons will have how many electrons? What element is it?

15. Use the periodic table to determine the number of protons in an atom of each of the following elements.
a. boron
b. sulfur
c. copper

16. Use the periodic table to determine the number of protons in an atom of each of the following elements.
a. magnesium   **b.** aluminum   **c.** zinc

17. How many electrons are there in each neutral atom of the elements listed in Question 15?

18. How many electrons are there in each neutral atom of the elements listed in Question 16?

19. Sketch a diagram of a sulfur-33 atom, with the correct number of electrons in its main shells, and the correct number of each type of nucleon.

20. Sketch a diagram of a chlorine-37 atom, with the correct number of electrons in its main shells, and the correct number of each type of nucleon.

21. What are the symbol, name, and atomic mass of the element with atomic number 76? You may use the periodic table.

22. What are the symbol, name, and atomic number of the element that has 40 protons in the nuclei of its atoms?

Nuclear Symbols and Isotopes

23. The following table describes four atoms.

	Atom A	Atom B	Atom C	Atom D
Number of protons	17	18	18	17
Number of neutrons	18	17	18	17
Number of electrons	17	18	18	17

(a) Are atoms A and B isotopes? **(b)** Are A and C isotopes? **(c)** Are A and D isotopes? **(d)** What element is A? **(e)** What element is B?

24. Look at the table in Problem 23. **(a)** Are atoms B and C isotopes? **(b)** Are C and D isotopes? **(c)** What element is C? **(d)** What element is D?

25. Which atoms in Problem 23 have about the same mass?

26. Which atoms in Problem 23 have masses that are different from those of any of the others?

27. An atom with an electric charge is
a. an ion
b. a molecule
c. a nucleus
d. radioactive

28. Describe the charge and mass of the three different kinds of radioactivity.

29. Give the symbol and name for an isotope with a mass number of 37 and an atomic number of 17.

30. Give the symbol and name for an isotope with 30 neutrons and 25 protons.

31. Fill in the table.

Element	Mass Number	Number of Protons	Number of Neutrons
Nickel	60		
	108	46	
		7	7
Iodine			74

32. Determine which element has 21 protons and 24 neutrons. What is its mass number and number of electrons?

33. How many different elements are listed below?
$$^{23}_{11}X \quad ^{22}_{10}X \quad ^{11}_{5}X \quad ^{24}_{11}X \quad ^{25}_{12}X$$

34. How many different isotopes of silver are listed below? (The X does not necessarily represent any specific element.)
$$^{108}_{47}X \quad ^{108}_{48}X \quad ^{110}_{47}X \quad ^{109}_{46}X \quad ^{107}_{47}X$$

35. What is the most likely mass number of an atom with an atomic number of 14?

36. A beryllium atom has four protons, four electrons, and five neutrons. What is the atom's mass number?

37. When an atom loses three electrons, the charge on its ion is _____.

38. When an atom gains two electrons, the charge on its ion is _____.

Subshell Notation for Electron Configuration

39. Which main electron shell can be occupied by a maximum of eight electrons?

40. How many electrons can the third main electron shell hold?

41. Without referring to the periodic table, give the atomic numbers of the elements with the following electron configurations.
a. $1s^22s^22p^4$
b. $1s^22s^22p^63s^23p^2$
c. $1s^22s^22p^63s^23p^63d^84s^2$

42. Without referring to the periodic table, give the atomic numbers of the elements with the following electron configurations.
a. $1s^22s^22p^5$
b. $1s^22s^22p^63s^23p^64s^1$
c. $1s^22s^22p^63s^23p^63d^{10}4s^2$

43. Indicate whether each electron configuration represents an atom in the ground state or one in a possible excited state, or is incorrect. In each case, explain why.
a. $1s^12s^1$ **b.** $1s^22s^22p^7$ **c.** $1s^22p^2$ **d.** $1s^22s^22p^2$

44. Indicate whether each electron configuration represents an atom in the ground state or one in a possible excited state, or is incorrect. In each case, explain why.
a. $1s^22s^23s^2$
b. $1s^22s^2p^23s^1$
c. $1s^22s^22p^62d^5$
d. $1s^22s^42p^2$

45. Draw a grid containing 16 squares, in two rows of 8, representing the 16 elements in the periodic table from lithium to argon. For each of the elements, give **(a)** the main-shell electron configuration and **(b)** the subshell notation for the electron configuration.

46. Write the electron configuration for the valence electrons of neutral atoms of F, Cl, and Br. Explain why those atoms probably react with other atoms in similar ways.

47. Referring only to the periodic table, tell how the electron configurations of silicon (Si) and germanium (Ge) are similar. How are they different?

48. Referring only to the periodic table, tell how the electron configurations of fluorine (F) and chlorine (Cl) are similar. How are they different? How do you expect the electron configurations of bromine (Br) and iodine (I) to be similar to those of fluorine and chlorine?

49. Refer to the periodic table to categorize each element as a metal or a nonmetal.
    a. manganese   b. strontium   c. cesium   d. argon

50. Using the periodic table, categorize each element as a metal, nonmetal, or semimetal.
    a. titanium   b. arsenic   c. sulfur   d. boron

Use the following list of elements to answer problems 51–54.

Mg, Cs, Ne, P, Kr, K, Ra, N, Fe, Ca, Mo

51. Which elements are alkali metals?
52. Which elements are noble gases?
53. Which elements are transition metals?
54. How many of the elements are nonmetals?

Expand Your Skills

55. An atom of an element has two electrons in the first shell, eight electrons in the second shell, and five electrons in the third shell. From this information, give the element's **(a)** atomic number, **(b)** name, **(c)** total number of electrons in each of its atoms, **(d)** total number of s electrons, and **(e)** total number of d electrons.

56. Without referring to any tables in the text, color the s, p, and d blocks and mark an appropriate location for each of the following in the blank periodic table provided: **(a)** the fourth-period noble gas, **(b)** the third-period alkali metal, **(c)** the fourth-period halogen, and **(d)** a metal in the fourth period and in group 3B.

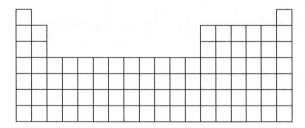

57. Suppose that two electrons are removed from the outermost shell of a magnesium atom. **(a)** What element's electron configuration would the atom then have? **(b)** Has that atom of magnesium been changed into that element? Why or why not?

58. Suppose three electrons are added to an arsenic atom. **(a)** Give the symbol for the element whose electron configuration matches the result. **(b)** Has that arsenic atom been changed into that element? Why or why not?

59. Refer to Figure 3.16 and write the electron configurations of **(a)** iron (Fe) and **(b)** tin (Sn).

60. Refer to Figure 3.16 and write the electron configuration of lead (Pb).

61. Atoms of two adjacent elements in the fourth period are in the ground state. An atom of element A has only s electrons in its valence shell. An atom of element B has at least one p electron in its valence shell. Identify elements A and B.

62. Atoms of two adjacent elements in the fifth period are in the ground state. An atom of element L has only s electrons in its valence shell. An atom of element M has at least one d electron in an unfilled shell. Identify elements L and M.

63. Atoms of two elements, one above the other in the same group, are in the ground state. An atom of element Q has two s electrons in its outer shell and no d electrons. An atom of element R has d electrons in its configuration. Identify elements Q and R.

64. Which of the following is true of the water-splitting reaction on page 88?
    a. It produces energy, for example, electricity or heat.
    b. It requires an energy input, for example, electricity or sunlight.
    c. It produces twice as much oxygen as hydrogen.
    d. It occurs inside a fuel cell.

65. What is produced when solar energy is converted into chemical-bond energy in the water-splitting reaction?
    a. hydrogen and nitrogen   b. oxygen and carbon
    c. hydrogen and oxygen   d. oxygen and calcium

66. Which of the following is not a benefit of using a solar fuel such as hydrogen?
    a. A solar fuel can store the energy of sunlight, which is intermittent.
    b. A solar fuel can replace some fossil fuels.
    c. Using a solar fuel such as hydrogen produces no carbon dioxide emissions.
    d. Electrodes used in electrolysis rely on platinum metal.

67. Isotopes played an important role in one of the most important experiments investigating photosynthesis; the chemical reaction that converts water (H_2O) and carbon dioxide (CO_2) into glucose ($C_6H_{12}O_6$) and oxygen (O_2). For most of the early twentieth century, scientists thought that the oxygen produced by photosynthetic plants and algae came from carbon dioxide they absorbed. To investigate this question, in the early 1940s, chemists at Stanford University used an isotope of oxygen to study the mechanism of the photosynthetic reaction. The researchers fed photosynthesizing algae with water enriched with oxygen-18 and discovered that the oxygen produced was enriched with oxygen-18. Did this result support or refute the hypothesis that the O_2 produced by photosynthesis comes from CO_2?

Critical Thinking Exercises

Apply knowledge that you have gained in this chapter and one or more of the FLaReS principles (Chapter 1) to evaluate the following statements or claims.

3.1 Suppose you read in the newspaper that a chemist in South America claimed to have discovered a new element with an atomic mass of 42. Extremely rare, it was found in a sample taken from the Andes Mountains. Unfortunately, the chemist has used all of the sample in his analyses.

3.2 Some aboriginal tribes have rain-making ceremonies in which they toss pebbles of gypsum up into the air. (Gypsum is the material used to make plaster of Paris by heating the rock to remove some of its water.) Sometimes it does rain several days after these rain-making ceremonies.

3.3 Some scientists think that life on other planets might be based on silicon rather than carbon. Evaluate this possibility.

3.4 You come across a website selling water made from only single isotopes of hydrogen and oxygen. A testimonial on the website claims that drinking only the isotopically pure water helps a person feel more refreshed throughout the day.

 Collaborative Group Projects

Prepare a PowerPoint, poster, or other presentation (as directed by your instructor) to share with the class.

1 Prepare a brief biographical report on one of the following.
a. Humphry Davy
b. Antoine Henri Becquerel
c. Wilhelm Roentgen
d. Marie Curie
e. Robert Millikan
f. Niels Bohr
g. Alessandro Volta
h. Ernest Rutherford
i. Michael Faraday
j. J. J. Thomson

2 Draw a grid containing 16 squares, in two rows of 8, representing the 16 elements in the periodic table from lithium to argon. For each of the elements, give **(a)** the main-shell electron configuration and **(b)** the subshell notation for the electron configuration.

3 Prepare a brief report on one of the **(a)** alkali metals, **(b)** alkaline earth metals, **(c)** halogens, or **(d)** noble gases. List sources and commercial uses.

4 Many different forms of the periodic table have been generated. Prepare a brief report showing at least three different versions of the periodic table, and comment on their usefulness.

LET'S EXPERIMENT! Birthday Candle Flame Test

Materials Needed

- Colorflame birthday candles (available in party supply stores or online)
- Matches or lighter
- Play-Doh or an aluminum pan of sand (container big enough to stand five candles in a row)

- Diffraction grating film (available as 8.5 × 11 inch or 12 × 6 inch sheets through online scientific education suppliers: 225 lines/mm or 500 lines/mm grating films work the best; the sheets can be cut into smaller pieces for use by multiple investigators)

What if you could test your ability to identify and describe elements by burning a birthday candle? Would you do it?

Traditionally, labs creating flame-emission spectra use solutions of hazardous metal salts and Bunsen burners. Even when used in tiny amounts, the metal ions in the salt solutions become volatized (airborne), creating a risk of inhalation. We can reduce the hazards by using birthday candles that burn at a much lower temperature than a Bunsen burner and eliminating the preparation and disposal of solutions of metal salts.

Your objective will be to compare and contrast the emission spectra from the colored birthday candles. Keep in mind that white light is composed of the multiple wavelengths of light that produce the whole rainbow. When individual atoms are heated, they absorb energy that causes transitions from lower to higher energy states. When the electrons return to the ground state, this energy is released as visible light. Every element has its own unique emission spectrum that can be used as its "fingerprint" for identification.

To perform this experiment, gather your materials and find a room that can easily be darkened. Place the five candles in the sand or Play-Doh about an inch apart, pressing them in about a quarter inch to steady them.

Create a small data table to record your findings. The order of the candles should match that of the data table—for instance, red, purple, green, blue, orange.

Light the candles using a match or lighter, and darken the room to observe them better. First observe the flames with your naked eye. Write down the color of each flame. Next, hold the diffraction film 6 to 12 inches away from the flames and observe the dominant color for each flame. Make a note about those colors. Blow out the candles. What did you see? How do the colors seen by the naked eye compare with the emission spectra emitted by each flame?

To validate your observations, check out emission-spectra websites such as http://webmineral.com/help/FlameTest.shtml

to identify the element in each flame. Which element is in each candle? Write down your answer in the data table.

Light individual candles again and compare your new and old observations. Are you able to justify your identifications of elements based on your observations? Why or why not? Brainstorm to find any mistake you might have made with this experiment.

Questions

1. How do the colors seen by the naked eye compare with the emission spectra emitted by each flame?
2. Why did the flames produce different colors?
3. Explain what happened within the atoms to produce the emission spectra.
4. Suggest potential sources of error in the analysis of the flames.
5. Justify your metal identifications based upon your observations.

Chemical Bonds

4

Have You Ever Wondered?

1 Why do atoms want to bond together anyway? Why are they not "happy" being alone? **2** Can you have chloride by itself? **3** How can metals be in salts, since metals are shiny and conductive, and salts are not? **4** If there is iron in blood, why aren't we magnetic? **5** Why does chlorine in a swimming pool have an odor but chlorine in salt does not? **6** Both oxygen and hydrogen are gases at room temperatures. Why is water—which combines them—a liquid?

THE TIES THAT BIND The element carbon is commonly found as soft, black soot formed by incomplete combustion or in chunks of coal. By heating that soot under tremendous pressure, we can make the hardest material known—diamond. We have not changed the carbon in soot to a different element in this process, so why is diamond so different from soot? The answer lies in the bonds that hold the carbon atoms together. About 18.5% of your body is carbon, but certainly not as soot or diamonds! The carbon in your body is *bonded* primarily to oxygen, hydrogen, and nitrogen atoms in myriad complex ways, and the reactions in which it participates in body chemistry are mind-boggling.

Atoms combine by forming chemical bonds. These bonds determine, to a great extent, the properties of the substance. We see carbon in the form of a diamond on the top of the previous page, but here we see carbon in graphite (pencil lead). The tetrahedral arrangement of essentially an infinite number of these bonds in a diamond makes it by far the hardest natural material in the world, but graphite is formed in slippery layers. The only difference between all the forms of carbon is the nature and arrangement of bonds between the carbon atoms.

We can now use what we have already learned about the structure of the atom to start understanding why these things happen. We are ready to consider chemical bonds, the ties that bind atoms together.

Chemical bonds are the forces that hold atoms together in molecules and hold ions together in ionic crystals. The vast number and incredible variety of chemical compounds—tens of millions are known—result from the fact that atoms can form bonds with many other types of atoms. Further, chemical bonds determine the 3-D shapes of molecules. Some important consequences of chemical structure and bonding include:

- whether a substance is a solid, liquid, or gas at room temperature
- the strengths of materials (as well as adhesives that hold materials together) used for building bridges, houses, and many other structures
- whether a liquid is low in density and volatile (like gasoline), or denser and viscous (like corn syrup)
- the optical properties of the substance (think of the "dusty" or dull appearance of soot versus that of a shiny diamond)
- the taste, odor, and drug activity of chemical compounds
- the structural integrity of skin, muscles, bones, and teeth
- the toxicity of certain molecules to living organisms

In this chapter, we will look at the different types of chemical bonds that can explain some of the unusual and interesting properties of many compounds.

4.1 The Art of Deduction: Stable Electron Configurations

Learning Objective • Determine the number of electrons in an ion and why it has a net positive or negative charge.

In our discussion of the atom and its structure (Chapter 2 and Chapter 3), we followed the historical development of some of the more important atomic concepts. Continuing to look at chemistry this way would require several volumes of print—and perhaps more of your time than you'd care to spend. We won't abandon the historical approach entirely but will emphasize another important aspect of scientific endeavor: deduction.

The art of deduction works something like this:

- **Fact**: Noble gases, such as helium, neon, and argon, are inert; they undergo few, if any, chemical reactions; they are "stable."
- **Theory**: The inertness of noble gases results from their electron configurations. Each gas (except helium) has an octet of electrons in its outermost shell.
- **Deduction**: An octet of electrons must be a very stable electron configuration. So other elements will likely seek to alter their electron configurations to become like those of noble gases in order to become less reactive (more stable).

We can use an example to illustrate this deductive argument. Sodium has eleven electrons, one of which is in the third shell. Recall that electrons in the outermost shell are called **valence electrons**, while those in all the other shells are lumped together as **core electrons**. The outermost shell is filled when it contains eight electrons: two electrons in an *s* orbital and six electrons in three *p* orbitals. (The first shell is an exception: It holds only two electrons in the $1s$ orbital.)

If the sodium atom were to get rid of its valence electron, its remaining core electrons would have the same electron configuration as an atom of the noble gas neon, a full octet. Using main-shell configurations (Chapter 3), we can represent this as

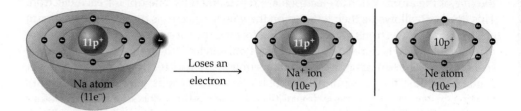

Similarly, if a chlorine atom were to gain an electron, it would have the same electron configuration as argon. The drive to have an electron configuration like a noble gas is strong. (We will see more in the subsection titled, "The Octet Rule: The Drive for Eight" on pages 107–108.)

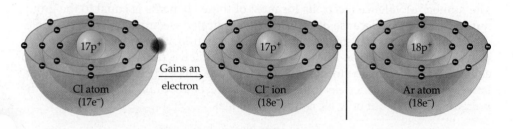

The sodium atom, having lost an electron, becomes positively charged. It has 11 protons (11+) and only 10 electrons (10−), for a net charge of 1^+. It is symbolized by Na^+ and is called a *sodium ion*. The chlorine atom, having gained an electron, has a net charge of 1^-. It has 17 protons (17+) and 18 electrons (18−). It is symbolized by Cl^- and is called a *chloride ion*. Note that a positive charge, as in Na^+, indicates that one electron has been lost. Similarly, a negative charge, as in Cl^-, indicates that one electron has been gained.

It is important to note that even though Cl^- and Ar are **isoelectronic** (i.e., they have the same electron configuration), they are *different* chemical species, because their nuclei are different. In the same way, a sodium atom does not *become* an atom of neon when it loses an electron; the sodium ion is simply isoelectronic with neon.

1 Why do atoms want to bond together anyway? Why are they not "happy" being alone?

Nature tends to make changes with a goal of "stability." A pile of teetering blocks is less stable than the blocks lying on the floor. Scientists have discovered that a filled octet of electrons in the outermost shell of an atom is very stable, so most atoms tend to give away, "steal," or share electrons in order to attain that octet.

SELF-ASSESSMENT Questions

1. Which group of elements in the periodic table is characterized by an especially stable electron arrangement?
   a. 2A
   b. 2B
   c. 8A
   d. 8B

2. The structural difference between a sodium atom and a sodium ion is that the sodium ion has
   a. one less proton than the sodium atom
   b. one less proton electron than the sodium atom
   c. more proton and one less electron than the sodium atom
   d. one more proton than the sodium atom

3. Which of the following are isoelectronic?
   a. S^{2-} and Ar
   b. Al^{3+} and Ar
   c. Kr and Cl^-
   d. Ne and Ar

4. Which of the following cations is isoelectronic with S^{2-}?
   a. Na^+
   b. Rb^+
   c. Mg^{2+}
   d. Ca^{2+}

Answers: 1, c; 2, b; 3, a; 4, d

▲ Electron-dot symbols are called *Lewis symbols*, after G. N. Lewis, the famous American chemist (1875–1946) who invented them. Lewis also made important contributions to our understanding of thermodynamics, acids and bases, and spectroscopy.

4.2 Lewis (Electron-Dot) Symbols

Learning Objective • Write the Lewis symbol for an atom or ion.

In forming ions, the *cores* (nucleus plus inner electrons) of the sodium and chlorine atoms do not change. It is convenient therefore to let the element symbol represent the *core* of the atom. Valence electrons are represented by one dot for each electron. This approach allows us then to focus on the valence electrons only, since the valence electrons are the electrons involved in bond formation. These electron-dot symbols are usually called Lewis dot symbols or **Lewis symbols** for short, since Gilbert N. Lewis (America, 1875–1946) first suggested that symbolism in 1916. In writing a Lewis dot symbol, consider that a chemical symbol has four sides (right, left, top, and bottom). Starting on *any side,* place single dots until each side has one dot before adding another one (if needed). There should be no more than two dots on any given side of the chemical symbol.

Lewis Dot Symbols and the Periodic Table

It is especially easy to write Lewis dot symbols for most of the main-group elements. The number of valence electrons for most of these elements is equal to the group number (Table 4.1). Because of their more complicated electron configurations, elements in the central part of the periodic table (the transition metals) cannot easily be represented by electron-dot symbols. The use of Lewis dot symbols in writing an equation representing the formation of an ion allows us to focus on the most important electrons, the valence electrons. The formation of a sodium ion was *drawn* in the preceding section, but now we can represent it as:

$$\text{Na} \cdot \longrightarrow \text{Na}^+ + 1\,e^-$$

and

$$\cdot \overset{..}{\underset{..}{\text{Cl}}}{:} \; + \; 1\,e^- \longrightarrow \overset{..}{\underset{..}{:\text{Cl}}}{:}^-$$

TABLE 4.1 **Lewis Dot Symbols for Selected Main Group Elements**

Group 1A	Group 2A	Group 3A	Group 4A	Group 5A	Group 6A	Group 7A	Noble Gases
H·							He:
Li·	·Be·	·B·	·C·	:N·	:O·	:F·	:Ne:
Na·	·Mg·	·Al·	·Si·	:P·	:S·	:Cl·	:Ar:
K·	·Ca·				:Se·	:Br·	:Kr:
Rb·	·Sr·				:Te·	:I·	:Xe:
Cs·	·Ba·						

EXAMPLE 4.1 Writing Lewis Symbols

Without referring to Table 4.1, but using the periodic table, write Lewis dot symbols for magnesium, oxygen, and phosphorus. In what respect was the periodic table useful?

Solution

According to the periodic table, magnesium is in group 2A, oxygen is in group 6A, and phosphorus is in group 5A. The Lewis dot symbols, therefore, have two, six, and five dots, respectively. They are

$$\cdot Mg \cdot \qquad :\overset{..}{\underset{..}{O}} \qquad :\overset{}{\underset{.}{P}}\cdot$$

> **Exercise 4.1A**

Without referring to Table 4.1, write Lewis symbols for each of the following elements. You may use the periodic table.

 a. Kr **b.** Ba **c.** I **d.** N **e.** K **f.** S

> **Exercise 4.1B**

Write Lewis symbols for each of the following *ions*. Remember that positive ions have lost electrons and negative ions have gained electrons. You may use the periodic table.

 a. N^{3-} **b.** Mg^{2+} **c.** K^+ **d.** Br^-

SELF-ASSESSMENT Questions

1. How many dots surround B in the Lewis dot symbol for boron?
   a. 2 **b.** 3 **c.** 5 **d.** 9

2. How many dots surround Br in the Lewis dot symbol for Br^-?
   a. 1 **b.** 8 **c.** 7 **d.** 9

3. Which of the following Lewis dot symbols is *incorrect*?
   a. $\cdot\overset{.}{\underset{.}{C}}\cdot$ **b.** $:\overset{..}{Cl}\cdot$ **c.** Li: **d.** $:\overset{.}{N}\cdot$

4. Which of the following Lewis dot symbols is *incorrect*?
   a. $\cdot\overset{.}{As}\cdot$ **b.** $:\overset{..}{O}\cdot$ **c.** Rb· **d.** $\cdot\overset{.}{Si}\cdot$

Answers: 1, b; 2, b; 3, c; 4, a

4.3 The Reaction of Sodium with Chlorine

Learning Objectives • Distinguish between an ion and an atom. • Describe the nature of the attraction that leads to formation of an ionic bond.

Sodium (Na) is a highly reactive metal. It is soft enough to be cut with a knife. When freshly cut, it is bright and silvery, but it dulls rapidly because it reacts with oxygen in the air. It reacts so readily in air that it is usually stored away from oxygen under oil. Sodium reacts violently with water, too, becoming hot from the reaction that tears water molecules apart, forming hydrogen gas and sodium hydroxide in a solution. The sodium metal melts and forms a spherical bead that races around on the surface of the water as it reacts. All other alkali metals react vigorously to violently with water, as shown on page 91.

Chlorine (Cl_2) is a greenish-yellow gas and is extremely irritating to the eyes and nose. In fact, it was used as a poisonous gas in World War I. Chlorine is a common disinfectant for drinking water and swimming pools, since it kills microorganisms. (The actual substance added is often a compound that reacts with water to form chlorine.) When a piece of sodium is dropped into a flask containing chlorine gas, a violent reaction ensues, producing beautiful white crystals of sodium chloride, table salt, that you might sprinkle on your food at the dinner table (Figure 4.1). Clearly, sodium chloride has very few properties in common with either of its parent materials, sodium or chlorine.

2 Can you have chloride ions only?

Formation of an anion, such as a chloride ion, is only possible when another atom or group of atoms gives up an electron and forms a cation. The presence of an anion requires the presence of a cation in the same jar. You can't have one without the other.

▶ **Figure 4.1** (a) Chlorine (a greenish gas) and sodium (a soft, silvery metal). (b) The two react violently to form sodium chloride (ordinary table salt), a white crystalline solid (c).

Q *What kinds of particles— ions or molecules—make up sodium chloride?*

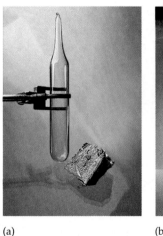

(a)   (b)   (c)

The Reaction of Sodium with Chlorine: Theory

A sodium atom achieves a filled valence shell by losing one electron. A chlorine atom achieves a filled valence shell by adding one electron. What happens when sodium atoms come into contact with chlorine atoms? The obvious answer: A chlorine atom takes an electron from a sodium atom.

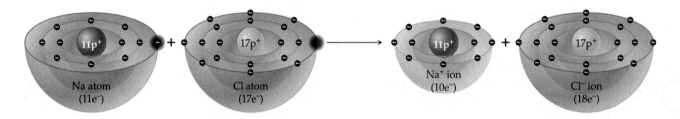

Na atom (11e⁻)   Cl atom (17e⁻)   Na⁺ ion (10e⁻)   Cl⁻ ion (18e⁻)

With Lewis symbols, this reaction is written as

$$\text{Na·} + \text{·}\overset{\cdot\cdot}{\underset{\cdot\cdot}{\text{Cl}}}\text{:} \longrightarrow \text{Na}^+\overset{\cdot\cdot}{\underset{\cdot\cdot}{\text{Cl}}}\text{:}$$

3 How can metals be in salts, since metals are shiny and conductive, and salts are not?

Metals contain neutral atoms of those elements. The packing of metal atoms often does reflect light, and they appear shiny. They conduct current because their outer electrons can easily be moved from one atom to the next, and so on, through a wire. When metal atoms form salts with non-metal atoms, their outermost electrons are "stolen" by the non-metal atoms; positive and negative ions are formed, which have completely different physical and chemical properties, compared with the original atoms.

But wait a minute! Chlorine gas is composed of diatomic molecules of Cl_2 (bonded together with *covalent bonds,* which we discuss later, in Section 4.6), and not of separate Cl atoms, so how can *one* chlorine atom combine with a sodium atom to form NaCl? It turns out that each chlorine atom in a diatomic chlorine molecule takes one electron from each of the two sodium atoms to fill its electron shell; two Cl⁻ and two Na⁺ ions are formed as a result, as shown in this overall chemical reaction:

$$Cl_2 + 2\,Na \longrightarrow 2\,Cl^- + 2\,Na^+$$

Since Cl⁻ and Na⁺ have different electronic configurations from chlorine and sodium atoms, the compound they form—sodium chloride—has very different properties from their parent atoms.

Ionic Bonds

The Na⁺ and Cl⁻ ions formed from sodium and chlorine atoms have opposite charges and are strongly attracted to one another. These ions arrange themselves in an orderly fashion in 3-D space. Each sodium ion attracts (and is attracted to) six chloride ions (above and below, front and back, and left and right), as shown in Figure 4.2a. Simultaneously, each chloride ion attracts (and is attracted to) six sodium ions (above and below, front and back, and left and right). The arrangement is repeated many times in all directions (Figure 4.2b). The result is a **crystal** of sodium chloride. The orderly microscopic arrangement of the ions is reflected in the macroscopic shape, a cube of a salt crystal (Figure 4.2c). Even the tiniest grain of salt has billions and billions of ions of each type. The forces holding the crystal together—the attractive forces between positive and negative ions—are called **ionic bonds**.

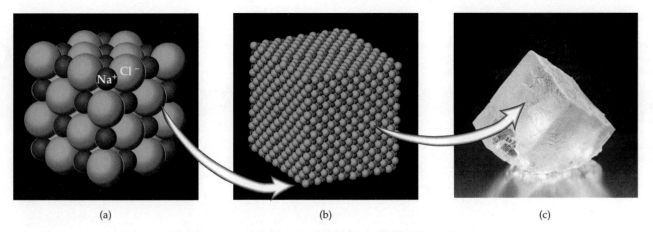

(a)   (b)   (c)

▲ **Figure 4.2** Molecular and macroscopic views of a sodium chloride crystal. (a) Each Na^+ ion (small purple spheres) is surrounded by six Cl^- ions (large green spheres), and each Cl^- ion by six Na^+ ions. (b) This arrangement repeats itself many, many times. (c) The highly ordered pattern of alternating Na^+ and Cl^- ions is observed in the macroscopic world as a crystal of sodium chloride.

Atoms and Ions: Distinctively Different

Ions are emphatically different from the atoms from which they are made, much as cars with and without an engine differ greatly. The names and symbols of an atom and its ion may look a lot alike—especially the positively charged ions, called cations—but the actual entities are very different (Figure 4.3). Unfortunately, the situation is confusing because people talk about needing iron to perk up "tired blood" and calcium for healthy teeth and bones. What they *really* need is iron(II) *ions* (Fe^{2+}) and calcium *ions* (Ca^{2+}). You wouldn't think of eating iron nails to get iron (although some enriched cereals do indeed contain powdered iron; the iron metal is readily converted to Fe^{2+} cations in the stomach). Nor would you eat highly reactive calcium metal.

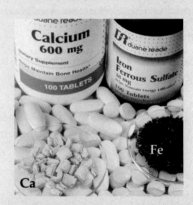

◀ **Figure 4.3** Forms of iron and calcium that are useful to the human body are the ions (usually Fe^{2+} and Ca^{2+}) in ionic compounds such as $FeSO_4$ and $CaCO_3$. The elemental forms (Fe and Ca) are chemically very different from the forms found in ionic compounds.

Similarly, if you are warned to reduce your sodium intake, your doctor is not concerned that you are eating too much sodium metal—talk about an upset stomach!—but that your intake of Na^+ *ions*—usually as sodium chloride—may be too high. Although the names are similar, the atom and the cation are quite different chemically. This is why it is important to make careful distinctions and use precise terminology. Thankfully, there is less room for confusion when it comes to naming the negatively charged ions, called anions. As you will see in Section 4.5, the names of anions, while based on the names of the atoms from which they come, are made to sound different through the use of suffixes (or endings), such as *-ide* in chlor*ide*.

SELF-ASSESSMENT Questions

1. A potassium ion and an argon atom have the same
 a. electron configuration   **b.** net charge
 c. nuclear charge   **d.** properties

2. How many dots are shown on the Lewis symbol for chloride ion, and what is the ion's charge?
 a. five dots, no charge
 b. six dots, positive charge
 c. seven dots, negative charge
 d. eight dots, negative charge

3. The bonding between Li^+ ions and F^- ions in lithium fluoride is
 a. covalent   **b.** ionic
 c. nonpolar   **d.** polar

4. Which group of elements will form stable ions with a 2– charge?
 a. 2A
 b. 2B
 c. 6A
 d. 8A

Answers: 1, a; 2, d; 3, b; 4, c

4.4 Using Lewis Symbols for Ionic Compounds

Learning Objectives • Write symbols for common ions and determine their charges. • Describe the relationship between the octet rule and the charge on an ion.

Potassium, a metal in the same family as sodium and therefore similar to sodium in properties, also reacts with chlorine. The reaction yields potassium chloride (KCl), a stable white crystalline solid, which is the primary ingredient in what is sold commercially as "substitute salt."

$$K\cdot \ + \ \cdot \ddot{\underset{..}{Cl}}: \longrightarrow K^+ \ + \ :\ddot{\underset{..}{Cl}}:^-$$

Potassium also reacts with bromine, a reddish-brown liquid in the same family as chlorine and therefore similar to chlorine in properties. The product, potassium bromide (KBr), is a stable white crystalline solid.

$$K\cdot \ + \ \cdot \ddot{\underset{..}{Br}}: \longrightarrow K^+ \ + \ :\ddot{\underset{..}{Br}}:^-$$

 EXAMPLE 4.2   Electron Transfer to Form Ions

Use Lewis dot symbols to show the transfer of electrons from sodium atoms to bromine atoms to form ions with noble gas configurations.

Solution
Sodium has one valence electron, and bromine has seven. Transfer of the single valence electron from sodium to bromine leaves each with a noble gas configuration. Sodium bromide is a white, solid compound, similar to sodium chloride.

$$Na\cdot \ + \ \cdot \ddot{\underset{..}{Br}}: \longrightarrow Na^+ \ + \ :\ddot{\underset{..}{Br}}:^-$$

> **Exercise 4.2A**
Use Lewis symbols to show the transfer of electrons from lithium atoms to fluorine atoms to form ions with noble gas configurations.

> **Exercise 4.2B**
Use Lewis dot symbols to show the transfer of electrons from cesium atoms to fluorine atoms. Does cesium now have a noble gas configuration?

Magnesium, a group 2A metal, is less reactive than sodium. When an atom of magnesium reacts with an atom of oxygen—a group 6A element that forms a colorless diatomic gas, O_2—a stable white crystalline solid called magnesium oxide (MgO) is formed:

$$\cdot Mg\cdot + \cdot \overset{\cdot\cdot}{\underset{\cdot\cdot}{O}}: \longrightarrow Mg^{2+} + :\overset{\cdot\cdot}{\underset{\cdot\cdot}{O}}:^{2-}$$

When an Mg atom loses two electrons, it becomes a Mg^{2+} ion and its second shell becomes its outermost shell. On the other hand, when an oxygen atom gains two electrons, it fills its current outermost shell. These are the *same* two electrons changing hands! Both the Mg^{2+} and O^{2-} ions become isoelectronic with the Ne atom and have the same stable main-shell electron configurations: 2, 8.

An atom such as oxygen, which needs two electrons to achieve a noble gas configuration, may react with potassium atoms, which have only one electron each to give. In this case, two atoms of potassium are needed for each one oxygen atom. The product is potassium oxide (K_2O).

$$\begin{matrix} K\cdot \\ \\ K\cdot \end{matrix} + \cdot\overset{\cdot\cdot}{\underset{\cdot}{O}}: \longrightarrow \begin{matrix} K^+ \\ \\ K^+ \end{matrix} + :\overset{\cdot\cdot}{\underset{\cdot\cdot}{O}}:^{2-}$$

Through this reaction, each potassium atom achieves the electron configuration of argon. As before, oxygen ends up with the neon configuration.

EXAMPLE 4.3   Electron Transfer to Form Ions

Use Lewis symbols to show the transfer of electrons from magnesium atoms to nitrogen atoms to form ions with noble gas configurations.

Solution

$$\begin{matrix} \cdot Mg\cdot \\ \cdot Mg\cdot \\ \cdot Mg\cdot \end{matrix} + \begin{matrix} \cdot\overset{\cdot\cdot}{\underset{\cdot}{N}}\cdot \\ \\ \cdot\overset{}{\underset{\cdot}{N}}\cdot \end{matrix} \longrightarrow \begin{matrix} Mg^{2+} \\ Mg^{2+} \\ Mg^{2+} \end{matrix} + \begin{matrix} :\overset{\cdot\cdot}{\underset{\cdot\cdot}{N}}:^{3-} \\ \\ :\overset{\cdot\cdot}{\underset{\cdot\cdot}{N}}:^{3-} \end{matrix}$$

Each of the three magnesium atoms gives up two electrons (a total of six), and each of the two nitrogen atoms acquires three electrons (a total of six). Notice that the total positive and negative charges on the products are equal (6+ and 6−). Magnesium reacts with nitrogen to yield magnesium nitride, Mg_3N_2.

❯ **Exercise 4.3A**
Calcium metal is quite reactive, combining with oxygen to form *quicklime* (calcium oxide). Use Lewis symbols to show the transfer of electrons from calcium atoms to oxygen atoms to form ions with noble gas configurations.

❯ **Exercise 4.3B**
Aluminum is a very reactive metal, but in the Earth's atmosphere, the outside surface of aluminum is converted to the rather unreactive aluminum oxide. Use Lewis symbols to show the transfer of electrons from aluminum atoms to oxygen atoms to form ions with noble gas configurations.

Generally speaking, metallic elements in groups 1A and 2A (those from the left side of the periodic table) react with nonmetallic elements in groups 6A and 7A (those from the right side) to form ionic compounds. These products are stable crystalline solids.

The Octet Rule: The Drive for Eight

Each atom of a metal tends to give up the electrons it has in its outer shell, leaving a filled shell just below it, and each atom of a nonmetal tends to take on enough electrons to *complete, or close, its valence shell* (s^2p^6). In other words, atoms are driven toward becoming

WHY IT MATTERS

Loss of electrons is known as *oxidation*, and gain of electrons is called *reduction*. Oxidation by oxygen gas is both useful and destructive. Oxygen gas oxidizes fuels to produce energy, but it also oxidizes iron metal into familiar orange or red rust (Fe_2O_3), which causes billions of dollars of damage each year. We explore the concepts of oxidation and reduction in greater detail in Chapter 8.

ions to have eight—an *octet* of—electrons in their valence shells. An octet of electrons is characteristic of all noble gases except helium. When atoms react with each other, they often tend to attain this very stable electron configuration of a noble gas closest to them. Thus, they are said to follow the **octet rule**, or the "rule of eight." In the case of helium, a maximum of two electrons can occupy its single electron shell. Consequently, such small atoms as hydrogen (in metal hydrides) or lithium often follow the "rule of two," or the **duet rule**, to obtain the electron configuration of their nearest noble gas—helium.

In following the octet rule, atoms of group 1A metals give up one electron to form 1+ ions, those of group 2A metals give up two electrons to form 2+ ions, and those of group 3A metals give up three electrons to form 3+ ions. Group 7A nonmetal atoms take on one electron to form 1− ions, and group 6A atoms tend to pick up two electrons to form 2− ions. Some group 5A atoms tend to take on three electrons to form 3− ions (most notably N and P). Atoms of B group metals can give up various numbers of electrons to form positive ions with various charges.

These periodic relationships are summarized in Figure 4.4.

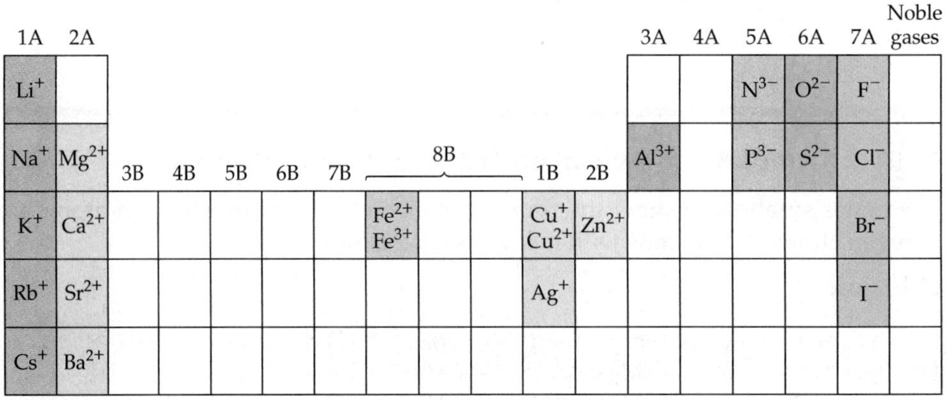

▶ **Figure 4.4** Periodic relationships of some simple ions. Many of the transition elements (B groups) can form multiple ions with different charges.

Q Can you write a formula for the simple ion formed from tellurium (Te)? For the simple ion formed from gallium (Ga)?

																	Noble gases
1A	2A										3A	4A	5A	6A	7A		
Li^+													N^{3-}	O^{2-}	F^-		
Na^+	Mg^{2+}	3B	4B	5B	6B	7B	8B	1B	2B	Al^{3+}			P^{3-}	S^{2-}	Cl^-		
K^+	Ca^{2+}						Fe^{2+} Fe^{3+}	Cu^+ Cu^{2+}	Zn^{2+}						Br^-		
Rb^+	Sr^{2+}							Ag^+							I^-		
Cs^+	Ba^{2+}																

4 **If there is iron in blood, why aren't we magnetic?**

Iron metal is magnetic, but compounds—such as the hemoglobin in blood—usually contain iron(II) or iron(III) ions. The ions differ chemically and physically from the metal, and are not magnetic.

EXAMPLE 4.4 Determining Formulas by Electron Transfer

What is the formula of the compound formed by the reaction of sodium and sulfur?

Solution

Sodium is in group 1A; the sodium atom has one valence electron. Sulfur is in group 6A; the sulfur atom has six valence electrons.

$$Na\cdot \quad \cdot \ddot{\underset{\cdot\cdot}{S}}\cdot$$

Sulfur needs two electrons to achieve an argon configuration, but sodium has only one electron to give. The sulfur atom therefore must react with two sodium atoms, stealing one electron from each of them:

$$\begin{array}{c} Na\cdot \\ + \quad \cdot\ddot{\underset{\cdot\cdot}{S}}\cdot \\ Na\cdot \end{array} \longrightarrow \begin{array}{c} Na^+ \\ + \quad :\ddot{\underset{\cdot\cdot}{S}}:^{2-} \\ Na^+ \end{array}$$

The formula of the compound, called sodium sulfide, is Na_2S.

> **Exercise 4.4A**

What are the formulas of the compounds formed by the reaction of **(a)** calcium with fluorine and **(b)** lithium with oxygen?

> **Exercise 4.4B**

Use Figure 4.4 to predict the formulas of the two compounds that can be formed from iron (Fe) and chlorine.

SELF-ASSESSMENT Questions

1. Which of the following pairs of elements would be most likely to form an ionic compound with each other?
   a. K and Ca
   b. Rb and Cl
   c. S and Se
   d. At and Ne

2. Oxygen forms a simple ion with a charge of
   a. 6−   **b.** 2−   **c.** 2+   **d.** 6+

3. Strontium forms a simple ion with a charge of
   a. 2−   **b.** 2+   **c.** 4+   **d.** 8+

4. Ca and P react to form Ca_3P_2, an ionic compound. How many electrons are there in the valence shell of the P^{3-} (phosphide) ion?
   a. 2   **b.** 8   **c.** 10   **d.** 16

5. The formula of the ionic compound formed by lithium and nitrogen is
   a. LiN
   b. Li_3N
   c. LiN_3
   d. Li_6N_2

6. The formula of the ionic compound formed by magnesium and bromine is
   a. MgBr
   b. Mg_2Br
   c. $MgBr_2$
   d. Mg_2Br_3

7. Only one of the following ions is likely to be formed in an ordinary chemical reaction; it is
   a. S^{6-}
   b. Br^-
   c. K^{2+}
   d. Ar^{3-}

Answers: 1, b; 2, b; 3, b; 4, b; 5, b; 6, c; 7, b

4.5 Formulas and Names of Binary Ionic Compounds

Learning Objective • Name and write formulas for binary ionic compounds.

Names of simple positive ions (*cations*) are derived from those of their parent elements and simply add the word "*ion*." For example, when a sodium atom loses an electron, it becomes a *sodium ion* (Na^+). A magnesium atom (Mg), on losing two electrons, becomes a *magnesium ion* (Mg^{2+}). When a metal can form more than one ion, the charges on the different ions are denoted by Roman numerals in parentheses. For example, Fe^{2+} is iron(II) ion and Fe^{3+} is iron(III) ion.

Names of simple negative ions (*anions*) end with *ide* and add the word *ion*. For instance, when a chlor*ine* atom gains an electron, it becomes a *chloride ion* (Cl^-). Oxygen becomes *oxide*, nitrogen becomes *nitride*, and sulfur is changed to *sulfide*.

Table 4.2 lists symbols and names for some ions formed when atoms gain or lose electrons. You can calculate the charge on the negative ions in the table by subtracting 8 from the group number. For example, by using Figure 4.4, you can see that the charge on the oxide ion (oxygen is in group 6A) is $6 - 8 = -2$. The nitride ion (nitrogen is in group 5A) has a charge of $5 - 8 = -3$.

5 **Why does chlorine in a swimming pool have an odor, but chlorine in salt does not?**

Chlorine used in water treatment is often the element Cl_2, while the chlorine in salt exists as chlor*ide* ions, Cl^-. An element is chemically quite different from its ions, and has different properties.

A Compound by Any Other Name . . .

As you read labels on foodstuffs and pharmaceuticals, you will see some names that are beginning to sound familiar and some that are confusing or mysterious. For example, why do we have two names for Fe^{2+}? Some names are historical: Iron, copper, gold, silver, and some other elements have been known for thousands of years. The first part of each of those names originates in the Latin words for those elements, such as "ferrum" (Latin) for iron. The ending *-ous* came to be used for the ion of smaller charge and the ending *-ic* for the ion of larger charge. The modern system uses Roman numerals to indicate ionic charge. Although the new system is more logical and easier to apply, the old names persist, especially in everyday life and in some of the biomedical sciences.

| POTASSIUM | mg | 108 |
| CHLORIDE | mg | 63 |

GETABLE OIL (PALM OLEIN, COCONUT, SOY AND OTEIN CONCENTRATE, GALACTOOLIGOSACCHA- N 1%: MORTIERELLA ALPINA OIL**, CRYPTHECO- ATE, POTASSIUM CITRATE, FERROUS SULFATE, DE, SODIUM CHLORIDE, ZINC SULFATE, CUPRIC SELENITE, SOY LECITHIN, CHOLINE CHLORIDE, NTOTHENATE, VITAMIN A PALMITATE, VITAMIN B12, CHLORIDE, VITAMIN B6 HYDROCHLORIDE, FOLIC

water powde

To M

2 fl o

4 fl o

8 fl o

▲ Labels for dietary supplements often use older, Latin-derived names.

TABLE 4.2 **Symbols and Names for Some Simple (Monatomic) Ions**

Group	Element	Name of Ion	Symbol for Ion
1A	Hydrogen	Hydrogen ion	H^+
	Lithium	Lithium ion	Li^+
	Sodium	Sodium ion	Na^+
	Potassium	Potassium ion	K^+
2A	Magnesium	Magnesium ion	Mg^{2+}
	Calcium	Calcium ion	Ca^{2+}
3A	Aluminum	Aluminum ion	Al^{3+}
5A	Nitrogen	Nitride ion	N^{3-}
	Phosphorus	Phosphide ion	P^{3-}
6A	Oxygen	Oxide ion	O^{2-}
	Sulfur	Sulfide ion	S^{2-}
7A	Fluorine	Fluoride ion	F^-
	Chlorine	Chloride ion	Cl^-
	Bromine	Bromide ion	Br^-
	Iodine	Iodide ion	I^-
1B	Copper	Copper(I) ion (cuprous ion)	Cu^+
		Copper(II) ion (cupric ion)	Cu^{2+}
	Silver	Silver ion	Ag^+
2B	Zinc	Zinc ion	Zn^{2+}
8B	Iron	Iron(II) ion (ferrous ion)	Fe^{2+}
		Iron(III) ion (ferric ion)	Fe^{3+}

Simple ions of opposite charge can be combined to form **binary** (two-element) **ionic compounds**. To get the correct formula for a binary ionic compound, (1) write each ion with its charge (positive ion on the left), (2) swap the charge numbers (but not the plus and minus signs), (3) write them as subscripts right after the element's symbol, and (4) confirm that the transposed (swapped) numbers have the simplest possible ratio, while maintaining the original atomic ratio. The process is best learned by practice, which is provided in the following examples, exercises, and end-of-chapter problems.

EXAMPLE 4.5 Determining Formulas from Ionic Charges

Write the formulas for **(a)** calcium chloride and **(b)** aluminum oxide.

Solution

a. First, we write the symbols for the ions. (We write the charge on chloride ion explicitly as "1−" to illustrate the method. You may omit the "1" when you are comfortable with the process.)

$$Ca^{2+} \; Cl^{1-}$$

We cross over the charge numbers (without the charges) as subscripts.

Then we write the formula. The formula for calcium chloride is

$$Ca_1Cl_2 \quad \text{or} \quad \text{(dropping the "1") simply } CaCl_2$$

b. We write the symbols for the ions.

$$Al^{3+}O^{2-}$$

We cross over the charge numbers as subscripts.

We write the formula for aluminum oxide as

$$Al_2O_3$$

> **Exercise 4.5A**
Write the formulas for **(a)** potassium oxide and **(b)** calcium nitride.

> **Exercise 4.5B**
Write the formulas for **(a)** magnesium phosphide and **(b)** aluminum phosphide.

The method just described, called the *crossover method*, works because it is based on the transfer of electrons and the conservation of charge. Two Al atoms lose three electrons each (a total of six electrons lost), and three O atoms gain two electrons each (a total of six electrons gained). The electrons lost must equal the electrons gained. Similarly, two Al^{3+} ions have six positive charges (three each), and three O^{2-} ions have six negative charges (two each). The net charge on Al_2O_3 is zero, just as it should be. Always remember that the formula should be the *simplest* ratio of elements. There are some elements that can have a 4+ charge as a cation, such as tin(IV). The correct formula for tin(IV) oxide is not Sn_2O_4; it is SnO_2.

Now that you are able to translate chemical names, such as aluminum oxide, into a chemical formula, such as Al_2O_3, you also can translate in the other direction, as shown in Example 4.6.

 EXAMPLE 4.6   Naming Ionic Compounds

What are the names of **(a)** MgS and **(b)** $FeCl_3$?

Solution

a. From Table 4.2, we can determine that MgS is made up of Mg^{2+} (magnesium ion) and S^{2-} (sulfide ion). The name is simply magnesium sulfide.

b. From Table 4.2, we can determine that the ions in $FeCl_3$ are

$$Fe^{3+} Cl^{-}$$

How do we know the iron ion in $FeCl_3$ is Fe^{3+} and not Fe^{2+}? Because there are three Cl^{-} ions, each with a 1− charge, the single Fe ion must have a 3+ charge, since the compound, $FeCl_3$, is neutral. The names of these ions are iron(III) ion (or ferric ion) and chloride ion. Therefore, the compound is iron(III) chloride (or, by the older system, ferric chloride).

> **Exercise 4.6A**
What are the names of **(a)** CaF_2 and **(b)** $CuBr_2$?

> **Exercise 4.6B**
What are the names of **(a)** Li_2S and **(b)** Fe_2S_3?

NaF in Medicine and Health

Sodium fluoride is on the World Health Organization's List of Essential Medicines, for its function to strengthen tooth enamel and prevent dental caries, and its utility in treating osteoporosis. It is an additive in many brands of toothpaste. Radioactive NaF (with fluorine-18) is used in positron emission tomography (PET) scans, to detect cancer, to determine whether a cancer has spread or treatment is working, to study blood flow to the heart, and to study Alzheimer's disease.

SELF-ASSESSMENT Questions

1. The formula for the binary ionic compound of barium and sulfur is
   a. BaS **b.** Ba_2S **c.** BaS_2 **d.** Ba_2S_3

2. Which of the following formulas is correct?
   a. ZnCl **b.** Zn_2Cl **c.** $ZnCl_2$ **d.** Zn_2Cl_5

3. Which of the following formulas is incorrect?
   a. $AlCl_3$ **b.** Al_3P_2 **c.** CaS **d.** Cs_2S

4. The correct name for KCl is
   a. krypton chloride
   c. potassium chloride
   b. krypton chlorite
   d. potassium chlorine

5. The correct name for Al_2S_3 is
   a. aluminum sulfide
   c. antimony sulfide
   b. dialuminum trisulfite
   d. antimony trisulfide

6. The correct name for LiBr is
   a. bromide lithide
   c. lithium bromium
   b. bromine lithide
   d. lithium bromide

7. The correct name for ZnI_2 is
   a. diiodomonozircon
   c. zincocide chloride
   b. zinc iodide
   d. zirconium iodide

Answers: 1, a; 2, c; 3, b; 4, c; 5, a; 6, d; 7, b

4.6 Covalent Bonds: Shared Electron Pairs

Learning Objectives • Explain the difference between a covalent bond and an ionic bond. • Name and write formulas for covalent compounds.

In Chapter 2, we stated that some elements form diatomic molecules. In Section 4.3 of this chapter, we mentioned that chlorine gas actually exists as a diatomic molecule. You now have the tools to understand what a diatomic molecule is, and how the atoms bond together, forming what is called a *covalent bond*.

Let's start with hydrogen. Because hydrogen follows the duet rule, it needs to have two electrons in order to complete its valence shell. The problem is that each hydrogen atom has only one electron. However, when two hydrogen atoms combine and share their electrons, each atom *gains access* to two electrons and a complete valence shell. Both hydrogen atoms come to this compromise because all hydrogen atoms have an equal attraction for electrons. This leads to the two hydrogen atoms' *sharing a pair* of electrons. Note that the two dots are drawn between the H atoms in the H_2 structure to the right of the arrow.

$$H\cdot + \cdot H \longrightarrow H:H$$

By sharing electrons, the two hydrogen atoms form a *diatomic* hydrogen molecule. The bond formed when atoms share electrons is called a **covalent bond**. When *one* pair of electrons is shared, the bond is a **single bond**.

H:H
└ covalent bond (shared pair of electrons)

A covalent bond is often represented as a *dash*, so that the Lewis dot structure for H_2 becomes H—H. The dash represents one shared pair of electrons, one from each atom in the bond.

Let's next consider the chlorine molecule, Cl_2. A chlorine atom readily takes an extra electron from anything willing to give one up. In this case, it is the octet rule that is at play so that the chlorine atom can fully fill its valence electron shell. Thus, two chlorine atoms can also attain a more stable arrangement by sharing a pair of electrons between them.

$$:\ddot{Cl}\cdot + \cdot\ddot{Cl}: \longrightarrow :\ddot{Cl}:\ddot{Cl}:$$

The shared pair of electrons in the chlorine molecule is another example of a covalent bond; the *two electrons* are called a **bonding pair**. The electrons that stay on one atom and are <u>not</u> shared are called *nonbonding pairs*, or **lone pairs**.

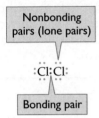

For simplicity, the *diatomic* hydrogen molecule is often represented as H_2, and the *diatomic* chlorine molecule as Cl_2. The word *diatomic* is usually dropped, and we refer to H_2 as simply a hydrogen molecule, while Cl_2 becomes a chlorine molecule, as already seen in Section 4.3. Sometimes the covalent bond is indicated by a dash between the bonding atoms, as in H—H and Cl—Cl.

Each chlorine atom in a chlorine molecule has eight electrons around it, an arrangement like that of the noble gas argon. Thus, the atoms involved in a covalent bond follow the octet rule by sharing electrons, just as those linked by an ionic bond follow it by giving up or accepting electrons. The formation of covalent bonds is by no means limited to diatomic molecules of one kind of element. Most covalent bonds are formed between atoms of different elements.

Multiple Bonds

In some cases, atoms may share more than one pair of electrons. An example would be oxygen. An oxygen atom has six valence electrons and seeks out two more electrons to complete its valence shell with an octet of electrons. When one oxygen atom finds another, they create a diatomic molecule by equally sharing two pairs of electrons, one pair from each atom:

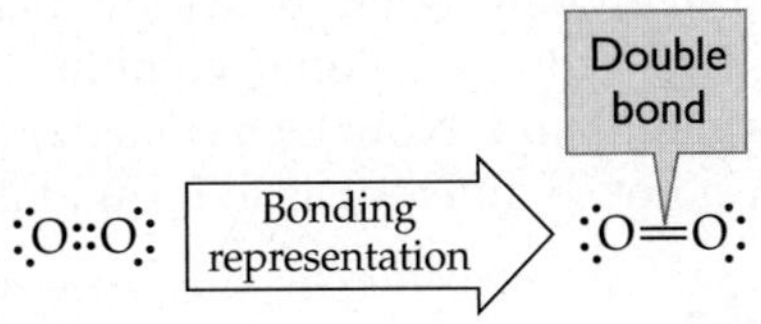

The covalent bond that forms from two pairs of electrons being shared between two atoms is called a **double bond**, and is indicated by a double dash between atoms.

Two atoms also can even share three pairs of electrons so that both atoms fill their valence shells. In the nitrogen (N_2) molecule, for example, each nitrogen atom equally shares three pairs of electrons with the other.

$$:N:::N:$$

The atoms are joined by a **triple bond**, a covalent linkage in which two atoms share three pairs of electrons (N≡N). Note that each of the nitrogen atoms has an octet of electrons around it.

The number of bonds between two atoms is known as the **bond multiplicity** (or **bond order**). The more bonds between the two atoms, the higher the bond multiplicity.

CONCEPTUAL EXAMPLE 4.7 Predicting Bond Multiplicity (Number of Bonds Between Two Atoms)

Predict the bond multiplicity between two atoms of the same element with the valence-shell electron configuration of 2, 4.

Solution

The valence-shell electron configuration of 2, 4 means that each atom has a total of six valence electrons and thus needs to obtain two more electrons to satisfy the octet rule. With two such atoms needing access to two additional electrons, and with both atoms being of the same element, a total of four electrons will be *equally* shared. Since each bond requires two shared electrons, four shared electrons will yield a double bond (and bond multiplicity of two).

SELF-ASSESSMENT Questions

1. A bond formed when two atoms share one pair of electrons is
   a. single covalent   **b.** double covalent
   c. triple covalent   **d.** ionic

2. What type of a bond forms when two atoms in group 7A combine to form a diatomic molecule?
   a. single covalent   **b.** double covalent
   c. triple covalent   **d.** ionic

3. A bond formed when two atoms share three pairs of electrons is
   a. single covalent   **b.** double covalent
   c. triple covalent   **d.** ionic

4. How many pairs of nonbonding electrons does an atom of oxygen have when it forms a double covalent bond with another atom of oxygen?
   a. 0   **b.** 1   **c.** 2   **d.** 4

Answers: 1, a; 2, a; 3, c; 4, c

4.7 Unequal Sharing: Polar Covalent Bonds

Learning Objectives • Classify a covalent bond as polar or nonpolar. • Use electronegativities of elements to determine bond polarity.

So far, we have seen that atoms combine in two different ways to fill their valence shells. Atoms that are quite different in electron configuration (from opposite sides of the periodic table) react by the complete transfer of one or more electrons from one atom to another to form an ionic bond. Some identical nonmetal atoms (such as two chlorine atoms or two hydrogen atoms) combine by equally sharing a pair of electrons to form a covalent bond. Now let's consider bond formation between atoms that are different, yet not so different as to form ionic bonds.

Hydrogen Chloride

Hydrogen and chlorine react to form a colorless gas called *hydrogen chloride* that stings the eyes and significantly irritates the respiratory tract. This reaction may be represented as

$$H\cdot + \cdot\ddot{\underset{\cdot\cdot}{Cl}}: \longrightarrow H\!:\!\ddot{\underset{\cdot\cdot}{Cl}}:$$

Keep in mind that the hydrogen atom and the chlorine atom each contribute an electron, to form a covalent bond between the atoms. Both hydrogen and chlorine achieved a noble gas configuration (filled valence shell)—a helium configuration for hydrogen and an argon configuration for chlorine—in forming the covalent bond.

In reality, when hydrogen and chlorine react to form hydrogen chloride, they do so as diatomic hydrogen and chlorine. For convenience and simplicity, the reaction is often represented as follows:

$$H_2 + Cl_2 \longrightarrow 2\,HCl$$

The bonds between the atoms and the lone pairs on the chlorine atoms are not shown explicitly when writing a simplified chemical reaction.

 EXAMPLE 4.8 Homonuclear and Heteronuclear Covalent Bonds with Lewis Symbols

Use Lewis symbols to show the formation of a covalent bond **(a)** between two fluorine atoms and **(b)** between a fluorine atom and a hydrogen atom.

Solution

$$:\overset{..}{\underset{..}{F}}\cdot \;+\; \cdot\overset{..}{\underset{..}{F}}: \;\longrightarrow\; :\overset{..}{\underset{..}{F}}:\overset{..}{\underset{..}{F}}: \quad\text{bonding pair}$$

$$\text{H}\cdot \;+\; \cdot\overset{..}{\underset{..}{F}}: \;\longrightarrow\; \text{H}:\overset{..}{\underset{..}{F}}: \quad\text{bonding pair}$$

❭ **Exercise 4.8A**
Use Lewis symbols to show how a covalent bond is formed between **(a)** two bromine atoms and **(b)** a fluorine atom and a chlorine atom.

❭ **Exercise 4.8B**
Use Lewis symbols to show how a covalent bond is formed between **(a)** two iodine atoms and **(b)** a hydrogen atom and an iodine atom.

You might ask why a chlorine atom would leave an equally attractive chlorine atom for a hydrogen atom, and the same for hydrogen. The answer lies in stability. There is stable, and there is *more* stable. The chlorine molecule represents a more stable arrangement than two separate chlorine atoms. However, given the opportunity, a chlorine atom selectively forms a bond with a hydrogen atom rather than with another chlorine atom.

Each molecule of hydrogen chloride consists of one atom of hydrogen and one atom of chlorine sharing a pair of electrons. *Sharing*, however, does not necessarily mean "equal" sharing. In fact, chlorine attracts the electron pair more strongly, making it the "electron hog" of the two atoms. The reason for this is that chlorine is more *electronegative*.

Electronegativity

The **electronegativity** of an element can be viewed as its ability to attract electrons that it shares with another atom. The more electronegative an atom is, the greater its tendency to pull the electrons toward its end of the covalent bond. Atoms of elements to the right in the periodic table are, in general, more electronegative than those to the left. The ones on the right (excluding atoms of the noble gas elements) are precisely the atoms that, in forming ions, tend to gain electrons and form negatively charged ions (anions). The ones on the left—metals—tend to give up electrons and become positively charged ions (cations). Within a column, electronegativity tends to be higher at the top and lower at the bottom. One reason for this trend is that atom size follows the same trend as electronegativity: smallest size at the top right and largest at the lower left. In smaller atoms, the electrons are physically closer to the positively charged nucleus, and thus the attraction for electrons is stronger.

The American chemist Linus Pauling (1901–1994) devised a scale of relative electronegativity values by arbitrarily assigning fluorine, the most electronegative element, a value of 4.0. Figure 4.5 displays the electronegativity values for some of the common elements that we will encounter in this text. Within the periodic table, electronegativity follows two trends: It increases as one goes from left to right along a period, and it decreases for the elements as one goes down the group.

▶ **Figure 4.5** Pauling electronegativity values are given for several common elements.

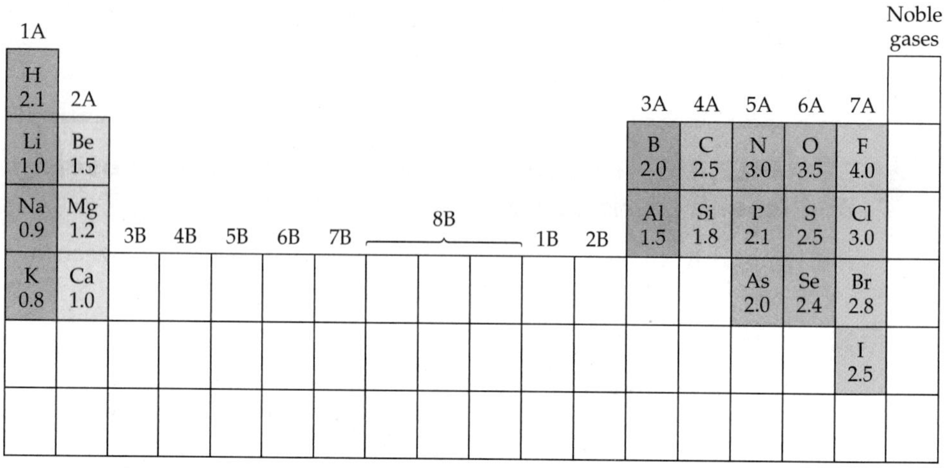

1A	2A								8B		1B	2B	3A	4A	5A	6A	7A	Noble gases
H 2.1																		
Li 1.0	Be 1.5												B 2.0	C 2.5	N 3.0	O 3.5	F 4.0	
Na 0.9	Mg 1.2	3B	4B	5B	6B	7B							Al 1.5	Si 1.8	P 2.1	S 2.5	Cl 3.0	
K 0.8	Ca 1.0														As 2.0	Se 2.4	Br 2.8	
																	I 2.5	

▲ Chlorine hogs the electron blanket, leaving hydrogen partially—but positively—exposed.

Let's now take a closer look at HCl. Chlorine has an electronegativity of 3.0, while hydrogen has an electronegativity of 2.1. This means that chlorine is more electronegative. As a result, the shared electron pair will be closer to chlorine, making the chlorine end of the bond more negative compared with the hydrogen end. An unequal sharing of an electron pair (or pairs) leads to bond polarity, and the bond is called a **polar covalent bond**. By comparison, the bonds where the electron pairs are shared equally—such as those you have already seen in the diatomic molecules of hydrogen, chlorine, oxygen, and nitrogen—are **nonpolar covalent bonds**.

A polar covalent bond is not an ionic bond. In an ionic bond, one atom completely loses an electron. In a polar covalent bond, the atom at the positive end of the bond (hydrogen in HCl) still has a partial share of the bonding pair of electrons (Figure 4.6). To describe this kind of bond from an ionic bond or a nonpolar covalent bond, we use the following notation:

$$\overset{\delta+}{H}-\overset{\delta-}{Cl}$$

The line between the atoms represents the covalent bond, a pair of shared electrons. The $\delta+$ and $\delta-$ (read "delta plus" and "delta minus") signify which end of the bond is partially positive and which is partially negative. (The word *partially* is used to distinguish these charges from the full charges on ions.)

Though not 100% foolproof, electronegativity values for the two atoms joined by a bond can be used to predict the type of bonding.

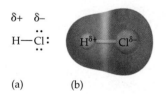

(a)   (b)

▲ **Figure 4.6** Representation of the polar hydrogen chloride molecule. (a) The electron-dot formula, with the symbols $\delta+$ and $\delta-$ indicating the partial positive and partial negative charge, respectively. (b) An *electrostatic potential diagram* depicting the unequal distribution of electron density in the hydrogen chloride molecule.

TABLE 4.3 How to Determine Bonding Polarity by Electronegativity Difference

Electronegativity Difference	Bonding Type
< 0.5	Nonpolar covalent
0.5–2.0	Polar covalent
> 2.0	Ionic

EXAMPLE 4.9 Using Electronegativities to Classify Bonds

Use data from Figure 4.5 and Table 4.3 to classify the bond between each of the following pairs of atoms as nonpolar covalent, polar covalent, or ionic:

a. H, H   **b.** C, F   **c.** C, H   **d.** C, O   **e.** K, O

Solution

a. Two H atoms have exactly the same electronegativity. The electronegativity difference is 0. The bond is nonpolar covalent.

b. The electronegativity difference is $4.0 - 2.5 = 1.5$. The bond is polar covalent.

c. The electronegativity difference is $2.5 - 2.1 = 0.4$. The bond is nonpolar covalent.

d. The electronegativity difference is $3.5 - 2.5 = 1.0$. The bond is polar covalent.

e. The electronegativity difference is $3.5 - 0.8 = 2.7$. The bond is ionic.

❭ Exercise 4.9A

Use data from Figure 4.5 and Table 4.3 to classify the bond between each of the following pairs of atoms as nonpolar covalent, polar covalent, or ionic:

a. H, Br   **b.** Na, O   **c.** C, C   **d.** B, F

❭ Exercise 4.9B

Use the periodic table to classify the following bonds as nonpolar covalent or polar covalent:

a. N—H   **b.** C—N   **c.** C—O   **d.** C=C

❭ Exercise 4.9C

Rank the bonds between each of the following pairs of atoms in terms of increasing polarity:

a. C, O   **b.** N, O   **c.** Si, F   **d.** As, O

Names of Covalent Compounds

Covalent compounds, or **molecular compounds**, are those in which electrons are shared, not transferred. Such compounds generally have molecules that consist of two or more nonmetal atoms. Many covalent compounds have common and widely used names. Examples are water (H_2O), ammonia (NH_3), and methane (CH_4).

However, when naming most binary covalent compounds, systematic rules apply. The prefixes *mono-*, *di-*, *tri-*, and so on are used to indicate the number of atoms of each element in the molecule. Prefixes for up to 10 atoms are given in Table 4.4. For example, the compound N_2O_4 is called *dinitrogen tetroxide*. (The ending vowel is often dropped from *tetra-* and other prefixes when they precede another vowel, to make them easier to pronounce.) We leave off the prefix *mono-* in front of the first atom in a formula but include it in front of the second atom. (For instance, NO is nitrogen monoxide.)

You can have fun with chemical names, as with water. Some compounds have strange—even funny—common names. Curious? Curium(III) chloride is *curious chloride*. Titanium(IV) chloride ($TiCl_4$) is *titanic chloride*. Even water is technically *dihydrogen monoxide*. A glass of dihydrogen monoxide sounds much more dangerous to the general public than a glass of water.

TABLE 4.4 Prefixes That Indicate the Number of Atoms of an Element in a Covalent Compound

Prefix	Number of Atoms
Mono-	1
Di-	2
Tri-	3
Tetra-	4
Penta-	5
Hexa-	6
Hepta-	7
Octa-	8
Nona-	9
Deca-	10

EXAMPLE 4.10   Naming Covalent Compounds

What are the systematic names of **(a)** $AsCl_3$, **(b)** P_4S_3, and **(c)** N_2O_5?

Solution

a. With one arsenic atom and three chlorine atoms, $AsCl_3$ is arsenic trichloride.

b. With four phosphorus and three sulfur atoms, P_4S_3 is tetraphosphorus trisulfide.

c. With two nitrogen atoms and five oxygen atoms, N_2O_5 is dinitrogen pentoxide.

❭ Exercise 4.10A

What are the names of **(a)** BrF_3, **(b)** BrF_5, **(c)** N_2O, and **(d)** NO_2?

❭ Exercise 4.10B

What are the names of **(a)** CO, **(b)** SF_6, **(c)** SO_3, and **(d)** Cl_2O_5?

EXAMPLE 4.11 **Formulas of Covalent Compounds**

Write the formula for triphosphorus pentaiodide.

Solution

The prefix *tri-* indicates three phosphorus atoms, and *penta-* specifies five iodine atoms. The formula is P_3I_5.

⟩ Exercise 4.11A

Write the formulas for **(a)** carbon dioxide and **(b)** dichlorine heptoxide.

⟩ Exercise 4.11B

Write the chemical formulas for **(a)** nitrogen triiodide and **(b)** disulfur decafluoride.

SELF-ASSESSMENT Questions

1. The bond in Br_2 is
   a. nonpolar double covalent   **b.** ionic
   c. nonpolar single covalent   **d.** polar triple covalent

2. Carbon monoxide possesses a _____ bond.
   a. polar double covalent   **b.** trisingle ionic
   c. nonpolar triple covalent   **d.** polar triple covalent

3. When HCl is formed
   a. a Cl atom gives one valence electron to an H atom
   b. an H atom gives one valence electron to a Cl atom
   c. a pair of valence electrons is shared, one each from the H atom and the Cl atom
   d. a pair of valence electrons is shared, none from the H atom and two from the Cl atom

4. Which of the following bonds is the most polar?
   a. C—Br   **b.** C—Cl   **c.** C—F   **d.** H—H

5. In which of the following pairs does carbon have the smallest share of electrons?
   a. C—C   **b.** C—F   **c.** C—O   **d.** C—N

6. In a polar covalent bond, the atom with the partial negative charge
   a. is less electronegative   **b.** has no electronegativity
   c. is more electronegative   **d.** is ionized

7. For an ionic bond to exist between two atoms, their difference in electronegativity should be
   a. between 0 and 0.5   **b.** between 0.5 and 2.0
   c. greater than 2.0   **d.** zero

8. The formula for phosphorus trichloride is
   a. KCl_3 **b.** K_3Cl **c.** PCl_3 **d.** P_3Cl

9. The formula for sulfur hexafluoride is
   a. S_5F **b.** SF   **c.** SF_6 **d.** S_6F_6

10. The correct name for N_2S_4 is
   a. dinitrogen disulfide   **b.** dinitrogen tetrasulfide
   c. tetrasulfodinitrogen   **d.** tetrasulfur dinitride

11. The correct name for I_2O_5 is
   a. diiodine pentoxide   **b.** diiodopentoxide
   c. iridium pentoxide   **d.** pentaoxodiiodine

Answers: 1, c; 2, d; 3, c; 4, c; 5, b; 6, c; 7, c; 8, c; 9, c; 10, b; 11, a

4.8 Polyatomic Molecules: Water, Ammonia, and Methane

Learning Objective • Predict the number of bonds formed by common nonmetals (the HONC rules).

To obtain an octet of electrons, an oxygen atom must share electrons with *two* hydrogen atoms, a nitrogen atom must share electrons with *three* hydrogen atoms, and a carbon atom must share electrons with *four* hydrogen atoms. In general, nonmetals tend to form a number of covalent bonds that is equal to eight minus the number of valence electrons they have. This equates to eight minus its group number on the periodic table. Oxygen, which is in group 6A, forms $8 - 6 = 2$ covalent bonds in most molecules. The same holds for sulfur because it also is in group 6A. Nitrogen, in group 5A, forms $8 - 5 = 3$ covalent bonds in most molecules. Carbon, in group 4A, forms $8 - 4 = 4$ covalent bonds in most molecules. This fact is central to a host of organic compounds (Chapter 9), including the compounds of life (Chapter 16).

The following simple guidelines—sometimes called the HONC rules—will enable you to write formulas for many molecules.

- Hydrogen forms one bond.
- Oxygen forms two bonds.
- Nitrogen forms three bonds.
- Carbon forms four bonds.

Water

Water has fascinated humankind for a long time, especially because it is essential to life on our planet. The electrolysis experiment of Nicholson and Carlisle (Section 3.1) and subsequent experiments have shown that the molecular formula for water is H_2O. Oxygen needs two electrons to complete its octet, but each hydrogen only has one electron to share. Therefore, for oxygen to satisfy its octet, it must share electrons with two hydrogen atoms. This means that oxygen fulfills its octet by sharing two pairs of electrons—one pair of electrons per hydrogen. This is how a *polyatomic* molecule is formed—a molecule containing more than two atoms.

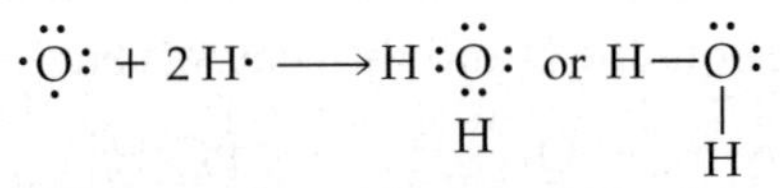

$$\cdot \ddot{O}: + 2\,H\cdot \longrightarrow H:\ddot{O}: \text{ or } H-\ddot{O}:$$
$$\qquad\qquad\qquad\quad H \qquad\qquad\; H$$

This arrangement completes the valence-shell octet of the oxygen atom, giving it the neon electron configuration. It also completes the outer shell of each hydrogen atom, giving each of these atoms the helium electron configuration. Oxygen is more electronegative (3.5) than hydrogen (2.1), so the H—O bonds formed are *polar covalent*.

Ammonia

A nitrogen atom has five electrons in its valence shell. It can attain the neon electron configuration by sharing three pairs of electrons with *three* hydrogen atoms. The result is the compound ammonia.

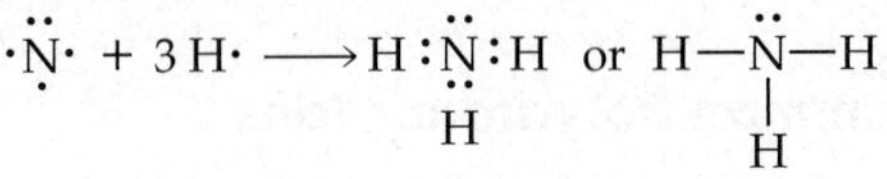

$$\cdot \ddot{N}\cdot + 3\,H\cdot \longrightarrow H:\ddot{N}:H \text{ or } H-\ddot{N}-H$$
$$\qquad\qquad\qquad\quad H \qquad\qquad\quad H$$

In ammonia, the bond arrangement resembles that of a tripod (see Figure 4.10, page 133), with a hydrogen atom at the end of each leg and the nitrogen atom with its unshared pair of electrons centrally at the top. (We will see why it has this shape in Section 4.11.) The electronegativity of N is 3.0, and that of H is 2.1, so all three N—H bonds are *polar covalent*.

Methane

A carbon atom has four electrons in its valence shell. It can achieve the neon configuration by sharing pairs of electrons with *four* hydrogen atoms, forming the compound methane.

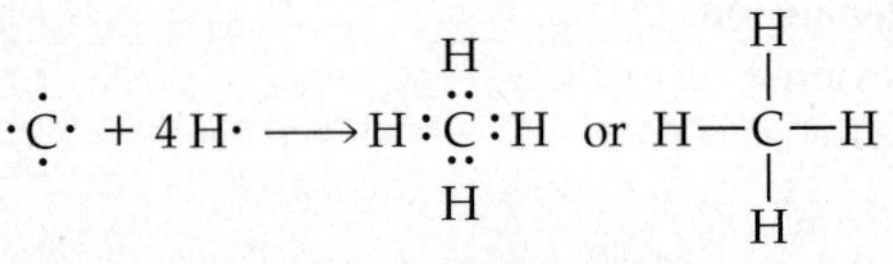

$$\qquad\qquad\quad H \qquad\qquad\qquad H$$
$$\cdot \dot{C}\cdot + 4\,H\cdot \longrightarrow H:\ddot{C}:H \text{ or } H-C-H$$
$$\qquad\qquad\qquad\quad H \qquad\qquad\quad H$$

The methane molecule, as shown above, appears to be planar (flat) but actually it is not—as we shall see in Section 4.11. The electronegativity difference between H and C is so small, less than 0.5, that the C—H bonds are considered nonpolar.

SELF-ASSESSMENT Questions

1. When two H atoms each share an electron with an O atom to form H₂O, the bonding is
   a. diatomic **b.** ionic
   c. nonpolar covalent **d.** polar covalent

2. How many electrons are in the valence shell of the C atom in a methane molecule?
   a. 2 **b.** 5 **c.** 6 **d.** 8

3. In which of the following molecules does hydrogen satisfy the duet rule?
   a. H₂ **b.** H₂O
   c. CH₄ **d.** all

4.9 Polyatomic Ions

Learning Objective • Recognize common polyatomic ions and be able to use them in naming and writing formulas for ionic compounds.

Many compounds contain both ionic and covalent bonds. Sodium hydroxide, commonly known as caustic soda or lye, consists of sodium ions (Na^+) and hydroxide ions (OH^-). A hydroxide ion contains an oxygen atom covalently bonded to a hydrogen atom, plus an "extra" electron, which gives it the negative charge. That extra electron is needed so that both the O and the H atoms have their valence shells filled.

$$e^- + \cdot \ddot{O} \cdot + \cdot H \longrightarrow [:\ddot{O}:H]^-$$

Where did the "extra" electron come from? The extra electron was "stolen" from the sodium atom, changing it into a sodium ion. The chemical formula for sodium hydroxide is NaOH. For each sodium ion, there is one hydroxide ion, so that NaOH is neutral. The bonds between the sodium and hydroxide ions are ionic, just like the ionic bonds you have already seen between sodium and chloride ions in Section 4.4.

There are many groups of atoms that (like hydroxide ions) remain together, as ions, through most chemical reactions. **Polyatomic ions** are charged particles containing two or more covalently bonded atoms. Table 4.5 lists some common polyatomic ions. By combining the ions listed in Table 4.2 and Table 4.5, you can determine formulas for many ionic compounds.

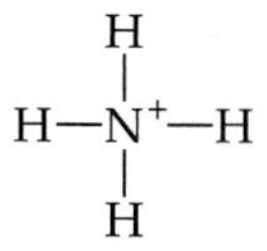

Ammonium ion

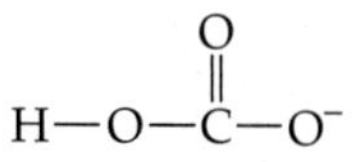

Hydrogen carbonate ion (bicarbonate ion)

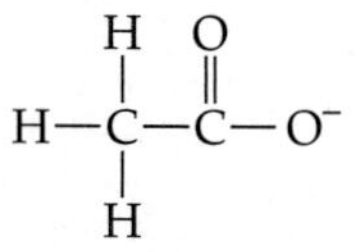

Acetate ion

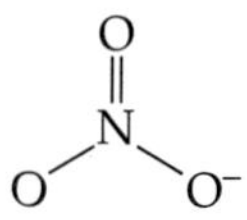

Nitrate ion 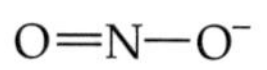   Nitrite ion

TABLE 4.5 Some Common Polyatomic Ions

Charge	Name	Formula
1+	Ammonium ion	NH_4^+
	Hydronium ion	H_3O^+
1−	Hydrogen carbonate (bicarbonate) ion	HCO_3^-
	Hydrogen sulfate (bisulfate) ion	HSO_4^-
	Acetate ion	$CH_3CO_2^-$ (or $C_2H_3O_2^-$)
	Nitrite ion	NO_2^-
	Nitrate ion	NO_3^-
	Cyanide ion	CN^-
	Hydroxide ion	OH^-
	Dihydrogen phosphate ion	$H_2PO_4^-$
	Permanganate ion	MnO_4^-
2−	Carbonate ion	CO_3^{2-}
	Sulfate ion	SO_4^{2-}
	Chromate ion	CrO_4^{2-}
	Hydrogen (monohydrogen) phosphate ion	HPO_4^{2-}
	Oxalate ion	$C_2O_4^{2-}$
	Dichromate ion	$Cr_2O_7^{2-}$
3−	Phosphate ion	PO_4^{3-}

EXAMPLE 4.12 **Formulas with Polyatomic Ions**

What is the formula for ammonium sulfide?

Solution

Ammonium ion is found in Table 4.5. Sulfide ion is a sulfur atom (group 6A) with two additional electrons. The ions are

$$NH_4^+ \quad S^{2-}$$

Crossing over gives

The formula for ammonium sulfide is $(NH_4)_2S$. (Note the parentheses to show that there are *two* ammonium ions for every one sulfide ion.)

❭ **Exercise 4.12A**

What are the formulas for **(a)** calcium carbonate, **(b)** potassium phosphate, and **(c)** ammonium sulfate?

❭ **Exercise 4.12B**

(a) How many hydrogen atoms are there in the formula of ammonium sulfide, and **(b)** how many oxygen atoms are there in calcium carbonate and potassium phosphate, respectively?

 EXAMPLE 4.13 **Naming Compounds with Polyatomic Ions**

What are the names of **(a)** NaCN, **(b)** $Fe(OH)_2$, and **(c)** $FePO_4$?

Solution

As with compounds containing only monatomic ions, we name the cation and then the anion.

  a. The cation is sodium ion, and the anion is cyanide ion (Table 4.5). The compound is sodium cyanide.

  b. The cation is an iron ion, and the anion is hydroxide ion (Table 4.5). There are two hydroxide ions, each with a charge of 1−, so the iron ion must be a 2+ ion [iron(II) or ferrous]. The compound is iron(II) hydroxide or ferrous hydroxide.

  c. Again, the cation is an iron ion and the anion is a phosphate ion. Phosphate has a charge of 3−. Since there is one iron ion in this neutral compound, the iron ion must have a 3+ charge [iron(III) or ferric]. The compound is iron(III) phosphate or ferric phosphate.

❭ **Exercise 4.13A**

What are the names of **(a)** $MgCO_3$, **(b)** $Ca_3(PO_4)_2$, and **(c)** $NH_4C_2H_3O_2$?

❭ **Exercise 4.13B**

What are the names of **(a)** $(NH_4)_2SO_4$, **(b)** KH_2PO_4, and **(c)** $CuCr_2O_7$?

SELF-ASSESSMENT Questions

1. Which of the following compounds incorporates a polyatomic ion?
   a. CO_2 **b.** $C_6H_{12}O_6$ **c.** K_2SO_3 **d.** $SrBr_2$

2. The formula for ammonium phosphate is
   a. NH_4PO_3 **b.** $NH_4(PO_4)_2$
   c. $(NH_4)_2PO_4$ **d.** $(NH_4)_3PO_4$

3. The formula for sodium hydrogen carbonate is
   a. $NaHCO_3$ **b.** Na_2HCO_3
   c. NaH_2CO_3 **d.** $NaHCO_4$

4. The formula for copper(I) hydrogen sulfate is
   a. $CuHSO_4$ **b.** Cu_2HSO_3
   c. Cu_2HSO_4 **d.** $Cu(HSO_4)_2$

5. The formula for ammonium dichromate is
   a. $NH_4Cr_2O_7$ **b.** NH_4CrO_4
   c. $(NH_4)_2Cr_2O_7$ **d.** $(NH_4)_2CrO_4$

6. The correct name for NH_4HCO_3 is
   a. ammonium acetate
   b. ammonium carbonate
   c. ammonium hydrogen carbonate
   d. ammonium cyanide

7. Sodium cyanide can be used in the extraction of gold according to the processes shown:
   $$4\ Au + 8\ NaCN + O_2 + 2\ H_2O \rightarrow$$
   $$4\ Na[Au(CN)_2] + 4\ NaOH$$

   Which compound (or compounds) contain a polyatomic ion?
   a. NaCN
   b. NaOH
   c. $Na[Au(CN)_2]$
   d. all

Answers: 1. c; 2. d; 3. a; 4. a; 5. c; 6. c; 7. d

4.10 Guidelines for Drawing Lewis Structures

. . . .

Learning Objectives • Draw Lewis structures for simple molecules and polyatomic ions. • Identify free radicals.

As we have seen, electrons are transferred or shared in ways that leave most atoms with octets (or a duet, for hydrogen) of electrons in their outermost shells. In this section, we describe how to draw **Lewis structures** for molecules and polyatomic ions. A Lewis structure shows how the atoms in a molecule are connected and also shows unshared (lone) pairs of electrons on atoms. Many methods have been devised for determining such an arrangement. We think this method is the easiest to use. This method will work *only* for molecules and polyatomic ions that follow the Octet Rule. (Yes, there are compounds that have central atoms that are stable with 4, 6, 10, or even 12 electrons in bonds. Beryllium seems "happy" with two pairs of electrons, as in BeH_2; boron seems "happy" with three pairs of electrons, as in BCl_3; phosphorus is "happy" with five pairs, as in PBr_5; and sulfur sometimes is "happy" with six pairs, as in SF_6.)

To draw a Lewis structure, we first decide on a *skeletal structure*, the arrangement of atoms in the molecule. One clue is that "nature loves symmetry." Keep the following in mind:

- Hydrogen atoms form only single bonds; they cannot be covalently bonded to two other atoms, on either side of that hydrogen atom. Hydrogen is often bonded to carbon, nitrogen, or oxygen.

- Oxygen tends to form two bonds, nitrogen usually forms three bonds, and carbon forms four bonds. (Recall the HONC rule.)

- Polyatomic molecules and ions often consist of a central atom surrounded by atoms of higher electronegativity. (Hydrogen is an exception; it is always on the outside, even when bonded to a more electronegative element.) The central atom of a polyatomic molecule or ion is often the *least* electronegative atom.

After choosing a skeletal structure for a polyatomic molecule or ion, we can use the following steps to draw the Lewis structure:

1. **Determine the total number of valence electrons *available*.** This total is the sum of the valence electrons for all the atoms in the molecule or ion. You must also account for the charge(s) on a polyatomic ion:

   a. For a polyatomic anion, *add* to its total number of valence electrons the number of negative charges.

b. For a polyatomic cation, *subtract* the number of positive charges from the total number of valence electrons.

Examples:

N_2O_4 has $[2 \times 5] + [4 \times 6] = 34$ valence electrons.

NO_3^- has $[1 \times 5] + [3 \times 6] + 1 = 24$ valence electrons.

NH_4^+ has $[1 \times 5] + [4 \times 1] - 1 = 8$ valence electrons.

2. **Determine the total number of electrons *needed* by all the atoms**, assuming that hydrogen atoms need two electrons and all others need eight electrons.

3. **Determine the number of bonding electrons by subtracting the number of electrons you have from the number of electrons you need**. (The short way to say this is "need"—"have.") You may divide that number by two to get the number of bonds to draw between the atoms.

4. **Fill in all the outer, lone-pair electrons to make sure all atoms have eight electrons**, except for H, which will have no lone pairs.

EXAMPLE 4.14 **Drawing Lewis Structures**

Draw Lewis structures for **(a)** carbon dioxide (CO_2), **(b)** sulfur dioxide (SO_2), and **(c)** the ammonium ion $(NH_4)^+$.

Solution

a. Carbon dioxide, CO_2

1. Since nature loves symmetry, the carbon atom should be in the middle between the two oxygen atoms.

$$O \quad C \quad O$$

2. How many electrons do you have? You have four from carbon and six each from the two oxygen atoms, for a total of sixteen electrons available.

3. How many electrons are needed to satisfy each atom's octet? $3 \times 8 = 24$ electrons.

4. Using [electrons needed − electrons available] = $[24 - 16] = 8$ electrons to be shared. The only way to do this would be to have a double bond (4 electrons) between the carbon and each of the oxygen atoms—in other words, 8 electrons/2 = 4 bonds. Draw them in.

$$O{=}C{=}O$$

5. Finally, add outer electrons to be sure all octets are satisfied. Each oxygen already has four in the double bond, so place two lone pairs on each oxygen atom. The carbon atom already has its eight electrons.

$$:\overset{..}{O}{=}C{=}\overset{..}{O}:$$

b. Sulfur dioxide, SO_2

1. For the skeletal structure, sulfur should be in the middle between two oxygen atoms.

$$O \quad S \quad O$$

2. How many electrons do you have? Six from sulfur and six each from the two oxygen atoms, for a total of eighteen electrons available ($6 \times 3 = 18$ electrons).

3. How many electrons are needed to satisfy each atom's octet? $3 \times 8 = 24$ electrons.

4. Using [electrons needed − electrons available] − $[24 - 18] = 6$ electrons to be shared. The only way to do this would be to have one double bond (4 electrons) and one single bond (2 electrons). (4 + 2) electrons/2 = 3 bonds.

Draw them in. Note: It does not matter whether the double bond is on the left or the right.

$$O=S—O$$

5. Finally, add outer electrons to be sure all electrons are satisfied. The oxygen on the left has four in the double bond, so place two lone pairs on that oxygen atom. The sulfur atom already has six electrons, so draw a lone pair on the S atom. The oxygen on the right has only two electrons in the single bond, so add three lone pairs on that atom.

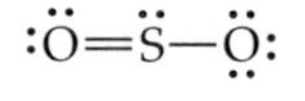

$$:\ddot{O}=\ddot{S}—\ddot{O}:$$

c. The Ammonium ion, NH_4^+

1. The skeletal structure would have the nitrogen atom in the center surrounded by four hydrogen atoms. (H can never be bonded to two other atoms, so a string of H's would be incorrect.)

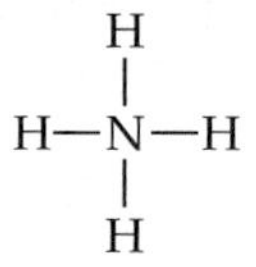

$$\begin{array}{c} H \\ | \\ H—N—H \\ | \\ H \end{array}$$

2. How many electrons do you have available? N has five and each H has one. However, for an ion with a +1 charge, one electron has been given away. $5 + 4 \times 1 - 1 = 8$ electrons.

3. How many electrons are needed to satisfy the octets? Nitrogen needs 8, plus 4 H atoms that are happy with a duet, so $8 + (4 \times 2) = 16$ electrons.

4. Subtracting electrons needed – electrons available, $16 - 8 = 8$ electrons to be shared. The only way to do this is to have four single bonds—in other words, 8 electrons/2 = 4 bonds. Draw them in. *Don't forget* to place a square bracket around the entire structure with the + charge on the outside. These are required for the polyatomic ions.

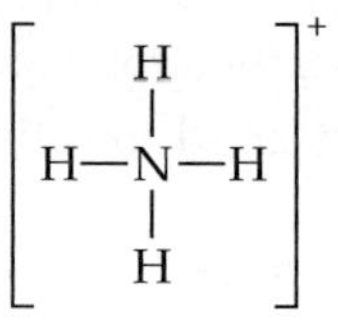

$$\left[\begin{array}{c} H \\ | \\ H—N—H \\ | \\ H \end{array}\right]^+$$

> **Exercise 4.14A**

Write Lewis structures for (a) hydrogen sulfide, H_2S, and (b) oxygen difluoride, OF_2.

> **Exercise 4.14B**

Write Lewis structures for (a) methyl chloride, CH_3Cl, and (b) the carbonate ion, $(CO_3)^{2-}$.

The rules we have used here lead to the results for selected elements that are summarized in Table 4.6. Note that the outermost (lone-pair) electrons are not drawn in the table, but are required for the Lewis structure to be correct. The ball-and-stick models represent what the molecules might look like if one uses wooden or plastic balls to represent atoms, and sticks or plastic connectors to represent the bonds. The molecular shapes may be more evident using that representation. Figure 4.7 relates the number of covalent bonds that a particular element can form to its position on the periodic table.

Odd-Electron Molecules: Free Radicals

Some molecules contain atoms that do not conform to the octet rule. For example, any compounds made with noble gases obviously will be exceptions, since noble gas atoms already have a full octet. These compounds are rare and were formed

TABLE 4.6 Number of Bonds Formed by Selected Elements

Lewis Symbol	Bond Picture	Number of Bonds	Representative Molecules	Ball-and-Stick Models
H·	H—	1	H—H H—Cl	HCl
He:		0	He	He
·C̈·	—C— (with vertical bonds)	4	H—C—H (with H above and below) H—C—F (with O above)	CH₄
·N̈·	—N—	3	H—N—H (with H below) N≡N	NH₃
·Ö:	—O—	2	H—O (with H below) O=O	H₂O
·F̈:	—F	1	H—F F—F	F₂
·C̈l:	—Cl	1	Cl—Cl H—C—Cl (with H above and below)	CH₃Cl

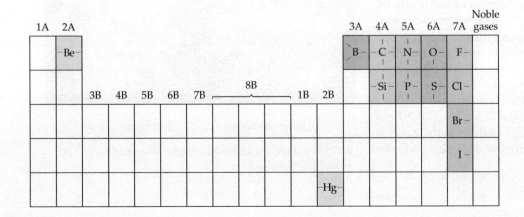

1A	2A												3A	4A	5A	6A	7A	Noble gases
	-Be-												B–	–C–	N–	O–	F–	
		3B	4B	5B	6B	7B	8B			1B	2B			–Si–	P–	S–	Cl–	
																	Br–	
																	I–	
										–Hg–								

◀ **Figure 4.7** Covalent bonding of representative elements on the periodic table.

Q *What is the relationship between an atom's position on the periodic table and the number of covalent bonds it tends to form? Why are the group 1A and most of the group 2A elements omitted from the table?*

with difficulty. Some beryllium, boron, sulfur, and phosphorus compounds may not follow the octet rules, either. BF₃—boron trifluoride—is an example in which boron seems satisfied with six electrons in three bonds.

A structure with an odd number of outermost electrons is a *free radical*. Molecules with odd numbers of valence electrons obviously cannot satisfy the octet rule. Remember that we only need to look at the valence electrons, because filled core shells and subshells contain only paired electrons, having placed two electrons in each orbital [Section 3.7]).

Examples of free radicals are nitrogen monoxide (NO, also called nitric oxide), with $5 + 6 = 11$ valence electrons; nitrogen dioxide (NO_2), with 17 valence electrons; and chlorine dioxide (ClO_2), with 19 valence electrons. Lewis structures of NO, NO_2, and ClO_2 each have one oxygen atom with an unpaired electron. The more electronegative the atom, the more likely it is to obey the octet rule. Therefore, it is usually the less electronegative atom that ends up with an unpaired electron and, as a consequence, remains short of the octet.

$$:\!\overset{..}{N}\!::\!\overset{..}{O}\!: \qquad :\!\overset{..}{O}\!:\!\overset{.}{N}\!::\!\overset{..}{O}\!: \qquad :\!\overset{..}{O}\!:\!\overset{..}{Cl}\!:\!\overset{..}{O}\!:$$

Useful Applications of Free Radicals

Moses Gomberg (1866–1947) discovered free radicals at the University of Michigan in 1900. The very reactive nature of free radicals does not preclude their importance and use in a variety of natural, medical, and industrial applications.

- Nitrogen monoxide (NO) has been shown to be a signaling molecule in the human cardiovascular system. American scientists Robert F. Furchgott (1916–2009), Louis J. Ignarro (1941–), and Ferid Murad (1936–) won the Nobel Prize in Physiology or Medicine in 1998 for that discovery. Interestingly, one of the physiological effects of sildenafil citrate (Viagra) is the production of small quantities of NO in the bloodstream.
- Small radical molecules are used in biochemistry to gain insight into how proteins—large biomolecules essential to life—function in our bodies. These studies also include biomedical research into drug design and other promising therapies.
- Hydroxyl radicals ($\cdot OH$) are produced in the body in the process of oxidation of foods and by radiation. The decomposition of hydrogen peroxide (HOOH) and other peroxides produces $\cdot OH$. These radicals inflict havoc on DNA and other essential substances, and have been implicated in the formation of cancerous cells due to their rapid reaction with DNA. Hydroxyl radicals also play a role in air pollution (Chapter 13). A beneficial application of $\cdot OH$ radicals is their use in the treatment of municipal wastewater and contaminated soils.

- Many plastics, including polyethylene and polyvinyl chloride (PVC), are made using free radicals to initiate the reactions. Liquid polyester resin is used to repair automobiles and to construct small boats and surfboards. The resin is converted to hard plastic by the addition of a free radical catalyst called *methyl ethyl ketone peroxide*.
- The free radical chlorine dioxide (ClO_2) is widely used to bleach paper and other products, including flour. Unbleached paper (left) is a pale cream color; bleached paper (right) is whiter and is considered more attractive to consumers. Chlorine radicals also play a role in ozone destruction, especially in the formation of the Antarctic ozone hole.

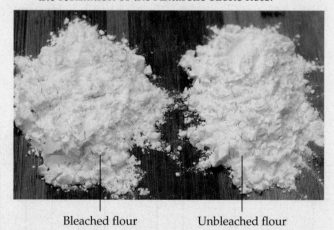

Bleached flour   Unbleached flour

▲ Chlorine dioxide is used to bleach flour. Bleached flour (left) and unbleached flour (right).

Most free radicals are highly reactive and exist for only a very brief time before reacting with another species. And unless they react with another free radical, the products of that reaction must include a free radical! (Odd number + even number = odd number.) Any atom or molecule with an odd number of electrons must have one unpaired electron, so such atoms or molecules are free radicals.

Nitrogen oxides are major components of smog (Chapter 13). Free chlorine atoms from the breakdown of chlorofluorocarbons in the stratophere lead to depletion of the ozone layer (Chapter 13). Some free radicals are quite stable, however, and have important functions in the body as well as in industrial processes.

SELF-ASSESSMENT Questions

1. How many lone pairs of electrons are there in the Lewis structure of CF_4?
   a. 4 **b.** 12 **c.** 14 **d.** 16
2. How many lone pairs of electrons are there in the Lewis structure of CF_2Cl_2?
   a. 4 **b.** 12 **c.** 14 **d.** 16
3. How many unshared electrons are there in the Lewis structure of H_2O_2?
   a. 3 **b.** 8 **c.** 14 **d.** 18
4. How many unpaired electrons are there on S in SCl_2?
   a. 4 **b.** 1 **c.** 2 **d.** 3
5. The Lewis structure for CCl_4 has
   a. 0 lone pairs **b.** 1 double bond
   c. 4 ionic bonds **d.** 24 unshared electrons
6. The correct Lewis structure for carbon dioxide is
   a. O—C—O **b.** :O—C—O:
   c. :Ö—C—Ö: **d.** :Ö=C=Ö:
7. The correct Lewis structure for the Cl_2O molecule is
   a. :C̈l—C̈l—Ö: **b.** :C̈l=O=C̈l
   c. :Ö—C̈l—Ö: **d.** :C̈l—Ö—C̈l:
8. In the Lewis structure for SO_2, the number of lone pairs of electrons in the outer shell of the central atom is
   a. 0 **b.** 1
   c. 2 **d.** 4
9. Which of the following species does not have a triple bond in its Lewis structure?
   a. N_2 **b.** CN^-
   c. CO **d.** O_2
10. The Lewis structure for the SO_3^{2-} ion that follows the octet rule is

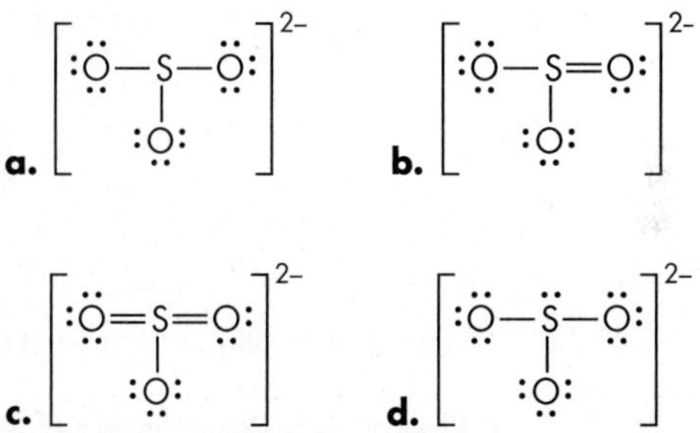

Answers: 1, b; 2, b; 3, b; 4, a; 5, d; 6, d; 7, d; 8, b; 9, d; 10, d

4.11 Molecular Shapes: The VSEPR Theory

Learning Objective • Predict the shapes of simple molecules from their Lewis structures.

It is now time to *lift* polyatomic molecules and ions—represented so far as being flat—off the two-dimensional (2-D) paper and give them three-dimensional (3-D) shapes. This third dimension is important in determining molecular properties. For example, the shapes of the molecules that make up gasoline determine its octane rating (Chapter 15), while drug molecules must have the right atoms in the right places to be effective (Chapter 18).

We can use Lewis structures as the initial step and then apply **valence-shell electron pair repulsion (VSEPR) theory** to predict the 3-D arrangement of atoms about a central atom in a molecule. In applying the VSEPR theory to predict molecular shapes, we will use the term *electron set* to refer to either a lone pair on a central atom or a bond—single, double, or triple—between the central atom and another atom. You can envision an *electron set* as a region in 3-D space that has a high electron density. The basis of the VSEPR theory is that electron sets, being negatively charged, arrange themselves as far away from each other as possible in 3-D space so as to minimize repulsions between these like-charged particles. To do this, molecules follow some simple geometric arrangements.

When there are only two electron sets on the central atom to consider, then their arrangement is *linear.* (See Figure 4.8a and Table 4.7 for the geometric representation and chemical examples, respectively.) This arrangement allows for the largest possible angle between electron sets, namely, $360°/2 = 180°$. This arrangement is

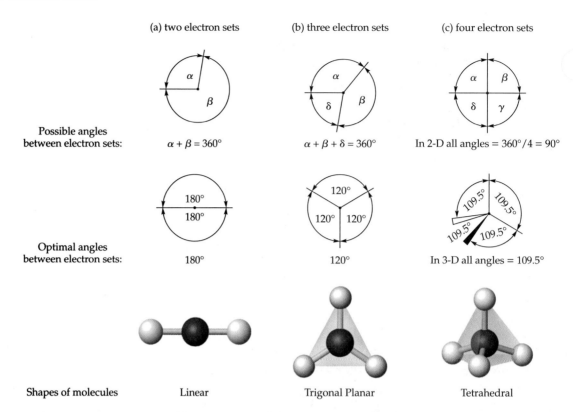

(a) two electron sets (b) three electron sets (c) four electron sets

Possible angles
between electron sets: $\alpha + \beta = 360°$ $\alpha + \beta + \delta = 360°$ In 2-D all angles = $360°/4 = 90°$

Optimal angles
between electron sets: 180°   120°   In 3-D all angles = 109.5°

Shapes of molecules   Linear   Trigonal Planar   Tetrahedral

▲ **Figure 4.8** Determination of the arrangement of electron sets in molecules based on maximum separation and minimum repulsion among them. (a) In a *linear* molecule, the bond angle is 180°. (b) In a *trigonal planar* molecule, bond angles are 120°. (c) In a *tetrahedral* molecule, bond angles are 109.5°. In the absence of lone pairs (LPs) of electrons on the central atom, the names for the shapes of the molecules are the same as the names of the three basic arrangements of electron sets—*linear*, *trigonal planar*, and *tetrahedral*.

one-dimensional (1-D)—that is, one in which all atoms are aligned. This arrangement would occur if there were *no* lone pairs of electrons on the central atom, or when there are two double bonds, or one single plus one triple bond on that central atom.

When three electron sets are in play, we must expand into 2-D space, where all atoms are arranged in one plane. In principle, the angles between electron sets could all be different as long as their sum was 360°; however, the three identical $360°/3 = 120°$ angles ensure the maximum separation of the electron sets and minimize repulsions among them. This geometric arrangement is called *trigonal planar*. (See Figure 4.8b and Table 4.7.) This could occur if there were one double bond and two single bonds on the central atom, and then the molecule or polyatomic ion has a trigonal planar shape. If one of the electron sets is simply a lone pair of electrons, then the molecule is "bent."

Using similar logic, we would expect four electron sets to be separated by $360°/4 = 90°$ angles, as shown in the top portion of Figure 4.8c. However, upon exploration of the 3-D space, it is possible for the four electron sets to be separated by even larger angles of 109.5°. The corresponding *tetrahedral* geometry is shown in the bottom portion of Figure 4.8c, and several chemical examples are listed in Table 4.7. If four atoms are bonded to the central atom, the molecule is, not surprisingly, tetrahedral in shape. But if one of those electron sets is simply a lone pair of electrons, the molecule has a "trigonal pyramid" shape (as in the ammonia molecule). If two of those electron sets are lone pairs of electrons, then the molecule is "bent" (as in the water molecule). If three of those electron sets are lone pairs of electrons, then the molecule is linear (as in hydrogen chloride). So you can see that the number and arrangement of the electron sets, whether bonded to other atoms or simply lone pairs, are the critical factors in determining the molecular shape of the molecule (or polyatomic ion).

TABLE 4.7   Bonding and the Shapes of Molecules

Number of Bonded Atoms	Number of Lone Pairs	Number of Electron Sets	Molecular Shape	Examples of Molecules	Ball-and-Stick Models
2	0	2	Linear	$BeCl_2$ $HgCl_2$ CO_2 HCN	 $BeCl_2$
3	0	3	Trigonal Planar	BF_3 $AlBr_3$ CH_2O	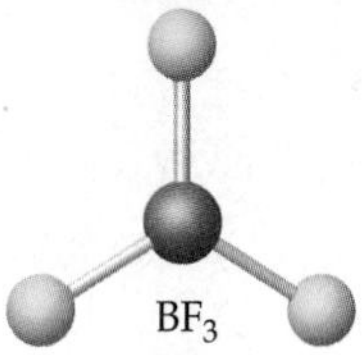 BF_3
4	0	4	Tetrahedral	CH_4 CBr_4 $SiCl_4$	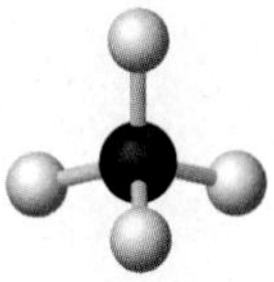 CH_4
3	1	4	Trigonal Pyramidal	NH_3 PCl_3	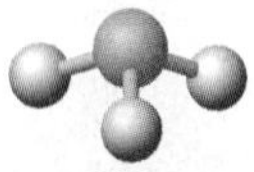 NH_3
2	2	4	Bent	H_2O H_2S SCl_2	 H_2O
2	1	3	Bent	SO_2 O_3	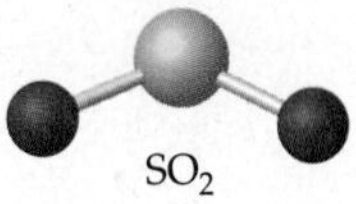 SO_2

We can determine and envision the shapes of many molecules (and polyatomic ions) by following this simple procedure:

1. Draw a Lewis structure in which a shared electron pair (bonding pair) is indicated by a line. Use dots to show any unshared pairs (lone pairs) of electrons.

2. To determine shape, count the number of electron sets around the *central* atom. Recall that a multiple bond counts as only one electron set. Examples are:

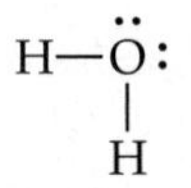

Four sets	Four sets	Four sets
(2 atoms, 2 LPs)	(3 atoms, 1 LP)	(4 atoms)

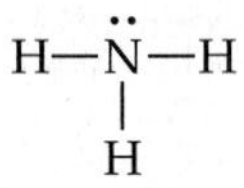

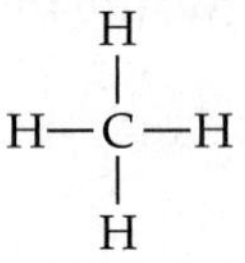

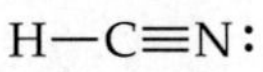 Two sets	Two sets	Three sets	Three sets
(2 atoms)	(2 atoms)	(3 atoms)	(2 atoms, 1 LP)

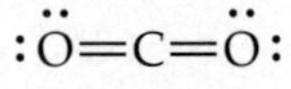

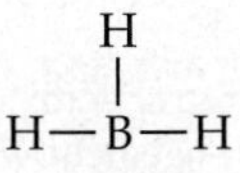

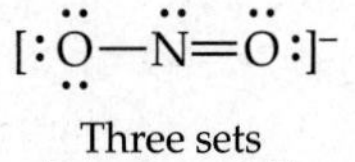

3. Using the number of electron sets determined in step 2, draw a shape *as if* all the sets were bonding pairs, and place these electron sets as far apart as possible (Table 4.7).

4. If there are *no* lone pairs, the shape from step 3 is the shape of the molecule. If there *are* lone pairs, remove them, leaving the bonding pairs exactly as they were. (This may seem strange, but it stems from the fact that *all* the sets determine the geometry, even though only the arrangement of bonded atoms is considered in the shape of the molecule.)

5. The presence of lone pairs of electrons on the central atom decreases the angles between the bonds, as lone pairs require more space than bonding pairs of electrons. Likewise, double and triple bonds command more space and push remaining bonds away. While somewhat distorted, most molecules maintain their predicted arrangement of electron sets (Table 4.7). The actual (distorted) angles must be measured experimentally, and you will *not* be asked to memorize or figure them out!

 EXAMPLE 4.15 Shapes of Molecules

What are the shapes of **(a)** the H_2CO molecule and **(b)** the SCl_2 molecule?

Solution

a. Formaldehyde, H_2CO
1. The Lewis structure for H_2CO is

$$\begin{array}{c} H \\ | \\ C{=}\overset{..}{\underset{..}{O}} \\ | \\ H \end{array}$$

2. There are three sets of electrons: two C—H single bonds and one C=O double bond.
3. A trigonal planar arrangement puts the three electron sets as far apart as possible.

$$\overset{H}{\underset{H}{\diagdown}}C{=}\overset{..}{O}:$$

4. All the electron sets are bonding pairs; the molecular shape is trigonal planar, the same as the arrangement of the electron sets.
5. Due to the presence of a double bond, the actual angles, while based on the ideal 120° angles, will be somewhat distorted.

$$\overset{H}{\underset{H}{\diagdown}} \overset{> 120°}{\underset{}{}} C{=}\overset{..}{O}:$$

b. Again, we follow the rules, starting with the Lewis structure of SCl_2.
1. The Lewis structure for SCl_2 is

$$\begin{array}{c} :\overset{..}{S}{-}\overset{..}{\underset{..}{Cl}}: \\ | \\ :\overset{}{\underset{..}{Cl}}: \end{array}$$

2. There are four sets of electrons on the sulfur atom.
3. A tetrahedral arrangement around the central atom puts the four sets of electrons as far apart as possible.

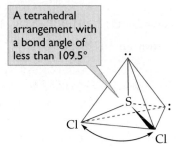

A tetrahedral arrangement with a bond angle of less than 109.5°

4. Two of the electron sets are bonding pairs and two are lone pairs. The molecular shape is *bent*.

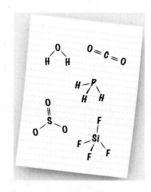

5. The two lone pairs of electrons require more space, which decreases the Cl—S—Cl bond angle from the ideal 109.5° to 103°. We will not expect students to determine these angle modifications.

› Exercise 4.15A

What are the shapes of **(a)** the PI_3 molecule and **(b)** $SiCl_4$?

› Exercise 4.15B

A student drew the following molecular shapes of several molecules. In the corresponding Lewis structure, which of the central atoms had **(a)** one and which had **(b)** two lone pairs of electrons?

SELF-ASSESSMENT Questions

1. Which of the following molecular shapes *cannot* be formed from the tetrahedral arrangement of the electron sets?
   a. linear   **b.** bent   **c.** trigonal planar

2. Which of the following molecular shapes *can* be formed from the trigonal planar arrangement of the electron sets?
   a. linear   **b.** bent
   c. tetrahedral   **d.** trigonal pyramidal

3. Which of the following has the smallest O—S—O bond angle?
   a. SO_2 **b.** SO_3^{2-} **c.** SO_3

Answer the next four questions by matching the formula in the left column with the molecular shape in the right column.

4. CS_2 **a.** bent
5. OF_2 **b.** linear
6. SiF_4 **c.** trigonal pyramidal
7. NH_3 **d.** tetrahedral

Answers: 1. c; 2. b; 3. b; 4. b; 5. a; 6. d; 7. c

4.12 Shapes and Properties: Polar and Nonpolar Molecules

Learning Objective • Classify a simple molecule as polar or nonpolar from its shape and the polarity of its bonds.

Section 4.7 discussed polar and nonpolar bonds. A diatomic *molecule* is nonpolar if its bond is nonpolar, as in H_2 or Cl_2, and polar if its bond is polar, as in HCl. Recall from Section 4.7 that the partial charges in a polar bond are indicated by $\delta+$ and $\delta-$. In the case of a **dipole**, a molecule with a positive end and a negative end, the polarity is commonly indicated with an arrow with a plus sign at the tail end ($\longleftrightarrow$). The plus sign indicates the part of the molecule with a partial positive charge, and the head of the arrow signifies the end of the molecule with a partial negative charge.

H – H and Cl – Cl $\overset{\delta+ \quad \delta-}{H—Cl}$ or $\overset{\longleftrightarrow}{H—Cl}$
     Nonpolar   Polar

For a molecule with three or more atoms, we must consider the polarity of the individual bonds as well as the overall geometry of the molecule to determine whether the molecule as a whole is polar. A **polar molecule** has separate centers of positive and negative charge, just as a magnet has north and south poles. Many properties of compounds—such as melting point, boiling point, and solubility—depend on the polarity of their molecules.

Methane: A Tetrahedral Molecule

There are four electron sets around the central carbon atom in methane, CH_4. Using the VSEPR theory, we would expect a tetrahedral arrangement and bond angles of 109.5° (Figure 4.9a). The actual bond angles are 109.5°, as predicted by the theory. All four electron sets are shared with hydrogen atoms, and therefore occupy identical volumes. Since the electronegativity difference between carbon and hydrogen is only 0.4 (< 0.5), carbon-to-hydrogen bonds are essentially nonpolar. Moreover, since the methane molecule is symmetric, these negligible bond polarities (Figure 4.9b) cancel out anyway, leaving the methane molecule, as a whole, nonpolar (Figure 4.9c).

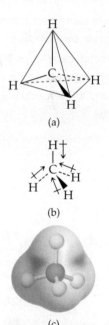

(a)

(b)

(c)

▲ **Figure 4.9** The methane molecule. In (a), black lines indicate covalent bonds; the red lines outline a tetrahedron; and all bond angles are 109.5°. The slightly polar C—H bonds cancel each other (b), resulting in a nonpolar, tetrahedral molecule (c).

Q *How does the electrostatic potential diagram in (c) look different from the one in Figure 4.6?*

Ammonia: A Trigonal Pyramidal Molecule

In ammonia, NH_3, the central nitrogen atom has three bonds and a lone pair around it. The N—H bonds of ammonia are more polar than the C—H bonds of methane. More important, the NH_3 molecule has a different geometry, as a result of the lone pair on the nitrogen atom. The VSEPR theory predicts a tetrahedral arrangement of the four sets of electrons, giving bond angles of 109.5° (Figure 4.10a). However, the lone pair of electrons occupies a greater volume than a bonding pair, pushing the bonding pairs slightly closer together and resulting in slightly decreased bond angles (in the case of ammonia, about 107°). The trigonal pyramidal geometry can be envisioned as a tripod with a hydrogen atom at the end of each leg and the nitrogen atom—with its lone pair—sitting at the top. Each nitrogen-to-hydrogen bond is polar (Figure 4.10b). The asymmetric structure makes the ammonia molecule polar. The partial negative charge is on the nitrogen atom and partial positive charges are on the three hydrogen atoms (Figure 4.10c). The existence of that negatively charged lone pair on the nitrogen atom is the most important clue that the N atom is the negative "pole" of the molecule, and that the H atoms constitute the positive "pole."

Water: A Bent Molecule

The O—H bonds in water are even more polar than the N—H bonds in ammonia, because oxygen is more electronegative than nitrogen. (Recall that electronegativity increases from left to right in the periodic table.) Just because a molecule contains polar bonds, however, does not mean that the molecule as a whole is polar. If the atoms in the water molecule were in a straight line (that is, in a linear arrangement, as in CO_2), the two polar bonds would cancel one another and the molecule would be nonpolar. In water, as in ammonia, the polar bonds do not cancel; both molecules have physical and chemical properties of polar molecules.

We can understand the dipole in the water molecule by using the VSEPR theory. The two bonds and two lone pairs should form a tetrahedral arrangement of electron sets (Figure 4.11a). Ignoring the lone pairs, we see that the molecular shape has the atoms in a bent arrangement, with a bond angle of about 104.5° (instead of 109.5°). As with ammonia, the actual angle is slightly smaller than the predicted angle because of the greater volume of space occupied by the two lone pairs compared with the bonding pairs. The larger space occupied by the lone pairs reduces the space in which the bonding pairs reside, pushing them closer together. The two polar O—H bonds do not cancel each other (Figure 4.11b), making the bent water molecule polar (Figure 4.11c). The lone pairs on the O atom logically would imply that the O atom is the negative "pole," and that the H atoms are the positive "poles."

6 Both oxygen and hydrogen are gases. Why is water—which combines them—a liquid at room temperature?

When atoms of different elements bond with each other to form compounds, they lose their elemental characteristics. They exhibit changes in chemical and physical properties, including state of matter.

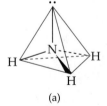
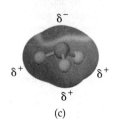
(a)
(b)
(c)

◀ **Figure 4.10** The ammonia molecule. In (a), black lines indicate covalent bonds. The red lines outline a tetrahedron. The polar N—H bonds do not cancel each other completely in (b), resulting in a polar, trigonal pyramidal molecule (c).

Q *How does the electrostatic potential diagram in Figure 4.10c resemble the one in Figure 4.6?*

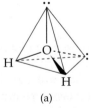
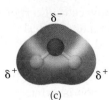
(a)
(b)
(c)

◀ **Figure 4.11** The water molecule. In (a), black lines indicate covalent bonds, and the red lines outline a tetrahedron. The polar O—H bonds do not cancel each other in (b), resulting in a polar, bent molecule (c).

A Chemical Vocabulary

Learning chemical symbolism is much like learning a foreign language. Once you have learned the basic "vocabulary," the rest is a lot easier. At first, the task is complicated, because different chemical species or definitions sound a lot alike (like sodium atoms versus sodium ions). Another complication is that we have several different ways to represent the *same* chemical species (Figure 4.12).

Ammonia ⟶ NH_3 ⟶ H—N̈—H ⟶

|Name|Chemical formula|Lewis formula|Molecular geometry|

◀ **Figure 4.12** Several representations of the ammonia molecule

In Chapter 1, we introduced symbols for the chemical elements. Chapter 3 discussed the structure of the nucleus and the symbolism for distinguishing different isotopes. This chapter covered chemical names and chemical formulas. Now you also know how to write Lewis structures and formulas using dots to represent valence electrons, and you can predict the shapes of thousands of different molecules with the VSEPR theory. In the next chapter, you will add the mathematics of chemistry to your toolbox.

CONCEPTUAL EXAMPLE 4.16 Polarity of Molecules

Which of these arrangements of water molecules is the most probable? Explain your answer.

- ● Oxygen
- ○ Hydrogen

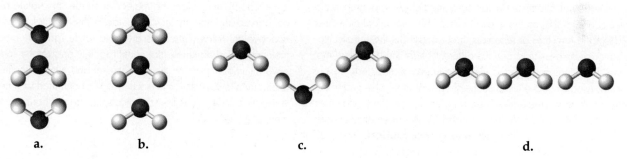

a.   b.   c.   d.

Solution

Due to electronegativity differences between the O and H atoms, bonds within water molecules are polar. Since the water molecule is bent, the bond polarities add up to render water molecules polar, with the partially negative oxygen atom at one end and the partially positive hydrogen atoms at the other. In liquid, water molecules arrange themselves so that the partially negative oxygen atom in one water molecule faces the partially positively charged hydrogen atoms in another water molecule. While greatly idealized, answer **b** is correct. The partially negative oxygen ends of water dipoles repel each other. Likewise, the partially positive hydrogen ends of water dipoles repel each other, making answers **a**, **c**, and **d** very unlikely.

SELF-ASSESSMENT Questions

1. Which of the following molecules is polar?
   a. H_2 b. CH_4 c. CO_2 d. NH_3

2. Which of the following molecules is nonpolar?
   a. HF   b. H_2O c. CCl_4 d. PF_3

3. Which of the following molecules is polar?
   a. CF_4 b. BF_3 c. HCN   d. F_2

4. Which of the following molecules has polar bonds but is nonpolar?
   a. SiF_4 b. Cl_2 c. NCl_3 d. OF_2

GREEN CHEMISTRY Green Chemistry and Chemical Bonds
John C. Warner, Warner Babcock Institute for Green Chemistry Amy S. Cannon, Beyond Benign

Principles 3, 5, 6, 8

Learning Objectives • Describe the concept of molecular recognition. • Explain the green chemistry advantages of using production methods based on molecular recognition.

You learned in this chapter that atoms bond together to form compounds and these new compounds have properties that differ greatly from the component elements. In molecular compounds (molecules), atoms join by forming covalent bonds that cause the molecules to adopt specific geometric shapes (Sections 4.6, 4.7, 4.12). Each atom is held in a specific position relative to other atoms in the molecule. In ionic compounds, atoms are held together through ionic bonds (Section 4.3).

Molecular shape and composition of compounds control reactivity and interactions with other substances. As a chemist manipulates the geometry and composition of molecules, the properties can change drastically. A wonderful example of how shape impacts the design of molecules is seen in medicine. Within our bodies, biological molecules have particular shapes that are recognized by enzymes and other receptors, in the same way that a key matches a lock (Chapter 16). The molecule and the receptor fit together to form complexes through noncovalent forces (intermolecular forces, Section 6.3).

Medicinal chemists design new medicines and drug molecules (Chapter 18) so that they will resemble biological molecules and bind to enzymes or receptors in our body. But these molecule–receptor complexes must behave differently from the biological molecule. The binding triggers a response in our bodies and helps us to heal. Medicinal chemists use the shapes of molecules and forces between them to create new life-saving or life-changing medicines. Interaction between molecules caused by the geometric orientations of their atoms is called **molecular recognition**. The 1987 Nobel Prize in Chemistry was awarded to Donald J. Cram, Jean-Marie Lehn, and Charles J. Pedersen for pioneering work in this area. Scientists are now learning to use molecular recognition to control molecular properties and create not only new medicines but also new materials.

What does this have to do with green chemistry? The processes that chemists rely on to make materials often use large amounts of energy or highly reactive and toxic chemicals. New production methods that take advantage of molecular recognition typically use lower energy, solvent-free processes that do not require harsh conditions. For example, the use of nanoparticles as delivery systems for tumor targeting.

This way of making materials takes advantage of several of the 12 principles of green chemistry. Principles 3, 5, 6, and 8 focus on the choices we make as chemists as we design new molecules and materials. For example, using less hazardous chemicals to make new medicines means we can reduce the environmental impact of the manufacturing process for these compounds. Doing so ensures that they can change the lives of sick patients as well as provide safer workplaces for the workers making the new medicines.

Typical reactions carried out in laboratories use solvents that dissolve the molecules (solutes) (Section 5.5) and allow them to react to form new molecules. Many solvents can be hazardous to human health and the environment. Processes for making new noncovalent complexes that rely on molecular recognition often do not require solvents. This drastically reduces the environmental impact of these processes.

Many traditional processes require high temperatures, which translates to high energy use. Noncovalent complexes created through molecular recognition function in much the same way as molecules interact within our bodies at moderate temperatures. Therefore, these new approaches use very little energy, which benefits the environment and reduces the manufacturing cost. Also, many processes that generate new molecules and materials are quite complex.

By understanding how molecules interact by shape and noncovalent forces, we can reduce the complexity in these processes and avoid the generation of waste and the reliance on hazardous materials. By understanding molecular shape, the nature of chemical bonds, and how molecules interact with other molecules through noncovalent forces, chemists are able to design new medicines, molecules, and materials that benefit society and, at the same time, have minimal impact on the environment and human health. Green chemistry allows chemists to use these skills combined with new innovative techniques to design chemicals that do not harm our environment in the process.

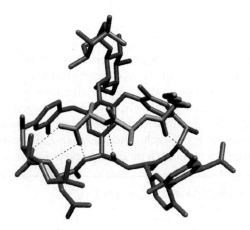

▲ The antibiotic *vancomysin* carries out its function using molecular recognition. Vancomysin undergoes strong, selective hydrogen bonding with *peptides* (small protein chains) of bacteria, preventing formation of the cell wall.

Summary

Section 4.1—Outermost-shell electrons are called **valence electrons**; those in all other shells are **core electrons**. The noble gases are inert because they have an octet of valence electrons. Helium is the exception; it is also inert, but it has a duet of valence electrons. Atoms of other elements become more stable by gaining or losing electrons to attain an electron configuration that makes them **isoelectronic** with (having the same electron configuration as) a noble gas.

Section 4.2—The number of valence electrons for most main-group elements is the same as the group number. Electron-dot symbols, or **Lewis symbols**, use dots singly or in pairs to represent valence electrons of an atom or ion.

Section 4.3—The element sodium reacts violently with elemental chlorine to give sodium chloride, or table salt. In this reaction, an electron is transferred from a sodium atom to a chlorine atom, forming ions. The ions formed have opposite charges and are strongly attracted to one another; this attraction results in an **ionic bond**. The positive and negative ions arrange themselves in a regular array, forming a **crystal** of sodium chloride. The ions of an element have very different properties from the atoms of that element.

Section 4.4—Metal atoms tend to give up their valence electrons to become positively charged ions, and nonmetal atoms tend to accept (8 − group number) electrons to become negatively charged ions. In either case, the ions formed tend to have eight valence electrons. Small atoms, such as lithium, are an exception and tend to form ions with two valence electrons. The formation of ions is thus said to follow either the **octet rule** or the **duet rule**. The symbol for an ion is the element's symbol with a superscript that indicates the number and type (+ or −) of charge. Some elements, especially the transition metals, can form ions of different charges. The formula of an ionic compound always has the same number of positive charges as negative charges.

Section 4.5—A **binary ionic compound** contains two different elements; it has a cation of a metal and an anion of a nonmetal. The formula of a binary ionic compound represents the numbers and types of ions in the compound. The formula is found by crossing over the charge numbers of the ions (without the plus and minus signs), giving the chemical formula Al_2O_3.

Binary ionic compounds are named by naming the cation first and then putting an *-ide* ending on the stem of the anion name. Examples include calcium chloride, potassium oxide, and aluminum nitride. Differently charged cations of a metal are named with Roman numerals to indicate the charge: Fe^{2+} is iron(II) ion and Fe^{3+} is iron(III) ion.

Section 4.6—Nonmetal atoms can bond by sharing one or more pairs of electrons, forming a **covalent bond**. A **single bond** is one pair of shared electrons; a **double bond** is two pairs; and a **triple bond** is three pairs. Shared pairs of electrons are called **bonding pairs**, and unshared pairs are called nonbonding pairs, or **lone pairs**.

Section 4.7—Bonding pairs are shared equally when the two atoms are the same, but may be unequally shared when the atoms are different. The **electronegativity** of an element is the attraction of an atom in a molecule for a bonding pair. Electronegativity generally increases toward the right and top of the periodic table; fluorine is the most electronegative element. When electrons of a covalent bond are shared unequally, the more electronegative atom takes on a partial negative charge ($\delta-$), the other atom takes on a partial positive charge ($\delta+$), and the bond is said to be a **polar covalent bond**. The greater the difference in electronegativity, the more polar is the bond. A bonding pair shared equally is a **nonpolar covalent bond**. Most binary covalent compounds are named by naming the first element in the formula, followed by the stem of the second element with an *-ide* ending. Prefixes such as *mono-*, *di-*, and *tri-* are used to indicate the number of atoms of each element.

Section 4.8—The HONC rules state one bond for hydrogen, two for oxygen, three for nitrogen, and four for carbon. A water molecule has two bonding pairs (O—H bonds) and two lone pairs, an ammonia molecule has three bonding pairs (N—H bonds) and one lone pair, and a methane molecule has four bonding pairs (C—H bonds).

Section 4.9—A **polyatomic ion** is a charged particle containing two or more covalently bonded atoms. Compounds containing polyatomic ions are named, and their formulas written, in the same fashion as compounds containing monatomic ions, except that parentheses are placed around the formula for a polyatomic ion if there is more than one of it in the compound. Names of polyatomic ions often end in *-ate* or *-ite*.

Section 4.10—A **Lewis structure** shows the arrangement of atoms, bonds, and lone pairs in a molecule or polyatomic ion. To draw a Lewis structure, after guessing a reasonable skeletal structure, we (1) determine the total electrons available, (2) determine the total electrons needed for all the atoms, (3) determine electrons in bonds by subtracting electrons available from electrons needed, and (4) fill in the bonds and the needed lone pairs. A multiple bond is formed if there are not enough electrons to give each atom an octet (except with hydrogen, which requires only two electrons). There are exceptions to the octet rule. Atoms and molecules with unpaired electrons are called **free radicals**. They are highly reactive and short-lived. Examples of free radicals are NO and ClO_2.

Section 4.11—The shape of a molecule can be predicted with the **VSEPR theory**, which assumes that sets of electrons (either lone pairs or electrons in a bond) around a central atom will get as far away from each other as possible. Two sets of electrons are 180° apart, three sets are about 120° apart, and four sets are about 109° apart. The presence of lone pairs of electrons on the central atom will somewhat decrease the angle between the bonds within the molecule. Once these angles have been established, we determine the shape of the molecule by examining only the bonded atoms. Simple molecules have shapes described as linear, bent, trigonal planar, trigonal pyramidal, or tetrahedral.

Section 4.12—Like a polar bond, a **polar molecule** has separation between its centers of positive charge and negative charge. A molecule with polar bonds is nonpolar if its shape causes the polar bonds to cancel one another. If the polar bonds do not cancel, the molecule is polar or is said to be a **dipole**. The water molecule is polar.

GREEN CHEMISTRY—Chemists manipulate the geometry and composition of molecules to change the properties of both molecules and materials. Typical methods for making these molecules and materials use highly reactive and toxic chemicals and high-energy processes. Greener approaches that take advantage of molecular recognition can instead use low energy, solvent-free processes that do not require harsh conditions.

Learning Objectives	**Associated Problems**
• Determine the number of electrons in an ion, and why it has a net positive or negative charge. (4.1)	1–3
• Write the Lewis symbol for an atom or ion. (4.2)	9–12, 15, 82, 83
• Distinguish between an ion and an atom. (4.3)	1–3, 15, 80, 81
• Describe the nature of the attraction that leads to formation of an ionic bond. (4.3)	53, 54
• Write symbols for common ions, and determine their charges. (4.4)	4, 9–12, 16–22
• Describe the relationship between the octet rule and the charge on an ion. (4.4)	4–6, 73, 89, 91
• Name and write formulas for binary ionic compounds. (4.5)	13, 14, 23–26, 75
• Explain the difference between a covalent bond and an ionic bond. (4.6)	7, 53, 54, 78
• Name and write formulas for covalent compounds. (4.6)	31–39, 84
• Classify a covalent bond as polar or nonpolar. (4.7)	31–37, 41, 42, 53–56
• Use electronegativities of elements to determine bond polarity. (4.7)	45–56
• Predict the number of bonds formed by common nonmetals (the HONC rules). (4.8)	8
• Recognize common polyatomic ions and be able to use them in naming and writing formulas for compounds. (4.9)	27–30, 86
• Draw Lewis structures for simple molecules and polyatomic ions. (4.10)	43, 44, 87
• Identify free radicals. (4.10)	69–72
• Predict the shapes of simple molecules from their Lewis structures. (4.11)	57–60, 76, 89, 90
• Classify a simple molecule as polar or nonpolar from its shape and the polarity of its bonds. (4.12)	61–68, 74
• Describe the concept of molecular recognition.	93–94
• Explain the green chemistry advantages of using production methods based on molecular recognition.	95–96

Conceptual Questions

1. How does sodium metal differ from sodium ions (in sodium chloride, for example) in properties?

2. Why is potassium metal so much more reactive than the potassium ion (in KBr, for example)?

3. What are the structural differences among chlorine atoms, chlorine molecules, and chloride ions? How do their properties differ?

4. What are the charges on simple ions formed from atoms of the following?
   a. group 1A elements   b. group 6A elements
   c. group 5A elements   d. group 2A elements

5. How many electrons are needed for the elements below to complete their octets?
   a. group 7A elements   b. group 6A elements
   c. group 5A elements   d. group 8A elements

6. In what group of the periodic table would elements that form ions with the following charges likely be found?
   a. 2−   b. 3+   c. 1+   d. 2+

7. How is a covalent bond different from an ionic bond?

8. How many covalent bonds does each of the following usually form? You may refer to the periodic table.
   a. H   b. Cl   c. S   d. F   e. N   f. P

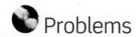 Problems

Lewis Symbols for Elements

9. Write Lewis symbols for each of the following elements. You may use the periodic table.
   a. calcium **b.** sulfur **c.** silicon

10. Write Lewis symbols for each of the following elements. You may use the periodic table.
    a. phosphorus **b.** fluorine **c.** boron

11. Write the Lewis symbol for each species in the following pairs.
    a. Na, Na^+ **b.** Cl, Cl^-

12. Write Lewis symbols for each species in the following pairs.
    a. O, O^{2-} **b.** Mg, Mg^{2+}

Lewis Structures for Ionic Compounds

13. Write Lewis structures for each of the following.
    a. potassium bromide **b.** sodium oxide
    c. magnesium fluoride **d.** aluminum chloride

14. Write Lewis structures for each of the following.
    a. lithium bromide **b.** strontium sulfide
    c. potassium oxide **d.** aluminum nitride

15. Which of the following statements regarding Lewis dot symbols of ions are false with respect to outermost electrons?
    a. In magnesium oxide, O^{2-} has eight electrons.
    b. Mg^{2+} always has one electron around it.
    c. In potassium chloride, K^+ has no electrons around it.
    d. In ionic compounds containing chloride ions, Cl^- is isoelectronic with Ar.

Names and Symbols for Simple Ions

16. Pair (match) the names and symbols for the ions below.
    a. nitride ion **b.** Fe^{2+}
    c. copper(I) ion **d.** N^{3-}
    e. iron(II) ion **f.** Cu^+
    g. silver ion **h.** Ag^+

17. Without referring to Table 4.2, supply a symbol (given the name) or a name (given the symbol) for each of the following ions.
    a. sulfide ion **b.** K^+
    c. Br^- **d.** fluoride ion
    e. Ca^{2+} **f.** iron(III) ion

18. Refer to the beginning paragraphs of Section 4.5, then use that information to name the following ions.
    a. Cr^{2+} **b.** Cr^{3+} **c.** Cr^{6+}

19. Refer to the beginning paragraphs of Section 4.5, then use that information to name the following ions.
    a. Co^{3+} **b.** Co^{6+}

20. Refer to the beginning paragraphs of Section 4.5, then write symbols for:
    a. vanadium(V) ion **b.** tin(II) ion
    c. titanium(IV) ion

21. Refer to the beginning paragraphs of Section 4.5, then write symbols for:
    a. molybdenum(II) ion **b.** molybdenum(IV) ion
    c. molybdenum(VI) ion

22. For copper and iron ions, we sometimes use the *-ic* and *-ous* endings. Which ending indicates the more positively charged ion?

Names and Formulas for Binary Ionic Compounds

23. Pair (match) the names and formulas for the following binary ionic compounds.
    a. iron(II) chloride **b.** $FeBr_3$
    c. silver fluoride **d.** $FeCl_2$
    e. iron(III) bromide **f.** Na_2O
    g. sodium oxide **h.** AgF

24. Write a formula to match the name or a name to match the formula of the following binary ionic compounds.
    a. $LiBr$ **b.** calcium iodide
    c. MgS **d.** aluminum chloride
    e. NiO **f.** copper(I) sulfide

25. There are two common binary ionic compounds formed from chromium and oxygen. One of them contains chromium(III) ions; the other contains chromium(VI) ions. Write the formulas for the two compounds, and name them.

26. One of two binary ionic compounds is often added to toothpaste. One of these compounds contains sodium and fluorine; the other contains tin(II) ions and fluorine. Write the formulas for these two compounds, and name them.

Names and Formulas for Ionic Compounds with Polyatomic Ions

27. Pair (match) the names and formulas for the following polyatomic ionic compounds.
    a. ammonium nitrate **b.** $FePO_4$
    c. calcium carbonate **d.** NH_4NO_3
    e. iron(III) phosphate **f.** $Ca(CH_3CO_2)_2$
    g. calcium acetate **h.** $CaCO_3$

28. Supply a formula to match the name or a name to match the formula for the following.
    a. potassium hydroxide **b.** $MgCO_3$
    c. iron(III) cyanide **d.** $CuSO_4$
    e. iron(II) oxalate **f.** $Na_2Cr_2O_7$

29. Supply a formula to match the name or a name to match the formula for the following.
    a. $AgNO_3$ **b.** sodium chromate
    c. $(NH_4)_2SO_3$ **d.** strontium hydrogen carbonate
    e. $Al(MnO_4)_3$ **f.** copper(II) phosphate

30. What is the formula of and charge on the following polyatomic ions?
    a. carbonate **b.** permanganate
    c. ammonium **d.** acetate

Molecules: Covalent Bonds

31. Use Lewis dot symbols to show the sharing of electrons between a hydrogen atom and an iodine atom.

32. Use Lewis dot symbols to show the sharing of electrons between two bromine atoms to form a bromine (Br_2) molecule. Label all electron pairs as bonding pairs (BPs) or lone pairs (LPs).

33. Use Lewis dot symbols to show the sharing of electrons between a phosphorus atom and hydrogen atoms to form a molecule in which phosphorus has an octet of electrons.

34. Use Lewis dot symbols to show the sharing of electrons between a silicon atom and hydrogen atoms to form a molecule in which silicon has an octet of electrons.

35. Use Lewis dot symbols to show the sharing of electrons between a carbon atom and fluorine atoms to form a molecule in which each atom has an octet of electrons.

36. Use Lewis dot symbols to show the sharing of electrons between a nitrogen atom and bromine atoms to form a molecule in which each atom has an octet of electrons.

37. When a hydrogen atom covalently bonds to another atom, how many electrons associated with this hydrogen atom become involved in such a bond? (Explain your answer.)

Names and Formulas for Covalent Compounds

38. Pair (match) the names and formulas for the following binary covalent compounds.
    a. dichlorine monoxide   b. ClO
    c. chlorine monoxide   d. Cl_2O_7
    e. dichlorine trioxide   f. Cl_2O
    g. dichlorine heptoxide   h. Cl_2O_3

39. Supply a formula for the name or a name for the formula for the following covalent compounds.
    a. dinitrogen tetroxide   b. bromine trichloride
    c. OF_2 d. nitrogen triiodide
    e. CI_4 f. N_2O_3

40. Supply a formula for the name or a name for the formula for the following covalent compounds.
    a. carbon disulfide   b. chlorine trifluoride
    c. PF_5 d. CBr_4
    e. sulfur trioxide   f. P_4S_3

Lewis Structures for Molecules and Polyatomic Ions

41. Draw Lewis structures that follow the octet rule for the following covalent molecules.
    a. SiH_4 b. N_2F_4 c. CH_5N
    d. H_2CO e. NOH_3 f. H_3PO_3

42. Draw Lewis structures that follow the octet rule for the following covalent molecules.
    a. NH_2Cl b. C_2H_4 c. H_2SO_4
    d. C_2N_2 e. $COCl_2$ f. SCl_2

43. Draw Lewis structures that follow the octet rule for the following ions.
    a. ClO^- b. HPO_4^{2-} c. BrO_3^-

44. Draw Lewis structures that follow the octet rule for the following ions.
    a. CN^- b. ClO_2^- c. HSO_4^{2-}

Electronegativity: Polar Covalent Bonds

45. If element A bonds covalently with element B, then what must the electronegativity difference between A and B be for this bond to be polar?

46. If element X bonds covalently with element Y, then what must the electronegativity difference between X and Y be for this bond to be nonpolar?

47. Classify the following covalent bonds as polar or nonpolar.
    a. $H-O$ b. $N-F$ c. $Cl-B$

48. Classify the following covalent bonds as polar or nonpolar.
    a. $H-N$ b. $O-Be$ c. $P-F$

49. Use the symbol $\leftrightarrow$ to indicate the direction of the dipole in each polar bond in Problem 47.

50. Use the symbol $\leftrightarrow$ to indicate the direction of the dipole in each polar bond in Problem 48.

51. Use the symbols $\delta+$ and $\delta-$ to indicate partial charges, if any, on the following bonds.
    a. $Si-O$ b. $F-F$ c. $F-N$

52. Use the symbols $\delta+$ and $\delta-$ to indicate partial charges, if any, on the following bonds.
    a. $O-H$ b. $C-F$ c. $C=C$

Classifying Bonds

53. Classify the bonds in the following as ionic or covalent. For bonds that are covalent, indicate whether they are polar or nonpolar.
    a. K_2O b. $BrCl$ c. CaF_2 d. I_2

54. Classify the bonds in the following as ionic or covalent. For bonds that are covalent, indicate whether they are polar or nonpolar.
    a. Na_2O b. $CaCl_2$ c. NBr_3 d. CS_2

55. Rank the following bonds in terms of increasing polarity.
    a. $H-F$ b. $N-F$ c. $B-F$ d. $Si-F$

56. Rank the following bonds in terms of decreasing polarity.
    a. $Si-Cl$ b. $H-Cl$ c. $P-Cl$ d. $Br-Cl$

VSEPR Theory: The Shapes of Molecules

57. Pair (match) the VSEPR predicted shape to the listed molecules.
    a. $AlCl_3$ b. linear
    c. CSe_2 d. trigonal planar
    e. SeH_2 f. bent
    g. SiI_4 h. tetrahedral

58. Use the VSEPR theory to predict the shape of each of the following molecules.
    a. silane (SiH_4)   b. hydrogen selenide (H_2Se)
    c. phosphine (PH_3)   d. silicon tetrafluoride (SiF_4)

59. Use VSEPR theory to predict the shape of each of the following molecules.
    a. chloroform ($CHCl_3$)
    b. carbon tetrafluoride (CF_4)
    c. sulfur difluoride (SF_2)
    d. hydrogen sulfide (H_2S)

60. Use VSEPR theory to predict the shape of each of the following molecules.
    a. nitrogen triiodide (NI_3)
    b. dichlorodifluoromethane (CCl_2F_2)
    c. sulfate ion (SO_4)$^{2-}$
    d. carbonate ion (CO_3)$^{2-}$

Polar and Nonpolar Molecules

61. Despite the large electronegativity difference between fluorine and beryllium atoms, BeF_2 forms linear molecules in gas phase. Are these molecules polar or nonpolar? Explain.

62. The molecule SF_2 is bent. Is it polar or nonpolar? Explain.

63. Look again at the molecules in Problem 58. For each one, are the bonds polar? What are the approximate bond angles? Is the molecule as a whole polar?

64. Look again at the molecules in Problem 59. For each one, are the bonds polar? What are the approximate bond angles? Is the molecule as a whole polar?

65. What makes the molecule of SO_2 polar?

66. What makes the molecule of SO_3 nonpolar?

67. Is it possible for a molecule of CH_2Cl_2 to be nonpolar? (C is the central atom.) Explain.

68. Is H_2O more or less polar than SCl_2?

Molecules That Are Exceptions to the Octet Rule

69. Which of the following species (atoms or molecules) are free radicals?
    a. Br   b. F_2 c. CCl_3

70. Which of the following species (atoms or molecules) are free radicals?
    a. S   b. NO_2 c. N_2O_4

71. Free radicals are one class of molecules in which atoms do not conform to the octet rule. Another exception involves atoms with fewer than eight electrons, as seen in elements of group 3 and in beryllium. In some covalent molecules, these atoms can have six or four electrons, respectively. Draw Lewis structures for the following covalent molecules.
    a. $AlBr_3$ b. BeH_2 c. BCl_3

72. Exceptions to the octet rule include molecules that have atoms with more than 8 valence electrons, most typically 10 or 12. Atoms heavier than Si can expand their valence shell, meaning that they can accommodate the "extra" electrons in unoccupied, higher-energy orbitals. Draw Lewis structures for the following molecules or ions.
    a. XeF_4 b. I_3^- c. SF_4 d. KrF_2

 Expand Your Skills

73. Why does radon tend not to form chemical bonds?

74. Use the symbols $\delta+$ and $\delta-$ to indicate the polarity of the H_2S molecule.

75. The gas phosphine (PH_3) is used as a fumigant to protect stored grain and other durable produce from pests. Phosphine is generated where it is to be used by adding water to aluminum phosphide or magnesium phosphide. Give formulas for these two phosphides.

76. Draw Lewis structures for the nitronium ion, NO_2^+ and for the nitrite ion, NO_2^-. Which has the smaller bond angle?

77. There are two different covalent molecules with the formula C_2H_6O. Draw Lewis structures for the two molecules. (Two molecules with the same formula but different arrangements of the atoms are called "isomers"; see Chapter 9.)

78. Solutions of iodine chloride (ICl) are used as disinfectants. Is the compound ICl ionic, polar covalent, or nonpolar covalent?

79. Consider the hypothetical elements X, Y, and Z, which have the following Lewis symbols:

    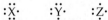

    a. To which group in the periodic table would each element belong?
    b. Write the Lewis formula for the simplest compounds X, Y, and Z would form with hydrogen.
    c. Draw Lewis structures for the ions that would be formed when X reacted with sodium and when Y reacted with sodium.

80. Potassium is a soft, silvery metal that reacts violently with water and ignites spontaneously in air. Your doctor recommends that you take a potassium supplement. Would you take potassium metal? If not, what would you take?

81. What is wrong with the phrase "just a few molecules of potassium iodide"?

82. Use subshell notation to write an electron configuration for the most stable simple ion formed by each of the following elements.
    a. Ca   b. Rb   c. S
    d. I   e. N   f. Se

83. Chemistry was simple in the early universe, with only hydrogen, helium, and a tiny bit of lithium, existing in part as ions. With enough helium atoms and enough H^+ ions wandering about, scientists think some of them combined to form HeH^+ molecules, the first molecule to form in the early universe. So far HeH^+ has eluded extraterrestrial observation. Write the Lewis structure for HeH^+ and the equation for its formation.

84. The halogens (F, Cl, Br, and I) tend to form only one single bond in binary molecules. Explain.

85. A science magazine for the general public contains this statement: "Some of these hydrocarbons are very light, like methane gas—just a single carbon molecule attached to three hydrogen molecules." Evaluate the statement, and correct any inaccuracies.

86. Sodium tungstate is Na_2WO_4. What is the formula for aluminum tungstate?

87. When metal atoms form monatomic ions, they do not necessarily lose them from the last subshell in the electron configuration. Instead, they lose electrons from the *highest-numbered* subshell. With that in mind, write electron configurations for copper(I) ion and vanadium(II) ion.

88. What is the formula for the compound formed by the imaginary ions Q^{2+} and ZX_4^{3-}?

89. Why are the 120° angles optimal in a trigonal planar geometry?

90. Why are the 109.5° angles optimal in a tetrahedral geometry?

91. Consider the statement "Germanium (Ge) can form ions with a 4+ charge, because Ge is in column 4A of the periodic table, and those elements tend to lose four electrons."
    a. The statement is correct, but the reason is wrong.
    b. The statement is wrong, but the reason is correct.
    c. Both the statement and the reason are correct.
    d. Neither the statement nor the reason is correct.

92. Consider the statement "Ions of tin and lead should both exist as 2+ and 4+, because elements in column 4A can lose varying numbers of electrons, usually two or four."
    a. The statement is correct, but the reason is wrong.
    b. The statement is wrong, but the reason is correct.
    c. Both the statement and the reason are correct.
    d. Neither the statement nor the reason is correct.

93. Give two design criteria that chemists use to design new medicines.

94. What is the term for the science that explains how molecules interact through geometric orientation of atoms?

95. Give three ways ways that molecular recognition approaches can support greener methods for making molecules and materials.

96. How can an understanding of enzymes and biological receptors guide medicinal chemists?

 # Critical Thinking Exercises

Apply knowledge that you have gained in this chapter and one or more of the FLaReS principles (Chapter 1) to evaluate the following statements or claims.

4.1 Some people believe that crystals have special powers. Crystal therapists claim that they can use quartz crystals to restore balance and harmony to a person's spiritual energy.

4.2 A "fuel-enhancer" device is being sold by an entrepreneur. The device contains a powerful magnet and is placed on the fuel line of an automobile. The inventor claims that the device "separates the positive and negative charges in the hydrocarbon fuel molecules, increasing their polarity and allowing them to react more readily with oxygen."

4.3 Sodium chloride (NaCl) is a metal–nonmetal compound held together by ionic bonds. A scientist has studied mercury(II) chloride ($HgCl_2$) and says that its atoms are held together by covalent bonds. The scientist says that since a solution of this substance in water does not conduct an electric current, it does not contain an appreciable number of ions.

4.4 Another scientist, noting that the noble gas xenon does not contain any ions, states that xenon atoms must be held together by covalent bonds.

4.5 A webpage claims, "Fifty years ago, the hydrogen bond angle in water was 108° and you rarely heard of anyone with cancer. Today, it's only 104° and, as a result, cancer is an epidemic!"

4.6 In the 1960s the concept of polywater came into fashion. The idea was that under the right circumstances, water could be made into a polymer—that is, a large molecule in which a number of water molecules would be covalently bonded with each other.

4.7 It was claimed that a compound containing a pentavalent carbon atom (a C atom with five bonds attached to it) was synthesized. This report was received with great skepticism by the scientific community.

4.8 A football coach claims that his players have more stamina, since they drink atomic O-saturated water rather than O_2-saturated water.

4.9 A fertilizer company is marketing a new form of garden fertilizer. It is made of thin flakes and is claimed to work better because it contains flat ammonia molecules, which have greater surface area and dissolve better in water.

4.10 A scientist claims to have detected HeH^+ and HeH_2.

 # Collaborative Group Projects

Prepare a PowerPoint, poster, or other presentation (as directed by your instructor) for presentation to the class. Projects 1 and 2 are best done by a group of four students.

1. Starting with 20 balloons, blown up to about the same size and tied, tie two balloons together. Next, tie three balloons together. Repeat this with four, five, and six balloons. (Suggestion: Three sets of two balloons can be twisted together to form a six-balloon set, and so on.) Show how these balloon sets can be used to illustrate the VSEPR theory.

2. Prepare a brief biographical report on one of the following:
   a. Gilbert N. Lewis
   b. Linus Pauling

3. There are several different definitions and scales for electronegativity. Search for information on these, and write a brief compare-and-contrast essay about two of them.

LET'S EXPERIMENT! Molecular Shapes: Please Don't Eat the Atoms!

Materials Needed
• Small bag of gumdrops or spice drops
• Toothpicks

Molecules can be written out as formulas or drawn in bond-line format. However, molecules are actually three-dimensional. Can you use household items to create your own 3-D models of molecules?

In this experiment, you will have the opportunity to visualize the geometric shapes defined by VSEPR theory, and you will be practicing writing Lewis formulas and predicting molecular geometry.

Using your gumdrops and toothpicks, you will create a series of molecules and observe their geometry.

a. First, connect two gumdrops of the same color with one toothpick. This geometry is called linear and is similar to any diatomic molecule, such as oxygen (O_2). To correctly represent the oxygen molecule, there should be two toothpicks between the gumdrops. Sketch the shape and write the Lewis formula.

b. Next, connect two gumdrops to one central gumdrop of a different color. Keep the gumdrops in one line. This geometry is also called linear and is similar to carbon dioxide (CO_2). To be correct for carbon dioxide, two toothpicks should be inserted on either side of the central atom. Sketch the shape and write the Lewis formula.

c. Now connect three gumdrops to one central gumdrop of a different color. Place them all in the same plane but spread them out equally. This geometry is called trigonal planar and is similar to a nitrate ion (NO_3^-). To be correct for the nitrate ion, there should be two toothpicks between the central gumdrop and one of the outer gumdrops. Sketch the 3-D shape and write the Lewis formula.

d. Remove one of the toothpicks and gumdrop. The remaining gumdrops define a "bent" or "angular" molecule similar to sulfur dioxide (SO_2). Sketch the shape and write the Lewis formula.

e. Now connect four gumdrops to one central gumdrop of a different color in the shape of a tetrahedron. This geometry is similar to methane (CH_4). Sketch the shape and write the Lewis formula.

f. Remove one of the gumdrops. Leave its toothpick in the structure. The remaining system is called trigonal pyramid. This shape is the same as that of an ammonia molecule (NH_3). Sketch the 3-D structure and write the Lewis formula.

g. Remove another gumdrop. This is another "bent" or "angular" geometry. Notice that it is different from the geometry in **c**. (In **c**, the angle was 120°; in this case, the single is about 109.5°, the tetrahedral angle.) This geometry is similar to that of water (H_2O). Sketch the structure and write the Lewis formula.

Questions

1. Why do the SO_2 and H_2O molecules have a bent or angular geometry? Why are they not linear geometries?

2. Using your gumdrops and toothpicks, create a titanium tetrachloride ($TiCl_4$) molecule. Write the Lewis formula to predict the geometry.

5

Chemical Accounting

Everything that burns fuel requires oxygen, and the human body is no exception. A SCUBA diver depends on oxygen in the air supply carried on his back. But how much oxygen does a diver need? A knowledge of chemistry allows this question to be answered, thus promoting the safety of the divers.

Have You Ever Wondered?

1 How can a car make more pollution than the gasoline it uses?

2 What's the difference between .925 silver and .800 silver?

3 Does 2% milk have just 2% of the calories of whole milk?

4 What does a toxicity rating (LD_{50}) of 100 ppm mean?

140

MASS AND VOLUME RELATIONSHIPS We can look at or represent chemical phenomena on three levels:

1. Macro and tangible: for example, a sample of yellow powder or a blue solution in a beaker.
2. Particle (atom, molecule, or ion) and invisible: for example, a sulfur atom, a carbon dioxide molecule, or a chloride ion.
3. Symbolic and mathematical: for example, S, CO_2, Cl, or $d = m/V$.

Macro-level representations of chemicals are familiar from everyday life. It is easy to visualize salt crystals in a shaker, gold in jewelry, baking soda in a box, vitamin C in supplement tablets, and many other common substances because they are tangible. We can see them and touch them. The other representations were introduced in earlier chapters. You may have recognized the mathematical equation for density from Chapter 1. We employed symbols when discussing atoms in Chapter 2 and Chapter 3, and formulas in our discussion of molecules and ions in Chapter 4. In this chapter, we consider symbolic and mathematical descriptions of chemical changes.

Why Chemical "Accounting"?

Much chemistry can be discussed and understood with little or no use of mathematics, but the quantitative aspects can shed a great deal of light on properties of matter. In this chapter, we will consider some of the basic calculations used in chemistry and related fields such as biology and medicine. This type of algebra is not limited to these fields but can also be applied to business, social sciences, and even art. Chemists use many kinds of mathematics, from simple arithmetic to sophisticated calculus and complicated computer algorithms, but our calculations here will require at most a bit of algebra.

The aspect of chemistry in this chapter, often called *stoichiometry*, is extremely important in the science of chemistry. The name comes from two words: *stoicheion*, which is the Greek word for element, and *metria*, referring to measurement. Literally, it means the measurement of elements. There is more to this field than just measuring, however. Like a financial accountant, who not only must count money but also must make an account of where the funds go and how they are used, a chemist must also account for the changes in amounts of matter involved in a chemical reaction. As we will see later in this chapter and in other parts of this text, Dalton's laws demand that we keep track of all the movements and locations of matter.

5.1 Chemical Sentences: Equations

Learning Objective • Identify balanced and unbalanced chemical equations, and balance equations by inspection.

Studying science, like studying a language, means that the student has to learn the vocabulary and grammar of that specific language. A beginner cannot launch directly into speaking that language. Chemistry is a study of matter and its changes as it interacts with energy. The symbols and formulas we have used to represent elements and compounds make up the letters (symbols) and words (formulas) of our chemical language. Using them, we can write sentences (chemical equations), but we must understand the rules for speaking the chemical language. A **chemical equation** uses symbols and formulas to represent the elements and compounds involved in the change.

At the macro level, we can describe a chemical reaction in words. For example,

Carbon reacts with oxygen to form carbon dioxide.

We can describe the same reaction symbolically.

$$C + O_2 \longrightarrow CO_2$$

The plus sign (+) indicates that carbon and oxygen are added together or combined in some way. The arrow ($\longrightarrow$) is read "yield(s)" or "react(s) to produce." Substances on the left of the arrow (C and O_2, in this case) are **reactants**, or *starting materials*. Those on the right (here, CO_2) are the **products** of the reaction. The order in which numbers are added makes no difference in a sum. (Do you recall the *commutative property* of addition from algebra? It applies here.) Likewise, reactants and products need not be written in any particular order in a chemical equation, as long as all reactants are to the left of the arrow and all products are to the right. In other words, we could also write the preceding equation as

$$O_2 + C \longrightarrow CO_2$$

At the atomic or molecular level, the equation means that one carbon (C) atom reacts with one oxygen (O_2) molecule to produce one carbon dioxide (CO_2) molecule. It is important to note that chemical substances sometimes can be found as elemental atoms, like the carbon, while others, like the oxygen, are only obtained in a molecular state.

Sometimes, we indicate the physical states of the reactants and products by writing the initial letter of the state immediately following the formula: (g) indicates a gaseous substance, (l) a liquid, and (s) a solid. The label (aq) indicates that the substance is in an aqueous (water) solution. Using state labels, we see that the above equation becomes

$$C(s) + O_2(g) \longrightarrow CO_2(g)$$

Balancing Chemical Equations

We can represent some reactions quite simply, like the previous reaction of carbon and oxygen to form carbon dioxide, but many other chemical reactions require more thought. For example, hydrogen reacts with oxygen to form water. Using formulas, we can represent this reaction as

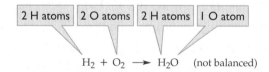

However, this representation shows two oxygen atoms in the reactants (as O_2), but only one in the product (in H_2O). Recall that matter is neither created nor destroyed in a chemical reaction (the law of conservation of mass, Section 2.2), so the equation must be *balanced* to represent the chemical reaction correctly. Because we are interpreting the equation to mean that each symbol represents multiples of a single atom or molecule, we cannot simply assume that we are using one-half of the oxygen molecule. Instead, we must maintain the relationships between the whole particles. Also, we must be sure that the same number of each of the different types of atom appears on both sides of the arrow. To balance the oxygen atoms, we need only place the coefficient 2 in front of the formula for water.

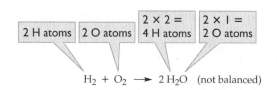

This coefficient means that two whole molecules of water are produced. As is the case with a subscript of 1, a coefficient of 1 is understood when there is no number in front of a formula. *Everything* in a formula is multiplied by the coefficient in front of it. In the preceding equation, adding the coefficient 2 before H_2O not only increases the total number of oxygen atoms on the product side to two but also increases the total number of hydrogen atoms to four.

But the equation is still not balanced. There are now four hydrogen atoms on the right (product) side of the equation but only two on the left (reactant) side of the equation. As we fixed the number of oxygen atoms in the product, we unbalanced the number of hydrogen atoms. To balance the hydrogen, we must now place a coefficient 2 in front of H_2.

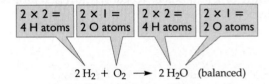

$$2 H_2 + O_2 \longrightarrow 2 H_2O \quad \text{(balanced)}$$

Now there are four hydrogen atoms and two oxygen atoms on each side of the equation. Atoms are conserved: The equation is balanced (Figure 5.1), and the law of conservation of mass is obeyed. Figure 5.2 illustrates two common pitfalls in the process of balancing equations, as well as the correct method. You cannot simply gain or lose substances on the product side of the equation, as it may represent an invalid reaction. Also, if you change the subscript of a specific chemical formula, you create either a different substance or an invalid formula.

○ = hydrogen
● = oxygen

▲ **Figure 5.1** To balance the equation for the reaction of hydrogen and oxygen to form water, the same number of each kind of atom must appear on each side. (Atoms are conserved.) When the equation is balanced, there are four H atoms and two O atoms on each side.

Q *Why can't you balance the equation by removing one of the oxygen atoms from the left side of the first balance?*

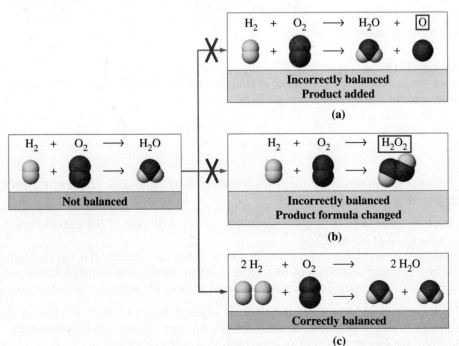

◀ **Figure 5.2** Balancing the equation for the reaction between hydrogen and oxygen to form water. (a) Incorrect. There is no atomic oxygen (O) as a product. Other products cannot be introduced simply to balance an equation. (b) Incorrect. The product of the reaction is water (H_2O), not hydrogen peroxide (H_2O_2). A formula can't be changed simply to balance an equation. (c) Correct. An equation can be balanced only by using the correct formulas and adjusting the coefficients.

EXAMPLE 5.1 Balancing Equations

Balance the following chemical equation, which represents the reaction that occurs when an airbag in a car deploys.

$$NaN_3 \longrightarrow Na + N_2$$

Solution

The sodium (Na) atoms are balanced, but the nitrogen (N) atoms are not. To balance this equation, we can use the concept of the least common multiple. There are three nitrogen atoms on the left (reactant side) and two on the right (product side). The least common multiple of 2 and 3 is 6. To get *six* nitrogen atoms on each side, we need *three* N_2 and *two* NaN_3:

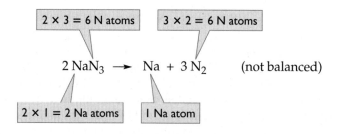

We now have *two* sodium atoms on the left. We can get two on the right by placing the coefficient 2 in front of Na.

$$2\,NaN_3 \longrightarrow 2\,Na + 3\,N_2 \quad \text{(balanced)}$$

Checking, we count two Na atoms and six N atoms on each side. The equation is balanced.

> **Exercise 5.1A**

The reaction between hydrogen and nitrogen to give ammonia, called the *Haber process*, is typically the first step in the industrial production of nitrogen fertilizers.

$$H_2 + N_2 \longrightarrow NH_3$$

Balance the equation.

> **Exercise 5.1B**

Balance the equation for the *smelting* of iron ore (which is mostly Fe_2O_3) with carbon to produce iron metal and carbon dioxide.

$$Fe_2O_3 + C \longrightarrow CO_2 + Fe$$

▲ Smelting iron ore.

Balancing equations by inspection means some trial and error, but a few strategies often help:

1. If an element occurs in just one substance on each side of the equation, try balancing that element *first*.
2. Balance any reactants or products that exist as the free element (Fe, O_2, etc.) *last*.
3. When you add a coefficient to a compound to correct one element, be aware that it will also change the number of atoms of other elements in the compound.

Perhaps the most important step in balancing any equation is to check that the result you obtain is indeed balanced. Remember that for each element, the same number of atoms of the element must appear on each side of the equation.

We have made the task of balancing equations appear easy by considering fairly simple reactions. It is more important at this point for you to understand the principle than to be able to balance complicated equations. You should know what is meant by a balanced equation and be able to balance simple equations. It is crucial that you understand the information that is contained in balanced equations.

SELF-ASSESSMENT Questions

1. Which of the following equations are balanced? (You need not balance the equations. Just determine whether they are balanced as written.)

   I. $Mg + 2 H_2O \longrightarrow Mg(OH)_2 + H_2$
   II. $4 LiH + AlCl_3 \longrightarrow 2 LiAlH_4 + 2 LiCl$
   III. $2 KOH + CO_2 \longrightarrow K_2CO_3 + H_2O$
   IV. $2 Sn + 2 H_2SO_4 \longrightarrow 2 SnSO_4 + SO_2 + 2 H_2O$

   a. I and II   **b.** I and III
   c. II and III   **d.** II and IV

2. Consider the equation for the reaction of phosphine with oxygen gas to form tetraphosphorus decoxide and water:

   $$PH_3 + O_2 \longrightarrow P_4O_{10} + H_2O \quad \text{(not balanced)}$$

   When the equation is balanced, how many molecules of water are produced for each molecule of P_4O_{10} formed?

   a. 1   **b.** 4   **c.** 6   **d.** 12

Items 3–5 refer to the following equation for the reaction of magnesium hydroxide [$Mg(OH)_2$] with phosphoric acid (H_3PO_4).

$$Mg(OH)_2 + H_3PO_4 \longrightarrow H_2O + Mg_3(PO_4)_2 \quad \text{(not balanced)}$$

3. When the equation is balanced, how many H_3PO_4 molecules must react to produce one $Mg_3(PO_4)_2$ formula unit? (Just as a molecule is the smallest unit of a molecular compound, a *formula unit* is the smallest unit of an ionic compound.)

   a. 1   **b.** 2   **c.** 3   **d.** 6

4. When the equation is balanced, how many H_2O molecules will be produced for each H_3PO_4 molecule that reacts?

   a. 1   **b.** 2   **c.** 3   **d.** 6

5. When the equation is balanced, how many $Mg(OH)_2$ formula units are needed to react with six H_3PO_4 molecules?

   a. 1   **b.** 1.5   **c.** 6   **d.** 9

Answers: 1, b; 2, c; 3, b; 4, c; 5, d

5.2 Volume Relationships in Chemical Equations

Learning Objective • Determine volumes of gases that react, using a balanced equation for a reaction.

One challenge in dealing with chemical reactions is that they seldom occur in reactions between just a few atoms or molecules. In the real world, chemists work with quantities of matter that contain billions upon billions of atoms.

John Dalton postulated that atoms of different elements have different masses. Therefore, equal masses of different elements must contain different numbers of atoms. By analogy, a kilogram of golf balls contains a smaller number of balls than a kilogram of Ping-Pong balls because a golf ball is heavier than a Ping-Pong ball. While you might determine the number of balls in each case simply by counting them, it is not possible to count individual atoms. The smallest visible particle of matter contains more atoms than you could count in 10 lifetimes! There are other ways, however, to determine the number of atoms in a substance, as you will learn in this chapter.

Ironically, it was studies of gases, which are also difficult to observe, that advanced our understanding of how to account for atoms involved in chemical reactions. These studies, which occurred in the 1700s, revealed the interesting nature and behavior of gases. (These behaviors are explored in Chapter 6.)

Here we focus on some historical experiments with gases that revealed how chemical equations can be balanced. French scientist Joseph Louis Gay-Lussac (1778–1850) performed experiments that led to an approach to quantifying atoms. In 1808, he announced the results of some chemical reactions that he had carried out with gases and summarized these experiments in a new law. The **law of combining volumes** states that when all measurements are made at the same temperature and pressure, the volumes of gaseous reactants and products are in small whole-number ratios.

One such experiment is illustrated in Figure 5.3. Hydrogen gas will react with nitrogen gas to create ammonia gas. In practice, three volumes of hydrogen will

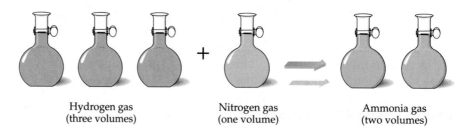

Hydrogen gas
(three volumes)

Nitrogen gas
(one volume)

Ammonia gas
(two volumes)

▲ **Figure 5.3** Gay-Lussac's law of combining volumes. Three volumes of hydrogen gas react with one volume of nitrogen gas to yield two volumes of ammonia gas, when measured at the same temperature and pressure.

Q *Can you sketch a similar figure for the reaction* $2\,SO_2(g) + O_2(g) \longrightarrow 2\,SO_3(g)$?

Amedeo Avogadro

▲ Amedeo Avogadro. The mineral *avogadrite*, named in his honor, is one of the few sources of cesium in the Earth's crust.

combine with one volume of nitrogen to yield two volumes of ammonia. The small whole-number ratio is 3:1:2.

Gay-Lussac thought there must be some direct relationship between the numbers of molecules and the volumes of gaseous reactants and products. It was Italian mathematician and scientist Amedeo Avogadro (1776–1856) who first explained the law of combining volumes in 1811. A consequence of this law is that equal volumes of all gases, when measured at the same temperature and pressure, contain the same number of molecules (Figure 5.4).

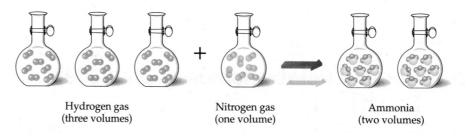

Hydrogen gas
(three volumes)

Nitrogen gas
(one volume)

Ammonia
(two volumes)

▲ **Figure 5.4** Avogadro's explanation of Gay-Lussac's law of combining volumes. Equal volumes of all the gases contain the same number of molecules.

Q *Can you sketch a similar figure for the reaction* $2\,NO(g) + O_2(g) \longrightarrow 2\,NO_2(g)$?

The equation then for the combination of hydrogen and nitrogen to form ammonia is

$$3\,H_2(g) + N_2(g) \longrightarrow 2\,NH_3(g)$$

Note that the coefficients of the molecules correlate with the combining ratio of the gas volumes in Figure 5.3, or 3:1:2. The equation says that three H_2 molecules react with one N_2 molecule to produce two NH_3 molecules. The balanced equation then provides equivalent ratios that can be scaled for the size of the reaction. If you had 3 million H_2 molecules, you would need 1 million N_2 molecules to produce 2 million NH_3 molecules. According to the equation, three volumes of hydrogen react with one volume of nitrogen to produce two volumes of ammonia because each volume of hydrogen contains the same number of molecules as that same volume of nitrogen.

EXAMPLE 5.2 *Volume Relationships of Gases*

What volume of oxygen is required to burn 2.00 L of propane in the following combustion reaction, assuming that both gases are measured at the same temperature and pressure?

$$C_3H_8(g) + 5\,O_2(g) \longrightarrow 3\,CO_2(g) + 4\,H_2O(g)$$

Solution

Looking at the balanced chemical equation, the coefficients indicate that each volume of $C_3H_8(g)$ requires five volumes of O_2 gas. Thus, we use $5\ L\ O_2(g)/1\ L\ C_3H_8(g)$ as a conversion factor to find the volume of oxygen required. (Conversion factors are explained in Section A.3 of the Appendix.)

$$?\,L\ O_2(g) = 2.00\ L\ \cancel{C_3H_8(g)} \times \frac{5\ L\ O_2(g)}{1\ L\ \cancel{C_3H_8(g)}} = 10.0\ L\ O_2(g)$$

> **Exercise 5.2A**

Using the equation in Example 5.2, calculate the volume in liters of $CO_2(g)$ produced when 0.553 L of propane is burned if the two gases are compared at the same temperature and pressure.

> **Exercise 5.2B**

If 4.00 L each of propane and oxygen are combined at the same temperature and pressure, which gas will be left over after reaction? What volume of that gas will remain?

SELF-ASSESSMENT Questions

1. In the reaction $2\ H_2(g) + O_2(g) \longrightarrow 2\ H_2O(g)$, with all substances at the same temperature and pressure, the ratio of volumes, for H_2, O_2, and H_2O, respectively, is
    a. 1:0:1 **b.** 1:1:1 **c.** 2:0:2 **d.** 2:1:2

2. How many CO_2 molecules are produced in the following reaction if 75 O_2 molecules react?

    $2\ C_8H_{18}(l) + 25\ O_2(g) \longrightarrow 16\ CO_2(g) + 18\ H_2O(l)$

    a. 50 molecules **b.** 100 molecules
    c. 48 molecules **d.** 32 molecules

3. In the reaction $N_2(g) + 3\ H_2(g) \longrightarrow 2\ NH_3(g)$, with all substances at the same temperature and pressure, what volume of ammonia is produced when 4.50 L of nitrogen reacts with excess hydrogen?
    a. 3.00 L
    b. 9.00 L
    c. 6.75 L
    d. 4.50 L

Answers: 1, d; 2, c; 3, b

5.3 Avogadro's Number and the Mole

Learning Objectives • Calculate the formula mass, molecular mass, or molar mass of a substance. • Use Avogadro's number to determine the number of particles of different types in a mass of a substance.

As stated earlier, equal volumes of gases at the same temperature and pressure contain equal numbers of molecules. This means that if we weigh equal volumes of different gases, the gases won't weigh the same. The ratio of their masses should be the same as the mass ratio of the molecules themselves. However, we want to extend this relationship in some fashion to solids and liquids, too.

Avogadro's Number: 6.02×10^{23}

Avogadro had no way of knowing how many molecules were in a given volume of gas. Scientists since his time have determined the number of atoms in various weighed samples of substances. Atoms are so incredibly small that these numbers are extremely large, even for tiny samples. Recall that the mass of a carbon-12 atom is exactly 12 atomic mass units (amu) (Section 2.4, page 50) because the carbon-12 atom sets the standard for atomic mass. The number of carbon-12 atoms in a 12 g sample of carbon-12 is called **Avogadro's number** and has been determined experimentally to be 6.0221367×10^{23}. We usually round this number to three significant figures: 6.02×10^{23}.

The Mole: "A Dozen Eggs and a Mole of Sugar, Please"

We buy socks by the pair (2 socks), eggs by the dozen (12 eggs), pencils by the gross (a dozen dozens, or 144 pencils), and paper by the ream (500 sheets of paper). We use these quantitative terms as constant measures and are comfortable with them. A dozen is the same number regardless of whether we are counting eggs or oranges. But a dozen eggs and a dozen oranges do not weigh the same. If an orange weighs five times as much as an egg, a dozen oranges will weigh five times as much as a dozen eggs. Furthermore, when we need to obtain large numbers of these items, we can do so easily. No one would want to purchase 2000 individual sheets of paper if it meant counting them out, but buying four reams of paper is easy. Likewise, a baker might need to make hundreds of cakes and therefore need many dozens of eggs, and can obtain them in multiples of a dozen or a gross.

The sellers of many of those items will not want to sell them in fragments, however, so even if staff in an office wanted to buy 500 pencils, they would not typically be able to buy the exact number. They would instead have to purchase 4 gross of pencils (576 pencils) and have a surplus. Similarly, sometimes we have to deal with an excess amount of chemical components due to the whole-number ratios used in chemical equations.

In the same way that an office manager may purchase paper by the ream and pencils by the gross, a chemist counts atoms and molecules by the *mole*. A single carbon atom is much too small to see or weigh, but a mole of carbon atoms fills a large spoon and weighs 12 g. A mole of carbon and a mole of titanium each contain the same number of atoms. But a titanium atom has a mass four times that of a carbon atom, so a mole of titanium has a mass four times that of a mole of carbon.

A **mole (mol)** is an amount of any substance or item that contains the same number of elementary units as there are atoms in exactly 12 g of carbon-12. That number is 6.02×10^{23}, Avogadro's number. The elementary units may be atoms (such as S or Ca), molecules (such as O_2 or CO_2), ions (such as K^+ or SO_4^{2-}), or any other kind of formula unit. A mole of NaCl, for example, contains 6.02×10^{23} NaCl formula units, which means that it contains 6.02×10^{23} Na^+ ions and 6.02×10^{23} Cl^- ions. The number must be extremely large because atoms, molecules, and ions are so very, very small, and it takes an enormous number of them to make an amount of matter that we can see and work with. The comparisons in Figure 5.5 may give you some feeling for the size of that number.

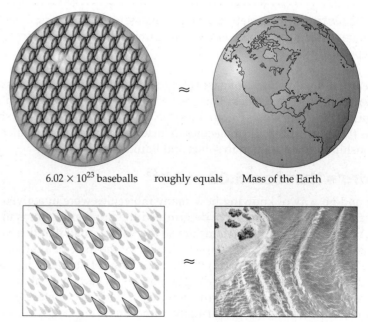

6.02×10^{23} baseballs   roughly equals   Mass of the Earth

6.02×10^{23} drops of water   roughly equals   Water in the oceans

▶ **Figure 5.5** Avogadro's number is so large that it is difficult to truly comprehend, even for scientists ... but these comparisons may help.

Formula Masses

Each element has a characteristic atomic mass. Because chemical compounds are made up of two or more elements, the masses of compounds are combinations of atomic masses. For any substance, the **formula mass** is the sum of the masses of the atoms represented in the formula. If the formula represents a molecule, the term **molecular mass** is often used instead of formula mass. For example, because the formula CO_2 specifies one carbon atom and two oxygen atoms per molecule of carbon dioxide, the formula (or molecular) mass of carbon dioxide is the atomic mass of carbon plus twice the atomic mass of oxygen.

$$\text{Formula mass } CO_2 = 1 \times \text{atomic mass of C} + 2 \times \text{atomic mass of O}$$

$$= (1 \times 12.0 \text{ amu}) + (2 \times 16.0 \text{ amu}) = 44.0 \text{ amu}$$

 EXAMPLE 5.3   Calculating Molecular Masses

Calculate **(a)** the molecular mass of sulfur dioxide (SO_2), a yellow-brown gas that is often associated with coal-burning power plants, and **(b)** the formula mass of ammonium sulfate [$(NH_4)_2SO_4$], a fertilizer commonly used by home gardeners.

Solution

a. We start with the molecular formula: SO_2. Then, to determine the molecular mass, we simply add the atomic mass of sulfur to twice the atomic mass of oxygen.

$$1 \times \text{atomic mass of S} = 1 \times 32.1 \text{ amu} = 32.1 \text{ amu}$$

$$2 \times \text{atomic mass of O} = 2 \times 16.0 \text{ amu} = 32.0 \text{ amu}$$

$$\text{Formula mass of } SO_2 = 64.1 \text{ amu}$$

b. The formula $(NH_4)_2SO_4$ signifies two nitrogen atoms, eight hydrogen atoms, one sulfur atom, and four oxygen atoms. Combining the atomic masses, we have

$$(2 \times 14.0 \text{ amu}) + (8 \times 1.01 \text{ amu}) + (1 \times 32.1 \text{ amu}) + (4 \times 16.0 \text{ amu})$$

$$= 132.2 \text{ amu}$$

▲ A bag of ammonium sulfate fertilizer.

AMMONIUM SULFATE

21–0–0–24 (S)

NET WEIGHT: 50.00KG

GROSS WEIGHT: 50.15KG

FOR AGRICULTURAL USE ONLY

❭ **Exercise 5.3A**
Calculate the formula mass of **(a)** sodium azide (NaN_3), used in automobile airbags, and **(b)** propylene glycol ($C_3H_8O_2$), used as a food preservative.

❭ **Exercise 5.3B**
Calculate the formula mass of **(a)** isopropyl alcohol [$(CH_3)_2CHOH$], better known as rubbing alcohol, and **(b)** calcium dihydrogen phosphate [$Ca(H_2PO_4)_2$], used as a leavening agent in baked goods and pancake mixes.

Percent Composition of a Compound from Formula Masses

A chemical formula represents the ratio of atoms as well as a ratio of mass. This relationship allows us to relate the mass to the number of particles and express the information in a meaningful way for a variety of applications. For example, the mass of 1.00 mol CO_2 is 44.0 g, of which 12.0 g is carbon and 32.0 g is oxygen. The composition of CO_2 can be expressed as percentages by mass of carbon and of oxygen:

$$\% \text{ by mass C} = \frac{\text{mass C}}{\text{mass } CO_2} \times 100\% = \frac{12.0 \text{ g}}{44.0 \text{ g}} \times 100\% = 27.3\%$$

We could find the percent by mass of oxygen in a similar way, but because oxygen is the only other element in CO_2, it is easier to find its percent by mass by subtraction:

$$100\% - 27.3\% = 72.7\% \text{ oxygen}$$

The idea of percent composition by mass is useful in several applications. Dietary sodium limits are usually indicated in milligrams of Na^+. We consume Na^+ mainly as sodium chloride. The American Heart Association (AHA) recommends a maximum intake of 2400 mg Na^+ per day. Example 5.4 shows how we can relate this amount to the NaCl in one teaspoon of table salt.

 EXAMPLE 5.4 Calculating Percent Composition from a Formula

What mass of Na^+ is present in one teaspoon (6.0 g) of NaCl? How does this compare with the maximum intake of 2400 mg Na^+ per day recommended by the AHA?

Solution

A mole of sodium chloride has 1 mole (23.0 g) Na^+ and 1 mole (35.5 g) Cl^-. The percent by mass of Na^+ is

$$\% \text{ by mass Na} = \frac{\text{mass Na}}{\text{mass NaCl}} \times 100\% = \frac{23.0 \text{ g}}{58.5 \text{ g}} \times 100\% = 39.3\%$$

So in 100 g of NaCl there are 39.3 g of Na^+. One teaspoon (6.0 g) of NaCl has

$$6.0 \text{ g NaCl} \times \frac{39.3 \text{ g Na}^+}{100 \text{ g NaCl}} = 2.4 \text{ g Na}^+$$

That is 2400 mg Na^+, the maximum recommended for one day's intake.

> **Exercise 5.4A**

Calculate the percent by mass of nitrogen in $(NH_4)_2S$.

> **Exercise 5.4B**

Nitrogen is an important element in plant growth, so fertilizers are marked with the percentage of nitrogen they contain. Which contains more nitrogen, a pound of ammonium sulfate, $(NH_4)_2SO_4$, or a pound of ammonium nitrate, NH_4NO_3? (*Hint*: You don't have to find the amount of nitrogen in one pound of each; the *percentage* of nitrogen tells you the same thing.)

Many environmental pollutants are reported as elements even when the actual pollutant is a compound or a mixture of compounds. For example, nitrates (NO_3^-) in groundwater (Chapter 14) are reported as parts per million nitrogen, and carbon dioxide emissions (Chapter 13) often are reported as tons of carbon. The idea of percent by mass is useful in understanding these reported quantities.

SELF-ASSESSMENT Questions

1. How many oxygen-16 atoms are there in exactly 16 g of oxygen-16?
   a. 1   b. 16   c. 256   d. 6.02×10^{23}

2. The mass of 1 mol of carbon-12 atoms is
   a. 12 g   b. 12.011 g   c. 12 amu   d. 12.011 amu

3. Which of the following has the smallest molecular mass?
   a. CH_4 b. HCl   c. H_2O d. PH_3

4. How many moles of sulfur atoms are in 1 mol $Cr(SO_4)_3$?
   a. 1 mol   b. 3 mol   c. 15 mol   d. 6.02×10^{23} mol

5. How many moles of hydrogen atoms are present in 2.00 mol of water H_2O?
   a. 2.00 mol   b. 4.00 mol
   c. 6.00 mol   d. 8.00 mol

6. Sodium chloride, by mass, is
   a. about 40% Na^+ and 60% Cl^-
   b. 50% Na^+ and 50% Cl^-
   c. about 60% Na^+ and 40% Cl^-
   d. 23.0% Na^+ and 35.5% Cl^-

Answers: 1. d; 2. a; 3. a; 4. b; 5. b; 6. a

5.4 Molar Mass: Mole-to-Mass and Mass-to-Mole Conversions

Learning Objectives • Convert from mass to moles and from moles to mass of a substance. • Calculate the mass or number of moles of a reactant or product from the mass or number of moles of another reactant or product.

By now, you should be able to see the usefulness of the relationship of moles to mass. Our ability to measure the mass of a substance and convert that value to a number of particles means that the chemical equation can be used not just as a guide to the reaction between equal volumes of gases or equal numbers of individual molecules. We can also scale the equation to any amount of a given substance. We just have to know how to apply the equations.

The **molar mass** of a substance is just what the name implies: the mass of one mole of that substance. The size of Avogadro's number makes the molar mass of a substance *numerically* equal to the atomic mass, molecular mass, or formula mass … but molar mass is expressed in the unit *grams per mole* (g/mol). The atomic mass of sodium is 23.0 amu; its molar mass is 23.0 g/mol. The molecular mass of carbon dioxide is 44.0 amu; its molar mass is 44.0 g/mol. The formula mass of ammonium sulfate is 132.2 amu; its molar mass is 132.2 g/mol. We can use these facts, together with the definition of the mole, to write the following relationships.

$$1 \text{ mol Na} = 23.0 \text{ g Na} = 6.02 \times 10^{23} \text{ Na atoms}$$

$$1 \text{ mol CO}_2 = 44.0 \text{ g CO}_2 = 6.02 \times 10^{23} \text{ CO}_2 \text{ molecules}$$

$$1 \text{ mol (NH}_4)_2\text{SO}_4 = 132.2 \text{ g (NH}_4)_2\text{SO}_4 = 6.02 \times 10^{23} \text{ (NH}_4)_2\text{SO}_4 \text{ formula units}$$

These relationships supply the conversion factors we need to convert between mass in grams and amount in moles, as illustrated in the following example. It is important to note that these conversion factors represent equalities, which means that we can invert them (see Example 5.5b) as needed in the equations so that the units cancel appropriately.

 EXAMPLE 5.5 Conversions Involving Moles and Mass

(a) What is the mass in grams of 0.400 mol CH_4? **(b)** Calculate the number of moles of Na in a 62.5 g piece of sodium metal.

Solution

a. The molecular mass of CH_4 is $(1 \times 12.0 \text{ amu}) + (4 \times 1.0 \text{ amu}) = 16.0$ amu. The molar mass of CH_4 is therefore 16.0 g/mol. Using this molar mass as a conversion factor, we have

$$? \text{ g CH}_4 = 0.400 \text{ mol CH}_4 \times \frac{16.0 \text{ g CH}_4}{1 \text{ mol CH}_4} = 6.4 \text{ g CH}_4$$

b. The molar mass of Na is 23.0 g/mol. To convert from mass to moles, we must use the *inverse* of the molar mass as a conversion factor (1 mol Na/23.0 g Na) to cancel units of mass. When we start with grams, we must have grams in the denominator of our conversion factor.

$$? \text{ mol Na} = 62.5 \text{ g Na} \times \frac{1 \text{ mol Na}}{23.0 \text{ g Na}} = 2.72 \text{ mol Na}$$

> **Exercise 5.5A**
Calculate the mass, in grams, of **(a)** 0.0728 mol carbon, **(b)** 55.5 mol H_2O, and **(c)** 0.0728 mol $Ca(HSO_4)_2$.

> **Exercise 5.5B**
Calculate the amount, in moles, of **(a)** 3.71 g Fe, **(b)** 165 g C_5H_{12}, and **(c)** 0.100 g $Mg(NO_3)_2$.

▲ **Figure 5.6** One mole of each of several familiar substances. From left to right on the table are sugar, salt, carbon, and copper. Behind them, the balloon contains helium. Each container holds Avogadro's number of elementary units of the substance. There are 6.02×10^{23} molecules of sugar, formula units of NaCl, and atoms of carbon, copper, and helium in the respective samples.

Q *How many Na^+ ions and how many Cl^- ions are there in the dish of salt? If the balloon were filled with oxygen gas (O_2), how many oxygen atoms would it contain?*

Figure 5.6 shows one-mole samples of several different substances. Each container holds Avogadro's number of formula units of the substance. That is, there are just as many sugar molecules in the largest dish as there are helium atoms in the balloon.

Mole and Mass Relationships in Chemical Equations

It is now time to put these important ideas together. We have learned the basics of balancing equations, and we have learned the importance of mole/mass conversions. Now, we can apply them in realistic situations.

Chemists, other scientists, and engineers are often confronted with questions such as "How much iron can be obtained from 1.0 ton of Fe_2O_3?" or "How much carbon dioxide is produced by burning 1000 g of propane?" There is no simple way to calculate the mass of one substance directly from the mass of another in the same reaction. Because chemical reactions involve atoms and molecules, quantities in reactions are calculated in moles of atoms, ions, or molecules. To do such calculations, it is necessary to convert grams of a substance to moles, as we did in Example 5.5. We also must write balanced equations to determine the relationship between moles.

We always start a problem of this type with our chemical equation. The balanced chemical equation is the recipe or guideline that we follow to solve the new and unknown equation. Recall that chemical equations provide ratios of atoms and molecules as well as ratios of moles. For example, the equation

$$C + O_2 \longrightarrow CO_2$$

tells us not only that one C atom reacts with one O_2 molecule (two atoms) to form one CO_2 molecule (one C atom and two O atoms). The equation also means that 1 mol (6.02×10^{23} atoms) of carbon reacts with 1 mol (6.02×10^{23} molecules) of oxygen to yield 1 mol (6.02×10^{23} molecules) of carbon dioxide. Because the molar mass in grams of a substance is numerically equal to the formula mass of the substance in atomic mass units, the equation can also tell us (indirectly) that 12.0 g (1 mol) C reacts with 32.0 g (1 mol) O_2 to yield 44.0 g (1 mol) CO_2 (Figure 5.7).

The crucial task is to maintain the *ratio* of masses. If the ratios are fixed, then we can always predict the outcomes. In the preceding reaction, the mass ratio of oxygen to carbon is 32.0:12.0, which could be simplified to 8.0:3.0 by reducing those values

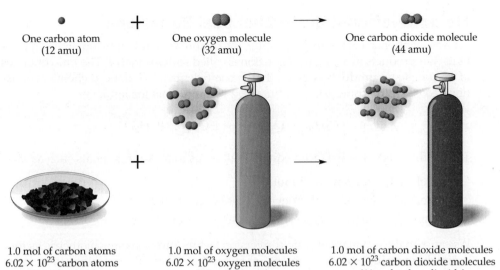

One carbon atom
(12 amu)

One oxygen molecule
(32 amu)

One carbon dioxide molecule
(44 amu)

▪ **Figure 5.7** We cannot weigh single atoms or molecules, but we can weigh equal numbers of these tiny particles.

1.0 mol of carbon atoms
6.02×10^{23} carbon atoms
(12 g of carbon)

1.0 mol of oxygen molecules
6.02×10^{23} oxygen molecules
(32 g of oxygen)

1.0 mol of carbon dioxide molecules
6.02×10^{23} carbon dioxide molecules
(44 g of carbon dioxide)

by a factor of 4. Likewise, the product would also be reduced by a factor of 4. Thus, 8.0 g O and 3.0 g C will produce 11.0 g CO_2. In fact, to calculate the amount of oxygen needed to react with a given amount of carbon, we need only multiply the amount of carbon by the factor 32.0:12.0. The following examples illustrate these relationships.

 CONCEPTUAL EXAMPLE 5.6 Molecular, Molar, and Mass Relationships

Nitrogen monoxide (nitric oxide), an air pollutant discharged by internal combustion engines, combines with oxygen to form nitrogen dioxide, a yellowish-brown gas that irritates our respiratory systems and eyes. The equation for this reaction is

$$2\,NO + O_2 \longrightarrow 2\,NO_2$$

State the molecular, molar, and mass relationships indicated by this equation.

Solution

The molecular and molar relationships can be obtained directly from the equation; no calculation is necessary. The mass relationship requires a little simple calculation.

Molecular: Two molecules of NO reacts with one molecule of O_2 to form two molecules of NO_2.

Molar: 2 mol NO reacts with 1 mol O_2 to form 2 mol NO_2.

Mass: 60.0 g NO (2 mol NO $\times$ 30.0 g/mol) reacts with 32.0 g O_2 (1 mol O_2 $\times$ 32.0 g/mol) to form 92.0 g NO_2 (2 mol NO_2 $\times$ 46.0 g/mol).

〉 Exercise 5.6A

Hydrogen sulfide, a gas that smells like rotten eggs, burns in air to produce sulfur dioxide and water according to the equation

$$2\,H_2S + 3\,O_2 \longrightarrow 2\,SO_2 + 2\,H_2O$$

State the molecular, molar, and mass relationships indicated by this equation.

〉 Exercise 5.6B

From the equation from Exercise 5.6A, which of the following are conserved? That is, for which of the following is the total of the reactants equal to the total of the products? **(a)** Number of atoms; **(b)** number of molecules; **(c)** number of moles; **(d)** mass in grams.

WHY IT MATTERS

If there is not enough oxygen gas to react completely with burning carbon, some of the carbon will be incompletely oxidized and deadly carbon monoxide will form. Over 400 deaths occur each year from unintentional carbon monoxide poisoning in the United States. Many of these are caused by improperly ventilated heat sources used indoors, including fireplaces and wood-burning stoves.

Molar Relationships in Chemical Equations

As was mentioned earlier in the chapter, the quantitative relationship between reactants and products in a chemical reaction is called **stoichiometry**. The ratio of moles of reactants and products is given by the coefficients in a balanced chemical equation. Consider the combustion of butane, a common fuel for lighters:

$$2\,C_4H_{10} + 13\,O_2 \longrightarrow 8\,CO_2 + 10\,H_2O$$

The coefficients in the balanced equation allow us to make statements such as

- 2 mol C_4H_{10} reacts with 13 mol O_2.
- 8 mol CO_2 are produced for every 2 mol C_4H_{10} that react.
- 10 mol H_2O are produced for every 8 mol CO_2 produced.

We can turn these statements into conversion factors known as stoichiometric factors. A **stoichiometric factor** is a conversion factor that relates the amounts, in *moles*, of any two substances involved in a chemical reaction. Note that stoichiometric factors, because they come from the coefficients in a balanced equation for a reaction, involve small whole numbers. In the examples that follow, stoichiometric factors are shown in color.

EXAMPLE 5.7 Molar Relationships

When 0.105 mol of propane is burned in a plentiful supply of oxygen, how many moles of oxygen are consumed?

$$C_3H_8 + 5\,O_2 \longrightarrow 3\,CO_2 + 4\,H_2O$$

Solution

The balanced equation tells us that 5 mol O_2 are required to burn 1 mol C_3H_8. From this relationship, we can construct conversion factors to relate moles of oxygen to moles of propane. The possible conversion factors are

$$\frac{1\ \text{mol}\ C_3H_8}{5\ \text{mol}\ O_2} \quad \text{and} \quad \frac{5\ \text{mol}\ O_2}{1\ \text{mol}\ C_3H_8}$$

Which one do we use? We're starting the problem with 0.105 mol C_3H_8, so we need the factor on the *right*, which will cancel "mol C_3H_8" and give the answer in the units requested (moles of oxygen gas).

$$?\ \text{mol}\ O_2 = 0.105\ \text{mol}\ C_3H_8 \times \frac{5\ \text{mol}\ O_2}{1\ \text{mol}\ C_3H_8} = 0.525\ \text{mol}\ O_2$$

❯ Exercise 5.7A

Consider the combustion (burning) of propane presented in Example 5.7. **(a)** How many moles of CO_2 are formed when 1.250 mol C_3H_8 is burned? **(b)** How many moles of CO_2 are produced when 0.059 mol O_2 is consumed?

❯ Exercise 5.7B

For the combustion of propane in Example 5.7, suppose that 25.0 mol O_2 and 12.0 mol propane are combined. After reaction, which reactant will be used up, and how many moles of the other reactant will be left over? (*Hint*: Calculate the number of moles of propane that would react with the given amount of oxygen, then the number of moles of oxygen that would react with the given amount of propane. Which reactant gets used up?)

Mass Relationships in Chemical Equations

A chemical equation defines the stoichiometric relationship in terms of moles, but because we cannot directly measure moles in the laboratory, problems are seldom presented in terms of moles. Typically, you are given the mass (in grams) of one substance and asked to calculate the mass of another substance that will react or can be produced. Such calculations involve several steps:

1. Write a balanced chemical equation for the reaction.
2. Determine the molar masses of the substances involved in the calculation.
3. Write down the given quantity and use the molar mass to convert this quantity to moles.
4. Use coefficients (stoichiometric relationship) from the balanced chemical equation to convert moles of the given substance to moles of the desired substance.
5. Use the molar mass to convert moles of the desired substance to grams of the desired substance.

If we know the quantity of any substance in an equation, we can determine the quantities of all the other substances. The conversion process is diagrammed in Figure 5.8. It is best learned by studying examples and working exercises.

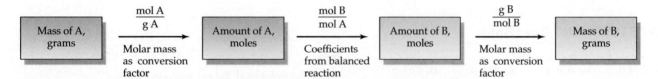

▲ **Figure 5.8** We can use a chemical equation to relate the number of *moles* of any two substances represented in the equation. These substances may be a reactant and a product, two reactants, or two products. We cannot directly relate the mass of one substance to the mass of another substance involved in the same reaction. To obtain *mass* relationships, we must convert the mass of each substance to moles, relate moles of one substance to moles of the other through a stoichiometric factor, and then convert moles of the second substance to its mass.

 EXAMPLE 5.8 Mass Relationships

Calculate the mass of oxygen needed to react with 5.00 g of carbon in the reaction that forms carbon dioxide.

Solution
Step 1. The balanced equation is

$$C + O_2 \longrightarrow CO_2$$

Step 2. The molar masses are 12.0 g/mol for C and $2 \times 16.0 = 32.0$ g/mol for O_2.

Step 3. We convert the given mass of carbon to an amount in moles.

$$5.0 \text{ g C} \times \frac{1 \text{ mol C}}{12.0 \text{ g C}} = 0.417 \text{ mol C}$$

Step 4. We use coefficients from the balanced equation to establish the stoichiometric factor (pink) that relates the amount of oxygen to that of carbon.

$$0.417 \text{ mol C} \times \frac{1 \text{ mol O}_2}{1 \text{ mol C}} = 0.417 \text{ mol O}_2$$

Step 5. We convert from moles of oxygen to grams of oxygen.

$$0.417 \text{ mol O}_2 \times \frac{32.0 \text{ g O}_2}{1 \text{ mol O}_2} = 13.3 \text{ g O}_2$$

1 How can a car make more pollution than the gasoline it uses?

A mole of carbon (12 g) combines with oxygen from the atmosphere to form a mole of carbon dioxide (44 g). Gasoline is mostly carbon by weight, so the mass of carbon dioxide formed is greater than the mass of gasoline burned. Carbon dioxide is a major greenhouse gas (Chapter 13).

We can also combine the five steps into a single setup. Note that the units in the denominators of the conversion factors are chosen so that each cancels the unit in the numerator of the preceding term.

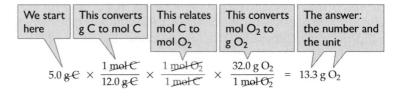

| We start here | This converts g C to mol C | This relates mol C to mol O_2 | This converts mol O_2 to g O_2 | The answer: the number and the unit |

$$5.0 \text{ g C} \times \frac{1 \text{ mol C}}{12.0 \text{ g C}} \times \frac{1 \text{ mol } O_2}{1 \text{ mol C}} \times \frac{32.0 \text{ g } O_2}{1 \text{ mol } O_2} = 13.3 \text{ g } O_2$$

> **Exercise 5.8A**

Calculate the mass of oxygen gas (O_2) needed to react with 1.505 g nitrogen gas (N_2) to form nitrogen dioxide. (*Hint*: Write the balanced equation first.)

> **Exercise 5.8B**

Calculate the mass of carbon dioxide formed by burning 355 g each of **(a)** methane (CH_4) and **(b)** pentane (C_5H_{12}) in oxygen gas. (*Hint*: The reaction in Example 5.7 may help in balancing the equations.)

SELF-ASSESSMENT Questions

1. The mass of 4.65 mol H_2SO_4 is
   a. 314 g   **b.** 456 g   **c.** 466 g   **d.** 564 g

2. For which of the following would a 10.0 g sample contain the most moles?
   a. Co   **b.** F   **c.** Se   **d.** P

3. How many moles of CO_2 are there in 1 lb. (453.6 g) CO_2?
   a. 0.0969 mol   **b.** 9.20 mol
   c. 10.3 mol   **d.** 14.2 mol

4. How many moles of benzene (C_6H_6) are there in a 7.80 g sample of benzene?
   a. 0.100 mol
   b. 1.00 mol
   c. 7.81 mol
   d. 610 mol

Items 5–7 refer to the equation $2 H_2 + O_2 \longrightarrow 2 H_2O$.

5. At the molar level, the equation means that
   a. 2 mol H reacts with 1 mol O to form 2 mol H_2O
   b. 4 mol H reacts with 2 mol O to form 4 mol H_2O
   c. 2 mol H_2 reacts with 1 mol O_2 to form 2 mol H_2O
   d. 2 mol O_2 reacts with 1 mol H_2 to form 2 mol H_2O

6. How many moles of water can be formed from 0.500 mol H_2?
   a. 0.500 mol   **b.** 1.00 mol   **c.** 2.00 mol   **d.** 4.00 mol

7. How many moles of oxygen (O_2) are required to form 0.222 mol H_2O?
   a. 0.111 mol
   b. 0.222 mol
   c. 0.444 mol
   d. 1.00 mol

Answers: 1, b; 2, b; 3, c; 4, a; 5, c; 6, c; 7, a

5.5 Solutions

Learning Objectives • Calculate the concentration (molarity, percent by volume, or percent by mass) of a solute in a solution. • Calculate the amount of solute or solution given the concentration and the other amount.

Many chemical reactions take place in solutions because the chemical components are more readily exchanged. A **solution** is a homogeneous mixture of two or more substances. The substance being dissolved is the *solute*, and the substance doing the dissolving is the *solvent*. The solvent is usually present in greater quantity than the solute. There are many solvents: Kerosene dissolves grease; ethanol dissolves many drugs; isopentyl acetate, a component of banana oil, is a solvent for model airplane glue.

Water is no doubt the most familiar solvent, dissolving as it does many common substances, such as sugar, salt, and ethanol. We focus our discussion here on **aqueous solutions (aq)**, those in which water is the solvent, and we take a more quantitative look at the relationship between solute and solvent.

Solution Concentrations

We say that some substances, such as sugar and salt, are *soluble* in water. Of course, there is a limit to the quantity of sugar or salt that we can dissolve in a given volume of water. Yet we still find it convenient to say that they are soluble in water because an appreciable quantity dissolves. Other substances, such as a copper penny or glass (silicon dioxide), we consider to be *insoluble* because their solubility in water is near zero. Such terms as *soluble* and *insoluble* are useful, but they are imprecise and must be used with care.

Two other roughly estimated but sometimes useful terms are *dilute* and *concentrated*. A **dilute solution** is one that contains a relatively small amount of solute in lots of solvent. For example, a pinch of sugar in a cup of tea makes a dilute, faintly sweet sugar solution. A **concentrated solution** has a relatively large amount of solute dissolved in a small amount of solvent. Pancake syrup, with lots of sugar solute in a relatively small amount of water, is a concentrated, and very sweet, solution.

Scientific work generally requires more precise concentration units than "a pinch of sugar in a cup of tea." Further, quantitative work often requires the *mole* unit because substances enter into chemical reactions according to *molar* ratios.

Molarity

A concentration unit that chemists often use is *molarity*. For reactions involving solutions, the amount of solute is usually measured in moles and the quantity of solution in liters or milliliters. The **molarity (M)** is the amount of solute, in moles, per liter of solution.

$$\text{Molarity (M)} = \frac{\text{moles of solute}}{\text{liters of solution}}$$

For example, we read 1.75 M NaCl as "1.75 molar sodium chloride." Be careful not to confuse molarity with moles. The molarity equation *includes* moles, but that does not mean that molarity *equals* moles, as can be seen in Example 5.9.

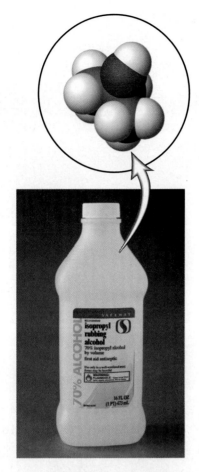

▲ Rubbing alcohol is a water–isopropyl alcohol solution that is usually 70% isopropyl alcohol [$(CH_3)_2 CHOH$] by volume.

EXAMPLE 5.9 Solution Concentration: Molarity and Moles

Calculate the molarity of a solution made by dissolving 3.50 mol NaCl in enough water to produce 2.00 L of solution.

Solution

$$\text{Molarity (M)} = \frac{\text{moles of solute}}{\text{liters of solution}} = \frac{3.50 \text{ mol NaCl}}{2.00 \text{ L solution}} = 1.75 \text{ M NaCl}$$

> Exercise 5.9A
Calculate the molarity of a solution that has 0.0400 mol NH_3 in 5.95 L of solution.

> Exercise 5.9B
Calculate the molarity of a solution made by dissolving 0.850 mol H_2SO_4 in enough water to produce 775 mL of solution.

Usually, when preparing a solution, we must *weigh* the solute. Balances do not display units of moles, so we usually work with a given mass and divide by the molar mass of the substance, as illustrated in Example 5.10.

EXAMPLE 5.10   Solution Concentration: Molarity and Mass

What is the molarity of a solution in which 333 g potassium hydrogen carbonate is dissolved in enough water to make 10.0 L of solution?

Solution

First, we must convert grams of $KHCO_3$ to moles of $KHCO_3$.

$$333 \text{ g KHCO}_3 \times \frac{1 \text{ mol KHCO}_3}{100.1 \text{ g KHCO}_3} = 3.33 \text{ mol KHCO}_3$$

Next, we use this value as the numerator in the defining equation for molarity. The solution volume, 10.0 L, is the denominator.

$$\text{Molarity} = \frac{3.33 \text{ mol KHCO}_3}{10.0 \text{ L solution}} = 0.333 \text{ M KHCO}_3$$

> **Exercise 5.10A**
Calculate the molarity of **(a)** 18.0 g H_3PO_4 in 2.00 L of solution and **(b)** 3.00 g KCl in 2.39 L of solution.

> **Exercise 5.10B**
Calculate the molarity of 0.309 g HF in 1752 mL of solution. HF is used for etching glass.

Often, we need to know the *mass* of solute required to prepare a given volume of a solution of a particular molarity. In such calculations, we can use molarity as a conversion factor between moles of solute and liters of solution. For instance, in Example 5.11, the expression "0.15 M NaCl" means 0.15 mol NaCl per liter of solution, expressed as the conversion factor

$$\frac{0.15 \text{ mol NaCl}}{1 \text{ L solution}}$$

 EXAMPLE 5.11   Solution Preparation: Molarity

What mass in grams of NaCl is required to prepare 0.500 L of typical over-the-counter saline solution (about 0.15 M NaCl)?

Solution

First, we use the molarity as a conversion factor to calculate moles of NaCl.

$$0.500 \text{ L solution} \times \frac{0.15 \text{ mol NaCl}}{1 \text{ L solution}} = 0.075 \text{ mol NaCl}$$

Then, we use the molar mass to calculate grams of NaCl.

$$0.075 \text{ mol NaCl} \times \frac{58.5 \text{ g NaCl}}{1 \text{ mol NaCl}} = 4.4 \text{ g NaCl}$$

> **Exercise 5.11A**
What mass in grams of potassium hydroxide is required to prepare **(a)** 1.500 L of 4.00 M KOH and **(b)** 200.0 mL of 1.00 M KOH?

> **Exercise 5.11B**
What mass in kilograms of sodium hydroxide is required to prepare 12.5 L of 19.1 M NaOH?

WHY IT MATTERS

The ethanol content of a beverage is almost always expressed by *volume*. (See Example 5.13A.) This may be done partly as a marketing ploy. Ethanol is about 20% less dense than water, so a beverage that is labeled "12% alcohol by volume" is only about 10% ethanol by mass. The higher percent by volume may make purchasers think they are getting "more for the money."

Quite often, solutions of known molarity are available commercially. For example, concentrated hydrochloric acid is 12 M. How would you determine the *volume* of this solution that contains a certain number of moles of solute? Since 12 M HCl means that 12 moles of HCl is equivalent to 1 L of solution, we can write *two* conversion factors:

$$\frac{12 \text{ mol HCl}}{1 \text{ L solution}} \quad or \quad \frac{1 \text{ L solution}}{12 \text{ mol HCl}}$$

The second conversion factor will take us from mol HCl to L HCl.

 EXAMPLE 5.12   Moles from Molarity and Volume

Concentrated hydrochloric acid is 12.0 M HCl. What volume in milliliters of this solution contains 0.425 mol HCl?

Solution

$$0.425 \text{ mol HCl} \times \frac{1 \text{ L solution}}{12 \text{ mol HCl}} = 0.0354 \text{ L solution}$$

In this case, 0.0354 L (35.4 mL) of the solution contains 0.425 mol of the solute. Remember that molarity is moles per liter of *solution*, not per liter of solvent.

⟩ Exercise 5.12A
What volume in milliliters of 18.0 M acetic acid ($HC_2H_3O_2$) solution contains 0.445 mol acetic acid?

⟩ Exercise 5.12B
What mass in grams of HNO_3 is in 500 mL of acid rain that has a concentration of 2.00×10^{-5} M HNO_3?

Percent Concentrations

For many practical applications, including those in medicine and pharmacy, solution concentrations are often expressed in percentages. There are different ways to express percentages, depending on whether mass or volume is being measured. If both solute and solvent are liquids, **percent by volume** is often used because liquid volumes are easily measured.

$$\text{Percent by volume} = \frac{\text{volume of solute}}{\text{volume of solution}} \times 100\%$$

In calculating percent by volume, we can express the solute and solution in any unit of volume as long as we use the same unit for both. For example, we could express both volumes in milliliters or cubic feet or gallons. Ethanol (CH_3CH_2OH) for medicinal purposes is 95% by volume. That is, it consists of 95 mL CH_3CH_2OH per 100 mL of aqueous solution.

 EXAMPLE 5.13   Percent by Volume

Two-stroke engines, commonly used in such equipment as chainsaws and lawnmowers, use a mixture of 120 mL of oil dissolved in enough gasoline to make 4.0 L of fuel. What is the percent by volume of oil in this mixture?

Solution

$$\text{Percent by volume} = \frac{120 \text{ mL oil}}{4000 \text{ mL solution}} \times 100\% = 3.0\%$$

2 What's the difference between .925 silver and .800 silver?

The purity of silver is expressed as a decimal, with "1" representing pure silver. Greater than .925 silver is often referred to as *sterling* silver. Sterling silver has at least .925 g of silver atoms for every 1.000 g of the alloy; .800 silver has 0.800 g of silver for every 1.000 g of the alloy. Items made of silver alloys are different from silver-plated items.

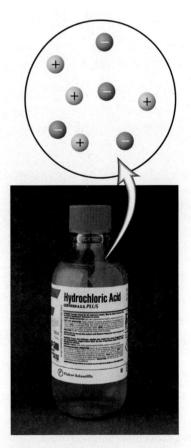

▲ "Reagent grade" concentrated hydrochloric acid is a 38% by mass solution of HCl (H^+ and Cl^- ions) in water.

3 Does 2% milk have just 2% of the calories of whole milk?

No. By law, whole milk must contain at least 3.25% butterfat by mass. Some of the butterfat is removed to make 2% milk, which is 2% butterfat by mass. Since the sugars (mostly lactose) and proteins in milk also provide energy, 2% milk has about 75% of the calories of whole milk. In contrast, skim milk has a maximum of 0.2% fat and 60% of the calories of whole milk. All three types of milk have the same calcium and protein content.

WHY IT MATTERS

An isotonic, or "normal," intravenous solution must have the proper concentration of solute to avoid damage to blood cells. High concentrations cause blood cells to shrivel (*crenation*) as water is drawn out of them by osmosis. Low concentrations cause the cells to swell and even burst (hemolysis).

> **Exercise 5.13A**

What is the volume percent of ethanol in 845 mL of an ethanol–water solution that contains 90.0 mL of water?

> **Exercise 5.13B**

Determine the volume percent of toluene ($C_6H_5CH_3$) in a solution made by mixing 35.5 mL of toluene with 80.0 mL of benzene (C_6H_6). Assume that the volumes are additive.[1]

Most commercial solutions are labeled with the concentration in **percent by mass**. For example, sulfuric acid is sold in several concentrations: 35.7% H_2SO_4 for use in storage batteries, 77.7% H_2SO_4 for the manufacture of phosphate fertilizers, and 93.2% H_2SO_4 for pickling steel. Each of these percentages is by mass: 35.7 g of H_2SO_4 per 100 g of sulfuric acid solution, and so on.

$$\text{Percent by mass} = \frac{\text{mass of solute}}{\text{mass of solution}} \times 100\%$$

As with percent by volume, we can express the masses of solute and solution in any mass unit, as long as we use the same unit for both.

 EXAMPLE 5.14 Percent by Mass

What is the percent by mass of NaCl if 25.5 g of it is dissolved in 425 g of water?

Solution

We use these values in the above percent-by-mass equation. Keep in mind that the solution mass is the sum of the masses of NaCl and water.

$$\text{Percent by mass} = \frac{25.5 \text{ g NaCl}}{(25.5 + 425) \text{ g solution}} \times 100\% = 5.66\% \text{ NaCl}$$

> **Exercise 5.14A**

Hydrogen peroxide solution for home use is usually 3.0% by mass of H_2O_2 in water. What is the percent by mass of a solution of 8.95 g H_2O_2 dissolved in 425 g of water?

> **Exercise 5.14B**

Sodium hydroxide (NaOH, lye) is used to make soap and is extremely soluble in water. What is the percent by mass of NaOH in a solution that contains 1.00 kg NaOH in 1025 g water?

EXAMPLE 5.15 Solution Preparation: Percent by Mass

Describe how to make 430 g of an aqueous solution that is 4.85% by mass $NaNO_3$.

Solution

We begin by rearranging the equation for percent by mass to solve for mass of solute.

$$\text{Mass of solute} = \frac{\text{percent by mass} \times \text{mass of solution}}{100\%}$$

---

[1]Masses are always additive. That is, adding 50 g of salt to 50 g sugar gives 100 g of the mixture. However, *volumes of liquids* are *not* always additive. For example, carefully measure 50 mL each of methanol (wood alcohol, found near paints in home improvement stores) and of water, and mix them together; the solution volume is *less than* 100 mL. This is because of differences in intermolecular forces (Section 6.3).

Substituting, we have

$$\frac{4.85\% \times 430 \text{ g}}{100\%} = 20.9 \text{ g}$$

We would weigh out 20.9 g NaNO₃ and add enough water to make 430 g of solution.

> **Exercise 5.15A**

Describe how you would prepare 250 g of an aqueous solution that is 4.50% glucose by mass.

> **Exercise 5.15B**

Describe how you would prepare 1.25 kg of *isotonic saline*, a commonly used intravenous (IV) solution that is 0.89% sodium chloride by mass.

Note that for percent concentrations, the mass or volume of the solute needed doesn't depend on what the solute is. A 10% by mass solution of NaOH contains 10 g NaOH per 100 g of solution. Similarly, 10% HCl, 10% $(NH_4)_2SO_4$, and 10% $C_{110}H_{190}N_3O_2Br$ each contain 10 g of the specified solute per 100 g of solution. For *molar* solutions, however, the mass of solute in a solution of specified molarity is different for different solutes. A liter of 0.10 M solution requires 4.0 g (0.10 mol) NaOH, 3.7 g (0.10 mol) HCl, 13.2 g (0.10 mol) $(NH_4)_2SO_4$, or 166 g (0.10 mol) $C_{110}H_{190}N_3O_2Br$.

4 **What does a toxicity rating (LD_{50}) of 100 ppm mean?**

The LD_{50} designation means the median lethal dose—the dosage of a substance that would result in fatality for 50% of the organisms tested. A 100 ppm or parts per million measurement means that there would only have to be 100 milligrams of the toxic substance per 1 kilogram of an organism's body weight.

SELF-ASSESSMENT Questions

1. In a solution of 85.0 g water and 5.0 g sucrose ($C_{12}H_{22}O_{11}$), the
   a. solute is sucrose and the solution is the solvent
   b. solute is sucrose and the solvent is water
   c. solute is water and sucrose is the solvent
   d. water and sucrose are both solutes

2. What is the molarity of a sodium hydroxide solution having 0.500 mol NaOH in 62.5 mL of solution?
   a. 0.00200 M   b. 0.200 M
   c. 2.00 M   d. 8.00 M

3. How many moles of sugar are required to make 2.00 L of 0.600 M sugar solution?
   a. 1.20 mol   b. 2.40 mol
   c. 24.0 mol   d. 6.67 mol

4. What mass of NaCl in grams is needed to prepare 2.500 L of a 0.800 M NaCl solution?
   a. 2.00 g   b. 4.00 g
   c. 58.5 g   d. 117 g

5. What volume of 6.00 M HCl solution contains 1.80 mol HCl?
   a. 250 mL   b. 300 mL
   c. 900 mL   d. 9.00 L

6. What is the percent by volume of ethanol in a solution made by adding 50.0 mL of pure ethanol to enough water to make 250 mL of solution?
   a. 16.7%   b. 20.0%
   c. 25.0%   d. 80.0%

7. A solution contains 15.0 g sucrose and 60.0 g of water. What is the percent by mass of sucrose in the solution?
   a. 15.0%   b. 20.0%
   c. 25.0%   d. 80.0%

8. To make 100.0 mL of 4.00 M NaCl solution, you would take 23.4 g NaCl and
   a. add 76.6 mL of water
   b. add 76.6 g of water
   c. add 100.0 mL of water
   d. add water until the solution volume reached 100.0 mL

Answers: 1. b; 2. d; 3. a; 4. d; 5. b; 6. b; 7. b; 8. d

GREEN CHEMISTRY Atom Economy
Margaret Kerr, Worcester State University

Principles 1, 2

Learning Objectives • Explain how the concept of atom economy can be applied to pollution prevention and environmental protection. • Calculate the atom economy for chemical reactions.

Imagine yourself in the future. Your job is related to environmental protection, which requires that you provide information to practicing chemists as they design new processes and reactions. Waste management is one of your top concerns. Although waste typically has been addressed only after production of desired commodities, you realize that a greener approach can mean minimizing the waste from the start. What methods would you use in this job? What topics are important? How can chemists provide new products while protecting the environment?

Intrinsic in greener approaches to waste management is the concept of *atom economy*, a calculation of the number of atoms conserved in the desired product rather than gone in waste. Atom Economy is an example of a "green chemistry metric," which helps to determine the "greenness" (or lack thereof) of the reaction of interest. In 1998, Barry Trost of Stanford University won a Presidential Green Chemistry Challenge Award for his work in developing this concept. By calculating the number of atoms that will not become part of the desired product and, therefore, will enter the waste stream, chemists can precisely determine the minimum amount of waste that will be produced by chemicals used in a reaction before even running the reaction.

You have learned how to write and balance chemical equations (Section 5.1), and you also can calculate molar mass, convert from mass to moles, and determine the amount of product formed from given amounts of reactants (Section 5.3 and Section 5.4). Other ways of calculating the efficiency of a reaction have traditionally been expressed using *theoretical yield, experimental yield,* and *percent yield.* The theoretical yield is the maximum amount of desired product that can be produced by the amounts of reactants used. Under laboratory conditions, the theoretical yield of a reaction is not usually obtained. The actual amount of desired product collected is called the experimental yield. Chemists often express their experimental results in percent yield, which is simply the experimental yield divided by the theoretical yield, then multiplied by 100%.

$$\% \text{ yield} = \frac{\text{experimental yield}}{\text{theoretical yield}} \times 100\%$$

A report of a 100% yield for a reaction seems good, but the percent yield considers only the desired product. Any other substance formed (for instance, a by-product) is not counted in the percent yield. By-products often become part of the chemical waste stream. Some reactions produce more by-products than desired products.

Atom economy is a better way to measure reaction efficiency. Percent atom economy (% A.E.) measures the proportion of the atoms in the reaction mixture that actually become part of the desired product. Reactant atoms that do not appear in the product are considered waste. The % A.E. given by the following equation is the theoretical maximum atom economy for a 100% yield of the desired product.

$$\% \text{ A.E.} = \frac{\text{molar mass of desired product}}{\text{molar masses of all reactants}} \times 100\%$$

A reaction can have a poor atom economy even when the percent yield is near 100%.

Consider the following two ways to make butene (C_4H_8), a compound that is an important chemical feedstock in the plastics industry. First, butene can be made using butylbromide (C_4H_9Br) and sodium hydroxide $(NaOH)$.

$$C_4H_9Br + NaOH \longrightarrow C_4H_8 + H_2O + NaBr$$

In this reaction, a Br atom and a H atom are eliminated from the butylbromide to form the final product. Generally, in reactions like this one, only one product is desired and all other products are not used.

We can calculate the atom economy for this reaction.

$$\% \text{ A.E.} = \frac{\text{molar mass } C_4H_8}{\text{molar mass } C_4H_9Br + \text{molar mass } NaOH \times 100\%}$$

$$\% \text{ A.E.} = \frac{56.11 \text{ g/mol}}{137.02 \text{ g/mol} + 40.00 \text{ g/mol}} \times 100\% = 31.7\%$$

These elimination reactions never have 100% atom economy, because there is always a by-product. This results in some atoms going into the waste stream.

Butene can also be made using butyne (C_4H_6) and hydrogen gas (H_2):

$$C_4H_6 + H_2 \longrightarrow C_4H_8$$

In this reaction, all of the atoms in both reactants appear in the desired product and none go into the waste stream. The percent atom economy is 100%.

$$\% \text{ A.E.} = \frac{\text{molar mass } C_4H_8}{\text{molar mass } C_4H_6 + \text{molar mass } H_2} = 100\%$$

$$\% \text{ A.E.} = \frac{56.11 \text{ g/mol}}{54.09 \text{ g/mol} + 2.02 \text{ g/mol}} \times 100\% = 100\%$$

Because the concept of atom economy offers a way to predict waste, it can lead to innovative design processes that eliminate or minimize waste. By focusing on waste from the beginning of the reaction, we reduce or eliminate cleanup at the end.

Summary

Section 5.1—A **chemical equation** is shorthand for a chemical change, using symbols and formulas instead of words: for example, $C + O_2 \longrightarrow CO_2$. Substances to the left of the arrow are starting materials or **reactants**; those to the right of the arrow are **products** or what the reaction produces. Physical states may be indicated with (s), (l), or (g). A chemical equation must be balanced, with the same numbers and types of atoms on each side. Equations are balanced by placing coefficients in front of the formula for each reactant or product. The balanced equation $2\,NaN_3 \longrightarrow 2\,Na + 3\,N_2$ means that two formula units of NaN_3 react to give two atoms of Na and three molecules of N_2.

Section 5.2—Gay-Lussac's experiments led him to the **law of combining volumes**: At a given temperature and pressure, the volumes of gaseous reactants and products are in small whole-number ratios (such as 1:1, 2:1, or 4:3). **Avogadro's law** states that equal volumes of all gases contain the same number of molecules at fixed temperature and pressure.

Section 5.3—**Avogadro's number** is defined as the number of ^{12}C atoms in exactly 12 g of ^{12}C—that is, 6.02×10^{23} atoms. A **mole (mol)** is the amount of a substance that contains Avogadro's number of elementary units—atoms, molecules, or formula units, depending on the substance. The **formula mass** is the sum of masses of the atoms represented in the formula of a substance. If the formula represents a molecule, the term **molecular mass** often is used. A formula represents a mass ratio as well as an atom ratio, and the composition by mass of a compound may be calculated from the formula.

Section 5.4—The **molar mass** is the mass of one mole of a substance. The numeric part of molar mass is the same as for formula mass, molecular mass, or atomic mass—but the units of molar mass are grams per mole (g/mol). It is used to convert from grams to moles of a substance, and vice versa. A balanced equation can be read in terms of atoms and molecules or in terms of moles. **Stoichiometry**, or mass

relationships in chemical reactions, can be found by using **stoichiometric factors**, which relate moles of one substance to moles of another. Stoichiometric factors are obtained directly from the balanced equation. To evaluate mass relationships in chemical reactions, molar masses and stoichiometric factors are both used as conversion factors.

Section 5.5—A **solution** is a homogeneous mixture of two or more substances, often consisting of a *solute* dissolved in a *solvent*. A substance is soluble in a solvent if some appreciable quantity of the substance dissolves in the solvent. Any substance that does not dissolve significantly in a solvent is insoluble in that solvent. An **aqueous solution (aq)** has water as the solvent. A **concentrated solution** has a relatively large amount of solute compared with solvent, and a **dilute solution** has only a little solute in a large amount of solvent. One quantitative unit of concentration is **molarity (M)**—moles of solute dissolved per liter of solution.

$$\text{Molarity (M)} = \frac{\text{moles of solute}}{\text{liters of solution}}$$

There are several ways to express concentration using percentages. Two important ones are **percent by volume** and **percent by mass**.

$$\text{Percent by volume} = \frac{\text{volume of solute}}{\text{volume of solution}} \times 100\%$$

$$\text{Percent by mass} = \frac{\text{mass of solute}}{\text{mass of solution}} \times 100\%$$

For each concentration unit, if we know two of the three terms in the equation, we can solve for the third.

GREEN CHEMISTRY—Atom economy is an important component of green chemistry and waste prevention. By strategically designing reactions to maximize the number of atoms used, a chemist has the ability to reduce waste production.

Learning Objectives

	Associated Problems
• Identify balanced and unbalanced chemical equations, and balance equations by inspection. (5.1)	4, 11–16, 51–53, 62
• Determine volumes of gases that react, using a balanced equation for a reaction. (5.2)	17–22, 55
• Calculate the formula mass, molecular mass, or molar mass of a substance. (5.3)	2, 3, 27–34
• Use Avogadro's number to determine the number of particles of different types in a mass of a substance. (5.3)	23–26
• Convert from mass to moles and from moles to mass of a substance. (5.4)	29–34, 71
• Calculate the mass or number of moles of a reactant or product from the mass or number of moles of another reactant or product. (5.4)	35–38, 54, 56–58, 62, 69
• Calculate the concentration (molarity, percent by volume, or percent by mass) of a solute in a solution. (5.5)	39, 40, 45, 46, 60, 61
• Calculate the amount of solute or solution given the concentration and the other amount. (5.5)	41–44, 47–50, 59, 64, 65
• Explain how the concept of atom economy can be applied to pollution prevention and environmental protection.	75
• Calculate the atom economy for chemical reactions.	73–77

Conceptual Questions

1. Define or illustrate each of the following.
   a. formula unit
   b. formula mass
   c. mole
   d. Avogadro's number
   e. molar mass
   f. molar volume (of a gas)

2. Explain the difference between the atomic mass of chlorine and the formula mass of chlorine gas.

3. Which best represents the amount of a mole of a typical substance like sugar or salt: a pinch (like a pinch of salt), a handful, a bucket full, or a truckload?

4. Referring to the law of conservation of mass, explain why we must work with balanced chemical equations.

5. Define or explain and illustrate the following terms.
   a. solution
   b. solvent
   c. solute
   d. aqueous solution

6. Define or explain and illustrate the following terms.
   a. concentrated solution
   b. dilute solution
   c. soluble
   d. insoluble

 Problems

Interpreting Formulas

7. How many oxygen atoms does each of the following contain?
   a. $Al(H_2PO_4)_3$
   b. $HOC_6H_4COOCH_3$
   c. $(BiO)_2SO_4$

8. How many carbon atoms does each of the following contain?
   a. $Fe_2(C_2O_4)_3$
   b. $Al(CH_3COO)_3$
   c. $(CH_3)_3CCH(CH_3)_2$

9. How many atoms of each element (N, P, H, and O) does the notation $3\ (NH_4)_2HPO_4$ indicate?

10. How many atoms of each element (Fe, C, H, and O) does the notation $4\ Fe(HOOCCH_2COO)_3$ indicate?

Interpreting Chemical Equations

11. Consider the following equation. **(a)** Explain its meaning at the molecular level. **(b)** Interpret it in terms of moles. **(c)** State the mass relationships conveyed by the equation.

$$2\ H_2O_2 \longrightarrow 2\ H_2O + O_2$$

12. Express each chemical equation in terms of moles.
    a. $2\ Mg + O_2 \longrightarrow 2\ MgO$
    b. $2\ C_2H_6 + 7\ O_2 \longrightarrow 4\ CO_2 + 6\ H_2O$

Balancing Chemical Equations

13. Balance the following equations.
    a. $Li + O_2 \longrightarrow Li_2O$
    b. $Mg + Co_2O_3 \longrightarrow MgO + Co$
    c. $Zr + H_2S \longrightarrow ZrS_2 + H_2$

14. Balance the following equations.
    a. $K + O_2 \longrightarrow K_2O_2$
    b. $FeCl_2 + Na_2SiO_3 \longrightarrow NaCl + FeSiO_3$
    c. $F_2 + AlCl_3 \longrightarrow AlF_3 + Cl_2$

15. Write balanced equations for the following processes.
    a. Nitrogen gas and oxygen gas react to form dinitrogen tetroxide (N_2O_4).
    b. Ozone (O_3) reacts with carbon to form tricarbon dioxide, C_3O_2.
    c. Uranium(VI) oxide reacts with hydrogen fluoride (HF) to form uranium(VI) fluoride and water.

16. Write balanced equations for the following processes.
    a. Aluminum metal reacts with oxygen gas to form aluminum oxide.
    b. Calcium carbonate (the active ingredient in many antacids) reacts with stomach acid (HCl) to form calcium chloride, water, and carbon dioxide.
    c. Hexane (C_6H_{14}) burns in oxygen gas to form carbon dioxide and water.

Volume Relationships in Chemical Equations

17. Three balloons are filled with equal volumes of carbon dioxide, chlorine gas, and nitrogen gas, respectively. **(a)** Which one contains the most molecules? **(b)** Which one has the greatest mass?

18. Three balloons are filled with equal masses of the noble gases neon, krypton, and argon, respectively. Which balloon is largest?

19. Consider the following equation, which represents the combustion (burning) of pentane (C_5H_{12}).

$$C_5H_{12}(g) + 8\ O_2(g) \longrightarrow 5\ CO_2(g) + 6\ H_2O(g)$$

    a. What volume, in liters, of carbon dioxide is formed when 20.6 L of pentane vapor is burned? Assume that both gases are measured under the same conditions.
    b. What volume, in milliliters, of pentane vapor must burn to react with 58.4 mL of oxygen gas? Assume that both gases are measured under the same conditions.

20. Consider the following equation, which represents the combustion of ammonia.

$$4\ NH_3(g) + 3\ O_2(g) \longrightarrow 2\ N_2(g) + 6\ H_2O(g)$$

    a. What volume, in liters, of nitrogen gas is formed when 3.58×10^4 L of ammonia is burned? Assume that both gases are measured under the same conditions.
    b. What volume, in liters, of oxygen gas is required to form 6.42 L of water vapor? Assume that both gases are measured under the same conditions.

21. Using the equation in Problem 19, determine the ratio of volume of carbon dioxide formed to volume of pentane vapor that reacts, assuming that both gases are measured under the same conditions.

22. Using the equation in Problem 20, determine the ratio of volume of water vapor formed to volume of ammonia that reacts, assuming that both gases are measured under the same conditions.

Avogadro's Number

23. How many **(a)** phosphorus *molecules* and how many **(b)** phosphorus *atoms* are there in 1.00 mol of white phosphorus, P_4?

24. How many **(a)** aluminum ions and how many **(b)** sulfide ions are there in 2.00 mol Al_2S_3?

25. Choose one of the following to complete this statement correctly: One *mole* of bromine (Br_2) gas _____.
  a. has a mass of 79.9 g.
  b. contains 6.02×10^{23} Br atoms.
  c. contains 12.04×10^{23} Br atoms.
  d. has a mass of 6.02×10^{23} g.

26. **(a)** How many *moles* of barium ions and how many *moles* of nitrate ions are there in 7.00 mol $Ba(NO_3)_2$? **(b)** How many *moles* of nitrogen atoms and how many *moles* of oxygen atoms are there in 7.00 mol $Ba(NO_3)_2$?

Formula Masses and Molar Masses

You may round atomic masses to one decimal place.

27. Calculate the molar mass of each of the following compounds.
  a. $CuSO_4$ **b.** $Sr(ClO_4)_2$
  c. $Cd(BrO_3)_2$ **d.** $(CH_3)_2CHCH_2OH$

28. Calculate the molar mass of each of the following compounds.
  a. Na_3PO_4 **b.** $(NH_4)_2SO_3$
  c. $Cr_2(Cr_2O_7)_3$ **d.** $(CH_3)_3CCOCH(C_2H_5)_2$

29. Calculate the mass, in grams, of each of the following.
  a. 3.15 mol $AgNO_3$ **b.** 0.0901 mol $CaCl_2$
  c. 11.86 mol H_2S

30. Calculate the mass, in kilograms, of each of the following.
  a. 194 mol $TiCl_4$ **b.** 2.25×10^7 mol Fe_3O_4
  c. 24.8 mol $Fe_3[Fe(CN)_6]_2$

31. Calculate the amount, in moles, of each of the following.
  a. 77.3 g Sb_2S_3 **b.** 321 g MoO_3
  c. 908 g $AlPO_4$

32. Calculate the amount, in moles, of each of the following.
  a. 16.3 g SF_6 **b.** 25.4 g $Pb(C_2H_3O_2)_2$
  c. 15.6 g $SrCO_3$

33. Calculate the percent by mass of N in **(a)** $NaNO_3$ and **(b)** NH_4Cl.

34. Calculate the percent by mass of C, H, and O in glucose ($C_6H_{12}O_6$).

Mole and Mass Relationships in Chemical Equations

35. Consider the reaction for the combustion of isopropanol (C_2H_5OH, rubbing alcohol):

$$2\,C_3H_7OH + 9\,O_2 \longrightarrow 6\,CO_2 + 8\,H_2O$$

  a. How many moles of CO_2 are produced when 0.845 moles of isopropanol are burned?
  b. How many moles of oxygen gas are required to burn 2.54 moles of isopropanol?

36. Consider the reaction of potassium superoxide, used in some breathing apparatus, with exhaled carbon dioxide to produce breathing oxygen:

$$4\,KO_2 + 2\,CO_2 \longrightarrow 2\,K_2CO_3 + 3\,O_2$$

  a. How many moles of oxygen gas are produced when 2.81 moles of KO_2 react?
  b. How many moles of CO_2 are consumed when 8.12 moles of oxygen gas are produced?

37. What mass in grams **(a)** of ammonia can be made from 250 g of H_2 and **(b)** of hydrogen is needed to react completely with 923 g N_2?

$$N_2 + H_2 \longrightarrow NH_3 \quad \text{(not balanced)}$$

38. Toluene (C_7H_8) and nitric acid (HNO_3) are used in the production of trinitrotoluene (TNT, $C_7H_5N_3O_6$), an explosive used for excavation and demolition.

$$C_7H_8 + HNO_3 \longrightarrow C_7H_5N_3O_6 + H_2O \quad \text{(not balanced)}$$

What mass in grams **(a)** of nitric acid is required to react with 454 g and **(b)** of TNT can be made from 829 g C_7H_8?

Molarity of Solutions

39. Calculate the molarity of each of the following solutions.
  a. 9.66 mol $LiNO_3$ in 8.83 L of solution
  b. 0.575 mol $FeCl_3$ in 955 mL of solution

40. Calculate the molarity of each of the following solutions.
  a. 8.82 mol H_2SO_4 in 7.50 L of solution
  b. 1.22 mol C_2H_5OH in 96.3 mL of solution

41. What mass in grams of solute is needed to prepare **(a)** 2.25 L of 0.288 M HCl and **(b)** 175 mL of 0.375 M K_2CrO_4?

42. What mass in grams of solute is needed to prepare **(a)** 0.250 L of 0.167 M $K_2Cr_2O_7$ and **(b)** 625 mL of 0.0200 M $KMnO_4$?

43. What volume in liters of **(a)** 6.00 M NaOH contains 2.50 mol NaOH and **(b)** 0.0500 M KH_2AsO_4 contains 8.10 g KH_2AsO_4?

44. What volume in liters of **(a)** 0.250 M NaOH contains 2.50 M NaOH and **(b)** 4.25 M $H_2C_2O_4$ contains 0.225 g $H_2C_2O_4$?

Percent Concentrations of Solutions

45. What is the percent by volume concentration of **(a)** 18.9 mL of hydrogen peroxide in 514 mL of a hydrogen peroxide–water solution and **(b)** 3.81 L of ethylene glycol in 7.55 L of an ethylene glycol–water solution?

46. What is the percent by volume concentration of **(a)** 35.0 mL of water in 725 mL of an ethanol–water solution and **(b)** 78.9 mL of acetone in 1550 mL of an acetone–water solution?

47. Describe how you would prepare 3375 g of an aqueous solution that is 8.2% NaCl by mass.

48. Describe how you would prepare 2.44 kg of an aqueous solution that is 16.3% KOH by mass.

49. Describe how you would prepare 2.00 L of an aqueous solution that is 2.00% acetic acid by volume.

50. Describe how you would prepare 600.0 mL of an aqueous solution that is 30.0% isopropyl alcohol by volume.

Expand Your Skills

51. Both magnesium and aluminum react with hydrogen ions in aqueous solution to produce hydrogen. Why is it that only one of the following equations correctly describes the reaction?

$$Mg(s) + 2\,H^+(aq) \longrightarrow Mg^{2+}(aq) + H_2(g)$$

$$Al(s) + 2\,H^+(aq) \longrightarrow Al^{3+}(aq) + H_2(g)$$

52. Which of the following correctly represents the decomposition of potassium chlorate to produce potassium chloride and oxygen gas?
    a. $KClO_3(s) \longrightarrow KClO_3(s) + O_2(g) + O(g)$
    b. $2\,KClO_3(s) \longrightarrow 2\,KCl(s) + 3\,O_2(g)$
    c. $KClO_3(s) \longrightarrow KClO(s) + O_2(g)$
    d. $KClO_3(s) \longrightarrow KCl(s) + O_3(g)$

53. Write a balanced chemical equation to represent (a) the reaction of solid iron(II) oxide with aluminum metal to form liquid iron metal and solid aluminum oxide, and (b) the reaction of aqueous potassium sulfide with aqueous copper(II) nitrate to form aqueous potassium nitrate and solid copper(II) sulfide.

54. Joseph Priestley discovered oxygen in 1774 by heating "red calx of mercury" [mercury(II) oxide]. The calx decomposed to its elements. The equation is

$$HgO \longrightarrow Hg + O_2 \quad \text{(not balanced)}$$

    What mass of oxygen is produced by the decomposition of 18.0 g HgO?

55. Consider 1.00 mol $H_2(g)$, 2.00 mol $He(g)$, and 0.50 mol $C_2H_2(g)$ at the same temperature and pressure. (a) Do the three samples have the same number of atoms? (b) Which sample has the greatest mass?

56. What mass in grams of the magnetic oxide of iron (Fe_3O_4) can be made from 50.0 g of pure iron and an excess of oxygen? The equation is

$$Fe + O_2 \longrightarrow Fe_3O_4 \quad \text{(not balanced)}$$

57. When heated above ~900 °C, limestone (calcium carbonate) decomposes to quicklime (calcium oxide, used to make cement) and carbon dioxide. What mass in grams of quicklime can be produced from 2.5×10^5 g of limestone?

58. Ammonia reacts with oxygen to produce nitric acid (HNO_3) and water. What mass of nitric acid, in grams, can be made from 549 g of ammonia?

59. Hydrogen peroxide solution sold in drugstores is 3.0% H_2O_2 by mass dissolved in water. How many moles of H_2O_2 are in a typical 16 fl. oz. bottle of this solution? (1 fl. oz = 29.6 mL; density = 1.00 g/mL)

60. What is the mass percent of (a) NaOH in a solution of 2.59 g NaOH in 100.0 g of water and (b) ethanol in a solution of 5.25 mL of ethanol (density = 0.789 g/mL) in 50.0 g of water?

61. What is the volume percent of (a) ethanol in 355 mL of an ethanol–water solution that contains 18.0 mL of water and (b) acetone in 1.55 L of an acetone–water solution that contains 4.00 mL of acetone?

62. Laughing gas (dinitrogen monoxide, N_2O; also called nitrous oxide) can be made by very carefully heating ammonium nitrate. (Heating carelessly causes an explosion, which is very poor lab technique.) The other product is water.
    a. Write a balanced equation for the process.
    b. Draw the Lewis structure for N_2O. (*Hint*: Nitrogen is the central atom.)
    c. What mass in grams of N_2O can be made from 150.0 g of ammonium nitrate?

63. In Table 1.6 we listed some multiplicative prefixes used with SI base units. In the 1990s, some new prefixes were recommended because of ever-smaller amounts of matter in advanced experiments:

    zetta- (Z, 10^{21})   *yotta-* (Y, 10^{24})

    zepto- (z, 10^{-21})   *yocto-* (y, 10^{-24})

    a. What is the mass of 1.50 ymol of uranium in yoctograms?
    b. How many atoms are in 1.20 zmol of uranium?

64. What volume of 0.859 M oxalic acid contains 31.7 g of oxalic acid ($H_2C_2O_4$)?

65. In tests for intoxication, blood alcohol levels are expressed as percentages by volume. A blood alcohol level of 0.080% by volume means 0.080 mL of ethanol per 100 mL of blood, and is considered proof of intoxication. If a person's total blood volume is 5.0 L, what volume of alcohol in the blood gives a blood alcohol level of 0.165% by volume?

66. Homeopathic "remedies," found in many pharmacies in the United States, are prepared by using a process of dilution, starting with a substance that causes symptoms of the illness being treated. Some homeopaths use a scale (X) in which the substance is diluted by a factor of ten at each stage. For example, for a 6X preparation, the original substance is diluted tenfold six times, for a total dilution of 10^6. (Many homeopathic preparations are diluted far more than this; some use a C scale, meaning dilution 100 times.)
    a. If the original substance is in the form of a 1.00 M solution, what would be the yoctomolar concentration of a 24X preparation? (Few starting solutions are 1 M; most are far less concentrated.)
    b. How many molecules of the original substance remain in 10.0 L (about 2.4 gallons) of a 24X "remedy"? (For comparison, there are about 400 *trillion* atoms of gold in 10 L of seawater.)
    c. What does this suggest about the effectiveness of such a "remedy"?

67. Evaluate this statement: "One cup of water has more molecules than there are cups of water in all of Earth's oceans." (1 cup = 236 mL = 236 cm³) For the volume of the oceans, use 3.50×10^8 cubic miles.

68. If a Neanderthal excreted 500 mL of urine into the ocean 50,000 years ago, how many molecules in 500 mL of water you drink today are from that urine? Refer to Problem 67, and assume complete mixing of the waters of Earth over 50,000 years.

69. A truck carrying about 50,000 kg of sulfuric acid (H_2SO_4) is involved in an accident, spilling the acid. What mass of sodium bicarbonate ($NaHCO_3$) is needed to react with and

neutralize the sulfuric acid? The products of the reaction are sodium sulfate (Na_2SO_4), water, and carbon dioxide.

70. Referring to Problem 69, determine the mass of sodium carbonate (Na_2CO_3) needed to neutralize the acid. (The products of the reaction are the same.)

71. How many moles of H_2O are in 2.00 L of water? ($d = 1.00$ g/mL)

72. A one-liter bottle of a typical sports drink contains 425 mg of sodium ions, 114 mg of potassium ions, and 53 grams of sugars, and provides 190 calories. If sodium chloride is the source of the sodium ions, potassium hydrogen phosphate (K_2HPO_4) is the source of the potassium ions, and sucrose ($C_{12}H_{22}O_{11}$) is the only sugar, calculate the molarity of sodium chloride, potassium hydrogen phosphate, and sucrose in the bottle. (*Hint*: One mole of sodium ion is contained in one mole of sodium chloride; *two* moles of potassium ions are in one mole of potassium hydrogen phosphate.)

73. A professor would like you to prepare 25 mL of 1.25 M silver nitrate ($AgNO_3$) solution for an experiment for 72 students in each of her three sections of general chemistry. Silver nitrate currently costs $85 for 100 grams. How much (to the nearest dollar) will it cost the chemistry department to have this solution prepared?

74. Consider the following equation for the reaction of sodium azide that forms nitrogen gas in airbags. **(a)** What is the atom economy for the production of N_2? **(b)** What is the mass in grams of N_2 that is formed if 5.74 g of NaN_3 are reacted?

$$2\ NaN_3 \longrightarrow 2\ Na + 3\ N_2$$

75. Ethanol (C_2H_5OH) is a very important chemical. In addition to being widely used as an industrial solvent, a gasoline additive, and an alternative fuel, it is well known as the alcohol in alcoholic beverages. A common name for ethanol is *grain alcohol* because it is formed by the fermentation of glucose ($C_6H_{12}O_6$) and other sugars in grains such as corn, wheat, and barley:

$$C_6H_{12}O_6 \longrightarrow C_2H_5OH + CO_2$$

The previous reaction has been carried out for many centuries and is one of the oldest manufacturing processes. In a more recent development, ethanol can be prepared by reacting ethylene (found in petroleum) with water:

$$C_2H_4 \longrightarrow H_2O + C_2H_5OH$$

a. Write the *balanced* equation for each of the two reactions above.
b. Calculate the % A.E. (for the product ethanol) for each of the two reactions.
c. Explain how you can determine which reaction has the higher atom economy without doing the calculation in part (b) above.
d. Is either method sustainable? Which one? Justify your answer.
e. Given the results from parts (b) and (d) above, select one of the two methods as the better candidate overall for preparing ethanol. Justify your selection.

76. Consider the following equation for the formation of carbon dioxide CO_2, a known greenhouse gas, from the combustion of hydrocarbons.

$$C_5H_{12} + O_2 \longrightarrow CO_2 + H_2O \quad \text{(not balanced)}$$

a. What is the percent atom economy based on carbon dioxide as the target product?
b. How many grams of CO_2 form if 12.0 g C_5H_{12} is burned?
c. What is the percent yield of the reaction if 15.2 g CO_2 are formed at the end of the reaction? (Hint: Use the information in the Atom Economy essay for this calculation.)

77. Balance each reaction and calculate the percent atom economy based on the underlined molecule as your desired product.
a. $CuCl(aq) + Na_2CO_3(aq) \longrightarrow \underline{CuCO_3}(s) + NaCl(aq)$
b. $NH_3(g) + O_2(g) \longrightarrow \underline{NO}(g) + H_2O(g)$
c. $KMnO_4(aq) + KOH(aq) + KI(aq) \longrightarrow \underline{K_2MnO_4}(aq) + KIO_3(aq) + H_2O(l)$

Critical Thinking Exercises

Apply knowledge that you have gained in this chapter and one or more of the FLaReS principles (Chapter 1) to evaluate the following statements or claims.

5.1. Suppose that someone has published a paper claiming to have established a new value for Avogadro's number. The author says that he has made some very careful laboratory measurements and his calculations indicate that the true value for Avogadro's number is 3.0187×10^{23}. Is this claim credible, in your opinion? What questions would you ask the author about his claim?

5.2. A chemistry teacher asked her students, "What is the mass, in grams, of a mole of chlorine?" One student answered 35; another said 70; and several others gave answers of 36, 37, 72, and 74. The teacher stated that all of these answers were correct. Do you believe her statement?

5.3. A website on fireworks provides directions for preparing potassium nitrate, KNO_3 (molar mass = 101 g/mol),

using potassium carbonate (138 g/mol) and ammonium nitrate (80 g/mol), according to the following equation:

$$K_2CO_3 + 2\ NH_4NO_3 \longrightarrow 2\ KNO_3 + CO_2 + H_2O + NH_3$$

The directions state: "Mix one kilogram of potassium carbonate with two kilograms of ammonium nitrate. The carbon dioxide and ammonia come off as gases, and the water can be evaporated, leaving two kilograms of pure potassium nitrate." Use information from this chapter to evaluate this statement.

5.4. A manufacturer of fuel cells claims that his hydrogen fuel cells are lighter than rechargeable batteries for use in electric cars because the batteries use metals like lead, nickel, and cadmium, and his fuel cell is powered by hydrogen, which has a lower atomic mass.

5.5. After working Problem 66 of Expand Your Skills, evaluate the claim that homeopathy is a valid alternative medical treatment.

Collaborative Group Projects

Prepare a PowerPoint, poster, or other presentation (as directed by your instructor) to share with the class.

1. Prepare a brief biographical report on Amedeo Avogadro.
2. In the view of many scientists, hydrogen holds great promise as a source of clean energy. A 2008 report by the U.S. National Research Council suggested that a government subsidy of roughly $4 billion per year through 2023 could make hydrogen cars competitive with petroleumpowered vehicles. Search the web to explore the proposals for using hydrogen as a fuel. Write a brief summary of what you find, and use an appropriate balanced equation to explain why hydrogen is considered a clean, renewable source. More discussion of hydrogen as a fuel is found in Chapter 15.

 ## LET'S EXPERIMENT! Cookie Equations

Materials Needed

- Four visually different kinds of sandwich cookies (chocolate, vanilla, chocolate chip, and peanut butter shown)
- Butter knife
- Napkin

Wait, wait, don't eat the molecules…yet!
The law of conservation of matter states that matter is neither created nor destroyed during a chemical reaction. Although atoms may be combined differently, the number and mass of the atoms in the reactants must equal the number and mass of the atoms in the product.

Let's practice writing chemical equations using cookies as manipulatives. Remember that chemical equations use symbols and formulas to represent the elements and compounds in the change. Whatever elements are on the left side of the equation will be in the final product on the right side of the equation. The combinations may be different, but the amount on the left must equal the amount on the right. "What goes in must come out."

First, twist apart each cookie. Separate the fillings from the cookie parts for several of each type of cookie, using the knife. Save the fillings so you can "bond" atoms together in different combinations, if you wish.

Use the cookie parts to construct the reactants in the following reactions. Follow the key to match up the cookies with the elements. Now take your "molecules" apart and re-assemble them into the products. How many "atoms" do you start with? How many atoms do you end up with? How many molecules do you start with and end up with?

$$Fe_2O_3 + C \longrightarrow CO_2 + Fe$$

> **Key**
> Cookie #1 = Fe
> Cookie #2 = C
> Cookie #3 = O

$$CH_4 + 2\,O_2 \longrightarrow CO_2 + 2\,H_2O$$

> **Key**
> Cookie #1 = H
> Cookie #2 = C
> Cookie #3 = O

Use the cookie parts to construct the two reactants in the following equation. Now take the reactants apart and try to construct the products. Finally, balance the equation, then repeat the construction of reactants and products.

$$NaOH + HCl \longrightarrow NaCl + H_2O$$

> **Key**
> Cookie #1 = H
> Cookie #2 = Na
> Cookie #3 = O
> Cookie #4 = Cl

Questions

1. In the three balanced equations from above, are atoms conserved? Are molecules conserved?
2. Why is it important to balance equations?

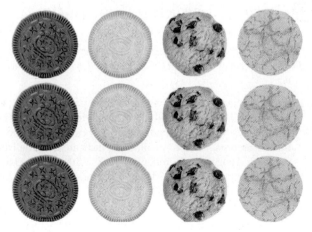

Gases, Liquids, Solids ... and Intermolecular Forces

6

Water has many unusual properties, It can exist as solid, liquid (both left), or gas under rather ordinary conditions. Interestingly, "steam" rising from the athlete's head (above) is actually billions of microscopic droplets of liquid water. Steam—gaseous water—is invisible. Both solid and liquid water are excellent cooling agents, and liquid water dissolves many substances, both ionic and covalent. These properties are largely due to the *intermolecular forces* in water. We will learn about those forces in this chapter.

Have You Ever Wondered?

1 Why does "slime" have such strange properties? **2** What does "5W30" on a bottle of motor oil mean? **3** What keeps oil and water from mixing? **4** Why doesn't creamy Italian dressing separate like ordinary Italian dressing? **5** How cold can it possibly get, and how hot can it possibly get? **6** Why is it recommended that tire pressure be measured when tires are cold?

WE INTRODUCED THE THREE STATES OF MATTER—gas, liquid, and solid—in Chapter 1. Water is the most familiar substance that we experience in all three states: as Earth's most common liquid, as a solid (ice), and as a gas (steam). Many other substances can exist in the three states, depending on conditions. An understanding of those conditions is important for our understanding of matter. The state that a substance exhibits depends on the physical properties of that substance, on the pressure, and especially on the temperature. When it is very cold, water exists as solid ice. When it is warmer, water is a liquid. At still higher temperatures, water boils and becomes steam.

Chapter 4 described the forces called *chemical bonds* that bind atoms together within molecules. Chemical bonds are **intramolecular forces** that determine such molecular properties as geometry and polarity. Forces *between* molecules—**intermolecular forces**—determine the macroscopic physical properties of liquids and solids. Intermolecular forces are also related to the kinds of bonds found in substances. Figure 6.1 compares intermolecular and intramolecular forces.

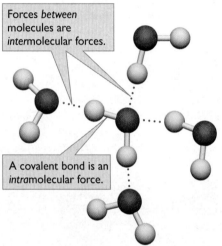

Forces *between* molecules are *intermolecular forces.*

A covalent bond is an *intramolecular force.*

▶ **Figure 6.1** Intramolecular and intermolecular forces. The gray "sticks" represent covalent bonds in water molecules—*intra*molecular forces, or pairs of shared electrons. The dotted red lines represent forces between water molecules—*inter*molecular forces.

▲ Most solids are made up of highly ordered particles. Glass is an exception. The atoms in glass (Chapter 12) are randomly arrayed.

Intermolecular forces are forces of attraction. In solids and liquids, molecules remain close together. If these forces did not exist, there would be no liquids or solids—everything would be in a gaseous state.

The shape, size, and polarity of molecules determine how they interact with each other. The physical state of a material depends on the strength of the intermolecular forces that hold the molecules or ions together, relative to the thermal energy (temperature) that acts to separate them.

6.1 Solids, Liquids, and Gases

Learning Objective • Explain how the different properties of solids, liquids, and gases are related to the motion and spacing of atoms, molecules, or ions.

Most solids are highly ordered assemblies of atoms, molecules, or ions in close contact with one another, as suggested by Figure 6.2a. This makes it difficult to compress solids. The motion of these particles is mainly vibration. Table salt (sodium chloride) is a typical crystalline solid. In this solid, ionic bonds hold the Na^+ and Cl^- ions in position and maintain the orderly arrangement. (Look again at Figure 4.2.)

In liquids, the particles are still in close contact, but they are randomly arranged and are freer to move about (Figure 6.2b). Like solids, liquids are difficult to compress, but they are able to flow and to conform to the shape of their container. To get a better image of a liquid at the molecular level, think of a box of marbles being

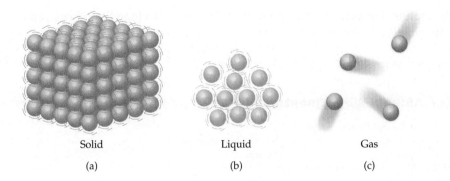

Solid   Liquid   Gas

(a)   (b)   (c)

◄ **Figure 6.2** Three states of matter. (a) Most solids have molecules (or atoms) that vibrate around fixed positions. (b) In liquids, the molecules are still close together but are free to move about like people in a crowded room. (c) Molecules of gases are widely spaced and move freely and randomly.

shaken continuously. The marbles move back and forth, rolling over one another. The particles of a liquid (like the marbles) are not held in place as rigidly as the particles in a solid are. Even so, there must be some force attracting the particles in a liquid to one another. Otherwise, they would not stay close together.

In gases, the particles are separated by relatively great distances and are moving quite rapidly in random directions (Figure 6.2c). In this state, the atoms or molecules have no interactions with one another. We will look more closely at gases in Section 6.5.

Most solids can be changed to liquids; that is, they can be *melted*. When a solid is heated, the thermal energy is absorbed by the particles of the solid. The energy causes the particles to vibrate more vigorously until, finally, the forces holding the particles in their regular arrangement are overcome. The solid becomes a liquid. The temperature at which this happens is the **melting point** of the solid. A high melting point indicates that the forces holding a solid together are very strong.

Many liquids can change to gases in a process called **vaporization**. Again, all that is needed to achieve this change is enough heat. Energy is absorbed by the liquid particles, which move faster and faster as a result. Finally, this increasingly violent motion overcomes the attractive forces holding the liquid particles close to one another, and one or more particles leave the liquid. The liquid becomes a gas. The temperature at which the pressure of the escaping vapor is the same as the outside pressure is the **boiling point** of the liquid.

Removing energy from the gas and slowing down the particles can reverse the sequence of changes. A gas changes to a liquid in a process called **condensation**; a liquid changes to a solid in a process called **freezing**. Figure 6.3 presents a diagram of the changes in state that occur as energy is added to or removed from a sample. Some substances go directly from the solid state to the gaseous state, a process called

1   Why does "slime" have such strange properties?

"Slime," readily made at home from household glue and laundry borax, has strange properties. It sags under its own weight, and flows like a very slow liquid. That's because it is just that: a slow-flowing liquid, not a solid. The borax reacts with the glue to form large, flexible molecules that do not easily form a regular arrangement. The molecules are so intertwined that slime flows very slowly.

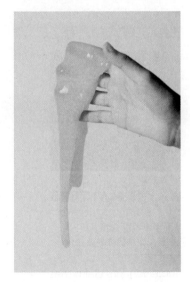

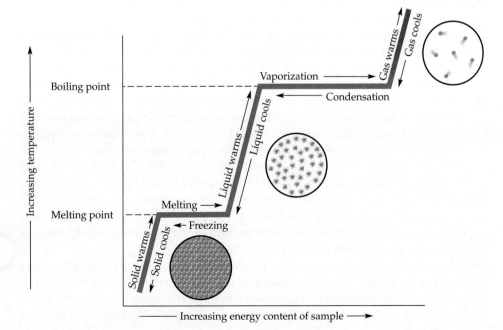

◄ **Figure 6.3** Diagram of changes in state of a substance on heating or cooling. Contrary to popular belief, the melting point and the freezing point of a pure substance are exactly the same temperature. Once the melting point is reached, additional heat does not raise the temperature, but instead causes more solid to melt. Once all the solid is melted, the temperature will rise when heat is added. Likewise the boiling point and condensation point are the same temperature.

▲ The reverse of sublimation is *deposition*, a process that occurs on cold mornings. Water vapor deposits directly onto the fence as solid frost.

sublimation. For example, ice cubes left in a freezer for a long time shrink, and solid mothballs slowly disappear in a drawer or closet. In each case, the solid undergoes sublimation. And sometimes a gas can go directly to the solid state, as when water vapor forms frost on the grass. That process is called **deposition**.

SELF-ASSESSMENT Questions

1. Molecules are farthest apart in a(n)
   a. ionic solid   **b.** covalent solid
   c. liquid   **d.** gas

2. Liquids are
   a. difficult to compress and have no definite shape
   b. difficult to compress and have no definite volume
   c. easy to compress and have a definite shape
   d. easy to compress and have a definite volume

3. When a liquid changes into a gas, the process is called
   a. condensation   **b.** sublimation
   c. gasification   **d.** vaporization

4. When a liquid changes into a solid, the process is called
   a. condensation   **b.** freezing
   c. fusion   **d.** melting

5. When a solid changes directly into a gas, the process is called
   a. evaporation   **b.** deposition
   c. sublimation   **d.** vaporization

Answers: 1, d; 2, a; 3, d; 4, b; 5, c

6.2 Comparing Ionic and Molecular Substances

Learning Objective • Identify some differences between ionic and molecular substances, and explain why these differences exist.

Recall that Figure 6.1 illustrates *intermolecular* forces. Ionic substances such as sodium chloride (see Figure 4.2) are made of ions, not molecules, so the term *intermolecular forces* really doesn't apply to them. Nevertheless, there are attractive forces between ions, and we can compare them with the forces between molecules. First, let's consider some differences between ionic and molecular substances.

- Almost all ionic compounds, such as NaCl, KBr, $CaCO_3$, and NH_4NO_3, are solids at room temperature. Some molecular substances are liquids or gases at room temperature, although many are solids. Ethane (CH_3CH_3), hydrogen sulfide (H_2S), and chlorine (Cl_2) are gases. Ethanol (CH_3CH_2OH), bromine (Br_2), and phosphorus trichloride (PCl_3) are liquids. Sulfur (S_8), glucose ($C_6H_{12}O_6$), and iodine (I_2) are solids.

- Ionic compounds generally have much higher melting points and boiling points than molecular compounds. High temperatures are necessary to overcome the strong ionic bonds (Section 4.3) in ionic compounds. Less energy is required to overcome the much weaker attractions between molecules (Section 6.3). For example, sodium chloride (NaCl) must be heated to 801°C before it melts, but the molecular compound ethane (CH_3CH_3) melts at −184 °C.

- It generally takes between 10 and 100 times as much energy to melt one mole of a typical ionic compound as it does to melt one mole of a typical molecular compound. To melt one mole of sodium chloride requires almost 30 kJ, while it takes only about 3 kJ to melt one mole of ethane.

- Many familiar ionic compounds dissolve in water and form solutions that conduct electricity. The dissociated ions carry the electric current through the solution. Most molecular compounds, even those that dissolve in water, do not dissociate into ions and thus do not form conducting solutions. A solution of

WHY IT MATTERS

A salt lamp has a light bulb mounted in a large piece of sodium chloride. Some vendors claim that the heat from the lamp causes negative ions to be emitted from the salt. But ionic compounds like sodium chloride have extremely high melting and boiling points, far above the temperature of an ordinary light bulb. No significant quantity of ions can be produced from such a lamp, and there is no conclusive evidence that such ions would be beneficial anyway.

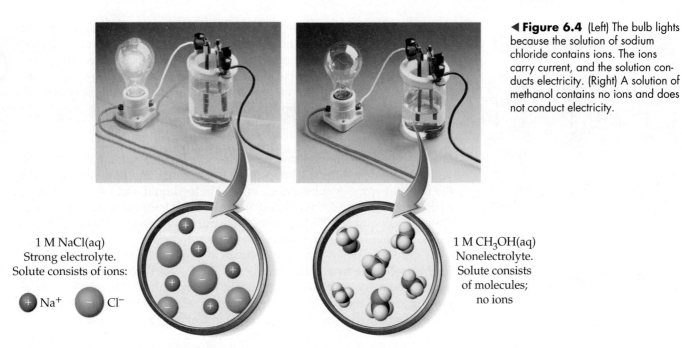

◀ **Figure 6.4** (Left) The bulb lights because the solution of sodium chloride contains ions. The ions carry current, and the solution conducts electricity. (Right) A solution of methanol contains no ions and does not conduct electricity.

1 M NaCl(aq)
Strong electrolyte.
Solute consists of ions:

⊕ Na⁺   ⊖ Cl⁻

1 M CH₃OH(aq)
Nonelectrolyte.
Solute consists
of molecules;
no ions

(ionic) sodium chloride conducts electricity, as seen in Figure 6.4, while a solution of methanol (CH$_3$OH, molecular) does not.

- As solids, most ionic compounds are crystalline, hard, and often quite brittle. Solid molecular compounds are usually much softer than ionic ones.

Next, we consider some other interactions that hold the particles of solids and liquids together.

SELF-ASSESSMENT Questions

1. Which of the following is not composed of particles that exhibit intermolecular forces?
   a. HCN   **b.** HF   **c.** NaBr   **d.** CH$_4$

2. Which of the following solids is likely to require the greatest input of energy to melt?
   a. C$_6$H$_{12}$O$_6$ **b.** N$_2$O   **c.** KBr   **d.** Kr

Answers: 1, c; 2, c

6.3 Forces between Molecules

Learning Objectives • Classify forces between molecules as dipole–dipole forces, dispersion forces, or hydrogen bonds. • Explain the effects that intermolecular forces have on melting and boiling points.

As we have noted, many substances, such as water and carbon dioxide, do not consist of ions but instead are molecular compounds. Molecular compounds are characterized by covalent bonds. For example, in a hydrogen chloride molecule, a covalent bond holds the hydrogen and chlorine *atoms* together. These atoms share a pair of electrons. But what makes one HCl molecule interact with another? In this section, we examine several types of *inter*molecular forces—forces *between* molecules.

Dipole–Dipole Forces

Recall that the hydrogen chloride molecule is a *dipole*: a molecule with a positive end and a negative end. (See Figure 4.6.) When oppositely charged ends of two dipoles are near each other, they attract one another. In solid HCl, the molecules line up so that the positive end of one molecule attracts the negative end of neighboring molecules, as suggested by Figure 6.5a. If we heat the solid sufficiently, the orderly arrangement is

WHY IT MATTERS

Pure water is a good insulator because it contains only the tiniest trace of ions, about one ion for every 10 million water molecules. It conducts so little electricity that the electrical resistance of water is used to determine its purity. The reason that toasters are poor bath toys is that water encountered in everyday life has enough ions from dissolved solids to conduct electricity quite well.

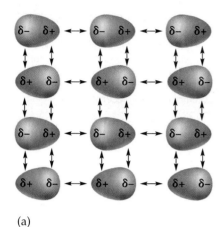

(a)

(b)

▲ **Figure 6.5** An idealized representation of dipole–dipole forces in (a) a solid and (b) a liquid. In a real liquid or solid, the interactions between particles are more complex.

undone, and the solid melts. In the liquid state, the oppositely charged dipoles in the liquid still attract one another, but more randomly, as shown in Figure 6.5b. These **dipole–dipole forces** occur between any two polar molecules. Generally, these forces are much weaker than ionic bonds, but they are stronger than the forces between nonpolar molecules of comparable size. The more polar the molecules, the stronger are the dipole–dipole forces.

Dispersion Forces

It is easy enough to understand how ions or polar molecules maintain contact with one another. After all, opposite charges attract. But how can we explain the fact that *nonpolar* substances can exist in the liquid and solid states? Even hydrogen (H_2) can exist as a liquid or a solid if the temperature is low enough. (Its melting point is $-259\ °C$.) Some force must hold nonpolar molecules close to one another in the liquid and solid states.

Figure 6.6a suggests that the electrons in a covalent bond are evenly distributed between the two atoms sharing the bond. Recall, however, that electrons are *not* stationary but move continuously. Within a covalent bond, electrons are moving around the nuclei. On *average*, the two electrons in a nonpolar bond are evenly distributed around the two nuclei. But at any given instant, it is possible for more electrons in a molecule to be at one end than at the other (Figure 6.6b), causing a temporary negative charge at that end. This causes electrons in an adjacent molecule to move toward the opposite end of that molecule, as in Figure 6.6c. Thus, an attractive force is created between the electron-rich end of one molecule and the electron-poor end of the next. This attractive force only lasts for an instant before the electrons redistribute themselves evenly.

These momentary, usually weak attractive forces between molecules are called **dispersion forces**. Because the electrons are in constant motion, dispersion forces can be thought of as dipole–dipole forces in which the dipoles are constantly shifting. Generally, the bigger the molecule, the stronger are the dispersion forces, because more electrons are available to cause these momentary dipole forces. Dispersion forces exist between *any* two particles containing electrons, whether polar, nonpolar, or ionic. However, they are the *only* force that exists between nonpolar molecules.

Hydrogen Bonds

Certain polar molecules exhibit stronger attractive forces than would be expected from ordinary dipole–dipole interactions. These forces are strong enough to be given a special name, *hydrogen bonds*. The name is a bit misleading because it emphasizes only one of the two atoms in the interaction. The name is also misleading because hydrogen bonds are not bonds at all; no electrons are shared in a hydrogen bond. Not all compounds containing hydrogen exhibit this kind of attractive force. In most cases, the hydrogen atom *must* be attached to a small, very electronegative atom—namely, nitrogen, oxygen, or fluorine. The high electronegativity of such atoms leaves the hydrogen

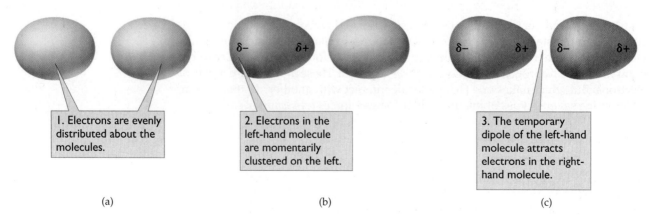

1. Electrons are evenly distributed about the molecules.

2. Electrons in the left-hand molecule are momentarily clustered on the left.

3. The temporary dipole of the left-hand molecule attracts electrons in the right-hand molecule.

(a)   (b)   (c)

▲ **Figure 6.6** The mechanism underlying dispersion forces.

with a large partial positive charge. A hydrogen bond forms when this hydrogen comes near a nitrogen, oxygen, or fluorine atom on another polar molecule. The positively charged hydrogen is strongly attracted to the negatively charged N, O, or F.

In hydrogen fluoride (HF), for example, the hydrogen–fluorine bond is strongly polar, with a negative fluorine end and a positive hydrogen end. Both hydrogen and fluorine are small atoms, so the electrons on the fluorine end of one molecule can approach very closely to the hydrogen end of a second molecule. This unusually strong interaction is the **hydrogen bond**. Hydrogen bonds are often explicitly represented by *dotted* lines (see Figure 6.7) to emphasize that their strength is much greater than that of other intermolecular forces.

Water has both an unusually high melting point (0 °C) and an unusually high boiling point (100 °C) for a compound with such small molecules. These abnormal values are attributed to the water molecules' ability to form hydrogen bonds. Remember, however, that during melting and boiling, only the hydrogen bonds *between water molecules* are overcome. The covalent bonds between the hydrogen atoms and the oxygen atom in each water molecule remain intact. No chemical change occurs when there is a change in state, because covalent bonds are much stronger than hydrogen bonds.

The unique properties of water resulting from hydrogen bonding are discussed in Chapter 14. Hydrogen bonding also plays an important role in biological molecules. The 3-D structures of proteins and enzymes and the arrangement of DNA (Chapter 16) are dependent on the presence and strength of hydrogen bonds.

The primary classification of intermolecular forces that exist between two molecules can be assessed by first asking "Are the molecules polar?" (Remember, ordinarily we must draw the Lewis structures and determine the molecular shapes to determine polarity.) If they are nonpolar, the only kind of force that can exist is dispersion force. For polar molecules, we then determine whether hydrogen bonding can occur. The requirements are: H covalently bonded to N, O, or F in one molecule, and an N, O, or F in a polar bond of a neighboring molecule. If hydrogen bonding can occur, it does so. If not, then dipole–dipole forces exist between the polar molecules. The process is illustrated in Example 6.1.

EXAMPLE 6.1 Determining Intermolecular Forces

What is the major kind of force that exists between **(a)** NH_2Cl molecules; **(b)** CF_4 molecules; and **(c)** an H_2O molecule and an H_2S molecule?

Solution

a. NH_2Cl molecules are similar to NH_3 molecules, in that they are both trigonal pyramidal and polar. They also have the requirements for hydrogen bonding: H covalently bonded to N in one molecule, and N in a polar bond (NH bond) in a neighboring molecule. The major force is therefore hydrogen bonding.

b. Despite the fact that CF_4 molecules contain highly electronegative fluorine atoms, they are nonpolar because the fluorine atoms are symmetrically arranged around the carbon atom, similar to CH_4. Therefore, the only forces that exist between CF_4 molecules are dispersion forces.

c. Both H_2O and H_2S are bent molecules with polar bonds, so both are polar. Water molecules have hydrogen atoms covalently bonded to oxygen. However, H_2S does not contain N, O, or F atoms in a polar bond. Therefore, the major force here is a dipole–dipole force.

> **Exercise 6.1A**
What is the major kind of force between **(a)** SiH_4 molecules and **(b)** SF_2 molecules?

> **Exercise 6.1B**
What is the major kind of force between a H_3CCHO molecule (H and O bonded to C) and a water molecule?

2 What does "5W30" on a bottle of motor oil mean?

Motor oil, whether synthetic or from crude oil, must have the proper *viscosity* or thickness—which depends on the strength of the oil molecules' intermolecular forces—to lubricate an engine properly. As the engine gets hot, the high temperature makes the oil flow more easily, so a thicker oil (viscosity of 30) is needed. However, when the engine is first started, especially on a cold winter morning, the engine needs a thinner oil (viscosity of 5) to work properly. Oil labeled "5W30" has molecules with intermolecular forces that are about the same whether the oil is cold or hot, and such oil can be used in both summer and winter.

Hydrogen bond

Hydrogen fluoride

Water

▲ **Figure 6.7** Hydrogen bonding in hydrogen fluoride and in water. See also Figure 6.1.

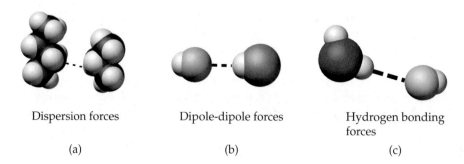

Dispersion forces   Dipole-dipole forces   Hydrogen bonding forces

(a)   (b)   (c)

▲ **Figure 6.8** A summary of three important intermolecular forces. (a) Dispersion forces exist between any two particles, such as butane and propane, shown here. Usually weak, their strength depends on size and shape of the particles. (b) Dipole–dipole forces occur between polar molecules, such as HCl and HBr. The opposite partial charges attract one another. Their strength depends on polarity. (c) Hydrogen bonding, such as between HF and H_2O, is a special, strong type of dipole–dipole force. It occurs when a hydrogen atom, bonded to O, N, or F, is attracted to O, N, or F in a neighboring polar bond or molecule.

Strength of Intermolecular Forces

For molecules of similar size and shape, intermolecular forces range in strength from dispersion forces (generally the weakest) to dipole–dipole forces to hydrogen bonds (the strongest), as indicated in Figure 6.8. Stronger attractions between molecules lead to higher melting points and boiling points. Melting requires weakening the forces between molecules, while boiling requires separating the molecules completely from one another. So the stronger the intermolecular forces, the more energy in the form of a higher temperature that must be applied before a compound can melt or boil. These effects can be seen in Figure 6.9. The boiling points of the group 4A compounds (CH_4 to SnH_4), which are all nonpolar, increase with mass as you go down the group. The same trend with mass is seen in the group 5A, 6A, and 7A compounds. However, since these molecules are polar, the boiling point trend lines for these compounds lie above the line for the group 4A compounds. The dipole–dipole forces present in these compounds give them higher boiling points than their group 4A analogs.

Three compounds in Figure 6.9 don't follow these trends. H_2O, HF, and NH_3 have much higher boiling points than their third-row equivalents. These three compounds can form hydrogen bonds with other molecules of themselves. In fact, the hydrogen bonds in water are so strong that they raise the boiling point by almost 200 °C!

▶ **Figure 6.9** Each color represents the boiling point of a covalent compound containing hydrogen and an element within a column of the periodic table. The boiling points of water, hydrogen fluoride, and ammonia are much higher than theory would predict, because these three compounds can hydrogen-bond with other molecules of themselves. Methane, CH_4, has a much lower boiling point than the three other compounds because it cannot hydrogen-bond with other molecules of itself.

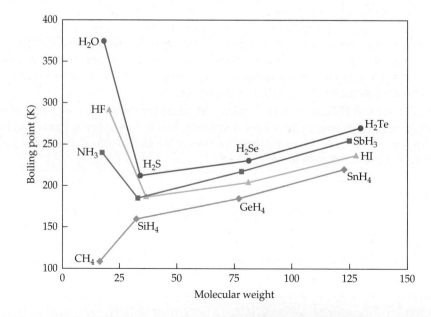

SELF-ASSESSMENT Questions

1. What intermolecular forces are most significant in accounting for the high boiling point of liquid water relative to other substances of similar molecular weight?
   a. dispersion forces
   b. dipolar attractions
   c. hydrogen bonds
   d. ionic bonds

2. Which one of the following interactions is the strongest?
   a. covalent bonds
   b. dipole–dipole forces
   c. dispersion forces
   d. hydrogen bonds

3. In ethanol (CH_3CH_2OH), the attractions between neighboring molecules are
   a. covalent bonds and dispersion forces only
   b. dipole–dipole forces and hydrogen bonds only
   c. covalent bonds only
   d. hydrogen bonds, dipole–dipole forces, and dispersion forces only

4. The molecules of liquid water in a beaker are held to each other by
   a. covalent bonds
   b. intramolecular forces
   c. intermolecular forces
   d. ionic bonds

5. Of CH_3OH, H_2, HF, and H_2O, which molecule(s) can form hydrogen bonds?
   a. H_2 only
   b. H_2O only
   c. CH_3OH, HF, and H_2O
   d. all four

6. Which one of the following molecules does *not* form hydrogen bonds?
   a. ammonia (NH_3)
   b. chloroform ($CHCl_3$)
   c. methyl alcohol (CH_3OH)
   d. ethylene glycol ($HOCH_2CH_2OH$)

Answers: 1, c; 2, a; 3, d; 4, c; 5, c; 6, b

6.4 Forces in Solutions

Learning Objective • Explain why nonpolar solutes tend to dissolve in nonpolar solvents and polar and ionic solutes tend to dissolve in polar solvents.

To complete our look at intermolecular forces, let's briefly examine the interactions that occur in solutions. A **solution** is an intimate, homogeneous mixture of two or more substances. Here *intimate* means that mixing occurs down to the level of individual ions and molecules, as suggested in Figure 6.10. For example, a sugar-in-water solution has separate sugar molecules randomly distributed among water molecules. *Homogeneous* means that all parts of the solution have the same distribution of components. A sugar solution is equally sweet at the top, bottom, and middle of the solution. The substance being dissolved, and present in a lesser amount, is called the **solute**. The substance doing the dissolving, and present in a greater amount, is the **solvent**. In a sugar-in-water solution, sugar is the solute and water is the solvent.

Ordinarily, solutions form most readily when the substances involved have *similar* strengths of forces between their molecules (or atoms or ions). An old chemical adage is "Like dissolves like." Nonpolar solutes dissolve best in nonpolar solvents. For example, oil and gasoline, both nonpolar, mix (Figure 6.11a), but oil and vinegar do not (Figure 6.11b). Vinegar is mostly water, and the strong hydrogen bonds between water molecules do not allow nonpolar oil molecules (which only have much weaker dispersion forces) to mix freely with them. On the other hand, both ethyl alcohol molecules and water molecules can form hydrogen bonds. Thus, ethyl alcohol readily dissolves in water (Figure 6.11c). In general, a solute dissolves when attractive forces between it and the solvent are at least almost as strong as the attractive forces operating in the pure solute and in the pure solvent.

• = Solvent molecule
• = Solute molecule

▲ **Figure 6.10** In a solution, solute molecules (orange spheres) are randomly distributed among solvent molecules (purple spheres).

3 What keeps oil and water from mixing?

Oil molecules are nonpolar, and water molecules are very polar. The weak forces between oil and water molecules aren't strong enough to overcome the strong forces between the water molecules. Thus, oil and water don't mix.

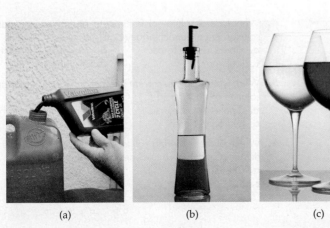

(a)   (b)   (c)

◄ **Figure 6.11** (a) Lawn mowers with two-cycle engines are fueled and lubricated with a solution of nonpolar lubricating oil in nonpolar gasoline. (b) In a salad dressing, polar vinegar and nonpolar olive oil are mixed. However, the two liquids do not form a solution, and separate on standing. (c) Wine is a solution of polar ethyl alcohol in polar water.

Q *What are some other examples of mixtures of polar and nonpolar substances?*

▶ **Figure 6.12** An ionic solid (sodium chloride) dissolving in a polar solvent (water). The hydrogen ends of the water molecules surround the negatively charged chloride ions, while the oxygen ends of the water molecules surround the positively charged sodium ions.

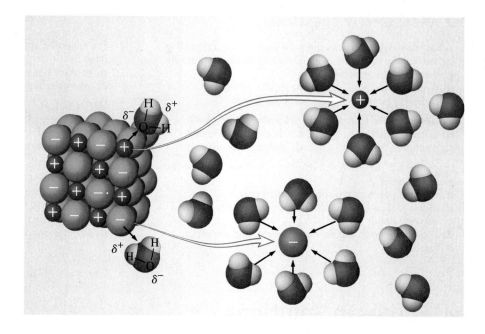

④ Why doesn't creamy Italian dressing separate like ordinary Italian dressing?

Both kinds of dressing contain oil and vinegar. However, proteins (from egg yolk or other sources) are mixed into the creamy dressing. Like soap molecules, the protein molecules have both polar parts and nonpolar parts and keep tiny droplets of oil suspended in the vinegar.

Why, then, does salt dissolve in water? Ionic solids are held together by strong ionic bonds. We have already seen that high temperatures are required to melt ionic solids and break these bonds. Yet if we simply place sodium chloride in water at room temperature, the salt dissolves. And when such a solid dissolves, its bonds *are* broken. The difference between the two processes is the difference between brute force and persuasion. In the melting process, we simply put in enough energy (as heat) to break the crystal down. In the dissolving process, we offer the ions an attractive alternative to their ionic interactions in the crystal.

When a salt crystal dissolves, water molecules first surround the crystal. Opposite charges attract one another, so the positive (hydrogen) ends of their dipoles point toward a negative ion, as shown in Figure 6.12. The negative (oxygen) ends of their dipoles point toward the positive ion. Although the attraction between a dipole and an ion is not as strong as that between two ions, several water molecules surround each ion, and the many *ion–dipole* interactions that are formed as the crystal dissolves overcome the ionic bonds.

In an ionic solid, the positive and negative ions are strongly bonded together in an orderly crystalline arrangement. In solution, cations and anions move about more or less independently, each surrounded by a cage of solvent molecules. Water—including the water in our bodies—is an excellent solvent for many ionic compounds. Water also dissolves many molecules that are polar covalent like itself. These solubility principles explain how nutrients reach the cells of our bodies (dissolved in blood, which is mostly water) and how so many kinds of pollutants get into our water supplies (as we will see in Chapter 14).

SELF-ASSESSMENT Questions

1. A solution is a mixture that is
  a. heterogeneous   **b.** homogeneous
  c. homologous   **d.** humectic

2. Which of the following pairs of substances is *least likely* to form a solution?
  a. an ionic compound in a nonpolar solvent
  b. an ionic compound in a polar solvent
  c. a nonpolar compound in a nonpolar solvent
  d. a polar compound in a polar solvent

3. A positive test for iodine is the purple color of a solution of I_2 in hexane (C_6H_{14}). What type of solute–solvent interaction is most important in a solution of nonpolar I_2 in nonpolar C_6H_{14}?
  a. dipole–dipole forces   **b.** dispersion forces
  c. ion–dipole interactions   **d.** hydrogen bonds

4. What type of solute–solvent interaction is most important in a solution of calcium chloride ($CaCl_2$) in water?
  a. dipole–dipole forces   **b.** ion–dipole interactions
  c. ionic bonds   **d.** dispersion forces

5. What type of solute–solvent interaction is most important in a solution of acetic acid (CH_3COOH) in water?
  a. dipole–dipole forces   **b.** dipole–hydrogen bonds
  c. hydrogen bonds   **d.** dispersion forces

6. Which of the following solutes is most likely to dissolve in octane (C_8H_{18}), a component of gasoline?
  a. $CH_3(CH_2)_4CH_3$ **b.** $CaCl_2$
  c. KCl **d.** MgO

Answers: 1, b; 2, a; 3, b; 4, b; 5, c; 6, a

6.5 Gases: The Kinetic-Molecular Theory

Learning Objective • List the five basic concepts of the kinetic–molecular theory of gases.

At the macroscopic level, gases may seem more difficult to understand than liquids and solids—perhaps because most gases are invisible, while we can see and feel liquids and solids. However, at the microscopic level, gases are more readily treated mathematically. Gas molecules are so far apart that intermolecular forces can be ignored. Therefore, the behavior of a gas normally doesn't depend on the composition of the gas, so the description of gases that appears in the rest of the chapter can be applied to almost any gas. On the other hand, in liquids and solids the molecules are much closer together, so intermolecular forces are stronger, and they vary depending on the type of molecules and the distance between them. This variation complicates any theoretical treatment.

Experiments with gases were instrumental in developing the concepts of Avogadro's number and of molar ratios in reactions. Let's look at the behavior of gases more closely, using a model known as the **kinetic–molecular theory** (Figure 6.13). There are five basic concepts of this theory.

1. Particles of a gas (usually *molecules*, but *atoms* in the case of noble gases) are in rapid, constant motion and move in straight lines.
2. The particles of a gas are tiny compared with the distances between them.
3. Because the particles of a gas are so far apart, there is very little interaction between them.
4. Particles of a gas collide with one another. Energy is *conserved* in these collisions; energy lost by one particle is gained by the other.
5. Temperature is a measure of the average kinetic energy (energy of motion) of the gas particles.

Section 6.3 described how intermolecular forces hold molecules together in solids and liquids. In gases, the particles are separated by relatively great distances and are moving about randomly. Although these particles can collide, they seldom interact. We can use the kinetic–molecular theory to explain the gas laws presented in the next section.

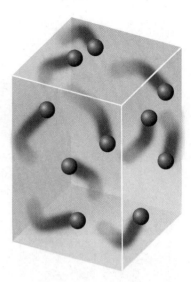

▲ **Figure 6.13** According to the kinetic–molecular theory, particles (molecules or atoms) of a gas are in constant, random motion. They move in straight lines and undergo collisions with each other and with the walls of the container.

SELF-ASSESSMENT Questions

1. Which of the following is *not* a concept of the kinetic–molecular theory of gases?
   a. Particles of a gas are in constant motion.
   b. The pressure and volume of a gas are inversely related.
   c. Temperature is a measure of the average kinetic energy of the gas particles.
   d. The particles are very small compared with the distances between them.

2. According to the kinetic–molecular theory of gases, a gas particle
   a. will collide with another gas particle with no loss of kinetic energy for either
   b. will stick to a wall of its container
   c. is tiny compared with the volume occupied by the gas
   d. repels another gas particle when the two are nearby each other

3. In collisions between gas particles, the total energy
   a. decreases slightly
   b. decreases considerably
   c. increases slightly
   d. remains the same

Answers: 1, b; 2, c; 3, d

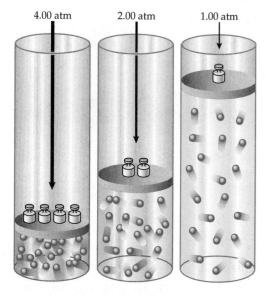

4.00 atm   2.00 atm   1.00 atm

▲ **Figure 6.14** The kinetic–molecular theory and Boyle's law. As the pressure is reduced from 4.00 atm to 2.00 atm and then to 1.00 atm, the volume of the gas doubles and then doubles again.

Q *What would happen to the volume of the gas if the pressure were changed to 0.500 atm?*

6.6 The Simple Gas Laws

Learning Objectives • State the four simple gas laws, by name and mathematically. • Use a gas law to find the value of one variable if the other values are given.

The behavior of gases can be described by mathematical relationships called *gas laws*. These equations use four variables to specify properties of a sample of gas: its amount in moles (n), its volume (V), its temperature (T), and its pressure (P). These variables are related through simple laws that show how one of the variables (for example, V) changes as a second variable (for example, P) changes and the other two (for example, n and T) remain constant.

Boyle's Law: Pressure and Volume

The first gas law was discovered by the English chemist and physicist Robert Boyle (1627–1691) in 1662. It describes the relationship between the pressure and the volume of a gas. **Boyle's law** states that *for a given amount of gas at a constant temperature, the volume of the gas varies inversely with its pressure*. That is, when the pressure in a closed container of gas increases, the volume decreases; when the pressure decreases, the volume increases.

A bicycle pump illustrates Boyle's law. When you push down on the plunger, you decrease the volume of the air in the pump. The pressure of the air increases, and the higher-pressure air flows into the bicycle tire.

Think of gases as depicted by the kinetic–molecular theory. A gas exerts a particular pressure because its molecules bounce against the container walls with a certain frequency and speed. If the volume of the container is increased while the amount of gas remains fixed, the number of molecules per unit volume decreases (Figure 6.14). The frequency with which molecules strike a unit area of the container walls decreases, and the gas pressure decreases. Thus, as the volume of a gas is increased, its pressure decreases.

Mathematically, for a given amount of gas at a constant temperature, Boyle's law is written

$$V \propto \frac{1}{P}$$

where the symbol $\propto$ means "is proportional to." This relationship can be changed to an equation by inserting a proportionality constant, a.

$$V = \frac{a}{P}$$

Multiplying both sides of the equation by P, we get

$$PV = a \text{ (at constant temperature and amount of gas)}$$

Another way to state Boyle's law, then, is that *for a given amount of gas at a constant temperature, the product of the pressure and the volume is a constant*. This is an elegant and precise way of summarizing a lot of experimental data. If the product $P \times V$ is constant, then when V increases, P must decrease, and *vice versa*. This relationship is illustrated in Figure 6.15 by a pressure–volume graph.

Boyle's law has a number of practical applications, which are perhaps best introduced through some examples. In Example 6.2, we see how to estimate an answer. Sometimes an estimate is all we need. Even when we need a quantitative answer, however, an estimate helps us determine whether the answer we get is reasonable.

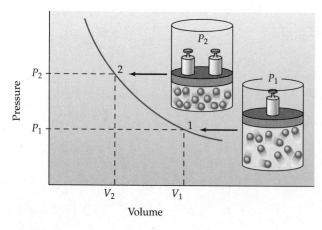

▲ **Figure 6.15** A graphic representation of Boyle's law. As the pressure is increased, the volume of a gas decreases. When the pressure is doubled ($P_2 = 2 \times P_1$), the volume of the gas decreases to one-half of its original value ($V_2 = \frac{1}{2} \times V_1$). The pressure–volume product is a constant ($PV = a$).

Q *What would happen to the volume of the gas if the pressure were quadrupled?*

EXAMPLE 6.2 Boyle's Law: Pressure–Volume Relationships

A gas is enclosed in a cylinder fitted with a piston. The volume of the gas is 2.00 L at 0.524 atm. The piston is moved to increase the pressure to 5.15 atm. Which of the following is a reasonable value for the volume of the gas at the greater pressure if the temperature of the gas stays the same?

0.20 L   0.4 L   1.00 L   16.0 L

Solution

The pressure increase from 0.524 atm to 5.15 atm is almost tenfold. The volume should drop to about one-tenth of the initial value. Thus, we estimate a volume of 0.20 L. (The calculated value is 0.203 L.)

〉 Exercise 6.2A

A syringe used for injecting large animals holds 100 mL. The air in the syringe is at 1.02 atm pressure. If the end is plugged with a finger, which of the following is a reasonable value for the volume when 3.3 atm of pressure is applied to the plunger?

0.030 mL   0.30 mL   3.0 mL   30 mL

〉 Exercise 6.2B

Argon gas is enclosed in a 10.2 L tank at 1208 mmHg. (*mmHg* is a pressure unit: 760 mmHg = 1 atm.) Which of the following is a reasonable value for the pressure when the argon is transferred to a 30.0 L tank at a constant temperature?

300 mmHg   400 mmHg   3600 mmHg   12,000 mmHg

Example 6.3 illustrates quantitative calculations using Boyle's law. For a given confined sample of a gas at a constant temperature, the initial pressure (P_1) times the initial volume (V_1) is equal to the final pressure (P_2) times the final volume (V_2). That is, Boyle's law can be written as the following equation.

$$P_1V_1 = a = P_2V_2$$

 EXAMPLE 6.3 Boyle's Law: Pressure–Volume Relationships

A cylinder of oxygen has a volume of 2.25 L. The pressure in the cylinder is 1470 pounds per square inch (psi) at 20 °C. What volume will the oxygen occupy at standard atmospheric pressure (14.7 psi), assuming no temperature change?

Solution

We find it helpful to first separate the initial and final conditions.

Initial	Final	Change
$P_1 = 1470$ psi	$P_2 = 14.7$ psi	The pressure goes down; therefore,
$V_1 = 2.25$ L	$V_2 = ?$	the volume goes up.

Next, we solve the equation $P_1V_1 = P_2V_2$ for the desired volume or pressure. In this case, we solve for V_2. Finally, we substitute the given values for P_1, V_1, and P_2.

$$V_2 = \frac{P_1V_1}{P_2}$$

$$V_2 = \frac{1470 \text{ psi} \times 2.25 \text{ L}}{14.7 \text{ psi}} = 225 \text{ L}$$

Note how the units for pressure cancel. It doesn't matter what units we use for pressure and volume, as long as we use the same ones throughout the calculation. Because the final pressure (14.7 psi) is *less than* the initial pressure (1470 psi), we expect the final volume (225 L) to be *larger than* the original volume (2.25 L), and we see that it is.

> **Exercise 6.3A**
A sample of air occupies 73.3 mL at 98.7 atm and 0 °C. What volume will the air occupy at 4.02 atm and 0 °C?

> **Exercise 6.3B**
A sample of helium occupies 535 mL at 988 mmHg and 25 °C. If the sample is transferred to a 1.05 L flask at 25 °C, what will the gas pressure in the flask be?

---

5 How cold can it possibly get, and how hot can it possibly get?

Quantum mechanics (Section 3.7) says that all atoms must always keep moving, at least a little bit. At 0 K (absolute zero, −273.15 °C), atomic motion is essentially nonexistent. Because atoms cannot move any more slowly than this, no lower temperature is possible. Theoretically, there is no upper limit on the amount of kinetic energy atoms can have. However, at such high values, temperature ceases to have much meaning.

Charles's Law: Temperature and Volume

In 1787, the French physicist Jacques Charles (1746–1823), a pioneer hot-air balloonist, studied the relationship between volume and temperature of gases. He found that when a fixed mass of gas is cooled at constant pressure, its volume decreases. When the gas is heated, its volume increases. Temperature and volume vary directly; that is, they rise or fall together. But this relationship is a bit more complex. If 1.00 L of gas is heated from 100 °C to 200 °C at constant pressure, the volume increases, but only to 1.27 L rather than to 2.00 L. The relationship between temperature and volume is not as simple as it may seem at first.

For most measurements and quantities, "zero" means "none." A measured pressure of 0 atm means no measurable pressure, and it is not possible to have a lower pressure than 0 atm. On the other hand, zero degrees Celsius (0 °C) does not signify the lowest possible temperature, but merely signifies the freezing point of water. The Celsius scale is a *relative* scale rather than an absolute scale; temperatures below 0 °C are often encountered. So ... is there a temperature that is actually zero, meaning "no temperature"?

Charles noted that for each degree Celsius that the temperature rises, the volume of a gas increases by $\frac{1}{273}$ of its volume at 0 °C. A plot of volume against temperature is a straight line (Figure 6.16). We can extrapolate (extend) the line to the temperature at which the volume of the gas should become zero. This temperature is −273.15 °C.

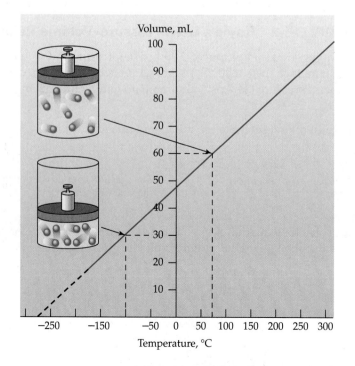

▶ **Figure 6.16** Charles's law relates gas volume to temperature at constant pressure. In the case shown, when the gas is at about 70 °C, its volume is 60 mL. As the temperature is lowered from about 70 °C to −100 °C, the volume drops to 30.0 mL. The volume continues to fall as the temperature is lowered. The extrapolated (dashed) line intersects the temperature axis (corresponding to a volume of zero) at about −273 °C.

Q *What would be the approximate volume of the gas at a temperature of −150 °C?*

In 1848, the Scots-Irish physicist William Thomson (Lord Kelvin; 1824–1907) made this temperature the zero point on an absolute temperature scale now called the Kelvin scale. (There are other absolute temperature scales, but the Kelvin scale is by far the most commonly used.) As noted in Chapter 1, the unit on this temperature scale is the kelvin (K).

Charles's law says that the volume of a fixed amount of a gas at a constant pressure is directly proportional to its absolute temperature. Mathematically, this relationship is expressed as

$$V = kT \quad \text{or} \quad \frac{V}{T} = k$$

where k is a proportionality constant. To keep V/T at a constant value, when the temperature increases, the volume must also increase. When the temperature decreases, the volume must decrease accordingly (Figure 6.17).

With a given sample of trapped gas at a constant pressure, the initial volume (V_1) divided by the initial absolute temperature (T_1) is equal to the final volume (V_2) divided by the final absolute temperature (T_2). We can therefore use the following equation to solve problems involving Charles's law.

$$\frac{V_1}{T_1} = \frac{V_2}{T_2}$$

The kinetic–molecular model readily explains the relationship between gas volume and temperature. When we heat a gas, the gas molecules absorb energy and move faster. These speedier molecules strike the walls of the container more forcefully and more often. For the pressure to stay the same, the volume of the container must increase so that the increased molecular motion will be distributed over a greater space.

 EXAMPLE 6.4 Charles's Law: Temperature–Volume Relationship

In a room at 27 °C, a balloon has a volume of 2.00 L. What would its volume be **(a)** in a sauna, where the temperature is 47 °C, and **(b)** outdoors, where the temperature is −23 °C? Assume that there is no change in pressure in either case.

Solution

First, and most important, we convert all temperatures to the Kelvin scale:

$$T(\text{K}) = t(°\text{C}) + 273$$

The initial temperature (T_1) in each case is 27 + 273 = 300 K, and the final temperatures are **(a)** 47 + 273 = 320 K and **(b)** −23 + 273 = 250 K.

a. We start by separating the initial from the final condition.

Initial	Final	Change
$T_1 = 300$ K	$T_2 = 320$ K	⇑
$V_1 = 2.00$ L	$V_2 = ?$	⇑

Solving the equation

$$\frac{V_1}{T_1} = \frac{V_2}{T_2}$$

for V_2, we have

$$V_2 = \frac{V_1 T_2}{T_1}$$

Substituting the known values gives

$$V_2 = \frac{2.00\ \text{L} \times 320\ \text{K}}{300\ \text{K}} = 2.13\ \text{L}$$

As expected, because the temperature increases, the volume must also increase.

(a)

(b)

▲ **Figure 6.17** A dramatic illustration of Charles's law. (a) Liquid nitrogen (boiling point −196 °C) cools the balloon and its contents to a temperature far below room temperature. (b) As the balloon warms back to room temperature, the volume of air increases proportionately (about fourfold).

b. We have the same initial conditions as in **(a)** but different final conditions.

Initial	Final	Change
$T_1 = 300$ K	$T_2 = 250$ K	⇓
$V_1 = 2.00$ L	$V_2 = ?$	⇓

Again, using Charles's law, we solve the equation for V_2 and substitute the known quantities.

$$V_2 = \frac{V_1 T_2}{T_1}$$

$$V_2 = \frac{2.00 \text{ L} \times 250 \text{ K}}{300 \text{ K}} = 1.67 \text{ L}$$

As expected, the volume decreases because the temperature decreases.

> **Exercise 6.4A**
A sample of ethane gas occupies a volume of 16.8 L at 25 °C. What volume will this sample occupy at 109 °C? (Assume no change in pressure.)

> **Exercise 6.4B**
At what Celsius temperature will the initial volume of ethane in Exercise 6.3A occupy 0.750 L? (Assume no change in pressure.)

In addition to Boyle's law, which describes the relationship between pressure and volume, and Charles's law, relating volume and temperature, another simple gas law—called Gay-Lussac's law—relates temperature and pressure. **Gay-Lussac's law** states that *the pressure of a fixed amount of a gas at a constant volume is directly proportional to its absolute temperature.* In other words:

$$P = kT \quad \text{or} \quad \frac{P_1}{T_1} = \frac{P_2}{T_2}$$

 EXAMPLE 6.5 **Gay-Lussac's Law: Temperature and Pressure Relationship**

If a 20.0 L cylinder containing a gas at 6.00 atm pressure at 27 °C is heated to 77 °C, what would the pressure of the gas be?

Solution
The cylinder's volume remains unchanged. Gay-Lussac's law can be expressed as

$$\frac{P_1}{T_1} = \frac{P_2}{T_2}$$

Whenever working with the gas laws, first convert all temperatures to the Kelvin scale:

The initial temperature (T_1) is $(27 + 273) = 300$ K and the final temperature is $(77 + 273) = 350$ K.

We start by separating the initial from the final condition.

Initial	Final	Change
$t_1 = 27$ °C	$t_2 = 77$ °C	⇑
$T_1 = 300$ K	$T_2 = 350$ K	⇑
$P_1 = 6.00$ atm	$P_2 = ?$	⇑

Solving the Gay-Lussac equation for P_2:

$$P_2 = \frac{P_1 T_2}{T_1}$$

$$P_2 = \frac{6.00 \text{ atm} \times 350 \text{ K}}{300 \text{ K}} = 7.00 \text{ atm}$$

❭ Exercise 6.5A

A CO_2 tank for use in a paintball is filled to 124 psi at 35 °C. What will the tank pressure be if the temperature drops to 5 °C?

❭ Exercise 6.5B

A weather balloon has a pressure of 1.00 atm at 23 °C. At what temperature will the pressure of the balloon reach 0.600 atm if the volume of the balloon doesn't change?

Molar Volume

In Section 5.2 we saw that equal numbers of molecules of different gases at the same temperature and pressure occupy equal volumes. Thus, the amount of a gas (number of molecules or of moles) is directly related to the gas volume when temperature and pressure remain constant. We call the simple gas law implied by this relationship **Avogadro's law**: At a fixed temperature and pressure, the volume of a gas is directly proportional to the amount of gas (that is, to the number of moles of gas, n, or to the number of molecules of gas). If we double the number of moles of gas at a fixed temperature and pressure, the volume of the gas doubles. Mathematically, we can state Avogadro's law as

$$V \propto n \quad \text{or} \quad V = cn \quad \text{(where c is a constant)}$$

A mole of gas contains Avogadro's number of molecules (or atoms, if it is a noble gas). Furthermore, a gas sample consisting of Avogadro's number of molecules occupies the same volume (at a given temperature and pressure) regardless of the size or mass of the individual molecules. The volume occupied by one mole of gas is the **molar volume** of a gas.

Because the volume of a gas is altered by changes in temperature or pressure, a particular set of conditions has been chosen as the standard set for reference purposes. Standard pressure is 1 atmosphere (atm), which is the normal pressure of the air at sea level. Standard temperature is 0 °C, which is the freezing point of water. A mole of any gas at **standard temperature and pressure (STP)** occupies a volume of about 22.4 L, slightly more than a five-gallon water bottle (Figure 6.18). This is known as the *standard molar volume* of a gas. At a temperature of 0 °C and 1 atm pressure, a 22.4 L container holds one mole (28.0 g) of N_2, one mole (32.0 g) of O_2, one mole (44.0 g) of CO_2, and so on.

We can readily calculate the density (g/L) of a gas at STP. We begin with the molar mass of the gas (g/mol) and use the conversion factor 1 mol gas = 22.4 L. Because the conversion factor can be inverted, we can also use it to calculate molar mass from density. Note that this relationship is only valid at STP.

▲ **Figure 6.18** An "empty" bottle from an office water cooler is not really empty. It holds five gallons of air, slightly less than 22.4 L—the volume of one mole of a gas at STP.

Q *About how many C_2H_6 molecules would the bottle hold at STP?*

EXAMPLE 6.6 Density of a Gas at STP

Calculate the density of **(a)** nitrogen gas and **(b)** methane (CH_4) gas, both at STP.

Solution

a. The molar mass of N_2 gas is 28.0 g/mol. We multiply by the conversion factor 1 mol N_2 = 22.4 L, arranged to cancel units of *moles*.

$$\frac{28.0 \text{ g } N_2}{1 \text{ mol } N_2} \times \frac{1 \text{ mol } N_2}{22.4 \text{ L } N_2} = 1.25 \text{ g/L}$$

6 **Why is it recommended that tire pressure be measured when tires are cold?**

Tire pressure is measured when the tires are cold because a tire has nearly constant volume. According to Gay-Lussac's law, as the temperature increases (by friction from the road), the pressure increases, too. If the tire pressure were adjusted right after a long drive, the tires might be soft and underpressurized the next morning. Air temperature also affects tire pressure. So it's a good idea to check your car's tire pressure any time the temperature changes significantly.

b. The molar mass of CH_4 gas is $(1 \times 12.0)\,\text{g/mol} + (4 \times 1.01)\,\text{g/mol} = 16.0\,\text{g/mol}$. Again, we use the conversion factor $1\,\text{mol}\;CH_4 = 22.4\,\text{L}$.

$$\frac{16.0\,\text{g }CH_4}{1\,\text{mol }CH_4} \times \frac{1\,\text{mol }CH_4}{22.4\,\text{L }CH_4} = 0.714\,\text{g/L}$$

❯ Exercise 6.6A
Calculate the density of xenon gas at STP.

❯ Exercise 6.6B
Estimate the density of air at STP (assume that air is 78% N_2 and 22% O_2), and compare this value with the value you calculated for xenon in Exercise 6.6A.

SELF-ASSESSMENT Questions

1. Gas pressure is caused by
 a. gas molecules colliding with each other
 b. gas molecules colliding with vessel walls
 c. gas molecules condensing to form a liquid
 d. measurement with a barometer

2. Boyle's law states that the pressure of a gas is inversely proportional to its
 a. amount **b.** mass
 c. temperature **d.** volume

3. A gas sample occupies 6.00 L at 2.00 atm pressure. At what pressure would the volume be 1.50 L if the temperature remained constant?
 a. 0.500 atm **b.** 8.00 atm
 c. 9.00 atm **d.** 12.0 atm

4. In an experiment, as the temperature is changed at constant pressure, the volume of a gas is halved. How is the final temperature related to the initial temperature?
 a. The Celsius temperature is doubled.
 b. The Celsius temperature is halved.
 c. The kelvin temperature is doubled.
 d. The kelvin temperature is halved.

5. One 1.00 L flask (Flask A) contains NO gas, and another 1.00 L flask (Flask B) contains NO_2 gas, both at STP. Flask A contains
 a. less mass and fewer molecules than Flask B
 b. less mass but the same number of molecules as Flask B
 c. less mass but more molecules than Flask B
 d. more mass and more molecules than Flask B

6. How many moles are there in 5.60 L of a gas at STP?
 a. 0.250 mol **b.** 0.500 mol
 c. 4.00 mol **d.** 125 mol

Answers: 1, b; 2, d; 3, b; 4, d; 5, b; 6, a

6.7 The Ideal Gas Law

Learning Objective • State the ideal gas law, and use it to calculate one of the quantities if the others are given.

Boyle's law, Charles's law, Gay-Lussac's law, and Avogadro's law are useful when two variables of the four (P, V, n, and T) are held constant. Often, however, the temperature, pressure, and volume all change at the same time. In such cases, the simple gas laws can be incorporated into a single relationship called the **combined gas law**. Mathematically, this law is written as

$$\frac{PV}{T} = k \quad or \quad PV = kT$$

where k is a constant. For comparing the same sample of gas under two different sets of conditions, the combined gas law can be written as

$$\frac{P_1 V_1}{T_1} = \frac{P_2 V_2}{T_2}$$

The combined gas law can also be used in place of Boyle's law, Charles's law, and Gay-Lussac's law, simply by cancelling the variable P, V, or T that is constant on both sides. Because the amount of gas, n, may change as well, incorporating this variable gives an expression called the **ideal gas law**, which involves all four variables.

$$\frac{PV}{nT} = R \quad \text{or} \quad PV = nRT$$

In this equation, R is a constant (called the *gas constant*), which can be calculated from the fact that one mole of gas occupies 22.4 L at 273 K (0 °C) and 1 atm (Section 6.6). If P is in atmospheres, V in liters, and T in kelvins, then R has a value of

$$R = 0.0821 \frac{\text{L} \cdot \text{atm}}{\text{mol} \cdot \text{K}}$$

Note that the equation is valid *only* if the units are atmospheres, liters, and kelvins. Otherwise we must first convert to those units before applying the equation. The ideal gas equation can be used to calculate the value of any of the four variables— P, V, n, or T—if the other three values are known.

EXAMPLE 6.7 The Combined Gas Law

A balloon used for underwater salvage is inflated to 50.0 L at a depth of 200 ft., where the pressure is 6.89 atm and the temperature is 3 °C. The balloon rises to the surface (22 °C and 0.988 atm). What is the new volume of the balloon?

Solution
We start by solving the combined gas law equation for V_2.

$$\frac{P_1 V_1}{T_1} = \frac{P_2 V_2}{T_2}$$

$$V_2 = \frac{P_1 V_1 T_2}{T_1 P_2}$$

We have $T_1 = 3 + 273 = 276$ K, and $T_2 = 22 + 273 = 295$ K.

$$V_2 = \frac{6.89 \ \cancel{\text{atm}} \times 50.0 \ \text{L} \times 295 \ \cancel{\text{K}}}{276 \ \cancel{\text{K}} \times 0.988 \ \cancel{\text{atm}}} = 373 \ \text{L}$$

> Exercise 6.7A
What will be the final volume of a 425 mL sample of a gas initially at 22.0 °C and 760 mmHg pressure when the temperature is changed to 30.0 °C and the pressure to 360 mmHg?

> Exercise 6.7B
What will be the final pressure of a 2.34 L sample of a gas initially at 28.0 °C and a pressure of 823 mmHg if the temperature is changed to 44.0 °C and the volume to 4.76 L?

EXAMPLE 6.8   Ideal Gas Law

Use the ideal gas law to calculate **(a)** the volume occupied by 1.00 mol N_2 gas at 244 K and 1.00 atm pressure and **(b)** the pressure exerted by 0.500 mol O_2 gas in a 15.0 L container at 303 K.

Solution

a. We solve the ideal gas equation for V and substitute the known quantities. It's important when using the ideal gas law to always to express volume in L, pressure in atm, and temperature in K.

$$V = \frac{nRT}{P}$$

$$V = \frac{1.00 \ \cancel{mol}}{1.00 \ \cancel{atm}} \times \frac{0.0821 \ \text{L} \cdot \cancel{atm}}{\cancel{mol} \cdot \cancel{K}} \times 244 \ \cancel{K} = 20.0 \ \text{L}$$

b. Here we solve the ideal gas equation for P and substitute the known quantities.

$$P = \frac{nRT}{V}$$

$$P = \frac{0.500 \ \cancel{mol}}{15.0 \ \cancel{L}} \times \frac{0.0821 \ \cancel{L} \cdot \text{atm}}{\cancel{mol} \cdot \cancel{K}} \times 303 \ \cancel{K} = 0.829 \ \text{atm}$$

❯ **Exercise 6.8A**
Determine **(a)** the pressure exerted by 0.0450 mol O_2 gas in a 22.0 L container at 313 K and **(b)** the volume occupied by 0.400 mol N_2 gas at 298 K and 0.980 atm.

❯ **Exercise 6.8B**
Determine the volume occupied by 132 g N_2 gas at 25 °C and 1.03 atm.

Table 6.1 summarizes the gas laws.

TABLE 6.1   The Gas Laws

Law	Description	Related Equation
Boyle's law	For a given amount of gas at a constant temperature, the volume of the gas varies inversely with its pressure.	$PV = k$ or $P_1V_1 = P_2V_2$ (where k is a constant)
Charles's law	The volume of a fixed amount of a gas at a constant pressure is directly proportional to its absolute temperature.	$\dfrac{V}{T} = k$ or $\dfrac{V_1}{T_1} = \dfrac{V_2}{T_2}$ (k is a constant)
Gay-Lussac's law	The pressure of a fixed amount of a gas at a constant volume is directly proportional to its absolute temperature.	$\dfrac{P}{T} = k$ or $\dfrac{P_1}{T_1} = \dfrac{P_2}{T_2}$ (k is a constant)
Avogadro's law	At a fixed temperature and pressure, the volume of a gas is directly proportional to the amount of gas (that is, to the number of moles of gas, n, or to the number of molecules of gas).	$\dfrac{V}{n} = k$ or $\dfrac{V_1}{n_1} = \dfrac{V_2}{n_2}$ (k is a constant)
Combined gas law		$\dfrac{PV}{T} = k$ or $\dfrac{P_1V_1}{T_1} = \dfrac{P_2V_2}{T_2}$ (k is a constant)
Ideal gas law		$\dfrac{PV}{nT} = k$ or $PV = nRT$ (R is the gas constant)

GREEN CHEMISTRY Supercritical Fluids
Doug Raynie, *South Dakota State University*

Learning Objectives • Describe how Green Chemistry Principle 3 must be considered in designing chemical reactions and processes. • Identify the properties that make supercritical fluids applicable in greener chemical processes.

Many chemical reactions take place in solution (Section 5.5), and most chemical processes involve a separation of components at some stage. Water is an excellent solvent for many substances, but some substances do not dissolve in water under ordinary conditions (Section 6.4). Chemists have traditionally used organic solvents such as benzene (C_6H_6), a cancer-causing aromatic hydrocarbon (Section 9.2), and methylene chloride (CH_2Cl_2), a toxic chlorinated hydrocarbon (Section 9.3), to dissolve nonpolar and slightly polar substances. Wouldn't it be nice to have a nontoxic, benign solvent that could be recycled readily?

Let's look again at the states of matter and changes of state (Section 6.1), particularly the change from liquid to gas. Have you ever used a pressure cooker, or tried to cook food while backpacking in the mountains? If you have, you've probably noticed that boiling point changes with pressure. In a pressure cooker, the increase in pressure lets us cook at higher temperatures. In the mountains, where the air pressure is lower, water boils at a lower temperature, so it takes longer to prepare a meal. This behavior is shown in the following graph called a *phase diagram*:

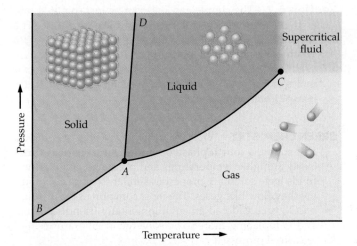

▲ A phase diagram shows the physical state of a material at given temperatures and pressures. As the temperature and pressure increase, the material shown in the graph changes from a solid to a liquid or gas. Above a certain temperature and pressure (the critical point C), the material is neither gas nor liquid but becomes a supercritical fluid.

The curve *AC* slopes upward, showing that the boiling point increases with increasing pressure. Notice that the line *AC* does not continue forever. Eventually, continued heating will take a substance

to point *C*, called the *critical point*. Above the critical point, matter exists as neither a gas nor a liquid but as a kind of a hybrid called a supercritical fluid. We can think of supercritical fluids as having properties of both gases and liquids. Most importantly, they can dissolve things as liquids do, but they flow rapidly like gases. We can vary these properties to make the fluid more liquid-like or more gas-like by changing the temperature or pressure.

The most commonly used supercritical fluid is carbon dioxide ($scCO_2$), an inexpensive, convenient-to-use, nonflammable, and nontoxic substance. Carbon dioxide becomes a supercritical fluid at temperatures greater than 31 °C and pressures greater than 73 atm. As a solvent, carbon dioxide (nonpolar) is a complement to water (polar), giving us two environmentally friendly solvents with different properties and uses. Individually or combined, these supercritical fluids can dissolve a wide range of chemical compounds. Also, because they can flow like gases, supercritical fluids can often make processes occur more quickly.

Supercritical water and $scCO_2$ are "generally regarded as safe" by the U.S. Food and Drug Administration and are often used in the food industry. (See Chapter 17.) For example, coffee is sometimes decaffeinated using $scCO_2$ extraction. After the extraction, the $scCO_2$ is decompressed and returned to a gaseous state, leaving the caffeine behind. The gaseous CO_2 is cooled and compressed to return it to a liquid form to be used again. This process is much greener than the older one using toxic methylene chloride, a liquid that had to be removed from the coffee by distillation.

Flavors and fragrances are often isolated using $scCO_2$. In the cosmetics industry, oils such as palm oil or jojoba oil are isolated from their plant sources using $scCO_2$. Supercritical fluids are used to clean or degrease precision parts in electronic instruments and medical devices. In more specialized cases, chemical reactions are carried out using supercritical fluid solvents. The main drawback to the use of supercritical fluids is that special equipment is needed to control the temperatures and high pressures required for these processes. The generation of specialty materials as nanoparticles (Section 12.2) or low-density aerogels, very strong insulators with nanoscale structures, is also favored in $scCO_2$.

Nearly all chemical processes require the separation of one chemical from a mixture of others by dissolving the compound of interest in a separate phase. Water and $scCO_2$ give us a pair of inexpensive and safe solvents that dissolve a wide range of chemical compounds—an excellent application of Green Chemistry Principle 5, which urges the use of safer reaction conditions and less toxic solvents.

SELF-ASSESSMENT Questions

1. In the ideal gas equation, $PV = nRT$, the variable n stands for
  a. a constant
  b. the number of atoms
  c. the number of moles
  d. the principal quantum number

2. At STP, 1.00 mol of an ideal gas occupies 22.4 L. What volume does 1.00 mol of an ideal gas occupy at 20 °C and 1.50 atm?
  a. 13.9 L   **b.** 22.4 L   **c.** 16.0 L   **d.** 672 L

3. The molar volume of an ideal gas at 2.00 atm and 546 K is
  a. 5.60 L   **b.** 22.4 L   **c.** 44.8 L   **d.** 89.6 L

Answers: 1, c; 2, c; 3, b

 # Summary

Section 6.1—When a substance is melted or vaporized, the forces that hold its particles (molecules, atoms, or ions) close to one another are overcome. The **melting point** of a solid is the temperature at which the forces holding the particles together in a regular arrangement are overcome. **Freezing** is the reverse process, a liquid changing to a solid. **Vaporization** is the conversion of a liquid to a gas; the reverse process is called **condensation**. The temperature at which the vapor leaving a liquid is at the same pressure as the air around it is called its **boiling point**. Some substances undergo **sublimation**, a conversion directly from the solid to the gaseous state; the reverse process is **deposition**.

Section 6.2—Ionic solids have very strong forces holding the ions to one another, so most ionic compounds have higher melting points and higher boiling points than molecular compounds.

Section 6.3—Polar molecules exhibit **dipole–dipole forces** from the attraction of the positive end of one dipole for the negative end of another dipole. Nonpolar molecules exhibit **dispersion forces** resulting from electron motions within molecules, which cause tiny, short-lived dipoles. Dispersion forces are generally weak. An especially strong type of force called a **hydrogen bond** occurs when a hydrogen atom attached to N, O, or F is attracted to a second N, O, or F atom. Generally, hydrogen bonds are much stronger than dipole–dipole forces, which are stronger than most dispersion forces.

Section 6.4—A **solution** is a homogeneous mixture of two or more substances. A substance that is dissolved is called a **solute**, and the dissolving substance is called the **solvent**. Nonpolar solutes dissolve in nonpolar solvents. Polar and ionic solutes dissolve in polar solvents.

Section 6.5—The behavior of gases is explained using **kinetic–molecular theory**, which describes gas particles: (1) They move rapidly and constantly and in straight lines;

(2) they are far apart; (3) there is little interaction among them; (4) when they collide, energy is conserved; and (5) temperature is a measure of their average kinetic energy.

Section 6.6—**Boyle's law** says that the volume of a fixed amount of a gas varies inversely with pressure at constant temperature, or $P_1V_1 = P_2V_2$. **Charles's law** says that the volume of a fixed amount of a gas varies directly with absolute (Kelvin) temperature at constant pressure, or $V_1/T_1 = V_2/T_2$. **Gay-Lussac's law** says that the pressure of a fixed amount of a gas varies directly with absolute (Kelvin) temperature at constant volume, or $P_1/T_1 = P_2/T_2$. **Avogadro's law** states that at constant temperature and pressure, the volume of a gas is directly proportional to the number of moles of gas, or $V_1/n_1 = V_2/n_2$. The **molar volume** of any gas is the volume occupied by 1 mol of the gas. At 0 °C and 1 atm, known as **standard temperature and pressure (STP)**, the molar volume is 22.4 L for any gas. We can use that molar volume to convert from moles of gas to volume, and vice versa.

Section 6.7—The simple gas laws can be incorporated into a single relationship called the **combined gas law**. The **ideal gas law** shows the relationship among pressure (P), volume (V), number of moles (n), and absolute temperature (T): $PV = nRT$. When P is in atmospheres, V is in liters, n is moles of gas, and T is in kelvins, the value of the constant R is 0.0821 liter·atm/(mol·K).

GREEN CHEMISTRY—While we are familiar with gases, liquids, and solids, a fourth phase of matter, **supercritical fluid**, exists at higher temperatures and pressures. Supercritical fluids have solubility properties similar to those of liquids, yet they flow like gases. The most common of these fluids, $scCO_2$, is considered an environmentally friendly solvent and is replacing more toxic solvents in laboratories and industrial processes.

Learning Objectives

Learning Objectives	Associated Problems
• Explain how the different properties of solids, liquids, and gases are related to the motion and spacing of atoms, molecules, or ions. (6.1)	1–4
• Identify some differences between ionic and molecular substances, and explain why these differences exist. (6.2)	8, 9
• Classify forces between molecules as dipole–dipole forces, dispersion forces, or hydrogen bonds. (6.3)	5, 10–15
• Explain the effects that intermolecular forces have on melting and boiling points. (6.3)	6, 13

• Explain why nonpolar solutes tend to dissolve in nonpolar solvents and polar and ionic solutes tend to dissolve in polar solvents. (6.4)	14, 15
• List the five basic concepts of the kinetic–molecular theory of gases. (6.5)	51, 52
• State the four simple gas laws, by name and mathematically. (6.6)	7, 54
• Use a gas law to find the value of one variable if the other values are given. (6.6)	16–38, 54–64
• State the ideal gas law, and use it to calculate one of the quantities if the others are given. (6.7)	39–50, 67
• Describe how Green Chemistry Principle 3 must be considered in designing chemical reactions and processes.	68, 69
• Identify the properties that make supercritical fluids applicable in greener chemical processes.	70, 71

 Conceptual Questions

1. In what ways are liquids and solids similar? In what ways are they different?

2. Define each of the following terms.
   a. melting   b. vaporization
   c. condensation   d. freezing

3. Label each arrow with the correct term from Question 2 that identifies the process.

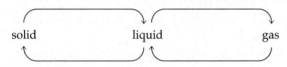

solid   liquid   gas

4. In which process is energy absorbed by the substance undergoing the change of state?
   a. melting or freezing
   b. condensation or vaporization

5. List four types of interactions between the particles of a substance in the liquid and solid states, from weakest to strongest. Give an example of each type.

6. What intermolecular forces or combinations of intermolecular forces are likely to lead to
   a. the highest melting point?
   b. the lowest boiling point?

7. In the combined gas law, is the volume of a gas directly proportional or inversely proportional to the pressure? Is the volume directly or inversely proportional to the absolute temperature?

 Problems

Intermolecular Forces

8. What type of intermolecular interaction predominates in each substance?
   a. HBr   b. SiF$_4$ c. NH$_2$OH

9. Which of the following have ionic bonds: N$_2$, NBr$_3$, and NaBr?

10. For which of the following would hydrogen bonding be an important intermolecular force?

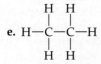

a. H—S | H
b. H—C—N—H (with H's on C and N)
c. H—C—F (with H's on C)
d. H—C—O (with H's)
e. H—C—C—H (with H's)

11. In which of the following are dispersion forces the only type of intermolecular force: F$_2$, H$_2$S, and CF$_4$?

12. In which of the following substances are dipole–dipole forces an important intermolecular force: Br$_2$, HBr, and NaBr?

13. Order the substances in Question 8 from lowest to highest expected melting point.

14. Hexane (C$_6$H$_{14}$) is a nonpolar solvent. Would you expect KNO$_3$ to dissolve in hexane? Explain.

15. Which of the following would you expect to dissolve in water and which in carbon tetrachloride (CCl$_4$)?
    a. CF$_4$ b. HF   c. C$_6$H$_{12}$ d. (NH$_4$)$_2$S

Boyle's Law

16. A sample of helium occupies 1820 mL at 719 mmHg. Assume that the temperature is held constant and determine (a) the volume of the helium at 752 mmHg and (b) the pressure, in mmHg, if the volume is changed to 345 mL.

17. A decompression chamber used by deep-sea divers has a volume of 10.1 m³ and operates at an internal pressure of 4.25 atm. What volume, in cubic meters, would the air in the chamber occupy if it were at 1.00 atm pressure, assuming no temperature change?

18. Oxygen used in respiratory therapy is stored at room temperature under a pressure of 150 atm in a gas cylinder with a volume of 60.0 L.
    a. What volume would the oxygen occupy at 0.925 atm? Assume no temperature change.
    b. If the oxygen flow to the patient is adjusted to 6.00 L/min. at room temperature and 0.925 atm, how long will the tank of gas last?

19. *Sealab II*, a habitat used for underwater research, had a volume of 2.2 × 10⁴ L, and in use it was held at an internal pressure of 5.4 × 10³ mmHg with a mixture of nitrogen, oxygen, and helium. What volume in liters of this mixture would be needed to fill *Sealab II* at 0.945 atm?

Sealab II

20. A gas sample has a volume of 0.954 L at a pressure of 764 mmHg. What would be the pressure of the gas if the volume were changed to 477 mL? Assume that the amount and the temperature of the gas remain constant.

21. A balloon filled with air at a pressure of 0.994 atm has a volume of 1.88 L. What will be the volume of the balloon if the pressure is changed to 0.497 atm? Assume that the temperature and the amount of the gas remain constant.

Charles's Law

22. A balloon is filled with helium. Its volume is 5.90 L at 26 °C. What will its volume be at 78 °C, assuming no pressure change?

23. A gas at a temperature of 100 °C occupies a volume of 2894 mL. What will the volume be at 19 °C, assuming no change in pressure?

24. A sample of gas at STP is to be heated at constant pressure until its volume triples. What will be the final gas temperature in kelvins?

25. A normal breath of air is about 1.00 L. If the air is drawn in at 18 °C and is warmed to 37 °C, what will be the new volume of the air? Assume that the pressure and amount of the gas remain constant.

26. A normal breath of air is about 1.00 L. If the air is drawn in at −10 °C and is warmed to 37 °C, what will be the new volume of the air? Assume that the pressure and the amount of the gas remain constant.

27. A 567 mL sample of a gas at 305 °C and 1.20 atm is cooled at constant pressure until its volume becomes 425 mL. What is the final gas temperature in °C ?

28. An air/gasoline vapor mix in an automobile cylinder has an initial temperature of 119 °C and a volume of 8.34 cm³. If the mixture is heated to 611 °C with the pressure and amount held constant, what will be the final volume of the gas in cubic centimeters?

Gay-Lussac's Law

29. A flask is filled with argon gas at 24 °C and 1.18 atm. What will be its pressure in atmospheres when cooled to liquid nitrogen temperature, 77 K?

30. A tank of neon gas is at 3.44 atm at a temperature of 25 °C. What will be its pressure in atmospheres at 100 °C?

31. A can containing Freon gas (CCl_2F_2) at 2.03 atm and 17 °C is thrown into a fire which is at 752 °C. The can will withstand a pressure of 5.0 atmospheres. Will the can burst?

32. A flask filled with krypton gas at 752 mmHg and 22 °C is cooled in dry ice until its pressure is 500 mmHg. What is the final temperature in °C?

Molar Volume and Gas Densities

33. What is the volume of each of the following samples of gas at STP?
    a. 1.00 mol C_2N_2 b. 8.12 mol SF_6
    c. 0.197 mol H_2

34. What is the mass in grams of one molar volume of each of the gases in Problem 33?

35. Calculate the density, in grams per liter, of xenon (Xe) gas at STP.

36. Calculate the density, in grams per liter, of carbon dioxide (CO_2) gas at STP.

37. Calculate the molar mass of **(a)** a gas that has a density of 2.12 g/L at STP and **(b)** an unknown liquid, whose vapor has a density of 2.97 g/L at STP.

38. Calculate the molar mass of **(a)** a gas that has a density of 1.98 g/L at STP and **(b)** an unknown liquid for which 3.33 L of its vapor at STP weighs 10.88 g.

The Ideal Gas Law

39. Will the volume of a fixed amount of a gas increase, decrease, or remain unchanged with
    a. an increase in pressure at constant temperature?
    b. a decrease in temperature at constant pressure?
    c. a decrease in pressure coupled with an increase in temperature?

40. Will the pressure of a fixed amount of a gas increase, decrease, or remain unchanged with
    a. an increase in temperature at constant volume?
    b. a decrease in volume at constant temperature?
    c. an increase in temperature coupled with a decrease in volume?

41. According to the kinetic–molecular theory, **(a)** what change in temperature occurs if the particles of a gas begin to move more slowly, on average, and **(b)** what change in pressure occurs when the particles of a gas strike the walls of the container less often?

42. For each of the following, indicate whether a given gas will have the same density or different densities in the two containers. If different densities, in which container is the density greater?
    a. Containers A and B have the same volume and are at the same temperature, but the gas in A is at a higher pressure.
    b. Containers A and B are at the same pressure and temperature, but the volume of A is greater than that of B.
    c. Containers A and B are at the same pressure and volume, but the gas in A is at a higher temperature.

43. Calculate (a) the volume, in liters, of 0.00600 mol of a gas at 31 °C and 0.870 atm and (b) the pressure, in atmospheres, of 0.0108 mol CH_4 (g) in a 0.265 L flask at 37 °C.

44. Calculate (a) the volume, in liters, of 1.12 mol H_2S(g) at 62 °C and 1.38 atm and (b) the pressure, in atmospheres, of 4.64 mol CO(g) in a 3.96 L tank at 29 °C.

45. How many moles of Kr(g) are there in 0.555 L of the gas at 0.918 atm and 25 °C?

46. What is the mass in grams of 185 mL of Freon gas, CCl_2F_2, at 0.944 atm and 17 °C?

47. How many moles of gas are there in a 660 mL sample at 298 K and a pressure of 0.154 atm?

48. A 0.334 mol sample of xenon gas has a volume of 20.0 L at a pressure of 0.555 atm. What is the temperature of the gas in degrees Celsius?

49. What must the volume be for 4.55 mol of a gas at 7.32 atm and 285 K?

50. A proposed car air bag uses a canister of CO_2 gas to inflate a 67 L bag at 18 °C and 1.04 atm pressure. What mass of CO_2 must the canister contain?

Expand Your Skills

51. What are the five basic postulates of the kinetic–molecular theory? Which postulate best explains why a gas can be compressed?

52. Which of the postulates of the kinetic–molecular theory explains why the same gas laws can be used to describe most gases?

53. Calculate the density of ethane, C_2H_6(g), at 17 °C and 743 mmHg. (*Hint*: Density is mass/volume. Find the mass and volume of one mole of ethane under these conditions.)

54. Graphing volume of a gas sample as the pressure is changed (at constant temperature) gives a curve called a hyperbola (see Figure 6.14). Graphing volume of a gas sample as temperature is changed gives a straight line (see Figure 6.15). What will be the appearance of a graph of volume of a gas as the number of moles of gas is changed? (*Hint*: Volume is inversely proportional to pressure, while volume is directly proportional to kelvin temperature.)

55. What mass of He(g) will occupy the same volume at STP as 0.75 mol H_2 at STP? (*Hint*: It is not necessary to find the volume of the H_2 at STP.)

56. From Chapter 1 you may recall that density determines whether a material will float in water. The same is true for floating in air. The average molar mass of air is about 28.6 g/mol. Which one of the gases CH_4, Cl_2, and C_4H_{10} might be used to fill a balloon that would float in air? Explain how this question may be answered *without* calculating the densities.

57. Look again at Figure 6.16. If the balloon has a volume of a 1.50 L at 20 °C, what will its final volume be in part (a), assuming that the air in the balloon reaches the temperature of the liquid nitrogen? You may ignore the stretching forces in the rubber.

58. Three 2.00 L flasks, labeled X, Y, and Z, each at 758 mmHg and 21 °C, contain neon (Flask X), argon (Flask Y), and krypton (Flask Z). (a) Which flask holds the most atoms of gas? (b) In which flask does the gas have the greatest density? (c) If Flask X is heated and Flask Y is cooled, which of the three flasks will have the highest pressure?

(d) If the temperature of Flask X is lowered and that of Flask Z is raised, which of the three flasks will contain the largest number of moles of gas?

59. A 14.4 g sample of an unknown gas has a volume of 8.00 L at 760 mmHg and 25 °C. What is the molar mass of the gas?

60. If a basketball is inflated to an internal pressure of 1.32 atm at 25 °C and then taken outside, where the temperature is 10 °C, what will be the final pressure in the ball?

61. Consider a room that is 4.61 m × 9.48 m × 2.63 m. What mass in kg of CO_2(g) at STP will the room hold?

62. A gas at 750 mmHg and 27 °C has a density of 2.32 g/L. Which of the following could it be: CO_2, Kr, H_2S, or C_4H_{10}?

63. The density (mass per unit volume) of air in the atmosphere is proportional to the atmospheric pressure. At sea level, mean atmospheric pressure is 760 mmHg. Boyle's law, which relates the pressure of a given amount of gas to its volume, indicates that density decreases with altitude in the atmosphere. As altitude increases, there is less mass of air above a given point; thus, less pressure is exerted. About half the mass of the atmosphere lies below an altitude of about 5.5 km, about 95% of the mass lies below 25 km, and 99% below 30 km. What is the approximate pressure at an altitude of (a) 5.5 km, (b) 25 km, and (c) 30 km?

64. How many liters of hydrogen gas (at STP) are produced from the electrolysis of 1.00 L H_2O(l)? (*Hint*: 1.00 L H_2O weighs 1.00 kg.)

65. Which of the following will *not* result in an increase in the volume of a gas?
    a. an increase in temperature
    b. an increase in pressure
    c. a decrease in temperature
    d. a threefold increase in pressure together with a twofold reduction in Kelvin temperature

66. Choose the answer that correctly completes the statement: At 0 °C and 0.500 atm, 4.48 L NH_3(g) (a) contains 0.20 mol NH_3; (b) has a mass of 3.40 g; (c) contains 6.02×10^{22} molecules; or (d) contains 0.40 mol NH_3.

67. Butane (C_4H_{10}), a gas at room temperature, is compressed to a liquid and used as a fuel in cigarette lighters and camping stoves. If the 5.00 mL of liquid butane ($d = 0.601$ g/mL at 20 °C) in a lighter vaporizes, about what volume will the gas occupy at STP?

68. Radiation called cosmic microwave background (CMB) was formed about 300,000 years after the Big Bang. CMB radiation fills the universe and can be detected in every direction. The temperature of the universe (and therefore of the radiation) at the time of the Big Bang was about 3000 K. Since that time, the universe has expanded by a factor of 1000. What is the temperature in kelvins of the radiation today?

69. Which of the following properties must be considered when selecting a solvent for a chemical process?

a. solute and solvent polarity
b. ease of isolation of solute
c. flammability
d. toxicity
e. all of the above

70. Many of the Presidential Green Chemistry Challenge Awards have involved the use of liquid or supercritical carbon dioxide. What properties of $scCO_2$ support its use in greener chemical processing?

71. Give two examples of solvents previously used in industrial processes that have been replaced by $scCO_2$.

72. In addition to Green Chemistry Principles 3 and 5, how else might the use of an environmentally friendly solvent like $scCO_2$ impact green chemistry?

 Critical Thinking Exercises

Apply knowledge that you have gained in this chapter and one or more of the FLaReS principles (Chapter 1) to evaluate the following statements or claims.

6.1 Some tire stores claim that filling your car tires with pure, dry nitrogen is much better than using plain air. They make the following claims: **(1)** The pressure inside N_2-filled tires does not rise or fall with temperature changes. **(2)** Nitrogen leaks out of tires much more slowly than air because the N_2 molecules are bigger. **(3)** Nitrogen is not very reactive, and moisture and O_2 in air cause corrosion that shortens tire life by 25% to 30%.

6.2 A researcher claims to have discovered why the gecko can walk on walls and ceilings. His claim is that the lizard's feet have many microscopic hairlike protrusions that get close enough to the wall or ceiling surface to allow intermolecular forces to "take over" and hold the gecko to the surface.

6.3 A microbrewery claims to have prepared "hydrogen beer." This beer is reportedly "carbonated" with H_2 gas instead of CO_2 gas. A chemist at the brewery claims that the H_2 gas molecules engage in hydrogen bonding with the water molecules in the beer, so that more gas can dissolve in the beer.

6.4 A student in an advanced chemistry course has been assigned to determine the density of the vapor from an unknown liquid. She claims that the density is 0.053 g/L.

 Collaborative Group Projects

Prepare a PowerPoint, poster, or other presentation (as directed by your instructor) to share with the class.

1. Look ahead to Table 9.3 and consider this series of compounds: methane (CH_4), ethane (CH_3CH_3), propane ($CH_3CH_2CH_3$), butane ($CH_3CH_2CH_2CH_3$), and pentane ($CH_3CH_2CH_2CH_2CH_3$). For each compound: **(a)** Indicate whether its molecules are polar or nonpolar. **(b)** Indicate the predominant type of intermolecular forces in the liquid state. **(c)** Calculate the molar mass. **(d)** Convert each boiling point from Celsius degrees to kelvins. **(e)** Plot a graph of boiling point (in K) versus molar mass. **(f)** Is there any discernible relationship between molar mass and boiling point?

2. Consider this series of compounds: methane (CH_4), fluoromethane (CH_3F), chloromethane (CH_3Cl), bromomethane

(CH3Br), and iodomethane (CH_3I). Repeat parts **(a)** through **(e)** of Project 1 for these compounds. Boiling points may be found in a printed reference work, such as the *CRC Handbook* or *Merck Index*, or on a website such as that of the National Institute for Standards and Technology (NIST). Is there any discernible relationship between molar mass and boiling point? How does the graph for these compounds differ from that in Project 1? Suggest a reason for the difference.

3. Most ionic compounds are solids, but *ionic liquids* have recently been made and have found many uses. Search the Web for information on ionic liquids. In particular, find out why they are liquids rather than solids, and report on some of their current and potential applications.

LET'S EXPERIMENT! Blow Up My Balloon

Materials Needed

- 1 small balloon
- 1 container with lid large enough to hold the balloon (for instance, a pot and a lid)
- Warm water

- Ice
- Measuring tape, or a piece of string and a ruler (*Hint:* Use the string to measure the circumference of the balloon, then measure the length using the ruler.)

For this exercise, we will blow up a small balloon and observe the volume of the gas upon heating and cooling. You will be applying your knowledge of Charles's law in this experiment.

First, inflate the balloon and secure it so that the air cannot escape. Measure and record the circumference of the balloon (the distance around the outside of the balloon).

Next, place the balloon in a container that is filled with warm tap water. Make sure your container has enough water to cover much of the balloon, but not so much water so that it spills out. Now, cover the container and allow the balloon to warm up to the temperature of the water (about 5 minutes).

Take the balloon out of the container, and measure and record the circumference of the hot balloon. Allow the balloon to cool to room temperature for about 2 minutes to 3 minutes. While the balloon is cooling, pour the warm water out of the container and replace it with ice water.

Then, place the balloon into the container with the ice water and cover the container with the lid. Allow the balloon to cool for about 5 minutes. Remove the balloon from the container, and measure and record the circumference of the cold balloon. What do you notice about the size of the balloon when cold and when warm? Now, place your hands on the cold balloon as it is warming up. What do you notice about the balloon?

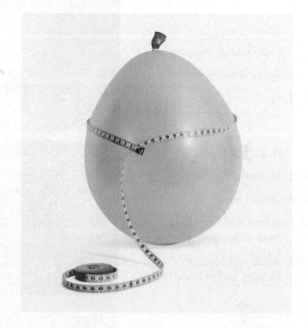

Questions

1. What was the overall change in the circumference of the balloon after heating?
2. What was the overall change in the circumference of the balloon after cooling?
3. Why did the balloon change in size when you heated and cooled the balloon?
4. What can you conclude about the interdependence of temperature and volume of a gas?
5. What would happen if you had a rigid container that did not expand or contract like the balloon? Would you expect the pressure to rise or fall if heated or cooled?

7 Acids and Bases

The chemical substances we call acids and bases are all around us. Acids are used in steel production and metal plating, as well as in chemical analysis. Acids are even found in the food we eat. Pancakes and muffins rise because of acids and bases in the ingredients, and the tart taste of fruit comes from the different acids in its flesh. If you get too much acid in your stomach, antacids (bases) can provide relief.

Have You Ever Wondered?

1 Are all acids corrosive? **2** What is an amino acid? **3** Lye isn't an acid, so why is it dangerous? **4** What is meant by "pH-balanced shampoo"?

PLEASE PASS THE PROTONS Are you fond of sour candy? Have you ever had the slippery feel of bleach on your skin? These are examples of acids (citric, malic, and tartaric acids in sour candy) and bases (sodium hypochlorite in bleach) that we encounter in our daily lives. Other familiar acids are vinegar (acetic acid), vitamin C (ascorbic acid), and battery acid (sulfuric acid). Some familiar bases are drain cleaners (sodium hydroxide), baking soda (sodium bicarbonate), and window cleaner (ammonia).

From "acid indigestion" to "acid rain," the word *acid* appears frequently in the news and in advertisements. Air and water pollution often involve acids and bases. Acid rain, for example, is a serious environmental problem. In arid areas, alkaline (basic) water is sometimes undrinkable.

Did you know that our senses recognize four tastes related to acid–base chemistry? Acids taste sour, bases taste bitter, and **salts** (ionic compounds formed when acids react with bases) taste salty. The sweet taste is more complicated. To taste sweet, a compound must have an acidic part and a basic part, plus just the right geometry to fit the sweet-taste receptors of our taste buds.

In this chapter, we discuss some of the chemistry of acids and bases. You use them every day. Your body processes them continuously. Proteins that you eat are broken down into amino acids, then to be reassembled into proteins that your body needs, with the rest used as fuel. Stomach acid helps digest your food. Basic compounds called amines give fish its "fishy" odor; partially neutralizing the amines with acidic malt vinegar or lemon juice makes the fish more palatable. You will probably hear and read about acids and bases as long as you live. What you learn here can help you gain a better understanding of these important classes of compounds.

7.1 Acids and Bases: Experimental Definitions

Learning Objectives • Distinguish between acids and bases using their chemical and physical properties. • Explain how an acid–base indicator works.

Acids and bases are chemical opposites, and so their properties are quite different or opposite. Let's begin by listing a few of these properties.

An *acid* is a compound that

- Tastes sour
- Causes litmus indicator dye to turn red
- Dissolves active metals such as zinc and iron, producing hydrogen gas
- Reacts with a base to form water and a salt

A *base* is a compound that

- Tastes bitter
- Causes litmus indicator dye to turn blue
- Feels slippery on the skin
- Reacts with an acid to form water and a salt

We can identify foods that are acidic, such as vinegar and lemon juice, by their sour taste. Vinegar is a solution of acetic acid (about 5%) in water. Lemons, limes, and other citrus fruits contain citric acid. Lactic acid gives yogurt its tart taste, and phosphoric acid is often added to carbonated drinks to impart tartness. The bitter taste of tonic water, on the other hand, comes in part from quinine, a base. Figure 7.1 shows some common acids and bases (and salts; see page 200).

▶ **SAFETY ALERT**
Although all acids taste sour and all bases taste bitter, a taste test is *not* an appropriate way to determine whether a substance is an acid or a base. Some acids and bases are poisonous, and some are quite corrosive unless greatly diluted. ***Never taste laboratory chemicals.*** Too many of them are toxic, and others might be contaminated.

◀ **Figure 7.1** Some common acids (left), bases (center), and salts (right). Acids, bases, and salts are components of many familiar consumer products.

Q *How would each of the three classes of compounds affect the indicator dye litmus? (See Figure 7.2.)*

WHY IT MATTERS

Hydrangea is one of many types of flowers that may show different colors depending on the acidity of the soil in which they are grown. Some varieties are blue (top) when planted in slightly acidic soil and pink (bottom) when planted in more basic soil.

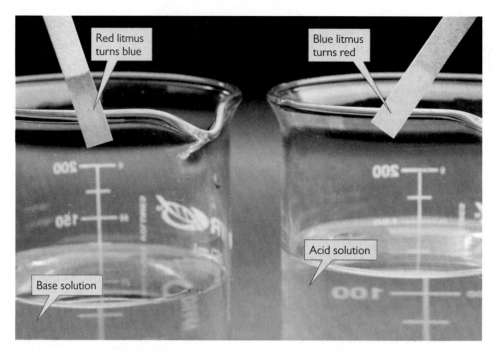

▲ **Figure 7.2** Strips of paper impregnated with litmus dye (extracted from a fungus) are often used to distinguish between acids and bases. The sample on the left turns litmus blue and is therefore basic. The sample on the right turns litmus red and is acidic.

A litmus test is a common way to identify a substance as an acid or a base. Litmus (Figure 7.2) is an **acid–base indicator**, a compound that has a different color in acid than it does in a base. Blue litmus paper that turns pink when dipped into a solution means that the solution is acidic. Red litmus that turns blue in a solution indicates that the solution is basic. (Neutral litmus is a red-violet color.) Many substances, such as those that give the colors to grape juice, red cabbage, blueberries, and many flower petals, are acid–base indicators.

SELF-ASSESSMENT Questions

1. Which of the following is *not* a property of acids?
   a. Feel slippery on the skin
   b. React with Zn to form hydrogen gas
   c. Taste sour
   d. Turn litmus red

2. Which of the following is *not* a property of bases?
   a. Feel slippery on the skin
   b. React with salts to form acids
   c. Taste bitter
   d. Turn litmus blue

3. In general, when an acid and a base are mixed,
   a. a new acid and a salt are formed
   b. a new base and a salt are formed
   c. no reaction occurs
   d. water and a salt are formed

4. A common substance that contains lactic acid is
   a. olive oil
   b. soap
   c. vinegar
   d. yogurt

5. The sour taste of grapefruit is due to
   a. acetic acid
   b. ammonia
   c. sodium chloride
   d. citric acid

▲ You can make your own indicator dye from red cxabbage; see the Let's Experiment! box on page 223. Other plant materials that contain indicators include blackberries, black raspberries, red radish peels, red rose petals, and turmeric.

Answers: 1, a; 2, b; 3, d; 4, d; 5, d

7.2 Acids, Bases, and Salts

Learning Objectives • Identify Arrhenius and Brønsted–Lowry acids and bases.
• Write a balanced equation for a neutralization or ionization reaction.

Acids and bases have certain characteristic properties. But why do they have these properties? We use several different theories to explain them.

The Arrhenius Theory

Swedish chemist Svante Arrhenius (1859–1927) developed the first successful theory of acids and bases in 1887. According to Arrhenius's concept, an **acid** is a substance that forms hydrogen ions (H^+) and anions in water. (Because a hydrogen ion is a hydrogen atom from which the sole electron has been removed, H^+ ions are also called *protons*.) The acid is said to *ionize*. For example, nitric acid ionizes in water.

$$HNO_3(aq) \longrightarrow H^+(aq) + NO_3^-(aq)$$

In water, the properties of acids are those of the H^+ ion. It is the hydrogen ion that turns litmus red, tastes sour, and reacts with bases and active metals. Table 7.1 lists some common acids. Notice that each formula contains one or more hydrogen atoms. Chemists often indicate an acid by writing the formula with the H atom(s) first. HCl, H_2SO_4, and HNO_3 are acids; NH_3 and CH_4 are not. The formula $HC_2H_3O_2$ (acetic acid) indicates that one H atom ionizes when this compound is in aqueous solution, and three do not.

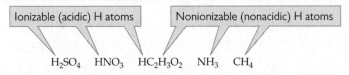

An Arrhenius **base** is a substance that forms hydroxide ions (OH^-) in aqueous solution. Some bases are ionic solids that contain OH^-, such as sodium hydroxide (NaOH) and calcium hydroxide [$Ca(OH)_2$]. These compounds simply release hydroxide ions into the solution when the solid is dissolved in water:

Other bases are molecular substances, such as ammonia, that ionize to produce OH^- when placed in water. (See page 205.)

$$NaOH(s) \xrightarrow{H_2O} Na^+(aq) + OH^-(aq)$$

Table 7.2 lists some common bases. Most of these are ionic compounds containing positively charged metal ions, such as Na^+ or Ca^{2+}, and negatively charged hydroxide ions. When these compounds dissolve in water, they all provide OH^- ions, and thus they are all bases. The properties of bases are those of hydroxide ions, just as the properties of acids are those of hydrogen ions.

▲ Swedish chemist Svante Arrhenius proposed the theory that acids, bases, and salts in water are composed of ions. He also was the first to relate carbon dioxide in the atmosphere to the greenhouse effect (Chapter 13).

1   Are all acids corrosive?

It is clear from Table 7.1 that acid does not necessarily mean "corrosive." Many acids are harmless enough to be included in foods we eat, and some are necessary for life.

TABLE 7.1   Some Common Acids

Name	Formula	Acid Strength	Common Uses/Notes
Sulfuric acid	H_2SO_4	Strong	Battery acid; ore processing, fertilizer manufacturing; oil refining
Hydrochloric acid	HCl	Strong	Cleaning of metals and bricks; removing scale from boilers
Phosphoric acid	H_3PO_4	Moderate	Used in colas and rust removers
Lactic acid	$CH_3CHOHCOOH$	Weak	Yogurt; acidulant (food additive to increase tartness); lotion additive
Acetic acid	CH_3COOH	Weak	Vinegar; acidulant
Boric acid	H_3BO_3	Very weak	Antiseptic eyewash; roach poison
Hydrocyanic acid	HCN	Very weak	Plastics manufacture; extremely toxic

TABLE 7.2 **Some Common Bases**

Name	Formula	Classification	Common Uses/Notes
Sodium hydroxide	NaOH	Strong	Acid neutralization; soap making
Potassium hydroxide	KOH	Strong	Liquid soap making; biodiesel fuels
Lithium hydroxide	LiOH	Strong	Alkaline storage batteries
Calcium hydroxide	$Ca(OH)_2$	Strong[a]	Plaster; cement; soil neutralizer
Magnesium hydroxide	$Mg(OH)_2$	Strong[a]	Antacid; laxative
Ammonia	NH_3	Weak	Fertilizer; household cleansers

[a]Although these bases are classified as strong, they are not very soluble in water. Calcium hydroxide is only slightly soluble, and magnesium hydroxide is even less so; it would take over 20 gallons of water to dissolve a single gram of $Mg(OH)_2$.

▲ Many skin-peel preparations used by cosmetologists contain alpha-hydroxy acids such as glycolic acid or lactic acid.

Arrhenius further proposed that the essential reaction between an acid and a base, **neutralization**, is the combination of H^+ and OH^- to form water. The cation that was originally associated with the OH^- combines with the anion that was associated with the H^+ to form a **salt** (an ionic compound formed in an acid–base reaction).

$$\text{An acid} + \text{a base} \longrightarrow \text{a salt} + \text{water}$$

 CONCEPTUAL EXAMPLE 7.1 **Ionization of Acids and Bases**

Write equations showing **(a)** the ionization of sulfuric acid (H_2SO_4) in water and **(b)** the ionization of solid potassium hydroxide (KOH) in water.

Solution

a. An H_2SO_4 molecule ionizes to form *two* hydrogen ions and a sulfate ion. Because this reaction occurs in water, we can use the label "(aq)" to indicate that the substances involved are in aqueous solution.

$$H_2SO_4(aq) \longrightarrow 2\,H^+(aq) + SO_4^{2-}(aq)$$

b. Potassium hydroxide (KOH), an ionic solid, simply dissolves in the water, forming separate $K^+(aq)$ and $OH^-(aq)$ ions.

$$KOH(s) \xrightarrow{H_2O} K^+(aq) + OH^-(aq)$$

> **Exercise 7.1A**
Write equations showing **(a)** the ionization of HI (hydroiodic acid) in water and **(b)** the ionization of solid calcium hydroxide in water.

> **Exercise 7.1B**
Historically, formulas for carboxylic acids (compounds containing a —COOH group; see Chapter 9) are often written with the ionizable hydrogen *last*. For example, instead of writing the formula for acetic acid as $HC_2H_3O_2$, it can be written as CH_3COOH. Write an equation showing the ionization of CH_3COOH in water.

2 What is an amino acid?

An amino acid is a molecule that contains both a carboxyl group (—COOH), which is acidic, and —NH_2 (an *amino group*; see Section 9.8 and Section 9.9), which is basic. Amino acids link together to form the proteins in living tissue.

Limitations of the Arrhenius Theory

The Arrhenius theory is limited in several ways.

▪ A simple free proton does not exist in water solution. The H^+ ion has such a high positive charge density that it is immediately attracted to a lone pair of electrons on an O atom of an H_2O molecule, forming a *hydronium ion*, H_3O^+.

$$\text{H}\!:\!\overset{\displaystyle ..}{\underset{\displaystyle H}{\text{O}}}\!: \;+\; \text{H}^+ \;\longrightarrow\; \left[\text{H}\!:\!\overset{\displaystyle ..}{\underset{\displaystyle H}{\text{O}}}\!:\!\text{H}\right]^+$$

Water   Hydronium ion

- It does not explain the basicity of ammonia and related compounds. Ammonia seems out of place in Table 7.2 because it contains no hydroxide ions.
- It applies only to reactions in aqueous solution.

Although the Arrhenius theory is still useful within its limitations, it has been largely supplanted by a broader, more useful theory based on newer data.

The Brønsted-Lowry Acid-Base Theory

The shortcomings of the Arrhenius theory were largely overcome by a theory proposed in 1923 by J. N. Brønsted (1879–1947) in Denmark and T. M. Lowry (1874–1936) in Great Britain, who were working independently. In the Brønsted–Lowry theory (often simply the Brønsted theory)

- An acid is a *proton* (H^+) *donor*
- A base is a *proton acceptor*

The theory portrays the ionization of hydrogen chloride in this way:

$$HCl(aq) + H_2O \longrightarrow H_3O^+(aq) + Cl^-(aq)$$

Here, water is a reactant. The acid molecules donate hydrogen ions (protons) to the water molecules, so the acid (HCl) acts as a proton donor. Water acts as a proton acceptor. This can be confusing, because we think of water as being neutral—and it is, by itself. However, when water interacts with a proton donor, the water *acts as* a Brønsted base.

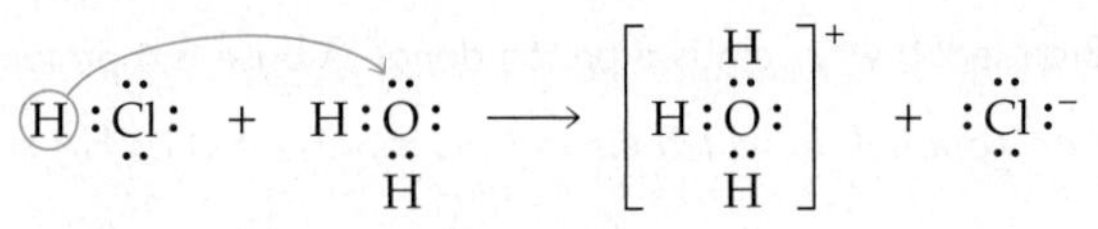

Acid
(proton donor)   Hydrochloric acid

The HCl molecule donates a proton to a water molecule, producing a hydronium ion and a chloride ion and forming a solution called *hydrochloric acid*. Other acids react similarly. They donate hydrogen ions to water molecules to produce hydronium ions. If we let HA represent any acid, the reaction is written as

$$HA(aq) + H_2O \longrightarrow H_3O^+(aq) + A^-(aq)$$

Even in solvents other than water, some compounds can still act as acids (proton donors), transferring H^+ ions to the solvent molecules.

> In water, an H^+ ion is associated with several H_2O molecules—for example, four H_2O molecules in the ion $H(H_2O)_4^+$ or $H_9O_4^+$. For many purposes, we simply use H^+ and ignore the associated water molecules. However, keep in mind that the notation H^+ is a simplification of the real situation; protons in water are always associated with water molecules.

CONCEPTUAL EXAMPLE 7.2 Brønsted-Lowry Acids

Write an equation showing the reaction of water with the Brønsted acid HNO_3. What is the role of water in the reaction?

Solution
As a Brønsted acid, HNO_3 donates a proton to a water molecule, forming a hydronium ion and a nitrate ion.

$$HNO_3(aq) + H_2O \longrightarrow H_3O^+(aq) + NO_3^-(aq)$$

Water acts as a Brønsted base in this reaction; it accepts the proton from the HNO_3.

〉 Exercise 7.2A
Write an equation showing the reaction of water with the Brønsted acid HBr.

〉 Exercise 7.2B
Write an equation showing the reaction of methanol (CH_3OH) with the Brønsted acid $HClO_4$.

Where does the OH^- come from when bases such as ammonia (NH_3) are dissolved in water? The Arrhenius theory proved inadequate in answering this question, but the Brønsted theory explains how ammonia acts as a base in water.

Ammonia is a gas at room temperature. When it is dissolved in water, some of the ammonia molecules react as shown by the following equation.

$$NH_3(aq) + H_2O \longrightarrow NH_4^+(aq) + OH^-(aq)$$

An ammonia molecule accepts a proton from a water molecule; NH_3 acts as a Brønsted base. (Recall that the N atom of ammonia has a lone pair of electrons, which it can share with a proton.) The water molecule acts as a proton donor—an acid. The ammonia molecule accepts the proton and becomes an ammonium ion. When a proton leaves a water molecule, it leaves behind the electron pair that it shared with the O atom. The water molecule becomes a negatively charged hydroxide ion.

In general, then, *a base is a proton acceptor* (Figure 7.3). This definition includes not only hydroxide ions but also neutral molecules such as ammonia and *amines* (organic compounds such as CH_3NH_2 that are derived from ammonia; more on this in Section 9.8). It also includes other negative ions such as the oxide (O^{2-}), carbonate (CO_3^{2-}), and bicarbonate (HCO_3^-) ions. The idea of an acid as a proton donor and a base as a proton acceptor greatly expands our concept of acids and bases.

Acid   Base

▲ **Figure 7.3** A Brønsted–Lowry acid is a proton donor. A base is a proton acceptor.

Q *Can you write an equation in which the acid is represented as HA and the base as B⁻?*

Salts

Salts, formed from the neutralization reactions of acids and bases, are ionic compounds composed of cations and anions. These ions can be simple ions, such as sodium ion (Na^+) or chloride ion (Cl^-), or they can be polyatomic ions, such as ammonium ion (NH_4^+), sulfate ion (SO_4^{2-}), or acetate ion (CH_3COO^-). Sodium chloride, ordinary table salt, is probably the most familiar salt, but many different salts exist.

Salts that conduct electricity when dissolved in water are called *electrolytes*. Various electrolytes, in certain amounts, are critical for many bodily functions, including nerve conduction, heartbeat, and fluid balance. Medical blood tests often check levels of Na^+, K^+, Cl^-, HCO_3^-, and other electrolyte ions. (See Section 17.4.) Sports drinks that boast of containing electrolytes are simply solutions of potassium phosphate, sodium chloride, and other salts along with sweeteners and flavorings.

There are many common salts with familiar uses. Sodium chloride and calcium chloride are used to melt ice on roads and sidewalks in winter. Copper(II) sulfate is used to kill tree roots in sewage lines. We will encounter many other examples in later chapters as dietary minerals (Section 17.2), fertilizers (Section 19.1), and more. Table 7.3 lists some salts used in medicine.

TABLE 7.3   Some Salts with Present or Past Uses in Medicine

Name	Formula	Uses
Silver nitrate	$AgNO_3$	Germicide and antiseptic
Stannous fluoride [tin(II) fluoride]	SnF_2	Toothpaste additive to prevent dental cavities
Calcium sulfate (plaster of Paris)	$2\ CaSO_4 \cdot H_2O^a$	Plaster casts
Magnesium sulfate (Epsom salts)	$MgSO_4 \cdot 7\ H_2O^a$	Laxative; foot baths
Potassium permanganate	$KMnO_4$	Cauterizing agent; antiseptic
Ferrous sulfate [iron(II) sulfate]	$FeSO_4$	Prescribed for iron deficiency (anemia)
Zinc sulfate	$ZnSO_4$	Skin treatment (eczema)
Barium sulfate	$BaSO_4$	Provides the contrast material in "barium cocktail" given for gastrointestinal X-rays
Mercurous chloride [mercury(I) chloride; calomel]	Hg_2Cl_2	Laxative; no longer used

[a]These compounds are *hydrates*, substances containing water molecules combined in a definite ratio as an integral part of the compound.

SELF-ASSESSMENT Questions

For Questions 1–5, match each formula with the compound's application.

1. CH_3COOH **a.** battery acid
2. H_3BO_3 **b.** soap making
3. HCl **c.** antiseptic eyewash
4. H_2SO_4 **d.** removal of boiler scale
5. $NaOH$ **e.** vinegar

6. Which of the following equations best represents what happens when hydrogen bromide dissolves in water?
   a. $2\ HBr \xrightarrow{H_2O} Br_2 + H_2$
   b. $HBr(g) \xrightarrow{H_2O} H(aq) + Br(aq)$
   c. $HBr(g) \xrightarrow{H_2O} H^+(aq) + Br^-(aq)$
   d. $HBr(g) + H_2O \longrightarrow Br^-(aq) + H_3O^+(aq)$

For Questions 7–10, match each term with the correct definition.

7. Arrhenius acid   **a.** proton acceptor
8. Arrhenius base   **b.** proton donor
9. Brønsted acid   **c.** produces H^+ in water
10. Brønsted base   **d.** produces OH^- in water

Answers: 1. e; 2. c; 3. d; 4. a; 5. b; 6. d; 7. c; 8. d; 9. b; 10. a

7.3 Acidic and Basic Anhydrides

Learning Objective • Identify acidic anhydrides and basic anhydrides, and write equations showing their reactions with water.

Certain metal and nonmetal oxides are well known for their ability to produce or to neutralize acids or bases. For example, nitrogen dioxide (NO_2) and sulfur dioxide (SO_2) are notorious for producing acid rain. Calcium oxide (quicklime, CaO) is widely used to neutralize acidic soils. In the Brønsted view, many metal oxides act directly as bases because the oxide ion can accept a proton. These metal oxides also react with water to form metal hydroxides, compounds that are bases in the Arrhenius sense. And many nonmetal oxides react with water to form acids.

Nonmetal Oxides: Acidic Anhydrides

Many acids are made by reacting nonmetal oxides with water. For example, sulfur trioxide reacts with water to form sulfuric acid.

$$SO_3 + H_2O \longrightarrow H_2SO_4$$

Similarly, carbon dioxide reacts with water to form carbonic acid.

$$CO_2 + H_2O \longrightarrow H_2CO_3$$

In general, nonmetal oxides react with water to form acids.

$$\text{Nonmetal oxide} + H_2O \longrightarrow \text{acid}$$

Nonmetal oxides that act in this way are called **acidic anhydrides**. *Anhydride* means "without water." These reactions explain why rainwater is acidic. (See Section 7.8 and Section 13.6.)

 CONCEPTUAL EXAMPLE 7.3   Acidic Anhydrides

Give the formula for the acid formed when sulfur dioxide reacts with water.

Solution

The formula for the acid, H_2SO_3, is obtained by adding the two H atoms and one O atom of water to SO_2. The equation for the reaction is simply

$$SO_2 + H_2O \longrightarrow H_2SO_3$$

WHY IT MATTERS

Slaked lime is inexpensive, easily made from cheap raw materials, and low in toxicity. These properties make it the most widely used base for neutralizing acids. It is an ingredient in such diverse materials as bricklayer's mortar, plaster, glues, and even pickles and corn tortillas. The whitewash made famous in Mark Twain's *Adventures of Tom Sawyer* was a simple mixture of slaked lime and water. Here, that same whitewash is being used to protect the tender bark of young trees.

> **Exercise 7.3A**
Give the formula for the acid formed when selenium dioxide (SeO_2) reacts with water.

> **Exercise 7.3B**
Give the formula for the acid formed when dinitrogen pentoxide (N_2O_5) reacts with water. (*Hint:* Two molecules of acid are formed.)

Metal Oxides: Basic Anhydrides

Just as acids can be made from nonmetal oxides, many common hydroxide bases can be made from metal oxides. For example, calcium oxide reacts with water to form calcium hydroxide ("slaked lime").

$$CaO + H_2O \longrightarrow Ca(OH)_2$$

Another example is the reaction of lithium oxide with water to form lithium hydroxide.

$$Li_2O + H_2O \longrightarrow 2\,LiOH$$

In general, metal oxides react with water to form bases (Figure 7.4). These metal oxides are called **basic anhydrides**.

$$\text{Metal oxide} + H_2O \longrightarrow \text{base}$$

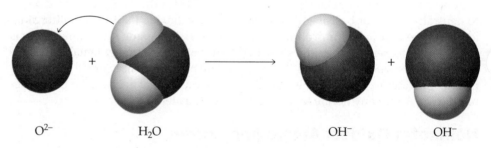

O^{2-} H_2O OH^- OH^-

▲ **Figure 7.4** Metal oxides are basic because the oxide ion reacts with water to form two hydroxide ions.

 Can you write an equation that shows how solid sodium oxide (Na_2O) reacts with water to form sodium hydroxide?

CONCEPTUAL EXAMPLE 7.4 **Basic Anhydrides**

Give the formula for the base formed by the addition of water to barium oxide (BaO).

Solution
Again, we add the atoms of a water molecule to BaO. Because a barium ion has a 2+ charge, the formula for the base has *two* hydroxide (1−) ions.

$$BaO + H_2O \longrightarrow Ba(OH)_2$$

> **Exercise 7.4A**
Give the formula for the base formed by the addition of water to strontium oxide (SrO).

> **Exercise 7.4B**
What base is formed by the addition of water to potassium oxide, K_2O? (*Hint:* Two moles of base are formed.)

SELF-ASSESSMENT Questions

1. Selenic acid (H_2SeO_4) is an extremely corrosive acid that, when heated, is capable of dissolving gold. It is quite soluble in water. The anhydride of selenic acid is
   a. SeO   **b.** SeO_2 **c.** SeO_3 **d.** SeO_4

2. Zinc hydroxide [$Zn(OH)_2$] is used as an absorbent in surgical dressings. The anhydride of zinc hydroxide is
   a. ZnO   **b.** ZnOH   **c.** ZnO_2 **d.** ZnH_2

7.4 Strong and Weak Acids and Bases

Learning Objective • Define and identify strong and weak acids and bases.

One of the most important acid–base reactions occurs between two water molecules. As shown in Figure 7.5, it is possible for one water molecule to transfer a proton to another water molecule.

$$2\ H_2O(l) \longrightarrow H_3O^+(aq) + OH^-(aq)$$

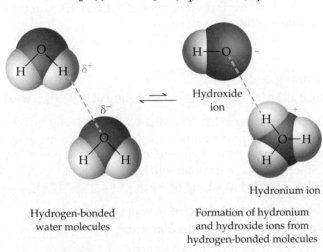

◄ Figure 7.5 A water molecule can transfer a proton to another water molecule to create a hydroxide ion and a hydronium ion. A hydronium ion can donate a proton back to a hydroxide ion to make two water molecules. Both reactions occur constantly, and the reactants and products are in equilibrium.

Hydroxide ion

Hydronium ion

Hydrogen-bonded water molecules

Formation of hydronium and hydroxide ions from hydrogen-bonded molecules

This is the opposite reaction of neutralization.

$$H_3O^+(aq) + OH^-(aq) \longrightarrow 2\ H_2O(l)$$

Both reactions occur simultaneously all the time. So, instead of writing both reactions, we use arrows pointing both ways to indicate that both the forward reaction and the backward reaction occur.

> Equilibrium: forward and reverse reactions both occur at the same rate

$$2\ H_2O(l) \rightleftharpoons H_3O^+(aq) + OH^-(aq)$$

Chemists read an equation like this, with the double arrows, as water is "in equilibrium" with hydronium and hydroxide ions. Consider a leaky boat; if you bail water out of the boat at the same rate at which it leaks in, the level of water in the boat does not change. Likewise, when a chemical system is in equilibrium, both the "forward" and the "reverse" chemical reactions are occurring, but they occur at the same rate, so the concentrations of reactants and products stay the same. In some equilibria, a lot of product is produced. For others, like the dissociation of water, very little product is produced. At any given time in neutral water, only about 1 in 500 million water molecules is dissociated into hydroxide ions and hydronium ions. In this reaction, one of the water molecules is acting as an acid by donating a proton; the other is acting as a base by accepting the proton. A substance such as water that can either donate a proton or accept a proton is said to be **amphiprotic**. (See also Problem 71.)

We can also express the autoionization of water as

$$H_2O(l) \rightleftharpoons H^+(aq) + OH^-(aq)$$

This equation focuses on the dissociation of a water molecule, and (aq) is used to indicate a solution in water. As mentioned earlier, a free hydrogen ion never exists in a water solution. However, we often represent hydronium ions with the simpler H^+ formula.

Many acids participate in similar equilibria. The poisonous gas hydrogen cyanide (HCN) ionizes in water to produce hydrogen ions and cyanide ions. HCN also ionizes only to a slight extent. In a solution that has 1 mol of HCN in 1 L of water, only 1 HCN molecule in 40,000 ionizes to produce a hydrogen ion. The rest remain as dissolved HCN molecules.

$$HCN(aq) \rightleftharpoons H^+(aq) + CN^-(aq)$$

Remember, the opposite-pointing arrows indicate that the reaction is reversible; the ions can also combine to form HCN molecules.

On the other hand, when gaseous hydrogen chloride (HCl) reacts with water, it reacts completely to form hydronium ions and chloride ions.

$$HCl(aq) \longrightarrow H^+(aq) + Cl^-(aq)$$

This time, only a single arrow is used because effectively 100% of the HCl molecules ionize. The opposite reaction doesn't occur.

Acids can be classified according to their extent of ionization.

- An acid such as HCl that ionizes completely in (reacts completely with) water is called a **strong acid**.
- An acid such as HCN that ionizes only slightly in water is a **weak acid**.

There are only a few strong acids. The first two acids listed in Table 7.1 (sulfuric and hydrochloric) are the most common ones. The other strong acids are nitric acid (HNO_3), hydrobromic acid (HBr), hydroiodic acid (HI), and perchloric acid ($HClO_4$). Virtually any other acid you encounter will be a weak acid.

Bases are also classified as strong or weak.

- A **strong base** is completely ionized in water.
- A **weak base** is only slightly ionized in water.

The word *strong* does not refer to the *amount* of acid or base in a solution. As mentioned in Section 5.5, a solution that contains a relatively large amount of an acid or a base, whether strong or weak, as the solute in a given volume of solution is called a *concentrated* solution. A solution with only a little solute in that same volume of solution is a *dilute* solution.

Neither does *strong* specifically mean *corrosive* or *dangerous*. In fact, all of the strong acids and most strong bases are corrosive and dangerous to a greater or lesser degree, but there are weak acids that are more corrosive and more dangerous. For example, a splash of hydrochloric acid on the arm, if washed off immediately, is unlikely to do more than redden the skin. By contrast, a splash of hydrofluoric acid, HF(aq), calls for immediate medical attention and can be a life-threatening situation. Also, the six strong acids mentioned previously are safely stored in glass bottles, while hydrofluoric acid will dissolve most materials—including glass—and must be kept in bottles made of, or coated with, certain plastics such as Teflon.

Perhaps the most familiar strong base is sodium hydroxide (NaOH), commonly called *lye*. It exists as sodium ions and hydroxide ions even as a solid. Other strong bases include potassium hydroxide (KOH) and the hydroxides of all the other group 1A metals. Except for $Be(OH)_2$, the hydroxides of group 2A metals are also strong bases. However, $Ca(OH)_2$ is only slightly soluble in water, and $Mg(OH)_2$ is nearly insoluble. The concentration of hydroxide ions in a solution of $Ca(OH)_2$ or $Mg(OH)_2$ is, therefore, not very high.

The most familiar weak base is ammonia (NH_3). It reacts with water to a slight extent to produce ammonium ions (NH_4^+) and hydroxide ions (Figure 7.6).

$$NH_3 + H_2O \rightleftharpoons NH_4^+ + OH^-$$

In its reaction with HCl (page 204), water acts as a base (proton acceptor). In its reaction with NH_3, water acts as an acid (proton donor).

3 **Lye isn't an acid, so why is it dangerous?**

Lye is solid sodium hydroxide. When it contacts moisture, Na^+ and OH^- ions are released. Hydroxide ions are different from hydrogen ions, but are equally corrosive in various chemical reactions; in your eyes it would blind you by denaturing some of the proteins.

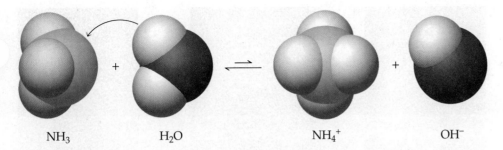

NH$_3$ H$_2$O   NH$_4^+$ OH$^-$

▲ **Figure 7.6** Ammonia is a base because it accepts a proton from water. A solution of ammonia in water contains ammonium ions and hydroxide ions. Only a small fraction of the ammonia molecules react, however; most remain unchanged. Ammonia is therefore a *weak* base.

Q *Amines are related to ammonia; in an amine, one or more of the H atoms of NH$_3$ are replaced by a carbon-containing group. Amines react in the same way that ammonia does. Can you write an equation that shows how the amine CH$_3$NH$_2$ reacts with water? One product is hydroxide ion.*

SELF-ASSESSMENT Questions

For Questions 1–7, select the correct classification. (More than one substance may fit into a given classification.)

1. Ca(OH)$_2$
2. HCN
3. HF
4. HNO$_3$
5. NH$_3$
6. KOH
7. CH$_3$NH$_2$

a. strong acid
b. strong base
c. weak acid
d. weak base

8. Acetic acid reacts with water to form
   a. CH$_3$COO$^-$ + H$_2$O
   b. CH$_3$COOH + OH$^-$
   c. CH$_3$COO$^-$ + H$_3$O$^+$
   d. CH$_3$COO$^+$ + OH$^-$

9. Ammonia reacts with water to form
   a. NH$_3$ + H$_2$O
   b. NH$_4^+$ + OH$^-$
   c. NH$_2^-$ + H$_3$O$^+$
   d. NH$_3$ + OH$^-$

Answers: 1. b; 2. c; 3. c; 4. a; 5. d; 6. b; 7. d; 8. c; 9. b

7.5 Neutralization

Learning Objective • Identify the reactants and predict the products in a neutralization reaction.

When an acid reacts with a base, the products are water and a salt. If a solution containing hydrogen ions (an acid) is mixed with another solution containing exactly the same amount of hydroxide ions (a base), the resulting solution does not change the color of litmus, dissolve zinc or iron, or feel slippery on the skin. It is no longer either acidic or basic. It is neutral. As was mentioned earlier, the reaction of an acid with a base is called *neutralization* (Figure 7.7).

$$H^+ + OH^- \longrightarrow H_2O$$

If sodium hydroxide is neutralized by hydrochloric acid, the products are water and sodium chloride (ordinary table salt).

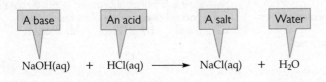

| A base | An acid | A salt | Water |

NaOH(aq) + HCl(aq) ⟶ NaCl(aq) + H$_2$O

(a)　　　　　　　　　　　(b)　　　　　　　　　　　(c)

▲ **Figure 7.7** The amount of acid (or base) in a solution is determined by careful neutralization. (a) Here a 5.00 mL sample of vinegar, some water, and a few drops of phenolphthalein (an acid–base indicator) are placed in a flask. (b) A solution of 0.1000 M NaOH is added to the flask slowly from a buret (a device for precise measurement of volumes of solutions). (c) As long as the acid is in excess, the solution is colorless. When the acid has been neutralized and a tiny excess of base is present, the phenolphthalein indicator turns pink.

Q *Can you write an equation for the reaction of acetic acid (CH_3COOH), the acid in vinegar, with aqueous NaOH?*

 EXAMPLE 7.5　Neutralization Reactions

Potassium nitrate, a component of black powder gunpowder and some fertilizers, was obtained from the late Middle Ages through the nineteenth century by precipitation from urine. Commonly called *saltpeter*, it can be prepared by the reaction of nitric acid with potassium hydroxide. Write the equation for this neutralization reaction.

Solution

The OH^- from the base and the H^+ from the acid combine to form water. The cation of the base (K^+) and the anion of the acid (NO_3^-) form a solution of the salt (potassium nitrate, KNO_3).

$$KOH(aq) + HNO_3(aq) \longrightarrow KNO_3(aq) + H_2O(l)$$

❯ **Exercise 7.5A**

Countertop spills of lye solutions (aqueous sodium hydroxide) can be neutralized with vinegar (aqueous acetic acid; see Table 7.1). Write the equation for the neutralization reaction.

❯ **Exercise 7.5B**

A toilet-bowl cleaner contains hydrochloric acid. An emergency first-aid treatment for accidental ingestion is a teaspoon of milk of magnesia (magnesium hydroxide). Write the equation for the neutralization reaction between magnesium hydroxide and hydrochloric acid. (*Hint:* Be sure to write the correct formulas for the reactants and the salt before attempting to balance the equation.)

SELF-ASSESSMENT Questions

1. When equal amounts of acids and bases are mixed
   a. the acid becomes more concentrated
   b. the base becomes more concentrated
   c. no reaction occurs
   d. they neutralize each other

2. How many moles of hydrochloric acid are needed to neutralize 1.5 mol of sodium hydroxide?
   a. 1.0 mol　　b. 1.5 mol　　c. 3.0 mol　　d. 4.5 mol

3. How many moles of hydrochloric acid are needed to neutralize 2.4 mol of calcium hydroxide?
   a. 1.2 mol　　b. 2.4 mol　　c. 3.6 mol　　d. 4.8 mol

4. How many moles of strontium hydroxide are needed to neutralize 2.0 mol of phosphoric acid? Be careful to write the balanced chemical equation first.
   a. 1.0 mol　　b. 2.0 mol　　c. 3.0 mol　　d. 6.0 mol

Answers: 1, d; 2, b; 3, d; 4, c

7.6 The pH Scale

Learning Objectives • Describe the relationship between the pH of a solution and its acidity or basicity. • Find the molar concentration of hydrogen ion, [H⁺], from a pH value, or the pH value from [H⁺].

In solutions, the concentrations of ions are expressed in moles per liter (molarity; see Section 5.5). Because hydrogen chloride is completely ionized in water, a 1 molar solution of hydrochloric acid (1 M HCl), for example, contains 1 mol H^+ ions per liter of solution. Likewise, 1 L of 3 M HCl contains 3 mol H^+ ions, and 0.500 L of 0.00100 M HCl contains 0.500 L × 0.00100 mol/L = 0.000500 mol H^+ ions.

We can describe the acidity of a particular solution in moles per liter of hydrogen ions: The hydrogen ion concentration of a 0.001 M HCl solution is 1×10^{-3} mol/L. However, exponential notation isn't very convenient. More often, the acidity of this solution is reported simply as pH 3.

We usually use the **pH** scale, first proposed in 1909 by the Danish biochemist S. P. L. Sørensen (1868–1939), to describe the degree of acidity or basicity of a solution. The numbers on the pH scale are directly related to the hydrogen ion concentration. In Section 7.4 we saw how, at any given time in pure water, about 1 out of every 500 million water molecules is split into H^+ and OH^- ions. This gives a concentration of hydrogen ions and of hydroxide ions in pure water of 0.0000001 mol/L, or 1×10^{-7} M each. Can you see why 7 is the pH of pure water? It is simply the power of 10 for the molar concentration of H^+, with the negative sign removed. (The H in pH stands for "hydrogen," and the p for "power.") Thus, pH is defined as the negative logarithm of the molar concentration of hydrogen ion (refer to Table 7.4):

$$pH = -\log[H^+]$$

The brackets around H^+ mean "molar concentration." The relationship between pH and [H⁺] is perhaps easier to see when the equation is written in the following form. $[H^+] = 10^{-pH}$

Most solutions have a pH that lies in the range of 0–14. The neutral point on the scale is 7, with values below 7 indicating increasing acidity and those above 7 increasing basicity. Thus, pH 6 is slightly acidic, whereas pH 12 is very basic (Figure 7.8). Although pH is an acidity scale, note that its value goes down when acidity goes up. Not only is the relationship an inverse one, but it is also logarithmic. A decrease of one pH unit represents a tenfold increase in acidity, and when pH goes down by two units, acidity increases by a factor of 100. This relationship may seem strange at first, but once you understand the pH scale, you will appreciate its convenience.

Table 7.4 summarizes the relationship between hydrogen ion concentration and pH. A pH of 4 means a hydrogen ion concentration of 1×10^{-4} mol/L, or 0.0001 M. If the concentration of hydrogen ions is 0.01 M, or 1×10^{-2} M, the pH is 2. Typical pH values for various common solutions are listed in Table 7.5.

TABLE 7.4 **Relationship between pH and Concentration of H⁺ Ion**

Concentration of H⁺(mol/L)	pH
1×10^{-0}	0
1×10^{-1}	1
1×10^{-2}	2
1×10^{-3}	3
1×10^{-4}	4
1×10^{-5}	5
1×10^{-6}	6
1×10^{-7}	7
1×10^{-8}	8
1×10^{-9}	9
1×10^{-10}	10
1×10^{-11}	11
1×10^{-12}	12
1×10^{-13}	13
1×10^{-14}	14

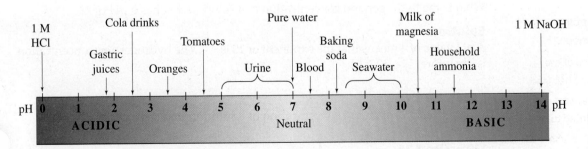

▲ **Figure 7.8** The pH scale. A change in pH of one unit means a tenfold change in the hydrogen ion concentration.

Q *Are tomatoes more acidic or less acidic than oranges? About how many times more (or less) acidic?*

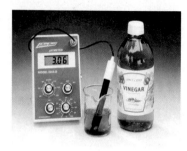

▲ A pH meter is a device for determining pH quickly and accurately.

Q *Is the solution in the beaker more basic or less basic than household ammonia? (Refer to Figure 7.8.)*

TABLE 7.5 Approximate pH Values of Some Common Solutions

Solution	pH
Hydrochloric acid (4%)	0
Gastric juice	1.6–1.8
Soft drink	2.0–4.0
Lemon juice	2.1
Vinegar (4%)	2.5
Urine	5.5–7.0
Rainwater[a]	5.6
Saliva	6.2–7.4
Milk	6.3–6.6
Pure water	7.0
Blood	7.4
Fresh egg white	7.6–8.0
Bile	7.8–8.6
Milk of magnesia	10.5
Washing soda	12.0
Sodium hydroxide (4%)	13.0

[a]Saturated with carbon dioxide from the atmosphere but unpolluted.

 EXAMPLE 7.6   pH from Hydrogen Ion Concentration

What is the pH of a solution that has a hydrogen ion concentration of 1×10^{-5} M?

Solution

The hydrogen ion concentration is 1×10^{-5} M. The exponent is -5; the pH is, therefore, the negative of this exponent: $-(-5)$ or 5.

❭ **Exercise 7.6A**

What is the pH of a solution that has a hydrogen ion concentration of 1×10^{-9} M?

❭ **Exercise 7.6B**

What is the pH of a solution that is 0.010 M HNO_3? (*Hint:* HNO_3 is a strong acid.)

 EXAMPLE 7.7   Finding Hydrogen Ion Concentration from pH

What is the hydrogen ion concentration of a solution that has a pH of 4?

Solution

A pH value of 4 means that the exponent of 10 is -4. The hydrogen ion concentration is, therefore, 1×10^{-4} M.

❭ **Exercise 7.7A**

What is the hydrogen ion concentration of a solution that has a pH of 2?

❭ **Exercise 7.7B**

What is the concentration of an HI solution that has a pH of 3? (*Hint:* HI is a strong acid.)

4 What is meant by "pH-balanced shampoo"?

Shampoo on either end of the pH scale would damage hair (and probably skin as well!). Most shampoos are formulated to be neutral (pH 7) or slightly basic. See Section 20.6 for more information on shampoo.

CONCEPTUAL EXAMPLE 7.8   Estimating pH from Hydrogen Ion Concentration

Which of the following is a reasonable pH for a solution that is 8×10^{-4} M in H^+?

   a. 2.9   **b.** 3.1   **c.** 4.2   **d.** 4.8

Solution

The $[H^+]$ is greater than 1×10^{-4} M, so the pH must be less than 4. That rules out answers **(c)** and **(d)**. The $[H^+]$ is less than 10×10^{-4} (or 1×10^{-3} M), so the pH must be greater than 3. That rules out answer **(a)**. The only reasonable answer is **(b)**, a value between 3 and 4.

   With a calculator you can calculate the actual pH. On many graphing calculators you would enter $\log(8 \times 10^{-4})$ then press = or ENTER. This gives a value of -3.1 for the log of 8×10^{-4}. Because pH is the *negative* log of the H^+ concentration, the pH value is 3.1.

> **Exercise 7.8A**

Which of the following is a reasonable pH for a solution that is 2×10^{-10} M in H^+?

   a. 2.0   **b.** 8.7   **c.** 9.7   **d.** 10.2

> **Exercise 7.8B**

What is the pH of 3.6×10^{-3} M HI? (*Hint*: See the solution to Example 7.8 above, and recall that HI is a strong acid.)

In a neutral solution, the hydrogen ion and the hydroxide ion concentrations are both 1×10^{-7} M. Multiplying the two together gives 1×10^{-14}. It turns out that in an aqueous solution at 25 °C, $[H^+] \times [OH^-]$ always equals 1×10^{-14}. So, if the hydrogen ion concentration is greater than 1×10^{-7} M, the hydroxide ion concentration must be less than 1×10^{-7} M. For example, a solution with a pH of 4 is 1×10^{-4} M in H^+ ions but only 1×10^{-10} M in OH^- ions. On the other hand, basic solutions have more than 1×10^{-7} M OH^- ions and, therefore, must have less than 1×10^{-7} M of H^+ ions. Hence, a basic solution with 1×10^{-2} M of OH^- ions must have 1×10^{-12} M of H^+ ions and a pH of 12. This is why a pH greater than 7 means the solution is basic.

CONCEPTUAL EXAMPLE 7.9   Hydrogen and Hydroxide Ions in Acids and Bases

Classify each of the three following aqueous solutions shown as either acidic, basic, or neutral. (Water molecules are not shown.)

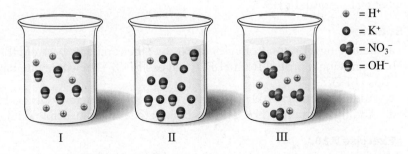

Solution

Beaker I contains equal numbers of OH^- ions and H^+ ions—six of each—so that solution is neutral. Beaker II contains more OH^- ions (seven) than H^+ ions (one), so its solution is basic. Beaker III contains six H^+ ions and one OH^- ion; the solution is acidic.

> **Exercise 7.9A**

What is the formula of the compound that was dissolved in beaker III of Example 7.9? What is the formula of the compound that was dissolved in beaker II?

> **Exercise 7.9B**

To represent an aqueous solution of calcium nitrate, a student draws a beaker with three Ca^{2+} ions and five H^+ ions. How many OH^- ions and how many NO_3^- ions should she draw?

SELF-ASSESSMENT Questions

1. The hydrogen ion concentration, $[H^+]$, of a 0.0010 M HNO_3 solution is
   a. 1.0×10^{-4} M **b.** 1.0×10^{-3} M
   c. 1.0×10^{-2} M **d.** 10 M

2. What is the pH of a solution that has a hydrogen ion concentration of 1.0×10^{-11} M?
   a. 1.0 **b.** 3 **c.** 10 **d.** 11

3. What is the pH of a solution that has a hydroxide ion concentration of 1.0×10^{-5} M?
   a. 1.0 **b.** 5 **c.** 9 **d.** 14

4. Swimming pool water with a pH of 8 has a hydrogen ion concentration of
   a. 8.0 M **b.** 8.0×10^{-8} M
   c. 1.0×10^{-8} M **d.** 1.0×10^8 M

5. The pH of pure water is
   a. 0 **b.** 1 **c.** 7 **d.** 10 **e.** 14

6. Which of the following is a reasonable pH for 0.015 M HCl?
   a. 0.015 **b.** 1.82 **c.** 8.24 **d.** 12.18

7. Which of the following is a reasonable pH for 0.015 M NaOH?
   a. 0.015 **b.** 1.82 **c.** 8.24 **d.** 12.18

8. Physiological pH (7.4) is the average pH of blood. Which of the following is a reasonable hydrogen ion concentration of a solution at physiological pH?
   a. -7.4 M **b.** 0.6 M
   c. 6×10^{-7} M **d.** 1×10^{-8} M
   e. 4×10^{-8} M

Answers: 1, b; 2, d; 3, c; 4, c; 5, c; 6, b; 7, d; 8, e

7.7 Buffers and Conjugate Acid–Base Pairs

Learning Objectives • Write the formula for the conjugate base of an acid or for the conjugate acid of a base. • Describe the action of a buffer.

In the Brønsted theory, a **conjugate acid–base pair** is a pair of compounds or ions that differ by one proton (H^+). For example, HF and F^- are a conjugate acid–base pair, as are NH_3 and NH_4^+. When a base (for example, NH_3) accepts a proton, it becomes the acid NH_4^+ because NH_4^+ now has an "extra" proton that it can donate. Similarly, when an acid (for example, HF) donates a proton, what remains (F^-) becomes a base—the conjugate base of HF—because F^- can now accept a proton.

 CONCEPTUAL EXAMPLE 7.10   Conjugate Acid–Base Pairs

What is the conjugate base **(a)** of HBr and **(b)** of H_3PO_4? What is the conjugate acid **(c)** of OH^- and **(d)** of HSO_4^-?

Solution

   a. Removing a proton from HBr leaves Br^-; the conjugate base of HBr is Br^-.
   b. Removing a proton from H_3PO_4 leaves $H_2PO_4^-$; the conjugate base of H_3PO_4 is $H_2PO_4^-$.
   c. Adding a proton to OH^- gives H_2O; the conjugate acid of OH^- is H_2O.
   d. Adding a proton to HSO_4^- gives H_2SO_4; the conjugate acid of HSO_4^- is H_2SO_4.

> **Exercise 7.10A**

What is the conjugate acid **(a)** of SO_4^{2-} and **(b)** of HCO_3^-?

> **Exercise 7.10B**

What is the conjugate base **(a)** of HCN and **(b)** of NH_3?

Buffer Solutions

A **buffer solution** maintains an almost constant pH when small amounts of a strong acid or a strong base are added to it. Buffer solutions have many important applications in industry, in the laboratory, and in living organisms because some chemical reactions consume acids, others produce acids, and many are catalyzed by acids. (A *catalyst* is a substance or mixture that speeds up a reaction and can be recovered unchanged after the reaction is complete. More on catalysts in Section 8.6.) A buffer solution consists of a weak acid and its conjugate base (for example, $HC_2H_3O_2$ and $C_2H_3O_2^-$) or a weak base and its conjugate acid (for example, NH_3 and NH_4^+).

Consider the equation for the ionization of acetic acid, in which the double arrow indicates the slight extent to which the acid ionizes.

$$HC_2H_3O_2(aq) \rightleftharpoons H^+(aq) + C_2H_3O_2^-(aq)$$

If we add sodium acetate to a solution of acetic acid, the sodium acetate dissociates completely into Na^+ ions and $C_2H_3O_2^-$ ions. Thus, we are adding the conjugate base of acetic acid—that is, acetate ion—forming a buffer solution. If we add a little strong base (OH^-) to this solution, it will react with the weak acid to produce a little of its weak conjugate base.

$$OH^- + HC_2H_3O_2 \longrightarrow H_2O + C_2H_3O_2^-$$

Do you see that the solution no longer contains the strong base? Instead, it has a little more weak base and a little less weak acid than it did before the strong base was added. The pH remains very nearly constant. Likewise, if a little strong acid (H^+) is added, it will react with the weak base to produce its conjugate weak acid.

$$H^+ + C_2H_3O_2^- \longrightarrow HC_2H_3O_2$$

The strong acid is consumed, a little weak acid is formed, and the solution pH decreases only slightly. This behavior of buffered solutions contrasts sharply with that of unbuffered solutions, in which any added strong acid or strong base changes the pH greatly.

A dramatic and essential example of the action of buffers is found in our blood. Blood must maintain a pH very close to 7.4, or it cannot carry oxygen from the lungs to cells. The most important buffer for maintaining acid–base balance in the blood is the carbonic acid–bicarbonate ion $\left(H_2CO_3/HCO_3^-\right)$ buffer.

SELF-ASSESSMENT Questions

1. Which of the following pairs is a conjugate acid–base pair?
   a. CH_3COOH and OH^-
   b. HCN and CN^-
   c. HCN and OH^-
   d. HCl and OH^-

2. Which of the following is *not* a conjugate acid–base pair?
   a. CH_3COO^- and CH_3COOH
   b. F^- and HF
   c. H_2O and H_3O^+
   d. NH_3 and H_3O^+

3. A buffer solution is made from formic acid ($HCOOH$) and sodium formate ($HCOONa$). Added acid will react with
   a. $HCOO^-$
   b. $HCOOH$
   c. Na^+
   d. OH^-

4. Which of the following pairs could form a buffer?
   a. C_6H_5COOH and C_6H_5COONa
   b. HCl and $NaCl$
   c. $NaCl$ and $NaOH$
   d. NH_3 and H_3BO_3

Answers: 1, b; 2, d; 3, a; 4, a

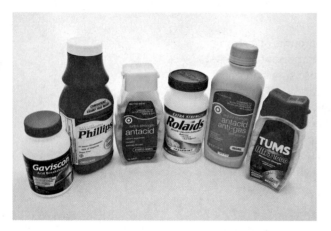

▲ **Figure 7.9** Claims that antacids are "fast-acting" are almost meaningless because most common acid–base reactions are almost instantaneous. Some tablets may dissolve a little more slowly than others. You can speed their action by chewing them.

You can make your own aspirin-free "Alka-Seltzer." Simply place half a teaspoon of baking soda in a glass of orange juice. (What is the acid and what is the base in this reaction?)

7.8 Acids and Bases in Industry and in Daily Life

Learning Objective • Describe everyday uses of acids and bases and how they affect daily life.

Acids and bases play an important role in our bodies, in medicine, in our homes, and in industry; they are useful products in many ways. However, as industrial by-products, they can damage the environment. Their use requires caution, and their misuse can be dangerous to human health.

Antacids: A Basic Remedy

The stomach secretes hydrochloric acid to aid in the digestion of food. Sometimes overindulgence or emotional stress leads to *hyperacidity*. (Too much acid is secreted.) Hundreds of brands of antacids (Figure 7.9) are sold in the United States to treat this condition. Despite the many brand names, there are only a few different antacid ingredients, all of which are bases. Common ingredients are sodium bicarbonate, calcium carbonate, aluminum hydroxide, magnesium carbonate, and magnesium hydroxide.

Sodium bicarbonate ($NaHCO_3$), commonly called *baking soda*, was one of the first antacids and is still used occasionally. It is the principal antacid in most forms of Alka-Seltzer for heartburn relief. The bicarbonate ions react with the acid in the stomach to form carbonic acid, which then breaks down to carbon dioxide and water.

$$HCO_3^-(aq) + H^+(aq) \longrightarrow H_2CO_3(aq)$$

$$H_2CO_3(aq) \longrightarrow CO_2(g) + H_2O(l)$$

The $CO_2(g)$ is largely responsible for the burps that bicarbonate-containing antacids produce. Overuse of sodium bicarbonate can make the blood too alkaline, a condition called **alkalosis**. Also, antacids that contain sodium ion can increase blood pressure in people with hypertension (high blood pressure).

Why Doesn't Stomach Acid Dissolve the Stomach?

We know that strong acids are corrosive to skin. Look back at Table 7.1 and you will see that the gastric juice in your stomach is extremely acidic. Gastric juice is a solution containing about 0.5% hydrochloric acid. Why doesn't the acid in your stomach destroy your stomach lining? The cells that line the stomach are protected by a layer of mucus, a viscous solution of a sugar–protein complex called *mucin* and other substances in water. The mucus serves as a physical barrier, but its role is broader than that. The mucin acts like a sponge that soaks up bicarbonate ions from the cells lining the stomach and hydrochloric acid from within the stomach. The bicarbonate ions neutralize the acid within the mucus.

It used to be thought that too much stomach acid was the cause of ulcers, but it is now known that the acid plays only a minor role in ulcer formation. Research has shown that most ulcers develop as a result of infection with a bacterium called *Helicobacter pylori* (*H. pylori*) that damages the mucus, exposing cells in the stomach lining to the harsh stomach acid. Other agents, such as aspirin and alcohol, may be contributing factors to the development of ulcers. Treatment usually involves antibiotics that kill *H. pylori*.

Another ingredient found in antacids, calcium carbonate ($CaCO_3$), is safe in small amounts, but regular use can cause constipation. Also, taking large amounts of calcium carbonate can actually result in increased acid secretion after a few hours.

Tums and many store brands of antacids have calcium carbonate as the only active ingredient.

Aluminum hydroxide [$Al(OH)_3$], like calcium carbonate, can cause constipation. There is also some concern that antacids containing aluminum ions can deplete the body of essential phosphate ions. Aluminum hydroxide is the active ingredient in *Amphojel*.

A suspension of magnesium hydroxide [$Mg(OH)_2$] in water is sold as *milk of magnesia*. Magnesium carbonate ($MgCO_3$) is also used as an antacid. These magnesium compounds act as antacids in small doses but as laxatives in large doses.

Many antacid products have a mixture of antacids. Rolaids® and Mylanta® contain calcium carbonate and magnesium hydroxide. Maalox® liquid has aluminum hydroxide and magnesium hydroxide. These products balance the tendency of magnesium compounds to cause diarrhea with that of aluminum and calcium compounds to cause constipation.

Although antacids are generally safe for occasional use, they can interact with other medications. Anyone who has severe or repeated attacks of indigestion should consult a physician. Self-medication can sometimes be dangerous.

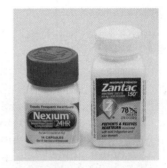

▲ Drugs such as ranitidine (Zantac®), famotidine (Pepcid AC®), cimetidine (Tagamet HB®) and esomeprazole (Nexium®) are not antacids. Rather than neutralizing stomach acid, these drugs act on cells in the lining of the stomach, reducing the amount of acid that is produced.

Acids and Bases in Industry and at Home

Sulfuric acid is the leading chemical product in the United States (average production of about 40 billion kg each year), with a worldwide production of over 200 billion kg annually. Most of it is used for making fertilizers and other industrial chemicals. Around the home, we use sulfuric acid in automobile batteries and in some special kinds of drain cleaners.

Hydrochloric acid is used in industry to remove rust from metal, in construction to remove excess mortar from bricks and to etch concrete for painting, and in the home to remove lime deposits from fixtures and toilet bowls. The product used in the home is often called *muriatic acid*, an old name for hydrochloric acid. Concentrated solutions (about 38% HCl) cause severe burns, but dilute solutions can be used safely in the home if handled carefully. Yearly production of hydrochloric acid is more than 4 billion kg in the United States and about 20 billion kg worldwide.

Lime (CaO) is the cheapest and most widely used commercial base. It is made by heating limestone ($CaCO_3$) to drive off CO_2.

$$CaCO_3(s) + heat \longrightarrow CaO(s) + CO_2(g)$$

Yearly production of calcium oxide is about 22 billion kg in the United States and about 230 billion kg worldwide. Calcium oxide is very corrosive. By adding water, calcium hydroxide [$Ca(OH)_2$, slaked lime] is formed; it is generally safer to handle. Slaked lime is used in agriculture and to make mortar and cement.

Sodium hydroxide (NaOH, lye) is the strong base most often used in the home. It is employed in products such as *Easy Off* for cleaning ovens, in products such as *Drano* for unclogging drains, and to make both commercial and homemade soaps. Yearly U.S. production of sodium hydroxide is about 9 billion kg.

Ammonia (NH_3) is produced in huge volume, mainly for use as fertilizer. Yearly U.S. production is nearly 11 billion kg. Ammonia is used around the home in a variety of cleaning products (Chapter 20).

Acids and Bases in Health and Disease

When they are misused, acids and bases can be damaging to human health. Concentrated strong acids and bases are corrosive poisons (Chapter 21) that can cause serious chemical burns. Once the chemical agents are removed, these injuries resemble burns caused by heat and are often treated in the same way. Besides being a strong acid, sulfuric acid is also a powerful dehydrating agent that can react with water in the cells.

WHY IT MATTERS

The proper pH is as important for plant growth as fertilizer is. Soil that is "sour," or too acidic, is "sweetened" by adding slaked lime (calcium hydroxide). A few plants, such as blueberries and citrus fruit, require acidic soil, and a little acid may have to be periodically added to the soil in which they grow.

Acid Rain

Carbon dioxide is the anhydride of carbonic acid (Section 7.3). Raindrops falling through the air absorb CO_2, which is converted to H_2CO_3 in the drops. Rainwater is thus a dilute solution of carbonic acid, a weak acid. Rain saturated with carbon dioxide is very slightly acidic, with a pH of 5.6. In many areas of the world, particularly those downwind from industrial centers, rainwater is much more acidic, with a pH as low as 3 or less. Rain with a pH below 5.6 is called **acid rain**.

Acid rain is due to acidic pollutants in the air. As we shall see in Chapter 13, several air pollutants are acid anhydrides. Sulfur dioxide (SO_2) and sulfur trioxide (SO_3) are formed mainly from the burning of high-sulfur coal in power plants and metal smelters, and nitrogen dioxide (NO_2) and nitrogen monoxide (NO) come from automobile exhaust fumes.

Some acid rain is due to natural pollutants, such as those resulting from volcanic eruptions and lightning. Volcanoes give off sulfur oxides and sulfuric acid, and lightning produces nitrogen oxides and nitric acid from nitrogen, oxygen, and water in the air.

Acid rain is an important environmental problem that involves both air pollution (Chapter 13) and water pollution (Chapter 14). It can have serious effects on plant and animal life.

Strong acids and bases—even in dilute solutions—break down, or **denature**, the protein molecules in living cells, much as cooking does. Generally, the fragments are not able to carry out the functions of the original proteins. If exposure to the acid or base is sustained, this fragmentation continues until the tissue has been completely destroyed.

Acids and bases affect human health in more subtle ways. A delicate balance between acids and bases must be maintained in the blood, other body fluids, and cells. If the acidity of the blood changes too much, the blood loses its capacity to carry oxygen. In living cells, proteins function properly only at an optimal pH. If the pH changes too much in either direction, the proteins can't carry out their functions. Fortunately, the body has a complex but efficient mechanism for maintaining a proper acid–base balance (Section 7.7).

SELF-ASSESSMENT Questions

1. Which of the following is a common ingredient in antacids?
   a. $CaCO_3$ b. $Ca(OH)_2$
   c. HCl   d. KOH

2. When a person with excess stomach acid takes an antacid, the pH of the person's stomach changes
   a. from a low value to a value nearer 7
   b. from 7 to a much higher value
   c. from a low value to an even lower value
   d. from a high value to a lower value

3. The leading chemical product of U.S. industry is
   a. ammonia   b. lime
   c. plastics   d. sulfuric acid

4. The acid used in automobile batteries is
   a. citric acid
   b. hydrochloric acid
   c. nitric acid
   d. sulfuric acid

5. The base often used in soap making is
   a. $Ca(OH)_2$
   b. KOH
   c. NaOH
   d. NH_3

Answers: 1, a; 2, a; 3, d; 4, d; 5, c

GREEN CHEMISTRY Acids and Bases–Greener Alternatives
Irvin J. Levy, Gordon College, Wenham, MA

Principles 1, 3, 7, and 12

Learning Objectives • Recognize carbon dioxide in water as a useful, safe source of acid. • Describe alternate choices for greener reaction conditions when using acids and bases.

Acids and bases are among the most common types of chemical substances. We find them everywhere. Although you may not identify acids and bases by their technical terms, we are in contact with them every moment of our lives. For example, a healthy person's blood is very slightly basic (pH ~ 7.4), also referred to as alkaline, while that person's stomach is rather acidic (pH ~ 2–3) (Section 7.6). Acids and bases also are found in many food ingredients. For instance, vinegar and orange juice are slightly acidic solutions, and baking soda is an alkaline solid.

Although acids and bases are important chemicals, concentrated solutions can be hazardous. Some industrial processes produce large quantities of highly basic or acidic waste. Those waste streams, if untreated, are dangerous to human health and the environment if unintended exposure occurs. Neutralization of acidic and alkaline waste is necessary as part of its proper disposal. Acids and bases also are used in many chemical reactions and, therefore, must be transported, stored, and used in large quantities. These corrosive substances can harm the equipment and storage containers where they handled and, potentially, the people who work with them. As described in the following text, chemical processes that include acids and bases provide opportunities for developing greener alternatives.

One example is the conversion of aqueous sodium chloride solution into important chemicals, including chlorine and sodium hydroxide. The process produces millions of tons of these useful chemicals every year but also results in a vast amount of alkaline waste. For the waste to be handled safely, an acid must be added to reduce the pH to between 6 and 9. Typically, mineral acid—hydrochloric or sulfuric acid—is added to the waste stream to neutralize the waste (Section 7.5). Yet that method has several disadvantages.

First, the amount of added acid is significant, and it represents a wasteful use of chemicals. Further, the amount must be very carefully controlled or the process could "over-neutralize" the waste (Section 7.5). This would convert the hazardous alkaline waste stream into a hazardous acidic waste stream, which would not be helpful. Also, the unwanted products produced when the neutralizing acid comes together with the alkaline waste can cause additional problems. Lastly, and maybe most importantly, mineral acids are hazardous substances that pose threats during production, transport, storage, and use at the end site. There must be a greener way to handle these waste streams.

Indeed, companies are beginning to neutralize alkaline waste with a much greener substance, carbon dioxide (CO_2). CO_2 often is seen as a danger to the environment because the large amounts that we release from burning fossil fuels contribute to global climate changes. However, there are many good uses for the CO_2 that is already in our atmosphere. For example, a weakly acidic solution is formed when carbon dioxide dissolves in water, so it is possible to neutralize alkaline waste streams by simply bubbling CO_2 through them. The solutions cannot be "over-neutralized" with CO_2 because it will not lower the pH of the solution below 7 in practical use, and excess CO_2 simply bubbles out of the water. This method replaces mineral acids, prevents waste, relies on a renewable resource, and leads to the same outcome with safer chemicals (Principles 1, 3, 7, and 12).

Chemistry is a very lively field and the Principles of Green Chemistry remind chemists of many opportunities to develop methods that are safer for human health and the environment, more cost-effective, and work as well as—or even better than—more hazardous alternatives. One area of research is replacing acids in chemical processes with naturally occurring safer substances. For example, a common chemical reaction that converts an alcohol (a molecule with a carbon bonded to an OH group) to a different chemical substance requires a high concentration of very strong mineral acid and high temperature:

$$\underset{\displaystyle -\overset{\displaystyle |}{\underset{\displaystyle |}{C}}-\overset{\displaystyle |}{\underset{\displaystyle |}{C}}-}{\overset{\displaystyle H \quad OH}{}} \xrightarrow[\text{high temp}]{\text{strong acid}} \quad \underset{/}{\overset{\backslash}{C}}=\underset{\backslash}{\overset{/}{C}} \quad + \quad H_2O$$

Recently, it has been learned that the sulfuric and phosphoric acids used in this chemical transformation can be replaced by a much safer substance—a type of clay called montmorillonite (Section 12.2, page 357). Clay catalysts are finding many uses in the chemical industry and significantly improve the safety of those processes.

Many methods that have been in place in the chemical industry for decades are now again areas of active research by chemists who are dedicated to using green chemistry's guiding principles to develop safer substances and methods. You might say that's a really "basic" idea. (And acidic, too!)

Summary

Section 7.1—Acids taste sour, turn litmus red, and react with active metals to form hydrogen. Bases taste bitter, turn litmus blue, and feel slippery to the skin. Acids and bases react to form **salts** and water. An **acid–base indicator** such as litmus has different colors in acid and in base and is used to determine whether solutions are acidic or basic.

Section 7.2—An Arrhenius acid produces hydrogen ions (H^+), also called protons, in aqueous solution, and an Arrhenius base produces hydroxide ions (OH^-). **Neutralization** is the reaction of H^+ and OH^- to form water. The anion and cation that were associated with H^+ and OH^- ions combine to form an ionic salt. In the more general Brønsted acid–base theory, an **acid** is a proton donor and a **base** is a proton acceptor. When a Brønsted acid dissolves in water, H_2O molecules pick up H^+ to form hydronium ions (H_3O^+). A Brønsted base in water accepts a proton from a water molecule, forming OH^-.

Section 7.3—Some nonmetal oxides (such as CO_2 and SO_3) are **acidic anhydrides** and react with water to form acids. Some metal oxides (such as Li_2O and CaO) are **basic anhydrides**; they react with water to form bases. **Amphiprotic** substances can act as either acids or bases.

Section 7.4—A **strong acid** is one that ionizes completely in water to form H^+ ions and anions. A **weak acid** ionizes only slightly in water; most of the acid exists as intact molecules. Common strong acids are sulfuric, hydrochloric, and nitric acids. Likewise, a **strong base** is completely ionized in water, and a **weak base** is only slightly ionized. Sodium hydroxide and potassium hydroxide are common strong bases.

Section 7.5—The reaction between an acid and a base is called neutralization. In aqueous solution, it is often the reaction of H^+ and OH^- to form water. The other anions and cations form an ionic salt.

Section 7.6—The **pH** scale indicates the degree of acidity or basicity; pH is defined as $pH = -\log[H^+]$, where $[H^+]$ is the molar concentration of hydrogen ion. A pH of 7 ($[H^+] = 1 \times 10^{-7}$ M) is neutral; pH values lower than 7 represent increasing acidity, and pH values greater than 7 represent increasing basicity. A change in pH of one unit represents a tenfold change in $[H^+]$.

Section 7.7—A pair of compounds or ions that differ by one proton (H^+) is called a **conjugate acid–base pair**. A **buffer solution** is a mixture of either a weak acid and its conjugate base or a weak base and its conjugate acid. A buffer maintains a nearly constant pH when a small amount of a strong acid or a strong base is added.

Section 7.8—An antacid is a base such as sodium bicarbonate, magnesium hydroxide, aluminum hydroxide, or calcium carbonate that is taken to relieve hyperacidity. Overuse of some antacids can make the blood too alkaline (basic), a condition called **alkalosis**. Sulfuric acid is the number one chemical produced in the United States and is used for making fertilizers and other industrial chemicals. Hydrochloric acid is used for rust removal and for etching mortar and concrete. Lime (calcium oxide) is made from limestone and is the cheapest and most widely used base. It is an ingredient in plaster and cement and is used in agriculture. Sodium hydroxide is used to make many industrial products, as well as soap. Ammonia is a weak base produced mostly for use as fertilizer. **Acid rain** is rain with a pH lower than 5.6. The acidity is due to sulfur oxides and nitrogen oxides from natural sources as well as industrial air pollution and automobile exhaust fumes. Acid rain can have serious effects on plant and animal life. Concentrated strong acids and bases are corrosive poisons that can cause serious burns and **denature** (destroy) proteins. Living cells have an optimal pH that is necessary for the proper functioning of proteins.

GREEN CHEMISTRY—Acids can be used for many purposes, including neutralizing basic waste streams or promoting chemical reactions. The Twelve Principles of Green Chemistry can guide us in the selection of acids and bases in chemical processes that are safer than the ones that were once considered standard choices. Replacements of hazardous mineral acids by carbon dioxide in water and montmorillonite clay are two examples.

Learning Objectives

Learning Objectives	Associated Problems
• Distinguish between acids and bases using their chemical and physical properties. (7.1)	1, 2, 57
• Explain how an acid–base indicator works. (7.1)	3, 58
• Identify Arrhenius and Brønsted acids and bases. (7.2)	4–6, 13–28, 59
• Write a balanced equation for a neutralization or ionization reaction. (7.2)	7, 39–42, 60–62
• Identify acidic and basic anhydrides, and write equations showing their reactions with water. (7.3)	8, 29, 30
• Define and identify strong and weak acids and bases. (7.4)	9–11, 31–38
• Identify the reactants and predict the products in a neutralization reaction. (7.5)	43–46
• Describe the relationship between the pH of a solution and its acidity or basicity. (7.6)	43, 44

• Find the molar concentration of hydrogen ion [H⁺] from a pH value or the pH value from [H⁺]. (7.6)	45–52, 63–65
• Write the formula for the conjugate base of an acid or for the conjugate acid of a base. (7.7)	53, 54
• Describe the action of a buffer. (7.7)	66, 67
• Describe everyday uses of acids and bases and how they affect daily life. (7.8)	12, 55, 56, 68–71
• Recognize carbon dioxide in water as a useful, safe source of acid.	72–74
• Describe alternate choices for greener reaction conditions when using acids and bases.	73, 75

 ## Conceptual Questions

1. Define the following terms, and give an example of each.
   a. acid   **b.** base   **c.** salt

2. List four general properties **(a)** of acidic solutions and **(b)** of basic solutions.

3. Describe the effect on litmus and the action on iron or zinc of a solution that has been neutralized.

4. Can a substance be a Brønsted–Lowry acid if it does not contain H atoms? Are there any characteristic atoms that must be present in a Brønsted–Lowry base?

5. What is meant by a proton in acid–base chemistry? How does it differ from a nuclear proton (Chapter 3)?

6. According to the Arrhenius theory, all acids have one element in common. What is that element? Are all compounds containing that element acids? Explain.

7. Describe the neutralization of an acid or a base.

8. What is an acidic anhydride? A basic anhydride? What is meant by the word *anhydride*?

9. Both strong bases and weak bases have properties characteristic of hydroxide ions. How do strong bases and weak bases differ?

10. Magnesium hydroxide is completely ionic, even in the solid state, yet it can be taken internally as an antacid. Explain why taking it does not cause injury although taking sodium hydroxide would.

11. What are the effects of strong acids and strong bases on the skin?

12. What is alkalosis? What antacid ingredient might cause alkalosis if taken in excess?

Problems

Acids and Bases: The Arrhenius Theory

13. Write equations that represent the action in water **(a)** of hydrobromic acid as an Arrhenius acid and **(b)** of cesium hydroxide as an Arrhenius base.

14. Write equations that represent the action in water **(a)** of nitrous acid, HNO_2, as an Arrhenius acid and **(b)** of strontium hydroxide as an Arrhenius base. (Be sure to include the correct charges for ions.)

15. When a particular Arrhenius acid is dissolved in water, one of the products is hypochlorite ion, ClO^-. Write the formula for the Arrhenius acid.

16. A particular Arrhenius base reacts with water, producing hydrogen phosphate ions. Write the equation for this reaction. (Be sure to include any charges on ions.)

Acids and Bases: The Brønsted–Lowry Acid–Base Theory

17. Use the Brønsted–Lowry definitions to identify the first compound in each equation as an acid or a base.
    a. $CH_3NH_2 + H_2O \longrightarrow CH_3NH_3^+ + OH^-$
    b. $H_2O_2 + H_2O \longrightarrow H_3O^+ + HO_2^-$
    c. $NH_2Cl + H_2O \longrightarrow NH_3Cl^+ + OH^-$

18. Use the Brønsted–Lowry definitions to identify the first compound in each equation as an acid or a base. (*Hint:* What is produced by the reaction?)
    a. $(CH_3)_2NH + H_2O \longrightarrow (CH_3)_2NH_2^+ + OH^-$
    b. $C_6H_5NH_2 + H_2O \longrightarrow C_6H_5NH_3^+ + OH^-$
    c. $C_6H_5CH_2NH_2 + H_2O \longrightarrow C_6H_5CH_2NH_3^+ + OH^-$

19. Write equations that represent the action in water **(a)** of formic acid (HCOOH) as a Brønsted–Lowry acid and **(b)** of pyridine (C_5H_5N) as a Brønsted–Lowry base.

20. Write equations that represent the action in water **(a)** of hypochlorous acid (HOCl) as a Brønsted–Lowry acid and **(b)** of diethylamine [$(CH_3CH_2)_2NH$] as a Brønsted–Lowry base.

21. Write the equation that shows how ammonia acts as a Brønsted–Lowry base in water.

22. Although hydroxylamine ($HONH_2$) contains an OH group, it does not dissociate into NH_2^+ and OH^- ions. However, it *is* a Brønsted–Lowry base. Write an equation that shows how hydroxylamine acts as a Brønsted–Lowry base in water.

Acids and Bases: Names and Formulas

23. For the following acids and bases, supply a formula to match the name or a name to match the formula.
    a. HCl   b. strontium hydroxide
    c. KOH   d. boric acid

24. For the following acids and bases, supply a formula to match the name or a name to match the formula.
    a. rubidium hydroxide
    b. $Al(OH)_3$
    c. hydrocyanic acid
    d. HNO_3

25. Name the following, and classify each as an acid or a base.
    a. H_3PO_4
    b. CsOH
    c. H_2CO_3

26. Name the following, and classify each as an acid or a base.
    a. $Mg(OH)_2$
    b. NH_3
    c. CH_3COOH

27. When an acid name ends in -ic acid, there is often a related acid whose name ends in -ous acid. The formula of the -ous acid has one fewer oxygen atom than that of the -ic acid. With this information, write formulas for (a) nitrous acid and (b) phosphorous acid.

28. Refer to Problem 27 and Table 7.1. Tellurium is in the same group as sulfur. Use this information to write the formulas for (a) telluric acid and (b) tellurous acid.

Acidic and Basic Anhydrides

29. Give the formula for the compound formed when (a) selenium dioxide reacts with water and (b) strontium oxide reacts with water. In each case, is the product an acid or a base?

30. Give the formula for the compound formed when (a) cesium oxide reacts with water and (b) tetraphosphorus decoxide reacts with water. (*Hint*: Six molecules of water react, and four molecules of the product are formed.) In each case, is the product an acid or a base?

Strong and Weak Acids and Bases

31. When 1.0 mol of hydrogen iodide (HI) gas is dissolved in a liter of water, the resulting solution contains 1.0 mol of hydronium ions and 1.0 mol of iodide ions. Classify HI as a strong acid, a weak acid, a weak base, or a strong base.

32. Thallium hydroxide (TlOH) is a water-soluble ionic compound. Classify TlOH as a strong acid, a weak acid, a weak base, or a strong base.

33. A solution made by adding 1.300 mol of methylamine (CH_3NH_2) gas to 1.00 L of water contains 0.00936 mol each of methylammonium ions $(CH_3NH_3^+)$ and hydroxide ions (OH^-) and 1.291 mol of CH_3NH_2 molecules. Classify CH_3NH_2 as a strong acid, weak acid, weak base, or strong base.

34. A solution made by adding 1.000 mol of HOCN to 1.00 L of water contains 0.055 mol each of H^+ ions and OCN^- ions and 0.945 mol of HOCN molecules. Classify HOCN as a strong acid, weak acid, weak base, or strong base.

35. Identify each of the following substances as a strong acid, a weak acid, a strong base, a weak base, or a salt.
    a. LiOH
    b. HBr
    c. HNO_2
    d. $CuSO_4$

36. Identify each of the following substances as a strong acid, a weak acid, a strong base, a weak base, or a salt.
    a. K_3PO_4
    b. $CaBr_2$
    c. $Mg(OH)_2$
    d. NH_4Cl

37. Place the following aqueous solutions in order from highest concentration of H^+ ions to lowest: 0.20 M NH_3; 0.10 M HClO; 0.10 M $HClO_4$.

38. Rank 0.10 M solutions of formic acid (HCOOH), ammonia, nitric acid, and lithium hydroxide in order of increasing pH.

Neutralization

39. Write equations for the reaction (a) of silver hydroxide (AgOH) with hydrochloric acid and (b) of rubidium hydroxide with nitric acid.

40. Write equations for the reaction (a) of 1 mol of barium hydroxide with 2 mol of hydrobromic acid and (b) of 3 mol of perchloric acid with 1 mol of aluminum hydroxide.

41. Write the equation for the reaction of 1 mol of sulfurous acid (H_2SO_3) with 1 mol of magnesium hydroxide.

42. Write the equation for the reaction of 2 mol of phosphoric acid with 3 mol of calcium hydroxide.

The pH Scale

43. Indicate whether each of the following pH values represents an acidic, basic, or neutral solution.
    a. 3   b. 11.4   c. 0.8   d. 9.6

44. Lime juice is quite sour. Which of the following is a reasonable pH for lime juice?
    a. 13   b. 7.4   c. 2.2   d. 6.5

45. What is the pH of a solution that has a hydrogen ion concentration of 1.0×10^{-8} M?

46. What is the pH of a solution that has a hydrogen ion concentration of 1.0×10^{-2} M?

47. What is the pH of a solution that has a hydroxide ion concentration of 1.0×10^{-3} M?

48. What is the pH of a solution that has a hydroxide ion concentration of 1.0×10^{-10} M?

49. What is the hydrogen ion concentration of a solution that has a pH of 6?

50. What is the hydrogen ion concentration of a solution that has a pH of 11?

51. Black coffee has a hydrogen ion concentration between 1.0×10^{-5} M and 1.0×10^{-6} M. The pH of black coffee is between which two whole numbers?

52. A window cleaning spray has a pH between 10 and 11. What two whole-number values of x should be used in 1.0×10^{-x} M to express the range of hydrogen ion concentration?

Conjugate Acid–Base Pairs

53. In the following reaction in aqueous solution, identify **(a)** which of the reactants is the acid and which is the base, **(b)** the conjugate base of the acid, and **(c)** the conjugate acid of the base.

$$NH_2OH + HCl \longrightarrow NH_3OH^+ + Cl^-$$

54. In the following reaction in aqueous solution, identify **(a)** which of the reactants is the acid and which is the base, **(b)** the conjugate base of the acid, and **(c)** the conjugate acid of the base.

$$HCO_3^- + HCOOH \longrightarrow HCOO^- + H_2CO_3$$

Antacids

55. Mylanta liquid has 200 mg of $Al(OH)_3$ and 200 mg of $Mg(OH)_2$ per teaspoonful. Write the equation for the neutralization of stomach acid [represented as HCl(aq)] by each of these substances.

56. What is the Brønsted–Lowry base in each of the following compounds? Which are ingredients in antacids?
    a. $NaHCO_3$
    b. $Mg(OH)_2$
    c. $MgCO_3$
    d. $CaCO_3$

Expand Your Skills

57. Most soaps have a bitter taste. What does this indicate about the pH of most soaps?

58. Which one(s) of the following could be the solute in the pictured solution at left?
    a. HNO_2 b. KOH   c. NH_3
    d. CH_3NH_2 e. CF_3COOH

59. Which one of the following could be the gas coming out of the tube in the drawing at the right?
    a. N_2 b. CO_2 c. NH_3 d. SO_2

60. According to the Arrhenius theory, is every compound that contains OH a base? Explain.

61. Lime deposits on brass faucets are mostly $CaCO_3$. The deposits can be removed by soaking the faucets in hydrochloric acid. Write an equation for the reaction that occurs.

62. Strontium iodide can be made by the reaction of solid strontium carbonate ($SrCO_3$) with an acid. Identify the acid, and write the balanced equation for the reaction.

63. When a well is drilled into rock that contains sulfide minerals, some of the minerals may dissolve in the well water, yielding "sulfur water" that smells like rotten eggs. The water from these wells feels slightly slippery, contains HS^- ions, and turns litmus paper blue. From this information, write an equation representing the reaction of HS^- ions with water.

64. The *pOH* is related to $[OH^-]$ just as pH is related to $[H^+]$. What is the pOH **(a)** of a solution that has a hydroxide ion concentration of 1.0×10^{-2} M? **(b)** of a 0.1 M KOH solution? **(c)** of a 0.001 M HCl solution?

65. Rank 0.1 M solutions of **(a)** acetic acid ($HC_2H_3O_2$), **(b)** ammonia (NH_3), **(c)** nitric acid (HNO_3), **(d)** sodium chloride (NaCl), and **(e)** sodium hydroxide (NaOH) from the highest pH to the lowest pH.

66. Calculate the pH of a 0.050 M $Ba(OH)_2$ solution.

67. Human blood contains buffers that minimize changes in pH. One blood buffer is the hydrogen phosphate ion (HPO_4^{2-}). Like water, HPO_4^{2-} is amphiprotic. That is, it can act either as a Brønsted–Lowry acid or as a Brønsted–Lowry base. Write equations that illustrate the reaction with water of HPO_4^{2-} as an acid and as a base.

68. Which is likely to be a better buffer, a solution of ammonia with ammonium acetate or a solution of sodium hydroxide with sodium acetate? Why?

69. Three varieties of Tums have calcium carbonate as the only active ingredient: Regular Tums tablets have 500 mg; Tums E-X, 750 mg; and Tums ULTRA, 1000 mg. How many regular Tums would you have to take to get the same quantity of calcium carbonate as you would get with two Tums E-X? With two Tums ULTRA tablets?

70. The active ingredient in the antacid Basaljel is a gel of solid aluminum carbonate. Write the equation for the neutralization of aluminum carbonate by stomach acid (aqueous HCl).

71. Liquid ammonia, $NH_3(l)$, is used as a solvent for certain unusual reactions. Like water, liquid ammonia can self-protonate to a very small extent. Write an equation representing the Brønsted–Lowry reaction of an ammonia molecule with another ammonia molecule.

72. Sulfuric acid is produced from elemental sulfur by a three-step process: (1) Sulfur is burned to produce sulfur dioxide. (2) Sulfur dioxide is oxidized to sulfur trioxide using oxygen and a vanadium (V) oxide catalyst. (3) Finally, the sulfur trioxide is reacted with water to produce 98% sulfuric acid. Write equations for the three reactions.

73. A buffer solution can be prepared by mixing one mole of sodium hydroxide with two moles of acetic acid. What is the conjugate acid–base pair in this buffer solution? Explain.

74. When carbon dioxide mixes with water, a weak acid called carbonic acid is formed. Will CO_2 reduce the pH of the water, increase the pH of the water, or leave the pH of the water unchanged?

75. Suppose a certain company has an alkaline waste stream that requires 1 mL of concentrated HCl to neutralize 1 liter of waste. If that company chooses to switch to CO_2 to neutralize their annual 1 million liter waste stream, how many liters of concentrated HCl are replaced annually by this choice?

76. Carbonated water (pure water in which CO_2 is dissolved) has a slightly sour taste. What does this hint indicate about the pH of carbonated water?

77. An important type of chemical substance called an ester can be prepared by heating two chemicals in the presence of a strong mineral acid. No water is used in this reaction. Suggest a greener alternative to the strong mineral acid.

 ## Critical Thinking Exercises

Apply knowledge that you have gained in this chapter and one or more of the FLaReS principles (Chapter 1) to evaluate the following statements or claims.

7.1 A television advertisement claimed that the antacid *Maalox* neutralizes stomach acid faster and, therefore, relieves heartburn faster than *Pepcid AC*, a drug that inhibits the release of stomach acid. To illustrate this claim, two flasks of acid were shown. In one, *Maalox* rapidly neutralized the acid. In the other, *Pepcid AC* did not neutralize the acid.

7.2 Testifying in a court case, a witness makes the following statement: "Although runoff from our plant did appear to contaminate a stream, the pH of the stream before the contamination was 6.4, and after contamination it was 5.4. So the stream is now only slightly more acidic than before."

7.3 A Jamaican fish recipe calls for adding the juice of several limes to the fish, but no heat is used to cook the fish. The directions state that the lime juice in effect "cooks" the fish.

7.4 An advertisement claims that vinegar in a glass-cleaning product will remove the spots left on glasses by tap water. The spots are largely calcium carbonate deposits.

7.5 A Web page on making biodiesel claims that any water in the waste oil must be removed before starting the reaction.

 ## Collaborative Group Projects

Prepare a PowerPoint, poster, or other presentation (as directed by your instructor) to share with the class.

1. Prepare a brief report on one of the following acids or bases. List sources (including local sources for the acid or base, if available) and commercial uses.
   a. ammonia
   b. hydrochloric acid
   c. phosphoric acid
   d. nitric acid
   e. sodium hydroxide
   f. sulfuric acid

2. Examine the labels of at least five antacid preparations. Make a list of the ingredients in each. Look up the properties (medical use, side effects, toxicity, and so on) of each ingredient on the Web or in a reference book such as *The Merck Index*.

3. Examine the labels of at least five toilet-bowl cleaners and five drain cleaners. Make a list of the ingredients in each. Look up the formulas and properties of each ingredient on the Web or in a reference book such as *The Merck Index*. Which ingredients are acids? Which are bases?

LET'S EXPERIMENT! Acids and Bases and pH, Oh My!

Materials Needed

- Red cabbage
- Water
- Container (for strained cabbage juice indicator)
- Measuring cup
- Lemon juice
- Vinegar
- Ammonia
- Orange juice
- Borax solution (1 tsp in $^1/_4$ cup water); stir until mostly dissolved

- Baking soda solution (1 tsp in $^1/_4$ cup water); stir until mostly dissolved
- Milk of magnesia
- Clear shampoo
- Seltzer water
- Clear plastic cups, 10
- Tablespoon or dropper

Think of all of the products that we use every day. Is everything that we use pH-neutral? How do we test if something is acidic or basic?

You can rank how acidic or basic something is by using the pH scale. Recall that the number of hydrogen (H^+) or hydroxide (OH^-) ions in a solution affects how acidic or basic (alkaline) it is. pH is measured using chemicals that change color when they bind to the extra hydrogen or hydroxyl ions in water. So the more acidic a solution, the more hydrogen ions there are, and the more the color will change. The more basic (alkaline) hydroxyl ions there are, the more basic (alkaline) a solution and the more the color will change.

In this lab, you will be using a natural indicator from red cabbage to test the pH of different substances that can be found in your local grocery store or pharmacy. The chemicals responsible for the red color in red cabbage and many other vegetables, fruits, and flower petals belong to the anthocyanin family. More than 300 kinds of anthocyanin have been discovered so far. They are also powerful antioxidants that have been shown to be very beneficial to human health. The red cabbage indicator has its own pH color change chart.

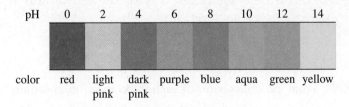

pH	0	2	4	6	8	10	12	14
color	red	light pink	dark pink	purple	blue	aqua	green	yellow

▲ Red Cabbage Color Changes with pH Chart.

Begin by preparing your indicator from red cabbage.
- First, peel off leaves from a head of red cabbage and finely chop 2 cups' worth of red cabbage. (You can also use a blender.)
- Meanwhile, boil 8 cups of water.
- Add the chopped cabbage leaves to the pot, lower the heat, and let it simmer for 20 minutes.
- Pour the mixture through a strainer into a large container. Let the liquid cool to room temperature, and use the cabbage juice as your indicator.

Once you have prepared your cabbage-juice indicator and gathered your supplies, you are ready for the experiment. Set out 10 clear plastic cups, and fill them halfway with indicator juice. For each household item, add one tablespoon of the liquid to the indicator juice and stir. Be sure to leave one cup with only red cabbage indicator and add tap water to another. This will give you 10 samples to examine. Be sure to record your observations.

Does the indicator change color? Create a data table to track the chemical tested, color change, and pH. Once you've finished, arrange the cups from most acidic to most basic.

Questions

1. Did the pH of any of the products surprise you?
2. Which substances were acidic?
3. Which substances were basic?
4. How would you go about neutralizing a particularly acidic solution?
5. What would account for any of your pH measurements not matching the actual pH?

8 Oxidation and Reduction

Plug-in electric cars such as the Tesla are seeing tremendous interest as zero-carbon-footprint vehicles. Their charging and discharging cycles are but an example of the thousands of ways that oxidation–reduction reactions affect our lives.

Have You Ever Wondered?

1 Why are copper roofs green? **2** How does an electric eel create electricity? **3** Why does a battery go dead? **4** Why are some batteries rechargeable but others are not? **5** Why do we not see more "rust bucket" ships?

BURN AND UNBURN From the conversion of sunlight to food, to charging a smartphone to view a recipe, to cooking the food, to metabolizing the food to gain energy so we can look up the next recipe, to the simple act of breathing oxygen in and carbon dioxide out, we depend on an important group of reactions called *redox reactions* (a shortening of *reduction–oxidation*). These reactions are extremely diverse: Charcoal burns; iron rusts; bleach removes stains; a hybrid automobile battery is charged or discharged. Redox reactions are employed in pollution remediation, wastewater treatment, and determination of blood sugar levels by diabetics. Even the generation of a shock of 600 volts by an electric eel is due to redox processes.

Oxidation and reduction always take place together. They represent opposite directions of a single process, a redox reaction. When one substance is oxidized, another is reduced. In other words, you can't have one without the other

▲ **Figure 8.1** Oxidation and reduction always occur together. Pictured on the left is a reaction called the ammonium dichromate volcano. In the reaction, the nitrogen in ammonium ion $\left(NH_4^+\right)$ is oxidized and the chromium in dichromate ion $\left(Cr_2O_7^{2-}\right)$ is reduced. Considerable heat and light are evolved. The equation for the reaction is $(NH_4)_2Cr_2O_7 \longrightarrow Cr_2O_3 + N_2 + 4\,H_2O$. The water is driven off as vapor, and the nitrogen gas escapes, leaving pure Cr_2O_3 as the visible product (right).

(Figure 8.1). However, it is sometimes convenient to discuss only a part of the process—the oxidation part or the reduction part.

Reduced forms of matter—foods, coal, and gasoline—are high in energy. *Oxidized* forms—carbon dioxide and water—are low in energy. The energy in foods and fossil fuels is released when these materials are oxidized. In this chapter, we examine the processes of oxidation and reduction in some detail to better understand the chemical reactions that keep us alive and maintain our civilization.

8.1 Oxidation and Reduction: Four Views

Learning Objectives • Identify an oxidation–reduction reaction. • Classify a particular change within a redox reaction as either oxidation or reduction.

One of the most important redox reactions is combustion—that is, the burning of a substance in oxygen. The term **oxidation** came from the observation that during combustion, oxygen was added. The opposite process would require the removal of oxygen, and thus it was called **reduction**.

Originally, the term *oxidation* was limited to reactions involving combination with oxygen. As chemists came to realize that combination with chlorine or another active nonmetal was quite similiar to combination with oxygen, they broadened the definition of oxidation to include reactions involving these other substances.

Let's take a look at the simplest combustion reaction, the combustion (oxygen present) of hydrogen to form water.

$$2\,H_2 + O_2 \longrightarrow 2\,H_2O$$

In this reaction, oxygen is added to hydrogen, oxidizing it. At the same time, the oxygen is reduced. Whenever oxidation occurs, reduction must also occur in an exactly equivalent amount, and vice versa.

Because oxidation and reduction are chemical opposites and constant companions, their definitions are linked. We can view oxidation and reduction in four different ways (Figure 8.2). As we will see shortly, the four ways are generally consistent with one another and we tend to use the most convenient one.

▶ **Figure 8.2** Four different views of oxidation and reduction.

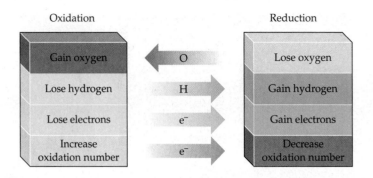

1. Redox as Gain or Loss of Oxygen

The first view of redox is the one used historically. *Oxidation* is defined as a gain of oxygen atoms, and *reduction* as a loss of oxygen atoms. For example, at high temperatures (such as those in automobile engines), nitrogen—normally quite unreactive—combines with oxygen to form nitrogen monoxide.

$$N_2 + O_2 \longrightarrow 2\,NO$$

Nitrogen gains oxygen atoms; there are no O atoms in the N_2 molecule and one O atom in each of the NO molecules. Therefore, nitrogen is oxidized.

Now consider what happens when methane is burned to form CO_2 and water. Both carbon and hydrogen gain oxygen atoms, and so both elements are oxidized.

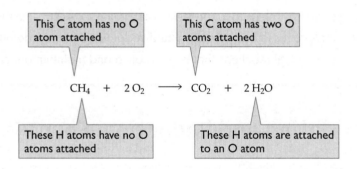

When lead dioxide is heated at high temperatures, it decomposes as follows:

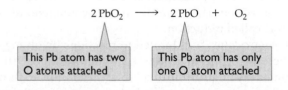

The lead dioxide loses oxygen, so it is reduced.

 CONCEPTUAL EXAMPLE 8.1 Redox Processes—Gain or Loss of Oxygen Atoms

In each of the following changes, is the reactant undergoing oxidation or reduction? (These are not complete chemical equations.)

a. $Pb \longrightarrow PbO_2$ b. $SnO_2 \longrightarrow SnO$

c. $KClO_3 \longrightarrow KCl$ d. $Cu_2O \longrightarrow 2\,CuO$

Solution

a. Lead gains oxygen atoms (it has none on the left and two on the right); it is oxidized.

b. Tin loses an oxygen atom (it has two on the left and only one on the right); it is reduced.

c. There are three O atoms on the left and none on the right. The compound loses oxygen; it is reduced.

d. The two copper atoms on the left share a single oxygen atom—that is, half an oxygen atom each. On the right, each Cu atom has an O atom all to itself. Cu has gained oxygen; it is oxidized.

> **Exercise 8.1A**

In each of the following changes, is the reactant undergoing oxidation or reduction? (These are not complete chemical equations.)

a. $3\,Fe \longrightarrow Fe_3O_4$ b. $NO \longrightarrow NO_2$ c. $CrO_3 \longrightarrow Cr_2O_3$

> **Exercise 8.1B**

In each of the following changes, is the reactant undergoing oxidation or reduction? (These are not complete chemical equations.)

a. $SO_2 \longrightarrow SO_3$ b. $NaBrO \longrightarrow NaBr$

c. $CH_3CH_2OH \longrightarrow CH_3CH_2COOH$

2. Redox as Gain or Loss of Hydrogen

When hydrogen is involved in a redox reaction, we can define *oxidation* as a loss of hydrogen atoms, and *reduction* as a gain of hydrogen atoms. Look once more at the burning of methane:

$$CH_4 + 2\,O_2 \longrightarrow CO_2 + 2\,H_2O$$

The oxygen gains hydrogen to form water, so the oxygen is reduced. Methane loses hydrogen and is oxidized. Do you see that the carbon and hydrogen atoms in CH_4 also gain oxygen? Our two views of oxidation and reduction are consistent with one another.

This view is particularly useful when oxygen gas is not involved. For example, methyl alcohol (CH_3OH), when passed over hot copper gauze, forms formaldehyde and hydrogen gas.

$$CH_3OH \longrightarrow CH_2O + H_2$$

The C and O atoms have four H atoms attached	The C and O atoms have only two H atoms attached

Because the methyl alcohol loses hydrogen, it is oxidized in this reaction.

A mixture of hydrogen gas and chlorine gas reacts explosively when ignited.

$$H_2 + Cl_2 \longrightarrow 2\,HCl$$

Because each chlorine atom gains a hydrogen atom, chlorine is reduced; hydrogen therefore must be oxidized.

Biochemists often find the gain or loss of hydrogen atoms a useful way to look at oxidation–reduction processes. For example, a substance called NAD^+ is changed to NADH in a variety of biochemical redox reactions. The actual molecules are rather complex, but we can write the equation for the oxidation of ethyl alcohol to acetaldehyde, one step in the metabolism of the alcohol, as follows.

$$CH_3CH_2OH + NAD^+ \longrightarrow CH_3CHO + NADH + H^+$$

We can see that ethyl alcohol is oxidized (loses hydrogen) and NAD^+ is reduced (gains hydrogen).

 CONCEPTUAL EXAMPLE 8.2 Redox Processes—Gain or Loss of Hydrogen Atoms

In each of the following changes, is the reactant undergoing oxidation or reduction? (These are not complete chemical equations.)

a. $C_2H_6O \longrightarrow C_2H_4O$ b. $C_2H_2 \longrightarrow C_2H_6$

Solution

a. There are six H atoms in the reactant on the left and only four in the product on the right. The reactant loses H atoms; it is oxidized.
b. There are two H atoms in the reactant on the left and six in the product on the right. The reactant gains H atoms; it is reduced.

〉 Exercise 8.2A

In each of the following changes, is the reactant undergoing oxidation or reduction? (These are not complete chemical equations.)

a. $C_6H_{12} \longrightarrow C_6H_6$
b. $2\,CH_3SH \longrightarrow CH_3SCH_3$

〉 Exercise 8.2B

In each of the following changes, is the reactant undergoing oxidation or reduction? (These are not complete chemical equations.)

a. $C_3H_6O \longrightarrow C_3H_8O$
b. $C_3H_6 \longrightarrow C_3H_4$

---

3. Gain or Loss of Electrons in Redox Reactions

A third view of oxidation and reduction involves gain or loss of electrons. (In fact, electron transfer occurs in all redox reactions, though the other views of redox are sometimes more convenient to use.) Oxidation is defined as a loss of electrons; the oxidized species becomes more positive. Reduction, in turn, involves a gain of electrons and the reduced species becomes more negative. For example, a strip of magnesium metal can be ignited in air, then placed in a container of chlorine gas, and it will continue to burn. A white powdery "soot" of magnesium chloride forms, with two chloride ions for each magnesium ion.

$$Mg + Cl_2 \longrightarrow Mg^{2+} + 2\,Cl^-$$

Because the magnesium atom loses electrons (two), magnesium metal is oxidized. Each chlorine atom gains one electron, so the chlorine is reduced.

There are two popular mnemonics for remembering the link between oxidation/reduction and electron loss/gain, seen in the margin.

LEO the lion says GER
  Loss of
    Electrons is
      Oxidation
  Gain of
    Electrons is
      Reduction

OIL RIG
  Oxidation
    Is
      Loss of electrons
  Reduction
    Is
      Gain of electrons

 CONCEPTUAL EXAMPLE 8.3 Redox Processes—Gain or Loss of Electrons

In each of the following changes, is the reactant undergoing oxidation or reduction? (These are not complete chemical equations.)

a. $Zn \longrightarrow Zn^{2+}$ b. $Fe^{3+} \longrightarrow Fe^{2+}$ c. $S^{2-} \longrightarrow S$ d. $F_2 \longrightarrow 2F^-$

Solution

a. In forming a 2+ ion, a Zn atom loses two electrons: $Zn \longrightarrow Zn^{2+} + 2\,e^-$, so zinc is oxidized.

b. To change from a 3+ ion to a 2+ ion, Fe^{3+} gains an electron: $Fe^{3+} + e^- \longrightarrow Fe^{2+}$ This means that each Fe(III) ion is reduced.

c. To change from an S^{2-} ion to an atom with no charge, S loses two electrons: $S^{2-} \longrightarrow S + 2e^-$. So, the sulfide ion is oxidized.

d. Here two F^- ions are formed from neutral diatomic F_2, so an electron is gained by each fluorine atom in F_2: $F_2 + 2e^- \longrightarrow 2\,F^-$. In other words, elemental fluorine is reduced.

We must be careful to use the correct terms in redox reactions. In (c) it would be incorrect to simply say that "sulfur is oxidized," as that phrase would suggest that elemental sulfur, S, has lost one or more electrons to change to a positive ion. "Sulfide ion is oxidized" is the correct statement for S^{2-}.

❯ Exercise 8.3A

In each of the following changes, is the reactant undergoing oxidation or reduction? (These are not complete chemical equations.)

a. $Co^{2+} \longrightarrow Co^{3+}$ b. $Hg^{2+} \longrightarrow Hg$ c. $H_2 \longrightarrow 2\,H^+$

❯ Exercise 8.3B

In each of the following changes, is the reactant undergoing oxidation or reduction? (These are not complete chemical equations.)

a. $Cr^{6+} \longrightarrow Cr^{3+}$ b. $2\,Cl^- \longrightarrow Cl_2$ c. $I_3^- \longrightarrow 3\,I^-$

4. Redox Reactions and Oxidation Numbers

At this stage, it will be useful to introduce the concept of *oxidation numbers* (or oxidation states) and their applications. An **oxidation number** is a hypothetical value that can be thought of as "the charge an atom would have in a formula if all bonds were ionic." This fourth view of oxidation and reduction involves an increase or decrease in oxidation number. (See Figure 8.3.)

When an oxidation number increases, the actual charge (of an ion) or perceived charge (of an atom) becomes more positive or less negative as a result of one or more electrons being lost. The opposite is true for a reduction: The charge of an atom becomes less positive or more negative as a result of an electron (or electrons) being gained.

The following set of rules *generally* yields correct oxidation numbers:

1. An atom in its elemental form has an oxidation number of 0. Ag, N_2, Fe, and S all have oxidation numbers of zero.

2. A monatomic ion has an oxidation number equal to its charge. $K^+ = +1$; $Br^- = -1$; $Cu^{2+} = +2$. In TiF_4, titanium is a 4+ ion, so its oxidation number is +4. Fluoride is a 1− ion, so its oxidation number is −1.

3. Hydrogen and oxygen in compounds have oxidation numbers of +1 and −2, respectively. For example, in $HClO_4$, H has an oxidation number of +1. Oxygen's oxidation number is −2. (Exceptions: −1 for oxygen in peroxides—e.g., H_2O_2; −1 for hydrogen in metal hydrides—e.g., CaH_2, NaH.)

4. The sum of the oxidation numbers of the atoms in a molecule is zero. The sum of the oxidation numbers of the atoms in a polyatomic ion equals the charge of the ion. This means that the oxidation number of Cl in $HClO_4$ (#3, above) is +7.

Note that by convention, the sign of an ion's charge is placed *after* the number, while the sign of an oxidation number is placed *before* the number. So iron(II) ion is written as Fe^{2+}, not Fe^{+2}, while its oxidation number is +2, not 2+.

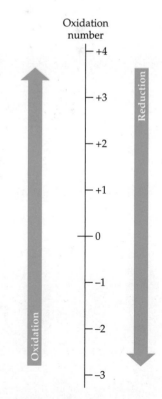

▲ **Figure 8.3** An increase in oxidation number means a loss of electrons and is therefore oxidation. A decrease in oxidation number means a gain of electrons and is therefore reduction.

CONCEPTUAL EXAMPLE 8.4 Determining Oxidation Numbers in Molecules and Polyatomic Ions

What are the oxidation numbers of atoms in the following molecules and ions?

a. SO_2 **b.** HNO_2 **c.** NH_3^- **d.** $Cr_2O_7^{2-}$

Solution

Finding an oxidation number in a molecule or polyatomic ion is a matter of setting up a simple equation, using the four rules above to determine what is known, and solving for the unknown.

a. Here's what we know: Rule 4 says that the sum of the oxidation number of S plus that of two O is zero. Rule 3 tells us that each O has an oxidation number of –2, so two O equals 2 × (–2) or –4. So to find the oxidation number of sulfur, we solve $S + (-4) = 0$, which means that S is +4.

b. For HNO_2, again the sum of the oxidation numbers of the three elements is zero. Rule 3 says that H is +1 and each O is –2. To find N's oxidation number we solve $(+1) + N + [2 \times (-2)] = 0$, giving +3 for N.

c. For NO_3^-, the sum of the oxidation numbers is the charge, –1. Oxygen has an oxidation number of –2. The equation to solve becomes $N + [3 \times (-2)] = -1$, which gives +5 for N. Note from this problem and (b) that the oxidation number for nitrogen (and many other elements) can vary.

d. The sum of the oxidation numbers in $Cr_2O_7^{2-}$ is the charge, –2. Oxygen has an oxidation number of –2. There are *two* chromium atoms, so the equation to solve becomes $(2 \times Cr) + [7 \times (-2)] = -2$, which gives +6 for Cr.

❯ Exercise 8.4A

What is the oxidation number of carbon in each of the following?

a. CO   **b.** C_2H_6 **c.** H_2CO

❯ Exercise 8.4B

In each of the following, determine the oxidation number of the chlorine atom.

a. Cl_2 **b.** HCl   **c.** ClO_3^- **d.** Cl_2O_3

 CONCEPTUAL EXAMPLE 8.5 Redox Processes—Oxidation Number as Applied to Reactions

In the following change, is the reactant undergoing oxidation or reduction? (This is not a complete chemical equation.)

$$C_2H_6O \longrightarrow C_2H_4O$$

Note that we have previously looked at this case in Conceptual Example 8.2A in terms of the hydrogen removal. This time, however, let's do so in terms of the oxidation numbers.

Solution

C_2H_6O: The molecule has no charge, so the sum of the oxidation numbers of all the atoms must equal zero. Each of the six hydrogen atoms has an oxidation number of +1, for a sum of +6. Oxygen has an oxidation number of −2. The sum of the oxidation numbers of hydrogen and oxygen is +6 + (−2) = +4. This means that the sum of the oxidation numbers for the carbon atoms must equal −4. There are two carbon atoms, so each carbon must have an oxidation number of −2.

C_2H_4O: Again, the molecule has no charge, so the sum of all the oxidation numbers must equal zero. Each of the four hydrogen atoms has an oxidation number of +1, for a total of +4. The oxidation number of the oxygen atom is −2, so the sum of the oxidation numbers of hydrogen plus oxygen is +4 + (−2) = +2. This means that the sum of the oxidation numbers of the two carbon atoms must be -2. Therefore, each of the two carbon atoms must have the oxidation number of -1.

This means that the oxidation number for each carbon has become less negative, changing from -2 to -1, and an electron was lost. Therefore, carbon was oxidized.

> **Exercise 8.5A**

In each of the following changes, is the reactant undergoing oxidation or reduction? (These are not complete chemical equations.)

   a. $C_2H_2 \longrightarrow CO_2$ **b.** $MnO_4^- \longrightarrow Mn^{2+}$ **c.** $C_2Cl_4 \longrightarrow C_2Cl_6$

> **Exercise 8.5B**

In each of the following changes, is the reactant undergoing oxidation or reduction? (These are not complete chemical equations.)

   a. $ClO_3^- \longrightarrow Cl^-$ **b.** $Br_2 \longrightarrow 2BrO_3^-$ **c.** $S_2O_8^{2-} \longrightarrow 2\,SO_4^{2-}$

Which of the Four Views Should We Use?

Why do we have different ways to look at oxidation and reduction? Oxidation as a gain of oxygen is historical and specific, but the definition in terms of electrons applies more broadly. Which one should we use? Ordinarily the definitions are consistent with one another, so we use the most convenient one.

Let's look at several reactions and see how to decide which approach to use. The following reaction is of great importance within the environment, as it changes the highly toxic $Cr_2O_7^{2-}$ to the much less toxic Cr^{3+}.

$$Cr_2O_7^{2-}(aq) + 3\,Sn^{2+}(aq) + 14\,H^+(aq) \longrightarrow 2\,Cr^{3+}(aq) + 3\,Sn^{4+}(aq) + 7\,H_2O(l)$$

In this case, the oxidation number approach is probably easiest, because tin is present as monatomic ions. The oxidation number of Sn^{2+} is $+2$, and that of Sn^{4+} is $+4$, so tin(II) is oxidized to tin(IV). The chromium product is Cr^{3+}, so its oxidation number is $+3$. The oxidation number of chromium in $Cr_2O_7^{2-}$ can be found to be $+6$, so chromium is reduced from $+6$ to $+3$. (In this case we can reason our way to the answer. We have already determined that tin is oxidized. The oxygen and hydrogen in the equation do not change their oxidation numbers. Therefore, the only element left—chromium—must be reduced.)

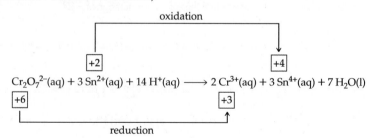

For the combustion (burning) of carbon,

$$C + O_2 \longrightarrow CO_2$$

it is most convenient to see that carbon is oxidized by gaining oxygen atoms. The only other reactant is oxygen gas, so it must be reduced. Similarly, for the reaction

$$CH_2O + H_2 \longrightarrow CH_4O$$

it is easy to see that the reactant CH_2O gains hydrogen atoms and is thereby reduced. Finally, in the case of the reaction

$$3\,Sn^{2+} + 2\,Bi^{3+} \longrightarrow 3\,Sn^{4+} + 2\,Bi$$

it is clear that tin is oxidized because it loses electrons, increasing its oxidation number from $+2$ to $+4$. Similarly, we see that bismuth is reduced because it gains electrons, decreasing its oxidation number from $+3$ to 0.

Viewing redox reactions through the prism of oxygen and hydrogen gain or loss is useful in organic chemistry and biochemistry (Chapter 9 and Chapter 16)

1 **Why are copper roofs green?**

Copper is a versatile roofing material, highly regarded for its appearance as well as its economical heat-conducting properties. One interesting aspect of copper roofs is that they turn green. The name for this desired green coating is *patina*. This change in copper's appearance is due to a number of chemical reactions. The first of these reactions is the oxidation of Cu to Cu^+ by oxygen:

$$4\,Cu + O_2 \longrightarrow 2\,Cu_2O$$

Copper(I) is then further oxidized to Cu^{2+}, forming black CuO and CuS. Further reactions with sulfur dioxide and carbon dioxide present in the air, aided by moisture, yield azurite (blue), $Cu_3(CO_3)_2(OH)_2$; brochantite (green), $Cu_4SO_4(OH)_6$; and malachite (green), $Cu_2CO_3(OH)_2$. These three minerals combine to give patina its unique green hues.

and for some simple reactions. However, it is the electron loss or gain analysis that is universally applicable to all redox processes, especially to those not involving any oxygen- or hydrogen-containing species. The oxidation number approach is the method of choice for analyzing and studying redox processes ranging from oxygen transport in our bodies to powering nearly all high-performance electronic devices, from cell phones to glucose meters.

SELF-ASSESSMENT Questions

1. Which of the following is *not* one of the ways we can view oxidation?
   a. electrons gained
   b. electrons lost
   c. hydrogen atoms lost
   d. oxygen atoms gained

2. Which of the following partial equations represents an oxidation of nitrogen?
   a. $NH_3 \longrightarrow NH_4^+$
   b. $N_2O_4 \longrightarrow NI_3$
   c. $NO_3^- \longrightarrow NO$
   d. $NO_2 \longrightarrow N_2O_5$

3. When CrO_4^{2-} reacts to form Cr^{3+}, the chromium in CrO_4^{2-} is
   a. reduced; it loses electrons
   b. reduced; it loses oxygen
   c. oxidized; its oxidation number increases
   d. oxidized; it gains positive charge

4. Which of the following equations best represents reduction of a molybdenum ion?
   a. $MoO_2 + H_2O \longrightarrow H_2MoO_3$
   b. $MoO_2 + 4 H_2 \longrightarrow Mo + 2 H_2O$
   c. $2 MoO_2 + O_2 \longrightarrow 2 MoO_3$
   d. $MoO_2 + 4 H^+ \longrightarrow Mo^{4+} + 2 H_2O$

5. According to the following equation . . .

   $$3 Cl_2 + 2 Al \longrightarrow 2 AlCl_3$$

   . . . Cl_2 is
   a. both oxidized and reduced
   b. neither oxidized nor reduced
   c. oxidized only
   d. reduced only

6. According to the following partial equation . . .

   $$Mn \longrightarrow Mn^{2+}$$

   . . . manganese
   a. gains electrons and is oxidized
   b. gains electrons and is reduced
   c. loses electrons and is oxidized
   d. loses electrons and is reduced

7. When an element is oxidized, its oxidation number
   a. decreases, as electrons are gained
   b. decreases, as electrons are lost
   c. increases, as electrons are gained
   d. increases, as electrons are lost

8. What are the oxidation numbers of sodium and oxygen atoms in Na_2O, respectively?
   a. 0, 0   **b.** +1, +1   **c.** +1, −2   **d.** +1, −1

9. What is the oxidation number of manganese in $KMnO_4$?
   a. +7   **b.** +8   **c.** +1   **d.** +6

10. Which of the following statements is correct with regard to assigning oxidation numbers?
    a. Alkali metals can have an oxidation number of either +1 or −1.
    b. Halogens can never have an oxidation number of −1.
    c. Oxygen must always have an oxidation number of −2.
    d. The oxidation number of alkaline earth metals in ionic compounds is always +2.

11. What is the charge of the PO_4^{n-} ion, in which phosphorus has the oxidation number of +5 and each of the oxygen atoms has the oxidation number of −2?
    a. 0   **b.** 1−   **c.** 2−   **d.** 3−

Answers: 1, a; 2, d; 3, b; 4, b; 5, d; 6, c; 7, d; 8, c; 9, a; 10, d; 11, d

8.2 Oxidizing and Reducing Agents

Learning Objective • Identify the oxidizing agent and the reducing agent in a redox reaction.

It is often convenient to focus attention on one reactant at a time to see how this reactant affects the entire redox reaction. An **oxidizing agent** is a reactant that causes another reactant to become oxidized; a **reducing agent** is a reactant that causes another reactant to become reduced. (By analogy, a lending agent does not get money; she causes another person to get money.) Stated differently, the reducing agent is the substance being oxidized; the oxidizing agent is the substance being reduced. For example, in the reaction

$$\boxed{+2} \xrightarrow{\text{reduction}} \boxed{0}$$
$$CuO + H_2 \longrightarrow Cu + H_2O$$
$$\boxed{0} \xrightarrow[\text{oxidation}]{} \boxed{+1}$$

we see that copper in copper(II) oxide is reduced and hydrogen is oxidized. Since CuO brought about the oxidation of H_2, CuO is the oxidizing agent. On the other hand, since H_2 brought about the reduction of CuO, H_2 is the reducing agent. Each oxidation–reduction reaction has an oxidizing agent and a reducing agent among the reactants.

 CONCEPTUAL EXAMPLE 8.6   Oxidizing and Reducing Agents

Identify the oxidizing and reducing agents in the following reactions.

a. $2\,C + O_2 \longrightarrow 2\,CO$

b. $N_2 + 3\,H_2 \longrightarrow 2\,NH_3$

c. $3\,Mg + N_2 \longrightarrow Mg_3N_2$

d. $2\,H_2S + 8\,NO_3^- \longrightarrow SO_4^{2-} + 4\,N_2 + 4\,H_2O + 2\,H^+$

Solution

We can determine the answers by applying the various definitions of oxidation and reduction.

a. C gains oxygen and is oxidized, so it must be the reducing agent. Therefore, O_2 is the oxidizing agent.

b. N_2 gains hydrogen and is reduced, so it is the oxidizing agent. Therefore, H_2 is the reducing agent.

c. The oxidation number of magnesium increases from 0 to +2 as a result of magnesium losing two electrons. In other words, Mg is oxidized and is the reducing agent. As a result, N_2 is the oxidizing agent, gaining electrons by decreasing the oxidation number of the nitrogen atom from 0 to −3. Note that the argument involving oxygen gain or hydrogen loss for oxidation would not have worked in this case! Neither of these elements is in this reaction.

d. S in H_2S increases its oxidation number from −2 to +6 by losing eight electrons and becoming oxidized, so H_2S is the reducing agent. The oxidation number of the nitrogen atom in the NO_3^- ion decreases from +5 to 0; N is reduced. Consequently, NO_3^- is the oxidizing agent.

❭ **Exercise 8.6A**

Identify the oxidizing agents and reducing agents in the following reactions.

a. $Mg + F_2 \longrightarrow MgF_2$ b. $4\,H_2 + Fe_3O_4 \longrightarrow 3\,Fe + 4\,H_2O$

c. $Ce^{4+} + Ni^{2+} \longrightarrow Ce^{3+} + Ni^{3+}$

❭ **Exercise 8.6B**

Identify the oxidizing agents and reducing agents in the following reactions.

a. $Se + O_2 \longrightarrow SeO_2$ b. $2\,K + Br_2 \longrightarrow 2\,K^+ + 2\,Br^-$

c. $4\,Fe + 3\,O_2 \longrightarrow 2\,Fe_2O_3$

SELF-ASSESSMENT Questions

1. A reducing agent
   a. gains electrons   b. gains protons
   c. loses electrons   d. loses protons

2. When an atom gains two electrons, it is
   a. an oxidizing agent   b. a reducing agent
   c. oxidized   d. a double-reducing agent

3. If an oxidation number of an atom in a compound increases from +2 to +4, the compound
   a. is a reducing agent   b. is an oxidizing agent
   c. gains electrons   d. gains protons

4. Which substance is the oxidizing agent in the following reaction?

   $$Cu(s) + 2\,Ag^+(aq) \longrightarrow Cu^{2+}(aq) + 2\,Ag(s)$$

   a. Ag   b. Ag^+ c. Cu   d. Cu^{2+}

5. In the following reaction . . .

   $$Al(s) + Cr^{3+}(aq) \longrightarrow Al^{3+}(aq) + Cr(s)$$

   . . . the reducing agent is

   a. Al(s)   b. $Al^{3+}(aq)$ c. $Cr^{3+}(aq)$ d. Cr(s)

6. What is the oxidizing agent in the following reaction?

   $$Zn(s) + 2\,Tl^+(aq) \longrightarrow Zn^{2+}(aq) + 2\,Tl(s)$$

   a. Tl(s)   b. $Tl^+(aq)$ c. Zn(s)   d. $Zn^2(aq)$

7. What is the reducing agent in the following reaction?

   $$12\,H^+(aq) + 2\,IO_3^-(aq) + 10\,Fe^{2+}(aq)$$
   $$\longrightarrow 10\,Fe^{3+}(aq) + I_2(s) + 6\,H_2O(l)$$

   a. $Fe^{2+}(aq)$ b. $H^+(aq)$ c. $I_2(s)$ d. $IO_3^-(aq)$

8.3 Electrochemistry: Cells and Batteries

Learning Objectives • Balance redox equations. • Identify and write the half-reactions in an electrochemical cell.

We saw in Chapter 3 that electricity can cause chemical change in the process called *electrolysis*. For example, an electric current passed through water produces hydrogen and oxygen gases. In Sections 8.1 and 8.2, we saw that oxidation–reduction reactions involve electron transfer. This electron transfer can be harnessed to produce electricity. An electric current in a wire is simply a flow of electrons.

The study of this natural phenomenon has a long history. It all started with the desire to understand how muscles work. An Italian scientist by the name of Luigi Galvani (1737–1798), a lecturer of anatomy, discovered that amputated frog legs suspended by a brass hook twitched when the hook was dragged along an iron rail or when lightning struck close by (or so the story goes). He concluded that the electricity came from the frog's leg.

Another Italian scientist, Alessandro Volta (1745–1827)—a very good professor of experimental physics—did not agree with Galvani's conclusion. Instead, Volta set up a number of experiments in which he tried to connect several metals through brine (saltwater) solutions. He used his tongue (the best detector of the day) to detect whether electricity was generated and, if so, how much. He settled on the combination of silver and zinc. To make his device as portable as possible, and to demonstrate his idea in the process, he piled up alternating silver and zinc discs separated by brine-soaked cardboard discs. This was the first battery, and is very similar to the technology that is used in nearly all modern-day portable electronics. The brine was the original electrolyte—a solution that can conduct electricity. All this came from the fundamental question of why muscles move.

When a coil of copper wire is placed in a solution of silver nitrate (Figure 8.4a), the copper atoms give up their outer electrons to the silver ions. (We can omit the nitrate ions from the equation because they do not change.) The copper metal dissolves, going into solution as copper(II) ions that color the solution blue. The silver ions come out of solution as beautiful needles of silver metal (Figure 8.4b). The copper is oxidized; the silver ions are reduced.

So far, so good, but how can we go about harnessing electricity? Because the reaction involves only the transfer of electrons, it can occur even when the silver ions are separated from the copper metal. We can place the copper metal and copper(II) ions in one compartment, the silver metal and silver ions in an adjoining compartment, and connect them with a wire. The electrons will flow through the wire from the copper side, where the electrons are lost due to the oxidation, to the silver side, where the electrons are simultaneously taken up in the reduction process. This flow of electrons is an electric current, and it can be used to run a motor or light a lamp.

In the **electrochemical cell** pictured in Figure 8.5, we see the two separate compartments. One contains copper metal in a blue solution of copper(II) sulfate, and the other contains silver metal in a colorless solution of silver nitrate. Copper atoms give up electrons much more readily than silver atoms do, so electrons flow away from the copper compartment and toward the silver. The copper bar slowly dissolves as copper atoms give up electrons to form Cu^{2+} ions. The electrons flow through the wire to the silver compartment, where silver ions pick them up to become silver atoms and add to the mass of the silver bar.

But hold on! Isn't the copper side getting more positive and the silver side more negative as the electrons flow from the copper side to the silver side? This is when the nitrate anions, originally present in the silver nitrate solution,

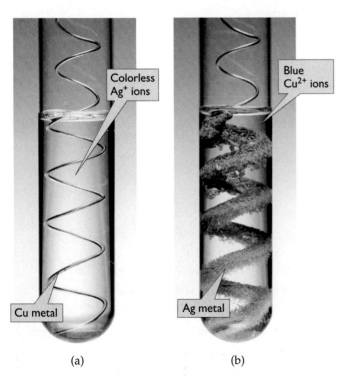

(a)   (b)

▲ **Figure 8.4** (a) A coil of copper wire is immersed in a colorless solution of Ag^+ ions. (b) As the reaction progresses, the more active copper displaces the less active silver from solution, producing needle-like crystals of silver metal and a blue solution of Cu^{2+} ions. The equation for the reaction is $2\ Ag^+(aq) + Cu(s) \longrightarrow 2\ Ag(s) + Cu^{2+}(aq)$.

come into play. In all electrochemical cells, the transfer of electrons is balanced by the transfer of anions in the opposite direction. For this anion transfer to take place, the two sides (or halves) of the electrochemical cell must be connected by something that allows for the transport of the anions, much as wire allows the transfer of electrons. This barrier is usually called a *salt bridge* and can take many forms, but it always has a medium (electrolyte) in which ions are present and can be transported. In Volta's electrochemical cell (battery), the salt bridges were the brine-soaked cardboard disks. In the electrochemical cell in Figure 8.5, each copper atom gives up two electrons, two silver ions pick up one electron each, and two nitrate ions move from the right compartment to the left compartment through the salt bridge.

The two pieces of metal where electrons are transferred are called **electrodes**. The electrode where oxidation occurs is called the **anode**. The one where reduction occurs is the **cathode**. In our example cell, copper gives up electrons, and it is oxidized, and the copper bar is therefore the anode. Silver ions gain electrons and are reduced, so the silver bar is the cathode.

Electrochemical reactions are often represented as two *half-reactions*. The following representation shows the two half-reactions for the copper–silver cell and their addition to give the overall cell reaction.

$$\textit{Oxidation:} \qquad Cu(s) \longrightarrow Cu^{2+}(aq) + 2e^-$$

$$\textit{Reduction:} \qquad 2Ag^+(aq) + 2e^- \longrightarrow 2\,Ag(s)$$

$$\overline{\textit{Overall reaction:} \quad Cu(s) + 2\,Ag^+(aq) \longrightarrow Cu^{2+}(aq) + 2\,Ag(s)}$$

Note that the electrons cancel when the two half-reactions are added.

CONCEPTUAL EXAMPLE 8.7 Oxidation and Reduction Half-Reactions

Represent the following reaction as two half-reactions, and label them as an oxidation half-reaction and a reduction half-reaction.

$$Sn^{2+} + SO_4^{2-} + 2\,H^+ \longrightarrow Sn^{4+} + SO_3^{2-} + H_2O$$

Solution

Tin is oxidized from Sn^{2+} to Sn^{4+} by losing two electrons. The oxidation half-reaction is therefore

$$\textit{Oxidation:} \quad Sn^{2+} \longrightarrow Sn^{4+} + 2\,e^-$$

The reduction half-reaction involves sulfur-containing ions, in which the sulfur atom gains two electrons. The reduction half-reaction is therefore

$$\textit{Reduction:} \quad SO_4^{2-} + 2\,H^+ + 2\,e^- \longrightarrow SO_3^{2-} + H_2O$$

Note that to balance all atoms in the reduction half-reaction, hydrogen ions and water had to be involved. This is not unusual. Water provides the medium in which the reaction takes place, and is plentiful. Hydrogen ions, on the other hand, must have been added to the solution to make it more acidic. The concept of supplying the reaction with the needed pH so that an oxidation–reduction reaction can take place is an extension of what we saw in our discussion of acid–base chemistry in Chapter 7.

> **Exercise 8.7A**

Represent the following reaction as two half-reactions. Label each as either an oxidation or a reduction half-reaction.

$$2\,Co + 3\,S \longrightarrow 2\,Co^{3+} + 3\,S^{2-}$$

> **Exercise 8.7B**

Represent the following reaction as two half-reactions, and label each as either an oxidation or a reduction half-reaction.

$$Mg + ZnCl_2 \longrightarrow Zn + MgCl_2$$

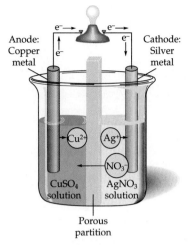

▲ **Figure 8.5** A simple electrochemical cell. The half-reactions are

Oxidation (anode):

$$Cu(s) \longrightarrow Cu^{2+}(aq) + 2\,e^-$$

Reduction (cathode):

$$Ag^+(aq) + e^- \longrightarrow Ag(s)$$

Q *The balanced equation for the overall reaction must have a coefficient of 2 for both Ag^+ and $Ag(s)$. Why?*

2 How does an electric eel create electricity?

The electric eel generates electricity by utilizing thousands of special cells, stacked like coins in a roll. Sodium pores in the cells' membranes are closed, while potassium ions pass through the membrane, creating a concentration *gradient* (gradual change). The eel generates a shock by opening the pores, producing about 120 mV per cell. A stack of 5000 cells thus yields about 600 volts—enough to knock a horse off its feet!

When balancing ordinary equations, we ensure that there are the same number and type of each atom on each side. For redox equations, we must also be careful to make the *total charge* on each side the same. That is, the number of electrons lost by the oxidizing agent must equal the number of electrons gained by the reducing agent, as illustrated in Example 8.8.

 EXAMPLE 8.8 Balancing Redox Equations

Balance the following half-reactions, and combine them to give a balanced overall reaction.

$$Sn^{2+} \longrightarrow Sn^{4+}$$
$$Bi^{3+} \longrightarrow Bi$$

Solution

Atoms are balanced in both half-reactions, but electric charge is not. To balance charge in the first half-reaction, we add two electrons on the right side.

$$Sn^{2+} \longrightarrow Sn^{4+} + 2\,e^-$$

The second half-reaction requires three electrons on the left side.

$$Bi^{3+} + 3\,e^- \longrightarrow Bi$$

Before we can combine the two, however, we must set electron loss equal to electron gain. Electrons lost by the substance being oxidized must be gained by the substance being reduced. To make electron loss and gain equal, we multiply the first half-reaction by 3 and the second by 2.

$$3 \times (Sn^{2+} \longrightarrow Sn^{4+} + 2e^-) = 3\,Sn^{2+} \longrightarrow 3\,Sn^{4+} + 6e^-$$
$$2 \times (Bi^{3+} + 3e^- \longrightarrow Bi) = 2\,Bi^{3+} + 6e^- \longrightarrow 2\,Bi$$
$$= 3\,Sn^{2+} + 2\,Bi^{3+} \longrightarrow 3\,Sn^{4+} + 2\,Bi$$

Note that both atoms and charges balance in the overall reaction.

〉 Exercise 8.8A

Balance the following half-reactions, and combine them to give a balanced overall reaction.

$$Zn(s) \longrightarrow Zn^{2+}$$
$$2\,H^+ \longrightarrow H_2(g)$$

〉 Exercise 8.8B

Balance the following half-reactions, and combine them to give a balanced overall reaction.

$$Pb \longrightarrow Pb^{2+}$$
$$Ag(NH_3)_2^+ \longrightarrow Ag + 2\,NH_3$$

Photochromic Glass

Eyeglasses with photochromic lenses darken when exposed to bright light, eliminating the need for sunglasses. This response is the result of oxidation–reduction reactions. Ordinary glass is a complex matrix of silicates (Chapter 12) that is transparent to visible light. Photochromic lenses have silver chloride (AgCl) and copper(I) chloride (CuCl) crystals uniformly embedded in the glass. Silver chloride is susceptible to oxidation and reduction by light. First, the light displaces an electron from a chloride ion.

$$\textit{Oxidation:} \quad Cl^- \longrightarrow Cl + e^-$$

The electron reduces a silver ion to a silver atom.

Reduction: $Ag^+ + e^- \longrightarrow Ag$

Clusters of silver atoms block the transmittance of light, causing the lenses to darken. This process occurs quite quickly. The degree of darkening depends on the intensity of the light.

   To be useful in eyeglasses, the photochromic process must be reversible. The darkening process is reversed by the copper(I) chloride. When the lenses are removed from light, the chlorine atoms formed by the exposure to light are reduced by the copper(I) ions, which are oxidized to copper(II) ions.

$$Cl + Cu^+ \longrightarrow Cu^{2+} + Cl^-$$

The copper(II) ions then oxidize the silver atoms.

$$Cu^{2+} + Ag \longrightarrow Cu^+ + Ag^+$$

The net effect is that the silver and chlorine atoms are converted to their original oxidized and reduced states, and the lenses become transparent once more. It should be noted that this reaction depends on temperature, making the darkening less effective at higher temperatures. Although this technology is commonly used to convert a pair of regular glasses into sunglasses and back, sunglass manufacturers have started recently to apply this technology to help adjust the tint of sunglasses depending on the intensity of light.

▲ Photochromic glass darkens in the presence of light.

Dry Cells

The familiar *dry cell* (Figure 8.6) is used in many flashlights, toys, and remote controls. The cylindrical zinc case serves as an anode, while the cathode is a carbon rod in the center of the cell. The space between the cathode and the anode contains a moist paste of graphite powder (carbon), manganese(IV) oxide (MnO_2), and ammonium chloride (NH_4Cl) in contact with the carbon cathode. This paste is separated from the zinc anode (case) by a porous spacer. The reaction on the anode is the oxidation of the zinc cylinder to zinc ions:

Oxidation (anode): $Zn(s) \longrightarrow Zn^{2+} + 2\,e^-$

The reaction on the cathode involves reduction of manganese(IV) oxide:

Reduction (cathode): $2\,MnO_2(s) + H_2O(l) + 2\,e^- \longrightarrow Mn_2O_3(s) + 2\,OH^-(aq)$

A simplified version of the overall reaction is

$$Zn + 2\,MnO_2 + H_2O \longrightarrow Zn^{2+} + Mn_2O_3 + 2\,OH^-$$

WHY IT MATTERS

Batteries labeled "heavy duty" or "super" may be ordinary dry cells, which will not have the longer shelf life and operating life of alkaline cells. Usually, alkaline cells are labeled as such and are only slightly higher in cost, making them a better buy.

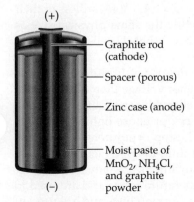

(+)

— Graphite rod (cathode)

— Spacer (porous)

— Zinc case (anode)

— Moist paste of MnO_2, NH_4Cl, and graphite powder

(−)

◀ **Figure 8.6** Cross section of a zinc–carbon cell.

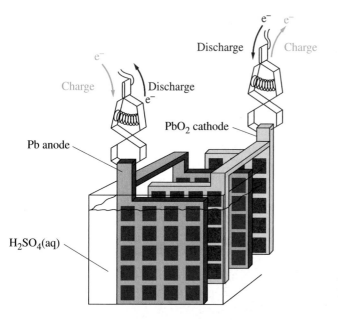

▲ **Figure 8.7** One cell of a lead–acid battery has two anode plates and two cathode plates. Six such cells make up the common 12-V car battery. The half-reactions are

Oxidation (anode):

$$Pb(s) + SO_4^{2-}(aq) \longrightarrow PbSO_4(s) + 2\ e^-$$

Reduction (cathode):

$$PbO_2(s) + 4\ H^+(aq) + SO_4^{2-}(aq) + 2\ e^- \longrightarrow PbSO_4(s) + 2\ H_2(l)$$

3 Why does a battery go dead?

When one or more of the reactants (zinc and MnO_2 in a dry cell) are used up, the cell can no longer provide useful energy.

4 Why are some batteries rechargeable but others are not?

A rechargeable battery uses a *reversible* redox reaction. The normal, forward reaction provides electrical energy. Applying electrical energy in the opposite direction using a charger reverses the reaction, and the starting materials are reformed. Not all redox reactions can be reversed in this manner.

A significant difference in *alkaline* cells is that potassium hydroxide (KOH) is used instead of NH_4Cl. While somewhat more expensive, alkaline cells are preferred because zinc-corroding ammonium ions are not present. They also last longer—both in storage and in use.

Lead-Acid Storage Batteries

Although we often refer to dry cells as batteries, a **battery** is actually a collection of electrochemical cells. (Recall Volta's battery.) The 12-volt (V) lead–acid battery used in automobiles, for example, is a series of six 2.1-V cells. Each cell (Figure 8.7) contains a pair of electrodes, one lead and the other lead(IV) oxide, in a chamber filled with sulfuric acid. Lead–acid batteries have been known for over 150 years.

An important feature of the lead–acid battery is that it can be recharged. A car battery discharges while supplying electricity for the ignition or when the engine is off but the lights or the radio is on. The net reaction during discharge is:

$$Pb + PbO_2 + 2\ H_2SO_4 \longrightarrow 2\ PbSO_4 + 2\ H_2O$$

The battery is recharged when the car is moving, and an electric current generated by the mechanical action of the car is supplied back to the battery. The reaction during recharge is the reverse:

$$2\ PbSO_4 + 2\ H_2O \longrightarrow Pb + PbO_2 + 2\ H_2SO_4$$

Lead–acid batteries are durable, but they are heavy, they contain corrosive sulfuric acid, and lead is a toxic environmental problem. (In fact, the vast majority of battery technologies are not friendly to the environment.) However, advances in cell technology have made lead–acid batteries last much longer than they did a few decades ago, and the batteries are routinely recycled.

Other Batteries

Rechargeable batteries are one of the most important classes of batteries in today's marketplace. Over their lifetime, these batteries are more economical and environmentally friendly than their non-rechargeable counterparts. They are also more convenient in most applications; you can have your cake and eat it too—with the correct science, of course. Several rechargeable-battery technologies are on the market today. They utilize different chemical combinations, with the most common being lead–acid, nickel–cadmium (NiCd), nickel–metal hydride (NiMH), lithium ion (Li-ion), and lithium-ion polymer (Li-ion polymer or LiPo).

When purchasing rechargeable cells, one issue to be considered is the cell voltage. Voltage depends on the cell reaction; as mentioned earlier, lead–acid cells produce 2.1-V per cell. Dry cells and alkaline cells produce 1.5-V regardless of their size. On the other hand, NiCd and NiMH cells, while the same physical size as comparable dry cells, produce only 1.2-V. Whether NiCd/NiMH cells will work in a particular application depends on the application.

Much of modern battery technology involves the use of lithium, which has an extraordinarily low density and yet provides higher voltage than other metals. Lithium batteries are very popular despite some well-publicized issues, including bursting and causing fires. In fact, nearly all high-performance lightweight electronic devices, such as smartphones, tablets, and laptop computers, were made possible due to their use of lithium-ion batteries. Lithium-ion polymer batteries are the current technology of choice, as they can be made into nearly any shape imaginable. Lithium–SO_2 cells operate over a wide temperature range and are used in military electronics. Lithium–iodine cells have long working lives and are used in pacemakers. Lithium–FeS$_2$ cells provide the same 1.5-V as alkaline cells, but have several times their working life.

Button batteries used in hearing aids and hand calculators formerly contained mercury (zinc anode, HgO cathode), but these cells are gradually being phased out and replaced with zinc–air cells. Tiny silver oxide cells (zinc anode, Ag$_2$O cathode) are used mainly in watches and cameras.

Fuel Cells

An interesting kind of battery is the fuel cell. In a **fuel cell**, the oxidizing agent (usually oxygen) and the reducing agent are supplied continuously to the cell rather than having a fixed amount within the cell. Unlike other cells, a fuel cell does not "go dead" as long as the reactants are supplied. Moreover, when fossil fuels such as coal are burned to generate electricity, at most 35% to 40% of the energy from the combustion is actually harnessed. In today's fuel cells, the fuel—usually hydrogen—is oxidized and oxygen is reduced with 70% to 75% efficiency. As we shall see in Chapter 15, most present-day fuel cells use platinum, nickel, or rhodium electrodes, in a solution of potassium hydroxide. Hydrogen–oxygen fuel cells are widely used in spacecraft; not only are they relatively small, but they can also provide drinking water from the reaction of hydrogen with oxygen. Research in designing cells that will work with natural gas and other more easily stored fuels is ongoing.

▲ Some carbon monoxide detectors use an electrochemical cell in which carbon monoxide is oxidized to carbon dioxide at the anode, and oxygen is consumed at the cathode. Sulfuric acid serves as the electrolyte. The cell produces current related to the CO concentration in the air, and the current triggers an alarm or produces a readout of the CO concentration.

Electrolysis

Dry cells and fuel cells produce electricity from chemical reactions. Sometimes, we can reverse the process. In electrolysis (discussed briefly in Section 3.1), electricity supplied externally causes a chemical reaction to occur. The charging reaction of the lead–acid cell shown in Figure 8.7 is one type of electrolysis. The car's alternator provides electric current, causing lead(II) sulfate to be converted back to lead metal and PbO$_2$. Other rechargeable cells undergo the reverse of their electricity-producing reaction when they are recharged.

Electrolysis is very important in industry. About 1% of all of the electricity in the United States is used to electrolyze aluminum oxide to the aluminum metal we use for pots, foil, and cans. Chrome plating is performed by passing electricity through a solution of chromium ions $\left(Cr^{6+}\right)$; electrons convert the ions to shiny chromium metal. Crude copper metal is purified by electrolysis, which converts the impure copper metal at the anode into copper(II) ions and then deposits nearly pure copper metal at the cathode. Many other materials—including sodium hydroxide, chlorine bleach, and many less abundant or more reactive metals—are produced by electrolysis.

SELF-ASSESSMENT Questions

1. In an electrochemical cell, electrons flow from the
   a. anode to the cathode
   b. cathode to the anode
   c. anode to the salt bridge
   d. salt bridge to the cathode

2. In the electrochemical cell whose overall reaction is represented by the following equation . . .
   $$Zn(s) + Cu^{2+}(aq) \longrightarrow Zn^{2+}(aq) + Cu(s)$$
   . . . the negative electrode is
   a. Cu(s) **b.** Cu^{2+}(aq) **c.** Zn(s) **d.** Zn^{2+}(aq)

3. In the zinc–carbon dry cell, the anode is _____ and the cathode is _____.
   a. Zn, MnO$_2$ **b.** Zn, C **c.** C, Zn **d.** C, MnO$_2$

4. In an electrochemical cell composed of two half-cells, ions flow from one half-cell to the other through
   a. electrodes **b.** air **c.** a salt bridge **d.** a wire

5. For the following reaction . . .
   $$Mg + CuO \longrightarrow MgO + Cu$$
   . . . the oxidation half-reaction is
   a. Cu^{2+} + 2 e$^-$ $\longrightarrow$ CuO
   b. CuO $\longrightarrow$ Cu^{2+} + 2 e$^-$

   c. Mg^{2+} + 2 e$^-$ $\longrightarrow$ MgO
   d. Mg $\longrightarrow$ Mg^{2+} + 2 e$^-$

6. In a lead–acid battery, the product of both cathode and anode half-reactions is
   a. H$_2$SO$_4$ **b.** Pb **c.** PbO$_2$ **d.** PbSO$_4$

7. Which statement is not true with regard to lithium batteries?
   a. They are low-density.
   b. They are used in laptop computers.
   c. They provide high voltage.
   d. They are non-rechargeable.

8. The overall electricity-producing reaction in the H$_2$/O$_2$ fuel cell is
   $$2\ H_2(g) + O_2(g) \longrightarrow 2\ H_2O(g)$$
   The half-reaction at the cathode is
   a. H$_2$ $\longrightarrow$ 2 H$^+$ + 2 e$^-$
   b. 2 H$^+$ + 2 e$^-$ $\longrightarrow$ H$_2$
   c. O$_2$ + 4 H$^+$ + 4 e$^-$ $\longrightarrow$ 2 H$_2$O
   d. 2 H$_2$O $\longrightarrow$ O$_2$ + 4 H$^+$ + 4 e$^-$

Answers: 1, a; 2, c; 3, b; 4, c; 5, d; 6, d; 7, d; 8, c

5 Why don't we see more "rust bucket" ships?

Rust is a pervasive problem because the $Fe(OH)_3$ that forms will flake off, exposing fresh metal. Most ship hulls are made of steel (mostly iron) and, according to Figure 8.8, when a ship sails the ocean (lots of electrolyte present!) it should simply rust away. So why is this not the case?

For rust to form, iron must give up its electrons and be oxidized. Some metals give up electrons more easily than others. (See page 249.) For instance, zinc is more easily oxidized than iron. When a block of zinc is simply attached to the steel hull under the water line, the zinc is oxidized instead of the iron. The zinc block acts as a so-called *sacrificial anode*. More recently, paints and coatings have been developed that react with rust to form a strongly adherent, protective layer. Some of these have been found to be useful for oil rigs and other steel-containing ocean structures.

8.4 Corrosion and Explosion

Learning Objectives • Describe the reactions that occur when iron rusts.
• Explain why an explosive reaction is so energetic.

You might think *corrosion* and *explosion* have little in common except that they rhyme. However, both processes involve redox reactions. Corrosion of metals is of great economic importance. Corrosion cost in the United States in 2013 was about $451 billion, or about 2.7% of the gross domestic product. Worldwide, the figure was about $2.5 trillion. The cost is even greater when safety and environmental consequences are included. Perhaps as much as 20% of all the iron and steel produced in the United States each year goes to replace corroded items.

The Rusting of Iron

In moist air, iron and steel are oxidized, particularly at a nick or scratch. (Steel is mostly iron with a small amount of carbon and other alloying elements.)

$$Fe(s) \longrightarrow Fe^{2+}(aq) + 2\,e^-$$

As iron is oxidized, (atmospheric) oxygen is reduced.

$$O_2(g) + 2\,H_2O(l) + 4\,e^- \longrightarrow 4\,OH^-(aq)$$

The net result, initially, is the formation of insoluble iron(II) hydroxide, which appears dark green or black.

$$2\,Fe(s) + O_2(g) + 2\,H_2O(l) \longrightarrow 2\,Fe(OH)_2(s)$$

This product is usually further oxidized to orange iron(III) hydroxide.

$$4\,Fe(OH)_2(s) + O_2(g) + 2\,H_2O(l) \longrightarrow 4\,Fe(OH)_3(s)$$

Iron(III) hydroxide [$Fe(OH)_3$], sometimes written as $Fe_2O_3 \cdot 3\,H_2O$, is the familiar iron rust.

Oxidation and reduction often occur at separate points on the metal's surface. Electrons are transferred through the iron metal. The circuit is completed by an electrolyte in aqueous solution. In the Snow Belt (a region near the Great Lakes in the United States), this solution is often the slush from road salt and melting snow. The metal is pitted in an anodic area, where iron is oxidized to Fe^{2+}. These ions migrate to the cathodic area, where they react with hydroxide ions formed by the reduction of oxygen.

$$Fe^{2+}(aq) + 2\,OH^-(aq) \longrightarrow Fe(OH)_2(s)$$

As noted previously, the iron(II) hydroxide is then oxidized to $Fe(OH)_3$, or rust. This process is diagrammed in Figure 8.8. Note that the anodic area is protected from oxygen by a water film, while the cathodic area is exposed to air.

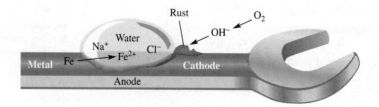

▲ **Figure 8.8** The corrosion of iron requires water, oxygen, and an electrolyte.

Protection of Aluminum

Aluminum is more reactive than iron, yet corrosion is ordinarily not a serious problem with aluminum. We use aluminum foil, we cook with aluminum pots, and we buy beverages in aluminum cans. Even after many years, they do not corrode. How is this possible? The aluminum at the surface reacts with oxygen in the air to form a thin layer of oxide:

$$4\,Al(s) + 3\,O_2(g) \longrightarrow 2\,Al_2O_3(s)$$

Instead of being porous and flaky like iron oxide in rust, aluminum oxide is hard and tough, and adheres strongly to the surface of the metal, protecting it from further oxidation.

Nonetheless, corrosion can sometimes be an issue with aluminum. Certain substances, such as some salts, can interfere with the protective oxide coating on aluminum, allowing the metal to oxidize. This problem can have some serious consequences. If the aluminum landing gear of an aircraft corrodes in this manner, the wheels can shear off.

Silver Tarnish

The tarnish on silver results from oxidation of the silver surface by hydrogen sulfide (H_2S) in the air or from food. It produces a film of black silver sulfide (Ag_2S) on the metal surface.

$$2\,Ag(s) + H_2S(g) \longrightarrow Ag_2S(s) + H_2(g)$$

You can use silver polish to remove the tarnish, but polishing off the silver sulfide means that a little of the silver has been removed. An alternative method involves the use of aluminum metal. The overall equation is:

$$3\,Ag_2S(s) + 2\,Al(s) \longrightarrow 6\,Ag(s) + Al_2S_3(s)$$

Since sulfur remains unchanged as sulfide ions (its oxidation number does not change) in the overall reaction, the equation may be written in terms of silver ions, which are reduced (oxidation number decreases from +1 to 0), and aluminum metal, which is oxidized (oxidation number increases from 0 to +3.)

$$3\,Ag^+(aq) + Al(s) \longrightarrow 3\,Ag(s) + Al^{3+}(aq)$$

This reaction also requires an electrolyte; sodium bicarbonate ($NaHCO_3$, baking soda) is usually used. The tarnished silver is placed in contact with aluminum metal (often in the form of foil) and covered with a hot solution of sodium bicarbonate. The silver ions are reduced to silver metal at the expense of the cheaper metal.

Explosive Reactions

Corrosion is often quite a slow reaction. A chemical explosion (that is, one not based on nuclear reactions) is an extremely rapid redox reaction, accompanied by a large increase in volume. This volume increase is due to the conversion of a solid or liquid explosive into gaseous products. Such a reaction often involves a compound of nitrogen, such as nitroglycerin (the active ingredient in dynamite), ammonium nitrate (NH_4NO_3, a common fertilizer), or trinitrotoluene (TNT). These compounds can decompose readily, yielding nitrogen gas as a product. As you may recall from Chapter 6, the molar volume of a gas at STP conditions is 22.4 L. For instance, when TNT detonates, it may decompose by one or both of the following reactions:

$$2\,C_7H_5N_3O_6(s) \longrightarrow 3\,N_2(g) + 5\,H_2O(g) + 7\,CO(g) + 7\,C(s)$$

$$2\,C_7H_5N_3O_6(s) \longrightarrow 3\,N_2(g) + 5\,H_2(g) + 12\,CO(g) + 2\,C(s)$$

This means that two moles of TNT, about 0.28 L, yields between 336 L (top reaction, 15 × 22.4 L) and 448 L (bottom reaction, 20 × 22.4 L) at STP. In any case the expansion is greater than 1000-fold!

Nitroglycerin has a very interesting history. The Swedish chemist Alfred Nobel (1833–1896), who established the Nobel Prize, made his fortune in nitroglycerin as an explosive, including inventing dynamite. The motivation behind establishing this prize was born out of a sense of guilt because commercial explosives afforded by Nobel's invention—intended mainly for mining and public works—had been used as a weapon, with very tragic consequences. Somewhat ironically, in the later part of his life, Nobel used nitroglycerin to treat his heart condition. This application of nitroglycerin is still in use to this day.

▲ Silver tarnish (Ag_2S) is readily removed from silver metal brought into contact with aluminum metal in a baking soda solution. Working in a sink, add about 1 cup of baking soda to 1 gal. of hot water. Immerse a piece of aluminum (wire or sheet aluminum generally works better than foil) and then the tarnished silver object so that it touches the aluminum.

Explosive mixtures are used for mining, earth-moving projects, and the demolition of buildings. Dynamite is a mixture of nitroglycerin and an absorbent material; the absorbent makes dynamite much less sensitive and far safer to transport and store than nitroglycerin alone. *ANFO* (*Ammonium Nitrate mixed with Fuel Oil*) is an inexpensive and effective explosive, widely used for mining salt, coal, and other minerals. When ANFO explodes, ammonium ion is the reducing agent and nitrate ion is the oxidizing agent. The fuel oil provides additional oxidizable material. Using $C_{17}H_{36}$ as the formula for a typical molecule in fuel oil, we can write an equation for the explosive reaction.

$$52\,NH_4NO_3(s) + C_{17}H_{36}(l) \longrightarrow 52\,N_2(g) + 17\,CO_2(g) + 122\,H_2O(g)$$

Once again, notice that 53 moles of solid and liquid reactants generate 191 moles (over 4000 L at STP) of gaseous products. This reaction produces a huge volume increase, which becomes part of the explosive force.

Unfortunately, ANFO has also been used in terrorist attacks around the world. Ammonium nitrate is a widely used fertilizer, and fuel oil is used in many home furnaces. The fact that both of these materials are readily accessible to potential terrorists is cause for some concern.

SELF-ASSESSMENT Questions

1. When iron rusts, it is being
   a. converted to gold   **b.** oxidized
   c. reduced   **d.** neutralized

2. Corrosion of aluminum is not as great a problem as corrosion of iron because aluminum
   a. forms an oxide layer that adheres to the metal's surface
   b. is less reactive than Fe
   c. needs water to corrode
   d. has a higher oxidation number

3. When silver is tarnished by H_2S, hydrogen is
   a. reduced   **b.** dissolved
   c. oxidized   **d.** evaporated

4. In a typical explosive reaction, which of the following usually occurs?
   a. Gases are converted from one form to another.
   b. Liquids or solids are converted to aqueous solutions.
   c. Liquids or solids are converted to gases.
   d. Nitrogen nuclei are split.

5. Black powder is a mixture of charcoal (carbon), potassium nitrate, and sulfur. The oxidizing agent in black powder is
   a. charcoal
   b. nitroglycerin
   c. potassium nitrate
   d. sulfur

Answers: 1, b; 2, a; 3, a; 4, c; 5, c

8.5 Oxygen: An Abundant and Essential Oxidizing Agent

Learning Objectives • Write equations for reactions in which oxygen is an oxidizing agent. • List some of the common oxidizing agents encountered in daily life.

Oxygen itself is the most common oxidizing agent. Many other oxidizing agents—sometimes called *oxidants*—are important in the laboratory, in industry, and in the home. They are used as antiseptics, disinfectants, and bleaches, and play a role in many chemical syntheses.

Oxygen: Occurrence and Properties

Certainly one of the most important elements on Earth, oxygen is one of about two dozen elements essential to life. Making up one-fifth of the air, it oxidizes the wood in our campfires and the gasoline in our automobiles. It is found in most compounds that are important to life. Foodstuffs—carbohydrates, fats, and proteins—all contain oxygen. The human body is approximately 65% water (by mass). Because water is 89% oxygen and many other compounds in your body also contain oxygen, almost two-thirds of your body mass is oxygen.

Oxygen comprises about half of the accessible portion of Earth by mass. It occurs in all three subdivisions of Earth's outermost structure. Oxygen occurs as O_2 molecules in the lower atmosphere, as a small amount of highly reactive ozone (O_3) in the stratosphere (about 30 km or 20 miles high), and as atomic oxygen (O) in higher altitudes. In the hydrosphere (oceans, rivers, etc.), oxygen is combined with hydrogen in water. In the lithosphere (outer solid portion of the Earth's crust), oxygen is combined with silicon (sand is SiO_2), aluminum (in clays), and other elements.

WHY IT MATTERS

Physical activity under conditions in which plenty of oxygen is available to body tissues is called *aerobic* exercise. When the available oxygen is insufficient, the exercise is called *anaerobic*, and it often leads to weakness and pain resulting from "oxygen debt." (See Chapter 17.)

The atmosphere is about 21% elemental oxygen (O_2) by volume. The rest is about 78% nitrogen (N_2) and 1% argon (Ar), both of which are quite unreactive. The free, uncombined oxygen in the air is taken into our lungs, passes into our bloodstreams, is carried to our body tissues, and reacts with the food we eat. This is the process that provides us with energy to move and think.

Fuels such as natural gas, gasoline, and coal also need oxygen to burn and release their stored energy. Oxidation, or combustion, of these fossil fuels currently supplies about 86% of the energy that turns the wheels of civilization.

Pure oxygen is obtained commercially by cooling air until it liquefies, and then letting the nitrogen and argon boil off. Nitrogen boils at −196 °C, argon at −186 °C, and oxygen at −183 °C. Over 400 million tons of oxygen is used annually worldwide. Almost half is used to burn off impurities in making steel. A small amount, about 1% of production, is used in welding, medicine, and similar purposes.

Not everything that oxygen does is desirable. Besides causing corrosion (Section 8.4), it promotes food spoilage and wood decay.

▲ Fuels burn more rapidly in pure oxygen than in air. Huge quantities of liquid oxygen are used to burn the fuels that blast rockets such as SpaceX's *Falcon 9* into orbit.

Oxygen: Reactions with Other Elements

Many oxidation reactions involving atmospheric oxygen are quite complicated, but we can gain some understanding by looking at some of the simpler ones. For example, as we have seen, oxygen combines with many metals to form metal oxides and with nonmetals to form nonmetal oxides.

EXAMPLE 8.9 **Writing Equations: Reaction of Oxygen with Other Elements**

Magnesium combines readily with oxygen when ignited in air. Write the equation for this reaction.

Solution

Magnesium is a group 2A metal. Oxygen occurs as diatomic molecules (O_2). The product, magnesium oxide, has the formula MgO. The reaction is

$$Mg(s) + O_2(g) \longrightarrow MgO(s) \text{ (not balanced)}$$

To balance the equation, we need 2 Mg and 2 MgO.

$$2\,Mg(s) + O_2(g) \longrightarrow 2\,MgO(s)$$

〉 **Exercise 8.9A**
Write equations for the following reactions.
  a. Zinc burns in air to form zinc oxide (ZnO).
  b. Diamond (C) burns in air at very high temperatures to form CO_2.

〉 **Exercise 8.9B**
Write equations for the following reactions.
  a. The surface of aluminum metal reacts with oxygen to form aluminum oxide.
  b. A cut piece of lithium metal quickly oxidizes to form lithium oxide.

Oxygen: Reactions with Compounds

Oxygen reacts with many compounds, oxidizing one or more of the elements in the compound. As we have seen, combustion of a fuel is oxidation, and produces oxides. There are many other examples. Hydrogen sulfide, a gaseous compound with a rotten-egg odor, burns to produce oxides of hydrogen (water) and of sulfur (sulfur dioxide).

$$2\,H_2S(g) + 3\,O_2(g) \longrightarrow 2\,H_2O(l) + 2\,SO_2(g)$$

EXAMPLE 8.10 Writing Equations: Reaction of Oxygen with Compounds

Carbon disulfide, a highly flammable liquid, combines readily with oxygen, burning with a blue flame. What products are formed? Write the balanced equation.

Solution

Carbon disulfide is CS_2. The products are oxides of carbon (carbon dioxide) and of sulfur (sulfur dioxide). The balanced equation is

$$CS_2(g) + 3\,O_2(g) \longrightarrow CO_2(g) + 2\,SO_2(g)$$

❭ Exercise 8.10A

When heated in air, lead(II) sulfide combines with oxygen to form lead(II) oxide and sulfur dioxide. Write a balanced equation for the reaction.

❭ Exercise 8.10B

Write a balanced equation for the combustion (reaction with oxygen) of ethanol (C_2H_5OH).

Ozone: Another Form of Oxygen

In addition to diatomic O_2, oxygen also has a triatomic form (O_3) called *ozone*. Ozone is a powerful oxidizing agent. In fact, it is a more powerful oxidizer than any chlorine-based disinfectant or antiseptic. This has made ozone a major chemical in water treatment, especially in Europe. Ozone can completely break down urea, through oxidation, as follows:

$$(NH_2)_2\,CO(aq) + O_3(g) \longrightarrow N_2(g) + CO_2(g) + 2\,H_2O(l)$$

In the reaction above, the oxidation number of nitrogen has changed from −3 to 0. This oxidation is caused by ozone.

In the following reaction

$$2\,Mn^{2+}(aq) + 2\,O_3(g) + 4\,H_2O(l) \longrightarrow 2\,MnO(OH)_2(s) + 2\,O_2(g) + 4\,H^+(aq)$$

ozone increases the oxidation number of manganese from +2 to +4 and, more importantly, removes manganese—a potential toxin causing psychiatric and motor disorders—from solution by forming a poorly soluble solid called manganite.

Ozone can also be a harmful air pollutant. It can be extremely irritating to both plants and animals, and is especially destructive to rubber. On the other hand, a layer of ozone in the upper stratosphere serves as a shield that protects life on Earth from ultraviolet radiation from the sun (Chapter 13). Ozone clearly illustrates that the same substance can be extremely beneficial or quite harmful, depending on the context.

Atomic Oxygen

Above the bulk of Earth's atmosphere, we find *atomic oxygen* (O), formed when O_2 or O_3 is split by high-energy ultraviolet light. Atomic oxygen is even more reactive than ozone. Although there is very little of this gas at orbital altitudes, (150–500 km), the speed of those satellites (typically 27,000 km/h) means that they encounter a lot of this corrosive element over many hundreds of orbits. Design of long-term satellites and spacecraft such as the International Space Station must include protection against atomic oxygen.

Other Common Oxidizing Agents

Oxidizing agents used around the home include antiseptics, disinfectants, and bleaches. Many others are used in workplaces and in laboratories.

Disinfectants and antiseptics are alike in that both are materials used to destroy microorganisms (that is, they are germicidal), but antiseptics are applied mainly to living tissue and disinfectants to nonliving things. A good example of a disinfectant is chlorine, which is used to kill disease-causing microorganisms in drinking water and some swimming pools. Roughly 1 part per million (0.0001% by mass chlorine) is sufficient to kill organisms in drinking water. Slightly higher concentrations are usually used in large swimming pools.

Smaller swimming pools are often chlorinated with solid calcium hypochlorite [$Ca(OCl)_2$], which is more convenient for the homeowner to handle than is gaseous chlorine. Because calcium hypochlorite is alkaline, it also increases the hydroxide ion concentration and raises the pH of the water.

$$OCl^-(aq) + H_2O(l) \longrightarrow HOCl(aq) + OH^-(aq)$$

When a pool becomes too alkaline, the pH is lowered by adding muriatic (hydrochloric) acid. Swimming pools are usually maintained at pH 7.2–7.8.

In recent years, concern has been raised about by-products that form when chlorine is used to disinfect water containing certain impurities. Although chlorine kills harmful microorganisms, it can also oxidize certain small molecules to form harmful by-products, such as chloroform.

Hydrogen peroxide (H_2O_2) is a common oxidizing agent that has the advantage of being converted to water in most reactions. Pure hydrogen peroxide is a syrupy liquid. It is available (in laboratories) as a dangerous 30% solution that has powerful oxidizing power, or in solutions of 3% to 12% (sold in stores) for various uses around the home. When combined with a specially designed catalyst, hydrogen peroxide can be used to oxidize colored compounds and water impurities.

As a dilute aqueous solution, hydrogen peroxide is used to bleach hair (which gave rise to the phrase "peroxide blonde"). Dilute H_2O_2 was often used medically to clean and deodorize wounds and ulcers, but it is falling out of favor because of the damage it does to healthy tissue. Hydrogen peroxide is the active ingredient in many tooth-whitening products. A solution of hydrogen peroxide, detergent, and baking soda is a remedy recommended for skunk spray. The peroxide oxidizes the smelly sulfur-containing compounds to less-smelly substances.

Iodine's oxidizing properties make it a useful antiseptic. An alcoholic solution (called *tincture of iodine*) can be found in drugstores. Povidone–iodine solution (common trade name Betadyne), containing about 10% available iodine, is widely used in surgery as well as for minor injuries, to prevent infection. Bacteria do not develop resistance to povidone–iodine.

An oxidizing agent sometimes used in the laboratory is potassium dichromate ($K_2Cr_2O_7$), although its use is falling out of favor due to its toxic chromium(VI) content. Potassium dichromate is an orange substance that turns green when it is reduced to chromium(III) compounds such as Cr_2O_3. One of the compounds that potassium dichromate oxidizes is ethyl alcohol. The early Breathalyzer test for intoxication made use of a dichromate solution. The exhaled breath of the person being tested was passed through dichromate coated on silica gel, and the degree of color change to green chromium(III) ions indicated the level of alcohol in the blood.

Potassium permanganate ($KMnO_4$) is a deep purple solid that sees use in some kinds of water treatment. It minimizes or removes the rotten-egg smell of "sulfur water" from wells that have been drilled in sulfur-containing rock, by oxidizing sulfides to sulfates. It is also used to treat fungal infections. In the laboratory, it is preferred to the more-toxic potassium dichromate, and finds use in testing for carbon–carbon double and triple bonds. A purple solution of $KMnO_4$ is oxidized to brown MnO_2 in the presence of compounds containing C=C or C≡C bonds. Highly diluted solutions of $KMnO_4$ are pale pink (called "pinky water" in India) and are used in developing countries to disinfect foods. Their use is limited because the manganese dioxide product leaves a brown stain on the treated substance.

Ointments for treating acne often contain 5% to 10% benzoyl peroxide, a powerful antiseptic and also a skin irritant. It causes old skin to slough off and be replaced by new, fresher-looking skin. When used on areas exposed to sunlight, however, benzoyl peroxide may promote skin cancer.

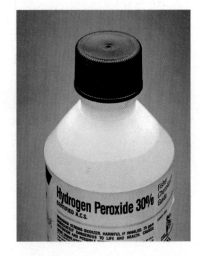

▲ Hydrogen peroxide is a powerful oxidizing agent. Dilute solutions are used as disinfectant and to bleach hair.

▶ **SAFETY ALERT**

Strong oxidizing agents such as 30% aqueous hydrogen peroxide can cause severe burns. In general, strong oxidizing agents are corrosive, and many are highly toxic. They can also initiate or increase the severity of a fire when brought into contact with combustible materials.

▲ Pure water (left) has little effect on a dried tomato sauce stain. Sodium hypochlorite bleach (right) removes the stain by oxidizing the colored tomato pigments to colorless products.

Bleaches are oxidizing agents, too. A **bleach** removes unwanted color from fabrics or other material. Most bleaches work by oxidizing carbon–carbon double bonds in the colored substance to single bonds. This renders the stain colorless, or nearly so. Almost any oxidizing agent could do the job, but some might be unsafe to use, or harmful to fabrics, or perhaps too expensive.

Bleaches (Chapter 20) are usually sodium hypochlorite (NaOCl) in aqueous solution (in laundry products such as Purex and Clorox)—often called *chlorine bleaches*—or calcium hypochlorite [Ca(OCl)$_2$], known as *bleaching powder*. The powder is usually preferred for large industrial operations, such as the whitening of paper or fabrics. Nonchlorine bleaches, often called *oxygen bleaches*, contain sodium percarbonate (a combination of Na$_2$CO$_3$ and H$_2$O$_2$) or sodium perborate (a combination of NaBO$_3$ and H$_2$O$_2$).

Stain removal is more complicated than bleaching. A few stain removers are oxidizing agents, but some are reducing agents, some are solvents or detergents, and some have quite different action. Stains often require rather specific stain removers.

SELF-ASSESSMENT Questions

1. When CH$_4$ is burned in plentiful air, the products are
   a. C + H$_2$ **b.** CH$_2$ + H$_2$O
   c. CO$_2$ + H$_2$ **d.** CO$_2$ + H$_2$O

2. When sulfur burns in plentiful air, the product is
   a. CO$_2$ **b.** H$_2$S   **c.** H$_2$SO$_4$ **d.** SO$_2$

3. Which of the following is a common oxidizing agent?
   a. HCl   **b.** H$_2$O$_2$ **c.** NaOH   **d.** NaCl

4. Which of the following is a common oxidizing agent?
   a. Cl$_2$ **b.** H$_2$ **c.** HCl   **d.** KOH

5. The main active ingredient in common household bleach, often called *chlorine bleach*, is
   a. HCl   **b.** NaOCl   **c.** NaOH   **d.** NaCl

6. Bleach for hair usually contains
   a. ammonia   **b.** hydrochloric acid
   c. hydrogen peroxide   **d.** sodium hypochlorite

7. Many disinfectants are
   a. oxidizing agents   **b.** reducing agents
   c. strong acids   **d.** strong bases

8. Potassium permanganate is a powerful oxidizing agent. In acidic solutions, the permanganate anion turns into Mn^{2+} cation, while in basic solution, it forms solid MnO$_2$. How many electrons are involved in these processes, respectively?
   a. 6, 2   **b.** 6, 3   **c.** 4, 5   **d.** 5, 3

Answers: 1. d; 2. d; 3. b; 4. a; 5. b; 6. c; 7. a; 8. d

8.6 Some Common Reducing Agents

Learning Objective • Identify some common reducing agents.

In every reaction involving oxidation, the oxidizing agent is reduced, and the substance undergoing oxidation acts as a reducing agent. Let's now consider reactions in which the purpose of the reaction is reduction.

Metals

Most metals occur in nature as compounds. To prepare the free metals, the compounds must be reduced. Metals are often freed from their ores by using coal or *coke* (elemental carbon obtained by heating coal to drive off volatile matter). Tin(IV) oxide is one of the many ores that can be reduced with coal or coke.

$$SnO_2(s) + C(s) \longrightarrow Sn(s) + CO_2(g)$$

Sometimes a metal can be obtained by heating its ore with a more active metal. Chromium(III) oxide, for example, can be reduced by heating it with aluminum.

$$Cr_2O_3(s) + 2\,Al(s) \longrightarrow Al_2O_3(s) + 2\,Cr(s)$$

One reason why aluminum was so expensive in earlier times was that it could only be produced by using (expensive, reactive) metals such as molten sodium or potassium:

$$Al_2O_3(s) + 6\,Na(l) \longrightarrow 3\,Na_2O(s) + 2\,Al(s)$$

The modern *electrolysis process* (see Chapter 12) reduced the cost of obtaining aluminum by a factor of a hundred or more.

WHY IT MATTERS

In making a black-and-white photograph, silver bromide in the film is exposed to light. The silver ions so exposed can then be reduced to black silver metal, forming a photographic negative.

Antioxidants

In food chemistry, certain reducing agents are called **antioxidants**. Ascorbic acid (vitamin C) can prevent the browning of fruit (such as sliced apples or pears) by inhibiting air oxidation. Vitamin C is water-soluble, but tocopherol (vitamin E) and beta-carotene (a precursor of vitamin A) are fat-soluble antioxidants. All these vitamins are believed to retard various oxidation reactions that are potentially damaging to vital components of living cells (Chapter 17). However, the health benefits of large oral doses of antioxidants such as vitamins C and E have not been demonstrated.

Hydrogen as a Reducing Agent

Hydrogen is an excellent reducing agent that can free many metals from their ores, but it is generally used to produce more expensive metals, such as tungsten (W).

$$WO_3(s) + 3\,H_2(g) \longrightarrow W(s) + 3\,H_2O(l)$$

Hydrogen can be used to reduce many kinds of chemical compounds. Ethylene, for example, can be reduced to ethane.

$$C_2H_4(g) + H_2(g) \longrightarrow C_2H_6(g)$$

The reaction requires a **catalyst**, a substance that increases the rate of a chemical reaction without itself being used up. Nickel is the catalyst used in this case. (Catalysts lower the **activation energy**—the minimum energy needed to get a reaction started; they are further explored in Chapter 15.)

Hydrogen also reduces nitrogen, from air, in the industrial production of ammonia.

$$N_2(g) + 3\,H_2(g) \longrightarrow 2\,NH_3(g)$$

Ammonia is the source of most nitrogen fertilizers in modern agriculture. The reaction that produces ammonia employs an iron catalyst.

A stream of pure hydrogen burns quietly in air with an almost colorless flame, but when a mixture of hydrogen and oxygen is ignited by a spark or a flame, an explosion results. The product in both cases is water.

$$2\,H_2(g) + O_2(g) \longrightarrow 2\,H_2O(l)$$

Certain metals, including platinum and palladium, have an unusual affinity for hydrogen. They absorb large volumes of the gas. Palladium can absorb up to 900 times its own volume of hydrogen. It is interesting to note that hydrogen and oxygen can be mixed at room temperature with no perceptible reaction. If a piece of platinum gauze is added to the mixture, however, the gases react violently. The platinum acts as a catalyst. The heat from the initial reaction heats the platinum, making it glow; it then ignites the hydrogen–oxygen mixture, causing an explosion.

Nickel, platinum, and palladium are often used as catalysts for reactions involving hydrogen. These metals have the greatest catalytic activity when they are finely divided and have a high active surface area. Hydrogen adsorbed on the surface of these metals is more reactive than ordinary hydrogen gas.

Catalysts are an important area of research in green chemistry. Chemists are able to develop more energy-efficient and sustainable processes by designing catalysts that enable chemical reactions to occur at lower temperatures or pressures. Using a catalyst can also allow a reaction to occur using milder reactants, such as oxygen or hydrogen peroxide instead of chlorine for oxidizing stains on clothes or destroying harmful microorganisms in drinking water.

A Closer Look at Hydrogen

Unlike oxygen, hydrogen is rarely found as a free, uncombined element on Earth. Most of it is combined with oxygen in water. Some is combined with carbon in petroleum and natural gas, which are mixtures of hydrocarbons. Nearly all compounds derived from plants and animals contain hydrogen.

▲ Figure 8.9 Hydrogen gas can be prepared in the laboratory by the reaction of zinc metal with hydrochloric acid.

Q *Is the reaction between HCl and Zn a redox reaction? Explain.*

Small amounts of elemental hydrogen can be made for laboratory use by reacting zinc with hydrochloric acid (Figure 8.9).

$$Zn(s) + 2\,HCl(aq) \longrightarrow ZnCl_2(aq) + H_2(g)$$

Commercial quantities of hydrogen are obtained as by-products of petroleum refining or from the reaction of natural gas with steam. About 8 billion kg of hydrogen is produced each year in the United States. At present, hydrogen is used mainly to make ammonia and methanol. Its possible use as a fuel in the future is discussed in Chapter 15.

SELF-ASSESSMENT Questions

1. Which of the following is a common reducing agent?
   a. Cl_2 **b.** F_2 **c.** H_2 **d.** I_2

2. Which of the following is a common reducing agent?
   a. C   **b.** O_2 **c.** CO_2 **d.** $KMnO_4$

3. All the following metals are used as catalysts for reactions involving hydrogen gas *except*
   a. aluminum   **b.** nickel   **c.** palladium   **d.** platinum

4. A catalyst changes the rate of a chemical reaction by
   a. changing the products of the reaction
   b. changing the reactants involved in the reaction
   c. increasing the temperature at the catalyst's surface
   d. lowering the activation energy

Answers: 1, c; 2, a; 3, a; 4, d

8.7 Oxidation, Reduction, and Living Things

Learning Objective • Write the overall equations for the metabolism of glucose and for photosynthesis.

Perhaps the most important oxidation–reduction processes are the ones that maintain life on this planet. We obtain energy for all our physical and mental activities by metabolizing food in a process called *cellular respiration*. The process has many steps, but eventually the food we eat is converted mainly into carbon dioxide, water, and energy. Essentially, the food is "burned" at body temperature, with catalysts (enzymes) to speed the reaction.

Bread and many of the other foods we eat are largely made up of carbohydrates, composed of carbon, hydrogen, and oxygen (Chapter 16). If we represent carbohydrates with the simple example of glucose ($C_6H_{12}O_6$), we can write the overall equation for their metabolism as follows.

$$C_6H_{12}O_6 + 6\,O_2 \longrightarrow 6\,CO_2 + 6\,H_2O + energy$$

This process occurs constantly in animals, including humans. The carbohydrate is oxidized in the process.

Meanwhile, plants need carbon dioxide, water, and energy from the sun to produce carbohydrates. The process by which plants synthesize carbohydrates is called **photosynthesis** (Figure 8.10). The overall chemical equation is

$$6\,CO_2 + 6\,H_2O + energy \longrightarrow C_6H_{12}O_6 + 6\,O_2$$

▲ Figure 8.10 Photosynthesis occurs in green plants. The chlorophyll pigments that catalyze the photosynthesis process give the green color to much of the land area of Earth.

Notice that this process in plant cells is exactly the reverse of the process going on inside animals. In food metabolism in animals, we focus on an oxidation process. In photosynthesis, we focus on a reduction process.

The carbohydrates produced by photosynthesis are the ultimate source of all our food because fish, fowl, and other animals either eat plants or eat other animals that eat plants. Note that the photosynthesis process not only makes carbohydrates but also yields free elementary oxygen (O_2). In other words, photosynthesis does not just provide the food we eat—it also provides the oxygen we breathe.

GREEN CHEMISTRY Green Redox Catalysis

Roger A. Sheldon, *Delft University of Technology, Netherlands*

Principles 1, 2, 3, and 9

Learning Objectives ● Use a balanced reaction to calculate the atom economy and E factor of the reaction.
● Describe the role of catalysts in minimizing waste in the chemical and allied industries.

Oxidation and reduction processes always occur together in what is called a redox reaction (Section 8.1). When one substance is oxidized, another is always reduced. Oxidation and reduction reactions are extremely important in the chemical and allied industries. The simplest and least expensive reducing agent and oxidizing agent are molecular hydrogen and oxygen, respectively. Hydrogen is used in catalytic reductions (hydrogenations), and oxygen is used in catalytic oxidations. Both types of reactions produce a variety of bulk industrial chemicals in the petrochemical and other large industries. The catalyst increases the reaction rate or the selectivity to the product. In contrast, the manufacture of specialty chemicals, such as pharmaceuticals, generally involves the use of complicated reducing and oxidizing agents that are usually more expensive and also generate large amounts of waste.

A main goal of green chemistry is to minimize or eliminate waste, which is defined as everything formed in the process except the desired product (Principle 1). Comparing different reactions can illustrate how greener redox reactions can reduce waste and improve efficiency. To compare the greenness of processes, it is necessary to use meaningful green metrics. Two such metrics were developed in the early 1990s. These are *atom economy* (see the essay "Atom Economy" in Chapter 5) and the *E factor* (kilograms of waste per kilogram of product). As discussed in the green chemistry essay in Chapter 5, the ideal atom economy is 100%, which means that all of the atoms end up in the product, and the ideal E factor is zero-no waste (Principle 2).

In Scheme 1, we compare the reduction of acetone to an alcohol using one equivalent of sodium borohydride or hydrogen in the presence of catalytic amounts of palladium-on-charcoal. Reduction with sodium borohydride produces the desired isopropyl alcohol product and one equivalent each of boric acid and sodium hydroxide as waste. In the catalytic hydrogenation, in contrast, no waste product is formed (Principle 9).

You might ask, however, what is important about the catalyst? Since a catalyst is defined as a substance that changes the rate of a reaction (by lowering its activation energy) without itself being changed in the process, it can, in principle, be recycled indefinitely and no waste is generated (Section 8.6).

As shown in Scheme 1, the atom economy of the sodium borohydride reduction is 70% and the theoretical E factor is 0.42. However, in practice, the E factor is generally higher than predicted by the atom economy because the chemical yield is not 100%. Reagents, including sodium borohydride, are generally used in excess amounts to obtain complete conversion of the starting material in a reasonable time. Also, some waste is not contained in the stoichiometric equation, for example, solvents. In stark contrast,

the atom economy of the catalytic hydrogenation is 100%, and the theoretical E factor is zero.

Similarly, in Scheme 2, we compare the reverse reaction, the oxidation of an alcohol to acetone. One option is to use a classical stoichiometric reagent, potassium dichromate. The other route is a catalytic oxidation with molecular oxygen in the presence of a palladium catalyst. The atom economies are 20% and 76%, and the

(1) Stoichiometric reduction of a ketone with sodium borohydride

$$4 \text{ acetone } (C_3H_6O) + NaBH_4 + 4 H_2O \longrightarrow$$

$$4 \text{ isopropanol } (C_3H_8O) + H_3BO_3 + NaOH$$

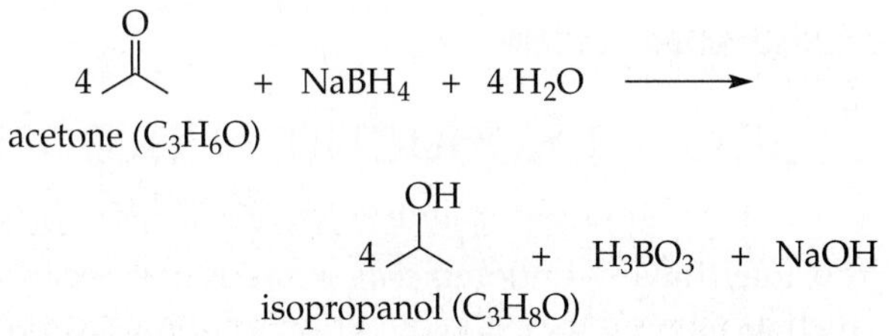

$$\text{atom economy} = \frac{\text{product MW}}{\text{total MW of all reactants}} = \frac{240}{232 + 38 + 72} = 70\%$$

$$\text{theoretical E factor} = \frac{\text{kg waste}}{\text{kg product}} = \frac{102}{240} = 0.42$$

(2) Palladium-on-charcoal catalyzed hydrogenation of a ketone

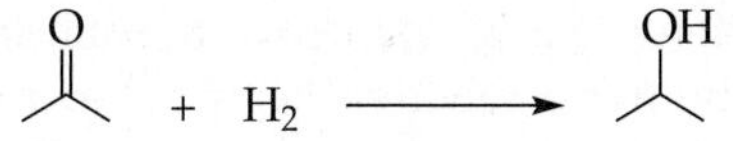

$$O + H_2 \longrightarrow OH$$

$$\text{atom economy} = \frac{78}{78} = 100\%$$

$$\text{theoretical E factor} = 0$$

▲ **Scheme 1** Atom economies and theoretical E factors of stoichiometric vs catalytic reduction.

theoretical E factors are 3.98 and 0.31, respectively. In practice, the E factor for the first reaction is substantially higher than 4 because dichromate and sulfuric acid are used in large excess.

Yet it is not only the amount of waste formed that is important. Chromium(VI) is a suspected carcinogen and, hence, chromium-containing waste is potentially toxic. In contrast, the byproduct in the catalytic oxidation is one equivalent of water, which is harmless (Principle 3). Replacement of chromium(VI) as the oxidant is important from another green chemistry viewpoint, avoiding the use of toxic or hazardous reagents. In short, catalysis is the key to green and sustainable redox chemistry.

There are many oxidation reactions that occur in nature (with oxygen being reduced in the process). The net photosynthesis reaction is unique in that it is a natural reduction of carbon dioxide (with oxygen being oxidized). Many processes in nature use oxygen. Photosynthesis is the only natural process that produces it.

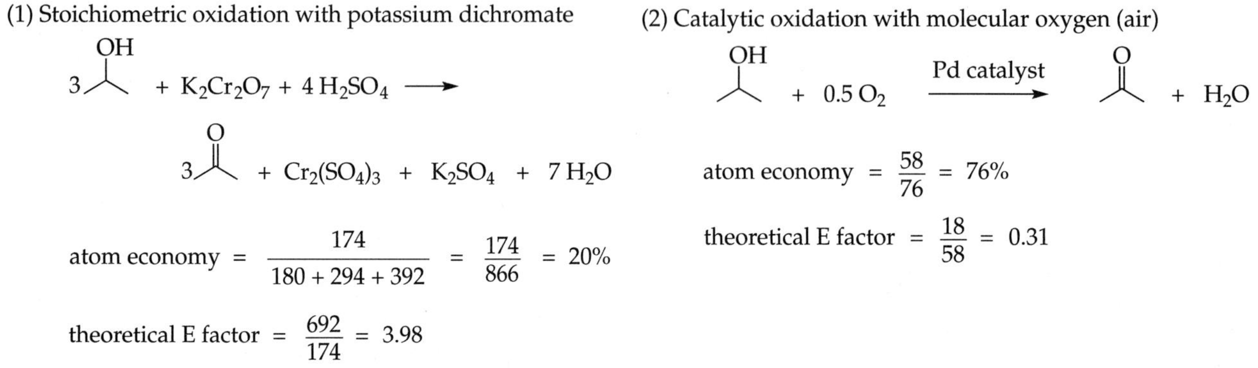

(1) Stoichiometric oxidation with potassium dichromate

$$3 \overset{OH}{\wedge} + K_2Cr_2O_7 + 4\,H_2SO_4 \longrightarrow$$

$$3 \overset{O}{\wedge} + Cr_2(SO_4)_3 + K_2SO_4 + 7\,H_2O$$

$$\text{atom economy} = \frac{174}{180 + 294 + 392} = \frac{174}{866} = 20\%$$

$$\text{theoretical E factor} = \frac{692}{174} = 3.98$$

(2) Catalytic oxidation with molecular oxygen (air)

$$\overset{OH}{\wedge} + 0.5\,O_2 \xrightarrow{\text{Pd catalyst}} \overset{O}{\wedge} + H_2O$$

$$\text{atom economy} = \frac{58}{76} = 76\%$$

$$\text{theoretical E factor} = \frac{18}{58} = 0.31$$

▲ **Scheme 2** Atom economies and theoretical E factors of stoichiometric vs catalytic oxidation.

Standard Reduction Potentials

You may have wondered *why* some redox reactions (such as silver ions with copper metal) proceed spontaneously, while others (copper ions with silver metal) do not. Every half-reaction has a particular *standard reduction potential* ($E°$), expressed in volts, that determines how easily the reactant will be reduced. Reactive nonmetals such as fluorine have high values of $E°$ and are very easily reduced. Cations of reactive metals such as lithium and potassium have low standard potentials and are very difficult to reduce to those reactive metals. The difference in $E°$ values for the half-reactions determines whether a reaction occurs. For example, silver ions have a higher $E°$ than do copper(II) ions, so Ag^+ ions react with copper metal to form less-reactive Cu^{2+} ions. In turn, the Ag^+ ions remove electrons from Cu and become silver metal.

The potential of a particular cell depends largely on the half-reactions that make up the cell. That is why an alkaline D cell produces the same voltage as an alkaline AA cell; both use the same half-reactions. However, the D cell, with more reactants in the larger cell, will last longer than the AA cell.

The technology that brought lithium to consumer batteries was an enormous breakthrough. Lithium is the lowest-density metal, so batteries made from it are very light. Furthermore, lithium ions have the lowest value of $E°$ and are the most difficult to reduce—which means that in a cell, lithium *metal* will produce a higher voltage than any other metal. And as a bonus, the cell reaction can be made reversible, so lithium-containing cells can be engineered to be rechargeable.

To recharge a cell—that is, to cause a cell reaction to go in reverse—the applied voltage must be greater than the cell's working potential. The lead–acid battery in a car produces slightly more than 12-V as it operates. The charging circuit uses a regulator that sets the charging voltage at 13-V to 14-V to reverse the cell reaction.

SELF-ASSESSMENT Questions

1. The main process by which animals obtain energy from food is
  a. acidification   **b.** cellular respiration
  c. photosynthesis   **d.** reduction of CO_2

2. The only natural process that produces O_2 is
  a. acidification of carbonates   **b.** photosynthesis
  c. oxidation of CO_2 **d.** reduction of metal oxide

3. In photosynthesis and cellular respiration, carbon is _____ and _____, respectively.
  a. reduced, reduced   **b.** reduced, oxidized
  c. oxidized, oxidized   **d.** oxidized, reduced

Answers: 1, b; 2, b; 3, b

Summary

Section 8.1—Oxidation and reduction are processes that occur together. They may be defined in four ways. **Oxidation** occurs when a substance gains oxygen atoms, loses hydrogen atoms, or loses electrons, or when its oxidation number increases. **Reduction** is the opposite of oxidation; it occurs when a substance loses oxygen atoms, gains hydrogen atoms, gains electrons, or when its **oxidation number** decreases. We usually use the definition that is most convenient for the situation.

Section 8.2—In an oxidation–reduction (or redox) reaction, the substance that causes oxidation is the **oxidizing agent**. The substance that causes reduction is the **reducing agent**. In a redox reaction, the oxidizing agent is reduced, and the reducing agent is oxidized.

Section 8.3—A redox reaction involves a transfer of electrons. That transfer can be made to take place in an **electrochemical cell**, which has two solid **electrodes** where the electrons are transferred. The electrode where oxidation occurs is the **anode**, and the one where reduction occurs is the **cathode**. Electrons are transferred from anode to cathode through a wire; their flow is an electric current. When the oxidation half-reaction is added to the reduction half-reaction, the electrons in the half-reactions cancel and we get the overall cell reaction. Knowing this, we can balance many redox equations.

A dry cell (flashlight battery) has a zinc case and a central carbon rod, and contains a paste of manganese dioxide with graphite powder and ammonium chloride. Alkaline cells are similar but contain potassium hydroxide instead of ammonium chloride. A **battery** is a series of connected cells. An automobile's lead–acid storage battery contains six cells. Each cell contains lead and lead(IV) oxide electrodes in sulfuric acid, and the battery can be recharged. A **fuel cell** generates electricity by oxidizing a fuel in a cell to which the fuel and oxygen are supplied continuously.

Section 8.4—Corrosion of metals is ordinarily an undesirable redox reaction. When iron corrodes, the iron is first oxidized to Fe^{2+} ions, and oxygen gas is reduced to hydroxide ions. The $Fe(OH)_2$ that forms is further oxidized to $Fe(OH)_3$, or rust. Oxidation and reduction often occur in separate places on the metal surface, and an electrolyte (such as salt) on the surface can intensify the corrosion reaction. Aluminum corrodes, but the oxide layer is tough and adherent, and protects the underlying metal. Silver corrodes in the presence of sulfur compounds, forming black silver sulfide, which can be removed electrochemically. An explosive reaction is a redox reaction that occurs rapidly and with a considerable increase in volume due to the formation of gases.

Section 8.5—Oxygen is an oxidizing agent that is essential to life. It makes up about one-fifth of the air and about two-thirds of our body mass. Our food is "burned" in oxygen as we breathe, and fuels such as coal react with oxygen when they burn and release most of the energy used in civilization. Pure oxygen is obtained from liquefied air, and most of it is used in industry. Oxygen reacts with metals and nonmetals to form oxides. It reacts with many compounds to form oxides of the elements in the compound. Ozone, O_3, is a form of oxygen that is an air pollutant in the lower atmosphere but provides a shield from ultraviolet radiation in the upper atmosphere. Other oxidizing agents include hydrogen peroxide, potassium dichromate, benzoyl peroxide, and chlorine. A **bleach** is an oxidizing agent that removes unwanted color from fabric or other material.

Section 8.6—Most metals are reducing agents. An **antioxidant** is a reducing agent that retards damaging oxidation reactions in living cells. Vitamins C and E and beta-carotene are antioxidants. Common reducing agents include carbon, hydrogen, and active metals. Carbon in the form of coal or coke is often used as a reducing agent to obtain metals from their oxides. Active metals such as aluminum are also used for this purpose. Hydrogen can free many metals from their ores. Some reactions of hydrogen require a **catalyst**, which speeds up a reaction without itself being used up **activation energy**. Formation of ammonia from hydrogen and nitrogen uses a catalyst.

Section 8.7—Oxidation of carbohydrates and other substances in foods produces the energy humans and other animals need to survive. The reduction of carbon dioxide in **photosynthesis** is the most important reduction process on Earth. We could not exist without it. Photosynthesis provides the food we eat and the oxygen we breathe.

GREEN CHEMISTRY—In organic synthesis, it is important to use methods where most of the materials end up in the product (high atom economy). This process minimizes the amount of waste generated (low E factor). It is also important to avoid, where possible, the use of hazardous or toxic reagents and solvents, thereby minimizing the environmental impact of reagents used and waste generated.

Learning Objectives	Associated Problems
• Classify a particular change within a redox reaction as either oxidation or reduction. (8.1)	18–20, 23, 37, 38, 43–45, 47, 49, 51, 72, 76, 82
• Identify an oxidation–reduction reaction. (8.1)	16, 17, 72
• Identify the oxidizing agent and the reducing agent in a redox reaction. (8.2)	3, 25–34, 42, 48, 50, 52, 54, 55, 60–62, 73, 75, 77, 79
• Balance redox equations. (8.3)	9, 10, 12–14, 21, 22, 35–41, 56, 57, 64–69, 77, 78

• Identify and write the half-reactions in an electrochemical cell. (8.3)	21, 22, 35–38, 65, 69
• Describe the reactions that occur when iron rusts. (8.4)	12, 55, 63
• Explain why an explosive reaction is so energetic. (8.4)	74
• Write equations for reactions in which oxygen is an oxidizing agent. (8.5)	56–59, 62, 67, 68
• List some of the common oxidizing agents encountered in daily life. (8.5)	13–15, 75
• Identify some common reducing agents. (8.6)	52, 53, 73
• Write the overall equations for the metabolism of glucose and for photosynthesis. (8.7)	15, 51, 69–71
• Use a balanced reaction to calculate the atom economy and E factor of the reaction.	83, 84
• Describe the role of catalysts in minimizing waste in the chemical and allied industries.	85, 86

 Conceptual Questions

1. What happens to the oxidation number of one of its elements when a compound is oxidized? When it is reduced?

2. What is an electrochemical cell? A battery?

3. What happens at the anode and the cathode of an electrochemical cell? Why do electrons flow from the anode to the cathode?

4. How does a fuel cell differ from other cells (dry cell, alkaline cell, etc.)?

5. What is the purpose of a porous partition between the two electrode compartments in an electrochemical cell?

6. What is a half-reaction? How are half-reactions combined to give an overall cell reaction?

7. From what material is the case of a dry cell made? What purpose does this material serve? What happens to it as the cell discharges?

8. How does an alkaline cell differ from a dry cell, and why are alkaline cells preferred?

9. How does a lead–acid battery become recharged?

10. Describe what happens when a lead–acid storage battery discharges.

11. Why are most modern high-performance electronic devices powered by lithium-based batteries?

12. Describe what happens when iron corrodes. Why does road salt intensify this process?

13. How does silver tarnish? How can the tarnish be removed without the loss of silver?

14. Describe how a bleaching agent, such as hypochlorite (ClO^-), works.

15. Relate the chemistry of photosynthesis to the chemical process that provides energy for your heartbeat.

 Problems

Recognizing Oxidation and Reduction

16. Which of the following reactions are redox reactions?
   a. $C_2H_2 + 2H_2 \longrightarrow C_2H_6$
   b. $Pb(NO_3)_2 + MgCl_2 \longrightarrow Mg(NO_3)_2 + PbCl_2$
   c. $2SO_2 + O_2 \longrightarrow 2SO_3$

17. Which of the following reactions are redox reactions?
   a. $2NaF + MgBr_2 \longrightarrow MgF_2 + 2NaBr$
   b. $C_4H_8 + 6O_2 \longrightarrow 4CO_2 + 4H_2O$
   c. $HCl + KOH \longrightarrow H_2O + KCl$

18. Indicate whether the reactant in each of the following partial equations is being oxidized or reduced. Explain.
   a. $Mg^{2+} \longrightarrow Mg$ b. $S^{2-} \longrightarrow S$
   c. $C_2H_4O_2 \longrightarrow C_2H_6O$ d. $Cl^- \longrightarrow ClO^-$

19. In the following reaction, identify whether carbon has undergone reduction or oxidation, and determine how the oxidation number of carbon has changed.
   a. $C + 2Cl_2 \longrightarrow CCl_4$ b. $2C + O_2 \longrightarrow 2CO$
   c. $C + O_2 \longrightarrow CO_2$ d. $C_2H_2 + I_2 \longrightarrow C_2H_2I_2$
   e. $C + 2H_2 \longrightarrow CH_4$

20. In which of the following changes is the reactant undergoing oxidation? Explain.
   a. $C_2H_4 \longrightarrow C_2H_6$
   b. $CrO_3 \longrightarrow Cr$
   c. $Fe^{3+} \longrightarrow Fe^{2+}$
   d. $C_{27}H_{33}N_9O_{15}P_2 \longrightarrow C_{27}H_{35}N_9O_{15}P_2$
   e. $2Cl^- \longrightarrow Cl_2$

21. Write a balanced oxidation half-reaction similar to $Na \longrightarrow Na^+ + e^-$ for each of the following metals.
   a. K   b. Mg   c. Fe (two different equations)

22. Write a balanced reduction half-reaction similar to $Cl_2 + 2e^- \longrightarrow 2Cl^-$ for each of the following nonmetals.
   a. F_2 b. P_4 c. O_3

23. In each reaction, identify the atom that is oxidized and the one that is reduced.
   a. $4Cu + O_2 \longrightarrow 2Cu_2O$
   b. $Mg + 2H^+ \longrightarrow Mg^{2+} + H_2$
   c. $Br_2 + H_2O_2 \longrightarrow O_2 + 2H^+ + 2Br^-$
   d. $4S_2O_3^{2-} + O_2 + 2H^+ \longrightarrow 2S_4O_6^{2-} + 2H_2O$

24. Discuss what happens to copper in photochromic lenses from a redox perspective.

Oxidizing Agents and Reducing Agents

25. Identify the oxidizing agent and the reducing agent in each reaction.
    a. $H_2 + CuO \longrightarrow Cu + 2 H_2O$
    b. $O_2 + 2 HCOH \longrightarrow 2 HCOOH$
    c. $2 Al_2O_3 + 3 Mn \longrightarrow 3 MnO_2 + 2 Al$
    d. $2 CrCl_2 + SnCl_4 \longrightarrow 2 CrCl_3 + SnCl_2$

26. Identify the oxidizing agent and the reducing agent in each reaction.
    a. $CuS + H_2 \longrightarrow Cu + H_2S$
    b. $4 K + CCl_4 \longrightarrow C + 4 KCl$
    c. $C_3H_4 + 2 H_2 \longrightarrow C_3H_8$
    d. $Fe^{3+} + Ce^{3+} \longrightarrow Fe^{2+} + Ce^{4+}$

27. Look again at Figure 8.4. What is the reducing agent?

28. What is the oxidizing agent in Figure 8.4?

29. Look again at Figure 8.6. Identify the oxidizing and reducing agents.

30. Look again at Figure 8.7. (a) Is SO_4^{2-} the oxidizing agent, the reducing agent, both, or neither? (b) When the cell is being charged, the discharge half-reactions are reversed. During charging, is Pb^{2+} being oxidized, reduced, both, or neither?

31. In an electrochemical cell, electrons flow from the anode to the cathode. How would you classify the anode and cathode in terms of oxidizing and reducing agents?

32. When aluminum is placed in contact with copper, electrons flow from the aluminum to the copper. Which of the two metals is the oxidizing agent?

33. The photo below shows the result of immersion of an iron nail in a solution of copper(II) sulfate. In this reaction
    a. copper(II) ions are oxidized and iron metal is reduced
    b. copper(II) ions are reduced and iron metal is oxidized
    c. copper(II) ions are oxidized and iron(II) ions are reduced
    d. copper metal is oxidized and iron(II) ions are reduced

34. In the photo at the top of the next column, a printed circuit board is being made from a copper-coated plastic board immersed in ammonium persulfate solution, $(NH_4)_2S_2O_8(aq)$. The persulfate ions are converted to

sulfate ions, and some copper metal is converted to blue copper(II) ions. Identify the oxidizing agent and the reducing agent.

Half-Reactions

35. Write the reduction half-reaction that occurs in the redox equation $K + Mg^{2+} \longrightarrow Mg + K^+$, then balance the overall equation so that the numbers of atoms/ions and the total charges are the same on both sides.

36. Write the oxidation half-reaction that occurs in the redox equation $H_2 + MnO_2 \longrightarrow H_2O + Mn$, then balance the overall redox equation so that the numbers of atoms/ions and the total charges are the same on both sides.

37. Separate the following redox reactions into half-reactions, and label each half-reaction as oxidation or reduction.
    a. $Mo^{6+}(aq) + 3 H_2(g) \longrightarrow 6 H^+(aq) + Mo(s)$
    b. $Cd(s) + Zr^{4+}(aq) \longrightarrow Cd^{2+}(aq) + Zr^{2+}(aq)$

38. Separate the following redox reactions into half-reactions, and label each half-reaction as oxidation or reduction.
    a. $Mn^{4+}(aq) + 2 Cr^{2+}(aq) \longrightarrow Mn^{2+}(aq) + 2 Cr^{3+}(aq)$
    b. $2 Ca(s) + Ti^{4+}(aq) \longrightarrow 2 Ca^{2+}(aq) + Ti(s)$

39. Label each of the following half-reactions as oxidation or reduction, and then combine them to obtain a balanced overall redox reaction.
    a. $Cu^{2+} + 2 e^- \longrightarrow Cu(s)$ and $Fe(s) \longrightarrow Fe^{2+} + 2 e^-$
    b. $2 H_2O_2 \longrightarrow 2 O_2 + 4 H^+ + 4 e^-$ and
    $$Fe^{3+} + e^- \longrightarrow Fe^{2+}$$
    c. $WO_3 + 6 H^+ + 6 e^- \longrightarrow W + 3 H_2O$ and
    $$C_2H_6O \longrightarrow C_2H_4O + 2 H^+ + 2 e^-$$

40. Label each of the following half-reactions as oxidation or reduction, and then combine them to obtain a balanced overall redox reaction.
    a. $2 I^- \longrightarrow I_2 + 2 e^-$ and $Cl_2 + 2 e^- \longrightarrow 2 Cl^-$
    b. $HNO_3 + H^+ + e^- \longrightarrow NO_2 + H_2O$ and
    $$SO_2 + 2 H_2O \longrightarrow H_2SO_4 + 2 H^+ + 2 e^-$$

41. The following half-reactions are of environmental importance. Label each of the reactions as oxidation or reduction, and then combine them to obtain a balanced overall redox reaction.
    a. $Ti \longrightarrow Ti^{4+} + 4 e^-$ and $Pb^{4+} + 2 e^- \longrightarrow Pb^{2+}$
    b. $S \longrightarrow S^{4+} + 4 e^-$ and $O + 2 e^- \longrightarrow O^{2-}$

Oxidation and Reduction: Chemical Reactions

42. In the following reactions, which substance is oxidized? Which is the oxidizing agent?
    a. $2 HNO_3 + SO_2 \longrightarrow H_2SO_4 + 2 NO_2$
    b. $Co_2S_3 + 2 HBr \longrightarrow H_2S + 2 CoS + Br_2$

43. In the following reactions, which element is oxidized and which is reduced?
   a. $CH_3CHO + H_2O_2 \longrightarrow CH_3COOH + H_2O$
   b. $5 C_2H_6O + 4 MnO_4^- + 12 H^+ \longrightarrow$
   $$5 C_2H_4O_2 + 4 Mn^{2+} + 11 H_2O$$

44. In which reaction is nitrogen being reduced, and in which is it being oxidized?
   a. $2 MnO_2 + NO_2^- + 2H^+ \longrightarrow Mn_2O_3 + NO_3^-$
   b. $3 MnO_2 + 2 NH_4^+ + 4 H^+ \longrightarrow 3 Mn^{2+} + N_2 + 6 H_2O$

45. Is manganese reduced or oxidized in equations **a** and **b** in Question 44?

46. Zinc metal reacts with vanadium(V) ions to form zinc(II) ions and vanadium(II) ions. **(a)** Is Zn oxidized or reduced? **(b)** Are electrons gained or lost by vanadium(V) ions? How many electrons in the balanced equation?

47. Unsaturated vegetable oils react with hydrogen to form saturated fats. A typical reaction is
   $$C_{57}H_{104}O_6 + 3 H_2 \longrightarrow C_{57}H_{110}O_6$$
   Is the unsaturated oil oxidized or reduced? Explain.

48. Tantalum, a metal used in electronic devices such as cellular phones and computers, can be manufactured by the reaction of its oxide with molten sodium metal.
   $$Ta_2O_5 + 10 Na \longrightarrow 2 Ta + 5 Na_2O$$
   Which substance is reduced? Which is the reducing agent?

49. To test for iodide ions (for example, in iodized salt), a solution is treated with chlorine to liberate iodine.
   $$2 I^- + Cl_2 \longrightarrow I_2 + 2 Cl^-$$
   Which substance is oxidized? Which is reduced?

50. In the Fukushima nuclear reactor that failed in March 2011, zirconium metal reacted with steam to produce hydrogen gas.
   $$Zr + 2 H_2O \longrightarrow ZrO_2 + 2 H_2$$
   What substance was oxidized in the reaction? What was the oxidizing agent?

51. Unripe grapes are exceptionally sour because of a high concentration of tartaric acid ($C_4H_6O_6$). As the grapes ripen, this compound is converted to glucose ($C_{14}H_{12}O_6$). Is the tartaric acid being oxidized or reduced?

52. Vitamin C (ascorbic acid) is thought to protect our stomachs from the carcinogenic effect of nitrite ions (NO_2^-) by converting the ions to nitrogen oxide (NO). Are the nitrite ions oxidized or reduced? Is ascorbic acid an oxidizing agent or a reducing agent?

53. In the reaction in Problem 52, ascorbic acid ($C_6H_8O_6$) is converted to dehydroascorbic acid ($C_6H_6O_6$). Is ascorbic acid oxidized or reduced in this reaction?

54. Bacteria are able to drive energy for growth utilizing Mn in the following manner:
   $$C_6H_{12}O_6 + 12 MnO_2 + 24 H^+ \longrightarrow$$
   $$6 CO_2 + 12 Mn^{2+} + 18 H_2O$$
   In the above equation, which element is being reduced and which element is being oxidized? Which compound is the oxidizing agent? What changes in oxidation numbers do carbon and manganese undergo?

55. Bacteria are able to obtain energy for growth utilizing Fe in the following manner:
   $$C_6H_{12}O_6 + 24 Fe(OH)_3 + 48 H^+ \longrightarrow$$
   $$6 CO_2 + 24 Fe^{2+} + 66 H_2O$$
   In the above equation, which element is being reduced and which element is being oxidized? What compound is the oxidation agent? What changes in oxidation numbers do carbon and iron undergo?

Combination with Oxygen

56. Write formula(s) for the product(s) formed when each of the following substances reacts with oxygen (O_2).
   a. S
   b. CH_3OH
   c. C_3H_6O

57. Write formula(s) for the product(s) formed when each of the following substances reacts with oxygen (O_2). (There may be more than one correct answer.)
   a. N_2
   b. CS_2
   c. C_5H_{12}

58. What is the oxidation number of oxygen on the reactant side in the following examples?
   a. $C + O_2 \longrightarrow CO_2$
   b. $O_2 + H_2 \longrightarrow H_2O_2$
   c. $O_2 + 2F_2 \longrightarrow 2F_2O$

🌑 Expand Your Skills

59. In the following compounds, oxygen and hydrogen do *not* have their usual oxidation numbers. Deduce the oxidation number of oxygen or hydrogen in each compound. What is unusual about the oxidation number of oxygen in CsO_2, cesium superoxide?
   a. CaH_2 b. Na_2O_2 c. CsO_2

60. Rank the following elements in terms of their ability to act as reducing agents.
   a. Na
   b. Ca
   c. O
   d. Cl

61. Rank the following elements in terms of their ability to act as oxidizing agents.
    a. F **b.** Cl
    c. Br **d.** I

62. The dye indigo (used to color blue jeans) is formed by exposure of indoxyl to air.

    $$2\,C_8H_7ON + O_2 \longrightarrow C_{16}H_{10}N_2O_2 + 2\,H_2O$$
    Indoxyl Indigo

    What substance is oxidized? What is the oxidizing agent?

63. Why do some mechanics lightly coat their tools with grease or oil before storing them?

64. When an aluminum wire is placed in a blue solution of copper(II) chloride, the blue solution turns colorless, and reddish-brown copper metal comes out of the solution. Write an equation for the reaction. (Chloride ion is not involved in the reaction.)

65. Researchers led by Elisabeth Bouwman, a chemist at Leiden University in the Netherlands, discovered a copper-based catalyst that can convert CO_2 in the air to solid oxalate compounds, such as sodium oxalate, $Na_2C_2O_4$. This could lead to a means to reduce the amount of that greenhouse gas in the atmosphere. Write the equation for the half-reaction of CO_2 to oxalate ion. Is the CO_2 oxidized or reduced?

66. When exposed to air containing hydrogen sulfide, lead-based paints turn black because the Pb^{2+} ions react with the H_2S to form black lead sulfide (PbS). This has caused the darkening of old oil-based paintings. Hydrogen peroxide lightens the paints by oxidizing the black sulfide (S^{2-}) to white sulfates (SO_4^{2-}). **(a)** Write the equation for the darkening reaction. **(b)** Write the equation for this reaction that lightens the paints.

67. The oxidizing agent that our bodies use to obtain energy from food is oxygen (from the air). If you breathe 15 times a minute (at rest), taking in and exhaling 0.5 L of air with each breath, what volume of air do you breathe each day? Air is 21% oxygen by volume. What volume of oxygen do you breathe each day?

68. To oxidize 1.0 kg of fat, our bodies require about 2000 L of oxygen. A healthy diet contains no more than about 80 g of fat per day. What volume of oxygen (at STP) is required to oxidize that fat?

69. The photosynthesis reaction (Section 8.7) can be expressed as the net result of two processes, one of which requires light and is called the *light reaction*. This reaction may be written as
    $$12\,H_2O \xrightarrow{\text{light}} 6\,O_2 + 24H^+ + 24\,e^-$$
    a. Is the light reaction an oxidation or reduction?
    b. Write the equation for the other half-reaction, called the *dark reaction*, for photosynthesis.

70. In cellular respiration, the average oxidation number of carbon changes from _____ to _____.

71. In photosynthesis, the oxidation number of carbon changes from _____ to _____.

72. Indicate whether the first-named substance in each change undergoes an oxidation, a reduction, or neither. Explain your reasoning.
    a. A violet solution of vanadium(II) ions, $V^{2+}(aq)$, is converted to a green solution of $V^{3+}(aq)$.
    b. Nitrogen dioxide converts to dinitrogen tetroxide when cooled.
    c. Carbon monoxide reacts with hydrogen to form methanol.

73. As a rule, metallic elements act as reducing agents, not as oxidizing agents. Explain. (*Hint*: Consider the charges on metal ions.) Do nonmetals act only as oxidizing agents and not as reducing agents? Explain your answer.

74. Why are TNT explosions so powerful?

75. If a stream of hydrogen gas is ignited in air, the flame can be immersed in a jar of chlorine gas and it will still burn. Write the balanced equation for the reaction that occurs in the jar. What is the oxidizing agent in the reaction?

76. Consider the following reaction of calcium hydride (CaH_2) with molten sodium metal. Identify the species being oxidized and the species being reduced according to **(a)** the second definition shown in Figure 8.2 and **(b)** the third definition in Figure 8.2. What difficulty arises?
    $$CaH_2(s) + 2\,Na(l) \longrightarrow 2\,NaH(s) + Ca(l)$$

77. Aluminum metal can react with water, but the oxide coating normally found on the metal prevents that reaction. In 2007, a Purdue University researcher found that adding gallium to aluminum prevented the formation of the oxide coating, allowing the aluminum to react with water to form hydrogen gas. This process might lead to a compact source of hydrogen gas for fuel. Write a balanced equation for the reaction of aluminum with water (aluminum hydroxide is the other product), and identify the oxidizing and reducing agents in the reaction.

78. The strength of the elemental halogens (F_2, Cl_2, etc.) as oxidizing agents, forming their corresponding halide ions, decreases from top to bottom of the periodic table. With this information, which of the following will react? Write a balanced equation for the reaction(s) that occur.
    a. Cl_2 with F^- **b.** Br_2 with I^-

79. A 3% solution of hydrogen peroxide as sold in drugstores is no longer recommended for treating minor cuts and scrapes. When it is applied to a cut, the enzyme *peroxidase* in blood causes rapid decomposition of the hydrogen peroxide:
    $$2\,H_2O_2(aq) \longrightarrow 2\,H_2O(l) + O_2(g)$$
    The high concentration of elemental oxygen kills bacteria, but also damages cells in the skin and underlying tissue. What are the oxidizing agent and the reducing agent in this reaction? (*Hint*: What are the oxidation numbers of oxygen in the reactant and products?)

80. Refer to Problem 77. Aluminum has a density of 2.70 g/cm³. What volume in cubic centimeters of aluminum metal would be needed to produce 2.0×10^4 L of hydrogen gas at STP? How many times smaller is this volume of aluminum than is the volume of hydrogen gas? (Recall that at STP, 1 mol of gas occupies 22.4 L.)

81. Consider the oxidation of ethanol (CH_3CH_2OH) to acetaldehyde (CH_3CHO) by NAD^+(page 230). **(a)** Write half-reactions for the process. **(b)** NADH is a biochemical reducing agent, in the reverse of the oxidation reaction. Write half-reactions for the reduction of pyruvic acid ($CH_3COCOOH$) to lactic acid ($CH_3CHOHCOOH$) by NADH, and then combine these two half-reactions to get the equation for the overall reaction.

82. Following are some organic chemistry processes. Classify each as an oxidation or a reduction. R represents a hydrocarbon group and does not change in the process.
    a. RCH_3 (alkane) $\longrightarrow$ RCH_2OH (alcohol)
    b. RCHO (aldehyde) $\longrightarrow$ RCH_2OH (alcohol)
    c. RCHO (aldehyde) $\longrightarrow$ RCOOH (carboxylic acid)
    d. RCOOH (carboxlic acid) $\longrightarrow$ RCHO (aldehyde)
    e. RCH_2OH (alcohol) $\longrightarrow$ RCHO (aldehyde)

83. The traditional process for producing propylene oxide involves reaction of propylene with hypochlorous acid (HOCl), which is formed by absorbing chlorine gas in water, to form propylene chlorohydrin. In a second step, the propylene chlorohydrin is treated with calcium hydroxide to form propylene oxide (see reaction below). Recently, an alternative process involving hydrogen peroxide as the oxidant, and titanium silicalite as the catalyst, has been commercialized.

Propylene

Propylene chlorohydrin   →   Propylene oxide

a. Write down the stochiometric equation for both processes.
b. Calculate the atom economy and the E factor of both processes.
c. Compare the two processes with regard to reagents used and waste generated.

84. A green catalyst is one that
    a. contains earth-abundant elements
    b. makes reagents react more efficiently
    c. decreases the cost of producing a chemical
    d. does all of the above

85. Give one reason why it can be difficult to design a long-lived oxidation catalyst.

86. Identify the oxidizing agent and the substance that is oxidized in the following reaction.
$$3\,C_2H_6O + Cr_2O_7^{2-} + 8\,H^+ \longrightarrow 3\,C_2H_4O + 2\,Cr^{3+} + 7\,H_2O$$

Critical Thinking Exercises

Apply knowledge that you have gained in this chapter and one or more of the FLaReS principles (Chapter 1) to evaluate the following statements or claims.

8.1. Over the years, some people have occasionally claimed that someone has invented an automobile engine that burns water instead of gasoline. They say that we have not heard about it because the oil companies have bought the rights to the engine from the inventor so that they can keep it from the public and people will continue to burn gasoline in their cars.

8.2. In January 2011, Italian scientists claimed to have constructed an apparatus to produce energy. The apparatus uses hydrogen and nickel and is supposed to produce several hundred times as much energy as would be obtained by merely burning the hydrogen.

8.3. A website claims that electricity can remove rust. The procedure is described as follows: Connect a wire to the object to be derusted, and immerse in a bath of sodium carbonate (washing soda) solution. Immerse a second wire in the bath. Connect the wires to a battery so that the object to be derusted is the cathode. The site claims that hydrogen gas is produced at the cathode and oxygen at the anode. The hydrogen then changes the rust back to iron metal.

8.4. You have just moved to a colder climate. A car salesperson tries to sell you zinc undercoating for your car. He claims that, due to salt being spread on the roads in winter months, the zinc undercoating will protect your car's iron frame from corrosion.

8.5. A co-worker states that the reason explosions are so destructive is the release of heat, which expands air.

8.6. A friend claims that an automobile can be kept free of rust simply by connecting the negative pole of the car's battery to the car's body. He reasons that this action makes the car the cathode, and oxidation does not occur at the cathode. He claims that for years he has kept his car from rusting by doing this.

8.7. A salesman claims that your tap water is contaminated. As evidence, he connects a battery to a pair of "special electrodes" and dips the electrodes in a glass of your tap water. Within a few minutes, white and brown "gunk" appears in the water. The salesman sells a special purifying filter that he claims will remove the contaminants and leave the water pure and clear.

Collaborative Group Projects

Prepare a PowerPoint, poster, or other presentation (as directed by your instructor) to share with the class.

1. Prepare a brief report on one of the following methods of protecting steel from corrosion. Identify the chemical reactions involved in the process, and tell how the process is related to oxidation and reduction.
    a. galvanization
    b. coating with tin
    c. cathodic protection

2. Batteries contain toxic substances and should be recycled rather than thrown in the trash. Prepare a brief report on battery recycling. Are there facilities in your area for recycling batteries?

3. Military personnel, outdoor recreation enthusiasts such as wilderness campers and trekkers, and others often have to drink water from untreated streams or lakes. People who temporarily lack safe drinking water after a natural disaster also need a way to purify water. A variety of water purification tablets are available to treat contaminated water. Research several such tablets, and list their ingredients.

LET'S EXPERIMENT! Light My Fruit

Materials Needed

- 2 lemons
- 2 kiwis
- Copper wire (or a penny)
- Zinc strip (or galvanized nail)

- Voltmeter
- 2 connection wires with alligator clips
- A 2 milliamp, 1.5-volt light-emitting diode (LED)

Can fruits make electricity? Do you think comparing lemons and kiwis will show how redox reactions work?

In this experiment, you will compare a lemon and a kiwi. Both the lemon and kiwi have juicy and acidic characteristics that make for the exact properties needed in a conductive solution, or electrolyte, which is any acid, base, or salt solution that is rich in ions. The chemical bath within these fruits allows the transfer of electrons from different elements, thereby making a single-cell electric battery. This means the charge "wants to" move from one metal to the other—creating electric potential.

Begin by gathering your supplies. Start by squeezing the fruit gently so the juice goes to the middle. Be careful not to burst the skin.

Poke your copper wire (or penny) into one side of the fruit. Push your zinc strip (or nail) in as close as you can to the copper without having them touch. If the two metals touch inside the fruit, the experiment will *not* work.

Clip your alligator clips to the wires. The red wire goes to the copper wire (or penny), and the black wire goes to the zinc (or nail). Record the voltage you have coming from your fruit battery.

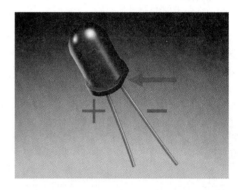

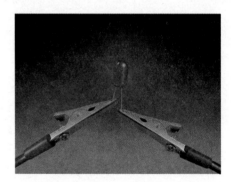

Connect the copper to the other wire of the diode, and connect the zinc to the side of the diode with the metal strip. (See following photo.)

Now clip your alligator clips from the battery to the diode. The diode has a small metal strip underneath one of the connection wires; this indicates the negative side of the diode.

Predict what will happen next. Record the amount of time it took to light and the amount of time it was on. Repeat for the other fruit and compare your results.

What happens when you connect the fruit in a series? Attach both the lemon and the kiwi together and see if you can light up the 1.5-volt LED.

Questions

1. Which metal is more reactive, zinc or copper? Why?
2. How many milliamps and volts did your lemon produce?
3. How many milliamps and volts did your kiwi produce?
4. Which fruit produced more milliamps and volts? Why?
5. Did your fruit produce enough voltage and current to power a 1.5-volt mini light bulb? Why or why not?

Organic Chemistry

Have You Ever Wondered?

1 What is the "petroleum distillate" that I see listed on some product labels? **2** Why are benzene-like compounds that have many multiple bonds called "aromatic"? **3** Why does fresh fruit smell so good and rotting fruit smell so bad? **4** Why do so many organic compounds have such strong tastes and odors? **5** How do fabric softeners, dryer sheets, and anti-static sprays work?

ORGANIC CHEMISTRY VERSUS ORGANIC FOODS We are constantly exposed to "organic" products on grocery shelves as well as ads for "organic" or other "natural" products in the media. Very often, people use the term "organic" to refer to foods that have been grown using "all-natural" methods and without the use of synthetic pesticides or herbicides. Many chemists find this concept ironic because of the modern chemical definition of organic compounds. Most of the materials that "organic" gardeners would never use are organic chemical substances. Organic farming and agriculture use techniques that include biological pest-control methods.

Ground-cover crops are planted that discourage the growth of weeds. Intercropping methods utilize plants that are naturally repellent to insects and animals to discourage their predation of the desired crops.

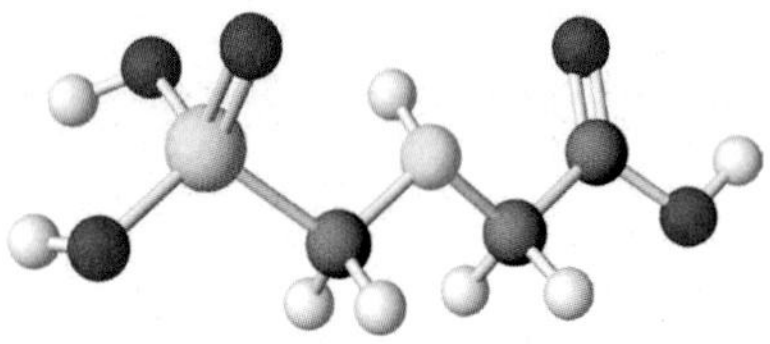

Glyphosate

Rotenone

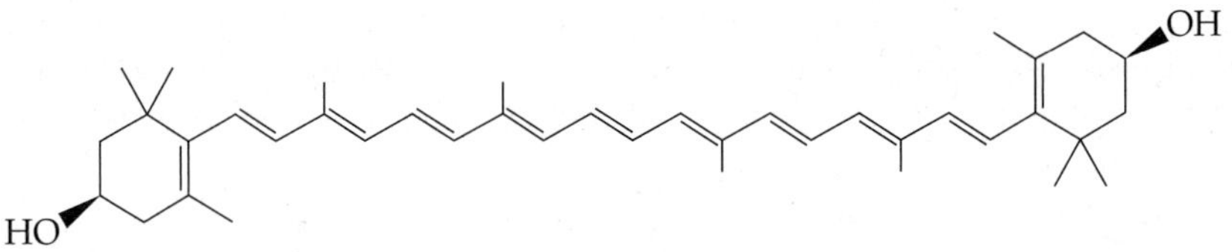

Xanthophyll

▶ Glyphosate is a common agricultural herbicide, and rotenone is a common agricultural pesticide. Both have been widely used in agriculture, but neither would be used by an "organic" farmer, because they are synthetic compounds. However, both fit the chemical definition of an organic compound. The overuse of pesticides and herbicides can be a serious environmental problem. Other "organic" methods of keeping insects and pests away from crops include using a spray from the oil of chili peppers, or planting garlic, marigolds, or other herbs around the crops. These natural materials are repellent to insects and other pests. Xanthophyll is a natural pesticide produced by Mexican marigolds. Planting crops that have insect-repelling properties interspersed with food crops can extend that protection to the food crops.

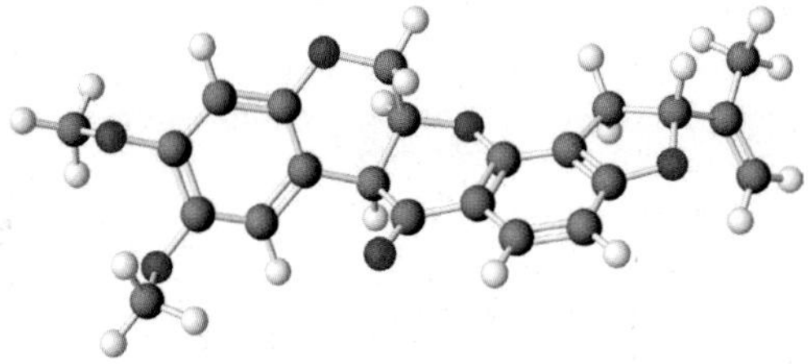

The definition of organic chemistry has changed over the years. The changes exemplify the dynamic character of science and illustrate how scientific concepts change in response to experimental evidence. Until the nineteenth century, chemists believed that *organic* chemicals originated only in tissues of living *organisms* and required a "vital force" for their production. All chemicals not manufactured by living tissue were once regarded as *inorganic*. Some chemists even believed that organic and inorganic chemicals followed different laws. This all changed in 1828, when Friedrich Wöhler (Germany, 1800–1882) synthesized urea, an organic compound found in urine, from ammonium cyanate, an inorganic compound. This important event led many chemists to attempt synthesis of other organic chemicals from inorganic ones and changed the very definition of organic chemicals.

Our modern definition of "organic" chemistry encompasses the field of carbon-containing compounds. This chapter will examine the wide variety of types of organic compounds that exist.

9.1 Organic Chemistry and Compounds

Learning Objective • Define *organic chemistry*, and identify differences between organic and inorganic compounds.

Organic chemistry is now defined as the chemistry of carbon-containing compounds. Most of these compounds do come from living things or from things that were once living, but some do not. Perhaps one of the most remarkable things about organic compounds is their sheer number. Of the tens of millions of known chemical compounds, more than 95% are compounds of carbon.

One reason for the huge number of organic compounds is the carbon atom itself. Recall that carbon is a group 4A element with four valence electrons, so it can form bonds with up to four other atoms. Carbon is unique in that its atoms bond readily to each other and form long chains. Carbon chains can also have branches or form rings of various sizes. (Silicon and a few other elements can also form chains, but only short ones.) Organic compounds can have as few as one carbon atom or many thousands of carbon atoms. When we consider that carbon atoms also bond strongly to other elements, particularly nonmetals such as hydrogen, oxygen, and nitrogen, and that these atoms can be arranged in many different ways, it soon becomes obvious why there are so many organic compounds. Almost all organic compounds also contain hydrogen, many contain oxygen and/or nitrogen, and quite a few contain sulfur or halogens. Nearly all of the common elements are found in at least a few organic compounds.

In addition to the millions of carbon compounds already known, new ones are being discovered every day. Carbon can form an almost infinite number of molecules of various shapes, sizes, and compositions. We use thousands of carbon compounds every day without even realizing it because they are silently carrying out important chemical reactions within our bodies. Many of these carbon compounds are so vital that we cannot live without them.

Given the large number of organic compounds that exist, it is not surprising that the various types of organic compounds have some differences from each other, which we will examine later in this chapter. However, organic compounds differ even more from the inorganic compounds we have previously seen.

Atoms in organic compounds generally form covalent bonds, while inorganic compounds often have ionic bonds. This means that organic compounds have much lower melting and boiling points and may exist as gases, liquids, or solids at room temperature. Inorganic compounds have very high melting points and are primarily solids at room temperature. Organic compounds generally are not soluble in water (except for very small molecules) and do not conduct electricity well, while many inorganic compounds are polar, dissolve readily in water, and are good conductors of electricity. Organic compounds react much more slowly than ionic compounds do, and they almost always undergo combustion. Inorganic compounds react very quickly, but they generally do not burn. Table 9.1 summarizes some of the different properties of organic and inorganic compounds. Table 9.2 illustrates the differences in some properties for two specific compounds: an organic compound, hexane (C_6H_{14}), and an inorganic compound, sodium chloride (NaCl).

TABLE 9.1   **General Properties of Organic and Inorganic Compounds**

Property	Organic Compounds	Inorganic Compounds
Bonding	Covalent	Ionic
State at room temperature	Liquid, gas, or solid	Solid
Boiling points	Low	High
Melting points	Low	High
Solubility in water	Most are not soluble in water	Frequently soluble in water
Solubility in inorganic solvents	Soluble	Not soluble
Conduct electricity in aqueous solution	No	Yes
Combustion	Most burn	Few burn
Reaction rates	Slow	Fast

TABLE 9.2   **Properties of Hexane and Sodium Chloride**

Property	Hexane (C_6H_{14}) (Organic)	Sodium Chloride (NaCl) (Inorganic)
Bonding	Covalent	Ionic (Na^+ and Cl^- ions)
State at room temperature	Liquid	Solid
Boiling point	68 °C	1413 °C
Melting point	−95 °C	801 °C
Solubility in water	0.95 mg/100 mL	35.7 g/100 mL

9.2 Aliphatic Hydrocarbons

Learning Objectives • Define *hydrocarbon*, and recognize structural features of alkanes, alkenes, and alkynes. • Identify hydrocarbon molecules as alkanes, alkenes, or alkynes, and name them.

Organic compounds can be subdivided into several main classes. Most organic compounds could be described as hydrocarbons, but a **hydrocarbon** is usually defined as containing only hydrogen and carbon.

The hydrocarbons can then be subdivided according to the type of bonding that exists between carbon atoms. We divide hydrocarbons into two main classes. The difference is based on the degree of multiple bonding between the atoms. Those discussed in this section are **aliphatic compounds** (the Greek word *aleiphar* means "fat" or "oil"), which have little or no multiple bonding between carbon atoms. The *aromatic compounds*, which have ring structures like benzene as well as extensive multiple bonds between the carbon atoms, are discussed in Section 9.3.

In this section, we will consider further subdivisions of the aliphatic compounds. These are the alkanes, cyclic structures, alkenes, and alkynes.

Alkanes

Each carbon atom forms four bonds, and each hydrogen atom forms only one bond, so the simplest hydrocarbon molecule possible is methane, CH_4. Methane is the main component of natural gas. Its structure is shown to the left.

Methane is the first member of a group of related compounds called **alkanes**, hydrocarbons that contain only single bonds. The next member of the series is ethane, which has two carbons and is also shown in the margin.

Ethane is a minor constituent of natural gas. It is seldom encountered as a pure compound in everyday life, but many common compounds are derived from it.

We saw in Chapter 4 that the methane molecule is tetrahedral, in accord with the VSEPR theory. In fact, the tetrahedral shape results whenever a carbon atom is connected to four other atoms, whether they are hydrogen, carbon, or other elements. Figure 9.1 shows models of methane and ethane. The ball-and-stick models show the bond angles best, but the space-filling models more accurately reflect the shapes of the molecules. Ordinarily, we use simple structural formulas such as the one shown previously for ethane because they are much easier to draw. A **structural formula** shows which atoms are bonded to each other, but it does not attempt to show the actual shape of the molecule.

Alkanes are often called **saturated hydrocarbons** because each carbon atom is *saturated* with hydrogen atoms; that is, each carbon is bonded to the maximum

(a)

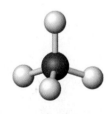
(b)

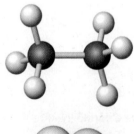

▲ **Figure 9.1** (a) Ball-and-stick and (b) space-filling models of methane (left) and ethane (right).

number of hydrogen atoms. In constructing a formula, we connect the carbon atoms to each other through single bonds, then we add enough hydrogen atoms to give each carbon atom four bonds. All alkanes have two more than twice as many hydrogen atoms as carbon atoms. That is, alkanes can be represented by a general formula C_nH_{2n+2}, in which n is the number of carbon atoms.

The three-carbon alkane is propane. Models of propane are shown in Figure 9.2. To draw its structural formula, we place three carbon atoms in a row.

$$C-C-C$$

Then we add enough hydrogen atoms (eight in this case) to give each carbon atom a total of four bonds. Therefore, the structural formula of propane is

$$H-\underset{\underset{H}{|}}{\overset{\overset{H}{|}}{C}}-\underset{\underset{H}{|}}{\overset{\overset{H}{|}}{C}}-\underset{\underset{H}{|}}{\overset{\overset{H}{|}}{C}}-H$$

Condensed Structural Formulas

The complete structural formulas that we have used so far show all the carbon and hydrogen atoms and how they are attached to one another. But these formulas take up a lot of space, and they can be a bit of trouble to draw, especially as the number of carbons increases. For these reasons, chemists usually prefer to use **condensed structural formulas**. These formulas show how many hydrogen atoms are attached to each carbon atom without showing the bond to each hydrogen atom. For example, the condensed structural formulas for ethane and propane are written CH_3-CH_3 and $CH_3-CH_2-CH_3$, respectively. These formulas can be simplified even further by omitting some (or all) of the remaining bond lines, resulting in CH_3CH_3 and $CH_3CH_2CH_3$.

Homologous Series

Note that the given formulas for methane, ethane, and propane form a pattern. We can build alkanes of any length simply by joining carbon atoms together in long chains and adding enough hydrogen atoms to give each carbon atom four bonds. Even the naming of these compounds follows a pattern (Table 9.3), with a stem name indicating the number of carbon atoms. For compounds of five carbon atoms or more, each stem is derived from the Greek or Latin name for the number. The compound names end in -ane, signifying that the compounds are saturated and are therefore *alkanes*. Table 9.4 gives condensed structural formulas and names for the continuous-chain (straight-chain, or unbranched) alkanes with up to 10 carbon atoms.

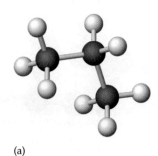

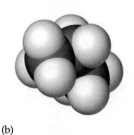

▲ **Figure 9.2** (a) Ball-and-stick and (b) space-filling models of propane.

TABLE 9.3 Stems for Organic Molecule Names

Stem	Number of Carbon Atoms
Meth-	1
Eth-	2
Prop-	3
But-	4
Pent-	5
Hex-	6
Hept-	7
Oct-	8
Non-	9
Dec-	10

TABLE 9.4 The First 10 Continuous-Chain Alkanes

Name	Molecular Formula	Condensed Structural Formula	Number of Possible Isomers
Methane	CH_4	CH_4	—
Ethane	C_2H_6	CH_3CH_3	—
Propane	C_3H_8	$CH_3CH_2CH_3$	—
Butane	C_4H_{10}	$CH_3CH_2CH_2CH_3$	2
Pentane	C_5H_{12}	$CH_3CH_2CH_2CH_2CH_3$	3
Hexane	C_6H_{14}	$CH_3CH_2CH_2CH_2CH_2CH_3$	5
Heptane	C_7H_{16}	$CH_3CH_2CH_2CH_2CH_2CH_2CH_3$	9
Octane	C_8H_{18}	$CH_3CH_2CH_2CH_2CH_2CH_2CH_2CH_3$	18
Nonane	C_9H_{20}	$CH_3CH_2CH_2CH_2CH_2CH_2CH_2CH_2CH_3$	35
Decane	$C_{10}H_{22}$	$CH_3CH_2CH_2CH_2CH_2CH_2CH_2CH_2CH_2CH_3$	75

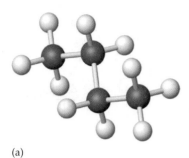

(a)

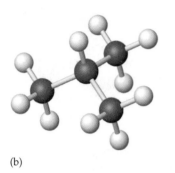

(b)

▲ **Figure 9.3** Ball-and-stick models of (a) butane and (b) isobutane.

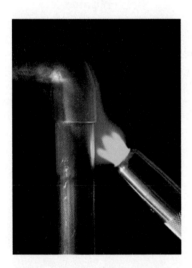

▲ A propane torch. Propane burns in air with a hot flame.

1 What is the "petroleum distillate" that is listed on some product labels?

Petroleum distillates are various components of crude oil, most of which are hydrocarbons. Paint thinner, lighter fluid, paraffin wax, petroleum jelly, and asphalt all are petroleum distillates.

Notice that the molecular formula for each alkane in Table 9.4 differs from the one preceding it by precisely one carbon atom and two hydrogen atoms—that is, by a CH_2 unit. Such a series of compounds has properties that vary in a regular and predictable manner. This principle, called *homology*, gives order to organic chemistry in much the same way that the periodic table gives organization to the chemistry of the elements. Instead of studying the chemistry of a bewildering array of individual carbon compounds, organic chemists study a few members—called *homologs*—of a **homologous series**, from which they can deduce the properties of other compounds in the series.

We need not stop at 10 carbon atoms as Table 9.4 does. Hundreds, thousands, even millions of carbon atoms can be linked together. We can make an infinite number of alkanes simply by lengthening the chain. But lengthening the chain is not the only option. Beginning with alkanes having four carbon atoms, chain branching is also possible.

Isomerism—Branched Molecules

When we extend the carbon chain to four atoms and add enough hydrogen atoms to give each carbon atom four bonds, we get $CH_3CH_2CH_2CH_3$. This formula represents butane, a compound that boils at about 0 °C. A second compound, which has a boiling point of −12 °C, has the same molecular formula, C_4H_{10}, as butane. The structure of this compound, however, is not the same as that of butane. Instead of having four carbon atoms connected in a continuous chain, this compound has a continuous chain of only three carbon atoms. The fourth carbon is branched off the middle carbon of the three-carbon chain. To show the two structures more clearly, let's use complete structural formulas, showing all the bonds to hydrogen atoms.

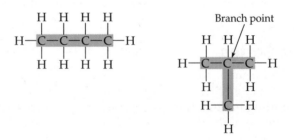

Compounds that have the same molecular formula but different structural formulas are called **isomers** or **constitutional isomers**. Because it is an isomer of butane, the branched four-carbon alkane is sometimes called *isobutane*. Condensed structural formulas for the two isomeric butanes are written as follows.

$$CH_3CH_2CH_2CH_3 \qquad CH_3CHCH_3$$
$$\qquad\qquad\qquad\qquad | $$
$$\qquad\qquad\qquad\qquad CH_3$$

Butane   Isobutane

Figure 9.3 shows ball-and-stick models of the two compounds.

The number of isomers increases rapidly with the number of carbon atoms (Table 9.4). There are three pentanes, five hexanes, nine heptanes, and so on. Isomerism is common in carbon compounds and provides another reason for the existence of millions of organic compounds.

Propane and the butanes are familiar fuels. Although they are gases at ordinary temperatures and under normal atmospheric pressure, they are liquefied under pressure and are usually supplied in tanks as liquefied petroleum gas (LPG). Gasoline is a mixture of hydrocarbons, mostly alkanes, with 5 to 12 carbon atoms.

Not all the possible isomers of the larger molecules have been isolated. Indeed, the task rapidly becomes more and more prohibitive going up the series. There are, for example, 4,111,846,763 possible isomers with the molecular formula $C_{30}H_{62}$.

EXAMPLE 9.1 **Hydrocarbon Formulas**

Without referring to Table 9.4, write the molecular formula, the complete structural formula, and the condensed structural formula for heptane.

Solution

The stem *hept-* means seven carbon atoms, and the ending *-ane* indicates an alkane. For the complete structural formula, we write a string of seven carbon atoms.

$$C—C—C—C—C—C—C$$

Then we attach enough hydrogen atoms to the carbon atoms to give each carbon four bonds. This requires three hydrogens on each end carbon and two hydrogens on each of the other carbons.

$$
\begin{array}{c}
\ \ \ \ H\ \ H\ \ H\ \ H\ \ H\ \ H\ \ H \\
\ \ \ \ |\ \ \ \ |\ \ \ \ |\ \ \ \ |\ \ \ \ |\ \ \ \ |\ \ \ \ | \\
H—C—C—C—C—C—C—C—H \\
\ \ \ \ |\ \ \ \ |\ \ \ \ |\ \ \ \ |\ \ \ \ |\ \ \ \ |\ \ \ \ | \\
\ \ \ \ H\ \ H\ \ H\ \ H\ \ H\ \ H\ \ H
\end{array}
$$

For the condensed form, we simply write each set of hydrogen atoms following the carbon atom to which they are bonded.

$$CH_3CH_2CH_2CH_2CH_2CH_2CH_3$$

For the molecular formula, we can simply count the carbon and hydrogen atoms to arrive at C_7H_{16}. Alternatively, we could use the general formula C_nH_{2n+2}, with $n = 7$, to get C_7H_{16}.

> **Exercise 9.1A**

Write the molecular, complete structural, and condensed structural formulas for **(a)** hexane and **(b)** octane.

> **Exercise 9.1B**

Write the molecular, complete structural, and condensed structural formulas for **(a)** pentane and **(b)** decane.

Properties of Alkanes

Note in Table 9.5 that after the first few members of the alkane series, the rest show an increase in boiling point of about 30 °C with each added CH_2 group. Note also that at room temperature alkanes that have 1 to 4 carbon atoms per molecule are gases. The compounds with 5 to about 16 carbon atoms per molecule are liquids, and those with more than 16 carbon atoms per molecule are solids.

The densities of liquid and solid alkanes are typically less than that of water. Recall that water has a density close to 1.00 g/mL at room temperature. All the alkanes listed in Table 9.5 have densities less than that. Alkanes are nonpolar molecules and are insoluble in water. The combination of these two properties means the alkanes float on top of water. Because these are nonpolar molecules, they will also dissolve other organic substances of low polarity, such as fats, oils, and waxes.

Alkanes undergo few chemical reactions, but one very important chemical property is that they undergo combustion and burn, producing a lot of heat. Alkanes have many uses, but they mainly serve as fuels.

The effects of alkanes on the human body vary. Methane appears to be physiologically inert. We probably could breathe a mixture of 80% methane and 20% oxygen without ill effect. This mixture would be flammable, however, and no fire or spark of any kind could be permitted in such an atmosphere. Even a spark from static electricity could set it off. Breathing an atmosphere of pure methane (the "gas"

 SAFETY ALERT

Hydrocarbons and most other organic compounds are flammable or combustible. Volatile ones may form explosive mixtures with air. Hydrocarbons are the main ingredients in gasoline, motor oils, fuel gases (natural gas and bottled gas), and fuel oils. They are also ingredients, identified on labels as "petroleum distillates" and "mineral spirits," in products such as floor cleaners, furniture polishes, paint thinners, wood stains and varnishes, and car waxes and polishes. Some products also contain flammable alcohols, esters, and ketones. Products containing flammable substances can be dangerous but may be used safely with proper precautions. They should be used only as directed and kept away from open flames.

TABLE 9.5 **Physical Properties, Uses, and Occurrences of Selected Alkanes**

Name	Molecular Formula	Melting Point (°C)	Boiling Point (°C)	Density^a at 20 °C (g/mL)	Use/Occurrence
Methane	CH_4	−183	−162	(Gas)	Natural gas (main component); fuel
Ethane	C_2H_6	−172	−89	(Gas)	Natural gas (minor component); plastics
Propane	C_3H_8	−188	−42	(Gas)	LPG (bottled gas); plastics
Butane	C_4H_{10}	−138	0	(Gas)	LPG; lighter fuel
Pentane	C_5H_{12}	−130	36	0.626	Gasoline component
Hexane	C_6H_{14}	−95	69	0.659	Gasoline component; extraction solvent for food oils
Heptane	C_7H_{16}	−91	98	0.684	Gasoline component
Octane	C_8H_{18}	−57	126	0.703	Gasoline component
Decane	$C_{10}H_{22}$	−30	174	0.730	Gasoline component
Dodecane	$C_{12}H_{26}$	−10	216	0.749	Gasoline component
Tetradecane	$C_{14}H_{30}$	6	254	0.763	Diesel fuel component
Hexadecane	$C_{16}H_{34}$	18	280	0.775	Diesel fuel component
Octadecane	$C_{18}H_{38}$	28	316	(Solid)	Paraffin wax component
Eicosane	$C_{20}H_{42}$	37	343	(Solid)	Paraffin wax and asphalt component

^{a}Densities of the gaseous alkanes vary with pressure. Densities of the solids are not available.

of a gas-fueled stove) can lead to death—not because of the presence of methane but because the body would be deprived of oxygen, a condition called *asphyxia*.

Light liquid alkanes such as those in gasoline will dissolve and wash away body oils when spilled on the skin. Repeated contact may cause dermatitis. While mineral oil, a mixture of long-chain liquid alkanes, has been used for many years as a laxative, ingestion of other alkanes can be extremely dangerous. Even a small amount could be aspirated into the lungs or cause serious damage to other internal tissues.

Heavier liquid alkanes act as emollients (skin softeners). Petroleum jelly (Vaseline® is one brand) is a semisolid mixture of hydrocarbons that can be applied as an emollient or simply as a protective film. (Skin lotions and creams are discussed in Chapter 20.) However, in the lungs, alkanes cause chemical pneumonia by dissolving fatlike molecules from the cell membranes in the alveoli, allowing the lungs to fill with fluid.

Cyclic Hydrocarbons: Rings and Things

The hydrocarbons we have encountered so far (alkanes) have open-ended chains of carbon atoms. Carbon atoms can also connect to form closed rings. The simplest possible ring-containing hydrocarbon, or **cyclic hydrocarbon**, has the molecular formula C_3H_6 and is called *cyclopropane* (Figure 9.4).

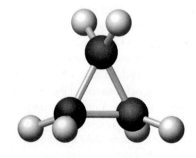

▲ **Figure 9.4** Ball-and-stick model of cyclopropane. Cyclopropane is a potent, quick-acting anesthetic with few undesirable side effects. It is no longer used in surgery, however, because it forms an explosive mixture with air at nearly all concentrations.

Names of *cycloalkanes* (cyclic hydrocarbons containing only single bonds) are formed by adding the prefix *cyclo-* to the name of the open-chain compound with the same number of carbon atoms as are in the cycloalkane's ring. Note that the formula for the cycloalkanes is different from that for the regular alkane. The carbon–carbon bond that is made to form a ring replaces bonds to two hydrogen atoms. The cycloalkane would fit a C_nH_{2n} formula, but this would still represent a saturated hydrocarbon, as there are only single bonds within the molecule.

Chemists often use geometric shapes to represent cyclic compounds (Figure 9.5). These are sometimes called **skeletal structures**, as the shape represents the carbon skeleton of the molecule. For example, a triangle is used to represent the cyclopropane ring, and a hexagon represents cyclohexane. Each corner of such a shape represents a carbon atom with enough associated hydrogen atoms to make four bonds to the carbon.

Cyclopropane   Cyclohexane   Cyclohexene

◀ **Figure 9.5** Structural formulas and symbolic representations of some cyclic hydrocarbons.

EXAMPLE 9.2   Structural Formulas of Cyclic Hydrocarbons

Write the structural formula for cyclopentane. What geometric figure is used to represent cyclopentane?

Solution
Cyclopentane has five carbon atoms arranged in a ring.

Each carbon atom needs two hydrogen atoms to complete its set of four bonds.

Cyclopentane can also be represented by a pentagon.

> **Exercise 9.2A**
Write the structural formula for cyclobutane and the geometric shape used to represent it.

> **Exercise 9.2B**
Explain why the formula for a cycloalkane does not fit that for a regular alkane.

Unsaturated Hydrocarbons: Alkenes and Alkynes

Two carbon atoms can share more than one pair of electrons. In ethylene (C_2H_4), the two carbon atoms share two pairs of electrons and are therefore joined by a double bond (Figure 9.6).

$$\ddot{\text{:}}C\text{::}C\ddot{\text{:}} \quad \text{or} \quad \overset{H}{\underset{H}{>}}C=C\overset{H}{\underset{H}{<}} \quad \text{or} \quad CH_2{=}CH_2$$

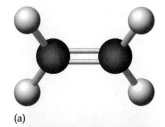

(a)

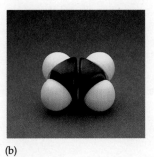

(b)

▲ **Figure 9.6** (a) Ball-and-stick and (b) space-filling models of ethylene.

▲ Common polyethylene products.

Ethylene is the common name; the systematic, or IUPAC, name is *ethene*. (The *-ene* ending shows that it contains a carbon–carbon double bond.) Ethylene is the simplest member of the alkene family. An **alkene** is a hydrocarbon that contains one or more carbon-to-carbon double bonds. Alkenes with one double bond have the general formula C_nH_{2n}, where n is the number of carbon atoms. This is the same general formula as that of a cycloalkane.

Ethylene is the most important organic chemical produced commercially. Annual U.S. production is over 25 billion kg. More than half goes into the manufacture of polyethylene, one of the most familiar plastics. Another 15% or so is converted to ethylene glycol, the major component of many formulations of antifreeze used in automobile radiators. Other commercially important alkenes are propylene (C_3H_6) and butylene (C_4H_8).

In acetylene (C_2H_2), the two carbon atoms share three pairs of electrons. The carbon atoms are joined by a triple bond (Figure 9.7).

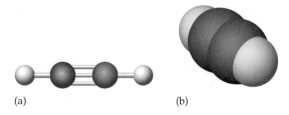

(a)   (b)

▲ **Figure 9.7** (a) Ball-and-stick and (b) space-filling models of acetylene.

$$H:C:::C:H \quad \text{or} \quad H-C\equiv C-H$$

Acetylene (IUPAC name ethyne, with the ending *-yne* indicating the carbon–carbon triple bond) is the simplest member of the alkyne family. An **alkyne** is a hydrocarbon that contains one or more carbon-to-carbon triple bonds. Alkynes with one triple bond have the general formula C_nH_{2n-2}, where n is the number of carbon atoms.

Acetylene is used in oxyacetylene torches for cutting and welding metals. Such torches can produce very high temperatures. Acetylene is also converted to a variety of other chemical products.

Collectively, alkenes and alkynes are called *unsaturated hydrocarbons*. Recall that a *saturated hydrocarbon* has the maximum number of hydrogen atoms attached to each carbon atom; it has no double or triple bonds. An **unsaturated hydrocarbon** can become saturated by having more hydrogen atoms added to it.

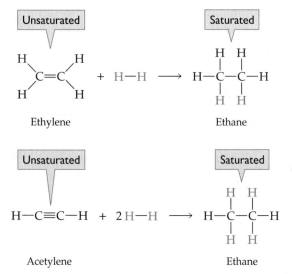

 EXAMPLE 9.3 **Molecular and Condensed Structural Formulas of Unsaturated Hydrocarbons**

What are the molecular and condensed structural formulas for 1-pentene? (The "1-" is a part of a systematic name that indicates the location of the double bond, which in this case connects the first carbon atom of the chain to the second.)

Solution

The stem *pent-* indicates five carbon atoms. The ending *-ene* tells us that the compound is an alkene. Using the general formula C_nH_{2n} for an alkene, with $n = 5$, we see that an alkene with five carbon atoms has the molecular formula C_5H_{10}. The condensed structural formula for 1-pentene is $CH_2=CHCH_2CH_2CH_3$.

❯ **Exercise 9.3A**

What are the molecular and condensed structural formulas for **(a)** 2-hexene and **(b)** 3-heptyne?

❯ **Exercise 9.3B**

What are the molecular and condensed structural formulas for **(a)** 4-nonene and **(b)** 2-pentyne?

Properties of Alkenes and Alkynes

The physical properties of alkenes and alkynes are much like those of the corresponding alkanes. Generally, unsaturated hydrocarbons with 2 to 4 carbon atoms per molecule are gases at room temperature, those with 5 to 18 carbon atoms are liquids, and most of those with more than 18 carbon atoms are solids. Like alkanes, alkenes and alkynes are insoluble in, and float on, water.

Both alkenes and alkynes burn, as alkanes do. However, these unsaturated hydrocarbons undergo many more chemical reactions than alkanes. An alkene or alkyne can undergo an **addition reaction** to the double or triple bond, in which all the atoms of the reactants are incorporated into a single product. In these reactions, one mole of hydrogen (H_2) adds to one mole of ethylene (C_2H_4) to form ethane (C_2H_6), and two moles of hydrogen ($2 H_2$) add to one mole of acetylene (C_2H_2) to form ethane. Chlorine, bromine, water, and many other kinds of small molecules also add to double and triple bonds.

One of the most unusual features of alkene (and alkyne) molecules is that they can add *to each other* to form large molecules called *polymers*. These interesting molecules are discussed in Chapter 10.

Geometric Isomers (*Cis–Trans* Isomers)

As with the alkanes, there is a homologous series of alkenes. The one with three carbon atoms (C_3H_6) is propylene (IUPAC: propene).

$$CH_2=CHCH_3$$

Propylene is also an important industrial chemical. It is converted to plastics (Chapter 10), isopropyl alcohol (Section 9.6), and a variety of other products. Although there is only one alkene with the formula C_2H_4 (ethene) and only one with the formula C_3H_6 (propene), there are several alkenes with the formula C_4H_8. The compound with a double bond between the first and second carbon atoms is called 1-butene.

$$CH_2=CHCH_2CH_3$$

Now consider the alkene with the formula $CH_3CH=CHCH_3$. We could name it "2-butene," but there are actually *two* such compounds. The double bond results in **geometric isomers** (*cis–trans isomers*), which are molecules with the same molecular formula but with a different arrangement of attached groups around the double bond (Figure 9.8).

We do not have free rotation about the double bond the way we would around the single bonds. As you might deduce from the models, the carbon-to-carbon single bonds in alkanes such as butane could function like an axle. The atoms at either end of the single bond are able to pivot about the shared electrons. In contrast, the structure of alkenes requires that the carbon atoms of a double bond and the two atoms bonded to each carbon all lie in a single plane. This part of the molecule's structure is rigid; rotation about doubly bonded carbon atoms is *not* possible without rupturing the bond. Now look at the two forms of 2-butenes in Figure 9.8. One isomer, called *cis*-2-butene, has both methyl groups attached to the carbons of the double bond on the same side of the molecule. In *trans*-2-butene, the methyl groups are on opposite sides of the molecule. Their structural formulas are:

CH₃⎯C=C⎯CH₃ with H, H   CH₃⎯C=C⎯H with H, CH₃

cis-2-butene   *trans*-2-butene

Note that each of the carbons in the double bond has two different groups attached to it. In this case each carbon has the same two groups, but they could be different.

An unsaturated hydrocarbon tends to burn with a smoky flame, especially if the oxygen supply is limited. Saturated hydrocarbons usually burn with a clean yellow flame. Burning of natural gas, propane, and butane is almost soot-free, but paint thinner, gasoline, and kerosene (all containing unsaturated hydrocarbons) burn with sooty flames.

WHY IT MATTERS

Fats are probably the best-known examples of compounds that can be saturated or unsaturated. In saturated fats, all of the carbon–carbon bonds are single bonds. These fats are found mostly in milk, meats, and eggs, and their consumption tends to increase low-density lipoproteins ("bad" cholesterol) in the human bloodstream. Most plant oils are unsaturated fats, with one or more carbon–carbon double bonds. Consumption of unsaturated fats tends to increase high-density lipoproteins ("good" cholesterol) in the blood.

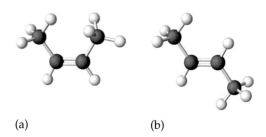

▲ Figure 9.8 Ball-and-stick models of (a) *cis*-2-butene and (b) *trans*-2-butene.

However, it is also possible to introduce the structural isomerization that was mentioned for alkanes. A three-carbon alkene chain, with a methyl group attached to the second carbon, would have the same molecular formula as the 2-butene molecule. This molecule would be named 2-methylpropene in the IUPAC system.

$$\begin{array}{c} H \\ \diagdown \\ \end{array} C = C \begin{array}{c} CH_3 \\ \diagup \\ \end{array}$$
$$H \diagup \qquad \diagdown CH_3$$

2-methylpropene

While this molecule is a structural isomer of the 2-butene molecule, it does not have a geometric isomer, because it has two identical groups on at least one of the doubly bonded carbon atoms.

The two requirements for geometric isomerism then are as follows:

1.   Rotation must be restricted in the molecule.
2.   There must be two different groups on *each* of the doubly bonded carbon atoms.

EXAMPLE 9.4   Geometric Isomers

Which of the following compounds can exist as geometric isomers? Explain.

   a. $CH_3CH{=}CHCH_2CH_3$ b. $CH_2{=}CBrCH_3$

Solution

Both structures have a double bond and thus meet rule 1 for geometric isomerism.

   a. Structure **(a)** meets rule 2; it has two different groups on *each* carbon (H and CH_3 on one and H and CH_2CH_3 on the other). It exists as *cis* and *trans* isomers:

$$\begin{array}{cc} H_3C & CH_2CH_3 \\ \diagdown \diagup \\ C = C \\ \diagup \diagdown \\ H & H \end{array} \quad \text{and} \quad \begin{array}{cc} H_3C & H \\ \diagdown \diagup \\ C = C \\ \diagup \diagdown \\ H & CH_2CH_3 \end{array}$$

   b. Structure **(b)** has two hydrogen atoms on one of its doubly bonded carbon atoms; it fails rule 2 and does not exist as *cis* and *trans* isomers.

❯ Exercise 9.4A

Are there *cis* and *trans* isomers of 1-butene? Explain.

❯ Exercise 9.4B

Which of the following compounds can exist as geometric isomers? Draw them.

   a. 1-pentene   **b.** 2-pentene   **c.** 3-pentene   **d.** 2-methyl-1-butene

---

SELF-ASSESSMENT Questions

1. Which of the following classes of organic compounds has the simplest composition?
   a. alcohols   **b.** amino acids
   c. carbohydrates   **d.** hydrocarbons

2. Successive members of the alkane series differ from the preceding member by one additional
   a. C atom and one H atom
   b. C atom and two H atoms
   c. C atom and three H atoms
   d. H atom and two C atoms

3. Structural isomers are compounds whose molecules have the same
   a. numbers of C atoms but different numbers of H atoms
   b. numbers of H atoms but different numbers of C atoms
   c. numbers and kinds of atoms but different structures
   d. kinds of atoms but different numbers of these atoms

4. As the molecular mass of the members of the alkane series increases, their boiling points
   a. decrease   **b.** increase
   c. remain the same   **d.** vary randomly

5. Addition reactions occur with alkenes but not with alkanes because alkenes have
   a. double bonds
   b. a greater molecular mass
   c. more C atoms per molecule
   d. a tetrahedral arrangement of bonds

6. Which of the following compounds will *not* form *cis* and *trans* isomers?
   a. $CH_3CH{=}CHCH_2CH_3$ **b.** $CH_2BrCH{=}CHCH_3$
   c. $CH_3CH{=}CHCH_3$ **d.** $CBr_2{=}CHBr$

9.3 Aromatic Compounds: Benzene and Its Relatives

Learning Objectives • Define *aromatic compound*, and recognize the structural features such compounds share. • Name simple aromatic hydrocarbons.

An important and interesting hydrocarbon is benzene. Discovered by Michael Faraday in 1825, benzene has the molecular formula C_6H_6. The structure of benzene puzzled chemists for decades. The formula seems to indicate an unsaturated compound, but benzene does not react as if it contains any double or triple bonds. It does not readily undergo addition reactions the way unsaturated compounds usually do.

Finally, in 1865, German chemist August Kekulé (1829–1896) proposed a structure with a ring of six carbon atoms, each attached to one hydrogen atom.

or

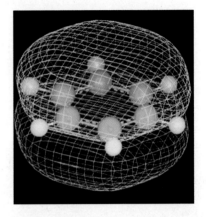

▲ **Figure 9.9** A computer-generated model of the benzene molecule. The six unassigned electrons occupy the yellow and blue areas above and below the plane of the ring of carbon atoms.

The two structures shown appear to contain double bonds, but in fact they do not. Both structures represent the same molecule, and the actual structure of this molecule is a *hybrid* of these two structures. The benzene molecule has six identical carbon-to-carbon bonds that are neither single bonds nor double bonds but something in between. The three pairs of electrons that would form the three double bonds are not tied down in one location but are spread around the ring (Figure 9.9). Today, we usually represent the benzene ring with a circle inside a hexagon.

The hexagon represents the ring of six carbon atoms, and the inscribed circle represents the six unassigned electrons. Because its ring of electrons resists being disrupted, the benzene molecule is exceptionally stable. Since benzene may cause leukemia after long exposure, its use has been restricted. Benzene compounds are classified as *carcinogens*, or compounds that can cause cancer. (Carcinogens will be discussed in greater detail in Chapter 21.) There are thousands of related compounds that contain the benzene ring, and many have also been labeled as carcinogenic. However, many benzene-related compounds are *not* carcinogenic, including common painkillers such as aspirin (Section 9.8) and some flavorings in foods, such as cinnamaldehyde (Section 17.5).

Benzene and similar compounds are called *aromatic hydrocarbons* because many of the first benzene-like substances discovered had strong aromas. Even though other compounds derived from benzene have turned out to be odorless, the name has stuck. Today, an **aromatic compound** is any compound that contains a benzene ring or has certain properties like those of benzene. All other compounds are said to be *aliphatic* (Section 9.2).

Structures of some common aromatic hydrocarbons are shown in Figure 9.10. Benzene, toluene, and the xylenes are all liquids that float on water. They are used mainly as solvents and fuels, but they are also used to make other benzene derivatives (compounds that contain the benzene ring). Naphthalene is a volatile, white, crystalline solid that has been widely used as an insecticide, especially as a moth repellant.

Monosubstituted benzenes—in which one hydrogen is replaced by another atom or group, called a **substituent**—are often represented by condensed formulas; for example, toluene is $C_6H_5CH_3$ and ethylbenzene is $C_6H_5CH_2CH_3$. Similar designations

2 Why are benzene-like compounds that have many multiple bonds called "aromatic"?

Many of these compounds were named before the details of the structure were known. Some of the first ones known have strong sweet smells and would have been described as "aromatic."

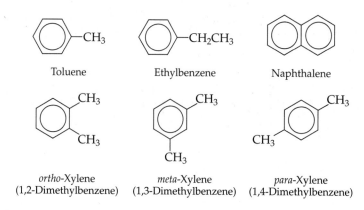

▲ **Figure 9.10** Some aromatic hydrocarbons, like benzene, toluene, and the three xylenes, are components of gasoline and also serve as solvents. All of these compounds are intermediates in the synthesis of polymers (Chapter 10) and other organic chemicals.

are sometimes used for disubstituted aromatic rings, indicating the relative positions of substituents with numbers, as in $1,4\text{-}C_6H_4(CH_3)_2$ for 1,4-dimethylbenzene.

A circle inside a ring indicates that a compound is aromatic, but it does not always mean six electrons. The structural formula for naphthalene (Figure 9.10), for example, has a circle in each ring, indicating that it is an aromatic hydrocarbon. The total number of unassigned electrons, however, is 10. Some chemists still prefer to represent naphthalene (and other aromatic compounds) with alternating double and single bonds.

SELF-ASSESSMENT Questions

1. Which of the following *cannot* be the molecular formula for an unsaturated hydrocarbon?
   a. C_4H_{10} **b.** C_5H_{10}
   c. C_7H_{12} **d.** C_8H_{16}

2. Which compound is *not* an aromatic hydrocarbon?
   a. benzene   **b.** cyclohexene
   c. naphthalene   **d.** toluene

3. A hydrocarbon with a benzene-like structure is called a(n) _____ hydrocarbon.
   a. aliphatic   **b.** aromatic
   c. saturated   **d.** unsaturated

4. Naphthalene is an
   a. alkane   **b.** alkene
   c. alkyne   **d.** aromatic hydrocarbon

5. The reactivity of the benzene ring reflects
   a. its aromatic odor
   b. the presence of carbon–carbon double bonds
   c. its saturated structure
   d. the spreading of six electrons over all six carbon atoms

Answers: 1, a; 2, b; 3, b; 4, d; 5, d

9.4 Halogenated Hydrocarbons: Many Uses, Some Hazards

Learning Objective • Name a halogenated hydrocarbon given its formula, and write the formula for such a compound given its name.

Many other organic compounds are derived from hydrocarbons by replacing one or more hydrogen atoms with another atom or group of atoms. For example, replacement of hydrogen atoms by chlorine atoms gives chlorinated hydrocarbons. When chlorine gas (Cl_2) is mixed with methane gas (CH_4) in the presence of ultraviolet (UV) light, a reaction takes place at a very rapid, even explosive, rate. The result is a mixture of products, some of which may be familiar to you. Several billion kilograms of chlorinated methanes are produced each year for use as solvents and as starting materials for manufacturing other commercially valuable compounds. The following equation shows the reaction in which one hydrogen atom of methane is replaced by a chlorine atom to give methyl chloride (chloromethane).

$$CH_4(g) + Cl_2(g) \longrightarrow CH_3Cl(g) + HCl(g)$$

 EXAMPLE 9.5   **Formulas of Chlorinated Hydrocarbons**

What is the formula for dichloromethane (also known as methylene chloride)?

Solution
The prefix *dichloro-* indicates two chlorine atoms. The rest of the name indicates that the compound is derived from methane (CH_4). We conclude that two hydrogen atoms of methane have been replaced by chlorine atoms; the formula for dichloromethane is therefore CH_2Cl_2.

⟩ Exercise 9.5A
What are the formulas for **(a)** trichloromethane (also known as chloroform) and **(b)** tetrachloromethane (carbon tetrachloride)?

⟩ Exercise 9.5B
What are the formulas for **(a)** tetrafluoromethane and **(b)** difluoromethane?

Methyl chloride is used mainly to make silicone polymers (Chapter 10). Methylene chloride (dichloromethane) is a solvent used, for example, as a paint remover. Chloroform (trichloromethane), also a solvent, was once used as an anesthetic, but such use is now considered dangerous. The dosage required for effective anesthesia is too close to a lethal dose. Carbon tetrachloride (tetrachloromethane) has been used as a dry-cleaning solvent and in fire extinguishers, but it is no longer recommended for either use. Exposure to carbon tetrachloride or any of many other chlorinated hydrocarbons can cause severe damage to the liver. The use of a carbon tetrachloride fire extinguisher in conjunction with water to put out a fire can be deadly. Carbon tetrachloride reacts with water at elevated temperatures to form phosgene ($COCl_2$), an extremely poisonous gas that was used during World War I.

A number of chlorinated hydrocarbons are of considerable interest. For example, vinyl chloride ($CH_2{=}CHCl$) is the starting material for manufacture of polyvinylchloride (PVC, Chapter 10).

Many chlorinated hydrocarbons have similar properties. Most are only slightly polar and do not dissolve in (highly polar) water. Instead, they dissolve (and dissolve in) fats, oils, greases, and other substances of low polarity. This is why certain chlorinated hydrocarbons make good dry-cleaning solvents; they remove grease and oily stains from fabrics. This is also why DDT and PCBs cause problems for fish and birds and perhaps for people; these toxic substances are concentrated in fatty animal tissues rather than being easily excreted in (aqueous) urine. DDT (dichlorodiphenyltrichloroethane) and other chlorinated hydrocarbon insecticides are discussed in Chapter 19. In Chapter 10, we will discuss polychlorinated biphenyls (PCBs).

Chlorofluorocarbons and Fluorocarbons

Compounds containing carbon and both fluorine and chlorine are called *chlorofluorocarbons (CFCs)*. CFCs have been used as the dispersing gases in aerosol cans, for making foamed plastics, and as refrigerants. Three common CFCs are best known by their industrial code designations:

$$CFCl_3 \qquad CF_2Cl_2 \qquad CF_2ClCF_2Cl$$
$$\text{CFC 11} \qquad \text{CFC 12} \qquad \text{CFC 114}$$

At room temperature, CFCs are gases or liquids with low boiling points. They are insoluble in water and inert toward most other substances. These properties make them ideal propellants in aerosol cans for deodorant, hair spray, and food products. Unfortunately, these properties also allow CFCs to migrate into the upper atmosphere, where they participate in chemical reactions that lead to depletion of the ozone layer that protects Earth from harmful UV radiation, and are therefore being phased out. We will look at this problem and some attempts to solve it in Chapter 13.

Fluorinated compounds have found some interesting uses. Some have been used as blood extenders. Oxygen is quite soluble in certain *perfluorocarbons*. (The prefix *per-* means that all hydrogen atoms have been replaced, in this case by fluorine atoms.) These compounds can therefore serve as temporary substitutes for hemoglobin, the oxygen-carrying protein in blood. Perfluoro compounds have been used to treat premature babies, whose lungs are often underdeveloped.

Polytetrafluoroethylene (PTFE, or Teflon®, discussed in Chapter 10) is a perfluorinated polymer. It has many interesting applications because of its resistance to corrosive chemicals and to high temperatures. It is especially noted for its unusual nonstick properties.

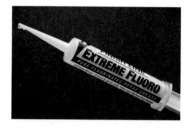

▲ Perfluorinated lubricants are used where highly corrosive or reactive conditions would cause most hydrocarbon greases to burst into flame.

SELF-ASSESSMENT Questions

1. The compound $CHCl_3$ is named
   a. chloroform
   b. methyl chloride
   c. methylene chloride
   d. methyl formate

2. The compound CF_2Cl_2 is
   a. a chlorofluorocarbon
   b. chloroform
   c. methyl chloride
   d. methylene chloride

3. Which of the following is an isomer of 2-chloropropane $(CH_3CHClCH_3)$?
   a. butane
   b. 2-chlorobutane
   c. 1-chloropropane
   d. propane

4. Which of the following is *not* a property of chlorofluorocarbons (CFCs)?
   a. They are inert.
   b. They have low boiling points.
   c. They react with ozone to help deplete the ozone layer.
   d. They are soluble in water.

Answers: 1, a; 2, a; 3, c; 4, d

9.5 Functional and Alkyl Groups

Learning Objectives • Classify an organic compound according to its functional group(s), and explain why the concept of a functional group is useful in the study of organic chemistry. • Recognize and write the formulas of simple alkyl groups.

While the study of organic chemistry focuses on the compounds containing carbon and hydrogen, we saw in Section 9.2 that saturated hydrocarbons do not undergo extensive chemical reactions. Compounds with double and triple carbon–carbon bonds and halogen substituents are examples of **functional groups**, atoms or groups of atoms that give a family of organic compounds its characteristic chemical and

TABLE 9.6 Selected Organic Functional Groups

Name of Class	Functional Group[a]	General Formula of Class
Alkane	None	R—H
Alkene	$\overset{\mid}{-}C=C\overset{\mid}{-}$	$R_2C=CR_2$
Alkyne	$-C\equiv C-$	$RC\equiv CR$
Alcohol	$-\overset{\mid}{\underset{\mid}{C}}-OH$	R—OH
Ether	$-\overset{\mid}{\underset{\mid}{C}}-O-\overset{\mid}{\underset{\mid}{C}}-$	R—O—R'
Thiol	$-\overset{\mid}{\underset{\mid}{C}}-SH$	R—SH
Aldehyde	$-\overset{O}{\overset{\parallel}{C}}-H$	$R-\overset{O}{\overset{\parallel}{C}}-H$
Ketone	$-\overset{O}{\overset{\parallel}{C}}-$	$R-\overset{O}{\overset{\parallel}{C}}-R'$
Carboxylic acid	$-\overset{O}{\overset{\parallel}{C}}-OH$	$R-\overset{O}{\overset{\parallel}{C}}-O-H$
Ester	$-\overset{O}{\overset{\parallel}{C}}-O-\overset{\mid}{\underset{\mid}{C}}-$	$R-\overset{O}{\overset{\parallel}{C}}-O-R'$
Amine	$-\overset{\mid}{\underset{\mid}{C}}-\overset{\mid}{N}-$	$R-\overset{H}{\overset{\mid}{N}}-H \qquad R-\overset{H}{\overset{\mid}{N}}-R' \qquad R-\overset{R'}{\overset{\mid}{N}}-R''$
Amide	$-\overset{O}{\overset{\parallel}{C}}-\overset{\mid}{N}-$	$R-\overset{O}{\overset{\parallel}{C}}-\underset{H}{\overset{\mid}{N}}-H \qquad R-\overset{O}{\overset{\parallel}{C}}-\underset{H}{\overset{\mid}{N}}-R' \qquad R-\overset{O}{\overset{\parallel}{C}}-\underset{R''}{\overset{\mid}{N}}-R'$

[a]Neutral functional groups are shown in green, acidic groups in red, and basic groups in blue.

physical properties. Table 9.6 lists a number of common functional groups found in organic molecules. Each of these groups is the basis for its own homologous series.

In many simple molecules, a functional group is attached as a substituent to a hydrocarbon stem called an *alkyl group*. An **alkyl group** is derived from an alkane by removing a hydrogen atom. The methyl group (CH_3—), for example, is derived from methane (CH_4), and the ethyl group (CH_3CH_2—) from ethane (CH_3CH_3). Propane yields two different alkyl groups, depending on whether the hydrogen atom is removed from an end carbon or from the middle carbon atom. (See Table 9.7.) Table 9.7 also lists the four alkyl groups derived from the two butanes. The names for alkyl groups are based on those of the original alkane, with the -*e* ending changed to -*yl*.

Often the letter R is used to stand for an alkyl group in general. Thus, ROH is a general formula for an alcohol, and RCl represents any alkyl chloride.

There is a major distinction between a functional group and an alkyl group. The functional group dictates the chemical reactions that a compound will undergo, while the molecule's alkyl group has little or no effect on how the molecule will react.

TABLE 9.7 Common Alkyl Groups

	Name	Structural Formula	Condensed Structural Formula
Derived from Propane	Propyl	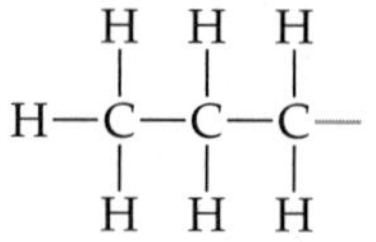	$CH_3CH_2CH_2-$
	Isopropyl	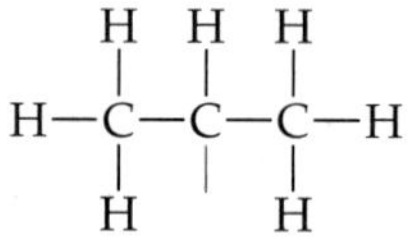	CH_3CHCH_3
Derived from Butane	Butyl	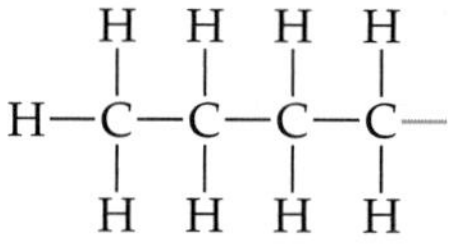	$CH_3CH_2CH_2CH_2-$
	Secondary butyl (*sec*-butyl)	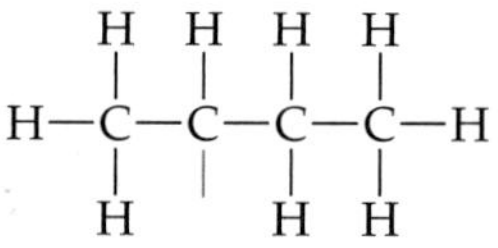	$CH_3CHCH_2CH_3$
Derived from Isobutane	Isobutyl	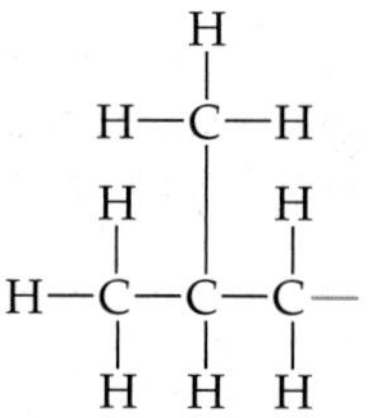	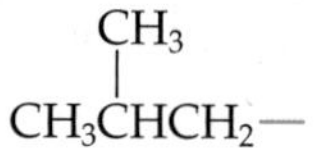 CH_3CHCH_2- with CH_3
	Tertiary butyl (*tert*-butyl)	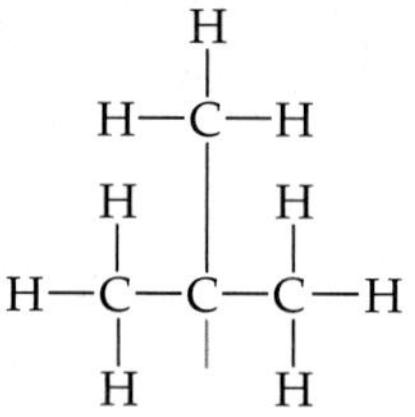	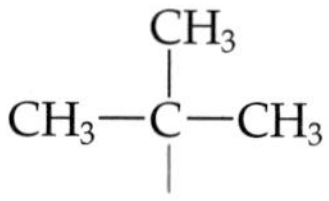 CH_3-C-CH_3 with CH_3

SELF-ASSESSMENT Questions

1. A specific structural arrangement of atoms or bonds that imparts characteristic properties to a molecule is called a(n)
   a. catalyst
   b. empirical formula
   c. functional group
   d. substituent

2. The formula $CH_3CH_2CH_2-$ represents
   a. an alkyl group
   b. a functional group
   c. propane
   d. propylene

3. How many carbon atoms are there in an ethyl group?
   a. 2
   b. 3
   c. 4
   d. 6

4. The formula $CH_3CH_2CH_2CH_2-$ represents a(n)
   a. butyl group
   b. isobutyl group
   c. isopropyl group
   d. propyl group

9.6 Alcohols, Phenols, Ethers, and Thiols

Learning Objectives • Recognize the general structure and properties for an alcohol, a phenol, an ether, and a thiol. • Name simple alcohols, phenols, ethers, and thiols.

When a hydroxyl group (—OH) is substituted for any hydrogen atom in an alkane (R—H), the molecule becomes an **alcohol**. Like many organic compounds, alcohols can be called by their common names or by their systematic IUPAC names. The IUPAC names for alcohols are based on those of alkanes, with the ending changed from *-e* to *-ol*. Methanol, ethanol, and 1-propanol are the first three members of the homologous series of straight-chain alcohols.

For many industries and businesses, the synthesis of alcohols represents the starting point for the production of important chemical products. Lower costs for these starting materials can lead to lower costs for consumers. Likewise, the use of alcohols as gasoline additives can lead to a smaller environmental impact.

A number designates the carbon to which the (—OH) group is attached when more than one location is possible. For example, the *1* in 1-propanol indicates that the (—OH) is attached to an end carbon atom of the three-carbon chain. The isomeric compound 2-propanol (also called isopropyl alcohol) has the (—OH) group on the second (middle) carbon atom of the chain.

			CH_3CHCH_3
$CH_3—OH$	$CH_3CH_2—OH$	$CH_3CH_2CH_2—OH$	$\overset{\|}{OH}$
Methanol	Ethanol	1–Propanol	2–Propanol

Common names for these alcohols are methyl alcohol, ethyl alcohol, propyl alcohol, and isopropyl alcohol. The IUPAC names are the ones used in most scientific literature.

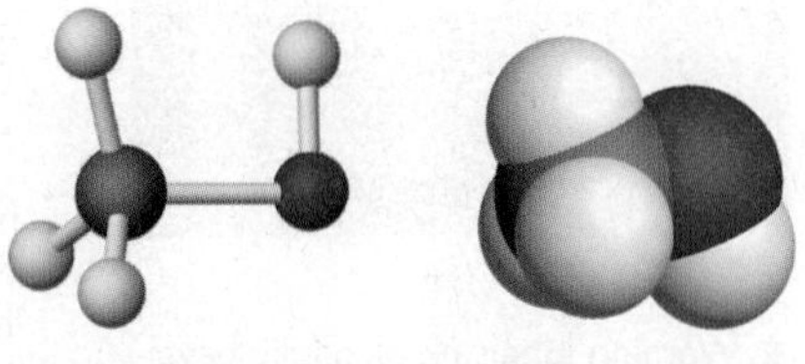

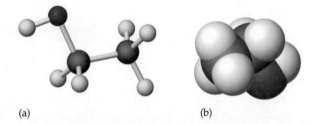

(a)   (b)

▲ **Figure 9.11** (a) Ball-and-stick and (b) space-filling models of methanol.

Methyl Alcohol (Methanol)

The simplest alcohol is *methanol,* or *methyl alcohol* (Figure 9.11). Methanol is sometimes called *wood alcohol* because it was once made from wood. The modern industrial process makes methanol from carbon monoxide and hydrogen at high temperature and pressure in the presence of a catalyst.

$$CO(g) + 2\,H_2(g) \longrightarrow CH_3OH(l)$$

Methanol is important as a solvent and as a chemical intermediate. It is also a gasoline additive and a potential replacement for gasoline in automobiles.

Ethyl Alcohol (Ethanol)

The next member of the homologous series of alcohols is *ethyl alcohol* (CH_3CH_2OH), also called ethanol or *grain alcohol* (Figure 9.12). Ethanol is made by fermentation of grain (or other starchy or sugary materials). If the sugar is glucose, the reaction is

▲ **Figure 9.12** (a) Ball-and-stick and (b) space-filling models of ethanol.

$$C_6H_{12}O_6(aq) \xrightarrow{\text{yeast}} 2\,CH_3CH_2OH(aq) + 2\,CO_2(g)$$

Ethanol for beverages and for automotive fuel is made in this way. Ethanol for industrial use is made by reacting ethylene with water.

$$CH_2{=}CH_2(g) + H_2O(l) \xrightarrow{H^+} CH_3CH_2OH(l)$$

This industrial alcohol is identical to that made by fermentation and is generally cheaper, but by law, it cannot be used in alcoholic beverages. Because it carries no excise tax, the law requires that noxious substances be added to this

The *proof* of an alcoholic beverage is twice the percentage of alcohol by volume. The term originated in a seventeenth-century English method for testing whiskey to ensure that a dealer was not increasing profits by adding water to the booze. A qualitative test was to pour some of the whiskey on gunpowder and ignite it. Ignition of the gunpowder after the alcohol had burned away was considered "proof" that the whiskey did not contain too much water.

▲ Warning labels on alcoholic beverages alert consumers to the hazards of consumption.

alcohol to prevent people from drinking it. The resulting *denatured alcohol* is not fit to drink. This is the kind of ethanol commonly found on the shelves in chemical laboratories.

Gasoline in many parts of the United States contains up to 10% ethanol that is produced by fermentation. Because the United States has a large corn (maize) surplus, there is a generous government subsidy for gasoline producers who use ethanol made this way. As a result, many factories now make ethanol by fermentation of corn.

Toxicity of Alcohols

Although ethanol is an ingredient in wine, beer, and other alcoholic beverages, alcohols in general are rather toxic. Toxicity depends on the nature of the substance, the amount, and the route by which it is taken into the body. (Toxicity is discussed in detail in Chapter 21.) Methanol, for example, is oxidized in the body to formaldehyde (HCHO). Drinking as little as 1 oz (about 30 mL or 2 tablespoonsful) can cause blindness and even death. Several poisonings each year result when people mistake methanol for the less toxic ethanol.

Ethanol is not as poisonous as methanol, but it is still toxic. About 50,000 cases of ethanol poisoning are reported each year in the United States, and there are over 2200 deaths from acute ethanol poisoning annually.

Generally, ethanol acts as a mild depressant; it slows down both physical and mental activity. Table 9.8 lists the effects of various doses. Although ethanol generally is a depressant, small amounts of it seem to act as a stimulant, perhaps by relaxing tensions and relieving inhibitions.

Ingesting too much ethanol over a long period of time can alter brain-cell function, cause nerve damage, and shorten life span by contributing to diseases of the liver, cardiovascular system, and practically every other organ of the body. Further, about half of all fatal automobile accidents involve at least one drinking driver. Babies born to mothers who abuse alcohol often are small and deformed, and exhibit cognitive impairment and developmental delays. Some investigators believe that this *fetal alcohol syndrome* can occur even if mothers drink only moderately. Ethanol, by far the most abused drug in the United States, is discussed more fully in Chapter 18.

TABLE 9.8 Approximate Relationship among Drinks Consumed, Blood Alcohol Level, and Behavior[a]

Number of Drinks[b]	Blood Alcohol Level (percent by volume)	Behavior[c]
2	0.05	Mild sedation; tranquility
4	0.10	Lack of coordination
6	0.15	Obvious intoxication
10	0.30	Unconsciousness
20	0.50	Possible death

[a]Data are for a 70 kg (154 lb.) moderate drinker.
[b]Rapidly consumed 30 mL (1 oz.) shots of 90 proof whiskey, 360 mL (12 oz.) bottles of beer, or 150 mL (5 oz.) glasses of wine.
[c]An inexperienced drinker would be affected more strongly, or more quickly, than someone who is ordinarily a moderate drinker. Conversely, an experienced heavy drinker would be affected less. However, it is important to remember that whether a person is driving while intoxicated is legally determined by one's blood alcohol level and not one's behavior.

Rubbing alcohol is a 70% solution of isopropyl alcohol. Because isopropyl alcohol is also more toxic than ethanol, it is not surprising that people become ill, and sometimes die, after drinking rubbing alcohol.

Multifunctional Alcohols

Several alcohols have more than one hydroxyl group. Examples are ethylene glycol, propylene glycol, and glycerol.

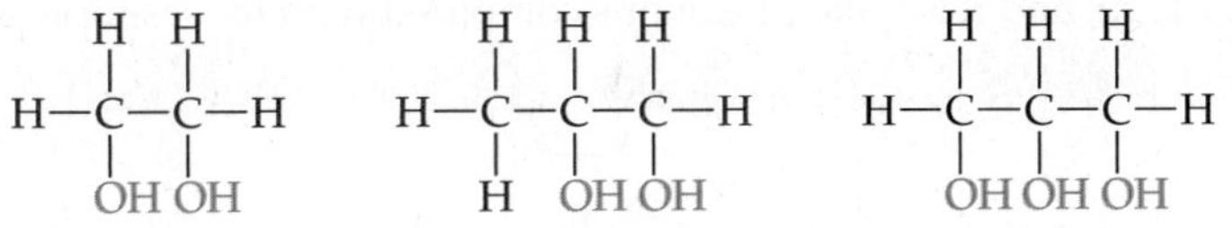

Ethylene glycol   Propylene glycol   Glycerol

Ethylene glycol is the main ingredient in many permanent antifreeze mixtures. Its high boiling point keeps it from boiling away in automobile radiators. Ethylene glycol is a syrupy liquid with a sweet taste, but it is quite toxic. It is oxidized in the liver to oxalic acid.

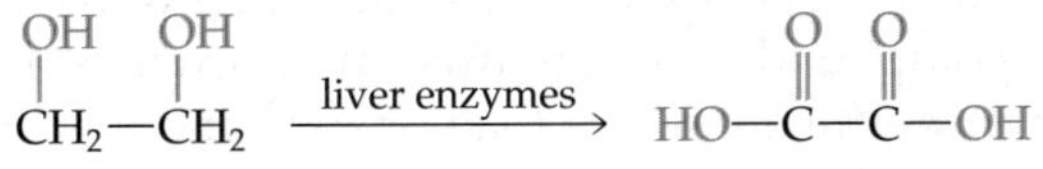

$$\underset{\text{Ethylene glycol}}{\overset{\overset{\displaystyle OH \quad OH}{|\qquad|}}{CH_2-CH_2}} \xrightarrow{\text{liver enzymes}} \underset{\substack{\text{Oxalic acid} \\ \text{(toxic)}}}{\overset{\overset{\displaystyle O \quad O}{\parallel \quad \parallel}}{HO-C-C-OH}}$$

Oxalic acid forms crystals of its calcium salt, calcium oxalate (CaC_2O_4), which can damage the kidneys, leading to kidney failure and death. Propylene glycol, a high-boiling substance that is not poisonous, is now marketed as a safer permanent antifreeze.

Glycerol (or glycerin) is a sweet, syrupy liquid that is a by-product from fats during soap manufacture (Chapter 20). It is used in lotions to keep the skin soft and as a food additive to keep cakes moist. It reacts with nitric acid to make nitroglycerin, the explosive material in dynamite. Nitroglycerin is also important as a vasodilator, a medication taken by heart patients to relieve angina pain.

WHY IT MATTERS

Antifreeze containing ethylene glycol should always be drained into a container and disposed of properly. Pets have been known to die from licking the sweet, but toxic, ethylene glycol mixture from a driveway or open container.

EXAMPLE 9.6   Structural Formulas of Alcohols

Write the structural formula for *tert*-butyl alcohol, sometimes used as an octane booster in gasoline.

Solution

An alcohol consists of an alkyl group joined to a hydroxyl group. From Table 9.7, we see that the *tert*-butyl group is

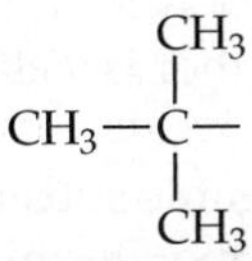

$$CH_3-\underset{\overset{\displaystyle |}{CH_3}}{\overset{\overset{\displaystyle CH_3}{|}}{C}}-$$

Connecting this group to an OH group gives *tert*-butyl alcohol.

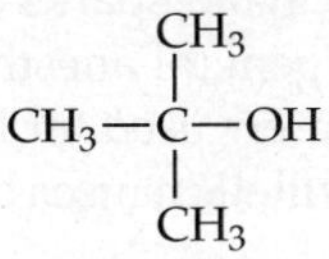

$$CH_3-\underset{\overset{\displaystyle |}{CH_3}}{\overset{\overset{\displaystyle CH_3}{|}}{C}}-OH$$

❯ **Exercise 9.6A**

Write the structural formulas for **(a)** butyl alcohol and **(b)** *sec*-butyl alcohol.

❯ **Exercise 9.6B**

Write the structural formulas for **(a)** propyl alcohol and **(b)** isopropyl alcohol.

Phenols

When a hydroxyl group is attached to a benzene ring, the compound is called a **phenol**. Although a phenol may appear to be an alcohol, it is not. The benzene ring greatly alters the properties of the hydroxyl group. Unlike alcohols, phenol is a weak acid (sometimes called *carbolic acid*), and it is quite poisonous compared with most simple alcohols.

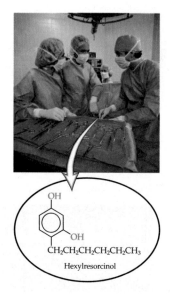

▲ Phenolic compounds such as hexyl-resorcinol help to ensure antiseptic conditions in hospital operating rooms.

Phenol was the first antiseptic used in an operating room—by Joseph Lister (England, 1827–1912) in 1867. Until that time, surgery was not antiseptic, and many patients died from infections following surgical operations. Although phenol has a strong germicidal action, it is far from an ideal antiseptic, because it causes severe skin burns and kills healthy cells along with harmful microorganisms.

Phenol is still sometimes employed as a disinfectant for floors and furniture, but other phenolic compounds are now used as antiseptics. Hexylresorcinol, for example, is a more powerful germicide than phenol, and it is less damaging to the skin and has fewer other side effects.

Many phenols also act as antioxidants by stabilizing free radicals and stopping the breakdown of long chain carboxylic acids (Section 9.8). Vitamin E is a natural antioxidant. Synthetic antioxidants, such as BHT (butylated hydroxytoluene) or BHA (butylated hydroxyanisole), are often added to products to retard spoilage.

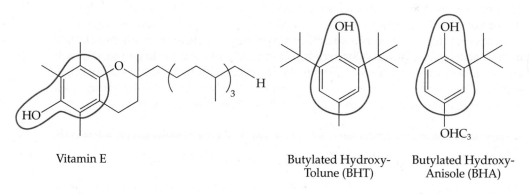

Vitamin E

Butylated Hydroxy-Tolune (BHT)

Butylated Hydroxy-Anisole (BHA)

▲ Some common antioxidants. The phenol rings, which help stabilize free radicals, are circled in red.

Ethers

Compounds with two alkyl or aromatic groups attached to the same oxygen atom are called **ethers**. The general formula is ROR (or ROR′, because the alkyl groups need not be alike). The best-known ether is *diethyl ether* ($CH_3CH_2OCH_2CH_3$), often called simply *ether*.

Diethyl ether dissolves many organic substances that are insoluble in water. It boils at 36 °C, which means that it evaporates readily, making it easy to recover dissolved materials. Although diethyl ether has little chemical reactivity, it is highly flammable. Great care must be taken to avoid sparks or flames when it is in use. Diethyl ether was introduced in 1842 as a general anesthetic for surgery and was once the most widely used anesthetic. It is rarely used for humans today because it has undesirable side effects, such as postanesthetic nausea and vomiting. We will discuss modern anesthesia in Chapter 18.

Another problem with ethers is that they can react slowly with oxygen to form unstable peroxides, which may decompose explosively. Beware of previously opened containers of ether, especially old ones.

The ether produced in the largest amount commercially is a cyclic compound called *ethylene oxide*. Its two carbon atoms and one oxygen atom form a three-membered ring. Like cyclic hydrocarbons, cyclic ethers can be represented by geometric shapes. Each corner of such a shape (except the O for an oxygen atom) stands for a carbon atom with enough hydrogen atoms attached to give the carbon four bonds. Ethylene oxide is represented as

Ethylene oxide is a toxic gas and is mainly used to make ethylene glycol.

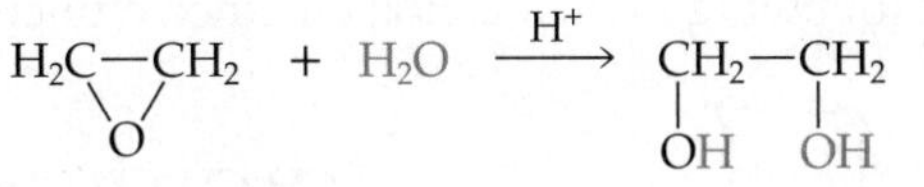

$$H_2C-CH_2 \ + \ H_2O \ \xrightarrow{H^+} \ CH_2-CH_2$$

Ethylene oxide	Water	Ethylene glycol

Ethylene oxide is also used to sterilize medical instruments and as an intermediate in the synthesis of some detergents (Chapter 20). Most ethylene glycol produced is used for making polyester fibers (Chapter 10) and antifreeze.

Thiols

When a sulfhydryl group (—SH) is substituted for any hydrogen atom in an alkane, the molecule becomes a **thiol.** Like many organic compounds, thiols can be called by their common names or by systematic IUPAC names. The IUPAC names for thiols are based on those of alkanes, with the ending -*thiol* added to the name of the alkane. For example, the IUPAC name for CH_3SH would be methanethiol. Many small thiols are commonly named as *mercaptans,* CH_3SH is methyl mercaptan.

Small thiols generally have very unpleasant odors, such as those emanating from raw sewage or sprayed by skunks. This property can be used in the detection of natural gas, which is normally odorless. A trace amount of ethanethiol added to natural gas makes it easier to detect by smell. Thiols can be oxidized to form disulfide groups (—S—S—), which play a major role in stabilizing protein structures. (See Chapter 16.)

 CONCEPTUAL EXAMPLE 9.7   Identifying Functional Groups

Classify each of the following as an alcohol, an ether, a phenol, or a thiol.

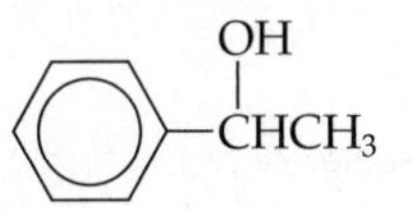

(a)

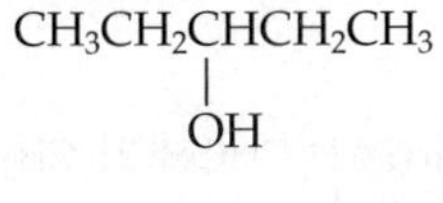

(b) $CH_3CH_2CHCH_2CH_3$ with OH below

(c) $CH_3CH_2OCH_2CH_3$

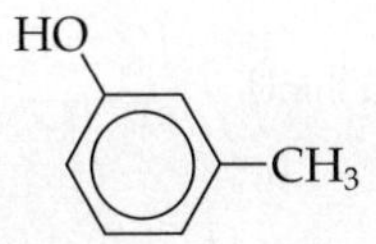

(d)

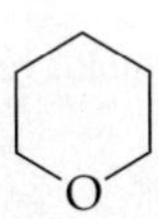

(e)

(f) $CH_3CH_2CH_2CH_2SH$

Solution

  a. The compound is an alcohol; the OH is not attached directly to the benzene ring.
  b. The OH group is attached to an alkyl group; the compound is an alcohol.
  c. The O atom is between two alkyl groups; the compound is an ether.
  d. The OH functional group is attached directly to the benzene ring; the compound is a phenol.
  e. The compound is an ether; the O atom is between two C atoms.
  f. The SH group is attached to an alkyl group; the compound is a thiol.

> **Exercise 9.7A**

Classify each of the following as an alcohol, an ether, a phenol, or a thiol.

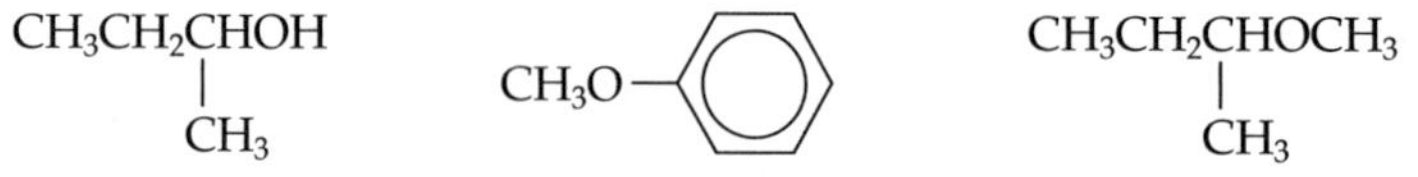

CH₃CH₂CHOH
     |
     CH₃

(a)

CH₃O—⬡

(b)

CH₃CH₂CHOCH₃
       |
       CH₃

(c)

> **Exercise 9.7B**

Classify each of the following as an alcohol, an ether, a phenol, or a thiol.

HO—⬡
    |
    Br

(a)

▭=O

(b)

CH₃CH₂CH₂CH₂CH₂SH

(c)

EXAMPLE 9.8 Formulas for Ethers

What is the formula for isopropyl methyl ether?

Solution

Isopropyl methyl ether has an oxygen atom joined to an isopropyl group (three carbons joined to oxygen by the middle carbon) and a methyl group. The formula is

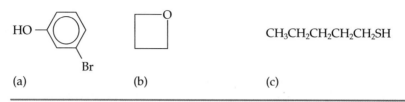

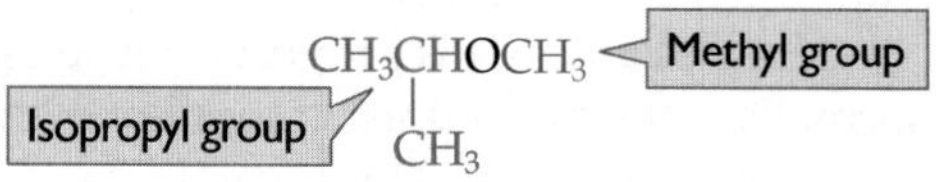

> **Exercise 9.8A**

Write formulas for **(a)** butyl methyl ether and **(b)** ethyl *tert*-butyl ether.

> **Exercise 9.8B**

Write formulas for **(a)** methyl pentyl ether and **(b)** hexyl propyl ether.

SELF-ASSESSMENT Questions

1. Which of the following formulas represents an alcohol?
  a. CH₃CHOHCH₂CH₃   **b.** CH₃COCH₃
  c. CdOH   **d.** CH₃CHO

2. The formula for ethylene glycol is
  a. CH₃CH₂OH   **b.** CH₃CH(OH)₂
  c. C₃H₅(OH)₃   **d.** HOCH₂CH₂OH

3. Which of these alcohols has the longest carbon chain?
  a. 1-butanol   **b.** 2-butanol
  c. 2-pentanol   **d.** 2-propanol

4. The general formula R—O—R′ represents a(n)
  a. ester   **b.** ether   **c.** ketone   **d.** phenol

5. The formula CH₃CH₂OCH₂CH₃ represents
  a. 2-butanol
  b. 2-butanone
  c. butyraldehyde
  d. diethyl ether

6. The compound CH₃CHOHCH₂CH₃ is an isomer of
  a. CH₃COCH₂CH₂CH₃
  b. CH₃CH₂CH₂COOH
  c. CH₃COOCH₂CH₃
  d. CH₃CH₂CH₂CH₂OH

7. Which of the following formulas represents a thiol?
  a. CH₃CHOHCH₂CH₃
  b. CH₃COCH₃
  c. CH₃CH₂SH
  d. CH₃CHO

8. An organic compound with a hydroxyl (OH) functional group on an aromatic carbon atom is called a(n)
  a. alcohol
  b. aldehyde
  c amine
  d. phenol

Answers: 1. a; 2. d; 3. c; 4. b; 5. d; 6. d; 7. c; 8. d

9.7 Aldehydes and Ketones

Learning Objectives • Recognize the general structure for simple aldehydes and ketones and list their important properties. • Name simple aldehydes and ketones.

Two families of organic compounds that contain the **carbonyl group** (C=O) are **aldehydes** and **ketones**. Aldehydes have at least one hydrogen atom attached to the carbonyl carbon (and are at the end of a hydrocarbon chain), whereas ketones have two other carbon atoms joined to the carbonyl carbon.

$-\overset{\displaystyle O}{\underset{\displaystyle \|}{C}}-$	$R-\overset{\displaystyle O}{\underset{\displaystyle \|}{C}}-H$	$R-\overset{\displaystyle O}{\underset{\displaystyle \|}{C}}-R'$
A carbonyl group	An aldehyde	A ketone

To simplify typing, these structures are often written on one line with the C=O double bond omitted.

—C(=O)— or —CO—	R—CH(=O) or R—CHO	R—C(=O)—R' or R—CO—R'
A carbonyl group	An aldehyde	A ketone

Models of three familiar carbonyl compounds are shown in Figure 9.13.

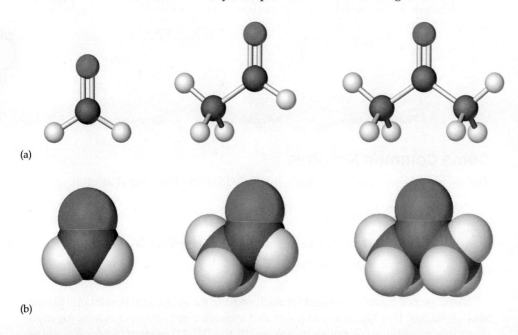

(a)

(b)

▲ **Figure 9.13** (a) Ball-and-stick and (b) space-filling models of formaldehyde (left), acetaldehyde (center), and acetone (right).

Some Common Aldehydes

The simplest aldehyde is *formaldehyde* (HCHO). It is a gas at room temperature but is readily soluble in water. As a 40% solution called *formalin*, it is used as a preservative for biological specimens and in embalming fluid. Systematic names for aldehydes are based on those of alkanes, with the ending changed from *-e* to *-al*. Thus, in the IUPAC system, formaldehyde is named methanal. IUPAC names are seldom used

WHY IT MATTERS

A formaldehyde solution is frequently used as a fixative to preserve biological specimens, such as those that you see in jars in biology labs. The aldehyde functional group reacts with proteins to form covalent bonds and crosslinks between protein molecules. This stops further biological activity and preserves the tissue, which can then be cut into very thin sections for microscopic examination and clinical diagnosis. Formaldehyde and glutaraldehyde (CHOCH₂CH₂CH₂CHO) are widely used in embalming fluid. Glutaraldehyde is also added to hydraulic fracking liquids as a biocide to prevent bacterial growth.

for simple aldehydes, because of possible confusion with the corresponding alcohols. For example, "methan*al*" is easily confused with "methan*ol*," when spoken or (especially) when handwritten.

Formaldehyde is used in making certain plastics (Chapter 10). It also is used to disinfect homes, ships, and warehouses. Commercially, formaldehyde is made by the oxidation of methanol. The same net reaction occurs in the human body when methanol is ingested, and accounts for the high toxicity of that alcohol.

Organic chemists often write equations that show only the organic reactants and products. Inorganic substances are omitted for the sake of simplicity.

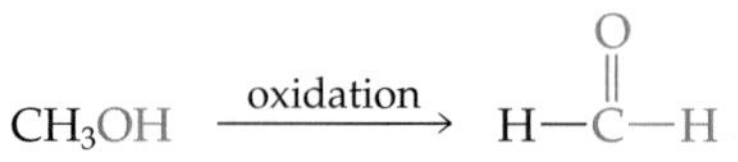

$$CH_3OH \xrightarrow{\text{oxidation}} \underset{\text{Formaldehyde}}{H-\overset{\overset{\displaystyle O}{\|}}{C}-H}$$

Methanol

The next member of the homologous series of aldehydes is acetaldehyde (ethanal), formed by the oxidation of ethanol.

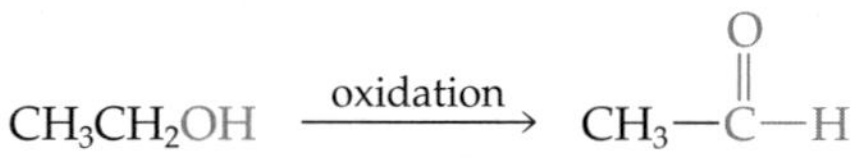

$$CH_3CH_2OH \xrightarrow{\text{oxidation}} \underset{\text{Acetaldehyde}}{CH_3-\overset{\overset{\displaystyle O}{\|}}{C}-H}$$

Ethanol

The next two members of the aldehyde series are propionaldehyde (propanal) and butyraldehyde (butanal). Both have strong, unpleasant odors. Benzaldehyde has an aldehyde group attached to a benzene ring. Also called synthetic oil of almond, benzaldehyde is commonly used in perfumes and is the flavoring ingredient in maraschino cherries.

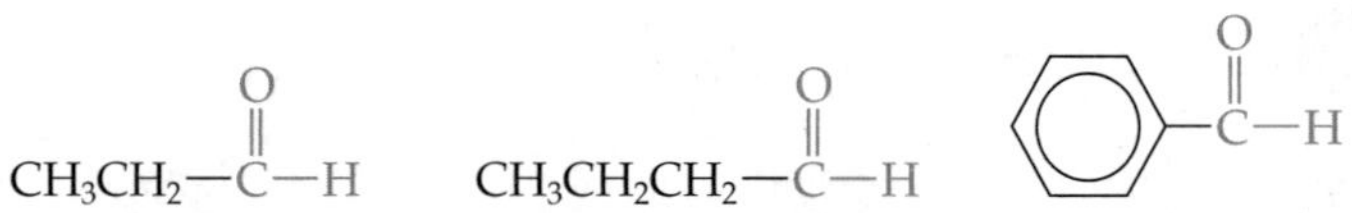

$$\underset{\text{Propionaldehyde}}{CH_3CH_2-\overset{\overset{\displaystyle O}{\|}}{C}-H} \qquad \underset{\text{Butyraldehyde}}{CH_3CH_2CH_2-\overset{\overset{\displaystyle O}{\|}}{C}-H} \qquad \underset{\text{Benzaldehyde}}{\text{⬡}-\overset{\overset{\displaystyle O}{\|}}{C}-H}$$

Some Common Ketones

The simplest ketone is *acetone*, made by the oxidation of isopropyl alcohol.

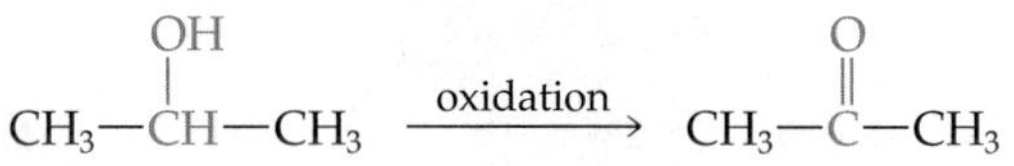

$$\underset{\text{Isopropyl alcohol}}{CH_3-\overset{\overset{\displaystyle OH}{|}}{CH}-CH_3} \xrightarrow{\text{oxidation}} \underset{\text{Acetone}}{CH_3-\overset{\overset{\displaystyle O}{\|}}{C}-CH_3}$$

Acetone is a common solvent for such organic materials as fats, rubbers, plastics, and varnishes. It is also used in paint and varnish removers and is a major ingredient in some fingernail polish removers. In the IUPAC system, acetone is named propanone. The systematic names for ketones are based on those of alkanes, with the ending changed from -*e* to -*one*. When necessary, a number is used to indicate the location of the carbonyl group. For example, $CH_3CH_2COCH_2CH_2CH_3$ is 3-hexanone.

Two other familiar ketones are ethyl methyl ketone and isobutyl methyl ketone, which, like acetone, are frequently used as solvents.

A ketone is a carbon chain with a carbonyl inside. If the carbonyl is on the end, then it's an aldehyde.

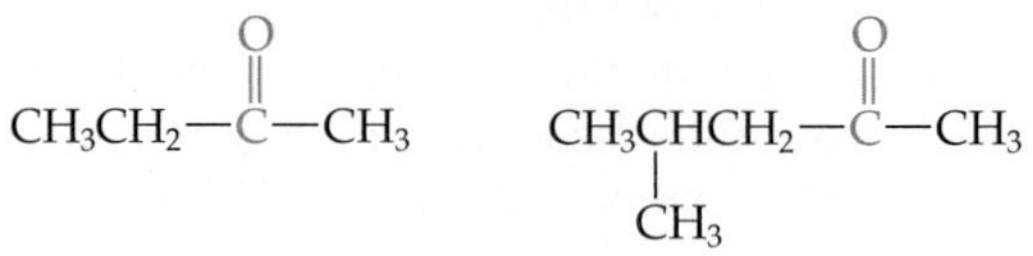

$$\underset{\text{Ethyl methyl ketone}}{CH_3CH_2-\overset{\overset{\displaystyle O}{\|}}{C}-CH_3} \qquad \underset{\text{Isobutyl methyl ketone}}{\underset{\overset{\displaystyle |}{CH_3}}{CH_3CHCH_2}-\overset{\overset{\displaystyle O}{\|}}{C}-CH_3}$$

CONCEPTUAL EXAMPLE 9.9 Aldehyde or Ketone?

Identify each of the following compounds as an aldehyde or a ketone.

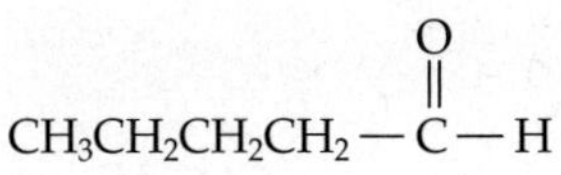

(a) $CH_3CH_2CH_2CH_2-\overset{\displaystyle O}{\overset{\|}{C}}-H$ (b) 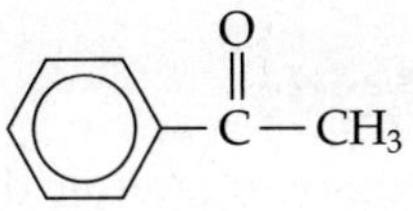   (c)

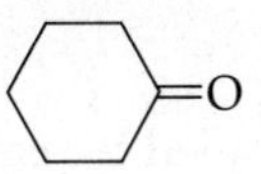

Solution

a. A hydrogen atom is attached to the carbonyl carbon atom; the compound is an aldehyde.
b. The carbonyl group is between two other (ring) carbon atoms; the compound is a ketone.
c. The compound is a ketone. (Remember that the corner of the hexagon stands for a carbon atom.)

› Exercise 9.9A

Identify each of the following compounds as an aldehyde or a ketone.

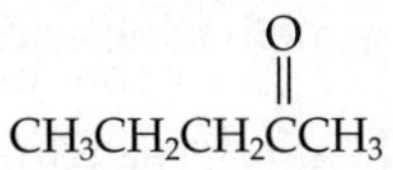

(a) $CH_3CH_2CH_2\overset{\displaystyle O}{\overset{\|}{C}}CH_3$ (b)

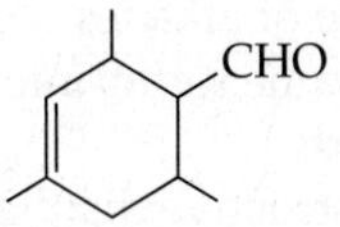

› Exercise 9.9B

Identify each of the following compounds as an aldehyde or a ketone.

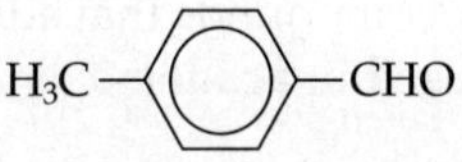

(a)   (b) $CH_3CH_2COCH_2OH$

SELF-ASSESSMENT Questions

1. The aldehyde functional group has a carbonyl group attached to
 a. an H atom
 b. an N atom
 c. an OH group
 d. two other C atoms

2. The ketone functional group has a carbonyl group attached to
 a. an OH group
 b. an OR group
 c. two other C atoms
 d. two OR groups

3. The general formula R—CO—R′ represents a(n)
 a. ester
 b. ether
 c. ketone
 d. phenol

4. Which of the following formulas represents an aldehyde?
 a. $CH_3CH_2CH_2OH$ **b.** CH_3COOCH_3
 c. $CH_2CH_2CH_2COOH$ **d.** $CH_3CH_2CH_2CHO$

5. A name for $CH_3COCH_2CH_3$ is
 a. butanone
 b. butyraldehyde
 c. propanone
 d. propionaldehyde

6. Which alcohol is oxidized to produce formaldehyde?
 a. CH_3OH
 b. CH_3CH_2OH
 c. $CH_3CH_2CH_2OH$
 d. $CH_3CHOHCH_3$

7. Which alcohol is oxidized to produce acetone?
 a. CH_3OH
 b. CH_3CH_2OH
 c. $CH_3CH_2CH_2OH$
 d. $CH_3CHOHCH_3$

9.8 Carboxylic Acids and Esters

Learning Objectives • Recognize the general structure for simple carboxylic acids and esters and list their important properties. • Name simple carboxylic acids and esters.

Carboxylic Acids

The functional group of organic acids is called the **carboxyl group (—COOH)**, and the acids are called **carboxylic acids**. The carboxyl group is always at the end of the hydrocarbon chain.

$$\overset{\overset{\textstyle O}{\|}}{-C}-OH \qquad \overset{\overset{\textstyle O}{\|}}{R-C}-OH$$

A carboxyl group   A carboxylic acid

As with aldehydes and ketones, these formulas are often written on one line:

$$-C(=O)OH \ \text{ or } \ -COOH \qquad R-C(=O)OH \ \text{ or } \ R-COOH$$

A carboxyl group   A carboxylic acid

The simplest carboxylic acid is formic acid (HCOOH). The systematic (IUPAC) names for carboxylic acids are based on those of alkanes, with the ending changed from *-e* to *-oic acid*. For example, the IUPAC name for formic acid is methanoic acid, and $CH_3CH_2CH_2CH_2COOH$ is pentanoic acid.

Formic acid was first obtained by the destructive distillation of ants. (The Latin word *formica* means "ant.") It smarts when an ant bites because the ant injects formic acid into the skin. The stings of wasps and bees also contain formic acid (as well as other poisonous compounds).

Acetic acid (ethanoic acid, CH_3COOH) can be made by the fermentation of a mixture of cider and honey in the presence of air. This produces a solution (vinegar) containing about 4% to 10% acetic acid plus a number of other compounds that add flavor. Acetic acid is probably the most familiar weak acid used in academic and industrial chemistry laboratories.

The third member of the homologous series of acids, propionic acid (propanoic acid, CH_3CH_2COOH), is seldom encountered in everyday life, but the fourth member is more familiar, at least by its odor. If you've ever smelled rancid butter, you know the odor of butyric acid (butanoic acid, $CH_3CH_2CH_2COOH$). It is one of the most foul-smelling substances imaginable. Butyric acid can be isolated from butterfat or synthesized in the laboratory. It is one of the ingredients of body odor, and extremely small quantities of this acid and other chemicals enable bloodhounds to track lost people and fugitives.

The acid with a carboxyl group attached directly to a benzene ring is called *benzoic acid*.

▲ The irritation from an ant bite is caused by the injection of formic acid.

▲ Acetic acid is a familiar weak acid in chemistry laboratories. It is also the principal active ingredient in vinegar.

$$\overset{\overset{\textstyle O}{\|}}{\underset{\bigcirc}{}C}-OH$$

Benzoic acid

Carboxylic acid salts—calcium propionate, sodium benzoate, and others—are widely used as food additives to prevent mold (Chapter 17).

 EXAMPLE 9.10 **Structural Formulas of Oxygen-Containing Organic Compounds**

Write the structural formula for each of the following compounds.

   a. propionaldehyde   **b.** ethanoic acid   **c.** ethyl methyl ketone

Solution

a. Propionaldehyde has three carbon atoms with an aldehyde functional group.

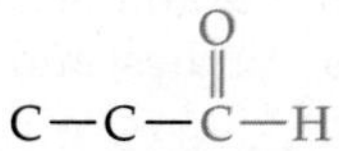

$$\underset{C-C-\overset{\displaystyle \overset{O}{\|}}{C}-H}{}$$

Adding the proper number of hydrogen atoms to the other two carbon atoms gives the structure

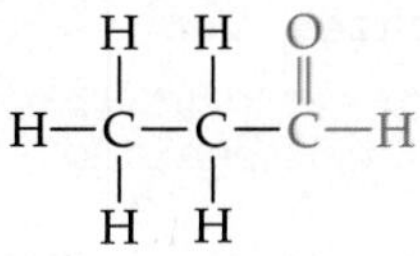

$$\underset{\displaystyle \underset{H}{\overset{H}{|}}\ \underset{H}{\overset{H}{|}}}{H-\overset{\overset{H}{|}}{C}-\overset{\overset{H}{|}}{C}-\overset{\overset{O}{\|}}{C}-H}\quad \text{or}\quad CH_3CH_2CHO$$

b. Ethanoic acid has two carbon atoms with a carboxylic acid functional group.

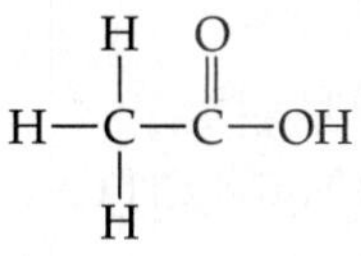

$$\underset{\displaystyle \underset{H}{\overset{H}{|}}}{H-\overset{\overset{H}{|}}{C}-\overset{\overset{O}{\|}}{C}-OH}\quad \text{or}\quad CH_3COOH$$

c. Ethyl methyl ketone has a ketone functional group between an ethyl and a methyl group.

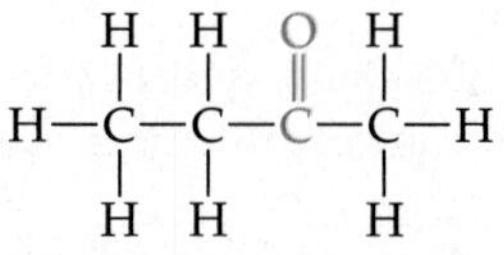

$$\underset{\displaystyle \underset{H}{\overset{H}{|}}\ \underset{H}{\overset{H}{|}}\ \ \ \underset{H}{\overset{H}{|}}}{H-\overset{\overset{H}{|}}{C}-\overset{\overset{H}{|}}{C}-\overset{\overset{O}{\|}}{C}-\overset{\overset{H}{|}}{C}-H}\quad \text{or}\quad CH_3CH_2COCH_3$$

> **Exercise 9.10A**
Write the structural formula for each of the following compounds.

  a. butyric acid   b. acetaldehyde   c. diethyl ketone

> **Exercise 9.10B**
Write the structural formula for each of the following compounds.

  a. 3-octanone   b. heptanal   c. hexanoic acid

Esters: Sweet-Smelling Compounds

Esters (RCOOR′) are derived from carboxylic acids and alcohols or phenols. The general reaction involves splitting out a molecule of water. Note that the two alkyl groups may be different and that they are represented as R and R′ in the general formula RCOOR′.

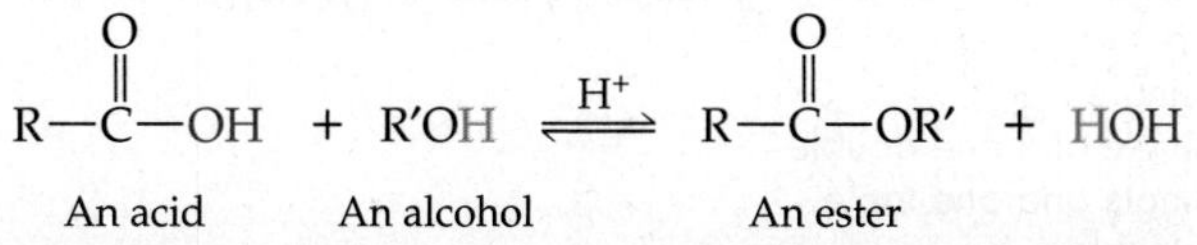

$$\underset{\text{An acid}}{R-\overset{\overset{O}{\|}}{C}-OH}\ +\ \underset{\text{An alcohol}}{R'OH}\ \underset{}{\overset{H^+}{\rightleftharpoons}}\ \underset{\text{An ester}}{R-\overset{\overset{O}{\|}}{C}-OR'}\ +\ HOH$$

The common name of an ester ends in *-ate* and is formed by naming the part from the alcohol first and the part from the carboxylic acid last. For example, the ester derived from butyric acid and methyl alcohol is methyl butyrate. In the IUPAC system, the part derived from the alcohol is also named first, but the part from the acid has the ending *-oate*, so the IUPAC name would be methyl butanoate.

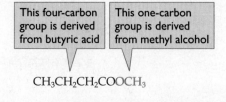

This four-carbon group is derived from butyric acid | This one-carbon group is derived from methyl alcohol

$$CH_3CH_2CH_2COOCH_3$$

3 Why does fresh fruit smell so good and rotting fruit so bad?

Esters are often the main components of fruity flavors. When fruit is broken down by bacteria, the esters are often separated into their corresponding alcohols and (unpleasant-smelling) carboxylic acids.

Although carboxylic acids often have strongly unpleasant odors, the esters derived from them are usually quite fragrant, especially when dilute. Many esters have fruity odors and tastes. Some examples are given in Table 9.9. Esters are widely used as flavorings in cakes, candies, and other foods and as ingredients in perfumes.

TABLE 9.9 Ester Flavors and Fragrances

Ester	Formula	Flavor/Fragrance
Methyl butyrate	$CH_3CH_2CH_2COOCH_3$	Apple
Ethyl butyrate	$CH_3CH_2CH_2COOCH_2CH_3$	Pineapple
Propyl acetate	$CH_3COOCH_2CH_2CH_3$	Pear
Pentyl acetate	$CH_3COOCH_2CH_2CH_2CH_2CH_3$	Banana
Pentyl butyrate	$CH_3CH_2CH_2COOCH_2CH_2CH_2CH_2CH_3$	Apricot
Octyl acetate	$CH_3COOCH_2CH_2CH_2CH_2CH_2CH_2CH_2CH_3$	Orange
Methyl benzoate	$C_6H_5COOCH_3$	Kiwifruit
Ethyl formate	$HCOOCH_2CH_3$	Rum
Methyl salicylate	$o\text{-}HOC_6H_4COOCH_3$	Wintergreen
Benzyl acetate	$CH_3COOCH_2C_6H_5$	Jasmine

SELF-ASSESSMENT Questions

1. Which of the following formulas represents a carboxylic acid?
   a. $CH_3CH_2CH_2OH$
   b. CH_3COOCH_3
   c. $CH_2CH_2CH_2COOH$
   d. $CH_3CH_2CH_2CHO$

2. Which of the following is the formula for butanoic acid?
   a. CH_3CH_2COOH
   b. $CH_3CH_2CH_2CH_2OH$
   c. $CH_3CH_2CH_2CHO$
   d. $CH_3CH_2CH_2COOH$

3. Alcohols can be oxidized with hot copper oxide to form carboxylic acids. If the alcohol is ethanol, the carboxylic acid will be
   a. CH_3COOH
   b. CH_3CH_2OH
   c. $CH_3CH_2CH_2OH$
   d. CH_3CH_2COOH

4. The general formula for an ester is
   a. ROR′
   b. RCOR′
   c. RCOOR′
   d. RCONHR′

5. The carbon atom in a carboxyl group is bonded to two oxygen atoms. The carbon–oxygen bonds are
   a. both single
   b. one single and one double
   c. one single and one triple
   d. both double

6. A name for the compound with the formula $CH_3CH_2CH_2CH_2COOCH_3$ is
   a. ethyl propanoate
   b. ethyl butanoate
   c. methyl pentanoate
   d. propyl propanoate

7. Which of the following formulas represents an ester?
   a. $CH_3CH_2CH_2OH$
   b. CH_3COOCH_3
   c. $CH_2CH_2CH_2COOH$
   d. $CH_3CH_2CH_2CHO$

Answers: 1, c; 2, d; 3, a; 4, c; 5, b; 6, c; 7, b

Salicylates: Pain Relievers Based on Salicylic Acid

Salicylic acid is both a carboxylic acid and a phenol. We can use it to illustrate some of the reactions of these two families of compounds.

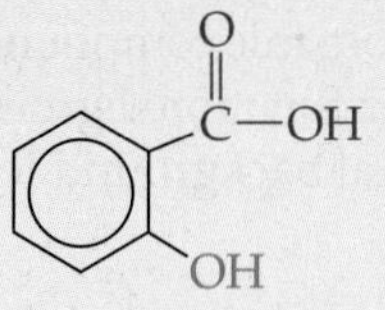

Salicylic acid

Since ancient times, various peoples around the world have used willow bark to treat fevers. Edward Stone, an English clergyman, reported to the Royal Society in 1763 that an extract of willow bark was useful in reducing fever.

Salicylic acid was first isolated from willow bark in 1838. It was first prepared synthetically in 1860 and soon was used in medicine as an *antipyretic* (fever reducer) and as an *analgesic* (pain reliever). However, it is sour and irritating when taken orally. Chemists sought to use chemical reactions to modify its structure so that these undesirable properties would be reduced while retaining or even enhancing its desirable properties. These reactions are summarized in Figure 9.14. The first such modification was simply to neutralize the acid (Reaction 1 in Figure 9.14). The resulting salt, sodium salicylate, was first used in 1875. It was less unpleasant to swallow than the acid but was still highly irritating to the stomach.

By 1886, chemists had produced another derivative (Reaction 2), the phenyl ester of salicylic acid, called *phenyl salicylate*, or *salol*. Salol was less unpleasant to swallow, and it passed largely unchanged through the stomach. In the small intestine, salol was converted to the desired salicylic acid, but phenol was formed as a by-product. A large dose could produce phenol poisoning.

Acetylsalicylic acid, an ester formed when the phenol group of salicylic acid reacts with acetic acid, was first produced in 1853. It is usually made by reacting salicylic acid with acetic anhydride, the acid anhydride of acetic acid (Reaction 3). (Anhydrides are made by combining two carboxylic acid molecules and splitting out a water molecule.) The German Bayer Company introduced acetylsalicylic acid as a medicine in 1899 under the trade name Aspirin. It soon became the best-selling drug in the world.

Another derivative, methyl salicylate, is made by reacting the carboxyl group of salicylic acid with methanol, producing a different ester (Reaction 4). Methyl salicylate, called *oil of wintergreen*, is used as a flavoring agent and in rub-on analgesics. When applied to the skin, it causes a mild warming sensation, providing some relief for sore muscles.

Aspirin and its use as a drug are discussed in more detail in Chapter 18.

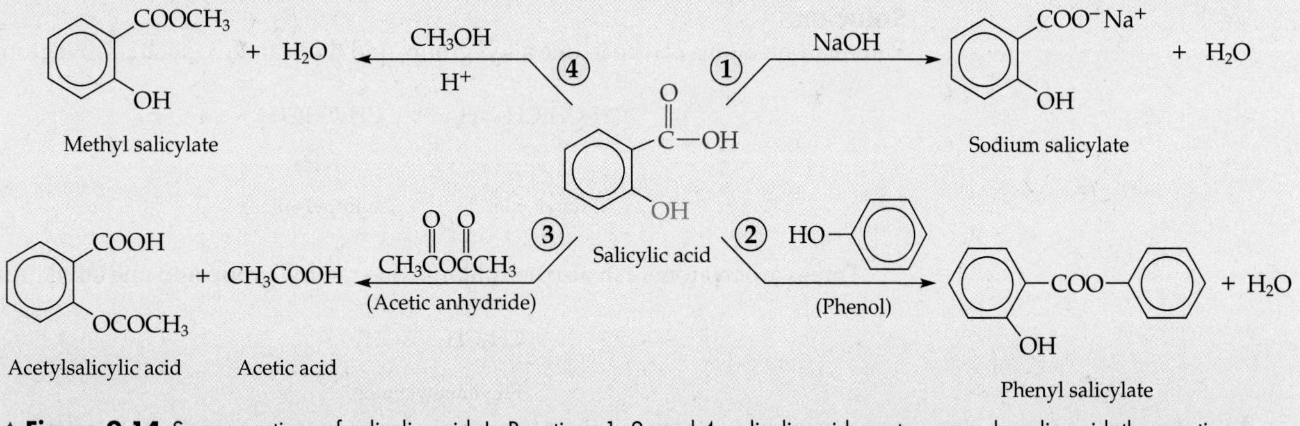

▲ **Figure 9.14** Some reactions of salicylic acid. In Reactions 1, 2, and 4, salicylic acid reacts as a carboxylic acid; the reactions occur at the carboxyl group. In Reaction 3, salicylic acid reacts as a phenol; the reaction takes place at the hydroxyl group.

Q *Can you identify and name the functional groups present in (a) salicylic acid, (b) methyl salicylate, (c) acetylsalicylic acid, and (d) phenyl salicylate?*

4 Why do organic compounds have such strong tastes and odors?

Organic compounds have many different complex structures. The chemoreceptors in our taste buds and smell receptors have a strong affinity for various carbon compounds. The aromatics have strong odors, the esters have sweet odors, the carboxylic acids have unpleasant odors, and the amides and amines have fishy or rotten odors.

Biologically, as we are creatures who consume organic compounds, our senses have evolved to recognize the chemical compounds that we either need or should avoid.

9.9 Nitrogen-Containing Compounds: Amines and Amides

Learning Objectives • Name and write the formulas for simple amines and amides. • Recognize a structure as that of a heterocyclic compound.

Many organic substances of interest to us in the chapters that follow contain nitrogen, which is the fourth-most-common element in organic compounds after carbon, hydrogen, and oxygen. The two nitrogen-containing functional groups that we consider in this section, amines and amides, provide a vital background for the material ahead.

Amines

Amines contain the elements carbon, hydrogen, and nitrogen. An **amine** is derived from ammonia by replacing one, two, or three of the hydrogen atoms with one, two, or three alkyl or aromatic groups:

$$H-\underset{\underset{H}{|}}{N}-H \qquad R-\underset{\underset{H}{|}}{N}-H \qquad R-\underset{\underset{R'}{|}}{N}-H \qquad R-\underset{\underset{R'}{|}}{N}-R''$$

Ammonia   Amines

The simplest amine is methylamine (CH_3NH_2). Amines with two or more carbon atoms can be isomers: Both ethylamine ($CH_3CH_2NH_2$) and dimethylamine (CH_3NHCH_3) have the molecular formula C_2H_7N. With three carbon atoms, there are several possibilities, including trimethylamine $\left[(CH_3)_3N\right]$.

CONCEPTUAL EXAMPLE 9.11   Amine Isomers: Structures and Names

Trimethylamine contains three carbons. Write structural formulas and names for the other three-carbon amines.

Solution
Three carbon atoms can be in one alkyl group, and there are two such propyl groups.

$$CH_3CH_2CH_2NH_2 \qquad\qquad CH_3\underset{\underset{NH_2}{|}}{C}HCH_3$$

Propylamine   Isopropylamine

Three carbon atoms can also be split into one methyl group and one ethyl group.

$$CH_3CH_2NHCH_3$$

Ethylmethylamine

❯ **Exercise 9.11A**
Write structural formulas for the following amines.
   a. butylamine
   b. diethylamine

❯ **Exercise 9.11B**
Write structural formulas for the following amines.
   a. methylpropylamine
   b. isopropylmethylamine

The amine with an −NH$_2$ group (called an **amino group**) attached directly to a benzene ring has the special name *aniline*. Like many other aromatic amines, aniline is used in making dyes. Aromatic amines tend to be toxic, and some are strongly carcinogenic.

Simple amines are similar to ammonia in odor, basicity, and other properties. The larger and aromatic amines are more interesting. Figure 9.15 shows a variety of these.

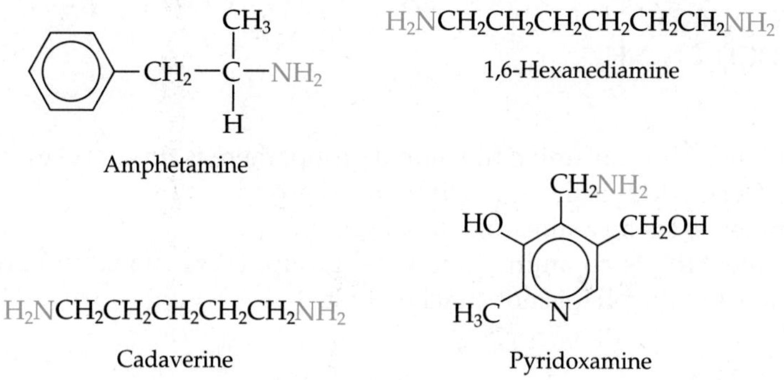

▲ **Figure 9.15** Some amines of interest. Amphetamine is a stimulant drug (Chapter 18). Cadaverine has the odor of decaying flesh. 1,6-Hexanediamine is used in the synthesis of nylon (Chapter 10). Pyridoxamine is a B vitamin (Chapter 17).

Q *1,6-Hexanediamine is the IUPAC name for a compound also known by the common name hexamethylenediamine. Cadaverine is a common name. What is the IUPAC name for cadaverine?*

Among the most important kinds of organic molecules are *amino acids*. As the name implies, these compounds have both amine and carboxylic acid functional groups. Amino acids are the building blocks of proteins. The simplest amino acid, H$_2$NCH$_2$COOH, is glycine. Amino acids and proteins are considered in detail in Chapter 16.

Amides

Another important nitrogen-containing functional group is the **amide group**, which contains a nitrogen atom attached directly to a carbonyl group.

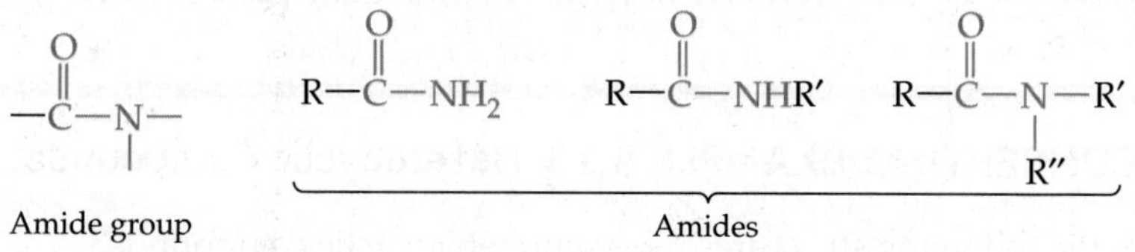

Like those of other compounds that have C=O groups, the formulas of amides are often written on one line.

$$RCONH_2 \quad RCONHR' \quad RCONR'R''$$

Note that urea (H$_2$NCONH$_2$), the compound that helped change the understanding of organic chemistry, is an amide.

Complex amides are of much greater interest than the simple ones considered here. Your body contains many kinds of proteins, all held together by amide linkages (Chapter 16). Nylon, silk, and wool molecules also contain hundreds of amide functional groups.

Names for simple amides are derived from those of the corresponding carboxylic acids. For example, HCONH$_2$ is formamide (IUPAC name: methanamide), and CH$_3$CONH$_2$ is acetamide (IUPAC name: ethanamide). Alternatively, we can also view amide names as based on those of alkanes, with the ending changed from -*e* to -*amide*. For example, CH$_3$CH$_2$CH$_2$CH$_2$CONH$_2$ is pentanamide.

5 How do fabric softeners, dryer sheets, and anti-static sprays work?

These products often contain quaternary ammonium salts that contain two methyl groups and two long hydrocarbon chains attached to an ammonium group, such as dioctadecyldimethylammonium chloride, which coat the surface of clothing. The long hydrocarbon chains lubricate the clothes and help to offset "static cling" that often results when clothes are laundered.

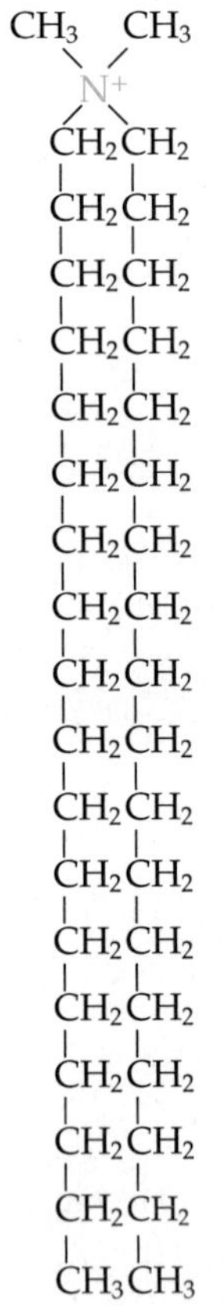

$$CH_3 \quad CH_3$$
$$N^+$$
$$CH_2 CH_2$$
$$CH_2 CH_2$$
$$CH_2 CH_2$$
$$CH_2 CH_2$$
$$CH_2 CH_2$$
$$CH_2 CH_2$$
$$CH_2 CH_2$$
$$CH_2 CH_2$$
$$CH_2 CH_2$$
$$CH_2 CH_2$$
$$CH_2 CH_2$$
$$CH_2 CH_2$$
$$CH_2 CH_2$$
$$CH_2 CH_2$$
$$CH_2 CH_2$$
$$CH_2 CH_2$$
$$CH_2 CH_2$$
$$CH_3 CH_3$$

CONCEPTUAL EXAMPLE 9.12   Amine or Amide?

Which of the following compounds are amides and which are amines? Identify the functional groups.

   a. $CH_3CH_2CH_2NH_2$
   b. CH_3CONH_2
   c. $CH_3CH_2NHCH_3$
   d. $CH_3COCH_2CH_2NH_2$

Solution

   a. An amine: NH_2 is an amine functional group; there is no $C{=}O$ group.
   b. An amide: $CONH_2$ is the amide functional group.
   c. An amine: NH is an amine functional group.
   d. An amine: NH_2 is an amine functional group; there is a $C{=}O$ group in the molecule, but the NH_2 is not attached to it.

❭ **Exercise 9.12A**

Which of the following compounds are amides and which are amines? Identify the functional groups.

   a. CH_3NHCH_3
   b. $CH_3CH_2COCH_2CH_2NHCH_3$

❭ **Exercise 9.12B**

Which of the following compounds are amides and which are amines? Identify the functional groups.

   a. $CH_3CH_2N(CH_3)_2$
   b. $CH_3CH_2CONHCH_2CH_3$

Heterocyclic Compounds: Alkaloids and Others

Cyclic hydrocarbons feature rings of carbon atoms. Now let's look at some compounds that have atoms other than carbon within the ring. These **heterocyclic compounds** usually have one or more nitrogen, oxygen, or sulfur atoms as a member of the ring.

CONCEPTUAL EXAMPLE 9.13 Heterocyclic Compounds

Which of the following structures represent heterocyclic compounds?

   (a)   (b)   (c)   (d)

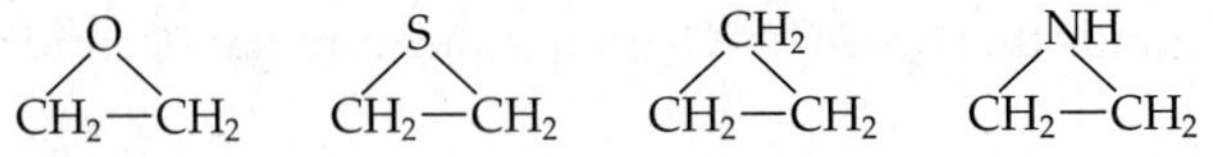

Solution

Compounds a, b, and d have an oxygen, a sulfur, and a nitrogen atom, respectively, in a ring structure; these are heterocyclic compounds.

❭ **Exercise 9.13A**

Which of the following structures represents a heterocyclic compound?

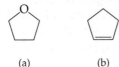

   (a)   (b)

> **Exercise 9.13B**

Which of the following structures represents a heterocyclic compound?

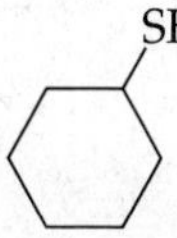

(a)    (b)

Many amines, particularly heterocyclic ones, occur naturally in plants. Like other amines, these compounds are basic. They are called **alkaloids**, which means "like alkalis." Among the familiar alkaloids are morphine, caffeine, nicotine, and cocaine. The actions of these compounds as drugs are considered in Chapter 18.

Other important heterocyclic amines are pyrimidine, which has two nitrogen atoms in a six-membered ring, and purine, which has four nitrogen atoms in two rings that share a common side. Compounds related to pyrimidine and purine are constituents of nucleic acids (Chapter 16).

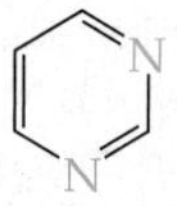

Pyrimidine 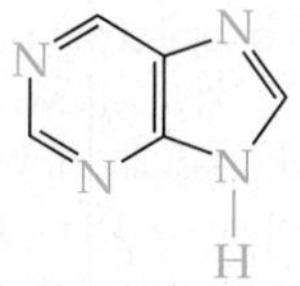   Purine

SELF-ASSESSMENT Questions

1. Which of the following is an amine attached to only one carbon group?
   a. dimethylamine
   b. ethyldimethylamine
   c. isopropylamine
   d. methylpropylamine

2. A name for $CH_3CH_2NHCH_3$ is
   a. butylamine
   b. ethylmethylamine
   c. propanamide
   d. propylamine

3. Which of the following amides is formed when ammonia reacts with acetic acid?
   a. CH_3CONH_2
   b. $CH_3CH_2CONH_2$
   c. $CH_3CH_2CH_2CONH_2$
   d. $C_6H_5CONH_2$

4. The amide functional group has a carbonyl group attached to
   a. an H atom
   b. an N atom
   c. an OH group
   d. two other C atoms

5. Which of the following is *not* an amine?
   a. $CH_3CH_2NCH_3$
   b. $CH_3CH_2NHCOCH_3$
   c. $N(CH_3)_3$
   d. $CH_3CH_2CH_2NH_2$

6. A cyclic organic compound that contains at least one atom in the ring that is not a carbon atom is called a(n)
   a. aromatic compound
   b. cyclic amide
   c. heterocyclic compound
   d. saturated compound

7. A basic organic heterocyclic compound that contains at least one nitrogen atom and is found in plants is called a(n)
   a. alkaloid
   b. amide
   c. cyclic ester
   d. phenol

Answer: 1. c; 2. b; 3. a; 4. b; 5. b; 6. c; 7. a

GREEN CHEMISTRY The Art of Organic Synthesis: Green Chemists
Find a Better Way
Thomas E. Goodwin, *Hendrix College*

Principles 1, 3, 5, 6, 7

Learning Objective ● Identify greener solvents, use renewable resources, less hazardous compounds, and more efficient procedures in organic synthesis, and use efficient energy sources for chemical reactions.

At one time or another, most of us have taken medicines purchased over the counter or obtained by prescription. Many of these are synthetic drugs prepared by an organic chemist at a pharmaceutical company or in a research lab at a college or university, not compounds that occur in nature. These drugs are synthesized via chemical transformations that make more complicated organic compounds from simpler ones. (You will learn more about drugs in Chapter 18.)

Chemical reactions generally proceed faster when there is a higher chance for reacting molecules to collide, such as when they are in heated, homogeneous solutions. In the past, many reactions were carried out in nonrenewable, petroleum-based solvents such as hydrocarbons, chlorinated hydrocarbons, esters, ketones, and ethers (Sections 9.2, 9.4, and 9.6–9.8). These solvents are often flammable, toxic, or both and they present serious disposal problems. Using the Twelve Principles of Green Chemistry (backside of front cover), chemists are now replacing these solvents with greener alternatives—or using no solvent at all. They also are seeking to improve the efficiency of reactions by reducing energy, time, and amounts of chemicals.

An increasing number of chemical reactions are run successfully in solvents from renewable resources or even without a solvent (Principle 5). Aqueous media have become popular because water is an ideal green solvent: cheap, abundant, and safe.

Unfortunately, many organic compounds are not very soluble in water. Solutions form readily when intermolecular forces in the solvent and solute are of similar strength (Section 6.4). Most organic compounds have relatively weak dispersion and dipole–dipole intermolecular forces, whereas water molecules are held together by a strong network of hydrogen bonds. Therefore, water molecules tend to associate with one another rather than exchanging some of their hydrogen bonds for weaker attractions to the nonpolar organic solutes.

Homogeneity is not always required, however. K. B. Sharpless (2001 Nobel Laureate) and his colleagues developed procedures in which the reactants are simply floated on the surface of water. In these "on water" reactions, rapid stirring produces suspensions of reactants with large surface contact areas that enhance reaction rates. Green benefits include reduced use of organic solvents, faster reaction times, and simplified isolation of hydrophobic products.

The EPA's Presidential Green Chemistry Challenge Awards recognize individuals, groups, and organizations that have developed chemical technologies that incorporate the principles of green chemistry. For example, the Pfizer pharmaceutical company developed an improved route to the antidepressant sertraline (Zoloft®) (Section 18.7). Four solvents (toluene, tetrahydrofuran, dichloromethane, and hexane) were replaced with

ethanol (Section 9.6). In addition, overall solvent use was reduced from 60,000 gallons to 6,000 gallons per ton of sertraline. Another award winner, Life Technologies Corporation, developed the green organic chemistry necessary to synthesize deoxyribonucleotide triphosphates (see an analogous deoxyribonucleotide monophosphate in Exercise 16.4, Section 16.9). These nucleotide triphosphates are the building blocks that are used in the very important polymerase chain reaction (PCR) (see Section 16.11). Compared to previous syntheses, solvent use was reduced by 95% and other waste was reduced by 65%, resulting in a total savings of 1.5 million pounds of hazardous waste per year.

A third award winner was Merck Research Laboratories which successfully applied green chemistry design principles to Prevymis (letermovir), an antiviral drug. The improvements to the way the drug was made, including use of a better catalyst, increased the overall yield by more than 60%, reduced raw material costs by 93%, and reduced water usage by 90%.

Green chemistry can also use microwaves to speed up reactions and reduce energy consumption. Microwaves are a form of electromagnetic radiation (Section 3.2) and are more energetic than radio waves but less energetic than X-rays (Section 11.1). In kitchen microwave ovens, the targets for microwave energy are the water, sugar, and fat molecules in food. Although home microwave ovens must not be used to carry out chemical reactions due to potential dangers, specially designed microwave ovens are available for laboratory use.

Additional innovations that can enhance organic chemistry reaction rates with greater energy efficiency include sonochemistry and mechanochemistry. Sonochemistry uses ultrasonic waves (ultrasound) to produce cavitation, a process that involves the creation of bubbles that collapse and liberate energy to drive chemical reactions. Mechanochemistry uses mechanical energy to carry out chemical reactions in the absence of a solvent. This can often be carried out on a small scale simply by grinding the solid reactants by hand using a mortar and pestle. The use of specially designed, automated ball mills is more efficient for larger scale or more difficult syntheses. Ball mills typically use stainless steel balls in a rotating cylinder to grind the reactants and enable the chemical reaction. In contrast to the mortar and pestle technique that derives energy from the chemist who does the grinding (that is, originally solar energy), ball mills require electricity. Their advantages, however, include the absence of solvents and, often, faster reactions than those using conventional energy sources.

Organic chemists today are actively designing ways to use greener solvents, renewable starting materials, and more energy-efficient techniques. These green chemistry innovations will provide sustainable solutions for us and for future generations.

 Summary

Section 9.1—**Organic chemistry** is the chemistry of carbon-containing compounds. More than 95% of all known compounds contain carbon. Carbon can form long chains, rings, and branches. The properties of organic compounds differ markedly from those of inorganic compounds. Generally, organic compounds react slowly, are not soluble in water, do not conduct electricity, have low melting and boiling points, often are liquids or gases at room temperature, and burn.

Section 9.2—**Hydrocarbons** contain only carbon and hydrogen. **Aliphatic compounds** form one main class of hydrocarbons. **Alkanes** are hydrocarbons that contain only single bonds and are **saturated hydrocarbons** because each of their carbon atoms is bonded to the maximum number of hydrogen atoms. **Structural formulas** show which atoms are bonded to one another. **Condensed structural formulas** omit the C—H bond lines and are often more convenient. A **homologous series** of compounds can be made by inserting CH_2 units into the chain. Its members are called *homologs*, and their properties vary in a regular and predictable manner.

Alkanes with more than three carbon atoms have **isomers**, compounds with the same molecular formula but different structures. Alkanes are nonpolar, are insoluble in water, undergo few chemical reactions, and are used mainly as fuels. **Cyclic hydrocarbons** have one or more closed rings and are often represented by geometric figures.

Alkenes are hydrocarbons with at least one carbon–carbon double bond. **Alkynes** are hydrocarbons with at least one carbon–carbon triple bond. Alkenes and alkynes are **unsaturated hydrocarbons** and can have more hydrogen atoms added to them. They undergo **addition reactions** in which hydrogen or another small molecule adds to the double or triple bond; they can also add to one another to form polymers. **Geometric isomers** have fixed rotation around a double bond that results in two unique versions of the molecule.

Section 9.3—Benzene (C_6H_6) has a ring of six carbon atoms with six pairs of bonding electrons between the carbon atoms and six unassigned electrons spread over the ring. It is often drawn as a hexagon with alternating double and single bonds or with a circle in it. Benzene and similar compounds with this type of stable ring structure are called **aromatic compounds**. Substituted benzenes have one or more hydrogen atoms replaced by another atom or group called a **substituent**. Aromatic hydrocarbons are used as solvents and fuels and to make other aromatic compounds.

Section 9.4—Chlorinated hydrocarbons are derived from hydrocarbons by replacing one or more hydrogen atoms with chlorine atom(s). Chlorofluorocarbons (CFCs) contain both chlorine and fluorine. Their use has been restricted because of their role in depleting the ozone layer. In perfluorinated compounds, all hydrogens have been replaced with hydrogens.

Section 9.5—A **functional group** is an atom or group of atoms that confers characteristic physical and reaction properties to a family of organic compounds. A functional group is often attached to an **alkyl group** (R—), an alkane with a hydrogen atom removed. A summary of different classes of functional groups is given in Table 9.10.

Section 9.6—A hydroxyl group (—OH) joined to an alkyl group produces an **alcohol** (ROH). Methanol (CH_3OH), ethanol (CH_3CH_2OH), and 2-propanol $[(CH_3)_2CHOH]$ are well-known and widely used alcohols. Alcohols with more than one —OH group include ethylene glycol, used in antifreeze, and glycerol, used in skin lotions and as a food additive. A **phenol** is a compound with an —OH group attached to a benzene ring. Phenols are slightly acidic, and some are used as antiseptics. Some phenols are antioxidants.

An **ether** has two alkyl or aromatic groups attached to the same oxygen atom (ROR′). Ethers are used as solvents, and they can react slowly with oxygen to form explosive peroxides. Diethyl ether ($CH_3CH_2OCH_2CH_3$) is a commonly used solvent. Ethylene oxide is a cyclic ether used to make ethylene glycol and to sterilize instruments.

A sulfhydryl group (—SH) joined to an alkyl group produces a **thiol** (RSH), also called a mercaptan.

Section 9.7—An **aldehyde** contains a **carbonyl group** ($C{=}O$) with a hydrogen atom attached to the carbonyl carbon. A **ketone** has two carbon atoms attached to the carbonyl carbon. Formaldehyde is used in making plastics and as a preservative and a disinfectant. Benzaldehyde is a flavoring ingredient. Acetone, the simplest ketone, is a widely used solvent.

Section 9.8—A carboxylic acid has a **carboxyl group** (—COOH) as its functional group. Formic acid (HCOOH) is the acid in ant, bee, and wasp stings. Acetic acid (CH_3COOH) is in vinegar. Butyric acid gives rancid butter its odor. Carboxylic acid salts are used as food additives to prevent mold.

The structure of an **ester** (RCOOR′) is similar to that of a carboxylic acid, but with an alkyl group replacing the hydrogen atom of the carboxyl group. Esters are made by reactions of a carboxylic acid with an alcohol or phenol. They often have fruity or flowery odors and are used as flavorings and in perfumes.

Section 9.9—An **amine** is an organic derivative of ammonia in which alkyl or aromatic groups replace one or more hydrogen atoms. Many amines consist of an alkyl group joined to an **amino group** (—NH$_2$). Like ammonia, amines are basic and often have strong odors. Amino acids, the building blocks of proteins, contain both amine and carboxylic acid functional groups. An **amide group** consists of a carbonyl group whose carbon atom is attached to a nitrogen atom. Proteins, nylon, silk, and wool contain amide groups.

A **heterocyclic compound** has at least one ring that contains one or more nitrogen, sulfur, or oxygen atoms. **Alkaloids** are (usually heterocyclic) amines that occur naturally in plants and include morphine, caffeine, nicotine, and cocaine. Heterocyclic structures are found in DNA and RNA.

GREEN CHEMISTRY—Organic chemistry plays an important role in enhancing the quality of human life, such as in the preparation of new life-saving drugs. Organic chemists are designing ways to use greener solvents, renewable starting materials, and more energy-efficient techniques.

TABLE 9.10 **Summary of Various Classes of Organic Compounds Containing O, N, or S Atoms**

Class	General Formula	Example	Name Common, IUPAC
Alcohols	ROH	CH_3CH_2OH	Ethyl alcohol, ethanol
Phenols	ROH (R is aromatic ring)	C_6H_5OH	Phenol
Thiols	RSH	CH_3SH	Methyl mercaptan, methanethiol
Ethers	ROR′	$CH_3CH_2OCH_2CH_3$	Diethyl ether
Aldehydes	RCHO	CH_3CHO	Acetaldehyde, ethanal
Ketones	RCOR	CH_3COCH_3	Acetone
Carboxylic acids	RCOOH	CH_3COOH	Acetic acid, ethanoic acid
Esters	RCOOR′	CH_3COOCH_3	Methyl acetate, methyl ethanoate
Amines	RNH_2 RNHR′ RR′R″N	CH_3NH_2 $CH_3NHCH_2CH_3$ $(CH_3)_3N$	Methylamine ethylmethylamine trimethylamine
Amides	$RCONH_2$	CH_3CONH_2	Acetamide, ethanamide

Learning Objectives

Learning Objectives	Associated Problems
• Define *organic chemistry*, and identify differences between organic and inorganic compounds. (9.1)	1, 3, 17–20,
• Define *hydrocarbon*, and recognize structural features and properties of alkanes, alkenes, and alkynes. (9.2)	2, 4, 9, 10, 65, 66, 73
• Identify hydrocarbon molecules as alkanes, alkenes, or alkynes, and name them. (9.2)	5, 6, 21–30
• Define *aromatic compound*, and recognize the structural feature such compounds share. (9.3)	7, 8, 31–34, 74
• Name simple aromatic hydrocarbons. (9.3)	24
• Name a halogenated hydrocarbon given its formula, and write the formula for such a compound given its name. (9.4)	35–38
• Classify an organic compound according to its functional group(s), and explain why the concept of a functional group is useful in the study of organic chemistry. (9.5)	15, 39, 40, 75
• Recognize and write the formulas of simple alkyl groups. (9.5)	41–44
• Recognize the general structure and properties for an alcohol, a phenol, an ether, and a thiol. (9.6)	12, 13, 45, 46, 48–52, 67, 68, 71, 72
• Name simple alcohols, phenols, ethers, and thiols. (9.6)	11, 45–47, 49–52
• Recognize the general structure for simple aldehydes and ketones and list their important properties. (9.7)	53, 54
• Name simple aldehydes and ketones. (9.7)	53–56
• Recognize the general structure for simple carboxylic acids and esters and list their important properties. (9.8)	14, 57–60, 70
• Name simple carboxylic acids and esters. (9.8)	57–60
• Name and write the formulas of simple amines and amides. (9.9)	61, 63
• Recognize the structure of a heterocyclic compound. (9.9)	16, 62, 64
• Identify greener solvents, use renewable resources in organic synthesis, and use efficient energy sources for chemical reactions.	76–82

 Conceptual Questions

1. Give three differences between organic and inorganic compounds.

2. How do structural formulas and condensed structural formulas differ? What are the advantages of each type of formula?

3. List three characteristics of the carbon atom that make possible the existence of millions of organic compounds.

4. Define or give an example for each of the following terms.
   a. hydrocarbon **b.** alkyne   **c.** alkane   **d.** alkene

5. What are isomers? How do geometric isomers differ from constitutional isomers?

6. What is an unsaturated hydrocarbon? Give examples of two types of unsaturated hydrocarbon.

7. What is the meaning of the circle inside the hexagon in the modern representation of the structure of benzene?

8. What is an aromatic hydrocarbon? How can you recognize an aromatic compound from its structure?

9. Which alkanes are gases at room temperature? Which are liquids? Which are solids? State your answers in terms of the number of carbon atoms per molecule.

10. Compare the densities of liquid alkanes with that of water. If you added octane to water in a beaker, what would you expect to observe?

11. What are the formulas and the systematic names for the alcohols known by the following familiar names?
    a. grain alcohol **b.** rubbing alcohol   **c.** wood alcohol

12. What are some of the long-term effects of excessive ethanol consumption?

13. State an important historical use of diethyl ether. What is its main use today?

14. How do carboxylic acids and esters differ in odor? In chemical structure?

15. Identify the family for each of the following general formulas.
a. ROH	**b.** RCOR'	**c.** RCOOR'
d. ROR'	**e.** RCOOH	**f.** RCHO
g. RSH	**h.** RCONH$_2$	**i.** RNH$_2$

16. What is an alkaloid? Name three common alkaloids.

 Problems

Organic Chemistry and Organic Compounds

17. Which of the following compounds are organic?
    a. $CH_3CH_2SCH_2CH_3$ **b.** H_2CrO_4
    c. $HONH_2$ **d.** $CH_3CH_2CCl_3$

18. Which of the following compounds are organic?
    a. $C_6H_{12}O$ **b.** H_2CS
    c. $KMnO_4$ **d.** $Co(NH_3)_6Cl_2$

19. Which of the following is *not* a characteristic of most organic compounds?
    a. They dissolve in ether but not in water.
    b. They are poor conductors of electricity.
    c. They react very quickly.
    d. They undergo combustion.

20. If you add decane ($C_{10}H_{22}$) to water and mix them, what would you expect to happen?
    a. They would form a homogeneous solution.
    b. They would undergo a rapid reaction, and heat and steam would be given off.
    c. They would separate into two layers, with the decane in the upper layer and the water in the lower layer.
    d. They would separate into two layers, with the decane in the bottom layer and the water in the top layer.

Aliphatic Hydrocarbons

21. How many carbon atoms are there in a molecule of each of the following?
    a. hexane   **b.** nonane
    c. cyclopentane   **d.** 2-pentene

22. The general formula for an alkane is C_nH_{2n+2}. Give the molecular formulas for the alkanes with **(a)** 8 carbon atoms and **(b)** 14 carbon atoms.

23. Name the following hydrocarbons.
    a. $CH_3CH_2CH_2CH_3$ **b.** $CH_2{=}CHCH_2CH_3$
    c. $HC{\equiv}CH$

24. Name the following hydrocarbons.

$$\begin{array}{l} CH_2{-}CH_2 \\ |\qquad | \\ CH_2{-}CH_2 \end{array}$$

   a.   **b.**   **c.**

25. Write the molecular formulas and condensed structural formulas for **(a)** pentane and **(b)** decane.

26. Write the structural formulas for the four-carbon alkanes (C_4H_{10}). Identify butane and isobutane.

27. Indicate whether each of the following compounds is saturated or unsaturated. Classify each as an alkane, alkene, or alkyne.

   a. $CH_3{-}C{\equiv}C{-}CH_3$ **b.**

28. Indicate whether the structures in each pair represent the same compound or isomers.

   a. CH_3CH_3 and $\begin{array}{c} CH_3 \\ | \\ CH_3 \end{array}$

   b. $\begin{array}{c} CH_3CH_2 \\ | \\ CH_3 \end{array}$ and $CH_3CH_2CH_3$

   c. $\begin{array}{c} CH_3CHCH_2CH_2CH_3 \\ | \\ CH_3 \end{array}$ and $\begin{array}{c} CH_3CH_2CHCH_3 \\ | \\ CH_3 \end{array}$

29. Indicate whether the structures in each pair represent the same compound or isomers.

    a. CH_3CHCH_2OH and $CH_3CHCH_2CH_3$
    | |
    CH_3 OH

    b. $CH_3CHCH_2CH_3$ and $CH_3CH_2CHNH_2$
    | |
    NH_2 CH_3

30. Classify the following pairs as homologs, the same compound, isomers, or none of these.

    a. $CH_3CH_2CH_3$ and $CH_3CH_2CH_2CH_3$

    b. 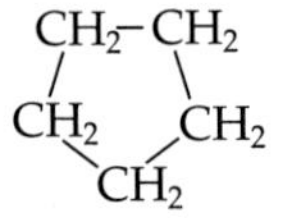 CH_2-CH_2 and $CH_3CH_2CH_2CH_2CH_3$
    /   \
    CH_2 CH_2
    \   /
    CH_2

Aromatic Compounds: Benzene and Relatives

31. Which of the following statements about aromatic compounds is true?
    a. They contain alternating single and double bonds.
    b. They can react easily to add small molecules like HCl to an aromatic ring.
    c. They may have substituents attached to the aromatic ring.
    d. They only contain one ring per molecule.

32. Which of the following is *not* true of aromatic compounds?
    a. Many aromatic compounds have distinctive odors.
    b. Aromatic compounds are not particularly stable compounds
    c. Some aromatic compounds are carcinogenic.
    d. Aromatic compounds are mainly used as solvents or as fuels.

33. Which of the following describe(s) structures that have been used to represent benzene?
    a. C_6H_6
    b. Hexagon with a circle
    c. Hexagon with alternating double and single bonds
    d. All of the above

34. Which of the following structures is aromatic?

a.

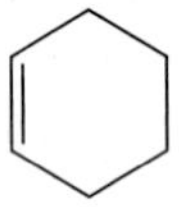

b.

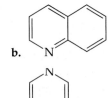

c.

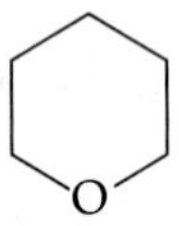

d.

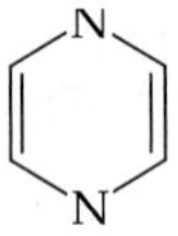

Halogenated Hydrocarbons

35. What is one possible product when ethane reacts with Cl_2 in the presence of light? $CH_3CH_3 + Cl_2 \xrightarrow{light}$
    a. CH_3CH_2Cl
    b. $CH_2=CH_2$
    c. $CH_2=CHCl$
    d. CH_3Cl

36. The formula for dibromomethane is
    a. CBr_4
    b. $CHBr_3$
    c. CH_2Br_2
    d. CH_3Br

37. Chlorinated hydrocarbons have been used for several purposes since they were first developed, but their use has been discontinued because of hazards. However, chlorinated hydrocarbons are still widely used
    a. as anesthetics
    b. in fire extinguishers
    c. as dry-cleaning agents
    d. as solvents

38. Hydrocarbons containing fluorine have been used for all of the following except
    a. aerosol propellants
    b. blood extenders
    c. ozone layer protection agents
    d. refrigerants

Functional and Alkyl Groups

39. Which of the following compounds would be a homolog of CH_3CO_2H?
    a. CH_3CH_2OH
    b. $CH_3CH_2CO_2H$
    c. CH_3CH_2CHO
    d. CH_3COCH_3

40. All of the following functional groups are neutral except
    a. aldehydes
    b. amides
    c. amines
    d. esters

41. Name the following alkyl groups.
    a. CH_3-
    b. CH_3CH_2CH-
    |
    CH_3

42. Name the following alkyl groups.
    a. CH_3-CH-
    |
    CH_3

    b. CH_3
    |
    CH_3-C-
    |
    CH_3

43. What is the difference between the propyl group and the isopropyl group? How are they alike?

44. How do the butyl group, the *sec*-butyl group and the *tert*-butyl group differ? How are they alike?

Alcohols, Phenols, Ethers, and Thiols

45. Give a structure to match the name or a name to match the structure for each of the following compounds.
    a. CH_3CH_2OH
    b. $CH_3CH_2CH_2OH$
    c. methanol
    d. 2-heptanol

46. Give a structure to match the name or a name to match the structure for each of the following compounds.
    a. $CH_3CH_2CHCH_3$
    |
    OH

    b. $CH_3CH_2CH_2CH_2CH_2OH$
    c. *sec*-butyl alcohol
    d. isobutyl alcohol

47. Phenols and other aromatic compounds are often named by numbering the carbon atoms of the benzene ring, starting with the carbon atom bounded to the major functional group. (In a phenol, this group is the —OH group.) To illustrate, the structure in Exercise 9.7b is named 3-bromophenol. With this information, write the structures for **(a)** 2-methylphenol and **(b)** 4-iodophenol.

48. How do phenols differ from alcohols? How are they similar?

49. Write the structure for each of the following.
    a. dipropyl ether
    b. butyl ethyl ether

50. Write the structure for **(a)** methyl propyl ether, an anesthetic known as Neothyl, and **(b)** dimethyl ether, used as a compressed gas to "freeze" warts from the skin. What alcohol is an isomer of dimethyl ether?

51. Which of the following compounds is a phenol?

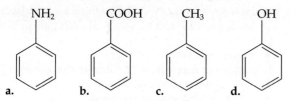

   a.   b.   c.   d.

52. Which of the following compounds is a thiol?
   a. CH_3CH_2OH **b.** CH_3OCH_3
   c. CH_3CH_2SH **d.** CH_3SCH_3

Aldehydes and Ketones

53. Give a structure to match the name or a name to match the formula of each of the following compounds.
   a. butanone
   b. formaldehyde
   c.
$$CH_3CH_2CH_2\overset{\displaystyle O}{\overset{\|}{C}}-H$$
   d.
$$CH_3CH_2-\overset{\displaystyle O}{\overset{\|}{C}}-CH_2CH_2CH_2CH_3$$

54. Give a structure to match the name or a name to match the formula of each of the following compounds.
   a. ethyl methyl ketone
   b. propionaldehyde
   c.
$$CH_3CH_2-\overset{\displaystyle O}{\overset{\|}{C}}-CH_2CH_3$$
   d.
$$\text{⬡}-\overset{\displaystyle O}{\overset{\|}{C}}-H$$

55. Which of the following is an aldehyde?
   a. $CH_3CH_2CH_2\overset{\displaystyle H}{\overset{|}{C}}=O$ **b.** $CH_3CH_2CH_2\overset{\displaystyle OH}{\overset{|}{C}}=O$
   c. $CH_3CH_2CH_2NH_2$ **d.** $CH_3CH_2CH_2OH$

56. Which of the following compounds is a ketone?
   a. $CH_3CH_2CH_2NH_2$ **b.** $CH_3CH_2OCH_3$
   c. $CH_3\overset{\displaystyle O}{\overset{\|}{C}}CH_2CH_3$ **d.** $CH_3CH_2CH_2OH$

Carboxylic Acids and Esters

57. Give a structure to match the name or a name to match the structure for each of the following compounds.
   a. CH_3CH_2COOH
   b. $CH_3CH_2CH_2CH_2COOH$
   c. formic acid
   d. benzoic acid

58. Give a structure to match the name or a name to match the structure for each of the following compounds.
   a. CH_3COOH
   b. $CH_3CH_2CH_2CH_2CH_2COOH$
   c. butanoic acid
   d. chloroacetic acid [*Hint*: To what atom must the chlorine atom (chloro group) be attached?]

59. Give a structure to match the name or a name to match the structure for each of the following.
   a. methyl acetate
   b. ethyl butyrate
   c.
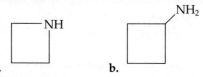

60. Give a structure to match the name or a name to match the structure for each of the following.
   a. methyl butyrate
   b. ethyl propionate
   c.
$$H-\overset{\displaystyle O}{\overset{\|}{C}}-OCH_2CH_2CH_2CH_3$$

Nitrogen-Containing Compounds: Amines and Amides

61. Give a structural formula to match the name or a name to match the structure for each of the following.
   a. methylamine
   b. isopropylamine
   c. $CH_3CH_2NHCH_2CH_3$
   d. ⬡$-NH_2$

62. Which of the following represents a heterocyclic compound? Classify each compound as an amine, ether, or cycloalkane.
   a. ☐$-NH$ **b.** ☐$-\overset{NH_2}{}$

63. Which of the following compounds represents an amide?
   a. $CH_3CH_2CONH_2$
   b. $CH_3CH_2NHCH_2CH_3$
   c. $CH_3COCH_2NHCH_3$
   d. $CH_3CH_2NH_2$

64. The compound acridine is often used in the manufacture of dyes and drugs. Acridine is

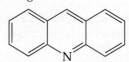

   a. both aromatic and heterocyclic
   b. aromatic but not heterocyclic
   c. heterocyclic but not aromatic
   d. neither aromatic nor heterocyclic

Expand Your Skills

65. As noted in Section 9.2, alkenes and alkynes can add hydrogen to form alkanes. These addition reactions, called *hydrogenations*, are carried out using H_2 gas and a catalyst such as nickel or platinum. Write equations, using condensed structural formulas, for the complete hydrogenation of each of the following.

    a. $CH_3CH_2C{\equiv}CCH_3$ b. $CH_3CH{=}CCH_2CH_2CH_2CH_3$
                                                    |
                                                    CH_3

66. As exemplified by the reaction of ethylene in Section 9.6, alkenes can add water to form alcohols. These addition reactions, called *hydrations*, are carried out using H_2O and an acid $\left(H^+\right)$ as catalyst. Write an equation, using structural formulas, for the hydration of each of the following.

    a.   b. $CH_3CH{=}CHCH_3$

67. As shown in Section 9.7, alcohols can be oxidized to aldehydes and ketones. Write the structure of the alcohol that can be oxidized to form each of the following.
    a. $CH_3CH_2CH_2CHO$ b. $CH_3CH_2COCH_2CH_3$
    c. $CH_3COCH(CH_3)_2$ d. C_6H_5CHO

68. Consider the following set of compounds. What concept does the set illustrate? Explain.

    CH_3OH CH_3CH_2OH
    $CH_3CH_2CH_2OH$ $CH_3CH_2CH_2\,CH_2\,OH$

69. Methanol is a possible replacement for gasoline. The complete combustion of methanol forms carbon dioxide and water.

    $$CH_3OH + O_2 \longrightarrow CO_2 + H_2O$$

    Balance the equation. What mass of carbon dioxide is formed by the complete combustion of 775 g of methanol?

70. Referring to Table 9.9, give the name and condensed structural formula of **(a)** the alcohol and the carboxylic acid that would be needed to make pineapple flavoring and **(b)** the alcohol and the carboxylic acid that would be needed to make apricot flavoring.

71. Refer back to Section 6.3. Describe the type of intermolecular forces that exists **(a)** between dimethyl ether molecules and **(b)** between ethanol molecules. Use this information to explain why ethanol is a liquid at room temperature but dimethyl ether is a gas.

72. Hydroquinone $(HO{-}C_6H_4{-}OH)$, a phenol with two hydroxyl groups in the 1 and 4 (or *para*) positions on the ring, is commonly used as a developer in black and white photography. **(a)** Write the structural formula for hydroquinone. (*Hint*: See Figure 9.10.) **(b)** In the developing process, hydroquinone is oxidized to *para*-benzoquinone $(C_6H_4O_2)$, a compound that is not aromatic and that has two ketone functions. Write the structural formula for *para*-benzoquinone.

73. Using data from Table 9.4, a printed reference such as the *CRC Handbook* or the *Merck Index*, or an online source such as the website of the National Institute for Standards and Technology (NIST), state which of each pair of isomers boils at a lower temperature.
    a. hexane or 2,3-dimethylbutane
    b. pentane or neopentane (2,2-dimethylpropane)
    c. heptane or 2,4-dimethylpentane
    d. octane or 2,2,4-trimethylpentane

    Use this information to form a generalization about the boiling points of branched-chain versus straight-chain alkanes.

74. When benzene and hydrogen are mixed, what will happen?
    a. Hydrogen will add to benzene, because benzene has double bonds.
    b. Hydrogen will not add to benzene, because benzene is already saturated.
    c. They will not react, because benzene has only single bonds.
    d. They will not react, because benzene does not have double bonds.

75. The compounds $CH_3CH_2CH_2OH$ and $CH_3CH_2CH_2CH_2OH$ will react
    a. differently because they have different numbers of carbon atoms
    b. differently because they are both alcohols
    c. similarly because they both have a hydrocarbon chain
    d. similarly because they are both alcohols

76. Give three reasons why water is often considered to be a green solvent for organic synthesis.

77. Which of the following is a potential drawback to using water as a solvent for organic synthesis?
    a. Water is derived from fossil fuels.
    b. Many organic molecules have limited solubility in water.
    c. The intermolecular forces in water are similar to those in most organic solvents.
    d. Reaction rates are always slower in water.

78. In an "on water" reaction:
    a. Reactants are floated on water.
    b. Rapid stirring leads to large surface contact areas.
    c. Reactions often occur faster than if the reactants were dissolved in an organic solvent.
    d. Insoluble products can be separated easily.
    e. All of the above are true.

79. Identify two aspects of the new synthesis of sertraline that make it greener than the traditional process.

80. Why is the improved synthesis of nucleotide triphosphates more environmentally benign?

81. What advantage does microwave heating provide in carrying out chemical reactions? Why does it have this advantage?

82. Name two advantages that sonochemistry and mechanochemistry have over conventional methods of organic synthesis.

Critical Thinking Exercises

Apply knowledge that you have gained in this chapter and one or more of the FLaReS principles (Chapter 1) to evaluate the following statements or claims.

9.1. A television advertisement claims that gasoline with added ethanol burns more cleanly than gasoline without added ethanol.

9.2. A news feature states that herbal remedies are more effective than those produced in the laboratory because they contain fewer harmful chemicals.

9.3. An environmental activist states that all toxic chemicals should be banned from the home.

9.4. An advertisement claims that synthetic amino acid additives in foods are less beneficial than amino acids made by living organisms because they don't contain a life force.

9.5. A Web page claims that acetylsalicylic acid (aspirin), polyester plastics, and polystyrene plastics are all carcinogens (cancer-causing agents). The reason given is that all three contain benzene rings in their structures, and benzene is a carcinogen.

9.6. A television program suggests that life on a distant planet could be based on silicon instead of carbon, because silicon (like carbon) tends to form four bonds. The narrator of the program proposes that long protein molecules and DNA molecules could be based on silicon chains.

9.7. The soot in chimney ash contains polycyclic aromatic hydrocarbons (PAHs), which are compounds that contain multiple aromatic rings. In the 1800s many children worked as chimney sweeps and later developed scrotal cancer. A medical historian claims that PAHs were responsible for the cancers that developed, and that we should not allow fireplaces in homes with children.

 # Collaborative Group Projects

Prepare a PowerPoint, poster, or other presentation (as directed by your instructor) to share with the class.

1. The theory that organic chemistry was the chemistry of living organisms, which was overturned by Wöhler's discovery in 1828, was known as *vitalism*. Search the Internet for information about this theory. What was the effect of this philosophy on areas outside chemistry?.

2. Many organic molecules contain more than one functional group. Look up the structural formula for each of the following in this text (use the index), by searching online, or in a reference work such as the *Merck Index*. Identify and name the functional groups in each.
   a. butesin **b.** estrone
   c. tyrosine **d.** morphine
   e. eugenol **f.** methyl anthranilate

3. Prepare a brief report on an alcohol or a carboxylic acid that has three or more carbon atoms per molecule. List sources and commercial uses.

4. Calculate the molar mass for (a) acetic acid, (b) acetone, (c) ethanol, (d) hexane, and (e) methyl acetate. Then use the Web or a printed reference such as the *CRC Handbook* or the *Merck Index* to find boiling points for these compounds. Is molar mass alone a good predictor of boiling point? Explain.

LET'S EXPERIMENT! Saturate This!

Materials Needed

- Tincture of iodine, 2% (Hint: Do not buy the colorless tincture of iodine. You will need the colored solution.)
- Tablespoon
- Small clear plastic cups
- Shallow pan with warm water

- Safflower oil
- Peanut oil
- Sesame oil (Hint: Use the lighter cold-pressed oil rather than toasted sesame oil.)

Why are saturated fats not good for your health? Why do doctors recommend unsaturated fats? How can you tell the difference?

Let's put your knowledge of saturated and unsaturated fats to the test. The terms *saturated* and *unsaturated* are often used when discussing fats and oils. There are many recommendations to avoid saturated fats because they contribute to "bad" cholesterol that clogs arteries in our bodies. Unsaturated fats are the "healthier" fats that contribute to "good" cholesterol levels. What is the difference?

Fats are made up of triglycerides, which are tri-esters of glycerol and long-chain fatty acids (hydrocarbon chains with a carboxylic acid on one end). Remember that saturated fatty acids do not have any double bonds and are therefore able to pack together closely with the long hydrocarbon chains, which results in their being solid at room temperature, contributing to "bad" cholesterol. Unsaturated fatty acids have one or more double bonds and do not pack as easily, due to the presence of the double bonds, which results in the fats' typically being liquid at room temperature. The double bonds create a "kink" in the structure, which disrupts the stacking (Figure 1).

In this experiment, you will determine the degree of unsaturation of the fatty acid hydrocarbon chains by doing a simple experiment with iodine. Carbon–carbon double bonds will react with iodine to produce a diiodide product (Figure 2). The iodine is a red-violet color (often looking brown), and the product is colorless. This test is commonly used to determine the degree of unsaturation of a triglyceride and is called the iodine number, which is the number of grams of iodine that will react with 100 grams of the triglyceride. The larger the value, the greater the unsaturation of the triglyceride (and the greater the number of carbon–carbon double bonds). A light color confirms unsaturation.

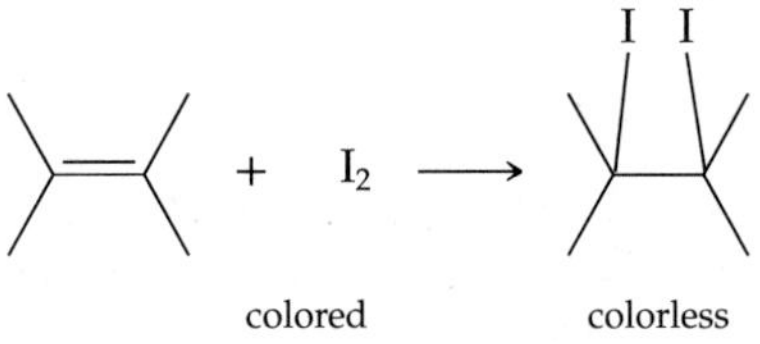

$$\mathrm{C{=}C} \quad + \quad I_2 \quad \longrightarrow \quad \mathrm{C{-}C}$$

colored   colorless

Figure 2

For this experiment, gather the following vegetable oils that will be used to determine which one has the greatest amount of unsaturation: safflower oil, peanut oil, and sesame oil.

Begin by measuring one tablespoon of each vegetable oil into the small cups (one oil in each) and label them appropriately. Place 2–3 drops of the tincture of iodine in each of the oils and gently swirl. Write down any initial observations about the color of each oil solution.

Place the oil solutions in a shallow pan filled with warm tap water to speed up the reaction. The iodine solution will not dissolve in water initially, so be sure to gently swirl the solutions occasionally as you are observing the reaction.

Observe the oil solutions over 10 minutes and note any color changes. What can you conclude about the degree of unsaturation of each oil? Compare the final colors of the oils to the original colors of the oils to observe the overall color change.

Questions

1. Look up the amount of saturated versus unsaturated fats in each of the three oils. Does this help to explain the color differences?
2. What would you expect the degree of unsaturation to be in animal fats compared with the oils that you tested? (Hint: Animal fats are solid at room temperature.)
3. What is the structural difference in *cis* versus *trans* fats? What is it about the structure of *trans* fats that makes them contribute to increasing the levels of bad cholesterol in our bodies?

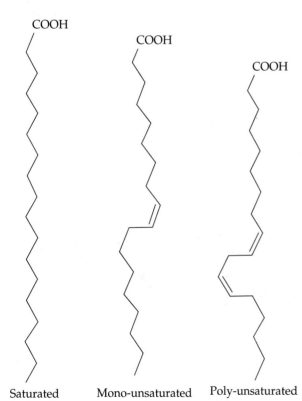

Saturated   Mono-unsaturated   Poly-unsaturated

Figure 1

Polymers

Have You Ever Wondered?

1 Are all polymers synthetic? **2** What is the source of most synthetic polymers? **3** Are there polymers in chewing gum? **4** Why can silicone be used for flexible baking pans? **5** What fabrics are made from natural polymers? **6** Why is it so important to recycle plastics?

GIANTS AMONG MOLECULES Look around you. Polymers are everywhere. Carpets, curtains, upholstery, towels, sheets, floor tile, books, furniture, most toys and containers (not to mention such things as telephones, toothbrushes, and piano keys), and even your clothes are made of polymers. Your car's dashboard, seats, tires, steering wheel, floor mats, and many parts that you cannot see are made of polymers. Much of the food you eat contains polymers, and many vital molecules in your body are polymers. You couldn't live without them. Some of the polymers that permeate our lives come from nature, but many are synthetic. Like other synthetic chemical compounds, they are made in chemical plants in an attempt to improve on nature.

As often happens, world events had an unexpected impact on the development of science. During World War II, the Japanese occupation of Southeast

Plastics—synthetic polymers—are used in most consumer items, including bottles, telephones, calculators, computers, cooking utensils, wire insulation, sunglasses, pens, and most types of packaging. The source of most of these polymers is crude oil, of which there is a limited supply. This means that it is important to recycle these polymers when possible. The bales of plastics shown are ready

303

to go to the recycling center, where they may be used directly to make toys, pipes, or building materials. Or the recycled material may be converted to shreds or beads of plastic (on previous page) that can be used by molding machines. In this chapter, we examine how polymers are made, review their useful properties, explain why they have those properties, and look at their long-term effects.

▲ Cotton is nearly pure cellulose, a natural polymer of the simple sugar glucose.

1 Are all polymers synthetic?

No. The total weight of natural polymers on Earth's surface far exceeds the weight of synthetic polymers. Starches, proteins, cellulose, latex, DNA, and RNA are all natural polymers.

Asia cut off most of the Allies' supply of natural rubber. The polymer industry, to a large extent, developed from the search for a rubber replacement, and many other synthetic products were also created. In this chapter we will look at some of the new polymers that have resulted from this work.

10.1 Polymerization: Making Big Ones Out of Little Ones

Learning Objectives • Define *polymer* and *monomer*. • List several natural polymers, including a chemically modified one.

Polymers are *macromolecules* (from the Greek *makros*, meaning "large" or "long"). Macromolecules are in fact not very large, as most such "giant" molecules are invisible to the human eye, but compared with other molecules, they are enormous.

A **polymer** (from the Greek *poly*, meaning "many," and *meros*, meaning "part") is made from much smaller molecules called *monomers* (from the Greek *monos*, meaning "one"). Hundreds, even thousands, of monomer units combine to make one polymer molecule. A **monomer** is a small-molecule building block from which a polymer is made. These monomers generally have a functional group, or groups, that undergo multiple reactions that link the many units. The process by which monomers are converted to polymers is called **polymerization**. A polymer is as different from its monomer as a long strand of spaghetti is from tiny specks of flour. For example, polyethylene, the familiar waxy material used to make plastic bags, is made from the monomer ethylene, a gas.

Natural Polymers

Polymers have served humanity for millennia by providing the starches and proteins in our food, the wood we use for shelter, and the wool, cotton, and silk we make into clothing. Starch is a polymer made up of glucose ($C_6H_{12}O_6$, a simple sugar) units. Cotton is made of cellulose, also a glucose polymer; wood is largely cellulose as well. Proteins are polymers made up of amino acid monomers. Wool and silk are two of the thousands of different kinds of proteins found in nature.

Living things could not exist without polymers. Each plant and animal requires many different specific types of polymers. Probably the most amazing natural polymers are *nucleic acids*, which carry the coded genetic information that makes each individual unique. The polymers found in nature are discussed in Chapter 16. In this chapter, we will focus mainly on macromolecules that are made in the laboratory.

Celluloid: Billiard Balls and Collars

The earliest synthetic attempt to improve on natural polymers involved chemical modification of a common macromolecule. The semisynthetic material **celluloid** was derived from natural cellulose (from cotton and wood, for example). When cellulose is treated with nitric acid, a derivative called *cellulose nitrate* is formed.

In response to a contest to find a substitute for ivory to use in billiard balls, American inventor John Wesley Hyatt (1837–1920) found a way to soften cellulose nitrate by treating it with ethyl alcohol and camphor. The softened material could be molded into smooth, hard balls. Thus, Hyatt brought the game of billiards within the economic reach of more people—and saved untold numbers of elephants. Celluloid was widely used as a substitute for more expensive substances, such as ivory, amber, and tortoiseshell.

The movie industry was once known as the "celluloid industry." Celluloid was used in movie film as well as for stiff shirt collars, which didn't require laundering and repeated starching. Unfortunately, the cellulose nitrate material is highly flammable (cellulose nitrate is also used as smokeless gunpowder), and as it becomes older, it starts to decompose and can even explode. The 1988 Italian movie *Cinema Paradiso* shows the evolution of film technology and includes a scene in which the heat from a bulb in the movie projector ignites film that becomes stuck in the projector. Celluloid film was removed from the market in the 1950s when safer substitutes became available. Because of the decomposition of the film stock and the resulting lack of preservation, the Library of Congress found that only 14% of the nearly 11,000 silent films released before 1929 are still available to watch. Today, movie film is made mainly from polyethylene terephthalate, a polyester. And high, stiff collars on men's shirts are definitely out of fashion, although clerical collars are still made of plastic.

It didn't take long for the chemical industry to recognize the potential of synthetic polymers. Scientists found ways to make macromolecules from small molecules rather than simply modifying large ones. The first such truly synthetic polymers were phenol–formaldehyde resins (such as Bakelite), initially made in 1909. These complex polymers are discussed in Section 10.5, but we'll look at some simpler ones first.

SELF-ASSESSMENT Questions

1. Cellulose is an example of a(n)
   a. fat   **b.** isomer   **c.** monomer   **d.** polymer

2. All of the following are polymers except
   a. celluloid   **b.** glucose   **c.** silk   **d.** wool

3. The first semisynthetic polymer was made from a natural polymer. It was
   a. Bakelite, based on bacon grease
   b. celluloid, based on cellulose from cotton
   c. polypropylene, based on isopropyl alcohol
   d. polypantene, based on human hair

4. The first truly synthetic polymer was
   a. Bakelite, made from phenol and formaldehyde
   b. cellulose, made from glucose
   c. polyethylene, made from ethanol
   d. rubber, made from isoprene

Answers: 1, d; 2, b; 3, b; 4, a

10.2 Polyethylene: From the Battle of Britain to Bread Bags

Learning Objectives • Describe the structure and properties of the two main types of polyethylene. • Use the terms *thermoplastic* and *thermosetting* to explain how polymer structure determines properties.

The prevalent plastic *polyethylene* is the simplest and least expensive synthetic polymer. It is familiar to us in the form of plastic bags used for packaging fruit and vegetables, garment bags for dry-cleaned clothing, garbage-can liners, and many other items. Worldwide annual production of polyethylene in 2018 was estimated to be 100 billion kg, with a commercial value of $164 billion. Polyethylene is made from ethylene ($CH_2{=}CH_2$), an unsaturated hydrocarbon (Chapter 9). Ethylene is produced in large quantities from the cracking of petroleum, a process by which large hydrocarbon molecules are broken down into simpler hydrocarbons.

With pressure and heat, and in the presence of a catalyst, ethylene monomers join together to form long chains. This process involves breaking *one* of the double bonds in the monomer, leaving lone electrons on each carbon atom of the original double bond as well as a single bond between the carbon atoms. Because electrons like to be paired in forming bonds, random collisions of those molecules cause bonds between carbons with lone electrons in two monomers to be formed, and this continues along the chain.

$$\cdots + \underset{\underset{H}{|}}{\overset{\overset{H}{|}}{C}}=\underset{\underset{H}{|}}{\overset{\overset{H}{|}}{C}} + \underset{\underset{H}{|}}{\overset{\overset{H}{|}}{C}}=\underset{\underset{H}{|}}{\overset{\overset{H}{|}}{C}} + \underset{\underset{H}{|}}{\overset{\overset{H}{|}}{C}}=\underset{\underset{H}{|}}{\overset{\overset{H}{|}}{C}} + \underset{\underset{H}{|}}{\overset{\overset{H}{|}}{C}}=\underset{\underset{H}{|}}{\overset{\overset{H}{|}}{C}} + \cdots \longrightarrow \sim \underset{\underset{H}{|}}{\overset{\overset{H}{|}}{C}}-\underset{\underset{H}{|}}{\overset{\overset{H}{|}}{C}}-\underset{\underset{H}{|}}{\overset{\overset{H}{|}}{C}}-\underset{\underset{H}{|}}{\overset{\overset{H}{|}}{C}}-\underset{\underset{H}{|}}{\overset{\overset{H}{|}}{C}}-\underset{\underset{H}{|}}{\overset{\overset{H}{|}}{C}}-\underset{\underset{H}{|}}{\overset{\overset{H}{|}}{C}}-\underset{\underset{H}{|}}{\overset{\overset{H}{|}}{C}}\sim$$

Using condensed formulas, this becomes

$$\cdots + CH_2{=}CH_2 + CH_2{=}CH_2 + CH_2{=}CH_2 + CH_2{=}CH_2 + \cdots \longrightarrow \sim CH_2CH_2CH_2CH_2CH_2CH_2CH_2CH_2\sim$$

Such equations can be tedious to draw, so we often use abbreviated forms like these:

$$n\,\underset{\underset{H}{|}}{\overset{\overset{H}{|}}{C}}{=}\underset{\underset{H}{|}}{\overset{\overset{H}{|}}{C}} \longrightarrow \left[\underset{\underset{H}{|}}{\overset{\overset{H}{|}}{C}}-\underset{\underset{H}{|}}{\overset{\overset{H}{|}}{C}}\right]_n$$

or

$$n\,CH_2{=}CH_2 \longrightarrow {+}CH_2CH_2{+}_n$$

The ellipses (. . .) and tildes (~) serve as *et ceteras*; they indicate, respectively, that the number of monomers is greater than the four shown and that the polymer structure extends for many more units in each direction. The molecular fragment within the square brackets is called the *repeat unit* of the polymer. In the formula for the polymeric product, the repeat unit is placed within brackets with bonds extending to both sides. The subscript n indicates that this unit is repeated many times (n in the equation above) in the full polymer structure.

The simplicity of the abbreviated formulas facilitates certain comparisons between the monomer and the polymer. Note that the monomer ethylene contains a double bond and polyethylene does not. Note also that each repeat unit in the polymer has the same composition (C_2H_4) as the monomer. (This is the case for addition polymers, but will not be true for condensation polymers, as we will see in Section 10.5.)

Molecular models provide 3-D representations. Figure 10.1 presents models of a tiny part of a very long polyethylene molecule, whose number of carbon atoms can vary from a few hundred to several thousand.

Polyethylene was invented in England shortly before the start of World War II. It proved to be tough and flexible, excellent as an electric insulator, and able to withstand both high and low temperatures. Prior to the use of polyethylene, wires were insulated using a cotton braid weave coated with rubber (both natural polymers). This material was much more susceptible to temperature changes, as the insulation aged and cracked, and bare wires were exposed. Polyethylene was used for insulating cables in radar equipment, a top-secret invention that helped British pilots detect enemy aircraft before the aircraft could be spotted visually. Without polyethylene, the British could not have had effective radar, and without radar, the Battle of Britain might have been lost. The invention of this simple plastic helped change the course of history.

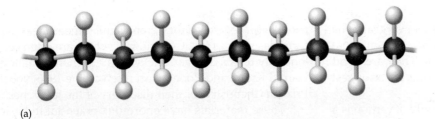

(a)

(b)

▲ **Figure 10.1** (a) Ball-and-stick and (b) space-filling models of a short segment of a polyethylene molecule.

2 **What is the source of most synthetic polymers?**

The raw materials for most synthetic polymers are petroleum and natural gas. These are not renewable sources, and we not only have a limited supply of these fossil fuels, but we are literally burning up over 95% for energy. Comparatively little is left for making other products such as polymers.

Types of Polyethylene

Today, there are three main types of polyethylene that are widely used, although other kinds do exist. *High-density polyethylene (HDPE)* has mostly linear molecules that pack closely together and can assume a fairly well-ordered, crystalline structure. HDPEs therefore are rather rigid and have good tensile strength. As the name implies, the densities of HDPEs ($0.94-0.96$ g/cm^3) are high compared with those of other polyethylenes. HDPEs are used for such items as threaded bottle caps, toys, detergent bottles, and milk jugs.

The Many Forms of Carbon

In many of the smaller molecules that we studied in Chapter 9 and the macromolecules that we encounter here, carbon atoms are bonded to other atoms, especially hydrogen. However, carbon atoms alone can form a variety of interesting structures, some of which have been known since ancient times. Diamond is pure crystalline carbon, in which each atom is bonded to four other atoms. In graphite, the black material of pencil lead, each carbon atom is bonded to three others. But the most intriguing forms of carbon were discovered only in the last few decades.

Andre Geim and Konstantin Novoselov of the University of Manchester, United Kingdom, used adhesive tape to lift one-atom-thick planar sheets of carbon atoms from graphite. This material, called *graphene*, is like molecular-scale chicken wire made of carbon atoms. Graphene has properties that make it suitable for applications in electronics, sensing devices, and touch screens, and for fundamental studies of electron flow in 2-D materials. Geim and Novoselov were awarded the 2010 Nobel Prize in Physics for their work on graphene.

Graphite is composed of many stacked graphene layers. A stack of 3 million graphene sheets would be only a millimeter thick. Graphene can also be thought of as an extremely large aromatic molecule, a system of fused benzene rings.

Many other structures are formed exclusively of carbon atoms. The tube-shaped carbon molecule called a *nanotube* can be visualized as a sheet of graphene rolled into a hollow cylinder. Nanotubes have unusual mechanical and electrical properties. They are good conductors of heat, and they are among the strongest and stiffest materials known.

Yet another structure consisting only of carbon is a roughly spherical collection of hexagons and pentagons like the pattern on a soccer ball. This carbon molecule has the formula C_{60}. Because it resembles the geodesic-dome structures pioneered by architect R. Buckminster Fuller, the molecule is called *buckminsterfullerene*. The general

name *fullerenes* is used for C_{60} and similar molecules with formulas such as C_{70}, C_{74}, and C_{82}, which are often colloquially referred to as "buckyballs." Richard Smalley, Harold Kroto, and Robert Curl received the 1996 Nobel Prize in Chemistry for their discovery of the fullerenes.

These materials have enormous potential in basic studies as well as in applied research. In the field of materials science, research is being done on polymers of carbon nanotubes as a type of radiation shielding for aerospace applications as well as for their use in drug delivery systems.

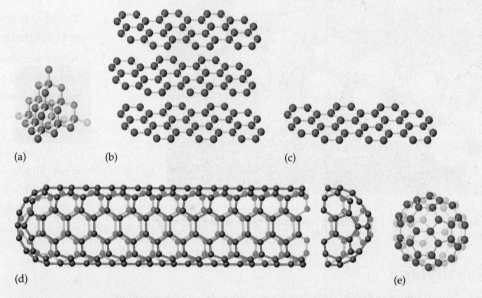

(a)   (b)   (c)

(d)   (e)

▲ Different forms of carbon have dramatically different properties. (a) Diamond is hard and transparent. (b) Graphite is soft and gray and conducts electricity. (c) Graphene is a single layer of graphite. (d) Rolling a sheet of graphene into a cylinder makes a carbon nanotube, the strongest material known. (e) Buckminsterfullerene, a C_{60} molecule, resembles a soccer ball.

▲ **Figure 10.2** Two bottles, both made of polyethylene, were heated in the same oven for the same length of time.

Q *Which of these bottles is made of HDPE and which of LDPE? Explain.*

Low-density polyethylene (LDPE), on the other hand, has many side chains branching off the polymer molecules. The branches prevent the molecules from packing closely together and assuming a crystalline structure. LDPEs are waxy, bendable plastics that have lower densities ($0.91-0.94 \text{ g/cm}^3$) and lower melting temperatures than high-density polyethylenes. Objects made of HDPE hold their shape in boiling water, whereas those made of LDPE are severely deformed (Figure 10.2). LDPEs are used to make plastic bags and film, squeeze bottles, electric wire insulation, and many common household products for which flexibility is important.

The third type of polyethylene, called *linear low-density polyethylene (LLDPE)*, is actually a **copolymer**, a polymer formed from two (or more) different monomers. LLDPEs are made by polymerizing ethylene with a branched-chain alkene such as 4-methyl-1-pentene.

$$n \, CH_2{=}CH_2 + m \, CH_2{=}CH \longrightarrow$$

Ethylene   4-Methyl-1-pentene   An LLDPE

LLDPEs are used to make such things as plastic films for use as landfill liners, trash cans, tubing, and automotive parts.

Thermoplastic and Thermosetting Polymers

Polyethylene is one of a variety of thermoplastic polymers. Because its molecules can slide past one another when heat and pressure are applied, a **thermoplastic polymer** can be softened and then reshaped. It can be repeatedly melted down and remolded. Total annual production of thermoplastic polymers in the United States is about 44 billion kg, of which 19 billion kg is polyethylene.

Not all polymers can be readily melted. About 180 billion kg of various plastics and 22 billion kg of rubber are manufactured globally each year. About 15% of U.S. production of polymers consists of **thermosetting polymers**, which harden permanently when formed. They cannot be softened by heat and remolded. Instead, strong heating causes them to discolor and decompose. The permanent hardness of thermosetting plastics is due to cross-linking (side-to-side connection) of the polymer chains, which we'll discuss later in this chapter.

WHY IT MATTERS

Polyethylene with a molecular mass from 2 million to 6 million, called *ultra-high-molecular-weight polyethylene (UHMWPE)*, is used to make fibers 10 times stronger than steel and around 40% stronger than Kevlar for use in body armor. Most modern skis are coated with UHMWPE on the bottom surface.

SELF-ASSESSMENT Questions

1. Each building-block molecule that joins to form a polymer is called a
   a. cross-link **b.** monomer
   c. polymer **d.** polypeptide
2. Compared with HDPE, LDPE has a lower melting point because it
   a. is made from fewer polar monomers
   b. is composed of molecules with a much lower molecular mass
   c. has fewer cross-links
   d. has more branches in its molecules, preventing them from packing closely
3. Which of the following is made of polyethylene?
   a. pre-1900 electrical wire insulation
   b. foamed coffee cups
   c. plastic grocery bags
   d. plastic plumbing pipes
4. A copolymer is made
   a. by blending two simple polymers
   b. from cobalt and an alkene
   c. from two identical monomers
   d. from two different monomers
5. A polymer that *cannot* be melted and reshaped is said to be
   a. highly crystalline **b.** rubbery
   c. thermoplastic **d.** thermosetting
6. A buckyball is a
   a. large carbon molecule (C_{60})
   b. carbon nanotube wound up like a ball of string
   c. long polymer molecule wound up like a ball of string
   d. circular ring of carbon atoms
7. A polymer that can be melted and reshaped is said to be
   a. highly crystalline **b.** rubbery
   c. thermoplastic **d.** thermosetting

Answers: 1, b; 2, d; 3, c; 4, d; 5, d; 6, a; 7, c

10.3 Addition Polymerization: One + One + One + . . . Gives One!

Learning Objective • Identify the monomer(s) of an addition polymer, and write the structural formula for a polymer from its monomer structure(s).

There are two general types of polymerization reactions: addition polymerization and condensation polymerization. In **addition polymerization** (also called *chain-reaction polymerization*), the monomer molecules add to one another in such a way that the polymeric product contains all the atoms of the starting monomers. The polymerization of ethylene to form polyethylene is an example. In polyethylene, as we noted in Section 10.2, the two carbon atoms and the four hydrogen atoms of each monomer molecule are incorporated into the polymer structure. In *condensation polymerization* (Section 10.5), some part of each monomer molecule is not incorporated in the final polymer.

Polypropylene

Most of the many familiar addition polymers are made from derivatives of ethylene in which one or more of the hydrogen atoms are replaced by another atom or group. Replacing one of the hydrogen atoms with a methyl group gives the monomer propylene (propene). Polypropylene molecules look much like polyethylene molecules, except that there is a methyl group ($-CH_3$) attached to every other carbon atom.

$$\sim CH_2-\underset{\underset{CH_3}{|}}{CH}-CH_2-\underset{\underset{CH_3}{|}}{CH}-CH_2-\underset{\underset{CH_3}{|}}{CH}-CH_2-\underset{\underset{CH_3}{|}}{CH}\sim \quad \text{or} \quad \left[CH_2-\underset{\underset{CH_3}{|}}{CH} \right]_n$$

Polypropylene

The chain of carbon atoms is called the polymer *backbone*. Groups attached to the backbone, such as the $-CH_3$ groups of polypropylene, are called *pendant groups*, shown in green.

Polypropylene is a tough plastic material that resists moisture, oils, and solvents. It is molded into hard-shell luggage, battery cases, and various kinds of appliance parts. It is also used to make packaging material, fibers for textiles such as upholstery fabrics and carpets, and ropes that float. Because of polypropylene's high melting point (121 °C), objects made from it can be sterilized with steam.

Polystyrene

Replacing one of the hydrogen atoms in ethylene with a benzene ring gives a monomer called *styrene*, which has the formula $CH_2=CHC_6H_5$, where C_6H_5 represents the benzene ring. Polymerization of styrene produces polystyrene, which has benzene rings as pendant groups, shown in green.

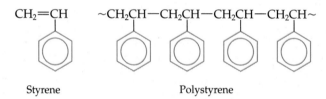

Styrene   Polystyrene

Polystyrene is the plastic used to make transparent disposable drinking cups. With color and filler added, it is the material of thousands of inexpensive toys and household items. When a gas such as air is blown into polystyrene liquid, it foams and hardens into the familiar material (called Styrofoam) of some disposable ice chests and coffee cups. The polymer can easily be formed into shapes as packing material for shipping instruments and appliances, and it is widely used for home insulation. Unfortunately, it is also one of the materials that most often gets discarded. Americans throw away about 25 billion coffee cups made of polystyrene foam each year.

Vinyl Polymers

Would you like a tough synthetic material that looks like leather at a fraction of the cost? Perhaps a clear, rigid material from which unbreakable bottles could be made? Do you need an attractive, long-lasting floor covering? Or lightweight, rustproof, easy-to-connect plumbing? Polyvinyl chloride (PVC) has all these properties—and more.

▲ Styrofoam insulation saves energy by reducing the transfer of heat from the warm interior of a house to the outside in winter and from the hot outside to the cooled interior in summer.

▶ Some furniture, window sashes, vinyl siding, and waterproof clothing are made from PVC. PVC can also be coated onto copper wire for insulation; made into colorful, resilient flooring; or formed into many other familiar consumer products.

Replacing one of the hydrogen atoms of ethylene with a chlorine atom gives vinyl chloride (CH_2=CHCl), a compound that is a gas at room temperature. Polymerization of vinyl chloride yields the tough thermoplastic material PVC. A segment of the PVC molecule is shown here.

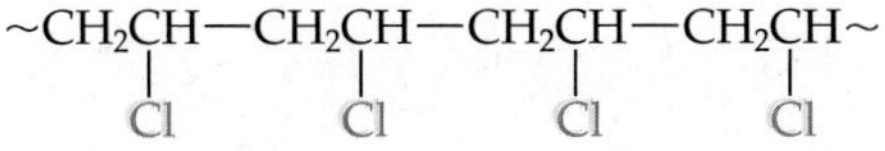

$$\sim CH_2CH\!-\!CH_2CH\!-\!CH_2CH\!-\!CH_2CH\sim$$
$$\underset{Cl}{\big|}\qquad\underset{Cl}{\big|}\qquad\underset{Cl}{\big|}\qquad\underset{Cl}{\big|}$$

Polyvinyl chloride (PVC)

PVC is readily formed into various shapes. The clear, transparent polymer is used in plastic wrap and clear plastic bottles. Adding color and other ingredients to a vinyl plastic yields artificial leather. Most floor tile and shower curtains are made from vinyl plastics, which are also widely used to simulate wood in home siding panels and window frames. About 40% of the PVC produced is molded into pipes.

Vinyl chloride, the monomer from which vinyl plastics are made, is a carcinogen. A number of people who worked closely with this gas later developed a kind of cancer called *angiosarcoma*. (Carcinogens are discussed in Chapter 21.)

PTFE: The Nonstick Coating

In 1938, a young American chemist at DuPont, Roy Plunkett (1910–1994), was working with the gas tetrafluoroethylene (CF_2=CF_2). He opened the valve on a tank of the gas—and nothing came out. Rather than discarding the tank, he decided to investigate. The tank was found to be filled with a waxy, white solid. He attempted to analyze the solid but ran into a problem: It simply wouldn't dissolve, even in hot concentrated acids. Plunkett had discovered polytetrafluoroethylene (PTFE), the polymer of tetrafluoroethylene, best known by its trade name, Teflon®. In this case, the fluorine atoms are pendant groups, and they are shown in green.

$$\sim CF_2\!-\!CF_2\!-\!CF_2\!-\!CF_2\!-\!CF_2\!-\!CF_2\!-\!CF_2\!-\!CF_2\sim$$

Teflon

Because its C—F bonds are exceptionally strong and resistant to heat and chemicals, PTFE is a tough, unreactive, non-flammable material. PTFE is used to coat the sole plates of irons used for pressing clothes and to make electrical insulation, bearings, and gaskets. Medical catheters are often coated with PTFE to help prevent bacteria from adhering to them and causing infections. However, it is most widely known as a coating for surfaces of cookware to eliminate sticking of food. PTFE begins to decompose at temperatures above 260 °C (500 °F), but most frying is done at much lower temperatures than that.

▲ Frying pans, baking pans, and other cookware often take advantage of the inert, nonstick nature of PTFE.

 CONCEPTUAL EXAMPLE 10.1 Repeat Units in Polymers

What is the repeat unit in polyvinylidene chloride? What is the monomer from which it was made? A segment of the polymer is represented as

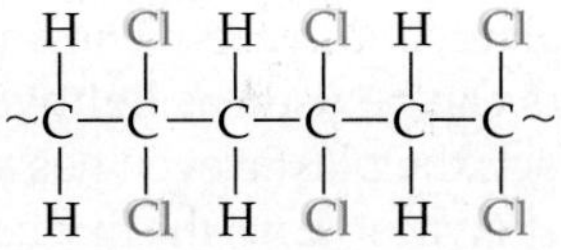

$$\overset{\displaystyle H\;\;Cl\;\;H\;\;Cl\;\;H\;\;Cl}{\underset{\displaystyle H\;\;Cl\;\;H\;\;Cl\;\;H\;\;Cl}{\sim C\!-\!C\!-\!C\!-\!C\!-\!C\!-\!C\sim}}$$

Solution
Inspection of the segment above reveals that it consists of three repeat units, each with two carbon atoms, two hydrogen atoms, and two chlorine atoms. Placing a double bond between the carbon atoms of one unit, we get

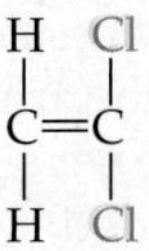

$$\overset{\displaystyle H\quad Cl}{\underset{\displaystyle H\quad Cl}{C\!=\!C}}$$

Joining hundreds of these units would yield a molecule of the polymer.

> **Exercise 10.1A**

What is the repeat unit in polyacrylonitrile? What is the monomer from which it was made? A segment of the polymer is represented as

$$\sim CH_2CHCH_2CHCH_2CHCH_2CHCH_2CHCH_2CH \sim$$
$$\quad\;\; | \qquad | \qquad | \qquad | \qquad | \qquad |$$
$$\quad\;\; CN \quad\; CN \quad\; CN \quad\; CN \quad\; CN \quad\; CN$$

> **Exercise 10.1B**

What is the repeat unit in polyvinyl bromide? What is the monomer from which it was made? A segment of the polymer is represented as

$$\sim CH_2CHCH_2CHCH_2CHCH_2CH \sim$$
$$\quad\;\; | \qquad | \qquad | \qquad |$$
$$\quad\;\; Br \quad\; Br \quad\; Br \quad\; Br$$

▲ Conducting polymers are used in products such as this backpack. Solar cells can capture energy that can be used to charge cell phones or other devices while you are walking outside.

EXAMPLE 10.2 Structure of Polymers

Write the structure of the polymer made from methyl vinyl ketone (CH_2=$CHCOCH_3$). Show at least four repeat units.

Solution

The carbon atoms bond to form a chain with only single bonds between carbon atoms. The methyl ketone is a pendant group on the chain. (Two of the electrons in the double bond of the monomer form the bond joining the units.) The polymer is

$$\sim CH_2CHCH_2CHCH_2CHCH_2CH \sim$$
$$\quad\;\; | \qquad\;\; | \qquad\;\; | \qquad\;\; |$$
$$\quad O=C \;\; O=C \;\; O=C \;\; O=C$$
$$\qquad\quad | \qquad\;\; | \qquad\;\; | \qquad\;\; |$$
$$\qquad\quad CH_3 \quad CH_3 \quad CH_3 \quad CH_3$$

> **Exercise 10.2A**

Write the structure of the polymer made from vinyl fluoride (CH_2=CHF). Show at least four repeat units of each polymer.

> **EXERCISE 10.2B**

Write the structure of the polymer made from vinyl acetate (CH_2=$CHOCOCH_3$). Show at least four repeat units of each polymer.

Conducting Polymers: Polyacetylene

Acetylene (H—C≡C—H) has a triple bond rather than a double bond, but it can still undergo addition polymerization, forming polyacetylene (Figure 10.3). Notice that, unlike polyethylene—which has a carbon chain containing only single bonds—every other carbon-to-carbon bond in polyacetylene is a double bond.

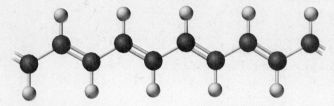

▲ **Figure 10.3** A ball-and-stick model of polyacetylene.

Q *Write an equation for the formation of polyacetylene similar to that for polyethylene on page 306, in which the repeat polymer unit is placed within brackets.*

The alternating double and single bonds form a *conjugated* system. Such a system makes it easy for electrons to travel along the chain, so polyacetylene is able to conduct electricity. Most other plastics are electrical insulators. Polyacetylene and similar conjugated polymers can be used as lightweight substitutes for metal. In fact, the plastic even has a silvery luster like metal.

Polyacetylene, the first conducting polymer, was discovered in 1970. Since then, a number of other polymers that conduct electricity have been made. More recently, paints and inks that contain conducting polymers have been made available on the open market. Artists can use the paints and inks as part of their art, incorporating electrical devices like light-emitting diodes (LEDs) and motors into the artwork. Similar uses of these materials in textiles are being investigated. Imagine using a jacket with embedded solar cells to keep your smartphone charged while walking around the city!

Processing Polymers

In everyday life, many polymers are called plastics. In chemistry, a *plastic* material is one that can be made to flow under heat and pressure. The material can then be shaped in a mold or in other ways. Plastic products are often made from granular polymeric material. In *compression molding*, heat and pressure are applied directly to such plastic grains in the mold cavity. In *transfer molding*, the polymer is softened by heating before being poured into molds to harden.

There also are several methods of molding molten polymers. In *injection molding*, the plastic is melted in a heating chamber and then forced by a plunger into cold molds to set. In *extrusion molding*, the melted polymer is extruded through a die in continuous form to be cut into lengths or coiled. Bottles and similar hollow objects often are *blow-molded*; a "bubble" of molten polymer is blown up like a balloon inside a hollow mold.

Table 10.1 lists some of the more important addition polymers, along with a few of their uses. The pendant groups are shown in green.

▲ Granular polymeric resins are the basic stock for many molded plastic goods.

TABLE 10.1 Some Addition Polymers

Monomer	Polymer	Polymer Name	Some Uses
$CH_2{=}CH_2$	$\left[\begin{array}{cc} H & H \\ -C-C- \\ H & H \end{array}\right]_n$	Polyethylene	Bags; bottles; toys; electrical insulation
$CH_2{=}CH-CH_3$	$\left[\begin{array}{cc} H & H \\ -C-C- \\ H & CH_3 \end{array}\right]_n$	Polypropylene	Carpeting; bottles; luggage
$CH_2{=}CH-\bigcirc$	$\left[\begin{array}{cc} H & H \\ -C-C- \\ H & \bigcirc \end{array}\right]_n$	Polystyrene	Simulated wood furniture; insulation; cups; toys; packing materials
$CH_2{=}CH-Cl$	$\left[\begin{array}{cc} H & H \\ -C-C- \\ H & Cl \end{array}\right]_n$	Polyvinyl chloride (PVC)	Food wrap; simulated leather; plumbing; garden hoses; floor tile
$CH_2{=}CCl_2$	$\left[\begin{array}{cc} H & Cl \\ -C-C- \\ H & Cl \end{array}\right]_n$	Polyvinylidene chloride (Saran)	Food wrap; seat covers
$CF_2{=}CF_2$	$\left[\begin{array}{cc} F & F \\ -C-C- \\ F & F \end{array}\right]_n$	Polytetrafluoroethylene (Teflon)	Nonstick coating for cooking utensils; electrical insulation
$CH_2{=}CH-C{\equiv}N$	$\left[\begin{array}{cc} H & H \\ -C-C- \\ H & CN \end{array}\right]_n$	Polyacrylonitrile (Acrilan; Creslan; Dynel)	Yarns; wigs; paints
$CH_2{=}CH-OCOCH_3$	$\left[\begin{array}{cc} H & H \\ -C-C- \\ H & O-C-CH_3 \\ & \parallel \\ & O \end{array}\right]_n$	Polyvinyl acetate	Adhesives; textile coatings; chewing gum resin; paints
$CH_2{=}C(CH_3)COOCH_3$	$\left[\begin{array}{cc} H & CH_3 \\ -C-C- \\ H & C-O-CH_3 \\ & \parallel \\ & O \end{array}\right]_n$	Polymethyl methacrylate (Lucite; Plexiglas)	Glass substitute; bowling balls

SELF-ASSESSMENT Questions

1. When addition polymers form, the monomer units are joined by
   a. delocalized electrons that flow along the polymer chain
   b. dispersion forces
   c. double bonds
   d. new bonds formed by electrons from the double bonds in the monomers

2. The monomer styrene (CH_2=CHC_6H_5, where C_6H_5 represents a benzene ring) forms a polymer that has chains with repeating
   a. eight-carbon units in the backbone
   b. six-carbon units in the backbone and two-carbon groups as pendants
   c. CH_2CH units in the backbone and C_6H_5 groups as pendants
   d. CH_2=CH units in the backbone and C_6H_5 groups as pendants

3. Water pipes, bottles, and raincoats are commonly made from
   a. polyethylene
   b. polypropylene
   c. polytetrafluoroethylene
   d. polyvinyl chloride

4. Which of the following is a conducting polymer?
   a. polyacetylene
   b. polypropylene
   c. polytetrafluoroethylene
   d. polyvinyl chloride

Answers: 1, d; 2, c; 3, d; 4, a

10.4 Rubber and Other Elastomers

Learning Objective • Define *cross-linking* and explain how it changes the properties of a polymer.

As noted in the introduction to this chapter, the search for a material to replace natural rubber when the supply was cut off during World War II was the basis for much of the development of the synthetic polymer industry.

Natural rubber can be broken down into simple hydrocarbon units called *isoprene*. Isoprene is a volatile liquid, whereas rubber is a semisolid, elastic material. Chemists can make polyisoprene, a substance identical to natural rubber, except that the isoprene comes from petroleum refineries rather than from the cells of rubber trees.

$$n\ CH_2=\underset{\underset{CH_3}{|}}{C}-CH=CH_2 \longrightarrow \left[CH_2-\underset{\underset{CH_3}{|}}{C}=CH-CH_2\right]_n$$

| Isoprene | Polyisoprene (rubber) |

Chemists have also developed several synthetic rubbers and devised ways to modify these polymers to change their properties.

Vulcanization: Cross-linking

The long-chain molecules that make up rubber are coiled and twisted and intertwined with one another. When rubber is stretched, its coiled molecules are straightened. Natural rubber is soft and tacky when hot. It can be made harder by reaction with sulfur. In this process, called **vulcanization**, sulfur atoms **cross-link** the hydrocarbon chains side-to-side (Figure 10.4). Charles Goodyear (United States, 1800–1860) discovered vulcanization and was issued U.S. Patent 3633 in 1844 for the process.

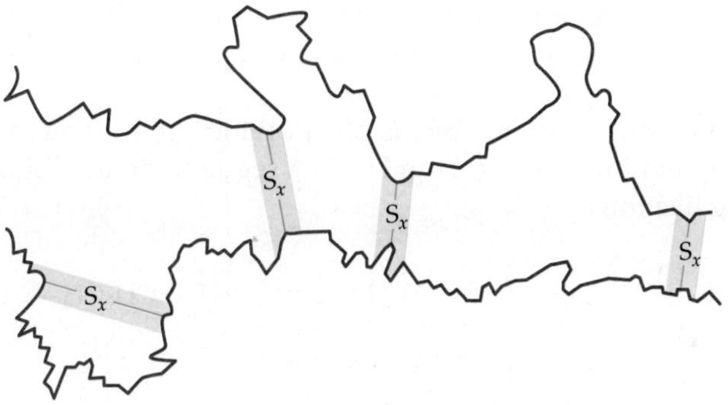

◀ **Figure 10.4** Vulcanized rubber has long hydrocarbon chains (represented here by red lines) cross-linked by short chains of sulfur atoms. The subscript x indicates a small, indefinite number of S atoms, usually not more than four.

Its cross-linked structure makes vulcanized rubber a harder, stronger substance that is suitable for automobile tires. Surprisingly, cross-linking also improves the elasticity of rubber. With just the right degree of cross-linking, the individual chains are still free to uncoil and stretch somewhat. When stretched vulcanized rubber is released, the cross-links pull the chains back to their original arrangement (Figure 10.5). Rubber bands owe their snap to this sort of molecular structure. Materials that can be extended and will return to their original size are called **elastomers**.

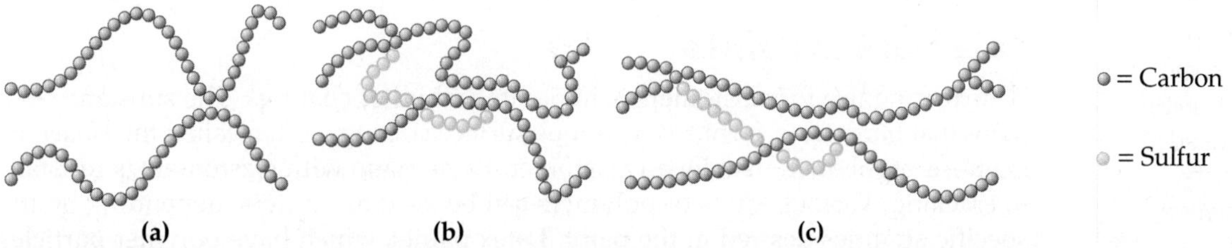

▲ **Figure 10.5** Vulcanization of rubber cross-links the hydrocarbon chains. (a) In unvulcanized rubber, the chains slip past one another when the rubber is stretched. (b) Vulcanization involves the addition of sulfur cross-links between the chains. (c) When vulcanized rubber is stretched, the sulfur cross-links prevent the chains from slipping past one another. Vulcanized rubber is stronger than unvulcanized rubber.

Synthetic Rubber

Natural rubber is a polymer of isoprene, and some synthetic elastomers are closely related to natural rubber. For example, polybutadiene is made from the monomer butadiene ($CH_2{=}CHCH{=}CH_2$), which differs from isoprene only in that it lacks a methyl group on the second carbon atom. Polybutadiene is made rather easily from this monomer.

$$n\,CH_2{=}CH{-}CH{=}CH_2 \longrightarrow {+}CH_2{-}CH{=}CH{-}CH_2{+}_n$$

However, this polymer has only fair tensile strength, and has poor resistance to gasoline and oils. These properties limit its value for automobile tires, the main use of elastomers.

Another synthetic elastomer, polychloroprene (Neoprene), is made from a monomer similar to isoprene but with a chlorine (shown in green) in place of the methyl group on isoprene.

$$n\,CH_2{=}\underset{\displaystyle Cl}{\underset{|}{C}}{-}CH{=}CH_2 \longrightarrow {+}CH_2{-}\underset{\displaystyle Cl}{\underset{|}{C}}{=}CH{-}CH_2{+}_n$$

3 Are there polymers in chewing gum?

Chewing gum has always contained polymers. Originally, it was the sap of the sapodilla tree, called "chicle," and an early brand of chewing gum was named Chiclets. Since World War II, the base of chewing gum has been synthetic rubbers, like polyethylene or polyvinyl acetate.

WHY IT MATTERS

A polymer related to SBR is poly(styrene-butadiene-styrene), or SBS. Called a *block copolymer*, SBS has molecules made up of three segments: One end is a chain of polystyrene repeat units; the middle is a long chain of polybutadiene repeat units; and the other end is another chain of polystyrene repeat units. SBS is a hard rubber used in items where durability is important, such as shoe soles and tire treads.

Neoprene is more resistant to oil and gasoline than other elastomers are. It is used to make gasoline pump hoses and similar items used at automobile service stations.

Styrene–butadiene rubber (SBR) is a copolymer of styrene (about 25%, and shown in green) and butadiene (about 75%). A segment of an SBR molecule might look something like this.

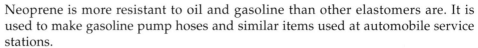

| Butadiene unit | Styrene unit | Butadiene unit | Butadiene unit |

SBR is more resistant to oxidation and abrasion than natural rubber, but it has poorer physical strength and resilience. Like those of natural rubber, SBR molecules contain double bonds and can be cross-linked by vulcanization. SBR accounts for about a third of the total U.S. production of elastomers, and is used mainly for making tires.

Polymers in Paints

A surprising use for elastomers is in paints and other coatings. The substance in a paint that hardens to form a continuous surface coating—often called the *binder*, or resin—is a polymer, usually an elastomer. Paint made with elastomers is resistant to cracking. Various kinds of polymers can be used as binders, depending on the specific qualities desired in the paint. Latex paints, which have polymer particles dispersed in water and thus avoid the use of organic solvents, are most common. Brushes and rollers can easily be cleaned in soap and water. This replacement of the hazardous organic solvents historically used in paints with water is a good example of green chemistry.

▲ Synthetic polymers serve as binders in paints. Pigments provide color and opacity to the paint. Titanium dioxide (TiO_2), a white solid, is the most widely used pigment.

SELF-ASSESSMENT Questions

1. Natural rubber is a polymer of
   a. butadiene
   b. isobutylene
   c. isoprene
   d. propylene

2. Which of the following is the structure of isoprene?
   a. $CH_2 = CH - C \equiv N$
   b. $CH_2 = C(CH_3) - CH_3$
   c. $CH_2 = CCl - CH - CH_2$
   d. $CH_2 = C(CH_3) - CH = CH_2$

3. A cross-linking agent is a(n)
   a. atom or group that bonds polymer chains together
   b. functional group on a polymer chain
   c. repeating unit in a polymer chain
   d. substituent group on a polymer chain

4. Charles Goodyear discovered that natural rubber could be cross-linked by heating it with
   a. a binder
   b. carbon
   c. isoprene
   d. sulfur

5. Styrene–butadiene rubber is an example of a
   a. copolymer
   b. natural elastomer
   c. natural cross-linked polymer
   d. polyamide

6. Polymers, usually elastomers, are used in paints as the
   a. binder
   b. initiator
   c. pigment
   d. solvent

7. Materials that can be extended and will then return to their original size are
   a. binders
   b. copolymers
   c. elastomers
   d. polyamides

Answers: 1, c; 2, d; 3, a; 4, d; 5, a; 6, a; 7, c

10.5 Condensation Polymers

Learning Objectives • Differentiate between addition and condensation polymerization. • Write the structures of the monomers that form polyesters and polyamides.

The polymers considered so far are all addition polymers. All the atoms of the monomer molecules are incorporated into the polymer molecules. In a *condensation polymer*, part of the monomer molecule is not incorporated in the final polymer. During **condensation polymerization**, also called *step-reaction polymerization*, a small molecule—usually water (shown in green in the following examples) but sometimes methanol, ammonia, or HCl—is formed as a by-product.

Nylon and Other Polyamides

As an example, let's consider the formation of nylon. The monomer in one type of nylon, called *nylon 6*, is a six-carbon carboxylic acid (magenta) with an amino (blue) group on the sixth carbon atom: 6-aminohexanoic acid ($HOOCCH_2CH_2CH_2CH_2CH_2NH_2$). (There are several different nylons, each prepared from a different monomer or set of monomers, but all share certain common structural features.)

In this polymerization reaction, a carboxyl group of one monomer molecule forms an amide bond with the amine group of another.

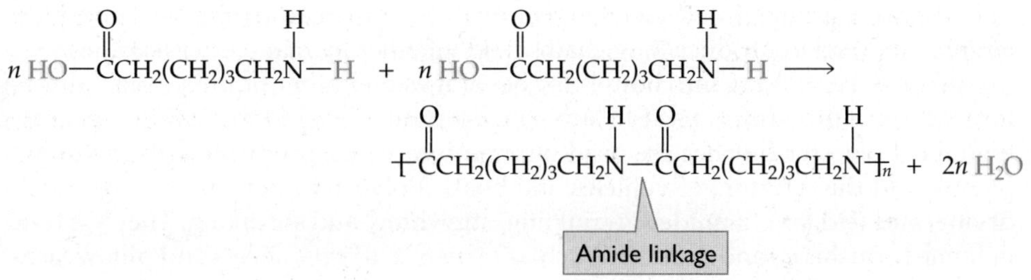

$$n \; HO-\overset{\overset{O}{\|}}{C}CH_2(CH_2)_3CH_2\overset{\overset{H}{|}}{N}-H \;+\; n \; HO-\overset{\overset{O}{\|}}{C}CH_2(CH_2)_3CH_2\overset{\overset{H}{|}}{N}-H \longrightarrow$$

$$\Big[\overset{\overset{O}{\|}}{C}CH_2(CH_2)_3CH_2\overset{\overset{H}{|}}{N}-\overset{\overset{O}{\|}}{C}CH_2(CH_2)_3CH_2\overset{\overset{H}{|}}{N}\Big]_n \;+\; 2n \; H_2O$$

Amide linkage

For each amide bond made, a water molecule is formed as a by-product. This formation of a nonpolymeric by-product distinguishes condensation polymerization from addition polymerization. Note that the formula of a repeat unit of a condensation polymer is *not* the same as that of the monomer.

Because the linkages holding the polymer together are amide bonds, nylon 6 is a **polyamide**. Another nylon is made by the condensation of two different monomers: 1,6-hexanediamine ($H_2NCH_2CH_2CH_2CH_2CH_2CH_2NH_2$) and adipic acid ($HOOCCH_2CH_2CH_2CH_2COOH$). Each monomer has six carbon atoms; this polymer is called *nylon 66*.

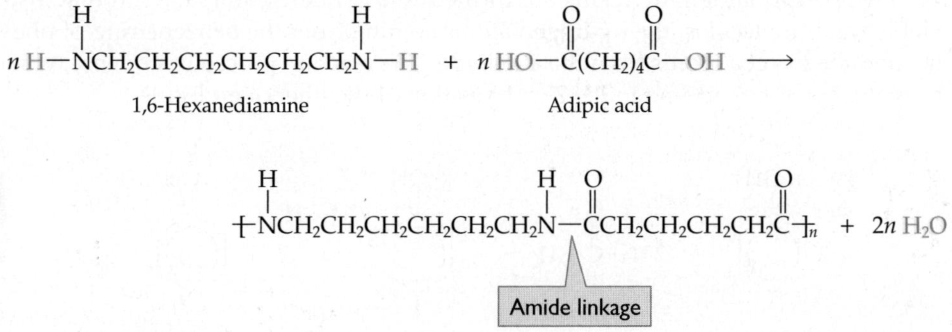

$$n \; H-\overset{\overset{H}{|}}{N}CH_2CH_2CH_2CH_2CH_2CH_2\overset{\overset{H}{|}}{N}-H \;+\; n \; HO-\overset{\overset{O}{\|}}{C}(CH_2)_4\overset{\overset{O}{\|}}{C}-OH \longrightarrow$$

1,6-Hexanediamine   Adipic acid

$$\Big[\overset{\overset{H}{|}}{N}CH_2CH_2CH_2CH_2CH_2CH_2\overset{\overset{H}{|}}{N}-\overset{\overset{O}{\|}}{C}CH_2CH_2CH_2CH_2\overset{\overset{O}{\|}}{C}\Big]_n \;+\; 2n \; H_2O$$

Amide linkage

This was the original nylon polymer discovered in 1937 by DuPont chemist Wallace Carothers (United States, 1896–1937). Note that one monomer has two amino groups and the other has two carboxyl groups, but the product is still a polyamide, quite similar to nylon 6. Silk and wool, which are protein fibers, are natural polyamides.

Although nylon can be molded into various shapes, most nylon is made into fibers. Some is spun into fine thread to be woven into silk-like fabrics, and some is made into yarn that is much like wool. Carpets were once made mainly from wool or cotton, but now at least 90% of carpets are made from nylon.

Polyethylene Terephthalate and Other Polyesters

A **polyester** is a condensation polymer made from molecules with alcohol and carboxylic acid functional groups. The most common polyester is made from ethylene glycol and terephthalic acid. It is called *polyethylene terephthalate (PET)*.

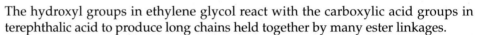

▲ Mylar balloons are very commonly used for celebrations and other festive occasions and events. They are made from colorful polyester films and are filled with helium.

The hydroxyl groups in ethylene glycol react with the carboxylic acid groups in terephthalic acid to produce long chains held together by many ester linkages.

PET can be molded into bottles for beverages and other liquids. It can also be formed into a film that is used to laminate documents and to make tough packaging tape. Polyester finishes are used on premium wood products such as guitars, pianos, and the interiors of vehicles and boats. Polyester fibers are strong, quick-drying, and resistant to mildew, wrinkling, stretching, and shrinking. They are used in home furnishings and products such as carpets, curtains, sheets and pillowcases, and upholstery. Because polyester fibers do not absorb water, they are ideal for outdoor clothing to be worn in wet and damp environments and for insulation in boots and sleeping bags. For other clothing, they are often blended with cotton for a more natural feel. A familiar use of polyester film (Mylar) is to make the shiny balloons that are filled with helium to celebrate special occasions.

Phenol-Formaldehyde and Related Resins

Let's go back to Bakelite, the original synthetic polymer. Bakelite, a phenol–formaldehyde resin, was first synthesized by Leo Baekeland (United States, 1863–1944), who received U.S. Patent 942,699 for the process in 1909. Phenolic resins are no longer important as industrial polymers, but they are used as a substitute for porcelain and in board and tabletop game pieces such as billiard balls, dominoes, and checkers.

Phenol–formaldehyde resins are formed in a condensation reaction that also yields water molecules, the hydrogen atoms coming from the benzene ring of phenol and the oxygen atoms from the aldehyde. The reaction proceeds stepwise, with formaldehyde first adding to the 2 or 4 position of the phenol molecule.

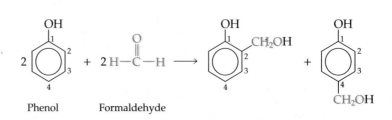

The substituted phenol molecules then link up as water molecules are formed. (Remember that there are hydrogen atoms at all the unsubstituted corners of a benzene ring.) The hookup of molecules continues until an extensive network is achieved.

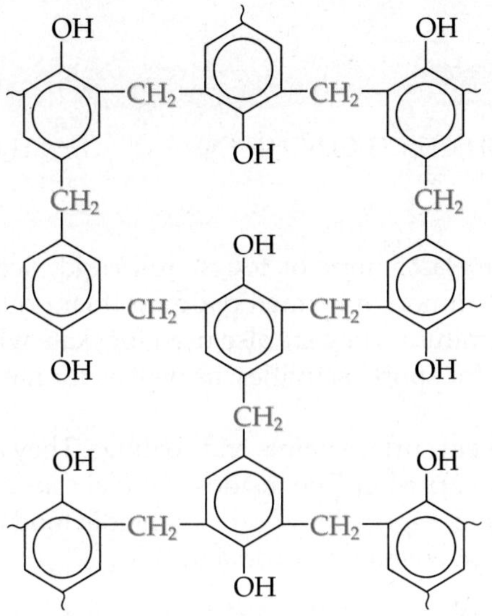

Phenol–formaldehyde resin

Water produced by the reaction is driven off by heat as the polymer sets. The structure of the polymer is an extremely complex three-dimensional network somewhat like the framework of a giant building. Note that the phenolic rings are joined together by CH_2 units from the formaldehyde. These network polymers are **thermosetting resins**; they cannot be melted and remolded. Instead, they decompose when heated to high temperatures.

Formaldehyde can also be condensed with urea $[H_2N(C=O)NH_2]$ to make urea–formaldehyde resins and with melamine to form melamine–formaldehyde resins. (Melamine is formed by condensation of three molecules of urea; see structure to the right.)

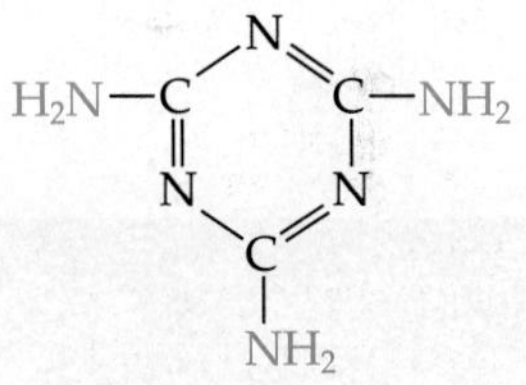

These resins, like phenolic resins, are thermosetting. The polymers are complex three-dimensional networks formed by the condensation reaction of formaldehyde ($H_2C=O$) molecules and amino ($-NH_2$) groups (in blue). Urea–formaldehyde resins are used to bind wood chips together in panels of particle board. Melamine–formaldehyde resins are used in plastic (Melmac) dinnerware and laminate countertops.

Melamine

Other Condensation Polymers

There are many other kinds of condensation polymers, but we will look at only a few.

Polycarbonates are "clear-as-glass" polymers tough enough to be used in bulletproof windows. They are also used in protective helmets, safety glasses, clear plastic water bottles, baby bottles, and even dental crowns. One polycarbonate is made from bisphenol-A (BPA) and phosgene ($COCl_2$). The products will have a carbonate group (shown in green).

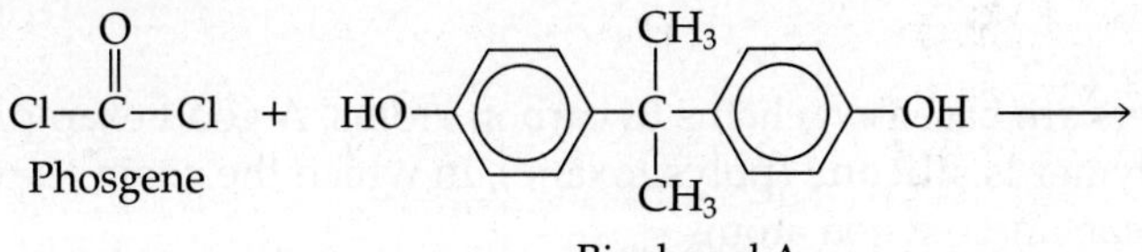

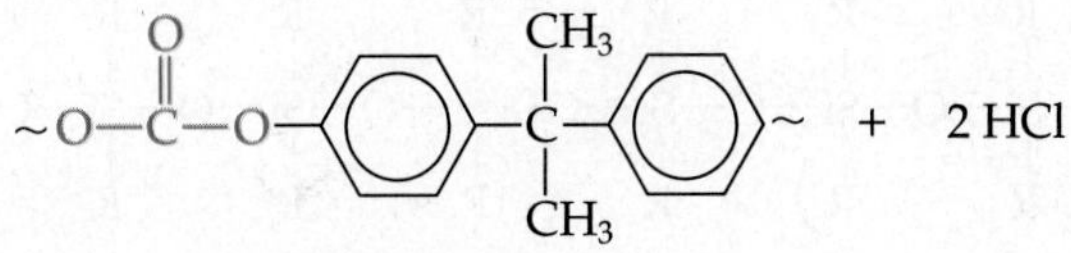

Phosgene   Bisphenol A

A polycarbonate

WHY IT MATTERS

Bisphenol-A (BPA) is an endocrine disruptor and can act like a hormone. A tiny amount of BPA leaches out of polycarbonate bottles, leading to long-term, low-dose exposure. Canada banned BPA in baby bottles in 2008. Many manufacturers in the United States and elsewhere have switched to making BPA-free bottles.

Polyurethanes are similar to nylon polymers in structure, except that they have an isocyanate group (—N=C=O) that reacts with an alcohol group (—OH) to form a —NHCOO— bond rather than an amide bond. The repeat unit in one common polyurethane is

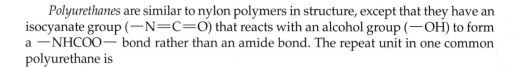

Polyurethanes may be elastomeric or tough and rigid, depending on the monomers used. They are common in foamed padding (foam rubber) in cushions, mattresses, and padded furniture. They are also used for skate wheels, in running shoes, and in protective gear for sports activities, as well as for hard lacquer-like coatings for wood.

Epoxies make excellent surface paints and coatings. They are used to protect steel pipes and fittings from corrosion. The insides of metal cans are often coated with an epoxy to prevent rusting, especially in cans for acidic foods like tomatoes. A common epoxy is made from epichlorohydrin and BPA.

An epoxy

Epoxies also make powerful adhesives. These adhesives usually have two components that are mixed just before use. The polymer chains become cross-linked, and the bonding is extremely strong.

Composite Materials

Composite materials are made up of high-strength fibers (of glass, graphite, synthetic polymers, or ceramics) held together by a polymeric matrix, usually a thermosetting condensation polymer. The fiber reinforcement provides the support, and the surrounding plastic keeps the fibers from breaking.

Among the most commonly used composite materials are polyester resins reinforced with glass fibers. These are widely used in boat hulls, molded chairs, automobile panels, and sports gear such as tennis rackets. A notable example is their use in poles used for pole vaulting. The world record is now over 75% greater than what could be achieved with wood poles. Some composite materials have the strength and rigidity of steel but with only a fraction of the weight of steel.

▲ Composite construction using carbon fiber and epoxies makes bicycle frames strong and light (only about 1 kg). The direction of the carbon fiber allows the maker to tailor the material for maximum strength where needed.

Silicon is in the same group (4A) as carbon and, like carbon, is tetravalent and able to form chains. However, carbon can form chains consisting only of carbon atoms (as in polyethylene), whereas the chains of silicone polymers have alternating silicon and oxygen atoms.

Silicones

Not all polymers are based on chains of carbon atoms. A good example of a different type of polymer is **silicone** (polysiloxane), in which the chains have a series of alternating silicon and oxygen atoms.

(In simple silicones, R represents a hydrocarbon group, such as methyl, ethyl, or butyl.)

Silicones can be linear, cyclic, or cross-linked networks. They are heat-stable and resistant to most chemicals, and are excellent waterproofing materials. Depending on chain length and amount of cross-linking, silicones can be oils or greases, rubbery compounds, or solid resins. Silicone oils are used as hydraulic fluids and lubricants. Other silicones are used in such products as sealants, auto polish, shoe polish, and waterproof sheeting. Fabrics for raincoats and umbrellas are frequently treated with silicone.

Perhaps the most remarkable silicones are the ones used for synthetic human body parts, ranging from finger joints to eye sockets. Artificial ears and noses are also made from silicone polymers. These can even be specially colored to match the surrounding skin.

4 Why can silicone be used for flexible baking pans?

The silicone material is temperature-resistant from the range of $-55°C$ to $300 °C$ ($-67°F$ to $570 °F$). It retains its flexibility, shape, and other properties. It is very stable and durable, though it is often reinforced to give it better strength and rigidity.

EXAMPLE 10.3 Condensed Structural Formulas for Polymers

Write the condensed structural formula for the polymer formed from dimethylsilanol, $(CH_3)_2Si(OH)_2$.

Solution
Let's start by writing the structural formula of the monomer.

$$
\begin{array}{c}
CH_3 \\
| \\
HO-Si-OH \\
| \\
CH_3
\end{array}
$$

▲ Cookware made of silicone is colorful, flexible, and nonstick.

Because there are no double bonds in this molecule, we do not expect it to undergo addition polymerization. Rather, we expect a condensation reaction, in which an OH group from one molecule and an H atom from another combine to form a molecule of water. Moreover, because there are two OH groups per molecule, each monomer can form bonds with two other monomers (one on each side). This is a key requirement for polymerization. We can represent the reaction as follows.

$$
\underset{\underset{CH_3}{|}}{\overset{\overset{CH_3}{|}}{HO-Si-OH}} + \underset{\underset{CH_3}{|}}{\overset{\overset{CH_3}{|}}{HO-Si-OH}} + \underset{\underset{CH_3}{|}}{\overset{\overset{CH_3}{|}}{HO-Si-OH}} + \cdots \longrightarrow \left[\underset{\underset{CH_3}{|}}{\overset{\overset{CH_3}{|}}{Si-O}}\right]_n + n\,H_2O
$$

> **Exercise 10.3A**

  a. Write the structural formula for the polymer formed from glycolic acid (hydroxyacetic acid, $HOCH_2COOH$), showing at least four repeat units.

  b. Write a condensed structural formula in which the repeat unit in part a is shown in brackets.

> **Exercise 10.3B**

  a. Write the structural formula for the polymer formed from 4-aminobutanoic acid ($NH_2CH_2CH_2CH_2COOH$), showing at least four repeat units.

  b. Write a condensed structural formula in which the repeat unit in part a is shown in brackets.

SELF-ASSESSMENT Questions

1. Condensation polymerization produces small molecules, usually water, and requires
   a. the presence of a carbon–carbon double bond
   b. the presence of a carbon–oxygen double bond
   c. two monomers with at least two functional groups each
   d. two monomers with only one functional group each

2. The backbone of a silicone polymer chain is composed of the repeating unit
   a. $\sim OCH_2CH_2OSi\sim$
   b. $\sim OCH_2CH_2Si\sim$
   c. $\sim SiO\sim$
   d. $\sim OSiSi\sim$

3. Which of the following molecules can serve as the sole monomer for a polyamide?
   a. $H_2NCH_2CH_2CH_2CH_2OH$
   b. $H_2NCH_2CH_2CH_2CH_2NH_2$
   c. $HOOCCH_2CH_2CH_2CH_2CH_2NH_2$
   d. $HOOCCH_2CH_2CH_2CONH_2$

4. Thermosetting polymers
   a. are easily recycled
   b. are always addition polymers
   c. cannot be melted and remolded
   d. usually melt at a lower temperature than do thermoplastic polymers

5. Which of the following pairs of molecules can form a polyester?
   a. $HOCH_2CH_2OH$ and $HOCH_2CH_2CH_2CH_2OH$
   b. $H_2NCH_2CH_2CH_2CH_2CH_2NH_2$ and $HOOCCH_2CH_2CH_2CH_2COOH$
   c. $H_2NCH_2CH_2CH_2CH_2CH_2NH_2$ and $HOCH_2CH_2CH_2CH_2CH_2OH$
   d. $HOOCCH_2CH_2CH_2CH_2COOH$ and $HOCH_2CH_2CH_2CH_2OH$

Answers: 1, c; 2, c; 3, c; 4, c; 5, d

10.6 Properties of Polymers

Learning Objectives • Explain the concept of the glass transition temperature.
• Explain how crystallinity affects the physical properties of polymers.

Polymers differ from substances consisting of small molecules in three main ways. First, the long chains can be entangled with one another, much like the strands in a dish of spaghetti. It is difficult to untangle the polymer molecules, especially at low temperatures. This property lends strength to many plastics, elastomers, and other materials.

Second, although intermolecular forces affect polymers just as they do small molecules, these forces are greatly multiplied for large molecules because the number of intermolecular interactions is so much greater. The larger the molecules are, the greater the intermolecular forces between them are. Even when ordinarily weak dispersion forces are the only intermolecular forces present, they can bind polymer chains together strongly. This property also provides strength in polymeric materials. For example, polyethylene is nonpolar, with only dispersion forces between its molecules. As noted in Section 10.2, though, ultra-high-molecular-weight polyethylene forms fibers so strong they can be used in bulletproof vests.

Third, large polymer molecules move more slowly than small molecules do. A group of small molecules (monomers) can move around more rapidly and more randomly when independent than they can when joined together in a long chain (polymer). The slower molecular speed makes a polymeric material different from one made of small molecules. For example, a polymer dissolved in a solvent will form a solution that is a lot more viscous than the pure solvent is and that flows more slowly than the pure solvent does.

Crystalline and Amorphous Polymers

Some polymers are highly crystalline; their molecules line up neatly to form long fibers of great strength. Other polymers are largely amorphous, composed of randomly oriented molecules that get tangled up with one another (Figure 10.6). Crystalline polymers tend to make good synthetic fibers, while amorphous polymers are often elastomers.

Sometimes the same polymer is crystalline in one region and amorphous in another. For example, scientists designed spandex fibers (used in stretch fabrics [e.g., Lycra] in ski pants, exercise clothing, and swimsuits) to combine the tensile

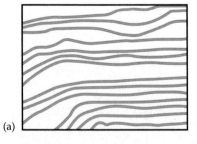
(a)

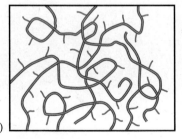
(b)

▲ **Figure 10.6** Organization of polymer molecules. (a) Crystalline arrangement. (b) Amorphous arrangement.

strength of crystalline fibers with the elasticity of amorphous rubber. Two molecular structures are combined in one polymer chain, with crystalline blocks alternating with amorphous blocks. The amorphous blocks are soft and rubbery, while the crystalline parts are quite rigid. The resulting polymer exhibits both sets of properties—flexibility and rigidity.

The Glass Transition Temperature

An important property of most thermoplastic polymers is the **glass transition temperature (T_g)**. Above this temperature, the polymer is rubbery and tough. Below it, the polymer is like glass: hard, stiff, and brittle. Each polymer has a characteristic T_g. We want automobile tires to be tough and elastic, so we make them from polymers with low T_g values. On the other hand, we want plastic substitutes for glass to look like glass. Therefore, these polymers have T_g values well above room temperature. We can apply the glass transition temperature in everyday life. For example, we can remove chewing gum from clothing by applying ice to lower the temperature of the polyvinyl acetate resin that gives the gum its "chewiness" below its T_g. The cold, brittle resin then crumbles readily and can be removed.

Fiber Formation

Not all synthetic polymers can be formed into useful fibers, but those that can often yield fibers with properties superior to those of natural fibers. More than half of the 60 billion kg of textile fibers produced each year in the United States is synthetic.

Silk fabrics are beautiful and have a luxurious "feel," but nylon fabrics are also attractive and feel much like silk. Moreover, nylon fabrics wear longer, are easier to care for, and are less expensive than silk.

Polyesters such as PET can substitute for either cotton or silk, but they outperform the natural fibers in many ways. Polyesters are not subject to mildew, as cotton is, and many polyester fabrics do not need ironing. Fabrics in which polyester fibers are blended with 35% to 50% cotton combine the comfort of cotton with the no-iron easy care of polyester.

Acrylic fibers, made from polyacrylonitrile, can be spun into yarns that look like wool. Acrylic sweaters have the beauty and warmth of wool, but they do not shrink in hot water, are not attacked by moths, and do not cause the allergic skin reaction wool causes in some people.

Exceedingly fine fibers—called *microfibers* because they measure less than 10 μm in diameter, half as thick as silk fibers—can be made from almost any fiber-forming polymer, but most are made of polyesters or polyamides, or a combination of the two. Varying the size, shape, and composition of the fibers results in products that are water repellent (for apparel) or absorbent (for a mop). Microfibers can be shaped for efficient trapping of dust or for wicking away liquids. These fibers already have many uses and are likely to have many more.

One downside to synthetic fibers is that they shed small microfibers when washed. Microfibers do not break down the way natural fibers do. They can accumulate in marine environments and present a hazard to aquatic animals and throughout the food chain.

5 What fabrics are made from natural polymers?

Cotton, silk, wool, and linen are all naturally occurring polymers that can be made into fabric. Other materials like rayon, which is made from naturally occurring cellulose but is synthetically modified and manufactured into thread, are considered semisynthetic.

▲ Formation of fibers by extrusion through a spinneret. A melted polymer is forced through the tiny holes to make fibers that solidify as they cool.

SELF-ASSESSMENT Questions

1. Crystalline polymers are
   a. elastic
   b. flexible
   c. highly cross-linked
   d. used in fibers

2. The temperature at which a polymer changes from soft and rubbery to glassy and hard is its
   a. T_c
   b. T_f
   c. T_g
   d. T_i

3. A material that is flexible in boiling water but rigid at room temperature would most likely have which of the following as its T_g?
   a. −20 °C
   b. 15 °C
   c. 75 °C
   d. 150 °C

Answers: 1, d; 2, c; 3, c

▲ Huge amounts of discarded plastics wind up in the oceans each year. Small plastic pieces carried by ocean currents migrate and accumulate in certain areas, such as the regions of the Pacific Ocean, which have much higher concentrations of plastic pieces than many other areas.

10.7 Plastics and the Environment

Learning Objectives • Describe the environmental problems associated with plastics and plasticizers. • Name two types of sustainable, nonpetroleum-derived polymers and give their sources.

An advantage of many plastics is that they are durable and resistant to environmental conditions. This can also be a disadvantage, as they are so resistant that they last almost forever. Once plastic objects are dumped, they do not go away. You see them littering our parks, our sidewalks, our highways, and our coasts, and if you looked into the middle of the ocean, you would see them there, too. Roughly 8 billion kg of plastic finds its way into the oceans each year. Small plastic pieces are carried by ocean currents and tend to accumulate in certain areas. In the Pacific Ocean, for example, there is a large area north of the Hawaiian Islands and another east of Japan that have much higher concentrations of plastic pieces than other areas. Many small fish have been found dead with their digestive tracts clogged by bits of plastic foam or microfibers ingested with their food. Scientists estimate that plastic bags and other plastic trash thrown into the ocean kill as many as a million sea creatures each year.

Plastics make up about 13% by mass of solid waste in the United States but about 25% by volume. Their bulk creates a problem because 55% of all solid waste goes into landfills, and it is increasingly difficult to find suitable landfill space. (Solid wastes are discussed in more detail in Chapter 12.) Americans use about 60 million plastic bottles each day, with only about 12% being recycled. Recycling just one plastic bottle conserves enough energy to burn a 60-W light bulb for six hours. Recycling plastics reduces our dependency on oil and decreases CO_2 emissions that result from the use of fossil fuels.

Another way to dispose of discarded plastics is to burn them. Most plastics have a high fuel value. For example, a pound of polyethylene has about the same energy content as a pound of fuel oil. Heat from community garbage incinerators can generate electricity, and some utility companies use powdered coal mixed with a small proportion of ground-up rubber tires as fuel, which also helps with the problem of tire disposal.

On the other hand, burning plastics and rubber can create new problems. For example, PVC produces toxic hydrogen chloride and vinyl chloride gases when it burns, while automobile tires give off soot and a stinking smoke. Incinerators can be corroded by acidic fumes and clogged by materials that are not readily burned.

Degradable Plastics

About half of the waste plastic generated in the United States is from packaging. One approach to the problem of disposal of plastics is to make plastic packages that are either biodegradable or photodegradable (broken down in the presence of bacteria or light, respectively). Of course, such packages must remain intact and not start to decompose while still being used. There has recently been an increase in research into new polymers, such as polylactic acid (PLA) made from corn or sugar cane or a number of polyhydroxyalkanoates (PHAs) made by microorganisms, and new plastic food containers and tableware that are degradable are being produced.

Recycling

Recycling is perhaps the best way to handle waste plastics. The plastics must be collected, sorted, chopped, melted, and then remolded. Collection works well when there is strong community cooperation. The separation step is simplified by code numbers stamped on plastic containers. Once the plastics have been separated, they can be chopped into flakes, melted, and remolded or spun into fibers.

6 Why is it so important to recycle plastics?

Polyethylene and most other polymers are made of petroleum. Since our supply of petroleum is limited, from a sustainability standpoint, we can make the best use of that resource by reusing those polymers. Most plastic items are imprinted with a code number so that they can be sorted for recycling.

Almost all plastics can theoretically be recycled, but it is not always feasible to do so. Use of "single-stream" recycling allows people to throw most of their recyclables in one container, but disposable plastic bags cannot be put into single-stream recycling because they clog up the sorters used for single-stream systems. Currently only two kinds of plastics are recycled on a large scale: PET (28% recycled) and HDPE (29%). Recycling is green because it keeps plastics out of landfills, and it increases sustainability by lessening the use of petroleum-derived monomers or hazardous chemicals. We have a long way to go; in 2016 less than 10% of the nearly 30 billion kg of plastic wastes in the United States was recycled.

▲ Recycled plastic can be used for many things. These Adirondack chairs are colorful, strong, and made from recycled milk jugs.

Plastics and Fire Hazards

The accidental ignition of fabrics, synthetic or other, has caused untold human misery. The U.S. Department of Health and Human Services estimates that fires involving flammable fabrics kill several thousand people annually and injure between 150,000 and 200,000.

Research has led to a variety of flame-retardant fabrics. Many incorporate chlorine and bromine atoms within the polymeric fibers. Federal regulations require that children's sleepwear be made of such flame-retardant materials. Another synthetic fabric, meta-aramid (or Nomex), is made of fibers that don't ignite or melt when exposed to flames or high heat. Nomex has such high heat resistance that it is used for protective clothing for firefighters and race-car drivers. It is also used in electric insulation and for machine parts exposed to high heat.

Burning plastics often produce toxic gases. Hydrogen cyanide is formed in large quantities when polyacrylonitrile and other nitrogen-containing polymers burn. Lethal amounts of cyanide found in the bodies of victims of plane crashes have been traced to burned plastics. Firefighters often refuse to enter burning buildings without gas masks, for fear of being overcome by fumes from burning plastics. Smoldering fires also produce lethal quantities of carbon monoxide. To make plastics safer, chemists are making new kinds of polymers that don't burn or that don't generate toxic chemicals when burned.

Plasticizers and Pollution

Chemicals used in plastics manufacture can also present problems. Plasticizers are an important example. Some plastics, particularly vinyl polymers, are hard and brittle, and thus difficult to process. A **plasticizer** can make such a plastic more flexible and less brittle by lowering its glass transition temperature. Unplasticized PVC is rigid and is used for water pipes. Raincoats, garden hoses, and seat covers for automobiles can be made from plasticized PVC. Plasticizers are liquids of low volatility, but they are generally lost by diffusion and evaporation as a plastic article ages. The plastic becomes brittle and then cracks and breaks.

Once used widely as plasticizers but now banned, polychlorinated biphenyls (PCBs) are derived from biphenyl ($C_{12}H_{10}$), a hydrocarbon that has two benzene rings joined at a corner. In PCBs, some of the hydrogen atoms of biphenyl are replaced with chlorine atoms (Figure 10.7, in green). Note that PCBs are structurally similar to the insecticide DDT.

PCBs were also widely used as insulating materials in electric transformers, because their stability and low polarity gave them high electrical resistance and the ability to absorb heat. However, they degrade slowly in nature, and their solubility in nonpolar substances—animal fat as well as vinyl plastics—leads to their becoming concentrated in the food chain. PCB residues have been found in fish, birds, water, and sediments. The physiological effect of PCBs is similar to that of DDT. Production of PCBs in the United States was discontinued in 1977, but the compounds still persist in the environment. Today, the most widely used plasticizers for vinyl plastics are phthalate esters, a group of diesters derived from phthalic acid (1,2-benzenedicarboxylic acid). Phthalate plasticizers (Figure 10.8) have low acute

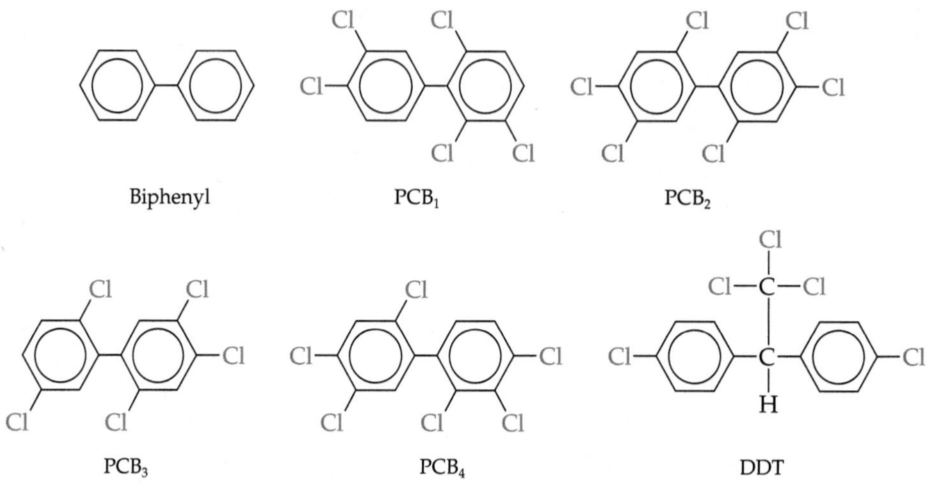

▶ **Figure 10.7** Biphenyl and some of the PCBs derived from it (only a few of the hundreds of possible PCBs). DDT is also shown for comparison.

toxicity, and although some studies have suggested possible harm to young children, the FDA has recognized these plasticizers as being generally safe. They pose little threat to the environment because they degrade fairly rapidly. Another approach of chemists is to make plastics that already have the right amount of flexibility and thus do not require plasticizers when they are manufactured.

Plastics and the Future

Widely used today, synthetic polymers are the materials of the future. New kinds of plastics and new uses will be discovered. We already have polymers that conduct electricity, amazing adhesives, and synthetic materials that are stronger than steel but much lighter in weight. Plastics present problems, but they have become such an important part of our daily lives that we would find it difficult to live without them.

In medicine, body replacement parts made at least partially from polymers have become common. There are about 300,000 total hip replacements, 700,000 knee replacements, and over 3.6 million cataract surgeries in the United States each year. Artificial lungs and artificial hearts are available, but they are very expensive, and their use is currently limited to periods of recovery from injury or illness, or while waiting for donor organs for transplantation. In the future, the cost of replacements will likely drop, while their efficacy will probably improve.

PVC water pipes, siding, window frames, plastic foam insulation, and polymeric surface coatings are used in home construction today. Some homes also contain lumber and wall panels of artificial wood made from recycled plastics.

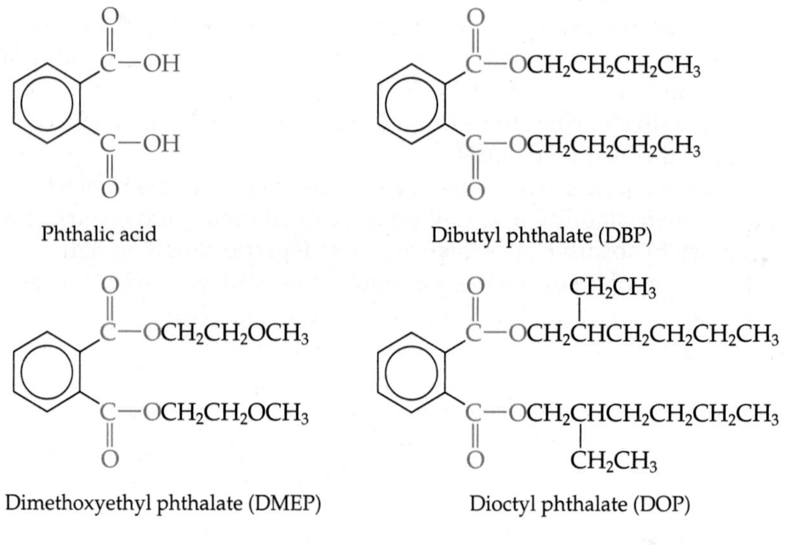

▶ **Figure 10.8** Phthalic acid (carboxylic acid groups in magenta) and some esters (ester groups in green) derived from it. Dioctyl phthalate is also called di-2-ethylhexyl phthalate.

GREEN CHEMISTRY Life-Cycle Impact Assessment of New Products
Eric J. Beckman, *University of Pittsburgh*

Principles 1, 2, 3, 4, 6, 7, 10

Learning Objective ▪ Explain how life-cycle impact analysis can be used to identify the environmental benefits and effects of newly designed products.

In any proposed design, a key question is, "How do we know that what we've made is an improvement?" The 12 principles of green chemistry provide strategies and outcomes to help chemical professionals create inherently safer products. Yet well-intentioned attempts to create safer products have sometimes led to unintended and undesirable effects. For example, mobile phones allowed developing countries to leapfrog to high-speed communication without grids of wire, which substantially reduced material use and energy. However, these devices rapidly became "disposable" products (average lifetime, 18 months), and they contain toxic and conflict metals including cadmium and tantalum (see Section 12.4). Can green chemistry principles help us avoid unfortunate trade-offs?

New products are assessed against specifications that depend on user requirements, so we can "grade" a new compound or product based on its environmental footprint. One way to do this is life-cycle impact analysis (LCIA), which creates a report card for products in which "grades" are associated with environmental aspects. The first step in an LCIA is to outline a product's full life cycle. If, for example, we examine a polyethylene terephthalate (PET) bottle (see Section 10.5), we would include the manufacturing process that turns PET pellets into bottles, the polymerization process that forms PET from the monomers (ethylene glycol and dimethyl terephthalate), and the synthesis processes used to make the monomers. The outline covers this "chemical family tree" of PET back to the oil well or plant matter that provides the raw materials. Finally, a full life cycle also covers the bottle's end of life – is it sent to a landfill or recycled? Learning about the life cycle provides opportunities to apply green chemistry principles 7 and 10.

Once the life cycle is outlined, the *inventory phase* of the LCIA totals all of the inputs (mass, water, energy) and outputs for each step. Also, because no process is 100% efficient, the waste during each step (emitted to air, land, and water) is tallied, too. The inventory is important when applying green chemistry principles 1, 2, and 6.

In the next phase, inventory is converted to *impact*. Although there are several ways to accomplish this conversion, one method pioneered by the EPA. and widely used in North America is midpoint analysis. In midpoint analysis, the potential for a given chemical to cause harm is calculated for each impact category by comparing the chemical with a compound whose behavior is well studied. For example, the climate-change potential is calculated relative to that of CO_2. As such, a simple bar graph can show the impacts (in 12 categories) of the full life cycle of a given product compared to previous versions. Once the impact is known, green chemistry principles 3 and 4 become relevant.

The report card for previous and "improved" products shows whether or not we've created any unfortunate trade-offs (some impacts declined yet others dramatically increased). For example, when petrochemical-based raw materials and polymers are replaced with those created from soybeans or corn, we often see a significant drop in fossil-fuel use and greenhouse-gas emissions for the agricultural products, but eutrophication impacts increase dramatically (such as damaging algae blooms in estuaries due to high phosphate and nitrogen runoff from fertilizer use).

While LCIA can show which trade-offs have been created during design of an alternative, it cannot make value judgments as to which is better. As in the case of a student's report card, where GPA summarizes performance, the impacts in an LCIA can be averaged to create a total score. While simple, reliance on a single score leads to a loss of detail. Many questions remain about how to conduct an effective LCIA practice. Researchers are trying to resolve how to weight the various measurements, how to include toxicological phenomena like endocrine disruption, and how to include all of a product's often-complex supply chain, among other factors.

Life cycle impact analysis provides a way to explore the question, "Has my use of green design principles led to an alternative with a smaller environmental footprint?" LCIA rarely produces a resounding *"Yes, this new product is greener!"* Yet LCIA leads to insight into the ways that chemical products make an impact on our environment, shows where improvement is needed, and informs the development of green chemistry solutions.

Synthetic polymers are used extensively in airplane interiors. The bodies and wings of some planes and many automobile bodies are made of lightweight composite materials. Electrically conducting polymers will aid in making lightweight batteries for electric automobiles. More electrically conducting thermoplastics are being used in miniaturized circuits.

As people develop new "improved" and safer products for use, it is important to look at both the intended and the unintended impacts of those products on the environment. One approach is to carry out a complete life-cycle impact assessment (LCIA), which considers not only the intended usefulness of the product, but also the requirements for raw materials and resources to manufacture it as well as the ultimate fate of the product once it is no longer in use.

And here is something else to think about: Most synthetic polymers are made from petroleum or natural gas. These natural resources are nonrenewable, and our fossil fuel supply is limited. We are likely to run out of petroleum during this century. You might suppose that we would be actively conserving this valuable resource, but unfortunately this is not the case. We are taking petroleum out of the ground at a rapid rate, converting most of it to gasoline and other fuels, and then simply burning it. There are other sources of energy, but is there anything that can replace petroleum as the raw material for making plastics? Yes! Several new types of plastics, such as polylactic acid and polyhydroxybutyrates, are being made from renewable resources such as corn, soybeans, and sugarcane rather than from fossil fuel sources. Developing polymers from renewable resources is an active but still fairly new area of green chemistry research.

SELF-ASSESSMENT Questions

1. In general, the best way to dispose of plastics is
   a. composting
   b. incineration
   c. in landfills
   d. recycling

2. Many plastics are fire hazards because they burn and produce
   a. CFCs
   b. nitrogen oxides
   c. solid wastes
   d. toxic gases

3. Once used as a plasticizer, polychlorinated biphenyls (PCBs) were discontinued because of
   a. their costs
   b. the nonsustainable petroleum source
   c. their environmental effect similar to DDT
   d. the biodegradability of the products

4. Many flame-retardant fabrics incorporate
   a. Br and Cl atoms
   b. CFCs
   c. PCBs
   d. phthalate plasticizers

5. Most synthetic polymers are made from
   a. coal
   b. cotton
   c. petroleum and natural gas
   d. sustainable sources

6. A polymer made from a renewable resource is
   a. LDPE
   b. polyacrylonitrile
   c. polylactic acid
   d. styrene–butadiene rubber

Answers: 1, d; 2, d; 3, c; 4, a; 5, c; 6, c

Summary

Section 10.1—A **polymer** is a giant molecule made from smaller molecules, building blocks called **monomers**. There are many natural polymers, including starch, cellulose, nucleic acids, and proteins. **Celluloid** was the first semisynthetic polymer and was made from natural cellulose treated with nitric acid.

Section 10.2—The simplest, least expensive, and highest-volume synthetic polymer is polyethylene, made from the monomer ethylene. High-density polyethylene molecules can pack closely together to yield a rigid, strong structure. Low-density polyethylene molecules have many side chains, resulting in more flexible plastics. Linear low-density

polyethylene is a **copolymer**, formed from two (or more) different monomers. Polyethylene is one of many **thermoplastic polymers** that can be softened and reshaped with heat and pressure. *Thermosetting* polymers decompose, rather than soften, when heated.

Section 10.3—In **addition polymerization**, the monomer molecules add to one another; the polymer contains all the atoms of the monomers. Addition polymers are made from monomers containing carbon-to-carbon double bonds. The properties and uses of addition polymers differ greatly, depending on the pendant groups in the monomer. Thermoplastic polymers can be molded in many different ways.

Section 10.4—Natural rubber is an **elastomer**, a material that returns to its original shape after being stretched. The monomer of natural rubber, isoprene, can be made artificially and polymerized. Polyisoprene is soft and tacky when hot. In **vulcanization**, polyisoprene is reacted with sulfur atoms that **cross-link** the hydrocarbon chains and make the rubber harder, stronger, and more elastic. Synthetic polymers similar in structure to polyisoprene include polybutadiene, polychloroprene, and styrene–butadiene rubber (SBR). These are elastomers that can coil and uncoil. Elastomers are used as binders in paints.

Section 10.5—In **condensation polymerization**, parts of the monomer molecules are not incorporated in the product. Small molecules such as water are formed as by-products. Nylon is a **polyamide**, with amide [—(CO)NH—] linkages joining the monomers. A **polyester** is made from monomers with alcohol and carboxylic acid functional groups. Polyethylene terephthalate (PET) is the most common polyester.

Various **thermosetting resins** can be prepared by condensing formaldehyde with urea, melamine, or phenol. Combining a polymer with high-strength fibers gives a composite material that exploits the best properties of both materials. **Silicones** have chains consisting of silicon and oxygen atoms instead of carbon atoms, and have varied structures and many uses.

Section 10.6—Properties of polymers differ greatly from those of their monomers. A polymer's strength arises partly because the molecules entangle with one another and partly because the large molecules have strong dispersion forces. Polymers may be crystalline or amorphous. Above the **glass transition temperature** (T_g), a polymer is rubbery and tough; below it, the polymer is brittle. Synthetic fibers made from crystalline polymers are strong because their long molecules can align neatly with one another.

Section 10.7—The durability of plastics means that they make up a large fraction of the content of landfills. Proper incineration is one way of disposing of discarded plastics, but it can produce toxic gases. Some polymers are engineered to be degradable. Recycling of plastics requires that they be collected and sorted according to the code numbers and is increasing. Flame-retardant fabrics for clothing are made of polymers that often incorporate chlorine or bromine atoms.

Plasticizers are added to make polymers more flexible. Polychlorinated biphenyls (PCBs) were once used as plasticizers but are now banned. Phthalate esters are the most widely used plasticizers today.

A life-cycle impact analysis is useful in determining the effects of new products on the environment. Most synthetic polymers come from petroleum, a limited and nonrenewable resource. This is an important reason for recycling plastics and for developing polymers that can be made from renewable sources.

GREEN CHEMISTRY—While application of the principles of green chemistry can lead to inherently safer products, it is important to analyze the improvements and understand any unintended and unfortunate impacts of the new products. Life-cycle impact analysis can help measure the environmental footprint and inform the greener design of products.

Learning Objectives	Associated Problems
• Define *polymer* and *monomer*. (10.1)	2, 13
• List several natural polymers, including a chemically modified one. (10.1)	7, 14
• Describe the structure and properties of the two main types of polyethylene. (10.2)	6, 15–17, 71
• Use the terms *thermoplastic* and *thermosetting* to explain how polymer structure determines properties. (10.2)	8, 18, 35, 36, 52, 72
• Identify the monomer(s) of an addition polymer, and write the structural formula for a polymer from its monomer structure(s). (10.3)	1, 3–5, 19–24, 49–51, 53, 54, 57
• Define *cross-linking*, and explain how it changes the properties of a polymer. (10.4)	25–30, 69
• Differentiate between addition and condensation polymerization. (10.5)	10, 45–48, 63, 65, 67, 68
• Write the structures of the monomers that form polyesters and polyamides. (10.5)	9, 31–34, 55, 56, 58, 60–64
• Explain the concept of the glass transition temperature. (10.6)	37
• Explain how crystallinity affects the physical properties of polymers. (10.6)	38–40, 59, 66
• Describe the environmental problems associated with plastics and plasticizers. (10.7)	11, 12, 41–44, 70
• Name two types of sustainable, nonpetroleum-derived polymers and give their sources. (10.7)	55
• Explain how life-cycle impact analysis can be used to identify the environmental benefits and effects of newly designed products.	73–75

Conceptual Questions

1. How does the structure of PVC differ from that of polyethylene? List several uses of polyethylene.

2. Define the following terms.
   a. macromolecule   b. elastomer
   c. copolymer   d. plasticizer

3. What is addition polymerization? What structural feature usually characterizes molecules used as monomers in addition polymerization?

4. What is Teflon? What special property does it have? What are some of its uses?

5. From what monomer are disposable foamed plastic coffee cups made?

6. What plastic is used to make (a) gallon milk jugs and (b) 2-L soft drink bottles?

7. What is celluloid? From what material is it made?

8. What is a thermoplastic polymer? Give an example.

9. Which type of fibers, natural or synthetic, is used most in the United States? Why?

10. How do addition and condensation polymerization differ?

11. What steps must be taken to recycle plastics?

12. What problems arise when plastics are (a) discarded into the environment, (b) disposed of in landfills, and (c) disposed of by incineration?

 # Problems

Polymerization

13. What is the difference between a polymer and a monomer?

14. Identify three polymers that occur in nature. What monomer(s) make up each polymer?

Polyethylene

15. Describe the structure of low-density polyethylene (LDPE). How does this structure explain the properties of this polymer?

16. Describe the structure of linear low-density polyethylene (LLDPE). How does this structure explain the properties of this polymer?

17. How do the structures of high-density polyethylene (HDPE) and low-density polyethylene (LDPE) differ? How do their properties differ?

18. What is the difference between a thermoplastic polymer and a thermosetting polymer?

Addition Polymerization

19. Write the structure of the monomer from which each of the following polymers is made.
    a. polyethylene   b. polystyrene

20. Write the structure of the monomer from which each of the following polymers is made.
    a. polypropylene   b. polyvinyl chloride

21. Write the structure of a chain segment that is at least eight carbon atoms long for the polymer made from each of the following monomers.
    a. acrylonitrile
    b. vinylidene fluoride ($H_2C\!=\!CF_2$)

22. Write the structure of a chain segment that is at least four repeat units long for the polymer formed from each of the following monomers.
    a. tetrafluoroethylene   b. methyl methacrylate

$$CH_2\!=\!\overset{\overset{\displaystyle CH_3}{\displaystyle |}}{C}COOCH_3$$

23. Write the structure of the polymer made from each of the following. Show at least four repeat units.
    a. 1-pentene ($CH_2\!=\!CHCH_2CH_2CH_3$)
    b. methyl cyanoacrylate

24. Write the structure of the polymer made from each of the following. Show at least four repeat units.
    a. vinyl acetate [$H_2C\!=\!CH\!-\!O(C\!=\!O)CH_3$]
    b. methyl acrylate [$H_2C\!=\!CH\!-\!(C\!=\!O)OCH_3$]

Rubber and Other Elastomers

25. Draw the structure of the monomer isoprene.

26. Describe the process of vulcanization. How does vulcanization change the properties of rubber?

27. Explain why rubber is elastic, while many other polymers are rigid.

28. How does polybutadiene differ from natural rubber in properties? In structure?

29. How is SBR made? From what monomer(s) is it made?

30. Name three synthetic elastomers and the monomer(s) from which they are made.

Condensation Polymers

31. Nylon 88 is made from the monomers $H_2N(CH_2)_8NH_2$ and $HOOC(CH_2)_6COOH$. Draw the structure of nylon 88, showing at least two repeat units from each monomer.

32. Kodel is a polyester fiber. The monomers used to make it are terephthalic acid (page 318) and 1,4-cyclohexanedimethanol (shown). Write a condensed structural formula for the repeat unit of the Kodel molecule.

$$HOCH_2\!-\!\langle\ \rangle\!-\!CH_2OH$$

1,4-Cyclohexanedimethanol

33. Draw the structure of a polymer made from glycolic acid (hydroxyacetic acid, $HOCH_2COOH$). Show at least four repeat units. (Hint: Compare with nylon 6 in Section 10.5.)

34. Kevlar, a polyamide used to make bulletproof vests, is made from terephthalic acid (page 318) and *para*-phenylenediamine (shown). Write a condensed structural formula for the repeat unit of the Kevlar molecule.

$$H_2N-\langle \bigcirc \rangle-NH_2$$

para-Phenylenediamine

Properties of Polymers

35. Rank the following types of matter from smallest to largest: atom, compound, element, polymer.

36. What does the word *plastic* mean **(a)** in everyday life and **(b)** in chemistry?

37. What is the glass transition temperature (T_g) of a polymer? For what uses do we want polymers with a low T_g? With a high T_g?

38. Give one advantage that each of the following synthetic fibers has over the natural fiber it has replaced: **(a)** nylon **(b)** acrylic **(c)** polyesters.

39. Identify three ways in which the properties of polymers differ from the properties of monomers.

40. What is the difference between crystalline and amorphous polymers?

Plastics and the Environment

41. How do plasticizers make polymers less brittle?

42. What are PCBs? Why are they no longer used as plasticizers?

43. What are the most commonly used plasticizers today?

44. One way to dispose of plastics is to burn them. Identify **(a)** advantages and **(b)** disadvantages of burning discarded plastics.

Expand Your Skills

45. In organic polymer chemistry, an addition reaction is one in which
   a. the molecular weight of the product is the sum of that of the starting materials
   b. one large molecule splits apart into two smaller ones
   c. two identical molecules come together to form a larger one
   d. two or more molecules combine to form a larger one

46. In condensation polymerization reactions
   a. all the atoms of reactants are incorporated into the polymer product
   b. one large molecule splits apart into many smaller ones
   c. the product polymer is highly cross-linked
   d. a small molecule such as water is split out as the product forms

47. In the following equation, identify the parts labeled a, b, and c as monomer, polymer, and repeat unit. What type of polymerization (addition or condensation) is represented?

$$\overset{a}{\overbrace{n\ CH_2{=}CHF}} \longrightarrow$$

$$\underset{c}{\underbrace{{\sim}CH_2{-}CHF{-}\overset{b}{\overbrace{CH_2{-}CHF}}{-}CH_2{-}CHF{-}CH_2{-}CHF{\sim}}}$$

48. Is nylon 88 (Problem 31) an addition polymer or a condensation polymer? Explain.

49. One type of Saran has the structure shown below. Write the structures of the two monomers from which it is made.

$${\sim}CH_2CCl_2{-}CH_2CHCl{-}CH_2CCl_2{-}CH_2CHCl{\sim}$$

50. From what monomers could the following copolymer, called poly(styrene-co-acrylonitrile) (SAN), be made?

$${\sim}CH_2CH{-}CH_2CH{-}CH_2CH{\sim}$$
$$\quad\ \ \ \underset{C{\equiv}N}{|}\qquad\quad |\qquad\quad \underset{C{\equiv}N}{|}$$

51. Cyanoacrylates, such as those made from methyl cyanoacrylate (Problem 23b), are used in instant-setting "super-glues." However, poly(methyl cyanoacrylate) can irritate tissues. Cyanoacrylates with longer alkyl ester groups are less harsh and can be used in surgery in place of sutures. A good example is poly(octyl cyanoacrylate). Write a polymerization reaction for the formation of poly(octyl cyanoacrylate) from octyl cyanoacrylate, using a condensed structural formula to show the repeat unit of the polymer.

52. What is the first vivid change that occurs **(a)** when a thermoplastic polymer and **(b)** when a thermosetting polymer are heated to increasingly higher temperatures?

53. Isobutylene [$CH_2{=}C(CH_3)_2$] polymerizes to form polyisobutylene, a sticky polymer used as an adhesive. Draw the structure of polyisobutylene. Show at least four repeat units.

54. Copolymerization of isoprene and isobutylene (Problem 53) forms butyl rubber. Write the structure of butyl rubber. Show at least three isobutylene repeat units and one isoprene repeat unit.

55. The bacteria *Alcaligenes eutrophus* produce a polymer called polyhydroxybutyrate whose structure is shown below. Write the structure of the hydroxy acid from which this polymer is made.

$$\begin{array}{cc} CH_3 & O \\ | & \| \\ \end{array}$$
$${+}O{-}CH{-}CH_2{-}C{+}_n$$

56. DSM Engineering Plastics makes two specialty polymers: Stanyl nylon 46 and polytetramethylene terephthalamide (PA$_4$T). (A methylene group is CH_2; tetramethylene consists of four methylene groups: $CH_2CH_2CH_2CH_2$.) Both polymers are made using 1,4-diaminobutane as the diamine. **(a)** What diacid is used for each? Write a condensed structural formula showing the repeat unit of **(b)** Stanyl nylon 46 and **(c)** PA$_4$T.

57. A student writes the formula for polypropylene as shown below. Identify the error(s) in the formula.

$${+}CH_2{-}CH{-}CH_3{+}_n$$

58. Based on the condensed structural formula of the repeat unit of poly(ethylene naphthalate) (PEN) shown below, **(a)** write the structure(s) of the monomer(s) and **(b)** state whether the polymer is a polyester or a polyamide.

59. A rubber ball dropped from 100 cm will bounce back up to about 60 cm. Balls made from polybutadiene, called SuperBalls, bounce back up to about 85 cm. Golf balls, made from cross-linked polybutadiene, bounce back up to 89 cm. What type of polymer are all three balls made of? Which kind of ball exhibits the property of this type of polymer to the greatest degree?

60. Draw a structure of a likely product if 2,3-butanediol (shown) is substituted for ethylene glycol in the reaction with terephthalic acid (page 318).

61. A student is asked to synthesize a polyester from terephthalic acid (page 318) and 1,4-butanediol ($HOCH_2CH_2CH_2CH_2OH$). By mistake, the student uses 2-butanol rather than 1,4-butanediol. What is the likely result of this substitution?

62. A student is asked to synthesize a polyester from terephthalic acid and 1,4-butanediol ($HOCH_2CH_2CH_2CH_2OH$). By mistake, the student uses 1,3-butanediol (shown below) rather than the 1,4-butanediol. What is the likely result of this substitution?

63. Draw the structure of the silicone polymer formed from the compound shown below, showing four repeating units.

64. The two alkenes shown below are different compounds, but when each undergoes addition polymerization, the same polymer is formed. Explain.

65. Each *amino acid* shown in Table 16.4 (Section 16.6) has an amino group ($-NH_3^+$) and a carboxylic acid group ($-COO^-$). (A proton has been removed from the carboxylic group and one has been added to the amino group.) Proteins are polymers of amino acids. Draw the structure of the protein formed by polymerization of glycine, showing four repeat units.

66. What are composite materials? What are some of the advantages of composite materials?

67. Which of the following pairs of monomers will form a condensation polymer? Explain.
    a. $NH_2CH_2CH_2CH_2CH_2NH_2$ and $NH_2CH_2CH_2CH_2NH_2$
    b. $NH_2CH_2CH_2CH_2CH_2NH_2$ and $COOHCH_2CH_2COOH$
    c. $COOHCH_2CH_2CH_2COOH$ and $HOCH_2CH_2CH_2OH$
    d. $NH_2CH_2CH_2CH_2COOH$ and $NH_2CHCH_2CH_2COOH$

68. Children's sleepwear must be made of flame-retardant fabrics. How do flame-retardant fabrics differ from normal fabrics?

69. LDPE has many small side chains on the main polymer chains, and vulcanization results in short sulfur chains attached to the main polymer chains. How do these side chains differ in their effect on the properties of polymers?

70. Which of the following statements about recycling of polymers is true?
    a. PET and HDPE are recycled at higher rates than Styrofoam because they are bulkier and are easier to handle.
    b. PET and HDPE are recycled at lower rates than Styrofoam because polymers that have an aromatic ring cannot be recycled.
    c. PET and HDPE are recycled at higher rates than Styrofoam because they are more cost-effective to recycle.
    d. PET and HDPE are recycled at lower rates than Styrofoam because they cannot be separated in "single-stream" recycling operations.

71. Assume that you have two monomers, $CH_2{=}CHA$ and $CH_2{=}CHB$, where A and B represent different pendant groups. Which of the following statements is true?
    a. The polymer made from each monomer will have a different carbon backbone, because the pendant groups are part of the backbone.
    b. The polymer made from each monomer will have the same carbon backbone, because the pendant groups are not part of the polymerization reaction.
    c. The polymer made from each monomer will have the same properties, because changing the pendant group will not affect the properties of the polymer.
    d. The polymer made from each monomer will not contain all of the atoms that were in the monomers, because a small molecule will be released.

72. Which of the following statements about thermoplastic and thermosetting materials is true?
    a. Thermoplastic materials are stronger than thermosetting materials because they cannot be softened by heating.
    b. Thermoplastic materials are easier to recycle than thermosetting materials because they will decompose when heated.
    c. Thermoplastic materials are easier to recycle than thermosetting materials because they can be softened by heating.
    d. Thermoplastic materials are harder to recycle than thermosetting materials because they cannot be softened by heating.

73. Which of the following processes would be included in the outline of the life cycle of a PET water bottle? Choose all that apply.
 a. Whether the bottle is sent to a landfill or recycled
 b. The PET polymerization process
 c. The source of the water inside the bottle
 d. The synthesis processes to make the monomers
 e. The source of the raw materials for the monomers

74. Give an example of an unintended and undesirable effect that has resulted from an "improved" product or technology.

75. How can LCIA help with the development of green chemistry solutions?
 a. It gives a total score that makes it obvious which product is greener.
 b. It makes sure that people will recycle PET bottles.
 c. It shows which trade-offs have been created during the design of an alternative and enables the comparison of the environmental footprints of products.
 d. It identifies that production of CO_2 is bad for the environment.

Critical Thinking Exercises

Apply knowledge that you have gained in this chapter text and one or more of the FLaReS principles (Chapter 1) to evaluate the following statements or claims.

10.1 An environmental activist states that all-natural materials should be used to replace synthetic plastics because they are more sustainable.

10.2 A news report states that incinerators that burn plastics are a major source of chlorine-containing toxic compounds called *dioxins* (Chapter 19).

10.3. Vinyl chloride is a carcinogen. Some people warn that children will get cancer from playing with toys made of PVC.

10.4. The Plastics Division of the American Chemistry Council says that, in comparison with paper bags, "plastic bags are . . . an environmentally responsible choice."

10.5 A manufacturer argues that more plastic than metal should be used in car engines because plastics are more resistant to oil and gas.

Collaborative Group Projects

Prepare a PowerPoint, poster, or other presentation (as directed by your instructor) to share with the class.

1. Prepare a brief report on possible sources of plastics when Earth's supplies of coal and petroleum become scarce.

2. Prepare a brief report on biobased polymers such as bio-HDPE, nylon 11 (polyundecylamide), biobased PET, and biobased polytrimethylene terephthalate (PTT). Identify the natural source from which each is derived, and describe intermediate steps from source to final product.

3. Do a risk–benefit analysis (Section 1.4) of the use of synthetic polymers for one or more of the following.
 a. grocery bags   **b.** building materials
 c. clothing   **d.** carpets
 e. food packaging   **f.** picnic coolers
 g. automobile tires   **h.** artificial hip sockets

4. To what extent is plastic litter a problem in your community? Survey one city block (or other area as directed by your instructor), and inventory the litter found. (You might as well pick it up while you are at it.) What proportion of the litter is plastics? What proportion of it is fastfood containers? To what extent are plastics recycled in your community? What factors limit further recycling?

5. Do some research on Kevlar, which is used in body armor. Report on its other uses, and explain why it is so strong. Describe the Kevlar Survivors' Club, jointly sponsored by the International Chiefs of Police and DuPont.

6. Compare the treatment of the topic of the environmental impact of plastics and other polymers at several websites, such as those of the American Plastics Council, an industry group, and an environmental group such as Greenpeace or the Sierra Club.

LET'S EXPERIMENT! Polymer Bouncing Ball

Materials Needed:

- 2 small plastic cups (4 oz)
- Measuring spoons
- Warm water
- Borax
- 2 wooden craft sticks

- White craft glue
- Cornstarch
- Food coloring (if desired)
- Plastic bag with zip lock (for storage)

Did you know that the earliest balls were made of wood and stone? What are most bouncy balls made of today?

Many bouncing balls are made out of rubber, but they can also be made out of leather or plastic and can be hollow or solid. This experiment will use common, inexpensive ingredients to make a ball that bounces.

Polymers are molecules made up of repeating chemical units. Glue is made up of the polymer polyvinyl acetate (PVA). In this experiment, borax (sodium borate, a salt of boric acid) is responsible for hooking the molecules together and cross-linking the molecules, providing the ball with its putty-like and bouncy properties.

To start, label the two cups *Borax Solution* and *Ball Mixture*.

- For the borax solution, pour 2 tablespoons of warm water and $1/2$ teaspoon of borax powder into the cup. Use a craft stick to stir the mixture to dissolve the borax. Add food coloring, if desired.
- For the ball mixture, pour 1 tablespoon of glue into the cup. Add $1/2$ teaspoon of the borax solution you just made and 1 tablespoon of cornstarch. Do not stir. Allow the ingredients to interact on their own for 10–15 seconds. Then use the other craft stick to stir them together to fully mix them. Once the mixture becomes too thick to stir, take it out of the cup and start molding the ball with your hands.

The ball will start out sticky and messy but will solidify as you knead it. Once the ball is less sticky, go ahead and bounce it! To keep your ball from drying out, store your ball in a plastic zip lock bag.

Questions

1. Does your ball bounce? How high?
2. Does making a polymer ball cause a chemical or physical reaction? Explain.
3. Describe how changing the amounts of each ingredient would affect the ball mixture.
4. Is this ball biodegradable? Why or why not?

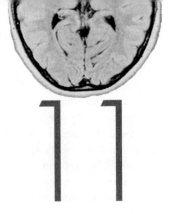

Nuclear Chemistry

11

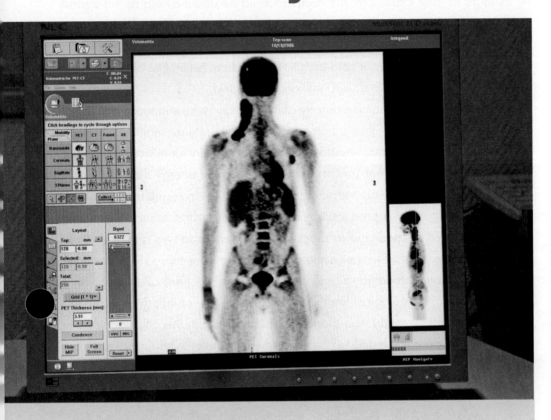

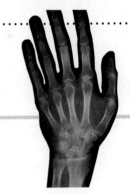

The photo to the left shows an image from PET (positron emission tomography). At the upper right of this page is an image from a CAT (computerized axial tomography). The picture directly above is an example of an X-ray, which is used for examining bones and internal organs. All of these diagnostic techniques use radiation or a radioactive isotope. With these techniques, we can now examine tissues and diagnose diseases that

Have You Ever Wondered?

1 Is radiation entirely a human-made problem? **2** How do we measure radioactivity and its effect on people? **3** Do irradiated foods contain radioactive material? **4** Are we exposed to dangerous radiation during X-rays and other medical procedures? **5** What causes radiation sickness, and how serious is it? **6** Can we minimize or recycle radioactive wastes to make nuclear power generation a more sustainable process?

THE HEART OF MATTER Many people associate the term *nuclear energy* with fearsome images of a mighty force: giant mushroom clouds from nuclear explosions that devastated cities and nuclear power plant accidents at Three Mile Island, Pennsylvania, in 1979; Chernobyl, Ukraine, in 1986; and Fukushima, Japan, in 2011 (discussed in Chapter 15). But some amazing stories can also be told about life-giving applications of nuclear energy. Indeed, medical uses of nuclear energy have undoubtedly saved more lives than have been lost as a result of nuclear bombs or accidents.

previously could only be detected by much more invasive procedures such as biopsy or exploratory surgery. The nuclear age has produced both serious problems and tremendous advantages, which will be examined in this chapter.

The discussion of atomic structure in Chapter 3 focused mainly on the electrons, because the electrons are the particles that determine an element's chemical and physical properties. In this chapter, we take a closer look at that tiny speck in the center of the atom—the atomic nucleus.

Atoms undoubtedly are small, but the even-tinier size of the atomic nucleus is almost beyond our imagination. The diameter of an atom is roughly 100,000 times greater than the diameter of its nucleus. If an atom could be magnified to be as large as your classroom, the nucleus would be about as big as the period at the end of this sentence. Yet this very tiny nucleus contains almost all the atom's mass. How dense the atomic nucleus must be! A cubic centimeter of water weighs 1 g, and a cubic centimeter of gold about 19 g. A cubic centimeter of pure atomic nuclei would weigh more than 100 million metric tons!

Even more amazing than the density of the nucleus is the enormous amount of energy that is contained within it. Some atomic nuclei undergo reactions that can fuel the most powerful bombs ever built or provide electricity for millions of people. Our sun is one huge nuclear power plant, supplying the energy that warms our planet and the light necessary for plant growth. The twinkling light from every star we see in the night sky is produced by powerful nuclear reactions. Radioactive isotopes are used in medicine to diagnose and treat diseases and save lives every day. Many applications of nuclear chemistry in science and industry have improved the human condition significantly. Nuclear reactions allow us to date archaeological and geological finds, to assess the quality of industrial materials, to provide electricity, and to be alerted to deadly fires. In this chapter, we will learn about the destructive and the healing power of the infinitesimally small and wonderfully dense heart of every atom—the nucleus.

11.1 Natural Radioactivity

Learning Objectives • Identify the major sources of radiation to which we are exposed. • List the sources and dangers of ionizing radiation.

Recall from Chapter 3 that most elements occur in nature in several isotopic forms, whose nuclei have the same number of protons but differ in their number of neutrons. Some of these nuclei are stable, but many are unstable and undergo **radioactive decay**. The nuclei that undergo such decay are called **radioisotopes**, and the process of decay produces one or more types of radiation. Stable isotopes have some common characteristics, which are identified on page 338.

Background Radiation

Humans have always been exposed to radiation. Even as you read this sentence, you are being bombarded by **cosmic rays**, which originate from the sun and outer space. Other radiation reaches us from natural radioactive isotopes in air, water, soil, and rocks. We cannot escape radiation, because it is part of many natural processes, including those in our bodies. A naturally occurring radioactive isotope of potassium, ^{40}K, exists in all our cells and in many foods. The average person absorbs about 4000 particles of radiation each second from ^{40}K and another 1200 particles per second from ^{14}C. (This may sound like a lot, but recall how small atoms are and how many atoms exist in even a tiny sample of matter.)

This ever-present natural radiation is called **background radiation**. Figure 11.1 shows that only about half of the average American's radiation exposure comes from background radiation. Most of the remainder comes from medical irradiation, including X-rays, CT scans, and nuclear medicine. In many other areas of the world, medical irradiation is not as widely used, and people living in those areas receive roughly three-fourths of their radiation exposure from background sources. Other sources, such as fallout from testing of nuclear bombs, releases from nuclear industry power plants, and occupational exposure, account for only a minute fraction of the total average exposure.

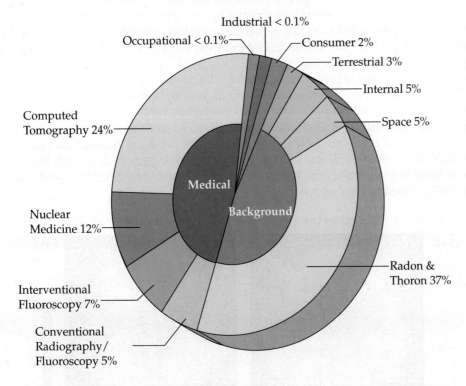

Sources of U.S. Radiation Exposure

◀ **Figure 11.1** About half of the radiation to which Americans are exposed comes from natural sources (green colors), and close to half comes from medical treatments (blue colors). A very small amount of this radiation comes from other human activities (other colors).

Q *What natural source of ionizing radiation contributes most to our exposure? What source provides the largest proportion of our exposure due to human activities?*

Harmful effects arise from the interaction of radiation with living tissue. Radiation with enough energy to knock electrons from atoms and molecules, converting them into ions (electrically charged atoms or groups of atoms), is called **ionizing radiation**. Nuclear radiation and X-rays are examples. Because radiation is invisible and because it has such great potential for harm, we are very much concerned about our exposure to it. However, it is worth noting that the total amount of background radiation to which the average person is exposed is less than 0.5% of the amount that causes symptoms of radiation sickness.

Radiation Damage to Cells

Radiation-caused chemical changes in living cells can be highly disruptive. Ionizing radiation can devastate living cells by interfering with their normal chemical processes. Molecules can be splintered into reactive fragments called *free radicals* (Section 4.10), which can disrupt vital cellular processes. White blood cells, the body's first line of defense against bacterial infection, are particularly vulnerable. High levels of radiation also affect bone marrow, causing a drop in the production of red blood cells, which results in anemia. Radiation has also been shown to induce leukemia, a disease of the blood-forming organs.

Ionizing radiation also can cause changes in the molecules of heredity (DNA) in reproductive cells. Such changes can show up as mutations in the offspring of exposed parents. Little is known of the effects of such exposure on humans.

1 Is radiation entirely a human-made problem?

Many people think that a large proportion of radiation comes from human-made sources. However, Figure 11.1 clearly shows that roughly half of the ionizing radiation to which the average American is exposed is background radiation that comes from natural sources while just under half of ionizing radiation results from human-made sources.

2 How do we measure radioactivity and its effect on people?

There are several units of radioactivity—and they measure different things. Two units that measure the actual amount of radiation emitted are the *curie*, which is the amount of a radioactive substance that corresponds to the decay of 3.7×10^{10} nuclei per second, and the *becquerel* (Bq), which is equal to 1 disintegration per second. Other units look at the amount of radiation actually absorbed by the body (radiation absorbed dose, or *rad*) or the effect of absorbed radiation on the body (*Roentgen equivalent man*, or *rem*). Clearly the *rem* was named in an earlier era.

However, many of the mutations that occurred during the evolution of present species may have been caused by background radiation.

Ionizing radiation from decaying atomic nuclei is highly energetic and can damage tissue, as discussed above. *Electromagnetic* radiation has many forms—visible light, radio waves, television broadcast waves, microwaves, ultraviolet light, military ULF (ultra-low-frequency), and others. Some types of electromagnetic radiation, such as X-rays and gamma rays, do have enough energy to ionize tissue, and ultraviolet light has been strongly implicated in *melanoma* (skin cancer).

But what about microwaves and radio waves, which don't have nearly as much energy? There is no conclusive evidence that low levels of microwaves pose a significant threat to human health. A 2006 study of nearly half a million Danish citizens and several smaller studies failed to show a relationship between cell phone use and cancer. While cell phones *do* give off radiation that affects parts of the brain, as can be seen in Figure 11.2, it is not ionizing radiation and thus does not cause cell damage. However, higher levels of radio frequencies over long periods of time *may* be a different matter. A 2011 World Health Organization (WHO) report classified heavy cell phone use—30 minutes of talking daily for ten years—as *possibly* posing an increased risk of *glioma*, a type of brain cancer. Although no conclusive evidence has been found, studies continue, and California has issued guidance on how cell phone users can limit their exposure to radiation. For low levels of exposure to nonionizing electromagnetic radiation, the hazard remains difficult to measure, let alone assess.

Because of the potentially devastating effects on living things, knowledge of radiation, radioactive decay, and nuclear chemistry in general is crucial. Next, we will learn how to identify the various types of radiation that are emitted and how to write balanced nuclear equations.

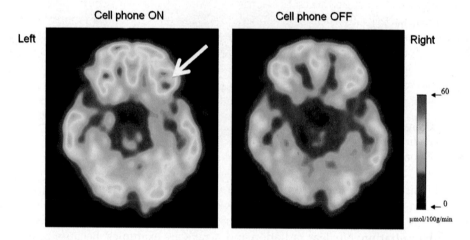

▶ **Figure 11.2** A brain scan shows slight differences when a cell phone held to the ear is on (left) and off (right).

Why Are Isotopes Stable—or Unstable?

Some isotopes are stable, but most are radioactive and not stable. What factors tend to make an atomic nucleus stable? Generally, stable isotopes tend to have:

1. *Even* numbers of either protons or neutrons or (especially) both. Of the 264 stable isotopes, 157 have even numbers of both protons and neutrons, and only 4 have odd numbers of both protons and neutrons. Elements with an even Z (protons) have more stable isotopes than do those with an odd Z.
2. So-called *magic numbers* of either protons or neutrons. (Magic numbers are 2, 8, 20, 28, 50, 82, and 126.)

3. An atomic number of 83 or less. All isotopes with Z > 83 are radioactive. (Bismuth, with Z = 83, was found in 2003 to be very weakly radioactive, but it exhibits no significant decay even over a billion years.)
4. Fewer protons than neutrons in the nucleus (or the same number of them), and a ratio of neutrons to protons close to 1:1 if the atomic number is 20 or below. As the atomic number gets larger, the stable n/p ratio also increases, up to about 1.5:1. There is a zone of stability within which the n/p ratio should lie for an atom with a given atomic number.

SELF-ASSESSMENT Questions

1. Approximately what percentage of background radiation for Americans comes from natural sources?
   a. 5%   **b.** 20%
   c. 50%   **d.** 80%

2. The largest source of natural background radiation is
   a. cosmic rays   **b.** isotopes in the body
   c. radon   **d.** isotopes in rocks and soil

3. At present levels of exposure, which of the following forms of radiation is most likely to give you cancer?
   a. fallout from the Fukushima nuclear power plant accident
   b. medical X-rays
   c. radiation from operating nuclear power plants
   d. sunlight

4. Which of the following isotopes is most likely to be a cause of natural radiation in our bodies?
   a. ^{12}C **b.** ^{125}I **c.** ^{40}K **d.** ^{238}U

11.2 Nuclear Equations

. . . .

Learning Objectives • Balance nuclear equations. • Identify the types of products formed by various nuclear decay processes.

Writing balanced equations for nuclear processes is relatively simple. Nuclear equations differ in two ways from the chemical equations discussed in Chapter 5. First, while chemical equations must have the same elements on both sides of the arrow, nuclear equations rarely do. Second, while we balance atoms in ordinary chemical equations, we balance the *nucleons* (protons and neutrons) in nuclear equations. What this really means is that we must balance both the atomic numbers (number of protons) and the nucleon numbers (number of protons plus neutrons) of the starting materials and products. For this reason, we must always specify the *isotope* of each element appearing in a nuclear equation. We use nuclear symbols (Section 3.5) when writing nuclear equations because they make the equations easier to balance.

In one example of a nuclear reaction, radon-222 atoms break down spontaneously in a process called **alpha decay**, giving off alpha (α) particles, as shown in Figure 11.3a. Because alpha particles are identical to helium nuclei, which contain two protons and two neutrons, this reaction can be summarized by the following equation:

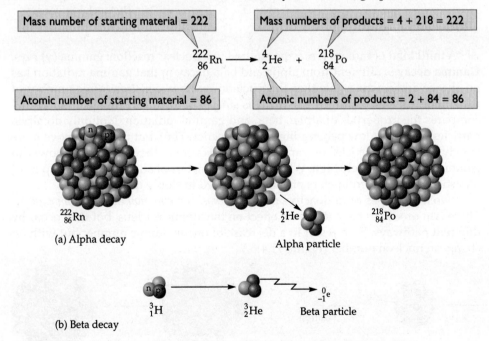

| Mass number of starting material = 222 | | Mass numbers of products = 4 + 218 = 222 |

$$^{222}_{86}\text{Rn} \longrightarrow \, ^{4}_{2}\text{He} \; + \; ^{218}_{84}\text{Po}$$

| Atomic number of starting material = 86 | | Atomic numbers of products = 2 + 84 = 86 |

$^{222}_{86}\text{Rn}$

(a) Alpha decay

$^{4}_{2}\text{He}$
Alpha particle

$^{218}_{84}\text{Po}$

$^{3}_{1}\text{H}$ $^{3}_{2}\text{He}$ $^{0}_{-1}e$
                                         Beta particle

(b) Beta decay

▲ **Figure 11.3** Nuclear emission of (a) an alpha particle and (b) a beta particle.

Q *What changes occur in a nucleus when it emits an alpha particle? A beta particle?*

We use the symbol $^{4}_{2}\text{He}$ (rather than α) for the alpha particle because it allows us to check the balancing of mass and atomic numbers more readily. The atomic number, $Z = 84$, identifies the element produced as polonium (Po). In nuclear chemistry, the mass number, A, is the number of nucleons in the starting material. The number of nucleons in the starting material must equal the total number of the nucleons in the products. The same is true for the atomic numbers. In alpha decay, the atomic number decreases by 2 and the number of nucleons decreases by 4.

The heaviest isotope of hydrogen, hydrogen-3, often called *tritium*, decomposes by a process called **beta decay** (Figure 11.3b). Because a beta (β) particle is identical to an electron, this process can be written as

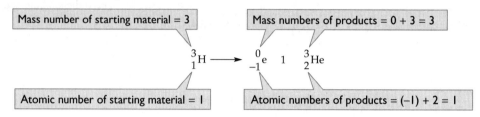

Mass number of starting material = 3

Mass numbers of products = 0 + 3 = 3

$$^{3}_{1}\text{H} \longrightarrow \ ^{0}_{-1}\text{e} \ + \ ^{3}_{2}\text{He}$$

Atomic number of starting material = 1

Atomic numbers of products = (−1) + 2 = 1

The atomic number, $Z = 2$, identifies the product isotope as helium.

Beta decay is a little more complicated than indicated by the preceding reaction. In beta decay, a neutron within the nucleus is converted into a proton (which remains in the nucleus) and an electron (which is ejected).

$$^{1}_{0}\text{n} \longrightarrow \ ^{1}_{1}\text{p} + \ ^{0}_{-1}\text{e}$$

With beta decay, the atomic number increases by one, but the total number of nucleons remains the same.

TABLE 11.1 Common Types of Radiation in Nuclear Reactions

Radiation	Mass (u)	Charge	Identity	Velocity[a]	Penetrating Power: (Can Be Stopped by)
Alpha (α)	4	2+	He^{2+}	$0.1c$	Very low (sheet of paper or outer skin)
Beta (β)	0.00055	1−	e^-	$<0.9c$	Moderate (glass or metal sheet)
Gamma (γ)	0	0	High-energy photon	c	Extremely high (1–4 inches of lead)

[a]c is the speed of light.

A third kind of radiation may be emitted in a nuclear reaction: gamma (γ) rays. **Gamma decay** is different from alpha and beta decay in that gamma radiation has no charge and no mass. Neither the nucleon number nor the atomic number of the emitting atom is changed; the nucleus simply become less energetic. Table 11.1 compares the properties of alpha, beta, and gamma radiation. Not only do alpha particles have very low penetrating power (Section 11.6), but they also move more slowly than beta particles or gamma rays. In contrast, the penetrating power of gamma rays is extremely high. While a few millimeters of aluminum will stop most β particles, several centimeters of lead are needed to stop γ rays.

Two other types of radioactive decay are *positron emission* and *electron capture*. These two processes have the same effect on the atomic nucleus, but they occur by different pathways. Both result in a decrease of one in atomic number but with no change in nucleon number (Figure 11.4).

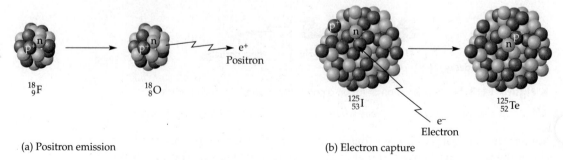

(a) Positron emission

(b) Electron capture

▲ **Figure 11.4** Nuclear change accompanying (a) positron emission and (b) electron capture.

Q *What changes occur in a nucleus when it emits a positron? When it undergoes electron capture?*

The **positron** (β^+) is a particle equal in mass but opposite in charge to the electron. It is represented as $_{+1}^{0}e$. Fluorine-18 decays by positron emission.

$$_{9}^{18}\text{F} \longrightarrow _{+1}^{0}e + _{8}^{18}\text{O}$$

We can envision positron emission as the change of a proton in the nucleus into a neutron and a positron, with the positron being emitted from the nucleus.

$$_{1}^{1}\text{p} \longrightarrow _{0}^{1}\text{n} + _{+1}^{0}e$$

After a positron is emitted, the original nucleus has one fewer proton and one more neutron than it had before. The nucleon number of the product nucleus is the same, but its atomic number has been reduced by one. The emitted positron quickly encounters an electron (there are many electrons in all kinds of matter), and both particles are changed into energy—in this case, two gamma rays.

$$_{+1}^{0}e + _{-1}^{0}e \longrightarrow 2_{0}^{0}\gamma$$

Electron capture (EC) is a process in which a nucleus absorbs an electron from an inner electron shell, usually the first or second. When an electron from a higher shell drops to the level vacated by the captured electron, an X-ray is released. Once inside the nucleus, the captured electron combines with a proton to form a neutron.

$$_{1}^{1}\text{p} + _{-1}^{0}e \longrightarrow _{0}^{1}\text{n}$$

Iodine-125, used in medicine to diagnose pancreatic function and intestinal fat absorption, decays by electron capture.

$$_{53}^{125}\text{I} + _{-1}^{0}e \longrightarrow _{52}^{125}\text{Te}$$

Note that unlike alpha, beta, gamma, or positron emission, electron capture has the electron as a reactant (on the left side) and not as a product. Conversion of a proton to a neutron (by the absorbed electron) yields a nucleus with the atomic number lowered by one but unchanged in atomic mass. Emission of a positron and absorption of an electron have the same effect on an atomic nucleus (lowering the atomic number by one), except that positron emission is accompanied by gamma radiation and electron capture by X-radiation. Positron-emitting isotopes and those that undergo electron capture have important medical applications (Section 11.5).

The five types of radioactive decay are summarized in Table 11.2.

TABLE 11.2 Radioactive Decay and Nuclear Change

Type of Decay	Decay Particle	Particle Mass (amu)	Particle Charge	Change in Nucleon Number	Change in Atomic Number
Alpha decay	α	4	2+	Decreases by 4	Decreases by 2
Beta decay	β	0	1−	No change	Increases by 1
Gamma radiation	γ	0	0	No change	No change
Positron emission	β^+	0	1+	No change	Decreases by 1
Electron capture (EC)	e^- absorbed	0	1−	No change	Decreases by 1

EXAMPLE 11.1 Balancing Nuclear Equations

Write balanced nuclear equations for the following processes.
  a. Plutonium-239 emits an alpha particle when it decays.
  b. Protactinium-234 undergoes beta decay.
  c. Carbon-11 emits a positron when it decays.
  d. Carbon-11 undergoes electron capture.

Solution

a. We start by writing the symbol for plutonium-239 and a partial equation showing that one of the products is an alpha particle (helium nucleus).

$$^{239}_{94}\text{Pu} \longrightarrow {}^{4}_{2}\text{He} + \text{?}$$

To balance this equation, the other product must have $A = 239 - 4 = 235$ and $Z = 94 - 2 = 92$. The atomic number 92 identifies the element as uranium (U).

$$^{239}_{94}\text{Pu} \longrightarrow {}^{4}_{2}\text{He} + {}^{235}_{92}\text{U}$$

b. We write the symbol for protactinium-234 and a partial equation showing that one of the products is a beta particle (electron).

$$^{234}_{91}\text{Pa} \longrightarrow {}^{0}_{-1}\text{e} + \text{?}$$

The other product must have a nucleon number of 234 and $Z = 92$ to balance the equation. The atomic number identifies this product as another isotope of uranium.

$$^{234}_{91}\text{Pa} \longrightarrow {}^{0}_{-1}\text{e} + {}^{234}_{92}\text{U}$$

c. We write the symbol for carbon-11 and a partial equation showing that one of the products is a positron.

$$^{11}_{6}\text{C} \longrightarrow {}^{0}_{+1}\text{e} + \text{?}$$

To balance the equation, a particle with $A = 11 - 0 = 11$ and $Z = 6 - 1 = 5$ (boron) is required.

$$^{11}_{6}\text{C} \longrightarrow {}^{0}_{+1}\text{e} + {}^{11}_{5}\text{B}$$

d. We write the symbol for carbon-11 and a partial equation showing it capturing an electron.

$$^{11}_{6}\text{C} + {}^{0}_{-1}\text{e} \longrightarrow \text{?}$$

To balance the equation, the product must have $A = 11 + 0 = 11$ and $Z = 6 + (-1) = 5$ (boron).

$$^{11}_{6}\text{C} + {}^{0}_{-1}\text{e} \longrightarrow {}^{11}_{5}\text{B}$$

As we noted previously, positron emission and electron capture result in identical changes in atomic number and, therefore, affect a given nucleus in the same way, as parts **(c)** and **(d)** illustrate for carbon-11. Also note that carbon-11 (and certain other nuclei) can undergo more than one type of radioactive decay.

> **Exercise 11.1A**

Write balanced nuclear equations for the following processes.

a. Bismuth-214 decays by alpha emission.
b. Niobium-94 undergoes beta decay.
c. Gallium-68 decays by positron emission.
d. Argon-37 undergoes electron capture.

> **Exercise 11.1B**

Write balanced nuclear equations for the following processes.

a. Argon-42 decays by beta emission.
b. Thorium-228 undergoes alpha decay.
c. Titanium-44 undergoes electron capture.
d. Aluminum-26 emits a positron.

We noted earlier that nuclear equations differ in two ways from ordinary chemical equations. Nuclear reactions also exhibit many differences from chemical reactions. Some important distinctions are summarized in Table 11.3. Some nuclear reactions also involve processes other than the five simple ones we have discussed here. Nevertheless, all nuclear equations must be balanced for both nucleon (mass) numbers and atomic numbers. When an unknown particle has an atomic number that does not correspond to an atom, that particle may be a subatomic particle. A list of symbols that have been used to represent subatomic particles in nuclear equations is given in Table 11.4.

TABLE 11.3 Some Differences Between Chemical Reactions and Nuclear Reactions

Chemical Reactions	Nuclear Reactions
Atoms retain their identity.	Atoms usually change their identity—from one element to another.
Reactions involve only electrons and usually only outermost electrons.	Reactions involve mainly protons and neutrons. It does not matter what the valence electrons do.
Reaction rates will be greater at higher temperatures.	Reaction rates are unaffected by temperature.
The energy absorbed or given off in reactions is comparatively small.	Reactions sometimes involve enormous changes in energy.
Mass is conserved. The mass of products equals the mass of starting materials.	Huge changes in energy are accompanied by measurable changes in mass ($E = mc^2$).*

*See Section 11.7.

TABLE 11.4 Symbols for Subatomic Particles

Particle	Symbol	Nuclear Symbol
Proton	p	$_1^1p$ or $_1^1H$
Neutron	n	$_0^1n$
Electron	e^- or β^-	$_{-1}^0e$ or $_{-1}^0\beta$
Positron	e^+ or β^+	$_{+1}^0e$ or $_{+1}^0\beta$
Alpha particle	α	$_2^4He$ or $_2^4\alpha$
Beta particle	β or β^-	$_{-1}^0e$ or $_{-1}^0\beta$
Gamma ray	γ	$_0^0\gamma$

SELF-ASSESSMENT Questions

1. What particle is needed to complete the following nuclear equation?

$$_4^9Be + ? \longrightarrow _6^{12}C + _0^1n$$

   a. alpha particle
   b. beta particle
   c. neutron
   d. proton

2. Emission of which of the following will result in a decrease of two in atomic number?
   a. alpha particle
   b. beta particle
   c. neutron
   d. proton

3. For which type of radioactive decay is an electron found on the left side of the equation?
   a. alpha emission
   b. beta emission
   c. electron capture
   d. positron emission

4. Which of the following is the smallest particle?
   a. alpha particle
   b. beta particle
   c. neutron
   d. proton

5. Which of the following has a mass of 1 and a charge of 0?
   a. alpha particle
   b. beta particle
   c. neutron
   d. proton

Answers: 1, a; 2, a; 3, c; 4, b; 5, c

11.3 Half-Life and Radioisotopic Dating

Learning Objectives • Solve simple half-life problems. • Use the concept of half-life to solve simple radioisotope dating problems.

Thus far, we have discussed radioactivity as applied to single atoms. In the laboratory, we generally deal with great numbers of atoms—numbers far larger than the number

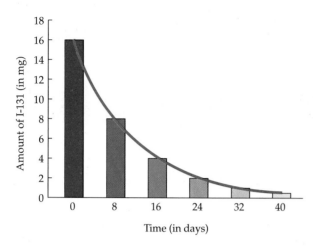

▲ **Figure 11.5** The radioactive decay of iodine-131, which has a half-life of 8 days.

 How much of a 32 mg sample of iodine-131 remains after five half-lives have passed?

of all the people on Earth. If we could see the nucleus of an individual atom, we could tell whether it would undergo radioactive decay by noting its composition. Certain combinations of protons and neutrons are unstable. However, we would not be able to predict *when* the atom would undergo a change. Radioactivity is a random process, generally independent of outside influences.

Half-life

With large numbers of atoms, the process of radioactive decay becomes more predictable. We can measure the *half-life,* a property characteristic of each radioisotope. The **half-life** of a radioactive isotope is the time it takes for one-half of the original number of atoms to undergo radioactive decay.

Suppose, for example, we had 16.00 mg of the radioactive isotope iodine-131. The half-life of iodine-131 is 8.0 days. This means that in 8.0 days, half the iodine-131, or 8.00 mg, will have decayed, and there will be 8.00 mg left. In another 8.0 days, half of the remaining 8.00 mg will have decayed. After two half-lives, or 16.0 days, one-quarter of the original iodine-131, or 4.00 mg, will remain. Note that two half-lives do not make a whole. The concept of half-life is illustrated by the graph in Figure 11.5. Half-lives of radioisotopes can differ enormously. The half-life of tellurium-128 is 2×10^{24} y, while that of beryllium-13 is 2.7×10^{-21} s.

The rate of decay is inversely related to half-life. An isotope with a long half-life decays slowly; an isotope with a short half-life decays more rapidly. We measure the rate of decay, also referred to as the isotope's activity, in disintegrations per second, a unit called a **becquerel (Bq)**. We can calculate the fraction of the original isotope that remains after a given number of half-lives from the relationship

$$\text{Fraction remaining} = \frac{1}{2^n}$$

where n is the number of half-lives.

EXAMPLE 11.2 Half-lives

A 4.00 mg sample of cobalt-60, half-life 5.271 y, is to be used for radiation treatment. How much cobalt-60 remains after 15.813 years (three half-lives)?

Solution
The fraction remaining after three half-lives is

$$\frac{1}{2^n} = \frac{1}{2^3} = \frac{1}{2 \times 2 \times 2} = \frac{1}{8}$$

The amount of cobalt-60 remaining is $\frac{1}{8} \times 4.00$ mg $= 0.50$ mg.

> **Exercise 11.2A**
The half-life of phosphorus-32 is 14.3 days. What percentage of a sample of phosphorus-32 remains after four half-lives?

> **Exercise 11.2B**
On April 4, 2011, a small fish caught about 50 mi. south of the Fukushima nuclear complex had an activity of 4000 Bq of iodine-131 per kilogram. The half-life of iodine-131 is 8.0 d. What was the activity of the isotope **(a)** on April 28, 2011 (three half-lives later)? **(b)** On June 23, 2011 (ten half-lives later)?

CONCEPTUAL EXAMPLE 11.3   Time and Radioactive Decay

We cannot determine how long it will take for *all* of a radioactive isotope to decay. For many isotopes, it is assumed that the activity is near zero after about ten half-lives. What mass of a 0.0260 mg sample of mercury-190, half-life 20 min., remains after ten half-lives?

Solution

The fraction remaining after ten half-lives is found in the usual manner.

$$\frac{1}{2^n} = \frac{1}{2^{10}} = \frac{1}{2 \times 2 \times 2 \times 2 \times 2 \times 2 \times 2 \times 2 \times 2 \times 2} = \frac{1}{1024}$$

The amount of mercury-190 remaining is $\frac{1}{1024} \times 0.0260$ mg $= 0.000025$ mg.

> **Exercise 11.3A**

The disposal of some radioactive materials is scheduled according to the rule stated in Conceptual Example 11.3. Would the rule work for a 1.00 g sample of rubidium-87 with an initial decay rate of 3200 Bq?

> **Exercise 11.3B**

The disposal of some radioactive materials is scheduled according to the rule stated in Conceptual Example 11.3. Would the rule work for a 1.00 g sample of cobalt-60, which has an initial activity of 4.1×10^{13} Bq? Explain.

Radioisotopic Dating

The half-lives of certain radioisotopes can be used to estimate the ages of rocks and archaeological artifacts. Uranium-238 decays with a half-life of 4.5 billion years. The initial products of this decay are also radioactive, and they and subsequent products (daughter nuclei) continue to decay until lead-206 is formed. By measuring the relative amounts of uranium-238 and lead-206, chemists can estimate the age of a rock. Some of the older rocks on Earth have been found to be 3.0–4.5 billion years old. Moon rocks and meteorites have been dated at a maximum age of about 4.5 billion years. Thus, the age of Earth (and of the solar system itself) is generally estimated to be about 4.5 billion years.

The dating of artifacts derived from plants or animals usually involves radioactive carbon-14. Of the carbon on Earth, about 99% is carbon-12 and 1% is carbon-13, and both of these isotopes are stable. However, in the upper atmosphere, radioactive carbon-14 is formed by the bombardment of ordinary nitrogen by neutrons from cosmic rays.

$$^{14}_{7}\text{N} + ^{1}_{0}\text{N} \longrightarrow ^{14}_{6}\text{C} + ^{1}_{1}\text{H}$$

This process results in a small but steady concentration of carbon-14 in the CO_2 molecules of Earth's atmosphere. Plants use CO_2, and animals consume plants and other animals; thus, living things constantly incorporate this isotope into their cells. When organisms die, the incorporation of carbon-14 ceases, and the isotope begins to decay—with a half-life of 5730 years—back to nitrogen-14. We can measure the carbon-14 activity remaining in an artifact of plant or animal origin to determine its age, using a technique called **carbon-14 dating**. For instance, a sample that has half the carbon-14 activity of new plant material is 5730 years old; it has been dead for one half-life. Similarly, an artifact with one-fourth of the carbon-14 activity of new plant material is 11,460 years old; it has been dead for two half-lives.

We have assumed that the formation of the carbon-14 isotope has been constant over the years, but this is not quite the case. For the last 7000 years or so, carbon-14 dates do correlate with those obtained from the annual growth rings of trees and with documents of known age. Calibration curves have been constructed from the accurate dates that can be determined. Generally, carbon-14 is reasonably accurate for dating objects from about 100 to 50,000 years old. Newer objects may not yet have seen measurable decay of carbon-14. Objects older than 50,000 years have too little of the isotope left for accurate measurement.

Charcoal from the fires of ancient peoples, dated by determining the carbon-14 activity, has been used to estimate the age of other artifacts found at the same archaeological sites. For example, carbon dating was used to confirm that ancient Hebrew writing found on a stone tablet unearthed in Israel dates from the ninth century B.C.E. Carbon dating has also been used to date the collection of Dead Sea Scrolls as being written between the third century B.C.E. and the first century C.E.

Tritium, the radioactive isotope of hydrogen, can also be used for dating materials. Its half-life of 12.26 years makes it useful for dating items up to about 100 years old. An interesting application of tritium dating is the dating of brandies. Alcoholic beverages that are 10–50 years old can be quite costly. Tritium dating can be used to check the veracity of advertising claims about the ages of the more expensive brandies.

Many other isotopes are useful for estimating the ages of objects and materials. Several of the more important ones are listed in Table 11.5.

It should also be noted that all isotopes of an element react the same way chemically. However, differences in isotopes can be quite important to some investigations. The variance in mass among isotopes can affect their rates of reaction, with lighter isotopes generally moving a bit faster than heavier ones. This difference can lead to a measurable characteristic called an *isotopic signature* in a material. For example, the ratio of carbon-13 to carbon-12 in materials formed by photosynthesis is less than the isotopic ratio in inorganic carbonates.

Water molecules with oxygen-18 atoms are heavier than water molecules with oxygen-16 atoms, and evaporate less readily than the latter do. This difference is less pronounced at higher temperatures. Oxygen is incorporated into calcium carbonate shells of aquatic organisms, whose fossils provide a chronological record of the temperature of the water in which the organisms lived. The atmospheric oxygen isotope ratio varies with the season and with geographic location, and allows scientists to determine the location in which a material originated.

The ratio of nitrogen-14 to nitrogen-15 is different for herbivores and carnivores because organisms higher in the food chain tend to concentrate the nitrogen-15 isotope in their tissues. This allows scientists to learn about the diets of people from ages past.

TABLE 11.5 Several Isotopes Used in Radioactive Dating

Isotope	Half-life (years)	Useful Range	Dating Applications
Carbon-14	5730	100–50,000 years	Charcoal; organic material
Hydrogen-3 (tritium)	12.26	1–100 years	Aged wines and brandies
Lead-210	22	1–75 years	Skeletal remains
Potassium-40	1.25×10^9	10,000 years to the oldest Earth samples	Rocks; Earth's crust; the moon's crust
Rhenium-187	4.3×10^{10}	4×10^7 years to the oldest samples in the universe	Meteorites
Uranium-238	4.51×10^9	10^7 years to the oldest Earth samples	Rocks; the Earth's crust

EXAMPLE 11.4   Radioisotopic Dating

An old wooden implement shows carbon-14 activity that is one-eighth that of new wood. How old is the artifact? The half-life of carbon-14 is 5730 y.

Solution
Using the relationship

$$\text{Fraction remaining} = \frac{1}{2^n}$$

we see that one-eighth is $\frac{1}{2^n}$, where $n = 3$; that is, the fraction $\frac{1}{8}$ is $\frac{1}{2^3}$. The carbon-14 has gone through three half-lives. The wood is therefore about $3 \times 5730 = 17{,}190$ y old.

▲ A body that has been preserved for centuries in the peat bogs of Europe.

> **Exercise 11.4A**
Calculate the approximate age of human remains have been preserved in the bogs of Northern Europe if they have a carbon-14 activity one-fourth that of living tissue. The half-life of carbon-14 is 5730 years.

> **Exercise 11.4B**
How old is a bottle of brandy that has a tritium activity one-sixteenth that of new brandy? The half-life of tritium (hydrogen-3) is 12.26 years.

SELF-ASSESSMENT Questions

1. A patient is given a 48 mg dose of technetium-99m (half-life 6.0 h) to treat a brain cancer. How much of it will remain in his body after 36 h?
   a. 0.75 mg
   b. 1.5 mg
   c. 3.0 mg
   d. 12 mg

2. What is the half-life of arsenic-74 if a 50.0 g sample decays to 6.25 g in 53.7 d?
   a. 8.9 d
   b. 17.9 d
   c. 27.8 d
   d. 53.7 d

3. Of a 64.0 g sample of strontium-90 (half-life 28.5 y) produced by a nuclear explosion, what mass in grams would remain unchanged after 285 y?
   a. 0.000 g
   b. 0.000625 g
   c. 0.00625 g
   d. 0.0625 g

4. Lead-210 (half-life 22.26 y) can be used to date skeletal remains. About how old is a bone that has an activity of 4.0 Bq if new bone has an activity of 16 Bq?
   a. 7.4 y
   b. 44.5 y
   c. 67 y
   d. 89 y

Answers: 1, a; 2, b; 3, d; 4, b

11.4 Artificial Transmutation

Learning Objective • Write a nuclear equation for a transmutation, and identify the product element formed.

During the Middle Ages, alchemists tried to turn base metals, such as lead, into gold. However, they were trying to do it chemically and were therefore doomed to failure because chemical reactions involve only the outer electrons of atoms. **Transmutation** (changing one element into another) requires altering the number of protons in the *nucleus*.

Thus far, we have considered only natural forms of radioactivity. Nuclear reactions can also be initiated by bombarding stable nuclei with alpha particles, neutrons, or other subatomic particles. These particles, given sufficient energy, can

▲ Ernest Rutherford carried out the first nuclear bombardment experiment. Element 104 was named rutherfordium (Rf) in his honor.

penetrate a stable nucleus and result in some kind of radioactive emission. Just as in natural radioactive processes, one element is changed into another. Because the change does not occur naturally, the process is called *artificial transmutation*.

In 1919, a few years after his famous gold-foil experiment (Section 3.4), Ernest Rutherford reported on the bombardment of a variety of light elements with alpha particles. One such experiment, in which he bombarded nitrogen, resulted in the production of protons, as shown in this balanced nuclear equation.

$$^{14}_{7}N + ^{4}_{2}He \longrightarrow ^{17}_{8}O + ^{1}_{1}H$$

(The hydrogen nucleus is simply a proton, which can be represented by the symbol $^{1}_{1}H$.) This provided the first empirical verification of the existence of protons in atomic nuclei, which Rutherford had first postulated in 1914.

Recall that Eugen Goldstein had produced protons in his gas-discharge tube experiments in 1886 (Section 3.1). The significance of Rutherford's experiment lay in the fact that he obtained protons from the *nucleus* of an atom other than hydrogen, thus establishing that protons are constituents of nuclei. Rutherford's experiment was the first induced nuclear reaction and the first example of artificial transmutation. His experiment not only produced protons but also changed nitrogen into oxygen!

 EXAMPLE 11.5 Artificial Transmutation Equations

When potassium-39 is bombarded with neutrons, chlorine-36 is produced. What other particle is emitted?

$$^{39}_{19}K + ^{1}_{0}n \longrightarrow ^{36}_{17}Cl + ?$$

Solution
To determine the particle, we need a balanced nuclear equation. To balance the equation, the unknown particle must have $A = 4$ and $Z = 2$. It is an alpha particle.

$$^{39}_{19}K + ^{1}_{0}n \longrightarrow ^{36}_{17}Cl + ^{4}_{2}He$$

❯ **Exercise 11.5A**
Technetium-97 is produced by bombarding molybdenum-96 with a deuteron (a hydrogen-2 nucleus). What other particle is emitted?

$$^{96}_{42}Mo + ^{2}_{1}H \longrightarrow ^{97}_{43}Tc + ?$$

❯ **Exercise 11.5B**
A team of scientists want to form uranium-235 by bombarding a nucleus of another element with an alpha particle. They expect a neutron to be produced along with the uranium-235 nucleus. What nucleus must be bombarded with the alpha particle?

SELF-ASSESSMENT Questions

1. What isotope is formed in the following artificial transmutation?

$$^{14}_{7}N + ^{1}_{0}n \longrightarrow ^{1}_{1}H + ?$$

a. carbon-12   **b.** carbon-14
c. nitrogen-15   **d.** oxygen-15

2. What isotope is formed in the following artificial transmutation?

$$^{9}_{4}Be + ^{1}_{1}H \longrightarrow ^{4}_{2}He + ?$$

a. boron-9   **b.** boron-10
c. lithium-6   **d.** lithium-7

3. What isotope is formed in the following artificial transmutation?

$$^{27}_{13}Al + ^{4}_{2}He \longrightarrow ^{1}_{0}n + ?$$

a. phosphorus-30   **b.** phosphorus-31
c. silicon-29   **d.** sulfur-31

Answers: 1, b; 2, c; 3, a

11.5 Uses of Radioisotopes

Learning Objective • List some applications of radioisotopes.

Most of the 3000 known radioisotopes are produced by artificial transmutation of stable isotopes. The value of both naturally occurring and artificial radioisotopes goes far beyond their contributions to our knowledge of chemistry.

Radioisotopes in Industry and Agriculture

Scientists in a wide variety of fields use radioisotopes as **tracers** in physical, chemical, and biological systems. Isotopes of a given element, whether radioactive or not, behave nearly identically in chemical and physical processes. Because radioactive isotopes are easily detected through their decay products, it is relatively easy to trace their movement, even through a complicated system. For example, tracing radioisotopes allows us to do the following:

- **Detect leaks in underground pipes.** Suppose there is a leak in a pipe that is buried beneath a concrete floor. We could locate the leak by digging up extensive areas of the floor, or we could add a small amount of radioactive material to liquid poured into the drain and trace the flow of the liquid with a Geiger counter (an instrument that detects radioactivity). Once we located the leak, only a small area of the floor would have to be dug up to repair it. A compound containing a short-lived isotope (for example, iodine-131, half-life 8.04 days) is usually employed for this purpose.

- **Measure the thickness of sheet material during production.** Radiation from a beta emitter is allowed to pass through sheet metal, paper, or plastic on the production line. The amount of radiation that passes through a sheet is related to its thickness.

- **Determine frictional wear in piston rings.** A ring is subjected to neutron bombardment, which converts some of the carbon in the steel to carbon-14. Wear in the piston ring is assessed by the rate at which the carbon-14 appears in the engine oil.

- **Determine the uptake of phosphorus and its distribution in plants.** This can be done by incorporating phosphorus-32, a β^- emitter with a half-life of 14.3 d, into phosphate fertilizers fed to plants.

Radioisotopes are also used to study the effectiveness of weed killers, compare the nutritional value of various feeds, determine optimal methods for insect control, and monitor the fate and persistence of pesticides in soil and groundwater. They have also been used to test the stability of buildings such as hospitals, schools, and structures of historical interest following natural disasters, such as the 2015 Nepal earthquake.

One of the most successful applications of radioisotopes in agriculture involves inducing heritable genetic alterations known as *mutations*. Exposing seeds or other parts of plants to neutrons or gamma rays increases the likelihood of genetic mutations. At first glance, this technique may not seem very promising. However, genetic variability is vital, not only to improve varieties but also to protect species from extinction. Some hybrid plants, such as seedless watermelons and bananas, are sterile. The lack of genetic variability among these plants may place the entire population at risk. Another use has been to irradiate insect populations to render them sterile. When those insects are released into the environment, they can mate with the native insects, but they cannot reproduce, limiting the need for insecticides. Sterile *Aedes aegypti* mosquitoes have been released to combat the Zika virus.

Radioisotopes are also used to irradiate foodstuffs as a method of preservation. Some 25% to 30% of harvested food can spoil before it is consumed. The radiation

▲ Scientists can trace the uptake of phosphorus by a green plant by adding a compound containing some phosphorus-32 to the applied fertilizer. When the plant is later placed on photographic film, radiation from the phosphorus isotopes exposes the film, much as light does. This type of exposure, called a *radiograph*, shows the distribution of phosphorus in the plant.

▲ Mosquitoes carry the Zika virus and other diseases. Irradiated mosquitoes will mate but will not reproduce, because they have been rendered sterile.

3 Do irradiated foods contain radioactive material?

Irradiated foods have absorbed energy in the form of gamma rays. But the foods themselves do *not* contain any radioactive material, so there is no danger to individuals who ingest irradiated food. Irradiated food must carry labels, but many people are wary of anything that is connected with radiation.

destroys microorganisms and enzymes that cause foods to spoil. Irradiated food shows little change in taste or appearance, and it has a much longer shelf life. While some people are concerned about possible harmful effects of chemical substances produced by the radiation, there has been no good evidence of harm to laboratory animals fed irradiated food or any known adverse effects in humans living in countries where food irradiation has been used for years. No residual radiation remains in food after irradiation, because gamma rays do not have nearly enough energy to change nuclei.

Radioisotopes in Medicine

Nuclear medicine involves two distinct uses of radioisotopes: therapeutic and diagnostic. In a therapeutic application, an attempt is made to treat or cure a disease with radiation. The diagnostic use of radioisotopes is aimed at obtaining information about the state of a patient's health.

Table 11.6 lists some radioisotopes in common use in medicine. The list is necessarily very incomplete, but it should give you an idea of their importance and the wide range of their uses. The claim that nuclear medicine has saved many more lives than nuclear bombs have taken is not an idle one.

Cancer is not one disease but many. Some forms are particularly susceptible to radiation therapy. The aim of **radiation therapy** is to destroy cancerous cells before too much damage is done to healthy tissue. Radiation is most lethal to rapidly reproducing cells, and rapid reproduction is the characteristic of cancer

TABLE 11.6 Some Radioisotopes and Their Medical Applications

Isotope	Name	Half-life[a]	Use
^{11}C	Carbon-11	20.39 min	Brain scans
^{15}O	Oxygen-15	2.05 min	Determining oxygen metabolism and blood flow and volume
^{51}Cr	Chromium-51	27.8 d	Blood volume determination
^{57}Co	Cobalt-57	270 d	Measuring vitamin B_{12} uptake
^{60}Co	Cobalt-60	5.271 y	Radiation cancer therapy
^{153}Gd	Gadolinium-153	242 d	Determining bone density
^{67}Ga	Gallium-67	78.1 h	Scan for lung tumors
^{131}I	Iodine-131	8.040 d	Thyroid diagnoses and therapy
^{192}Ir	Iridium-192	74 d	Breast cancer therapy
^{59}Fe	Iron-59	44.496 d	Detection of anemia
^{32}P	Phosphorus-32	14.3 d	Detection of skin cancer or eye tumors
^{238}Pu	Plutonium-238	86 y	Provision of power in pacemakers
^{226}Ra	Radium-226	1600 y	Radiation therapy for cancer
^{75}Se	Selenium-75	120 d	Pancreas scans
^{24}Na	Sodium-24	14.659 h	Locating obstructions in blood flow
^{99m}Tc	Technetium-99m	6.0 h	Imaging of brain, liver, bone marrow, kidney, lung, or heart
^{201}Tl	Thallium-201	73 h	Detecting heart problems during treadmill stress test
^3H	Tritium	12.26 y	Determining total body water
^{133}Xe	Xenon-133	5.27 d	Lung imaging

[a]Abbreviations: y—years; d—days; h—hours; min—minutes.

cells that allows radiation therapy to be successful. Radiation is carefully aimed at cancerous tissue while minimizing the exposure of normal cells. If the cancer cells are killed by the destructive effects of the radiation, the malignancy is halted.

Patients undergoing radiation therapy may show signs of radiation sickness during treatment. Nausea and vomiting are the usual early symptoms of radiation sickness, as rapidly growing cells, such as those in the gastrointestinal tract, are affected most. Radiation therapy can also interfere with the replenishment of white blood cells and thus increase patients' susceptibility to infection.

Radioisotopes used for diagnostic purposes provide information about the functioning of some part of the body or about the type or extent of an illness. For example, radioactive iodine-131 is used to determine the size, shape, and activity of the thyroid gland as well as to treat cancers in this gland and to control its hyperactivity. Small doses are used for diagnostic purposes, and large doses for treatment of thyroid cancer. After the patient drinks a solution of potassium iodide incorporating iodine-131, the iodide ions become concentrated in the thyroid. A detector showing the differential uptake of the isotope is used in diagnosis. The resulting *photoscan* can pinpoint the location of tumors or other abnormalities in the thyroid. In cancer treatment, radiation from therapeutic (large) doses of iodine-131 kills the thyroid cells in which the radioisotope has become concentrated.

The radioisotope gadolinium-153 is used to determine bone mineralization. Its widespread use is an indication of the large number of people, mostly women, who suffer from osteoporosis (reduction in bone density) as they grow older. Gadolinium-153 gives off two types of radiation: gamma rays and X-rays. A scanning device compares these types of radiation after they pass through bone. Bone density is then determined from the difference in absorption of the rays.

Technetium-99m is used in a variety of diagnostic tests (Table 11.6). The *m* stands for *metastable*, which means that this isotope gives up some energy when it changes to a more stable version of technetium-99 (which has the same atomic number and the same nucleon number). The energy it gives up is in the form of a gamma ray.

$$^{99m}_{43}\text{Tc} \longrightarrow {}^{99}_{43}\text{Tc} + \gamma$$

Note that the decay of technetium-99m produces no alpha or beta particles that could cause unnecessary damage to the body. Technetium-99m also has a short half-life (6.0 h), so the radioactivity does not linger very long in the body after the scan has been completed. With such a short half-life, use of the isotope must be carefully planned. In fact, medical labs do not purchase technetium-99m. Instead they buy molybdenum-99, which decays to form technetium-99m.

$$^{99}_{42}\text{Mo} \longrightarrow {}^{99m}_{43}\text{Tc} + {}^{0}_{-1}\text{e} + \gamma$$

The decay product, technetium-99m, is "milked" from the container as needed.

Using modern computer technology, *positron emission tomography (PET)* can measure dynamic processes occurring in the body, such as blood flow or the rate at which oxygen or glucose is being metabolized. For example, PET scans have shown that the brain of a person diagnosed with schizophrenia metabolizes only about one-fifth as much glucose as a normal brain. These scans can also reveal metabolic changes that occur in the brain during tactile learning (learning by the sense of touch) as well as identify brain damage that triggers severe epileptic seizures.

Compounds incorporating a positron-emitting isotope, such as carbon-11 or oxygen-15, are inhaled or injected before the scan. Before the emitted positron can travel very far in the body, it encounters an electron (numerous in any ordinary matter), and two gamma rays are produced, exiting from the body in exactly opposite directions.

$$^{11}_{6}\text{C} \longrightarrow {}^{11}_{5}\text{B} + {}^{0}_{+1}\text{e}$$

$$^{0}_{+1}\text{e} + {}^{0}_{-1}\text{e} \longrightarrow 2\gamma$$

WHY IT MATTERS

The difference between a radioactive isotope and the alpha, beta, gamma, or positron radiation that it produces is very important and often misunderstood. A radioisotope such as iodine-131 or radon-222, when absorbed by the human body, can cause great harm because the radiation continues to be released as the isotope decays. However, matter exposed to ionizing radiation simply absorbs that energy, causing a chemical change. A patient undergoing gamma-radiation treatment does *not* become radioactive.

4 Are we exposed to dangerous radiation during X-rays and other medical procedures?

A number of medical procedures do use ionizing radiation or radioisotopes, but the amount of radiation is kept to a minimum. The hazard of the ionizing radiation is ordinarily much less than the risk that would occur because of a lack of diagnosis or treatment!

5 What causes radiation sickness, and how serious is it?

Alpha, beta, and gamma rays; X-rays; and positrons that are absorbed by the body do cause cell damage. By and large, they are *not* energetic enough to create radio-active isotopes. Individuals who undergo radiation treatments often develop radiation sickness, and may require some time to recover once treatment ends. However, individuals who are exposed to very large amounts of radiation, such as from a bomb or a nuclear power plant disaster, will develop very serious radiation sickness or even die from acute radiation exposure.

Detectors, positioned on opposite sides of the patient, record the gamma rays. An image of an area in the body is formed using computerized calculations of the points at which annihilation of positrons and electrons occurs.

Yet a third role for radiation in medicine is the use of gamma rays or other very-high-energy electromagnetic radiation to sterilize medical instruments and equipment without subjecting them to the high temperatures and pressures that are encountered when materials are autoclaved.

Many people wonder just how much radiation is absorbed in a dental X-ray or other procedure, and how much must be absorbed to make a person ill. There are no clear-cut answers to these questions, because the effect of radiation depends to some extent on the timespan over which the radiation is absorbed. A single dose of 100 rem (*R*oentgen *e*quivalent *m*an, a measure of the effect of ionizing radiation on humans) is likely to have a significantly greater effect than two rem per month over a 50-month period. Table 11.7 shows the amount of radiation absorbed in various situations and the effects of higher doses.

TABLE 11.7 Radiation Dosages

Type/Effect of Exposure	Typical Dose
Living near a nuclear power plant	0.001 rem
One dental X-ray	0.005 rem
One chest X-ray	0.010 rem
One head CT scan	0.20 rem
Exposure from radon in average U.S. home	0.29 rem
Whole-body CT scan	1 rem
Dose (short-term) that usually gives symptoms of radiation poisoning (nausea, vomiting, hair loss)	40 rem
Dose (short-term) that usually is fatal	400–600 rem

EXAMPLE 11.6   Positron Emission Equations

One of the isotopes used to perform PET scans is oxygen-15, a positron emitter. What other isotope is formed when oxygen-15 decays?

Solution
First, we write the nuclear equation

$$^{15}_{8}O \longrightarrow {}^{0}_{+1}e + ?$$

The nucleon number, A, does not change, but the atomic number, Z, does: $8 - 1 = 7$. The product isotope is nitrogen-15.

$$^{15}_{8}O \longrightarrow {}^{0}_{+1}e + {}^{15}_{7}N$$

❭ **Exercise 11.6A**
Phosphorus-30 is a positron-emitting radioisotope suitable for doing PET scans. What other isotope is formed when phosphorus-30 decays?

❭ **Exercise 11.6B**
Copper-64 (a positron emitter) is used to study Wilson's disease, a genetic disease in which copper accumulates in the body. What other isotope is formed when copper-64 emits a positron?

Other Uses of Radioisotopes

Much of what we know about the chemistry of biological systems has been learned by using radioactive tracers in experiments. These include carbon-14, hydrogen-3 (tritium), and phosphorus-32, radioactive isotopes of elements that are found in the four major types of macromolecules in biological systems: carbohydrates, lipids, proteins, and nucleic acids. By adding a labeled compound, it is possible to follow the radioactive atoms through the metabolic pathways of living systems.

Another use of radioisotopes is in forensics. Nuclear activation analysis (NAA) involves placing a sample in a nuclear reactor and bombarding it with neutrons. The targeted atoms will then emit a gamma ray that can be detected. For example, NAA has been used to analyze bullet fragments used in the assassination of President John F. Kennedy, and the results support the findings of the Warren Commission that only two bullets struck Kennedy.

SELF-ASSESSMENT Questions

1. Iodine-131 is sometimes used for detecting leaks in underground pipes because
   a. it has a short half-life
   b. it is highly penetrating
   c. it is nontoxic
   d. its radiation is harmless

2. What is used to irradiate foodstuffs to extend shelf life?
   a. alpha particles
   b. beta particles
   c. gamma rays
   d. microwaves

3. Which of the following isotopes is used for detecting heart problems during a treadmill stress test?
   a. cobalt-60
   b. gadolinium-153
   c. iodine-131
   d. thallium-201

4. Which of the following isotopes is used in thyroid diagnosis and therapy?
   a. cobalt-60
   b. gadolinium-153
   c. iodine-131
   d. tritium

5. Which of the following isotopes is used to follow the uptake of fertilizer in plants?
   a. cobalt-60
   b. iodine-131
   c. lithium-7
   d. phosphorus-32

6. Which of the following isotopes has been used to provide power in pacemakers?
   a. chromium-51
   b. gallium-67
   c. iron-59
   d. plutonium-238

Answers: 1. a; 2. c; 3. d; 4. c; 5. d; 6. d

11.6 Penetrating Power of Radiation

Learning Objective • Describe the nature of materials needed to block alpha, beta, and gamma radiation.

The danger of radiation to living organisms comes from its potential for damaging cells and tissues. The ability to inflict injury relates to the penetrating power of the radiation. The two aspects of nuclear medicine just discussed (therapeutic and diagnostic) are also dependent on the penetrating power of various types of radiation.

All other things being equal, the more massive the particle, the less its penetrating power. *Alpha particles*, which are helium nuclei with a mass of 4 amu, are the least penetrating of the three main types of radiation. *Beta particles*, which are identical to the almost massless electrons, are somewhat more penetrating. *Gamma rays*, like X-rays, have no mass; they are considerably more penetrating than the other two types.

But all other things are not always equal. The faster a particle moves or the more energetic the radiation is, the more penetrating power it has.

It may seem counterintuitive that the biggest particles make the least headway. But keep in mind that penetrating power reflects the ability of radiation to make its way through a sample of matter. It is as if you were trying to roll some rocks through a field of boulders. The alpha particle acts as if it were a boulder itself. Because of its size, it cannot get very far before it bumps into and is stopped by other boulders. The beta particle acts as if it were a small stone. It can sneak between and perhaps ricochet off boulders until it makes its way farther into the field. The gamma ray can be compared with a grain of sand that can get through the smallest openings.

People have used different means of shielding to protect themselves while they were conducting experiments. Researchers using phosphorus-32, which emits beta particles, often wore lead aprons and were quite weary by the end of the day! Plexiglass shields both provided protection and let workers see what they were doing more easily.

The danger of a specific type of radiation to human tissue depends on the location of the radiation's source as well as on its penetrating power. If the radioactive substance is outside the body, alpha particles are the least dangerous; they have low penetrating power and are stopped by the outer layer of skin. Beta particles are also usually stopped before they reach vital organs. Gamma rays readily pass through tissue, so an external gamma source can be quite dangerous. People working with radioactive materials can protect themselves through the following actions:

- Move away from the source. The intensity of radiation decreases with distance from the source.
- Use shielding. A sheet of paper can stop most alpha particles. A block of wood or a thin sheet of aluminum or Plexiglas can stop most beta particles. Yet it takes a meter of concrete or several centimeters of lead to stop most gamma rays (Figure 11.6).

When the radioactive source is *inside* the body, as is the case for many medical applications, the situation is reversed. The nonpenetrating alpha particles can do great damage. All such particles are trapped within the body, which must then absorb all of the energy the particles release. Alpha particles inflict all their damage in a tiny area because they do not travel far. Therefore, getting an alpha-emitting therapeutic radioisotope close to the targeted cells is vital.

Beta particles distribute their damage over a somewhat larger area because they travel farther. Tissue may recover from limited damage spread over a large area; it is less likely to survive concentrated damage.

Many diagnostic applications rely on the highly penetrating power of gamma rays. In most cases, the radiation created inside the body must be detected by instruments outside the body, so a minimum of absorption is desirable.

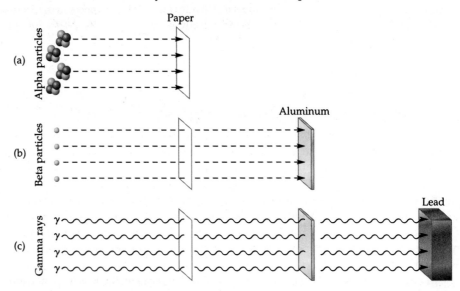

▲ **Figure 11.6** The relative penetrating power of alpha, beta, and gamma radiation. (a) Alpha particles are stopped by a sheet of paper. (b) Beta particles will not penetrate a sheet of aluminum. (c) It takes several centimeters of lead to block gamma rays.

 CONCEPTUAL EXAMPLE 11.7   Radiation Hazard

Most modern smoke detectors contain a tiny amount of americium-241, a solid element that is an alpha emitter. The alpha particles ionize air molecules, and the ions flow between charged plates and create a small electrical current. In a fire, the alpha particles hit smoke particles instead of air particles, the current drops, and an alarm is set off. The americium is in a chamber that includes thin aluminum shielding, and the device is usually mounted on a ceiling. Rate the radiation hazard of this device in normal use as (a) high, (b) moderate, or (c) very low. Explain your choice.

Solution

Alpha particles have very little penetrating ability—they are stopped by a sheet of paper or a layer of skin. The emitted alpha particles cannot exit the detector's chamber because even thin metal is more than sufficient to absorb them. Even if alpha particles could escape, they would likely be stopped by the air after a short distance or by the dead cells of your skin. Therefore, the radiation hazard is rated **(c)**—very low.

> **Exercise 11.7A**
Radon-222 is a gas that can diffuse from the ground into homes. Like americium-241, it is an alpha emitter. Is the radiation hazard from radon-222 likely to be higher or lower than that from a smoke detector? Explain.

> **Exercise 11.7B**
Polonium-210 is an alpha emitter that can be dissolved in water. Is it more likely to be a radiation hazard if it is in a glass standing on the counter or if a person drinks the water? Explain.

SELF-ASSESSMENT Questions

1. Which of the following pairs represents the *most and least* (most/least) penetrating types of radiation?
   a. alpha/gamma
   b. beta/gamma
   c. gamma/alpha
   d. X-rays/beta

2. Which form of radiation requires a thin sheet of aluminum to be stopped?
   a. alpha particles
   b. beta particles
   c. gamma rays
   d. all of these

3. A Geiger counter registered 50 Bq of activity from an unknown type of radiation. A piece of paper inserted between the source and the counter caused the reading to drop to about 2 Bq. The radiation was mostly
   a. alpha particles
   b. beta particles
   c. gamma rays
   d. X-rays

Answers: 1, c; 2, b; 3, a

11.7 Energy from the Nucleus

Learning Objectives • Explain where nuclear energy comes from. • Describe the difference between fission and fusion.

We have seen that radioactivity—quiet and invisible—can be beneficial or dangerous. A much more dramatic—and equally paradoxical—aspect of nuclear chemistry is the release of nuclear energy by either *fission* (splitting of heavy nuclei into smaller nuclei) or *fusion* (combining of light nuclei to form heavier ones).

Einstein and the Equivalence of Mass and Energy

The potential power in the nucleus was established by German physicist Albert Einstein (1879–1955), a famous and most unusual scientist. Whereas most scientists work with glassware and instruments in laboratories, Einstein worked with a pencil and a notepad. By 1905, at the age of 26, he had already worked out his special theory of relativity and developed his famous **mass–energy equation**, in which mass (m) is multiplied by the square of the speed of light (c) to determine the amount of energy (E) released.

$$E = mc^2$$

The equation suggests that mass and energy are equivalent—just two different aspects of the same thing—and that a little bit of mass can yield enormous energy. The atomic bombs that destroyed Hiroshima and Nagasaki (Section 11.8) in World War II converted less than an ounce of matter into energy.

A chemical reaction that gives off heat must also lose mass in the process, but the change in mass is far too small to measure. Reaction energy must be enormous—at

▲ Albert Einstein. Element 99 was named einsteinium (Es) in his honor.

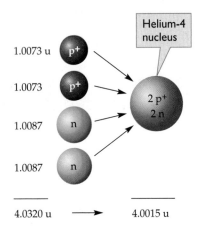

▲ Figure 11.7 Nuclear binding energy in $^4_2$He. The mass of a helium-4 nucleus is 4.0015 amu, which is 0.0305 amu less than the masses of two protons and two neutrons. The missing mass is equivalent to the binding energy of the helium-4 nucleus.

the level of the energy given off by nuclear explosions—for the mass loss to be measurable. (If every atom in a 1 kg lump of coal became energy, it could produce 25 billion kWh of electricity—enough to keep an 8-W LED bulb going for over 360 million years. Burning 1 kg of coal in a conventional power plant produces only enough energy to keep the bulb shining for about 5 weeks.)

Binding Energy

Nuclear fission releases the tremendous amounts of energy produced by atomic bombs and in nuclear power plants. Where does all this energy come from? It is locked inside the atomic nucleus. When protons and neutrons combine to form atomic nuclei, a small amount of mass is converted to energy. This is the **binding energy** that holds the nucleons together in the nucleus. For example, the helium nucleus contains two protons and two neutrons. The masses of these four particles add up to 2×1.0073 amu $+ 2 \times 1.0087$ amu $= 4.0320$ amu (Figure 11.7). However, the actual mass of the helium nucleus is only 4.0015 u, and the missing mass—called the *mass defect*—amounts to 0.0305 amu. Using Einstein's equation $E = mc^2$, we can calculate (see Problem 50) a value of 28.3 million electron volts (28 MeV; 1 MeV $= 1.6022 \times 10^{-13}$ J) for the binding energy of the helium nucleus. This is the amount of energy it would take to separate one helium nucleus into two protons and two neutrons.

When the binding energy per nucleon is calculated for all the elements and plotted against nucleon number, the graph in Figure 11.8 is obtained. The elements with the highest binding energies per nucleon have the most stable nuclei. They include iron and elements with nucleon numbers close to that of iron. When uranium atoms undergo nuclear fission, they split into atoms with higher binding energies. In other words, the fission reaction converts large atoms into smaller ones with greater nuclear stability.

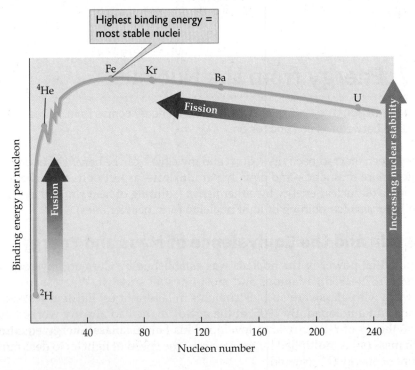

▲ Figure 11.8 Nuclear stability is greatest for iron and elements near iron in the periodic table. Fission of very large nuclei or fusion of very small ones results in greater nuclear stability.

Q *Which process, fission of uranium nuclei or fusion of hydrogen nuclei, releases more energy?*

We can also see from Figure 11.8 that even more energy can be obtained by combining small atoms, such as hydrogen or deuterium, to form larger atoms with more stable nuclei. This kind of reaction is called **nuclear fusion**. It is what happens when a hydrogen bomb explodes, and it is also the source of the sun's energy.

Nuclear Fission

In 1934, the Italian scientists Enrico Fermi (1901–1954) and Emilio Segrè (1905–1989) bombarded uranium atoms with neutrons. They were trying to make elements with higher atomic numbers than uranium, which had the highest atomic number then known. To their surprise, they found four radioactive species among the products. One was presumably element 93, formed by the initial conversion of uranium-238 to uranium-239, which then underwent beta decay.

$$^{238}_{92}U + ^{1}_{0}n \longrightarrow ^{239}_{92}U$$

$$^{239}_{92}U \longrightarrow ^{239}_{93}Np + ^{0}_{-1}e$$

However, they were unable to explain the remaining radioactivity.

German chemists Otto Hahn (1879–1968) and Fritz Strassman (1902–1980) repeated the Fermi–Segrè experiment in 1938 and were perplexed to find isotopes of barium among the many reaction products. Hahn wrote to Lise Meitner (1878–1968), an Austrian physicist who had worked with Hahn in Berlin for many years, to ask what she thought about these strange results. Because she was Jewish, Meitner had fled to Sweden when the Nazis took over Austria in 1938.

On hearing about Hahn's work, she noted that barium atoms were roughly half the size of uranium atoms. Was it possible that the uranium nuclei might be splitting into fragments? She made some calculations that convinced her that the uranium nuclei had indeed been split apart. Her nephew, Austrian physicist Otto Frisch (1904–1979), was visiting for the winter holidays, and they discussed this discovery with great excitement. Frisch later coined the term **nuclear fission** (Figure 11.9).

▲ Enrico Fermi. Element 100 was named fermium (Fm) in his honor.

▲ Lise Meitner. Element 109 was named meitnerium (Mt) in her honor.

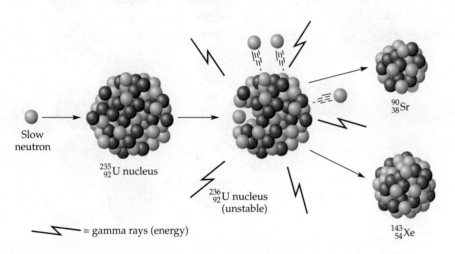

Slow neutron

$^{235}_{92}U$ nucleus

$^{236}_{92}U$ nucleus (unstable)

$^{90}_{38}Sr$

$^{143}_{54}Xe$

〰〰 = gamma rays (energy)

▲ **Figure 11.9** One possible way a uranium nucleus can undergo fission. The neutrons produced by the fission reactions can split other uranium nuclei, thus sustaining a chain reaction.

Q *If each of the neutrons emitted by the ^{236}U nucleus caused another ^{235}U nucleus to split in the way shown here, how many more neutrons would be released?*

Frisch was working with Niels Bohr at the University of Copenhagen, and when Frisch returned to Denmark, he told Bohr of the fission reaction. Shortly afterward, Bohr attended a physics conference in the United States. The discussions in the corridors about this new reaction would be the most important talks to take place at that meeting.

Meanwhile, Fermi had just received the 1938 Nobel Prize in Physics. Fermi's wife, Laura, was Jewish, and the fascist Italian dictator Benito Mussolini was a close ally of Hitler's. Fermi accepted the award in Stockholm and then immediately traveled to the United States with his wife and children. Thus, by 1939, the United States had received news about the German discovery of nuclear fission and had also acquired from Italy one of the world's foremost nuclear scientists.

Nuclear Chain Reaction

Leo Szilard (1898–1964), another Jewish refugee, had been born in Hungary and had come to the United States in 1937. Szilard was one of the first scientists to realize that neutrons released in the fission of one atom could trigger the fission of other uranium atoms, thus setting off a **chain reaction** (Figure 11.10). Because massive amounts of energy could be obtained from the fission of uranium, such a chain reaction might be used in a bomb with tremendous explosive force.

Aware of the destructive forces that could be produced, and concerned that Germany might develop such a bomb, Szilard prevailed on Einstein, who had moved to the United States, to sign a letter to President Franklin D. Roosevelt that described the importance of the discovery. It was critical that the U.S. government act quickly.

▶ **Figure 11.10** Schematic representation of a nuclear chain reaction. Neutrons released in the fission of one uranium-235 nucleus can strike other uranium-235 nuclei, causing them to split and release more neutrons as well as a variety of other nuclei. For simplicity, some fission fragments are not shown.

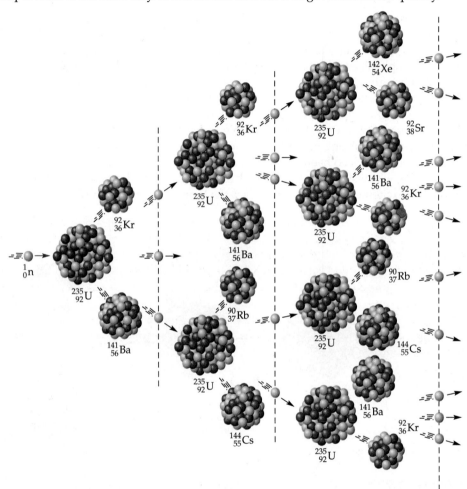

SELF-ASSESSMENT Questions

Refer to the graph in Figure 11.8 to answer questions 1–3.

1. Helium has a particularly high binding energy per nucleon, which means that it
   a. is especially unstable compared with H and Li
   b. is more stable than H and Li
   c. will readily fuse into larger nuclei
   d. will readily split into hydrogen-2 atoms

2. The most stable nuclei are those of
   a. Ba
   b. He
   c. Fe
   d. U

3. Which equation explains why the products of fission have less mass than the original substances?
   a. $E = hv/\lambda$
   b. $E = hv$
   c. $E = mc^2$
   d. $PV = nRT$

4. For a nuclear reaction to become self-sustaining, there must be a release of which particles?
   a. alpha particles
   b. beta particles
   c. neutrons
   d. protons

5. Combination of two smaller atomic nuclei into one larger nucleus is called
   a. a chain reaction
   b. fission
   c. fusion
   d. radioactive decay

6. Splitting an atomic nucleus into two or more smaller nuclei is called
   a. a chain reaction
   b. fission
   c. fusion
   d. radioactive decay

7. A self-sustaining nuclear reaction is called
   a. a chain reaction
   b. fission
   c. fusion
   d. radioactive decay

11.8 Nuclear Bombs

Learning Objectives • Describe the goals of the Manhattan Project and how uranium and plutonium bombs were made. • Identify the most hazardous fallout isotopes and explain why they are particularly dangerous.

In 1939, President Franklin D. Roosevelt launched a highly secret research project for the study of atomic energy. Called the *Manhattan Project*, it eventually became a massive research effort involving more scientific brainpower than had ever been devoted to a single project. Amazingly, it was conducted under such extreme secrecy that even Vice President Harry Truman did not know of its existence until after Roosevelt's death.

The Manhattan Project included four separate research teams that focused on:

- sustaining a nuclear fission chain reaction
- enriching uranium so that it contained about 90% of the fissionable isotope ^{235}U
- making plutonium-239 (another fissionable isotope)
- constructing a bomb based on nuclear fission

Sustainable Chain Reaction

By 1939, it had been established that neutron bombardment could initiate the fission reaction, but many attributes of this reaction were unknown. Fermi and his group, working in a lab under the bleachers at Stagg Field on the campus of the University of Chicago, worked on the fission reaction and how to sustain it. They found that the neutrons used to trigger the reaction had to be slowed down to increase the probability that they would actually hit a uranium nucleus. Because graphite slows down neutrons, a large "pile" of graphite was built to house the reaction. Then the amount of uranium "fuel" was gradually increased. The major goal was to determine the **critical mass**—the amount of uranium-235 needed to sustain the fission reaction. There had to be enough fissionable nuclei in the "fuel" for the neutrons released in one fission event to have a good chance of being captured by another fissionable nucleus before escaping from the pile.

On December 2, 1942, Fermi and his group achieved the first sustained nuclear fission reaction. The critical mass of uranium (enriched to about 94% uranium-235) for this reactor turned out to be about 16 kg.

◄ Graphite pile constructed under the bleachers at Stagg Stadium at the University of Chicago. This was the site of the first sustained nuclear fission reaction.

Isotopic Enrichment

Natural uranium is 99.27% uranium-238, which does not undergo fission by itself. Uranium-235, the fissionable isotope, makes up only 0.72% of natural uranium.

Because making a bomb required **enriching** the uranium to about 90% uranium-235, it was necessary to separate the uranium isotopes. This was the job of the Manhattan Project research team in Oak Ridge, Tennessee.

Chemical separation was almost impossible; uranium-235 and uranium-238 are chemically almost identical. The separation method used converted uranium to gaseous uranium hexafluoride, UF_6. Molecules of UF_6 containing uranium-235 are slightly lighter and therefore move slightly faster than molecules containing uranium-238. Uranium hexafluoride was passed through a series of thousands of pinholes, and the molecules containing uranium-235 gradually outdistanced the others. Enough enriched uranium-235 was finally obtained to make a small explosive device.

Depleted Uranium

Naturally occurring uranium contains 0.72% ^{235}U, the fissile isotope that is used in nuclear reactors, although the amount may vary slightly depending on the source of the ore. However, nuclear power plants require that the fissile isotope be increased to a concentration of 3.5% to 5.0% to support a controlled chain reaction. Enrichment to approximately 90% is needed for nuclear weapons.

The enrichment process concentrates the ^{235}U isotope. The remaining uranium ore from which most ^{235}U has been removed is known as *depleted uranium* (DU). After enrichment, the ^{235}U concentration in DU is only ~0.2% of the total.

The military has used armor plating and ammunition made from depleted uranium. A major product has been 30 mm projectiles that can be directed against military vehicles. Uranium's density of 19.1 g/cm^3 is much greater than that of most other armored materials, so a bullet made of depleted uranium rather than lead (with a density of only 11.34 g/cm^3) will be heavier but smaller in diameter and will be able to pierce armor plating made from steel, which has a density of only 7.75–8.05 g/cm^3. A direct hit will also result in converting some of the uranium into dust particles, which can then be inhaled by people in the area.

There have been myriad studies of the health effects of exposure to depleted uranium. These generally show that the two organs most affected are the lungs and the kidneys. Small particles of uranium that result from impact of DU weapons can be inhaled and can settle in the lungs. This can lead to radiation damage or even lung cancer if the concentration is high enough. Other means of exposure are through ingestion of contaminated food and water or from metal shrapnel fragments embedded in the body after impact. The overall toxic effects of DU seem to be less related to the radioactivity in the material than to toxicity to uranium itself. There are numerous reports of increased numbers of birth defects or other health problems in areas where DU has been used.

▲ Glenn T. Seaborg. Element 106 was named seaborgium (Sg) in his honor. Seaborg is the only person to have had an element named for him while still alive. He is shown here pointing to his namesake element on the periodic table of elements.

Synthesis of Plutonium

While the tedious work of separating uranium isotopes was under way at Oak Ridge, another research team, led by American scientist Glenn T. Seaborg (1912–1999), approached the challenge of obtaining fissionable material by another route. Although uranium-238 would not undergo fission when bombarded by neutrons, this more common uranium isotope *would* decay to form a new element, neptunium (Np), which quickly decayed to another new element, plutonium (Pu).

$$^{238}_{92}U + ^{1}_{0}n \longrightarrow ^{239}_{92}U$$

$$^{239}_{92}U \longrightarrow ^{0}_{-1}e + ^{239}_{93}Np$$

$$^{239}_{93}Np \longrightarrow ^{239}_{94}Pu + ^{0}_{-1}e$$

Plutonium-239 was found to be fissionable and thus was suitable material for the making of a bomb. A group of large reactors was built near Hanford, Washington, to produce plutonium.

Bomb Construction

The actual building of the nuclear bombs was carried out at Los Alamos, New Mexico, under the direction of American physicist J. Robert Oppenheimer (1904–1967). In a top-secret laboratory at a remote site, a group of scientists planned and then constructed what would become known as atomic bombs. Two different models were developed, one based on ^{235}U and the other on ^{239}Pu.

The critical mass of uranium-235 could not be exceeded prematurely or the bomb would detonate early, so it was important that no single piece of fissionable material in the bomb be large. The bomb was designed to contain pieces of uranium of subcritical mass plus a neutron source to initiate the fission reaction. Then, at the chosen time, all the pieces would be forced together using an ordinary high explosive, thus triggering a runaway nuclear chain reaction.

The synthesis of plutonium turned out to be easier than the isotopic separation, and by July 1945, enough fissionable material had been made for three bombs to be assembled—two using plutonium and one using uranium. The first atomic bomb (one of the plutonium devices) was tested in the desert near Alamogordo, New Mexico, on July 16, 1945. The heat from the explosion vaporized the 30 m steel tower on which the bomb was placed and melted the sand for several acres around the site. The light produced was the brightest anyone had ever seen.

Some of the scientists were so awed by the force of the blast that they argued against using the bombs on Japan. A few, led by Leo Szilard, suggested a demonstration of the bombs' power at an uninhabited site. But fear of a well-publicized "dud" and the desire to avoid millions of American casualties during an invasion of Japan led President Truman to order the dropping of the bombs on Japanese cities. The lone uranium bomb, called "Little Boy" (Figure 11.11), was dropped on Hiroshima on August 6, 1945, and caused over 100,000 casualties. Three days later, the other plutonium bomb, called "Fat Man," was dropped on Nagasaki with comparable results (Figure 11.12). World War II ended with the surrender of Japan on August 14, 1945.

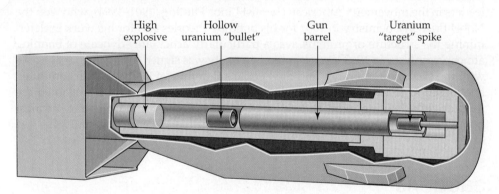

High explosive   Hollow uranium "bullet"   Gun barrel   Uranium "target" spike

◀ **Figure 11.11** An internal schematic of the atomic bomb "Little Boy" dropped on Hiroshima. Each of the cylinders of uranium-235 had less than the critical mass. The high explosive shot the cylindrical "bullet" down the gun barrel and onto the target spike. The result was about two critical masses of uranium-235, and a nuclear explosion occurred about a millisecond later.

Radioactive Fallout

When a nuclear explosion occurs in the open atmosphere, radioactive materials can rain down on parts of Earth thousands of miles away, days and weeks later, in what is called **radioactive fallout**. The uranium atom can split in several different ways, including those shown here.

$$^{235}_{92}U + {}^{1}_{0}n \longrightarrow {}^{90}_{38}Sr + {}^{143}_{54}Xe + 3{}^{1}_{0}n$$
$$\longrightarrow {}^{102}_{39}Y + {}^{131}_{53}I + 3{}^{1}_{0}n$$
$$\longrightarrow {}^{95}_{37}Rb + {}^{137}_{55}Cs + 4{}^{1}_{0}n$$

▲ **Figure 11.12** The mushroom cloud over Nagasaki, following the detonation of "Fat Man" on August 9, 1945.

The primary (first) fission products are radioactive. These decay to daughter isotopes, many of which are also radioactive. In all, over 200 different fission products are produced, with half-lives that vary from less than a second to more than a billion years. Also, the neutrons produced in the explosion act on molecules in the atmosphere to produce carbon-14, tritium, and other radioisotopes. Fallout is therefore exceedingly complex. We will consider three of the more worrisome isotopes here.

Of all the isotopes released in an atomic explosion, strontium-90 (half-life 28.5 y) presents the greatest hazard to people. Strontium-90 reaches us primarily through dairy products and vegetables. Because of its similarity to calcium (both are group 2A elements), strontium-90 is incorporated into bone. There it remains an internal source of radiation for many years. Strontium-90 has been tied to an increase in bone cancer and leukemia.

Iodine-131 may present a greater threat immediately after a nuclear explosion or reactor meltdown. Its half-life is only 8 days, but it is produced in relatively large amounts. Iodine-131 is readily transferred up the food chain. In the human body, it is concentrated in the thyroid gland, and it is precisely this characteristic that makes a trace of iodine-131 so useful for diagnostic scanning. However, for a healthy individual, larger amounts of radioactive iodine offer only damaging side effects. To minimize the absorption of radioactive iodine, many people in the areas of Chernobyl and Fukushima were given large amounts of potassium iodide, which effectively diluted the amount of radioactive iodine absorbed by their thyroid glands.

Cesium-137 (half-life 30.2 y) is, like strontium-90, capable of long-term effects. Because of its similarity to potassium (both are group 1A elements), it is taken up by living organisms as part of body fluids. It can be obtained from sources in the environment and reconcentrated in living organisms. Elevated levels of cesium-137 were perhaps the greatest health concern following the Fukushima Daiichi disaster.

By the late 1950s, radioactive isotopes from atmospheric testing of nuclear weapons were detected in the environment. Concern over radiation damage from nuclear fallout led to a movement to ban atmospheric testing. Many scientists were leaders in the movement. American chemist Linus Pauling (1901–1994), who won the Nobel Prize in Chemistry in 1954 for his bonding theories and for his work in determining the structure of proteins, was a particularly articulate advocate of banning atmospheric tests. In 1963, a nuclear test ban treaty was signed by the major nations—with the exception of France and the People's Republic of China, which continued aboveground tests. Since the signing of the treaty, other countries have joined the nuclear club. Pauling, who had endured being called a communist and a traitor because of his outspoken position, was awarded the Nobel Peace Prize in 1962.

Thermonuclear Reactions

In the chapter opening we noted that the sun is a nuclear power plant vital to life on Earth. However, the nuclear reactions that take place in the sun are somewhat different from the ones we have discussed so far in this chapter. These **thermonuclear reactions** require enormously high temperatures (millions of degrees) to initiate them. The intense temperatures and pressures in the sun cause nuclei to fuse and thus release enormous amounts of energy. Instead of large nuclei being split into smaller fragments (fission), small nuclei are fused into larger ones (fusion). The main reaction in the sun is thought to be the fusion of four hydrogen nuclei to produce one helium nucleus and two positrons.

Fusion of 1 g of hydrogen into helium releases an amount of energy equivalent to the burning of nearly 20 tons of coal. Every second, the sun fuses 600 tons of hydrogen, producing millions of times more energy than has been produced on Earth in the entire history of humankind. Much current research is aimed at reproducing such a reaction in the laboratory by using ultrapowerful magnets to contain the intense heat required for ignition. Fusion technology is discussed further in Chapter 15. To date, however, fusion reactions on Earth have been limited to the uncontrolled reactions in explosions of hydrogen (thermonuclear) bombs and to small amounts of energy produced by very expensive experimental fusion reactors.

$$4 {}_{1}^{1}\text{H} \longrightarrow {}_{2}^{4}\text{He} + 2 {}_{+1}^{0}\text{e}$$

SELF-ASSESSMENT Questions

1. The isotope used in uranium fission reactions is
   a. uranium-232   **b.** uranium-234
   c. uranium-235   **d.** uranium-238

2. Strontium-90 substitutes for calcium in bones because Sr and Ca
   a. are in the same group of the periodic table
   b. are in the same period of the periodic table
   c. have identical electron configurations
   d. have identical half-lives

3. Many people in the Chernobyl and Fukushima areas were given large doses of potassium iodide to limit the uptake of iodine-131 and its concentration in the
   a. bones   **b.** kidneys   **c.** lungs   **d.** thyroid

4. Which of the following is *not* one of the isotopes produced by fission that present the greatest hazard to people?
   a. cesium-137
   b. iodine-131
   c. iron-59
   d. strontium-90

5. Which of the following radioisotopes in fallout is dangerous mainly because it becomes concentrated in bones?
   a. cesium-137
   b. iodine-131
   c. strontium-90
   d. radon-222

Answers: 1, c; 2, a; 3, d; 4, c; 5, c

11.9 Uses and Consequences of Nuclear Energy

Learning Objective • List some uses and consequences of nuclear energy.

A significant portion of the electricity we use today is generated by nuclear power plants. In the United States, one-fifth of all the electricity produced comes from nuclear power plants. Europeans rely even more on nuclear energy. France, for example, obtains 72% of its electric power from nuclear plants, while Belgium, Spain, Switzerland, and much of Eastern Europe generate about one-third of their power from nuclear reactors.

Ironically, the same nuclear chain reaction that occurred in the detonation of the bomb dropped on Hiroshima is used extensively today under the familiar concrete containment tower of a nuclear power plant. The key difference is that the power plant employs a *slow*, *controlled* release of energy from the nuclear chain reaction, rather than an explosion. The slowness of the process is due to the use of uranium fuel that is much less enriched (about 3.5% to 5% ^{235}U rather than the 90% enrichment for weapons-grade uranium).

While the generation of electricity is currently the major use for uranium, nuclear reactors are also used to propel a small number of seagoing vessels. These are mainly military rather than civilian ships. The first of these vessels was the submarine U.S.S. *Nautilus*, which was launched in 1955. Nuclear power enabled the submarine to remain submerged or at sea for long time periods without refueling. The *Nautilus*, for example, was able to reach the North Pole by navigating beneath the Polar Ice Pack. There are currently about 150 marine vessels that carry nuclear reactors, including much larger ships with several reactors, such as the aircraft carrier U.S.S. *John C. Stennis*.

Currently there are some 450 reactors that generate electricity worldwide. There are also roughly 225 small reactors that are important in the manufacturing of radioisotopes for medical or other purposes. These include a wide range of isotopes, such as Mo-99 (a precursor to the short-lived Tc-99m, which is used in medical diagnoses); Sr-82, which is used for heart imaging; and Ge-68, which is used to calibrate PET equipment for medical diagnostic procedures.

One of the main problems with the production of nuclear power arises from the products of the nuclear reactions. As in nuclear fallout, most of the daughter nuclei produced by the fission of uranium-235 are also radioactive, including some with very long half-lives. We will discuss the problems associated with nuclear waste further in Chapter 15. Another problem is the potential transformation of spent nuclear fuel into weapons-grade material. A third problem occurs when nuclear plants malfunction.

There have been three serious malfunctions at nuclear plants. Three Mile Island near Harrisburg, Pennsylvania, experienced a partial meltdown of the core in 1978, but relatively little radioactivity was released into the environment. Chernobyl involved a fire and the release of radioactivity into the surrounding area. Over 330,000 people were relocated to other areas after the 1986 explosion, and much of the forest and agricultural area around the plant still has high levels of cesium-137 contamination. More recently, the Fukushima Daiichi reactors malfunctioned following a 2011 earthquake that created tsunami waves that were much higher than the sea walls protecting the facility. Although the plants survived the earthquake, flooding caused failure of the cooling system, a partial core meltdown, and the release of radioactive material. Cleanup and decommissioning of the Fukushima units are progressing, but are expected to take several decades.

In addition to plant malfunctions, problems resulting during the production of nuclear materials have taken a serious toll on the environment. For example, mining of uranium in the American Southwest from the 1940s through the 1980s resulted in many open mines with uranium tailings that were not sealed properly. Testing on members of the Navajo nation who reside in the region has shown a significant increase both in blood uranium levels and in the rates of kidney and urinary system cancers. Hanford, Washington, saw the initial production of plutonium followed by several decades of subsequent work on radioactivity. It now has many leaky casks of radioactive wastes. Hanford is the location of the largest environmental cleanup effort in the United States, which is expected to take decades.

When power is generated in nuclear plants, important changes occur in the fuel. Neutron bombardment converts uranium-235 to radioactive daughter products. Eventually, the concentration of uranium-235 becomes too low to sustain the nuclear chain reaction. The fuel rods (Figure 11.13) must be replaced about every three years, and the rods then become high-level nuclear waste.

▶ **Figure 11.13** The core of a nuclear reactor contains hollow rods filled with uranium pellets. The heat generated by the fission reaction is used to boil water, which drives turbines to generate electricity in the same way as in a coal- or gas-fired generating plant. Because the pellets are only about 3.5% to 5% uranium-235, a nuclear explosion cannot occur, though loss of coolant may lead to a reactor meltdown.

One potential problem that can arise from fuel rods in nuclear reactors is the production of fissile plutonium-239, which could be diverted to use in plutonium bombs. As uranium-235 nuclei undergo fission, the nonfissionable (and very concentrated) uranium-238 nuclei absorb neutrons and are converted to plutonium-239, a transmutation described in Section 11.8. In time, some plutonium-240 is also formed. However, if the fuel rods are removed after only about three months, the fissionable plutonium-239 can be easily separated from the other fission products by chemical means. Plutonium bombs require much less fissile material to produce than is needed to make nuclear weapons from uranium. Plutonium is thus of greater concern with respect to weapons proliferation, because operation of a nuclear plant can produce materials suitable for use in a bomb. North

Korea has nuclear power plants and a plutonium separation facility capable of producing enough plutonium-239 for several weapons each year (Figure 11.14). Iran is also reported to have a facility capable of enriching uranium. In addition, there are a number of other countries, such as Pakistan, which now have nuclear weapon capabilities. The activities of these and other developing countries raise the fear of a dangerous proliferation of nuclear weapons.

The amount of nuclear waste from reactors and from isotopic enrichment is quite large, though not nearly as large as the amount of ash and waste from coal-fired power plants. Ways to deal with the waste and to minimize the amount of material have not been well developed. In some countries, most notably in Europe, reprocessing used fuel has been carried out for years. Recovery of unused uranium and plutonium can result in 25% to 30% more energy coming from the original fuels. Recently, interest in recovering actinides with plutonium has grown. This would lower the amount of long-term radioactivity in the waste as well as the risk of having plutonium used for illegal activities. However, the U.S. government has prohibited reprocessing by civilian companies. Much effort and money have been spent on finding a suitable repository for nuclear waste. Funding for development of the most promising site, Yucca Mountain in Nevada, has been suspended, although there is discussion about looking at that site once again.

▲ **Figure 11.14** A satellite photograph of part of the plutonium processing plant at Yongbyon, North Korea.

The Nuclear Age

The splitting of the atom made the Chinese curse "May you live in interesting times" seem quite appropriate. The goal of the alchemists, to change one element into another, has been achieved through the application of scientific principles. New elements have been formed, and the periodic table has been extended well beyond uranium (Z = 92). This modern alchemy produces plutonium by the ton; neptunium (Z = 93), americium (Z = 95), and curium (Z = 96) by the kilogram; and berkelium (Z = 97) and einsteinium (Z = 99) by the milligram.

As we have seen, radioactive isotopes have many uses, from killing tiny but deadly cancer cells and harmful microorganisms in food to serving as tracers in a variety of biological experiments and imaging technologies, and generating fearsome weapons. Figure 11.15 displays a number of other constructive uses of nuclear energy. We live in an age in which the extraordinary forces present in the atom have been unleashed as a true double-edged sword. The threat of nuclear war—and nuclear terrorism—has been a constant specter for some 75 years, yet it is hard to believe that the world would be a better place if we had not discovered the secrets of the atomic nucleus.

6 Can we minimize or recycle radioactive wastes to make nuclear power generation a more sustainable process?

Processes that involve the production of radioisotopes or that use radioisotopes create a large amount of radioactive waste, some of which has a very long half-life. The main strategy so far in the United States has been to store waste materials until they can be moved to a permanent long-term storage facility, but a suitable site has still not been established. Just the sheer volume of waste is overwhelming. If some of the waste could be reused, or if the volume could be reduced by separating out radioactive metals, it would minimize the volume of radioactive waste, increase the U.S. energy supply, and make nuclear power a more sustainable source of energy.

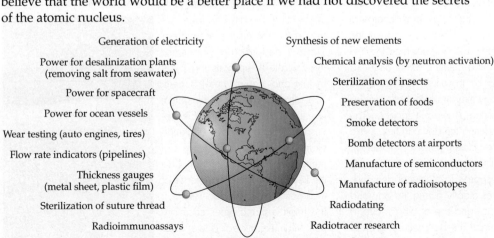

Generation of electricity

Power for desalinization plants (removing salt from seawater)

Power for spacecraft

Power for ocean vessels

Wear testing (auto engines, tires)

Flow rate indicators (pipelines)

Thickness gauges (metal sheet, plastic film)

Sterilization of suture thread

Radioimmunoassays

Medical diagnosis

Synthesis of new elements

Chemical analysis (by neutron activation)

Sterilization of insects

Preservation of foods

Smoke detectors

Bomb detectors at airports

Manufacture of semiconductors

Manufacture of radioisotopes

Radiodating

Radiotracer research

Cross-linking of polymers

▲ **Figure 11.15** Some constructive uses of nuclear energy.

GREEN CHEMISTRY Can Nuclear Power Be Green?
Galen Suppes and Sudarshan Loyalka University of Missouri

Principles 1, 3, 6, 10, and 12

Learning Objective • Explain how applications of Green Chemistry Principles can be applied to make nuclear power safer and more cost competitive with other power sources.

For decades, nuclear power has provided about 20% of the U.S. electricity with a proven performance and safety record (Section 11.9). Building upon this tradition and making the use of nuclear power greener and more sustainable may have economic and environmental benefits.

Like all technologies, nuclear power has advantages and challenges that are not easily compared. Among the advantages advocates of nuclear power claim are (1) reduced CO_2 emissions compared to coal and natural gas; (2) a lower volume of waste; (3) lower fuel cost (~0.7 ¢/kWh of electricity versus ~ 1.5 to 4 ¢/kWh for coal and natural gas); (4) the best safety history (about one-fifth the annual occupational health deaths per kWh of electrical power versus coal); and (5) reliability (nuclear power plants operate at the industry's highest capacity factors).

Challenges people point to include (1) the need for specialized spent-fuel handling and waste storage; (2) base load mismatch to peak demand; (3) potential for environmental contamination and broad-scale lethal exposure; (4) potential use in weapons proliferation; and (5) high capital costs.

If society is going to use nuclear power, it's probably a good idea to see how green chemistry and improved safety standards can be used to meet these challenges:

Radioactive Waste – Chemists have developed greener approaches to waste processing and storage. Nuclear waste typically contains a small amount of radioactive material that requires handling the entire mixture as radioactive waste. The first pass on handling radioactive waste is to store the waste for several months before attempting to handle it. Short half-life isotopes undergo natural and quick degradation to safer materials (Green Chemistry Principle 10) These radioactive parts of the waste can be removed using known chemistry (opportunities for Green Chemistry Principle 3). France and Japan partially concentrate nuclear waste today and produce a reprocessed fuel that can be used to generate energy at a low cost (0.9 ¢/kWh). If all of the nuclear fuel used to date in the United States were fully reprocessed and useful isotopes recycled, the remaining radioactive waste would fit in a building about one third the size of a small house. Also, it is possible to convert that waste into harmless isotopes, and future technology may make that affordable.

Peak Power Demand – Storing excess electrical energy produced at night followed by its use during the day can help make nuclear energy easier to use and more efficient (Green Principle 6). For both safety and economy, nuclear reactors operate continuously and do not change their output in response to electrical demand. However, the need for electricity varies greatly during the day and night. Use of large grid-based batteries and use of plug-in hybrid electric vehicles (being charged at night) are both topics of great interest—solutions to both can be attained through green chemistry.

Safety – Nuclear plants employ multiple safety features and also use the philosophy of "defense in depth" against accidental radioactivity releases (Green Chemistry Principle 12). Yet accidents with large consequences have occurred. The three most costly accidents have been at Three Mile Island, Chernobyl, and Fukushima Daiichi. Chemical processes during accidents include build-up of pressure, formation of hydrogen, combustion of hydrogen with air, and reactions and transport of radioactive fission products and other substances. Use of green adsorbents that will remove these radioactive components from the vented gases is a green chemistry solution.

Security (Weapons Proliferation and Physical Protection) – It is difficult to make a nuclear bomb from low-enriched uranium (used in all power reactors) because the process involves separating uranium isotopes. It is easier to separate plutonium from the spent fuel (mainly uranium-238), because plutonium is a different element with different chemical properties, and hence plutonium should be removed from the spent fuel for recycling or safeguarding. The process used for this separation (known as PUREX) is a first-generation commercial process with ample opportunity to be improved upon with green chemistry. The objective is to remove the plutonium from spent fuel and to use it in new fuel rods thereby minimizing the amount on site—this represents a "recycle to extinction process," which is an example Green Chemistry Principle 1.

High Power-Plant Costs – The high capital cost of nuclear power plants is caused in part by the delays between capital investment and startup due to permitting and licensing processes. The approach taken by the United States to address this concern is the use of small modular nuclear reactors (SMNR), with a $450 million investment announced in 2012. Solutions to higher plant costs are not so much a matter of green chemistry but rather of good practices that improve the economics so that green chemistry can solve the issues related to other topics.

The processing of nuclear waste is performed at places like the Sellafield reprocessing plant in England (shown here). Chemistry allows recycling of plutonium (weapons proliferation), recycling of unspent fuel, and removal of benign elements from the radioactive elements. Control of chemistry is also the key to preventing the release of radiation to the environment (safety). It is improved and green chemistry that will make nuclear power safer and sustainable for millennia.

SELF-ASSESSMENT Questions

1. Nuclear fission has been used for all of the following except
   a. development of new elements
   b. excavation of an underground nuclear waste repository
   c. generation of electricity
   d. powering of ships and submarines

2. A major problem with nuclear power is
   a. it does not generate enough electricity
   b. it results in large amounts of nuclear waste
   c. power plants can only be built where there is access to water for cooling
   d. large amounts of radiation are released into the air.

3. Which of the following isotopes presents the greatest risk for the development of nuclear weapons?
   a. plutonium-239   b. radon-222
   c. radium-228   d. uranium-238

Summary

Section 11.1—Some isotopes of elements are **radioisotopes**, unstable nuclei that undergo changes in nucleon number, atomic number, or energy during the **radioactive decay** process. We are constantly exposed to naturally occurring **background radiation**. **Ionizing radiation** (including nuclear radiation and X-rays) causes harm by forming ions, disrupting cellular processes, and damaging DNA.

Section 11.2—Nuclear equations are used to represent nuclear processes. These equations are balanced when the sums of nucleon numbers and the sums of atomic numbers on each side are the same. Four types of radioactive decay are **alpha ($_2^4$He) decay**, **beta ($_{-1}^0$e) decay**, **gamma ($_0^0\gamma$) decay**, and emission of a **positron ($_{+1}^0$e)**. **Electron capture (EC)** is a fifth type of decay, in which a nucleus absorbs one of the atom's electrons. There are a number of important differences between nuclear reactions and chemical reactions.

Section 11.3—The unit of radioactive decay is the **becquerel (Bq)** which is equal to 1 disintegration/s.

The **half-life** of a radioactive isotope is the time it takes for half of a sample to decay. The fraction of a radioisotope remaining after n half-lives is given by

$$\text{Fraction remaining} = \frac{1}{2^n}$$

Half-lives of certain isotopes can be used to estimate the ages of various objects. **Carbon-14 dating** is the best-known of the radioisotopic dating techniques.

Section 11.4—**Transmutation**, the conversion of one element into another, cannot be carried out by chemical means but can be accomplished by nuclear processes. Bombarding a stable nucleus with energetic particles can cause artificial transmutation.

Section 11.5—Radioisotopes have many uses. A radioisotope and a stable isotope of an element behave nearly the same chemically, so radioisotopes can be used as **tracers** in physical and biological systems. Radiation can be used to produce useful mutations in agriculture and to preserve foodstuffs.

Radiation therapy to destroy cancer cells depends on the fact that radiation is more damaging to those rapidly reproducing cells than to healthy cells. Radioisotopes are used in the diagnosis of various medical disorders.

Section 11.6—Different types of radiation have different penetrating abilities. *Alpha particles* (helium nuclei) are relatively slow and have low penetrating power. *Beta particles* (electrons) are much faster and more penetrating. *Gamma rays* (high-energy photons) travel at the speed of light and have great penetrating power. How hazardous radiation is depends on the location of the source.

Section 11.7—Einstein's mass–energy equation, $E = mc^2$, relates mass and energy. The total mass of the nucleons in a nucleus is greater than the actual mass of the nucleus. The missing mass, or the mass defect, is equivalent to the **binding energy** holding the nucleons together. Binding energy can be released either by breaking down heavy nuclei into smaller ones, a process called **nuclear fission**, or by joining small nuclei to form larger ones, called **nuclear fusion**.

Bombarding uranium atoms with neutrons produced radioactive species, including lighter nuclei among the products. It was hypothesized that the uranium nuclei had undergone fission and that neutrons released in the fission reaction of one nucleus could initiate a **chain reaction**.

Section 11.8—The Manhattan Project had four goals: (1) to determine the **critical mass** of fissionable material required to achieve a sustained nuclear fission reaction; (2) to extract fissionable uranium-235 from ordinary uranium; (3) to synthesize plutonium-239; and (4) to construct a nuclear fission bomb before the Germans did so.

In addition to the devastation at the site of a nuclear explosion, much radioactive debris, or **radioactive fallout**, is produced, including strontium-90, iodine-131, and cesium-137, which are particularly hazardous.

Thermonuclear reactions (fusion reactions) combine small nuclei into larger ones. Fusion is the basis of the hydrogen bomb and the source of the sun's energy and potentially will produce more energy than fission.

Section 11.9—Nuclear power plants use the same nuclear chain reaction as atomic bombs, but fuel rods have a much lower concentration of fissionable material and the reaction can be controlled. Disposal of the products of nuclear fission in power plants and the potential conversion of nuclear fuel into weapons are important problems facing us.

GREEN CHEMISTRY—Chemistry is important in nuclear power, and improved application of green chemistry could make nuclear power the safest and most sustainable source of power for humankind. Improved green chemistry can be applied to allowing safe release of pressures that can build up in reactors and to reduce risks associated with spent fuel rods from nuclear reactors.

Learning Objectives	Associated Problems
• Identify the major sources of radiation to which we are exposed. (11.1)	1, 2, 17–19
• List the sources and dangers of ionizing radiation. (11.1)	20, 73
• Balance nuclear equations. (11.2)	21–23, 60
• Identify the types of products formed by various nuclear decay processes. (11.2)	5, 6, 24–26, 61–65
• Solve simple half-life problems. (11.3)	27–29, 31–33, 66, 67, 69, 70
• Use the concept of half-life to solve simple radioisotopic dating problems. (11.3)	7–9, 30, 34, 68, 71
• Write a nuclear equation for a transmutation and identify the product element formed. (11.4)	4, 35–38, 72
• List some applications of radioisotopes. (11.5)	10, 39–42
• Describe the nature of materials needed to block alpha, beta, and gamma radiation. (11.6)	11, 12, 15, 43–46
• Explain where nuclear energy comes from. (11.7)	13, 47–50
• Describe the difference between fission and fusion. (11.7)	49
• Describe the goals of the Manhattan Project and how uranium and plutonium bombs were made. (11.8)	14, 51–53
• Identify the most hazardous fallout isotopes, and explain why they are particularly dangerous. (11.8)	54, 59
• List some uses and consequences of nuclear energy. (11.9)	16, 55–58
• Explain how applications of Green Chemistry Principles can be applied to make nuclear power safer and more cost competitive with other power sources.	74, 75

Conceptual Questions

1. Which of the following examples of radiation comes from naturally occurring sources?
   a. rocks and soil   b. nuclear medicine
   c. nuclear power plants   d. radon
   e. cosmic rays

2. Which of the following examples of radiation comes from human-made sources?
   a. rocks and soil   b. nuclear medicine
   c. cosmic rays   d. radon
   e. internal radiation from within our bodies

3. Match each description with the type of change.
   i. A new compound   a. nuclear
      is formed.
   ii. A new element is   b. chemical
       formed.
   iii. Size, shape, appearance,   c. physical
        or volume is changed
        without changing the
        composition.

4. Define or describe each of the following.
   a. half-life   b. positron
   c. background radiation   d. radioisotope

5. What changes occur in the nucleon number and the atomic number of a nucleus during emission of each of the following?
   a. alpha particle   b. gamma ray   c. proton

6. What changes occur in the nucleon number and atomic number of a nucleus during emission of each of the following?
   a. beta particle   b. neutron   c. positron

7. True or false: A half-life is the time it takes for 50% of a radioactive sample to decay, so it takes two half-lives for all of the radioactivity to decay. Explain.

8. How many days does it take for 1.00 mg of lead-103 (half-life 17.0 d) to decay to 0.125 mg?

9. Fluorine-20 has a half-life of 11.0 s. If a sample initially contains $5.00\ \mu g$ of this isotope, how much remains after 33.0 s?

10. What are some of the characteristics that make technetium-99m such a useful radioisotope for diagnostic purposes?

11. (a) From which type of radiation would a pair of gloves be sufficient to shield the hands: heavy alpha particles or massless gamma rays? (b) From which type of radiation would heavy lead shielding be necessary to protect a worker: alpha, beta, or gamma?

12. List two ways in which workers exposed to radioactive materials can protect themselves from radiation hazard.

13. If you add together the masses of all of the particles (protons and neutrons) in a nucleus, the mass will be greater than the actual mass of the nucleus. Explain why this occurs. What happened to the "missing" mass?

14. Identify the four goals of the Manhattan Project.

15. Plutonium is especially hazardous when inhaled or ingested because it emits alpha particles. Why do alpha particles cause more damage to tissue than beta particles when their source is inside the body?

16. Identify three uses of nuclear fission.

 Problems

Natural Radioactivity

17. Approximately what percentage of radiation to which the average person is exposed is background radiation?
    a. < 10% **b.** 25% **c.** 50% **d.** > 75%

18. Radioactivity in the environment
    a. comes only from human activity
    b. comes from many human and natural sources
    c. comes mainly from rocks
    d. is harmless because it is natural

19. The radioactivity of a substance can be neutralized by
    a. burying it in salt mines **b.** electrolysis
    c. incineration **d.** no known method

20. What is ionizing radiation?

Nuclear Equations

21. Write a balanced equation for emission of **(a)** an alpha particle by californium-250, **(b)** a beta particle by bismuth-210, and **(c)** a positron by iodine-117.

22. Write a balanced equation for **(a)** alpha decay of gold-173, **(b)** beta decay of iodine-138, and **(c)** capture of an electron by cadmium-104.

23. Complete the following equations.
    a. $^{179}_{79}Au \longrightarrow ^{175}_{77}Ir + ?$
    b. $^{12}_{6}C + ^{2}_{1}H \longrightarrow ^{13}_{6}C + ?$
    c. $^{154}_{62}Sm + ^{1}_{0}n \longrightarrow ? + 2^{1}_{0}n$

24. When a magnesium-24 nucleus is bombarded with a neutron, a proton is ejected. What element is formed? (*Hint:* Write a balanced nuclear equation.)

25. A radioisotope decays to give an alpha particle and a protactinium-231 nucleus. What was the original nucleus?

26. A nucleus of astatine-210 decays by beta emission, forming nucleus A. Nucleus A also decays by beta emission to form nucleus B. Write the nuclear symbol for B.

Half-life and Isotopic Dating

27. Niobium was first isolated from a rock that was found in a Connecticut farm field and sent to the British Royal Society in 1743 by John Winthrop, the grandson of the first colonial governor of Connecticut and the great-grandson of the first colonial governor of Massachusetts. One isotope, niobium-94, has a half-life of 20,000 years. How long will it take for a 2.00 μg sample of niobium-94 to decay to 0.0625 μg?

28. The half-life of a radioactive isotope is
    a. the amount that combines with 1 mol hafnium
    b. half its lifetime
    c. half the time it takes for one isotope to decay
    d. the time it takes for half of a given sample to decay

29. A lab worker reports an activity of 80,000 counts/s for a sample of magnesium-21, whose half-life is 122 ms. Exactly 5.00 min later, another worker records an activity of 10 counts/min on the same sample. What best accounts for this difference? Magnesium-21
    a. gives off only neutrinos
    b. has a very short half-life, so almost every radioactive atom in the sample has decayed
    c. is an alpha emitter
    d. is a gamma emitter

30. The ratio of carbon-14 to carbon-12 in a piece of charcoal from an archaeological excavation is found to be one-half the ratio in a sample of modern wood. Approximately how old is the charcoal? How old would it be if the isotopic ratio were 25% of that in a sample of modern wood?

31. Gallium-67 is used in nuclear medicine (Table 11.6). After treatment, a patient's blood shows an activity of 20,000 counts per minute (counts/min). How long will it be before the activity decreases to about 5000 counts/min?

32. How long will it take a radioactive sample of iodine-131 (half-life 8.07 d) to lose 87.5% of its original radioactivity?
    a. 8.0 d **b.** 16 d **c.** 24 d **d.** 32 d

33. If it takes 75 days for 10.0 mg of a particular radioactive isotope to decay to 1.25 mg, what is its half-life?

34. Living matter has a carbon-14 activity of about 16 counts/min per gram of carbon. What is the age of an artifact for which the carbon-14 activity is 8 counts/min per gram of carbon?

Artificial Transmutation

35. In 2010, Russian and American scientists produced a few atoms of element 117 by shooting an intense beam of ions of the rare isotope calcium-48 at a target of berkelium-247. Two isotopes of element 117 were formed, one having 176 neutrons and the other 177 neutrons. Write two separate nuclear equations for the formation of the two isotopes. How many neutrons were released in each process?

36. Meitnerium-278 undergoes alpha decay to form element 107, which in turn also emits an alpha particle. What are the atomic number and nucleon number of the isotope formed by these two steps? Write balanced nuclear equations for the two reactions.

37. Scientists have tried to make element 120. Three different combinations of projectile and target were used to try to produce the nucleus $^{302}_{120}X$: **(a)** nickel-64 and uranium-238, **(b)** iron-58 and plutonium-244, and **(c)** chromium-54 and curium-248. However, after 120 days, no sign of element 120 was found. Write nuclear equations for the three expected reactions.

38. How does artificial transmutation differ from naturally occurring transmutation?

Uses of Radioisotopes

39. There are several technological applications for the transuranium elements (Z > 92). An important one is in smoke detectors, which can use the decay of a tiny amount of americium-241 to neptunium-237. What subatomic particle is emitted from that decay process?

40. Which of the following is true of the use of radioisotopes as tracers to follow the movement of various compounds through complicated biochemical systems?
    a. They react at a slower rate than stable isotopes do.
    b. Radioisotopes are smaller than stable isotopes, so they can move through the reactions more quickly.
    c. Compounds with radioisotopes will react chemically in the same way that compounds without radioactive labels do.
    d. Compounds with radioactive isotopes are taken up by plants and cells more easily than non-labeled compounds are.

41. Radioisotopes are widely used in medicine for all of the following except
    a. cancer treatments   **b.** diagnosis
    c. positron emission   **d.** treatment of radiation
    tomography (PET)   sickness

42. Which of the following is *not* a use of radioisotopes?
    a. detecting leaks in underground pipes
    b. determining the stability of buildings following natural disasters, such as earthquakes
    c. food irradiation
    d. stabilizing concrete structures

Penetrating Power of Radiation

43. Assume that samples that emit radiation are placed on a table across the room. Which sample would produce the greatest damage to you?
    a. alpha emitter   **b.** beta emitter   **c.** gamma emitter
    d. They would all cause the same damage to an individual.

44. Assume that samples that emit radiation are placed on a table in front of you. Which sample would produce the least damage to you?
    a. alpha emitter   **b.** beta emitter   **c.** gamma emitter
    d. They would all cause the same damage to an individual.

45. Barium-140 is a gamma emitter. Bismuth-210 is a beta emitter. Radium-223 is an alpha emitter. Rate these isotopes in order from the most penetrating to least penetrating.
    a. barium-140, bismuth-210, radium-223
    b. bismuth-210, radium-223, barium-140
    c. radium-223, barium-140, bismuth-210
    d. barium-140, radium-223, bismuth-210

46. Alpha emitters are much more likely to cause damage to tissue than gamma emitters are. Why are most alpha emitters less dangerous to living systems than gamma emitters are?

Energy from the Nucleus

47. Nuclear energy is created through splitting of
    a. atomic nuclei   **b.** ionic compounds
    c. molecules   **d.** petroleum products

48. In the first step of the chain reaction of a nuclear explosion, a ^{235}U nucleus absorbs a neutron. The resulting ^{236}U nucleus is unstable and can fission into ^{92}Kr and ^{141}Ba nuclei as shown in the figure. What are the other products of this reaction? After assessing this reaction, explain how the chain reaction can continue.

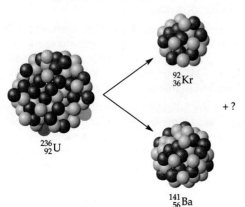

$^{92}_{36}$Kr

$+$?

$^{236}_{92}$U

$^{141}_{56}$Ba

49. Compare nuclear fission and nuclear fusion. Why is energy liberated in each case?

50. Einstein's mass–energy equation is $E = mc^2$, where mass is in kilograms and the speed of light is 3.00×10^8 m/s. The unit of energy is the joule (4.184 J = 1 cal; 1000 cal = 1 kcal; 1 J = 1 kg · m^2/s^2).
    a. Calculate the energy released, in calories and kilocalories, when 1 g of matter is converted to energy.
    b. A bowl of cornflakes supplies 110 kcal. How many bowls of cornflakes would supply the same amount of energy as you calculated for 1 g of matter in part **a**?

Nuclear Bombs

51. The compounds ^{235}UF$_6$ and ^{238}UF$_6$ are nearly chemically identical. How were they separated to enrich the percentage of ^{235}U?

52. How was plutonium-239 synthesized from uranium-238?

53. Describe the logic behind the designs for the uranium-235 bomb and the plutonium-239 bombs that were built at Los Alamos during World War II.

54. Which of the following isotopes is *not* one of the three most hazardous radioisotopes in the fallout that follows a nuclear explosion?
    a. cesium-137
    b. iodine-131
    c. potassium-40
    d. strontium-90

Uses and Consequences of Nuclear Energy

55. Fuel rods in nuclear power plants are generally replaced after roughly three years of use, when the amount of ^{235}U remaining falls below the enrichment that is needed to generate electricity efficiently. What happens to the original ^{235}U isotope during nuclear fission?

56. If fuel rods are used and then removed after only three months, why can that lead to the production of nuclear weapons?

57. The same fission reaction is used for power plants that generate electricity and for atomic bombs. Give two differences between the operation of a power plant and the explosion of a bomb.

58. Which of the following isotopes has the greatest potential for ultimate conversion to weapons-grade material?
    a. polonium-210
    b. plutonium-239
    c. radium-226
    d. thallium-205

Expand Your Skills

59. Which of the following isotopes released after an atomic explosion is considered to be the biggest *immediate* threat?
    a. cesium-137   b. iodine-131
    c. potassium-40   d. strontium-90

60. Complete the following equations.
    a. $^{10}_{5}B + ^{1}_{0}n \longrightarrow ? + ^{4}_{2}He$ b. $^{23}_{10}Ne \longrightarrow ^{23}_{11}Na + ?$
    c. $^{121}_{51}Sb + ? \longrightarrow ^{121}_{52}Te + ^{1}_{0}n$

61. In 2006, Soviet dissident Alexander Litvinenko was murdered when polonium-210 was added to his tea; he died three weeks later. Polonium-210 is an alpha emitter. What other product is formed?

62. A proposed method of making fissionable nuclear fuel is to bombard the relatively abundant isotope thorium-232 with a neutron. The product X of this bombardment decays quickly by beta emission to nucleus Y, which then emits a beta to form nucleus Z. Write the nuclear symbol for Z.

63. Radium-223 nuclei usually decay by alpha emission. Once in every billion decays, a radium-223 nucleus emits a carbon-14 nucleus. Write a balanced nuclear equation for each type of emission.

64. A radionuclide undergoes alpha emission, and thorium-234 is produced. What was the original radionuclide?

65. In 1932, James Chadwick discovered a new subatomic particle when he bombarded beryllium-9 with alpha particles. One of the products was carbon-12. What particle did Chadwick discover?

66. Krypton-81m is used for lung ventilation studies. Its half-life is 13 s. How long does it take the activity of this isotope to reach one-quarter of its original value?

67. Radium-223 has a half-life of 11.4 days. Approximately how long would it take for the activity of a sample of ^{223}Ra to decrease to 1% of its initial value?

68. You are offered a great price on a case of brandy supposedly bottled during the lifetime of Napoleon (1769–1821). Before buying it, you insist on testing a sample of the brandy and find that its tritium content is 12.5% that of newly produced brandy. How long ago was the brandy bottled? Is it likely to be authentic Napoleon-era brandy?

69. Radioactive copper ($^{64}_{29}Cu$, half-life 12.7 h) is found in quantities exceeding the pollution standard in the sediments of a reservoir during a routine check on Monday at noon, when 112 ppm/m³ is measured. The standard allows up to 14 ppm/m³. About when will the level of copper-64 return to 14 ppm/m³?

70. An unidentified corpse was discovered on April 21 at 7:00 a.m. The pathologist discovered that there were 1.24×10^{17} atoms of $^{32}_{15}P$ remaining in the victim's bones, and so he placed the time of death sometime on March 15. The half-life of $^{32}_{15}P$ is 14.28 days. How much $^{32}_{15}P$ was present in the bones at the time of death?

71. Several different radioisotopes can be used much like carbon-14 for determining the age of rocks and other matter. In the geosciences, the age of rocks may be determined using potassium-40, which decays to argon-40, with a half-life of 1.2 billion years. (a) Some rocks brought back from the moon were dated as being 3.6 billion years old. What percentage of the potassium-40 has decayed after that time? (b) Give a reason why uranium-238 (half-life of 4.6 billion years) is more useful for confirming this age than is carbon-14.

72. Write balanced nuclear equations for (a) the bombardment of $^{121}_{51}Sb$ by alpha particles to produce $^{124}_{53}I$, followed by (b) the radioactive decay of $^{124}_{53}I$, by positron emission.

73. Which of the following isotopes is most likely to be radioactive?
    a. 18 protons and 22 neutrons, because it has even numbers of protons and neutrons
    b. 29 protons and 35 neutrons, because it has odd numbers of protons and neutrons
    c. 50 protons and 69 neutrons, because it has an odd number of neutrons
    d. 82 protons and 126 neutrons, because it has a magic number of both protons and neutrons

74. Which of the following is *not* an advantage to using nuclear fuels to generate electricity?
    a. smaller volume of waste produced than with coal plants
    b. lower fuel cost of nuclear fuel
    c. ease of extracting weapons-grade plutonium from nuclear waste
    d. lower CO_2 emissions than from coal or natural gas plants.

75. All of the following are challenges to using nuclear power to generate electricity except
    a. need for specialized nuclear waste storage and handling
    b. low cost of building nuclear power plants
    c. potential for extracting weapons-grade plutonium from nuclear waste
    d. generation of electricity is not easily increased during peak demand hours

 Critical Thinking Exercises

Apply knowledge that you have gained in this chapter and one or more of the FLaReS principles (Chapter 1) to evaluate the following statements or claims.

11.1 Brazil nuts are known to absorb and concentrate the element barium. A nationally known laboratory announces that Brazil nuts contain tiny traces of radium.

11.2 Carbon-14 analyses of the Shroud of Turin have indicated that the shroud dates back only to 1260–1390 C.E. Some people have claimed that these analyses are flawed, because the shroud was subjected to high temperatures during a fire in Chambery in 1532 and this affected the carbon-14 samples.

11.3 A scientist reports that he has found an artifact that has been shown to be 80,000 years old by carbon-14 dating.

11.4 The americium in a smoke detector (see Conceptual Example 11.7) is encased in a chamber with aluminum shielding, about as thick as the wall of a soft-drink can. A woman who purchases a smoke detector mounts it at the highest point on the ceiling. She claims that this minimizes her exposure to the radiation.

11.5 A young girl refuses to visit her grandmother, who has recently undergone radiation therapy for breast cancer. The teenager is afraid that she will catch radiation sickness from her grandmother.

Collaborative Group Projects

Prepare a PowerPoint, poster, or other presentation (as directed by your instructor) to share with the class.

1. Write a brief report on the impact of nuclear science on one of the following.
   a. war and peace
   b. industrial progress
   c. medicine
   d. agriculture
   e. human, animal, and plant genetics

2. Write an essay on radioisotopic dating using one or more of the isotopes in Table 11.5 other than carbon-14 or tritium.

3. Write a brief biography of one of the following scientists.
   a. Otto Hahn
   b. Enrico Fermi
   c. Glenn T. Seaborg
   d. J. Robert Oppenheimer
   e. Lise Meitner
   f. Albert Einstein

4. Find a website that is strongly in favor of nuclear power plants and one that is strongly opposed. Note the sites' sponsors, and analyze their viewpoints. Try to find a website with a balanced viewpoint.

5. In 2002 and again in 2006, researchers at Dubna in Russia and at Lawrence Livermore Laboratory in California reported the synthesis of a total of four atoms of element 118. However, in July 1999, researchers at a different California laboratory had also reported creation of element 118, a report that was found to be fraudulent two years later. Report on either (a) the scientific ethical questions raised by the fraudulent report or (b) the difficulties involved in confirming the synthesis of a new human-made element.

6. Find the location of the nuclear power plant closest to where you live. Try to determine the risks and benefits of the plant. To how many houses does it provide power? What are the environmental impacts of the plant under normal operating conditions?

LET'S EXPERIMENT! The Brief Half-Life of Candy

Materials Needed

- Milk chocolate M&Ms or Skittles (or similar candy with print on one side)
- Resealable plastic bag
- Paper and pencil

How do scientists know when an atom will decay? What is the process of radioactive decay? Let's explore and find out more.

Radioactive decay is a random process, much like flipping a coin or a piece of candy. It is not possible to predict the radioactive behavior of one individual atom. One atom can survive for only a single half-life, or it can survive many half-lives. If we are observing not one atom but many, though, we can try to calculate and predict the radioactive decay for the group.

To understand this concept better, we will use candy, such as M&Ms or Skittles, to simulate radioactive decay. First, place 100 pieces of candy into a small plastic bag. Shake the bag gently to mix the candy.

Gently pour the candy out on the table. The candies with their printed side up have decayed, while the candies with their printed side down have not. Count and record the number of decayed pieces of candy. Push these aside (or eat them!) and return only the pieces of candy with the print side down to the bag.

Reseal the bag and repeat the pouring, counting, recording, and setting aside of the candy until all of the candy has decayed. Based on your collected data, what is the half-life of the candy? Plot the "radioactive decay" of your candy.

Typical Results

Pour	Trial 1: Number of Decayed Candies	Trial 2: Number of Decayed Candies	Average Number of Non-decayed Candies
1	54	49	48.5
2	54 + 20 = 74	49 + 27 = 76	25
3	74 + 10 = 84	76 + 10 = 86	15
4	84 + 10 = 94	86 + 7 = 93	6.5
5	94 + 2 = 96	93 + 2 = 95	4.5
6	96 + 2 = 98	95 + 3 = 98	2
7	98 + 1 = 99	98 + 1 = 99	1
8	99 + 0 = 99	99 + 1 = 100	0.5
9	99 + 1 = 100		0

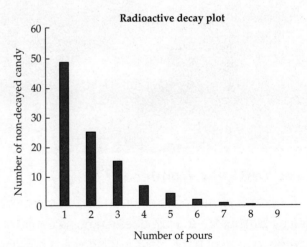

▲ Chart: Radioactive decay of the candy (number of non-decayed candies versus number of pours).

Questions

1. If one period of shaking of the bag of candy equals 1000 years of time, what is the half-life in years for your candy?
2. If you came across a sample of candy containing 85 pieces of "decayed" candy, how "old" would you expect the candy to be? (Again, assume one period of shaking = 1000 years.)
3. What would you expect your radioactive decay plot to look like if you started the experiment with 1000 pieces of candy instead of 100?
4. If you ran this experiment with only one piece of candy, how long would you expect that piece of candy to last (how many half-lives)?

Chemistry of Earth

12

Have You Ever Wondered?

1 If there is so much aluminum in Earth's crust, why can't we find any native aluminum? **2** How does clumping cat litter work? **3** Is there any difference between the glass in windows, a drinking glass, and car windshields? **4** Isn't cement the same as concrete? **5** Why are knives stiff and sharp and paper clips bendable? **6** What are gemstones made of?

Did your mother or grandmother ever tell you to gargle with salt water to relieve a sore throat? Did you know that the most important use of salt throughout history was to preserve meat and fish, and to prevent infections in wounds? This chapter will provide insight into the chemical composition of Earth and the important metals and minerals—including table salt—that are not only useful, but critical for human society.

Spaceship Earth is a tiny blue-green jewel in the vastness of space, about 40,000 km in circumference, with a surface area of 50 million km², about three-quarters covered in water. It is self-contained and provides, along with the

▲ Many areas of Earth are full of minerals, many of which are useful to humankind. Copper mining started in Mesopotamia at least 6000 years B.C.E., and possibly earlier. The modern Chino copper mine east of Silver City, N.M., is the third-oldest active open pit mine in the world. The copper it has produced is primarily used for electrical wires and cables.

sun, all that is needed for life on it, ranging from the structures we live in to all the objects we buy and use, from the food we eat to the salt we sprinkle on the food, from our porcelain plates to the metal utensils used to prepare and consume that food. The fact that this spaceship is self-contained does mean, however, that its resources are finite and must be managed as such—in other words, sustainably.

12.1 Spaceship Earth: Structure and Composition

Learning Objectives • Describe the structure of Earth and the regions of Earth's surface. • List the most abundant elements in Earth's crust and the common compounds in which each are found.

What kinds of materials do we have aboard this spaceship? Are they sufficient for this enormous load of passengers? Let's begin by looking at the composition of Earth. Earth's structure can be broken down into three parts (Figure 12.1).

1. **Core:** Its diameter is ~6800 km. It has two physical states: solid (inner; ~2600 km in diameter) and liquid (outer; ~2100 km thick). The core is composed mainly of iron with a smaller percentage of nickel.
2. **Mantle:** Though the thickness varies considerably, the mantle averages ~2800 km thick and consists of a lower (~2000 km thick), transition (~500 km thick), and upper (~360 km thick) mantle. It is composed mainly of high-density magnesium and iron silicates.
3. **Crust:** This outermost layer is between 8 km (oceans) and 40 km (continents) thick, accounting for less than 1% of Earth's volume. The oceanic crust consists mainly of magnesium and iron silicates in rocks such as basalt. The continental crust is composed mainly of sodium, potassium, and aluminum silicates, and

▲ **Figure 12.1** Diagram showing the structural regions of Earth (not to scale).

Q *Earth is roughly 12,000 km in diameter, and the crust averages 35 km thick. Why is the crust not shown to scale?*

TABLE 12.1 **Elemental Composition of the Earth's Surface**

Element	Atom Percent	Percent by Mass
Oxygen	53.3	49.5
Silicon	15.9	25.7
Hydrogen	15.1	0.9
Aluminum	4.8	7.5
Sodium	1.8	2.6
Iron	1.5	4.7
Calcium	1.5	3.4
Magnesium	1.4	1.9
Potassium	1.0	2.4
All others	3.7	1.4
Total	100.0	100.0

quartz, which is silicon dioxide. Metal oxides, carbonates, sulfides, and sulfates are also present in lesser amounts.

The *crust* is the outer solid shell of Earth, often called the *lithosphere*. We will get to the gaseous and watery parts, known as the *atmosphere* and *hydrosphere*, in Chapter 13 and Chapter 14, respectively. Based on extensive sampling, the elemental composition of the Earth's surface (including the *atmosphere*, *hydrosphere*, and *lithosphere*) has been determined and is summarized in Table 12.1.

More than half the atoms at Earth's surface are oxygen atoms, which occur in the atmosphere as molecular oxygen (O_2) and as ozone (O_3); in the hydrosphere bonded to hydrogen as water; and in the lithosphere bonded to silicon (in SiO_2 or in silicates), bonded with metals as metal oxides such as Fe_2O_3, or in carbonates or sulfates. Silicon is the second most abundant element, and hydrogen is third, with most of the H atoms combined with oxygen in water. Because hydrogen is the lightest element, it makes up only 0.9% of Earth's crust by mass. Just three elements—oxygen, silicon, and hydrogen—make up 84.3% of the atoms at Earth's surface, and the top nine elements account for 96.3% of the atoms, leaving only 3.7% of the atoms representing all the other elements.

1 If there is so much aluminum in Earth's crust, why can't we find any native aluminum?

Essentially all aluminum atoms in the lithosphere are bonded to oxygen atoms in specific, stable structures in rocks, or entrapped in complex polymeric structures, called aluminum silicates, not as free atoms.

The Lithosphere: Its Importance to Humans

In terms of mass, the lithosphere is almost completely composed of inorganic minerals. However, when discussing the lithosphere's importance to humans and to life on Earth, we should include more than rocks and minerals. The lithosphere is composed of three equally important components, namely, inorganic (rocks and minerals), organic (soil and fossil fuels), and biological (flora and fauna). In this chapter, we will focus mainly on the inorganic components.

The predominant rocks and minerals in the lithosphere are:

- *silicate minerals* (compounds of metals with silicon and oxygen)
- *carbonate minerals* (metals combined with carbon and oxygen)
- *oxide minerals* (metals combined with oxygen only)
- *sulfide minerals* (metals combined with sulfur only)

Some typical nonsilicate minerals are listed in Table 12.2.

Although much, much smaller in quantity, the *organic* portion of Earth's outer layers includes soil (mostly waste and decomposition products of the *biological* component) and and the fossil fuels materials (such as coal, natural gas, petroleum, and oil shale) that were living organisms millions of years ago. This organic material always contains carbon, nearly always hydrogen, and often oxygen, nitrogen, and other elements. In Chapter 15 we will discuss the energy content of those fossil fuels.

TABLE 12.2 **Some Nonsilicate Minerals of Economic Importance**

Mineral Type	Name	Chemical Formula	Source and/or Use
Oxide	Hematite	Fe_2O_3	Ore of iron; pigment
	Magnetite	Fe_3O_4	Ore of iron
	Corundum	Al_2O_3	Gemstone; abrasives
Sulfide	Galena	PbS	Ore of lead
	Chalcopyrite	$CuFeS_2$	Ore of copper
Carbonate	Calcite	$CaCO_3$	Lime; cement; glass; cave structures

The Lithosphere's Bounty: From Materials to Energy

Early people obtained food by hunting and gathering. After the agricultural revolution (about 10,000 years ago), people were no longer forced to search for food. Domesticated animals and plants supplied human needs for food and clothing. A reasonably assured food supply enabled them to live in villages and created a demand for more sophisticated building materials. People learned to convert natural materials into products with superior properties. For example, adobe bricks, which serve well in arid areas, could be made simply by drying a mixture of clay and straw in the sun.

Fire was one of the earliest agents of chemical change. When people learned to control fire, they learned that cooking improved the flavor and digestibility of meat and grains. People learned that heating adobe bricks made stronger ceramic bricks. Similarly, firing clay at high temperatures produced ceramic pots, making cooking and storage of food much easier. As they became able to produce higher temperatures, people learned to make glass and to extract metals from ores. They made tools from bronze and, later, from iron.

SELF-ASSESSMENT Questions

1. The regions of the Earth, listed in order from the center outward, are
   a. core, crust, mantle, hydrosphere   **b.** core, mantle, crust
   c. core, mantle, crust, hydrosphere   **d.** mantle, crust, core

2. Earth's core is composed mainly of
   a. lead   **b.** iron
   c. radioactive elements   **d.** sulfur

3. The Earth's lithosphere can be broken down into how many components?
   a. one   **b.** two
   c. three   **d.** four

4. Which of the following is not a component of Earth's surface?
   a. lithosphere   **b.** hydrosphere
   c. crust   **d.** mantle

5. Three elements account for more than 80% of the Earth's crust by atom percent. In decreasing order of abundance, they are
   a. H, O, Si   **b.** H, Si, O
   c. O, H, Si   **d.** O, Si, H

Answers: 1, b; 2, b; 3, c; 4, d; 5, d

WHY IT MATTERS

One of the more common rocks on Earth is limestone, which is calcium carbonate ($CaCO_3$). In nature, calcium carbonate is found in many different physical forms, including marble, seashells, eggshells, coral, pearls, and chalk. Striking examples of limestone formations are the stalagmites and stalactites in caves, such as those seen here from the Carlsbad Caverns National Park.

12.2 Silicates and the Shapes of Things

Learning Objectives • Describe the arrangement of silicate tetrahedra in common silicate minerals. • Describe how glass differs in structure from other silicates.

We will consider only a few representative minerals of the thousands that occur in the lithosphere. First, let's look at some silicates. The basic unit of silicate structure is the SiO_4 tetrahedron (Figure 12.2). SiO_4 tetrahedra can be connected in a number of ways to make an amazing array of arrangements and unique mineral phases. Some of the possibilities are illustrated in Figure 12.3, with some real mineral phases presented in Table 12.3.

Quartz is pure silicon dioxide, SiO_2. The ratio of silicon to oxygen atoms is 1:2. However, because each silicon atom is surrounded by *four* oxygen atoms, the basic unit of quartz is the SiO_4 tetrahedron. These tetrahedra are arranged in a complex 3-D structure. Crystals of pure quartz are colorless, but impurities produce a variety of quartz crystals sometimes used as gems (Figure 12.4).

Micas are composed of SiO_4 tetrahedra arranged in 2-D sheetlike arrays and can be easily cleaved into thin, transparent sheets (Figure 12.5). Micas are used as insulators for electronics and in high-temperature furnaces.

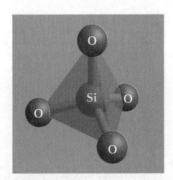

▲ **Figure 12.2** The silicate tetrahedron has a silicon atom at the center and an oxygen atom at each of the four corners. The shaded area (green) shows how the SiO_4 tetrahedron is typically represented in a mineral structure.

Connecting SiO$_4$ Tetrahedra

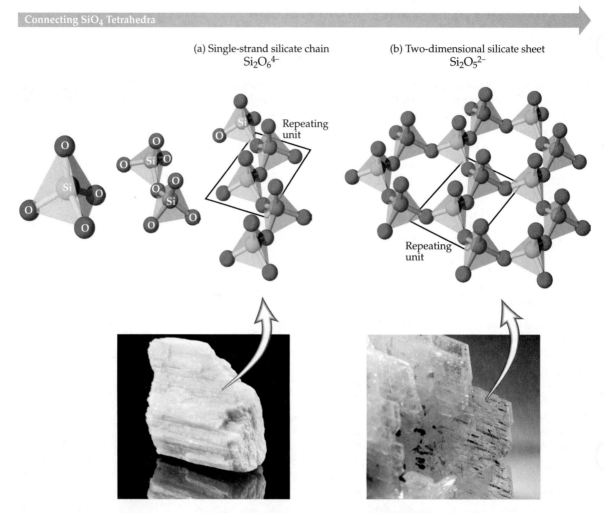

(a) Single-strand silicate chain
$Si_2O_6^{4-}$

(b) Two-dimensional silicate sheet
$Si_2O_5^{2-}$

▲ **Figure 12.3** SiO$_4$ tetrahedra bridged by oxygen atoms and some resulting minerals (wollastonite, above left, and muscovite, above right).

TABLE 12.3 **Silicate Tetrahedra in Some Minerals**

Mineral(s)	Silicate Arrangement	Formula	Uses
Zirconium silicate	Simple anion (SiO$_4^{4-}$)	ZrSiO$_4$	Ceramics; gemstones (zircon)
Spodumene	Long chains	LiAl (SiO$_3$)$_2$	Source of lithium and its compounds
Chrysotile asbestos	Double chains	Mg$_3$(Si$_2$O$_5$)(OH)$_4$	Fireproofing (now banned)
Muscovite mica	Sheets	KAl$_2$(AlSi$_3$O$_{10}$)(OH)$_4$	Insulation; lustrous paints; packing (vermiculite)
Quartz	3-D array	SiO$_2$	Sand; making glass; amethyst; agate; citrine

► **Figure 12.4** The collection of quartz crystals includes (counterclockwise from upper right) citrine (yellowish quartz), colorless quartz, amethyst (purple quartz), and smoky quartz. The chemical structure of quartz (left) shows that each silicon atom is bonded to four oxygen atoms and each oxygen atom is bonded to two silicon atoms.

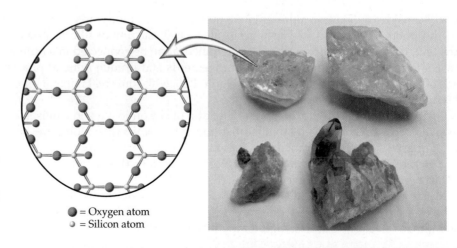

● = Oxygen atom
○ = Silicon atom

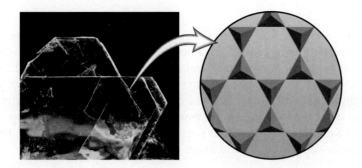

◀ **Figure 12.5** A sample of mica, showing cleavage into thin, transparent sheets. The chemical structure of mica shows sheets of SiO_4 tetrahedra. The sheets are linked by bonds between O atoms and cations, mainly Al^{3+} (not shown). Mica is used as transparent "window" material in industrial furnaces.

Asbestos is a generic term for a variety of fibrous silicates. Perhaps the best known of these is *chrysotile*, a magnesium silicate (Figure 12.6). Chrysotile has double chains of SiO_4 tetrahedra. The oxygen atoms that have only one covalent bond also bear a negative charge. Magnesium ions (Mg^{2+}) balance these negative charges.

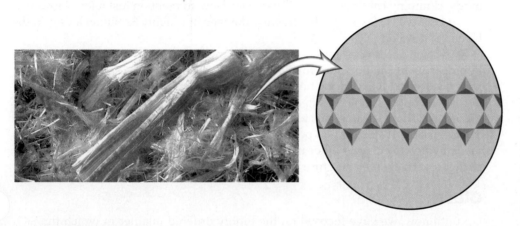

◀ **Figure 12.6** A sample of chrysotile, a type of asbestos. The chemical structure of chrysotile shows double chains of SiO_4 tetrahedra. The two chains are joined to each other through O atoms. The double chains are linked by bonds between O atoms and cations, mainly Mg^{2+} (not shown).

Asbestos: Benefits and Risks

Asbestos is an excellent fireproof thermal insulator. It was once used widely to insulate furnaces, heating ducts, and steam pipes. A city in Russia called Asbest, on the eastern slope of the Ural Mountains, is the site of a significant deposit of that mineral that has been mined since the 1880s. It is still a major producer of that material. An open-pit asbestos mine in Asbestos, Quebec, however, was closed in 2011.

The health hazards to those who work with asbestos are well known. Asbestos fibers are extremely tiny and have very sharp edges. Inhalation of fibers 5–50 μm long over a period of 10–20 years can cause *asbestosis* and may lead to lung cancer or *mesothelioma*, a rare and incurable cancer of the linings of body cavities.

Long-term occupational exposure to asbestos increases the risk of lung cancer by a factor of two. Cigarette smoking causes a tenfold increase in the risk of lung cancer. We might expect asbestos workers who smoke to have a risk of developing lung cancer that is 20 times that of nonsmokers who do not work with asbestos, but instead, their risk is increased by a factor of 90. Cigarette smoke and asbestos fibers act in such a way that each increases the effect of the other. Such a reinforcing interaction is called a **synergistic effect**.

Aluminum Silicates: From Clays to Ceramics

When 2-D sheets of connected SiO_4 tetrahedra are stacked, we have the start of a 3-D mineral phase called *clay*. Use of clay by humans has a long history, dating back to pottery making in 14,000 B.C.E. Today, we use the same basic concept and techniques to manufacture a range of useful items, including clay coffee cups and roofing tiles. Clay owes its usefulness and versatility to the manner in which Al, Si, and O atoms are bonded. There are two major classes of clays, based on how the tetrahedral silicate sheets are assembled with the octahedral, mainly aluminum hydroxide, sheets.

2 **How does clumping cat litter work?**

The secret to superabsorbent clumping cat litters is that they contain montmorillonite, a 2:1 clay that swells. When urine (mostly water) hits a sodium montmorillonite–based clay, it causes the clay particles to swell, stick together, and, ultimately, create a clump that can easily be scooped out. Importantly, only the soiled portion of litter needs to be removed. This makes day-to-day maintenance of a litter box easier.

Gold in glass? The medieval glass artisans used salts of gold to manufacture red glass using conditions that would produce gold nanoparticles smaller than 100 nm by reducing the gold. Gold nanoparticles interact with light differently than do large gold objects. Different hues of red, due to differences in the exact size of gold nanoparticles, could be obtained by slightly varying the conditions used to make the glass. Skilled glass artisans were sought after and could demand high salaries.

3 **Is there any difference between the glass in windows, a drinking glass, and car windshields?**

Car windshields have a layer of plastic laminated between two curved layers of glass, so in a car crash the glass does not break apart into potentially lethal sharp pieces. Both ordinary windows and drinking glasses are essentially soda-lime glass—amorphous mixtures of silicates, sodium, and calcium ions. Some window glass has minor amounts of additives, usually iron ions, for UV absorption.

1:1 Clays

In these clays, a tetrahedral silicate sheet and an octahedral aluminum hydroxide sheet are connected by oxygen atoms bound to both the Si atoms of the tetrahedral sheet and the Al atoms of the octahedral sheet, to form a layer. These 1:1 layers are then connected by hydrogen bonding to each other, alternating the silicate and aluminum hydroxide sheets as they pile one on top of the other. It also means that these clays do not expand or swell. A well-known example of this type of clay is *kaolinite*, named after Kaoling, a mountain in Jiangxi Province in China, where the mineral was used for thousands of years to make porcelain plates and bowls.

2:1 Clays

These clays consist of layers composed of one octahedral aluminum hydroxide sheet sandwiched between two tetrahedral silicate sheets, bridged by oxygen atoms. If the space between the 2:1 layers contains H_2O, this type of clay contracts when it is dried and expands when it is wetted. Expanding clays have found applications in drilling muds, clumping cat litter, and pollutant clean-up, to mention just a few. However, expandability also leads to infrastructure damage of roughly $2 billion a year in the United States alone.

Adding water to clay makes it malleable, and a clay/water mixture can be made into nearly any shape. Once formed, the clay object becomes solid as the excess water is driven away. Bricks, flowerpots, and tiles are made in this manner. When porosity is not desirable—as in cooking pots or water jugs—the clay object can be glazed by adding various salts to the surface. Heat then turns the entire surface into a glass-like matrix. Bricks and pottery are examples of **ceramics**, inorganic materials made by heating clay or other mineral matter to a high temperature at which the particles partially melt and fuse together.

Glass

Up until now, we have focused on the highly ordered manner in which the SiO_4 tetrahedra are connected in sand and clays. However, this high degree of order can be broken by heating certain silicate minerals to very high temperatures, which was discovered about 5000 years ago in northern coastal Syria, Mesopotamia, and Egypt. The resulting material was **glass**. Decorative glass beads in ancient times were so rare and precious that they were placed in the tombs of the pharaohs, alongside items of gold.

Nature has created its own version of glass, in areas with a history of volcanism. Obsidian is a black, naturally occurring glass mineral, made from molten magma; it has no ordered crystal structure. One of the most famous sites of its occurrence is in Yellowstone National Park, where its occurrence is called "obsidian cliff."

The raw materials for ordinary window and container glass manufactured nowadays are sand (SiO_2), "soda ash" (Na_2CO_3), and limestone ($CaCO_3$). Glassmakers, however, typically describe the composition of "soda-lime" glass in terms of $\%SiO_2$, $\%Na_2O$ (traditionally called "soda"), and $\%CaO$ (traditionally called "lime"). An ordinary recipe may be described as 75% SiO_2, 15% Na_2O, and 10% CaO.

At the high temperatures (up to 1500 °C) of the glass-making facility, the carbonates break down into CaO and Na_2O, releasing carbon dioxide, according to the reactions:

$$CaCO_3 \longrightarrow CaO + CO_2 \quad \text{and} \quad Na_2CO_3 \longrightarrow Na_2O + CO_2$$

After several days of heating, glass is formed, then slowly cooled. Basic glass is essentially colorless and can be considered to be a mixture of sodium silicate and calcium silicate, according to the following reactions:

$$CaO + SiO_2 \longrightarrow CaSiO_3 \quad \text{and} \quad Na_2O + SiO_2 \longrightarrow Na_2SiO_3$$

On the molecular level, what is really happening is that the bridging interactions within the crystalline matrix of silicates are broken, leading to an irregular 3-D arrangement of the SiO_4 tetrahedra, as seen in Figure 12.7. In this new *phase*, called *amorphous,* chemical bonds linking the SiO_4 tetrahedra differ in strength. Thus, when

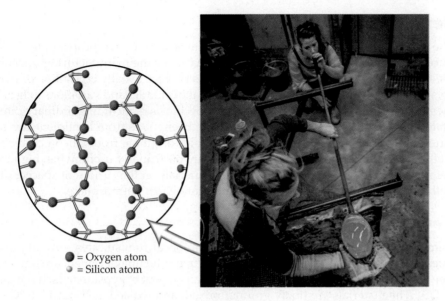

= Oxygen atom
= Silicon atom

▲ **Figure 12.7** Heat-softened glass can be shaped by blowing or molding. Compare the irregular arrangement of atoms in glass with the regular arrangement of those in quartz (Figure 12.4).

glass is heated, the weaker bonds break first and the glass softens gradually. This explains why amorphous substances melt *over a range* of temperatures, instead of *at* one specific temperature. Many amorphous substances are quite pliable, such as some kinds of rubber, Play-Doh, and various synthetic polymers. (See Chapter 10.)

Oxides of various metals can be added to or substituted for some or all of the sodium or calcium oxides, to change their colors or give them new properties. Many special types of glass can be made, including those used in lenses in microscopes, eyeglasses, and telescopes (made with lead(II) oxide). Glass that is highly resistant to breaking, even when subjected to sudden changes in temperature, as when a glass baking dish is removed from the oven and cooled, is made with boron oxides (sometimes called borosilicate glass). If glass being produced is meant for windows or patio doors, a small amount of iron oxide is added, in order to absorb some ultra-violet radiation striking them. Some examples of glass additives including colorants are given in Table 12.4. Safety glass in automobile windows and windshields is manufactured with a thin plastic layer laminated between two curved glass sheets.

WHY IT MATTERS

Optical fibers are hair-thin threads of glass that carry messages as intermittent bursts of light. Sounds are translated into electric signals, which are converted into pulses of laser light that are transmitted by the glass fibers. At the end of the line, the light pulses are converted back to sound waves. A bundle of optical fibers can carry several hundred times as many messages as a copper cable of the same size, and these fiber bundles have replaced many old telephone lines and computer cables. They are also used in healthcare, in devices that provide visual access inside the body and that guide surgical instruments.

TABLE 12.4 Compositions and Properties of Various Glasses

Type	Special Ingredient(s)	Special Properties and/or Uses
Soda-lime glass	None	Ordinary glass for windows, bottles, drinking glasses
Borosilicate glass	Boron oxide (instead of lime)	Heat-resistant, for laboratory glassware and ovenware
Leaded glass	Lead oxide (instead of lime)	Highly refractive (bends light); optical glass for microscopes, telescopes, table crystal
Colored glass	Selenium compounds	Red (ruby glass)
	Cobalt compounds	Blue (cobalt glass)
	Chromium compounds	Green
	Manganese compounds	Violet
	Carbon and iron oxide	Brown (amber glass)
Photochromic glass	Silver chloride or bromide	Darkens when exposed to light; for sunglasses, hospital windows

Manufacture of ordinary glass uses no vital raw materials, but the furnaces used to melt and shape the glass expend energy. Glass is easily recycled. It can be melted and formed into new objects, using considerably less energy than manufacturing new glass.

WHY IT MATTERS

Research has led to the development of amazing new ceramic materials. Ceramics have been used to make high-temperature industrial furnaces, rocket nose cones, heat-resistant tiles on the Space Shuttle, maglev trains, automobile engines, jet engine turbine blades, and memory elements in computers. *Superconducting ceramics* exhibit virtually no resistance to the flow of electricity; they make possible high-speed magnetically levitated trains.

▲ Concrete for sidewalks, highways, basements, and many other applications is a mixture. An important ingredient is Portland cement, made by roasting a mixture of limestone and clay and then crushing it to make a powder. This cement is the "glue" that holds sand, water, and fine gravel together. Some concrete can be stamped with a design before it hardens, or crushed recycled glass can be added for strength. Some high-quality concrete can be polished to create a very smooth concrete floor, or even colored with other additives.

Ceramics

The process of heating and then cooling to cause a change in property, just introduced in the formation of glass, is also the basis for the manufacturing of ceramics. **Ceramics** are inorganic nonmetallic solids that are usually crystalline or partly crystalline. The variety of materials used to make ceramics and variations in heating them mean that ceramics can be tailored to specific uses, including ordinary dinner plates and coffee mugs. Many ceramics have been made from kaolinite clays and aluminum oxide. The more advanced ceramics originate from silicon carbide as well as tungsten carbide. Ceramic materials are particularly useful when high temperatures are present, such as the ceramic materials on the surface of spacecraft to withstand the high temperatures of re-entry into Earth's atmosphere.

Cement and Concrete

Cement is a complex mixture mainly made up of calcium silicates, aluminum silicates, and carbonates. The raw materials for the production of cement are limestone (see the following Section 12.3), clay, and a small percent of gypsum, which is mainly $CaSO_4$. The materials are finely ground, mixed, and roasted at about 1500 °C in a rotary kiln heated by burning natural gas or powdered coal. The finished product is crushed to a fine powder and then mixed with sand, gravel, and water to form **concrete**. Besides sidewalks, driveways, and highways, many large structures have been made of concrete, such as parking ramps, the Hoover Dam, and the Panama Canal. Such structures have embedded steel reinforcements for strength.

Concrete is used widely in construction because it is inexpensive, strong, chemically inert, durable, and tolerant of a wide range of temperatures. However, its production is environmentally damaging in terms of raw material mining, from the destruction of entire mountains to the emission of CO_2, particulate matter, and other pollutants, such as sulfur dioxide. On the positive side, tires and toxic organic wastes can be destroyed when they are used as fuel for the kiln. For disposal, concrete can be broken up and used as rock fill.

SELF-ASSESSMENT Questions

1. All silicate minerals contain silicon and
    a. carbon   **b.** hydrogen   **c.** iron   **d.** oxygen

2. The mineral composed of pure silicon dioxide (SiO_2) is
    a. asbestos   **b.** calcite   **c.** quartz   **d.** zircon

3. The silicate tetrahedron consists of one _____ atom at the center and _____ atoms at the corners.
    a. O; 4 Si   **b.** O; 2 Si and 2 Al
    c. Si; 4 O   **d.** Si; 2 O and 2 Al

For items 4–9, match each substance with the correct structure.

4. asbestos   **a.** double chains

5. mica   **b.** sheets

6. quartz   **c.** simple anions

7. zircon   **d.** three-dimensional array

8. clay   **e.** amorphous

9. glass   **f.** stacked sheets

10. 2:1 clays are made up of
    a. tetrahedral/octahedral/tetrahedral sheets   **b.** octahedral/tetrahedral/octahedral sheets
    c. tetrahedral/tetrahedral/octahedral sheets   **d.** tetrahedral/tetrahedral/tetrahedral sheets

11. The red color in medieval stained glass was due to
    a. blood   **b.** iron nanoparticles
    c. iron cations   **d.** gold nanoparticles

12. One type of material that is hard and durable but brittle is a
    a. ceramic   **b.** metal
    c. polymer   **d.** semiconductor

13. The main ingredients of ordinary glass are
 a. boron oxide, sand, and limestone
 b. clay and limestone
 c. coke, sand, and limestone
 d. sand, sodium carbonate, and limestone

14. The main ingredients of ordinary cement are
 a. clay and limestone
 b. coke, sand, and limestone
 c. sand and limestone
 d. sand, sodium carbonate, and limestone

Answers: 1, d; 2, c; 3, c; 4, a; 5, b; 6, d; 7, c; 8, f; 9, e; 10, a; 11, d; 12, d; 13, d; 14, a

4 **Isn't cement the same as concrete?**

No. Concrete is a mixture of sand, gravel, water, and Portland cement. Portland cement is a fine powder made from limestone and clay, often containing some gypsum, which is calcium sulfate. To make a concrete sidewalk, that powdered cement is mixed with sand, gravel, and water to make a mixture that hardens into a solid, long-lasting artificial stone.

12.3 Carbonates: Caves, Chalk, and Limestone

Learning Objective • Explain the importance, abundance, and reactions of calcium and other carbonates.

Limestone is the most commonly found sedimentary rock on earth. It consists mainly of calcium carbonate and is found in clear, warm, shallow ocean waters—particularly where coral reefs and other mollusks are abundant. The Caribbean Sea, the Gulf of Mexico, the Pacific Ocean, the Indian Ocean, and the Indonesian Archipelago are active building grounds for limestone. In regions where primeval oceans covered the continents, limestone cliffs, outcroppings, and underground caves are evidence of such ancient waters. Limestone structures have been built throughout the world, and many have lasted hundreds of years; college campus and government buildings made of limestone are attractive and long-lasting.

In its purest form, $CaCO_3$ is known as calcite. One type of calcite called Iceland spar is a clear, crystalline mineral that exhibits the unusual property of *birefringence*, in which a double image is seen when viewing an object through it.

When limestone was melted eons ago, it was metamorphosed into a very hard mineral called marble, the material from which statuary is carved, kitchen countertops are made, and stately stone buildings have been constructed. Notre-Dame Cathedral in Paris; the Washington Monument and Lincoln Memorial in Washington, DC; and the Taj Mahal in Agra, India, are all made mainly of marble.

Underground cave systems are found on every inhabited continent on Earth. Some of the better known in North America are Carlsbad Caverns in New Mexico (see photo p. 377) and Mammoth Cave in Kentucky. Stalactites and stalagmites are just some of the fascinating forms of limestone formed over eons by evaporation of water in caves. Jewel Cave, near Rapid City, South Dakota, is the third-longest cave system in the world (over 310 mapped km of passageways); its formations are mainly calcite.

Calcium carbonate, as are all other carbonates, is attacked by acids, forming carbon dioxide, slowly dissolving the solid. Geologists often have a small container of hydrochloric acid packed amongst their supplies when in the field. A few drops of acid will form bubbles of CO_2 when dropped on any carbonate material, as in the reaction:

$$CaCO_3(s) + 2\,H^+(aq) \longrightarrow H_2O\,(l) + CO_2(g) + Ca^{2+}(aq)$$

Similarly, limestone and marble structures are susceptible to slow degradation by acidic rainfall. (See Section 13.6.)

Calcium carbonate is found in many antacid formulations (see Section 7.8), taking advantage of the reaction of a carbonate with an acid. Calcium carbonate is also used in the making of glass (see previous Section 12.2) and in blast furnaces to help extract metals from their ores (see the following Section 12.4), and is found in ordinary chalk, though some brands of sidewalk chalk may be made of calcium sulfate.

Besides calcium, Mg^{2+}, Ba^{2+}, Fe^{2+}, Cu^{2+}, Mn^{2+}, and many other cations form carbonate minerals, often in crystal form and in varying colors. Both azurite (deep blue) and malachite (bluish green), for example, are important compounds of copper that are carbonates. Rhodochrosite ($MnCO_3$) is a beautiful bright pink mineral.

▲ Clear calcite ("Iceland spar") exhibits the unusual property called *birefringence*. This double-refraction phenomenon results in a double image of the object viewed through the crystal.

SELF-ASSESSMENT Questions

1. Which of the following statements about carbonates is NOT true?
   a. The charge on the carbonate ion is 2−.
   b. Sodium carbonate is the most common carbonate found in the lithosphere.
   c. CaCO₃ is a primary component of mollusk shells.
   d. Calcium carbonate is formed mainly in shallow, warm oceanic waters.

2. Copper carbonates
   a. are never found in nature
   b. are usually coppery-colored
   c. are the most common form of copper in nature
   d. are surprisingly blue

3. Cave formations may consist of all the following except
   a. Ca(OH)₂
   b. CaSO₄
   c. CaCO₃

4. Acidic rain has a deleterious effect on
   a. sandstone
   b. marble statuary
   c. gypsum
   d. quartz

Answers: 1, b; 2, d; 3, a; 4, b

12.4 Metals and Their Ores

Learning Objectives • List the most important metals with their principal ores, and explain how they are extracted and their uses. • Describe some of the environmental costs associated with metal production.

Progress through the ages is often described in terms of the materials, especially metals, used for making tools. We speak of the Stone Age, the Bronze Age, and the Iron Age.

Copper and Bronze

Copper is sometimes found in the *native state* (as the element, not in compounds), usually in polycrystalline form, and it was probably also the first metal to be used, since it is found free in nature. Ancient records from the Middle East show that copper was known as far back as 10,000 years ago. Copper is usually found associated with sulfur, as *chalcopyrite* ($CuFeS_2$) and to a lesser extent as *chalcocite* (Cu_2S). Malachite and azurite are less commonly found, and are made of copper, carbonate, and hydroxide groups. Copper can be isolated from chalcocite by heating, and the two stepwise reduction reactions that take place are:

$$\text{(oxidation of sulfur): } 2\,Cu_2S + 3\,O_2 \rightarrow 2\,Cu_2O + 2\,SO_2$$

$$\text{(reduction of copper): } 2\,Cu_2O \rightarrow 4\,Cu + O_2$$

Copper gained its historic importance as the main component of **bronze**. Classic bronze is an **alloy** (a mixture of two or more elements) of ∼90% copper and ∼10% tin. Another form, called mild bronze, contains only 6% tin. Classic bronze is harder and was used for blades, while mild bronze was softer and used for helmets and armor. Bronze alloys are harder than pure copper and, thus, better for a variety of uses, such as weapons or boat and ship fittings. Societies that learned bronze technology moved into more dominant roles in the world.

Today, copper is still very valuable due to its excellent electrical conductivity. Most of the electrical wiring in buildings around the world is now made of copper. Copper is sometimes sought out as a roofing material due to its thermal conductivity as well as its aesthetic value. It is also a very popular material for plumbing piping. A minor use of copper has been in coinage, starting in Roman times, when bronze coins with images of Roman emperors were in use. Copper-plated one-cent coins are still in use in the United States.

Iron and Steel

Iron is barely harder than copper in its natural form. However, when alloyed with carbon to form **steel**, the result is much different. Steel is much harder than any bronze alloy, and a steel weapon will nearly always be superior to a bronze weapon. Nevertheless, the technology for production of iron and steel lagged far behind that for copper and bronze for two reasons. Iron reacts so readily with oxygen and sulfur that it is not found uncombined in nature. It also melts at a much higher temperature (1536 °C) than does copper (1084 °C) or bronze (600 °C to 950 °C).

▲ Examples of a bronze sword (*left*) and a steel sword (*right*).

Mixing crushed coal with reddish iron ore in a very hot fire to form molten iron was discovered ~1500 B.C.E. in Mesopotamia. In modern times, iron ore blast furnaces (smelters) use a purified form of carbon called "coke" as the reducing agent, mixed with concentrated iron ore pellets. Limestone is added to combine with any silicate impurities to form "slag."

First, carbon is converted to carbon monoxide by limiting the amount of air allowed into the blast furnace:

$$2\,C(s) + O_2(g) \rightarrow 2\,CO(g)$$

Then, the carbon monoxide reduces the iron oxide in the ore to iron metal.

$$Fe_2O_3(s) + 3\,CO(g) \rightarrow 2\,Fe(l) + 3\,CO_2(g)$$

The liquid iron is poured into molds and solidified (then called *cast iron*), and sent to the steel mills to be changed into useful objects. Usually the cast iron is remelted and purified. Other metals can be added to form alloys for a variety of purposes.

Aluminum: An Abundant, Low-Density Metal

Aluminum is the most abundant metal in Earth's crust, but much of it is thinly distributed in aluminum-containing clays. The metal is tightly bound in other compounds, and considerable energy is required to extract it from its ores. Nevertheless, aluminum has replaced iron for many purposes. Worldwide production of aluminum—about 63 million tons/year—is second only to that of iron production, at about 1630 million tons/year.

The main ore of aluminum is *bauxite*, impure aluminum oxide (Al_2O_3). The challenge of figuring out a method to extract pure aluminum from the oxide was significant, since aluminum oxide does not melt until well over 2000 °C, so the use of carbon as a reducing agent simply would not work well. Few manufactured objects made of aluminum metal existed before the 1880s. It was Charles Martin Hall, a young graduate of Oberlin College in Ohio, who discovered that process in 1886, and soon became wealthy as one of the co-founders of the ALCOA company. The impure Al_2O_3 is reduced by electrolysis of aluminum ions in a solution of molten cryolite, Na_3AlF_6, at about 1000 °C. Think of the myriad common items made of aluminum today, from pots, pans, and cooking foil to fishing boats, canoes, beverage cans, and machine parts.

To produce 1 ton of aluminum, it takes 2 metric tons (t) (1 t = 1000 kg) of aluminum oxide and 17,000 kilowatt hours (kWh) of electricity. If this electricity is generated from coal, about 16 t of CO_2 is released into the atmosphere. Further smelting yields aluminum ingots—the starting material for all aluminum products.

Aluminum is both light and strong. An aluminum object has a density less than 3 g/cm³, which is about one-third the density of a steel object. Although it is considerably more reactive than iron, aluminum corrodes much more slowly. Freshly prepared aluminum metal reacts rapidly with oxygen to form a hard, transparent film of aluminum oxide over its surface. This film protects the metal from further oxidation. Iron, on the other hand, forms an oxide coating that is porous and flaky. Instead of protecting the metal, the coating flakes off, allowing further oxidation. (Review the discussion of the corrosion of metals in Chapter 8.)

The Environmental Costs of Iron and Aluminum Production

Although steel and aluminum play essential roles in the modern industrial world, both are produced at significant environmental cost.

Both steel mills and aluminum plants produce considerable waste. Environmental damage caused by such wastes was demonstrated when 1 million cubic meters of aluminum waste known as "red mud" was released in 2010 into the surrounding environment of Kolontár, Hungary. The spill led to the death of 10 people and killed all life in the nearby Marcal River. The "red mud" also flowed to the Danube River, causing further damage. Pollution from metal production in developing nations is still a major problem.

5 Why are knives stiff and sharp, but paper clips bendable?

Stainless steel knives have at least 13% chromium for strength; some knives use a manganese/iron alloy. Paper clips are made of a simple steel, consisting of mainly iron, with a small amount of carbon.

Aluminum production from ore is more energy intensive than steel production, but the low density of aluminum means energy savings down the road. An automobile made largely from aluminum is so much lighter than one made mainly from steel that it takes a good deal less energy to operate it.

In addition, aluminum is highly recyclable. Take an aluminum beverage can. Recycling it requires 5% of the energy needed to make a new one. In terms of energy costs, 20 recycled cans are equivalent to one brand-new can. Thus, recycling is a highly sustainable activity, both environmentally and economically. Clearly, throwing an aluminum can or almost any other aluminum product in the trash is both environmentally harmful and wasteful. So drink up and recycle!

"Thar's Gold in Them Thar Hills!"

As we stated previously, relatively few metals are found in the pure state (in which they are called *native metals*) in nature. They are generally very difficult to find, usually in smaller amounts that preclude commercial development. Gold, silver, copper, and platinum are the best known, but many of the metals found in the middle section of the periodic table have been found in pure form in nature—but only rarely.

The presence or absence of precious metals has been an important factor in the rise and fall of nations throughout history. Egypt was a dominant culture in ancient times. Thousands of gold artifacts have been uncovered in Egyptian ruins. The tomb of King Tutankhamun contained at least 1000 kg of gold, estimated as a significant portion of the gold in active use at that time (~1300 B.C.E.) in the entire world. His coffin alone, made of 110 kg of pure gold, is worth about $4.7 million U.S. dollars (at $1328/ounce).

The discovery of new regions where gold could be found has sparked significant migrations of people. Just think of the California gold rush in 1848–1849, when hundreds of thousands of people swarmed central California in an attempt to strike it rich. Besides gold, silver, platinum, palladium, and other precious metals are actively mined, since they are important commodities in today's world, in commerce, electronics, and jewelry.

Coinage

Coins have been in use since several hundred years B.C.E. Ancient Greek and Roman coins still exist in collections and museums, usually made of copper or copper alloys such as brass (Cu and Zn) or bronze (Cu and Sn), featuring images of emperors or military leaders. Gold and silver coins have been dated to the times of the Romans. The Roman silver denarius was worth a day's labor, and the widow's mite ("pruta") was the smallest Jewish coin, both of which were mentioned in the Christian New Testament. Chinese coins have been in active use since at least the 8th century C.E., and European and Middle Eastern coins have been used since the Middle Ages.

Modern coinage consists of metal alloys, making bi-metallic or layered coins in a variety of shapes and sizes, utilizing primarily copper, zinc, nickel, and aluminum. U.S. pennies currently consist of zinc with a thin layer of copper, since zinc is much cheaper than copper. Unfortunately, it still costs nearly 2 cents to mint a penny. Canada ceased minting pennies in 2013, and the United States may cease minting them at some point in the future. Gold and silver coins have not been in common use for nearly a century. In 1933 more than 445,000 double eagle gold coins ($20 gold pieces) were struck by the U.S. Mint, but most were surrendered by Executive Order 6102 and melted down during the Great Depression. The United States and some other countries, including South Africa, mint precious metal coins in gold and silver for collectors, not for everyday use in commerce. Numismatics is a popular hobby.

▲ Examples of coins from around the world made of copper, coins made from brass (Cu/Zn alloys), two-part coins, and coins made of various copper/nickel alloys. The tiniest Chinese coins are made of aluminium.

Other Important Metals

Many other important metals are essential to modern technology. Some that have high-tech applications are shown in Table 12.5. The United States has depleted most of its high-grade ores and is now dependent on imports of many such essential minerals, also shown in Table 12.5. The metal reserves in other countries are also being depleted.

TABLE 12.5 Technologically Important Metals

Metal	Source(s)	Typical Use(s)	Annual World Production (t)
Indium	China; Canada	Transparent, conducting coatings on LCD displays and photovoltaic cells	655
Lithium	Chile; Argentina	Batteries for tools, laptops, digital cameras, and cell phones	35,074
Palladium	Russia	Catalytic converters; energy-storing capacitors in electronic devices	208
Platinum	South Africa; Russia	Catalytic converters; computer hard drives; glass for LCD displays; fuel cells	200
Rare earths (lanthanides)	China	Electronics; energy-efficient technologies	126,000
Europium		Red phosphor in CRT displays	
Cerium		Commercial catalysts; glass polishing; self-cleaning ovens	
Neodymium		Ingredient in super magnets and lasers	
Yttrium	China	Fluorescent light bulbs; ceramics; lasers	5000
Tantalum	Democratic Republic of the Congo; Rwanda	Electronic capacitors in computers, digital cameras, and cell phones	1100

How can we ever run short of a metal? Aren't atoms conserved? Yes, there is as much iron on Earth as there was 100 years ago, but in using it, we scatter the metal throughout the environment. This is why sustainable practices, such as recycling, are important beyond energy savings.

SELF-ASSESSMENT Questions

1. The metal most likely to be found in its native state is
   a. aluminum   **b.** copper   **c.** iron   **d.** sodium

2. When copper is produced from Cu_2S, the by-product is
   a. CuO   **b.** CO_2 **c.** S   **d.** SO_2

3. Bronze is an alloy of Cu and
   a. Fe   **b.** Pb   **c.** Sn   **d.** Zn

4. In steelmaking, carbon acts as
   a. a reducing agent and hardener
   b. an oxidizing agent and a softener
   c. a reducing agent and softener
   d. an oxidizing agent and a hardener

5. Complete the following sentences using the following:
   (a) electrolysis, (b) heat, and (c) carbon. Use each term once.
   An ore of aluminum is reduced by _____.
   An ore of copper is reduced by _____.
   An ore of iron is reduced by _____.

6. Bauxite is impure
   a. Al_2O_3 **b.** Cu_2S
   c. Fe_2O_3 **d.** gold

7. In which of the following reactions does the oxidation number of aluminum change?
   a. $2\ Al(OH)_3 \longrightarrow Al_2O_3 + 3H_2O$
   b. $Al_2O_3 + 2\ NaOH + 3\ H_2O \longrightarrow 2\ NaAl(OH)_4$
   c. $NaAl(OH)_4 \longrightarrow Al(OH)_3 + NaOH$
   d. none

8. How much more energy does it take to make a new aluminum can than to recycle one?
   a. 2 times   **b.** 10 times
   c. 20 times   **d.** 50 times

9. The _____ the tin content of bronze, the _____ it is.
   a. lower; harder
   b. higher; harder
   c. higher; softer
   d. none of the above

10. Uses for copper include all the following except
    a. electrical wires and cables
    b. galvanizing steel
    c. layered coinage
    d. bronze statues

Answers: 1, b; 2, d; 3, c; 4, a; 5, a/b/c; 6, a; 7, d; 8, c; 9, b; 10, b

12.5 Salts and "Table Salt"

Learning Objectives • Explain why sodium chloride has been one of the most important minerals throughout human history. • Describe how salt kills microorganisms. • Explain how freezing point depression works with rock salt in winter.

▲ The otherwise barren Salar de Uyuni in Bolivia, the largest salt flats in the world, holds enormous amounts of sodium chloride, a substance widely used in both industry and the home. Most sodium chloride is obtained from salt lakes and from the oceans. The water is evaporated, and the solids are heaped in piles. After drying, purifying, and packaging, the product finds its way into our kitchens as table salt.

We have already learned that salts consist of positively charged metal ions and negatively charged non-metal ions or polyatomic ions in an orderly geometric arrangement (Sections 4.3–4.5). Ionic salts are the reaction products of acids reacting with bases (Section 7.5). The simplest salts are those of the alkali metal (group I) ions with halide ions, Cl^-, F^-, Br^-, and I^-.

The most abundant alkali halide on Earth—and the most important one throughout human history, up to and including the 21st century—is NaCl, sodium chloride. The most important use for sodium chloride for thousands of years has *not* been to flavor food.

Although early peoples had no idea that microorganisms were responsible for decomposing fresh meat and fish, or that bacteria caused infections in wounds, most early cultures realized that salting meat would help preserve those foods, and cleaning wounds with salt would prevent infections. Our word "salary" comes from "salt," since part of the salaries of many Roman soldiers was an allowance to purchase salt. Salt was traded for slaves in ancient Greece and salt played a crucial role in European and American history, particularly in its wars, from the American Revolution to the settling of the American west. Whole settlements were built up near underground saltwater sources, near solid salt deposits, or near oceans where salt water evaporation ponds could be tended. Salt was carried along major trade routes to areas where no salt was readily available, from ancient times up until the present. From its use in religious rites in ancient Egypt to its use in leather tanning to government taxation as a major source of revenue, salt is inextricably intertwined with human history.

Salt kills microorganisms by a process called *osmosis*. This process results in a dehydration of cells, since water molecules migrate through cell walls toward the salt or salt solution on its surface, in order to "equalize" the concentrations of ions. In modern society, people in developed nations have refrigeration and many commercially available antiseptics, but salt is still important in developing nations for preserving meat and preventing infections.

Nowadays, rock salt—which is primarily NaCl with other impurities—constitutes a major expenditure for municipalities and highways in the winter, because it is used (usually mixed with sand for traction) to melt ice and snow on roads. Over a million tons of rock salt is used in New York state on its roads in an average year! And that figure does not count over a million gallons of brine (salt) solution applied to roads in that state, per year, as well.

Salting roads and sidewalks is effective because the presence of ions *in solution* results in a lowering of any liquid's freezing point. Scientists label this phenomenon *freezing point depression*. When rock salt is dissolved in water, Na^+ and Cl^- ions are freed in the solution. One way to envision why this works is to consider that the ions present "get in the way" of water molecules forming the ice structure. There is a limit, however, to the effectiveness of salting roads, since there is a solubility limit of any salt in water. The solubility of NaCl in water is about 35 g/100 g water. Calculations (beyond the scope of this text book) show that if the outside temperature is colder than about −22 °C (about −8 °F), ice won't melt.

Table salt is obviously used in many foods, and salt shakers seem to have taken a permanent place on most dining tables, at least in western cultures. More will be discussed about salt and other minerals in the diet in Chapter 17.

SELF-ASSESSMENT Questions

1. Primary uses of NaCl throughout history include all the following except
   a. to make food taste good
   b. to melt ice
   c. to improve digestion
   d. to prevent infections in wounds

2. Rock salt is used on roads in the winter because
   a. salt dissolves better in ice than water
   b. salt molecules affect the freezing point
   c. salt helps traction
   d. ions lower the freezing point of water

12.6 Gemstones and Semi-Precious Stones

Learning Objective • *Name and describe the primary components and properties of gemstones.*

Although relatively rare, gemstones and semi-precious stones have played a role in history, dating back to ancient Egypt and China. Kings and queens have worn elaborate jewelry and crowns festooned with jewels throughout history. Today, gemstones have an important niche in our modern economy and business. Diamond saws are used to cut the hardest of materials. In much of western society, an engagement ring with a diamond or other precious stone is the most common symbol of a commitment to marriage. Surprisingly, some diamonds are colored, such as the blue Hope Diamond, currently on display at the National Museum of Natural History in Washington, DC.

Precious stones include diamond, ruby, sapphire, and emerald, due to their relative rarity, high degree of hardness (diamonds are hardest, and the other three are harder than quartz), translucency, and capability of being cut or faceted to produce a highly reflective surface ("sparkle"). Other stones that are less hard, more common, and have less brilliant colors are labeled semi-precious and generally cost less, though naturally the size of the stone is important to the cost, as are any flaws that may be apparent. Semi-precious stones include topaz, aquamarine, amethyst, and tanzanite (originally found in Tanzania).

Gemstones were formed under conditions of high temperature and pressure and slow cooling to form the crystal structure inherent in those minerals, which results in their rarity. For example, in Russia, such conditions occurred in a small area of the Ural Mountains, where the emerald "belt" is only 25 km wide and less than 1 km long. Emeralds are also found in Colombia, Zambia, and Brazil. Rubies are found in Asia, particularly India and Thailand, and sapphires in Madagascar, Sri Lanka, and other locations.

Surprisingly, the elemental composition of gemstones seems quite ordinary. Diamonds are pure carbon; however, every carbon atom in a diamond is covalently bonded to four other carbon atoms in continuous tetrahedra, which are very stable structures (Figure 12.8), formed in nature under conditions of high heat and pressure. Graphite is also pure carbon, but consists of layers of carbon atoms bonded in hexagonal patterns. Graphite is used in pencils and as a lubricant, since the layers can easily slide past one another. Diamonds left in a burning home will likely disappear, as they are pure carbon and will be oxidized into some very expensive carbon dioxide.

Sapphires and rubies are made of a mineral called *corundum*, which is aluminum oxide (Al_2O_3), but in a very hard form consisting of slightly distorted octahedral shapes. The difference between the two gemstones is the presence of specific trace elements. Only diamonds are harder than sapphires and rubies. Rubies contain about 1% Cr^{3+} ions that replace Al^{3+} ions. Energy transitions that chromium undergoes are different from those of aluminum. The trace chromium ions can absorb most of the yellow-green and violet light that strikes it, and reflect all the red and some of the blue, resulting in a pinkish red color seen by the eye. Rubies fluoresce a brilliant bright red when ultraviolet light strikes them.

Sapphires, on the other hand, are blue due to the presence of a trace amount of Ti^{4+} and Fe^{2+} ions, located in pairs right next to each other in the distorted crystal structure of aluminum oxide. When photons in the yellow part of the electromagnetic spectrum are absorbed, that energy helps transfer electrons from the iron ions to the titanium ions, so only blue is reflected to the eye.

▲ An array of beautiful faceted gemstones. Order from left: blue sapphire, red ruby, green emerald, diamond, blue aquamarine, yellow topaz and pink sapphire.

6 What are gemstones made of?

Diamonds are simply carbon atoms in an essentially infinite structure with each C atom bonded to four others. Other precious stones are crystallized aluminium oxide or have Be atoms incorporated into the structure, with trace impurities to provide the colors. Beryllium is a well-known element in the gem world, since it is an essential constituent of many gemstones, including emerald, beryl, aquamarine and chrysoberyl.

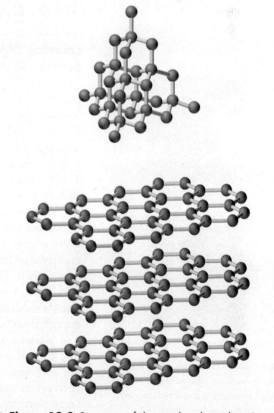

▲ **Figure 12.8** Structures of diamond and graphite. In diamond, every carbon atom is covalently bonded to four other carbon atoms in an essentially "infinite" repetition of tetrahedra in 3-D space. In graphite, layers of carbon atoms are loosely bonded together to form repeated hexagons.

Emeralds are different, and consist of a beryllium aluminum silicate with very low trace amounts of chromium. In this case, the beryllium atoms weaken the possible electron transitions, and photons in the yellow-to-red part and some of the violet part of the electromagnetic spectrum are absorbed, but all of the green light is reflected to the eye.

Semi-precious stones are often made into jewelry. Quartz with impurities includes amethyst found near Lake Superior shorelines into Canada, citrine, aquamarine, and topaz, which are either quartz or corundum with metal impurities. Pearls are mainly calcium carbonate. Many believe that semi-precious stones have healing or metaphysical properties. For example, orange calcite ($CaCO_3$ with iron impurities) is claimed to be an energy amplifier, "enhancing all the gifts you possess and getting rid of the pollution that is present in your body, and increasing the positive energies in your body and dispelling the negative energies," according to one website. There is no scientific evidence that atoms rubbed off from stones can enter the skin and cause such effects.

SELF-ASSESSMENT Questions

1. Which statement is untrue?
   a. Diamonds are made entirely of carbon atoms.
   b. Rubies are made of carbon with chromium ions as an impurity.
   c. Sapphires and rubies are both aluminum oxide, with different trace metals incorporated into the crystal structure.
   d. Pearls are mostly calcium carbonate.

2. Which of the following is false with respect to diamonds?
   a. They have carbon atoms bonded in layers.
   b. They are used to cut rubies and emeralds into faceted shapes.
   c. They are the hardest mineral known on Earth.
   d. They are sometimes colored.

3. Semi-precious stones
   a. are more expensive than diamonds and rubies
   b. are, like diamonds, made primarily of carbon
   c. are usually made of aluminum oxides or quartz with metal ion impurities
   d. have metaphysical powers

Answers: 1, b; 2, a; 3, c

12.7 Earth's Dwindling Resources

Learning Objectives • List the main components of solid waste. • Name and describe the three Rs of garbage.

▲ When the Europeans first came to North America, native copper (shown here) was readily available and was used by Native Americans. In the United States today, ore with less than 0.6% copper is mined, including chalcopyrite ($CuFeS_2$), bornite (Cu_5FeS_4), and chalcocite (Cu_2S).

For decades, international politics has created threats of shortages and high prices for various commodities, including metals and minerals. The United States has exhausted its high-grade ores of metals such as copper and iron. Miners once found nuggets of pure gold and silver, but the "glory holes" of western North America are long gone. Now we mine ores yielding a fraction of an ounce of gold per ton of ore. There remain very few high-grade deposits of any metal ores that are readily accessible to the industrialized nations.

What's wrong with low-grade ores? Because more material must be mined to obtain the same final amount of metal, there is more environmental disruption, and more energy is required to concentrate the ores. Both environmental cleanup and energy cost money. Consider an analogy. A bag of popcorn is useful. You can pop it and eat it. The same popcorn, scattered all over your room, would be less useful. Of course, you could gather it all up and then pop it, but that would take a lot of energy—perhaps more energy than you would care to expend and perhaps more energy than you would get from eating the corn after popping it.

Modern society is heavily dependent on chemicals, and we are rapidly depleting known reserves of many key elements that are used to make the large number and variety of chemicals utilized for everyday products. For example, catalytic converters used on automobiles use platinum, rhodium, and palladium. While a large portion of these valuable elements is recycled when a car is scrapped, some of these metals are lost through the exhaust pipes of automobiles. Another example is indium. Why should you care about indium, you ask? Surely you have a cell phone—or perhaps a tablet, laptop, or some other electronic device—that has a visual display. Or perhaps you've seen solar panels on roofs or in a field of panels.

All of these contain indium. Now multiply your devices by a few billion and add yearly replacements on the order of 500 million. Even if only a small amount of indium is in these devices, it adds up to metric tons of indium per year. We currently don't recycle indium, and doing so would be a major challenge. Eventually, getting indium is going to be harder than it is already.

How can we move from the chemicals we currently use to new kinds of chemicals that don't have all the problems we have now? It's possible, but we need to get moving on this sooner rather than later. Many of those answers lie in chemical research.

How Crowded Is Our Spaceship?

Each day, people die and others are born, but the number born is about 200,000 more than the number who die. Every five days, the world population goes up by a million people, and it reached more than 7.5 billion in 2018.

Figure 12.9 is a graphic picture of population growth. For tens of thousands of years, the human population was never more than a few hundred million. It did not reach the 1 billion mark until about 1800, after which the rate of growth skyrocketed. World population quadrupled during the twentieth century, with growth peaking in the 1960s at about 2.05% a year. The rate of growth in 2017 was 1.10% per year. The Population Division of the United Nations Department of Economic and Social Affairs estimates a world population of 9.8 billion by 2050, and 11.2 billion by 2100. How many people can Spaceship Earth accommodate? Many believe that we have already gone beyond the optimal population for this planet. They predict increasing conflicts over resources and increasing pollution of the environment.

▲ Where will we get metals in the future? The sea is one possible source. Nodules rich in manganese cover vast areas of the ocean floor. These nodules also contain copper, nickel, and cobalt. Questions about who owns them and how they can be mined without major environmental disruption remain to be resolved.

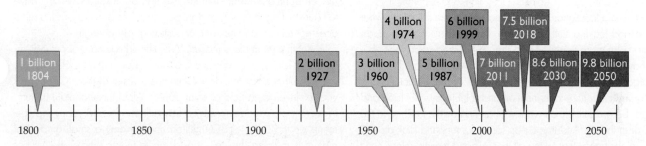

▲ **Figure 12.9** Two thousand years ago, Earth's population was about 300 million—roughly the same as the population of the United States today. Population changed little over the next thousand years and is estimated to have been only 310 million at the end of the first millennium. It reached 1 billion in about 1804 and stood at nearly 7.6 billion in 2018. This graph shows the intervals in which an additional billion people were added to Earth's population, with projections for 2030 and 2050. *Source:* United Nations, *World Population Prospects: The 2017 Revision.* New York: 2017.

Why did Earth's population start increasing so rapidly after centuries of little change? Population growth depends on both the birth rate and the death rate. If they are equal, population growth is zero. After thousands of years of nearly equal rates, scientific progress in the nineteenth century led to a decline in the death rate in developed countries. Yet the birth rate changed little, and the population grew rapidly. These changes reached the less developed countries in the 1950s, when the death rate fell dramatically but the birth rate remained high—creating a population explosion.

Population growth has slowed considerably in recent years, especially in developed countries. Scientific progress continues to cause a decline in death rates, but birth rates have also declined, especially in Europe and East Asia. The decrease in birth rates in the less developed countries has come more slowly. With more people comes more waste to be processed. We must continue to develop technologies and methods to deal with our ever-increasing wastes.

SELF-ASSESSMENT Questions

1. The three Rs of garbage are
   a. reduce, recycle, and redeploy   b. reuse, redeploy, and recycle
   c. reuse, reprocess, and recycle   d. reduce, reuse, and recycle

2. Which kind of material is the largest component of the garbage we put in landfills?
   a. food   b. metals   c. paper   d. plastic

GREEN CHEMISTRY Critical Supply of Key Elements
David Constable, ACS Green Chemistry Institute

Principles **2, 3, 6, 7**

Learning Objectives • Identify a few elements critical to products used every day by modern society that are not abundant that have to be replaced by earth-abundant materials. • Describe why recycling and reuse of critical elements is not sustainable.

You may not have thought about it this way, but chemists are a lot like people who started out thinking about how to build shelters. Early build-ers started with what was readily available – sticks, stones, logs, mud, clay – whatever they could use. People who were interested in chemistry began in a similar way. They looked around and started to think about how they could use those same kinds of materials to make different things. They wondered about what the stuff around them, like rocks, was made of and whether or not they could make something of greater value from these materials. Sometimes, they boiled things or put them in water to see what would happen. At one point, they happily discovered that the juice from berries or grapes would sometimes ferment, and when they drank it, it changed the way they felt. From all the things around them – rocks, plants, blood, feathers – they started to experiment.

Fast forward thousands of years, and people are still taking what is at hand and making stuff. Of course, the materials that are made today are a lot more sophisticated than a hundred years ago, but people haven't changed all that much. The holes in the ground have gotten bigger and deeper, the fires are hotter or more efficient, and thankfully, we know how to ferment all kinds of things to get food and chemicals. But we still rely on what we can find on the land, or in the oceans, or in the air. And if you stop for a moment and look around you, you might realize that taking all that stuff and making things can't go on forever in the same way we do it now.

Modern society is heavily dependent on chemicals, and we are rapidly depleting known reserves of many key elements that are used to make the large number and variety of chemicals we use for everyday products. For many elements that are of critical importance to making chemicals and products, we are taking a large amount of mass (think mountains of rock, for example), concentrating the desired elements, and dispersing them into products in forms that are difficult to recover and reuse. As we use up more and more of these materials, it will be harder and harder to extract the desired elements.

At some point, we'll have to figure out how to obtain them from the ocean or from lower-grade ores, which will come at increasing human and environmental costs.

Let's go through a few examples where critical elements are used. There are several in catalytic converters used on automobiles, including the transition metals platinum, rhodium, and palladium. While a large portion of these valuable elements is recycled when a car is scrapped, some of these metals are lost through the exhaust pipes of automobiles. As long as we drive cars that use gasoline and employ these metals in catalytic converters to ensure better air quality, the growing number of automobiles means that more metals are lost to the environment. We will have to develop catalytic converters that rely on earth-abundant elements like iron or nickel, or we will have to figure out another way to transport ourselves and our products that does not rely on an internal combustion or diesel engine.

Another example is indium. Why should you care about indium, you ask? Surely, you have a cell phone, or perhaps a tablet, laptop, or some other electronic device that has a visual display. Or perhaps you've driven by a field of solar panels. All of these contain indium. Now multiply your devices by a few billion and add yearly replace-ments on the order of 500 million. Even if only a small amount of indium is in these devices, it adds up to metric tons of indium per year. We currently don't recycle indium, and doing so would be a major challenge. Eventually, getting indium is going to be harder than it is already. So what do we do?

This is where green chemistry comes in. In the case of visual displays, we can move toward the use of things like earth abundant materials and organic light-emitting diodes to replace the current displays. That is what the world needs to start thinking about. How can we move from the chemicals we currently use to new kinds of chemicals that don't have all the problems we have now? It's pos-sible, but we need to get moving on this sooner rather than later.

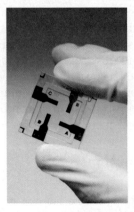

Summary

Section 12.1—Earth is divided into the core, the mantle, and the crust. The core is thought to be mainly iron. The crust consists of the solid lithosphere, the liquid (water) hydrosphere, and the gaseous atmosphere. Nine elements make up 96% of the atoms in the crust. The lithosphere is made up largely of rocks and minerals, including silicates, carbonates, oxides, and sulfides. The lithosphere is composed of an inorganic component as well as organic and biological components.

Section 12.2—Silicates contain SiO_4 tetrahedra in various arrangements. **Quartz** is a crystalline silicate that consists of a three-dimensional network of these tetrahedra. Sand is mainly quartz. The **micas** consist of sheets of SiO_4 tetrahedra. **Asbestos** is a fibrous silicate mineral made of chains of SiO_4 tetrahedra. Asbestos exposure has a **synergistic effect** with smoking; that is, the two together produce a higher risk of lung cancer than their simple combination indicates. Clays are aluminum silicates made up of tetrahedral silica and octahedral aluminum sheets. They come in two types: the 1:1 and 2:1 clays, with the 2:1 type being potentially water expandable. Ceramics, glass, and cement are modified silicates. **Glass** is a noncrystalline solid having no fixed melting point. Most glass is made from sand, limestone, and sodium carbonate with additives to impart desired properties such as heat resistance, color, and light sensitivity. Glass does not degrade in the environment, but it is easy to recycle. **Ceramics** are made by heating clay and other minerals to fuse them. **Cement** is a complex mixture of calcium and aluminum silicates made from limestone and clay. **Concrete** is a mixture of cement, sand, gravel, and water. Producing the energy needed in the cement-baking step has a negative environmental impact.

Section 12.3—Calcium carbonate is an important compound in the lithosphere, found in limestone, most cave formations, cliffs, and outcroppings created under ancient oceans. It is found in chalk and antacid formulations. Marble is a metamorphosed limestone, often used in statuary and large structures that last hundreds of years. Carbonates are susceptible to degradation by acids, especially acidic rainfall.

Section 12.4—Copper was one of the first metals to be obtained from its ore (chalcopyrite or chalcocite), but it is too soft to be used for tools. **Bronze** is an **alloy**—a mixture of a metal and one or more other elements. Bronze contains copper and tin. The more tin bronze contains, the harder it is. Bronze is harder than copper, and tools made of it were used for thousands of years. Iron-making technology was developed much later. Iron is produced in a furnace from iron ore (hematite) and carbon (coke). **Steel** is an alloy of iron and carbon, and is the form in which iron is normally used. The higher the carbon content, the harder the steel is. Aluminum is less dense than steel and does not rust. Aluminum is isolated from its ore, bauxite, by a reduction process that uses electricity. Both the energy required and the waste produced in metal extraction have a heavy environmental impact. Recycling reduces this impact. The discoveries of precious metal deposits such as of gold and silver have impacted history, and modern technological society has made use of the rarer minerals in nature, such as Os, Ir, and Ta. Coinage has been used for over 2000 years, and has been made from gold, silver, copper, and now cheaper metals with copper or other metal coatings.

Section 12.5—Sodium chloride is the most important alkali halide found in nature. Preserving meat and fish and cleaning wounds have been the most important uses of salt throughout history—so important that many people were paid in salt, not money or coins. Salt kills microorganisms by osmosis. Rock salt is used in the winter to melt snow and ice by the process known as freezing point depression, and is very expensive for communities in northern climates.

Section 12.6—Gemstones are an important commodity, usually for jewelry, but are actually made of common elements: C, Al, and O, with just the right amount of other trace metal ions to provide color. They are rare, because the conditions of very high temperature and pressure followed by slow cooling are so uncommon. A diamond is made of carbon atoms, each bonded to four other carbon atoms on nearly an infinite level, and is the hardest mineral on Earth.

Section 12.7—We face potential shortages of many metals and minerals because most of the high-grade ores have been used up. The United States now imports some metals and ores that it used to mine. It takes much less energy to recycle most metals than it does to produce them from ores.

Although birth rates have declined somewhat, Earth is crowded and is going to become more so in years to come. We must develop technologies and methods to deal with ever-increasing wastes.

GREEN CHEMISTRY—Green chemistry helps us to think about ways we might make the things we need without using rare elements. It helps us to think about recycling, new kinds of materials, or using more materials from waste and turning them into useful products.

Learning Objectives

Learning Objectives	Associated Problems
• Describe the structure of Earth and the regions of Earth's surface. (12.1)	1, 20–22, 28
• List the most abundant elements in Earth's crust and the common compounds in which they are found. (12.1)	23–27
• Describe the arrangement of silicate tetrahedra in common silicate minerals. (12.2)	29–36
• Describe how glass differs in structure from other silicates. (12.2)	2, 3, 37–39
• Explain the importance, abundance, and reactions of calcium and other carbonates. (12.3)	5, 43–50

• List the most important metals with their principal ores, and explain how they are extracted and their uses. (12.4)	6–9, 18, 51–64, 72–74
• Describe some of the environmental costs associated with metal production. (12.4)	10–12, 70, 71
• Explain why sodium chloride has been one of the most important minerals throughout human history. (12.5)	15
• Describe how salt kills microorganisms. (12.5)	13
• Explain how freezing point depression works with rock salt in winter. (12.5)	14
• Name and describe the primary components and properties of gemstones. (12.6)	16, 17, 67–69
• List the main components of solid waste. (12.7)	17
• Identify a few elements critical to products used every day by modern society that are not abundant that have to be replaced by earth-abundant materials.	76–77
• Describe why recycling and reuse of critical elements is not sustainable.	78–80

Conceptual Questions

1. Where are essentially 100% of the materials found that humankind needs? Compare with the mass of the Earth.

2. Compare an amorphous substance with a crystalline substance.

3. List at least three metal oxides that can be added to a glass mixture and explain what properties of the glass have been changed.

4. What environmental problems are associated with the manufacture of cement?

5. Describe where and from what limestone is formed. Why can it be found in the interior of continents?

6. Why was copper one of the first metals used?

7. Why did the Bronze Age precede the Iron Age in most places?

8. What is steel? How does it differ from pure iron in composition and in properties?

9. Aluminum is the most abundant metal in Earth's crust, yet aluminum ores are not very plentiful. Explain.

10. What are the environmental costs and benefits of aluminum?

11. Of Al, Cu, and Fe, which metal requires the most energy to extract and purify, and why?

12. Compare the energy cost of production of new aluminum from bauxite extraction with that from recycled aluminum.

13. Explain how salt kills microorganisms via osmosis.

14. What is freezing point depression? Explain how sprinkling rock salt on a sidewalk will melt ice in winter.

15. What have been the most important uses for sodium chloride throughout human history?

16. List at least three reasons why gemstones are prized.

17. List three methods of solid waste disposal. Give the advantages and disadvantages of each.

18. If matter is conserved, how can we ever run out of a metal?

19. Explain why metals can be recycled fairly easily. What factors limit the recycling of metals?

Problems

Composition of Earth

20. Which statement about Earth's structure is incorrect?
    a. The crust is between 8 km and 40 km thick.
    b. The mantle is composed mainly of silica.
    c. The inner core is liquid.
    d. The outer core is liquid.

21. The Earth's surface consists of:
    a. the atmosphere   b. the hydrosphere
    c. the lithosphere   d. all of the above

22. Define *lithosphere*, *hydrosphere*, and *atmosphere*.

23. Name four kinds of minerals found in Earth's crust, and give an example of each.

24. Rank the following elements in terms of their abundance on the Earth's surface by atom percent: calcium, hydrogen, silicon, magnesium, potassium, iron, and sodium.

25. Rank the following elements in terms of their abundance on the Earth's surface by percentage by mass: calcium, hydrogen, silicon, magnesium, potassium, iron, and sodium.

26. Why are the rankings in Problems 24 and 25 different?

27. What materials make up the organic portion of the lithosphere?

28. Referring to Table 12.2, write a balanced equation for the dissolving of galena in hydrochloric acid. The products are hydrogen sulfide gas and lead(II) chloride.

Silicate Minerals

29. What is the chemical composition of quartz? How are the basic structural units of quartz arranged?

30. In silica, four O atoms surround each Si atom, yet the formula for silica is SiO_2. Explain how this can be.

31. How are the SiO_4 tetrahedra connected? How does this compare with how carbon atoms are connected in diamond?

32. Draw the Lewis dot structure for SiO_4^{4-}, and then draw the Lewis dot structures for two connected SiO_4 tetrahedra. (Hint: Look at it as $O_3Si-O-SiO_3^{6-}$.)

33. Why does mica occur in sheets? How are the basic structural units of mica arranged?

34. Why does asbestos occur as fibers? How are the basic structural units of chrysotile asbestos arranged?

35. In clays, silica is found
    a. only in the tetrahedral sheet
    b. only in the octahedral sheet
    c. in both tetrahedral and octahedral sheets
    d. in sheets other than tetrahedral or octahedral

36. Which of the following clays would be expandable, and how?
    a. a 1:1 clay
    b. a 2:1 clay with no water between 2:1 layers
    c. a 2:1 clay with water between 2:1 layers
    d. There are no expandable clays.

Modified Silicates

37. Which of the following elements is the most abundant in a typical glass window, by atom percent? Does this change when percentage by mass is used?
    a. Si   b. O   c. Ca   d. Na

38. How does heating destroy long-range order in crystalline materials?

39. Focusing on the bonds, explain how glass becomes softer when heated.

40. What are the two basic raw materials required for making cement?

41. What is the difference between cement and concrete?

42. What raw materials are used to make concrete?

Carbonates

43. In what kind of environment is limestone created?

44. What is the difference between limestone and marble?

45. List several specific uses of marble.

46. List several specific uses of limestone.

47. What are the correct chemical formulas for (a) barium carbonate and (b) potassium carbonate?

48. What are the correct chemical formulas for (a) iron(II) carbonate and (b) iron(III) carbonate?

49. Write a balanced chemical reaction for when sulfuric acid (H_2SO_4) reacts with calcium carbonate.

50. Write a balanced chemical reaction for when hydrochloric acid (HCl) reacts with barium carbonate.

Metals and Ores

51. What are the main ores from which aluminum, copper, and iron are extracted?

52. What modern-day uses do copper and iron have?

53. For each of the following chemical reactions, discuss the change in oxidation number of the metal atom:
    a. $2\,Cu_2O \rightarrow 4\,Cu + O_2$
    b. $Fe_2O_3 + 3\,CO \rightarrow 2\,Fe + 3\,CO_2$

54. For each of the following chemical reactions, discuss the change in oxidation number of the metal atom:
    a. $2\,Al_2O_3 \rightarrow 4\,Al + 3\,O_2$
    b. $2\,FeO + CO \rightarrow CO_2 + 2\,Fe$

55. How many electrons are involved in each of the reactions in Problem 53?

56. How many electrons are involved in each of the reactions in Problem 54?

57. By what kind of chemical process (see Chapter 8) is a metal obtained from its ore? Give an example.

58. By what chemical process does a metal corrode? Give an example.

59. What is the advantage of adding tin to copper?

60. What alloy is formed between copper and tin? Which copper tin alloy is harder, one with 6% or 10% of tin content?

For Problems 61–64, first write (if it is not given) or complete and balance the equation for the reaction. Next, answer the following: (a) *What substance is reduced?* (b) *What is the reducing agent?* (c) *What substance is oxidized?* (d) *What is the oxidizing agent? (These questions require some knowledge of material in Chapter 8.)*

61. Vanadium metal can be prepared by reacting vanadium(V) oxide with calcium; the other product is calcium oxide.

62. Erbium metal (Er) can be prepared by reacting erbium(III) fluoride with magnesium; the other product is magnesium fluoride.

63. Europium (Eu) was used in older television screens to give a red color. The metal can be prepared by electrolysis of molten europium(III) chloride; chlorine gas is a by-product.

64. Crushed carbon was mixed with crushed iron ore (Fe_2O_3) by early cultures and heated strongly in ceramic pots to create iron to make swords and shields. What is the second product of the reaction? Write a balanced equation for that process.

Salt

65. What is the general definition of "salt"? Give an example other than NaCl and list the reactants that would be used to form that salt.

66. Write the chemical formulas of at least four important salts and why they are important.

Gemstones

67. List the four most important gemstones.

68. Explain the difference between diamond and graphite in terms of chemical structure.

69. Which two important gemstones have the same basic chemical formula, and differ only on the grounds of the presence of different trace minerals? What is that basic chemical formula?

Expand Your Skills

70. An aluminum plant produces 72 million kg of aluminum per year. How much aluminum oxide is required? How much bauxite is required? (It takes 2.1 kg of crude bauxite to produce 1.0 kg of aluminum oxide.)

71. It takes 17 kWh of electricity to produce 1.0 kg of aluminum. How much electricity does the plant in Problem 70 use for aluminum production in one year?

    Problems 72, 73, and 74 require some knowledge of material in Chapter 5.

72. Neodymium metal, a vital component of the magnets used in gasoline–electric hybrid automobiles, is prepared by reduction of neodymium(III) fluoride with calcium metal.

    $$2\,NdF_3 + 3\,Ca \longrightarrow 2\,Nd + 3\,CaF_2$$

    What mass of neodymium is obtained from 100.0 g of NdF_3?

73. Indium is a vital component of the electrically conducting transparent coatings on LCD screens. Indium metal can be prepared by electrolysis of salts such as indium(III) chloride. What mass in grams of indium(III) chloride is required to produce 20.0 g of indium?

    $$2\,InCl_3 \longrightarrow 2\,In + 3\,Cl_2$$

74. The equation for the reaction by which a sodium cyanide solution dissolves gold from its ore is

    $$4\,Au + 8\,NaCN + 2\,H_2O + O_2 \longrightarrow 4\,NaAu(CN)_2 + 4\,NaOH$$

    What is the minimum mass of NaCN required to dissolve the 10 g of Au in a ton of gold ore?

75. What do the three Rs of garbage stand for?

76. Find out what minerals contain indium and where these are mined.
    a. What is the process used to get pure indium?
    b. What are some of the environmental problems from mining indium?

77. Describe the process of obtaining platinum.
    a. Chemists like to use platinum in chemical processes. Describe one form of platinum they use.
    b. Why do chemists like to use platinum and what might some alternatives be?

78. Explain why platinum, palladium, and rhodium might be easier to recycle than indium.

79. What makes recycling an element like indium so difficult?

80. Give an example of a different approach to electiric lighting that wouldn't use a rare element like indium.

 Critical Thinking Exercises

Apply knowledge that you have gained in this chapter and one or more of the FLaReS principles (Chapter 1) to evaluate the following statements or claims.

12.1. An economist has said that we need not worry about running out of copper, because it can be made from other metals.

12.2. A citizen testifies against establishing a landfill near his home, claiming that the landfill will leak substances into the groundwater and contaminate his well water.

12.3. A citizen lobbies against establishing an incinerator near her home, claiming that plastics burned in the incinerator will release hydrogen chloride into the air.

12.4. An environmental activist claims that we could recycle all goods, leaving no need for the use of raw materials to make new ones.

12.5. A salesperson tells you that *ceramics* is just a fancy word for glass.

 Collaborative Group Projects

Prepare a PowerPoint, poster, or other presentation (as directed by your instructor) to share with the class.

1. Prepare a brief report on one of the metals listed in Table 12.6. List the main ores, and describe how the metal is obtained from one of those ores. Describe some of the uses of the metal, and explain how its properties make it suitable for those uses.

2. Prepare a brief report on recycling one of the following types of municipal solid waste. List the advantages of recycling, and identify problems involved in the process.
   a. glass   b. plastics   c. yard waste
   d. paper   e. lead   f. aluminum

3. Discuss the environmental effects **(a)** of use of cleaning cloths, mop heads, and sponges that can be cleaned and reused versus use of similar disposable products and **(b)** of use of cloth diapers that need to be washed versus use of disposable diapers.

4. Search the Internet to find the current estimated population of the United States and of the world. What is the current rate of population growth, as a percentage per year, in **(a)** France, **(b)** Bolivia, **(c)** Uganda, and **(d)** the United States.

LET'S EXPERIMENT! Fizzy Flintstones, Crumbling Calcium Carbonate

Materials Needed
- Vinegar
- Lemon juice
- Water
- Clear plastic cups

- Dropper
- Several different rock types (limestone, chalk, quartz, granite, or others)

Why do stone statues dissolve over time? Which rock types are sensitive to acids?

Limestone or marble statues are common in our society. Many are found in parks, such as those in the Gettysburg National Battlefield and in Central Park in New York City.

Unfortunately, these statues are suffering damage due to acid rain. Limestone ($CaCO_3$) is water insoluble, but it will dissolve if it is in an acidic environment. Acids in rain such as sulfuric acid (H_2SO_4) contribute to the dissolving of the limestone. Sulfuric acid will dissolve the calcium carbonate to make a moderately soluble form of calcium, $CaSO_4$ (called gypsum), along with carbon dioxide and water. The equation is listed below.

$$CaCO_3 + H_2SO_4 \rightarrow CaSO_4 + CO_2 + H_2O$$

In this experiment, you will collect at least three rocks of three different types. Limestone, chalk, and granite are available outdoors in some regions, or at your hardware store in others areas. You will use lemon juice and vinegar to test the presence of carbonate in rocks.

Prepare three sets of clear plastic cups, three cups for each of the rock types. Label appropriately.

- Put 5–10 drops of lemon juice on one piece of each of the three rock types. What happens when you put lemon juice on each rock?

- Put 5–10 drops of vinegar on a second piece of each of the three rock types. What happens when you put vinegar on each rock?
- Put 5–10 drops of water on the third piece of each of the three rock types. What happens when you put water on each rock?

Look and listen carefully each time you add the lemon juice, vinegar, and water to the rocks.

Questions
1. Did the lemon juice and vinegar act the same way on each rock?
2. Why did some of the rocks react differently?
3. What does this experiment have to do with weathering?
4. Write the chemical equations for the reaction between acetic acid ($C_2H_4O_2$) and limestone ($CaCO_3$).

13 Air

Have You Ever Wondered?

1 Is air less polluted today than it was in the past? **2** Is it practical to collect automobile emissions and convert them back to gasoline? **3** How do greenhouse gases keep the Earth warm? **4** Is global warming caused by the ozone hole? **5** What does smog testing of vehicles actually test?

THE BREATH OF LIFE We could live about four weeks without food. We could live for four days without water. But without air we cannot live even four minutes. Most of us cannot hold our breath for very long, even if we had to do so in an emergency. The thin blue boundary at the edge of Earth in the photograph actually exaggerates the thickness of the atmosphere that keeps us alive. Although the atmosphere has no definite boundary, 99% of its gas is less than 30 km from the surface of Earth. If the Earth were the size of an apple, the atmosphere would be thinner than the apple's skin.

Sally Ride, the first American woman in space, said the following after her experience: "I saw the blackness of space, and then the bright blue Earth. And then it looked as if someone had taken a royal blue crayon and traced along Earth's horizon. And then I realized that blue line, that really thin royal blue line, was Earth's atmosphere, and that was all there was of it. And it's so clear from that perspective how fragile our existence is."

The concept of living today in such a way that does not compromise the lives of future generations is one way of viewing sustainability. How humans interact with our atmosphere is clearly a determining factor in the sustainability of the Earth. If large parts of our planet do not have air that is healthy to breathe, or if the condition of the overall atmosphere causes too much of the sun's energy to be retained, with negative consequences, we have not fulfilled our obligation to those who come after us. As you read about the topics related to air in this chapter, keep in mind the concept of sustainability as a measuring tool that can indicate the need for action.

A community may run short of food or fresh water, but we cannot run out of air, given the size of Earth's atmosphere. On the other hand, we *can* pollute the air so badly that it becomes unpleasant or even unhealthy to breathe. In some areas, the air is already of such poor quality that people become sick from breathing it, even to the point of dying in some cases. The World Health Organization (WHO) estimates (2017) that 9 million people die each year from air pollution. This corresponds to 1 in 6 deaths worldwide—almost twice the rate of 2012. Most of these deaths occur in developing countries.

This chapter first discusses the natural chemistry of the atmosphere and then presents some of the effects human intervention is having on the natural balance in the components of our air. Some possible ways to reduce these negative effects are also considered.

13.1 Earth's Atmosphere: Divisions and Composition

Learning Objectives • List and describe the layers of the atmosphere. • Give the approximate proportions of N_2, O_2, Ar, and CO_2 in Earth's atmosphere.

We passengers on Spaceship Earth live under a thin blanket of air called the **atmosphere**. We divide the atmosphere into four main layers based on the properties of the gases in those regions (Figure 13.1). The temperature in each layer varies with altitude. The layer nearest Earth, called the *troposphere*, contains nearly all living things and nearly all human activity. Although other planets in our solar system have atmospheres, Earth's atmosphere is unique in its ability to support higher forms of life, such as us humans. Many people suppose that atmospheric temperature decreases as the altitude increases, but this is mainly true only for the troposphere and *mesosphere*. In the *stratosphere* and *thermosphere*, the temperature increases as you go higher.

The altitudes of the various atmospheric layers and the composition of dry air are summarized in Table 13.1. The amount of water in the air varies markedly from almost none up to about 4%. The atmosphere has a number of minor constituents, the most important of which is CO_2. The concentration of carbon dioxide in the atmosphere increased from about 280 parts per million (ppm) in the preindustrial world to 408 ppm in 2018, and it continues to rise at 2–3 ppm each year.

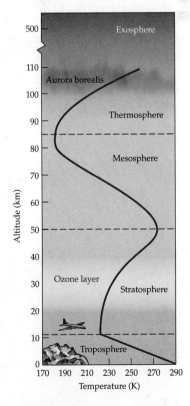

▲ **Figure 13.1** Approximate altitudes and temperature variations of the layers of the atmosphere.

WHY IT MATTERS

Although the increase in atmospheric CO_2 over the last two centuries—from 0.028% to 0.040%—appears to be small in absolute terms, that increase of almost a factor of 1.5 represents a significant change with major consequences. This change has been for the entire volume of the atmosphere above every part of the Earth, which should give us pause to consider the impact of humans on this grand scale. Most of the increased CO_2 comes from carbon that was stored in fossil fuels for tens of millions of years but has recently been burned in air. About 30 billion tons of CO_2 are emitted into our atmosphere each year from this burning.

TABLE 13.1 Approximate Composition of Clean Dry Air near Sea Level

Component	Percent by Volume
Nitrogen (N_2)	78.08
Oxygen (O_2)	20.94
Argon (Ar)	0.93
Carbon dioxide (CO_2)	0.04
Trace gases[a]	<0.01

[a]In order of decreasing abundance: neon (Ne), helium (He), methane (CH_4), krypton (Kr), hydrogen (H_2), dinitrogen monoxide (N_2O), xenon (Xe), ozone (O_3), sulfur dioxide (SO_2), nitrogen dioxide (NO_2), ammonia (NH_3), carbon monoxide (CO), and iodine (I_2), plus some others.

SELF-ASSESSMENT Questions

1. Nearly all human activity occurs in the
   a. mesosphere **b.** stratosphere **c.** thermosphere **d.** troposphere

2. The ozone layer is found in the
   a. mesosphere **b.** stratosphere **c.** thermosphere **d.** troposphere

3. Earth's atmosphere
   a. blends into outer space **b.** ends abruptly at 30 km altitude
   c. ends abruptly at 50 km altitude **d.** ends abruptly at the moon

4. The approximate percentage of nitrogen in the atmosphere is
   a. 18% **b.** 28% **c.** 48% **d.** 78%

5. After nitrogen and oxygen, and ignoring water vapor concentration, the most abundant gas in the Earth's atmosphere is
   a. argon **b.** carbon dioxide **c.** hydrogen **d.** methane

6. The present carbon dioxide concentration in the Earth's atmosphere is about
   a. 280 ppm **b.** 400 ppm **c.** 1% **d.** 4%

7. The stratosphere occurs in what range of altitude?
   a. 0–11 km **b.** 11–20 km **c.** 11–50 km **d.** 50–85 km

Answers: 1, d; 2, b; 3, a; 4, d; 5, a; 6, b; 7, c

13.2 Chemistry of the Atmosphere

Learning Objectives • Describe the nitrogen and oxygen cycles. • Describe the origin and effects of temperature inversions.

Under ordinary conditions, nitrogen and oxygen gases coexist in the atmosphere as a simple mixture, with little or no reaction between them. Under certain circumstances, however, the two elements can combine in a reaction that has consequences for life itself.

Nitrogen makes up 78% of the atmosphere in the form of diatomic N_2 molecules. Nitrogen is essential to life, but most animals and plants cannot incorporate N_2 directly into their tissues. How then has nitrogen become a component of so many vital compounds, such as the proteins and nucleic acids of living organisms? The answer is that organisms first combine nitrogen with another element, a process called **nitrogen fixation**. This process does not occur readily, because the nitrogen atoms in N_2 are held together by a strong triple bond ($N≡N$), which must be broken for fixation to take place.

The Nitrogen Cycle

Several natural phenomena convert nitrogen gas to more usable forms. Energy from lightning and from the heat of vehicle and industrial engines fixes nitrogen by causing it to combine with oxygen, forming first nitrogen monoxide (NO), commonly called nitric oxide, and then nitrogen dioxide (NO_2).

$$N_2 + O_2 \xrightarrow{\text{energy}} 2\,NO$$

$$2\,NO + O_2 \longrightarrow 2\,NO_2$$

(These oxides of nitrogen, NO and NO_2, are collectively designated as NO_x.) Nitrogen dioxide reacts with water to form nitric acid (HNO_3) and nitrogen monoxide.

$$3\,NO_2 + H_2O \longrightarrow 2\,HNO_3 + NO$$

Nitric acid in rainwater falls on Earth's surface, adding to the supply of available nitrates in the oceans and the soil.

Nitrogen is fixed industrially by combining nitrogen with hydrogen to form ammonia, a procedure called the *Haber–Bosch process* (described in Section 19.1), which is facilitated with a catalyst.

$$N_2 + 3\,H_2 \longrightarrow 2\,NH_3$$

This technology has greatly increased our food supply because the availability of fixed nitrogen is often the limiting factor in the production of food. Not all consequences of this intervention have been favorable, however. Excessive runoff of nitrogen fertilizer has led to serious water pollution problems in some areas (which we discuss in Chapter 14).

In an amount comparable to the production from the Haber–Bosch process, nitrogen fixation is performed by bacteria in the roots of legumes (peas, beans, clover, and the like). Certain types of bacteria reduce N_2 to ammonia (NH_3). Other bacteria oxidize ammonia to nitrites (NO_2^-), and still others oxidize nitrites to nitrates (NO_3^-). Plants are then able to take up the nitrates and ammonia from the soil. Animals in turn can get their required nitrogen compounds from eating plants, both directly and by eating animals that consume plants. The **nitrogen cycle** (Figure 13.2) is completed by the action of other types of microbes, which can use nitrate ions as their oxygen source for the decomposition of organic matter and release N_2 gas back to the atmosphere in a process called denitrification.

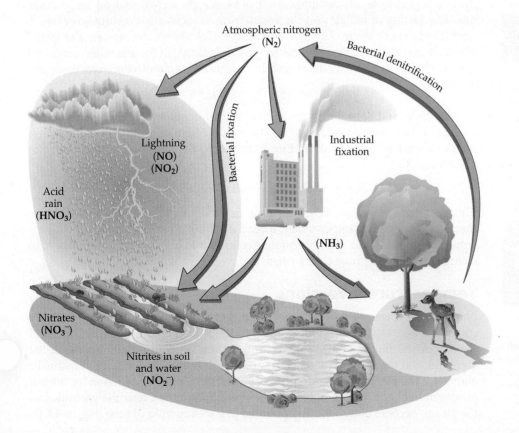

Atmospheric nitrogen
(**N₂**)

Bacterial denitrification

Lightning
(**NO**)
(**NO₂**)

Bacterial fixation

Industrial fixation

Acid rain
(**HNO₃**)

(**NH₃**)

Nitrates
(**NO₃⁻**)

Nitrites in soil and water
(**NO₂⁻**)

◄ **Figure 13.2** The nitrogen cycle. Atmospheric nitrogen is *fixed*, or converted to water-soluble forms, both naturally and industrially. Animal wastes, and dead plants and animals are converted back to atmospheric nitrogen by certain bacteria.

▶ **Figure 13.3** The oxygen cycle. Animals and people use oxygen gas and produce carbon dioxide. Plants, in turn, consume the carbon dioxide, converting it to oxygen gas and glucose for food. In the stratosphere, atmospheric oxygen is involved in ozone production.

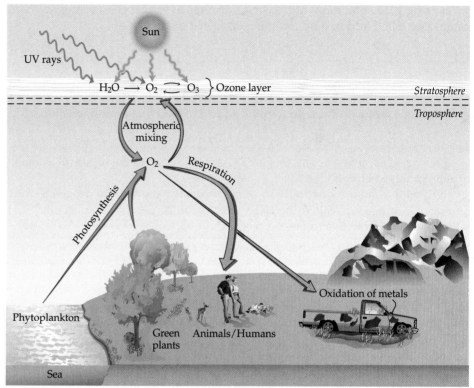

(a) Normal

(b) Temperature inversion

▲ **Figure 13.4** (a) Ordinarily, the air is colder at increasing altitudes. (b) During a temperature (thermal) inversion, a cooler layer is trapped near the surface, remains beneath a warmer layer, and can readily be polluted.

The Oxygen Cycle

Oxygen makes up about 21% of Earth's atmosphere. As with nitrogen, a balance is maintained between consumption and production of oxygen in the troposphere. Both plants and animals use oxygen in the metabolism of the food that they eat. The decay and combustion of plant and animal materials consume oxygen and produce carbon dioxide, as does the burning of fossil fuels such as coal, oil, and natural gas. The rusting of metals and the weathering of rocks also consume oxygen. A simplified **oxygen cycle**, illustrated in Figure 13.3, shows the influence of green plants, including one-celled organisms (phytoplankton) in the sea, which consume carbon dioxide in photosynthesis and, in the process, replenish the oxygen supply.

$$6\,CO_2 + 6\,H_2O \longrightarrow C_6H_{12}O_6 + 6\,O_2$$

Another vital balancing act occurs in the stratosphere far above the Earth's surface. There, oxygen is formed by the action of ultraviolet (UV) radiation on water molecules. Perhaps more important, some oxygen is converted to ozone (Section 13.8).

$$3\,O_2(g) + energy\,(UV\,radiation) \longrightarrow 2\,O_3(g)$$

This ozone absorbs high-energy UV radiation that might otherwise make higher forms of life on Earth impossible if it reached the surface. We discuss the atmosphere in more detail in subsequent sections, paying particular attention to changes wrought by human activities.

Temperature Inversions

Normally, air is warmest near the ground, due to the heat released by land, and gets cooler at moderate altitudes. On clear nights, the ground cools rapidly, but the wind usually mixes the cooler air near the ground with the warmer air above it. If the air is still, and the lower layer of cold air becomes trapped by the layer of warm air above it, a meteorological condition known as a **temperature inversion** (or *thermal inversion*) happens. Pollutants in the cooler air are trapped near the ground, and the air can become quite seriously polluted in a short period of time (Figure 13.4).

A temperature inversion can also occur when a warm front collides with a cold front. The less-dense warm air mass slides over the cold air mass, producing a condition of atmospheric stability. Because there is little vertical air movement, the air near the ground stagnates and air pollutants accumulate.

SELF-ASSESSMENT Questions

1. Lightning can convert atmospheric N_2 to
   a. nitroglycerin   **b.** ammonia
   c. nitrogen oxides   **d.** bacteria

2. Conversion of atmospheric N_2 to biologically useful forms is called nitrogen
   a. addition   **b.** fixation
   c. reduction   **d.** denitrification

3. The starting materials in the industrial process for making ammonia are
   a. N_2 and H_2 **b.** N_2 and O_2
   c. NO_3^- and H_2 **d.** NO_3^- and O_2

4. Which of the following processes removes carbon dioxide from the atmosphere?
   a. photosynthesis   **b.** respiration
   c. temperature inversions **d.** burning of fossil fuels

5. Which of the following processes consumes oxygen?
   a. breakdown of ozone   **b.** corrosion of metals
   c. Haber–Bosch process   **d.** photosynthesis

6. A temperature inversion is a meteorological condition that
   a. makes people put on thermal underwear
   b. commonly occurs over flat countryside
   c. concentrates air pollutants
   d. occurs when warm air is below cold air

Answers: 1, c; 2, b; 3, a; 4, a; 5, b; 6, c

13.3 Pollution through the Ages

Learning Objectives • List some natural sources of air pollution. • List the main pollutants formed by burning coal, and describe some technologies used to clean up these pollutants.

Air pollution has always been with us. Wildfires, windblown dust, and volcanic eruptions added pollutants to the atmosphere long before humans existed, and continue to do so. Kilauea volcano in Hawaii emits 1000–2000 mt[1] of sulfur dioxide each day, the upper figure representing the production when the volcano is active. (This corresponds to 5% to 10% of the entire SO_2 production of China.) Sulfur dioxide produces acid rain downwind that has helped create the barren Kau Desert. Dust from the Sahara Desert often reaches the Caribbean and South America. Dust and pollution from China blanket Japan and even reach western North America. Nature isn't always benign.

The Air Our Ancestors Breathed

The problem of air pollution probably first occurred when fires were built in poorly ventilated caves or other dwellings. People cleared land, leading to larger dust storms. They built cities, and the soot from their hearths and the stench from their wastes filled the air. The Roman author Seneca wrote in 61 C.E. of the stink, soot, and "heavy air" of the imperial city. The Industrial Revolution brought even more air pollution, as coal was burned in large amounts to power factories. Soot, smoke, and sulfur dioxide filled the air.

Air pollution today is much more complex than that affecting our ancestors. When society changes its activities, the nature of its waste materials changes, including the ones it dumps into the atmosphere. The phenomenon is not new, but today's rapid rate of change is unprecedented. Our environment may not be able to absorb the changes without irreversible damage to living systems.

Pollution Goes Global

Air pollution knows no political boundaries. Pollution from midwestern power plants leads to acid rain in the Northeast. Los Angeles smog drifts to Colorado and beyond. The United States generates pollution that contributes to acid rain in

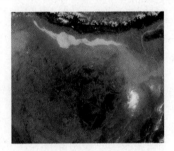

WHY IT MATTERS

A vast plume of polluted air called the *Asian brown cloud (ABC)* drapes over South Asia and the Indian Ocean during the winter. It is so huge that it may cool the area as much as or more than the greenhouse gases warm it. The cloud is composed of soot and ash, sulfates, nitrates, and mineral dust. Burning of biomass in forest fires, cooking, and heating are responsible for half of the pollution or more. Motor vehicles and factories also contribute.

----
[1]mt = metric ton; 1 mt = 1000 kg = 2200 lb.

▲ (Top) Burning of coal in electric power plants and other facilities is the main source of industrial smog, which is characterized by high levels of sulfur dioxide and particulate matter. (Bottom) Scanning electron microscope image of fly ash. Once inhaled, such particulate matter generally remains in the lungs.

1 **Is air less polluted today than it was in the past?**

The answer depends on when and where. As noted in the text, air quality varies markedly by location, such as urban areas compared with rural areas. The Industrial Revolution caused major degradation of air quality, but in the latter half of the twentieth century, many governments enforced environmental laws that led to a significant improvement in the air in many locations. For some parts of the world, the air was much cleaner in 2000 than it was in 1900, but in many places in the developing world, the negative trend has not been reversed. Some cities in China and India have recently reached record levels of air pollutants.

Canada. Contaminants from England and Germany pollute the snow in Norway. There is evidence that the shift of manufacturing from the eastern United States to Asia has led to cleaner New England air—but more pollutants on the West Coast.

As the number of people increases, more of the world's population lives in cities. Especially in developing countries, industrialization takes precedence over the environment. Cities in China, Iran, Mexico, Indonesia, and many other countries have experienced frightening episodes of air pollution. Large metropolitan areas are troubled the most by air pollution, but rural areas are affected, too. Neighborhoods around smoky factories have seen evidence of increased rates of miscarriage and decreased wool quality in sheep, decreased egg production and high mortality in chickens, and increased feed and care requirements for cattle. Plants are stunted, deformed, and even killed. Our agricultural operations for plants and animals are a key part of the need for sustainability, so it is important that we consider impacts to this industry.

What is a **pollutant**? It is too much of any substance in the wrong place or at the wrong time. A chemical may be a pollutant in one place and helpful in another. For example, ozone is a natural and important constituent of the stratosphere, where it shields Earth from life-destroying UV radiation. In the troposphere, however, ozone is a dangerous pollutant (Section 13.4).

Coal + Fire ⟶ Industrial (Sulfurous) Smog

The word **smog** is a blend of the words *smoke* and *fog*. There are two basic types of smog (Table 13.2). Polluted air associated with industrial activities is often called **industrial smog** or sulfurous smog. It is characterized by the presence of smoke, sulfur dioxide, and particulate matter such as ash and soot. The burning of coal—especially high-sulfur coal such as that found in the eastern United States, China, and Eastern Europe—causes most industrial smog. (*Photochemical smog* is discussed in Section 13.5.)

The chemistry of industrial smog is fairly simple. Coal is a complex combination of organic and inorganic materials. The organic materials are mainly carbon, and burn when the coal is combusted. The carbon is oxidized to carbon dioxide and heat is given off. The inorganic materials—minerals—wind up as ash, mostly metal oxides.

$$C(s) + O_2(g) \longrightarrow CO_2(g) + \text{heat}$$

Not all of the carbon is completely oxidized, however; some winds up as carbon monoxide.

$$2\,C(s) + O_2(g) \longrightarrow 2\,CO(g)$$

And some carbon, unburned, ends up as soot.

The sulfur in coal also burns, forming sulfur dioxide, a choking, acrid gas that gives rise to the term *sulfurous pollution*.

$$S(s) + O_2(g) \longrightarrow SO_2(g)$$

Sulfur dioxide is readily absorbed in the respiratory system. It is a powerful irritant known to aggravate the symptoms of people who suffer from asthma, bronchitis, emphysema, and other lung diseases.

TABLE 13.2 Types of Smog

	Industrial Smog	Photochemical Smog
Alternative names	Winter smog; sulfurous smog; "London smog"	Summer smog; "LA smog"
Main initial components	SO_2; particulates	NO_x; hydrocarbons
Secondary components	SO_3; H_2SO_4	O_3; aldehydes; peroxyacetyl nitrate (PAN)
Main source(s)	Electric power plants; factories	Automobiles
Typical weather	Cold; damp; foggy	Warm; dry; sunny

The next step in this chemistry makes the situation even worse. Some of the sulfur dioxide reacts further with oxygen in the air to form sulfur trioxide.

$$2\,SO_2(g) + O_2(g) \longrightarrow 2\,SO_3(g)$$

Sulfur trioxide then reacts with water to form sulfuric acid.

$$SO_3(g) + H_2O(l) \longrightarrow H_2SO_4(l)$$

Collectively, the two oxides of sulfur are often designated as SO_x.

In air, sulfuric acid forms an **aerosol**, a dispersion of tiny particles (solid) or droplets (liquid) in a gas. This aerosol is corrosive and even more irritating to the respiratory tract than sulfur dioxide.

Industrial smog usually has high levels of **particulate matter (PM)**, solid and liquid particles of greater-than-molecular size. The largest particles are often visible in the air as dust and smoke. PM consists mainly of soot and the mineral matter that occurs in coal.

Minerals do not burn, even in the roaring fire of a huge factory or power plant boiler, because they are made of very stable, ionically bonded species. Some mineral matter is left behind as *bottom ash*, but much of it is carried aloft in the tremendous draft created by the fire. This *fly ash* settles over the surrounding area, covering everything with dust. It is also inhaled, contributing to respiratory problems in animals and humans.

Small particulates—less than 12 μm in diameter, called PM12—are believed to be especially harmful. They contribute to respiratory and heart disease. The U.S. Environmental Protection Agency (EPA) also monitors even finer particulates, called PM2.5. EPA studies indicate that tens of thousands of premature deaths a year are linked to PM and that about 20,000 of them are due to PM2.5. Particulate concentrations have been lowered significantly since nationwide monitoring began in 1999. In 2010, three out of ten people in the United States lived in counties with particulate pollution levels above the EPA's standards for PM2.5, PM12, or both.

Health and Environmental Effects of Industrial Smog

Minute droplets of liquid sulfuric acid, as well as PM12 and smaller solids, are easily trapped in the lungs. Interaction can considerably magnify the harmful effects of pollutants. A certain level of sulfur dioxide, without the presence of PM, might be reasonably safe. A particular level of PM might be fairly harmless without sulfur dioxide around, but the effect of both together can be deadly. PMs can be very effective carriers of other pollutants into the lungs. *Synergistic effects* such as this are quite common whenever chemicals act together. Two examples are the synergistic effects of asbestos and cigarette smoke (discussed in Section 12.2) and the synergistic effects of some drugs (discussed in Section 18.7).

When the pollutants in industrial smog come into contact with the alveoli of the lungs, the alveoli lose their resilience, making it difficult for them to expel carbon dioxide. Such lung damage contributes to pulmonary emphysema, a condition characterized by increasing shortness of breath.

Sulfuric acid and SO_x pollutants also damage plants. Leaves become bleached and splotchy when exposed to SO_x. Yield and quality of farm crops can be severely affected. These compounds are also major culprits in the production of acid rain.

What to Do about Industrial Smog

An aspect of sustainability is the effort to prevent and alleviate industrial smog. How industrial plants operate is a significant factor in the impact they have on our atmosphere and therefore on humans as well as other animals and plants.

PM can be removed from smokestack gases by means of several devices.

- An **electrostatic precipitator** (Figure 13.5) induces electrical charges on the particles, which are then attracted to oppositely charged plates and deposited.

▲ Industrial smog was once so common a problem in London that it was called *London smog*. A notorious smog episode began in London on Thursday, December 4, 1952, lasted 5 days, and may have killed more than 4000 people.

▲ **Figure 13.5** A cross section of a cylindrical electrostatic precipitator for removing particulate matter from smokestack gases. Electrons from the negatively charged discharge electrode (in the center) become attached to the particles of fly ash, giving them a negative charge. These charged particles are then attracted to and deposited on the outer, positively charged collector plate, which permits their removal.

- *Bag filtration* works much like the bag in a vacuum cleaner. Particle-laden gases pass through filters in a bag house. These filters can be cleaned by shaking and periodically blowing air through them in the opposite direction.
- A *cyclone separator* causes the stack gases to spiral upward through it in a circular motion. The particles hit the outer walls, settle out, and are collected at the bottom.
- A **wet scrubber** removes PM by passing the stack gases through a fine mist of water. After use, the wastewater must be treated to remove the particulates.

The choice of device depends on the type of coal being burned, the size of the plant, and other factors. All require energy—electrostatic precipitators use about 10% of the plant's output—and the collected ash has to be disposed somewhere. Some of the ash is used to make concrete, as a substitute for aggregate in road base, as a soil modifier, and for backfilling mines. The rest has to be stored. Most of it goes into ponds, and the rest into landfills.

Sulfur and SO_x are harder to get rid of than is PM. Sulfur can be removed by processing coal before burning, but both the flotation method and the gasification or liquefaction process (Chapter 15) are expensive. Another way to get rid of sulfur is to scrub sulfur dioxide out of stack gases after the coal has been burned.

The most common scrubber uses the limestone–dolomite process. Limestone ($CaCO_3$) and dolomite (a mixed calcium–magnesium carbonate) are pulverized and heated. Heat drives off carbon dioxide to form calcium oxide (lime), a basic oxide that reacts with sulfur dioxide to form solid calcium sulfite ($CaSO_3$).

$$CaCO_3(s) + heat \longrightarrow CaO(s) + CO_2(g)$$

$$CaO(s) + SO_2(g) \longrightarrow CaSO_3(s)$$

This by-product presents a sizable disposal problem. Removal of 1 t of sulfur dioxide produces almost 2 t of solids. Some modified scrubbers oxidize the calcium sulfite to calcium sulfate ($CaSO_4$).

$$2\, CaSO_3(s) + O_2(g) \longrightarrow 2\, CaSO_4(s)$$

Calcium sulfate is much more useful than calcium sulfite; the sulfate is used to make commercial products such as plasterboard. In nature, calcium sulfate occurs as gypsum ($CaSO_4 \cdot 2H_2O$), which has the same composition as hardened plaster. It should be noted that the first step in this process produces additional amounts of CO_2, which creates problems of its own in terms of affecting the Earth's climate (Section 13.9).

SELF-ASSESSMENT Questions

1. The air in cities in eighteenth-century England
   a. smelled of the sea
   b. smelled of roses
   c. was smoky from coal burning
   d. was pristine

2. Which of the following is *not* usually released from burning fossil fuels?
   a. N_2 b. NO_x c. CO_2 d. SO_2

3. The pollutant usually described by size rather than composition is
   a. acid rain   b. CFCs
   c. particulates   d. photochemicals

4. Which of the following pairs of pollutant act synergistically in causing lung damage?
   a. O_3 and soot   b. O_3 and PM
   c. SO_2 and NO_x d. SO_2 and PM

5. A device that cleans stack gases by means of electric charges is called a(n)
   a. catalytic converter
   b. electric generator
   c. electrostatic precipitator
   d. particle accelerator

6. Which pollutant is removed from stack gases by scrubbers that use lime?
   a. CO_2 b. CO
   c. NO_x d. SO_2

7. A common by-product of sulfur dioxide removal from stack gases is
   a. CaO   b. $CaSO_4$
   c. $CaCO_3$ d. solid S

Answers: 1, c; 2, a; 3, c; 4, d; 5, c; 6, d; 7, b

13.4 Automobile Emissions

Learning Objectives • List the main gases in automobile emissions, and describe how catalytic converters reduce these gaseous pollutants. • Explain how carbon monoxide acts as a poison.

When a hydrocarbon burns in sufficient oxygen, the products are carbon dioxide and water. For example, consider the combustion of octane, one of the hundreds of hydrocarbons that make up the mixture we call gasoline:

$$2\,C_8H_{18}(l) + 25\,O_2(g) \longrightarrow 18\,H_2O(g) + 16\,CO_2(g)$$

Thus, the main components of automotive exhaust are water vapor, carbon dioxide, and unreacted nitrogen and oxygen from the atmosphere. Water vapor and nitrogen are innocuous, but CO_2 contributes to global warming. Furthermore, the combustion process is never quite complete, and side reactions also occur. These factors give rise to smaller quantities of more harmful products, including carbon monoxide, nitrogen oxides (NO_x), and volatile organic compounds (VOCs). NO_x contribute to smog and acid rain. VOCs from unburned fuel will react with oxygen atoms to form ozone (O_3), aldehydes (RCHO), and PAN (RCOOONO$_2$, page 409).

Carbon Monoxide: The Quiet Killer

When insufficient oxygen is present during combustion, carbon monoxide is formed. In the United States, CO makes up more than 60% (by mass) of all air pollutants entering the atmosphere, with more than half of all CO emissions coming from transportation sources. Total emissions of CO in the United States decreased by almost 50% over a period of 20 years but have plateaued since 2009.

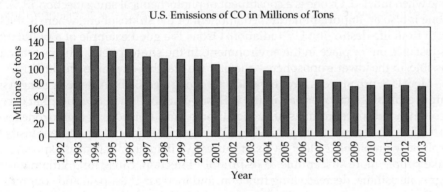

U.S. Emissions of CO in Millions of Tons

▲ Annual emissions of carbon monoxide in the United States.

Other sources of CO include industrial activities, residential wood burning, and natural sources such as forest fires. The EPA has set danger levels for CO at 9 ppm (average) over 8 h and 35 ppm (average) over 1 h. On city streets, these danger levels are exceeded much of the time. Even in off-street urban areas, levels often average 7–8 ppm. Such levels do not cause immediate death, but exposure over a long period can cause physical and mental impairment.

Because it is colorless and odorless, we can't tell that carbon monoxide is present except by using CO detectors or test reagents. (Automobile exhaust gets its odor from unburned hydrocarbons, not carbon monoxide.) Drowsiness, usually the only early symptom, is not always unpleasant.

Carbon monoxide acts by tying up the hemoglobin in the blood. Normally, hemoglobin transports oxygen throughout our body (Figure 13.6), but CO binds to hemoglobin much more strongly than oxygen does, which prevents the binding of oxygen. The symptoms of carbon monoxide poisoning are similar to those of oxygen deprivation, except that the skin may turn bright red from the CO–hemoglobin. All except the most severe cases of acute carbon monoxide poisoning are reversible, but prolonged hospital stays with oxygen therapy are sometimes necessary. Carbon monoxide poisoning is discussed further in Chapter 21.

2 Is it practical to collect automobile emissions and convert them back to gasoline?

This is hypothetically possible, though it would be difficult and would require much energy to accomplish. In fact, it would in practice require *more* energy to convert emissions back to gasoline than burning the original gasoline would have provided!

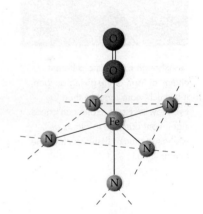
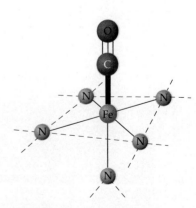

▲ **Figure 13.6** Schematic representations of a portion of the hemoglobin molecule. (See Figure 16.22.) Carbon monoxide bonds much more tightly than oxygen, as indicated by the heavier bond line.

Nitrogen Oxides: Some Chemistry of Amber Air

Nitrogen oxides are formed when N_2 reacts with O_2 at high temperatures. Because air contains both nitrogen and oxygen, NO_x can be formed during the combustion of any fuel, regardless of whether nitrogen is present in the fuel. Automobile exhaust and power plants that burn fossil fuels are major sources of NO_x. The main product of the reaction of nitrogen and oxygen is nitrogen monoxide (NO).

$$N_2(g) + O_2(g) \longrightarrow 2\,NO(g)$$

Nitrogen monoxide is then oxidized by atmospheric oxygen to nitrogen dioxide, an amber-colored gas that causes eye irritation and creates a brownish haze.

$$2\,NO_2(g) + O_2(g) \longrightarrow 2\,NO_2(g)$$

The two nitrogen oxides play a vital but nasty role in atmospheric chemistry.

At typical atmospheric levels, nitrogen oxides don't seem particularly dangerous. At high concentrations, however, NO reacts with hemoglobin, leading to oxygen deprivation. Such high levels seldom, if ever, result from ordinary air pollution, but they might be reached in areas close to industrial sources. More serious is the role of NO_x in smog formation. These gases also contribute to the fading and discoloration of fabrics and to acid rain when they form nitric acid (Section 13.6).

Ozone as an Air Pollutant

▲ Although ozone is a different chemical form of life-giving oxygen, it is much more reactive than O_2. Even low levels of ozone can cause significant damage to plants as well as to animal tissue, such as the human lung.

The ordinary oxygen in the air we breathe is made up of O_2 molecules. As was noted earlier, ozone is a form of oxygen that consists of O_3 molecules. Ozone and oxygen are **allotropes**, different forms of the same element, with different properties. Although one might think that having a third oxygen atom in the molecule would be something positive, the opposite is true. Ozone is quite reactive, with the result that it is highly toxic when inhaled. Ozone is a constituent of photochemical smog (Section 13.5). Yet ozone is also an important natural component of the stratosphere, where it shields Earth from life-destroying UV radiation. Ozone is a good example of a pollutant—a substance out of place in the environment. In the stratosphere, it helps make life possible. In the lower troposphere—the part we breathe—it makes life difficult.

Most U.S. urban areas have experienced ozone alerts. The American Lung Association estimates that over half of the U.S. population lives in counties that exceed recommended concentrations of ozone at some time. Episodes of high ozone levels correlate with increased hospital admissions and Emergency Department visits for respiratory problems. At low levels, it causes eye irritation. Repeated exposure can make people more susceptible to respiratory infection, cause lung inflammation, aggravate asthma, decrease lung function, and increase chest pain and coughing.

Ozone causes economic damage beyond its adverse effect on health. Its oxidizing properties cause rubber to harden and crack, shortening the life of automobile tires and other rubber items. Ozone also causes extensive damage to crops, especially cotton, peanuts, and soybeans.

Volatile Organic Compounds

Volatile organic compounds (VOCs) are organic substances that vaporize significantly at ordinary temperatures and pressures. They are major contributors to smog formation. There are many sources of VOCs, including trees, gasoline vapors and spills, incomplete combustion of fuels, and consumer products such as paints and aerosol sprays. Many VOCs are hydrocarbons.

Hydrocarbons are released from a variety of natural sources, such as the decay of plant matter in swamps. Only about 15% of all hydrocarbons found in the atmosphere are put there by people. In most urban areas, however, the processing and use of gasoline are the major sources of air-polluting hydrocarbons. Gasoline can evaporate anywhere it is used, contributing substantially to the total amount of hydrocarbons in urban air. This is why it is important to have a good seal on your gas tank cap and to not overfill and spill the liquid. Hydrocarbons from automobiles come largely from those vehicles with no pollution control devices or those with devices that are not operating properly. The automobile's internal combustion engine also contributes by exhausting unburned and partially burned hydrocarbons.

Hydrocarbons can act as pollutants even when they aren't burned. Certain hydrocarbons, particularly alkenes (Section 9.1), combine with oxygen atoms or ozone molecules to form aldehydes. Many aldehydes have foul, irritating odors. Another series of reactions involving hydrocarbons, oxygen, and nitrogen dioxide leads to the formation of peroxyacetyl nitrate (PAN).

$$CH_3\overset{\overset{\displaystyle O}{\|}}{C}-O-ONO_2$$
PAN

Ozone, aldehydes, and PAN are responsible for much of the destruction wrought by smog. They make breathing difficult and cause the eyes to sting and itch. People who already have respiratory ailments may be severely affected. The very young and the very old are particularly vulnerable.

The incidence of asthma has increased dramatically in recent years. Although it is unlikely that air pollution is the direct cause, medical studies show that pollutants at moderate levels can exert a small but measurable effect on the lung function and level of symptoms of asthmatic individuals. It seems more likely that pollutants and allergens interact to enhance the severity of asthma attacks rather than that they cause attacks.

A broad question to ask is how many individual automobiles can be used sustainably in a country or region. The efficiency of the auto is a factor in this discussion. Changes in U.S. standards for average mileage for cars were made in 2013. These include a fleet average of 35.5 mpg in 2016, which was slated to increase to 54.5 mpg in 2025 until the target value was reduced in 2018. Consider these data in the context of the number of vehicles, which is 821 per 1000 people in the United States, but only 587 in the United Kingdom and just 118 for China. Related to this is use of public transportation, which is much lower in the United States than many other countries. Buses and trains also produce the emissions discussed in this section, but the amount per person or per mile is usually much less.

SELF-ASSESSMENT Questions

1. Which of these is the largest source of carbon monoxide?
   a. transportation vehicles   b. volcanoes
   c. coal-burning power plants   d. home furnaces

2. Which of the following is *not* a problem caused by NO_x emissions?
   a. addiction
   b. formation of nitric acid, leading to acid rain
   c. formation of smog   d. eye irritation

3. Many volatile organic compounds (VOCs) have molecules composed of
   a. C and H only   b. C and N only
   c. N and H only   d. N and O only

4. Except in urban atmospheres, volatile organic compounds (VOCs) come mainly from
   a. automobiles   b. dry cleaning plants
   c. electric power plants   d. natural sources

5. Alkenes react with oxygen atoms to form
   a. alcohols   b. aldehydes   c. alkanes   d. ozone

6. Carbon monoxide is formed as a product of
   a. animal respiration
   b. removing SO_x from flue gas
   c. combustion of fossil fuels with excess O_2
   d. partial combustion of fossil fuels

7. Automobiles emit
   a. CFCs that cause ozone depletion
   b. O_3 directly
   c. CO_2, which reacts with hydrocarbons in sunlight to produce ozone
   d. NO_x and VOCs, which form other pollutants including ozone.

8. Which pollutant is likely to make breathing problems worse?
   a. CO   b. CO_2 c. NO   d. CH_4

9. Between 2003 and 2013, by approximately what percent did CO emissions decrease?
   a. 1%   b. 5%   c. 10%   d. 25%

Answers: 1. a; 2. a; 3. a; 4. d; 5. b; 6. d; 7. d; 8. a; 9. d

13.5 Photochemical Smog: Making Haze While the Sun Shines

Learning Objectives • Distinguish the origin of photochemical smog from the origin of sulfurous smog. • Describe the technologies used to alleviate photochemical smog.

Individually, hydrocarbons and nitrogen oxides are worrisome and have dramatic environmental impacts, including acid rain. Collectively, in the presence of sunlight, these pollutants can undergo a complex series of reactions that produces a very significant pollutant, **photochemical smog**, which is visible as a brownish haze.

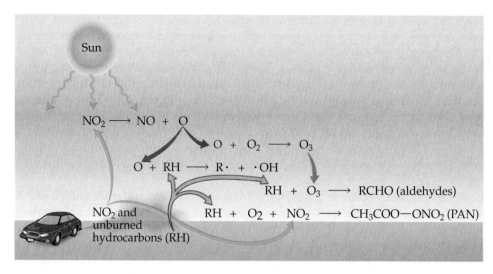

▲ **Figure 13.7** Some chemical processes involved in the formation of photochemical smog. Many other reactive intermediates have been omitted from this simplified scheme.

▲ Photochemical smog results from the action of sunlight on nitrogen oxides emitted from automobiles and other high-temperature combustion sources. A maroon or brownish haze, like that shown here, characterizes this type of smog.

Unlike sulfurous smog, which accompanies cold, damp air, photochemical smog usually occurs during dry, sunny weather. The "California climate" that has drawn so many people to the Los Angeles area is also the perfect setting for photochemical smog at any time of year. The principal culprits are unburned hydrocarbons and nitrogen oxides from automobiles. The chemistry of photochemical smog is exceedingly complex (Figure 13.7).

The development of air pollutants over time on a typical sunny summer day is shown in Figure 13.8. As traffic builds up in the early morning, NO enters the air in auto exhaust and reacts with O_2 in the air to form NO_2. Hydrocarbons enter the air in exhaust as unburned components of fuel and from fueling leaks and spills. Sunlight splits the NO_2 into NO and O atoms. The very reactive O atoms combine with O_2 to form O_3, ozone. Ozone levels continue to rise all day and then decrease after the sun sets because there are fewer reactant gases and little solar energy.

▶ **Figure 13.8** Concentrations of several urban air pollutants at different times during a typical sunny day. Hydrocarbons and NO are produced first. As they interact with sunlight, NO_2 and then ozone form.

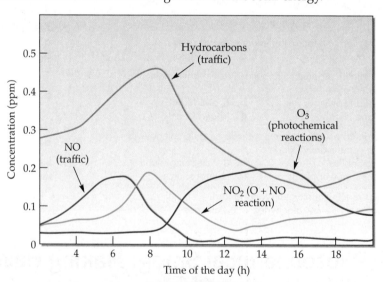

Solutions to Photochemical Smog

Photochemical smog requires the action of sunlight on nitrogen oxides and hydrocarbons. Reducing the quantity of any of these would diminish the amount of smog. It isn't likely that we would want to reduce the amount of sunlight, so let's focus on the other two.

Hydrocarbons (commonly called HC in air pollution reports) have many important uses as solvents. But if we could reduce the quantity of hydrocarbons entering the atmosphere, the amounts of aldehydes and PAN formed would also be reduced.

Improved design of storage and dispensing systems has decreased HC emissions from gasoline stations. Modified gas tanks and crankcase ventilation systems have reduced evaporative emissions from automobiles. Most of the reduction, however, has resulted from the use of **catalytic converters** to lower hydrocarbon and carbon monoxide emissions in automotive exhausts in two stages:

- The first stage employs a *reduction catalyst* made of platinum and rhodium to reduce NO_x emissions. Some of the carbon monoxide is used to reduce nitric oxide to N_2.

$$2\,NO(g) + 2\,CO(g) \longrightarrow N_2(g) + 2\,CO_2(g)$$

- The second stage in the catalytic converter employs an *oxidation catalyst*. It uses some of the remaining oxygen in the exhaust gas to oxidize the unburned HC and CO over a platinum and palladium catalyst to produce water and carbon dioxide.

Lowering the operating temperature of an engine also helps to reduce the quantity of nitrogen oxides emitted, but the engine then becomes less efficient in producing energy to propel the car. Running an engine on a richer (more fuel, less air) mixture lowers NO_x emissions but tends to raise CO and hydrocarbon emissions. A control system is needed to monitor the exhaust stream for optimum operation.

Catalytic converters convert more than 90% of HC, CO, and NO_x into less harmful CO_2, N_2, and water vapor. Over the past 40 years, their use has prevented more than 12 billion tons of harmful exhaust gases from entering Earth's atmosphere.

Many people now drive *hybrid vehicles* to save gasoline and to benefit the environment. A hybrid vehicle uses both an electric motor and a gasoline engine. In one variation, a small gasoline engine drives the vehicle most of the time and recharges the batteries that run the electric motor. The electric motor provides additional power when it is needed. Hybrid vehicles are much more efficient than conventional gasoline vehicles. Although these vehicles do provide a reduction in air pollution, a complete evaluation of their environmental impact must include the production, replacement, and recycling of the batteries. Researchers are working to develop more compact, more efficient batteries that generate less pollution in manufacture.

There are a few electric-only vehicles in production; more are being developed. At present, they have several limitations. Most electric vehicles have less range than cars with gasoline engines. They also take hours to recharge and have poorer acceleration than gasoline-powered vehicles. To attain widespread use, the batteries must be improved to provide a longer range and faster recharging. Standardization of charging hardware, more recharging stations, and more power plants to provide the electricity will also be required.

A sustainability factor for electric vehicles is the source of energy used to produce the electricity. For states such as California, which has a high percentage of renewable energy, the electric vehicle produces much less carbon dioxide than an internal combustion vehicle. For states with high use of coal, such as Ohio, there is still an advantage for electric vehicles, but the difference is small. For electric vehicles to be a net positive in terms of emissions, electricity produced by renewable methods must increase significantly.

SELF-ASSESSMENT Questions

1. Nitrogen oxides and hydrocarbons combine in the presence of sunlight to cause
   a. acid rain
   b. global warming
   c. an ozone hole
   d. photochemical smog

2. An amber haze over a city indicates the presence of
   a. HNO_3
   b. NO
   c. NO_2
   d. SO_2

3. Photochemical smog occurs mainly in
   a. cold, wet weather
   b. dry, sunny weather
   c. flat, dry areas
   d. snowy, mountainous areas

4. The main source of photochemical smog in most areas is
   a. automobiles
   b. electric power plants
   c. chemical factories
   d. solid waste burning

5. Removing NO_x from automobile exhaust requires a(n)
   a. bag filter
   b. electrostatic precipitator
   c. reduction catalyst
   d. wet scrubber

6. How long does it take for ozone concentrations to drop to one-half of their maximum value according to the time progression in Figure 13.8?
   a. 2.5 hours   b. 4 hours   c. 6 hours   d. 18 hours

Answers: 1, d; 2, c; 3, b; 4, a; 5, c; 6, a

13.6 Acid Rain: Air Pollution ⟶ Water Pollution

Learning Objectives • Name the air pollutants that contribute to acid rain. • List the major industrial and consumer sources of acid rain–producing pollutants.

We have seen how sulfur oxides are converted to sulfuric acid (Section 13.3) and nitrogen oxides to nitric acid (Section 13.2). If they are in the atmosphere, these acids fall on Earth as acid rain or acid snow, or are deposited from acid fog or adsorbed on particulates. **Acid rain** is defined as any form of precipitation having a pH less than 5.6. Rain with a pH as low as 2.1 and fog with a pH of 1.8 have been reported (Figure 13.9). These values are lower than the pH of vinegar or lemon juice.

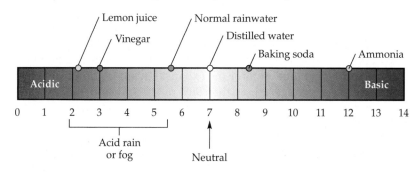

▲ **Figure 13.9** The pH of acid rain in relation to that of other familiar substances. Normal rainwater, if saturated with dissolved CO_2, has a pH of 5.6 because carbon dioxide reacts with water to make it acidic and not neutral. During thunderstorms, the pH of rainwater can be much lower because of nitric acid formed by lightning. Recall (from Chapter 7) that a solution is ten times more acidic than a second solution that has a pH value one unit higher. A decrease in pH from 5.6 to 4.6, for example, means an increase in acidity by a factor of 10. Two pH units lower means 100 times more acidic, and three pH units lower means 1000 times more acidic.

Acid rain is mainly due to sulfur oxides emitted from power plants and smelters and nitrogen oxides discharged from power plants and automobiles. These substances form acids that are often carried great distances before falling as rain or snow (Figure 13.10).

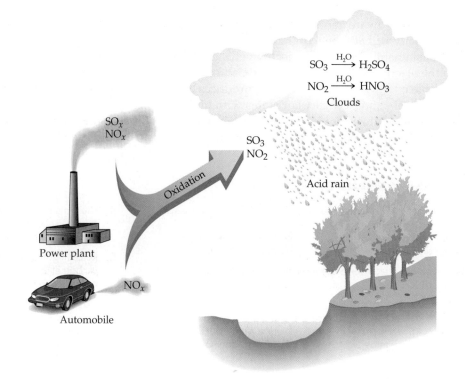

◀ **Figure 13.10** Acid rain. The two main substances responsible for acid rain are sulfur dioxide (SO_2) from power plants and nitric oxide (NO) from power plants and automobiles. These oxides are converted to SO_3 and NO_2, respectively, which then react with water to form H_2SO_4 and HNO_3. The acids then fall in rain, often hundreds of kilometers from their sources.

Acids corrode metals and can even erode stone buildings and statues. Sulfuric acid dissolves metals to form soluble salts and hydrogen gas.

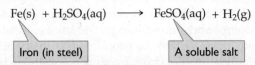

$$Fe(s) + H_2SO_4(aq) \longrightarrow FeSO_4(aq) + H_2(g)$$

Iron (in steel) A soluble salt

▲ (Left) As seen here, acid rain can damage artworks and decoration, especially those made of marble (calcium carbonate, which is readily soluble in acid). (Right) This shows the nineteenth-century decorations after they were restored to their original appearance.

The reaction shown here is oversimplified. For example, in the presence of water and oxygen (air), the iron is converted to rust (Fe_2O_3). One effect of these reactions is the increased concentration of metal ions in natural waters. This can lead to additional environmental problems when these ions are taken up by plants. The EPA has a website with "Quick Finder" links that will lead you to more information on topics such as acid rain, air, the Clean Air Act, climate change, ozone, and radon.

Marble or limestone buildings and statues disintegrate through a similar reaction with sulfuric acid that forms calcium sulfate, a slightly soluble, crumbly compound.

$$CaCO_3(s) + H_2SO_4(aq) \longrightarrow CaSO_4(aq) + H_2O(l) + CO_2(g)$$

Marble or limestone

SELF-ASSESSMENT Questions

1. Acid rain is defined as precipitation with a pH
  a. above 7.4
  b. above 7.0
  c. below 5.6
  d. below 7.0

2. Which of these reactions leads to acid rain?
  a. $C + O_2 \longrightarrow CO_2$
  b. $Cl_2 + H_2 \longrightarrow 2\,HCl$
  c. $N_2 + 3\,H_2 \longrightarrow 2\,NH_3$
  d. $S + O_2 \longrightarrow SO_2$

Answers: 1, c; 2, d

13.7 The Inside Story: Indoor Air Pollution

Learning Objectives • List the main indoor air pollutants and their sources.
• Explain where radon comes from and why it is hazardous.

Indoor air pollution can pose a major health concern. It can be serious in the United States but is often much worse in developing countries where people use wood, charcoal, coal, or dried dung as fuel for heating and cooking.

EPA studies on human exposure to air pollutants indicate that indoor air levels of many pollutants may be 2–5 times higher, and on occasion more than 100 times higher, than outdoor levels. Near high-traffic areas, the air indoors has the same carbon monoxide level as the air outside. In some areas, office buildings, airport terminals, and apartments have indoor CO levels that exceed government safety standards. Woodstoves, gas stoves, cigarettes, and unvented gas and kerosene space heaters are also sources of CO indoors.

A kitchen with a gas range often has levels of nitrogen oxides above U.S. government standards. Freestanding kerosene stoves produce levels of NO_x up to 20 times greater than those permitted by federal regulations for outdoor air. (These regulations do not apply to indoor air.)

Wood Smoke

Many people like the smell of smoke from burning wood or leaves, but wood smoke is far from benign. Most of the output is carbon dioxide and water vapor, but burning a kilogram of wood can yield 80–370 g of carbon monoxide plus smaller amounts of VOCs, aldehydes, acetic acid, particulates, hydrocarbons, and carcinogenic polycyclic aromatic hydrocarbons (PAHs). The odor of wood smoke may evoke memories of autumns past or singing around a campfire, but inhaling the smoke can lead to asthma attacks and possible long-term harm to the lungs.

Cigarette Smoke

Only about 15% of cigarette smoke is inhaled by the smoker; the rest remains in the air for others to breathe. More than 40 carcinogens have been identified among the 4000 or so chemical compounds found in cigarette smoke. In addition to the PAHs found in wood smoke, burning tobacco produces aromatic amines, nitrosamines, and even radioactive polonium-210 (from traces of naturally occurring uranium and thorium in tobacco fertilizers). The EPA even classifies secondhand smoke as a Class A carcinogen—known to cause cancer in humans. Smoke from just one burning cigarette in a room can raise the level of particulates above government standards.

The risk to nonsmokers from cigarette smoke is also well established. Medical research has shown that nonsmokers who regularly breathe secondhand smoke suffer many of the same diseases as active smokers, including lung cancer and heart disease. Women who have never smoked but live with a smoker have almost twice the risk of heart disease and of dying from lung cancer. Women who are exposed to secondhand smoke during pregnancy have a higher rate of miscarriages and stillbirths. Their children have decreased lung function and a greater risk of sudden infant death syndrome (SIDS). Children exposed to secondhand smoke are more likely to experience middle ear and sinus infections, asthma, colds, bronchitis, pneumonia, and other lung diseases. These health problems and others have caused many states and municipalities to ban smoking in most public places.

Radon and Its Dirty Daughters

A noble gas, radon is colorless, odorless, tasteless, and unreactive chemically—but it is radioactive. Radon-222 decays by alpha emission (Section 11.2) with a half-life of 3.8 days. The decay products, called **daughter isotopes**, are the major problem.

$$^{222}_{86}\text{Rn} \longrightarrow {}^{218}_{84}\text{Po} + {}^{4}_{2}\text{He}$$

When radon is inhaled, polonium-218 and other radioactive isotopes are trapped in the lungs, where they decay further to damage the tissues.

Radon is released naturally from soils and rocks, particularly granite and shale. The ultimate source is uranium atoms found in these materials. Radon is only one of several radioactive materials formed during the multistep decay of uranium. However, radon is unique in that it is a gas and escapes into the atmosphere. Outdoors, radon dissipates and presents no problems. However, a house built on a solid concrete slab or over a basement can trap the gas inside. Radon levels can reach several times the maximum safe level established by the EPA (0.15 Bq per liter of air). At five times this level, the hazard is thought to equal that of smoking two packs of cigarettes a day. Although the hazard has not been precisely determined, scientists at the National Cancer Institute estimate that radon may cause 20,000 lung cancer deaths a year.

Other Indoor Pollutants

Some people use kerosene heaters or unvented natural gas heaters to reduce heating costs in the winter. Heating only the occupied areas of a home seems like a good idea. However, when such heaters are not properly adjusted, carbon monoxide can be generated. Carbon monoxide from gas stoves, gas furnaces, gasoline generators, even from automobile exhaust in attached garages can contribute to the high levels of CO found in some houses. The EPA recommends a maximum of 9 ppm CO over an eight-hour period. Carbon monoxide from a poorly adjusted gas heater can easily exceed 30 ppm—continuously. Proper ventilation is key to preventing high levels of carbon monoxide.

Mold is an often-ignored pollutant in many homes. Leaky pipes or cool areas where water vapor condenses can lead to moisture problems, and mold will grow where there is moisture. Mold spores are a form of PM that can worsen asthma, bronchitis, and other lung diseases. Generally, mold is controlled by controlling moisture.

Electronic air cleaners are widely advertised, and may seem a convenient solution to indoor air pollution. Unfortunately, some of these cleaners *produce* pollution—ozone in particular. Ozone is quite reactive and can indeed reduce the levels of some pollutants by reacting with them. But in doing so, it can generate other pollutants, such as formaldehyde. The EPA advises the public "to use proven methods of controlling indoor air pollution. These methods include eliminating or controlling pollutant sources, increasing outdoor air ventilation, and using proven methods of air cleaning."

SELF-ASSESSMENT Questions

1. Even properly operated gas ranges in kitchens are a source of the indoor air pollutant
  a. CO   **b.** NO_x **c.** PAHs   **d.** VOCs

2. What pollutant can be generated by advertised indoor air electronic air cleaners?
  a. CO   **b.** ozone   **c.** VOCs   **d.** radon

3. Radon in homes comes from
  a. gas cooking ranges   **b.** smoke detectors
  c. leaking heater exhausts   **d.** soil and rocks

4. Which of the following is a radioactive radon daughter?
  a. cobalt-60   **b.** helium-4
  c. polonium-218   **d.** uranium-238

Answers: 1, b; 2, b; 3, d; 4, c

13.8 Stratospheric Ozone: Earth's Vital Shield

Learning Objectives • Explain the link between CFCs and depletion of the ozone layer. • Describe the consequences of stratospheric ozone depletion.

In the mesosphere (see Figure 13.1), some ordinary oxygen molecules are split into oxygen atoms by short-wavelength, high-energy UV radiation.

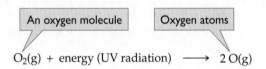

$$O_2(g) + \text{energy (UV radiation)} \longrightarrow 2\,O(g)$$

Some of these highly reactive atoms diffuse down to the stratosphere, where they react with O_2 molecules to form ozone molecules.

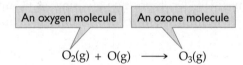

$$O_2(g) + O(g) \longrightarrow O_3(g)$$

The ozone in turn absorbs longer-wavelength—but still-damaging—UV radiation, shielding us from these harmful rays. In absorbing these photons, the ozone molecules are converted back to oxygen molecules and oxygen atoms in a reversal of the previous reaction.

$$O_3(g) + \text{energy (UV radiation)} \longrightarrow O_2(g) + O(g)$$

Undisturbed, the concentration of ozone in the stratosphere remains fairly constant due to this cyclic process. Over 300 billion tons of ozone are destroyed and created every day in this way in a very diffuse layer of the stratosphere about 20 km thick. However, the actual number of ozone molecules is relatively small. If the ozone were compressed to normal sea-level pressure, that layer would be only 3 mm thick! This delicate barrier protects all living things from harmful UV radiation. In recent decades, human activity has upset the balance, decreasing the protection from the ozone layer in some areas.

Chlorofluorocarbons and the Ozone Hole

The less ozone there is in the stratosphere, the more harmful UV radiation reaches Earth's surface. For humans, increased exposure to UV radiation can lead to skin cancer, cataracts, and impaired immune systems. The U.S. National Research Council

▲ In 1986, Susan Solomon, a chemist at the Oceanic and Atmospheric Administration in Boulder, Colorado, proposed a mechanism that explained the reason for greater ozone depletion over the poles by CFCs. Later experiments confirmed the mechanism and led to an international ban on the use of CFCs. Solomon was awarded a 1999 National Medal of Science for her work.

predicts a 2% to 5% increase in skin cancer for each 1% depletion of the ozone layer. Increased UV also reduces yields of many crops and can also decrease the population of phytoplankton in the oceans, disrupting the web of sea life and the oxygen cycle. (See Figure 13.3.)

The thickness of the ozone layer changes with latitude and also with the seasons. The cycling of ozone concentration is most prevalent over Antarctica, in part because the ozone-destroying reactions occur much more rapidly on the surfaces of ice crystals. During the extreme Antarctic winter (June–August), more clouds containing ice crystals are formed and ozone levels decrease. The ozone levels are lowest in September, when the increasing sunlight of the southern spring maximizes the decomposition reaction. In the 1970s, serious concerns were raised because the ozone layer was not recovering properly during the warm months. In 1974, Mexican chemist Mario Molina (1943–) and American chemist F. Sherwood Rowland (1927–2012) described a mechanism for the enhanced ozone depletion that implicated the presence of chlorofluorocarbons (CFCs) in the stratosphere. Their work was awarded the Nobel Prize in Chemistry in 1995.

CFCs are not found in nature but were developed to be used as the dispersing gases in aerosol cans and as refrigerants. At room temperature, CFCs are either gases or liquids with low boiling points. They are nearly insoluble in water and nearly chemically inert, making them ideal for many uses. However, once released into the atmosphere, the low reactivity means that they persist and diffuse to the upper stratosphere. There, above the protective ozone layer, energetic UV radiation cleaves a C—Cl bond and initiates a series of reactions that results in the net decomposition of ozone molecules:

$$CF_2Cl_2 + energy\,(UV\;light) \longrightarrow CF_2Cl\cdot + Cl\cdot$$

$$Cl\cdot + O_3 \longrightarrow ClO\cdot + O_2$$

$$ClO\cdot + O \longrightarrow Cl\cdot + O_2$$

Both products of the initial reaction are highly reactive free radicals (Section 4.10), species with an unpaired electron (indicated here by a dot). A chlorine atom reacts with an ozone molecule to produce O_2, ultimately forming another chlorine atom, which can break down yet another O_3 molecule. The second and third steps are repeated many times. The decomposition of a single CFC molecule in this catalytic *chain reaction* can result in the destruction of *thousands* of molecules of ozone. (See Section 9.3.)

International Cooperation

The United Nations has addressed the problem of ozone depletion through the 1987 Montreal Protocol, an international agreement enforcing the reduction and eventual elimination of the production and use of ozone-depleting substances. As a result, CFCs have been banned in the United States and many other countries, and the search for effective substitutes has been successful.

CFCs and other ozone-destroying compounds have been largely replaced by more benign substances. Hydrofluorocarbons (HFCs), such as CH_2FCF_3, and hydrochlorofluorocarbons (HCFCs), such as $CHCl_2CF_3$, are some of these alternatives. The C—H bonds in these compounds are susceptible to reactive species, so the molecules generally break down before reaching the stratosphere. For every solution, it seems there is a new problem; CFCs, HFCs, and HCFCs are all greenhouse gases (Section 13.9). Also, HFCs and HCFCs can break down while in use, corroding refrigeration components.

Due to the Montreal Protocol, the concentration of CFCs in the stratosphere peaked in 1998. Levels have declined slowly since then, but they still remain high enough to cause significant ozone destruction until 2025 or later. The size of the ozone hole over Antarctica (Figure 13.11) varies from year to year with temperature and other meteorological factors. A complete recovery from CFC-caused damage is not expected for decades.

October 2017

▲ **Figure 13.11** The ozone hole (blue) over Antarctica is at its largest in September. This is an image from the hole's largest value in October of 2017. The area is approximately equal to that of North America. This is 20% smaller than the record low from 2000 but still far from pre-CFC days.

SELF-ASSESSMENT Questions

1. In the stratosphere, some O_2 is converted to
   a. acid rain
   b. ammonia
   c. ozone
   d. water

2. Ozone is an allotrope of
   a. carbon
   b. chlorine
   c. nitrogen
   d. oxygen

3. The pollutant that is harmful at Earth's surface yet protects life when in the stratosphere is
   a. carbon monoxide
   b. chlorine
   c. oxygen
   d. ozone

4. In the mesosphere, oxygen atoms are produced by
   a. breakdown of O_3 molecules
   b. breakdown of H_2O molecules
   c. breakdown of H_2O_2 molecules
   d. splitting of O_2 molecules

5. Which portion of the sun's radiation spectrum is absorbed by ozone?
   a. blue
   b. infrared
   c. violet
   d. ultraviolet

6. Chlorofluorocarbons harm the ozone layer by
   a. adding more ozone
   b. blocking UV radiation
   c. destroying oxygen molecules
   d. destroying ozone molecules

Answers: 1, c; 2, d; 3, d; 4, d; 5, d; 6, d

13.9 Carbon Dioxide and Climate Change

Learning Objectives • List the important greenhouse gases, and describe the mechanism and significance of the greenhouse effect. • Describe some strategies for reducing the amount of CO_2 released into the atmosphere.

Every engine and factory that burns coal or petroleum products produces carbon dioxide. Analysis of air bubbles trapped in ice cores, which date back 800,000 years, indicated that the concentration of CO_2 in the atmosphere did not exceed 300 ppm in that time period. Values rose above 300 ppm in the early twentieth century, exceeded 400 ppm for the first time in 2013, and reached 408 ppm in 2018. Careful analysis indicates that the main cause is the burning of fossil fuels. We are turning huge amounts of solid and liquid carbon-containing materials, stored in the Earth for millions of years, into a gaseous form with serious consequences.

Carbon dioxide is a natural component of the environment and is not considered toxic. In fact, we exhale this gas with every breath. But its effect on humans, other animals, and plants is starting to become significant, and projections of the future are cause for great concern. In 2007, the U.S. Supreme Court decided that carbon dioxide should be considered an air pollutant for the purposes of environmental regulations.

The sun radiates many different types of radiation, largely visible, infrared, and ultraviolet. About half of this energy is either reflected or absorbed by the atmosphere. The light that gets through (mostly visible) acts to heat the surface of Earth.

The three main constituents of the atmosphere—oxygen, nitrogen, and argon—do not absorb infrared energy. However, carbon dioxide and some other gases, including water, produce a **greenhouse effect** (Figure 13.12). These gases are transparent, so visible light from the sun passes through the atmosphere to warm the Earth's surface. Warm surfaces give off infrared photons. This infrared energy radiates toward space, but greenhouse gases in the atmosphere absorb these photons, trapping some of the energy near the surface. This makes the air, water, and land warmer than they would be without these gases.

Some greenhouse effect is necessary to life. Without an atmosphere, all of the radiated heat would be lost to space, and the Earth would be much colder. In fact, based on the distance from the sun, it is estimated that the Earth's average temperature would be a cold −18°C. which means that ice would cover much of the surface.

▶ **Figure 13.12** The greenhouse effect. Sunlight passing through the atmosphere is absorbed by the surface of the Earth, causing it to warm. The warm surface reemits energy as infrared radiation. Some of this radiation is absorbed by CO_2, H_2O, CH_4, and other gases and therefore retained in the atmosphere.

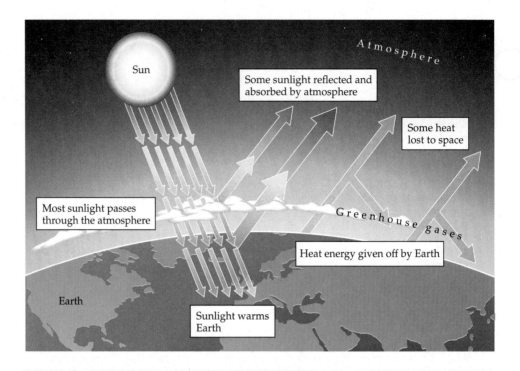

 EXAMPLE 13.1 CO_2 Stoichiometry

Many reports on emissions of carbon dioxide quantify the amount of the emissions in gigatons of carbon. This value does not include the mass of oxygen in the emitted gas.

If 4.0×10^{10} g of carbon from all sources are released by human actions in a year, how many grams of carbon dioxide are produced?

$$C(s) + O_2(g) \longrightarrow CO_2(g)$$

Solution

We have the mass of carbon and want to find the mass of CO_2 produced from it, so this is a stoichiometry problem (See Section 5.4)

We must convert the mass of carbon to moles.

$$4.0 \times 10^{10} \, \cancel{g \, C} \times \frac{1 \text{ mole C}}{12.0 \, \cancel{g \, C}} = 3.3 \times 10^9 \text{ mol C}$$

For the reaction, the mole relationship is 1 mol of C is equivalent to 1 mol of CO_2.

$$3.3 \times 10^9 \, \cancel{\text{mol C}} \times \frac{1 \text{ mol } CO_2}{1 \, \cancel{\text{mol C}}} = 3.3 \times 10^9 \text{ mol } CO_2$$

Finally we use the molar mass of CO_2 to determine the mass of CO_2 produced.

$$3.3 \times 10^9 \, \cancel{\text{mol } CO_2} \times \frac{44 \text{ g } CO_2}{1 \, \cancel{\text{mol } CO_2}} = 1.5 \times 10^{11} \text{ g } CO_2$$

The mass of CO_2 is almost four times the mass of C that was emitted, which is reasonable; one CO_2 is almost four times the mass of a C atom.

⟩ **Exercise 13.1A**
If 0.10% of the carbon produced by humans annually forms CO instead of CO_2, what mass in kg of CO is formed?

⟩ **Exercise 13.1B**
Without doing detailed calculations, determine whether CH_4 or CH_3OH will produce more CO_2 when 1 kg is burned. Explain your reasoning.

EXAMPLE 13.2 CO$_2$ Emissions from a Car

This example provides a perspective for the emissions from driving a car. What mass of CO$_2$ is produced by burning 2670 g (1 gal) of gasoline? Use C$_8$H$_{18}$ (114.0 g/mol) to represent gasoline.

$$2\,C_8H_{18}(l) + 25\,O_2(g) \longrightarrow 18\,H_2O(g) + 16\,CO_2(g)$$

Solution

The problem gives us a mass of a reactant, and we are to find the mass of a product, so this is a second stoichiometry problem but with a few more aspects to consider.

We begin by converting the mass of the given substance, C$_8$H$_{18}$, to moles.

$$2670\ \text{g}\ C_8H_{18} \times \frac{1\ \text{mol}\ C_8H_{18}}{114.0\ \text{g}\ C_8H_{18}} = 23.4\ \text{mol}\ C_8H_{18}$$

Next, the balanced equation tells us that 2 mol of C$_8$H$_{18}$ produces 16 mol of CO$_2$.

$$23.4\ \text{mol}\ C_8H_{18} \times \frac{16\ \text{mol}\ CO_2}{2\ \text{mol}\ C_8H_{18}} = 187\ \text{mol}\ CO_2$$

Finally, we use the molar mass of CO$_2$ to find the mass of CO$_2$ produced.

$$187\ \text{mol}\ CO_2 \times \frac{44.0\ \text{g}\ CO_2}{1\ \text{mol}\ CO_2} = 8240\ \text{g}\ CO_2$$

The mass of CO$_2$ is greater than the mass of gasoline consumed because the carbon from the gasoline combines with oxygen (much heavier than hydrogen).

> **Exercise 13.2A**
Calculate the mass in grams of water vapor produced from burning 2670 g (1 gal) of gasoline.

> **Exercise 13.2B**
Calculate the mass in kilograms of carbon dioxide produced by a class of 60 students in one week if each student burns 10 gal of gasoline in a week.

Greenhouse Gases and Global Warming

The present problem is an *enhanced* greenhouse effect caused by rapidly increasing concentrations of carbon dioxide and other greenhouse gases in the atmosphere. This enhancement is leading to **global warming** (Figure 13.13), a rise in Earth's average temperature. Indeed, periods of high CO$_2$ concentration in Earth's past have correlated with increased global temperatures.

About 36.6 billion mt of carbon dioxide were emitted into the atmosphere in 2017, a total that has risen almost every year since 1960. Most of this comes from the burning of fossil fuels as determined by the isotopic analysis of the carbon. About 57% of this CO$_2$ is removed, mostly by plants, soil, and the oceans. The remaining 43% causes a net increase of about 2 ppm CO$_2$ per year.

Methane, CH$_4$, also contributes to the greenhouse effect. The concentration of CH$_4$ in the atmosphere increased from about 0.7 ppm in 1750 to about 1.8 ppm today. Although this is much lower than CO$_2$ concentrations, CH$_4$ is about 25 times more effective than CO$_2$ at trapping heat. Most atmospheric methane comes from natural wetlands, where CH$_4$-producing bacteria decompose organic material. Similar reactions in landfills are the largest human-related source of CH$_4$ in the United States. Other sources include cattle and termites, which produce methane during their normal digestion. Recently, the process of fracking has produced large quantities of methane. One effect has been a decrease in the price of this gas used to heat many homes. However, there is concern that the drilling process leaks CH$_4$ to the atmosphere.

3 How do greenhouse gases keep the Earth warm?

Some polar molecules (recall Section 4.12) act as greenhouse gases because they strongly absorb infrared radiation (IR). Most of the radiation absorbed from the sun is in the form of visible light, and most of the radiation emitted by the Earth is IR. The more polar the molecule, the more readily it can absorb IR and cause warming of the Earth. This allows our planet to have a temperature that is reasonable for living systems.

▶ **Figure 13.13** Carbon dioxide content of the atmosphere and surface temperature variations, which are clearly correlated, as seen in this figure. The significant increase in carbon dioxide corresponds with the impact of the Industrial Revolution. Ice cores indicate that the atmospheric concentration of carbon dioxide did not exceed 300 ppm for at least 800,000 years.

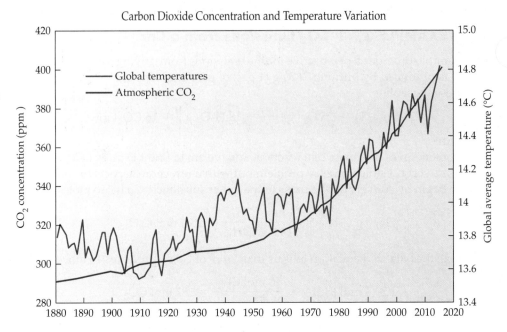
Carbon Dioxide Concentration and Temperature Variation

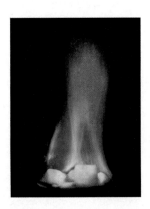

▲ "Methane ice" or methane clathrate consists of methane molecules trapped in spaces between water molecules in ice. As the ice melts, the methane is liberated and can be burned. Methane ice trapped in permafrost may give rise to a serious global warming problem in the future. (See text.)

Another source of methane is "methane ice" (methane clathrate), in which CH_4 molecules are physically trapped in the spaces in solid ice. There are large deposits of methane ice, mainly in some ocean floors. It is thought that global warming could cause some of this ice to melt and release its methane. This would increase global temperatures further, with more release of methane and even greater temperature increases. Although unlikely within our lifetimes, there is evidence that such a catastrophic event contributed to the Permian–Triassic extinction event of 250 million years ago, when over 90% of lifeforms became extinct.

Molecules that are polar or have moderate-to-strong dispersion forces are very effective greenhouse gases. CFCs are 5000–14,000 times and HCFCs 140–11,700 times as effective as CO_2 at holding heat in Earth's atmosphere. Water is only about one-tenth as efficient at trapping the heat, but when water is present in significant amounts, it has a noticeable effect. For instance, cloudy winter nights tend to be warmer than clear winter nights due to the presence of significant amounts of water in the clouds, which act to trap the heat close to the Earth's surface.

Predictions and Consequences

A massive study by the United Nations' Intergovernmental Panel on Climate Change (*Fifth Assessment Report: Climate Change 2013–14*) provided a comprehensive review of past changes and predictions of future changes. The global average surface temperature increased about 0.94 °C from 1880 to 2016.

Some areas, such as Russia and other northern countries, may well benefit from a warmer climate. However, many of the regions facing the greatest risks are among the world's poorest nations.

Predictions of future warming are complicated by the many natural variables (clouds, volcanoes, El Niño weather patterns, etc.) as well as the varying estimates of future greenhouse gas emissions, but the IPCC study predicts an increase of 1 °C to 6 °C by 2100. This may seem small, but the ramifications of such warming can be severe because so many of our physical and biological systems are attuned to early-twentieth-century conditions. This includes the locations of populations, which tend to congregate near bodies of water. Additionally, plants and animals have life cycles that are acutely sensitive to environmental factors such as temperature.

Another important factor is that the increases will not be uniform around the Earth. For example, in just the last 30 years, the temperatures of states in the U.S. Northeast have increased by 1.4 °C, while some states in the Southwest have experienced temperature changes of as much as 1.9 °C.

So far, some predictions made by models have underestimated the actual changes, particularly at high latitudes. Losses from the ice sheets of Greenland and Antarctica have exceeded expectations. March 2017 showed the lowest extent of Arctic ice ever seen, with 2018, 2016, and 2015 immediately following. If the overall trend continues, the Arctic Ocean may be ice-free in summers by 2030.

Sea-level rise is a consequence of the warming and expansion of ocean waters, which has caused about 40% of the change. The remaining increase is from the melting of land-based ice from Greenland, Antarctica, and glaciers. However, the melting of Arctic Sea ice is not a factor in sea-level rise.

An important effect of less Arctic ice is the reduced reflection of sunlight back into space. Arctic waters absorb more solar energy than does ice, which causes warming. This in turn leads to more ice melting and creates what is called a positive feedback effect.

The polar ice caps need not melt completely for global warming to create problems. The oceans have risen 19 cm since 1901, at an average rate of 0.17 cm/year. Again, this rise may seem trivial, but even slight rises in ocean level increase tides and result in higher, more damaging storm surges. The change in sea level is now 0.32 cm/year—almost double the long-term average. The IPCC report predicts an additional rise of 26–98 cm by 2100. The IPCC earned a share of the 2007 Nobel Peace Prize for its work on climate change.

Mitigation of Global Warming

Scientists know in broad outline what we need to do to alleviate global warming: Technologically advanced countries must make quick and dramatic cuts in emissions. They must also transfer technology to the rest of the world so that countries can develop their economies without heavy use of coal-fired power plants. The problems involved in implementing these ideas are also well known: Can we make such rapid cuts? Are governments and individuals willing to tackle these issues?

We discuss a variety of alternatives to fossil fuels for energy production in Chapter 15. However, most scientists realize that no single alternative technology—solar power, nuclear power, wind power, or biofuels—will solve the problem of increasing CO_2 emissions. We must pursue several strategies—especially conservation, which pays for itself in lower costs.

More energy-efficient cars, appliances, and home heating and cooling systems will help somewhat, as will general energy conservation. Perhaps as much as one-third of the needed reductions in emissions can come from such measures. Large-scale plans to capture and sequester CO_2 are under development, but they are quite costly and energy-intensive. **Carbon sequestration** is the collection and transportation of CO_2 from large emission sources, and storing it in underground reservoirs. Sequestration also involves removal of CO_2 from the atmosphere by agricultural practices such as planting trees and stopping deforestation. All these alternatives are difficult, but without effective carbon sequestration it will be almost impossible to keep CO_2 levels at values necessary to avoid major negative impacts.

A few nations have moved forward in reducing or capping CO_2 emissions, but overall there has been little or no progress. China is building coal-fired power plants at a breakneck rate—in fact, since 2008, China has become the largest emitter, replacing the United States. But to make a realistic comparison, the United States emits 2.5 times as much CO_2 per person than does China.

One promising factor has been the greater availability and reduced cost of methane. For each kilogram of CO_2 emitted, methane can produce 50% more electricity than coal. It also burns much cleaner than coal. The greater availability of methane is primarily due to fracking, which does have limitations in terms of leaks and some local reports of increases in earthquakes. (See Chapter 15.)

To consider perhaps the most important consequences of climate change, the WHO estimates that Earth's warming climate contributes to more than 150,000 deaths and 5 million illnesses each year and predicts that this toll could double by 2030.

You can do your own experiment to simulate melting of the floating Arctic ice cap. Place some water and two ice cubes in a glass. Mark the water level. Has the level changed after the ice has melted? To simulate melting of the Antarctic or Greenland glaciers, place some rocks in a glass and fill the glass with water to about a cm below the top of the rocks. Place two ice cubes on top of the rocks and mark the water level. Has the water level in the glass changed after the ice has melted?

4 Is global warming caused by the ozone hole?

The loss of some ozone in the stratosphere has an overall cooling effect on the Earth, but the amount is quite small. Most global warming is due to the recent large increases in greenhouse gas emissions.

According to data published in the journal *Nature*, a warmer climate is driving up rates of malaria, malnutrition, and diarrhea. These effects are in addition to other emissions, such as particulates and mercury, that come from burning coal.

SELF-ASSESSMENT Questions

1. What type of radiation is trapped on Earth's surface by the greenhouse effect?
   a. gamma   **b.** infrared
   c. ultraviolet   **d.** visible

2. During the twentieth century, the global average air temperature near the Earth's surface increased about
   a. 0.5 °C   **b.** 0.85 °C
   c. 1.25 °C   **d.** 2.25 °C

3. The main cause of recent increased global temperatures is
   a. CO_2 from factories, power plants, and automobiles
   b. Earth's orbital eccentricities
   c. variations in the sun's output
   d. water vapor

4. When CO_2 dissolves in the oceans, it
   a. decreases the acidity of the seawater
   b. forms an acid that dissolves shells of sea creatures
   c. raises the temperature of the ocean
   d. reacts with Ca^{2+} ions to form $CaCO_3$

5. To combat global warming, people should
   a. drain wetlands
   b. heat with charcoal
   c. pave dirt roads
   d. reduce the use of fossil fuels

6. If there were no greenhouse gases in our atmosphere, the average temperature of the Earth would be
   a. below the freezing point of water
   b. between the freezing point and the boiling point of water
   c. above the boiling point of water

7. In addition to CO_2 and methane, which gas is also an important greenhouse gas?
   a. Ar   **b.** NO   **c.** N_2O **d.** HCl

8. Atmospheric carbon dioxide concentrations increased by approximately what percent in the twentieth century?
   a. 1%   **b.** 10%   **c.** 25%   **d.** 50%

Answers: 1. b; 2. b; 3. a; 4. b; 5. d; 6. a; 7. c; 8. c

13.10 Who Pollutes? Who Pays?

Learning Objective • List the EPA's criteria pollutants and the major air pollutants that come mainly from automobiles and mostly from industry.

The EPA lists six *criteria pollutants*, so called because scientific criteria are employed to determine their health effects. Since the first Earth Day in 1970, nationwide emissions of the six criteria pollutants have declined dramatically (Table 13.3). These improvements in air quality have occurred even though the U.S. population increased by one-third and both gross domestic product and vehicle miles traveled tripled.

TABLE 13.3 Ambient Air Quality Standards for Six Criteria Air Pollutants

Pollutant	Limit
Carbon monoxide	
8 h average	9 ppm
1 h average	35 ppm
Nitrogen dioxide	
Annual average	0.053 ppm
Ozone[a]	
8 h average	0.075 ppm
Particulate matter	
PM12, 24 h average	150 $\mu g/m^3$
PM2.5, 24 h average	35 $\mu g/m^3$
Sulfur dioxide	
3 h average	0.50 ppm
1 h average	0.075 ppm
Lead	
3 month average	0.15

[a]Formed from precursor volatile organic compounds (VOCs) and NO_x.

5 **What does smog testing of vehicles actually test?**

Tests vary depending on the location, but most measure unburned hydrocarbons and carbon monoxide emissions concentrations. Unburned hydrocarbons may come from the tailpipe or may arise from evaporation or leakage elsewhere in the fuel system. Some states and cities also test for nitrogen oxides or particulates.

Table 13.4 lists the main sources of seven major air pollutants along with their health and environmental effects. Note that motor vehicles are the source of nearly one-half (by mass) of all air pollutants. Transportation accounts for about 85% of urban CO emissions, 40% of hydrocarbon emissions, and 40% of NO_x emissions. From the early twentieth century through the mid-1970s, cars were also a major source of lead in the air in the United States. The compound *tetraethyllead* [TEL, $(CH_3CH_2)_4Pb$] was added to gasoline during those years. In 1976, the United States began phasing out the use of TEL in automobile fuels. Now the major sources of lead in the air are industrial processes and nonroad equipment. Nonetheless, because most transportation in the United States is by private automobile, cars are a major source of air pollution.

TABLE 13.4 Seven Major Air Pollutants

Pollutant	Formula or Symbol	Major Sources	Health Effects	Environmental Effects
Carbon monoxide	CO	Motor vehicles	Interferes with oxygen transport; contributes to heart disease	Slight
Hydrocarbons	C_nH_m	Motor vehicles; industry; solvents	Narcotic at high concentrations; some aromatics are carcinogens	Precursors of aldehydes, PAN
Sulfur oxides	SO_x	Power plants; smelters	Irritates respiratory system; aggravates lung and heart diseases	Reduces crop yields; precursors of acid rain, SO_4^{2-} particulates
Nitrogen oxides	NO_x	Power plants; motor vehicles	Irritates respiratory system	Reduces crop yields; precursors of ozone and acid rain; produce brownish haze
Particulate matter	PM	Industry; power plants; dust from farms and construction sites; mold spores; pollen	Irritates respiratory system; synergistic with SO_2; contains adsorbed carcinogens and toxic metals	Impairs visibility
Ozone	O_3	Secondary pollutant from NO_2	Irritates respiratory system; aggravates lung and heart diseases	Reduces crop yields; kills trees (synergistic with SO_2); destroys rubber and paint
Lead	Pb	Motor vehicles; smelters	Toxic to nervous system and blood-forming system	Toxic to all living things

On the other hand, most PM comes from power plants (about 40%) and industrial processes (about 45%). Similarly, more than 80% of SO_x emissions come from power plants, with an additional 15% coming from other industries. Power plants alone contribute about 55% of NO_x emissions. Who uses electricity from power plants? We all do. A coal-fired plant must burn 275 kg of coal to light an ordinary 100 W bulb for one year.

What is the worst pollutant? Carbon monoxide is produced in huge amounts and is quite toxic. However, it is deadly only in concentrations approaching 4000 ppm. Its contribution to cardiovascular disease, by increasing stress on the heart, is difficult to measure. The WHO rates SO_x as the worst pollutants. They are powerful irritants, and according to WHO, people with respiratory illnesses are more likely to die from exposure to SO_x than from exposure to any other kind of pollutant. Sulfur oxides and acids formed from them have been linked to more than 50,000 deaths per year in the United States.

We all share the responsibility for pollution. In the words of Walt Kelly's comic strip character Pogo, "We have met the enemy, and he is us." But we can all be part of the solution by conserving fuel and electricity. Many utility companies now offer suggestions and incentives for saving energy. Some even give free analyses of home energy use.

GREEN CHEMISTRY Putting Waste CO$_2$ to Work
Philip Jessop and Jeremy Durelle, *Queen's University*

Principles 1, 5, and 7

Learning Objective • Describe how to use waste CO$_2$ to lessen the environmental impact of industrial processes.

The alarming rate of carbon dioxide (CO$_2$) emission into the atmosphere is causing the average temperature of the oceans, and even the air, to steadily increase. Because humans have burned roughly 550 billion tons of carbon since the Industrial Revolution, the concentration of atmospheric CO$_2$ is 100 parts per million (ppm) higher than it was 200 years ago. Despite this, energy demand is on the rise, which means we are burning more carbon faster. While CO$_2$ isn't the only greenhouse gas produced when burning carbon, it gets the most attention because it is responsible for 60% of the greenhouse effect (see Section 13.9).

To minimize CO$_2$ emissions, power plants can separate the CO$_2$ from the waste gas before it is released into the atmosphere. This is known as scrubbing and involves cooling the gas and treating it with a liquid solvent that binds to CO$_2$ (such as aqueous solutions of ammonia or ethanolamine). As you can imagine, scrubbing the world's power plants would produce a lot of waste CO$_2$. While most of that CO$_2$ would have to be sequestered somewhere, such as in depleted oil or gas reservoirs, could some of it be used to benefit society and the environment? While waste CO$_2$ is renewable (Principal 7) and is already used for making products such as urea or carbonated beverages, could some of it be used to make industrial processes greener?

As it turns out, we can use waste CO$_2$ to make a number of processes more environmentally friendly. For example, a very new kind of solvent, called switchable-hydrophilicity solvents (SHS), can exist in either of two forms that can be interconverted by the addition of CO$_2$. SHS are amines. In the absence of CO$_2$, they are hydrophobic solvents, meaning that, like hexane, they won't mix with water. However, in the presence of CO$_2$, SHS become hydrophilic, meaning that, like methanol, they do mix with water. (Equation 1 shows how this interconversion happens.)

$$NR_3 + H_2O + CO_2 \rightleftharpoons [NR_3H^+][HCO_3^-]$$

Hydrophobic   Hydrophilic
form of SHS   form of SHS

▲ **Equation 1.** This chemical reaction interconverts the hydrophobic and hydrophilic forms of the SHS. Adding CO$_2$ shifts the equilibrium to the hydrophilic form, while removing CO$_2$ shifts the equilibrium to the hydrophobic form.

SHS can be used, for example, to extract oily materials from insoluble solids such as removing oil from oil sands (a naturally occurring mixture of sand, clay, and heavy oil), extracting

vegetable oil from soybeans, and separating and recycling the oil and plastic in used bottles of motor oil. Most extractions of oily materials are done with a conventional (non-switchable) organic solvent. The mixture to be extracted (such as soybeans) is soaked in the conventional solvent. After the solids are removed by filtration, you are left with a solution of oil in solvent. The solvent is then removed by distillation. Unfortunately, that last step will only work if the solvent is volatile, which means it is also likely to be flammable, smog-forming, and an inhalation risk to workers. Switchable-hydrophilicity solvents, because they can be removed from product and recycled without distillation, do not need to be volatile; they therefore offer a potentially safer alternative (Figure 1) and Principle 5. The mixture to be extracted is soaked in the SHS. After the solids are removed by filtration, the SHS is extracted from the product by carbonated water. Removing the CO$_2$ from the water causes the SHS to come out of the water, so it can be recovered and reused.

While use of CO$_2$ isn't a solution to global warming, the utilization of waste CO$_2$ enables us to modify industrial processes in unconventional ways that can decrease environmental impact. That is, after all, the purpose of green chemistry.

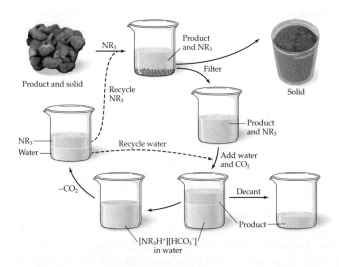

▲ **Figure 1.** The extraction of product (non-switchable oil) from a product-solid mixture using an SHS. Waste carbon dioxide from power plants can be used to make industrial processes such as product extraction from a product-solid mixture greener.

Paying the Price

Air pollution costs us tens of billions of dollars each year. It wrecks our health by causing or aggravating bronchitis, asthma, emphysema, and lung cancer. It destroys crops and sickens and kills livestock. It corrodes machines and blights buildings. Replacing these items takes more energy and produces more pollution—along with carbon dioxide emissions that have climate-change effects.

Elimination of air pollution will be neither cheap nor easy. It is especially difficult to remove the last fractions of pollutants. Cost curves are exponential, soaring toward infinity as pollutants approach zero. For example, if it costs $200 per car to reduce emissions by 50%, it usually costs about $400 to reduce them by 75%, $800 to reduce them by 87.5%, and so on. It would be extremely expensive to reduce emissions by 99%. We can have cleaner air, but how clean depends on how much we are willing to pay.

What would we gain by getting rid of air pollution? Certainly it is nice to see the clear blue sky and stars at night, and to breathe clean, fresh air. But in the long run, the stakes are much higher: possibly the survival of *homo sapiens*.

SELF-ASSESSMENT Questions

1. The U.S. agency responsible for gathering and analyzing air pollution data is the
   a. EPA   **b.** FAA   **c.** FCC   **d.** FDA

2. Air pollutants have decreased significantly since 1970 because
   a. there are fewer electric power plants
   b. there are fewer people
   c. there are more pollution controls
   d. we are planting more trees

3. The largest source of carbon monoxide pollution is
   a. electric power plants   **b.** coal mining
   c. motor vehicles   **d.** oil refineries

4. Electric power plants are the largest source of
   a. CO pollution   **b.** ozone
   c. SO_2 pollution   **d.** VOCs

5. According to the World Health Organization, the worst pollutant in terms of health effects is
   a. CO   **b.** ozone   **c.** NO_x **d.** SO_x

6. The largest sources of NO_x pollution are
   a. electric power plants and motor vehicles
   b. electric power plants and industry
   c. industry and motor vehicles
   d. smelters and motor vehicles

7. The largest sources of particulate pollutants are
   a. electric power plants and motor vehicles
   b. electric power plants and industry
   c. industry and motor vehicles
   d. smelters and motor vehicles

Answers: 1, a; 2, c; 3, c; 4, c; 5, d; 6, a; 7, b

Summary

Section 13.1—The layers of Earth's **atmosphere** differ in temperature, pressure, and composition. At sea level dry air is a mixture of about 78% nitrogen, 21% oxygen, and 1% argon by volume, plus small amounts of carbon dioxide and trace gases. Moist air can contain up to 4% water vapor.

Section 13.2—Animals and most plants cannot use atmospheric nitrogen unless it has undergone **nitrogen fixation** (combination with other elements). Nitrogen goes from the air into plants and animals and eventually back to the air via the **nitrogen cycle**. Oxygen from the air is involved in oxidation (of plant and animal materials) and is eventually returned to the air via the **oxygen cycle**. In the stratosphere, diatomic oxygen—O_2—absorbs UV radiation and is converted to ozone (O_3) and then changed back to diatomic oxygen. A **temperature inversion** occurs when a layer of cooler air is trapped under a layer of warmer air. The cooler air can become quite polluted in a short time.

Section 13.3—A **pollutant** is a substance that is in the wrong place or time and causes problems. Natural air pollution has existed for millions of years and includes dust storms, noxious gases from swamps, and ash and sulfur dioxide from erupting volcanoes. Today air pollution is more complex. Air pollution is a global problem because pollutants from one geographic area often migrate to another. The main air pollutants introduced by human activity are the smoke and gases produced by the burning of fuels and the fumes and particulates emitted by factories. **Smog** (*smoke* + *fog*) is a general term for this type of pollution. **Industrial smog** is produced in cold, damp air by excessive burning of fossil fuels. It consists of sulfur dioxide and an **aerosol** or suspension of **particulate matter (PM)**, or solid or liquid particles of greater-than-molecular size, such as sulfuric acid, soot, and fly ash from coal. The pollutants that make up industrial smog can act synergistically to cause severe damage to living tissue. PM can be removed from smokestack gases using an **electrostatic precipitator**, which charges the particles, or with a **wet scrubber**, which passes the gases through water that contains chemicals to trap or destroy the pollutants. Sulfur dioxide can be scrubbed from stack gasses using limestone to form calcium sulfite, which can then be oxidized to useful calcium sulfate.

Section 13.4—Complete combustion of gasoline produces carbon dioxide and water, but combustion is always incomplete. Carbon monoxide is one product of incomplete combustion, and three-fourths of the carbon monoxide produced results from transportation. Carbon monoxide ties up hemoglobin in the blood so that it cannot transport oxygen. Nitrogen oxides come from automobile exhaust gases and from power plants that burn fossil fuels; they contribute greatly to smog. The amber color of nitrogen dioxide causes the brownish tint to the haze often seen over certain large cities. Ozone is an **allotrope** or different chemical form of oxygen, formed in a complex reaction in the troposphere. It is very hazardous to breathing and is a significant component of urban pollution. **Volatile organic compounds (VOCs)** from evaporation of gasoline and incomplete combustion of fuels are major contributors to smog. Hydrocarbons can be pollutants in themselves and can react to form other pollutants such as aldehydes and peroxyacetyl nitrate (PAN), both of which are very irritating to tissues.

Section 13.5—**Photochemical smog** results when hydrocarbons and nitrogen oxides are exposed to bright sunlight. A complex series of reactions produces a variety of pollutants, including ozone. **Catalytic converters** in automobiles reduce photochemical smog by oxidizing unburned hydrocarbons and decreasing the emission of nitrogen oxides. Hybrid vehicles are more efficient than conventional vehicles and emit fewer pollutants.

Section 13.6—Nitrogen oxides and sulfur oxides (mainly from power plants) can dissolve in atmospheric moisture, forming **acid rain**. Acid rain often falls far from the original sources of the acids, and it can corrode metals and erode marble or limestone buildings and statues.

Section 13.7—Indoor pollution can be worse than outdoor pollution. Cigarette smoke can raise levels of both PM and carbon monoxide above the allowed EPA safety standards.

Because of the effects of secondhand smoke, most cities and states restrict tobacco smoking in public places. The radioactive gas radon can accumulate indoors in some cases. It produces **daughter isotopes** that are also radioactive and can accumulate in the lungs. Inhaled radon can produce a significant risk of lung cancer.

Section 13.8—While ozone is harmful in the troposphere, it forms a protective layer in the stratosphere that shields Earth against UV radiation from the sun. Chlorofluorocarbons (CFCs) have been linked to the destruction of ozone—termed a "hole" in the ozone layer—and are now widely banned. This ban should stop the expansion of the ozone hole, but it will take many years for complete recovery.

Section 13.9—Carbon dioxide and other gases contribute to the **greenhouse effect**, which occurs when infrared radiation emitted from the Earth's surface cannot escape the atmosphere. The result is **global warming**, an increase in the Earth's average temperature, which is predicted to continue to cause undesirable climate change. **Carbon sequestration**—the storage or chemical change of atmospheric carbon dioxide—is a possible partial solution to the greenhouse effect.

Section 13.10—The EPA has listed six criteria pollutants, which all have been reduced in concentration in the atmosphere during recent years. Motor vehicles are a prominent source of carbon monoxide, hydrocarbons, and nitrogen oxides, while most PM and sulfur oxides come from industrial processes. Carbon monoxide and sulfur oxides are the most problematic pollutants. We all share responsibility for pollution—and the responsibility for reducing it.

GREEN CHEMISTRY—We can recycle the waste greenhouse gas, CO_2, and help our environment at the same time. Applying CO_2 technology to industrial processes can reduce the waste emitted from plants and factories and lead to greener applications.

Learning Objectives

Learning Objectives	Associated Problems
• List and describe the layers of the atmosphere. (13.1)	12, 16
• Give the approximate proportion of N_2, O_2, Ar, and CO_2 in Earth's atmosphere. (13.1)	12
• Describe the nitrogen and oxygen cycles. (13.2)	10, 15
• Describe the origin and effects of temperature inversions. (13.2)	17, 70
• List some natural sources of air pollution. (13.3)	7, 18, 59
• List the main pollutants formed by burning coal, and describe some technologies used to clean up these pollutants. (13.3)	3, 4, 8, 20, 23
• List the main gases in automobile emissions, and describe how catalytic converters reduce these gaseous pollutants. (13.4)	8, 28, 29, 32, 35, 36
• Explain how carbon monoxide acts as a poison. (13.4)	32
• Distinguish the origin of photochemical smog from the origin of sulfurous smog. (13.5)	9, 20, 21, 25, 26
• Describe the technologies used to alleviate photochemical smog. (13.5)	28
• Name the air pollutants that contribute to acid rain. (13.6)	20, 43–46, 67
• List the major industrial and consumer sources of acid rain–producing pollutants. (13.6)	43, 45

• List the main indoor air pollutants and their sources. (13.7)	47, 48, 51, 52
• Explain where radon comes from and why it is hazardous. (13.7)	49–51
• Explain the link between CFCs and depletion of the ozone layer. (13.8)	1, 2, 40–42
• Describe the consequences of stratospheric ozone depletion. (13.8)	7, 10, 41, 42
• List the important greenhouse gases, and describe the mechanism and significance of the greenhouse effect. (13.9)	5, 6, 53–55, 57, 63, 68
• Describe some strategies for reducing the amount of CO_2 released into the atmosphere. (13.9)	56, 58, 68, 69
• List the EPA's criteria pollutants and the major air pollutants that come mainly from automobiles and mostly from industry. (13.10)	66
• Describe how to use waste CO_2 to lessen the environmental impact of industrial processes.	72, 73, 74

Conceptual Questions

1. List two (former) uses of CFCs.

2. Name one replacement for CFCs. What problems are associated with the replacements?

3. What is bottom ash? What is fly ash? Give two uses for fly ash.

4. Describe how **(a)** a bag filter and **(b)** a cyclone separator remove particulates from stack gases.

5. Describe the action of a greenhouse gas in terms of how it interacts with sunlight. Give three examples of such gases. What molecular feature of the gas is responsible for this effect?

6. What are the main sources of the three major greenhouse gases that have increased significantly over the past 100 years?

7. What are the health effects of **(a)** ground-level ozone as an air pollutant and **(b)** depletion of stratospheric ozone?

8. What is smog?

9. What makes photochemical smog different from industrial smog?

10. A student says that the hole in the ozone layer is the primary reason for warming of the Earth, because it allows more energy to enter the lower atmosphere. Explain why this is incorrect.

Problems

The Atmosphere: Composition and Cycles

11. What is nitrogen fixation? Why is it important?

12. What are the approximate proportions of the four main components of dry air?

13. Look again at Figure 13.1. The X-15 was an experimental aircraft tested from 1959 to 1968. One test-flew to an altitude of 67 miles. What layer of the atmosphere did the X-15 reach?

14. Referring to Figure 13.1, consider a small airplane flying at an altitude of 12,000 ft. In what layer of the atmosphere is the plane flying?

15. Figure 13.3 shows oxidation of metals as one aspect of the oxygen cycle. Write the balanced equations for **(a)** the oxidation of iron metal by atmospheric oxygen to form iron(III) oxide and **(b)** the oxidation of chromium metal to form chromium(III) oxide.

16. Trace gases in the atmosphere tend to be concentrated at different levels according to molar mass, with the heavier gases closer to the Earth's surface and the lighter gases at higher altitudes. Refer to Figure 13.1 and Table 13.1, and answer the following: **(a)** Which of the trace gases listed in Table 13.1 are likely to be found at altitudes above about 40 km? **(b)** Which ones are likely to be found below about 20 km? **(c)** Which gas is most likely to be found in the troposphere? [*Hint:* Consider the location of the ozone layer to answer parts (a) and (b).]

Industrial Smog

17. What weather conditions are associated with industrial smog?

18. Name two elements that are reactants in the production of industrial smog.

For Problems 19–23, write the balanced equation for each reaction described.

19. Sulfur dioxide is oxidized to sulfur trioxide by atmospheric oxygen.

20. Sulfur trioxide reacts with water vapor to form sulfuric acid.

21. Sulfur dioxide reacts with dihydrogen sulfide to form elemental sulfur and water.

22. Calcium sulfite is oxidized to calcium sulfate by atmospheric oxygen.

23. Nitrogen dioxide reacts with water vapor to form nitric acid and nitrogen oxide, NO.

24. How many PM2.5 particles, placed in a row, would be 1.0 inches long?

Photochemical Smog

25. Under what conditions do nitrogen and oxygen combine? Give the equation for the reaction.

26. What is the formula for the oxide of nitrogen that initiates the formation of photochemical smog?

27. What is PAN? From what is it formed? What are its health effects?

28. Describe two ways in which the level of nitrogen oxide emissions from an automobile can be reduced.

29. About 144 billion gallons of gasoline were used in the United States in 2017. Assume that the formula for gasoline is C_8H_{18}, its density is 0.77 g/mL, and 1 gal = 3785 mL.
    a. Write the balanced equation for the complete combustion of gasoline to form carbon dioxide and water.
    b. What mass in kg of carbon dioxide could be produced from the complete combustion of that gasoline?

30. An electrostatic precipitator could be useful for reducing soot emitted from a shoe factory but not for reducing noxious vapors from adhesives used in that factory. Explain.

Automobile Emissions and Carbon Monoxide

31. Of CO, SO_3, NO_2, and NO, which is/are free radicals?

32. A concentration of carbon monoxide of 4000 ppm (0.400% by mass) in air is lethal in about half an hour. What mass in grams of carbon monoxide in a car with an internal volume of 2.3 m^3 is sufficient to provide that concentration? Assume the density of air is 1.29 g/L; 1 m^3 = 1000 L.

33. Refer to Problem 32. If a person in the car takes in 1.0 L of air in each breath, and breathes eight times per minute, what mass in grams of carbon monoxide could be absorbed in half an hour?

34. Write the equation for the reaction by which carbon monoxide reduces nitrogen monoxide to nitrogen gas.

35. Describe the change in CO emissions over time in the United States. Give a plausible explanation for this pattern.

36. Which pollutants are reduced in concentration by a catalytic converter?

The Ozone Layer

37. Oxygen and ozone are the same element in two different forms. What term describes this phenomenon?

38. Seawater contains roughly 30 g of dissolved sodium chloride per liter. Suggest a reason why seawater is not a potent source of ozone-depleting chlorine atoms.

39. Ozone supports combustion much more vigorously than O_2 does. A mixture of ozone and methane, CH_4, is unstable and reacts immediately and violently. Write a balanced equation for the complete combustion of methane in ozone to form carbon dioxide and water.

40. How are free radicals related to the ozone layer?

41. A single chlorine atom (free radical) in the stratosphere can destroy up to 50,000 ozone molecules before it is rendered harmless. With what kind of particle must a free radical react to be made unreactive?

42. Give a medical condition that becomes more prevalent with a decreased ozone layer.

Acid Rain

43. What two acids are mainly responsible for acid rain?

44. Write the equations for the reaction of each acid in Problem 43 with marble.

45. Write the equation for the reaction of nitric acid (from acid rain) with iron to form iron(III) nitrate and a gaseous product.

46. Acid rain can be formed by the reaction of water with sulfur dioxide to form *sulfurous* acid. Write the equation for this reaction.

Indoor Air Pollution

47. Name an indoor air pollutant that is also an outdoor pollutant. What is the major source of that pollutant in each environment?

48. List three risks associated with secondhand cigarette smoke.

49. Explain why a house built over a crawl space is likely to have less of a radon problem than a house built on a concrete slab.

50. How does better insulation of buildings make indoor air pollution worse?

51. What is the physical state of radon, and why does its physical state make it especially hazardous?

52. What kind of particulate matter is often found inside homes?

Carbon Dioxide and Climate

53. What is the greenhouse effect?

54. List the major greenhouse gases in the atmosphere.

55. How much does human respiration affect the CO_2 levels of the atmosphere? (*Hint*: What is the ultimate source of the carbon in exhaled CO_2?)

56. If your neighbor asks how she can reduce her greenhouse gas emissions, what suggestions would you give to her?

57. By approximately what percent has the concentration of atmospheric carbon dioxide increased over the past 130 years?

58. What substance can contain CO_2 from the atmosphere for the longest period of time?

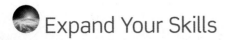

Expand Your Skills

59. Why is zero pollution not possible?

60. The atmosphere contains about 5.2×10^{15} t of air. What mass in metric tons of carbon dioxide is in the atmosphere if the concentration of CO_2 is 402 ppm?

61. A person who exercises vigorously several hours a day might take in about 22 m^3 of air per day. What mass in milligrams of particulates would the person inhale in a day if the particulate level averages 312 $\mu g/m^3$?

62. A tree 33 m tall and 0.55 m in diameter at its base produces about 84,000 L of oxygen per year. How many trees are needed to provide oxygen to 1000 people breathing as described in Problem 61?

63. Water vapor is a greenhouse gas that is present in significant concentrations in the Earth's atmosphere. Water is also a product of the combustion of fossil fuels. Why is there little concern about increasing atmospheric water vapor and its contribution to global warming? There is a correlation between water temperature and water vapor concentration; does this new information affect your analysis?

64. Evaluate the claim "There are more molecules in one breath of air than there are breaths in Earth's entire atmosphere." Use the following information: The total mass of the atmosphere is about 5.2×10^{21} g. The average molecular mass of air is about 29 amu. An average breath has a volume of about 0.50 L and a density of about 1.3 g/L.

65. Evaluate the claim "The breath that you just took contains at least one molecule of air that was in the last breath of the Buddha, Siddhartha Gautama, who died in about 400 B.C.E." See Problem 64 and assume complete mixing of Earth's atmosphere over 2400 years.

66. What is a criteria pollutant? Which of the pollutants in Table 13.3 come mainly from automobiles, and which come mostly from industry?

67. The reaction for formation of nitric acid is more complex than that shown in Figure 13.10. In the first step, three molecules of nitrogen dioxide react with a water molecule to form nitric acid and nitrogen monoxide. In the second step, the nitrogen monoxide is oxidized by atmospheric oxygen to nitrogen dioxide, and then the cycle continues. Write balanced equations representing these two steps.

68. Although carbon dioxide is the actual greenhouse gas, emissions from the burning of fossil fuels are often reported as gigatons (Gt) of carbon. In 2017, China emitted 10.5 Gt of carbon dioxide while the United States released 5.3 Gt of CO_2. To what mass in gigatons of carbon do these amounts correspond?

69. Given the information in Problem 68 and an estimate of population, compare the tons of carbon dioxide released per person in China to that figure for the United States. Suggest how you might approach U.S. citizens to convince them to reduce their emissions even though China's total emissions are greater than those of the United States.

70. Explain why a thermal inversion causes pollution problems.

71. The *specific heat* of a substance is the amount of heat required to raise the temperature of 1 g of the substance by 1 °C. The specific heat of air varies with temperature and humidity, but for ordinary calculations a value of 1.0 kJ/(kg °C) is sufficient. The average temperature of Earth according to NASA figures is 15 °C, and the mass of Earth's atmosphere is 5.1×10^{18} kg. **(a)** How much heat, in kilojoules, is added to the atmosphere for a 1.0 °C increase in temperature? How does this value compare with **(b)** the total world annual energy consumption of about 5.0×10^{20} J? **(c)** a hurricane that releases about 2×10^{20} J?

72. Why is it necessary for solvents to be volatile in conventional separation processes? Why isn't it necessary when employing CO_2 switching technology?

73. List 3 disadvantages of volatile solvents. Do switchable-hydrophilicity solvents (SHS) also have these problems?

74. Which of the following components is not recycled and used in multiple extractions using SHS technology: water, amine solvent, or CO_2?

Critical Thinking Exercises

Apply knowledge that you have gained in this chapter and one or more of the FLaReS principles (Chapter 1) to evaluate the following statements or claims.

13.1. A councilmember claims that the air pollution problem in the Los Angeles area could be solved by allowing only zero-emission vehicles to be sold and licensed there.

13.2. Ozone in aerosol cans was once used as a room deodorizer. An inventor proposes that such preparations be reintroduced because any ozone entering the environment would help to restore the protective ozone layer in the stratosphere.

13.3. Some people claim that aerosols will cool the planet, offsetting the enhanced greenhouse effect.

13.4. Some experts believe that global warming is mostly due to methane produced by such natural sources as plant decay in swamps and cattle flatulence. They say that industrial activities that produce carbon dioxide raise the standard of living and should be allowed to continue without interference because they are not the only sources of greenhouse gases.

13.5. An activist claims that if catalytic converters can remove 90% of pollutants from automobile exhaust, it would be easy to remove the other 10%.

13.6. A scientist claims that the increase in the atmosphere concentration of carbon dioxide from 0.02% to 0.04% is not important in terms of global warming. His reasoning is that water vapor makes up about 2% to 3% of the atmosphere and though water vapor is about one-tenth as effective as CO_2 as a greenhouse gas, still, the 0.02% increase in CO_2 is equivalent to a trivial 0.2% increase in water vapor.

Collaborative Group Projects

Prepare a PowerPoint, poster, or other presentation (as directed by your instructor) to share with the class.

1. What are the average and peak concentrations of each of the following in your community or in a nearby large city? How can you find out?

    a. carbon monoxide   **b.** ozone
    c. nitrogen oxides   **d.** sulfur dioxide
    e. particulate matter

2. The American Lung Association publishes a list of metropolitan areas with the worst ozone air pollution each year. Look up that list and others with other worst air problems. Find a list of areas with the best air quality.

3. Write a brief essay comparing the atmospheres of Venus and Mars to Earth's atmosphere and explaining the differences.

4. Air quality standards are maximum allowable amounts of various pollutants. The federal government and some state governments issue these standards. Using the Internet, find the current standards from by the EPA. Does your state issue such standards? (*Hint*: Try the EPA site and your own state's environmental protection site.)

 LET'S EXPERIMENT! Let the Sun Shine

Materials Needed

- UV-sensitive color-changing beads (40)
- 4 small plastic zipper bags
- Permanent marker

- Baking sheet or tray
- Spray sunscreens with three different SPFs: 4, 15, and 30 (or 50)

Have you ever wondered how well your sunscreen really works? And is there a difference between SPF 4 and SPF 50? Does one really protect more than the other from harmful UV rays?

As we all hear news about the depletion of the ozone layer and the resulting increase of UV light reaching the Earth's surface, sunscreen is becoming increasingly important to protect against skin cancer. The highest rates of skin cancer have been recorded in Australia, where approximately 1,200 people die from skin cancer each year. Australia has higher UV levels than other parts of the world because it is very close to the ozone hole over the Antarctic. Australia also gets closer to the sun than most other countries do in the summer.

Sunscreens are provided in different SPFs, or skin protection factor ratings. SPF 15 sunscreen provides 15-times-longer protection than unprotected skin, and blocks harmful UV rays. In this experiment, you will be using UV-sensitive color-changing beads as indicators so that you can visually determine what SPF sunscreen is most effective at protecting your skin. The first part of the experiment is done indoors, away from open windows or areas withstrong sunlight, while the second part of the experiment is performed outside in direct sunlight.

First, while indoors, place 12–15 UV-sensitive beads in each of the four plastic bags and seal them. (*Hint*: Use the same distribution of colors of beads in each bag for easy comparison.) Label one bag of beads as "control" and place on the tray.

Next, spray each bag of beads with your three different SPF sunscreens with one brief spray (1–2 seconds). As you spray, keep the nozzle at a distance of about 4–5 inches from the bag. Also try to keep the length of each spray the same, for consistency among the sets of beads. Label each bag with the appropriate SPF value and place the bags on the tray. Allow the sunscreen to dry for about 5 minutes.

Finally, bring your samples outdoors into the sun and immediately observe any color changes. What do you notice about the colors of the different sets of beads? Leave the samples in the sun for only about 1 minute, and then bring them back indoors to look at the changes more closely. What do you notice when you bring the beads back indoors?

Questions

1. What trends did you see in your observations? What does this tell you about the different SPF values of sunscreens (if anything)?
2. What SPF sunscreen do you typically use? Do you think this protects you from harmful UV rays?
3. What do you think would happen if you were to rinse the sunscreen off the bags?

Water

Have You Ever Wondered?

1 Why does water form droplets on plastic surfaces, yet there are no giant raindrops? **2** Why is the air temperature often warmer on the coast of Alaska than it is in North Dakota in winter? **3** Why isn't there a filter that removes all pollutants from water? **4** About how much water do we really use each day? **5** How much water is wasted when a faucet drips? **6** How do you get fresh water from saltwater?

Water is truly a unique substance; its most common state is as a liquid, which covers nearly three-quarters of Earth's surface, most in oceans. Water is sometimes called the "universal solvent." Water does dissolve most inorganic and polar substances, which is the reason the oceans are salty. Unfortunately, water dissolves many substances that are water pollutants. In this chapter we will look at the nature of water, and the methods currently used for purifying water.

RIVERS OF LIFE, SEAS OF SORROWS   Earth is a watery world. Most of its surface is covered with oceans and seas, lakes, and rivers: the hydrosphere. People contain a lot of water, too. About two-thirds of our body weight is water, and water makes up 60% to 90% of a typical cell. The water in our blood is much like ocean water, containing a variety of dissolved ions. You might even say that we are walking sacks of seawater. We need to drink about 2 L of water daily to replenish what is lost.

The presence of large quantities of water on Earth makes our planet unique in the solar system, the only planet capable of supporting higher forms of life. Water is the only substance on Earth that commonly exists in large amounts in all three physical states. Go outside on a snowy day and you are likely to experience all three forms at once. Gaseous water vapor is evident when you exhale and you can "see your breath." Tiny droplets of liquid water form clouds. Solid water falls to the ground as snow and melts into puddles of liquid water.

The early Greek philosopher Thales (ca. sixth century B.C.E.)—whom some consider the first scientist—held water in the highest regard, believing that it was the "primordial substance" from which all other things were made. Little wonder that a man from a nation with many islands would think so. Water's unique nature makes it essential to life.

Polar water can dissolve more substances than any other liquid, resulting in most water sources containing many dissolved ions and molecules. Many substances are easily dispersed throughout the environment because they are soluble in water. Once dissolved, it is not easy to remove substances from water. Furthermore, many strains of bacteria and other microorganisms that are harmful to humans thrive in water. In this chapter we will look at various methods that can be used to purify water, depending on the nature of the pollutant.

From a global perspective, a significant portion of Earth's population faces two major challenges with respect to water: the distribution of water (some regions have drought and others severe flooding) and the quality of the water available. Even when water is available in developing countries, that water may contain disease-causing substances. We should neither underestimate the importance of clean drinking water nor take it for granted. We could live for several weeks without food, but without water, we would last a few days at most.

14.1 Water: Some Unique Properties

Learning Objectives • Relate water's unique properties to polarity of the water molecule and to hydrogen bonding. • Explain how water on the surface of Earth acts to moderate daily temperature variations.

Water is the most common molecule on Earth, yet is a molecule with very unusual and unique properties. Why? The bent geometry of a water molecule (Chapter 4) and the large difference in electronegativities between oxygen and hydrogen together result in water molecules being highly polar. The central oxygen atom in water has two sets of nonbonding (lone) electron pairs. These lone pairs, in combination with the positive poles on the two hydrogen atoms, result in a single water molecule's being able to interact with four other water molecules through hydrogen bonding (Chapter 6), which is the reason for the unique properties of water. Let's take a look at some of those properties.

Water's High Boiling Point for a Small Molecule

Water is a small molecule, with a mass of 18 amu. Most small molecules with masses less than 50 amu are gases at room temperature (~25 °C); consider nitrogen, N_2

1 Why does water form droplets on plastic surfaces, yet there are no giant raindrops?

Plastic surfaces consist of huge, nonpolar molecules, and thus are hydrophobic (water-hating). Water tends to minimize its interactions with the plastic surface and maximize its interactions within itself by exploiting the hydrogen bonding to the maximum. The geometric shape with the lowest surface-to-volume ratio is a sphere. Water molecules on a flat, hydrophobic surface form hemispheric shapes. In the air, spherical raindrops are formed. So, why don't we see raindrops the size of a beach ball? Enter gravity. Gravity overcomes the strength of the hydrogen bonding network in a water droplet, which breaks into smaller, though still spherical, droplets.

(28 amu); oxygen, O_2 (32 amu); carbon dioxide, CO_2 (44 amu); propane, C_3H_8 (44 amu); and ozone, O_3 (48 amu). Yet water is a liquid at room temperature, due to the strong hydrogen bonding between water molecules, with a boiling point of 100 °C—significantly higher than the boiling points of the gases we just mentioned. The average air temperature on Earth is about 16 °C, well under the boiling point of water and higher than its freezing point. If water molecules did not exhibit strong hydrogen bonding, water would be gaseous on planet Earth, and life as we know it would not exist. Strong hydrogen bonding among the bent water molecules supports and is critical to life.

Solid Water Is Less Dense than Liquid Water

You have surely seen ice cubes (solid water) floating on top of liquid water and other water-based drinks. Most substances other than water contract when they cool and solidify, so most solids are slightly denser than they are in their liquid form. Why is water different?

Water is unique in that at about 4 °C, liquid water molecules are packed together as closely as they can be packed. As water is cooled from 4 °C down to 0 °C, water molecules actually spread apart from each other so that a highly ordered structure of repeating hexagons is formed, with hydrogen bonding playing a major role. Oxygen atoms are at the apices of the hexagons, with hydrogen atoms in the middle of each side (Figure 14.1). Each hydrogen is bonded to one oxygen by a covalent bond in its "own" molecule, and to a second oxygen in another molecule by a hydrogen bond. What is in the center of the hexagons in the solid ice structure? Nothing. Empty space. This is the reason that ice is less dense than liquid water and why ice floats in water.

(a)

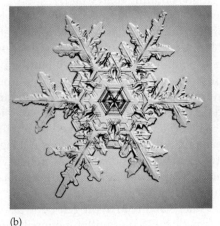
(b)

▲ **Figure 14.1** Hydrogen bonds in ice. (a) Oxygen atoms are stacked in layers of distorted hexagonal rings. Hydrogen atoms lie between pairs of oxygen atoms, closer to one (in the covalent bond) than to the other (in the hydrogen bond). The yellow dashed lines indicate hydrogen bonds. The structure has large "holes." (b) The hexagonal arrangement of water molecules in the crystal structure of ice is revealed at the macroscopic level in the hexagonal shapes of snowflakes.

▲ Water is the only substance that expands when cooled from 4 °C to 0 °C, when it freezes. If molecules are farther apart from each other in the solid state than they are in the liquid state, then the solid is less dense than the liquid. Ice cubes with a lower density than that of cold water float. A layer of ice forms in winter on the surface of lakes and ponds for the same reason, preserving aquatic life below it.

When ice melts, the hydrogen bonding network is perturbed and weakened, making water molecules more mobile. Melting ice allows for more water molecules to be packed together per unit volume, giving rise to the higher density of liquid water versus that of ice. Actually, water is denser than most liquids, including alcohols, oils, and gasoline. Its density as a liquid is very close to 1.00 g/cm^3, and the **specific gravity** of many substances is defined as the density of that substance compared with water. The density of gold, for example, is 19.3 g/cm^3; its specific gravity is 19.3.

The fact that ice is less dense than liquid water is essential to life on Earth. We know that a layer of ice forms on the tops of lakes and ponds in the winter, leaving water in the liquid state under that ice, preserving aquatic life in that water.

▲ One liter of oil can create an oil slick on the ocean surface with an area of 2.5 hectares (6.3 acres). It is sometimes difficult to estimate the size of a spin, as when the Deepwater Horizon rig sank in the Gulf of Mexico, following an explosion and the leak was a mile below the surface. Oil spills are reminders that hydrocarbons and water don't mix, since oil molecules are non-polar and water molecules are polar

If the opposite were true, when the air temperature falls to less than 0 °C, ice formed at the surface of natural waters would sink to the bottom of the lakes and rivers. Those natural bodies of water would completely freeze from the bottom upward. If this were to happen, all life in the lakes and rivers would die. The same would have happened during the ice ages our planet has gone through. In short, if ice did not float, all life on this planet would have become extinct a long time ago, and humans would never have lived.

The fact that water expands when it freezes can also be life-threatening. The majority of cells in our bodies can be envisioned as water-filled balloons (the balloon being the outer cell membrane). When water expands—by as much as 10% when it becomes frozen—the balloon or cell membrane may burst. Frostbite can kill cells and destroy tissue in this way.

Solvation Ability of Water

Water has sometimes been called "the universal solvent." Obviously this is not true, because sand, homes, metals, plants, and human beings do not dissolve in water! Nonpolar molecules and substances consisting of covalent bonding on an "infinite" level, like sand (SiO_2), will not dissolve in polar water. Oil spills in the ocean will not dissolve in the ocean water, because polar liquids *repel* nonpolar liquids like oils. Oils are typically less dense than water, so fortunately the oil slick will float.

Nonetheless, water does dissolve more substances than any other liquid. The main reason for water's being an extremely good solvent is that it is strongly polar and exhibits strong hydrogen bonding. For instance, NaCl readily dissolves in water. In fact, 1 L of water dissolves 359 g of NaCl, but 1 L of methanol only dissolves 14.9 g of NaCl. Why is this so? The answer is that methanol exhibits much weaker hydrogen bonding than water.

How does dissolving actually happen on the molecular level? One way to envision the process of dissolving NaCl in water is to imagine the positive poles of water molecules strongly pulling each Cl^- ion from the outer part of the crystal lattice by ion–dipole forces (Section 6.4). Simultaneously, the negative poles of other water molecules strongly pull each Na^+ ion from the outer part of the crystal lattice. This continues until all the Cl^- and Na^+ ions are separated and entrapped within the liquid network of hydrogen-bonded water molecules.

The Cl^- and Na^+ ions fit well within the free spaces among water molecules. This means that dissolving NaCl does not significantly disturb the hydrogen bonding network in water, and hence, little energy is required to dissolve NaCl. Yet size does matter! KBr, consisting of the larger potassium and bromide ions, is somewhat less soluble (119 g/L) in water.

Water's ability to hydrogen-bond also explains why water is such a good solvent for organic molecules with functional groups capable of hydrogen bonding, such as alcohols, carboxylic acids, and amides. (Refer to Section 9.4). An example of this can be seen if we compare the solubility of benzene (a nonpolar molecule with no functional groups) with that of phenol (a single alcohol functional group on the benzene ring), which in water are 1.8 g/L and 83 g/L, respectively.

High Specific Heat Capacity of Water

Specific heat capacity (sometimes referred to as "specific heat") is the quantity of heat required to raise the temperature of 1 g of a substance by 1 °C. Compared with other substances, water absorbs a relatively large amount of heat energy before its temperature increases by 1 °C. Conversely, water must release a relatively large amount of heat energy when its temperature drops by 1 °C. Table 14.1 gives the specific heat capacity for several familiar substances in the old units—calories per gram per degree (cal/g °C)—and in the SI units—joules per gram per kelvin (J/g K). The definition of a calorie dictates that the specific heat capacity of water is 1.00 cal/g °C, while in SI units, it is 4.184 J/g K.

Note that metals have much lower values for specific heat capacity than does water. It takes almost ten times as much heat to raise the temperature of 1 g of water

TABLE 14.1 **Specific Heat Capacity of Some Familiar Substances at 25 °C**

Substance	Specific Heat Capacity	
	(cal/g °C)	(J/g K)
Aluminum (Al)	0.216	0.902
Copper (Cu)	0.0920	0.385
Ethanol (CH_3CH_2OH)	0.588	2.46
Iron (Fe)	0.107	0.449
Ethylene glycol ($HOCH_2CH_2OH$)	0.561	2.35
Magnesium (Mg)	0.245	1.025
Mercury (Hg)	0.0332	0.139
Lead (Pb)	0.0306	0.128
Silver (Ag)	0.0562	0.235
Water (liquid)	1.00[a]	4.184

[a]Note that in cal/ g°C, the specific heat capacity of water is 1.00. As the metric system was established, the properties of liquid water were often taken as the standard.

by 1 °C as to raise the temperature of 1 g of iron by the same amount. A metal mug held over a campfire will heat up very rapidly, in contrast to water in the same mug heated over the fire. Because water stores heat so well, heated water in a mug will also stay hot a lot longer than will a hot, empty mug. The vast amounts of water on the surface of Earth thus act as a giant heat reservoir to moderate daily temperature variations. Cities located next to an ocean experience a moderate climate, where the temperature variations are much smaller than for inland cities. The extreme temperature changes on the surface of the waterless moon, ranging from 100 °C at noon to −173 °C at night, dramatically illustrate this important temperature-moderating property of water.

We can use the following equation, in which ΔT is the change in temperature (in either °C or kelvins), to calculate the quantity of heat absorbed or released by a system.

$$\text{Heat absorbed or released} = \text{mass} \times \text{specific heat capacity} \times \Delta T$$

2 **Why is the air temperature often warmer on the coast of Alaska than it is in North Dakota in winter?**

The average temperature in January in Juneau, Alaska, is −2.1 °C, and in Fargo, North Dakota, is −12.7 °C. Water has a high heat capacity, so it takes a lot more energy than most other substances to raise its temperature or vaporize it. During the day, a considerable amount of heat energy absorbed from the sun's radiation may result in a small temperature increase in ocean water. When ocean water cools at night—even with a small change in temperature—considerable heat is released. For this reason, the air temperature near coasts does not experience as much temperature variance as it does in areas farther from large bodies of water.

EXAMPLE 14.1 Heat Absorbed

How much heat—in calories, kilocalories, and kilojoules—does it take to raise the temperature of 225 g (about a glassful) of water from 25.0 °C to 100.0 °C?

Solution

Let's list the quantities we need for the calculations.

Specific heat capacity of water = 1.00 cal/g °C

Temperature change = 100 °C − 25 °C = 75 °C

Then we use the equation

$$\text{Heat absorbed} = \text{mass} \times \text{specific heat capacity} \times \Delta T$$

$$= 225 \text{ g} \times 1.00 \text{ cal/g °C} \times 75.0 \text{ °C}$$

$$= 16{,}900 \text{ cal}$$

We can then convert the unit cal to the units kcal and kJ.

$$16{,}900 \text{ cal} \times \frac{1 \text{ kcal}}{1000 \text{ cal}} = 16.9 \text{ kcal}$$

$$16.9 \text{ kcal} \times \frac{4.184 \text{ kJ}}{1 \text{ kcal}} = 70.7 \text{ kJ}$$

> **Exercise 14.1A**

How much heat, in calories, kilocalories, and kilojoules, is released by 975 g of water as it cools from 100.0 °C to 18.0 °C?

> **Exercise 14.1B**

Using information from Table 14.1, identify the substance if 148 J of heat energy is needed to heat 32.0 g of the substance by 12 K.

High Heat of Vaporization of Water

Heat of vaporization is the amount of heat required to evaporate a substance in units of kcal/mol or kJ/mol. A large amount of the heat needed to evaporate even a small amount of water goes toward breaking all of the existing intermolecular forces, of which hydrogen bonding is the strongest. This fact is of enormous importance to us because large amounts of body heat can be dissipated by the evaporation of small amounts of water (perspiration) from the skin.

The high heat of vaporization of water (40.8 kJ/mol) also accounts in part for the climate-modifying property of lakes and oceans. A large portion of the heat that would otherwise raise the temperature of the air and land instead vaporizes water from the surfaces of lakes and seas. Thus, in summer, it is cooler near a large body of water than it is in interior land areas.

All these fascinating properties of water depend on the unique structure of the highly polar water molecule and its unique dense hydrogen-bonding network.

SELF-ASSESSMENT Questions

1. Which of the following is *not* a unique property of water?
   a. Solid water floats on liquid water.
   b. Water has a boiling point a lot higher than those of most gases.
   c. Water has three states (phases): solid, liquid, and gas.
   d. Water has a dense hydrogen-bonding network.

2. In ice and water, each water molecule forms hydrogen bonds with how many water molecules?
   a. one   b. three   c. four   d. five

3. Water's unique properties arise because water molecules
   a. have hydrogen atoms that do not have stable electron configurations
   b. attract each other through hydrogen bonds
   c. can form covalent bonds with ionic substances
   d. can form covalent bonds with nonpolar substances

4. In a collection of water molecules, hydrogen bonds form between
   a. an H atom in one H_2O molecule and the O atom in another H_2O molecule
   b. the O atoms in different H_2O molecules

   c. the two H atoms in a single H_2O molecule
   d. two H atoms in different H_2O molecules

5. Frozen water has an open structure with
   a. an amorphous molecular arrangement
   b. large hexagonal holes
   c. molecules that are in constant motion
   d. molecules that tend to repel each other

6. How much heat (in cal) is required to raise the temperature of 50.0 g of water from 20.0 °C to 50.0 °C?
   a. 30.0 cal
   b. 50.0 cal
   c. 1500 cal
   d. 2500 cal

7. How much heat (in kJ) is required to raise the temperature of 131 g of iron from 15.0 °C to 95.0 °C?
   a. 0.058 kJ
   b. 1.11 kJ
   c. 4.71 kJ
   d. 4710 kJ

Answers: 1, c; 2, c; 3, b; 4, a; 5, b; 6, c; 7, c

14.2 Water in Nature

Learning Objectives • Explain why humans can only make use of less than 1% of all the water on Earth. • Explain the water cycle on Earth. • Identify natural sources of contaminants in rain and in natural bodies of water.

Saltwater and Fresh Water

Three-fourths of the surface of Earth is covered with water, but nearly 98% of it is salty seawater, or is found in inland lakes with few or no outlets, such as the Dead

Sea—unfit for drinking and not suitable for most industrial purposes. The main dissolved ions in seawater are sodium and chloride ions, but calcium, magnesium, and sulfate are also important. The overall salinity of oceans is highest in the tropics—about 35 parts per thousand (ppt)—since evaporation occurs more rapidly in hotter regions, and is lowest near the polar ice caps (about 30 ppt).

Fresh water is defined as natural water with a much lower concentration of ions than salty water. *No* water in nature is absolutely pure and free of dissolved material. Much of the fresh water on Earth is frozen in ice caps, leaving less than 1% of all Earth's water available for human use. Of that, most is underground, and obtained from wells. Lakes and streams account for only 0.01% of the fresh water on the planet, and are not usable for direct water consumption. While rain falls onto Earth in enormous amounts, most of it falls into the sea or on inaccessible land areas. A more detailed breakdown of the water distribution on planet Earth is presented in Figure 14.2.

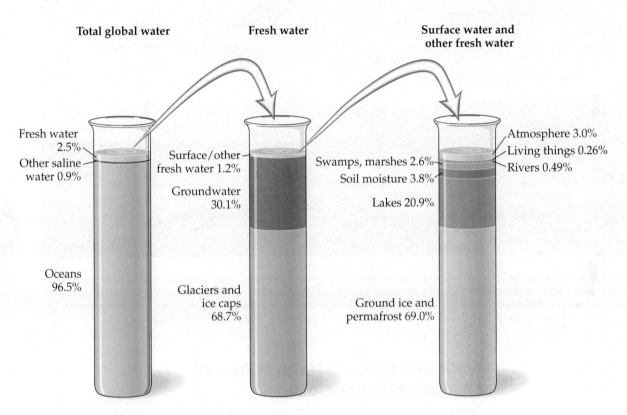

▲ **Figure 14.2** Where is Earth's water?

The underground temperature of groundwater from wells around 10–15 m underground is usually about 2 °C above the average air temperature at that locality. Water from a well decreases in temperature about a degree for every 20 m of depth. In some localities where the crust is thinner—such as Yellowstone Park, New Zealand, and Iceland—underground water may be superheated by very hot rocks, and may burst forth in geysers or create hot springs.

Water is constantly cycled on planet Earth through its evaporation (powered by the sun), precipitation, and flow, due to gravity. As a whole, the planet's water cycle is at a steady state (Figure 14.3), meaning that the percentages of water apportioned to the oceans; the ice caps; and the freshwater rivers, lakes, and streams remain fairly constant. The melting of Arctic and Antarctic ice due to climate change may, however, change that constancy within several generations. Water constantly evaporates from both water and land surfaces, and water vapor condenses into clouds and returns to Earth as rain, sleet, and snow. This fresh water becomes part of the ice caps, runs off in streams and rivers, and fills both lakes and underground pools of water in rocks and sand called *aquifers*.

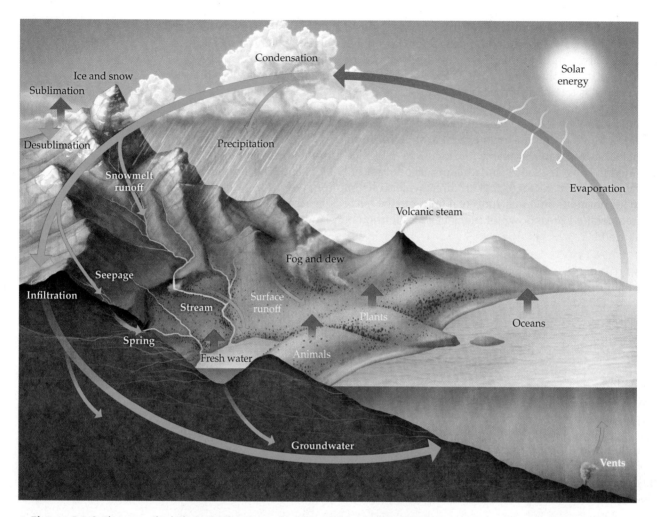

▲ **Figure 14.3** The water (hydrologic) cycle. Powered by the sun, evaporation of water from oceans, lakes, and land occurs. As moist air rises, it cools, and the water vapor condenses to form clouds. Water returns to the surface as precipitation. Some of this water becomes groundwater and recharges aquifers. Some of it enters rivers and streams and is carried back to the oceans. Evapotranspiration is the process by which water is discharged to the atmosphere from soils and transpiration by plants.

Acids and Bases in Natural Waters and Contaminants in Rain

Ocean water typically has a pH of around 8.0, which is slightly alkaline, due to the presence of carbonate and hydrogen carbonate ions. Recall that mollusk shells are made primarily of calcium carbonate (Section 12.3). The chemistry of ocean water is actually quite complex, but scientists are concerned about the slow acidification of ocean waters by the dissolving of carbon dioxide from the air, forming carbonic acid, and its effect on the fauna that live in oceans:

$$CO_2(g) + H_2O(l) \longrightarrow H_2CO_3(aq)$$

"Pristine" rainwater should have a pH of about 5.65 due to the dissolving of atmospheric CO_2 in raindrops to form carbonic acid (H_2CO_3), as in the chemical reaction above. Lightning can also make rainwater acidic by causing nitrogen, oxygen, and water vapor to combine into nitric acid. A natural contaminant in rain is dust in the air, consisting of extremely small particles of $CaSO_4$, SiO_2, aluminum silicates, NaCl, and KCl.

Other constituents within rainwater are caused by human activities. The burning of fossil fuels has led to acidic rain, fog, and snow (Section 13.6), with a pH as low as 4.3 in some regions, mainly due to sulfur dioxide in the air. Acid rain has caused the destruction of a large number of statues and other architectural treasures around the world. It has corroded metals and even ruined the finishes on automobiles. Despite the reduction of SO_x emissions (Section 13.6) in recent years, and the subsequent decreases in the severity and frequency of acid rain, thousands of bodies of water in eastern North America remain acidified. Acids often flow into these water bodies from abandoned mines, which is another factor.

Acidic water is neutralized when limestone, $CaCO_3$, is present (Section 12.3):

$$CaCO_3(s) + 2H^+(aq) \longrightarrow Ca^{2+}(aq) + CO_2(g) + H_2O(l)$$

Limestone   Acid

Without the abundance of limestone—as is the case in areas where rock is mainly granite—the acidic water is detrimental to aquatic life, causing the release of aluminum from the aluminum-silicate minerals (Section 12.2). Even at very low concentrations, aluminum ions are deadly to young fish. Ironically, lakes destroyed by excess acidity are often quite beautiful—clear and sparkling due to lack of life.

Acidic waters and acidic rain have also been linked to declining crop yields and to the destruction of forests and their inhabitants. In fact, acid rain has been linked to lower maple syrup production. Not a sweet situation.

Dissolved Minerals

As stated above, water is a wonderful solvent. It moves along or beneath the surface of Earth, dissolving minerals from rocks and soils, albeit slowly, as demonstrated by the formation of the Grand Canyon. Recall that minerals (mostly salts) are ionic and that ions are either positively charged (cations) or negatively charged (anions). The main cations in natural water, shown in red in Table 14.2, are ions of sodium, potassium, calcium, magnesium, and iron (as Fe^{2+} or Fe^{3+}). The anions (shown in blue) are usually sulfate, bicarbonate, and chloride ions.

Water for domestic use from wells or municipal water supplies is usually classified as either soft (containing mostly Na^+ and K^+ ions) or hard (containing higher concentrations of Ca^{2+}, Mg^{2+}, and $Fe^{2+/3+}$ ions). **Hard water** is not a health hazard. However, when soap is used in hard water, it forms a white solid soap scum, which causes bathroom fixtures to look dingy. More about water softening and the use of soaps and detergents will be discussed in Section 20.1.

Groundwater also contains dissolved gases, including oxygen, nitrogen, carbon dioxide, and the naturally occurring gas radon, but at such low concentrations that it is not a concern.

TABLE 14.2 Some Substances Found in Natural Waters

Substance	Formula	Source
Carbon dioxide	CO_2	Atmosphere
Dust	—	Atmosphere
Nitrogen	N_2	Atmosphere
Oxygen	O_2	Atmosphere
Nitric acid	HNO_3	Atmosphere (thunderstorms)
Sand and soil particles	—	Soil and rocks
Sodium ions	Na^+	Soil and rocks
Potassium ions	K^+	Soil, rocks, and fertilizer
Calcium ions	Ca^{2+}	Limestone rocks
Magnesium ions	Mg^{2+}	Dolomite rocks
Iron(III) ions	Fe^{3+}	Soil and rocks
Chloride ions	Cl^-	Soil, rocks, and fertilizer
Sulfate ions	SO_4^{2-}	Soil, rocks, and fertilizer
Bicarbonate ions	HCO_3^-	Soil and rocks
Radon	Rn	Radioactive decay

As Figure 14.3 shows, the water cycle replenishes our supply of fresh water. When water evaporates from the sea, salts are left behind. When water moves through the ground, many impurities are filtered out and trapped in the rock, gravel, sand, and clay. This capacity to purify is not infinite. Yet, globally, the water cycle cleans and purifies over 16 million liters of water a second, and it does so for free!

Organic Matter

Rainwater dissolves matter from decaying plants and animals. As part of the natural cycle in a forest, small quantities of this organic matter enrich the soil. As you will see in the next section, not all organic inputs are good for the environment. Lubricants, fuels, some fertilizers, and pesticides all contaminate water. Bacteria, other microorganisms, and animal wastes are all potential water contaminants, and often cause serious health problems.

SELF-ASSESSMENT Questions

1. Energy to power the water cycle comes from
   a. electric power plants
   b. hurricanes
   c. the sun
   d. thunderstorms

2. Most of the Earth's water is in
   a. groundwater
   b. ice and permafrost
   c. lakes and rivers
   d. oceans

3. Rainwater is naturally acidic because water reacts with
   a. CO_2
   b. N_2
   c. O_2
   d. SO_2

4. Which of the following is not commonly found in rainwater?
   a. HNO_3
   b. SiO_2
   c. H_2CO_3
   d. Rn (radon)

5. The main cations in water are
   a. Na^+, Al^{3+}, Ca^{2+}, and Mg^{2+}
   b. Na^+, K^+, Ca^{2+}, and Li^+
   c. Na^+, K^+, Ca^{2+}, and Mg^{2+}
   d. Ra^{2+}, Sr^{2+}, Ca^{2+}, and Mg^{2+}

6. Hard water may be characterized by the presence of all of the following ions except
   a. Ca^{2+}
   b. Fe^{2+}
   c. K^+
   d. Mg^{2+}

7. Water in a lake made acidic by acid rain or mine drainage is often
   a. clear
   b. green with algae
   c. rich in game fish
   d. rich in lime

Answers: 1, c; 2, d; 3, a and d; 4, d; 5, c; 6, c; 7, a

14.3 Organic Contamination; Human and Animal Waste

Learning Objectives • Describe how human and animal waste affect water quality. • Give an example of a biological water contaminant. • Identify the sources of nitrates in groundwater and problems they can cause.

Human waste, a euphemism for human excrement and urine, constitutes a significant disposal problem, especially in large metropolitan areas. Flushing a toilet does not make the contents simply disappear. Historically, the easiest manner of disposal was to apply the waste to land or release it into the nearest body of water. These forms of disposal led to environmentally damaging effects, some deadly. Farm animal wastes need to be considered, as well.

Too Much Organic Matter Means Too Little Oxygen

Organic matter from human and farm-animal waste is still very rich in nutrients such as nitrates and phosphates, protein, fats, and bacteria biomass. An increase in biomass continues as this organic material is metabolized by microorganisms in the environment to grow and multiply. If oxygen is required for this breakdown, it is called *aerobic* (as opposed to *anaerobic*) *oxidation*. When this process takes place in a body of water, the amount of **dissolved oxygen** is being depleted. If too much organic matter is present, too much oxygen will be removed from the water. When this happens, there is not enough oxygen in the water for the more advanced organisms, such as fish, potentially causing death by suffocation. A measure of the amount of oxygen needed for the degradation to occur is the **biochemical oxygen demand (BOD)**.

Streams and rivers can regenerate themselves by reintroducing oxygen from the atmosphere. Those with rapids soon come alive again as the swirling water helps with this process. Lakes with little or no flow can remain dead for years. Small human-made ponds may be oxygenated by percolating air through them using special pumping equipment.

Introducing human or animal waste to water also introduces nitrates (NO_3^- ions) and phosphates (PO_4^{3-} ions), which may serve as nutrients for the growth of algae. When this extra biomass dies and becomes organic matter itself, the cycle resumes, amplified by the biomass added due to the newly incorporated CO_2, further increasing the BOD. This process is called **eutrophication**. The eutrophication of a lake is a natural process, but the action can be greatly accelerated by human waste, by phosphates from detergents and from the runoff of fertilizers from farms and lawns, all of which stimulate algal bloom (Section 19.1) with resultant low oxygen content. A high-profile example of the ecological consequences of eutrophication is the **dead zone** in the northern Gulf of Mexico that encompasses over 8700 square miles (2017 estimate). Fertilizer runoff into the Mississippi River from Minnesota south to Louisiana has been the primary reason that the water there has such a low oxygen content that very little marine life there is possible.

Interestingly enough, some cities—such as Minneapolis (the "City of Lakes") and St. Paul, Minnesota—have banned the use of fertilizer that contains phosphates for lawn use, because there is enough phosphate in the soil already in those cities.

When too much organic matter—whether from sewage, dying algae, or other sources—depletes the dissolved oxygen in a body of water, **anaerobic decay** processes take over. Instead of oxidizing the organic matter, anaerobic bacteria reduce it. Methane (CH_4) is formed. Sulfur is converted to hydrogen sulfide (H_2S) and other foul-smelling organic compounds. Nitrogen is reduced to ammonia and odorous amines (Section 9.8). Foul odors are a good indication that the water is overloaded with organic wastes. Only anaerobic microorganisms can survive in such water.

▲ The eutrophication of a lake is a natural process, but the process can be greatly accelerated by human wastes, phosphates from detergents, and the runoff of fertilizers from farms, lawns, and golf courses.

Health Considerations from Human and Animal Waste

Water runoff from animal feedlots on farms, as well as extensive use of fertilizers to increase crop production or enhance golf-course greens, increases the concentration of nitrates (NO_3^- ions) in farm water wells. (Read more in Section 19.1). In the 1950s, bottle-feeding of infants became popular, replacing breast-feeding, and many women on farms used the water from the family farm well to prepare baby formulas from powdered concentrates. Unfortunately, high concentrations of nitrate ions in the well water (more than 10 mg/L) can cause *methemoglobinemia*, or blue baby syndrome. In an infant's delicate digestive tract, nitrate ions are reduced to nitrite ions. If not treated, methemoglobinemia can be fatal. Once the medical community realized the connection between increased nitrates and blue baby syndrome, bottled water was recommended for mixing with baby formula powders.

Discharge of human and animal waste into the environment raises several concerns other than those discussed above. One of these issues is the release of hormonally active and endocrine-disrupting chemicals. A second concern stems from antibiotics introduced to the environment in human and animal waste. Antibiotics have been overprescribed to humans, especially in some developed nations, and have been overused in farming to prevent livestock illnesses. If a small fraction of organisms are not killed by an antibiotic and mutate into stronger individuals, they can multiply into new generations of drug-resistant *super bugs*.

Contamination of water supplies by *pathogenic* (disease-causing) microorganisms from human wastes has been and still is a devastating problem throughout the world. In 1900, for example, there were more than 35,000 deaths from typhoid in the United States. Even in the twenty-first century, waterborne diseases are still quite common in much of the developing world. At any given time, about half the world's hospital beds are occupied by people suffering from waterborne diseases.

Whenever there is severe flooding, due to major hurricanes or even during the rainy season in underdeveloped nations, outbreaks of cholera or other gastrointestinal diseases become major public health challenges. Severe flooding contaminates the usual clean water supply; spreads disease-causing bacteria and viruses to levels that constitute a major public health catastrophe; and provides breeding grounds for mosquitoes that can carry malaria, or the Zika, West Nile, or chikungunya virus. WHO (the World Health Organization) reports, for example, that each year there are well over a million cases and 100,000 deaths from cholera, mainly through water contamination. In 2017, an outbreak of cholera infected half a million persons in Yemen. In 2010 waterborne diseases such as severe diarrhea affected over a quarter of a million people in Bangladesh in the wake of severe flooding. In the fall of 2017, hurricanes Irma and Harvey resulted in a major increase in disease in areas where those hurricanes inflicted major damage and caused significant flooding.

Agriculture and gardening contribute pesticides and herbicides to groundwater. (Section 19.2 and Section 19.3 provide further information about these organic compounds.) Industries such as tanning, wood treatment, pharmaceutical manufacturing, petrochemical manufacturing, and transportation have introduced a long list of organic chemicals into the environment. Household products and cosmetics also contribute to water pollution.

Today, as a result of chemical treatment, municipal water supplies in developed nations are generally safe. Municipal water utilities are constantly improving their treatment systems to deal with biological contamination. However, millions of people in the United States are still at risk because of bacterial contamination of drinking water. Ongoing threats include *Cryptosporidium* and *Giardia*, protozoans excreted in human and animal feces that resist standard chemical disinfection. Biological contamination also lessens the recreational value of water when it makes swimming, fishing, and other recreational activities hazardous.

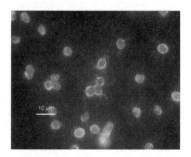

▲ *Cryptosporidium*, shown here in a fluorescent image, sickened 400,000 people and killed about 100 in the Milwaukee area in 1993. Since then, water utilities have generally improved their treatment systems.

SELF-ASSESSMENT Questions

1. It is estimated that half of the world's hospital beds are occupied by people with waterborne diseases. The main causes of these are
   a. microorganisms
   b. DDT and dioxins
   c. lead and mercury
   d. pesticides

2. The biochemical oxygen demand (BOD) of a water sample indicates the
   a. dissolved oxygen level
   b. organic matter level
   c. pH level
   d. salt levels (salinity)

3. How are dissolved oxygen and BOD related?
   a. BOD has no impact on dissolved oxygen.
   b. High BOD means a high level of dissolved oxygen.
   c. High BOD reduces dissolved oxygen as the organic matter decays.
   d. Low BOD means a low level of dissolved oxygen.

4. Anaerobic reduction produces
   a. CO_2, H_2O, and inorganic ions
   b. CO, NH_3, and H_2O
   c. CH_4, NH_3, and H_2S
   d. sewage sludge

5. Excess nitrates or phosphates in a lake or river cause
   a. aerobic decay
   b. eutrophication
   c. an increase in game fish
   d. nitrogen fixation

6. Which of the following are released into water by human and animal waste?
   a. fertilizers
   b. pesticides
   c. detergents
   d. antibiotics

7. Which of the following is most likely a source of safe drinking water?
   a. filtered pond water
   b. a municipal water supply
   c. a mountain lake
   d. a wilderness stream

8. Which ion, from fertilizer runoff and wastewater discharges, can lead to methemoglobinemia in young children?
   a. Ca^{2+} b. K^+ c. NO_3^- d. PO_4^{3-}

Answers: 1, a; 2, b; 3, c; 4, c; 5, b; 6, d; 7, b; 8, c

14.4 The World's Water Crisis

Learning Objective • Become aware of the global potable water crisis, including the inequitable distribution of water around the planet, and the quality of water available.

In developed nations, *potable* water safe for drinking is just a water faucet, a public restroom, or a home bathroom away. Yet from a global perspective, a large part of our planet already suffers from three problems: the distribution of water (either too scarce or in excess, as in severe flooding), the quality of the water, and the lack of adequate sanitation facilities.

Nearly a billion of the poorest people on Earth—most of them in Africa or Asia—subsist on less than 5 gallons of water a day, usually carried by women in "jerry cans" on their heads for an average of 9 km. WHO estimates that about 850 million people in the world have *no* access to "safe" drinking water. Experts predict that by 2025, ~1.8 billion people will live in regions of severe water scarcity, and some believe that water shortage will be the norm in the future, resulting in regional conflicts or even wars, simply over water rights.

Water shortages are not limited to the poorest in the world. In mid-January 2018, residents of Cape Town, South Africa, were being urged to use less than 19 gallons of water a day, and to take fewer and shorter showers, since the city's fresh water supply "could run out in less than two months!"* Three years of very low rainfall and a growing population were cited as contributing factors. In the United States, an increasing number of large cities risk running short of fresh, potable water, especially cities in California, Texas, and Arizona.

The opposite side of water scarcity is too much water. Some areas of the world are subject to periodic, occasionally severe flooding due to weather extremes or other factors. Section 14.3 mentioned the deleterious results of severe flooding in terms of human health.

Besides poor access to quality water, over 2.3 billion people in the world live without basic sanitation, defined as a toilet or latrine. Poor sanitation is a major factor in transmission of dysentery, hepatitis A, typhoid, polio, cholera, and diarrhea, as well as many tropical diseases, including intestinal worms. WHO estimates that inadequate sanitation causes 280,000 deaths annually. This global water crisis ranks second only to climate change in terms of overall threat to humans on planet Earth.

If you consider how important fresh, safe water is to you and how little of it there is, you should think about using water sustainably, especially given the ever-increasing human population. Ask yourself, no matter where you live, how long was your shower today? Can you limit yourself to a 90-second shower?

▲ Millions of women in developing nations worldwide must walk many km each day to obtain a few gallons of water for their families. This amount is a fraction of the amount of water used each day, per person, in developed nations. An article in National Geographic (April 2010) asserted that if women had faucets outside their doors, whole societies could be transformed.

SELF-ASSESSMENT Questions

1. All of the following are major aspects of the world's water crisis except
   a. bacterial contamination
   b. lack of an adequate quantity of potable water that is easily accessed
   c. lead and mercury in the water
   d. severe flooding that spreads bacterial and viral infections

2. Essentially every large city and town in the United States
   a. may not be able to provide adequate clean water to its inhabitants in the near future
   b. is fully capable of finding an unlimited amount of water, using modern technology

Answers: 1, c; 2, a

14.5 Tap Water and Government Standards for Drinking Water

Learning Objective • List some groundwater contaminants.

Surface water in the United States supplies about half the drinking water, and the other half is supplied by groundwater (from wells). In rural areas in particular, 97% of people drink groundwater. Due to differences in the physical nature of various rocks, water does not move within all rock in the same way. An aquifer forms when water-bearing rock readily transmits water to springs. Wells are drilled into an aquifer, and water is pumped out. Precipitation recharges the aquifer. In many instances, water is being pumped out of the aquifer much faster than it can be recharged, causing the water table to drop. In such areas, deeper and deeper wells have to be drilled. Clearly, this situation is not sustainable. Using an already depleted aquifer to irrigate crops today will cause crop losses tomorrow.

Safe Drinking Water Act

Over the last half-century, concerns have been expressed with respect to groundwater contamination. Stronger environmental laws and manufacturing methodology improvements have decreased those potential problems. The Safe Drinking Water Act was passed in the U.S. in 1974 and amended in 1986 and 1996. The act gives the Environmental Protection Agency (EPA) power to set, monitor, and enforce national health-based standards for a variety of contaminants in municipal water supplies. As our ability to identify smaller and smaller concentrations of potentially harmful substances has improved, the number of regulated substances with *maximum contaminant levels (MCLs)* has increased, from 22 in 1976 to 97 in 2016. The current list includes 12 microbial contaminants. Table 14.3 lists the MCLs for some inorganic and organic contaminants, including pesticides or herbicides used in agriculture. (See Section 19.2 and Section 19.3.)

TABLE 14.3 **U.S. Environmental Protection Agency Drinking Water Standards for Selected Substances[a]**

Inorganic Substance	Maximum Contaminant Level (mg/L)	Organic Substance	Maximum Contaminant Level (mg/L)
Arsenic	0.010	Atrazine	0.003
Barium	2	Benzene	0.005
Copper	1.3	*p*-Dichlorobenzene	0.075
Cyanide ion	0.2	Dichloromethane	0.005
Fluoride ion	4.0	Heptachlor	0.0004
Lead	0.01	Lindane	0.0002
Nitrate ion	10[b]	Toluene	1
Mercury	0.002	Trichloroethylene	0.005

[a]A much more extensive list and detailed explanation of the rules can be found on the EPA website.
[b]Measured as N. When measured as nitrate ion, the level is 45 mg/L.

Other Sources of Water Pollutants

In addition to being air pollutants (Section 13.4), VOCs (volatile organic chemicals) are also water pollutants. They add undesirable odor to water, and many are suspected carcinogens. VOCs are used as solvents, cleaners, and fuels, and they are components of gasoline, spot removers, oil-based paints, inks, dry cleaning solvents, and degreasers. Common VOC's are hydrocarbon solvents such as benzene (C_6H_6) and toluene (C_6H_5—CH_3), chlorinated hydrocarbons such as trichloroethylene (CCl_2=$CHCl$), as well as carbon tetrachloride (CCl_4), chloroform ($CHCl_3$) and methylene chloride (CH_2Cl_2), When spilled or discarded, VOCs enter the soil and, eventually, the groundwater. One well-documented case is the Love Canal site in Niagara Falls, New York, where people built schools and houses on or near old dump sites. The use of many of those VOCs in university laboratories has been significantly decreased or discontinued in many states.

Underground storage tanks, such as those at gas stations, that contain petroleum or hazardous chemicals last an average of about 15 years before they rust through and begin to leak. In the 1980s, the EPA estimated that there were at least 2.5 million leaking tanks. About 1,750,000 of all known tank sites have been cleaned up, but as of 2017, there were still 550,000 sites that had not been cleaned up.

Aboveground storage tanks are not immune to leakage either. In January 2014, 28,000 L of crude 4-methylcyclohexanemethanol (MCHM) leaked from a small hole in a tank capable of storing 150,000 L of the chemical. In spite of the existence of a containment area, the spill reached the groundwater and eventually contaminated the Elk River, which supplies water to the city of Charleston, West Virginia, and the surrounding areas. The "do-not-use" advisory for drinking water affected about 300,000 residents for several days.

▲ Half of all Chinese cities have significantly polluted groundwater, and most of those cities are facing a water crisis. Many industries pour wastewater directly into streams. More than 400 Chinese cities are threatened with shortages of potable water, with a third of them experiencing severe shortages. Here a water company worker takes a sample from the Songhua River in Songyuan in Jilin Province.

3 **Why isn't there a filter that removes all pollutants from water?**

Some water pollutants are present at extremely low levels, and the removal techniques available to us just aren't efficient enough to remove those pollutants. Distillation techniques will remove essentially dissolved ions, but this is an expensive method of purification. Filters attract some larger ions, but not every pollutant.

A much longer-lasting example of tainted tap water began in Flint, Michigan, in 2014, when the drinking water source for the city was changed from Detroit-treated water to the Flint River. Excess amounts of chlorine were added to kill bacteria, but the chlorine reacted with the extensive lead delivery pipes in the city, exposing hundreds of thousands of children to lead at levels up to 10 times the EPA maximum level. A federal state of emergency was declared in January 2016, and Flint residents were instructed to use only bottled or filtered water for drinking, cooking, cleaning, and bathing. Replacement of all the lead pipes will not be complete until 2020. This crisis certainly brought to the fore the problem of using lead pipes as water conduits.

SELF-ASSESSMENT Questions

1. Which of the following relationships between the withdrawal from and recharge of an aquifer is of concern?
   a. withdrawal > recharge   **b.** withdrawal < recharge   **c.** withdrawal = recharge

2. "V" in VOC stands for
   a. volcano   **b.** volatile   **c.** venomous   **d.** virulent

3. Which of the following is a VOC that contaminates groundwater?
   a. acetaldehyde from polluted air   **b.** acetic acid from fermenting fruit
   c. PAN from polluted air   **d.** toluene, a gasoline component

4. What is a possible source of VOCs?
   a. concrete   **b.** fracking liquid   **c.** glass bottles   **d.** rust

5. About how many substances are regulated under EPA drinking water standards?
   a. 25   **b.** 45   **c.** 100   **d.** 180

6. A value of 3 ppm of dissolved oxygen means
   a. 3 mg O_2/L H_2O **b.** 3 mg O_2/mL H_2O **c.** 3 g O_2/mL H_2O **d.** 3 g O_2/L H_2O

7. Which of the following corresponds to the *lowest* concentration?
   a. 3 ppt   **b.** 3 ppm   **c.** 3 ppb

Answers: 1, a; 2, b; 3, d; 4, b; 5, c; 6, a; 7, c

14.6 Water Consumption: Who Uses It and How Much?

Learning Objective • List the major users and uses of water.

Total *direct* daily use of water averages about 400 L per person in the United States, but ranges from less than 200 to more than 600 L/person. While only about 2 L are needed for drinking, such activities as cooking food, personal hygiene, laundry and dishwashing, watering the lawn and garden, and car washing consume the rest. Environmentally conscious individuals can cut their water use in half by carefully monitoring and controlling how much water goes down the drain.

A typical automobile takes several hundred kilograms of steel to produce. To make a metric ton (t) of steel requires about 100 t of water. About 4 t of water are lost through evaporation. The remainder is contaminated with acids, grease and oil, lime, and iron salts. Given the environmental and economic costs of producing elastomers for tires, fabrics for upholstery, glass for windows, and so on, it is easy to see that the private automobile is an ecological problem even before it hits the road.

About 70% of the world's freshwater supply is used in agriculture, almost all of it for irrigation. It takes up to 800 L of water to produce 1 kg of vegetables and 13,000 L of water to produce about 1 kg of beefsteak, including water used to grow animal feed, plus water for farm animals to drink. We also use water for recreation (for example, swimming, boating, and fishing). Most other industries and activities also contribute to water pollution. For most of these purposes, we need water that is free of bacteria, viruses, and parasitic organisms.

WHY IT MATTERS

Traditional paper production from virgin wood fibers uses enormous quantities of water. Worldwide water use for producing paper is about 315 billion kg each year, or about 100 lb of water for a single ream of paper. Wood is about half cellulose, but the cellulose fibers are held together by lignin, a resinous material that must be removed to make paper. Recycling newspapers, magazines, junk mail, and other paper reduces water usage by about 60% and energy consumption by about 40%. The EPA says that recycling results in 35% less water pollution and 74% less air pollution. An even more effective measure is to reduce paper use: Print only what you must!

4 About how much water do we really use each day?

The direct average daily water use per person in the United States is about 400 L, of which about 7 L is used for drinking and cooking, with the remainder used for personal hygiene, laundry, flushing toilets, washing dishes and cars, and other tasks. However, Table 14.5 shows that the indirect use of water is even more significant, most of it going toward energy production and agriculture. People can cut that use in half by using water more efficiently.

TABLE 14.5 Use of Water in the United States

Use	% of Total Usage
Coolant for electric power plants	48
Irrigation	34
Public water supplies	11
Industrial	5
Miscellaneous[a]	2

[a]Includes mining, livestock, aquaculture, and other uses.

Table 14.4 lists the estimated quantities of water required for the production of a variety of industrial and consumer products. Industries in the United States have substantially reduced their contribution to water pollution, and most are in compliance with the Federal Water Pollution Control Act as amended as of 1972, which requires that industries use the best practicable technology. Table 14.5 shows that significantly more water is used indirectly in agriculture and industry to produce food and products (34% of water usage) than is used by households via the public water supplies (about 11%).

TABLE 14.4 Estimated Quantities of Water Required[a] to Produce Various Materials

Industrial Products	Water Required[b] (t)	Consumer Products	Water Required (L)
Steel	100	Laptop computer	10,600
Paper	20	1 hamburger	2400
Copper	400	1 bowl rice	525
Rayon	800	1 kg bananas	860
Aluminum	1280	1 cup coffee	140
Synthetic rubber	2400	1 pair blue jeans	11,000

[a]Includes irrigation water, rainwater, water consumed directly, and water used in processing, cleaning, and waste disposal.
[b]Per metric ton of product.

SELF-ASSESSMENT Questions

1. The human activity that uses the most water is
   a. energy production   b. manufacturing electronics
   c. making elastomers   d. producing paper

2. Most of the water used by the steel industry is
   a. discharged untreated   b. lost through evaporation
   c. recycled   d. included in the final product

3. Most of the water used in agriculture is used for
   a. cooling   b. irrigation   c. drinking   d. cleanup

4. How much water does the average person in the United States use each day (for all direct uses)?
   a. 70 L   b. 100 L   c. 200 L   d. 400 L

Answers: 1, a; 2, c; 3, b; 4, d

14.7 Making Water Fit to Drink

Learning Objective • Describe how water is purified for drinking and cooking.

The per capita direct and indirect use of water in the United States is about 2.8 million L per year, slightly less than the volume of an Olympic pool. This value is more than twice the per capita consumption in Europe. Although we often take drinking water for granted, significant energy and resources are spent every year to ensure the safety and quality of drinking water supplied by the 155,000 public water systems in the United States.

Water Treatment

Since both surface water and groundwater are subject to contamination, most cities in developed nations treat their water supply before it flows into homes (Figure 14.4). The water to be purified is usually placed in a settling basin, where it is treated with slaked lime (calcium hydroxide) and a **flocculent** (a substance that causes particles

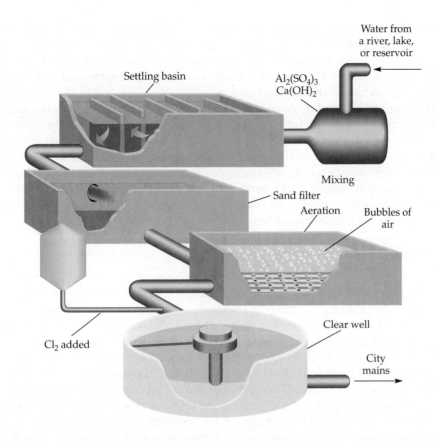

Water from
a river, lake,
or reservoir

Settling basin

$Al_2(SO_4)_3$
$Ca(OH)_2$

Mixing

Sand filter

Aeration

Bubbles of
air

Cl_2 added

Clear well

City
mains

◀ **Figure 14.4** Diagram of a municipal water purification plant. The plant takes water from a river, lake, or waterway and removes suspended matter, kills microorganisms, and adds dissolved air before piping the water to consumers.

Q *Why are aluminum sulfate and calcium hydroxide added before the water enters the settling basin?*

to clump together), such as alum (aluminum sulfate). These materials react to form a gelatinous mass of aluminum hydroxide that carries dirt particles and bacteria to the bottom of the settling tank:

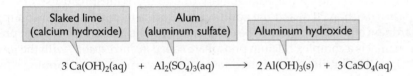

| Slaked lime (calcium hydroxide) | Alum (aluminum sulfate) | Aluminum hydroxide |

$$3\, Ca(OH)_2(aq) \;+\; Al_2(SO_4)_3(aq) \;\longrightarrow\; 2\, Al(OH)_3(s) \;+\; 3\, CaSO_4(aq)$$

The water above the settled solids is then filtered through sand and gravel, and sometimes through activated charcoal to remove colored and odorous compounds. The filtered water is usually further treated by *aeration*. It is sprayed into the air to remove odors and improve its taste. Water without dissolved air tastes flat.

Chemical Disinfection

In the final step of water treatment, chlorine is added to kill any remaining bacteria. In some communities that use river water, a lot of chlorine is needed to kill all the bacteria, and you can taste the chlorine in the water.

Analyses of the drinking water of several cities that take their water from rivers have found chlorinated hydrocarbons, including such known carcinogens as chloroform and carbon tetrachloride, formed by the conversion of dissolved organic compounds reacted with the chlorine. The concentrations are in the parts-per-billion range, probably posing only a small threat, but worrisome nonetheless. The presence of trace amounts of chlorinated hydrocarbons is not nearly as worrisome as the waterborne diseases that prevail in much of the world where adequate water treatment is not available.

Increasingly, ozone (O_3) is being used to disinfect drinking water. Adding ozone is more expensive than adding chlorine, but less of it is needed. An important advantage of ozonation is that it kills viruses, on which chlorine has little if any effect. For example, ozone is 100 times more effective than chlorine in killing polioviruses. Ozone acts by transferring its "extra" oxygen atom to the contaminant.

Oxygenated contaminants are generally less toxic than chlorinated ones. In addition, ozone imparts no chemical taste to water. Unlike chlorine, however, ozone does not provide residual protection against microorganisms. Some systems therefore use a combination of disinfectants: ozone for initial treatment and subsequent addition of chlorine to provide residual protection.

If you are on a camping trip, or are somewhere with no municipal water supply, heating water to a full, rolling boil for at least 1 minute often provides a quick and effective method of ridding water of pathogens. A variety of water purification tablets are available in many sporting goods stores, and are useful for back-country hiking where only surface waters are available for consumption. They usually contain chlorine dioxide or iodine, and take about 15–30 minutes to kill microorganisms in the water. Such products have been distributed in developing nations in Africa and Asia for purifying smaller quantities of water.

UV Irradiation

Water contaminated with a variety of microorganisms can be purified by irradiation with ultraviolet (UV) light. UV works rapidly, and it can be cost effective in small-scale applications. For instance, pouring clear water that has been previously filtered of solid impurities into a clear plastic bottle and exposing it to direct sunlight for at least 6 hours can purify a small amount of drinking water in an emergency. This methodology has been utilized all over Africa, where hundreds of large, clear plastic bottles of water are left on top of metal-roofed structures for at least 6 hours. No generation, storage, or handling of chemicals is required. UV is effective against *Cryptosporidium*, and no by-products are formed at levels that cause concern. The disadvantages include lack of residual protection for drinking water and of taste and odor control. UV has limited effectiveness in turbid water and generally costs more than chlorine treatment on a large-scale basis.

Fluoridation

Dental caries (tooth decay) was once considered the leading chronic disease of childhood. This is no longer true, mainly due to the widespread use of fluoride toothpastes (Section 20.6) and the addition of fluoride to municipal water supplies. The hardness of tooth enamel can be correlated with the amount of fluoride present. Tooth enamel is a complex calcium phosphate called *hydroxyapatite*, with the formula $Ca_5(PO_4)_3OH$. Fluoride ions replace some of the hydroxide ions, forming a harder mineral called *fluorapatite*, with the formula $Ca_5(PO_4)_3F$.

$$Ca_5(PO_4)_3OH(s) + F^-(aq) \longrightarrow Ca_5(PO_4)_3F(s) + OH^-(aq)$$

Early studies showed reductions in the incidence of dental caries by 50% to 70% because of fluoridation of water. More recent studies show a much smaller effect, perhaps because of the common practice of using fluoride treatments for teeth in children, and the presence of fluoride in food products prepared with fluoridated water.

Some people object to water fluoridation, and some cities either are considering or already have ceased fluoridation of their municipal water supply. The fluoride concentration in the drinking water of many communities has been adjusted to 0.7–1.0 ppm (by mass) by adding fluoride, usually as H_2SiF_6 or Na_2SiF_6. Fluoride salts are acute poisons in moderate to high concentrations. Indeed, sodium fluoride (NaF) is used as a poison for roaches and rats. Small amounts of fluoride ion, however, seem to contribute to our well-being through strengthening bones and teeth. There is some concern about the cumulative effects of consuming fluorides in drinking water, in the diet, in toothpaste, and from other sources. Excessive fluoride consumption during early childhood can cause mottling of tooth enamel. The enamel becomes brittle in certain areas and gradually discolors. Fluorides in high doses also interfere with calcium metabolism, kidney action, thyroid function, and the actions of other glands and organs.

▲ UV disinfection is a relatively simple way to destroy pathogenic bacteria in water, by laying large, clear plastic bottles filled with water on top of corrugated metal platforms above ground. In the hot sun of Africa, UV rays penetrate the plastic and kill bacteria, particularly at higher temperatures of the metal roof. This methodology is one answer to remediating or at least lessening incidence of gastrointestinal diseases in developing nations in Africa and Asia.

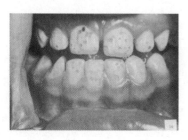

▲ Excessive fluoride consumption in early childhood can cause mottling of tooth enamel. The severe case shown here was caused by continuous consumption during childhood of water from a supply that had an excessive natural concentration of fluoride.

SELF-ASSESSMENT Questions

1. Public water supplies are aerated to
   a. improve the taste   **b.** kill bacteria
   c. remove suspended matter   **d.** remove trihalomethanes

2. Chlorine is added to public water supplies to
   a. improve the taste   **b.** kill bacteria
   c. remove suspended matter   **d.** remove trihalomethanes

3. What advantage does treatment of water with O_3 have over chlorination?
   a. O_3 is cheaper than Cl_2.
   b. O_3 provides residual protection; Cl_2 does not.
   c. O_3 kills viruses.
   d. Sedimentation is eliminated.

4. The chemical reaction responsible for the disinfecting properties of ozone is
   a. flocculation
   b. precipitation
   c. reduction
   d. oxidation

5. Fluoridation of drinking water strengthens tooth enamel by converting hydroxyapatite to
   a. $CaCO_3$
   b. CaF_2
   c. fluorapatite
   d. fluorotartrate

Answers: 1, a; 2, b; 3, c; 4, d; 5, c

The Soft Drink in Vogue: Bottled Water

Bottled water is an extremely profitable segment of the beverage industry. Though bottled water is not their exclusive source of water, more than half of all Americans drink it, even though it costs 240–10,000 times as much per liter as tap water. In 2016, bottled water consumption per capita was nearly 150 L.

In the United States in general, both bottled water and tap water are quite safe. Yet some people will continue to buy bottled water for its convenience, taste, and assumed (but not necessarily real) health benefits.

Because bottled water is considered food by the U.S. government, it is regulated by the Food and Drug Administration (FDA) rather than by the EPA. Even though the FDA has adopted the EPA standards for tap water as the standards for bottled water, testing of the latter is typically less rigorous than testing of municipal water. Most people are unaware that about 25% of bottled water sold in the United States comes from municipal water supplies. And just because water comes in a bottle with a glacier or a mountain spring on the label does not mean that it is *purer*. Mineral water is, in fact, likely to have *more* dissolved ions (typically Ca^{2+} and Mg^{2+}) than normal tap water. Read the label of your bottled water to find out what ions have been added.

Consider the environmental impact of bottled water. It takes about 17 million barrels of oil to make the polyethylene terephthalate (PET) water bottles that Americans use each year, enough to fuel 1,000,000 cars for a year. Only about 13% of those bottles are recycled. Add the fuel used for transporting bottled water to stores, and its environmental cost is even higher. Perhaps filling a cleaned plastic bottle with water from a drinking fountain would make much more sense than buying bottled water, both economically (for you!) and environmentally (for the Earth). Some airports and college campuses have even installed special taps for filling water bottles.

14.8 Wastewater Treatment

Learning Objective • Describe primary, secondary, and tertiary treatment of wastewater.

The wastes generated by nearly two-thirds of the U.S. population are gathered in sewer systems and carried to treatment plants by more than 50 billion L of water each day. Cities have to treat that wastewater before discharging it back into the environment. Treatment and purification of wastewater render it suitable for return to the water cycle.

Sewage Treatment Methods

For decades, most communities treated sewage simply by holding it in settling ponds for a while before discharging the resulting *effluent* into a stream, lake, or ocean, a process now called **primary sewage treatment** (Figure 14.5). Primary treatment removes 40% to 60% of suspended solids as *sludge* and about 30% of organic matter. The effluent still has a BOD level two-thirds of the original level, and nearly

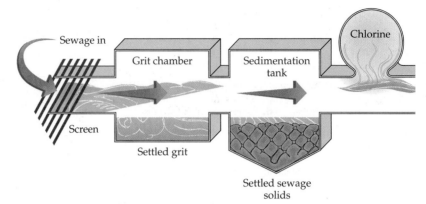

▲ **Figure 14.5** Diagram of a primary sewage treatment plant. Such treatment minimizes the harmful effect of wastewater on waterways.

Q *Why isn't primary treatment alone sufficient for wastewater purification?*

all its nitrates and phosphates. All the dissolved oxygen in the settling pond may be used up, and anaerobic decomposition—with its resulting odors—takes over.

In **secondary sewage treatment**, the effluent from primary treatment is passed through sand and gravel filters. There is some aeration in this step, and aerobic bacteria convert much of the organic matter to inorganic materials. In the **activated sludge method** (Figure 14.6), a combination of primary and secondary treatment methods, the sewage is placed in tanks and aerated with large blowers. This causes the formation of large, porous biological clumps called *flocs*, which filter and absorb contaminants. Aerobic bacteria further convert the organic material to sludge. Part of the sludge is recycled to keep the process going, but huge quantities must be removed for disposal. This sludge is stored on land (where it requires large areas), dumped at sea (where it pollutes the ocean), or burned in incinerators (which requires energy—such as natural gas—and can contribute to air pollution). Some of the sludge is used as fertilizer.

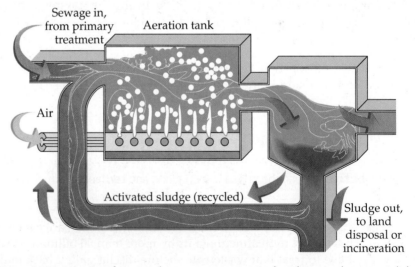

▲ **Figure 14.6** Diagram of a secondary sewage treatment plant that uses the activated sludge method.

Q *After organic matter is converted to sludge, why is some of the sludge recycled to the aeration tank?*

Secondary treatment of wastewater lowers the BOD by about 90% but often does not adequately reduce nitrates and phosphates. Increasingly, federal mandates require **advanced treatment** (sometimes called *tertiary treatment*). Several advanced processes are in use, and most are quite costly. Finding the money to finance adequate sewage treatment will be a major political problem for years to come.

Other water purification methods, used in homes or in industries using water in consumer products such as soft drinks and soups, include the following:

- In **activated charcoal filtration**, charcoal adsorbs certain organic molecules, such as chloroform ($CHCl_3$), that are difficult to remove by any other method. The organic molecules are adsorbed on the surface of the charcoal and thus removed from the water. After a period of time, the charcoal becomes saturated and is no longer effective. It can be regenerated by heating it to drive off the adsorbed substances. A home water-filtration system that attaches to a faucet usually contains activated charcoal.
- In **reverse osmosis**, pressure forces the wastewater through a semipermeable membrane, leaving contaminants behind. Bottled water and water for the preparation of soft drinks usually have been purified through reverse osmosis.
- **Phytoremediation** involves passage of the effluent into large natural or constructed lagoons for storage, allowing plants, such as reeds, to remove metals and other contaminants.

The effluent from sewage plants is usually treated with chlorine to kill any remaining pathogenic microorganisms before it is returned to a waterway. Chlorination has been quite effective in preventing the spread of waterborne infectious diseases such as typhoid fever. Further, some chlorine remains in the water, providing residual protection against pathogenic bacteria.

Water Pollution and the Future

For most of human history, our bodily wastes were an integral part of Earth's natural recycling system. The wastes provided food for microorganisms that degraded them, returning nutrients to soil and water. But with the growth of cities, human waste became disconnected from the cycle. In an effort to keep water clean and safe for our use, we also rid it of nutrients that could fertilize the land.

Many communities dry and sterilize sludge and then transport it to farmlands for return of nutrients to the soil. Others pump wet, suspended sludge directly to the fields. The water in the mixture irrigates the crops, and the sludge provides nutrients and humus. One concern is that the sludge is often contaminated with toxic metals that could be taken up by plants and eventually end up in our food.

A few communities treat their sewage by first holding it in sedimentation tanks, where the solids settle out as a sludge that is removed and processed for use as fertilizer. The effluent is then diverted into a marshy area, where the marsh plants filter the water and use nutrients from the sewage as fertilizer. Some plants even remove toxic metals. Effluent from the marsh is often as clean as the water in municipal reservoirs.

Other solutions are possible. Septic tanks have been used for home sewage treatment for decades, but their use is generally limited to rural and some suburban areas. They also require significant upkeep to remain effective. Toilets have been developed that compost wastes and use no energy or water. The mild heat of composting drives off water from the wastes. The system is ventilated to keep the process aerobic, and no odors enter the house from a properly installed system. Dried waste is removed about once a year. The initial cost is much higher than that of a flush toilet.

Each person in the United States flushes about 35,000 L of drinking-quality water each year. While newer toilet models use an average of 6 L of water, many households still use toilets that require about 15 L to flush.

We're the Solution to Water Pollution

We generally take our drinking water for granted. Perhaps we shouldn't. Thousands of cases of waterborne illnesses are reported in the United States each year. About 10% of public water supplies do not meet one or more of the EPA standards. Twenty million people in the U.S. have no running water at all. Another estimated 30 million people tap individual wells or springs, whose water is often of unknown quality. Much remains to be done before we can all be assured of safe drinking water.

5 How much water is wasted when a faucet drips?

When an average kitchen or bathroom faucet drips at an average rate of 15 drips a minute and an average faucet drop has a volume of 0.25 mL, over the span of one year, nearly 2000 L of water go straight down the drain. If in a medium-size city (200,000 homes) there is one dripping faucet per household, each day homeowners pay a utility company for over 1 million liters of water that are going to waste.

6 How do you get fresh water from saltwater?

With saltwater comprising more than 97% of our planet's water, desalination holds much promise. There are several ways of removing salts from water. Water can be evaporated (to separate it from the salt) and the resulting steam condensed back to liquid form. In reverse osmosis, water is allowed to pass through a membrane, leaving salt behind. Another method uses the difference in conductivity of saline versus that of fresh water to concentrate salt and isolate the salt-depleted water. Finally, careful freezing and thawing of saltwater may capture salt-depleted ice with higher salt concentration, which will melt first. Desalination can be quite costly in terms of energy consumption, since at least 1 kWh/m^3 is required to desalinate water. It takes only 0.2 kWh/m^3 or less to process fresh water.

GREEN CHEMISTRY Fate of Chemicals in the Water Environment
Alex S. Mayer, Michigan Technological University

Principles 1, 4, 5, and 10

Learning Objectives • Connect the properties of chemicals to their fate in the environment. • Describe how chemicals migrate between environmental compartments, including how chemicals' effects can magnify as they travel through food webs.

A key part of green chemistry is understanding how chemicals move and change when they enter the environment. Important questions include how chemicals move, how and why they react in the environment, and how these chemicals affect organisms in the environment like fish and other animals, plants, and people. Here, we focus on the movement of organic chemicals in water—lakes, rivers, and oceans—but the principles we are presenting also apply to other classes of chemicals and where they end up in the environment.

To make things easier, scientists and engineers like to track chemicals by dividing the larger environment into smaller parts or compartments such as air, water, and solids. Solids include sediments (found at the bottom of rivers, lakes, and oceans) and soils or other materials, such as layers of sand or rock that occur below the earth's surface. Once a chemical enters the environment, it can move into these different compartments through a process known as partitioning. Partitioning is best understood as the amount of a chemical that ends up in each compartment.

One way to think about partitioning is to think about salad dressing made of oil and vinegar. You may have seen how oil and vinegar will separate into distinct layers unless you vigorously mix them together. When you do mix them, most of the oil and vinegar will once again settle into different layers, but a little bit of the oil will dissolve or partition into the vinegar and a little bit of the vinegar will dissolve in the oil. The same phenomenon happens when a chemical in water comes into contact with air or sediment; some of the chemical will stay in the water, but some will partition into the sediment or into the air. (See Section 14.6 on groundwater contamination.)

How much of the chemical remains in the water and how much goes into the solid depends on the properties of the water and the solid and on the properties of the chemical. To understand this process better, scientists use water and oil as a model system for predicting how much of a chemical will remain in water, and how much will go into a solid. For example, a pesticide like DDT is about 10 million times more likely to go from water into oil (or fats, sediments, or soils) than a solvent like acetone. Why? DDT is very soluble in oil and is a nonpolar compound, whereas acetone (Section 9.6) is very soluble in water (less soluble in oil) and polar. Compounds like DDT tend to "hate" water, which is the most polar compound on earth (Section 4.12). Therefore, compounds such as DDT are usually found in smaller amounts in water than in other environmental compartments. More polar compounds, such as acetone, "prefer" to stay in the water compartment.

You might think that, in terms of environmental impact, it is better to have chemicals that like to stay in the solid (or in the oil) because less of the chemical will be found in water. However, chemicals that partition more strongly into solids tend to stay in the environment for

a longer time. Once they go into or onto the solids, only minute quantities are released into surrounding water over time, although potentially at toxic levels. Therefore, we can say that DDT is 10 million times more persistent than acetone. (See Section 1.1 and the work of Rachel Carson on pesticides, especially DDT.) Understanding the properties of chemicals can help us prevent pollution and use and design safer products through green chemistry (Principles 1, 4, and 5).

Organisms exposed to organic chemicals in the environment may break down or change the chemicals in some way before they excrete them. Green chemistry can help to design chemicals that degrade into benign products (Principle 10). However, sometimes organisms absorb or retain the chemicals. Many organic chemicals tend to be absorbed into fatty deposits or tissues in organisms. As discussed above, some chemicals like DDT tend to be more highly absorbed in the fatty tissues. These chemicals can lead to toxic effects including neurological damage and birth defects if present at high enough levels.

Chemical absorption in animals usually leads to the progressive increase in the concentration of some organic chemicals in animals that eat other animals. This progressive increase in concentration is known as *biomagnification* (Section 20.2). For example, if the smallest fish species in a food chain is exposed to a chemical, then absorbs and stores the chemical, a larger fish that eats many of those fish will store a much larger amount of the chemical. This phenomenon is bad news for animals at the top of the food web, such as predatory birds or humans.

Principles like the one we just described help scientists understand that the properties of chemicals determine where they end up in the environment. Thus, as we make new chemicals that are supposed to benefit humans, we should apply green chemistry principles and design these chemicals so that they are less persistent in the environment with less likelihood to absorb solids and biomagnify (Principles 4, 5, and 10). Furthermore, it makes sense to synthesize chemicals so that when their useful lives are over, they will break down quickly in the environment or we have an easy way to recover and reuse them. Testing chemicals before they are introduced is critical so we can avoid unintended consequences like biomagnification. Finally, we should minimize the generation of waste chemicals that are likely to persist in the environment.

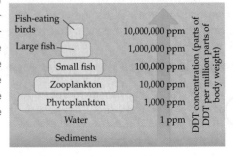

The Clean Water Act (a common name for the Federal Water Pollution Control Act) has drastically reduced water pollution from industrial sources. Yet the EPA's "National Water Quality Inventory: Report to Congress, 2017" stated that 46% of U.S. rivers and streams and 21% of lakes, ponds, and reservoirs are in poor biological condition, with nitrates, phosphates, and sediments too high. Complete elimination of pollution is not possible—to use water is to pollute it. All of us can do our share by conserving water and by minimizing our use of products that require vast amounts of water to make. As our population grows, it will cost a lot just to maintain present water quality. To clean up our water, and then keep it clean, will cost more. However, the costs of unclean water are even higher in terms of loss of recreation, aesthetics, and significant human health consequences.

SELF-ASSESSMENT Questions

1. At what stage in sewage treatment do bacteria consume dissolved organic matter?
   - **a.** aeration
   - **b.** primary
   - **c.** secondary
   - **d.** screening and sand removal

2. Trihalomethanes are formed when chlorine reacts with
   - **a.** bacteria
   - **b.** dissolved organic compounds
   - **c.** ethanol
   - **d.** phenols

3. Trihalomethanes can be removed from water by
   - **a.** charcoal filtration
   - **b.** chlorination
   - **c.** flocculation
   - **d.** sand filtration

4. Reverse osmosis is a way to purify water by using pressure to force the water through a
   - **a.** charcoal filter
   - **b.** sand filter
   - **c.** semipermeable membrane
   - **d.** trickling filter

5. Phytoremediation utilizes _____ to remove metals and other contaminants from wastewater.
   - **a.** plants
   - **b.** UV irradiation
   - **c.** charcoal
   - **d.** ozone

6. A wastewater treatment plant treats 500,000 gal. of sewage per day, using 16 lb of chlorine per day. Assume that 1 gal. of sewage weighs about 8 lb. The chlorine concentration, in parts per million, of the wastewater is about
   - **a.** 1 ppm
   - **b.** 4 ppm
   - **c.** 8 ppm
   - **d.** 16 ppm

7. Approximately how much water is wasted when a bathtub faucet drips for 24 hours at the rate of 10 times a minute? Assume the volume of a drip to be 0.25 mL.
   - **a.** 1 L
   - **b.** 3 L
   - **c.** 4 L
   - **d.** 15 L

Answers: 1, c; 2, b; 3, a; 4, c; 5, a; 6, b; 7, c

 Summary

Section 14.1—Water has unusual properties based primarily on its polarity and strong hydrogen bonding. It is the only common liquid on Earth; its solid form is less dense than the liquid. **Specific heat capacity** is the amount of heat needed to raise the temperature of 1 g of a substance by 1 °C (or 1 K). For liquid water, it is quite high: 1 cal/g °C (or 4.18 J/g K). Water also has a high **heat of vaporization**, the amount of heat required to vaporize a fixed amount of a liquid. Water is also an excellent solvent, especially for many other polar and ionic solutes.

Section 14.2—Water covers 75% of Earth's surface, but less than 1% of it is available as fresh water for human use. Fresh water has much lower concentrations of dissolved solutes; fresh does not equal pure. Water dissolves many ionic substances, which is why the oceans are salty. Potable water is readily available in most parts of the United States, but many other countries do not have enough fresh water. In the water (hydrologic) cycle, water evaporates from oceans and lakes, condenses into clouds, and returns to Earth as rain or snow. During this cycle, water can be contaminated by various chemicals and microorganisms from both natural sources and human activities. **Hard water** is formed when minerals (salts of calcium, magnesium, and iron) dissolve in groundwater. Acid rain created by the release of sulfur oxides and nitrogen oxides when coal and other fuels are burned has caused acidification of thousands of bodies of water. Besides corrosive effects, acid rain is detrimental to both plant and animal life.

Section 14.3—Waterborne diseases were a severe problem in wealthier countries until about 100 years ago, but are still common in many parts of the world. When sewage is dumped into waterways, the organic matter is biodegraded by microorganisms. Aerobic oxidation occurs in the presence of dissolved oxygen, and the **biochemical oxygen demand (BOD)** is a measure of the amount of oxygen needed. Sewage increases the BOD. If the BOD is high enough, only **anaerobic decay**, or decay in the absence of oxygen, can occur. High levels of nitrates and phosphates can accelerate **eutrophication**, in which algae grow and die, thereby increasing the BOD of the water. Other new substances introduced into the water cycle may cause new problems.

Section 14.4—A significant portion of the world's population subsists on about 5 gallons of water per day, particularly in developing nations in Africa and Asia. Many areas of the world are subject to severe flooding, spreading disease-causing bacteria. Waterborne diseases run rampant in many such countries, though they also are found more rarely in developed nations. Access to safe sanitation in those regions is a contributing factor to the world's water crisis.

Section 14.5—About half of the U.S. population drinks surface water, and half drinks groundwater. Use of groundwater is especially common in rural areas. The aquifers into which wells are drilled are being consumed faster than they are being recharged by precipitation, and deeper wells are now required in many places. Well water may be contaminated with various organic chemicals, including VOCs, nitrates, and phosphates. Nitrates can be deadly to babies in high enough concentrations, and phosphates cause eutrophication (rapid algae growth) in ponds and lakes. The Safe Drinking Water Act sets and enforces standards for water quality in the United States. In many cases, concentrations of contaminants must be expressed in tiny units, such as parts per million (ppm) or parts per billion (ppb).

Section 14.6—Producing industrial and consumer products uses a great deal of water, which is polluted in the process and must be cleaned up and recycled. Each person needs only about 2.0 L of water per day for drinking, but in the United States, each person directly uses a total of about 400 L per day, which includes various household uses. Indirect water use is much higher.

Section 14.7—Treatment of municipal water supplies to make the water safe to drink usually includes settling, treatment with a **flocculent** (to cause particles to clump together and settle out), filtration, aeration, and chemical disinfection with chlorine or ozone. Fluorides may also be added to prevent tooth decay. Boiling, or adding water-purifying tablets, may be used for small-scale water purification.

Section 14.8—Treatment of wastewater (used water) begins with **primary sewage treatment**, allowing the wastes to settle and removing sludge before discharging the water into the environment. In **secondary sewage treatment**, the effluent is filtered through sand and gravel. The **activated sludge method**, in which the sewage is aerated and the biological sludge is removed, may be used instead. **Advanced treatment** (tertiary treatment) may involve **activated charcoal filtration** to adsorb organic compounds, **reverse osmosis** to remove contaminants by forcing the wastewater through a semipermeable membrane, or **phytoremediation**, to allow plants to remove metals and other contaminants by storing the wastewater in lagoons. These advanced processes are expensive. Maintaining water quality is expensive and will become more so, but not maintaining it would cost much more in terms of our health and comfort.

GREEN CHEMISTRY—Chemicals can be released into the environment and have toxic effects on humans and other species. The properties of chemicals determine where and for how long they stay in the environment. It makes sense to design chemicals that will persist for shorter periods and occur in smaller amounts in organisms and to minimize the release of chemicals that are persistent and tend to magnify in the food web.

Learning Objectives

Learning Objectives	Associated Problems
• Relate water's unique properties to polarity of the water molecule and to hydrogen bonding. (14.1)	1–4, 18, 19–21, 63
• Explain how water on the surface of Earth acts to moderate daily temperature variations. (14.1)	5, 6, 22, 23
• Explain why humans can only make use of less than 1% of all the water on Earth. (14.2)	27
• Explain the water cycle on Earth. (14.2) • Identify natural contaminants in rain and natural bodies of water. (14.2)	8, 64, 65 7, 24–26, 34–36
• Describe how human and animal waste affect water quality. (14.3) • Give an example of a biological water contaminant. (14.3) • Identify the sources of nitrates in groundwater and problems they can cause. (14.3)	28–31, 70 10, 13, 28, 29 9
• Become aware of the global potable water crisis, including the inequitable distribution of water around the planet, and the quality of water available. (14.4)	11, 14
• List some groundwater contaminants. (14.5)	12, 15–17, 32, 33
• List the major users and uses of water. (14.6)	37, 41–44
• Describe how water is purified for drinking and cooking. (14.7)	39, 51, 66–67
• Describe primary, secondary, and tertiary treatment of wastewater. (14.8)	45–50, 52
• Connect the properties of chemicals to their fate in the environment.	74
• Describe how chemicals migrate between environmental compartments, including how chemicals' effects can magnify as they travel through food webs.	75–76

Conceptual Questions

1. Explain on a molecular level why ice is less dense than liquid water. Be sure to use the concept of hydrogen bonding in your explanation.

2. What consequences does the lower density of ice compared with that of liquid water have for life in northern lakes?

3. A tanker filled with light crude oil sinks and breaks open. Explain on the molecular level why the oil will not mix into the ocean water.

4. Explain why oil from an oil tanker spill floats on the surface of the ocean.

5. Why do inland regions have colder winters than regions near oceans, even if they are closer to the equator?

6. Why does your skin feel cool when you step out of a swimming pool into a breeze?

7. Why are oceans salty? Are they getting saltier? Explain on a molecular level.

8. Summarize the water cycle on Earth and describe one way that nature itself can begin to purify water.

9. Name several major sources of plant nutrients that enter waterways.

10. What are pathogenic microorganisms?

11. List some waterborne diseases that are common in developing nations. Why are these diseases no longer common in developed countries?

12. List some ways in which groundwater is contaminated with organic substances.

13. Discuss how sewage is related to biological oxygen demand (BOD), dissolved oxygen (DO), and eutrophication.

14. What are the consequences of significantly varied distribution of water around the globe, in terms of health challenges?

15. Acidic water that is neutralized by natural rock often becomes hard water. Explain.

16. What problems do leaking underground storage tanks cause?

17. Why do chlorinated hydrocarbons remain in groundwater for such a long time?

Problems

Properties of Water

18. Define *specific heat capacity*. Why is the high specific heat capacity of water important to Earth?

19. Refer to Table 14.1. How much heat, in calories, does it take to warm 750 g of water from 12.0 °C to 45.0 °C?

20. How much heat, in calories, does it take to warm 750 g of iron from 12.0 °C to 45.0 °C?

21. Compare your answers to Problem 19 and Problem 20. Both involve the same increase in temperature. Explain why.

22. Define *heat of vaporization*. Explain why water's value is exceptionally high.

23. Why is the high heat of vaporization of water important to our bodily functions?

Natural Waters

24. What impurities are present in rainwater and where do they come from?

25. Why is the pH of rainwater not 7.0?

26. What is the average pH of the ocean? Explain which ions or molecules regulate that pH.

27. Why do humans make use of less than 1% of the natural waters on Earth?

28. What are the products of the breakdown of organic matter by anaerobic decay?

29. What is the difference between aerobic and anaerobic bacteria?

30. Define *eutrophication*.

31. Eutrophication is amplified by
    a. heavy metals from landfills
    b. plant nutrients from fertilizer runoff and wastewater
    c. radioactive waste from nuclear power plants
    d. toxic chemicals from factories

32. Water containing more than 3.0 g of dissolved solids per gallon is considered to be hard. What is this "hardness limit" in milligrams of dissolved solids per liter?

33. Refer to Problem 32. A large container holds 18 liters of tap water, which is allowed to evaporate. The calcium- and magnesium-containing solid material left behind has a mass of 2.1 g. Can the tap water be considered to be hard?

Acidic Waters

34. List two ways in which lakes and streams become acidified.

35. Surface water is most likely to be acidified by acid rain in areas underlain by
    a. dolomite
    b. granite
    c. limestone
    d. magnesite ($MgCO_3$)

36. Why is acidic water with a pH under 4 especially harmful to fish?

Municipal Water Supplies

37. What is the optimal level of fluoride in drinking water?

38. What are some health effects of too much fluoride in the diet?

39. Can water be disinfected without adding chemicals? Explain.

40. What is *methemoglobinemia*? What is one cause?

41. How much water does a person need per day for drinking?

42. Explain why it takes many times as much water to produce a single serving of steak as it takes to produce a single serving of rice.

43. What are the pros and cons of using bottled water?

44. Which federal agency regulates drinking water standards from municipal water supplies? Which federal agency regulates bottled water?

Wastewater Treatment

45. Describe primary sewage treatment. What impurities does it remove?

46. Describe secondary sewage treatment. What impurities does it remove?

47. What substances remain in wastewater after effective secondary treatment?

48. Describe the activated sludge method of sewage treatment.

49. Identify each of the following as a primary, secondary, or tertiary method of wastewater treatment.
    a. charcoal filtration
    b. settling pond
    c. sand and gravel filters

50. What are the advantages and disadvantages of spreading sewage sludge on farmland?

Chemical Equations

51. Write the balanced equation for the neutralization of acid rain (assume HNO_3 is dissolved in water) by limestone (calcium carbonate).

52. Write the balanced equation for the reaction of slaked lime (calcium hydroxide) with alum (aluminum sulfate) to form aluminum hydroxide used in wastewater treatment.

53. Write the balanced equation for the reaction in which fluoride ions replace hydroxide ions in tooth enamel.

54. Write the balanced equation for the reaction that shows why natural water has an acidic pH.

Parts per Million and Parts per Billion

55. Express a concentration of 9 μg of benzene per liter of water, as ppb benzene.

56. Express 22 g Br_2 in 1.00×10^4 L of water as ppm Br_2.

57. Express 0.011% by mass $Ba(NO_3)_2$ as ppm $Ba(NO_3)_2$.

58. Express 36 μg chloroform in 65 L of water as ppb chloroform.

59. A 4.0 L sample of drinking water is found to contain 36 μg of copper and 32 mg of nitrate ion. Do either of these concentrations exceed the standards in Table 14.3?

60. A 0.50 L sample of drinking water is found to contain 4.0 μg of toluene. Does this exceed the standard in Table 14.3?

61. Using Table 14.3, and assuming a mass of 1 mg of each of the following, determine which substance/L must be more toxic.
    a. arsenic
    b. cyanide
    c. benzene
    d. heptachlor

62. Which sample of fluoride ions is more concentrated?
    a. 4 ppm
    b. 3×10^{-4} mol/L

 Expand Your Skills

63. Which of the following incorrectly describes the charge distribution in a molecule of water?
    a. Evenly; the molecule is nonpolar.
    b. There is a negative partial charge on the H atoms and a positive partial charge on the O atom.
    c. There is a negative partial charge on one H atom and a positive partial charge on the other.
    d. There is a positive partial charge on the H atoms and a negative partial charge on the O atoms.

64. Which of the following processes is *not* a part of the water cycle?
    a. eutrophication   b. condensation
    c. evaporation   d. precipitation

65. Most of the fresh water on Earth is
    a. above in the atmosphere
    b. frozen in the polar ice caps
    c. in motion in rivers, lakes, and streams
    d. underground in aquifers

66. Which of the following processes is *not* used in a conventional method of water treatment?
    a. aeration   b. coagulation
    c. filtration   d. percolation

67. Wastewater disinfected with chlorine must be dechlorinated before it is returned to sensitive bodies of water. The dechlorinating agent is often sulfur dioxide. The reaction is

$$Cl_2 + SO_2 + 2 H_2O \longrightarrow 2 Cl^- + SO_4^{2-} + 4 H^+$$

Is the chlorine oxidized or reduced? Identify the oxidizing agent and reducing agent in the reaction.

68. A large mug containing 550 g of hot water at 92°C is left sitting on the table, and gradually cools to 25°C. How much heat was released to the environment (air, mug, and table), in kilocalories? What scientific principle is illustrated?

69. The text indicates that radioactive radon gas is slightly soluble in water. What are the intermolecular forces responsible for the poor solubility?

70. Consider the statement "BOD (biological oxygen demand) indirectly measures the levels of sewage in water, because microorganisms in sewage release oxygen into the water." Which of the following is true?
    a. The statement is correct, but the reason is wrong.
    b. The statement is wrong, but the reason is correct.
    c. Both the statement and the reason are correct.
    d. Neither the statement nor the reason is correct.

71. A typical gas-fired home water heater holds 50 gal. of water and has the thermostat set to 60 °C.
    a. If the inlet water supply is at 10 °C, how much heat, in kilocalories, is needed to bring the contents of the water heater to the set temperature? (1 gal = 3.8 L)
    b. Burning 1.00 mol of methane (CH_4), the major component of natural gas, produces 890 kJ of energy. What mass (in grams) of methane must be burned to produce the amount of heat determined in part **a**?
    c. What volume in cubic feet (at STP) of methane does the mass from part **b** correspond to? (1 ft^3 = 28.3 L)
    d. A *therm* of natural gas is about 100 ft^3 and costs about $1.50 for residential customers. Use the results from parts **a–c** to calculate how much it costs to take a 10-minute shower at a usage rate of 3.0 gal. of hot water per minute.

72. How much calcium nitrate has to be added to 1000 L of solution to provide 180 ppm Ca^{2+} if the water supply already contains 40 ppm Ca^{2+}? How much NO_3^-, in parts per million, does this quantity of calcium nitrate add to the solution?

73. Consider the statement "60 calories of heat will cause 24 g of copper to warm up about 10 °C, and 60 calories of heat will cause 24 g of water to warm up only around 2 °C, because copper has a higher heat capacity." Which of the following is true?
    a. The statement is correct but the reason is wrong.
    b. The statement is incorrect, but the reason is correct.
    c. Both the statement and reason are correct.
    d. Neither the statement nor the reason is correct.

74. If a chemist were charged with developing a new pesticide and that chemist were concerned with the how the pesticide might persist in the environment, should the new pesticide be more polar or more nonpolar?

75. An experiment is conducted with two chemicals, X and Y, where equal masses of chemicals are exposed to a sample of water mixed with soil. At the end of the experiment, chemical X is found at a concentration in the soil of 1/100th that of chemical Y. Which chemical is more polar?

76. The Michivannia State Department of Public Health is setting fish advisories for two different fish, red fish and blue fish, that may be exposed to toxic organic chemicals. Red fish and blue fish have fat contents of 5% and 15%, respectively. Red fish is typically consumed by blue fish. Which fish would you recommend can be eaten more often than the other?

Critical Thinking Exercises

Apply knowledge that you have gained in this chapter and one or more of the FLaReS principles (Chapter 1) to evaluate the following statements or claims.

14.1 A friend states that ice cubes made of saltwater or heavy water will float on normal "light" water.

14.2 A company states that its special sports water, through a secret molecular process utilizing advanced laser technology, contains 500 mg of NaCl per 1 g of water.

14.3 A mother discovers that tests have found 1.0 ppb of trichloroethylene in the well water her family uses for drinking, bathing, cooking, and so on. The family has used the well for 2 years. The mother decides that the whole family should be tested for cancer.

14.4 A group of citizens wants to ban fluoridation of a community water supply because fluorides are toxic.

14.5 A company claims that its "superoxygenated water can boost athletic performance." (Note that under pressure, water can contain at most 0.3 g O_2/L H_2O, the amount in a 1 L breath.)

14.6 A company claims that its "oxygenated and structured water has smaller molecules that penetrate more quickly into the cells, and therefore hydrate your body faster and more efficiently."

14.7 A friend states, "The water we drink in the present day may be some of the same water that flowed in the time of the dinosaurs."

Collaborative Group Projects

Prepare a PowerPoint, poster, or other presentation (as directed by your instructor) to share with the class.

1. Consult online or print references, and prepare a report on the desalination of seawater by one of the following methods.
    a. distillation   b. freezing
    c. electrodialysis   d. reverse osmosis
    e. ion exchange

2. Call your local water utility's office or consult its website and obtain a chemical analysis of your drinking water. What substances are monitored? Are any of these substances considered problems? (If you use water from a private well, has the water been analyzed? If so, what were the results?)

3. Poll your class on consumption of bottled water. About how many bottles do they drink each year? Why do your classmates use it? Are there concerns regarding its purity?

4. Using the website of the Natural Resources Defense Council (NRDC) or another environmental group, investigate pollution from urban stormwater runoff, which some say rivals sewage plants and factories as a source of water contamination. Write a brief report on your findings.

LET'S EXPERIMENT! Disappearing Dilution

Materials Needed

- 6 plastic cups
- Measuring cups
- Water
- Large bowl or pitcher

- Spoon
- Colored powdered drink mix
- Sugar

Is dilution the solution to pollution? Let's use powdered drink mix and sugar as our contaminants to experiment with. How dilute does our mixture have to be before we can no longer see or taste them? Does that mean that our mixture is no longer polluted?

During the Industrial Revolution and up until the 1970s, it was common practice to dilute known hazardous substances in water. Environmental health research discovered that even small concentrations of certain chemicals can and do have very damaging effects on the environment. It is far better to reduce or eliminate the use of hazardous chemicals in the first place than to dilute the contaminant or try to clean up or remediate afterward.

During this experiment, you will compare pollution amounts using powdered drink mix and sugar as your "contaminants." Begin the experiment by labeling cups 1 through 6. Then, create a chart to record your observations for each cup. [Note that the diagram below shows an example dilution experiment, not the dilution ratios that are described below.]

Prepare the powdered drink mix using the recommended amount of sugar. Measure $1/2$ cup of this prepared drink with sugar into cup 1 and $1/4$ cup of water into each of cups 2–6. Taste the drink in cup 1 using a teaspoon. Now cup 1 is contaminated.

Place $1/4$ cup of the "polluted water" from cup 1 into cup 2 using the measuring cup. Make observations and notes.

Is this water less polluted than water in cup 1? What color differences do you notice? What is the difference in sweetness? Remember to record descriptions in your chart. Predict how dark the color will be and how it will taste in cups 3–6.

Slowly add $1/4$ of the "polluted water" from cup 2 to cup 3. Mix and record observations. Repeat this procedure for cups 4–6. When these observations are recorded, compare cup 1 with cup 3, and then compare it with cup 6. Place a white sheet of paper underneath each cup to emphasize the color differences. Note the differences in taste.

Questions

1. Was there evidence that pollution still remained in the water even when the solution was diluted?
2. How many more times do you think the polluted water would have to be diluted before color or taste changes ceased?
3. Do you think dilution is a good solution for pollution? Why or why not?
4. Does pollution always remain in the water? If not, where does it go?

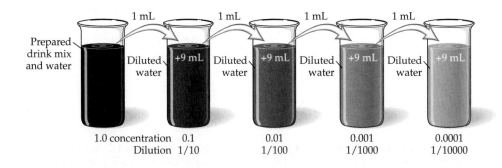

Energy

Have You Ever Wondered?

1 How can you measure energy? **2** How can chemical reactions give off heat? **3** Do we really have an energy shortage? **4** What is "frac sand," and how is it different from ordinary beach sand? **5** Which energy source is the best answer to the energy crisis?

Nearly all Earth's energy comes indirectly from the sun, a nuclear fusion reactor about 150 million km away. The wind that drives wind turbines is a result of solar heating. Biofuels such as biodiesel and biomass come from plants, which use sunlight as their source

FIRE! We were tempted to call this chapter "Fire" so that the titles of Chapter 12, Chapter 13, Chapter 14, and Chapter 15 would be "Earth," "Air," "Water," and "Fire" — the four elements of the ancient Greeks.

Indeed, "Fire" *is* an appropriate title for this chapter. Our ultimate energy source is the sun, an extremely dense and hot (about 10 million degrees on the surface), fiery ball of hydrogen and helium undergoing nuclear reactions that give off energy as infrared radiation, visible light, and ultraviolet radiation. We obtain most of our energy for heating homes, running our vehicles and our industries, and cooking by burning "fossil" fuels, which certainly involves fire. The plants we eat grew using energy from the sun, and the animals and fish we eat (who have generally subsisted on plants or other animals) provide us with energy to move, grow, think, and keep warm. Many sources of energy do not involve burning of energy. Even fossil fuels are derived from the sun. An increasingly popular method of directly harnessing solar energy is the use of solar cells that convert sunlight to electricity. Today, it is possible to buy enough solar cells to run an entire house, but at considerable expense. In this chapter, we will look at the general nature and properties of energy and at various ways of generating energy or converting it from one form to another. Only tiny fractions of the energy we use come from Earth's internal heat, visible in geysers and tapped as geothermal energy.

anything, however, and we will depend more heavily on these other sources as our reserves of fossil fuels are depleted.

Everything that we consume or use—our food and clothes, our homes and their contents, our cars and roads—requires energy to produce, package, distribute, operate, and discard. China is the largest consumer of energy in the world. The United States ranks second as an energy consumer. With less than 5% of the world's population, the United States uses about one-fifth of all the energy generated on the planet—making the US the largest energy consumer/capita. Abundant energy provides people in the United States with a high standard of living. But much of this energy is wasted. Australia, Switzerland, France, and the Scandinavian countries achieve higher standards of living with lower per capita use of energy. Our high energy consumption results in costs: to our health and to the environment, as well as our pocketbooks. It pollutes the air we breathe and the water we drink.

Industry uses about 22% of all the energy produced in the United States. This energy is used to convert raw materials to the many products our society needs and wants. Transportation uses about 29%, powering automobiles, trucks, trains, airplanes, and buses. Private homes use about 6%, commercial spaces consume about 4%, and electric power–generating plants consume 39%, from petroleum, natural gas, coal, renewable energy, and nuclear power, all of which we will discuss in this chapter.

Energy lights our homes, heats and cools our living spaces, and makes us the most mobile society in the history of the human race. It powers the factories that provide us with abundant material goods. Indeed, energy is the basis of modern civilization.

15.1 Our Sun, a Giant Nuclear Power Plant

Learning Objective • Perform power and energy calculations.

All living things on Earth—including us—depend on nuclear energy for survival, since most of the energy available to us on this planet comes from the giant nuclear reactor we call the sun. Although the sun is about 150 million km away, it has supplied Earth with most of its energy for billions of years, and is likely to continue to do so for billions more (Table 15.1).

Recall from Chapter 11 that the sun is a nuclear fusion reactor that steadily converts hydrogen to helium. Earth receives only about one billionth of the sun's energy output. Even so, this tiny fraction is equivalent to over 100 million nuclear power plants. In three days, Earth receives energy from the sun equivalent to all our fossil fuel reserves!

Only a small fraction of the energy that our biosphere (atmosphere, water, and soil in which all life exists) receives from the sun is used to support life. About 30% of the incident solar radiation is immediately reflected back into space as ultraviolet and visible light. Nearly half is converted to heat, making our planet a warm and habitable place. About 23% of solar radiation powers the water cycle (Section 14.2), evaporating water from land and seas. The radiant energy of the sun is converted to the potential energy of water vapor, water droplets, and ice crystals in

TABLE 15.1 Earth's Energy Ledger

Item	Energy (TW)[a]	Approximate Percentage
Energy in		
Solar radiation	174,000	99.97
Internal heat	46	0.025
Tides	3	0.002
Waste heat[b]	13	0.007
Energy out		
Direct reflection	52,000	30
Direct heating[c]	81,000	47
Water cycle[c]	40,000	23
Winds[c]	370	0.2
Photosynthesis[c]	40	0.02

[a]TW (terawatt) = 10^{12} W.
[b]From fossil fuel use.
[c]This energy is eventually returned to space as infrared radiation (heat).

the atmosphere. This potential energy is converted to the kinetic energy of falling rain and snow and of flowing rivers.

A tiny but most important fraction—less than 0.02%—of solar energy is absorbed by green plants, which use it to power **photosynthesis**. In the presence of chlorophyll and a number of enzymes, the energy converts carbon dioxide and water in a series of reactions to glucose, a simple sugar rich in energy.

$$6\,CO_2 \; + \; 6\,H_2O \; + \; energy \; \xrightarrow{\text{chlorophyll}} \; C_6H_{12}O_6 \; + \; 6\,O_2$$

Sunlight

Glucose (a simple sugar)

Photosynthesis also replenishes oxygen in the atmosphere. Glucose can be stored, or it can be converted to more complex foods, such as starch, and to structural materials, such as cellulose. All animals depend on the stored energy of green plants for survival.

Energy Units

A familiar unit for measuring the energy content of foods is the Calorie (1 Calorie = 1000 calories = 1 kilocalorie) (Section 1.9). Most adults need between 2000 and 2500 Calories per day to maintain all body functions. In SI units, energy is measured in joules (J). One calorie = 4.18 J, so one kilocalorie is 4,180 kJ.

Power is the *rate* at which energy is used (just as speed is the *rate* at which distance is covered). The SI unit for power is the *watt* (W). One watt is 1 joule per second (J/s).

$$1\,W = 1\frac{J}{s}$$

A unit of more convenient size is the kilowatt: 1 kW = 1000 W. The sun has a power output of 4×10^{26} W, a nearly incomprehensible amount.

To express energy consumption, the watt is combined with a unit of time. For example, electricity use is usually measured (and billed for) in terms of the *kilowatt hour* (kWh), the quantity of energy used by a 1 kW device in 1 h. There are 3600 seconds in an hour, so 1 kWh = 3600 kJ.

1 How can you measure energy?

Temperature, a measure of the *average* kinetic energy of molecules, is measured in degrees C, with a thermometer. Energy is not so simple. It is measured in terms of how much "work" it can accomplish, such as move a piano, provide light from a light bulb, or bake a cake. It is measured in units of calories or joules. Heat is thermal energy that is transferred from one object to another. The rate of consumption of energy is the watt, and your electric bill reflects the length of time you used energy at a certain rate.

WHY IT MATTERS

Consumers often have misconceptions about energy usage in the home. Appliances that generate or transfer heat usually consume a lot of energy. Two major energy "sinks" in most homes are the heating-and-cooling system and the water heater. Refrigerators, stoves, toasters, and similar appliances are also energy hogs. Lighting is often responsible for a surprisingly small fraction of the monthly energy bill. An ordinary toaster oven heating a slice of pizza for 10 minutes consumes as much energy as a 20 W fluorescent light bulb uses in 10 hours.

 EXAMPLE 15.1   Power and Energy Conversion

How much electrical energy, in joules, is consumed by a 75 W bulb left on for 1.0 h?

Solution

One watt is 1 J/s, so we know that 75 W = 75 J/s. The bulb is turned on for 1 h so

$$1\,\text{h} \times \frac{60\,\text{min}}{1\,\text{h}} \times \frac{60\,\text{s}}{1\,\text{min}} = 3600\,\text{s}$$

$$3600\,\text{s} \times \frac{75\,\text{J}}{1\,\text{s}} = 270{,}000\,\text{J}$$

❯ Exercise 15.1A

How much electrical energy, in joules, does a 650 W microwave oven consume in heating a cup of coffee for 1.0 min.?

❯ Exercise 15.1B

Compute the energy consumed by a 13 W compact fluorescent bulb burning for 1.0 h, and compare it with the energy consumed by a 60 W bulb that produces the same amount of light energy. What percent savings would this amount to, per hour?

❯ Exercise 15.1C

At 11.0 cents/kWh, how much would use of that 13 W compact fluorescent bulb save you in costs for a month in which you burned that bulb for 240 hours, compared with a regular 60 W bulb used for 240 hours? Both are supposed to provide the same intensity of light.

Kinetic and Potential Energy

Energy exists as potential and kinetic energy. Energy due to position or arrangement is called **potential energy**. Energy of motion is **kinetic energy**.

The water at the top of a dam has potential energy because of gravitational attraction. When the water is allowed to flow through a turbine to a lower level, the potential energy is converted to kinetic energy (energy of motion). As the water falls, it moves faster. Its kinetic energy increases as its potential energy decreases. The turbine can convert part of the kinetic energy of the water into electrical energy. The electricity thus produced can be carried by wires to homes and factories, where it can be converted to light energy, to heat, or to mechanical energy.

SELF-ASSESSMENT Questions

1. The rate of using or producing energy, in units of J/second, is the
 a. energy **b.** watt **c.** kilowatt hour **d.** volt

2. What sector of the U.S. economy uses the largest fraction of the nation's energy?
 a. residential
 b. electric power–generation industry
 c. industrial
 d. transportation

3. About how much of the world's energy does the United States use?
 a. 3% **b.** 5% **c.** 10% **d.** 20%

4. An eagle in flight has
 a. force energy **b.** kinetic energy
 c. potential energy **d.** rotational energy

5. About a third of Earth's incident solar radiation
 a. powers the water cycle **b.** is converted to heat
 c. powers photosynthesis **d.** is reflected back to space

6. About what percentage of Earth's incident solar radiation is used in photosynthesis?
 a. 0.02% **b.** 2% **c.** 5% **d.** 10%

15.2 Energy and Chemical Reactions

Learning Objectives • Classify chemical reactions and physical processes as exothermic or endothermic. • Explain how the heat of a reaction is calculated using bond energies.

We usually discuss energy changes in terms of an object or region's losing energy and another object or region's gaining it. When we warm our cold hands over a campfire, the burning wood gives off energy (as heat) and our hands gain energy, raising their temperature. In science, we define the **system** as the part of the universe under consideration. In the case of the campfire, the system is the burning wood. The **surroundings** are everything else—the rest of the universe (in theory)—but we usually limit the surroundings to those parts of the universe that exchange energy or matter or both with the system. In the case of the campfire, the surroundings are your hands and the air near the campfire.

Chemical reactions that result in the release of heat from the system to the surroundings are **exothermic**. The molecules producing the heat are the "system," and other things near those molecules are the "surroundings." The burning of methane, gasoline, or coal (Figure 15.1) is an exothermic reaction; chemical energy is converted to heat energy, which is lost by molecules but is absorbed by the surroundings. Because heat is released, the amount is written in the reaction on the right, as a "product" of the reaction. The numerical value of that energy is called the **heat of reaction** or the **enthalpy of reaction**. An example is the reaction for the burning of methane, the main component of "natural gas," the fuel that is used in most gas furnaces in people's homes:

$$CH_4(g) + 2\,O_2(g) \longrightarrow CO_2(g) + 2\,H_2O(g) + 803\text{ kJ}$$

An important thing to remember is that the heat of reaction—in this case, 803 kJ—is the amount of heat energy given off when one mole of methane is burned (since the stoichiometric coefficient for CH_4 in the reaction is a "one"), which would be 16.0 g.

Chemical reactions in which energy must be supplied to the system from the surroundings are **endothermic** (Figure 15.2). The decomposition of water into its components is an example. We can write the amount of heat required for this process as a reactant in the chemical equation, since it must be absorbed by the "system" molecules, as in

$$2\,H_2O(g) + 573\text{ kJ} \longrightarrow 2\,H_2(g) + O_2(g)$$

So decomposing 2 moles of gaseous water requires 573 kJ of energy (since the stoichiometric coefficient for water is 2).

Physical processes can also be either exothermic or endothermic. Melting, for example, requires heat, and freezing requires that heat energy be removed. Table 15.2 lists several examples of physical and chemical processes and classifies each as either exothermic (giving off heat) or endothermic (requiring heat).

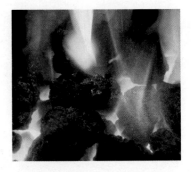

▲ **Figure 15.1** Coal burns in a highly exothermic reaction. The heat released can convert water to steam that can turn a turbine to generate electricity.

▲ **Figure 15.2** A striking endothermic reaction occurs when barium hydroxide octahydrate reacts with ammonium thiocyanate to produce barium thiocyanate, ammonia gas, and water.

Heat + $Ba(OH)_2 \cdot 8\,H_2O(s)$ + $2\,NH_4SCN(s) \rightarrow Ba(SCN)_2(s)$ + $2\,NH_3(g) + 10\,H_2O(l)$

Here the reaction is carried out in a flask placed on a wet block of wood. The temperature drops well below the freezing point of water, freezing the flask to the block.

TABLE 15.2 Some Exothermic and Endothermic Processes

Exothermic Processes	Endothermic Processes
Freezing of water	Melting of ice
Condensation of water vapor	Evaporation of water
Metabolism in animals	Photosynthesis in plants
Forming chemical bonds	Breaking chemical bonds
Discharging a battery	Charging a battery

For both chemical processes and physical changes, an important thing to remember is that the heat of reaction is for the number of moles of reactants in the balanced chemical reaction. Obviously, much more heat would be emitted by burning a ton of coal than by burning one charcoal briquet. Consider Example 15.2.

EXAMPLE 15.2 Energy Release in Chemical Reactions

Experiments show that burning 1.00 mol (16.0 g) methane to form carbon dioxide and water releases 803 kJ (192 kcal) energy as heat. How much heat is given off if we burn 2.00 mol methane?

Solution

$$2.00 \text{ mol} \times \frac{803 \text{ kJ}}{1 \text{ mol}} = 1606 \text{ kJ of heat released}$$

EXAMPLE 15.3 Energy Changes in Chemical Reactions

How much energy, in kilojoules, is released when 225 g propane is burned? The heat of reaction for burning propane is 2201 kJ/mol propane, considerably larger than for methane.

$$C_3H_8(g) + 5 O_2(g) \longrightarrow 3 CO_2(g) + 4 H_2O(g) + 2201 \text{ kJ}$$

Solution

For this problem, we are given grams, not moles. The formula mass of propane, whose molecule consists of 3 C atoms and 8 H atoms, is

$$(3 \times 12.0 \text{ amu}) + (8 \times 1.0 \text{ amu}) = 36.0 \text{ amu} + 8.0 \text{ amu} = 44.0 \text{ amu}$$

The molar mass is therefore 44.0 g. Next, we use the molar mass to convert grams of propane to moles of propane.

$$225 \text{ g } C_3H_8 \times \frac{1 \text{ mol } C_3H_8}{44.0 \text{ g } C_3H_8} = 5.11 \text{ mol } C_3H_8$$

Now we need the heat of combustion, 2201 kJ/mol, to complete our computation:

$$5.11 \text{ mol } C_3H_8 \times \frac{2201 \text{ kJ}}{1 \text{ mol } C_3H_8} = 11,300 \text{ kJ}$$

> Exercise 15.3A

How much energy in kilojoules is absorbed when 10.1 mol N_2 reacts with plenty of O_2 to form NO? The reaction of nitrogen and oxygen to form nitrogen monoxide requires an input of energy, so we write the energy as a reactant on the left.

$$N_2(g) + O_2(g) + 18.07 \text{ kJ} \longrightarrow 2 NO(g)$$

> Exercise 15.3B

How much energy in kilojoules is absorbed when 450.0 g of N_2 reacts with plenty of O_2 to form NO? The heat of reaction is 18.07 kJ, as written in the reaction of Exercise 15.3A.

2 How can molecules give off heat?

In reality, energy must be put into molecules first to break the bonds in the molecules. When the fragments recombine in a different arrangement, energy is released. It is the net result of those energy interactions that makes the overall process either give off heat or require heat to occur.

Bond Energies

Why do some processes, either physical or chemical, require heat to happen and other processes emit heat? What is happening with the molecules? It's fairly simple: First, all the bonds between atoms must break, which requires energy input, and then new bonds form from the molecular fragments, making new molecules, releasing energy to the surroundings. The net result is the amount of heat released or lost to the surroundings (if the calculation gives a negative number) or the amount of heat that has been absorbed (if the calculation gives a positive number). The energy inherent in every atom-to-atom bond can be measured, and is called the bond energy. The units are kJ/mol.

A simple example of how we determine if a chemical process is exothermic or endothermic is forming hydrogen chloride molecules from hydrogen molecules and chlorine molecules:

$$H_2(g) + Cl_2(g) \longrightarrow 2\,HCl(g)$$

For that reaction to happen, the bonds between the two hydrogen atoms in H_2 must first be broken and the bonds between the two chlorine atoms in Cl_2 must be broken. The free H atoms (each with one electron) and the free chlorine atoms (each with 7 electrons) now bond together, forming single covalent bonds between the H and Cl atoms, so the octet rule can be satisfied. Note that two HCl molecules are formed.

To calculate the heat of this reaction, we first find the bond energies from a reference book. We find that it takes 432 kJ to split up one mole of H—H bonds in hydrogen molecules and 243 kJ of energy to split up 1 mole of Cl—Cl bonds in chlorine molecules. We also find that 427 kJ of energy is released when 1 mole of HCl molecules is formed. Once we have the bond energies, we can compute the "enthalpy change" of the reaction:

enthalpy change =

sum of energy needed to break all the bonds − sum of energy released in forming bonds

For this reaction, the enthalpy of the reaction is calculated by

$$(1\,mol\,H—H\,bonds \times 432\,kJ/mol + 1\,mol\,Cl—Cl\,bonds \times 243\,kJ/mol)$$
$$-(2\,mol\,H—Cl\,bonds \times 427\,kJ/mol) = -179\,kJ/mol$$

The negative sign of the answer means that more energy was released to the surroundings than was put into the molecules to break them apart. This reaction is, therefore, exothermic, and heat can be considered a product of the reaction. We can now write a thermochemical reaction as

$$H_2(g) + Cl_2(g) \longrightarrow 2\,HCl(g) + 179\,kJ$$

Writing it this way reminds us that 179 kJ of heat is given off if one starts with 1 mole of hydrogen and 1 mole of chlorine.

In the event the sum of the bond energies *is greater than* the sum of the heat evolved in creating bonds, the resultant number of kJ of heat energy is written on the *left side* of the equation and the reaction is endothermic. In calculating the heat of reaction, it is usually best to draw all the Lewis structures for every molecule in the reaction, considering the stoichiometric coefficients in the balanced chemical equation. In this way, you can easily count the number and types of bonds that need to be broken and formed. Then find the bond energies and perform the calculation.

SELF-ASSESSMENT Questions

1. Which of the following is an exothermic process?
   a. $H_2O(g) \longrightarrow H_2O(l)$ b. $H_2O(s) \longrightarrow H_2O(g)$
   c. $SiO_2(g) + 3\,C(s) + 624.7\,kJ \longrightarrow SiC(s) + 2\,CO(g)$
   d. $H_2O(l) \longrightarrow H_2O(g)$

2. Consider a burning candle. Which of the following is a correct statement?
   a. The entire room in which the candle is burning is considered the "system."
   b. The burning process is endothermic.
   c. Energy is needed to break bonds in the wax, but more energy than that is released when carbon dioxide and water vapor molecules are formed.
   d. The numerical value of the heat of reaction must be positive.

3. Melting an ice cube
   a. is an exothermic process
   b. releases heat to the air
   c. requires the same amount of energy as melting a large block of ice
   d. is an endothermic process

4. 400 g of propane is burned. Which of the following is true?
   a. About 2000 kJ of heat is released.
   b. About 2000 kJ of heat is needed.
   c. About 20000 kJ of heat is released.
   d. About 800000 kJ of heat arise needed.

15.3 Reaction Rates

Learning Objective • List factors that affect the rates of chemical reactions, and explain how they affect those rates.

WHY IT MATTERS

On a daily basis we use the effect of temperature on chemical reactions. For example, we refrigerate milk or freeze vegetables to retard the chemical reactions that lead to spoilage. When we want to stir-fry meat or vegetables faster, we turn up the temperature under the frying pan.

Chemical **kinetics** is concerned with the rates at which chemical reactions occur. Reaction rates depend on several factors, including temperature, concentrations of reactants, and the presence or absence of catalysts. Because reaction rates are related to temperature, they are closely related to energy, the focus of this chapter.

Reactions generally proceed faster at higher temperatures because molecules move faster and collide more frequently at higher temperatures, increasing the chance that they will react. For example, at room temperature, coal (carbon) reacts so slowly with oxygen from the air that the change is imperceptible. However, if coal is heated to several hundred degrees, it reacts rapidly. The heat evolved in the reaction keeps the coal burning smoothly. The increase in temperature also supplies more energy to break chemical bonds.

The kinetics of reactions is important both in everyday life and in industrial processes. In manufacturing a medicinal product, for example, we usually want the chemical synthesis to happen rather quickly, rather than take hours or days on the production line. We heat water to boiling to cook noodles in a few minutes rather than let them sit in water for a few hours. Sometimes we want to slow down reactions, not speed them up. Refrigeration slows down spoilage reactions in foods.

The concentrations of reactants affect the rate of most reactions because when we crowd more molecules into a given volume of space, the molecules collide more often. More collisions each second means more reactions each second. For example, if you light a wood splint and then blow out the flame, the splint continues to glow as the wood reacts slowly with oxygen in the air. If the glowing splint is placed in pure oxygen—five times more concentrated than O_2 in air—the splint bursts into flame because of the increase in reaction rate.

Catalysts also affect the rates of chemical reactions, but are not used up in those reactions. A carefully selected catalyst can increase the rate of a reaction that otherwise would be so slow as to be impractical. Catalysts may provide an alternative method, called a pathway, to form the same products or lower the *activation energy*, the initial energy needed to get the reaction started in the first place. For example, the platinum in your automobile's catalytic converter causes the rapid breakdown of NO pollutant molecules coming from the engine and changes them back to nitrogen and oxygen molecules. Those oxygen molecules further oxidize carbon monoxide to carbon dioxide. This means that less CO and NO are emitted into our atmosphere.

Catalysts are of great importance in the chemical industry and in biology, where they are usually termed enzymes. Enzymes are critical to every biochemical process that happens in the human body (see Chapter 16). The body cannot function without enzymes to speed up reactions involved in digestion, for example. A lack of a specific enzyme can cause many genetic disorders, or excess of enzyme activity can cause cancer.

SELF-ASSESSMENT Questions

1. The rate of a chemical reaction usually decreases when
   a. a catalyst is added
   b. a chemist is watching
   c. the concentration of one of the reactants is decreased
   d. the temperature is increased

2. If a chemical process that is occurring in aqueous solution in a vessel of some sort seems to be proceeding far too rapidly, what practical thing can be done to slow the process and bring the reaction under control?
   a. Add water to the container to dilute the contents.
   b. Turn up the heat.
   c. Nothing, just let the reaction continue until the reactants are used up.

3. A catalyst speeds up the rate of a chemical reaction by
   a. decreasing the favorability of product formation
   b. increasing the pressure of reactants and thus favoring product formation
   c. lowering the activation energy required for the reaction
   d. raising the temperature at which the reaction occurs

4. Hydrogen peroxide breaks down over time into water and oxygen gas. Manganese dioxide will increase that decomposition rate by at least a factor of 100. Manganese dioxide is a(n)
   a. catalyst   **b.** inhibitor
   c. reactant   **d.** supplement

15.4 The Laws of Thermodynamics

Learning Objective • State the first and second laws of thermodynamics, and discuss their implications for energy production and use.

We will take a look at our present energy sources and our future energy prospects shortly. To do so scientifically, we first examine some natural laws. Recall that natural laws merely summarize the results of many experiments and observations. We won't recount all those experiments here. We will merely state the laws and some of their consequences.

Energy and the Laws of Thermodynamics

The **first law of thermodynamics** (*thermo* refers to heat; *dynamics* to motion) states that energy can be neither created nor destroyed. Energy can be changed from one form to another, however. This law is also called the **law of conservation of energy**. Energy cannot be made from nothing, which means that a machine cannot be made that produces more energy than it takes in. Energy does not just appear or disappear.

From the first law alone, we might conclude that we can't possibly run out of energy, because energy is conserved. This is true enough, but it doesn't mean that we don't have problems. One problem is that energy transformation may be into a form that we cannot "harness" for useful work. Recall from the margin note on page 461 that energy is the "ability to do work." There is another long-armed law from which we cannot escape, as well.

Energy and the Second Law: You CAN'T Break Even

Despite innumerable attempts (and even the granting of several patents), no one has ever built a successful perpetual-motion machine. You can't make a machine that runs indefinitely without consuming energy. Even if an engine isn't doing any work, it loses energy (as heat) because of the friction of its moving parts. In fact, in any real engine, it is impossible to get as much useful energy out as you put in. In other words, you can't break even.

If energy is neither created nor destroyed, why do we always need more? Won't the energy we have now last forever? The answer lies in these facts:

- Energy can be changed from one form to another.
- Not all forms are equally useful.
- More useful forms of energy are constantly being degraded into less useful forms or distributed in a way that makes the energy nearly impossible to recover.

Consider a cup of hot coffee sitting on the kitchen table. The coffee tends to cool off, dissipating its thermal energy to the ceramic cup into which it was poured, to molecules in the air, and to other objects in the kitchen. Thus, *there is a tendency toward an even distribution of energy.* If all the thermal energy from that hot coffee is spread out among the molecules in the air, the objects in the kitchen, and the walls, that thermal energy no longer is useful.

The second law of thermodynamics can be stated in many ways. One of them is "Heat always flows from a hot object to a cooler one" (Figure 15.3). The reverse does not occur spontaneously. (The hot coffee does not get hotter and the air around it does not simultaneously get cooler.) Another way to state the second law is "The forms of energy available for *useful* work are continually decreasing; energy spontaneously tends to distribute itself among the objects in the universe." When using machinery, mechanical energy is eventually changed into heat energy by friction and loss of energy/heat to the surroundings. The energy still exists, but we cannot easily use it.

It is true that we can make energy flow from a cold region to a hot one—refrigerators and freezers do so—but we cannot do so without an input of energy

WHY IT MATTERS

A steaming-hot cup of coffee or soup contains a lot of heat energy, which can be transferred to cold hands by wrapping one's fingers around the hot mug or bowl. But when the mug is left sitting on the table, the heat energy is soon transferred to the air molecules that are distributed throughout the room. When that happens, the heat energy is so dispersed that it is no longer *useful* for heating the hands. This is a good example of the second law of thermodynamics; nature simply wants to disperse energy.

▲ **Figure 15.3** Energy always flows spontaneously from a hot object to a cold one, never the reverse. It flows from a hot fire to the cooler marshmallow.

or without producing changes elsewhere. We can reverse a spontaneous process only at a price. The price, in the case of a refrigerator, is the consumption of electricity; that is, you must use energy to cause energy flow from a cold space to a warmer one.

When we change energy from one form to another, we can't concentrate all the energy in a particular source to do the job we want it to do. For example, we use the energy in a fuel to push a piston in a car engine, or we use water rushing down from the top of a dam to run dynamos to generate electricity. In either case—indeed, in all cases—some of the energy is converted to heat and thus is not available to do *useful* work.

Entropy

The energy to run an engine usually comes from the concentrated chemical energy inside molecules of oil or coal. In biochemistry, food molecules are the concentrated energy source. In either case, energy is spread out during the process. The spread-out energy in the product gases—CO_2 and H_2O, in each case—is less useful. A car won't run on exhaust gases, nor can organisms obtain energy for life processes from respiratory products.

Yet another way to look at the second law is in terms of *entropy*. Scientists use entropy as a measure of the dispersal of energy in a system. The more the energy is spread out, the higher the entropy of the system and the less likely it is that this energy can be harnessed to do useful work. Spontaneous processes tend toward greater entropy or are exothermic, or both. The total entropy always increases for an isolated system.

As was noted in Chapter 6, molecules move about constantly. Their motion in solids is limited to rapid but tiny back-and-forth movements much like vibrations about a nearly fixed point. The molecules in liquids move more freely but still over only short distances. Those in gases move still more freely and over much greater distances. Any process that changes a solid to a liquid or a liquid to a gas involves an increase in entropy. Let's delve a little deeper.

In Chapter 3, we saw that the energy of light occurs in little packets called *photons*. Similarly, the energy of molecular motion is described using a concept called *microstates*. If you could take a snapshot capturing the positions of all molecules in a sample of matter at a given instant, you would have captured an image of a microstate. The number of microstates in which a given sample of matter can exist differs depending on whether that matter is in the solid, liquid, or gaseous state. The number of possible energy microstates will be lowest for a solid because the molecules in a solid are limited mainly to vibrational motion. Because molecules in the liquid state can move more freely, the number of possible energy microstates is much larger for liquids than for solids. When converted to a gas, the molecules can move still more freely and for much greater distances, so even more microstates are available to molecules in the gaseous state.

As any form of matter is heated, the energy in it is spread out among more accessible microstates. (See Figure 15.4.) Technically, entropy is the dispersal of energy among these microstates. In practice, entropy is often thought of as being analogous to "molecular disorder." Of course, we can often reverse the tendency toward greater entropy—but only through an input of energy.

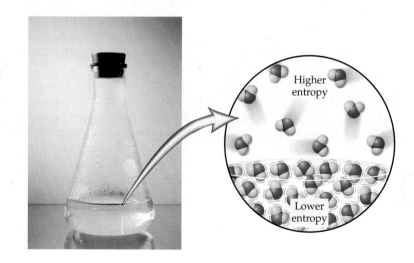

▶ **Figure 15.4** The photograph depicts the vaporization of water; a sample of liquid water [$H_2O(l)$] at room temperature spontaneously changes to $H_2O(g)$ through the process of evaporation. At the macroscopic level, nothing appears to be taking place. However, in the molecular view, we see that the molecules are in motion and are much more widely spaced in the gaseous state than in the liquid state. Vaporization is spontaneous because the gas has greater *entropy* than the liquid. Molecules in the gas can be "arranged" in many more ways in their spread-out spacing than can molecules of the liquid. We can make liquid water from water vapor but only by compressing the gas and/or lowering its temperature.

SELF-ASSESSMENT Questions

1. The first law of thermodynamics is also called the law of
   a. conservation of energy
   b. conservation of mass
   c. entropy
   d. perpetual motion

2. That energy always goes spontaneously from more useful forms to less useful forms is a statement of the
   a. first law of thermodynamics
   b. law of unintended consequences
   c. second law of thermodynamics
   d. standard law of energy conversion

3. According to the second law of thermodynamics, the
   a. entropy of a system always decreases
   b. entropy of a system always increases
   c. total entropy always decreases for an isolated system
   d. total entropy always increases for an isolated system

4. In the process represented by the equation
   $H_2O(l) \longrightarrow H_2O(g)$, the entropy
   a. decreases
   b. depends on the catalyst used
   c. increases
   d. remains the same

5. If you place a cool hand on a warm cup of tea
   a. heat spontaneously is transferred from the hand to the cup
   b. heat spontaneously is transferred from the cup to the hand
   c. both are in thermal equilibrium and no energy is transferred, because of the first law of thermodynamics
   d. the first law is relevant but not the second

Answers: 1, a; 2, c; 3, d; 4, c; 5, b

15.5 Fuels and Energy: People, Horses, and Fossils

Learning Objective • List the common fossil fuels, and describe how modern society is based on their use.

A **fuel** is a substance that burns readily with the release of significant amounts of energy, ideally in a controllable manner. Substances that explode are not usually satisfactory as fuels, even though they release a large amount of energy.

Early people obtained their energy (food and fuel) by collecting wild plants and hunting wild animals. They expended this energy in hunting and gathering. Domestication of horses and oxen increased the availability of energy only slightly. The raw materials used by these work animals were natural, replaceable plant materials. Plant materials kept early fires burning. As late as 1760, wood was almost the only fuel in use. Wood, dried dung, and crop residues still remain the main sources of heat for about one-third of the world's people.

One of the first mechanical devices used to convert energy to useful work was the waterwheel. The Egyptians first used waterpower about 2300 years ago, mainly for grinding grain. Later, waterpower was used for sawmills, textile mills, and other small factories. Windmills were introduced into western Europe during the Middle Ages, mainly for pumping water and grinding grain. In the 1800s, small windmills were seen on every small farm in the U.S. Midwest. More recently, large-scale wind power has been used to generate electricity, and wind turbine "farms" are likely to be a major option among renewable energy sources in the future. (See Section 15.10.)

Windmills and waterwheels are fairly simple devices for converting the kinetic energy of blowing wind and flowing water, respectively, to mechanical energy. They were sufficient to power the early part of the Industrial Revolution. The development of the steam engine allowed factories to be located away from waterways. Since 1850, the invention of turbines turned by water, steam, and gas; the internal-combustion engine; and a variety of other energy-conversion devices have boosted the energy available for society's use by an estimated factor of 10,000.

The Industrial Revolution was fueled, in effect, by fossils—the buried remains of living organisms. Combustion of this material initiated modern industrial civilization. Today, more than 85% of the energy used to support our way of life comes from **fossil fuels**—coal, petroleum, and natural gas that formed during Earth's Carboniferous period around 300 million years ago. In the following sections, we consider the origin and chemical nature of these fuels that release the energy captured by ancient plants from rays of sunlight.

▲ Waterwheels have been used to obtain mechanical energy from flowing water since antiquity. Some waterwheels at Hama, Syria are more than 2000 years old.

3 Do we really have an energy shortage?

There is no shortage of energy itself, but *not all* energy is equally useful to us. The supply of materials like fossil fuels that provide convenient, *useful* energy is dwindling.

Energy *production* refers to conversion of some form of energy into a more useful form. For example, production of petroleum means pumping, transporting, and refining it. Energy *consumption* involves using it in a way that changes it to a less useful form. In either case, energy is neither created nor destroyed.

Fuels are *reduced* forms of matter, and the burning process is an oxidation (Chapter 8). If an atom already has its maximum number of bonds to oxygen atoms (or to other electronegative atoms, such as chlorine or bromine), the substance cannot serve as a fuel. Indeed, some such substances can be used to put out fires. Figure 15.5 shows some representative fuels and nonfuels.

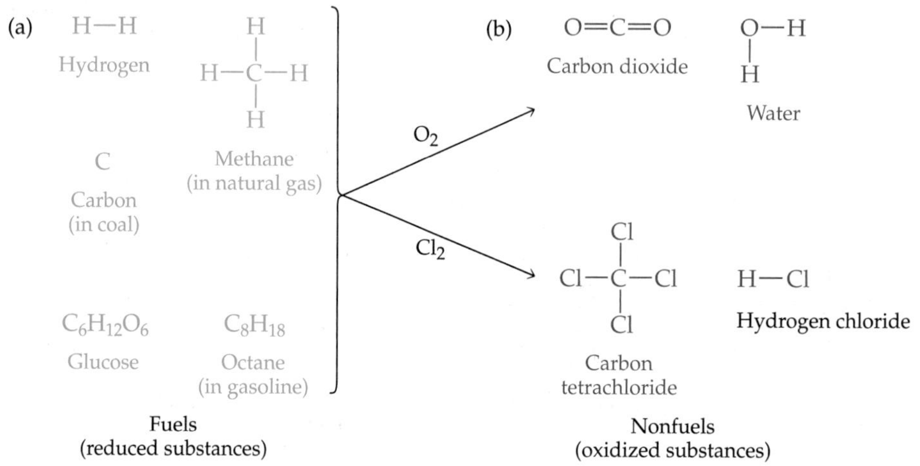

▶ **Figure 15.5** (a) Fuels are reduced forms of matter that release relatively large quantities of heat when burned. (b) Nonfuels are compounds that are oxidized forms of matter.

Reserves and Consumption Rates of Fossil Fuels

Earth has a limited supply of fossil fuels. Estimates of reserves vary greatly, depending on the assumptions made, and the numbers in Table 15.3 imply far greater certainty than there is in reality. The units for measuring those resources are not consistent, either, making comparisons difficult. Liquid fuels are usually estimated in barrels or gallons, gases in cubic feet, and electricity in kWh or British thermal units (Btu). However, the U.S. Energy Information Administration (EIA) has estimated that the total annual primary energy consumption in the United States is about 100 quadrillion (10^{16}) Btu, or 30 trillion (10^{12}) kWh. Estimated U.S. and world reserves and annual U.S. and world consumption for the various types of fossil fuels are given in Table 15.3, taken from the BP Statistical Review of World Energy June 2017. Even the most optimistic estimates, however, lead to the conclusion that these nonrenewable energy resources are being depleted rapidly. Indeed, in just a century, we will have used up more than half the fossil fuels that were formed over millions of years. In only a few hundred more years, we will have removed from Earth and burned virtually all the remaining recoverable fossil fuels. Of all that ever existed, about 90% will have been used in a period of 300 years.

TABLE 15.3 **Estimated U.S. and World Reserves of Economically Recoverable Fuels and Annual Consumption of Fossil Fuels**

Fuel	Reserves		Annual Consumption	
	United States	World	United States	World
Coal, billion tons	252	1139	0.37	3.66
Petroleum, billion barrels	48	1707	7.16	35.2
Natural gas, trillion cubic feet	308	6589	34.3	126

Data: BP Statistical Review of World Energy June 2017

Within the lifetime of today's 18-year-olds, natural gas and petroleum will likely become so scarce and so expensive that they won't be used much as fuels. At the current rate of consumption, world reserves of petroleum and natural gas will be substantially depleted sometime during this century. Coal reserves could last perhaps 300 years, but coal usage is now decreasing as renewable energy sources are increasing in the United States. However, in developing nations, the rate of use of all fossil

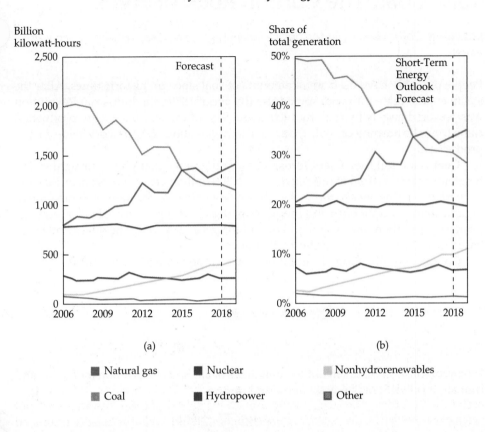

U.S. Electricity Generation by Energy Source (2006-2019)

(a)

(b)

- ■ Natural gas
- ■ Nuclear
- ■ Nonhydrorenewables
- ■ Coal
- ■ Hydropower
- ▨ Other

◀ **Figure 15.6** The U.S. Energy Information Administration predicts that coal use for electric power generation will continue to fall, and natural gas use will increase somewhat due to new sources discovered with fracking technology. Wind power use will increase significantly, but nuclear and hydropower should remain stable, at least in the near future.

Source: U.S. Energy Information Administration, *Short-Term Energy Outlook.*

fuels is increasing. Presumably, fossil fuels are still being formed in nature, a process that is perhaps most evident in peat bogs. The rate of formation is extremely slow, estimated to be only one fifty-thousandth of the rate at which fuels are being used.

In comparing the major energy sources available to us in the twenty-first century, it is noteworthy that the various sectors of users (transportation, industry, residential and commercial, and electric power–generation plants) each use different portions of those types of energy sources. Figure 15.6 includes nuclear power plants and renewable energy sources, besides fossil fuels, as significant contributors to our energy tally, both in terms of energy generation in billions of kWh (Figure 15.6a) and percent of total energy produced (Figure 15.6b).

Let's take a look at each of our major energy sources in some more detail.

SELF-ASSESSMENT Questions

1. Coal, petroleum, and natural gas are called fossil fuels because they
   a. are burned to release energy
   b. are nonrenewable and will run out
   c. cause air pollution
   d. were formed over millennia from the remains of ancient plants and animals

2. Which fossil fuel will likely last the longest, based on current rate of consumption?
   a. natural gas   b. crude oil
   c. coal   d. peat

3. Which of the following is not a fuel?
   a. C   b. $CH_3CH_2CH_3$
   c. CH_3CH_2OH d. CO_2

4. All fuels are
   a. carbon compounds   b. hydrocarbons
   c. oxidized forms of matter   d. reduced forms of matter

5. What fraction of the world's coal reserves is found in the United States?
   a. 2%   b. 10.7%   c. 31%   d. 25%

6. Which of the following is a false statement?
   a. The least important user of petroleum is electric power–generating plants.
   b. Natural gas supplies are used in about equal amounts by industry, residential, and electric power generation.
   c. Coal is used by all of the four major sectors of users.
   d. Nuclear power is primarily used for electric power generation.

15.6 Coal: The Carbon Rock of Ages

Learning Objective • List the origins, advantages, and disadvantages of coal as a fuel.

People probably have used small amounts of coal since prehistoric times. After the steam engine came into widespread use (by about 1850), the Industrial Revolution was powered largely by coal. By 1900, about 95% of the world's energy production came from the burning of coal. Today, coal supplies about 27% of worldwide energy production.

Coal is a fossil fuel. Giant ferns, reeds, and grasses that grew during the Carboniferous period (from about 360 million to 300 million years ago) were buried and converted over the ages to the coal we burn today. Coal is a complex combination of organic materials that burn and inorganic materials that produce ash. Coal consists primarily of carbon, plus water and other inorganic materials. The quality of coal is based on carbon content and those other elements.

Complete combustion of carbon produces carbon dioxide.

Carbon (from coal)   Oxygen (from air)

$$C(s) + O_2(g) \longrightarrow CO_2(g)$$

When combustion occurs in limited quantities of air, however, carbon monoxide and soot are formed. Soot is mostly unburned carbon.

$$2C(s) + O_2(g) \longrightarrow 2CO(g)$$

Coal is ranked by carbon content, from low-grade peat and lignite to high-grade anthracite (Table 15.4). The energy obtained from coal is roughly proportional to its carbon content. Soft (bituminous) coal is much more plentiful than hard coal (anthracite). Lignite and peat have become increasingly important as the supplies of higher grades of coal have been depleted.

TABLE 15.4 Major Components of Various Types of Coal (Minor Components Include Volatile Materials and Sulfur)

Fuel Type	Carbon	Water	Ash (Minerals)
Lignite	25% to 35%	35% to 66%	6% to 20%
Sub-bituminous	35% to 65%	13% to 30%	10% to 15%
Bituminous	65% to 85%	5% to 10%	3% to 12%
Anthracite	85% to 95%	< 5%	2% to 8%

Source: Diessel, F. K. *Coal-Bearing Depositional Systems.* New York: Springer-Verlag, 1992.

Coal deposits that exist today are less than 300 million years old. For millions of years, Earth was much warmer than it is now, and plant life flourished. Most plants lived, died, and decayed—playing their normal role in the carbon cycle (Figure 15.7). But some plant material became buried under mud and water. There, in the absence of oxygen, it decayed only partially. The structural material of plants is largely cellulose [$(C_6H_{10}O_5)_n$]. Under increasing pressure, as the material was buried more deeply, the cellulose molecules broke down. Small molecules rich in hydrogen and oxygen escaped (mostly as water), leaving behind a material increasingly rich in carbon.

Thus, peat is an immature coal, only partly converted, with plant stems and leaves clearly visible. Anthracite, on the other hand, has been almost completely carbonized. Most coal has a measurable water content.

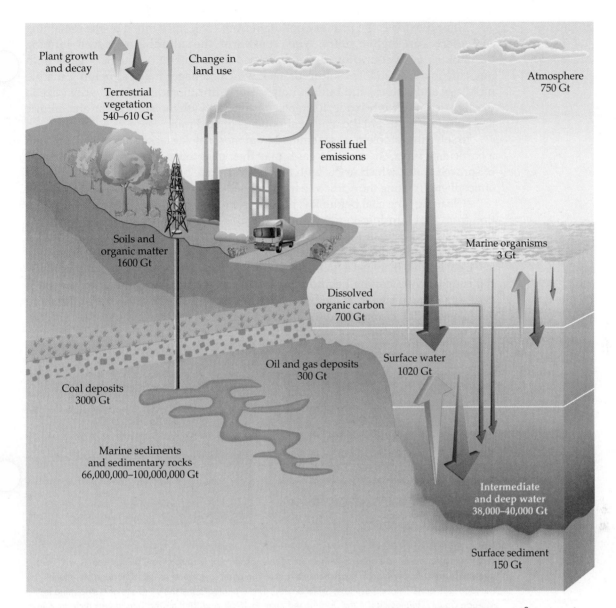

▲ **Figure 15.7** The carbon cycle. The numbers indicate the quantity of carbon, in gigatons (1 Gt = 10^9 t), in each location. The arrows indicate the general direction of travel of carbon during its cycle.

Abundant but Inconvenient Fuel

Coal is by far the most plentiful fossil fuel. The United States has about a quarter of the world's reserves. Electric utilities in the United States burn over half a billion metric tons of coal each year, generating 1.2 trillion kWh of electricity (less than 30% of the total).

Coal, a solid, is an inconvenient fuel to use and is dangerous to obtain. Coal mining is one of the most dangerous occupations in the world, accounting for over 100,000 deaths in the United States in the twentieth century alone, including accidents and black lung disease. Strip mining, though efficient, can devastate large areas. Coal is hauled in trains, barges, and trucks to the power plants and factories where it is used.

Curbing Pollution from Coal Burning

When coal burns, the carbon is changed to gaseous carbon dioxide. However, unlike natural gas and liquid petroleum, solid coal contains minerals that are left as ash when the coal is burned. As noted in Chapter 13, some of the ash may enter the air as particulate matter, constituting a major air pollution problem, including airborne mercury.

Sulfur dioxide is the primary pollutant from coal, since much of our remaining coal reserves are high in sulfur. When coal burns, the sulfur burns as well, forming the choking gas sulfur dioxide (as discussed in Chapter 13). The SO_2 reacts with moisture in the air to form sulfurous acid, which damages metal structures, marble buildings and statues, and human lungs. The sulfur dioxide can further oxidize in the air to form sulfur trioxide, forming sulfuric acid with moisture in the air. Sulfuric and sulfurous acids are the main components of acid rain.

To solve this problem, gases emitted from most smokestacks are first sent through a particulate filter. The residual gases may then pass through a scrubber, a fine mist of sprayed water, which reacts with the sulfur dioxide before being released into the atmosphere. Treating the acidic water with limestone is the final neutralization step.

"Cleaning" the coal before it is burned is another option for reducing air pollution from coal burning. Pyrite (FeS_2), the major source of sulfur in coal, is three to four times denser than the coal itself, so a flotation method is used to separate the heavier minerals from the coal after it is crushed.

Coal is more than just a fuel. When it is heated in the absence of air, volatile material is driven off, leaving behind a product—mostly carbon—called *coke*, which is used as a reducing agent in the production of iron and steel (Section 12.4). The more volatile material is condensed to liquid coal oil and a sticky mixture called *coal tar*, both of which are sources of organic chemicals for medical and industrial use.

(a)

(b)

▲ Strip mining bares vast areas of vegetation. The exposed soil washes away, filling streams with mud and silt. U.S. laws now require the expensive restoration of most stripped areas. Unfortunately, much damage was done before these laws were passed. These photographs show the site of a strip mine in Morgan County, Tennessee: (a) the abandoned mine in 1963 and (b) the area after it was reclaimed in a demonstration project in 1971.

SELF-ASSESSMENT Questions

1. Coal comes from
   a. microscopic ocean creatures
   b. compressed minerals under oceans
   c. decomposed plants
   d. other minerals

2. Coal is mainly carbon because
   a. plants have carbon dioxide in their cells
   b. plants are mainly cellulose, a compound of C, H, and O
   c. plants are mainly proteins
   d. all living things have carbon in them

3. Analysis of a solid fuel shows it to be 87% carbon. It is most likely
   a. anthracite coal
   b. bituminous coal
   c. peat
   d. wood

4. When coal is burned, the inorganic constituents end up as
   a. flue gases
   b. solid sulfur compounds
   c. ashes
   d. organic matter

5. Which fossil fuel contributes the most to atmospheric sulfur oxides that are major air pollutions?
   a. coal
   b. natural gas
   c. petroleum
   d. propane

6. On what physical property is the flotation method of coal cleaning based?
   a. density
   b. electrostatic precipitation
   c. melting point
   d. temperature

Answers: 1, c; 2, b; 3, b; 4, c; 5, a; 6, a

15.7 Natural Gas and Petroleum

Learning Objective • List the characteristics, advantages, and disadvantages of natural gas and petroleum.

Both natural gas and petroleum consist of hydrocarbons formed by the anaerobic decomposition of remains of living microorganisms. These remains settled to the bottom of the sea millions of years ago. The organic matter was buried under layers of mud and converted over the ages by high temperature and pressure to gaseous and liquid hydrocarbons.

Natural gas and petroleum (often called "crude oil") are frequently found close together in nature. Natural gas is trapped in geological formations capped by impermeable rock. It is removed through wells drilled into the gas-bearing formations. About 90% of today's natural gas in the United States is obtained by **hydraulic fracturing**, or *fracking*, of shale formations up to 5000 feet below ground. Up until recently, fracking has mainly involved forcing chemically treated water under high pressure into the layers of rock. The fractured rock releases the gas.

A recently developed method for propping open the cracks in the shale formations is using fine silica sand (often called *frac sand*) so that fluids (both liquid and gaseous) can easily flow out of the cracks. One major oil or natural gas well can require a few thousand tons of sand, so mining the specialized sand has become a major industry in recent years in several Midwestern states. This industrial boom has brought with it much controversy in small-town America, where energy industries seek to establish large mining operations. Usage of frac sand hit a peak at the beginning of 2017, and now is decreasing slightly, due in part to increasing costs of the sand and to the increase in use of renewable energy sources, such as wind power. Fracking apparently causes an increase in seismic activity in those regions, adding to the controversy, particularly in Midwestern states such as Oklahoma and North Dakota and in British Columbia, Canada.

Natural Gas

Natural gas, composed principally of methane, is the cleanest of the fossil fuels, used in a significant portion of homes in the United States. Minor components of this gaseous fossil fuel vary greatly. In North America, a pipeline natural gas supply might contain 83% to 95% methane, 2% to 6% ethane, 1% to 2% propane, and smaller amounts of butanes and pentanes. Natural gas burns with a relatively clean flame, and the products are mainly carbon dioxide and water.

$$CH_4(g) + 2\,O_2(g) \longrightarrow CO_2(g) + 2\,H_2O(g) + \text{heat}$$

When natural gas is burned in insufficient air, carbon monoxide and soot can be major products.

$$2\,CH_4(g) + 3\,O_2(g) \longrightarrow 2\,CO(g) + 4\,H_2O(g)$$
$$CH_4(g) + O_2(g) \longrightarrow C(s) + 2\,H_2O(g)$$

The gas, as it comes from the ground, often contains nitrogen (N_2), sulfur compounds, and other substances as impurities. As in any combustion in air, some nitrogen oxides are formed.

Most natural gas is used as fuel, supplying about 23% of worldwide energy consumption, but it is also an important raw material. In North America, a portion of the two- to four-carbon alkanes is separated from the rest of the natural gas. Ethane and propane can be *cracked*—decomposed by heating with a catalyst—to form ethylene and propylene. These alkenes are the raw materials used to make polyethylene and polypropylene plastics (Chapter 10) and many other useful commodities. Natural gas is also the raw material from which methanol and many other organic compounds are made. Like other fossil fuels, the supply of natural gas is limited (refer to Table 15.3), but its use in the United States is increasing, whereas coal use is decreasing.

▲ Ordinary sand and fracking sand. The sand on the left consists of ordinary, multivariant particles, but the frac sand on the right consists of small, rounded quartz particles, found extensively in North Dakota, Minnesota, Wisconsin, and Iowa. Grains in this image are about 0.50 mm in size.

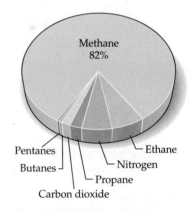

▲ A rough guide to the composition of typical natural gas. Natural gas is mostly methane, with small amounts of other hydrocarbons, nitrogen, and a few other gases.

▲ Natural gas burns with a relatively clean flame.

4 What is "frac sand," and how is it different from ordinary beach sand?

All sand is not the same! Usually sand is a mixture of small particles of varying color, size, and composition. "Frac sand" is 95% silicon dioxide (quartz), with rounded particles about 0.4–0.8 mm in diameter.

Petroleum: Liquid Hydrocarbons

Petroleum is an extremely complex, black, viscous liquid mixture of organic compounds—as many as 17,000 compounds were separated from a sample of Brazilian crude oil. Most are hydrocarbons: alkanes, cycloalkanes, and aromatic compounds. Complete combustion of these compounds forms carbon dioxide and water vapor.

Obtaining and Refining Petroleum

Crude oil is liquid, a convenient form for being transported. Petroleum is pumped easily through pipelines or hauled across the oceans in giant tankers. Pumping petroleum from the ground requires little energy at first. As petroleum is pumped out of the ground, the remaining oil becomes increasingly difficult to extract.

As it comes from the ground, crude oil is of limited. To make it better suited to our needs, we separate it into fractions by heating it in a fractional distillation column (Figure 15.8). Petroleum deposits nearly always have associated natural gas, part or all of which also goes through a separation process. The lighter hydrocarbon molecules come off at the top of the column because they have lower boiling points, and the heavier ones with the higher boiling points (Table 15.5) come off at the bottom.

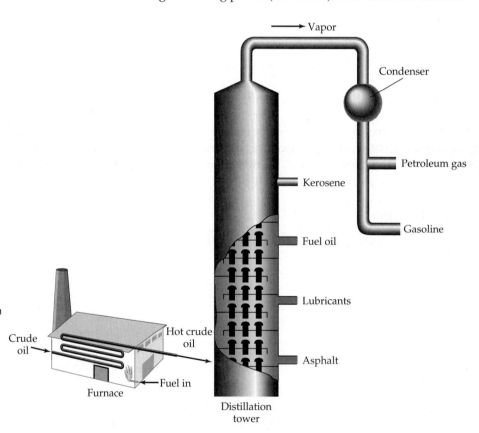

▶ **Figure 15.8** The fractional distillation of petroleum. Crude oil is vaporized, and the column separates the components according to their boiling points. The lower–boiling point constituents reach the top of the column, and the higher–boiling point components come off lower in the column. A nonvolatile residue collects at the bottom.

TABLE 15.5 Typical Petroleum Fractions

Fraction	Typical Size Range of Hydrocarbons	Approximate Range of Boiling Points (°C)	Typical Uses
Gas	CH_4 to C_4H_{10}	Less than 40	Fuel; starting materials for plastics
Gasoline	C_5H_{12} to $C_{12}H_{26}$	40–200	Fuel; solvents
Kerosene	$C_{12}H_{26}$ to $C_{16}H_{34}$	175–275	Diesel fuel; jet fuel; cracking to make gasoline
Heating oil	$C_{15}H_{32}$ to $C_{16}H_{34}$	250–400	Industrial and home heating; cracking to make gasoline
Lubricating oils; greases	$C_{17}H_{36}$ and up	Above 300	Lubricants
Residue	$C_{20}H_{42}$ and up	Above 350	Paraffin; asphalt

Since gasoline is so important today, fractions of crude oil from the fractionating column that boil at higher temperatures are often converted to gasoline by cracking. This process, with $C_{14}H_{30}$ as an example, is illustrated in Figure 15.9.

$$CH_3CH_2CH_2CH_2CH_2CH_2CH_2CH_2CH_2CH_2CH_2CH_2CH_2CH_3 \xrightarrow[\text{catalyst}]{\text{heat}}$$

$$CH_3CH_2CH_2CH_2CH_2CH_2CH_2CH_2CH_2CH_2CH_2CH_3 + CH_2{=}CH_2$$

and

$$CH_3CH_2CH_2CH_2CH_2CH_2CH_2CH_2CH_2CH_2CH_3 + CH_3CH{=}CH_2$$

and

$$CH_3CH_2CH_2CH_2CH_2CH_2CH_2CH_3 + CH_3CH_2CH_2CH_2CH{=}CH_2$$

and so on

◀ **Figure 15.9** Formulas of a few of the possible products formed when $C_{14}H_{30}$, a typical molecule in kerosene, is cracked. In practice, a wide variety of hydrocarbons (most of which have fewer than 15 carbon atoms), hydrogen gas, and char (mostly elemental carbon) are formed. Cyclic and branched-chain hydrocarbons are also produced.

Cracking not only converts some of the large molecules to those in the gasoline range (C_5H_{12} through $C_{12}H_{26}$) but also produces a variety of useful by-products, a dazzling array of substances with a wide variety of properties, including plastics, pesticides, herbicides, perfumes, preservatives, painkillers, antibiotics, stimulants, depressants, dyes, and detergents. In the future, as gasoline-powered vehicles are phased out, petroleum will still be important for the variety of other consumable materials that can be made from petroleum components.

Gasoline

Gasoline is usually the product from petroleum that is currently most in demand. With the development of the internal combustion engine, petroleum became increasingly important and, by 1950, had replaced coal as the principal fuel. Modern culture was forever changed by the invention of the automobile. Well over a billion gasoline-powered vehicles are in use in the world today.

Gasoline (often called "petrol" outside the United States), like the petroleum from which it is derived, is mainly a mixture of hydrocarbons. Commercial gasoline typically contains more than 150 different compounds, and up to 1000 have been identified in some blends. Among the hydrocarbons, a typical gasoline sample might include straight-chain alkanes, branched-chain alkanes, alkenes, cyclic hydrocarbons, and aromatic hydrocarbons. Gasoline also contains a variety of additives, including anti-knock agents, antioxidants, antirust agents, anti-icing agents, upper-cylinder lubricants, detergents, and dyes. Typical alkanes in gasoline range from C_5H_{12} to $C_{12}H_{26}$. There are also small amounts of some sulfur- and nitrogen-containing compounds present.

Gasoline-powered engines are rather inefficient, and the combustion of gasoline contributes greatly to air pollution (Section 13.4). Burning gasoline is a major source of CO_2 emissions that are causing global climate change. As with natural gas, complete combustion of these substances yields mainly carbon dioxide and water. Incomplete burning yields carbon monoxide and soot. Combustion of gasoline also produces nitrogen oxides, so cities with a high population density often record high CO and NO_2 levels in the air (Section 13.4).

Air pollution isn't the only problem with gasoline. Burning up our petroleum reserves will leave us without a ready source of many familiar materials, including plastics, synthetic fibers, solvents, and many other consumer products mentioned above.

Worldwide, petroleum still appears to be fairly abundant. (See Table 15.3.) U.S. petroleum reserves have declined since 1972, but with the development of fracking techniques, and the discovery of more oil shale reserves in the United States, about 77% of the oil the United States consumes is generated domestically, decreasing the country's dependence on foreign oil. Further tapping of offshore deposits and deeper drilling would produce more oil, but would require a great deal more energy (and money) than drilling on land. Political events in the oil-rich Middle East have also caused oil prices to fluctuate over the years. The large oil reserves in the Middle East will likely still be a factor in energy availability in the world's future.

The Octane Ratings of Gasolines

In an internal combustion engine, the gasoline–air mixture sometimes ignites before the spark plug "fires." This is called *knocking* and can damage the engine. Early on, scientists learned that some types of hydrocarbons, especially those with branched structures, burned more evenly and were less likely to cause knocking than straight-chain hydrocarbons. An arbitrary performance standard, called the **octane rating**, was established in 1927. Isooctane was assigned an octane rating of 100. An unbranched-chain compound, heptane, was given an octane rating of zero. A gasoline rated 90 octane was one that performed the same as a mixture that was 90% isooctane and 10% heptane.

$$CH_3-\underset{\underset{CH_3}{|}}{\overset{\overset{CH_3}{|}}{C}}-CH_2-\overset{\overset{CH_3}{|}}{CH}-CH_3 \qquad CH_3CH_2CH_2CH_2CH_2CH_2CH_3$$

Isooctane
(Octane rating 100)

Heptane
(Octane rating 0)

During the 1930s, chemists discovered that the octane rating of gasoline could be improved by heating it in the presence of a catalyst such as sulfuric acid (H_2SO_4) or aluminum chloride. This *isomerizes* some of the unbranched molecules to highly branched molecules. Chemists also can combine small hydrocarbon molecules (below the size range of gasoline) into larger ones more suitable for use as fuel, a process called *alkylation*. Petroleum refineries use **catalytic reforming** to convert low-octane alkanes to high-octane aromatic compounds. Hexane (with an octane number of 25) can be converted to benzene (octane number 106).

Unleaded Gasoline

Why is one of the pumps at a gasoline filling station labeled "unleaded"? The reason is that decades ago, tetraethyllead [$Pb(CH_2CH_3)_4$] was found to substantially improve the anti-knock quality of a gasoline mixture, hence "leaded" gasoline. As little as 1 mL of tetraethyllead per liter of gasoline (1 part per 1000) increases the octane rating by 10 or more. The problem, of course, is that lead atoms would be emitted from the exhaust pipes of automobiles when the leaded gas was burned, and so for many years lead was a major air pollutant. Lead, unfortunately, also fouls the catalytic converters used in modern automobiles. In the United States, *un*leaded gasoline became available in 1974, and leaded gasoline was phased out as automotive fuel. Since then, the amount of lead, as an air pollutant, has decreased at least 99% in the United States. In Mexico and many countries in Central America and Africa, however, leaded gas is still an option.

Octane boosters that have replaced tetraethyllead include ethanol, methanol, *tert*-butyl alcohol, and methyl *tert*-butyl ether (MTBE), which was phased out in 2014 because it was listed as a hazardous substance under the federal Superfund Act and is considered a potential human carcinogen by the Environmental Protection Agency (EPA). None of these is nearly as effective as tetraethyllead in boosting the octane rating.

Ethanol has become a common and important gasoline additive. The EPA allows up to 15% ethanol in gasoline (E15 fuel) for cars built in 2001 or later. Some states require all gasoline to contain 10% ethanol. Most ethanol comes from corn, so the use of ethanol in gasoline is controversial, because much of the corn grown in the central part of the United States is used to feed cattle rather than to be made into ethanol to be burned in car engines, a competing use. Brazil makes extensive use of ethanol made by fermentation of sucrose from sugarcane.

Unlike gasoline, which is made up of hydrocarbons, the alcohols used as octane boosters all contain oxygen and therefore are sometimes called *oxygenates*. Not only do these additives improve the octane rating, but they also decrease the amount of carbon monoxide in auto exhaust gas. A disadvantage of oxygenates is that they are partially oxidized. That makes their energy content lower than that of pure hydrocarbon fuels, so the distance a car can travel per tankful is somewhat shorter.

Alternative Fuels for Vehicles

An automobile engine can be made to run on nearly any liquid or gaseous fuel. There are cars on the road today powered by natural gas, by propane, by diesel fuel, by fuel cells, by hydrogen, and even by used fast-food restaurant grease.

Diesel fuel for vehicles overlaps the kerosene fraction of petroleum (refer to Table 15.5); it consists mainly of C_9 to C_{20} hydrocarbons. Diesel fuel has a greater proportion of straight-chain alkanes than gasoline does. The standard for performance, called the *cetane number*, is based on hexadecane ($C_{16}H_{34}$). A renewable fuel called *biodiesel* can be used in some unmodified diesel engines. Biodiesel is made by reacting ethanol with vegetable oils and animal fats.

In the United States, flexible fuel vehicles (FFVs) are designed to run on E85 gasoline, a fuel that is 85% ethanol and 15% gasoline. The price of E85 fuel is generally less than that of ordinary gasoline, but the mileage obtained is worse. Ethanol alone or blended with gasoline is not likely to be an answer to our energy problems.

Hybrid cars using both gasoline and electricity have been popular for nearly a decade. All-electric cars that can be plugged into the power grid may be the cars of the future. At the present time, some disadvantages of all-electric vehicles are their initial cost (higher than that of an otherwise similar gasoline-powered vehicle), the very high cost of replacing the lithium batteries after about 100,000 miles, the time required to recharge the batteries after driving 100–200 miles, and not enough convenient recharging stations. However, with engineering improvements, the cost of an all-electric vehicle is decreasing, the miles before needing recharging increasing, and the number of charging stations increasing. Electric-powered motorcycles are becoming much more common, particularly in high-density cities, where parking an ordinary vehicle becomes both challenging and expensive. In Taiwan, electric-powered motorcycles are very popular; batteries need recharging about once a week, and those batteries can be readily exchanged in minutes at a battery exchange facility.

SELF-ASSESSMENT Questions

1. The main constituent in natural gas is
   a. CH_4
   b. $CH_3CH_2CH_3$
   c. C_6H_6
   d. CH_3OH

2. Which of the fossil fuels is the cleanest and simplest in composition?
   a. coal
   b. It depends on the source.
   c. petroleum
   d. natural gas

3. Cracking is used to
   a. convert heavier fractions to lighter ones
   b. improve the octane rating
   c. remove sulfur
   d. separate various fractions

4. An alkylation process
   a. converts lighter fractions to heavier ones
   b. improves the asphalt yield
   c. removes sulfur
   d. separates various fractions

5. Isomerization is used to convert
   a. alcohols to ethers
   b. branched alkane molecules to straight-chain ones
   c. hexane to isooctane
   d. straight-chain molecules to branched ones

6. Catalytic reforming converts
   a. branched molecules to straight-chain ones
   b. convert low octane to high octane fuels
   c. straight-chain molecules to aromatic ones
   d. straight-chain molecules to branched ones

7. The octane rating for gasoline is related to the fuel's
   a. concentration of octane (C_8H_{16})
   b. energy content in kilocalories
   c. power rating
   d. tendency to cause knocking

8. Most gasoline stations offer, or are required to sell, automobile fuel that is
   a. 15% ethanol, 85% gasoline
   b. 10% ethanol, 90% gasoline
   c. 90% isooctane, 10% ethanol
   d. 10% ether, 90% octane

9. Hydraulic fracturing technology, including the use of fracking sand, has caused at least a 30% increase in reserves in the last few years of
   a. coal
   b. biomass
   c. gasoline
   d. natural gas

Answers: 1, a; 2, d; 3, a; 4, a; 5, d; 6, c; 7, d; 8, b; 9, d

15.8 Convenient Energy

Learning Objective • Explain why gaseous and liquid fuels are more convenient to use than solid fuels.

The convenience of a fuel depends on its physical state. Gases and liquids are convenient, while solids are much less so. Perhaps the most convenient form of energy is electricity. We can use electricity to provide light and hot water and to run motors of all sorts. We can use it to heat and cool our homes and workplaces. When looking at future energy sources, then, we look to a large degree at ways of generating electricity. But what is the advantage to electricity if we make it by burning coal? There is no simple answer.

Any fuel can be burned to boil water, and the steam produced can turn a turbine to generate electricity. Figure 15.10 shows a coal-fired steam power plant. At present,

▶ **Figure 15.10** A diagram showing how a coal-burning power plant generates electricity.

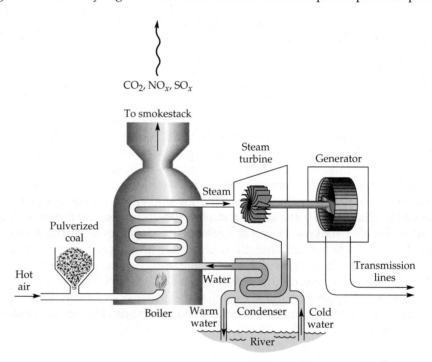

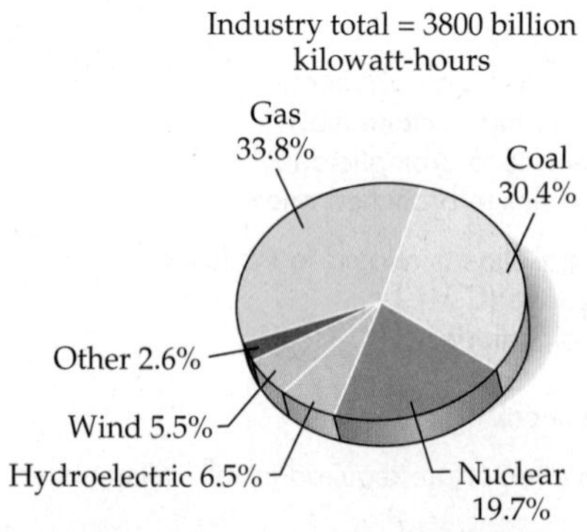

Industry total = 3800 billion kilowatt-hours

Gas 33.8%
Coal 30.4%
Other 2.6%
Wind 5.5%
Hydroelectric 6.5%
Nuclear 19.7%

▲ **Figure 15.11** Percentages of electric power generation in the United States from various energy sources, reported in 2017 (Source: U.S. Energy Information Administration).

about 30% of U.S. electric energy comes from coal-burning plants (Figure 15.11). Such facilities are at best only about 40% efficient. Thus, about 60% of the energy of the fossil fuel is wasted as heat, though some power installations use waste heat to warm buildings.

Coal Gasification and Liquefaction

Coal can be converted to gas or oil. When we run short of gas and petroleum, why not make them from coal? The technology has been around for years. Gasification and liquefaction do have some advantages: Gases and liquids are easy to transport, and the process of conversion leaves much of the sulfur and minerals behind, thus mitigating a serious disadvantage of coal as a fuel.

The basic process for converting coal to a synthetic gaseous fuel involves reduction of carbon by hydrogen. Passing steam over hot charcoal produces **synthesis gas**, a mixture of hydrogen and carbon monoxide.

$$C(s) + H_2O(g) \longrightarrow CO(g) + H_2$$

The hydrogen can be used to change the carbon in coal or other substances to methane. Coal can also be converted to methanol by way of synthesis gas, a multistep process that requires catalysts at every step. Every step adds to the cost.

Both gasification and liquefaction of coal require a lot of energy; up to one-third of the energy content of the coal is lost in the conversion. Liquid fuels from coal are high in unsaturated hydrocarbons and in sulfur, nitrogen, and arsenic compounds. Their combustion products require large amounts of water, yet the large coal deposits that would be used are in arid regions. Furthermore, these processes are messy. Without stringent safeguards, conversion plants would seriously pollute both air and water. Coal conversion also produces a lot of particulate matter.

SELF-ASSESSMENT Questions

1. About what percentage of the electricity produced in the United States comes from coal-burning plants?
   a. 20%   **b.** 30%   **c.** 40%   **d.** 80%

2. Natural gas is transported mainly by
   a. ocean-going tanker   **b.** pipelines
   c. railroad car   **d.** truck

3. About what percentage of the energy from coal burned in a power plant reaches customers as electricity?
   a. 20%   **b.** 40%   **c.** 60%   **d.** 80%

4. Coal gasification produces mainly
   a. coke   **b.** H_2
   c. CH_4 **d.** $CH_3CH_2CH_3$

5. The convenience of a fuel depends on its
   a. carbon content   **b.** density
   c. odor   **d.** physical state

Answers: 1, b; 2, b; 3, b; 4, c; 5, d

15.9 Nuclear Energy

Learning Objectives • List the advantages and disadvantages of nuclear energy.
• Describe how a nuclear power plant generates electricity.

In Chapter 11, we explored nuclear reactions, radioactive decay, medical uses for radioactive isotopes for both diagnosis and therapy, and both nuclear fission and fusion. So far, nuclear fusion reactions cannot be controlled for producing energy here on Earth, only in the stars, including our own sun. Nuclear fission reactions, however, can be controlled in a **nuclear reactor**. The energy released during fission can be used to generate steam, which can turn a turbine to generate electricity (Figure 15.12).

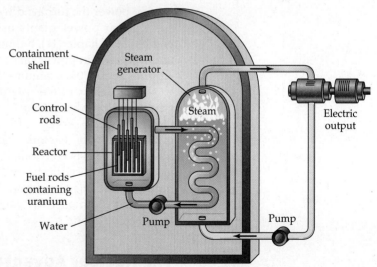

▲ **Figure 15.12** A diagram showing how a nuclear power plant generates electricity. Fission of uranium heats the water in the reactor, which generates steam that drives the turbines, producing electricity. Control rods absorb neutrons to slow the fission reaction as needed. Nuclear reactors are housed in containment buildings made of steel and reinforced concrete, designed to withstand nuclear accidents and prevent the release of radioactive substances into the environment.

At the dawn of the nuclear age, some people envisioned nuclear power as likely to fulfill the biblical prophecy of a fiery end to our world. Others saw it as a source of unlimited energy. During the late 1940s, some claimed that electricity

TABLE 15.6 Electricity Generation Using Nuclear Power Plants (Selected Countries)

Country	Total Electricity from Nuclear Power Plants (%)	Number of Operating Nuclear Power Plants	Number of Nuclear Power Plants under Construction
Canada	14.6	19	0
China	3.9	39	19
France	71.6	58	1
India	3.2	22	6
Japan	3.6	42	2
South Korea	27.1	25	3
Russia	17.8	35	7
Sweden	39.6	9	0
Ukraine	55.1	15	2
United Kingdom	19.3	15	0
United States	20.0	99	2
Other countries	NA	72	13
World total	NA	448	57

from nuclear power plants would become so cheap that it eventually would not have to be metered. Nuclear power has not yet brought either paradise or perdition, but it is still controversial.

Since about 1990 and continuing to this day, nuclear power has supplied about 20% of U.S. electricity. The Eastern Seaboard and upper Midwestern states, many of which have minimal fossil fuel reserves, are heavily dependent on nuclear power for electricity. The United States is currently the world's largest producer of nuclear energy, with 99 nuclear plants in 30 states. Over the last decade a number of aging power plants have been decommissioned, and two new nuclear power plants are now under construction. It takes about 10 years to build such a plant.

The United States could do as other nations do and rely more on nuclear power (Table 15.6), but the public is generally fearful of nuclear power plants. This apprehension was exacerbated in 2011 by the accident at Fukushima, Japan. There will have to be a dramatic change in public attitude if nuclear power is to play a big role in our energy future.

Nuclear Power Plants

There are several types of nuclear power plants, but we won't discuss the different designs. The one illustrated in Figure 15.12 is a pressurized water reactor. Earlier models were mainly boiling water reactors, in which the steam from the reactor was used to power the turbine directly.

Nuclear power plants use the same fission reactions employed in nuclear bombs (see Chapter 11), but nuclear power plants cannot blow up like bombs. The uranium used in power plants is enriched to only 3% to 4% uranium-235. A bomb requires about 90% uranium-235.

In a nuclear power plant, a *moderator* is used to slow down the neutrons produced by the fission reaction so that they can be absorbed by uranium-235 atoms. Ordinary water serves as the moderator in 75% of reactors worldwide, graphite is used in 20% of reactors, and heavy water (^2H_2O or D_2O) in 5% of reactors. The fission reaction is controlled by the insertion of boron steel or cadmium *control rods*. Boron and cadmium absorb neutrons readily, preventing them from participating in the chain reaction. These rods are installed when the reactor is built. Removing them partway starts the chain reaction; pushing them in all the way stops the reaction.

The Nuclear Advantage: Minimal Air Pollution

The main advantage of nuclear power plants over those that burn fossil fuels is in what they do not do. Unlike power plants that burn fossil fuels, nuclear power plants produce no carbon dioxide to add to the greenhouse effect, and they add no sulfur oxides, nitrogen oxides, soot, or fly ash to the atmosphere. They contribute almost nothing to global warming, air pollution, or acid rain. They reduce our dependence on foreign oil and lower our trade deficit. If the 99 nuclear plants in the United States were replaced by coal-burning plants, airborne pollutants would increase by 19,000 tons per day! Nuclear power plants run 24 hours a day, 7 days a week, and need to be refueled only every 1.5–2 years.

Problems with Nuclear Power

Nuclear power plants have some disadvantages. Elaborate and expensive safety precautions must be taken to protect plant workers and the inhabitants of surrounding areas from radiation. The reactor must be heavily shielded and housed inside a containment building of metal and reinforced concrete. Because loss of coolant water can result in a meltdown of the reactor core, backup emergency cooling systems are required.

Despite what proponents call utmost precautions, some opponents of nuclear power still fear a runaway nuclear reaction in which a containment building is breached and massive amounts of radioactivity escape into the environment. The chance of such an accident is very small, but one did happen at Fukushima. Yet the benefit of nuclear power—abundant electric energy—is clear, so the small probability of a serious accident causes scientists and others to endlessly debate its desirability.

Another problem is that the fission products are highly radioactive and must be isolated from the environment for centuries. Again, scientists disagree about the feasibility of disposal of nuclear waste. Proponents of nuclear power say that such wastes can be safely stored in old salt mines or other geologic formations. Opponents fear that the wastes may rise from their "graves" and eventually contaminate the groundwater. It is impossible to do a million-year experiment in a few years to determine who is right. Federal funding for an underground repository at Yucca Mountain, Nevada, was terminated in 2011 after billions of dollars were spent on the project. This leaves the United States with no long-term storage site for high-level radioactive waste. These wastes are currently stored at 126 sites around the country. It is important to note that high-level waste from nuclear power plant operations in the United States totals about 3000 tons per year. The U.S. Department of Energy estimates that the total volume of nuclear waste material from the last 60 years would fill a football field to a depth of under 10 meters. Compare this number with the coal ash produced in the United States—over 100 million tons per year! It is also important to realize that most of the nuclear wastes awaiting disposal are from nuclear weapons production, *not* energy production.

Mining and processing of uranium ore produce wastes called *tailings*. Over 200 million tons of tailings now afflict 10 western states. These tailings are mildly radioactive, giving off radon gas and gamma radiation.

Any power plant has another problem that is unavoidable—thermal pollution. As the energy from any material is converted to heat to generate electricity, some of the energy is released into the environment as waste heat. Nuclear power plants generate more thermal pollution than do plants that burn fossil fuels, but this provides an opportunity for engineers to design a mechanism to use the hot wastewater in some sustainable way.

Nuclear Accidents

In 1979, a loss-of-coolant accident at the Three Mile Island nuclear power plant near Harrisburg, Pennsylvania, released a tiny amount of radioactivity into the environment. Although no one was killed or seriously injured, the accident heightened public fear of nuclear power.

A 1986 accident at Chernobyl, Ukraine, was much more frightening. There, a reactor core meltdown killed several people outright. Others died from radiation sickness in the following weeks and months, and 135,000 people were evacuated. The Ukraine Radiological Institute estimated that the accident caused more than 2500 deaths overall. A large area will remain contaminated for decades. Radioactive fallout spread across much of Europe. The thyroid cancer rate among Ukrainians increased tenfold, and the overall risk of cancer rose because of exposure to radiation. At Three Mile Island, a containment building kept nearly all the radioactive material inside. The Chernobyl plant had no such protective structure.

In 2011, a tsunami caused by an earthquake destroyed the external power supply to a nuclear power plant in Fukushima. Thousands of people were killed by the tsunami, and large areas were rendered uninhabitable by released radiation, perhaps for centuries. The lack of power led to partial core meltdowns in three reactors, with widespread release of radiation. The Fukushima accident is rated as the second worst in history, after Chernobyl. Thousands were killed by the tsunami, but so far only a few by radiation.

There is considerable controversy over most aspects of nuclear power. Although scientists may be able to agree on the results of laboratory experiments, they don't always agree on what is best for society.

Breeder Reactors: Making More Fuel Than They Consume

The supply of the fissionable uranium-235 isotope is limited, as it makes up less than 1% of naturally occurring uranium. Separation of uranium-235 leaves behind large quantities of uranium-238, which is not fissionable. However, uranium-238 can be converted to fissionable plutonium-239 by bombardment with neutrons. Unstable uranium-239 is formed initially, but it rapidly decays to neptunium-239, which then decays to plutonium.

$$^{238}_{92}U + ^{1}_{0}n \longrightarrow ^{239}_{92}U \longrightarrow ^{239}_{93}Np + ^{0}_{-1}e$$

$$^{239}_{93}Np \longrightarrow ^{239}_{94}Pu + ^{0}_{-1}e$$

If a reactor is built with a core of fissionable plutonium-239 surrounded by uranium-238, neutrons from the fission of plutonium convert the uranium-238 shield to more plutonium. In this way, the reactor, called a **breeder reactor**, produces more fuel than it consumes. There is enough uranium-238 to last a few centuries, so one of the disadvantages of nuclear plants could be overcome by the use of breeder reactors. Molten sodium metal, not water, is used in the primary cooling loop, so these reactors are often called *liquid-metal fast breeder reactors*.

Breeder reactors have some problems of their own, however. Plutonium is fairly low melting (640 °C), and a plant is therefore limited to fairly cool, inefficient operation. Another problem is that plutonium is highly toxic and has a half-life of about 24,000 years. It emits alpha particles, making it especially dangerous if ingested. An estimated 1 μg in the lungs of a human is enough to induce lung cancer.

Only 11 breeder reactors with a production capacity of more than 100 MW have ever been built, none in the United States. Two are operational in Russia; the second has been in operation since November 2016 in the Sverdlovsk region of Russia, generating about 880 MW of power. China began construction of a breeder reactor in December 2017.

Nuclear Fusion: The Sun in a Magnetic Bottle

Chapter 11 discussed the thermonuclear reactions that power the sun and that occur in the explosion of hydrogen bombs. Control of these fusion reactions to produce electricity would give us nearly unlimited power. To date, fusion reactions have been useful only for making bombs, but research on the control of nuclear fusion is progressing (Figure 15.13).

Great technical difficulties have to be overcome before a controlled fusion reaction can be used to produce energy. A sustainable fusion reaction would require attainment of a critical ignition temperature between 100 million °C and 200 million °C,

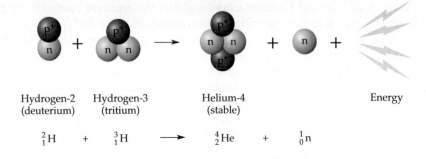

▶ **Figure 15.13** A promising fusion reaction is the deuterium–tritium reaction. A hydrogen-2 (deuterium) nucleus fuses with a hydrogen-3 (tritium) nucleus to form a helium-4 nucleus. A neutron is released, along with a considerable quantity of energy.

Hydrogen-2 (deuterium) Hydrogen-3 (tritium) Helium-4 (stable) Energy

$$^{2}_{1}H + ^{3}_{1}H \longrightarrow ^{4}_{2}He + ^{1}_{0}n$$

a confinement time of 1–2 s, and an ion density of between 2×10^{20} and 3×10^{20} ions per cubic meter. No molecule could hold together at the fusion temperature. No material on Earth can withstand more than a few thousand degrees. Even atoms are unstable under these conditions. Atoms are stripped of their electrons, and the nuclei and free electrons form a mixture called a *plasma*. The plasma, made of charged particles (nuclei and electrons), can be contained by a strong magnetic field (Figure 15.14).

Controlled fusion would have several advantages over nuclear fission as an energy source. The main fuel, deuterium $\left(^2_1H\right)$, is plentiful, and is obtained from electrolysis (splitting apart by means of electricity) of water. The problem of radioactive wastes would be minimized. The end product, helium, is stable and biologically inert. Escape of tritium $\left(^3_1H\right)$ might be a problem, because this hydrogen isotope would be readily incorporated into organisms. Also, neutrons are emitted in most fusion reactions, and neutrons can convert stable isotopes into radioactive ones. Finally, we would still be concerned with thermal pollution—the unavoidable loss of part of the energy as heat.

Nuclear fusion may well be our best hope for producing relatively clean, abundant energy in the future, but much work remains to be done. Once controlled fusion is achieved in a laboratory, it will still be decades before it becomes a practical source of energy.

▲ **Figure 15.14** A giant, doughnut-shaped electromagnet called a *tokamak* is designed to confine plasma at the extremely high temperatures and pressures required for nuclear fusion.

SELF-ASSESSMENT Questions

1. In a nuclear fission reactor
   a. carbon-14 decays to nitrogen-14
   b. hydrogen nuclei fuse to form helium
   c. uranium-235 absorbs a neutron and splits into two smaller nuclei
   d. uranium-238 decays to plutonium-238

2. The control rods in a nuclear reactor
   a. absorb neutrons, allowing them to initiate fission
   b. absorb neutrons, preventing them from causing fission
   c. slow the neutrons down so the moderator can stop them
   d. speed the neutrons up so they will initiate fission

3. What percentage of the electricity used in the United States is produced by nuclear power plants?
   a. 5%   b. 10%   c. 20%   d. 50%

4. Nuclear reactors that produce more nuclear fuel than they consume
   a. are called breeder reactors
   b. are called fusion reactors
   c. are called transuranic reactors
   d. violate the law of conservation of energy

5. Approximately how many nuclear power plants are in operation in the United States?
   a. 50   b. 100   c. 200   d. 400

6. Which of the following is a disadvantage of nuclear power plants?
   a. They produce greenhouse gases.
   b. They produce more radioactive waste than wastes left from nuclear weapons development.
   c. It is difficult to build safe reactors.
   d. They produce thermal pollution of nearby water sources.

7. What is the missing product in the following nuclear reaction?
$$^2_1H + {}^3_1H \longrightarrow {}^1_0n + ?$$
   a. ^5_2He b. ^3_2He c. ^4_2He d. ^3_1H

8. Fission, not fusion, is used in nuclear power plants because fusion
   a. uses expensive transuranium elements as fuel
   b. requires temperatures of millions of degrees, making the reaction difficult to contain
   c. produces less heat than fission
   d. produces waste products that are radioactive for centuries

9. Which country has the most nuclear power plants?
   a. France   b. United States
   c. England   d. Germany

Answers: 1. c; 2. b; 3. c; 4. a; 5. b; 6. d; 7. c; 8. b; 9. b

15.10 Renewable Energy Sources

Learning Objective • List important characteristics, advantages, and disadvantages of various kinds of renewable energy sources.

Burning fossil fuels leads to air pollution and to the depletion of vital resources. Using nuclear power also presents problems. Nuclear fuel is not unlimited, and the problem of waste disposal remains unsolved. The utilization of renewable energy sources, including wind energy, solar energy, fuel cells, and hydroelectric power, has experienced a very significant upswing in the last three years, as they now account for about 11% of the energy production in the United States. This section discusses these sustainable energy sources.

Solar Energy for Heating

We noted in Section 15.1 that most of the energy available on Earth comes from the sun. With all that energy from our celestial power plant, why are we currently dependent on fossil fuels or nuclear power? The answer lies in the fact that solar energy is thinly spread out and difficult to capture directly. Of all the solar energy that arrives on the surface of the Earth, 30% is reflected back into space and is thus unavailable for our use. About half of the solar energy that is available to Earth is converted into heat. We can increase the efficiency of this conversion rather easily. A black surface absorbs heat better than a light-colored one. To make a simple solar collector, we only have to cover a black surface with a glass plate. The glass is transparent to the incoming solar radiation, but it partially prevents the heat from escaping back into space. The hot surface is used to heat water or other liquids, and the hot liquids are usually stored in an insulated reservoir. Simple solar heating devices for individual family use are becoming more readily available in developing nations, particularly in Africa.

Water heated this way can be used directly for bathing, dish washing, and laundry, or it can be used to heat a building. Air is passed around the warm reservoir, and the warmed air is then circulated through the building (Figure 15.15). Even in cold northern climates, solar collectors could meet about 50% of home heating requirements. These installations are expensive but can supposedly pay for themselves, through fuel savings, in 8 to 10 years, but estimating such savings is quite difficult, because too many factors affect costs such as the size of the home, the external temperature fluctuations, and the temperature setting on the thermostat.

▼ **Figure 15.15** (a) Energy in sunlight is absorbed by solar collectors and used to heat water. The hot water can be used directly or used to warm air that is circulated to partially heat a building. This diagram shows how a solar collector furnishes hot water and warm air for heating a building. (b) A variety of collectors can be used to absorb solar energy.

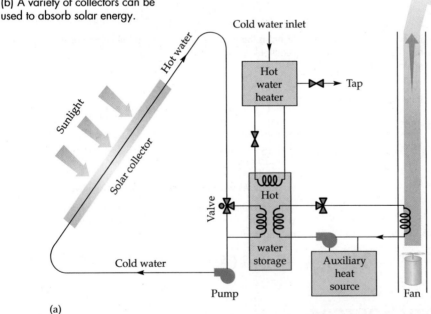

(a)

(b)

Solar Cells: Electricity from Sunlight

Sunlight also can be converted directly to electricity by devices called **photovoltaic cells,** or *solar cells.* These devices can be made from a variety of substances, but most are made from elemental silicon. In a crystal of pure silicon, each silicon atom has four valence electrons and is covalently bonded to four other silicon atoms (Figure 15.16). To make a solar cell, extremely pure silicon is "doped" with small amounts of specific impurities and formed into crystals.

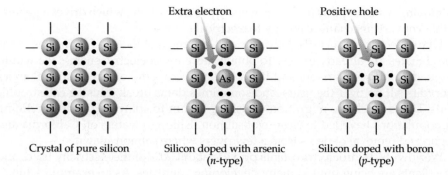

Crystal of pure silicon

Silicon doped with arsenic (*n*-type)

Silicon doped with boron (*p*-type)

◀ **Figure 15.16** Models of silicon crystals. Crystals doped with impurities such as arsenic and boron are more conductive than pure silicon and are used in solar cells.

One type of silicon crystal has about 1 ppm of arsenic added. Arsenic atoms have five valence electrons, four of which are used to form bonds to silicon atoms. The fifth electron is relatively free to move around. Because this material has extra electrons, and electrons are negatively charged, it is called an *n-type* (negative) *semiconductor*. Adding about 1 ppm of boron to silicon forms a different type of material. Boron has three valence electrons, producing a shortage of one electron and leaving a *positive hole* in the crystal. This boron-doped silicon is called a *p-type semiconductor*.

Joining the two types of crystals (*n*-type and *p*-type) forms a photovoltaic cell (Figure 15.17). Electrons flow from the *n*-type region, which has a high concentration of them, to the *p*-type region. However, the holes near the junction are quickly filled by nearby mobile electrons, and the flow ceases.

When sunlight hits the photovoltaic cell, an electric current is generated. The energetic photons knock electrons out of the Si—Si bonds, creating more mobile electrons and more positive holes. Because of the barrier at the junction between the two semiconductors, electrons cannot move through the interface. When an external circuit connects the two crystals, electrons flow from the *n*-type region around the circuit to the *p*-type region and that current can be used directly.

An array of solar cells, combined to form a solar battery, can produce about 200 W/m² of surface. That is, it takes 1 m² of cells to power two 100 W light bulbs. Solar batteries have been used for years to power artificial Earth satellites. They are now widely used to power small devices such as electronic calculators and highway signage in rural areas, and to provide electricity for weather instruments in remote areas.

The generation of enough energy to meet a significant portion of our demands requires covering vast areas of land with solar cells, in very sunny climates. For an example, a 1.5-mile-long area in the Nevada desert called "Crescent Dunes" consists of 10,000 solar panels, each 115 m², and powers about 75,000 homes, night and day, since it includes a central tower for power storage.

For individual home use, solar panels are not yet very practical in northern climates. Several companies sell home-sized solar electric systems that can be connected to the normal supply line and are claimed to reduce electric bills to nearly zero. Currently most commercially available solar cells are about 15% to 19% efficient, but at least two companies claim 22% efficiency. The best panels provide about 325 W of energy for a 1.5 m² panel, so an average home would need much of the roof covered with panels. The state of California passed a law in May 2018 requiring all new homes to have solar photovoltaic cells installed, starting in 2020. This would add an average of $9,500 to the cost of the home.

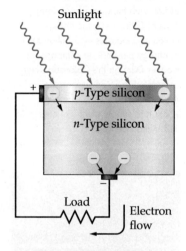

Sunlight

p-Type silicon

n-Type silicon

Load

Electron flow

▲ **Figure 15.17** Schematic diagram of the operation of a solar cell. Electrons flow from the *n*-type region (lower layer) to the *p*-type region (upper layer) through the external circuit.

Hydroelectric Power

Using the sun's energy seems to be a good idea, but it would be difficult to meet all our needs with solar energy. The sun, by heating Earth, causes winds to blow and water to evaporate and rise into the air, later to fall as rain. The kinetic energy of blowing wind and flowing water can be used as sources of energy—as they have been for centuries. Waterpower provides almost 7% of current electricity production in the United States, most of it in the mountainous West. In a modern hydroelectric plant, water is held behind huge dams. The potential energy of the stored water is converted to the kinetic energy of water flowing over the blades of giant turbines.

▲ The Three Gorges Dam on the Yangtze River was completed in 2009, with the goal of providing 10% of China's electric energy needs. It generates ~85 billion kWh of energy per year. In comparison, Hoover Dam on the Colorado River, completed in 1936, generates 4 billion kWh a year, which is enough to supply electrical energy to 1.3 million people.

▲ These willow trees are grown specifically as fuel. The biomass is harvested by cutting the trees near ground level, which stimulates regrowth. Willow is especially suitable for biomass because of its rapid growth.

The moving water imparts mechanical energy to the turbine, which drives a generator that converts mechanical energy to electrical energy.

Hydroelectric plants are relatively clean, but most of the good dam sites in the United States are already in use. To obtain more hydroelectric energy, we would have to dam up scenic rivers and flood valuable cropland and recreational areas. Reservoirs silt up over the years, and sometimes dams break, causing catastrophic floods. Dams block the migration of fish upstream to spawning beds. Declining fish populations have led to closure of salmon fishing in waters off California and Oregon and removal of dams from a few rivers where salmon spawn.

Worldwide, hydroelectric plants provide about 20% of the electricity used, and more plants are being built in many developing countries. As an example, China's huge Three Gorges project on the Yangtze River, completed in 2009, includes a 22,500 MW hydroelectric plant. It was projected to provide about 10% of China's electricity. Unfortunately, the project displaced 1.3 million people and may cause significant environmental damage. Water quality in the Yangtze's tributaries is already deteriorating rapidly because the dammed river disperses pollutants less effectively, and algae blooms flourish. The rising water caused extensive soil erosion, riverbank collapses, and landslides. Even renewable hydroelectric power has its problems.

Biomass: Photosynthesis for Fuel

Why bother with solar collectors and photovoltaic cells to capture energy from the sun when green plants do it every day? Indeed, dry plant material, called **biomass** when used as a fuel, burns quite well—consider a campfire that produces plenty of heat. The problem is that a lot of smoke and ash are produced when plant matter is burned. Biomass could possibly be used to fuel a power plant for the generation of electricity. The emissions are water vapor and the carbon dioxide the plants took from the air in the first place, plus smoke and ash. "Energy plantations" could grow plants for use as fuel. However, most available land is needed for the production of food. Even where productive land is available, plants have to be planted, harvested, and transported to the power plant.

In addition, we should realize that we don't have to burn plant material directly. Starches and sugars from plants can be fermented to form ethanol. Wood can be distilled in the absence of air to produce methanol. Both alcohols are liquids and thus convenient to transport, and both are excellent fuels that burn quite cleanly.

- Bacterial breakdown of plant material produces methane. Under proper conditions, this process can be controlled to produce a clean-burning fuel similar to natural gas.
- Oils and fats can be converted to a fuel called *biodiesel*. The triacylglycerols are converted into the methyl esters of the constituent fatty acids. A typical biodiesel molecule from palm oil is methyl palmitate, $CH_3(CH_2)_{14}COOCH_3$. Biodiesel from waste, algae, or nonfood plants can be profitable.

All these conversions, however, result in the loss of a portion of the useful energy. The laws of thermodynamics tell us that we can get the most energy by burning the biomass directly rather than converting it to a more convenient liquid or gaseous fuel.

Biomass and biofuels currently provide less than half of the renewable energy used in the United States, about 4% of the country's overall energy needs. We could make greater use of these sources by converting appropriate materials to biodiesel, fermenting some agricultural wastes to ethanol, and fermenting human and animal wastes to produce methane. The technology for all these processes is available now and can be further improved.

Wind Power

Wind power currently supplies nearly 6% of U.S. energy production, and that amount is increasing more rapidly than any other classification of energy sources,

20% to 30% per year. Large wind turbine "farms," consisting of hundreds to even thousands of turbines over several square miles, are being built in the United States every year, particularly in the Midwestern states, from North Dakota to Indiana all the way south to Texas. One large wind turbine can provide 1.65 megawatts of electrical energy. The up-front cost is enormous, but they pay for themselves in just a few years. More than 80 countries use on-land and offshore wind turbines for electric power generation, and at least 300,000 wind turbines are producing electrical energy in the world today.

Wind is clean, free, and abundant. However, it does not blow constantly, and some means of energy storage or an alternative source of energy is needed. Not all regions have enough wind to make wind power feasible. Some environmentalists oppose wind power because the rotating blades kill thousands of birds each year. Some recommend that the amount of land required for windmills could be used for farming or grazing.

▲ Wind turbine farms may consist of thousands of large turbines. By the end of 2017, wind turbines with a generating capacity of almost 200,000 MW were operating worldwide, producing about nearly 5% of the world's electricity. In 20 years, wind may produce upwards of 30% of the electricity generated in the world.

Geothermal Energy

The interior of Earth is heated by immense gravitational forces and by natural radioactivity. This heat comes to the surface in some areas through geysers and volcanoes. **Geothermal energy** has long been used in Iceland, New Zealand, Japan, and Italy. In Iceland, for example, geothermal energy provides heat to nearly 90% of the homes, and provides about 30% of the electricity used in that country. Geothermal energy in the United States is used in Hawaii, California, Nevada, and Utah; those energy plants provide as much electrical energy as three nuclear power plants, well over 3000 MW of energy (Figure 15.18). In general, this potential can be realized in areas where steam or hot water is at or near the surface of the Earth.

Even in areas where steam or hot water is *not* near the surface, installation of a geothermal heating system for a home is now possible. Though expensive up front, the system may pay for itself in 10–12 years, even in northern climates. Natural gas, the most commonly used fuel for home heating, may become much more expensive in the next few decades, so this home heating method for new home construction may be a viable option. (See Figure 15.19.) Four to six feet below the surface of the earth, the temperature is fairly constant, year-round. A geothermal heating (and cooling system) consists of an indoor handling unit; a buried system of pipes, called an "earth loop"; and a pump. In the winter the warmed water is used to warm the house directly, or used to warm air that is sent through ductwork. In the summer the water is used to cool the house.

▲ **Figure 15.18** Geothermal energy contributes to energy supplies in some areas. A geothermal field at Wairakei, New Zealand.

Oceans of Energy

The oceans that cover three-fourths of Earth's surface are an enormous reservoir of potential energy. The use of *ocean thermal energy* was first proposed in 1881 and was shown to be workable in the 1930s. The difference in temperature between the surface and the depths is 20 °C or more, enough to evaporate a liquid and use the vapor to drive a turbine. The liquid is then condensed by the cold from the ocean depths, and the cycle is repeated.

Other ways of using ocean energy have been tested. In some areas of the world, the daily rise and fall of the tide can be harnessed, in much the same way that the energy of a river is harnessed by a hydroelectric plant. At high tide, water fills a reservoir or bay. At low tide, the water escapes through a turbine to generate electricity. Even the energy of the waves crashing to shore can be used with the appropriate technology.

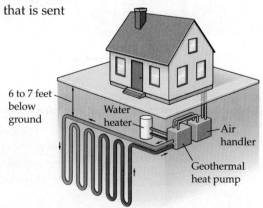

6 to 7 feet below ground

Water heater

Air handler

Geothermal heat pump

▲ **Figure 15.19** Geothermal system for heating a home. The earth loop of pipes buried underground brings water into the house. Electrical energy is used to operate the unit's fan, compressor, and pump.

Hydrogen: Light and Powerful

Natural gas is sent through pipes to wherever it is needed, and other gaseous fuels could be sent through the same pipes. One such gas is hydrogen.

When hydrogen burns, it produces water and gives off energy.

$$2\,H_2(g) + O_2(g) \longrightarrow 2\,H_2O(l) + 527\ kJ$$

Gram for gram, hydrogen yields more energy than any other chemical fuel. One problem is control of that energy, as hydrogen gas can be explosive. It is also a clean fuel, yielding only water as a chemical product. Although hydrogen is the most abundant element in the universe, elemental hydrogen (H_2) is almost nonexistent on Earth. There is a lot of hydrogen on Earth, but it is tied up in chemical compounds, mainly water, and releasing it requires more energy than the hydrogen produces when it is burned. Hydrogen can be made from seawater, so the supply is almost inexhaustible, but the production process involving electrolysis is costly.

Hydrogen can be used as a fuel for cars, either directly or in fuel cells. A small fleet of very expensive hydrogen-powered cars exists in southern California, but only 43 hydrogen-filling stations existed as of the writing of this publication in 2018. Hydrogen gas can be liquefied, but only at a temperature a few degrees above absolute zero, so storage, transportation, and dispensing the fuel are problematic. Heavy-walled fuel tanks in the vehicles are necessary. Fuel cell technology is becoming more efficient and costs are coming down, but there is still the problem of producing hydrogen gas economically. Hydrogen has to be made using another energy source: fossil fuels, nuclear power, or—preferably—some renewable source such as solar energy or wind power. It may be possible to store hydrogen in other ways. Scientists have found that various metals can absorb up to a thousand times their own volume of hydrogen gas. Special forms of carbon, such as ultrathin carbon nanotubes, may also hold large quantities of the gas.

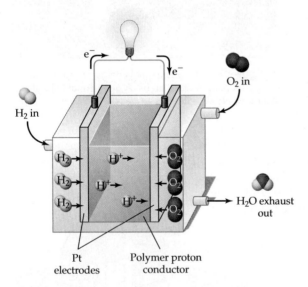

▲ **Figure 15.20** A hydrogen–oxygen fuel cell. The fuel cell continues to provide electricity as long as reactants are supplied.

Fuel Cells

A **fuel cell** is a device in which fuel is oxidized in an electrochemical cell (Section 8.3) to produce electricity directly. Fuel cells differ from the usual electrochemical cell in two ways:

- The fuel and oxygen are fed into the cell continuously. As long as fuel is supplied, current is generated.
- The electrodes are made of an inert material, such as platinum, that does not react during the process.

Most of today's fuel cells use hydrogen and oxygen (Figure 15.20). At the platinum anode, H_2 is oxidized to yield hydrogen ions and electrons.

$$2\,H_2(g) \longrightarrow 4\,H^+ + 4\,e^-$$

The electrons produced at the anode travel through the external circuit and arrive at the cathode, where they combine with the hydrogen ions and oxygen molecules to form water.

$$4\,e^- + O_2(g) + 4\,H^+ \longrightarrow 2\,H_2O(g)$$

The overall reaction is identical to direct combustion of hydrogen, but not all the chemical energy is converted to heat. About 40% to 55% of the chemical energy is converted directly to electricity, making fuel cells much more efficient than internal combustion engines.

$$2\,H_2(g) + O_2(g) \longrightarrow 2\,H_2O(l)$$

Fuel cells are often used on spacecraft to produce electricity, mainly because they have a weight advantage over storage batteries—an important consideration when launching a spaceship—and the water produced can be used for drinking. Because they have no moving parts, fuel cells are usually tough and reliable.

On Earth, research is under way to reduce costs and design long-lasting cells. Perhaps fuel cells will someday provide electricity to meet peak needs in large power plants, or run most of our vehicles. Unlike huge boilers and nuclear reactors, they can be started and stopped simply by turning the fuel on or off.

Energy Return on Energy Invested

We have discussed essentially all of the energy sources currently available to humankind. When considering major investments in energy production, how should we or our government choose? The answer is rarely clear-cut. Since the 1970s, considering the *energy return on energy invested* (EROEI) has been advocated, comparing the energy we can get from a project with how much it costs in terms of energy input. EROEIs are difficult to evaluate, with myriad factors to consider, and the results often have a high degree of uncertainty. However, if the energy and financial support needed far outweigh the energy obtained, a project becomes an energy "sink" rather than a source. In short, it is impractical.

In the early days of the petroleum industry, the EROEI for Texas crude oil may have been as much as 100:1; investing the equivalent of 1 barrel of oil could produce 100 barrels. However, as the primary energy supplies are depleted, it becomes more difficult to efficiently tap what is left, and the EROEI for crude oil has fallen over time. The EROEI for wind turbines is actually quite high, at least 30:1, despite expensive installation, since wind energy is "free" once the turbines are constructed and infrastructure created, and maintenance costs are not excessive. Solar energy panels work well in very sunny climates (high EROEI), but not when a foot of snow covers the roof in Minnesota (much lower EROEI). Hydroelectric power has a high EROEI rating, once the dams have been constructed. The EROEI for ethanol is the subject of much debate, with proposed values ranging from under 1:1 to about 1.7:1. There are two reasons for the low EROEI for ethanol: Much energy is needed to distill or evaporate the ethanol after fermentation, and ethanol has a lower energy content than petroleum fuels.

In addition to the EROEI, we have to consider the environmental consequences of exploiting an energy source. For example, getting oil from tar sands or oil shale is a messy process. Thus, our quest for plentiful energy will be costly and controversial.

The Future of Energy

Our long-term energy future is difficult to predict. For the near future, the U.S. Energy Information Administration (EIA) predicts that by 2019, the percentage of U.S. energy production from coal will decrease to 28%, that from natural gas will increase to 35%, nuclear will be steady at 20%, hydropower steady at 7%, and other renewables (mainly wind and solar) up to 11% (Figure 15.6b). Xcel Energy, Colorado's largest energy company, plans to replace two huge coal-burning electric power–generating plants with new facilities that use renewable energy (wind and solar) and possibly natural gas—and predicts that the cost for customers will actually decrease, while carbon emissions decrease. This change will affect people in seven states, and is a 'win–win' situation for customers and for the future of planet Earth.

World energy use will probably increase by 30% in the next 20 years, with much of the new demand from developing countries. Each of the energy sources discussed in this chapter has advantages and disadvantages. Our fossil fuels will become exceedingly scarce sometime in our future. Each of the renewable resources should be used wherever practical and available, and no one energy source is the ultimate answer. How do we choose the best way to deal with energy problems? Wise choices require informed citizens who examine the process from beginning to end. We must realize what is involved in the construction of power plants, the production of fuels, and the ultimate use of energy in our homes and factories.

We could significantly reduce energy consumption in the United States by greater use of public transportation. In Europe, where there is much wider use of public transit, per capita energy use is considerably less (Table 15.7), partly because the average price of gasoline is usually more than twice that in the United States. Even in Canada, the price of gasoline is much higher than in the United States. More and more people in large cities have given up personal vehicles where extensive

5 **Which energy source is the best answer to the energy crisis?**

No single energy source is the answer. We should use whatever we have that is convenient for the region we are in. Massive areas of solar cells are being built in deserts, and giant wind farms in the Great Plains provide renewable energy. Geothermal and hydropower are advantageous in the appropriate areas. Use of coal in electric energy–generating plants is decreasing, and natural gas use is increasing. None of those sources can directly power buses or cars, and since the availability of gasoline will likely be severely limited in the twenty-first century, the all-electric vehicle powered by batteries may be its substitute.

TABLE 15.7 Total Primary Energy Consumption Per Capita (Quadrillion BTUs in 2015)

Country	Consumption/ Capita
China	119.613
United States	92.896
Russia	29.625
India	25.27
Canada	14.361
Germany	13.203
Brazil	12.687
Saudi Arabia	10.657
France	10.261
United Kingdom	8.111
Mexico	7.62
Thailand	5.07
Nigeria	1.369
Guatemala	0.239
Zimbabwe	0.164

Source: U.S. Energy Information Administration

public transportation is available, but in small towns, a personal vehicle seems to be a necessity. Yet in the far-reaching future, increased construction of public transportation, such as subways and above-ground train systems in and between cities, may be critically important as the population continues to increase and space on Earth is limited.

Electric vehicles may be the car of the future, but as of yet most can travel only between 100 and 200 miles before they need recharging. Yet in January 2018, over 47,000 electric car charging outlets and over 17,000 electric car stations existed in the United States, making longer-distance travel more feasible. Most all-electric vehicles are still quite expensive, but prices are falling. One problem is that the very heavy lithium batteries may be recharged only a finite number of times. Replacement of the batteries is a daunting cost, equivalent to the cost of a modest new car, but battery development research may decrease that cost. With extensive computer sensors and controls, the self-driving electric car may be common in the future.

What can we do as individuals now? We can conserve. We can walk more and use cars less. We can reduce our wasteful use of electricity. We can buy more efficient appliances, and we can reduce our use of energy-intensive products (leaf blowers, lawn tractors, and larger automobiles).

Over the past three decades, we have made significant progress in energy conservation. Appliances are more energy efficient. New furnaces have efficiencies of 90% to 95%, as compared with the 60% efficiency of 20-year-old furnaces. New refrigerators and air conditioners use about 35% less energy than did earlier models. Incandescent light bulbs are being replaced by "LED" (light-emitting diode) bulbs, which theoretically can last 20 years.

The simple facts are that our population on Spaceship Earth is increasing and our fuel resources are decreasing. People in developing countries want to raise their standard of living, and that will require more energy. Energy is truly one of the great challenges for modern society.

SELF-ASSESSMENT Questions

1. Which of the following is an *n*-type semiconductor?
   a. arsenic doped with silicon
   b. arsenic doped with germanium
   c. silicon doped with arsenic
   d. silicon doped with boron

2. *p*-type semiconductors feature crystal sites with
   a. photons
   b. positrons
   c. positive holes
   d. protons

3. Which of the following is *not* a renewable energy source?
   a. biomass
   b. solar power
   c. hydropower
   d. petroleum

4. Biodiesel cannot be made from
   a. sugar
   b. olive oil
   c. soybean oil
   d. waste cooking fat

5. The major product from burning hydrogen gas as a fuel is
   a. CO_2
   b. H_2
   c. H_2O
   d. O_2

6. Which of the following fuels has the lowest EROEI?
   a. biodiesel
   b. hydrogen
   c. natural gas
   d. petroleum

7. Which of the following energy technologies does not use energy that originated in the sun?
   a. biomass
   b. hydroelectric
   c. nuclear
   d. wind

8. The fuel used by almost all current fuel cells is
   a. electricity
   b. hydrogen
   c. natural gas
   d. propane

9. A disadvantage of wind turbines is
   a. high maintenance costs
   b. production of greenhouse gases
   c. high up-front costs
   d. competition from other energy sources

10. The renewable energy source that provides the most electrical energy in the United States today is
   a. geothermal
   b. hydroelectric
   c. solar
   d. wind

11. The energy that comes from the internal heat of the Earth is
   a. bioenergy
   b. geothermal energy
   c. hydroelectric
   d. solar energy

12. Problems to overcome in the development of electric vehicles include all the following except
   a. yielding more miles before recharging
   b. the excessive cost of replacing batteries
   c. insufficient number of recharging stations
   d. designing engines to run at highway speeds

Answers: 1. c; 2. c; 3. d; 4. a; 5. c; 6. b; 7. c; 8. b; 9. c; 10. b; 11. b; 12. d

GREEN CHEMISTRY Where Will We Get the Energy?
Michael Heben, *University of Toledo*

Principles 1, 5, 6, 12

Learning Objective ● Explain how applying the green chemistry principles can help identify improved methods for energy generation.

Energy permeates every aspect of our existence. Think about your heart pumping blood through your body, the power of a rocket lifting explorers to Mars, a family drive to Grandma's house for Thanksgiving, a plant absorbing CO_2 and sunlight to make food, an actor, musician, artist, or athlete doing their thing, or a company making chemical compounds for industrial or pharmaceutical use. Energy is required anytime anything does something.

Is there anything that we do that does not require energy? No—and even thinking about the question can be exhausting—because thinking requires energy!

Chapter 15 considers sources of energy that power our world. Yet what we commonly consider to be energy sources, such as natural gas, coal, and oil, are not really energy sources at all. Rather, these are fuels. They contain stored energy in carbon-carbon and carbon-hydrogen chemical bonds. As we learned (Sections 15.6, 15.7), these bonds react with oxygen during combustion. The released energy can provide heat to homes or industrial processes or can be used to power a mechanical engine to do work or generate electricity.

Today's fossil fuels are the remains of ancient plant and animal matter that were produced by the energy from our own star, the Sun. Though fossil fuels have been built up over hundreds of millions of years by photosynthesis, our society is consuming the easily obtained portion very quickly. It's like putting a small amount of money in the bank each week for a long time and then spending it all at once.

Besides the limitations on the amount of stored energy still available in fossil fuels, we have to recognize broader concerns. Green chemistry principles help us to do so, by explicitly including the impacts of waste generation, toxic substances, transportation costs, feedstocks, and overall energy efficiency. To generate energy efficiently, we minimize the inputs and maximize the energy output. As principle 1 reminds us, clearly, waste must be reduced, and the process should produce other products only if needed.

As demonstrated by principle 6, processes should not limit the main product—in this case, energy—but should reduce negative environmental and economic impacts. It is only within the past 50 years that environmental pollution—undesired outputs from energy production—has become a big concern. The Clean Air Act of 1970, signed by President Nixon, has helped reduce dangerous smog in our cities and improve the health and economic vitality of our country. In 2007, the U.S. Supreme Court upheld a regulation that treated CO_2 as a pollutant. Unfortunately, the combustion of oil, coal, and natural gas produce CO_2 as a byproduct, and CO_2 is the big player in the world's urgent climate-change problem.

From a green chemistry point of view and particularly relating to principles 1, 5, and 12, hydraulic fracturing for production of natural gas makes an interesting case. Once considered a nonconventional method, "fracking" uses fluids to move the natural gas more easily from the ground to the surface and now accounts for approximately 67% of U.S. natural gas production. The benefit of this process is that it allows access to natural-gas resources that were previously not available, but many chemicals, some of which are carcinogens, may be added to the fracture fluids. These and other species released by fracturing (including naturally occurring radioactive species) may reach aquifers or be diverted to municipal water treatment facilities. Clearly, a great deal of waste is not accounted for in the process. Also, besides CO_2 emissions associated with combustion, considerable amounts of CH_4, which is 25 times more active as a greenhouse gas than CO_2, leak into the atmosphere during production.

Difficulties obtaining fuels from the dwindling supply of stored fossil energy and the impacts these fuels have on the Earth itself mean that we need to create a balance between energy supply and consumption by using the energy from the Sun. Direct methods include technologies like photovoltaics, wind power, and bioenergy. Luckily, there's more than enough solar energy available. For example, if we covered ~0.1% of the land area in the contiguous 48 U.S states with solar panels that are 15% efficient, all of our country's energy needs could be met. For comparison, the land needed is less than one fourth of the area covered by roads.

The biggest hurdle to developing clean, renewable energy has to do with cost. If we do the complete accounting and measure all inputs and outputs and impacts of a given process correctly, the true costs to society will guide us in making good choices. New science and technology and new mindsets, including those based on green chemistry, will be required. Perhaps you will want to help in this grand challenge?

Summary

Section 15.1—Energy is the ability to do work. The United States has about 5% of the world's population but uses about one-fifth of all the energy generated. Most of the energy used on Earth comes directly or ultimately from the sun. The SI unit of energy is the joule (J), and the unit of power is the watt (W), which is 1 joule per second. Energy can be classified as **potential energy** (energy of position) or **kinetic energy** (energy of motion). In the process called **photosynthesis**, solar energy is absorbed by plants and used to produce glucose, and oxygen is generated.

Section 15.2—The study of energy changes that occur during chemical reactions and physical processes is very important. In a chemical reaction, all the chemical bonds must be broken in the reactants; then the fragments combine in a different way to form the products. The first process requires an input of energy, and the second process releases energy. The net result of both is the heat of reaction. A chemical reaction or physical process may be either **exothermic** (giving off energy to the surroundings) or **endothermic** (requiring energy from the surroundings). Reactions involving burning of fuels are all exothermic.

Section 15.3— Kinetics is the study of reaction rates and mechanisms. The rate of a chemical reaction is affected by temperature, by concentrations of the reactants, and by the presence of catalysts. The rate increases if the temperature is increased or the concentrations of solutions are increased. Catalysts speed up reactions.

Section 15.4 The **first law of thermodynamics (law of conservation of energy)** says that energy can be neither created nor destroyed. The second law of thermodynamics states that in a spontaneous process, energy is degraded from more useful forms to less useful forms. **Entropy** is a measure of the dispersal of energy among the possible states of a system. There is a tendency for a system to increase in entropy.

Section 15.5—Ancient societies used human and animal power to do work. The main fuel was wood. A **fuel** is a substance that will burn readily with the release of significant energy, but in a controlled manner. The first mechanical devices for doing work were the waterwheel and the windmill. Energy from water, steam, gas, the internal combustion engine, and other sources fueled the Industrial Revolution. **Fossil fuels** are natural fuels derived from once-living plants and animals. They include coal, petroleum, and natural gas.

Section 15.6—**Coal** is a solid fuel that is also a source of chemicals. The main element in coal is carbon. The quality of coal is determined mostly by its carbon and energy content, with anthracite being highest in rank and peat lowest. Coal is the most plentiful fossil fuel but is inconvenient to use and produces more pollution than other fuels.

Section 15.7—**Natural gas** is mainly methane and burns relatively cleanly. It is important both as a fuel and as a raw material for synthesis of plastics and other products. **Petroleum**, or crude oil, is a thick liquid mixture of organic compounds, mainly hydrocarbons. Petroleum products can be burned fairly cleanly, but inefficient combustion produces pollution. **Hydraulic fracturing** using tiny grains of silica sand ("frac sand") has increased the availability of both natural gas and petroleum in the US. Petroleum is refined by separating it, by distillation, into different fractions. Gasoline is the fraction consisting of hydrocarbons

from about C_5 to C_{12}. Branched-chain hydrocarbons such as isooctane burn more smoothly than do straight-chain hydrocarbons. The **octane rating** of gasoline compares its anti-knock performance with that of pure isooctane. To raise the octane rating of gasoline, refineries use isomerization (to increase branching), **catalytic reforming** (to convert straight-chain hydrocarbons to aromatic hydrocarbons), and alkylation (to convert small hydrocarbons to larger branched molecules). Octane boosters such as methanol, ethanol, and *tert*-butyl alcohol can be added, since they burn cleaner with less carbon monoxide. Tetraethyllead was used to improve octane rating for more than 50 years but has been discontinued in the United States. It is still used in many other countries, however. Electric-powered vehicles are likely the vehicles of the future.

Section 15.8—Electricity is the most convenient form of energy. Less than half of the electricity in the United States is generated from coal-burning steam power plants, the rest from natural gas, nuclear energy, hydropower, wind, and solar. Coal can be converted to more convenient gaseous or liquid fuels, but at an energy cost.

Section 15.9—Nuclear fission reactions can be controlled in a **nuclear reactor**, which uses a low concentration (3% to 5%) of uranium-235. The reactions are controlled with rods of cadmium or boron steel. The energy released generates steam that turns a turbine to generate electricity. Approximately 20% of the United States' electricity comes from nuclear fission reactors. A nuclear power plant cannot explode like a nuclear bomb, it produces minimal air pollution, and its wastes are collected rather than dispersed. But elaborate safety precautions are needed to prevent escape of radioactive material, the needed fuel is limited in availability, and ultimate disposal of radioactive waste is an ongoing problem. A **breeder reactor** converts other isotopes to useful nuclear fuel and can make more fuel than it consumes. Nuclear fusion cannot yet be controlled, as there are great technical difficulties. Temperatures in the millions of degrees are needed to generate a plasma consisting of free electrons and nuclei, for controlled fusion. But fusion has very important potential advantages over most other energy sources.

Section 15.10—Simple solar collectors can produce hot water for heating. Solar cells, or **photovoltaic cells**, produce electricity directly from sunlight, using *n*-type and *p*-type semiconductors. The cells are becoming less expensive and more efficient, but photovoltaic systems require storage systems or connection to the local electricity grid; huge solar energy collectors have been built in deserts. **Biomass** such as wood can be burned directly or converted to liquid or gaseous fuels, but it requires much land area to produce and has a very low overall efficiency. Hydrogen is an energy storage and transport method rather than an energy source. It burns cleanly but requires more energy to make than it produces. One advantage of hydrogen is that it can be used in **fuel cells**, which consume fuel and oxygen continuously to generate electricity efficiently. Wind power and hydroelectric power are renewable energy sources that can be harnessed only in certain locations; the number of wind turbines in use around the world is increasing dramatically. Both sources have high initial costs. Heat energy from the interior of Earth is **geothermal energy**. We can also make use of the temperature difference between the ocean's surface and deep waters; tidal energy; and the energy of the waves. Each energy source has advantages, limitations, and consequences. We can all conserve energy by using more efficient appliances and simply by using less energy.

GREEN CHEMISTRY—Energy is needed for everything we do, and there are many ways we can get it. An accounting of all inputs, outputs, and impacts of an energy-generating technology should use green chemistry principles to assess its true value.

Learning Objectives	Associated Problems
• Perform power and energy calculations. (15.1)	9–12, 30, 85–88
• Classify chemical reactions and physical processes as exothermic or endothermic. (15.2)	79–84
• Explain how to calculate the heat of a reaction using bond energies. (15.2)	25–32, 89, 90
• List factors that affect the rates of chemical reactions, and explain how they affect those rates. (15.3)	3, 5
• State the first and second laws of thermodynamics, and discuss their implications for energy production and use. (15.4)	2, 4, 33–37
• List the common fossil fuels, and describe how modern society is based on their use. (15.5)	17–22, 41–50
• List the origins, advantages, and disadvantages of coal as a fuel. (15.6)	39, 40
• List the characteristics, advantages, and disadvantages of natural gas and petroleum. (15.7)	46, 49
• Explain why gaseous and liquid fuels are more convenient to use than solid fuels. (15.8)	46, 92
• List advantages and disadvantages of nuclear energy. (15.9)	51, 52, 56–58
• Describe how a nuclear power plant generates electricity. (15.9)	7, 8, 53–55
• List important characteristics, advantages, sources, and disadvantages of various kinds of renewable energy sources. (15.10)	65–76, 98
• Explain how applying the green chemistry principles can help identify improved methods for energy generation.	78–80

Conceptual Questions

1. Explain the difference between temperature and heat, in terms of molecular motion.
2. Explain why ice cream melts in a bowl at room temperature, in terms of energy transfer and the second law of thermodynamics.
3. How does temperature affect the rate of a chemical reaction? Explain why.
4. Besides the energy factor (the heat of reaction), what other factor is important in determining if a process is spontaneous or not?
5. Why does wood burn more rapidly in pure oxygen than in air?
6. Explain why tetraethyllead was discontinued as an anti-knock agent in gasoline in the 1970s.
7. Explain why a nuclear power plant cannot explode like a nuclear bomb.
8. Explain why the energy released in nuclear reactions is so much greater, by orders of magnitude, than the energy released in combustion reactions, in terms of what is happening on the atomic or molecular level.

Problems

Power and Energy

9. List two units that are commonly used to measure energy and explain how they are mathematically related to each other.
10. What is the relationship between watts and joules? Which represents energy and which represents power?
11. How much electrical energy is used by a 150 W outdoor bulb that is turned on for 10 hours a night? Express the answer in kJ.
12. How much electrical energy is used by a 1300 W electric heater that is turned on for 3 hours in the evening? Express the answer in kJ.

13. Describe the transformation of energy as water flows over a waterfall, in terms of kinetic and potential energy.

14. Describe the transformation of energy as an electric fan is turned on in terms of kinetic and potential energy.

Fuels and Combustion

15. What was the main fuel used in the United States before 1800?

16. What kind of reaction powers the sun?

17. Write a generalized word equation for the complete combustion of any hydrocarbon.

18. Write a generalized word equation from the incomplete combustion of any hydrocarbon.

19. Write a balanced chemical reaction for the complete combustion of propane, C_3H_8, in oxygen gas.

20. Write a balanced chemical reaction for the complete combustion of butane, C_4H_{10}, in oxygen gas.

21. Write a balanced chemical reaction for the incomplete combustion of propane, C_3H_8, in oxygen gas, with carbon monoxide as one of the products.

22. Write a balanced chemical reaction for the incomplete combustion of butane, C_4H_{10}, in oxygen gas, with carbon monoxide as one of the products.

23. Write a balanced chemical reaction for the complete combustion of octane, C_8H_{18}, in oxygen gas.

24. Write the balanced chemical reaction for the complete combustion of nonane, C_9H_{20}, in oxygen gas.

Energy and Chemical Reactions

25. Using bond energies, calculate the amount of energy, in kilojoules, that is released when 1.0 mol of methane is burned in plenty of oxygen gas, using the following steps:
    a. Draw the Lewis structures of all the molecules involved in the reaction (remember the balancing numbers):

    $$CH_4 + 2\,O_2 \longrightarrow CO_2 + 2\,H_2O$$

    b. If the bond energy for the C—H bond is 413 kJ/mol, how much energy is needed to break all the bonds in 1 mol of methane molecules?
    c. If the bond energy for the O=O bond is 498 kJ/mol, how much energy is needed to break all the bonds in 2 mol oxygen gas?
    d. If the bond energy for the C=O bond is 798 kJ/mol, how much energy is released when 1 mol carbon dioxide gas is formed?
    e. If the bond energy for the O—H bond is 467 kJ/mol, how much energy is released when 2 mol water molecules are formed?
    f. What is the net result of energy needed to break the bonds, less the energy released when new bonds are formed? How do you interpret the sign of the answer?

26. Calculate the amount of energy, in kilojoules, that is involved when 1 mol of nitrogen gas is reacted with 3 mol hydrogen gas to form 2 mol ammonia gas, using the following steps:
    a. Draw the Lewis structures of all the molecules involved in the reaction:

    $$N_2(g) + 3H_2(g) \longrightarrow 2\,NH_3(g)$$

    b. If the bond energy for the N≡N bond is 946 kJ/mol, how much energy is needed to break all the bonds in 1 mol nitrogen molecules?

    c. If the bond energy for the H—H bond is 432 kJ/mol, how much energy is needed to break all the bonds in 3 mol hydrogen gas?
    d. If the bond energy for the N—H bond is 391 kJ/mol, how much energy is released when 1 mol ammonia is formed?
    e. What is the net result of energy needed to break the bonds, less the energy released when new bonds are formed? How do you interpret the sign of that answer?

27. Burning 1.00 mol of methane releases 803 kJ of energy. How much energy is released by burning 24.5 mol of methane?

    $$CH_4(g) + 2\,O_2(g) \longrightarrow CO_2(g) + 2\,H_2O(g) + 803\ kJ$$

28. It takes 572 kJ of energy to decompose 2.00 mol of liquid water. How much energy does it take to decompose 30.0 mol of water?

    $$2\,H_2O(l) + 572\ kJ \longrightarrow 2\,H_2(g) + O_2(g)$$

29. When burned, 1.00 g of gasoline yields 1060 cal. What is this quantity in kilojoules?

30. Express 42,000 J in units of kcal.

31. How much heat, in kilojoules, is released when 10.0 g $H_2(g)$ reacts with plenty of $O_2(g)$ to form steam [$H_2O(g)$] according to the following equation?

    $$2\,H_2(g) + O_2 \longrightarrow 2\,H_2O(g) + 483.6\ kJ$$

32. Refer to Problem 31, and determine how much energy, in kilojoules, must be supplied to convert 1 mol of water vapor into hydrogen gas and oxygen gas.

Energy and the Laws of Thermodynamics

33. State the first law of thermodynamics.

34. State the second law of thermodynamics in terms of energy flow and in terms of distribution of energy.

35. What is entropy? Does entropy increase or decrease when a fossil fuel is burned?

36. Does entropy increase or decrease when water is frozen in an ice cube tray?

37. Which has higher entropy, nitrogen in the atmosphere or nitrogen in solid ammonium nitrate?

38. Which has higher entropy, potassium in solid potassium metal or potassium in KCl, in commercial fertilizer?

Fossil Fuels

39. What are the advantages and disadvantages of coal as a fuel?

40. Why did the United States shift from coal to petroleum and natural gas when we have much larger reserves of coal than we do of the other two fuels?

Estimates in Problems 41–44 are from U.S. Energy Information Administration data from 2017; see data in Table 15.3.

41. The estimated proven coal reserves in 2017 were 1139 billion tons. The world annual rate of use was 3.66 billion tons. How long will these reserves last if this rate of use continues? (Realize that the rate of use is actually likely to change.)

42. The estimated proven U.S. oil reserves in 2017 were 33.4 billion barrels. **(a)** How long will these reserves last if there are no imports or exports and if the U.S. annual rate

of use of 6.7 billion barrels continues? **(b)** Using the mean projection for oil reserves in the Arctic National Wildlife Refuge—10.4 billion barrels—how long would exploiting that resource extend our reserves at the current U.S. annual rate of use?

43. The estimated world natural gas reserves in 2013 were 6845 trillion cubic feet. The world annual rate of use was 120 trillion cubic feet. How long will these reserves last if this rate of use continues?

44. The U.S. proven natural gas reserves in 2013 were 323 trillion ft^3. How long will these reserves last if there are no imports or exports and if the U.S. annual rate of use of 24.5 trillion ft^3 continues?

45. What is thought to be the origin of petroleum? Compare the origins of petroleum with the origins of coal.

46. What are the advantages and disadvantages of petroleum as a source of fuels?

47. Describe the process of fractional distillation and how it is able to separate the various "fractions."

48. What does the term *octane rating* mean?

49. Consult Table 15.6, and then suggest a reason why asphalt is a relatively inexpensive way to pave highways and driveways (compared with concrete, for example).

50. Which should have a higher octane rating, octane ($CH_3CH_2CH_2CH_2CH_2CH_2CH_2CH_3$) or 2,3,3-trimethylpentane [$CH_3CH(CH_3)C(CH_3)_2CH_2CH_3$]? Both are isomers of octane. Give a reason for your answer.

Nuclear Energy

51. What proportion of U.S. electricity is generated by nuclear power plants?

52. What proportion of electricity in France is generated by nuclear power plants?

53. Write nuclear equations showing how uranium-238 is converted to fissionable plutonium-239. Write in words what the equations mean.

54. Can a nuclear bomb be made from reactor-grade uranium? Explain.

55. How does a breeder reactor produce more fuel than it consumes? Does doing so violate the law of conservation of energy?

56. What are some of the advantages and disadvantages of breeder reactors?

57. List some possible advantages of a nuclear fusion reactor over a fission reactor. What is plasma?

58. What are the main challenges in the development of nuclear fusion reactors?

59. Write nuclear equations showing how thorium-232 is converted to fissionable uranium-233.

60. Write the nuclear equation that shows how deuterium and tritium fuse to form helium and a neutron.

61. The fission of 1 mol of uranium-235 can be represented by

$$^{235}_{92}U + ^{1}_{0}n \longrightarrow 2^{1}_{0}n + ^{139}_{54}Xe + ^{95}_{38}Sr + 1.8 \times 10^{10} kJ$$

See the equation in Problem 25 for the combustion of methane. How many moles of methane must be burned to generate the energy produced by fission of 1.00 mol (235 g) of uranium-235?

62. From your answer in Problem 61, calculate the mass of methane, in metric tons (1 t = 1000 kg), that must be burned to release the equivalent amount of energy from the fission of 1 mol of uranium-235.

63. Coal is mostly carbon. Burning 1 mol (12.0 g) of carbon to form carbon dioxide releases 393 kJ of energy. Based on the reaction $C + O_2 \rightarrow CO_2$, determine how many moles of carbon must be burned to generate the energy produced by fission of 1.00 mol (235 g) of uranium-235.

64. Considering your answer to Problem 63, calculate the mass of carbon, in metric tons, that must be burned to release the equivalent amount of energy to that from the fission of 1 mol of uranium-235.

Renewable Energy Sources

65. What is a photovoltaic cell?

66. What are some problems associated with the use of solar cells for providing electrical energy, both for whole cities and for a single home?

67. The average power demand of an American household that uses gas for heating, cooking, and a water heater is 1.3 kW. Solar cells can produce about 100 W/m^2. What area of solar cells is needed to meet this demand?

68. Large wind turbines can produce 1.5 MW of electrical power. How many of the homes described in Problem 67 could be powered by one big wind turbine?

69. What are the advantages and disadvantages of using wind turbines for electrical energy?

70. What is biomass? What are the problems and advantages of biomass as an energy source?

71. List two ways in which fuel cells differ from electrochemical cells.

72. List the advantages, disadvantages, and limitations of geothermal power and of a geothermal home heating system as an energy source.

73. List the advantages, disadvantages, and limitations of hydroelectric power as an energy source.

74. List the advantages, disadvantages, and limitations of biodiesel as a fuel.

75. List the advantages and disadvantages of the all-electric car.

76. What are some of the reasons why the gasoline-powered car may, in your lifetime, become a thing of the past?

Synthetic and Converted Fuels

77. Write the chemical equations for the basic processes by which coal (carbon) is converted to **(a)** methane and **(b)** carbon monoxide and hydrogen.

78. Write the chemical equations for **(a)** the conversion of carbon monoxide and hydrogen to methanol and **(b)** the reaction that occurs in a hydrogen–oxygen fuel cell.

🌑 Expand Your Skills

79. A cold pack works by dissolving ammonium nitrate in water. Cold packs are carried by athletic trainers when transporting ice is not possible. Is this process endothermic or exothermic? Explain.

80. One type of hot-pack hand warmer contains a supersaturated and unstable solution of sodium acetate. It is activated by compressing a metal piece in the solution, which stimulates the crystallization. Hot packs are used by hunters and other outdoor sportsmen and workers in places where access to heaters or fires is not possible. Is this process endothermic or exothermic? Explain.

81. Consider the statement "Striking a match is an endothermic process because the air near the burning match is absorbing heat." Which of the following analyses of this statement is true?
   a. The statement is true, but the reason is false.
   b. The statement is false, but the reason is true.
   c. Both the statement and reason are true.
   d. Both the statement and reason are false.

82. Which of the following is a true statement about the process of freezing water?
   a. The process is exothermic.
   b. The process is endothermic.
   c. Water molecules absorb energy in the process.
   d. No energy is involved, because freezing occurs at the same temperature (0 °C) as melting.

83. Decide if the following are exothermic or endothermic—or neither.
   a. boiling water to cook vegetables
   b. changing water to steam

84. Decide if the following are exothermic or endothermic processes—or neither.
   a. baking a cake
   b. condensing water from the air on the outside of a glass of cold lemonade

85. With the invention of the steam engine, inventors tried to compare the output of steam engines with that of the horses the engines would replace. Over a day of sustained work, an average workhorse expends energy at a rate of—you guessed it—one horsepower (hp), originally defined as 550 foot pounds/second. This means that the average horse could pull 550 pounds a distance of one foot in one second. In modern units, one hp = 745 W.
   a. If the average human expends about 2100 kcal of energy per day, what is this output in watts?
   b. How does one human power compare to one horsepower?

86. A 1920 Model T Ford had an engine with 20 horsepower, or the power output of 20 horses. What is the power output of that early automobile in units of kilowatts? (See Question 85.)

87. A modern race car with a 2.4 L V8 engine can develop about 740 hp, or the power output of 740 horses. What is the power output of the race car in units of kilowatts?

88. Both a banana and a hand grenade have about 170 kcal of energy. (a) What is the power output in watts of a banana if its energy is expended over 2.0 h by a person lying on a couch watching television? (b) What is the power output

in watts of a hand grenade if its energy is expended in 0.0012 s? How do the banana and hand grenade compare in energy and in power?

89. Consider the combustion reactions of two gaseous fuels, hydrogen and isobutane:

$$H_2 + \frac{1}{2} O_2 \longrightarrow H_2O \qquad \text{Heat of combustion: 268.6 kJ}$$

$$CH_3CH(CH_3)_2 + \frac{13}{2} O_2 \longrightarrow 4\,CO_2 + 5\,H_2O$$
$$\text{Heat of combustion: 2686 kJ}$$

Which of those two combustion reactions yields more energy *per kilogram of the fuel?*

90. Consider the combustion reactions of the two fuels shown in Problem 89. Which of the two combustion reactions yields more energy per liter of the fuel when the volume is measured under the same conditions of temperature and pressure?

91. The United States leads the major developed countries in per capita emission of $CO_2(g)$ with a rate of 19.0 t per person per year (1 t = 100 kg). What mass, in metric tons of methane, would yield this quantity of CO_2?

92. Explain why and for what applications gaseous and liquid fuels are more convenient than solid fuels.

93. a. Write the chemical reaction for the burning of sulfur in coal to form sulfur dioxide.
   b. A large coal-burning power plant burns 2500 metric tons (1 t = 1000 kg) of coal per day. The coal contains 0.65% S by mass. Assume that all the sulfur is converted to SO_2. What mass of SO_2 is formed and released into the atmosphere?

94. Refer to Problem 93. If a thermal inversion traps all this SO_2 in a volume of air that is 45 km by 60 km by 0.40 km, will the level of SO_2 in the air exceed the primary national air quality standard of 365 $\mu g\ SO_2/m^3$ air?

95. The contribution of the combustion of various fuels to the buildup of CO_2 in the atmosphere can be assessed in different ways. One way relates the mass of CO_2 formed to the mass of fuel burned; another relates the mass of CO_2 to the quantity of heat evolved in the combustion. Which of the three fuels—C(graphite), $CH_4(g)$, or $C_4H_{10}(g)$—produces the smallest mass of CO_2 (a) per gram of fuel and (b) per kilojoule of heat evolved? The heat released per mole of each substance is C(graphite), 393.5 kJ; $CH_4(g)$, 803 kJ; and $C_4H_{10}(g)$, 2877 kJ.

96. Estimates of recoverable energy are notoriously uncertain, but some scientists think that a total of only 9.92×10^6 terawatt-hours (TWh) of recoverable energy is stored in the Earth (in the form of oil, coal, natural gas, and uranium). Worldwide use is about 1.32×10^5 TWh per year. (a) If demand for energy remains the same as today, approximately how many years will it be before all the energy stored in the Earth is gone? (b) Per capita consumption in the United States (population of about 313,000,000) is about four times that of the rest of the world (population of about 7.0 billion). If all the countries throughout the world consumed energy at the same per capita rate as the United

States, approximately how many years would it be before all the energy stored in the Earth was used up? **(c)** Compare and comment on your answers to parts **(a)** and **(b)**.

97. Production of gasoline from tar sands requires natural gas. If the ratio between the energy needed to generate the gasoline and the energy content of the gasoline itself is very low, why is this production method pursued?

98. Suppose you live in a typical three-bedroom home in a northern climate where your gas bill for one month in the wintertime states that 212 "therms" were used, and your electricity bill is for 490 kWh. You are considering installing solar cells on your roof to provide that amount of energy for the home. One company states that $1.0\,m^2$ of solar cells is needed to power *two* 100 W light bulbs. You want to calculate the total number of m^2 of solar cells you will theoretically need to provide enough energy for your home. (One practical problem, of course, is that in the winter, snow may cover the solar cells!)

(a) First you need to calculate the total kJ of energy needed to both heat your home and power electrical devices in that home, in one month, using the data from your utility bill. One "therm" = 100,000 BTU, and one BTU = 1.05506 kJ. One kWh = 3600 kJ.

(b) Next, determine the kWh of energy needed to power two 100 W light bulbs, for a 30-day month.

(c) Finally, determine the number of m^2 of solar cells needed by using your answers to parts **(a)** and **(b)** of this question. Watch the units!

 Critical Thinking Exercises

Apply knowledge that you have gained in this chapter and one or more of the FLaReS principles (Chapter 1) to evaluate the following statements or claims.

15.1 A futurist and inventor predicts that advances in nanotechnology will result in solar energy powering the world in just 16 years.

15.2 In 1996, Indian inventor Ramar Pillai claimed to have discovered a secret herbal formula that converted water to a hydrocarbon fuel.

15.3 An inventor claims to have discovered a catalyst that speeds the conversion of linear alkanes, such as hexane, to branched-chain isomers.

15.4 While watching TV during the summer, you hear someone proclaim that a way to cool your kitchen is to keep the refrigerator door open.

15.5 A fuel cell in every house can give each household energy independence.

15.6 A company claims that it can use a metal such as magnesium as a fuel for automobiles. A coil of the metal would be fed into a chamber, where it would react with superheated steam to produce hydrogen gas, which would power a fuel cell that would run the car.

 Collaborative Group Projects

Prepare a PowerPoint, poster, or other presentation (as directed by your instructor) to share with your class.

1. Find appliances such as a water heater, television set, computer monitor, or electric frying pan and determine the wattage of the appliance. Examine your electric bill to determine the cost in your area per kilowatt hour for electrical energy. Determine the cost for running that device for one hour. Determine the cost for 40 hours.

2. Compare the quantity of electricity used for two appliances that are alike except for different efficiencies. Extend this comparison to CO2 emissions from the appliances, assuming that the electricity is generated by burning coal.

3. Compare qualitatively the efficiency of heating a home with natural gas versus heating it with electricity produced from burning natural gas at a power plant.

4. All of the following are used for heating homes. What problems with supply, use, and waste products are involved in each case?
   a. natural gas　　b. electricity
   c. biomass/wood　　d. coal
   e. fuel oil　　f. geothermal energy

5. Use Internet sources to find directions and construct **(a)** a simple solar heater or **(b)** a solar oven. These devices re important in underdeveloped nations, particularly in areas with sunny climates such as much of Africa.

LET'S EXPERIMENT! Some Like It Hot and Some Like It Cool!

Materials Needed

- 2 small clear plastic cups
- 1 tablespoon water
- 1 tablespoon hydrogen peroxide 3% solution
- 3 grams powdered yeast
- Wooden splint (wood coffee stirrer)

- Matches
- 3 grams Pixy Stix or Fun Dip (candies containing citric acid)
- Kitchen scale
- Spoon
- Thermometer

How do you determine if a physical or a chemical change has occurred? What are the key differences between an exothermic and an endothermic reaction?

For this experiment, you will perform two different chemical reactions using safe materials and conditions. One will be an exothermic reaction that results in the release of heat from the system to the surrounding area. The other will be an endothermic reaction that results in energy being absorbed as heat.

Measure 1 tablespoon of water. Pour the water into a cup and label it. Record the initial starting temperature with the thermometer. Create a data table to record the starting temperature and the temperature at 10-second intervals for 60 seconds. Weigh out 3 grams of Pixy Stix granules using the kitchen scale. Quickly add all of the candy to the water and stir using the spoon. (The candy does not need to fully dissolve in the water.) Collect temperature data for 60 seconds.

What happens to the temperature of the water? What type of reaction occurred? (Figure 1.)

Figure 1

Measure 1 tablespoon of hydrogen peroxide. Pour the hydrogen peroxide into a cup and label it. Record the starting temperature in your data table. Create a data table to record the temperature at 10-second intervals for 60 seconds. Weigh out 3

grams of yeast using the kitchen scale. Add the yeast to the hydrogen peroxide and stir using the spoon. What happens when you add the yeast? Collect temperature data for 60 seconds (Figure 2).

Figure 2

Use matches to light the end of the wooden coffee stir stick, and let it burn for a moment. Then blow out the flame. You'll want it to remain glowing red but not flaming. Carefully bring the glowing end of the splint up to the larger bubbles in the foam. What happens next?

Questions

1. Were both reactions chemical or physical changes? Explain.
2. Which reaction caused an endothermic reaction? Why?
3. Which reaction caused an exothermic reaction? Why?
4. What gas was produced in the yeast-and-hydrogen-peroxide experiment? How did you know?
5. If you were to repeat the candy-and-water experiment, would you be able to relight the wooden splint? Why or why not?

Biochemistry

16

Life exists in many forms, including plants and animals. All of these forms, including those that are too small to see, utilize the same kinds of biochemical reactions to build molecules and then use them to provide the energy and the building blocks that support cells and maintain life in its many forms.

Have You Ever Wondered?

1 What is high-fructose corn syrup, and why is it used in so many products? **2** Since fruits are sweet, do they cause tooth decay, like candy? **3** Is high blood cholesterol mainly a problem for older people? **4** Is it true that there are more than just 20 amino acids? **5** Are enzymes different from proteins?

A MOLECULAR VIEW OF LIFE   The human body is an incredible chemical factory, far more complex than any industrial plant. To stay in good condition and perform its varied tasks, it needs many specific chemical compounds—most of which it manufactures in an exquisitely organized network of chemical production lines.

Every minute of every day, thousands of chemical reactions take place in the cells of living systems, including each of the 100 trillion tiny cells in your body. The study of these reactions and the chemicals they produce is called *biochemistry*. Every form of life is chemical in nature. The substances and reactions that occur in living organisms are often more complex than the ones we have already studied, but basic chemical concepts such as acid–base reactions, intermolecular forces, and the reactivity of organic functional groups also apply to living organisms.

In this chapter, we will examine acid–base reactions that govern the behavior of proteins in our bodies. We will look at intermolecular forces that influence the properties and behavior of DNA and RNA. We will see that the functional groups on biochemical molecules are the same as those on other organic compounds we have studied. We will also look at how foods provide the molecular building blocks from which our bodies are made and the energy that is needed for our life activities. That energy comes ultimately from the sun through photosynthesis.

16.1 Energy and the Living Cell

Learning Objectives • List the major parts of a cell, and describe the function of each part. • Name the primary source of energy for plants and the three classes of substances that are sources of energy for animals.

Biochemistry is the chemistry of living things and life processes. The structural unit of all living things is the **cell**. Every cell is enclosed in a **cell membrane**, a double layer of lipid molecules through which it gains nutrients and gets rid of wastes. Plant cells (Figure 16.1) also have walls made of cellulose. Animal cells (Figure 16.2) do not have cell walls. Cells have a variety of interior structures that serve a multiplicity of functions. We will consider only a few of them here.

The largest interior structure of a cell is usually the *nucleus*, which contains the DNA molecules. These molecular materials control heredity. Protein synthesis takes

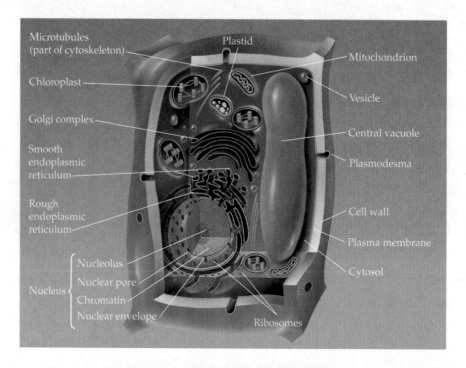

▶ **Figure 16.1** The general form of a plant cell. Not all the structures shown here occur in every type of plant cell.

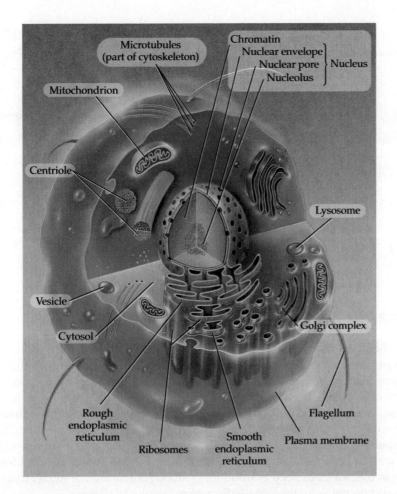

◀ **Figure 16.2** An animal cell. The entire range of structures shown here seldom occurs in a single cell, and only a few of them are discussed in this text. Each kind of plant or animal tissue has cells specific to the function of that tissue. Muscle cells differ from nerve cells, nerve cells differ from red blood cells, and so on.

place on the *ribosomes*. The *mitochondria* are the cell's "batteries," where energy is produced. Plant cells (but not animal cells) also contain *chloroplasts*, which convert energy from the sun to chemical energy that is stored in the plant in the form of carbohydrates.

Energy in Biological Systems

Life requires energy. Living cells are inherently unstable, and only a continued input of energy keeps them from falling apart. Living organisms are restricted to using certain forms of energy. Supplying a plant with heat energy by holding it in a flame will do little to prolong its life. On the other hand, a green plant is uniquely able to use sunlight, the richest source of energy for Earth. Chloroplasts in green plant cells capture radiant energy from the sun and convert it to chemical energy, which is then stored in carbohydrate molecules. The photosynthesis of glucose is represented by the following equation, which we have seen before:

$$6\,CO_2(g) \,+\, 6\,H_2O(l) \longrightarrow C_6H_{12}O_6(aq) \,+\, 6\,O_2(g)$$

Plant cells can also convert the carbohydrate molecules to fat molecules and, given the proper inorganic nutrients, to protein molecules.

Animals cannot directly use the energy of sunlight. They must get their energy by eating plants or by eating other animals that eat plants. Animals obtain energy from three major types of substances: carbohydrates, fats, and proteins. The use of carbohydrates, fats, and proteins as foods is discussed later in this chapter. However, first we must look at the synthesis and structure of each of these vital materials.

Once digested and transported to a cell, a food molecule can be used as a building block to make new cell parts or to repair old ones, or it can be "burned" for energy. The entire series of coordinated chemical reactions that keeps cells alive is called **metabolism**. In general, metabolic reactions are divided into two classes: the degrading of molecules to provide energy is called **catabolism**, and the process of building up, or synthesizing, the molecules of living systems is termed **anabolism**.

SELF-ASSESSMENT Questions

1. The three major types of substances from which we obtain energy are
   a. amino acids, proteins, and carbohydrates
   b. amino acids, proteins, and fats
   c. proteins, nucleic acids, and oils
   d. carbohydrates, fats, and proteins

2. The process of synthesizing the molecules to make cell structures within a living cell is called
   a. anabolism   b. catabolism
   c. metabolism   d. transcription

3. The overall set of chemical reactions that keeps cells alive is called
   a. anabolism   b. metabolism
   c. enzymology   d. proteolysis

4. The plant cell structure that captures sunlight and converts it to chemical energy is the
   a. chloroplast   b. mitochondrion
   c. nucleus   d. ribosome

Answers: 1, d; 2, a; 3, b; 4, a

16.2 Carbohydrates: A Storehouse of Energy

Learning Objective • Compare and contrast starch, glycogen, and cellulose.

It is difficult to give a simple formal definition of *carbohydrates*. Chemically, **carbohydrates** are polyhydroxy aldehydes or ketones, or compounds that can be hydrolyzed (split by water) to form such aldehydes and ketones. Composed of the elements carbon, hydrogen, and oxygen, carbohydrates include sugars, starches, and cellulose. Usually, the atoms of these elements are present in a ratio expressed by the formula $C_n(H_2O)_n$, which suggests that the compound could be described as a combination of carbon and water. Glucose, a simple sugar, has the formula $C_6H_{12}O_6$, which we can write as $C_6(H_2O)_6$. The term *carbohydrate* is derived from formulas written in this way, but keep in mind that carbohydrates are not in fact hydrates of carbon and do not actually contain discrete molecules of water.

Simple Sugars

Sugars are sweet-tasting carbohydrates. The simplest sugars are **monosaccharides**, carbohydrates that cannot be further hydrolyzed. Three familiar dietary monosaccharides are shown in Figure 16.3. They are *glucose* (also called *dextrose*), *galactose* (a component of lactose, the sugar in milk), and *fructose* (fruit sugar). Glucose and galactose are **aldoses**, monosaccharides with an aldehyde functional group (Table 9.6). Fructose is a **ketose**, a monosaccharide with a ketone functional group. Pure fructose is about 25% sweeter than table sugar (sucrose, shown in Figure 16.5). Glucose is about 25% less sweet than table sugar.

▶ **Figure 16.3** Three common monosaccharides. All have hydroxyl groups. Glucose and galactose have aldehyde functional groups, and fructose has a ketone group. Glucose and galactose differ only in the arrangement of the H and OH on the fourth carbon from the top (in green). Living cells use glucose as a source of energy. Fructose is fruit sugar, and galactose is a component of milk sugar.

Q *What is the molecular formula of each of the three monosaccharides? How are the three compounds related?*

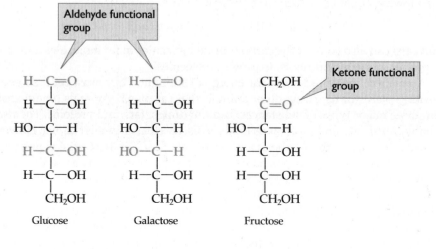

EXAMPLE 16.1 **Classification of Monosaccharides**

Shown below are structures of (left to right) erythrulose, used in some sunless tanning lotions; mannose, a sugar found in some fruits, including cranberries; and ribose, a component of ribonucleic acid (RNA; Section 16.9).

Classify erythrulose and ribose as an aldose or a ketose.

Solution
Erythrulose is a ketose. The carbonyl (C=O) group is on the second C atom from the top; it is between two other C atoms, an arrangement that defines a ketone. Ribose is an aldose. The carbonyl group includes a chain-ending carbon atom, as in all aldehydes.

❯ Exercise 16.1A
How does the structure of mannose (shown above) differ from that of galactose?

❯ Exercise 16.1B
How does the structure of mannose differ from that of fructose (Figure 16.3)?

The monosaccharides glucose, galactose, and fructose are represented in Figure 16.3 as open-chain compounds to show the aldehyde or ketone functional groups. However, these sugars actually exist in solution mainly as cyclic molecules. (See Figure 16.4.)

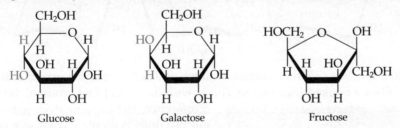

Glucose   Galactose   Fructose

◀ **Figure 16.4** Cyclic structures for glucose, galactose, and fructose. A corner with no letter represents a carbon atom. Glucose and galactose differ only in the arrangement of the H and OH on the fourth carbon (in green). Some sugars exist in more than one cyclic form, but only one form of each is shown here for simplicity.

Sucrose and lactose are examples of **disaccharides**, carbohydrates whose molecules can be hydrolyzed to yield the two monosaccharide units they contain (Figure 16.5).

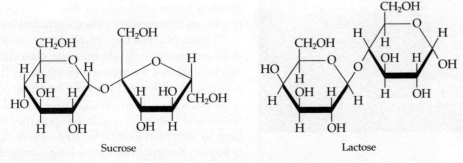

Sucrose   Lactose

▲ **Figure 16.5** Sucrose and lactose are disaccharides. On hydrolysis, sucrose yields glucose and fructose, whereas lactose yields glucose and galactose. Sucrose is cane or beet sugar, and lactose is milk sugar.

Q What is the molecular formula of each of these disaccharides? How are the two compounds related? Label the two monosaccharide units in each as fructose, galactose, or glucose.

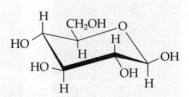

▲ Glucose is shown in a chair conformation here. Although we have represented the cyclic monosaccharides as flat hexagons, they are really three-dimensional. Most assume a conformation (shape) called a *chair conformation* because it somewhat resembles a reclining chair.

Sucrose is split into glucose and fructose. Hydrolysis of lactose yields glucose and galactose.

$$\text{Sucrose} + H_2O \longrightarrow \text{Glucose} + \text{Fructose}$$

$$\text{Lactose} + H_2O \longrightarrow \text{Glucose} + \text{Galactose}$$

Polysaccharides: Starch and Cellulose

Polysaccharides are composed of large molecules that yield many monosaccharide units on hydrolysis. Polysaccharides include starches and cellulose. Starches comprise the main energy-storage system of many plants, while cellulose provides their structural material. Figure 16.6 shows short segments of starch and cellulose molecules. Notice that both are polymers of glucose. Starch molecules generally have from 100 to about 6000 glucose units. Cellulose molecules are composed of 1800–3000 or more glucose units.

(a) Segment of a starch molecule

(b) Segment of a cellulose molecule

▶ **Figure 16.6** Both starch and cellulose are polymers of glucose. They differ in that the glucose units are joined by alpha linkages (in blue) in starch and by beta linkages (in red) in cellulose.

Q *The formulas for most polymers can be written in condensed form (Section 10.2). Write a condensed formula for starch. How would the condensed formula for cellulose differ from that of starch?*

An important difference between starch and cellulose is the way that the glucose units are hooked together. In starch, with the —CH₂OH group at the top as a reference, the oxygen atom joining the glucose units is pointed to the side *opposite* the —CH₂OH group. This arrangement is called an *alpha linkage*. In cellulose, the oxygen atom connecting the glucose units is pointed to the *same* side as the —CH₂OH group, an arrangement called a *beta linkage*.

This subtle but important difference in linkage determines whether the material can be digested by humans (Section 16.3), because the different linkages also result in different three-dimensional forms for cellulose and starch. In starch, the polymer is bent so the chain looks a bit like a helix. In cellulose, the polymer forms long, straight chains. Cellulose in the cell walls of plants is arranged in *fibrils*, bundles of parallel chains. As shown in Figure 16.7a, these fibrils lie parallel to each other in each layer

(a)

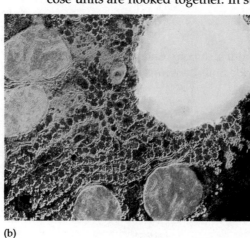

(b)

▲ **Figure 16.7** Cellulose molecules form fibers, whereas starch molecules (glycogen) form granules. Electron micrographs of (a) the cell wall of an alga, made up of successive layers of cellulose fibers in parallel arrangement, and (b) glycogen granules in a liver cell of a rat.

of the cell wall. In alternate layers, the fibrils are perpendicular, an arrangement that imparts great strength to the cell wall.

There are two kinds of starches found in plants. One, called *amylose*, has glucose units joined in a continuous chain like beads on a string. The other kind, *amylopectin*, has branched chains of glucose units. These starches are perhaps best represented schematically, as in Figure 16.8, where each glucose unit is represented by an open circle.

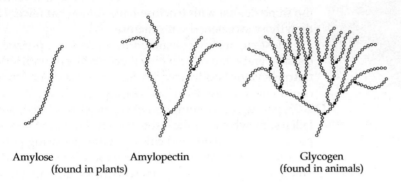

Amylose
(found in plants)

Amylopectin

Glycogen
(found in animals)

◀ **Figure 16.8** Schematic representations of amylose, amylopectin, and glycogen. Amylose is unbranched. Amylopectin has a branch approximately every 25 glucose units. Glycogen branches about every 10 glucose units.

Animal starch is called *glycogen*. Like amylopectin, it is composed of branched chains of glucose units. In contrast to cellulose, we see in Figure 16.7b that glycogen in muscle and liver tissue is arranged in small granules or clusters. Plant starch, on the other hand, forms large granules. When cornstarch is heated at about 400 °F for an hour, the starch polymers are broken down into shorter chains to form a product called *dextrin*, which has properties intermediate between those of sugars and starches. Mixed with water, dextrin makes a slightly thick, sticky mixture. Starch and dextrin adhesives account for about 58% of the U.S. packaging market.

Granules of plant starch rupture in boiling water to form a paste, which gels when it cools. Potatoes and cereal grains form this type of starchy broth. All forms of starch are hydrolyzed to glucose during digestion.

SELF-ASSESSMENT Questions

1. An aldose is a monosaccharide
   a. that is a diet aid
   b. that forms polyhydroxy cellulose
   c. that forms aldosterols
   d. with an aldehyde group

2. When most monosaccharides dissolve in water, they form
   a. branched polysaccharides   b. disaccharides
   c. ring structures   d. polysaccharides

3. The two monosaccharide units that make up a lactose molecule are
   a. both glucose   b. fructose and galactose
   c. glucose and fructose   d. glucose and galactose

4. Which of these polysaccharide molecules consists of branched chains of glucose units?
   a. amylose and amylopectin
   b. amylose and cellulose
   c. amylose and glycogen
   d. amylopectin and glycogen

5. Which of the following carbohydrates has beta linkages?
   a. amylose   b. amylopectin
   c. cellulose   d. glycogen

Answers: 1, d; 2, c; 3, d; 4, d; 5, c

16.3 Carbohydrates in the Diet

Learning Objective • Identify dietary carbohydrates, and state their sources and function.

Dietary carbohydrates include sugars and starches (Figure 16.9). Sugars are mainly monosaccharides and disaccharides; starches are polysaccharides.

Sugars have been used since ancient times to make food sweeter and more pleasant tasting. The underlying sweetness of fruits is due to the presence of fructose,

▲ **Figure 16.9** The main sugars in our diet are sucrose (cane or beet sugar), glucose (corn syrup), and fructose (fruit sugar, often in the form of high-fructose corn syrup).

Q *What is the molecular formula for glucose? For fructose? How are the two compounds related? How is sucrose related to glucose and fructose?*

▲ Sugarcane is a major agricultural crop in tropical regions. It is the source of about 45% of the refined sugar used in the United States.

▲ **Figure 16.10** Infants born with galactosemia can thrive on a milk-free substitute formula.

from the Latin word for fruit, *fructus*. The most ancient sweetener is honey, consisting of a mixture of the monosaccharides fructose and glucose. Common table sugar is sucrose, a disaccharide usually obtained from sugarcane or sugar beets. Ordinary corn syrup, a viscous liquid, consists mainly of glucose made by splitting long cornstarch chains. *High-fructose corn syrup (HFCS)* is made by treating corn syrup with enzymes to convert much of the glucose to fructose. Foods sweetened to the same degree with fructose have somewhat fewer calories than those sweetened with sucrose.

Most sugars are consumed in soft drinks, presweetened cereals, candy, and other highly processed foods with little or no nutritive value besides calories. Sugars in sweetened foods also contribute to tooth decay and obesity. From the 1960s through 2008, per capita consumption of sugars skyrocketed, but it has fallen somewhat over the last few years. Public awareness campaigns in some states and cities and the increasing popularity of "diet" soft drinks, with artificial sweeteners instead of the HFCS usually added to regular soft drinks, may account for the slight decrease in total sugar consumption among the general public in the recent years. A typical 12-ounce can of ordinary soda contains about 40 g (about three tablespoons) of sugar in the form of HFCS, but diet soft drinks contain none. Artificial sweeteners will be discussed in Section 17.5.

Digestion and Metabolism of Carbohydrates

Glucose and fructose are absorbed directly into the bloodstream from the digestive tract. Sucrose is hydrolyzed during digestion to glucose and fructose.

$$\text{Sucrose} + \text{H}_2\text{O} \longrightarrow \text{Glucose} + \text{Fructose}$$

The disaccharide lactose occurs in milk. During digestion, it is hydrolyzed to two simpler sugars, glucose and galactose.

$$\text{Lactose} + \text{H}_2\text{O} \longrightarrow \text{Glucose} + \text{Galactose}$$

Nearly all human babies have the enzyme necessary to accomplish this breakdown, but many adults do not. People who lack the enzyme get digestive upsets from drinking milk, a condition called *lactose intolerance*. When milk is cooked or fermented, the lactose is at least partially hydrolyzed. People with lactose intolerance may be able to enjoy cheese, yogurt, or cooked foods containing milk with little or no discomfort. Lactose-free milk, made by treating milk with an enzyme that hydrolyzes lactose, is available in most grocery stores. Lactose intolerance is particularly common in people from Asia and Africa, but much less common among people of European descent.

Although it has long been thought that all monosaccharides are converted to glucose during metabolism, research shows that fructose may remain in the liver, stimulating the formation of triglycerides (fats). This has focused more attention on the extensive use of HFCS. Glucose is the sugar used by the cells of our bodies for energy. Because it is the sugar that circulates in the bloodstream, it is often called **blood sugar**. Some babies are born with *galactosemia*, a deficiency of the enzyme that catalyzes the conversion of galactose to glucose. For proper nutrition, they must be fed a synthetic formula (Figure 16.10) in place of milk.

Complex Carbohydrates: Starch and Cellulose

Starch and cellulose are both polymers of glucose, but the connecting links between the glucose units are different (Figure 16.6). Human digestive enzymes are able to easily split **starch**, a polymer of glucose in which the units are joined by alpha linkages, resulting in a Slinky-like shape. On the other hand, those same enzymes cannot hydrolyze **cellulose**, because the beta linkages result in a straight chain of glucose units. Starch is needed in the diet, but it should provide only about a quarter of our daily caloric intake (Figure 16.11), and half of the starch we consume should be in the form of whole grains. Cellulose is an important component of dietary fiber.

▲ **Figure 16.11** Bread, flour, cereals, and pasta are rich in starches. Various grains for sale in Oshakati, Namibia, (left) and the assortment of pasta and baked goods available in the United States (right) show the range of carbohydrates eaten throughout the world.

When digested, starch is hydrolyzed to glucose, as represented by the following equation:

$$(C_6H_{10}O_5)_n + n\,H_2O \xrightarrow{\text{Carbohydrases}} n\,C_6H_{12}O_6$$

Starch   Glucose

The body then metabolizes the glucose, using it as a source of energy. Glucose is broken down through a complex set of more than fifty chemical reactions that result in the production of carbon dioxide and water and in the release of energy.

$$C_6H_{12}O_6 + 6\,O_2 \longrightarrow 6\,CO_2 + 6\,H_2O + \text{Energy}$$

Note that the net reaction is the reverse of photosynthesis. In this way, animals are able to make use of the energy from the sun that was captured by plants using the process of photosynthesis. However, it should be noted that the specific reactions for photosynthesis and for the metabolism of glucose are quite different.

Carbohydrates, which supply about 4 kcal of energy per gram, are our bodies' preferred fuels. When we eat more than our bodies can use, small amounts of carbohydrates can be stored in the liver and in muscle tissue as **glycogen** (animal starch), a highly branched polymer of α-glucose. Large excesses of carbohydrates are converted to fat for storage. Most health authorities recommend obtaining carbohydrates from a diet rich in whole grains, fruits, vegetables, and legumes (beans). We should minimize our intake of the simple sugars and refined starches found in many prepared foods.

Cellulose is the most abundant carbohydrate in nature. It is present in all plants, forming their cell walls and other structural features, such as stems, bark, leaves, and seeds. Wood is about 50% cellulose, and cotton is almost pure cellulose. Unlike starch, cellulose cannot be digested by humans and many other animals. Cellulose does, however, play an important role in human digestion, providing *dietary fiber* that absorbs water and helps move food through the digestive tract (Section 17.4).

We get no caloric value from dietary cellulose, because its glucose units are joined by beta linkages, and most animals lack the enzymes needed to break this kind of bond. Certain bacteria that live in the guts of termites and in the digestive tracts of grazing animals such as cows or deer do produce the enzymes, so that these animals can convert cellulose to glucose.

SELF-ASSESSMENT Questions

1. The most abundant carbohydrate found in nature is
   a. cellulose   **b.** lactose
   c. starch   **d.** sucrose

2. The monomer units that make up cellulose and starch, respectively, are
   a. glucose and fructose   **b.** α-glucose and β-glucose
   c. β-glucose and α-glucose   **d.** fructose and sucrose

2 **Since fruits are sweet, do they cause tooth decay, like candy?**

No. The main bacteria in the mouth that cause decay of tooth enamel, *Streptococcus mutans*, feed primarily on sucrose, the disaccharide that is found in candy and table sugar, and do not attack free glucose or fructose.

You can do your own carbohydrate digestion experiment. Chew an unsalted saltine cracker for several minutes, and you will notice a slight sweet taste. Your saliva contains enzymes that begin the digestion process. Explain the sweet taste.

WHY IT MATTERS

With today's emphasis on low-carbohydrate foods, labels on some diet foods give the mass in grams of carbohydrate that can be readily digested. "Only 1 carb" means that a serving contains 1 g of readily digestible carbohydrate. It is important to read the labels of such foods carefully. They may contain a large amount of fat, including saturated fat. They may also contain relatively large amounts of additives (such as glycerin or xylitol) that are digested as carbohydrate—though much more slowly than sugars or even starches.

3. Ordinary corn syrup
   a. consists of glucose and fructose   **b.** consists mainly of glucose
   c. is mainly sucrose, just like table sugar   **d.** is another name for high-fructose corn syrup

4. Small quantities of carbohydrates are stored in liver and muscle tissue as
   a. amylose **b.** cellulose **c.** glucose **d.** glycogen

5. Humans cannot digest cellulose, because they lack enzymes for the hydrolysis of
   a. α-amino acid linkages   **b.** α-glucose linkages   **c.** β-glucose linkages   **d.** sucrose

Answers: 1, a; 2, c; 3, b; 4, d; 5, c

16.4 Fats and Other Lipids

Learning Objectives • Describe the fundamental structure of a fatty acid and of a fat. • Classify fats as saturated, monounsaturated, or polyunsaturated.

Fats are the predominant form of a class of compounds called *lipids*. These substances are not defined by a functional group, as are most organic compounds. Rather, lipids all have common solubility properties. A **lipid** is a cellular component that is insoluble in water but soluble in organic solvents of low polarity, such as hexane, diethyl ether, and chloroform. In addition to fats, the lipid family includes **fatty acids** (long-chain carboxylic acids), steroids such as cholesterol and sex hormones (Chapter 18), fat-soluble vitamins (Chapter 17), and other substances. Figure 16.12 shows three representations of palmitic acid, a typical fatty acid.

A **fat** is an ester of fatty acids and the trihydroxy alcohol glycerol (Figure 16.13). A fat has three fatty acid chains joined to glycerol through ester linkages (which is why fats are often called **triglycerides**, or triacylglycerols). Related compounds are classified according to the number of fatty acid chains they contain: A *monoglyceride* has one fatty acid chain joined to glycerol, and a *diglyceride* has two.

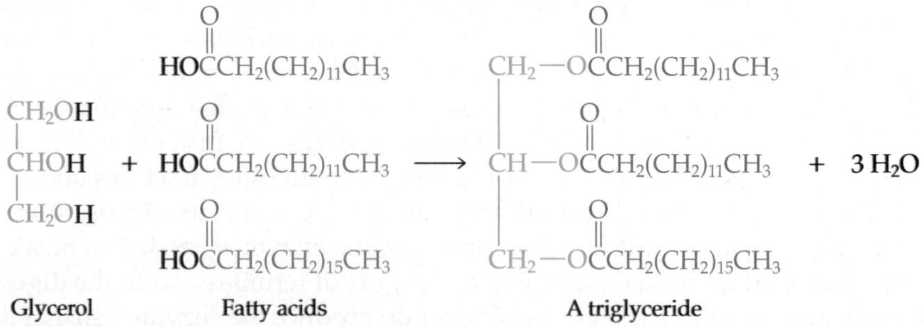

$CH_3CH_2CH_2CH_2CH_2CH_2CH_2CH_2CH_2CH_2CH_2CH_2CH_2CH_2CH_2COOH$
(a)

(b)

(c)

▲ **Figure 16.12** Three representations of palmitic acid: (a) Condensed structural formula. (b) Line-angle formula, in which the lines denote bonds and each intersection of lines or end of a line represents a carbon atom. (c) Space-filling model.

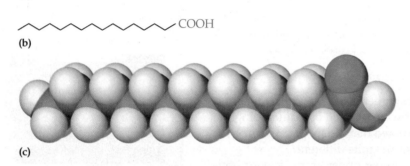

$$\begin{array}{c} CH_2OH \\ | \\ CHOH \\ | \\ CH_2OH \end{array} + \begin{array}{c} O \\ \| \\ HOCCH_2(CH_2)_{11}CH_3 \\ O \\ \| \\ HOCCH_2(CH_2)_{11}CH_3 \\ O \\ \| \\ HOCCH_2(CH_2)_{15}CH_3 \end{array} \longrightarrow \begin{array}{c} O \\ \| \\ CH_2-OCCH_2(CH_2)_{11}CH_3 \\ O \\ \| \\ CH-OCCH_2(CH_2)_{11}CH_3 \\ O \\ \| \\ CH_2-OCCH_2(CH_2)_{15}CH_3 \end{array} + 3\,H_2O$$

Glycerol   Fatty acids   A triglyceride
(a)

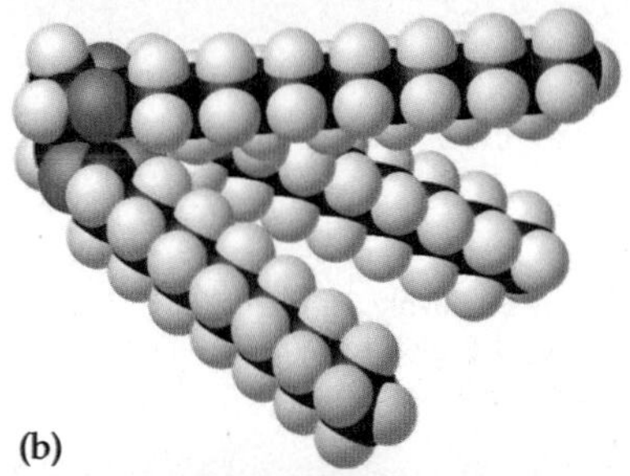

▶ **Figure 16.13** Triglycerides (triacylglycerols) are esters in which the trihydroxy (having three OH groups) alcohol glycerol is esterified with three fatty acid groups. (a) The equation for the formation of a triglyceride. (b) Space-filling model of a triglyceride.

(b)

TABLE 16.1 Some Fatty Acids in Natural Fats

Number of Carbon Atoms	Condensed Structural Formula	Name	Source
4	$CH_3CH_2CH_2COOH$	Butyric acid	Butter
6	$CH_3(CH_2)_4COOH$	Caproic acid	Butter
8	$CH_3(CH_2)_6COOH$	Caprylic acid	Coconut oil
10	$CH_3(CH_2)_8COOH$	Capric acid	Coconut oil
12	$CH_3(CH_2)_{10}COOH$	Lauric acid	Palm kernel oil
14	$CH_3(CH_2)_{12}COOH$	Myristic acid	Oil of nutmeg
16	$CH_3(CH_2)_{14}COOH$	Palmitic acid	Palm oil
18	$CH_3(CH_2)_{16}COOH$	Stearic acid	Beef tallow
18	$CH_3(CH_2)_7CH{=}CH(CH_2)_7COOH$	Oleic acid	Olive oil
18	$CH_3(CH_2)_4CH{=}CHCH_2CH{=}CH(CH_2)_7COOH$	Linoleic acid	Soybean oil
18	$CH_3CH_2(CH{=}CHCH_2)_3(CH_2)_6COOH$	Linolenic acid	Fish oils
20	$CH_3(CH_2)_4(CH{=}CHCH_2)_4CH_2CH_2COOH$	Arachidonic acid	Liver

Naturally occurring fatty acids typically have an even number of carbon atoms. Representative ones are listed in Table 16.1. Animal fats are generally rich in saturated fatty acids (fatty acids with no carbon-to-carbon double bonds) and have a smaller proportion of unsaturated fatty acids. At room temperature, most animal fats are solids. Liquid fats, called **oils**, are obtained mainly from vegetable sources. Oils typically have a higher proportion of unsaturated fatty acid units than fats do. Fats and oils feel greasy. They are less dense than water and will float on it.

Fats are often classified according to the degree of unsaturation of the fatty acids they incorporate. A *saturated fatty acid* contains no carbon-to-carbon double bonds, a *monounsaturated fatty acid* has one carbon-to-carbon double bond per molecule, and a *polyunsaturated fatty acid* has two or more carbon-to-carbon double bonds. A **saturated fat** contains a high proportion of saturated fatty acids; these fat molecules have relatively few carbon-to-carbon double bonds. A **polyunsaturated fat** (oil) incorporates mainly unsaturated fatty acids; these fat molecules have many double bonds.

The iodine number is a measure of the degree of unsaturation of a fat or oil. The **iodine number** is the mass in grams of iodine that is consumed by 100 g of fat or oil. Iodine, like other halogens, adds to a carbon-to-carbon double bond.

$$\text{C{=}C} + I_2 \longrightarrow -\overset{|}{\underset{I}{C}}-\overset{|}{\underset{I}{C}}-$$

The more double bonds a fat contains, the more iodine is required for the addition reaction. Thus, a high iodine number means a high degree of unsaturation. Representative iodine numbers are listed in Table 16.2. Note the generally lower values for animal fats (butter, tallow, and lard) compared with those for vegetable oils. Coconut oil, which is highly saturated, and fish oils, which are relatively unsaturated, are notable exceptions to the general rule.

▲ Many salad dressings are made of an oil and vinegar. In the bottle shown here, olive oil is floating on wine vinegar, an aqueous solution of acetic acid.

TABLE 16.2 Typical Iodine Numbers for Some Fats and Oils[a]

Fat or Oil	Iodine Number	Fat or Oil	Iodine Number
Coconut oil	8–10	Cottonseed oil	100–117
Butter	25–40	Corn oil	115–130
Beef tallow	30–45	Fish oils	120–180
Palm oil	37–54	Canola oil	125–135
Lard	45–70	Soybean oil	125–140
Olive oil	75–95	Safflower oil	130–140
Peanut oil	85–100	Sunflower oil	130–145

[a]Oils shown in blue are from plant sources.

SELF-ASSESSMENT Questions

1. A substance may be classified as a lipid because of its
   a. chemical reactivity   **b.** functional group
   c. plant source   **d.** solubility properties

Questions 2–4 refer to the structures S, T, U, and V.

$$S: CH_3(CH_2)_{10}COOH \qquad T: CH_3(CH_2)_{16}COOH$$

$$U: CH_3(CH_2)_7CH=CH(CH_2)_7COOH$$

$$V: CH_3CH_2(CH=CHCH_2)_3(CH_2)_6COOH$$

2. Which are saturated fatty acids?
   a. S and T   **b.** S and U   **c.** T and U   **d.** U and V

3. Which is a monounsaturated fatty acid?
   a. S   **b.** T   **c.** U   **d.** V

4. Polyunsaturated fats (oils) often have a high proportion of
   a. S   **b.** T   **c.** U   **d.** V

5. Fats are classified by their
   a. chemical reactivity   **b.** functional group
   c. function in the body   **d.** degree of unsaturation

6. The iodine number of a fat or oil expresses its
   a. calories per gram   **b.** degree of acidity
   c. degree of unsaturation   **d.** digestibility

7. In general, animal fats differ from vegetable oils in that the animal molecules
   a. are higher in calories   **b.** are shorter
   c. have fewer C=C bonds   **d.** have more C=C bonds

Answers: 1, d; 2, a; 3, c; 4, d; 5, d; 6, c; 7, c

16.5 Fats and Cholesterol

Learning Objective • Identify dietary lipids, and state their function.

In the previous section, we looked at the structure of fats, which are high-energy foods that yield about 9 kcal of energy per gram, more than twice as much as carbohydrates. Some fats are "burned" as fuel for our activities. Others are used to build and maintain important constituents of our cells, such as cell membranes. Fats help produce *prostaglandins*, which act as temperature regulators and inflammation mediators, and serve many other functions in the body. Fats can store vitamins and make many foods more palatable. The fat in our diet comes from a variety of sources, some of which are shown in Figure 16.14.

Digestion and Metabolism of Fats

Dietary fats are mainly triacylglycerols, commonly called triglycerides. Fats are digested by enzymes called *lipases* and are ultimately broken down into fatty acids and glycerol (Figure 16.15). Some fat molecules are hydrolyzed to a monoglyceride (*monoacylglycerol*) or a diglyceride (*diacylglycerol*), which contains a glycerol and either one or two fatty acids, respectively. Once absorbed, these products of fat digestion are reassembled into triglycerides, which are attached to proteins for transportation through the bloodstream.

▲ **Figure 16.14** Cream, butter, margarine, cooking oils, and foods fried in fat are rich in fats. The average American diet contains too much fat. Americans get 37% of their calories from fat, while the recommendation of the FDA is 20% to 30% fat calories.

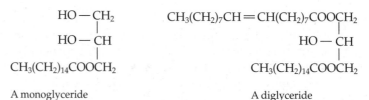

A monoglyceride   A diglyceride

Fats are stored throughout the body, principally in **adipose tissue**, in locations called **fat depots**. Fat depots around vital organs, such as the heart, kidneys, and

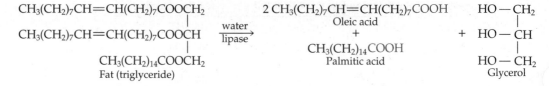

Fat (triglyceride)   Oleic acid   Palmitic acid   Glycerol

▲ **Figure 16.15** In the digestion of fats, a triglyceride is hydrolyzed to fatty acids and glycerol in a reaction catalyzed by the enzyme lipase.

Q *What diglyceride is formed by removal of the palmitic acid part of the triglyceride? What monoglyceride is formed by removal of both oleic acid parts?*

spleen, cushion and help prevent injury to these organs. Fat is also stored under the skin, where it helps to insulate against temperature changes.

When fat reserves are called on for energy, fat molecules are hydrolyzed back to glycerol and fatty acids. The glycerol can be burned for energy or converted to glucose. The fatty acids enter a process called the *fatty acid spiral* that removes carbon atoms two at a time. The two-carbon fragments can be used for energy or for the synthesis of new fatty acids.

Fats, Cholesterol, and Human Health

Dietary *saturated fats* and cholesterol have been implicated in *arteriosclerosis* ("hardening of the arteries"). Incidence of cardiovascular disease is strongly correlated with diets rich in saturated fats, which contain a large proportion of fatty acids, such as stearic or palmitic acids. As the disease develops, deposits form on the inner walls of arteries. Eventually these deposits harden, and the vessels lose their elasticity (Figure 16.16). Blood clots tend to lodge in the narrowed arteries, leading to a heart attack (if the blocked artery is in heart muscle) or a stroke (if the blockage occurs in an artery that supplies the brain).

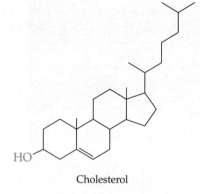

Cholesterol

▲ Structural formula of cholesterol.

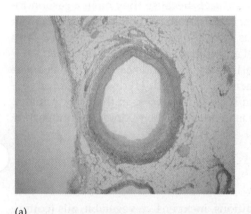
(a)

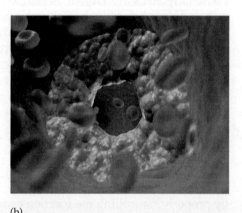
(b)

▲ **Figure 16.16** Cross-sections of (a) a normal artery and (b) a "hardened" artery, which shows deposits of plaque that contain cholesterol.

The plaque in clogged arteries is rich in cholesterol, a steroid alcohol found in animal tissues and various foods from animals. Cholesterol is normally synthesized by the liver, and is important as a constituent of cell membranes and as a precursor to steroid hormones. High blood levels of cholesterol, like those of triglycerides, correlate closely with the risk of cardiovascular disease. Like fats, cholesterol is insoluble in water, as you would expect from its molecular formula ($C_{27}H_{45}OH$) and structure. Cholesterol is transported in blood by water-soluble proteins. The cholesterol–protein combination is an example of a **lipoprotein**, any of a group of proteins combined with a lipid, such as cholesterol or a triglyceride.

Lipoproteins are usually classified according to their density (Table 16.3). Very-low-density lipoproteins (VLDLs) serve mainly to transport triglycerides, whereas low-density lipoproteins (LDLs) are the main carriers of cholesterol. LDLs carry cholesterol from the liver to the cells for use, and these lipoproteins are also the ones

3 **Is high blood cholesterol mainly a problem for older people?**

The effects of high blood cholesterol are seen mainly in older people because the buildup of plaque in the blood vessels takes time. Those effects are often the result of years of an unhealthy diet and high levels of LDLs. To minimize the long-term effects of high cholesterol, you should consider preventive measures early. Reducing saturated fat in your diet is a good start.

TABLE 16.3 **Lipoproteins in the Blood**

Class	Abbreviation	Protein (%)	Density (g/mL)	Main Function
Very-low-density	VLDL	5	1.006–1.019	Transport triglycerides
Low-density	LDL	25	1.019–1.063	Transport cholesterol to the cells for use
High-density	HDL	50	1.063–1.210	Transport cholesterol to the liver for processing and excretion

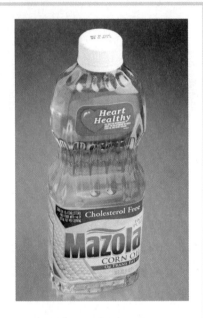

WHY IT MATTERS

Half of the total fat, three-fourths of the saturated fat, and all the cholesterol in a typical human diet come from animal products such as meat, milk, cheese, and eggs. Advertising that a vegetable oil (for example) contains no cholesterol is silly. No vegetable product contains cholesterol.

that deposit cholesterol in arteries, leading to cardiovascular disease. High-density lipoproteins (HDLs) also carry cholesterol, but they carry it from the cells to the liver for processing and excretion. Exercise is thought to increase the levels of HDLs, the lipoproteins sometimes called "good" cholesterol. High levels of LDLs, called "bad" cholesterol, increase the risk of heart attack and stroke. The American Heart Association recommends that people limit their intake of saturated fatty acids between 5% to 6% of their total daily calorie intake.

Fats differ in their effects on blood-cholesterol levels. Many nutritionists advise us to use olive oil and canola oil as our major sources of dietary lipids because they contain a high percentage of monounsaturated fatty acids, which have been shown to lower LDL cholesterol. There is also statistical evidence that fish oils can prevent heart disease. For example, Greenlanders eat a lot of fish but have a low risk of heart disease, despite a diet that is high in total fat and cholesterol. This is probably due to polyunsaturated fatty acids such as eicosapentaenoic acid (EPA) and docosahexaenoic acid (DHA):

$$CH_3(CH_2CH{=}CH)_5(CH_2)_3COOH \qquad CH_3(CH_2CH{=}CH)_6(CH_2)_2COOH$$
$$\text{EPA} \qquad\qquad \text{DHA}$$

These fatty acids are known as *omega-3 fatty acids* because they have a carbon-to-carbon double bond that begins at the *third* carbon from the end farthest away from the COOH group—the *omega* end. Studies have shown that adding omega-3 fatty acids to the diet leads to lower cholesterol and triglyceride levels in the blood.

Fats and oils containing carbon-to-carbon double bonds can undergo *hydrogenation*, addition of an H_2 molecule to each double bond. Hydrogenation of vegetable oils to produce semisolid fats is an important process in the food industry. The chemistry of this conversion process is identical to the hydrogenation reaction of alkenes described in Section 9.2.

$$CH_3(CH_2)_7CH{=}CH(CH_2)_7COOH \xrightarrow[Ni]{H_2} CH_3(CH_2)_7CH_2CH_2(CH_2)_7COOH$$
$$\text{Oleic acid (monounsaturated)} \qquad\qquad \text{Stearic acid (saturated)}$$

By properly controlling the reaction conditions, inexpensive vegetable oils (cottonseed, corn, and soybean) can be partially hydrogenated to yield soft, spreadable fats suitable for use in margarine or fully hydrogenated to produce harder fats like shortening. The consumer would get much higher unsaturation by using the oils directly, but most people would rather spread margarine than pour oil on their toast.

Concern about the role of saturated fats in raising blood cholesterol and clogging arteries has caused many consumers to switch from butter to margarine. However, partial hydrogenation converts some of the unsaturated fats into *trans* fats, which have structures similar to saturated fats and can also raise cholesterol levels and increase the risk of coronary heart disease. Figure 16.17 shows molecular models of different types of fatty acids that can be attached to glycerol via ester links to form a triglyceride. Saturated fatty acids have only single bonds connecting the chain of carbon atoms (Figure 16.17a), while unsaturated fatty acids have at least one double bond. Most naturally occurring unsaturated fatty acids have a *cis* arrangement about the double bond (with the two hydrogen atoms on the same side of the bond) and have a bend in the chain (Figure 16.17c). During hydrogenation, the arrangement in some of the molecules is changed so that the hydrogen atoms are on opposite sides of the double bond, a *trans* arrangement (Figure 16.17b).

Note that both saturated fatty acids (such as stearic acid) and *trans* fatty acids are more or less straight and can stack neatly like logs (Figure 16.17a and Figure 16.17b). This maximizes the intermolecular attractive forces, making saturated and *trans* fatty acids more likely to be solids than *cis* fatty acids. On the other hand, *cis* fatty acids (Figure 16.17c) have a bend in their structure, fixed in position by the double bond. Thus, they have weaker intermolecular forces and are more likely to be liquids.

Because *trans* fatty acids also resemble saturated fatty acids in their tendency to raise blood levels of LDL cholesterol, the FDA requires manufacturers to provide the *trans* fat content of foods on labels and is considering banning *trans* fats from

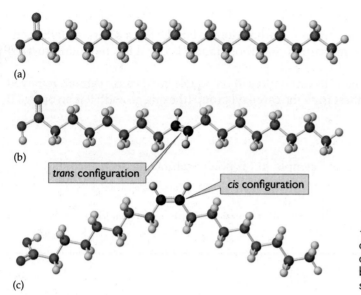

(a)

(b)

trans configuration

cis configuration

(c)

◀ **Figure 16.17** The structures of (a) stearic acid (a saturated fatty acid) and (b) the *trans* isomer of oleic acid (a *trans* unsaturated fatty acid) are similar in shape, and these acids behave similarly in the body; (c) oleic acid (a *cis* unsaturated fatty acid) has a very different shape.

prepared foods. Many fast-food establishments have eliminated the use of *trans* fats due to public pressure.

Emulsions

When oil and water are vigorously shaken together, the oil is broken up into microscopic droplets and dispersed throughout the water, a mixture called an **emulsion**. Unless a third substance has been added, the emulsion usually breaks down rapidly, as the oil droplets recombine and float to the surface of the water.

Emulsions can be stabilized by adding a type of gum, a soap, or a protein that can form a protective coating around the oil droplets and prevent them from coming together. Lecithin in egg yolks keeps mayonnaise from separating, while casein in milk keeps fat droplets suspended. Compounds called *bile salts* keep tiny fat droplets suspended and greatly aid the digestive process. The tiny emulsified droplets provide a much greater surface area for the water-soluble lipase enzymes to break down the triacylglycerides to glycerol and fatty acids.

▲ Many foods are emulsions. Milk is an emulsion of butterfat in water. The stabilizing agent is a protein called *casein*. Mayonnaise is an emulsion of vegetable oil in water, stabilized by egg yolk.

 EXAMPLE 16.1 Fatty Acids

Following are line-angle structures of three fatty acids (see Section 9.2). Each corner and each end of a line is a carbon atom, and each carbon atom is attached to enough hydrogen atoms to give each carbon atom four bonds. **(a)** Which is a saturated fatty acid? **(b)** Which is a monounsaturated fatty acid? **(c)** Which is an omega-3 fatty acid?

I.  COOH

II. COOH

III. COOH

Solution

   a. Fatty acid II has no carbon-to-carbon double bonds; it is a saturated fatty acid.

   b. Fatty acid III has one carbon-to-carbon double bond; it is a monounsaturated fatty acid.

   c. Fatty acid I has one of its carbon-to-carbon double bonds three carbons removed from the end farthest from the carboxyl group (the omega end); it is an omega-3 fatty acid.

› **Exercise 16.1A**

Which of the fatty acids in Example 16.1 is polyunsaturated?

› **Exercise 16.1B**

Is fatty acid III likely to be more similar physiologically to fatty acid I or to fatty acid II? Explain.

SELF-ASSESSMENT Questions

1. Fats are
  a. mixtures of fatty acids
  b. esters of glycerol and fatty acids
  c. esters of cholesterol and fatty acids
  d. polymers of amino acids

2. Among lipoproteins, LDLs are
  a. bad because they can block arteries
  b. the form that carries cholesterol to the liver for processing and excretion
  c. the good form, because they can clear arteries
  d. the kind increased by exercise

3. Which of the following structures represents an omega-3 fatty acid?.
  a. $CH_3(CH_2)_{16}COOH$
  b. $CH_3(CH_2)_5CH{=}CH(CH_2)_7COOH$
  c. $CH_3(CH_2)_3(CH_2CH{=}CH)_2(CH_2)_7HOOH$
  d. $CH_3(CH_2CH{=}CH)_3(CH_2)_7COOH$

4. The general shape of *trans* fatty acid molecules is similar to
  a. cholesterol
  b. disaccharides
  c. omega-3 fatty acids
  d. saturated fatty acids

5. Triglycerides are made up of
  a. one glycerol and three carbohydrate molecules
  b. one carbohydrate and three glycerol molecules
  c. one glycerol and three fatty acid molecules
  d. one fatty acid and three glycerol molecules

6. HDLs primarily carry
  a. cholesterol from the liver to the cells
  b. cholesterol from the cells to the liver
  c. triglycerides from the liver to the cells
  d. triglycerides from the cells to the liver

Answers: 1, b; 2, a; 3, d; 4, d; 5, c; 6, b

16.6 Proteins: Polymers of Amino Acids

Learning Objective • Draw the fundamental structure of an amino acid, and show how amino acids combine to make proteins.

Proteins are vital components of all life. No living part of the human body—or of any other organism—is completely without protein. There is protein in blood, muscles, brain, and even tooth enamel. The smallest cellular organisms—bacteria—contain protein. Viruses, so small that they make bacteria look like giants, are made up of little more than proteins and nucleic acids.

    Each type of cell makes its own kinds of proteins. Proteins serve a wide variety of functions. They provide structure (bones, insect exoskeletons, collagen, skin and nails), are involved in muscular movement (actin and myosin), transport chemicals (hemoglobin), catalyze chemical reactions, provide protection (antibodies and blood clotting), and even act as hormones (insulin). All proteins contain the elements carbon, hydrogen, oxygen, and nitrogen, and most also contain sulfur. The structure of a short segment of a typical protein molecule is shown in Figure 16.18.

(a)

(b)

▲ **Figure 16.18** (a) Space-filling model and (b) structural formula of a short segment of a protein molecule. In the structural formula, hydrocarbon side chains (in green), an acidic side chain (in red), a basic side chain (in blue), and a sulfur-containing side chain (with amber shading) are highlighted. The dashed lines crossing the formula indicate where two amino acid units join.

Q *The formulas for most polymers can be written in condensed form, while those for natural proteins cannot. Explain.*

Like starch and cellulose, *proteins* are polymers. They differ from other polymers that we have studied in that the monomer units include some 20 different amino acids, which are shown in Table 16.4. Each **amino acid** has two functional groups, an amino group ($-NH_2$) and a carboxyl group ($-COOH$) attached to the same carbon atom (called the *alpha carbon*). The amino acids differ in their side chains (R groups shown in green).

$$R$$
$$|$$
$$H_2N-C-COOH$$
$$|$$
$$H$$

In the formula shown above, the R can be a hydrogen atom (as it is in glycine, the simplest amino acid) or any of the groups shown in green in Table 16.4. Note that some R groups are nonpolar, some are acidic, some are basic, and others are polar but not charged. The formula shows the proper placement of the groups, but it is not really correct. Under physiological pH conditions, the amino and carboxylic acid groups are actually charged and can react to form salts. Within the amino acid, the carboxyl group has a negative change, and the amino group is positively charged. The resulting product is an inner salt, or **zwitterion**, a compound in which the negative charge and the positive charge are on different parts of the same molecule.

$$R$$
$$|$$
$$H_3N^+-C-COO^-$$
$$|$$
$$H$$

A zwitterion

TABLE 16.4 **The 20 Amino Acids Specified by the Human Genetic Code**

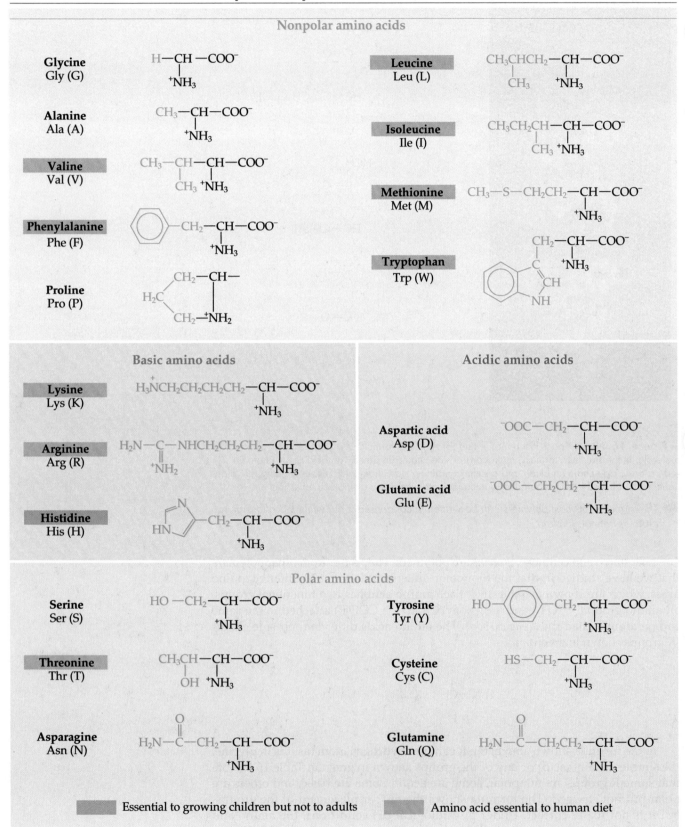

Nonpolar amino acids

Glycine Gly (G)

Alanine Ala (A)

Valine Val (V)

Phenylalanine Phe (F)

Proline Pro (P)

Leucine Leu (L)

Isoleucine Ile (I)

Methionine Met (M)

Tryptophan Trp (W)

Basic amino acids

Lysine Lys (K)

Arginine Arg (R)

Histidine His (H)

Acidic amino acids

Aspartic acid Asp (D)

Glutamic acid Glu (E)

Polar amino acids

Serine Ser (S)

Threonine Thr (T)

Asparagine Asn (N)

Tyrosine Tyr (Y)

Cysteine Cys (C)

Glutamine Gln (Q)

Essential to growing children but not to adults

Amino acid essential to human diet

Plants can synthesize amino acids and proteins from carbon dioxide, water, and minerals such as nitrates (NO_3^-) and sulfates (SO_4^{2-}). Animals need to consume proteins in their foods. Humans can only synthesize some of the amino acids shown in Table 16.4. Others are *essential amino acids*, which must be acquired from the diet and which will be discussed in Section 16.8.

The Peptide Bond: Peptides and Proteins

The human body contains tens of thousands of different proteins. Each of us has a tailor-made set. Proteins are polyamides. When an amide linkage joins two amino acids, it is called a **peptide bond** (shaded part in the following illustration).

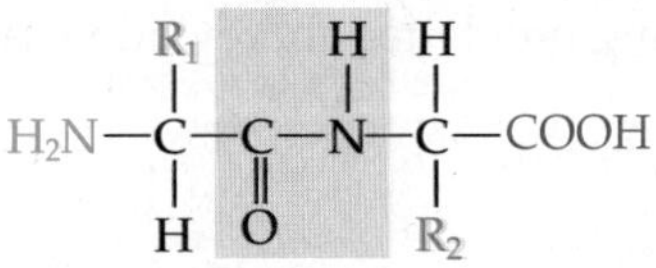

$$\text{H}_2\text{N}-\overset{\displaystyle R_1}{\underset{\displaystyle H}{\overset{|}{\underset{|}{C}}}}-\overset{\displaystyle H}{\underset{\displaystyle O}{\overset{|}{\underset{||}{C}}}}-\overset{\displaystyle H}{\underset{|}{N}}-\overset{\displaystyle H}{\underset{\displaystyle R_2}{\overset{|}{\underset{|}{C}}}}-\text{COOH}$$

Note that this molecule still has a reactive amino group on the left and a carboxyl group on the right that can react with other amino acid units. This process can continue until hundreds or even thousands of units have joined to form a giant molecule—a polymer called a protein. We will examine the structure of proteins in the next section, but first let's look at some smaller molecules called *peptides* that have only a few amino acids.

When only two amino acids are joined, the product is a *dipeptide*.

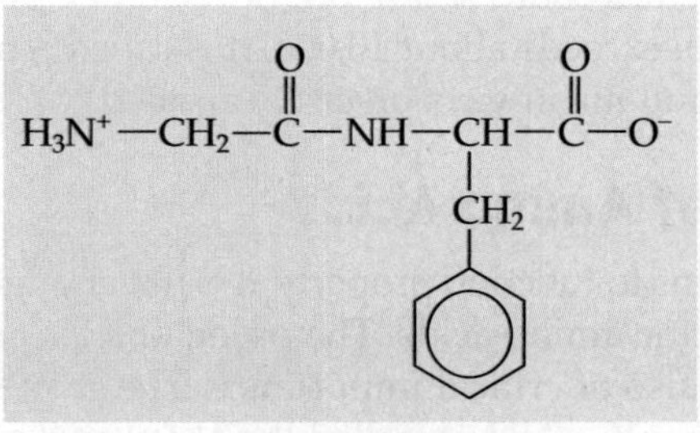

Glycylphenylalanine
(a dipeptide)

Three amino acids combine to form a *tripeptide*.

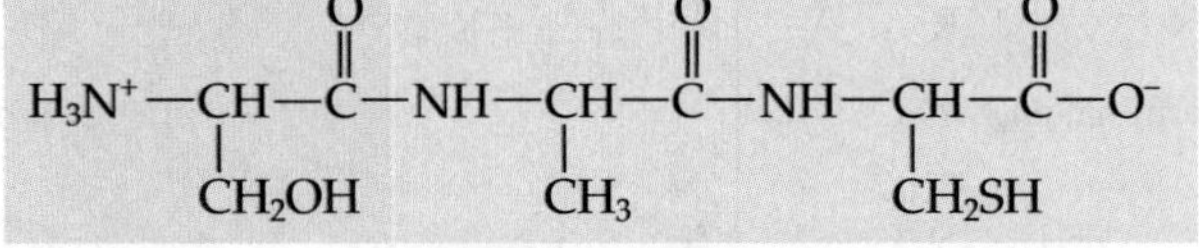

Serylalanylcysteine
(a tripeptide)

In describing peptides and proteins, scientists find it simpler to identify the amino acids in a chain by using the abbreviations given in Table 16.4. The three-letter abbreviations are more common, but the single-letter ones are now used quite frequently. Thus, glycylphenylalanine can be abbreviated as either Gly-Phe or G-F, and serylalanylcysteine as either Ser-Ala-Cys or S-A-C. Example 16.2 further illustrates peptide names and their abbreviations.

4 Is it true that there are more than just 20 amino acids?

Besides the 20 common amino acids, which are found in all living systems, two other rare ones have been found. They are selenocysteine and pyrrolysine. Selenocysteine, a selenium-containing analog of cysteine, is found in selenoproteins in all kingdoms of life. Pyrrolysine is found in methanogenic archaea and other bacteria, and it seems to be part of the enzyme that allows them to produce methane.

EXAMPLE 16.2 Names of Peptides

Give the **(a)** designation using one-letter abbreviations and **(b)** full name of the pentapeptide Met-Gly-Phe-Ala-Cys. You may use Table 16.4.

Solution

a. M-G-F-A-C

b. The endings of the names for all except the last amino acid are changed from *-ine* to *-yl*. The name is therefore methionylglycylphenylalanylalanylcysteine.

> **ExercisE 16.2A**

Give the **(a)** designation using one-letter abbreviations and **(b)** full name of the tetrapeptide His-Pro-Asn-Ala.

> **Exercise 16.2B**

Use **(a)** three-letter abbreviations and **(b)** one-letter abbreviations to identify all the amino acids in the peptide threonylglycylalanylalanylleucine.

A molecule with more than ten amino acid units is often simply called a **polypeptide**. Some protein molecules are enormous, with molecular weights in the tens of thousands. The molecular formula for hemoglobin, the oxygen-carrying protein in red blood cells, is $C_{3032}H_{4816}O_{780}N_{780}S_8Fe_4$, corresponding to a molar mass of 64,450 g/mol. Although they are huge compared with ordinary molecules, a billion average-sized protein molecules could still fit on the head of a pin. When the molecular weight of a polypeptide exceeds about 10,000, it is called a **protein**. These size distinctions are arbitrary and not always precisely applied.

The Sequence of Amino Acids

For peptides and proteins to function properly, it is not enough that they incorporate certain *amounts* of specific amino acids. The order, or *sequence*, in which the amino acids are connected is also of critical importance. The sequence is written starting with the free amino group ($-NH_3^+$), called the *N-terminal end*, on the left and continuing to the free carboxyl group ($-COO^-$), called the *C-terminal end*, on the right.

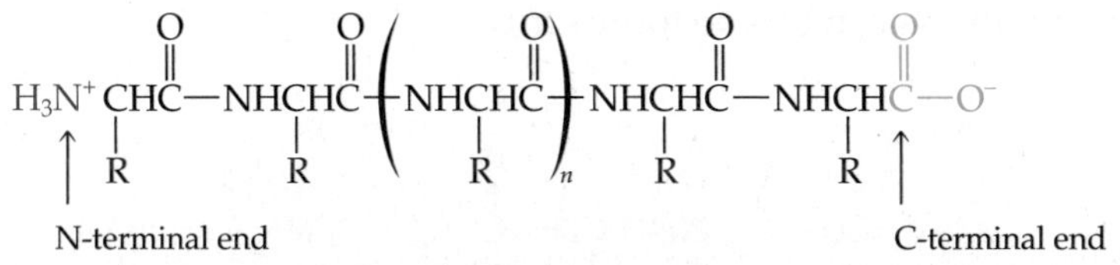

$$H_3N^+ \text{ CHC—NHCHC} \left(\text{NHCHC} \right)_n \text{NHCHC—NHCH}\overset{O}{\overset{\|}{C}}—O^-$$

N-terminal end   C-terminal end

Glycylalanine (Gly-Ala) is therefore different from alanylglycine (Ala-Gly). Although the difference seems minor, the structures are different, and the two substances behave differently in the body.

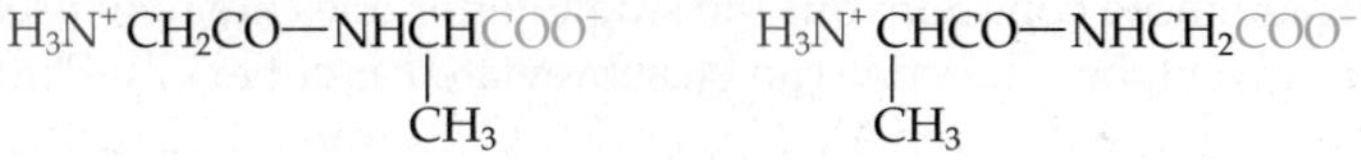

$$H_3N^+ \text{ CH}_2\text{CO—NHCHCOO}^- \qquad H_3N^+ \text{ CHCO—NHCH}_2\text{COO}^-$$
$$\underset{\text{CH}_3}{|} \qquad\qquad\qquad\qquad \underset{\text{CH}_3}{|}$$

Glycylalanine   Alanylglycine

As the length of a peptide chain increases, the number of possible sequence variations becomes enormous. Just as millions of different words can be made from the 26 letters of the English alphabet, millions of different proteins can be made from the 20 amino acids. And just as we can write gibberish with the English alphabet, nonfunctioning proteins can be formed by putting together the wrong sequence of amino acids.

 EXAMPLE 16.3 Numbers of Peptides

How many different tripeptides can be made from one unit each of the three amino acids methionine, valine, and phenylalanine? Write the sequences of these tripeptides, using the three-letter abbreviations in Table 16.4.

Solution
We write the various possibilities. Two begin with Met, two start with Val, and two with Phe:

Met-Val-Phe	Val-Met-Phe	Phe-Val-Met
Met-Phe-Val	Val-Phe-Met	Phe-Met-Val

There are six possible tripeptides.

❯ Exercise 16.3A
How many different tripeptides can be made from two methionine units and one phenylalanine unit?

❯ Exercise 16.3B
How many different tetrapeptides can be made from one unit each of the four amino acids methionine, valine, tyrosine, and glycine?

Although the correct sequence is ordinarily of utmost importance, it is not always absolutely required. Just as you can sometimes make sense of incorrectly spelled English words, a protein with a small percentage of "incorrect" amino acids may function, just not as well. And sometimes a seemingly minor difference can have a disastrous effect. Some people have hemoglobin with one incorrect amino-acid unit out of about 300. That "minor" error is responsible for sickle cell anemia, an inherited condition that often proves fatal.

SELF-ASSESSMENT Questions

1. The 20 common amino acids that make up proteins differ mainly in their
   a. alpha carbon atoms
   b. amino groups
   c. carboxyl groups
   d. side chains

2. Which of the following is *not* an alpha amino acid?
   a. $CH_3C(CH_3)(NH_2)COOH$
   b. $(CH_3)_2CHCH(NH_2)COOH$
   c. H_2NCH_2COOH
   d. $H_2NCH_2CH_2CH_2COOH$

3. In a dipeptide, two amino acids are joined through a(n)
   a. amide linkage
   b. dipolar interaction
   c. ionic bond
   d. hydrogen bond

4. The N-terminal amino acid and the C-terminal amino acid of the peptide Met-Ile-Val-Glu-Cys-Tyr-Gln-Trp-Ile are, respectively,
   a. Ile and Met
   b. Met and Asp
   c. Met and Ile
   d. Trp and Asp

5. The tripeptides represented as Ala-Val-Lys and Lys-Val-Ala are
   a. different only in name
   b. different in molar mass
   c. exactly the same
   d. two different peptides

6. Which of the following amino acids has a basic side chain?
   a. glutamic acid
   b. glycine
   c. lysine
   d. phenylalanine

7. Which of the following is *not* one of the types of side chains (R groups) on an amino acid?
   a. acidic
   b. basic
   c. nonpolar
   d. salt bridge

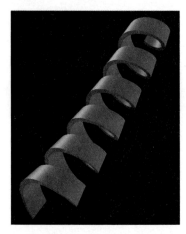

▲ The protein strands in an alpha helix are often represented as coiled ribbons.

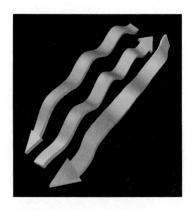

▲ The protein strands in a pleated sheet are often represented as ribbons.

16.7 Structure and Function of Proteins

Learning Objectives • Describe the four levels of protein structure, and give an example of each. • Describe how enzymes work as catalysts.

The structures of proteins have four organizational levels:

- **Primary structure.** Amino acids are linked by peptide bonds to form polypeptide chains. The primary structure of a protein molecule is simply the order of its amino acids. By convention, this sequence is written from the amino (N-terminal) to the carboxyl (C-terminal) end.
- **Secondary structure.** Polypeptide chains can fold into regular structures such as the alpha helix and the beta pleated sheet, which will be described in this section.
- **Tertiary structure.** Protein folding creates spatial relationships between amino acid units that are relatively far apart in the same protein chain.
- **Quaternary structure.** Two or more polypeptide chains can assemble into multiunit structures.

To specify the primary structure, we write out the sequence of amino acids. For even a small protein molecule, this sequence can be quite long. For example, it takes about two pages, even using the one-letter abbreviations, to give the sequence of the 1927 amino acids in lactase, the enzyme that catalyzes the hydrolysis of the disaccharide lactose into galactose and glucose. The primary structure of angiotensin II, a peptide that causes powerful constriction of blood vessels and is produced in the kidneys, is

Asp-Arg-Val-Tyr-Ile-His-Pro-Phe

This sequence specifies an octapeptide with aspartic acid at the N-terminal end, followed by arginine, valine, tyrosine, isoleucine, histidine, proline, and then phenylalanine at the C-terminal end.

The secondary structure of a protein refers to the arrangement of chains about an axis. The two main types of secondary structure are a *pleated sheet*, as in silk, and a *helix*, as in wool. In both cases, the structures are stabilized by hydrogen bonds between carbonyl groups (C=O) and N—H groups which are part of the *backbone* of the protein. (The protein backbone consists of the peptide bonds and the alpha carbons for all amino acids in the protein chain.)

In the pleated-sheet conformation (Figure 16.19), protein chains are arranged in an extended zigzag arrangement, with hydrogen bonds holding adjacent chains together. The appearance gives this type of secondary structure its name, the **beta (β) pleated sheet**. This structure, with its multitude of hydrogen bonds, makes silk strong and flexible.

▶ **Figure 16.19** Beta pleated–sheet structure consisting of protein chains. (a) Ball-and-stick model. (b) Model emphasizing the pleats. The side chains extend above or below the sheet and alternate along the chain. The protein chains are held together by interchain hydrogen bonds.

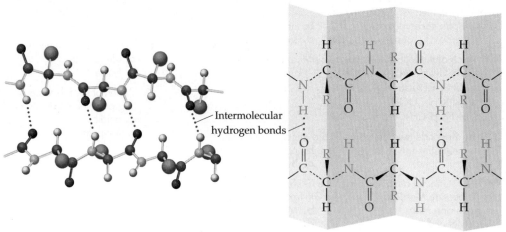

Intermolecular hydrogen bonds

(a)   (b)

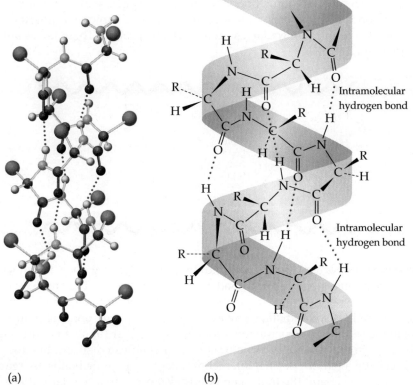

(a)　　　　　　　　　(b)

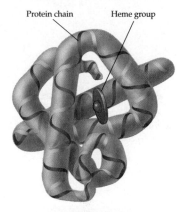

◀ **Figure 16.20** Two representations of the α-helical conformation of a protein chain. (a) Intrachain hydrogen bonding between turns of the helix is shown in the ball-and-stick model. (b) The skeletal representation better shows the helix.

▲ **Figure 16.21** The tertiary structure of myoglobin. The protein chain is folded to form a globular structure, much as a string can be wound into a ball. The disk shape represents the heme group, which binds the oxygen carried by myoglobin (or hemoglobin).

The protein molecules in wool, hair, and muscle contain large segments arranged in the form of a right-handed helix or **alpha (α) helix** (Figure 16.20). Each turn of the helix requires 3.6 amino-acid units. The N—H group at one turn forms a hydrogen bond to a carbonyl (C=O) group from another turn. These helices can wrap around one another in threes or sevens like strands of a rope. Unlike silk, wool can be stretched, much like stretching a spring by pulling the coils apart.

The tertiary structure of a protein refers to folding that affects the spatial relationships between amino-acid units that are relatively far apart in the chain. An example is the protein chain in a **globular protein**. Figure 16.21 shows the tertiary structure of myoglobin, which is folded into a compact, spherical shape.

A quaternary structure exists only if there are two or more polypeptide chains, which can then form an aggregate of subunits. Hemoglobin is the most familiar example. A single hemoglobin molecule contains four polypeptide units—each of which is roughly comparable to a myoglobin molecule—that are arranged in a specific pattern (Figure 16.22) and that form the quaternary structure of hemoglobin.

Four Ways to Link Protein Chains

As noted earlier, the primary structure of a protein is the order of the amino acids, which are held together by peptide bonds. What forces determine the secondary, tertiary, and quaternary structures? Four kinds of forces operate between protein

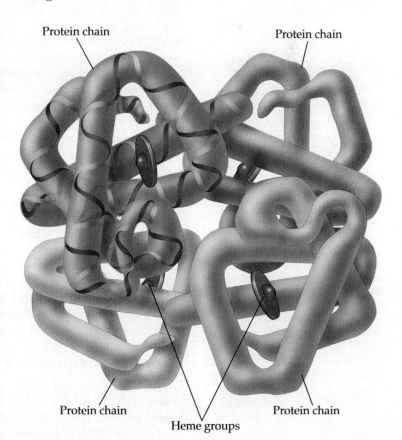

▲ **Figure 16.22** The quaternary structure of hemoglobin has four coiled protein chains, each analogous to myoglobin (Figure 16.21), grouped in a nearly tetrahedral arrangement.

chains: hydrogen bonds, ionic bonds, disulfide linkages, and dispersion forces. Hydrogen bonds and dispersion forces, introduced in Chapter 6 as intermolecular forces, can also be intramolecular, as occurs *within* a protein. These forces are illustrated in Figure 16.23.

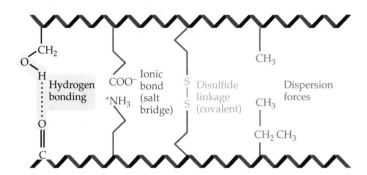

▶ **Figure 16.23** The tertiary structure of proteins is maintained by four different types of forces.

The most important hydrogen bonding in a protein involves an interaction between the atoms of one peptide bond and those of another. The amide hydrogen $(N\!-\!H)$ of one peptide bond can form a hydrogen bond to a carbonyl $(C\!=\!O)$ oxygen located (a) some distance away on the same chain, an arrangement that occurs in the secondary structure of wool (alpha helix), or (b) on an entirely different chain, as in the secondary structure of silk (beta pleated sheet). In either case, there is usually a pattern of such interactions. Because peptide bonds are regularly spaced along the chain, the hydrogen bonds also occur in a regular pattern, both intramolecularly (in wool) and intermolecularly (in silk). Hydrogen bonds can also form between side chains of amino acids.

Ionic bonds (also called salt bridges) occur when an amino acid with a basic side chain is located near one with an acidic side chain (Section 4.3). Under physiological conditions, these have opposite charges, which attract one another. These interactions can occur between groups that ordinarily are relatively distant but that happen to come in contact because of folding or coiling of a single chain. They can also occur between groups on adjacent chains.

A **disulfide linkage** is formed when the —SH groups on two cysteine units (either on the same chain or on two different chains) are oxidized. A disulfide linkage $(-S\!-\!S-)$ is a covalent bond and thus is much stronger than a hydrogen bond. Although far less numerous than hydrogen bonds, disulfide linkages are critically important in determining the shape of some proteins (for example, many of the proteins that act as enzymes) and the strength of others (such as the fibrous proteins in connective tissue and hair).

Dispersion forces are the only kind of forces that exists between nonpolar side chains. Recall from Section 6.3 that dispersion forces are relatively weak. These interactions can be important, however, when there are a large number of them or when other types of forces are missing or are minimized. Dispersion forces are increased by the cohesiveness of the water molecules surrounding the protein. Nonpolar side chains minimize their exposure to water by clustering together on the inside folds of the protein in close contact with one another (Figure 16.24). Dispersion forces become fairly significant in structures such as that of silk, in which a high proportion of amino acids in the protein has nonpolar side chains.

▲ **Figure 16.24** A protein chain often folds with its nonpolar groups on the inside, where they are held together by dispersion forces. The outside has polar groups, visible in this space-filling model of a myoglobin molecule as oxygen atoms (in red) and nitrogen atoms (in blue). The carbon atoms (in black) are mainly on the inside.

Enzymes: Exquisite Precision Machines

Enzymes are biological catalysts produced by cells. They have enormous catalytic power and are usually highly specific, catalyzing only one reaction or a closely related group of reactions. Nearly all known enzymes are proteins.

Enzymes enable reactions to occur at much higher rates and lower temperatures than they otherwise would. They do this by changing the reaction path. The reacting substance, called the **substrate**, attaches to an area on the enzyme called the **active site**, to form an enzyme–substrate complex. Biochemists often use the model pictured in Figure 16.25, called the *induced-fit* model, to explain enzyme action. In this model, the shapes of the substrate and the active site are not perfectly complementary, but the active site adapts to fit the substrate, much as a glove molds to fit the hand that is inserted into it. The enzyme–substrate complex reacts to change the substrate into the products, which are then released from the complex, and the enzyme is regenerated. We can represent the process as follows.

Enzyme + Substrate $\longrightarrow$ Enzyme − substrate complex $\rightleftharpoons$ Enzyme + Products

An enzyme–substrate complex is held together by electrical attraction. Charged groups on the enzyme complement either charged or partially charged groups on the substrate. The formation of new bonds between enzyme and substrate weakens bonds within the substrate, and these weakened bonds can then be more easily broken to form the products.

Portions of the enzyme molecule other than the active site may be involved in the catalytic process. Some interactions at a position remote from the active site can change the shape of the enzyme and thus change its effectiveness as a catalyst. In this way, it is possible to slow or stop the catalytic action. Figure 16.26 illustrates a model for the inhibition of enzyme catalysis. An inhibitor molecule attaches to the enzyme at a position (called the *allosteric site*) away from the active site where the substrate is bound. The enzyme changes shape as it accommodates the inhibitor, and the substrate is no longer able to bind to the enzyme. This is one of the mechanisms by which cells "turn off" enzymes when their work is done. A similar type of mechanism allows an inactive enzyme to be "turned on" and to begin catalyzing a reaction.

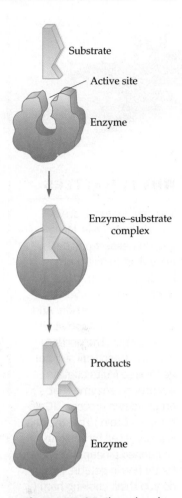

▲ **Figure 16.25** The induced-fit model of enzyme action. In this case, a single substrate molecule is broken into two product molecules.

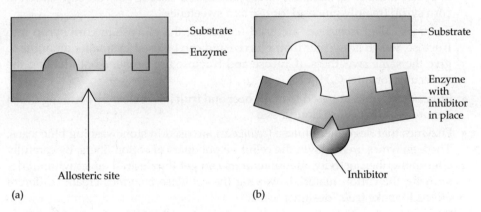

(a)   (b)

▲ **Figure 16.26** A model for the inhibition of an enzyme. (a) The substrate fits into the active site of the enzyme. (b) With the inhibitor bound to the allosteric site, the active site of the enzyme is distorted, and the enzyme cannot bind to its substrate.

Some enzymes consist entirely of protein chains. In others, another chemical component, called a **cofactor**, is necessary for proper function of the enzyme. This cofactor may be a metal ion such as zinc (Zn^{2+}), manganese (Mn^{2+}), magnesium (Mg^{2+}), iron(II) (Fe^{2+}), or copper(II) (Cu^{2+}). An organic cofactor is called a **coenzyme**. By definition, coenzymes are nonprotein. The pure protein part of an enzyme is called the **apoenzyme**. Both the coenzyme and the apoenzyme must be present for enzymatic activity to take place.

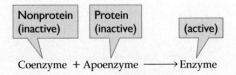

Coenzyme   + Apoenzyme $\longrightarrow$ Enzyme

Many coenzymes are vitamins or are derived from vitamin molecules. Enzymes are essential to the function of every living cell.

WHY IT MATTERS

You can demonstrate protein hydrolysis. Mix a gelatin-dessert powder according to directions. Divide into two parts. Add fresh pineapple, kiwi, or papaya to one portion. Add cooked fruit or none at all to the other. The portion with the fresh fruit will not gel! These fruits contain a protease enzyme that breaks down proteins. (Gelatin is a protein.) The enzyme is inactivated by cooking, so canned pineapple is perfectly fine in gelatin. (Why do you think cooking ham with pineapple tenderizes the ham, and eating too much raw pineapple can "burn" the lips?)

5 Are enzymes different from proteins?

Enzymes are biological catalysts. By far the greatest number of enzymes are also proteins. However, there are a very few instances when a ribozyme, which is made of RNA, can also act to catalyze a reaction. These reactions include aiding in peptide bond formation in ribosomes and making or breaking bonds in DNA and RNA.

Applications of Enzymes

Enzymes find many uses in medicine, industry, and everyday life. Enzymes have replaced the traditional synthetic route for making many chemicals. Enzymes are more efficient and specific catalysts that produce more of the desired product and less waste.

One widespread clinical use is in the test strips used by diabetics. Each strip contains *glucose oxidase*, an enzyme that catalyzes the oxidation of glucose at an electrode. The reaction generates an electrical current that is proportional to the amount of glucose in the blood that has reacted with the enzyme. The result is displayed directly on an external meter.

Clinical analysis for enzymes in body fluids or tissues is a common diagnostic technique in medicine. For example, enzymes normally found only in the liver may leak into the bloodstream from a diseased or damaged liver. The presence of these enzymes in blood confirms liver damage. Enzymes are also used to break up blood clots after a heart attack. If these medicines are given immediately after an attack, the patients' survival rate increases significantly. Other enzymes are used to do just the opposite: Blood-clotting factors are used to treat hemophilia, a disease in which the blood fails to clot normally.

Enzymes in intact microorganisms have long been used to make bread, beer, wine, yogurt, and cheese. More recently the use of enzymes has been broadened to include such things as the following:

- Enzymes that act on proteins (called *proteases*) are used in the manufacture of baby foods to predigest complex proteins and make them easier for infants to digest.
- Enzymes that act on starches (*carbohydrases*) are used to convert corn starch to corn syrup (mainly glucose) for use as a sweetener.
- Enzymes called *isomerases* are used to convert the glucose in corn syrup into fructose, which is sweeter than glucose and can be used in smaller amounts to give the same sweetness. (Glucose and fructose are isomers, both having the molecular formula $C_6H_{12}O_6$.)
- Protease enzymes are used to make beer and fruit juices clear by breaking down the proteins that cause cloudiness.
- Enzymes that degrade cellulose (*cellulases*) are used in stonewashing blue jeans. These enzymes break down the cellulose polymers of cotton fibers. By carefully controlling their activity, manufacturers can get the desired effect without destroying the cotton material. Varying the cellulase enzymes creates different effects to make true "designer jeans."
- Proteases and carbohydrases are added to animal feed to make nutrients more easily absorbed and improve the animals' digestion.

Enzymes are important components in some detergents. Enzymes called *lipases* attack the fats and grease that make up many stains. Proteases attack the proteins in certain types of stains, such as those from blood, meat juice, and dairy products, breaking them down into smaller, water-soluble molecules. Total global production of enzymes is more than $5 billion per year.

Microorganisms that contain enzymes that can degrade pollutants such as polychlorinated biphenyls (PCBs) have been isolated and thus have the potential to destroy these toxic pollutants. However, much research is still needed to make them cost-effective.

SELF-ASSESSMENT Questions

1. Hydrogen bonds form between the N—H group of one amino acid and the C=O group of another, giving a protein molecule the form of an alpha helix or a beta pleated sheet. What level of protein structure do those forms represent?
   a. primary **b.** quaternary **c.** secondary **d.** tertiary

2. Which level of protein structure is maintained in part by dispersion forces?
   a. primary structure **b.** quaternary structure
   c. secondary structure **d.** tertiary structure

3. A metal ion which much be added to an apoenzyme to make it into an active enzyme is a(n)
   a. activator   **b.** allosteric
   c. coenzyme   **d.** cofactor

4. An acidic side chain of an amino acid unit on a protein chain reacts with the basic side chain of an amino acid unit on another chain or at some distance away on the same chain to form a
   a. dispersion linkage   **b.** disulfide linkage
   c. hydrogen bond   **d.** salt bridge

5. The molecule on which an enzyme acts is called the enzyme's
   a. catalyst   **b.** coenzyme
   c. cofactor   **d.** substrate

6. According to the induced-fit model of enzyme action, the
   a. action of an enzyme is enhanced by an inhibitor
   b. active site is the same as the allosteric site

   c. shapes of active sites are rigid and only fit substrates with exact complementary shapes
   d. shape of the active site can change somewhat to fit a substrate

7. An inorganic metal ion working with an enzyme molecule is called a(n)
   a. cofactor **b.** apoenzyme **c.** coenzyme **d.** stimulator

8. An enzyme is a catalyst, or substance that
   a. binds to other molecules
   b. breaks down other molecules
   c. is made from smaller molecules
   d. increases reaction rate

9. Chemical reactions in the body require _____ in order to react rapidly and easily enough to sustain life.
   a. enzymes   **b.** inhibiters
   c. nucleotides **d.** saturated solutions

Answers: 1, c; 2, d; 3, d; 4, d; 5, d; 6, d; 7, a; 8, d; 9, a

16.8 Proteins in the Diet

Learning Objective • List the essential amino acids, and explain why we need proteins.

As we saw earlier, proteins are copolymers of amino acids. Genes carry the blueprints for specific proteins, and each protein serves a particular purpose. We require protein in our diet to provide the amino acids needed to make muscles, hair, enzymes, and many other cellular components vital to life. Proteins are mainly found in meats, poultry, fish, and eggs. Smaller amounts are found in all types of grains, nuts, and beans. Fruits and vegetables contain little protein.

Protein Metabolism: Essential Amino Acids

Proteins are broken down in the digestive tract into their component amino acids.

$$\text{Proteins} + n\ H_2O \xrightarrow{\text{Proteases}} \text{Amino acids}$$

From these amino acids, our bodies synthesize proteins for growth and repair of tissues. When a diet contains more protein than is needed for the body's growth and repair, the excess protein is used as a source of energy, providing about 4 kcal per gram.

The adult human body can synthesize all but nine of the amino acids needed for making proteins. These nine—isoleucine, lysine, phenylalanine, tryptophan, leucine, methionine, threonine, arginine, and valine (see Table 16.4)—are called **essential amino acids** and must be included in our diet. Each of the essential amino acids is a **limiting reactant** in protein synthesis. When the body is deficient in one of them, it can't make proper proteins.

An *adequate* (or *complete*) *protein* supplies all the essential amino acids in the quantities needed for the growth and repair of body tissues. Most proteins from animal sources contain all the essential amino acids in adequate amounts. Lean meat, milk, fish, eggs, and cheese supply adequate protein. Gelatin, a component of jellied candies and desserts, is one of the few animal proteins that does not. It contains almost no tryptophan and has only small amounts of threonine, methionine, and isoleucine.

In contrast, most plant proteins are deficient in one or more amino acids. Corn has insufficient lysine and tryptophan, and people who subsist chiefly on corn may suffer from malnutrition, even though they get adequate calories. Even soy protein, one of the best nonanimal proteins, is deficient in the essential amino acid methionine. Quinoa, consumed mainly in South America but becoming more popular in the United States, is the only grain that contains complete protein. Quinoa has no gluten and is an alternative to wheat for those who require a gluten-free diet.

Our daily requirement for protein is about 0.8 g per kilogram of body weight. Nutrition is especially important during a child's early years. This is readily apparent from the fact that the human brain reaches nearly full size by the age of two.

Early protein deficiency leads to both physical and mental disabilities. In the United States, protein deficiency occurs mainly among elderly persons in nursing homes.

SELF-ASSESSMENT Questions

1. Gram for gram, protein foods provide the same number of kilocalories as
   a. fats **b.** vitamins **c.** fatty acids **d.** carbohydrates

2. Proteins are copolymers of
   a. amino acids **b.** fatty acids
   c. monosaccharides **d.** peptides

3. An essential amino acid is one that
   a. has more calories than other amino acids
   b. is needed in greater quantities than some others
   c. is synthesized in the body
   d. must be included in the diet

4. The daily protein requirement for a 65 kg female is
   a. 0.8 g **b.** 52 g **c.** 65 g **d.** 100 g

16.9 Nucleic Acids: Structure and Function

Learning Objectives • Describe the two types of nucleic acids and the components from which they are made. • Explain complementary base pairing, and describe how a copy of DNA is synthesized.

Life on Earth has a fantastic range of forms, but all life arises from the same molecular ingredients: five nucleotides that serve as the building blocks for nucleic acids, and twenty common amino acids (Section 16.6) that are the building blocks for proteins. These components limit the chemical reactions that can occur in cells and thus determine what Earth's living things are like.

Nucleic acids serve as the information and control centers of the cell. There are two kinds of nucleic acids. **Deoxyribonucleic acid (DNA)** provides a mechanism for heredity and serves as the blueprint for all the proteins of an organism. **Ribonucleic acid (RNA)** directs protein assembly. DNA is a coiled threadlike molecule found mainly in the cell nucleus. RNA is found in all parts of the cell, where different forms do different jobs. Both DNA and RNA are chains of repeating units called **nucleotides**. Each nucleotide in turn consists of three parts (Figure 16.27): a pentose

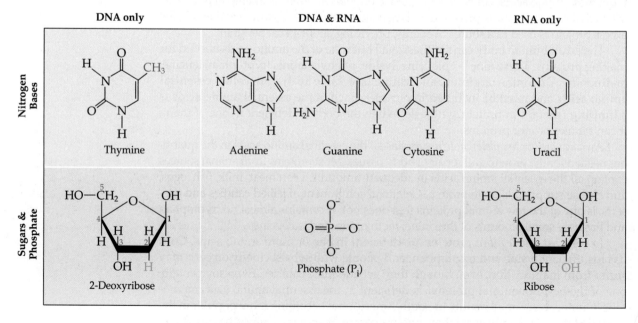

▲ **Figure 16.27** The components of nucleic acids. The sugars are 2-deoxyribose (in DNA) and ribose (in RNA). Note that deoxyribose differs from ribose in that it lacks an oxygen atom on the second carbon atom. *Deoxy* indicates that an oxygen atom is "missing." Phosphate units are often abbreviated as P_i ("inorganic phosphate"). Heterocyclic bases in nucleic acids are adenine, guanine, and cytosine in both DNA and RNA. Thymine occurs only in DNA, and uracil only in RNA. Note that thymine has a methyl group (in red) that is lacking in uracil.

(five-carbon sugar), a phosphate unit, and a heterocyclic amine (Section 9.9) base. The sugar is either ribose (in RNA) or deoxyribose (in DNA). Looking at the sugar in the nucleotide, note that the hydroxyl group on the first carbon atom is replaced by one of five bases. The bases with two fused rings, adenine and guanine, are classified as **purines**. The **pyrimidines**—cytosine, thymine, and uracil—have only one ring.

The hydroxyl group on the fifth carbon of the sugar unit is converted to a phosphate ester group. *Adenosine monophosphate (AMP)* is a representative nucleotide. In AMP, the base (blue) is adenine and the sugar (black) is ribose.

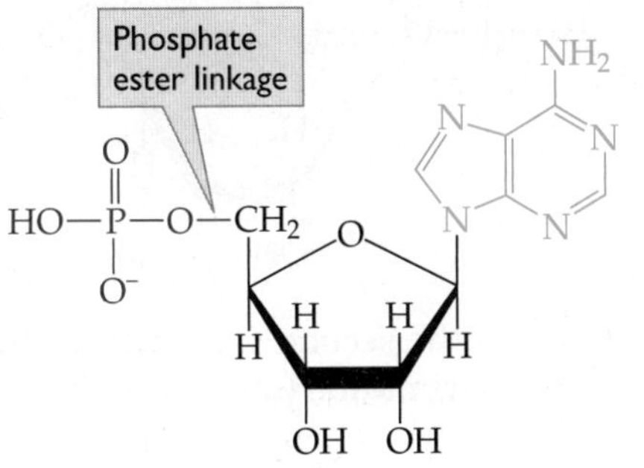

Adenosine monophosphate

Nucleotides can be represented schematically as shown in the margin, where P_i is the biochemists' designation for "inorganic phosphate." A general representation is shown on the left and a specific schematic for AMP on the right.

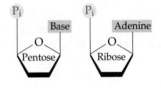

Nucleotides are joined to one another through the phosphate group to form nucleic acid chains. The phosphate linked to the hydroxyl group on carbon five of the sugar in one nucleotide forms a second ester linkage with the hydroxyl group on the third carbon atom of the sugar unit of a second nucleotide. (The prime numbers refer to position on the pentose sugar ring in nucleotides as shown in Figure 16.27. The 3' end has a hydroxyl group and the 5' end has a phosphate group.)

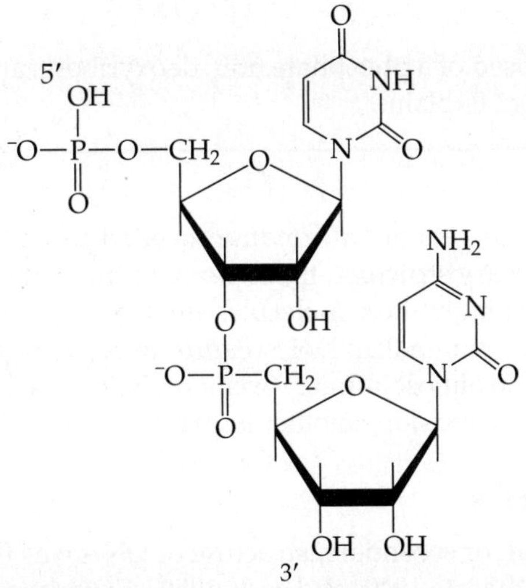

This dinucleotide is in turn joined to another nucleotide, and the process is repeated to build up a long nucleic acid chain. The backbone of the chain consists of alternating phosphate and sugar units, and the heterocyclic bases are branches off this backbone.

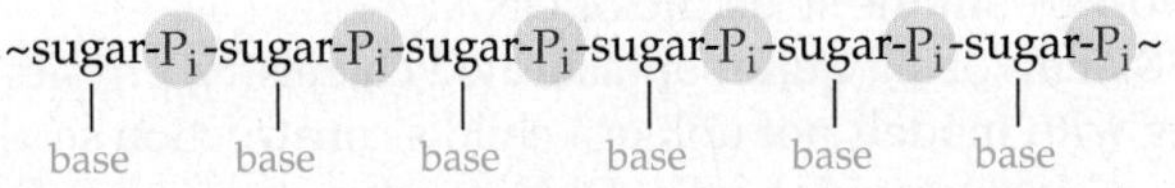

~sugar-P_i-sugar-P_i-sugar-P_i-sugar-P_i-sugar-P_i-sugar-P_i~

base   base   base   base   base   base

If the preceding diagram represents DNA, the sugar is deoxyribose and the bases are adenine, guanine, cytosine, and thymine. In RNA, the sugar is ribose and the bases are adenine, guanine, cytosine, and uracil.

▲ Francis H. C. Crick (1916–2004) (seated) and James D. Watson (b. 1928) used data obtained by British chemist Rosalind Franklin (1920–1958) to propose a double helix model of DNA in 1953. Franklin used a technique called X-ray diffraction, which showed DNA's helical structure. Franklin's colleague Maurice Wilkins shared her work with Watson and Crick. The data helped Watson and Crick decipher DNA's structure. Wilkins, Watson, and Crick were awarded the Nobel Prize for this discovery in 1962, a prize Franklin might have shared had she lived.

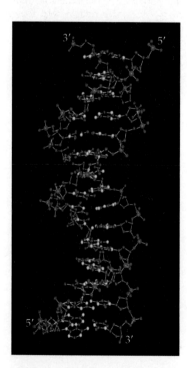

▲ **Figure 16.28** A DNA molecule. Each repeating unit is composed of a sugar, a phosphate unit, and a base. The sugar and phosphate units form backbones on the outside. The bases are attached to the sugar units. Pairs of bases make "steps" on the inside of the spiral staircase.

EXAMPLE 16.4 Nucleotides

Consider the following nucleotide. Identify the sugar and the base. State whether it is found in DNA or RNA.

Solution
The sugar has two H atoms on the second C atom and therefore is deoxyribose, found in DNA. The base is a pyrimidine (one ring). The amino group ($-NH_2$) helps identify it as cytosine.

› Exercise 16.4A
Consider the following nucleotide. Identify the sugar and the base. State whether it is found in DNA or RNA.

› Exercise 16.4B
A nucleotide is composed of a phosphate unit, deoxyribose, and uracil. Is it found in DNA, RNA, or neither? Explain.

The vast amount of genetic information needed to build living organisms is stored in its DNA primary structure, the *sequence* of the four bases along the nucleic acid strand. Not surprisingly, these molecules are huge, with molecular masses ranging into the billions for mammalian DNA (Figure 16.28). Along these chains, the four bases can be arranged in almost infinite variations. Before we look at what those variations mean, we will first consider another important feature of nucleic acid structure.

The Double Helix

The three-dimensional, or secondary, structure of DNA was the subject of an intensive research effort in the late 1940s and early 1950s. Experiments had shown that the molar amount of adenine (A) in DNA corresponds to the molar amount of thymine (T). Similarly, the molar amount of guanine (G) is the same as that of cytosine (C). How the DNA structure maintained this balance was a question that many illustrious scientists considered, but two who were relatively unknown announced in 1953 that they had worked out the structure of DNA.

Using data that involved quite sophisticated chemistry, physics, and mathematics, and working with models not unlike a child's construction set, James D. Watson and Francis H. C. Crick determined that DNA must contain two helices wound about one another to form a double helix (Figure 16.28). The two strands are antiparallel; they run in opposite directions. The phosphate and sugar backbones of the polymer chains form the outside of the structure, much like a spiral staircase. The heterocyclic bases are paired on the inside, with guanine always opposite cytosine and adenine always opposite thymine. The base pairs are the steps of the staircase (Figure 16.29).

Why do the bases pair in this precise pattern, always A to T and T to A, and always G to C and C to G? The answers lies in hydrogen bonding and a truly elegant molecular arrangement. Figure 16.30 shows the two sets of base pairs. You should notice two things. First, a pyrimidine is always paired with a purine in each case, and the distance across each of the pairs is identical (1.085 nm). Other pairings would distort the sugar–phosphate backbone.

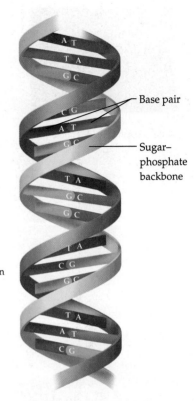

▲ **Figure 16.29** A model of a portion of a DNA double helix is shown in Figure 16.28. This schematic representation of the double helix shows the sugar–phosphate backbones as ribbons and the complementary base pairs as steps on the spiral staircase.

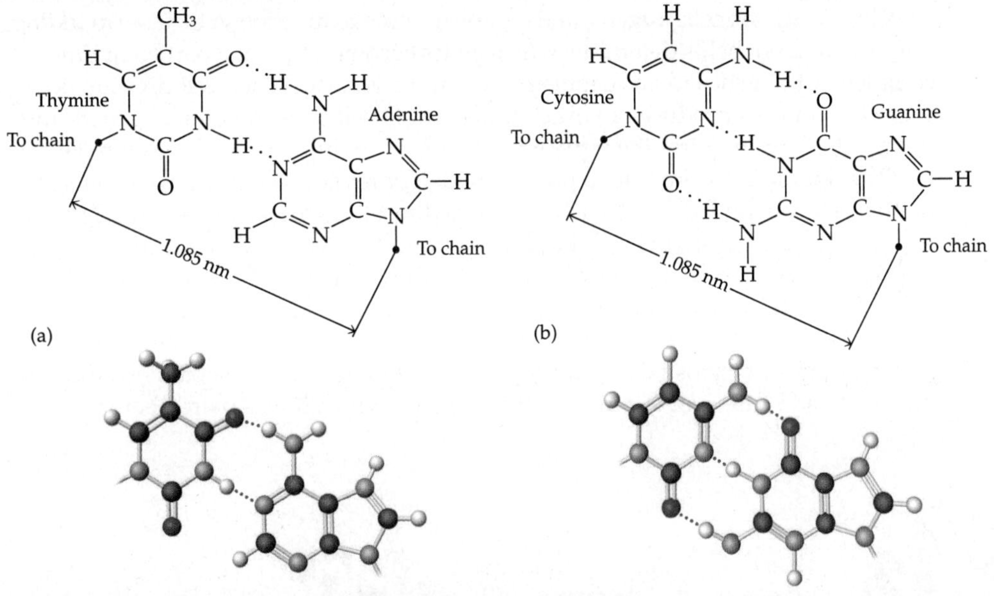

▲ **Figure 16.30** Pairing of the complementary bases (a) thymine and adenine and (b) cytosine and guanine. The pairing involves hydrogen bonding.

Second, notice the hydrogen bonding between the bases in each pair. When an adenine–thymine pair is formed, there are two hydrogen bonds, and when a guanine and a cytosine form a base pair, there are *three* hydrogen bonds between the bases. No other pyrimidine–purine pairing permits such extensive interaction.

Other scientists around the world quickly accepted the Watson–Crick structure because it answered so many crucial questions: how cells are able to divide and go on functioning, how genetic data are passed on to new generations, and even how proteins are built to required specifications. All of these processes depend on the base pairing, as we will see shortly.

Structure of RNA

Most RNA molecules consist of a single strand of nucleic acids. Some internal (intramolecular) base pairing can occur in sections where the molecule folds back on itself. Portions of some RNA molecules exist in double-helical form (Figure 16.31).

DNA: Self-Replication

How is it that each species reproduces its own kind? How does a fertilized egg "know" that it should develop into a kangaroo and not a koala?

The physical basis of heredity has been known for a long time. Higher organisms reproduce sexually. A sperm cell from a male unites with an egg cell from a female. The fertilized egg that results must carry all the information needed to make the various cells, tissues, and organs necessary for the functioning of a new individual. Further, if the species is to survive, information must be passed along in the germ cells—both sperm and egg—for the production of new individuals.

The hereditary material is found in the nuclei of all cells, concentrated in elongated, threadlike bodies called **chromosomes**. Chromosomes form compressed

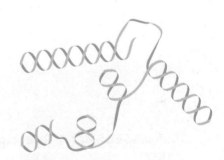

▲ **Figure 16.31** RNA occurs as single strands that can form double-helical portions through internal pairing of bases.

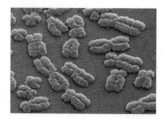

▲ Before cell division, human chromosomes have a distinctive X-shape.

X-shaped structures when strands of DNA coil up tightly just before a cell divides. The number of chromosomes varies with the species. In sexual reproduction, chromosomes come in pairs, with one member of each pair provided by each parent. All humans inherit 23 pairs and have 46 chromosomes in their bodies' cells. Thus, the entire complement of chromosomes is achieved only when the egg's 23 chromosomes combine with a like number from the sperm.

Chromosomes are made of DNA and proteins. Arranged along the chromosomes are the basic units of heredity, the genes. Structurally, a **gene** is a section of the DNA molecule, although some viral genes contain only RNA. Genes control the synthesis of proteins, which tell cells, organs, and organisms how to function in their surroundings. The environment helps determine which genes become active at a particular time. The complete set of genes in an organism is called its *genome*. When cell division occurs, each chromosome produces an exact duplicate of itself. Transmission of genetic information therefore requires the **replication** (copying or duplication) of DNA molecules.

The double helix provides a precise model for replication. If the two chains of the double helix are pulled apart, each chain can direct the synthesis of a new DNA chain using nucleotides from the cellular fluid surrounding the DNA. Each of the separating chains serves as a template, or pattern, for the formation of a new complementary chain. The two DNA strands are *antiparallel*. Synthesis begins with a base on a nucleotide pairing with its complementary base on the DNA strand—adenine with thymine and guanine with cytosine (Figure 16.32). Each base unit in the separated strand can only pick up a base unit identical to the one with which it was paired before.

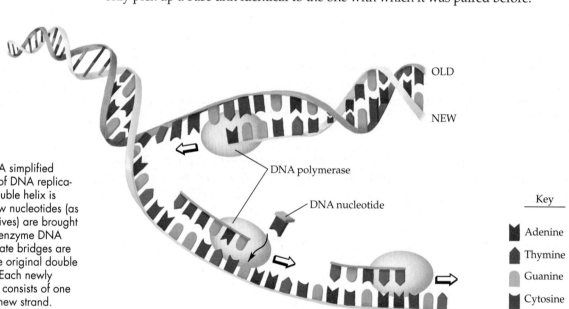

OLD

NEW

DNA polymerase

DNA nucleotide

Key

▮ Adenine

◢ Thymine

▯ Guanine

▮ Cytosine

▶ **Figure 16.32** A simplified schematic diagram of DNA replication. The original double helix is "unzipped," and new nucleotides (as triphosphate derivatives) are brought into position by the enzyme DNA polymerase. Phosphate bridges are formed, restoring the original double helix configuration. Each newly formed double helix consists of one old strand and one new strand.

As the nucleotides align, enzymes connect them to form the new chain. In this way, each strand of the original DNA molecule forms a duplicate of its former partner. All information that was encoded in the original DNA double helix is now contained in each of the replications. When the cell divides, each daughter cell gets one of the DNA molecules and thus has all the information that was available to the parent cell. We will look at how that information is used in the next section.

SELF-ASSESSMENT Questions

1. The monomer units of nucleic acids are
  a. amino acids   **b.** monosaccharides
  c. nucleosomes   **d.** nucleotides

2. A DNA nucleotide could contain
  a. A, P_i, and ribose   **b.** G, P_i, and glucose
  c. T, P_i, and deoxyribose   **d.** U, P_i, and deoxyribose

3. A possible base pair in DNA is
  a. A-G   **b.** A-T   **c.** A-U   **d.** T-G

4. Which of the following bases is a purine?
  a. A   **b.** C   **c.** T   **d.** U

5. Nucleic acid base pairs are joined by
  a. amide bonds   **b.** ester bonds
  c. glycosidic bonds   **d.** hydrogen bonds

6. Replication of a DNA molecule involves breaking hydrogen bonds between the
  a. base pairs
  b. pentose sugars
  c. phosphate units
  d. bases and pentoses

7. The second step of DNA replication involves
  a. bonding of free nucleotides in the correct sequence
  b. formation of a single-stranded RNA molecule

  c. linking of a phosphate group to an amino acid
  d. unzipping of the DNA molecule when the hydrogen bonds break

8. Chromosomes contain
  a. DNA only
  b. DNA and protein
  c. RNA and protein
  d. RNA only

Answers: 1, d; 2, c; 3, b; 4, a; 5, d; 6, a; 7, a; 8, b

16.10 RNA: Protein Synthesis and the Genetic Code

Learning Objectives • Explain how mRNA is synthesized from DNA and how a protein is synthesized from mRNA. • List important characteristics of the genetic code.

DNA carries a message that must somehow be transmitted and then acted on in a cell. Because DNA does not leave the cell nucleus, its information, or "blueprint," must be transported by something else, and it is. In the first step, called **transcription**, a segment of DNA called the *template strand* transfers its information to a special RNA molecule called **messenger RNA (mRNA)**. The base sequence of DNA specifies the base sequence of mRNA. Thymine in DNA calls for adenine in mRNA, cytosine specifies guanine, guanine calls for cytosine, and adenine requires uracil (Table 16.5). However, in RNA molecules, adenine will call for uracil rather than DNA's thymine. Notice the similarity in the structure of these two bases. (See Figure 16.27.) When transcription is completed, the RNA is released, and the DNA helix re-forms. The newly synthesized mRNA now contains the bases that will code for a protein.

Genes have functional regions called *exons* interspersed with inactive portions called *introns*. During protein synthesis, introns are snipped out and not translated. There are some genes, called *housekeeping genes*, that are expressed in all cells at all times and are essential for the most basic cellular functions. However, other genes are expressed only in specific types of cells or at certain stages of development. For example, the genes that encode brain nerve cells are expressed only in brain cells, not in the liver.

The next step in creating a protein involves deciphering the code copied by mRNA, followed by **translation** of that code into a specific protein structure. The decoding occurs when the mRNA travels from the nucleus and attaches itself to a ribosome in the cytoplasm of the cell. Ribosomes contain roughly 65% RNA and 35% proteins.

Another type of RNA molecule, called **transfer RNA (tRNA)**, carries amino acids from the cell fluid to the ribosomes. A tRNA molecule has the looped-clover-leaf structure shown in Figure 16.33. At the head of the molecule is a set of three base units, a *base triplet* called the *anticodon*, that pairs with a set of three complementary bases on mRNA, called the **codon**. The mRNA codon triplet determines which amino acid is carried at the tail of the tRNA. In Figure 16.33, for example, the codon GUA on the segment of mRNA pairs with the anticodon base triplet CAU on the tRNA molecule. All tRNA molecules with the anticodon base triplet CAU always carry the amino acid valine. Once the tRNA has paired with the mRNA codon, its amino acid binds to the growing polypeptide chain, and the now free tRNA returns to the cell fluid to pick up another amino acid molecule. Figure 16.34 provides an overall summary of protein synthesis.

Errors can occur at each step in the replication–transcription–translation process. In replication alone, each time a human cell divides, 4 billion bases are copied to make a new strand of DNA, and there may be up to 2000 errors. Most such errors are corrected, and others are unimportant, but some can result in genetic disease or even death.

Each of the 61 different tRNA molecules carries a specific amino acid into place on a growing peptide chain. The protein chain gradually built up in this way is released from the tRNA as it is formed. A complete dictionary of the genetic code has been compiled (Table 16.6). It shows which amino acids are specified by all the possible mRNA base triplets. There are 64 possible triplets and only 20 amino acids, so there is some redundancy in the code. Three amino acids (serine, arginine,

TABLE 16.5 DNA Bases and Their Complementary RNA Bases

DNA Base	Complementary RNA Base
Adenine (A)	Uracil (U)
Thymine (T)	Adenine (A)
Cytosine (C)	Guanine (G)
Guanine (G)	Cytosine (C)

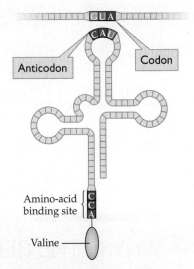

▲ **Figure 16.33** A transfer RNA (tRNA) molecule doubles back on itself, forming a clover-leaf structure with three loops that have intermolecular hydrogen bonding between complementary bases. The anticodon triplet at the head of the molecule base pairs with a complementary codon triplet on mRNA. Here the base triplet CAU in the anticodon specifies the amino acid valine.

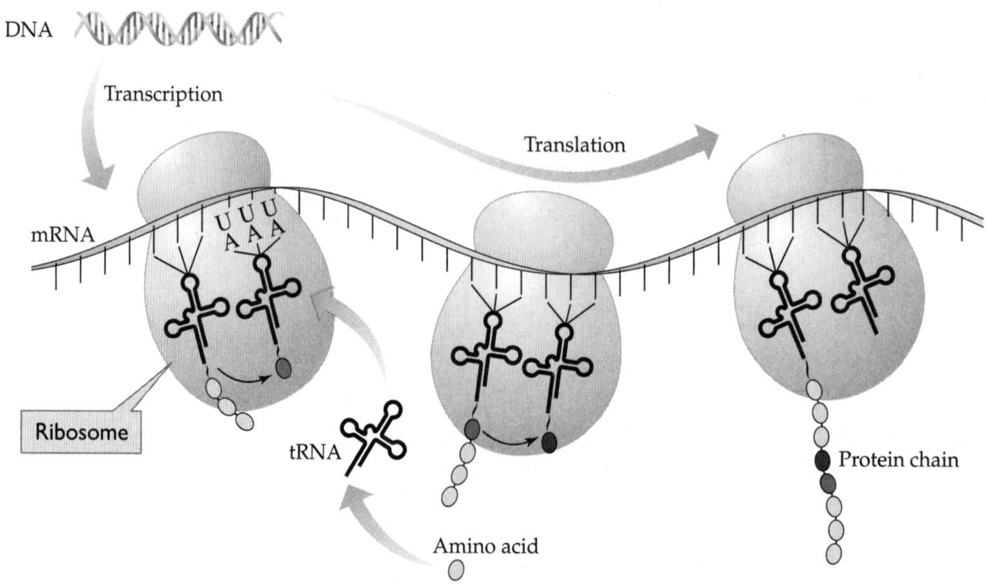

▶ **Figure 16.34** Protein synthesis requires the transcription from DNA to mRNA and then translation on the ribosomes to make a protein. The amino acids coded for in mRNA are carried to the ribosome by tRNA molecules.

TABLE 16.6 The Genetic Code

FIRST BASE	SECOND BASE: U	SECOND BASE: C	SECOND BASE: A	SECOND BASE: G	THIRD BASE
U	UUU=Phe UUC=Phe UUA=Leu UUG=Leu	UCU=Ser UCC=Ser UCA=Ser UCG=Ser	UAU=Tyr UAC=Tyr UAA=Termination UAG=Termination	UGU=Cys UGC=Cys UGA=Termination UGG=Trp	U C A G
C	CUU=Leu CUC=Leu CUA=Leu CUG=Leu	CCU=Pro CCC=Pro CCA=Pro CCG=Pro	CAU=His CAC=His CAA=Gln CAG=Gln	CGU=Arg CGC=Arg CGA=Arg CGG=Arg	U C A G
A	AUU=Ile AUC=Ile AUA=Ile AUG=Met	ACU=Thr ACC=Thr ACA=Thr ACG=Thr	AAU=Asn AAC=Asn AAA=Lys AAG=Lys	AGU=Ser AGC=Ser AGA=Arg AGG=Arg	U C A G
G	GUU=Val GUC=Val GUA=Val GUG=Val	GCU=Ala GCC=Ala GCA=Ala GCG=Ala	GAU=Asp GAC=Asp GAA=Glu GAG=Glu	GGU=Gly GGC=Gly GGA=Gly GGG=Gly	U C A G

and leucine) are each specified by six different codons. Two others (tryptophan and methionine) have only one codon each. Three base triplets on mRNA are "stop" signals that call for termination of the protein chain. The codon AUG signals "start" as well as specifying methionine in the chain.

WHY IS THE GENETIC CODE A TRIPLET CODE?

How is information stored in DNA? The code for the directions for building all the proteins that comprise an organism and enable it to function resides in the *sequence* of bases along the DNA chain. Just as *cat* means one thing in English and *act* means another, the sequence of bases CGT means one amino acid and GCT means another. Although there are only four "letters"—the four bases—in the genetic code, their sequence along the long strands can vary so widely that unlimited information storage is available. Each cell carries in its DNA all the information that determines its hereditary characteristics.

Why is a code that must specify 20 different amino acids based on a triplet of bases? A "doublet" code of four bases would have only 16 different amino acids (4 × 4)—not enough—and a "quadruplet" code would give 256 possibilities—far too many. A triplet code specifies 20 different amino acids with some redundancy (duplication).

Molecular Basis for Slowing or Reversing Aging

People have long been interested in finding a way to slow or even reverse aging. We must first understand the mechanism of aging so that some means can be devised to accomplish this. Scientists in the United States and in France have identified such a mechanism by studying progeria, a rare genetic disease which accelerates aging so that a person dies of "old age" at the average age of only 13 years old. A point mutation responsible for quickened aging in progeric patients was identified in 2003. This mutation, in which cytosine is replaced with thymine, appears in a gene that codes for the nuclear membrane protein LMNA—or lamin A—and leads to a defective protein called progerin.

After this defect was found, scientists at the National Cancer Institute (NCI) showed that much of the cellular damage accompanying normal old age is caused by irregular use of the same defect site in the lamin A gene.

The genetic defect can be masked by using an *antisense agent*—a strand of DNA or RNA, or a nucleic acid analog, that binds tightly to the messenger RNA (mRNA) produced by the gene (Figure 16.35). (This synthesized nucleic acid is termed *antisense* because its base sequence is complementary to the mRNA of the gene—the *sense* sequence.) The antisense molecule physically obstructs the translation machinery, effectively turning the gene off. Analogs with altered backbone linkages are used. This makes

▲ **Figure 16.35** (a) Segment of normal DNA, and (b) antisense analog.

the agent resistant to digestion by nuclease enzymes. Such an agent has been shown to reverse age-related defects in cultured cells taken from progeric patients and in cultured cells taken from normal persons of advanced age.

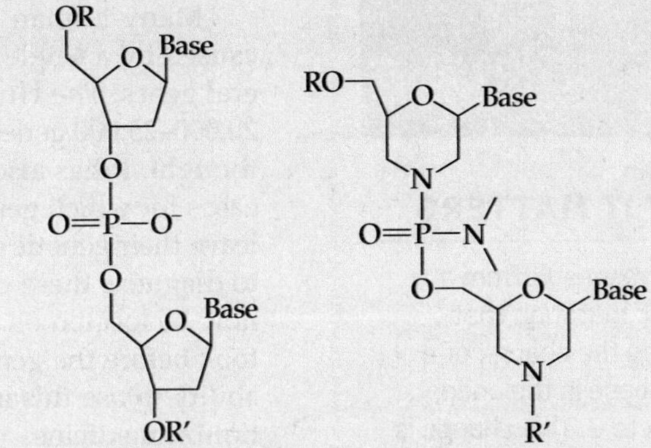

Use of antisense drugs is currently limited by a lack of efficient delivery into defective tissues and cells and by safety aspects such as general immune system activation. If these problems can be overcome, antisense technology may help us to attain the long-sought fountain of youth.

SELF-ASSESSMENT Questions

1. The conversion of mRNA to a sequence of amino acids in a protein is called
   a. transaction **b.** transcription **c.** transition **d.** translation

2. The anticodon base triplet is found on the
   a. original DNA **b.** ribosome **c.** mRNA **d.** tRNA

3. How many amino acids and how many codons are involved in the genetic code?
   a. 20 amino acids, 20 codons **b.** 20 amino acids, 64 codons
   c. 64 amino acids, 20 codons **d.** 64 amino acids, 64 codons

4. Each amino acid except tryptophan and methionine is coded for by more than one codon, indicating that the genetic code has
   a. evolved **b.** redundancy
   c. sequence assurance **d.** translation equability

5. Refer to Table 16.6; the codon UGA
   a. codes for Leu and signals "start"
   b. codes for Leu and signals "stop"
   c. signals "stop" only
   d. codes for Leu only

Answers: 1. d; 2. d; 3. b; 4. b; 5. c

16.11 The Human Genome

Learning Objective • Describe new DNA technologies, and explain how they are used.

In the nearly seven decades since Watson and Crick proposed their structure for DNA, scientists have made huge advances in technology and thus learned much more about DNA and how to alter or use DNA to benefit both individuals and humankind. These include the ability to sequence DNA, recombinant DNA, DNA profiling, and the very initial use of gene therapy. These technologies hold great promise. We will look at some of them in this section.

WHY IT MATTERS

Blue eyes result from a genetic mutation that involved the change of a single base in the code, from A to G. This change is thought to have occurred between 6000 and 10,000 years ago; before that time, no humans had blue eyes. The mutation turned off a gene involved in the production of melanin, the pigment that gives color to hair, eyes, and skin. Only people who inherit this gene from both parents have blue eyes.

In 1990, scientists set out to determine the sequence of all 3 billion base pairs in the human genome. The Human Genome Project was completed in 2003 at a cost of about $2.7 billion. We can now read the complete genetic blueprint for a human. This project has driven a worldwide revolution in biotechnology—and today a genome can be sequenced in a few days at a very low cost.

Many human diseases have clear genetic components. Some are directly caused by a single defective gene, and others arise through involvement of several genes. The Human Genome Project found that there are actually only about 20,000–25,000 genes in the human genome and not 80,000, as had previously been thought. It has also led to the discovery of more than 2000 genes related to diseases for which genetic tests have been developed. These tests enable patients to learn their genetic risk for certain diseases and also help health care professionals to diagnose these diseases. A gene suspected of causing an inherited disease can now be located on the human genome map in a few days, instead of the years it took before the genome sequence was known. Once the genes are identified, the ability to use this information to diagnose and cure genetic diseases will revolutionize medicine.

DNA is sequenced by using enzymes to cleave it into smaller segments of a few to several hundred nucleotides each. These fragments differ in length from each other by a single base that is identified in a later step. The pieces are duplicated and amplified by a technique called the **polymerase chain reaction (PCR)**. PCR employs enzymes called *DNA polymerases* to amplify the small amount of DNA by making millions of copies of each fragment. By repeatedly heating DNA to separate the chains in the double helix, synthesizing complementary strands to form new copies of the DNA, and then cooling the new DNA, large quantities of DNA can be made from a minuscule sample, thus providing enough DNA for sequencing. The DNA fragments can be sorted by length in an electric current, and the base at the end of each piece is identified. Computers are then used to compile the short sequences into longer segments and ultimately into the entire DNA base sequence. There are also other methods for sequencing DNA available, and research into faster and cheaper methods continues.

For many years people have tried to introduce changes into DNA. The methods have included exposing the organism to radiation or to mutagens that would alter DNA. These methods were very much happenstance, very time consuming, and expensive. A revolutionary new technology is *CRISPR*, a gene-editing system that results in very specific alterations to the DNA, even the replacement of a single base in a gene. Researchers design an RNA that binds to the gene that will be edited. The RNA guides an enzyme called *Cas9* to the gene, where the enzyme snips the DNA. Depending on the goal, CRISPR/Cas9 can disable or repair a gene, or paste a new gene into the specified location. Researchers can create a usable CRISPR/Cas9 system in days!

Recombinant DNA: Using Organisms as Chemical Factories

All living organisms (except some viruses) have DNA as their hereditary material. *Recombinant DNA* is DNA that has been created artificially when DNA from two different sources was incorporated into a single *recombinant* DNA molecule. The first step is to cut the DNA with an enzyme obtained from bacteria called an *endonuclease*. An endonuclease is a *restriction enzyme,* which cleaves DNA only at specific short sequences of DNA called *restriction sites*. The enzyme cuts two DNA molecules at the same sequence and produces DNA fragments with "sticky ends" that are very short, single-stranded sequences that are complementary. DNA pieces from both sources can form bonds to each other, thus creating a recombinant DNA. We will consider some examples of how recombinant DNA has been used next.

After scientists have identified the gene that codes for a particular protein, they can isolate it and amplify it by PCR. The gene can then be spliced (recombined) into a special kind of bacterial DNA called a *plasmid*. The recombined plasmid is

DNA Profiling

About 99.9% of DNA sequences are the same in every human, yet the 0.1% difference is enough to distinguish one person from another (except for identical twins). DNA profiling does not require sequencing of an entire genome. Rather, it uses *short tandem repeats*, or *STRs*, at 13 particular locations on the human genome. An STR is typically from two to five base units long and is repeated several times. An example is ACGT-ACGT-ACGT. The number of such repeats at the key points differs from one person to another. This is the key to a DNA fingerprint. When all 13 points are considered, each person has a different profile.

British scientist Alec Jeffreys invented DNA fingerprinting in 1985 and gave it its name. Like fingerprints, each person's DNA is unique. Any cells—skin, blood, semen, saliva, and so on—can supply the necessary DNA sample.

DNA fingerprinting is used to screen newborn babies for inherited disorders, such as cystic fibrosis, sickle cell anemia, Huntington's disease, and hemophilia. Early detection of such disorders enables doctors and parents to choose and initiate proper treatment of the child.

Because children inherit half their DNA from each parent, DNA fingerprinting is also used when there is a question about paternity. The odds of identifying the father through DNA fingerprinting are at least 99.99%, and they are 100% for excluding a man as a possible father.

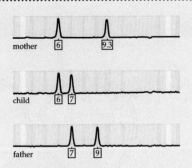

▲ This DNA test included samples from the mother (top), the child (middle), and the alleged father (bottom). The maternal marker 6 has been passed to the child. This means that marker 7 of the child must have come from the father, and the alleged father does indeed have a marker 7.

DNA fingerprinting was a major advance in criminal investigation, allowing thousands of criminal cases to be solved. DNA samples from evidence found at a crime scene are amplified by PCR and compared with DNA obtained from suspects and other people known to have been at the scene.

This technique has led to many criminal convictions, but perhaps more important, it can readily prove someone innocent. If the DNA does not match, the suspect could not have left the biological sample. More than 350 people have been shown to be innocent and freed, some after spending years in prison—including 20 who had been condemned to death. The technique is a major advance in the search for justice.

then inserted into the host organism, the bacteria from which the plasmids came (Figure 16.36). Bacterial cells carrying plasmids with the desired sequence are selected (cloned) and grown. Inside the host bacteria, the plasmids replicate, making multiple exact copies of themselves. As the engineered bacteria multiply, they become effective factories for producing the desired protein.

What kinds of proteins might scientists wish to produce using biotechnology? Insulin, a protein coded by DNA, is required for the proper use of glucose by cells. People with diabetes, an insulin-deficiency disease, formerly had to use insulin extracted from pigs or cattle. This was expensive, and over time many people developed an immune response and were no longer able to use the animal insulin. Human insulin is now made using recombinant DNA technology. Scientists insert the human gene for insulin production into the DNA of *Escherichia coli*, a bacterium commonly found in the human digestive tract, or into yeast. Expression factors, signals that provide instructions for transcription and translation of the gene by the cell, are also added. The recombinant cells multiply rapidly, making billions of copies of themselves, and each new *E. coli* cell carries in its DNA a replica of the gene for human insulin. The economic value of insulin and other recombinant products is huge. Worldwide revenue from insulin alone, for example, is projected to be nearly $40 billion by 2020. The hope for the future is that a functioning gene for insulin can be incorporated directly into the cells of clients with insulin-dependent diabetes.

A number of commercial companies have sprung up in recent years that offer to test your DNA. Some services look at potential health issues and markers for genetic diseases. Others focus primarily on tracing a person's ancestry and compare test results with information already in their databases, which exceeds many million people in some cases. Still others look at both. These companies do not sequence the genome but instead focus on specific areas of DNA that show differences among people.

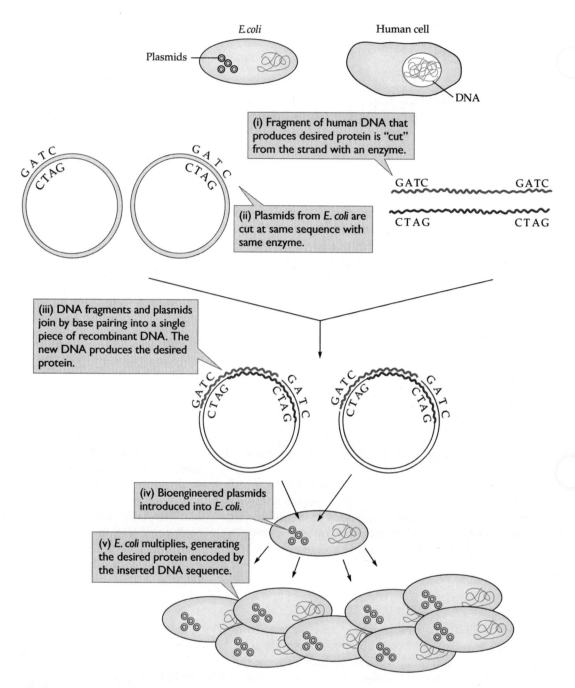

▲ **Figure 16.36** Recombinant DNA: cloning a human gene into a bacterial plasmid. Once the human gene is incorporated into the plasmid and inserted into bacteria such as *E. coli*, the bacteria multiply and make multiple copies of the recombinant gene. This process was initially used to make recombinant insulin, which was first marketed in 1982. Potentially, a gene from any organism—microbe, plant, or animal—can be incorporated into any other organism.

Besides insulin, many other valuable materials that are used in human therapy and are difficult to obtain in any other way can now be made using recombinant DNA technology. Some examples are

- *Human growth hormone* (HGH), which replaced cadaver-harvested HGH for treating children who fail to grow properly
- *Erythropoietin* (EPO) for treating anemia resulting from chronic kidney disease and from treatment for cancer
- *Factor VIII* (formerly harvested from blood), a clotting agent for treating males with hemophilia

- *Tissue plasminogen activator* (TPA) for dissolving blood clots associated with diseases such as heart attack and stroke
- *Interferons*, which assist the immune response and induce the resistance of host cells to viral infection, and are promising anticancer agents
- *Recombinant vaccines*, such as that for hepatitis B

Gene Therapy

In gene therapy, a functioning gene is introduced into a person's cells to correct the action of a defective gene. Viruses are commonly used to carry DNA into cells. Current gene therapy is experimental, and human gene transplants so far have had only modest success.

- In 1999, ten children with severe combined immunodeficiency disease (SCID) were injected with working copies of a gene that helps the immune system develop. Almost all the children improved and most are healthy today, although in 2002 two of them developed leukemia.
- Four severely blind young adults had a curative gene injected into their eyes. They can now see more light and have enough vision to walk without help. Two of them can read several lines of an eye chart.
- Genetically altered cells injected into the brain of patients with Alzheimer's disease appear to nourish ailing neurons and may slow cognitive decline in these people.
- In 2006, researchers at the National Cancer Institute (NCI) successfully reengineered immune cells to attack cancer cells in patients with advanced melanoma, the first successful use of gene therapy to treat cancer in humans. A considerable amount of gene therapy work focuses on developing cancer treatments.
- Researchers have genetically engineered the formation of new beta cells in the livers of mice. They also added a gene that inhibits activity of the T cells around the new beta cells in the liver but not in the rest of the body. (The T cells would otherwise destroy the new beta cells.) If this works in humans, clients with diabetes could produce their own insulin. Many problems remain, including getting the gene in the right place and targeting it so that it inserts itself only into the cells where it is needed. The stakes in gene therapy are huge. We all carry some defective genes. About one in ten people either has developed or will develop an inherited genetic disorder. Molecular genetics has already resulted in some impressive achievements. Its possibilities are mind-boggling—elimination of genetic defects, a cure for cancer, and who knows what else? Knowledge gives power, but it does not necessarily give wisdom. The greatest problem we are likely to face in bioengineering is deciding how this technology will be used.

SELF-ASSESSMENT Questions

1. Recombinant DNA cloning involves
   a. combining DNA from two different organisms to make a single new DNA
   b. creating a genetically identical organism from a skin cell
   c. isolation of a genetically pure colony of cells
   d. splitting a fertilized egg into two cells to generate identical twins

2. In producing recombinant DNA, the DNA from the two sources is cut with a
   a. DNA ligase b. phage c. restriction enzyme d. vector

3. Gene therapy is a process where scientists
   a. create new species of animals
   b. insert a functioning gene in place of a defective one

   c. increase the frequency of mutations
   d. use a DNA profile to identify an organism

4. Some geneticists suggest transferring some genes that direct photosynthesis from an efficient crop plant to a less efficient one, thus producing a more productive plant variety. This project would most likely involve
   a. genetic profiling   b. genetic engineering
   c. genetic screening   d. transcription

5. A technology that can be used to replace a single base or a base sequence in DNA is
   a. CRISPR   b. gene splicing
   c. PCR   d. recombination

Answers: 1, *a*; 2, *c*; 3, *b*; 4, *b*; 5, *a*

GREEN CHEMISTRY Green Chemistry and Biochemistry
David A. Vosburg, *Harvey Mudd College*

Principles 1–10, 12

Learning Objectives ● Name some advantages of using biochemistry to create useful molecules. ● Give examples of the use of biochemistry for energy production and other applications.

Chemistry has gone on constantly for millions of years in plants, animals, and bacteria. These living systems perform many chemical reactions, and plants have evolved to harness energy from the sun using photosynthesis. Living things store energy in the form of carbohydrates (like glucose, sucrose, and cellulose; Section 16.2) and lipids (fats and oils; Section 16.4). These renewable feedstocks (Principle 7) hold great potential for supplying human energy needs in the future.

Chemists are devising ways of converting these natural energy reservoirs into substitutes for nonrenewable petroleum fuels, and biochemists are engineering existing biochemical pathways in various microorganisms to directly produce usable fuels. In 2009 and 2010, three Presidential Green Chemistry Challenge Awards were given for work in this area to Virent Energy Systems, James Liao, and LS9. Virent's process transforms plant carbohydrates into gasoline (see graphic below), which could reduce dependence on fossil fuels. Liao and coworkers (UCLA and Easel Biotechnologies) have engineered bacteria to turn carbohydrates and carbon dioxide into alcohols that can be used as fuels. The LS9 company developed microorganisms that directly produce hydrocarbon fuels. All of these achievements use the products or processes of biochemistry to address the energy needs of tomorrow.

Using biochemistry to make molecules has several distinct advantages over traditional chemical synthesis and invokes many of the green chemistry principles. The reactions normally occur in water, a ubiquitous, nontoxic, and nonflammable solvent (Principle 5). When reaction products are nonpolar, they can often be separated easily from the more polar water. In contrast to many current industrial reactions that are performed at very high temperatures and pressures, most biochemical reactions are extremely rapid and occur at mild temperatures and atmospheric pressure. These reactions reduce both energy costs and potential hazards (Principles 6 and 12). The atom economy of biochemical reactions is very high, as protecting groups and auxiliaries are not necessary (Principles 2 and 8), and toxic reagents and products also can be avoided in most cases (Principles 3 and 4). Biological reactions are frequently catalytic, using a single enzyme (Section 16.7) for thousands or millions of repeated reactions that selectively generate products with very little waste produced (Principles 1 and 9). Furthermore, the products formed (and the biological catalysts themselves) are generally biodegradable and innocuous (Principle 10). When compounds are not themselves biodegradable,

other biochemical pathways may be employed to degrade molecules that otherwise would persist in the environment. This bioremediation of environmental waste by cleaning up toxins and purifying water is a very important advantage of using biochemistry.

Biochemical reactions can reduce waste and simplify processes, in part due to their impressive selectivity. Buckman International found enzymes that can strengthen paper fibers, which reduces the amount of wood pulp used and saves energy, money, and trees. Yi Tang (at UCLA) and Codexis developed an enzymatic route to the cholesterol-lowering drug Zocor that eliminated several synthetic steps and reduced waste. Both of these achievements were recognized with Presidential Green Chemistry Challenge Awards in 2012. Codexis also had received a Presidential Green Chemistry Challenge Award in 2010, with Merck, for their enzymatic production of the diabetes drug Januvia. All of these biochemical innovations benefit companies, customers, the environment, and society at large.

The self-replicating nature of the cell and its DNA and the self-assembling (folding) properties of proteins (Section 16.7) make biological systems ideal candidates for the discovery and implementation of novel chemical processes. One of the greatest strengths of biochemical reactions is also the greatest challenge: specificity. Biological catalysts have evolved to have exquisite selectivity for their natural substrate(s), so adapting these catalysts to have more general functions for a wider range of (potentially unnatural) substrates is difficult. Yet this is precisely what Merck and Codexis did for the synthesis of Januvia, so there is hope that similar approaches could lead to other successes.

▲ Genetic engineering of bacteria may transform natural energy sources into materials that will be energy-rich for human use. *Synechococcus elongatus*, shown here, may be particularly useful because of its small genome and its aptitude for photosynthesis.

 Summary

Section 16.1—Biochemistry is the chemistry of life processes. Life requires chemical energy. **Metabolism** is the set of chemical reactions that keep cells alive, and includes both **anabolism** (building up of molecules) and **catabolism** (breaking down of molecules). Substances that provide animals with energy are carbohydrates, fats, and proteins.

Section 16.2—A **carbohydrate** is a compound whose formula can be written as a "hydrate" of carbon. Sugars, starches, and cellulose are carbohydrates. The simplest sugars are **monosaccharides** and cannot be further hydrolyzed. A monosaccharide is either an **aldose** (with an aldehyde functional group) or a **ketose** (with a ketone functional group). A **disaccharide** can be hydrolyzed to yield two monosaccharide units. **Polysaccharides** contain many monosaccharide units linked together.

Section 16.3—Glucose, also called **blood sugar**, is the sugar used directly by cells for energy. Other sugars include fructose, sucrose (table sugar), and lactose (milk sugar). Sucrose and lactose are broken down during digestion. **Starch** is a polymer of α-glucose, and **cellulose** is a β-glucose polymer. Human beings can hydrolyze starch to glucose but cannot hydrolyze cellulose. The liver and muscles store small amounts of carbohydrates as **glycogen**.

Section 16.4—A **lipid** is a cellular component that is insoluble in water and soluble in solvents of low polarity. A fat is an ester of fatty acids and glycerol (a **triglyceride**). Lipids include solid fats and liquid oils (both are triglycerides) as well as steroids, hormones, and **fatty acids** (long-chain carboxylic acids). Fats may be **saturated** (only single carbon-to-carbon bonds), monounsaturated (one C=C), or **polyunsaturated** (more than one C=C). The iodine number of a fat or oil is the mass in grams of iodine that reacts with 100 g of the fat or oil.

Section 16.5—Dietary fats are triglycerides, which are broken down by enzymes to form fatty acids and glycerol. Fats are stored in **adipose tissue** located in **fat depots** and are hydrolyzed when needed. A **lipoprotein** is a cholesterol–protein combination transported in the blood. Low-density lipoproteins (LDLs) carry cholesterol to the cells for use and increase the risk of heart disease, while high-density lipoproteins (HDLs) carry it to the liver for excretion. Polyunsaturated fats (mostly from plants), *cis* fatty acids, and omega-3 fatty acids tend to lower LDL levels. Saturated fats (mostly from animals) and *trans* fatty acids made by hydrogenation of polyunsaturated fats have been implicated in cardiovascular disease.

Section 16.6—An **amino acid** contains an amino ($-NH_2$) group and a carboxyl ($-COOH$) group. An amino acid exists as a **zwitterion** in which both the carboxyl group and the amino group are charged. Proteins are polymers of amino acids, with 20 amino acids forming millions of different proteins. The bond that joins two amino acids in a protein is called a **peptide bond**. Two amino acids form a dipeptide, three form a tripeptide, and more than ten form a **polypeptide**. When the molecular weight of a polypeptide exceeds about 10,000, it is called a **protein**. The sequence of the amino acids in a protein is of great importance to its function.

Section 16.7—Protein structure has four levels. The **primary structure** of a protein is the sequence of its amino acids. The **secondary structure** is the arrangement of polypeptide chains that are formed by hydrogen bonding between atoms in the protein's backbone. Two such arrangements are the **beta pleated sheet**, in which the arrays of chains form zigzag sheets, and the **alpha helix**, in which protein chains coil around one another. The **tertiary structure** of a protein is its folding pattern; **globular proteins** such as myoglobin are folded into compact shapes. Some proteins have a **quaternary structure** consisting of an aggregate of subunits arranged in a specific pattern. There are four types of forces between amino acid side chains that determine the tertiary and quaternary structures of proteins: (a) hydrogen bonds; (b) salt bridges (ionic bonds) between acidic and basic side chains; (c) **disulfide linkages**, which are S—S bonds formed between cysteine groups; and (d) dispersion forces arising between nonpolar side chains.

Enzymes are biological catalysts. The reacting substance, or **substrate**, attaches to the **active site** of the enzyme to form a complex, which then forms and releases the products. An enzyme is often made up of an **apoenzyme** (a protein) and a **cofactor** that is necessary for proper function, such as a metal ion. An organic cofactor (such as a vitamin) is a **coenzyme**.

Section 16.8—Proteins, polymers of amino acids, are broken down to amino acids during digestion. **Essential amino acids** must be included in the diet. Each essential amino acid is a **limiting reactant** in protein synthesis; a deficiency leads to an inability to make the proper proteins. Most animal proteins have all of the essential amino acids, but plant proteins usually do not provide complete nutrition.

Section 16.9—**Nucleic acids** are the information and control centers of a cell. DNA (deoxyribonucleic acid) and RNA (ribonucleic acid) are polymers of **nucleotides**. Each nucleotide contains a pentose sugar unit, a phosphate unit, and a pyrimidine or purine base. In DNA, the sugar is deoxyribose and the bases are adenine, guanine, cytosine, and thymine. In RNA, the sugar is ribose and the bases are adenine, guanine, cytosine, and uracil. In DNA, the bases thymine and adenine are always paired, as are the bases cytosine and guanine. Pairing occurs via hydrogen bonding between the bases and gives rise to the double helix structure of DNA. RNA consists of single strands of nucleic acids, some parts of which may hydrogen bond to form double helixes. **Chromosomes** are made of DNA and proteins. A **gene** is a section of DNA. Genetic information is stored in the base sequence. Transmission of genetic information requires **replication**, or copying, of the DNA molecules.

Section 16.10—In transcription, a segment of DNA transfers its information to **messenger RNA (mRNA)**. The next step is the **translation** of the code into a specific protein structure, which occurs on the ribosomes. **Transfer RNA (tRNA)** delivers amino acids to the growing protein chain. The **genetic code** is made of base triplets that code for particular amino acids. The tRNA has an anticodon that pairs with a triplet of complementary bases on mRNA, called the **codon**, and carries the correct amino acid to be incorporated into a protein.

Section 16.11—DNA may be amplified by the polymerase chain reaction (PCR). DNA may also be edited using CRISPR technology. In genetic testing, DNA is cleaved and duplicated. The patterns of DNA fragments are compared with those of individuals with genetic diseases to identify and predict the occurrence of such diseases. A similar technique is used to compare DNA from different sources in DNA fingerprinting. In recombinant DNA technology, the base sequence for the gene that codes for a desired protein is determined, amplified and then spliced into a plasmid. The recombined plasmid is inserted into a host organism, which generates the protein as it reproduces.

GREEN CHEMISTRY—Biochemistry has important applications in renewable energy and offers several advantages over traditional chemical synthesis, including the use of water as the solvent and the low levels of waste and energy use. Drugs and fuels are examples of products that may be made using biochemical methods.

Learning Objectives

Learning Objectives	Associated Problems
• List the major parts of a cell, and describe the function of each part. (16.1)	2, 16
• Name the primary source of energy for plants and the three classes of substances that are sources of energy for animals. (16.1)	1, 4, 13–15, 79, 80, 86
• Compare and contrast starch, glycogen, and cellulose. (16.2)	17–22
• Identify dietary carbohydrates, and state their sources and function. (16.3)	23–26, 85
• Describe the fundamental structure of a fatty acid and of a fat. (16.4)	27–30
• Classify fats as saturated, monounsaturated, or polyunsaturated. (16.4)	31, 32
• Identify dietary lipids, and state their function. (16.5)	33–36, 87
• Draw the fundamental structure of an amino acid, and show how amino acids combine to make proteins. (16.6)	5, 6, 37, 39–44, 81
• Describe the four levels of protein structure, and give an example of each. (16.7)	45, 47–51
• Describe how enzymes work as catalysts. (16.7)	46, 52–57
• List the essential amino acids, and explain why we need proteins. (16.8)	3, 38, 55–58
• Describe the two types of nucleic acids and the components from which they are made. (16.9)	9, 10, 59, 60, 82, 83
• Explain complementary base pairing, and describe how a copy of DNA is synthesized. (16.9)	7, 8, 61–67
• Explain how mRNA is synthesized from DNA and how a protein is synthesized from mRNA. (16.10)	8, 68–71, 73, 84
• List important characteristics of the genetic code. (16.10)	12, 72, 74
• Describe new DNA technologies, and explain their uses. (16.11)	11, 75–78
• Name some advantages of using biochemistry to create useful molecules.	88, 91
• Give examples of the use of biochemistry for energy production and other applications.	89, 90

Conceptual Questions

1. What is the importance of photosynthesis to plants and animals?

2. Briefly identify and state a function of each of the following parts of a cell.
   a. cell membrane   b. cell nucleus
   c. chloroplasts   d. mitochondrial
   e. ribosomes

3. In what parts of the body are proteins found? What tissues are largely proteins?

4. What provides most of the energy for plants? Write the equation for the reaction that is driven by that energy.

5. What is the chemical nature of proteins?

6. Amino acid units in a protein are linked by peptide bonds. What is another name for this kind of bond?

7. What kind of intermolecular force is involved in base pairing?

8. Describe how replication and transcription are similar and how they differ.

9. How do DNA and RNA differ in structure?

10. What is the relationship among the cell parts called chromosomes, the units of heredity called genes, and the nucleic acid DNA?

11. List the steps in recombinant DNA technology.

12. Explain what is meant by redundancy in the genetic code. Does redundancy exist for every amino acid?

 Problems

Energy and the Living Cell

13. The overall set of chemical reactions that keeps cells alive and functioning is
    a. anabolism   b. catabolism
    c. metabolism   d. reductionism

14. The processes by which molecules are broken down to yield energy are collectively known as
    a. anabolism   b. catabolism
    c. metabolism   d. reductionism

15. All of the following are substances that provide animals with energy except
    a. carbohydrates   b. fats
    c. nucleic acids   d. protein

16. Which of the following structures is found in plant cells but not in animal cells?
    a. chloroplasts   b. mitochondria
    c. nucleus   d. ribosomes

Carbohydrates: A Storehouse of Energy

17. Which of the following compounds are carbohydrates?

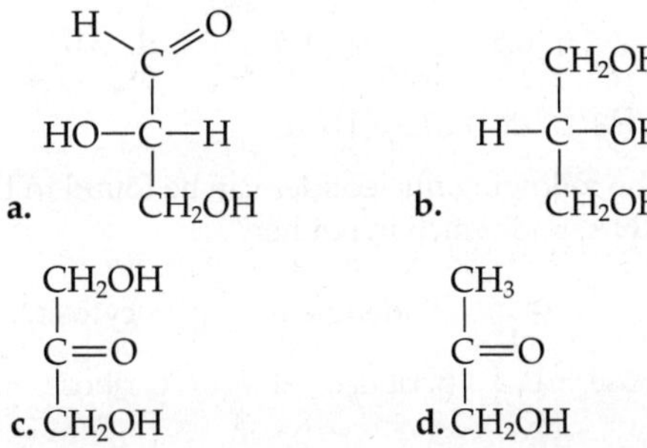

18. In what way are amylose and cellulose similar? What is the main structural difference between starch and cellulose?

19. Classify each of the following as a monosaccharide, a disaccharide, or a polysaccharide.
    a. amylose   b. cellulose   c. mannose   d. sucrose

20. What are the hydrolysis products of each of the following?
    a. amylopectin   b. cellulose

21. What functional groups are present in the formula for the open-chain form of lactose?

22. Mannose differs from glucose only in having the positions of the H and OH on the second carbon atom reversed. Write the formula for the open-chain form of mannose. What functional groups are present?

Carbohydrates in the Diet

23. The two monosaccharide units of a disaccharide are joined through a
    a. C-to-O-to-C linkage   b. C-to-O-to-H linkage
    c. C-to-O-to-O linkage   d. H-to-O-to-H linkage

24. What sugar is formed when starch is digested?

25. What is the role of carbohydrates in the diet?

26. Lactose intolerance results from the lack of an enzyme that will
    a. convert glucose to fructose
    b. hydrolyze lactose to form glucose and galactose
    c. metabolize galactose
    d. metabolize glucose

Fats and Other Lipids

27. How do fats and oils differ in structure? In properties?

28. A triacylglycerol contains fatty acids and
    a. glucose   b. glycerol
    c. glycogen   d. trehalose

29. Classify each of the following fatty acids as saturated, monounsaturated, or polyunsaturated.
    a. palmitic acid   b. linoleic acid
    c. oleic acid   d. stearic acid

30. How many carbon atoms are in a molecule of each of the following fatty acids?
    a. linolenic acid   b. palmitic acid   c. stearic acid

31. Which of the two foodstuffs shown below would you expect to have the higher iodine number? Explain your reasoning.

32. Which of the two fats shown below is likely to have the lower iodine number? Explain your reasoning.

Fats and Cholesterol

33. What is a *trans* fat? How are *trans* fats made?

34. How do animal fats differ from vegetable oils?

35. All of the following will tend to lower LDL levels except
    a. *cis* fatty acids   b. omega-3 fatty acids
    c. polyunsaturated fatty acids   d. *trans* fatty acids

36. The role of low-density lipoproteins (LDLs) is to
    a. carry cholesterol to the cells from the liver
    b. carry cholesterol to the liver from the cells
    c. carry triglycerides to the cells from the liver
    d. carry triglycerides to the liver from the cells

Proteins: Polymers of Amino Acids

37. Amino acids are the building blocks of
  a. carbohydrates   **b.** fats
  c. nucleic acids   **d.** proteins

38. Of the amino acids aspartic acid, lysine, serine, methionine, and glycine, which are essential to the human diet?

39. Which of the following structures for glycine is a zwitterion?
  a. $NH_2CH_2COO^-$ **b.** $^+NH_3CH_2COOH$
  c. NH_2CH_2COOH **d.** $^+NH_3CH_2COO^-$

40. Identify the following amino acids.

$$HOOC-CH_2CH_2-CH-COO^- \quad HS-CH_2-CH-COO^-$$
$$\qquad\qquad\qquad\; ^+NH_3 \qquad\qquad\qquad\qquad\; ^+NH_3$$

  a.   **b.**

41. Write structural formulas for the following dipeptides.
  a. glycylarginine   **b.** alanylcysteine

42. Identify the following dipeptides.

$$HO-CH_2 \qquad\qquad CH_2CH(CH_3)_2$$
$$^+NH_3-CH-CONH-CH-COO^-$$

  a.

$$HS-CH_2$$
$$^+NH_3-CH_2-CONH-CH-COO^-$$

  b.

43. How many different tripeptides can be made if you have one molecule each of three different amino acids?
  a. 2   **b.** 4   **c.** 6   **d.** 8

44. The sweetener *aspartame* has the structure shown below, in which a CH_3O- group (at right) is attached to a dipeptide. Identify the dipeptide.

$$H_2N-CH-\overset{O}{\overset{\|}{C}}-NH-CH-\overset{O}{\overset{\|}{C}}-OCH_3$$
$$\qquad CH_2 \qquad\qquad CH_2$$
$$\qquad COOH \qquad\qquad \bigcirc$$

Structure and Function of Proteins

45. List the four different forces that link protein chains.

46. Describe **(a)** the induced-fit model of enzyme action and **(b)** how an inhibitor deactivates an enzyme.

47. An alpha helix is an example of _____ protein structure.
  a. primary **b.** quaternary **c.** secondary **d.** tertiary

48. Hydrogen bonds help to stabilize all of the following levels of protein structure except
  a. primary **b.** quaternary **c.** secondary **d.** tertiary

49. A protein with more than one subunit has what level of struture?
  a. primary **b.** quaternary **c.** secondary **d.** tertiary

50. Which of the following pairs of amino acids can form salt bridges?
  a. lysine and glutamic acid   **b.** cysteine and lysine
  c. glutamic acid and glycine   **d.** cysteine and lysine

51. Which of the following amino acids can form disulfide bridges?
  a. cysteine **b.** glycine **c.** methionine **d.** serine

52. A substrate fits into an enzyme's
  a. active site   **b.** allosteric site
  c. coenzyme site   **d.** inhibitor site

53. A vitamin or other organic molecule that must be added to the protein part of an enzyme to activate the enzyme is a(n)
  a. activator **b.** coenzyme **c.** cofactor **d.** substrate

54. Describe the induced-fit enzyme model.

Protein in the Diet

55. What is the limiting reactant in any protein synthesis?

56. How do dietary proteins differ from dietary carbohydrates and fats in elemental composition?

57. All of the following provide adequate amounts of all essential amino acids except
  a. corn   **b.** eggs   **c.** fish   **d.** meat

58. The daily protein requirement is _____ g/kg body weight.
  a. 0.3   **b.** 0.8   **c.** 1.5   **d.** 3.0

Nucleic Acids: Parts and Structure

59. Which of the following nucleotides can be found in DNA, which in RNA, and which in neither?

  thymine   adenine   cytosine
  a. deoxyribose$-P_i$ **b.** ribose$-P_i$ **c.** ribose$-P_i$

60. Identify the sugar and the base in the following nucleotide.

61. In DNA, which base would be paired with the base listed?
  a. cytosine   **b.** adenine
  c. guanine   **d.** thymine

62. In RNA, which base would be paired with the base listed?
  a. adenine   **b.** guanine
  c. uracil   **d.** cytosine

63. In replication, a DNA molecule produces two daughter molecules. What is the fate of each strand of the original double helix?

64. DNA controls protein synthesis, yet most DNA resides within the cell nucleus, and protein synthesis occurs outside the nucleus. How does DNA exercise its control?

65. Which of the following is not a feature of a DNA double helix?
    a. A sugar–phosphate backbone is on the outside of the double helix.
    b. The two strands run parallel to each other.
    c. Each base pair contains a purine and a pyrimidine.
    d. It can be replicated to produce two identical daughter helices.

66. The Watson–Crick structure of the DNA molecule is
    a. a double helix
    b. globular
    c. a monolayer
    d. a tetrahedron

67. The process by which a DNA molecule produces an exact copy of itself is known as
    a. duplication
    b. replication
    c. transcription
    d. translation

68. RNA is made up of a _____ strand of nucleotides.
    a. bent
    b. double
    c. single
    d. straight

RNA: Protein Synthesis and the Genetic Code

69. Which nucleic acid(s) is/are involved in (a) transcription and (b) translation?

70. Explain the role of (a) mRNA and (b) tRNA in protein synthesis.

71. The process in which DNA unzips and makes a strand of mRNA is
    a. pairing
    b. replication
    c. transcription
    d. translation

72. Which of the following point mutations in DNA would be most likely to affect protein function? You may consult Table 16.6.
    a. AGG to AGA
    b. CAA to TAA
    c. CTT to CTC
    d. TAA to TGA

73. The process by which a protein is made from mRNA is known as
    a. duplication
    b. replication
    c. transcription
    d. translation

74. What amino acid would each of the following codons on mRNA code for? You may consult Table 16.6.
    a. UUG
    b. CAC

The Human Genome

75. The procedure by which large amounts of DNA can be made very quickly from a very small amount of DNA is
    a. CRISPR
    b. gene splicing
    c. polymerase chain reaction (PCR)
    d. transcription

76. It is estimated that the human genome contains approximately _____ genes.
    a. 500–1000
    b. 5000–10,000
    c. 20,000–25,000
    d. 80,000–85,000

77. The DNA used to make recombinant DNA must first be cut by a
    a. lipase
    b. polymerase
    c. protease
    d. restriction enzyme

78. DNA fingerprinting has been used to
    a. determine the best drug to treat an illness
    b. exonerate individuals falsely accused of a crime
    c. identify the base(s) which must be changed in order to use gene therapy
    d. provide a complete sequence of a person's genome

Expand Your Skills

79. Which of the following provide the greatest amount of energy per gram?
    a. carbohydrates
    b. lipids
    c. nucleic acids
    d. proteins

80. Which of the following provide the least amount of energy per gram?
    a. carbohydrates
    b. lipids
    c. nucleic acids
    d. proteins

81. Write the abbreviated versions of the following structural formula using (a) the three-letter abbreviations and (b) the one-letter abbreviations.

82. Answer the following questions for the molecule shown.
    a. Is the base a purine or a pyrimidine?
    b. Would the compound be incorporated in DNA or in RNA?

83. Answer the questions posed in Problem 82 for the following compound.

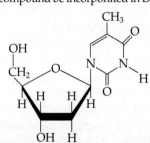

84. With what mRNA codon would the tRNA in the diagram form a codon–anticodon pair?

85. Which of the following statements about the ability of humans to digest cellulose is correct?
   a. We can digest cellulose, because it made from glucose molecules.
   b. We can digest cellulose, because it has no branches.
   c. We cannot digest cellulose, because it has β linkages between glucose units.
   d. We cannot digest cellulose, because it is highly branched.

86. Maintaining life requires both building up molecules and breaking them down. Which of the following statements about the use of energy to maintain life is correct?
   a. Photosynthesis provides energy that can be used to build molecules because photosynthesis can directly transfer that energy into new molecules.
   b. Photosynthesis provides energy that can be used to build molecules, because photosynthesis converts the sun's energy to chemical energy.
   c. Anabolic processes provide energy that can be used to build molecules, because they can harness radiant energy from the sun.
   d. Catabolic processes provide energy that can be used to build molecules, because they can convert radiant energy from the sun into useful energy.

87. Which of the following statements about lipids is correct?
   a. LDLs have low density, because they have more saturated fatty acids than HDLs do.
   b. HDLs are described as the "good" cholesterol, because they carry cholesterol to the cells from the liver.
   c. *Trans* fatty acids pack more closely together, because they have bends in their chains.
   d. Oils tend to be liquids at room temperature, because they contain more unsaturated fatty acids.

88. What solvent is typically used for biochemical reactions?

89. What types of molecules are used to store energy in biological systems?
   a. hydrocarbons
   b. carbohydrates and lipids
   c. amino acids
   d. water and carbon dioxide

90. Which is an example of a way to produce a renewable fuel using biochemical processes or products?
   a. turn carbohydrates into fuels
   b. use bacteria to convert carbohydrates or CO_2 into biofuels
   c. engineer microbes to make gasoline
   d. all of the above

91. Which of the following is NOT an advantage of using biochemical methods compared to traditional chemical synthesis?
   a. Water is often used as a solvent.
   b. Separation of products is difficult.
   c. Reaction conditions are mild.
   d. Reactions are usually catalytic.

 # Critical Thinking Exercises

Apply knowledge that you have gained in this chapter and one or more of the FLaReS principles (Chapter 1) to evaluate the following statements or claims.

16.1. The label on an energy drink claims, in large print, "Contains no artificial amino acids!"

16.2. DNA is an essential component of every living cell. An advertisement claims that DNA is a useful diet supplement.

16.3. A restaurant menu states, "Other restaurants use saturated beef fat for frying. We use only pure vegetable oil to fry our fish and French fries, for healthier eating."

16.4. An agricultural company claims that none of the plants or animals used in its products has ever been genetically modified.

16.5. A dietician states that a recommendation of 60 g of protein per day for an adult is somewhat misleading. She claims that the right kinds of proteins, with the proper amino acids in them, must be consumed; otherwise, even a much larger amount of protein per day may be insufficient to maintain health.

16.6. A man convicted of rape claims that he is innocent and that DNA fingerprinting of a semen sample taken from the rape victim showed that it contains DNA from another man.

16.7 A new bread that contains cellulose is marketed as a diet aid.

 # Collaborative Group Projects

Prepare a PowerPoint, poster, or other presentation (as directed by your instructor), to share with the class.

1. Prepare a brief report on (a) a carbohydrate such as fructose, glycogen, lactose, maltose, sucrose, or starch or (b) a lipid such as beef tallow, corn oil, lard, oleic acid, or palmitic acid. List principal sources and uses of the material.

2. Prepare a brief report on one of the following products of genetic engineering. List some advantages and disadvantages of the product.
   a. golden rice
   b. herbicide-resistant transgenic soybeans
   c. transgenic cotton that produces Bt toxin
   d. transgenic tobacco that produces human serum albumin

e. transgenic tomatoes that have improved resistance to viruses

f. transgenic *E. coli* that produce human interleukin-2, a protein that stimulates the production of T-lymphocytes that play a role in fighting selected cancers

3. Genetic testing carries both promise and peril, and its use is an important topic in bioethics today. After some online research, write a brief essay on one of the following—or on a similar question assigned by your instructor.

a. In what cases might a person not want to be tested for a disease his or her parent had?

b. When is genetic testing most valuable? What privacy issues must be addressed in this area?

c. Does testing negative for the breast cancer genes mean that a woman doesn't have to have mammograms?

LET'S EXPERIMENT! DNA Dessert

Materials Needed

- 1 resealable plastic bag
- Strawberries (fresh or frozen)
- 2 teaspoons dish detergent
- 1 teaspoon salt
- 1 cup of water

- 2 clear plastic cups (one cup will be used for the filtering apparatus below)
- Filtering apparatus: coffee filter and plastic cup
- Ice-cold 90% rubbing alcohol
- 1 wooden popsicle stick or plastic coffee stirrer

Ever wonder how scientists extract DNA from a cell? How many scientific fields use human DNA samples in their research?

Cells are the basic unit of life and make up all plants, animals, and bacteria. Deoxyribonucleic acid, or DNA, is the molecule that controls everything that happens in the cell. DNA contains instructions that direct the activities of cells and, ultimately, the body. To extract the DNA from a cell, we will add a detergent to lyse (pop open) the cell so that the DNA is released into solution. Alcohol added to the solution then causes the DNA to precipitate out.

In this activity, strawberries will be used, because each strawberry cell has eight copies of the genome (an octoploid genome), giving them a lot of DNA per cell. Most organisms only have one genome copy per cell.

Begin by pulling off any green leaves or by thawing frozen strawberries. Put the strawberry into the plastic bag, seal it, and gently smash it for about two minutes. Completely crush the strawberry. This starts to break open the cells and release the DNA.

In a plastic cup, make your DNA extraction liquid: Mix together 2 teaspoons of detergent, 1 teaspoon of salt, and 1 cup of water. Add 2 teaspoons of the DNA extraction liquid to the bag with the strawberry. This will further break open the cells. Reseal the bag and gently smash for another minute. Avoid making too many soap bubbles.

Place the coffee filter inside the other plastic cup. Open the bag and pour the strawberry liquid into the filter. You can twist the filter just above the liquid and gently squeeze the remaining liquid into the cup. Next, pour down the side of the cup an amount of cold rubbing alcohol equal to the strawberry liquid. Do not mix or stir.

You have just separated the DNA from the rest of the material contained in the cells of the strawberry. What happens to the mixture? Within a few seconds, watch for the development of a white, cloudy substance (DNA) in the top layer above the strawberry-extract layer.

Tilt the cup and pick up the DNA using a plastic coffee stirrer or wooden stick. Can you spool it onto the stick?

▲ Pouring into a coffee filter.

▲ The pulp in a glass with DNA floating on it.

Questions

1. Were you able to see the DNA clearly?
2. Were you able to pick up clumps of the DNA?
3. How do DNA samples help detectives solve crimes?
4. Why was the discovery of the structure of DNA so important?

17 Nutrition, Fitness, and Health

Exercise and a proper diet as parts of a healthy lifestyle are essential to good health, reducing the risk of diseases of the mind and body. Through chemistry, we gain a better understanding of the effects of diet and exercise on our bodies and minds.

Have You Ever Wondered?

1 How much fat in the diet is "too much"? **2** The doctor said my mother should cut her salt intake. Why? **3** Are large doses of vitamins beneficial? **4** What's the difference between natural and artificial sweeteners? **5** If I stop exercising, will the muscle I've developed turn to fat? **6** What is the difference between being overweight and being obese?

THE CHEMISTRY OF WELLNESS: FOOD, NUTRITION, EXERCISE, AND FITNESS For much of human history, the main concern of most people was obtaining enough food to stay alive. The primary purpose of food, after all, is to supply the nutrients and energy needed to sustain life. Food supplies all the molecular building blocks from which our bodies are made, enables children to grow, and provides all the energy for our life activities. That energy comes ultimately from the sun through photosynthesis.

Especially in developed countries, people now have scores of labor-saving devices and abundant food. Many people earn their daily bread with little physical exertion. So much food is available, and so little physical effort is required to obtain it, that many people eat more than they should and get much less exercise than they need.

In many places on Earth, however, some people never have enough food. According to *The Hunger Project*, over 800 million people (one in nine) are always hungry. Wars make the news, but ten times as many people die of chronic malnutrition than die in war. The world's population was 7.6 billion in 2018 and it continues to grow. Feeding this growing world population is a challenging task, especially in countries that are experiencing the highest birthrates and the deepest poverty. The population of Niger, for example, is growing at a rate of over 3% per year; most of the population lives in the direst poverty with little food to support them.

While millions starve, other locales have a surplus, and others in developed countries regularly eat the wrong kinds of food. The average American diet, often too rich in saturated fats, sugar, and alcohol, has been linked in part to at least five of the ten leading causes of death in the United States: heart disease, cancer, stroke, diabetes, and kidney disease. The U.S. Centers for Disease Control and Prevention (CDC) found that 70.7% of the adults in the United States are overweight, and 37.9% of them are considered obese. The International Obesity Task Force estimates that 1.5 billion people worldwide are overweight, with 500 million of those being obese. This contributes to poor health (increased risk of diabetes, heart attacks, strokes, and some forms of cancer) and an increased susceptibility to other diseases.

We have already looked at the types and sources of many foodstuffs (Chapter 16). In this chapter, we will extend our discussion of food to consider the requirements for proper nutrition: balanced proportions of carbohydrates, fats, and proteins as well as vitamins, minerals, water, electrolytes, and fiber. We will also examine muscle action, exercise, and other topics related to well-being and physical fitness.

17.1 Calories: Quantity and Quality

Learning Objectives • List the recommendations (sources and percentages) for calories in the American diet. • Describe the special dietary requirements of athletes.

Generally we talk about the calories that foods contain and that we consume. Total calorie intake is important, but the distribution of calories is even more important. The *2015 Dietary Guidelines for Americans* from the U.S. Department of Agriculture (USDA) and the U.S. Department of Health and Human Services (HHS) include some recommendations, presented as a food plate in Figure 17.1. This document is updated every five years and can be found online.

A food plan based on these general recommendations should be tailored to the individual. For the average 2000-calorie-a-day diet, the guidelines suggest the following: Avoid oversized portions and overeating. Half of the plate should be fruits and vegetable. At least half of total grains should be whole grains. Switch to fat-free or low-fat (1%) milk, and drink water instead of sugary drinks. Check the

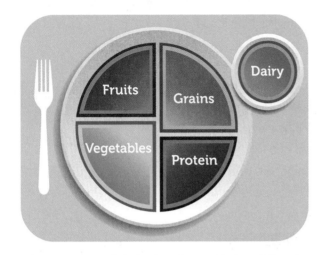

▲ **Figure 17.1** A daily food guide presented as a plate of food. More detailed recommendations in *Dietary Guidelines for Americans* are based on an individual's age, sex, activity level, and general health.

1 How much fat in the diet is "too much"?

Not more than about 35% of total daily calories should come from fat, and not more than 10% should come from saturated fat. For a 2000 kcal daily allowance, this means less than 700 kcal from fat and less than 200 kcal from saturated fat.

sodium content of foods like soup, bread, and frozen goods, and choose foods with lower sodium levels.

The 2000-calorie-per-day diet is an average. The recommended ranges for energy nutrients for most people are 20% to 35% of calories from fat, 45% to 65% of calories from carbohydrates, and 10% to 35% of calories from protein. (Recall from Chapter 16 that each gram of carbohydrate or protein provides about 4 kcal; a gram of fat provides about 9 kcal.) Your diet should take into account your weight, age, and level of activity. Then you should adjust it to include fewer calories if you want to lose weight or more calories if you want to gain weight. Calorie-intake calculators available online may be helpful.

An important point to remember when considering caloric intake is that the calorie listed on a food item is not the same as the calorie calculated in chemical energy. The food calorie is 1000 times larger than an energy calorie, and the food calorie is often expressed as a kcal (kilocalorie). This means that the average 2000-calorie-a-day recommendation would actually be 2000 kilocalories, or 2 million calories. In this chapter we refer mainly to food calories but we use the kcal notation when referring to food calories as energy calories.

A key recommendation for adults is to keep fat intake between 20% and 35% of all calories ingested. Most fats should be made up of polyunsaturated and monounsaturated fatty acids. Only 7% to 10% of calories should come from saturated fats, and intake of synthetic *trans* fatty acids should be kept to a minimum.

The "good" fats come from sources such as fish, nuts, and vegetable oils. Olive oil and canola oil are rich in monounsaturated fatty acids, and fish oils contain beneficial omega-3 fatty acids. A study of people living on the Greek island of Crete showed that they had an amazingly low incidence of cardiovascular disease, despite a diet that averaged about 40% fat. This result is thought to be due to the high percentage of olive oil in their diet.

More than half the fats in the typical American diet are animal fats, and 70% of saturated fat comes from animal products, such as butter, lard, red meats, and fast-food hamburgers. Saturated fats and cholesterol are implicated in the clogging of arteries. We should choose meat, poultry, milk, and milk products that are lean, low in fat, or fat-free.

Trans fats, produced when vegetable oils are hydrogenated (for example, to produce margarine), are high in *trans* fatty acids (Section 16.5), which behave more like saturated fatty acids. Unless margarine has been specially treated to remove the *trans* fat, it has about the same health effects as butter.

The *percentage* of fat in the American diet has dropped from 40% in 1990 to about 34% today, but the absolute quantity of fat consumed has increased because Americans are eating more total calories. Also, 34% is an average figure; for some people, the percentage is as high as 45%, which greatly exceeds the recommendation for fat consumption.

 EXAMPLE 17.1 Nutrient Calculations

A single-serving pepperoni pan pizza has 38 g of fat and 780 total food calories. (Recall that a food calorie is a kilocalorie.) Estimate the percentage of the total calories from fat.

Solution

First, we calculate the calories from fat. Fat furnishes about 9 kcal/g, so 38 g of fat furnishes

$$38 \text{ g fat} \times \frac{9 \text{ kcal}}{1 \text{ g fat}} = 340 \text{ kcal}$$

Now, we divide the calories from fat by the total calories. Then multiply by 100% to get the percentage (parts per 100).

$$\text{Percentage of calories from fat} = \frac{340 \text{ kcal}}{780 \text{ kcal}} \times 100\% = 44\%$$

This answer is only an estimate for two reasons: The value of 9 kcal/g is approximate, and the 38 g of fat is known only to the nearest gram. A proper answer is *about* 44%.

> **Exercise 17.1A**

A serving of French fries has 12 g of fat and furnishes 240 kcal. What percentage of the total calories comes from fat?

> **Exercise 17.1B**

What percent of one's recommended maximum intake of fats is provided by that single-serving pepperoni pizza and the serving of French fries from Example 17.1 and Exercise 17.1A? How many additional grams of fat could one ingest on that day while remaining within those guidelines? Do you think that would be an easy or a difficult goal?

Vegetarian Diets

Green plants trap a small fraction of the energy that reaches them from the sun. They use some of this energy to convert carbon dioxide, water, and mineral nutrients (including nitrates, phosphates, and sulfates) to proteins. Cattle eat plant protein, digest it, and convert a small portion of it to animal protein. It takes 100 g of protein feed to produce 4.7 g of edible beef or veal protein, an efficiency of only 4.7%. People eat this animal protein, digest it, and reassemble some of the amino acids into human protein.

Some of the energy originally transformed by green plants is lost as heat at every step. If people ate the plant protein directly, one highly inefficient step would be skipped. A vegetarian diet conserves energy. Pork production, at a protein conversion efficiency of 12.1%, and chicken or turkey production (at 18.2%) are more efficient than beef production. Milk production (22.7%) and egg production (23.3%) are even more efficient but still do not compare well with eating plant protein directly.

Vegetarians generally are less likely than meat eaters to have high blood pressure. Vegetarian diets that are low in saturated fat can help us avoid or even reverse coronary artery disease. These diets also offer protection from some other diseases. However, although complete proteins can be obtained by eating a carefully selected mixture of vegetable foods, total vegetarianism (vegan) can be dangerous, especially for young children. Even when the diet includes a wide variety of plant materials, an all-vegetable diet is usually short in several nutrients, including vitamin B_{12} (a nutrient not found in plants), calcium, iron, riboflavin, and vitamin D (required by children not exposed to sunlight). A modified vegetarian (*lacto-ovo*) diet that includes eggs, milk, and milk products without red meat or poultry can provide excellent nutrition.

A variety of traditional dishes supply relatively good protein by combining foods from plant sources, usually a cereal grain with a legume (beans, peas, peanuts, and so on). The cereal grain is deficient in tryptophan and lysine, but it has sufficient methionine. The legume is deficient in methionine, but it has enough tryptophan and lysine. A few such combinations are listed in Table 17.1. Peanut butter sandwiches are a popular American example of a legume–cereal grain combination.

TABLE 17.1 Traditional Foods That Combine a Cereal Grain with a Legume

Group	Food (Cereal listed first)
Mexican	Corn tortillas and beans
Japanese	Rice and soybean curd (tofu)
English	Toasted bread with baked beans
Native American	Corn and beans (succotash)
Western African	Rice and peanuts (groundnuts)
Cajun (Louisiana)	Rice and red beans
Middle Eastern	Pita bread and hummus

A healthy diet can help us avoid disease. Harvard Medical School studies indicate that correct dietary choices, regular exercise, and avoidance of smoking would prevent about 82% of heart attacks, about 70% of strokes, over 90% of type 2 diabetes, and over 70% of colon cancer. In contrast, the most effective drugs against cholesterol buildup, called *statins*, only reduce heart attacks by about 20% to 30%.

Nutrition and the Athlete

Athletes generally need more calories because they expend more energy than the average sedentary individual does. Those extra calories should come mainly from carbohydrates such as starches, which are the preferred source of energy for the healthy body.

Fat- and protein-rich foods also supply calories, but protein metabolism produces more toxic wastes that tax the liver and kidneys. The pregame steak dinner consumed by some athletes in the past was based on a myth that protein builds muscle. It doesn't. Although athletes do need the **Dietary Reference Intake (DRI)** quantity of protein (0.8 g protein/kg body weight), with few exceptions, they do not need an excess. DRIs are nutrient-based reference values established by the Food and Nutrition Board of the U.S. National Academy of Sciences for use in planning and assessing diets. They have replaced the Recommended Dietary Allowances (RDAs) that were published by the National Academy of Sciences for 75 years. Protein consumed in amounts greater than that needed for synthesis and repair of tissue only makes an athlete fatter (as a result of excessive calorie intake)—not more muscular.

Muscles are built through exercise, not through eating excess protein. When a muscle contracts against a resistance, an organic acid called *creatine* is released. Creatine stimulates production of the protein myosin, thus building more muscle tissue. If the exercise stops, the muscle begins to shrink after about two days. After about two months without exercise, muscle built through an exercise program is almost completely gone.

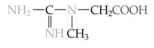

Creatine

SELF-ASSESSMENT Questions

1. The two major requirements for good health and fitness are
   a. food and entertainment
   b. food and medicine
   c. good nutrition and exercise
   d. nutrition and steroids

2. Which of the following is not a source of "good" fats?
   a. fish   b. peanuts   c. olives   d. lard

3. Strict vegetarian diets are often deficient in
   a. B vitamins and vitamin C
   b. magnesium and folic acid
   c. vitamin B_{12} and iron
   d. vitamin B_{12} and folic acid

4. Dietary Reference Intakes (DRIs) are nutrient-based reference values that are
   a. appropriate for undernourished and ill people
   b. guarantees of good health for everyone
   c. useful for planning and assessing diets
   d. minimum daily requirements

5. About how much protein does a 66 kg athlete require per day?
   a. 6.6 g   b. 8.3 g   c. 53 g   d. 2.5 kg

6. Which of the following foods do(es) NOT furnish complete protein?
   a. corn and beans   b. eggs   c. meat   d. rice

Answers: 1, c; 2, d; 3, c; 4, c; 5, c; 6, d

17.2 Minerals

Learning Objective • Identify the bulk dietary minerals, and state their functions.

In addition to the three major foodstuffs—carbohydrates, fats, and proteins—humans require a variety of minerals (and vitamins) for good nutrition. According to the World Health Organization (WHO), one of every three people in developing countries is affected by mineral and vitamin deficiencies. Let's start by considering some of the most important minerals.

Dietary Minerals

Thirty elements, listed in Table 17.2, are known to be essential to one or more living organisms. Among these are six elements found in organic compounds such as carbohydrates, fats, and proteins, as well as in water, which make up about two-thirds the weight of the human body. The other elements in the table, called **dietary minerals**, serve a variety of functions and are vital to life. They exist primarily as charged particles (ions) and not as solid, neutral metallic elements. Minerals represent about 4% of the weight of the human body, the main ones being calcium and phosphorus. There are nineteen trace elements, including iron, copper, zinc, and sixteen others called *ultratrace elements*.

The role of iodine is quite dramatic. A small amount of iodine is necessary for proper thyroid function; 0.15 mg of iodine per day is recommended for adults. According to the National Institutes of Health, nearly 1.6 billion people are at risk of iodine deficiency, and over 650 million of them have some degree of goiter, an enlargement of the thyroid that accompanies low thyroid activity (Figure 17.2). Greater deficiencies can have dire effects. Pregnant women deficient in iodine can experience stillbirths or spontaneous abortions, or have babies with congenital abnormalities and mental impairment. Iodine occurs naturally in seafood, but to guard against iodine deficiency, a small amount of potassium iodide (KI) is often added to table salt. "Iodized" salt was first put on the market in the United States in 1924. Many countries have been working to increase the consumption of iodine among their populations.

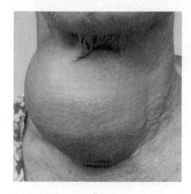

▲ **Figure 17.2** A person with goiter. The swollen thyroid gland in the neck results from a dietary deficiency of the trace element iodine.

TABLE 17.2 Elements Essential to Life

Element	Symbol	Form Used	Element	Symbol	Form Used
Bulk Structural Elements			**Ultratrace Elements**		
Hydrogen	H	Covalent	Manganese	Mn	Mn^{2+}
Carbon	C	Covalent	Molybdenum	Mo	Mo^{2+}
Oxygen	O	Covalent	Chromium	Cr	?
Nitrogen	N	Covalent	Cobalt	Co	Co^{2+}
Phosphorus[a]	P	Covalent	Vanadium	V	?
Sulfur[a]	S	Covalent	Nickel	Ni	Ni^{2+}
			Cadmium	Cd	Cd^{2+}
Macrominerals			Tin	Sn	Sn^{2+}
Sodium	Na	Na^+	Lead	Pb	Pb^{2+}
Potassium	K	K^+	Lithium	Li	Li^+
Calcium	Ca	Ca^{2+}	Fluorine	F	F^-
Magnesium	Mg	Mg^{2+}	Iodine	I	I^-
Chlorine	Cl	Cl^-	Selenium	Se	SeO_4^{2-}?
Phosphorus[a]	P	$H_2PO_4^-$	Silicon	Si	?
Sulfur[a]	S	SO_4^{2-}	Arsenic	As	?
			Boron	B	H_3BO_3
Trace Elements					
Iron	Fe	Fe^{2+}			
Copper	Cu	Cu^{2+}			
Zinc	Zn	Zn^{2+}			

[a]Note that phosphorus and sulfur each appear twice; they are structural elements and are also components of the macrominerals phosphate and sulfate, respectively.

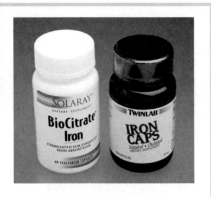

WHY IT MATTERS

Iron is an example of a substance that can be both essential and toxic. The National Institutes of Health recommends a daily intake of 8 mg for adult males and 18 mg for adult females but also points out that a 200 mg dose of iron in children can cause death.

2 **The doctor said my mother should cut her salt intake. Why?**

A low-sodium diet (no more than 1500 mg per day) is often recommended for anyone suffering from high blood pressure and for people whose family history includes early heart attack or stroke. Your mother may also be at risk for kidney stones, and lowering her sodium intake may avoid the use of prescription medication. In brief, it is better healthwise to cut salt.

Iron(II) ions (Fe^{2+}) are necessary for proper functioning of the oxygen-transporting compound hemoglobin. Without sufficient iron, the oxygen supply to body tissues is reduced, and anemia, a general weakening of the body, results. According to the World Health Organization (WHO), more than 30% of people worldwide (with a higher percentage in developing countries) suffer from anemia, which is often made worse by malaria and parasitic worm infections. Foods especially rich in iron compounds include red meat and liver. It appears that adult males need very little dietary iron, because iron is retained by the body, but women need higher amounts because of monthly blood loss. Iron "dust" is added to many dry cereals so that the body can digest it more efficiently than it does iron compounds found in iron supplements.

Calcium and phosphorus are necessary for the proper development of bones and teeth. Growing children need about 1.5 g of each of these minerals each day, which they can get from milk or milk products. The calcium and phosphorus needs of adults are less precisely known but just as real. *Osteoporosis*, a condition characterized by decreasing bone mass and an increased likelihood of bone fractures, is a problem especially among older women, who should continue to consume milk products throughout their lifetime. Calcium ions are necessary for the coagulation of blood (to stop bleeding) and for maintenance of the rhythm of the heartbeat. Phosphorus occurs in phosphate units in adenosine triphosphate (ATP), the "energy currency" of the body, which is necessary for the body to obtain, store, and use energy from foods. Phosphate units are also an important part of the backbone of the DNA and RNA chains (Section 16.9).

Sodium ions and chloride ions make up sodium chloride (salt). In moderate amounts, salt is essential to life. The sodium ions are important to the exchange of fluids between cells and plasma, so excess salt increases water retention. A high volume of retained fluids can cause swelling (called *edema*) and contribute to high blood pressure (*hypertension*). Over 100 million people in the United States suffer from hypertension, the leading risk factor for stroke, heart attack, kidney failure, and heart failure. Antihypertensives are among the most widely prescribed drugs in the United States.

Most Americans eat far more than the recommended 1500 mg of salt per day, and it is not a result of using the salt shakers on everyone's kitchen table. Convenience foods, such as canned soups, vegetables, pasta dishes, rice mixtures, and instant noodles, all contain a large amount of salt. Fortunately, many convenience food manufacturers are now offering "low-salt" or "no-salt-added" options for the consumer. Sodium has even been implicated in the creation of kidney stones, as it is involved in the sodium/calcium "pump" in the kidneys. Excess sodium results in the retention of too much calcium in the kidneys, which can then form kidney stones.

Iron, copper, zinc, cobalt, manganese, molybdenum, calcium, and magnesium are *cofactors* (Section 16.7) essential to the proper functioning of many life-sustaining enzymes. Copper, for example, is involved in the production of melanin, the pigment in skin and hair. Cobalt is in the center of the large vitamin B-12 molecule, nicknamed "cobalamin" because of the presence of that metal ion in the vitamin.

A well-balanced diet should supply an adequate amount of minerals for a healthy person. Table 17.3 summarizes the Dietary Reference Intake values for some minerals

TABLE 17.3 **Dietary Reference Intakes of Minerals for Young Adults**

Minerals	Females	Males
Calcium	1200 mg	1200 mg
Phosphorus	700 mg	700 mg
Magnesium	320 mg	420 mg
Iron	15 mg	10 mg
Zinc	12 mg	15 mg
Iodine	150 μg	150 μg
Fluoride	3 mg	4 mg
Selenium	55 μg	55 μg
Potassium	4.7 g	4.7 g

necessary for young adults. However, some people also choose to take a vitamin and mineral supplement during pregnancy, lactation, and periods of stress or recovery from disease. There is also evidence that for people over 65, a daily supplement of modest amounts of minerals and vitamins seems to strengthen the immune system and reduce infections. A great deal remains to be learned about the role of inorganic chemicals in our bodies, and bioinorganic chemistry is a flourishing area of research.

SELF-ASSESSMENT Questions

1. The mineral necessary for the healthy functioning of the thyroid is
   a. calcium   **b.** iodine   **c.** iron   **d.** sodium

2. High intake of which of the following contributes to hypertension?
   a. calcium   **b.** iodine   **c.** iron   **d.** sodium

3. The mineral needed for hemoglobin is
   a. calcium   **b.** iodine   **c.** iron   **d.** phosphorus

4. The mineral needed for the development of strong bones and teeth is
   a. calcium   **b.** iodine   **c.** iron   **d.** sodium

5. About what percentage of our body weight is minerals?
   a. 1%   **b.** 4%   **c.** 10%   **d.** 20%

Answers: 1. b; 2. d; 3. c; 4. a; 5. b

17.3 Vitamins

Learning Objective • Identify the vitamins and state their functions.

Why were British sailors called "limeys"? And what does this term have to do with food? Sailors on long voyages had been plagued since early times by *scurvy*, a disease whose symptoms include swollen and bleeding gums and lethargy. In 1747, Scottish naval surgeon James Lind showed that this disease could be prevented by including fresh fruit and vegetables in the diet. Fresh fruits that could conveniently be carried on long voyages were limes, lemons, and oranges. British ships put to sea with barrels of limes aboard, and sailors ate a lime or two every day.

The Vitamins: Vital, but Not All Are Amines

In 1897, the Dutch scientist Christiaan Eijkman (1858–1930) showed that polished rice lacked something found in the hull of whole-grain rice. Lack of that substance caused the disease *beriberi,* which was a serious problem in the Dutch East Indies (modern Indonesia) at that time. A British scientist, F. G. Hopkins (1861–1947), found that rats fed a synthetic diet of carbohydrates, fats, proteins, and minerals were unable to sustain healthy growth. Again, something was missing.

In 1912, Polish biochemist Casimir Funk (1884–1967) coined the word *vitamines* (from the Latin word *vita*, meaning "life") for these missing factors. Funk thought all these factors contained an amine group. In the United States, the final *e* was dropped after it was found that not all the factors were amines. The generic term became *vitamin.* Eijkman and Hopkins shared the 1929 Nobel Prize in Physiology or Medicine for their discoveries relating to vitamins.

Vitamins are specific organic compounds that are required (in addition to the usual proteins, fats, carbohydrates, and minerals) to prevent specific diseases. Unlike hormones and enzymes, which the body can synthesize, our bodies cannot make vitamins in the quantities we need, so they must be included in the diet. The role of vitamins in the prevention of deficiency diseases, such as *rickets* and *pellagra* (shown in Figure 17.3), has been well established. Some vitamins, along with their sources and deficiency symptoms, are listed in Table 17.4 (page 558).

Vitamins do not share a common chemical structure. They can, however, be divided into two broad categories: *fat-soluble vitamins* (A, D, E, and K) and *water-soluble vitamins* (B vitamins and vitamin C). The fat-soluble vitamins contain a high proportion of hydrogen and carbon, and usually only one or two oxygen atoms, so they generally are only slightly polar as a whole. They can therefore dissolve in the fatty tissue of the body, where reserves can be stored for future use.

One measure of the biological activity of many vitamins, hormones, and drugs is the International Unit (IU). An international agreement specifies the biological effect expected from a dose of 1 IU. The IU is also used to measure comparative potency

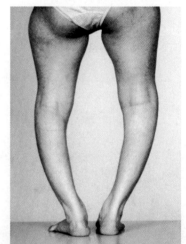
(a)

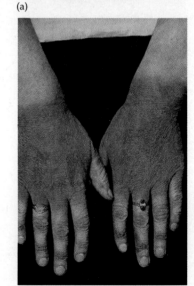
(b)

▲ **Figure 17.3** (a) Softened bones characterize rickets, caused by vitamin D deficiency.
(b) Inflammation and abnormal pigmentation, especially in skin that is exposed to the sun, characterize pellagra, caused by niacin deficiency.

of a substance such as a vitamin because the vitamin may exist in more than one form. For example, 1 IU of vitamin A is the biological equivalent of 0.3 μg retinol or 0.6 μg β-carotene.

Fat-Soluble Vitamins

An adult can store months' or even years' supply of vitamin A. If the diet becomes deficient in vitamin A, these reserves are mobilized, and the adult remains free of the deficiency disease for quite a while. However, a small child who has not built up a store of the vitamin exhibits deficiency symptoms much sooner. The WHO estimates that more than 250 million children in developing countries are deficient in vitamin A, and a quarter million to a half million become blind each year. Half of these children die within a year of losing their sight.

Vitamin A is a fat-soluble vitamin stored in the fatty tissues of the body and especially in the liver. Large doses can be toxic. The recommended upper limit for intake of vitamin A is 3000 μg/day. Studies in affluent countries have shown that ingesting vitamin A from supplements and in fortified foods at levels even slightly above the DRI leads to an increased risk of bone fractures later in life.

Larger quantities of β-carotene can be taken safely because excess β-carotene is not converted to vitamin A. Some nutritionists argue that β-carotene is needed in quantities greater than those needed to meet the vitamin A requirement because it may have important functions other than as a precursor of vitamin A. Consuming foods that contain β-carotene is probably the safest way to obtain the proper quantity of vitamin A. A single half-cup serving of carrots is sufficient to provide the DRI of vitamin A.

Vitamin D is a steroid hormone that protects children against rickets. Vitamin D is not strictly an essential dietary vitamin, because most mammals produce enough of it when they are exposed to sunlight. It promotes the absorption of calcium and phosphorus from foods to produce and maintain healthy bones. However, too much vitamin D can lead to excessive calcium and phosphorus absorption and subsequent formation of calcium deposits in various soft body tissues, including those of the heart. The DRI for vitamin D increases with age, from 200 international units (IU) for adults aged 20–50 to 600 IU for those over 70. The recommended upper limit is 2000 IU. Since the primary source of vitamin D comes from our exposure to sunlight, it should be noted that use of sunscreens may limit the capacity of the body to synthesize vitamin D (Chapter 20). It also suggests that exercise in the outdoors has specific benefits for vitamin D production, though the amount of vitamin D that the body can produce is self-limiting, to avoid toxic levels of the vitamin. Foods that supplement the availability of vitamin D, such as fatty fish or cheese, can prevent problems associated with vitamin D deficiency.

Vitamin E is a mixture of *tocopherols*, phenols with hydrocarbon side chains. Its antioxidant activity may have value in maintaining the cardiovascular system, and it has been used to treat coronary heart disease, angina, rheumatic heart disease, high blood pressure, arteriosclerosis, varicose veins, and a number of other cardiovascular problems. As an antioxidant, vitamin E can inactivate free radicals. It is generally believed that much of the physiological damage from aging results from the production of free radicals, and vitamin E has been called the "antiaging vitamin." However, studies of these effects are often contradictory and far from conclusive. Vitamin E is also an anticoagulant that has been useful in preventing blood clots after surgery.

Rats deprived of vitamin E become sterile. Vitamin E deficiency can also lead to muscular dystrophy, a disease of the skeletal muscles. A lack of vitamin E can lead to a deficiency in vitamin A, because vitamin A can be oxidized to an inactive form when vitamin E is not present to act as an antioxidant. Vitamin E also collaborates with vitamin C in protecting blood vessels and other tissues against oxidation. Vitamin E is a fat-soluble antioxidant vitamin, while vitamin C is a water-soluble one. Oxidation of unsaturated fatty acids in cell membranes can be prevented or reversed by vitamin E, which is itself oxidized in the process. Vitamin C can then restore vitamin E to its unoxidized form. The recommended upper limit for vitamin E is generally 400 IU/day, although there is some disagreement on this. And, according to a study published in October 2011, even 400 IU/day may increase the risk of prostate cancer in men. LDL cholesterol is oxidized before it is deposited in arteries, and vitamin E prevents the oxidation of cholesterol. Some think this is the mechanism by which vitamin E helps to protect against cardiovascular disease.

Because they are efficiently stored in the body, overdoses of fat-soluble vitamins can have adverse effects, although these are rarely encountered. Large excesses of vitamin A cause irritability, dry skin, and a feeling of pressure inside the head. Too much vitamin D can cause bone pain, hard deposits in the joints, nausea, diarrhea, and weight loss. Vitamin K is also fat-soluble, but it is metabolized and excreted. It is not stored to the extent that vitamins A and D are, and excess intake seldom causes problems.

Retinol

Vitamin D$_2$ (calciferol)

Fat-soluble vitamins

Water-Soluble Vitamins

The B family of vitamins has eight members. Because they are all water-soluble, excess intake is excreted in the urine. Little or no toxicity is connected with B vitamins, with the possible exceptions of vitamin B$_6$ and folic acid, both of which apparently can cause neurological damage in some people if taken in extremely large daily doses. Several B vitamins serve as coenzymes (Section 16.7). Besides preventing the skin lesions of pellagra, niacin (vitamin B$_3$) offers some relief from arthritis and helps in lowering the blood cholesterol level.

Vitamin B$_6$ (pyridoxine) has been found to help people with arthritis by shrinking the connective tissue membranes that line the joints. Vitamin B$_6$ is a coenzyme for more than 100 different enzymes.

Vitamin B$_{12}$ (cyanocobalamin) is not found in plants, and vegetarians are apt to be deficient in this vitamin. That deficiency can lead to pernicious anemia. Vitamin B$_{12}$ is a very large and complicated molecule with a molecular formula of $C_{63}H_{88}CoN_{14}O_{14}P$. Its structure was determined by British chemist Dorothy Hodgkin (1910–1994) of Oxford University in 1956 using X-ray crystallography, a technique in which an X-ray beam is used to bombard a crystal. The beam is scattered in a definite pattern determined by the crystal structure. No one had ever attempted to establish the structure of a molecule of this size and complexity before. Hodgkin received the 1964 Nobel Prize in Chemistry for her work.

Another B vitamin, folic acid, is critical in the development of the nervous system of a fetus. Its presence in a pregnant woman's diet prevents spina bifida in her baby. Folic acid also helps prevent cardiovascular disease. An upper limit of 1000 μg daily has been set; more than that can cause nerve damage.

Vitamin C is ascorbic acid, the component in citrus fruits that combats scurvy. About 60–80 mg daily will prevent scurvy, but vitamin C has many other roles. It promotes the healing of wounds, burns, and lesions such as gastric ulcers. It also seems to play an important role in maintaining collagen, the body's major structural protein. Like vitamin E, it is an antioxidant, and these two vitamins, along with β-carotene, are included in many antioxidant formulations. Antioxidants may also act as anticarcinogens. About 200 mg of vitamin C per day is probably optimal, and an upper limit of 2000 mg per day has been set.

Vitamin C seems to be essential for efficient functioning of the immune system. An increased level of vitamin C has been shown to increase the body's production of *interferons*, large molecules formed by the action of viruses on their host cells. Interferons are agents in the immune system that enable one virus to interfere with the growth of another virus. People with low levels of vitamin C are also more likely to develop cataracts and glaucoma, as well as gingivitis and periodontal disease. Some

3 Are large doses of vitamins beneficial?

Although there is some evidence that large doses of antioxidants such as vitamin C may have beneficial effects, "megadoses" of most vitamins have not generally been shown to have any significant benefit. In fact, some of the fat-soluble vitamins are toxic in large amounts. There are well-documented examples of polar explorers who suffered poisoning from eating the livers of polar bears or seals, which contain extremely high levels of vitamin A.

▲ Dorothy Crowfoot Hodgkin used X-rays to determine the structures of several important organic compounds, including vitamin B$_{12}$, penicillin, and cholesterol.

suggest that very large doses of vitamin C may cure or prevent colds, but there is little scientific evidence for this.

The body has a limited capacity to store water-soluble vitamins (the B-complex and vitamin C, ascorbic acid), since the human body will excrete anything over the amount that can be used immediately. Water-soluble vitamins are needed frequently, every day or so. Some foods lose their vitamin content when they are cooked in water and then drained. The water-soluble vitamins go down the drain with the water.

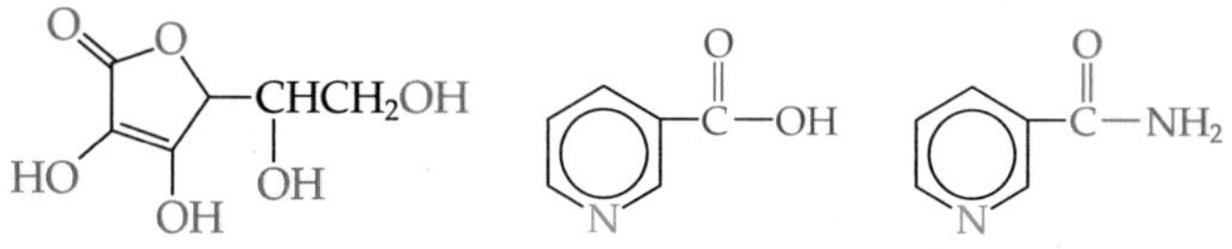

Ascorbic acid   Nicotinic acid   Nicotinamide

Water-soluble vitamins (sites for hydrogen bonding in color)

Many skin creams and lotions contain added vitamin C or vitamin E. Vitamin E has been demonstrated to accelerate healing of wounds and surgical incisions, as well as very dry and inflamed skin, but the addition of vitamin C has not been proved to be beneficial in skin creams.

Table 17.4 summarizes the sources of vitamins and their uses in preventing deficiency diseases. Table 17.5 gives the DRI for vitamins for young adults.

TABLE 17.4 Vitamin Sources and Deficiency Diseases

Vitamin[a]	Name	Sources	Deficiency Disease or Conditions
Fat-Soluble Vitamins			
A	Retinol	Fish; liver; eggs; butter; cheese; also (as β-carotene, a vitamin precursor) in carrots and other orange or red vegetables	Blindness in children; night blindness in adults
D_2	Calciferol	Cod liver oil; mushrooms; irradiated ergosterol (milk supplement)	Soft bones: rickets (children); osteomalacia (adults)
E	α-Tocopherol	Wheat germ oil; green vegetables; egg yolks; meat	Sterility; muscular dystrophy
K_1	Phylloquinone	Spinach; other green leafy vegetables	Hemorrhage
Water-Soluble Vitamins			
B_1	Thiamine	Germ of cereal grains; legumes; nuts; milk; brewer's yeast	Beriberi—polyneuritis resulting in muscle paralysis, enlargement of heart, and ultimately heart failure
B_2	Riboflavin	Milk; red meat; liver; egg white; green vegetables; whole wheat flour (or fortified white flour); fish	Dermatitis; glossitis (tongue inflammation)
B_3	Niacin	Red meat; liver; collards; turnip greens; yeast; tomato juice	Pellagra—skin lesions, swollen and discolored tongue, loss of appetite, diarrhea, and various mental disorders (Figure 17.3)
B_6	Pyridoxine	Eggs; liver; yeast; peas; beans; milk	Dermatitis, apathy, irritability, and increased susceptibility to infections; convulsions in infants
B_9	Folic acid	Liver; kidney; mushrooms; yeast; green leafy vegetables	Anemias (folic acid is used to treat megaloblastic anemia, a condition characterized by giant red blood cells); neural tube defects in fetuses of deficient mothers
B_{12}	Cyanocobalamin	Liver; meat; eggs; fish (not found in plants)	Pernicious anemia
C	Ascorbic acid	Citrus fruits; tomatoes; green peppers; broccoli; strawberries	Scurvy

[a]Some vitamins exist in more than one chemical form. We name a common form of each here.

TABLE 17.5 Dietary Reference Intakes of Vitamins for Young Adults

Vitamins	Females	Males
Vitamin A	700 μg^a	900 μg^a
Vitamin C	75 mg^b	90 mg^b
Vitamin D	200 IU^c	200 IU^c
Vitamin E	30 IU	30 IU
Thiamine (B_1)	1.1 mg	1.2 mg
Riboflavin (B_2)	1.1 mg	1.3 mg
Niacin	14 mg	16 mg
Pyridoxine (B_6)	1.5 mg	1.7 mg
Cyanocobalamin (B_{12})	3 μg	3 μg
Folacin (folic acid)	400 μg	400 μg
Pantothenic acid	5 mg	5 mg
Biotin	100–200 μg	100–200 μg

[a]To the extent that the vitamin A requirement is met by β-carotene, multiply these by 6.
[b]Smokers should add 35 mg.
[c]IU (for "International Unit") is a measure of the biological activity of many vitamins, hormones, and drugs.

SELF ASSESSMENT Questions

1. Which of the following cannot be synthesized by the body and therefore must be included in the diet?
   a. enzymes   **b.** hormones
   c. unsaturated fats   **d.** vitamins

2. Which of the following is a precursor of vitamin A?
   a. β-carotene   **b.** cholesterol
   c. pyridoxine   **d.** tryptophan

3. For which of the following are large doses most likely to be harmful?
   a. β-carotene   **b.** niacin   **c.** vitamin A   **d.** vitamin C

4. Which of the following vitamins provides the most benefit to bone health and development?
   a. vitamin D   **b.** vitamin B   **c.** vitamin C   **d.** vitamin E

5. Which of the following vitamins is fat-soluble?
   a. A   **b.** B-12   **c.** C   **d.** niacin

6. Which of the following vitamins is water-soluble?
   a. A   **b.** C   **c.** D   **d.** E

Answers: 1, d; 2, a; 3, c; 4, a; 5, a; 6, b

17.4 Fiber, Electrolytes, and Water

Learning Objective • Identify the roles of fiber, electrolytes, and water in maintaining health.

We need carbohydrates, proteins, fats, minerals, and vitamins in our diets, but some other items are also important to maintain health. These include *fiber*, *electrolytes*, and *water*.

Dietary fiber may be soluble or insoluble. Insoluble fiber is usually cellulose; soluble fiber generally consists of sticky materials called *gums* and *pectins*, often used for gelling, thickening, and stabilizing foods such as jams, jellies, and dairy products such as yogurt. High-fiber diets prevent constipation and are an aid to dieters. Fiber has no calories because the body can't absorb it. Therefore, high-fiber foods such as fruits and vegetables are low in fat and often low in calories. Fiber takes up space in the stomach, making us feel full and, therefore, we eat less food. Soluble fiber lowers cholesterol levels, perhaps by removing bile acids that digest fat. It may also help control blood sugar by delaying the emptying of the stomach, thus slowing sugar absorption after a meal. This may reduce the amount of insulin needed by a diabetic. These properties make dietary fiber beneficial to people with high blood pressure, diabetes, heart disease, or diverticulitis.

Another aspect of the relationship of chemistry and nutrition is the balance between fluid intake and electrolyte intake. An **electrolyte** is a substance that conducts electricity when dissolved in water. In the body, electrolytes are ions required by cells to maintain their internal and external electric charge and thus control the flow of water molecules across cell membranes. The main electrolytes are sodium ions (Na^+), potassium ions (K^+), and chloride ions (Cl^-). Others include calcium ions (Ca^{2+}), magnesium ions (Mg^{2+}), sulfate ions (SO_4^{2-}), hydrogen phosphate ions (HPO_4^{2-}), and bicarbonate ions (HCO_3^-).

Water is also an essential nutrient, a fact obvious to anyone who has been deprived of it. Many of the foods we eat are mainly water. It should come as no surprise that tomatoes are 90% water or that melons, oranges, and grapes are largely water. But water is one of the main ingredients in practically all foods, from roast beef and seafood to potatoes and onions.

In addition to the water we get in our food, we need to drink about 1.0–1.5 L of water each day. Some urologists recommend drinking 2.0 L of water per day. We could satisfy this need by drinking plain water, but we often choose other beverages, such as milk, coffee, tea, soft drinks, beer, fruit juices, or energy drinks. Some of these beverages actually interfere with the body's use of the water that they contain.

One problem with counting coffee and soft drinks as part of your water intake is that caffeine has a diuretic effect on the kidneys, which results in urine formation and consequent water loss when consumed in large amounts. Small amounts, such as a single cup of coffee or one soft drink, have little effect on urination volumes. Bottled water containing added vitamins and minerals is becoming quite popular, but should not be a substitute for fresh fruits and vegetables.

Many people also drink beverages that contain ethanol, made by fermenting grains or fruit juices. Beer is usually made from malted barley, and wine from grape juice. Even these alcoholic beverages are mainly water, but they should not be considered "nutrients." Alcoholic drinks promote water loss by blocking the action of *antidiuretic hormone (ADH)*, also known as *vasopressin*. Fruit juices and plain water are better alternatives.

The best way to replace water lost through respiration, sweat, tears, and urination is to drink water (Figure 17.4). Unfortunately, thirst is often a *delayed* response to water loss, and it may be masked by such symptoms of dehydration as exhaustion, confusion, headache, and nausea. Sweat is about 99% water, but a liter of sweat typically contains 1.15 g Na^+, 1.48 g Cl^-, 0.02 g Ca^{2+}, 0.23 g K^+, 0.05 g Mg^{2+}, and minute amounts of urea, lactic acid, and body oils. Most people already consume too much sodium chloride (the main electrolytes lost), so replacing the water component with plain water makes the most sense.

Commercial beverages such as sports and energy drinks usually contain one or more ingredients that are more effective as advertising gimmicks than as aids to athletic performance. Sports drinks usually contain sweeteners and electrolytes. Energy drinks usually have sugars and often contain caffeine as a mild stimulant. Although these drinks are quite popular with both serious and weekend athletes, they are so concentrated (a hazard that can lead to diarrhea) that they are of marginal value except for endurance athletes.

How do you know if you are drinking enough water and are in a proper state of hydration? The urine is a good indicator of hydration. When you are properly hydrated, your urine is clear and almost colorless. When you are dehydrated, your urine gets cloudy or dark because your kidneys are trying to conserve water so as to keep the blood volume from shrinking and to prevent shock.

Dehydration can be quite serious, even deadly. We become thirsty when total body fluid volume decreases by 0.5% to 1.0%. Beyond that, things get worse (Table 17.6). Muscles tire and cramp. Dizziness and fainting may follow, and brain cells shrink,

▲ **Figure 17.4** The best replacement for fluids lost during exercise generally is plain water. Sports drinks help little, except for those athletes engaging in endurance activities lasting for 2 h or more. At that point, carbohydrates in the drinks delay the onset of exhaustion by a few minutes.

TABLE 17.6 **Effects of Varying Degrees of Dehydration**

Percentage of Body Weight Lost as Sweat	Physiological Effect
2%	Performance impaired
4%	Muscular work capacity declines
5%	Heat exhaustion
7%	Hallucinations
10%	Circulatory collapse and heat stroke

resulting in mental confusion. Finally, the heat-regulatory system fails, causing *heat stroke*, which can be fatal without prompt medical attention.

It's a Drug! No, It's a Food! No, It's . . . a Dietary Supplement!

Almost every day, we see advertisements for products that claim to promote rapid weight loss, improve stamina, aid memory, or provide other seemingly miraculous benefits. Although these products may appear to act as drugs, many are not classified as such. Legally, they are *dietary supplements*.

The Dietary Supplement Health and Education Act, enacted by the U.S. Congress in 1994, changed the law so that dietary supplements became regulated as foods, not as drugs. Manufacturers are permitted to describe some specific benefits that may be attributed to use of a supplement. However, they must also include a disclaimer, such as: "This statement has not been evaluated by the Food and Drug Administration. This product is not intended to diagnose, treat, cure, or prevent any disease."

Some supplements are simply combinations of various vitamins. Others contain minerals; amino acids; other nutrients; herbs or other plant materials; or extracts of animal or plant origin. Should you take a particular supplement? Maybe; maybe not. A supplement containing omega-3 oils might well aid in reducing LDL cholesterol. But a tablespoon a day of a special

preparation of vitamins and minerals is not likely to reverse the effects of aging, no matter how brightly colored the bottle, or how attractive the person in the advertisement. In considering the use of a dietary supplement, using the FLaReS principles to evaluate the manufacturer's claims is certainly recommended.

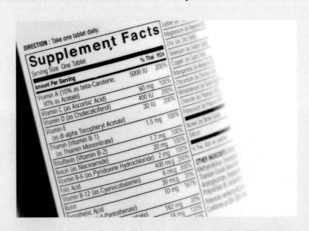

▲ Most dietary supplements do not go through the rigorous testing that is required by law for all prescription drugs and over-the-counter drugs.

SELF-ASSESSMENT Questions

1. Eating foods high in fiber
   a. can cause constipation
   b. helps build muscle
   c. helps to fill you up
   d. provides quick energy

2. The major electrolytes in body fluids are
   a. Ca^{2+}, Fe^{2+}, and Cl^-
   b. Ca^{2+}, Na^+, and HSO_4^-
   c. Na^+, K^+, and Cl^-
   d. Na^+, K^+, and HCO_3^-

3. The best replacement for water lost except in prolonged exercise is
   a. beer   **b.** non-fat milk   **c.** plain water   **d.** soft drinks

4. What is the minimal percentage of body weight lost in fluids that can cause heat exhaustion?
   a. 2%   **b.** 4%   **c.** 5%   **d.** 10%

5. Insoluble fiber usually is primarily
   a. cellulose   **b.** fats   **c.** lipids   **d.** protein

Answers: 1, c; 2, c; 3, c; 4, c; 5, a

17.5 Food Additives

Learning Objectives • Describe the historic impact of our desire for a variety of flavors and aromas, and of our search for sugar substitutes. • Identify the beneficial additives in foods and those that are somewhat controversial.

The search for new flavors to enhance food has changed human history. One well-known example is that Christopher Columbus was sent west with three small ships at considerable expense by Queen Isabella of Spain over five hundred years ago in an attempt to find a shorter route to East Asia to enhance the spice trade.

Spices, Herbs, and Flavorings

Spice cake, soft drinks, gingerbread, sausage, and many other foods depend on spices and other additives for most of their flavor. Cloves, for example, contain a unique molecule called eugenol that contributes to the captivating smell of a roast ham. Cinnamon, used on toast and in apple pie, contains an aldehyde called, appropriately, cinnamaldehyde. Ginger, turmeric, and nutmeg are examples of natural spices that come from the seeds, roots, or bark of those plants. Basil, marjoram, thyme, and rosemary are widely used herbs. Natural flavors are also extracted from fruits and other plant materials.

Some compounds that serve as flavors are listed in Table 9.9. Other flavor compounds are shown in Figure 17.5. Chemists can analyze natural flavors and then synthesize components and make mixtures that resemble those flavors. Major components of natural and artificial flavors are often identical. For example, both vanilla extract and imitation vanilla owe their flavor mainly to a compound called vanillin. The natural flavor is often more complex because vanilla extract contains a wider variety of chemicals than imitation vanilla does. Flavor additives, whether natural or synthetic, probably present little hazard when used in moderation, and they contribute considerably to our enjoyment of food.

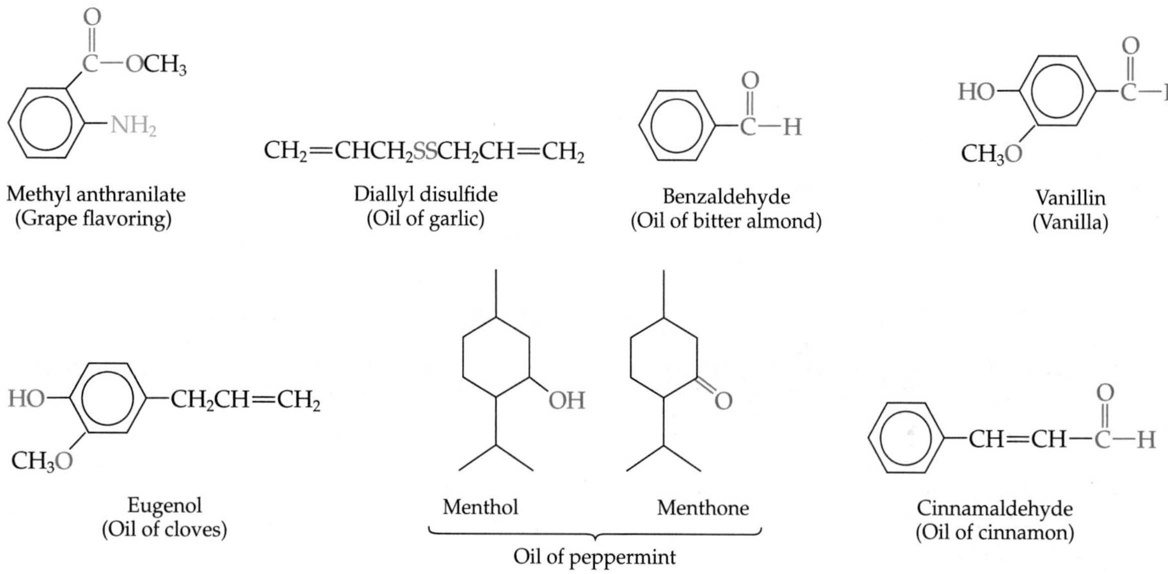

▲ **Figure 17.5** Some molecular flavorings. See Table 9.9 for others.

Ⓠ *Which of the compounds shown here have aldehyde functional groups? Which are phenols? Ethers? Alkenes? Which is an ester? A ketone? An alcohol?*

Other Natural and Artificial Sweeteners

Obesity is a major problem in most developed countries. Presumably, people could reduce their intake of calories by replacing sugars with noncaloric sweeteners, but there is little evidence that artificial sweeteners are of value in controlling obesity, since people tend to eat more of other foods when consuming "diet" soft drinks.

A natural product of note is *stevia*, a South American herb long used as a sweetener by the Guarani Indians of Paraguay. Leaves of this herb (*Stevia rebaudiana*) contain glycosides that provide a sweetness at least 30 times that of sucrose. The purified glycosides are considered to have GRAS (Generally Recommended As Safe) status, and steviol glycosides are now found in some soft drinks and other products in the United States.

For many years, the major artificial sweeteners were saccharin and cyclamates. Cyclamates were banned in the United States in 1970 after studies showed that they caused cancer in laboratory animals. (Although subsequent studies have failed to confirm these findings, the FDA has not lifted the ban.)

In 1977, saccharin was shown to cause bladder cancer in laboratory animals. However, the move by the FDA to ban it was blocked by Congress, because at that time saccharin was the only approved artificial sweetener. Its ban would have meant

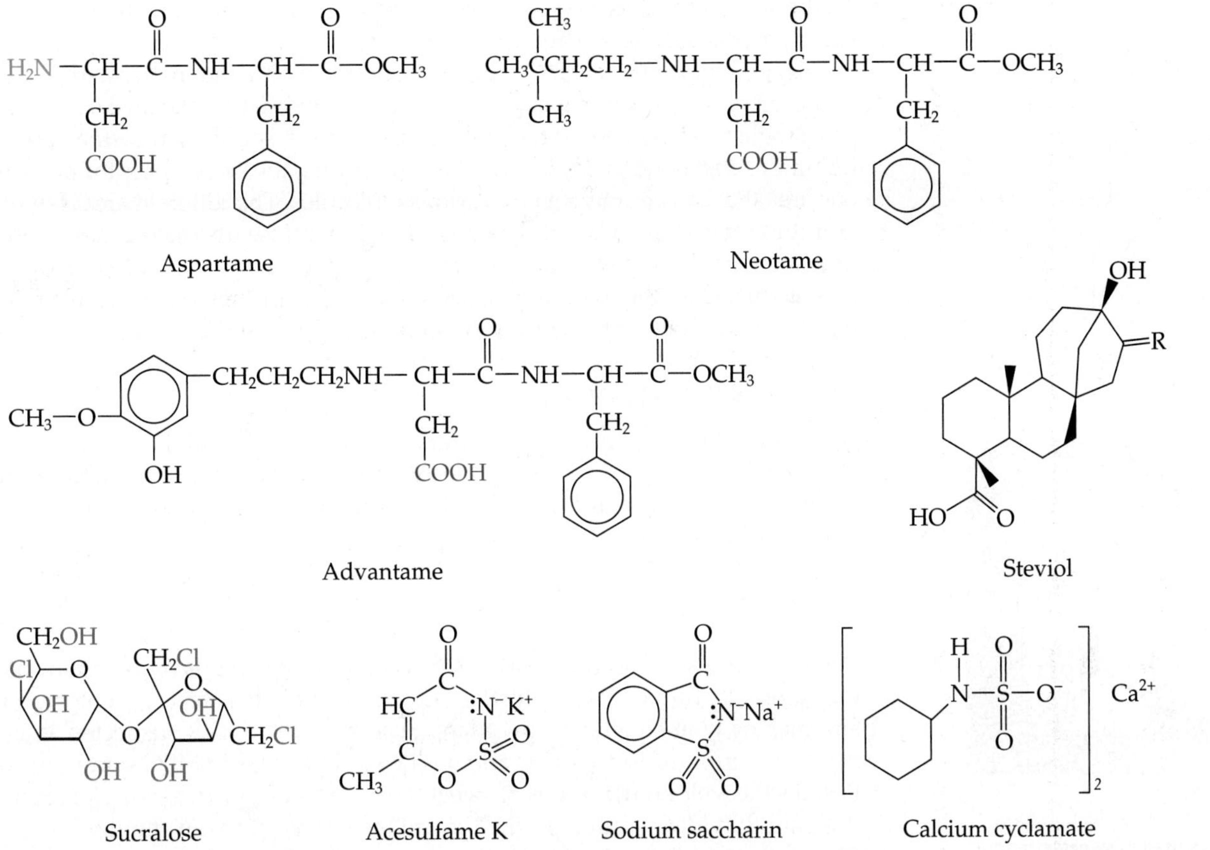

▲ **Figure 17.6** Eight artificial sweeteners. Note that sucralose (Splenda®) is a chlorinated derivative of sucrose. It is the only artificial sweetener actually made from sucrose. Only six of these sweeteners have actual FDA approval, although steviol is on the GRAS list. Cyclamates have been banned since 1970.

the end of diet soft drinks and other low-calorie products. Repeated studies did not show the same results as in 1977, so in 2000, warnings disappeared from saccharin packets and products containing saccharin.

There are now six FDA-approved artificial sweeteners (Figure 17.6). *Aspartame* (the methyl ester of the dipeptide aspartylphenylalanine) was approved in 1981. Two closely related compounds derived from aspartame that have also been approved are *neotame* and *advantame*. There are anecdotal reports of problems with aspartame, but repeated studies have shown it to be generally safe, except for people with phenylketonuria, an inherited condition in which phenylalanine cannot be metabolized properly. Other artificial sweeteners approved for use in the United States include acesulfame K (Sunette®) and sucralose (Splenda®); these sweeteners can survive the high temperatures of cooking processes, whereas aspartame is broken down by heat.

Table 17.7 compares the relative sweetness of a variety of substances. What makes a compound sweet? There is little structural similarity among the compounds that taste sweet, so the question remains unanswered. Most bear little resemblance to sugars. Recall from Chapter 16 that sugars are polyhydroxy compounds. Like sugars, many compounds with hydroxyl groups on adjacent carbon atoms are sweet. For example, ethylene glycol ($HOCH_2CH_2OH$) is sweet, although it is quite toxic. Glycerol, obtained from the hydrolysis of fats (Section 16.5), is also sweet, and is approved as a food additive. It is used principally as a **humectant** (moistening agent), however, and only incidentally as a sweetener.

Other polyhydroxy alcohols (polyols) used as sweeteners are *sorbitol*, made by the reduction of glucose, and *xylitol*, which has five carbon atoms with a hydroxyl group on each. These compounds occur naturally in foods such as fruits and berries. They are used as sweeteners in sugar-free chewing gums and candies. Unlike sugars, polyols do not cause sudden increases in blood sugar, and thus they can be used in

TABLE 17.7 Relative Sweetness of Some Compounds

Compound	Sweetness[a]
Lactose	0.16
Maltose	0.33
Glucose	0.74
Sucrose	1.00
Fructose	1.73
Steviol glycoside	30
Cyclamate	45
Aspartame	180
Acesulfame K	200
Saccharin	300
Sucralose	600
Neotame	13,000
Advantame	20,000

[a]Sweetness is relative to sucrose at a value of 1.

CH₂OH
|
CHOH   CH₂OH
| |
CHOH   CHOH
| |
CHOH   CHOH
| |
CHOH   CHOH
| |
CH₂OH   CH₂OH

Sorbitol   Xylitol

▲ Polyhydroxy alcohols (polyols) are used to sweeten sugar-free products such as chewing gum. These products aid in control of dental caries because the polyols are metabolized slowly if at all in dental plaque.

4 What's the difference between natural and artificial sweeteners?

Except for sucralose—which structurally looks very much like sucrose, but with chlorine atoms substituted for some —OH groups—the molecules that are commonly used as artificial sweeteners are completely different from natural sugars. Why they seem to taste "sweet" is somewhat of a mystery. Many people say that some artificial sweeteners have a bitter aftertaste, which may be due to the —NH₂ groups on the molecule. We do know that artificial sweeteners have almost no calories, because they are used in such small quantities.

moderation by people with diabetes. Large amounts—more than about 10 g—can cause gastrointestinal distress.

Several thousand sweet-tasting compounds have been discovered. They belong to more than 150 chemical classes. We now know that all the sweet substances act on a single taste receptor in the tongue. (In contrast, we have more than 30 receptors for bitter substances.) Unlike other taste receptors, the sweet receptor has more than one area that can be activated by various molecules. The different areas have varying affinities for certain molecules. For example, sucralose fits the receptor more tightly than sucrose does, partly because its chlorine atoms carry more negative charge than is carried by the oxygen atoms in the OH groups they replace. Neotame fits the receptor so tightly that it keeps the receptor firing repeatedly.

Flavor Enhancers

Some chemical substances, though not particularly flavorful themselves, are used to enhance other flavors. Common table salt (sodium chloride) is a familiar example. In addition to its being a necessary nutrient, salt seems to increase sweetness and helps to mask bitterness and sourness.

Another popular flavor enhancer is monosodium glutamate (MSG). MSG is the sodium salt of glutamic acid, a naturally occurring amino acid that is added to many convenience foods. Although glutamates are found in proteins, there is evidence that huge excesses can be harmful. MSG can numb portions of the brains of laboratory animals. It may also be *teratogenic*, causing birth defects when eaten in large amounts by women who are pregnant (Section 21.6). MSG is suspected as a cause of headaches in some individuals and has been implicated in MSG symptom complex. Besides the well-known tastes of sweet, sour, and salty, umami is a protein-like taste. It is said that MSG stimulates the umami taste sensors in our mouths.

$$HOOC-CH_2CH_2CH-\overset{\overset{\displaystyle O}{\|}}{C}-O^-$$
$$\underset{^+NH_3}{}$$

Glutamic acid

$$HOOC-CH_2CH_2CH-\overset{\overset{\displaystyle O}{\|}}{C}-O^- Na^+$$
$$\underset{NH_2}{}$$

MSG

In addition to the flavorings discussed here, more than 3000 substances appear in a list called "Substances Added to Food Inventory" on the website of the Food and Drug Administration (FDA). Most of the substances listed on a label are **food additives**—substances other than basic foodstuffs that are put into food for various reasons related to production, processing, packaging, or storage. Other substances are added to enhance color and flavor, to retard spoilage, to provide texture, to sanitize, to bleach, to ripen or prevent ripening, to control moisture levels, or to control foaming. Sugar, salt, and corn syrup are used in the greatest amounts. These three, plus citric acid, baking soda, vegetable colors, mustard, and pepper, make up more than 98% of all additives by weight.

In the United States, food additives are regulated by the FDA. The original Food, Drug, and Cosmetic Act was passed by Congress in 1938. Under this act, the FDA had to prove that an additive was unsafe before its use could be prevented. The Food Additives Amendment of 1958 shifted the burden of proof to the food industry. A company that wishes to use a food additive must first furnish proof to the FDA that the additive is safe for the intended use. The FDA can also regulate the quantity of additives that can be used.

Additives to Improve Nutrition

The first nutrient supplement approved by the Bureau of Chemistry (which later became the FDA) of the U.S. Department of Agriculture was potassium iodide (KI), added to table salt in 1924 to reduce the incidence of goiter.

Several other chemicals are added to foods to prevent deficiency diseases. Addition of vitamin B₁ (thiamine) to polished rice is essential in the Far East, where beriberi is still a problem. The replacement of the B vitamins thiamine, riboflavin, folic acid, and niacin (which are removed in processing) and the addition of iron, usually ferrous carbonate (FeCO₃), to flour are referred to as **enrichment**. Enriched bread or pasta made from this flour still isn't as nutritious as bread or pasta made from whole wheat. It lacks vitamin B$_6$, pantothenic acid, zinc, magnesium, and fiber, nutrients usually provided by whole-grain flour. Despite these shortcomings, the enrichment of bread, corn meal, and cereals has almost eliminated pellagra, a disease that once plagued the southern United States.

The GRAS List

Some food additives have been used for many years without apparent harmful effects. In 1958, the U.S. Congress established a list of additives *generally recognized as safe (GRAS)*. Many of the substances on the **GRAS list** are familiar spices, flavors, and nutrients. Some of the substances on the 1958 list have since been removed, including cyclamate sweeteners and some food colors, but other substances have been added to the original list. About 7000 substances currently have GRAS status.

Some deficiencies in the original testing procedures have come to light. The FDA has since reevaluated several of them based on new research findings—and greater consumer awareness. Improved instruments and better experimental designs have revealed the possibility of harm, though slight, from some additives previously thought safe. Most of the newer experiments involve feeding massive doses of additives to small laboratory animals, and these studies have been criticized for that reason.

Vitamin C (ascorbic acid) is frequently added to fruit juices, flavored drinks, and other beverages. Although our diets generally contain enough ascorbic acid to prevent scurvy, some scientists recommend a much larger intake than minimum daily requirements. Vitamin D is added to milk in developed countries, and consumption of fortified milk has led to the almost total elimination of rickets. Similarly, β-carotene, which changes to vitamin A in the body, is added to margarine so that the substitute more nearly matches the nutritional quality of butter, since vitamin A occurs naturally in butter.

If we were to eat a balanced diet of fresh foods, we probably wouldn't need nutritional supplements. But many people eat mainly processed foods and convenience foods. If their usual diet is composed mainly of highly processed foods, people may need the nutrients provided by vitamin and mineral food additives.

Additives That Retard Spoilage

Food spoilage can result from the growth of molds, yeasts, or bacteria. Substances that prevent such growth are often called *antimicrobials*, and they include certain carboxylic acids and their salts. Propionic acid and its sodium and calcium salts are used to inhibit molding of bread and cheese. Sorbic acid, benzoic acid, and their salts are also used (Figure 17.7) in some soft drinks and a wide variety of other prepared foods.

Some inorganic compounds can also be added to inhibit spoilage. Sodium nitrite ($NaNO_2$) is used to cure meat and to maintain the pink color of smoked hams, frankfurters, and bologna. It also contributes to the tangy flavor of processed meat products. Nitrites are particularly effective as inhibitors of *Clostridium botulinum*, the bacterium that produces botulism poisoning. However, only about 10% of the amount used to keep meat pink is needed to prevent botulism. Nitrites have been investigated as possible causes of cancer of the stomach. In the presence of the hydrochloric acid (HCl) in the stomach, nitrites are converted to nitrous acid.

$$NaNO_2 + HCl \longrightarrow HNO_2 + NaCl$$

This acid may then react with secondary amines (amines with two alkyl groups on nitrogen) to form *nitroso* compounds.

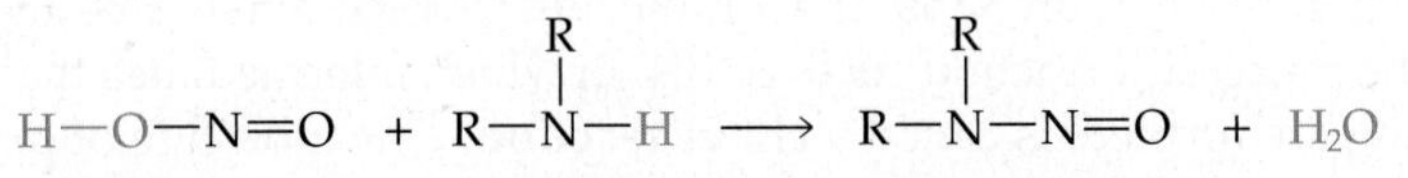

$$\underset{\text{Nitrous acid}}{H-O-N=O} + \underset{\substack{\text{A secondary}\\\text{amine}}}{R-\overset{R}{\underset{|}{N}}-H} \longrightarrow \underset{\substack{\text{A nitroso}\\\text{compound}}}{R-\overset{R}{\underset{|}{N}}-N=O} + H_2O$$

CH_3CH_2COOH
Propionic acid

$CH_3CH_2COO^-Na^+$
Sodium propionate

$CH_3CH=CHCH=CHCOOH$
Sorbic acid

$CH_3CH=CHCH=CHCOO^-K^+$
Potassium sorbate

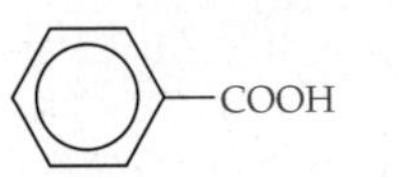

⬡—COOH
Benzoic acid

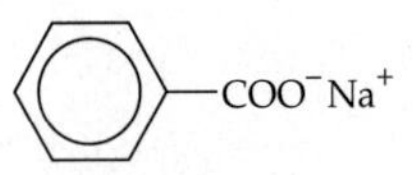

⬡—COO⁻Na⁺
Sodium benzoate

▲ **Figure 17.7** Most spoilage inhibitors are carboxylic acids or salts of carboxylic acids.

The R groups of nitroso compounds can be simple alkyl groups such as methyl (CH_3—), or ethyl (CH_3CH_2—), or they can be more complex. These compounds are among the most potent carcinogens known. The rate of stomach cancer is higher in countries where people use prepared meats than in developing nations, where people eat little or no cured meat. However, the incidence of stomach cancer is *decreasing* in the United States, perhaps because ascorbic acid (vitamin C) inhibits the reaction between nitrous acid and amines to form nitrosamines, and many people have orange juice with their breakfast bacon or sausage. Nevertheless, possible problems with nitrites have led the FDA to approve sodium hypophosphite (NaH_2PO_2) as an alternative meat preservative.

Other inorganic food preservatives include sulfur dioxide and sulfite salts. A gas at room temperature, sulfur dioxide (SO_2) serves as a disinfectant and preservative, particularly for dried fruits such as peaches, apricots, and raisins. It is also used as a bleach to prevent the browning of wines, corn syrup, jellies, dehydrated potatoes, and other foods. Sulfur dioxide seems safe for most people when ingested with food, but it is a powerful irritant when inhaled and is a damaging ingredient of polluted air in some areas (Chapter 13). Sulfur dioxide and sulfite salts cause severe allergic reactions in some people. The FDA requires food labels to indicate the presence of these compounds in a food.

Antioxidants: BHA and BHT

Another class of preservatives is the **antioxidants**, which are substances that inhibit the chemical spoilage of food that occurs in the presence of oxygen. A good example of chemical spoilage is a cut apple turning brown within a short time when exposed to air. Antioxidants are added to foods (or their packaging) to prevent fats and oils from forming rancid products that make foods unpalatable. Antioxidants also minimize the destruction of some essential amino acids and vitamins. Packaged foods that contain fats or oils (such as bread, potato chips, sausage, and breakfast cereal) often have antioxidants added. Compounds commonly used as antioxidants include butylated hydroxytoluene (BHT), butylated hydroxyanisole (BHA), propyl gallate, and *tert*-butylhydroquinone (Figure 17.8).

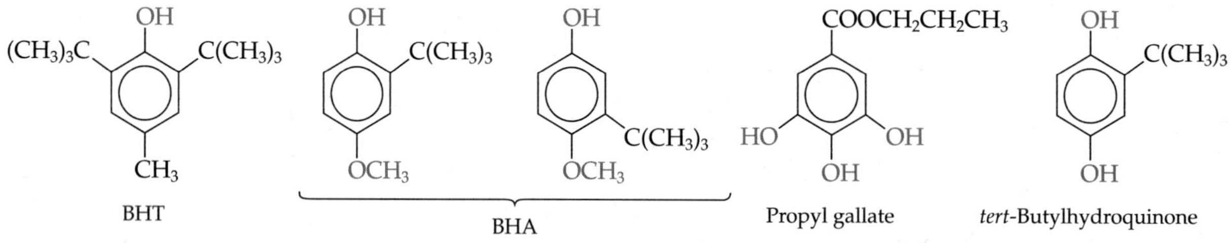

▲ **Figure 17.8** Four common antioxidants. BHA is a mixture of two isomers.

Q *What functional group is common to all four compounds? What other functional group is present in BHA? In propyl gallate?*

Fats turn rancid, in part as a result of oxidation, a process that occurs through the formation of molecular fragments called **free radicals**, which have an unpaired electron as a distinguishing feature. Without specifying the structures of radicals, we can summarize the process. First, a fat molecule reacts with oxygen to form a free radical.

$$Fat + O_2 \longrightarrow Free\ radical$$

The radical then reacts with another fat molecule to form a new free radical that can repeat the process. A reaction such as this, in which intermediates that keep the reaction going are formed, is called a **chain reaction**. One molecule of oxygen can lead to the decomposition of many fat molecules.

To preserve foods containing fats, processors package the products to exclude air. However, it cannot be excluded completely, so chemical antioxidants such as BHT are used to stop the chain reaction by reacting with the free radicals.

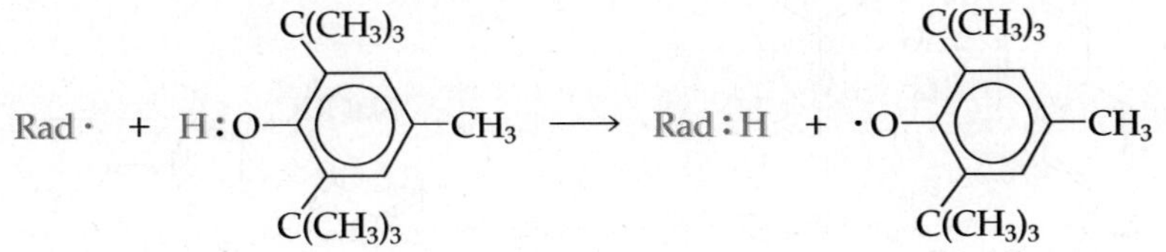

BHT

The new radical formed from BHT is rather stable. The unpaired electron doesn't have to stay on the oxygen atom but can move around in the electron cloud of the benzene ring. The BHT radical doesn't react with fat molecules, and the chain is broken.

Why are the butyl groups important? Without them, two phenol molecules (Section 9.6) would simply couple when exposed to an oxidizing agent.

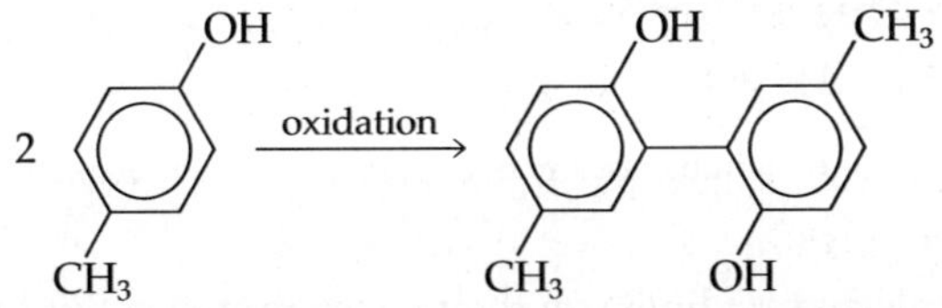

With the bulky butyl groups attached, the rings can't get close enough together for coupling. They are free, then, to trap free radicals formed by the oxidation of fats.

Many food additives have been criticized as being harmful, and BHA and BHT are no exceptions. They have been reported to cause allergic reactions in some people. In one study, pregnant mice fed diets containing 0.5% BHA or BHT gave birth to offspring with brain abnormalities. On the other hand, when relatively large amounts of BHT were fed to rats daily, their life spans were increased by a human equivalent of 20 years. One theory about aging is that it is caused in part by the formation of free radicals. BHT retards this chemical breakdown in cells in the same way that it retards spoilage in foods.

BHA and BHT are synthetic chemicals, but antioxidants such as vitamin E and vitamin C occur naturally. Vitamin E (like BHT) is a phenol, with several substituents on the benzene ring. Presumably, its action as an antioxidant is quite similar to that of BHT. However, recent studies suggest that large doses of vitamin E may also be somewhat harmful.

Color Additives

Some foods are naturally colored, especially fruits and vegetables. For example, the yellow compound β-carotene occurs in carrots and is used as a color additive in foods such as butter and margarine. Our bodies convert β-carotene to vitamin A by cutting the molecule in half at the center double bond. Thus, it is a vitamin additive as well as a color additive, especially in margarines.

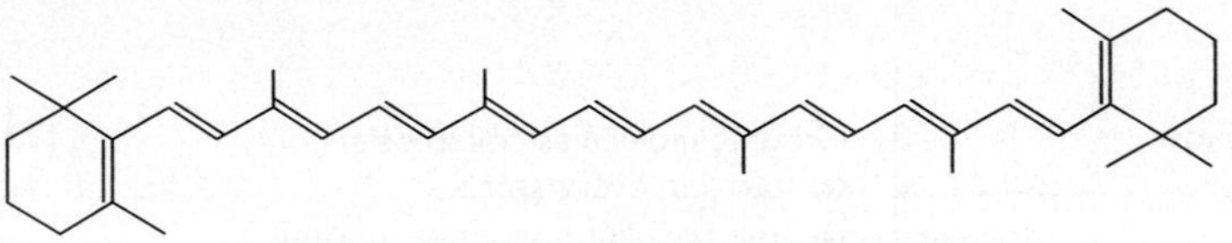

β-Carotene

Other natural food colors include beet juice, grape-hull extract, and saffron (from autumn-flowering crocus flowers).

We expect many foods to have characteristic colors. To increase the attractiveness and acceptability of its products, the food industry has used synthetic food colors for decades. Since the Food and Drug Act of 1906, the FDA has regulated the use of these

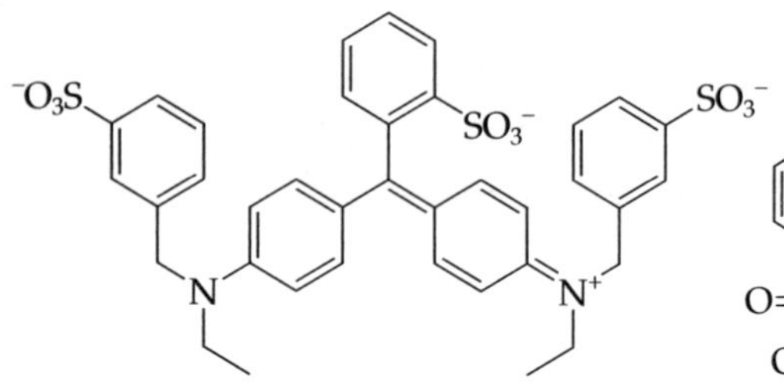

FD&C Blue No. 1 ("brilliant blue")

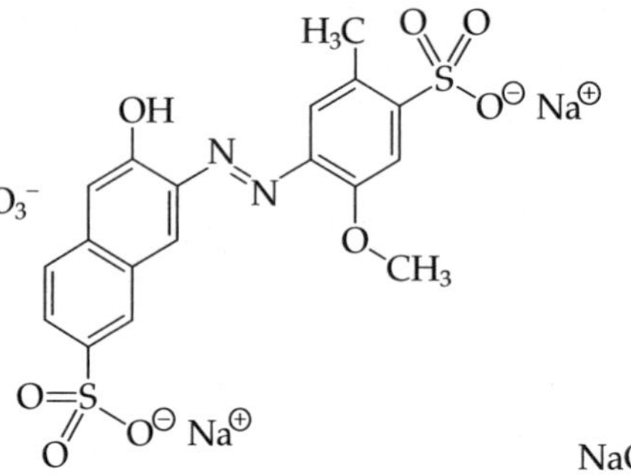

FD&C Red No. 40 ("allura red")

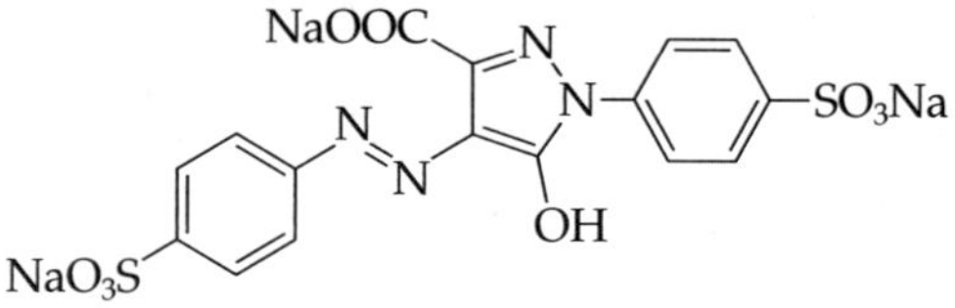

FD&C Yellow No. 5 ("tartrazine")

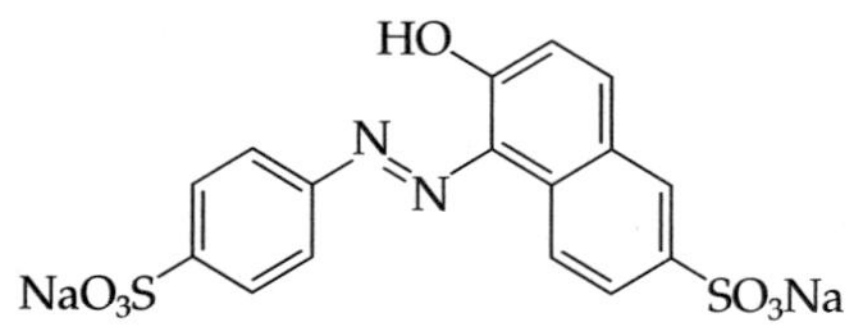

FD&C Yellow No. 6 ("sunset yellow")

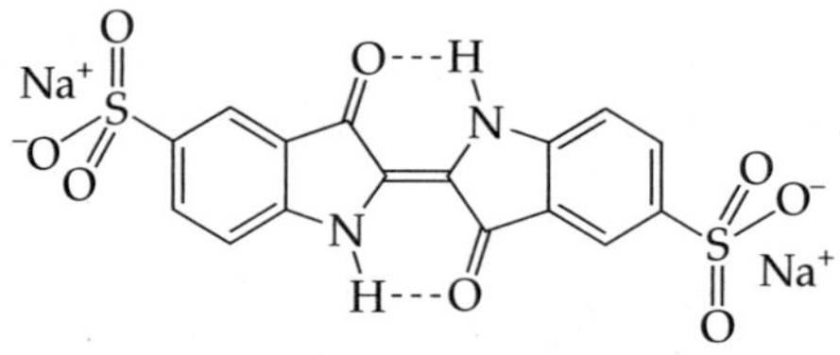

FD&C Blue No. 2 ("indigo carmine")

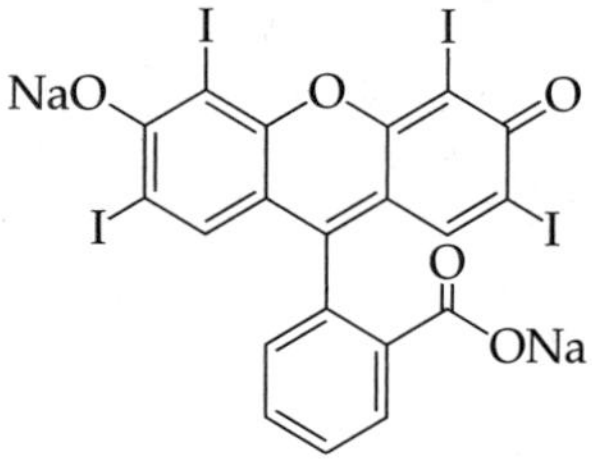

FD&C Red No. 3 ("erythrosine")

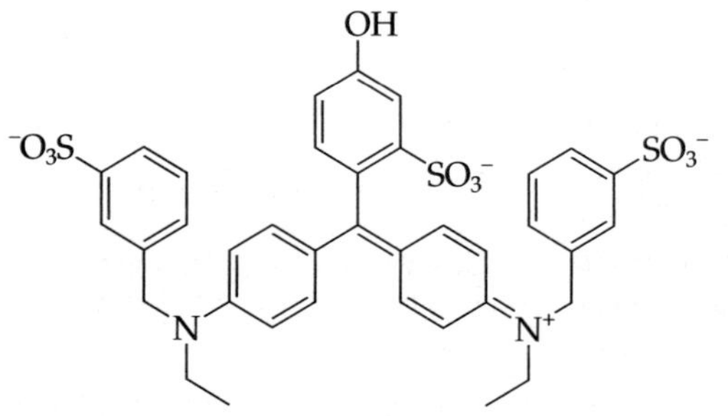

FD&C Green No. 3 ("fast green")

▲ **Figure 17.9** Food colorants accepted for use in foods.

chemicals and set limits on their concentrations. But the FDA is not infallible. Some colors once on the approved list were later shown to be harmful and were removed from the list. In 1950, Food, Drug, and Cosmetic (FD&C) Orange No. 1 in pumpkin-colored Halloween candy caused gastrointestinal upsets in several children and was then banned by the FDA.

In following years, other dyes were banned. FD&C Yellow No. 3 and No. 4 were found to contain small amounts of β-naphthylamine, a carcinogen that causes bladder cancer in laboratory animals. Furthermore, they reacted with stomach acids to produce more β-naphthylamine. FD&C Red No. 2 was also shown to be a weak carcinogen in laboratory animals.

Currently, six artificial colorants are commonly found in foods, including candies, prepared foods such as "mac and cheese," gelatin and pudding mixes, and a surprising number of prepared foods and beverages, such as "orange"-flavored soft drinks. The most common colorants include FD&C Red No. 40, Blue No. 1, and Yellow No. 5. Some candies also contain Red No. 3 (pink), Blue No. 2, and Yellow No. 6 (orange). Green No. 3, though approved, is rarely found in foods; some mouthwash products contain that color. The structures of those seven dyes are shown in Figure 17.9.

Food colors, even those that have been used for years with apparent safety and are normally used in tiny amounts, can present problems for some people. Their benefit is largely aesthetic. Kids like colorful candies! Any foods that contain artificial colors must say so on the label, so you can avoid them if you want to. If you see the word "lake" on a candy label, it means that the dye is entrapped in an aluminum oxide matrix, so the dyes will not rub off onto skin or clothing as easily.

SELF-ASSESSMENT Questions

1. All of the following compounds will taste sweet to most people except
  a. xylitol
  b. neotame
  c. saccharin
  d. MSG

2. Artificial vanilla
  a. contains exactly the same vanillin molecule as does natural vanilla
  b. contains a molecule different from vanillin but tastes similar to it
  c. contains every type of molecule found in natural vanilla
  d. is enhanced by sugars added to it

3. Organic molecules called esters are
  a. found in every spice
  b. sour or bitter flavorings in foods
  c. unique molecules found in highly aromatic and unique plant seeds or bark
  d. often found in vitamin supplements

4. MSG
  a. is safe to consume in any amount
  b. is the sodium salt of an amino acid
  c. is a sweetener with a structure similar to that of sucrose
  d. is safe to consume by those who wish to lower their blood pressure

5. To use a new food additive, a company must
   a. notify the FDA that it plans to use the additive
   b. pay the FDA to test the additive for safety
   c. provide the FDA with evidence that the additive is safe for the intended use
   d. submit an application to the FDA, but does not have to prove it safe and effective

6. The food additive that inhibits reproduction of botulism-causing bacteria in processed meats, such as bologna, ham, and bacon, is
   a. sodium benzoate   **b.** sodium ascorbate
   c. sodium sorbate   **d.** sodium nitrite

7. Calcium propionate is added to bread to
   a. act as an antioxidant   **b.** enhance the flavor
   c. inhibit mold growth   **d.** replace Ca^{2+} removed in processing

8. Antioxidants
   a. aid in lipid transport   **b.** aid in oxygen transport
   c. prevent color fading   **d.** trap free radicals

9. Which of the following is an antioxidant used to prevent rancidity?
   a. ascorbic acid **b.** BHT **c.** sodium sulfite **d.** sugar

Answers: 1, d; 2, a; 3, c; 4, b; 5, c; 6, d; 7, c; 8, d; 9, b

17.6 Starvation, Fasting, and Malnutrition

Learning Objective • Describe the effects of starvation, fasting, and malnutrition.

People in many parts of the world do not have access to adequate food for nutrition. Many of these people live in developing areas of the globe and many live in areas that have suffered natural disasters, such as droughts, or changing weather patterns that have affected agricultural productivity. Still others have had their normal sources of food interrupted because of wars, hostile actions, or forced migration. This has resulted in famines and consequent malnutrition or starvation. (See Section 19.5.)

When totally deprived of food, whether voluntarily or involuntarily, the human body suffers **starvation**. Involuntary starvation is a serious problem in much of the world. Although starvation is seldom the sole cause of death, those weakened by malnutrition succumb readily to disease.

Metabolic changes like those that accompany starvation also occur during fasting. During total fasting, the body's glycogen stores are depleted in less than a day, and the body calls on its fat reserves. Fat is first taken from around the kidneys and the heart. Then it is removed from other parts of the body, eventually even from the bone marrow.

Increased dependence on stored fats as an energy source leads to *ketosis*, a condition characterized by the appearance of compounds called *ketone bodies* in the blood and urine (Figure 17.10). Ketosis rapidly develops into *acidosis*; the blood pH drops, and oxygen transport is hindered. Oxygen deprivation leads to depression and lethargy.

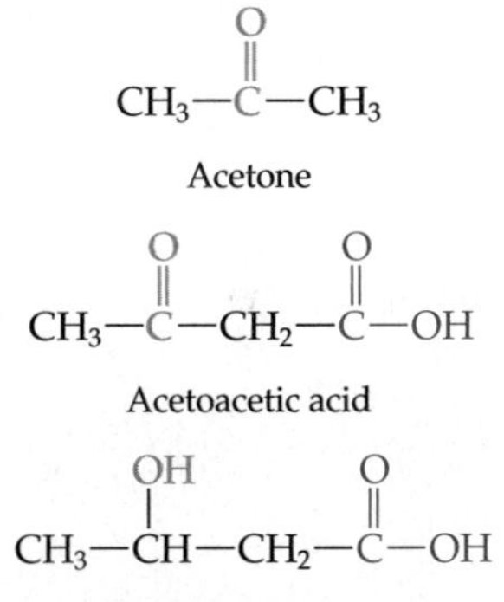

Acidosis is also associated with the insulin-deficiency disease diabetes. Insulin enables the body's cells to take up glucose from the bloodstream. A lack of insulin causes the liver to act as though the cells are starving, and they start burning fat for energy while blood sugar levels rise. This fat metabolism leads to the production of ketone bodies and subsequent acidosis.

In the early stages of a *total* fast, body protein is metabolized at a relatively rapid rate. After several weeks, the rate of protein breakdown slows considerably as the brain adjusts to using the breakdown products of fatty acid metabolism for its energy source. When fat reserves are substantially depleted, the body must again draw heavily on its structural proteins for its energy requirements. The emaciated appearance of a starving individual is due to the depletion of muscle proteins and swelling of the abdomen (Figure 17.11).

Processed Food: Less Nutrition

While we generally think about malnutrition as being caused by starvation or a severe shortage of food, it need not be due to starvation or dieting. It can also result from eating a diet that provides enough calories but that is missing some essential nutrients, such as an essential amino acid. There are a number of very common pairings of cereals with legumes to provide complete protein. (See Table 17.1.) Malnutrition can even occur in the midst of plenty, by eating too much highly processed food.

▲ Figure 17.10 The three ketone bodies produced in fat metabolism.

Acetone

Acetoacetic acid

β-Hydroxybutyric acid

Q *Identify the functional groups in each of the compounds. Which ketone body is not a ketone?*

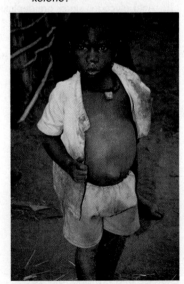

▲ Figure 17.11 Diets with inadequate protein are common in some parts of the world. A protein-deficiency disease called *kwashiorkor* is rare in developed countries but is common during times of famine in areas of the world where grains with incomplete protein are the only food staple available.

Whole wheat is an excellent source of vitamin B_1 and other vitamins. To make white flour, the wheat germ and bran are removed from the grain. This greatly increases the storage life of the flour, but the remaining material has few minerals or vitamins and little fiber. We eat the starch and use much of the germ and bran for animal food. Thus, cattle and hogs often get better nutrition than we do! Similarly, polished rice has had most of its protein and minerals removed, and it has almost no vitamins. The disease beriberi became prevalent when polished rice was introduced into Southeast Asia.

When many fruits and vegetables are peeled, they lose most of their vitamins, minerals, and fiber. The heat used to cook food also destroys some vitamins. If water is used in cooking, some of the water-soluble vitamins (vitamin C and B vitamins) and minerals are often drained off and discarded with the water.

It is estimated that the average family in the United States spends over 70% of its food budget on processed foods. A diet of hamburgers, potato chips, and soft drinks is lacking in many essential nutrients. Highly processed convenience foods are making many people in developed nations obese but poorly nourished despite their abundance of food.

Malnutrition and starvation can also occur in areas that have no shortage of food when individuals have serious eating disorders, such anorexia or bulimia, which frequently affect young women, although men can suffer from them as well. In anorexia, people believe that their weight is too high and that it is necessary to cut back on food or to exercise very strenuously in order to achieve the "desired" weight, which is well below their healthy weight. Frequently, they eat almost nothing, lose weight, have a very gaunt look, and develop symptoms, such as loss of menses or delayed puberty. In some cases, their weight loss even becomes life-threatening. In bulimia, individuals will frequently indulge in "binge eating" and then get rid of the ingested calories by vomiting or using laxatives. Sometimes a person will exhibit symptoms of both bulimia and anorexia.

SELF-ASSESSMENT Questions

1. During fasting or starvation, glycogen stores are depleted in about
   a. 1 hour   **b.** 1 day   **c.** 1 week   **d.** 1 month

2. After glycogen stores are depleted, the body draws on its reserves of
   a. amino acids   **b.** fat   **c.** glycogen   **d.** protein

3. When fat reserves have been used up, the body uses for its energy source
   a. adipose tissue   **b.** amino acid reserves
   c. glycogen stores   **d.** structural proteins

4. The average family in the United States spends what portion of its food budget on processed foods?
   a. 10%   **b.** 25%   **c.** 50%   **d.** 70%

5. Which of the following statements about malnutrition is *not* true?
   a. It results from an inadequate number of calories.
   b. It can result from eating too much processed food.
   c. It occurs only in developing countries.
   d. It can result from eating disorders such as anorexia or bulimia.

Answers: 1, b; 2, b; 3, d; 4, d; 5, c

17.7 Weight Loss, Diet, and Exercise

Learning Objectives • Explain how weight is lost through diet and exercise.
• Calculate weight loss due to calorie reduction and to exercise.

Exercise and a proper diet as parts of a healthy lifestyle are essential to good health, reducing the risk of diseases of the mind and body. Through chemistry, we gain a better understanding of the effects of diet and exercise on our bodies and minds.

With obesity at epidemic levels and so many people overweight in the United States, dieting is a major industry. It is possible to lose weight through dieting alone, but it isn't easy. One pound of adipose (fatty) tissue stores about 3500 kcal of energy. If you reduce your intake by 100 kcal/day and keep your activity level constant, you will burn off a pound of fat in 35 days. Unfortunately, people are seldom patient enough, and they resort to more stringent diets. To achieve their goals more rapidly, they exclude certain foods and reduce the amounts of others. Such diets can be harmful. Diets with fewer than 1200 kcal/day are likely to be deficient in necessary nutrients, particularly in B vitamins and iron. Furthermore, dieting slows down metabolism. Weight lost through dieting is quickly regained when the dieter resumes old eating habits.

EXAMPLE 17.2 **Weight Loss through Dieting**

If you ordinarily expend 1800 kcal/day and you go on a diet limited to 1500 kcal/day, about how long will it take to lose 1.0 lb of fat?

Solution

You will use 1800 − 1500 = 300 kcal/day more than you consume. There are about 3500 kcal in 1.0 lb of fat, so it will take

$$3500 \ \text{kcal} \times \frac{1 \ \text{day}}{300 \ \text{kcal}} = 11.66 \ \text{or about 12 days}$$

Keep in mind, however, that your weight loss will not be all fat. You will probably lose more than 1 lb, but it will be mostly water with some protein and a little glycogen.

> **Exercise 17.2A**

A person who expends 2000 kcal/day goes on a diet limited to 1200 kcal/day without a change in activities. Estimate how much fat she will lose if she stays on this diet for 3 weeks.

> **Exercise 17.2B**

A person who expends 2200 kcal/day goes on a diet limited to 1800 kcal/day and adds exercise activities that use 220 kcal/day. Estimate how much fat he will lose if he stays on this program for 6 weeks.

Biochemistry of Hunger

Scientists are just beginning to understand the complex biochemistry of hunger mechanisms. Several molecules that regulate body weight have been identified. Some determine whether we want to eat now or stop eating. Others control long-term fat balance. Much of what we now know is disappointing for people who want to lose weight: The hunger mechanisms protect against weight loss and favor weight gain.

Two peptide hormones, known as *ghrelin* and *peptide YY (PYY)*, are produced by the digestive tract and are linked to short-term eating behaviors. Ghrelin, a modified peptide, is an appetite stimulant produced by the stomach. PYY acts as an appetite suppressant. Studies at Imperial College, London, found that people ate about 30% less after they were given a dose of PYY. The research also found that obese people had lower natural levels of PYY, which may explain why these people feel hungrier and overeat.

The hormone insulin and a substance called *leptin*, which is produced by fat cells, determine longer-term weight balance. When insulin levels go up, glucose levels go down, and we experience hunger. Conversely, when glucose levels are high, the activity of brain cells sensitive to glucose is lessened, and we feel satiated.

Leptin, a protein consisting of 146 amino acids, is produced by fat cells. It causes weight loss in mice by decreasing their appetite and increasing their metabolic rates. Levels of leptin tell the hypothalamus (a regulatory center in the brain) how much fat is in the body. Humans also produce leptin, and scientists had hoped it would be the route to a cure for the ever-growing obesity problem, but that hope is largely unrealized. Only a few cases of severe human obesity, caused by defects in leptin production, have been helped by leptin treatment. As it turns out, most obese humans have *higher*-than-normal blood levels of leptin and are resistant to its actions. Leptin's main role seems to be to protect against weight loss in times of scarcity rather than against weight gain in times of plentiful food.

Other substances involved in weight control include *cholecystokinin (CCK)*, a peptide formed in the intestine that signals that we have eaten enough food, and a class of compounds called *melanocortins*, which act on the brain to regulate food intake.

It remains to be seen whether this new knowledge about the body's weight-control systems will pay off in better obesity treatments. A complication is that many social factors are also involved in eating behavior. Some people are motivated by environmental cues to eat. A family gathering, the sight or smell of food, and a stressful situation can all trigger the hunger mechanism.

Crash Diets: Quick = Quack

Any weight-loss program that promises a loss of more than a pound or two a week is likely to be dangerous quackery. Most quick-weight-loss diets depend on factors other than fat metabolism to hook prospective customers. The diets often include a **diuretic**, such as caffeine, to increase the output of urine. Weight loss is water loss, and that weight is regained when the body is rehydrated.

Other quick-weight-loss diets depend on depleting the body's stores of glycogen. On a low-carbohydrate diet, the body draws on its glycogen reserves, depleting them in about 24 hours. Recall that glycogen is a polymer of glucose (Section 16.2). Glycogen molecules have lots of hydroxyl (—OH) groups that can form hydrogen bonds to water molecules. We can store at most about 1 lb of glycogen. Each pound of glycogen carries about 3 lb of water held to it by these hydrogen bonds. Depleting the pound of glycogen results in a weight loss of about 4 lb (1 lb glycogen + 3 lb water). However, no fat is lost, and the weight is quickly regained when the dieter resumes eating carbohydrates.

If your normal energy expenditure is 2400 kcal/day, the most fat you can theoretically lose by *total fasting* for a day is 0.69 lb (2400 kcal/day divided by 3500 kcal/lb of adipose tissue). This assumes that your body burns nothing but fat—which does not occur. The brain runs on glucose, and if that glucose isn't supplied in the diet, it is obtained from protein. Any diet that restricts carbohydrate intake results in a loss of muscle mass as well as fat. However, when you gain the weight back (as 90% of all dieters do), you gain mostly fat. People who diet without exercising will replace metabolically active tissue (muscle) with inactive fat when they gain back the lost weight. Weight loss becomes harder with each subsequent attempted diet.

There is little evidence that commercial weight-loss programs are effective in helping people drop excess pounds and keep them off. Almost no rigorous studies of the programs have been carried out, and U.S. Federal Trade Commission officials say that companies are unwilling to conduct such studies. Many crash diets are also deficient in minerals such as iron, calcium, and potassium. A deficiency of these minerals can disrupt nerve-impulse transmissions to muscles, which would impair athletic performance. Nerve-impulse transmission to vital organs may also be impaired in cases of severe restriction, and death from cardiac arrest can result.

Exercise for Weight Loss

Studies consistently show that people who exercise regularly live longer. They are sick less often and have fewer signs of depression. They can move faster, and they have stronger bones and muscles. Although there are dozens of good reasons for regular exercise, many people actually begin exercise programs for one simple reason: They want to lose weight.

People who do not increase their food intake when they begin an exercise program lose weight. Contrary to a common myth, exercise (for up to an hour a day) does not cause an increase in appetite. Most of the weight loss from exercise results from an increase in metabolic rate during the activity, but the increased metabolic rate continues for several hours after completion of exercise. Exercise helps us maintain both fitness and proper body weight. On a weight-loss diet without exercise, about 65% of the weight lost is fat and about 11% is protein (muscle tissue). The rest is water and a little glycogen.

EXAMPLE 17.3 Weight Loss through Exercise

A 200 lb person doing high-impact aerobics burns about 11.2 kcal/min. How long does a person have to do such exercise to lose 1.0 lb of adipose (fat) tissue?

Solution

One pound of adipose tissue stores 3500 kcal of energy. To burn it at 11.2 kcal/min requires

$$3500 \text{ kcal} \times \frac{1 \text{ min}}{11.2 \text{ kcal}} = 313 \text{ min}$$

It takes about 313 min (about 5.25 h) to burn 1 lb of fat, even doing high-impact aerobics.

❯ Exercise 17.3A

Walking a mile uses about 100 kcal. About how far do you have to walk to burn 2.0 lb of fat?

❯ Exercise 17.2B

A moderately active person can calculate the calories needed each day to maintain a desired weight by multiplying the desired weight (in pounds) by 15 kcal/lb. How many calories per day does such a person need to maintain a weight of 180 lb?

Fad Diets

Weight-loss diets are often lacking in balanced nutrition and can be harmful to one's health. In the more extreme low-carbohydrate diets, ketosis is deliberately induced, and possible side effects include depression and lethargy. In the early stages of a diet deficient in carbohydrates, the body converts amino acids to the glucose that the brain requires. If there are enough adequate proteins in the diet, tissue proteins are spared.

Even low-carbohydrate diets high in adequate proteins are hard on the body, which must rid itself of the nitrogen compounds—ammonia and urea—formed by the breakdown of proteins. This puts an increased stress on the liver, where the waste products are formed, and on the kidneys, where they are excreted.

Contrary to a popular notion, fasting does not "cleanse" the body. Indeed, quite the reverse occurs. A shift to fat metabolism produces ketone bodies (Section 17.6), and protein breakdown produces ammonia, urea, and other wastes. You can lose weight by fasting, but the process should be carefully monitored by a physician.

The most sensible approach to weight loss is to adhere to a balanced low-calorie diet that meets the DRI for essential nutrients and to engage in a reasonable, consistent, individualized exercise program. This approach applies the principles of weight loss by decreasing intake and increasing output.

Weight loss or gain is based on the law of conservation of energy (Section 15.4). When we take in more calories than we use up, the excess calories are stored as fat. When we take in fewer calories than we need for our activities, our bodies burn some of the stored fat to make up for the deficit. One pound of adipose tissue requires 200 miles of blood capillaries to serve its cells. Excess fat therefore puts extra strain on the heart.

5 If I stop exercising, will the muscle I've developed turn to fat?

Muscle does not turn into fat, although without regular exercise, muscles become flabby and will lose mass.

SELF-ASSESSMENT Questions

1. To lose 1.0 lb per week, how many more kilocalories must you burn each day than you take in?
  a. 100   **b.** 300   **c.** 500   **d.** 1000

2. Which of the following hormones acts as an appetite suppressant?
  a. ADH   **b.** ghrelin   **c.** leptin   **d.** PYY

3. Blood glucose levels are lowered by the hormone
  a. cholecystokinin   **b.** glucagon
  c. insulin   **d.** oxytocin

4. Which of the following seems to prevent weight loss in humans during times when the body cannot get food?
  a. cholecystokinin (CCK)   **b.** ghrelin
  c. leptin   **d.** oxytocin

5. Crash diets
  a. are a good way to lose weight and keep it off
  b. depend on glycogen depletion for quick weight loss
  c. result in a decrease in the number of fat cells
  d. lead to a depletion of adipose tissue

Answers: 1, c; 2, d; 3, c; 4, c; 5, b

17.8 Fitness and Muscle

Learning Objectives • Describe several ways to measure fitness and percent body fat. • Differentiate between aerobic exercise and anaerobic exercise, and describe the chemistry that occurs during each. • Describe how muscles are built and how they work.

▲ **Figure 17.12** When submerged in water, a body displaces its own volume of water. The difference between a person's weight in air and when submerged in water corresponds to the *mass* of the displaced water. Because the density of water is 1.00 g/mL, the mass of water displaced (in grams) equals the person's volume (in milliliters).

Q *What is the volume, in liters, of a person who displaces 48.5 kg of water?*

Measuring fitness is roughly the same as measuring body fat. How much fat is enough? The male body requires about 3% body fat, and the average female body needs 10% to 12%. It is not easy to measure percent body fat accurately. Weight alone does not indicate degree of fitness. A tall 200 lb man may be much more fit than a 160 lb man with a smaller frame. Skinfold calipers are sometimes used, but they are quite inaccurate and measure water retention as well as fat.

One way to estimate body fat is by measuring a person's density. Mass is determined by weighing, and volume is calculated with the help of a dunk tank like the one shown in Figure 17.12. The difference between a person's weight in air and when submerged in water corresponds to the *mass* of the displaced water. When this mass is divided by the density of water and then corrected for air in the lungs, it gives a reasonable value for the person's volume. Mass divided by volume yields the person's density, although results can vary with the amount of air in the lungs. Fat is less dense (0.903 g/mL) than the water (1.00 g/mL) that makes up most of the body's mass. The higher the proportion of body fat a person has, the lower the density and the more buoyant that person is in water.

A simpler way to estimate percent body fat is by measuring the waist and the hips. The waist should be measured at the narrowest point and the hips should be measured where the circumference is the largest. The waist measurement should then be divided by the hip measurement. This ratio should be less than 1 for men and should be 0.8 or less for women. (Note that the units will cancel out, as long as the same units are used for both measurements.)

Some modern scales provide a measure of body fat content, using *bioelectric impedance analysis*. The person stands on the scale in bare feet, and a small electric current is sent through the body. Fat has greater impedance (resistance to varying current) than does muscle. By measuring the impedance of the body, the percentage of body fat can be calculated based on height and weight. However, many other variables, such as bone density, water content, and location of fat, can affect the reading. Although such scales are not very accurate, they may be useful for monitoring *changes* in body fat content.

Body Mass Index

Body mass index (BMI) is a commonly used measure of body fat. It is defined as weight (in kilograms) divided by the square of the height (in meters). For a person who is 1.6 m tall and weighs 62 kg, the body mass index is $62/(1.6)^2 = 24$. A BMI below 18.5 indicates that a person is underweight, while a BMI of 18.5 to <25 is in the normal range. A BMI from 25 to <30 indicates that a person is overweight, and a

BMI of 30 or more indicates obesity. When measurements are in pounds and inches, the equation is

$$BMI = \frac{705 \times body\ weight\ (lb)}{[height\ (in)]^2}$$

In other words, a person who is 5 ft, 10 in tall (70 in) and weighs 140 lb has a BMI of

$$\frac{705 \times 140}{70 \times 70} = 20$$

which is within the ideal range.

 EXAMPLE 17.4 Body Mass Index

What is the BMI for a person who is 6 ft, 3 in tall and weighs 350 lb?

Solution
The person's height is $(6 \times 12) + 3 = 72 + 3 = 75$ in

$$BMI = \frac{705 \times 350}{75 \times 75} = 43.9$$

The body mass index is 44.

> Exercise 17.4A
What is the BMI for a person who is 6.0 ft tall and weighs 120 lb?

> Exercise 17.4B
What is the maximum weight, in pounds, that a person who is 5 ft, 10 in. tall can maintain and have a BMI that does not exceed 25.0?

V_{O_2} max: A Measure of Fitness

As we increase exercise intensity, our uptake of oxygen must also increase. For example, the faster we run, the more oxygen we need to sustain the pace. However, the body reaches a point at which it simply cannot increase the amount of oxygen it consumes even if the intensity of exercise increases. The V_{O_2} max is the maximum amount of oxygen (in milliliters of oxygen per kilogram of body weight) that a person can use in 1 minute.

V_{O_2} max is therefore a measure of fitness. The higher the V_{O_2} max, the greater an athlete's fitness is. A person with a high V_{O_2} max can exercise more intensely than one who is less fit. The V_{O_2} max can be increased by working out at an intensity that raises the heart rate to 65% to 85% of its maximum for at least 20 minutes 3–5 times a week. Limitations on V_{O_2} max include the ability of muscle cells to use oxygen in metabolizing fuels and the ability of the cardiovascular system and lungs to transport oxygen to the muscles.

Direct testing of V_{O_2} max requires expensive equipment, including a gas analyzer to measure O_2 taken in and exhaled. Percent V_{O_2} max can be estimated indirectly from percent maximum heart rate (% MHR), where MHR is the highest number of beats the heart makes in a minute of exercising.

$$MHR = (0.64 \times \%V_{O_2}\ max) + 37$$

The relationship holds quite well for both males and females of all ages and activity levels. For example, 80% MHR corresponds to a % V_{O_2} max of

$$80 = (0.64 \times \%V_{O_2}\ max) + 37 \qquad \%V_{O_2}\ max = 67$$

WHY IT MATTERS

In Europe during the Middle Ages—and even later—people accused of crimes were often tried by some kind of ordeal. In trial by water, the innocent sank and the guilty floated. Suspected witches were tied hand to foot and thrown into the water. The guilty floated and were fished out, dried off, and executed. In those days, body density was *really* important!

6 **What is the difference between being overweight and being obese?**

While the human body needs some minimal fat stores to function normally, any amount more than the optimum is considered unhealthy. Anyone with a body mass index (BMI) greater than 25 kg/m² is considered overweight. A BMI over 30 kg/m² is defined as making one obese. Being overweight can lead to increased health risks, but being obese dramatically increases the likelihood of type 2 diabetes, heart disease, osteoarthritis, obstructive sleep apnea, and certain types of cancer.

Equations for determining MHR values and tables for evaluating fitness levels can be found online.

Some Muscular Chemistry

The human body has about 600 muscles. Exercise makes these muscles larger, more flexible, and more efficient in their use of oxygen. Exercise strengthens the heart because the heart is an organ composed mainly of muscle. With regular exercise, resting pulse and blood pressure usually decline. A person who exercises regularly is able to do more physical work with less strain. Exercise is an art, but it is also increasingly a science—a science in which chemistry plays a vital role.

Energy for Muscle Contraction: ATP

When cells metabolize glucose or fatty acids, only part of the chemical energy in these substances is converted to heat. Some of the energy is stored in the high-energy phosphate bonds of adenosine triphosphate (ATP) molecules.

Adenosine triphosphate

Stimulation of muscles causes them to contract. This contraction is *work*, and it requires energy that the muscles get from the molecules of ATP. The energy stored in ATP powers the physical movement of muscle tissue. Two proteins, actin and myosin, play important roles in this process. Together they form a loose complex called *actomyosin*, the contractile protein that makes up muscles (Figure 17.13).

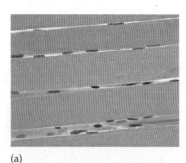

(a)

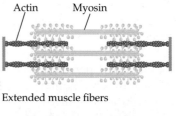

Extended muscle fibers

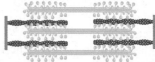

Resting fibers

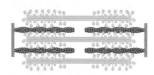

Partially contracted fibers

(b)

▲ **Figure 17.13** (a) Skeletal muscle tissue has a banded, or *striated*, appearance, shown here in a micrograph at 180× magnification. (b) The diagram of the actomyosin complex in muscle shows extended muscle fibers (top), resting fibers (middle), and partially contracted fibers (bottom).

When ATP is added to actomyosin in the laboratory, the protein fibers contract. Myosins turn the chemical energy of ATP into motion. Myosin molecules "walk" along the actin filaments. It takes about 2×10^{12} myosin molecules acting in unison in the arm to lift a 40 g spoonful of sugar. Besides serving as part of the structural complex in muscles, myosin also acts as an enzyme for the removal of a phosphate group from ATP. Thus, it is directly involved in liberating the energy required for muscle contraction.

In a resting person, muscle activity (including that of the heart muscle) accounts for only about 15% to 30% of the body's energy requirements. Other activities, such as cell repair, transmission of nerve impulses, and maintenance of body temperature, account for the remaining energy needs. During intense physical activity, the energy requirements of muscle may increase to more than 200 times the resting level.

Aerobic Exercise: Plenty of Oxygen

The ATP in muscle tissue is sufficient for activities lasting at most a few seconds. Fortunately, muscles have a more extensive energy supply: glycogen, a form of stored glucose.

When muscle contraction begins, glycogen is converted to pyruvic acid by muscle cells, in a series of steps.

$$(C_6H_{10}O_5)_n \longrightarrow 2n \ CH_3\overset{\overset{O}{\|}}{C}\!-\!\overset{\overset{O}{\|}}{C}OH$$

Glycogen   Pyruvic acid

Then, if sufficient oxygen is readily available, the pyruvic acid is oxidized to carbon dioxide and water in another series of steps. Because acids in biological systems are usually ionized at the pH of cellular fluids, biochemists usually refer to them by the names of their anions. For example, lactic acid becomes lactate, and pyruvic acid becomes pyruvate.

$$2 \ CH_3\overset{\overset{O}{\|}}{C}\!-\!\overset{\overset{O}{\|}}{C}OH + 5 O_2 \longrightarrow 6 CO_2 + 4 H_2O$$

Muscle contractions that occur under these circumstances—that is, in the presence of oxygen—constitute **aerobic exercise** (Figure 17.14).

Anaerobic Exercise and Oxygen Debt

When sufficient oxygen is not available, pyruvic acid is reduced to lactic acid.

$$CH_3COCOOH + [2 H] \longrightarrow CH_3CHOHCOOH$$

[2 H] represents hydrogen ions (H^+) transferred from one of several biochemical reducing agents.

If **anaerobic exercise** (muscle activity with insufficient oxygen) persists, an excess of lactic acid builds up in the muscle cells. This lactic acid ionizes, forming lactate ions and hydronium ions.

$$CH_3CHOHCOOH + H_2O \longrightarrow CH_3CHOHCOO^- + H_3O^+$$

Muscle fatigue correlates well with lactate levels, which were long thought to be its cause. However, studies have shown that muscle tiredness is actually related to calcium ion flow. Ordinarily, muscle contractions are managed by the ebb and flow of Ca^{2+} ions. As muscles grow tired, tiny channels in muscle cells start leaking Ca^{2+}, which weakens contractions. At the same time, the leaked Ca^{2+} ions stimulate an enzyme that breaks down muscle fibers, further contributing to the muscle fatigue. Levels of several other products of muscle metabolism—including phosphocreatine (the phosphorylated form of creatine, Section 17.1), ATP, and ions such as Na^+, K^+, and $H_2PO_4^-$—change during fatigue. These changes may also contribute to muscle tiredness.

$$H_2O_3PNH\!-\!\overset{\overset{}{\underset{\underset{NH}{\|}}{C}}}{}\!-\!\overset{\underset{CH_3}{|}}{N}\!-\!CH_2COOH$$

Phosphocreatine

▲ **Figure 17.14** Aerobic exercise is performed at a pace that allows us to get enough oxygen to our muscle cells to oxidize pyruvic acid to carbon dioxide and water. Aerobic dance is a popular form of aerobic exercise.

▲ **Figure 17.15** After a race, the metabolism that fueled the effort continues. The athlete gulps air to repay her oxygen debt.

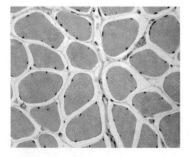

▲ **Figure 17.16** Slow-twitch muscle fibers have more mitochondria (elongated violet dots) than fast-twitch fibers do.

Well-trained athletes can continue for a while after muscle fatigue sets in, but most people quit. At this point, the oxygen debt starts to be repaid (Figure 17.15). After exercise ends, the cells' demand for oxygen decreases, making more oxygen available to oxidize the lactic acid resulting from anaerobic metabolism back to pyruvic acid, which is then converted to carbon dioxide, water, and energy.

Most athletes emphasize one type of exercise (anaerobic or aerobic) over the other. For example, an athlete training for a 60 m dash does mainly anaerobic exercising, but one planning to run a 10 km race does mainly aerobic exercising. Sprinting and weightlifting are largely anaerobic activities. A marathon run (42.195 km) is largely aerobic. During a marathon, athletes must set a pace to run for more than 2 hours. Their muscle cells depend on slow, steady aerobic conversion of carbohydrates to energy. During anaerobic activities, however, muscle cells use almost no oxygen. Rather, they depend on the quick energy provided by anaerobic metabolism.

After glycogen stores are depleted, muscle cells can switch over to fat metabolism. Fats are the main source of energy for sustained activity of low or moderate intensity, such as the last part of a marathon run.

Muscle Fibers: Twitch Kind Do You Have?

Muscles are tools with which we do work. The quality and type of muscle fibers that an athlete has affect that athlete's athletic performance, just as the quality of machinery influences how work is done. For example, we can remove snow from a driveway in more than one way. We can do it in 10 min with a snowblower, or we can spend 50 min shoveling. Muscles are classified according to the speed and effort required to accomplish this work. The two classes of muscle fibers are **fast-twitch fibers**, which are stronger, larger, and more suited for anaerobic activity, and **slow-twitch fibers**, which are better for aerobic work (Figure 17.16). Table 17.8 lists some characteristics of these two types of muscle fibers.

Type I (slow-twitch) fibers are called on during activity of light or moderate intensity. The respiratory capacity of these fibers is high, which means that they can provide a large amount of energy via aerobic pathways. The myoglobin level is also high. Myoglobin is a red, heme-containing protein in muscle that stores oxygen (similar to the way hemoglobin stores and transports oxygen in the blood). Aerobic oxidation requires oxygen, and muscle tissue rich in slow-twitch fibers is supplied with high levels of oxygen.

TABLE 17.8 A Comparison of Two Types of Muscle Fibers

Characteristic	Type I	Type IIB[a]
Category	Slow twitch	Fast twitch
Color	Red	White
Respiratory capacity	High	Low
Myoglobin level	High	Low
Catalytic activity of actomyosin	Low	High
Capacity for glycogen use	Low	High

[a]There is a type IIA fiber (not considered here) that resembles type I in some respects and type IIB in others.

Type I muscle fibers have a low capacity to use glycogen and thus are not geared to anaerobic generation of energy. Their action does not require the hydrolysis of glycogen. The catalytic activity of the actomyosin complex is low. Remember that actomyosin not only is the structural unit in muscle that undergoes contraction but also is responsible for catalyzing the hydrolysis of ATP to provide energy for the contraction. Low catalytic activity means that the energy is parceled out more slowly. Slow energy output is not good if you want to lift 200 kg, but it is great for a 15 km run.

Type IIB (fast-twitch) fibers have characteristics opposite those of type I fibers. Low respiratory capacity and low myoglobin levels do not bode well for aerobic oxidation. A high capacity for glycogen use and a high catalytic activity of actomyosin allow tissue rich in fast-twitch fibers to generate ATP rapidly and also to hydrolyze this ATP rapidly during intense muscle activity. Thus, this type of muscle tissue has the capacity to do short bursts of vigorous work but fatigues rather quickly. A period of recovery, during which lactic acid is cleared from the muscle, is required between the brief periods of activity.

Building Muscles

Endurance exercise increases myoglobin levels in skeletal muscles. The increased myoglobin provides faster oxygen transport and increased respiratory capacity (Figure 17.17), and these changes usually are apparent within 2 weeks. Endurance training does not necessarily increase the size of muscles.

If you want larger muscles, try weightlifting. Weight training (Figure 17.18) develops fast-twitch muscle fibers. These fibers increase in size and strength with repeated anaerobic exercise. Weight training does *not* increase respiratory capacity.

Some bodybuilders use supplements to try to increase muscle mass. *Creatine* is an organic acid that is naturally produced in the liver and kidneys (Section 17.1). As such, it is not an essential nutrient. Creatine is also acquired through the consumption of meat. A normal diet including protein can provide an adequate amount of creatine. An athlete or bodybuilder, though, might switch to a high-protein diet to increase muscle mass. Creatine, when used as a supplement for those seeking to build muscle mass, is often found in amounts 2 or 3 times that available from a high-protein diet.

Muscle-fiber type seems to be inherited. Research shows that world-class marathon runners may possess as high a proportion as 80% to 90% slow-twitch fibers, while championship sprinters may have up to 70% fast-twitch muscle fibers. Some exceptions have been noted, however, and factors such as training and body composition are important in athletics, as are nutrition, fluid and electrolyte balance, and drug use or misuse.

▲ **Figure 17.17** The Boston Marathon attracts close to 30,000 runners each year.

Q *What is the main muscle type in the elite runners in this race? Do their muscles have high or low aerobic capacity?*

▲ **Figure 17.18** Weightlifting is a popular activity in many gyms and fitness centers.

Q *What is the predominant muscle type in elite weightlifters? Do their muscles have high or low aerobic capacity?*

SELF-ASSESSMENT Questions

1. Body mass index (BMI) is calculated as
   a. percentage of body fat × waist circumference (in)
   b. (height ÷ weight) × 100
   c. [705 × weight (lb)] ÷ [height (in)]
   d. weight (kg) ÷ [height (m)]^2

2. The maximum amount of oxygen used in a minute of exercise is the
   a. aerobic threshold   **b.** lactate threshold
   c. oxygen debt   **d.** V_{O_2} max

3. Which of the following is a molecule used by the cells to energize the ATP molecule?
   a. an amino acid   **b.** actomycin
   c. glycogen   **d.** glucose

4. During aerobic exercise, pyruvic acid is converted to
   a. CO_2 and H_2O **b.** glycogen
   c. lactic acid   **d.** phosphocreatine

5. During anaerobic exercise, pyruvic acid is converted to
   a. CO_2 and H_2O **b.** glycogen
   c. lactic acid   **d.** phosphocreatine

6. When produced during exercise, which of the following results in oxygen debt?
   a. ATP   **b.** glycogen
   c. lactic acid   **d.** pyruvic acid

7. The reddish color of slow-twitch muscle fibers is due to the presence of
   a. actomyosin   **b.** creatine
   c. myoglobin   **d.** pyruvic acid

Answers: 1, d; 2, d; 3, d; 4, a; 5, c; 6, c; 7, c

GREEN CHEMISTRY The Future of Food Waste—A Green Chemistry Perspective

Katie Privett, *Green Chemistry Centre of Excellence, York, United Kingdom*

Principles 3, 6, 7

Learning Objectives • Explain how food supply chain waste can be used as a resource in the chemical industry.
• Identify the advantages of using food supply chain waste instead of fossil fuels as a source of chemicals, materials, and fuels.

A study by the United Nations Food and Agriculture Organization discovered that, globally, we generate 1.3 billion metric tons (t) (1 t = 1000 kg) of food waste every year—one-third of all the food we grow. Much of this waste can be avoided by improving the storage and processing of food before it reaches the shelves and by better educating shoppers to minimize post-consumer waste. However, some food waste within the supply chain is simply unavoidable—straw from farms, seafood shells, and citrus peels all are by-products that were never intended for human consumption.

▲ A simple food-waste bin in use.

In most cases, these food supply chain wastes (FSCW) currently go to landfills, where their decomposition affects soil quality, pollutes local water supplies, and releases the potent greenhouse gas methane. With the cost of disposal rising and the social acceptability of landfill sites decreasing, we have to find new ways to deal with this waste.

This is where green chemistry steps in. Green Chemistry Principle 7 states that resources should be renewable whenever technically and economically practicable. FSCW is a great example of an economically viable renewable material—instead of paying landfill taxes and fees to dispose of it as waste, the material can be used as a cheap feedstock for chemical processes. A relatively simple approach to gaining value out of FSCW is to turn it into a fuel. Anaerobic digestion is a technique that allows mixed biomass waste to be converted through microbial processes into a fuel and a fertilizer. Both of these products can replace alternatives derived from fossil fuels. However, these are relatively low-value products, and the efficiency of the process has to be very high to match the prices offered by the well-developed fossil-fuel industry.

When particular types of supply chain wastes are concentrated, such as agricultural wastes at large farms, citrus peels at juicing

plants, and spent grains at breweries, it can become technically practicable to collect and process them at large scales. Such concentrated wastes contain many higher-value chemical compounds that have industrial uses and that can be extracted economically.

One example of an industry that would benefit from utilizing its supply chain waste is orange juice. During juicing, half of the orange goes to waste, mostly as peel. Over 45 million t of oranges are juiced annually, so 22 million t of peel become waste. The usual disposal process is to bury the peels far underground because some of the components are toxic to vital soil microbes. Yet citrus peels contain a vast array of useful chemicals including limonene, a fragrant molecule used in detergents and personal care products, and pectin, used as a thickening agent in foods such as jam. These components are complex chemicals that nature has spent time and energy to make. It is a waste to break these chemicals down to a hydrocarbon mixture and try to rebuild similar compounds using synthetic organic techniques. Instead, by using mild, selective extraction techniques, these chemicals can be taken straight from the peel.

Innovative green chemistry also can be used to make bio-based material from FSCW for safer consumer products. Chemists have used waste straw from agricultural processes to make a bioboard that can replace traditional woodchip boards in furniture. The straw replaces the wood within the board, and the binder that holds the material together is made from waste biomass ash. This silicate binder replaces widely used and toxic chemical binders and makes the bioboards carbon neutral and nontoxic.

By using FSCW as a feedstock for chemical processes, we can avoid using fossil-based resources, minimize chemical processing, and reduce pollution. Green chemistry is a key to a bio-based economy that treats waste as a resource.

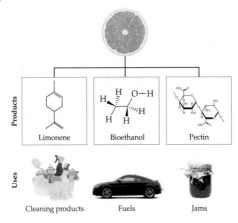

▲ Valuable chemicals from orange peel

Summary

Section 17.1—Good health and fitness require a nutritious diet and regular exercise. Today's dietary guidelines emphasize nutrient-dense foods and a balanced eating program. A balanced diet consists largely of fruits, vegetables, whole grains, and dairy products in the proper proportions. Total fats intake should not exceed 35% of calories; saturated fats (mostly of animal origin) should account for less than 10% of calories. Athletes generally need more calories, which should come mostly from carbohydrates—especially starches. Exercise, not extra protein, builds muscle tissue. The **Dietary Reference Intakes (DRIs)** for protein and other nutrients are established by the U.S. National Academy of Sciences for planning and assessing diets.

Section 17.2—**Dietary minerals** are inorganic substances vital to life. Bulk structural elements include carbon, hydrogen, and oxygen. Macrominerals include sodium, potassium, and calcium. We also need trace elements such as iron and ultratrace elements such as manganese, chromium, and vanadium. Minerals serve a variety of functions: Iodine is needed by the thyroid; iron is part of hemoglobin; and calcium and phosphorus are vital to the bones and teeth.

Section 17.3—Organic substances (vitamins) that the body cannot produce must be supplied by the diet. **Vitamins** are organic compounds that are required to prevent certain diseases. The B vitamins and vitamin C are water-soluble and are needed frequently, whereas vitamins A, D, E, and K are all fat-soluble and can be stored. Vitamin A is essential for vision and skin maintenance. There are several B vitamins. B_3 and B_6 may help arthritis patients; B_{12} is not found in plants, and a lack can cause anemia. Vitamin C prevents scurvy, promotes healing, is needed for the immune system, and is an antioxidant. Vitamin D promotes absorption of calcium and phosphorus. Vitamin E is an antioxidant and an anticoagulant, and a deficiency can lead to muscular dystrophy.

Section 17.4—Dietary fiber may be soluble (gums and pectins) or insoluble (cellulose). It prevents digestion problems such as constipation. An **electrolyte** is a substance that conducts electricity in a water solution. The main electrolytes in the body are ions of sodium, potassium, chlorine, calcium, and magnesium, as well as sulfate, hydrogen phosphate, and bicarbonate ions. Water is an essential nutrient.

Section 17.5—**Food additives** are substances other than basic foodstuffs that are added to food to aid nutrition, inhibit spoilage, enhance color and flavor, provide texture, and so on. There are thousands of additives. The FDA regulates which additives, and how much of them, can be used. The list of additives generally recognized as safe is the **GRAS list**, which is periodically reevaluated.

Spices and herbs come from plants, and make food more interesting, aromatic, and flavorful. Synthetic flavorings, such as vanilla, contain the same primary molecule as the original plants but are less expensive. A variety of artificial sweeteners has been developed to reduce sugar intake, but this does not appear to aid in controlling obesity. Glycerol is a sweet-tasting compound that is used mainly as a **humectant**, or moistening agent. Salt and MSG are flavor enhancers.

Enrichment includes the replacement of B vitamins and the addition of iron to flour. Vitamin C is added to many beverages, and vitamin D to milk. Propionate, sorbate, benzoate, and nitrite salts inhibit spoilage and bacterial growth.

Fats may turn rancid partly from formation of **free radicals** with unpaired electrons. One free radical can cause a **chain reaction** that leads to decomposition of many fat molecules. BHT and BHA are **antioxidants** that react with free radicals and prevent rancidity.

Natural food colorings, such as β-carotene, and synthetic food colorings improve the appearance of food. Seven artificial colorants have been approved for use in foods, but only six of them are easily found in foods.

Section 17.6—**Starvation** is deprivation of food that causes metabolic changes. The body first depletes its glycogen and then metabolizes fat and muscle tissue. Malnutrition can result from starvation, dieting, or eating too much highly processed food.

Section 17.7—Drastic diets are unlikely to be permanently successful, and the weight lost is usually regained. One pound of adipose tissue stores about 3500 kcal of energy. A variety of chemical substances that are produced in the body are regulators of body weight. Quick-weight-loss diets often include a **diuretic** to increase urine output, which results in temporary weight loss. Low-carbohydrate diets cause quick weight loss by depleting glycogen and water.

Exercise aids in weight loss, because the increase in metabolic rate that it causes continues after it ends. The most sensible approach to weight loss involves a balanced low-calorie diet and a consistent exercise program.

Section 17.8—The male body requires about 3% body fat; the female body, 10% to 12%. The percentage of body fat can be estimated by several methods. Density measurement is probably the most accurate. **Body mass index (BMI)** indicates percent body fat, and V_{O_2} max is a measure of fitness.

Exercise makes muscles larger, more flexible, and more efficient. Energy for muscular work comes from ATP. A protein complex called *actomyosin* contracts when ATP is added to it. ATP provides energy for just a few seconds, after which glycogen is metabolized. If metabolism occurs with plenty of oxygen—during **aerobic exercise**—CO_2 and H_2O are formed. With insufficient oxygen—during **anaerobic exercise**—lactic acid is formed and muscle fatigue occurs. When exercise ends, the **oxygen debt** is repaid; cell oxygen demand decreases, and the lactic acid can be oxidized to CO_2 and H_2O. There are two classes of muscle fibers. **Fast-twitch fibers** are larger, stronger, and good for short bursts of vigorous exercise, while **slow-twitch fibers** are geared for steady exercise of long endurance. Weight training develops fast-twitch muscles but does not increase respiratory capacity. Long-distance running develops slow-twitch muscles and increases respiratory capacity.

GREEN CHEMISTRY—Food supply chain waste (FSCW) can be used as a renewable source of valuable chemicals, materials, and fuels. Diverting food waste from other polluting endpoints such as landfills and incinerators reduces its environmental impact. More importantly, economically and technically practicable green chemistry routes to the creation of higher-value products from FSCW are being developed. Therefore, bio-based materials made from food waste can replace fossil-fuel-derived alternatives and facilitate the movement toward a bio-based economy.

Learning Objectives	Associated Problems
• List the recommendations (sources and percentages) for calories in the American diet. (17.1)	6, 8, 9, 13, 14, 16–18, 59, 62, 65, 68, 74
• Describe the special dietary requirements of athletes. (17.1)	3, 15
• Identify the bulk dietary minerals, and state their functions. (17.2)	19–22
• Identify the vitamins, and state their functions. (17.3)	1, 5, 23–28, 61, 75
• Identify the roles of fiber, electrolytes, and water in maintaining health. (17.4)	29–32, 37
• Describe the historic impact of our desire for a variety of flavors and aromas, and of our search for sugar substitutes. (17.5)	4, 11, 34, 36, 69
• Identify the beneficial additives in foods and those that are somewhat controversial. (17.5)	10, 33, 35, 38, 61
• Describe the effects of starvation, fasting, and malnutrition. (17.6)	2, 39–42
• Explain how weight is lost through diet and exercise. (17.7)	12, 43–46, 60, 66, 72
• Calculate weight loss due to calorie reduction and to exercise. (17.7)	47–50
• Describe several ways to measure fitness and percent body fat. (17.8)	7, 51, 63, 67
• Differentiate between aerobic exercise and anaerobic exercise, and describe the chemistry that occurs during each. (17.8)	52–58, 64, 73, 76
• Describe how muscles are built and how they work. (17.8).	70, 71
• Explain how food supply chain waste can be used as a resource in the chemical industry.	77, 78
• Identify the advantages of using food supply chain waste instead of fossil fuels as a source of chemicals, materials, and fuels.	79

Conceptual Questions

1. Is an excess of a water-soluble vitamin or an excess of a fat-soluble vitamin more likely to be dangerous? Why?

2. What is starvation?

3. How do the nutritional needs of an athlete differ from those of a sedentary individual? How are these extra needs best met?

4. Name three artificial sweeteners. Are any of them approved for current use in the United States?

5. What fat-soluble vitamin serves as an antioxidant? What water-soluble vitamin serves as an antioxidant?

6. Why are the federal dietary guidelines (Figure 17.1) not the same for everyone?

7. List two ways to determine percent body fat. Describe a limitation of each method.

8. What kinds of foods are the main dietary source of saturated fats?

9. List some problems that result from high-fat diets.

10. List five functions of food additives.

11. What U.S. government agency regulates the use of food additives? What must be done before a company can use a new food additive?

12. How might you lose 5 lb in just 1 week? Would it be a useful way to lose weight and keep it off?

Problems

Calories: Quality and Quantity

13. How do dietary proteins differ from dietary carbohydrates and fats in elemental composition?

14. The average energy yield per gram of carbohydrate and of fat is
    a. 4 kcal each
    b. 9 kcal each
    c. 4 kcal and 9 kcal, respectively
    d. 9 kcal and 4 kcal, respectively

15. What are the common dietary needs for athletes? Explain how athletes' needs differ from the typical diet.

16. Which essential amino acids are likely to be lacking in corn? In beans?

17. What is an adequate protein? List some foods that contain adequate proteins.

18. The recommended range for daily caloric intake is that a person get 45% to 65% of calories from
    a. carbohydrates   b. fats   c. fiber   d. protein

Minerals

19. Name the mineral ion that matches the biological function.
  a. thyroid (regulating metabolism)
  b. oxygen transport in the blood
  c. bones and teeth; heartbeat rhythm
  d. water regulation in the cells

20. Of the minerals Ca, Cl, Co, Mo, Na, P, and Zn, which is likely to be found in relatively large amounts in the human body? Explain.

21. A deficiency of which of the following minerals will result in anemia?
  a. calcium **b.** copper
  c. iron **d.** phosphorus

22. A major cause of high blood pressure is excessive intake of
  a. calcium **b.** iodine
  c. iron **d.** salt

Vitamins

23. What are the benefits of vitamin D? Is taking a daily mega-dose of vitamin D a good idea?

24. What is the safest way to supplement your vitamin A intake?

25. Of the vitamins calciferol, riboflavin, retinol, and cyano-cobalamin, which are most likely to be needed on a daily basis? Explain.

26. Vitamins C and E are antioxidants. What does this mean?

27. Match the compound with its designation as a vitamin.

Compound	Designation
Ascorbic acid	vitamin A
Calciferol	vitamin B_{12}
Cyanocobalamin	vitamin C
Retinol	vitamin D
Tocopherol	vitamin E

28. What is scurvy, and what vitamin prevents it?

Fiber, Electrolytes, and Water

29. What is a diuretic? Describe the function of antidiuretic hormone (ADH).

30. How is the appearance of urine related to dehydration?

31. Heat exhaustion occurs when what percentage of the body's fluid volume is lost?
  a. 3% **b.** 5% **c.** 7% **d.** 10%

32. Which of the following is helpful in avoiding constipation?
  a. carbohydrates **b.** electrolytes
  c. fiber **d.** protein

Food Additives

33. BHA and BHT help preserve the quality of packaged food by
  a. enabling radiation to sterilize the food
  b. killing pathogenic bacteria
  c. preventing the food from turning brown
  d. scavenging free radicals, consequently stopping fats from becoming rancid

34. Which of the following sugar substitutes contains chlorine?
  a. aspartame **b.** cyclamate
  c. stevia **d.** sucralose

35. What is the purpose of each of the following food additives?
  a. BHA **b.** FD&C Blue No. 2 **c.** saccharin

36. Which of the following is a flavor enhancer?
  a. iodine **b.** MSG
  c. saccharine **d.** sucralose

37. Coffee, tea, and cola drinks all contain the stimulant
  a. amphetamine **b.** caffeine
  c. fructose **d.** cocaine

38. What are antioxidants? Name a natural antioxidant and two synthetic antioxidants.

Starvation, Fasting, and Malnutrition

39. When the body is deprived of food, which source of energy does it utilize first?
  a. glycogen **b.** fat **c.** muscle **d.** protein

40. Ketosis results when the body begins to metabolize _____ as its main energy source.
  a. carbohydrate **b.** fat **c.** muscle **d.** protein

41. Malnutrition can result from all of the following except
  a. eating processed foods **b.** incomplete protein
  c. anorexia **d.** inadequate fiber

42. All of the following lower the nutrition of foods except
  a. boiling in water
  b. chopping fruits and vegetables
  c. peeling fruits and vegetables
  d. polishing rice

Weight Loss, Diet, and Exercise

43. Describe the role of each of the following substances involved in the control of body weight.
  a. leptin **b.** ghrelin

44. Describe the role of each of the following substances involved in the control of body weight.
  a. melanocortins **b.** PYY

45. What are some popular diets, and in what ways do they stimulate quick weight loss? Why are the effects sometimes short-lasting?

46. Why does a diet that restricts carbohydrate intake lead to loss of muscle mass as well as of fat?

47. A quarter-pound burger provides 420 kcal of energy. How long would a 180 lb person have to walk to burn those calories if 1 h of walking uses about 210 kcal?

48. How much protein is required each day by a 110 lb gymnast? (The DRI for protein is about 0.8 g/kg body weight.)

49. A 70 kg man can store about 2000 kcal as glycogen. How far could such a man run on this stored starch if he expends 100 kcal/km while running and the glycogen is his only source of energy?

50. A 70 kg man can store about 100,000 kcal of energy as fat. How far could such a man run on this stored fat if he expends 100 kcal/km while running and the fat is his only source of energy?

Fitness and Muscle

51. What is the body mass index (BMI) for a 5 ft, 10 in baseball player who weighs 186 lb?

52. What is aerobic exercise? What is anaerobic exercise?

53. Which type of metabolism (aerobic or anaerobic) is primarily responsible for providing energy for **(a)** intense bursts of vigorous activity and **(b)** prolonged low levels of activity?

54. What are the two functions of the actomyosin protein complex?

55. Explain why high levels of myoglobin are appropriate for muscle tissue geared to aerobic oxidation.

56. Categorize type I and type IIB muscle fibers as suited to aerobic oxidation or to anaerobic use of glycogen.

57. Why does the high catalytic activity of actomyosin in type IIB fibers suggest that these are the muscle fibers engaged in brief, intense physical activity?

58. Describe the differences between fast-twitch and slow-twitch muscle fibers. Can you change which type you develop?

Expand Your Skills

59. Ezekiel 4:9 describes a bread made from "wheat, and barley, and beans, and lentils, and millet, and spelt . . .." Would the bread supply complete protein?

60. What are two good reasons for not going on a high-protein diet?

61. What is one advantage of using beta-carotene as a yellow colorant in foods?

62. A diet that includes a lot of meat makes less efficient use of the energy originally captured by plants through photosynthesis than a vegetarian diet does. Explain.

63. A 100 mL sample of blood contains 15 g of hemoglobin (Hb). Each gram of Hb can combine with 1.34 mL of O_2(g) at body temperature and pressure. How much O_2 is in **(a)** 100 mL of blood and **(b)** the approximately 6.0 L of blood in an average adult?

64. Birds use large, well-developed breast muscles for flying. Pheasants can fly at 80 km/h, but only for short distances. Great blue herons can fly at about 35 km/h but can cruise great distances. What kind of fibers predominates in the breast muscles of each?

65. An adult female goes on a diet that provides 1200 kcal per day with no more than 25% from fat and no more than 30% of that fat being saturated fat. How much saturated fat, in grams, is permitted in this diet?

66. An athlete gains 40 lb after retiring. What is the likely explanation?

67. Fat tissue has a density of about 0.90 g/mL, and lean tissue a density of about 1.1 g/mL. Calculate the density of a person who has a body volume of 110 L and weighs 90 kg. Is the person fat or lean?

68. A 3 oz grilled lamb chop provides about 170 kcal, 23.5 g protein, and 6 g fat. How much carbohydrate does the lamb provide?

69. What structural differences are there between sucrose (Figure 16.5) and sucralose (Figure 17.6)?

70. Describe the biochemical process by which muscles are built.

71. Muscle is protein. Does an athlete need extra protein (above the DRI) to build muscles? What is the best way to build muscles?

72. If you are moderately active and want to maintain a weight of 160 lb, about how many calories do you need each day?

73. An athlete can run a 400 m race in 45 s. Her maximum oxygen intake is 4 L/min, but working muscles at their maximum exertion requires about 0.2 L of oxygen per minute for each kilogram of body weight. If the athlete weighs 50 kg, what oxygen debt will she incur?

74. The label on a can of "Healthy Request" chicken noodle soup indicates that each 1 cup portion supplies 110 kcal and has 7 g of protein, 14 g of carbohydrate, and 2.5 g of fat. Calculate the percentages of calories that come from carbohydrate, fat, and protein.

75. To maintain health, large doses of vitamins are
    a. necessary, because it is not possible to get the necessary vitamins from the diet
    b. necessary, because cooking destroys vitamins
    c. not necessary, because the average diet contains enough vitamins
    d. not necessary, because processed foods contain all of the vitamins of raw food

76. Oxygen debt can result from
    a. aerobic exercise, because there is not enough oxygen to convert pyruvic acid to carbon dioxide and water
    b. aerobic exercise, because there is not enough oxygen to convert lactic acid to carbon dioxide and water
    c. anaerobic exercise, because there is not enough oxygen to convert pyruvic acid to carbon dioxide and water
    d. anaerobic exercise, because there is not enough oxygen to convert glycogen to lactic acid

77. Match each source of food waste along the supply chain with an example usable product.
    a. Agricultural waste   1. Orange peel
    b. Processing waste   2. Canteen leftovers
    c. Post-consumer waste   3. Straw

78. Name three chemicals that can be derived from citrus peel and give a potential use for each.

79. Describe how each of the following benefits of using FSCW instead of fossil fuels can be realized.
    a. Reduced dependence on fossil fuels
    b. Minimized energy use and derivative use
    c. Saved money
    d. Improved business opportunities

Critical Thinking Exercises

Apply knowledge that you have gained in this chapter and one or more of the FLaReS principles (Chapter 1) to evaluate the following statements or claims.

17.1. A website states that natural vitamins may be worth the extra cost because those vitamins occur within a family of "things like trace elements, enzymes, cofactors, and other unknown factors that help them absorb into the human body and function to their full potential," whereas "synthetic vitamins are usually isolated into that one pure chemical, thereby leaving out some of the additional benefits that nature intended."

17.2. A friend states that he is going to use the "no-carb" diet to lose weight. He eats plenty of fruits and vegetables because he says they don't contain carbohydrates. His protein intake is high, since he eats meat at every meal, usually fried in butter or other fats. What is erroneous in this thinking?

17.3. A website states that a chemical in marijuana, delta 1-tetrahydrocannabinol (delta 1-THC), is a fat-soluble vitamin called "vitamin M." The site claims that vitamin status is justified "because many of the properties of delta 1-THC are similar to those of the fat-soluble vitamins."

17.4. Proponents of the Paleolithic diet suggest that we would be healthier if we only ate foods like those consumed by primitive pre-agricultural peoples, mainly fats and proteins, with few carbohydrates.

17.5. A natural products company claims that consuming large doses of its vitamins and minerals will lead to significant increases in health and fitness.

17.6. An online advertisement says, "I cut down 47 lb of stomach fat in a month by obeying this one old rule."

 Collaborative Group Projects

1. Prepare a brief report on the history of NaCl, where it is found in nature, important uses for sodium chloride throughout history, and dangers of overuse. Examine the labels on three convenience foods (a rice or spaghetti dish, soup), and compare the sodium content.

2. Prepare a brief report on one of the vitamins listed in Table 17.4. Identify the principal dietary sources and uses of the vitamin, as well as the history of its discovery, dangers (if any) of excessive ingestion, and physiological results of lack or low ingestion of that vitamin.

3. Examine the label on each of the following. Make a list of the food additives in each. Try to determine the function of each additive. Do all the labels provide this information?
   a. a can of soft drink   **b.** a can of beer
   c. a box of dried soup mix **d.** a can of soup
   e. a can of fruit drink   **f.** a cake mix box

4. Challenge: Can you eat on a $5.00 a day budget for food? Go to a local grocery store and investigate the costs of a variety of foods that you (theoretically) would buy to supply you with three meals a day for a week, assuming that you have only $35 and some change in your pocket. Those foods should supply you with at least 1800 kcal per day, provide minerals, vitamins, and enough protein and carbohydrates for a balanced diet, and leave you feeling reasonably "full." Create a detailed menu for each meal, as well as a shopping list and cost of each item. Assume you have enough condiments, spices, margarine or cooking oil and water available.

LET'S EXPERIMENT! Pumping Iron for Breakfast

Materials Needed

- Box of iron-fortified breakfast cereal (Total works well.)
- Box of breakfast cereal not fortified with iron (Hint: Check that trace iron is not listed among ingredients.)
- Measuring cup

- Super-strong magnet
- Quart-size resealable bags (2)
- Water

Are you getting enough iron in your diet? Does your breakfast cereal contain iron? Do you taste any difference between cereals fortified with iron and not? Can you see the difference?

A well-balanced diet of vegetables, fruits, grains, protein, and dairy along with adequate water intake provides average young adults with the necessary vitamins and minerals. There are slight variations in the amounts of vitamins and minerals needed by men and women.

Recommendations for iron, in particular, vary by 10 mg per day. The recommendation is 18 mg of iron for women but only 8 mg of iron for men. Iron(II) ions are needed for the proper functioning of the oxygen transport of hemoglobin. Women are more susceptible to anemia, a condition that can result in fatigue and reduced resistance to diseases. Many breakfast cereals are fortified with food-grade iron particles (metallic iron) as a mineral supplement. Generally, this form of iron cannot be seen or tasted.

Start by tasting each of the two cereals, one iron-fortified and other not, in a side-by-side test. Does the iron-fortified cereal taste metallic? It shouldn't. Otherwise, people probably would not eat it for breakfast.

Continuing your experiment, prepare two batches of cereal "soup." One bag will be filled with your iron-fortified cereal and the other with non-iron-fortified cereal. Measure 1 cup of cereal into each quart-size resealable bag. Fill each bag half full with warm water. Carefully seal the bag, leaving an air pocket inside.

Mix the cereal and the water by squeezing the bag until the contents become a brown, soupy mixture. Allow the mixture to sit for at least 20 minutes.

Begin by placing your super-strong magnet on your palm. See Figure 1:

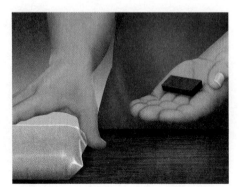

Figure 1

Place the non-iron-fortified cereal soup bag on top of the super-strong magnet. Sandwich the bag and the magnet between your two hands as in Figure 2:

Figure 2

Slowly shake the contents of the bag, using a circular motion, for 15 or 20 seconds. The idea is to attract any free moving bits of metallic iron in the cereal to the magnet.

Use both hands again and flip the bag and magnet over so the magnet is on top. Gently squeeze the bag to lift the magnet a little above the cereal soup. Don't move the magnet just yet. Look closely at the edges of the magnet where it's touching the bag. What do you see?

Repeat with the iron-fortified cereal soup. Now what do you see? What happens when you keep the magnet touching the bag and move it in circles? See Figure 3.

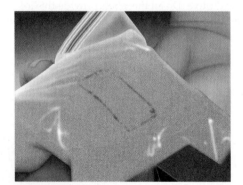

Figure 3

Questions

1. Were you able to detect any iron filings in the cereal not fortified with iron?
2. Why are there iron filings in your cereal?
3. Where else can you get iron in your diet? List at least three foods that are high in iron.

●Drugs

18

Have You Ever Wondered?

1 Why is aspirin not recommended for children? **2** What are the dangers of acetaminophen (Tylenol®)? **3** Why must I continue taking an antibiotic even after I am feeling better? **4** What kinds of drugs work for colds or the flu? **5** Do larger tablets contain more medicine? **6** What is the difference between a name-brand drug and a generic drug?

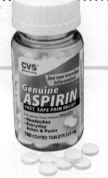

Willow bark and the herb meadowsweet (shown in the main image above left) contain salicylates, which became the basis for aspirin, the most widely used drug in the world. French chemist Charles Frédéric Gerhardt made crude acetylsalicylic acid (o-$CH_3COOC_6H_4COOH$) by reacting sodium salicylate (o-HOC_6H_4COONa) with acetyl chloride (CH_3COCl). *Aspirin*, Bayer's brand name, came from *acetylated Spirsäure*, the German name for salicylic acid, with the typical drug-naming ending *-in* added.

CHEMICAL CURES, COMFORTS, AND CAUTIONS The word *drug* originally referred to a dried plant (or plant part) used as a medicine, either directly or after extracting active ingredients as a tea. Today, a **drug** is defined as a chemical substance that affects the functioning of living things. Drugs are used to relieve pain, to treat illnesses, and generally to improve one's health or well-being. Many people literally owe their lives to drugs. Drugs can kill bacteria, lower blood pressure, prevent seizures, and relieve allergies. Some types of drugs can do more harm than good, however, especially if used in the wrong amounts or for the wrong reasons.

Humans have used drugs since prehistoric times, and drug use occurs in almost all cultures. Most societies have used alcohol, and the use of cannabis goes back to at least 3000 B.C.E. The narcotic effect of the opium poppy was known to the ancients. As early as 1400 B.C.E., the Egyptians grew poppies and the Minoans on Crete had terracotta statues of a poppy goddess. Indigenous people of the Andes Mountain region have long chewed leaves of the coca plant for the stimulating effect of cocaine.

In this chapter, we will look at natural and synthetic drugs that are intended to combat a wide range of ills, as well as those that are used illicitly.

18.1 Drugs from Nature and the Laboratory

Learning Objectives • Classify common drugs as natural, semisynthetic, or synthetic. • Define *chemotherapy*, and explain its origin.

Some drugs are made from molecules that come from natural sources. While there is no legal definition for *natural*, the word is commonly used to describe materials that are extracted directly from plant or animal sources by physical processes, as opposed to materials that are synthesized by chemical reactions.

We still obtain many drugs from natural sources. The following are just a few examples of the many drugs that have been obtained from plants.

▲ Opium poppy.

- Quinine, the antimalarial drug, from the bark of *Cinchona* species
- Morphine, the analgesic, from the opium poppy
- Digoxin, for heart disorders, from *Digitalis purpurea*
- Vinblastine and vincristine, anticancer agents from the Madagascar periwinkle (*Catharanthus roseus*)
- Tubocurarine, the muscle relaxant, from *Chondrodendron tomentosum*, used by Amazon natives in the arrow poison curare

▲ Madagascar periwinkle.

Many other drugs are obtained from microorganisms, including the following important ones.

- Antibacterial agents (penicillins) and cholesterol-lowering agents (statins) from *Penicillium* species
- Immunosuppressants, such as cyclosporins, from *Streptomyces* species

A potentially vast resource for new drugs is marine organisms. The following are examples of a few drugs from the sea.

- The anticancer drug dolastatin-10 from a sea hare
- An anti-inflammatory agent manoalide from the sponge *Luffariella variabilis*
- Ziconotide, a drug for treatment of severe pain, derived from cone snail venom

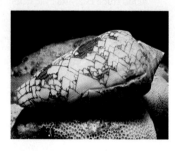

▲ Cone snail *(Conus textile)*.

A major source for naturally occurring drugs has been tropical rainforests— warm, moist, and fertile areas that contain more different kinds of plants and animals than anyplace else on Earth. The rainforests are living chemical factories full of valuable and irreplaceable natural resources, but they are rapidly disappearing, as people clear the trees and convert the land to agricultural uses. Thus we are losing a fertile source of potential drugs.

Other drugs are made by modifying molecules derived from natural sources to improve the substances' properties. These drugs are said to be *semisynthetic*. Many of the earliest science-based drugs were semisynthetic. Examples include the following.

- Salicylic acid (first obtained from willow bark) was converted to acetylsalicylic acid (aspirin).
- Morphine (from opium poppies) was converted to heroin.
- Lysergic acid (from ergot fungus) was converted to LSD.

There are now many completely synthetic drugs on the market. These substances are represented in nearly all categories, from antibacterial drugs to sleeping pills to narcotics such as methadone, fentanyl, and oxycodone.

In 1904, Paul Ehrlich (1854–1915), a German chemist, found that certain dyes used to stain bacteria to make them more visible under a microscope could also be used to kill the bacteria. He used dyes to combat the organism that causes African sleeping sickness. He also prepared an arsenic compound that proved somewhat effective against the organism that causes syphilis. Ehrlich realized that certain chemicals were more toxic to disease organisms than to human cells and could therefore be used to control or cure infectious diseases. He coined the term **chemotherapy** (from "chemical therapy"). Ehrlich was awarded the Nobel Prize in Physiology or Medicine in 1908.

The use of drugs is not new, but never before has there been such a vast array of them. There are dozens of pharmaceutical companies producing thousands of drugs. Their research laboratories spend huge sums to develop new drugs each year. According to a recent study, it costs up to an estimated $4.5 billion to develop and bring a new drug to market. (The cost varies substantially depending on the kind of drug.) That great cost means that companies often charge high prices for the products they develop, at least for a while, making them unaffordable for many of the people who need them. After the patent rights expire, other companies can make generic versions of the same products and sell them at lower prices because they have incurred no research costs. Generic drugs must act identically to the original products (Section 18.8).

Because many drugs are sold under generic names and several different brand names, there are thousands of prescription medicines on the market. Several hundred other medicines are available over-the-counter (OTC) under thousands of brand names and in many combinations. Some of these drugs are vital to the health of the people who take them. Drug-industry sales worldwide in 2017 were estimated to be nearly $2 trillion. In 2017, Americans had close to 4.5 billion prescriptions filled, including refills, an average of about 14 per person, and spent about $450 billion for prescription drugs.

SELF-ASSESSMENT Questions

For questions 1–3, state whether the drug is
a. natural   **b.** semisynthetic   **c.** synthetic

1. Quinine

2. Cyclosporins

3. Aspirin

4. Chemotherapy is
   a. improving one's health by avoiding chemicals
   b. improving one's mind by studying chemistry
   c. treating cancer with radiation
   d. treating a disease with chemicals

Answers: 1. a; 2. a; 3. b; 4. d

18.2 Pain Relievers: From Aspirin to Oxycodone

Learning Objectives • List common over-the-counter analgesics, antipyretics, and anti-inflammatory drugs, and describe how each works. • Name several common narcotics, describe how each functions, and state its potential for addiction.

Acetylsalicylic acid, commonly called *aspirin* in the United States, is widely used around the world. About 45 million kg are consumed annually—enough to make well over 100 billion standard tablets. Its history goes back at least 2400 years to the time of Hippocrates in ancient Greece; he knew that sick people could ease their pain and lower their fever by chewing willow leaves.

It was not until 1835 that chemists isolated the active ingredient, salicylic acid, an effective **analgesic** (pain reliever), **antipyretic** (fever reducer), and **anti-inflammatory** that, unfortunately, also caused considerable stomach distress and bleeding. Acetylsalicylic acid is a modified version of salicylic acid that had the desired medicinal properties but was gentler on the stomach. It was introduced in 1893 by the Bayer Company in Germany (Figure 9.14). Aspirin tablets are currently sold under a variety of trade names and store brands. A standard aspirin tablet contains 325 mg of acetylsalicylic acid, while "extra strength" formulations usually contain 500 mg.

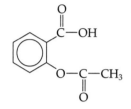

▲ Acetylsalicylic acid (aspirin).

Nonsteroidal Anti-Inflammatory Drugs (NSAIDs)

Aspirin is one of several substances called **nonsteroidal anti-inflammatory drugs (NSAIDs)**, a designation that distinguishes these drugs from the more potent steroidal anti-inflammatory drugs, such as cortisone or prednisone (Section 18.5). Other NSAIDs include ibuprofen (Advil®; Motrin®) and naproxen (Aleve®) (Figure 18.1). Acetaminophen (Tylenol), like aspirin, relieves minor aches and reduces fever but is not anti-inflammatory and is therefore not an NSAID.

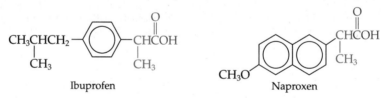

Ibuprofen   Naproxen

▲ **Figure 18.1** Two NSAIDs. Ibuprofen and naproxen are derivatives of propionic acid (CH_3CH_2COOH).

NSAIDs act to relieve pain and reduce inflammation by inhibiting the production of **prostaglandins**, hormone-like lipids that are derived from a fatty acid and are involved in sending pain messages to the brain (Section 18.5). Inflammation is caused by an overproduction of prostaglandin derivatives, and inhibiting prostaglandin synthesis reduces the inflammatory effect. Regrettably, prostaglandins also affect the blood platelets, the kidneys, and the stomach lining, and inhibiting them can leads to serious side effects such as excessive bleeding and stomach pains. NSAIDs have an **anticoagulant** effect (inhibition of the clotting of blood), and people who regularly take NSAIDs are at greater risk for bleeding. For example, about 1.5% of patients with rheumatoid arthritis who use NSAIDs for a year have a serious stomach complication. This is a small proportion, but it translates to a large number of people because so many patients use NSAIDs.

Because NSAIDs act as anticoagulants, small daily doses also seem to lower the risk of heart attack and stroke, presumably by the same anticoagulant action that causes bleeding in the stomach. Adult low-strength (usually 81 mg) aspirin tablets are now routinely used by people at risk for heart attack and stroke. NSAIDs should not be used by people facing surgery, childbirth, or some other hazard involving the possible loss of blood for at least a week before the event.

WHY IT MATTERS

Almost every single-ingredient analgesic product sold today uses one of four active ingredients: acetaminophen, aspirin, ibuprofen, or naproxen. These drugs are available in generic forms, under dozens of brand names, and in many combinations, including cough/cold and allergy/sinus products. You can search the Internet for brand names, using the generic names as key words.

NSAIDs also reduce body temperature. Fevers are induced by substances called *pyrogens*, compounds produced by and released from leukocytes (white blood cells) and other circulating cells. Pyrogens usually use prostaglandins as secondary mediators. Fevers therefore can be reduced by aspirin and other prostaglandin inhibitors. Pyrogens that do not work through prostaglandins are not affected by NSAIDs.

The body tries to fight off infection by elevating its temperature. Mild fevers in adults (those below 39 °C, or 102 °F) are usually best left untreated. High fevers, however, can cause brain damage and require immediate treatment.

How NSAIDs Work

Prostaglandins are produced in the body from a cell-membrane component called *arachidonic acid* (Section 18.5). Scientists have identified enzymes called *cyclooxygenases* (COX) that catalyze the conversion. Two of these are COX-1, found in stomach and kidney tissues where NSAID side effects occur, and COX-2, found in tissues where inflammation occurs. The older NSAIDs, such as aspirin, ibuprofen, and naproxen, inhibit both COX-1 and COX-2, providing relief from pain and inflammation but also producing undesirable side effects. Newer NSAIDs preferentially inhibit COX-2. Three COX-2 inhibitors came onto the market as anti-inflammatory agents but two, including Vioxx®, were quickly removed from the market because they led to a higher risk of stroke and heart attacks.

Celecoxib (Celebrex) (Figure 18.2) remains on the market, and a recent ten-year study shows that it does not carry the same risks as the removed drugs. The U.S. Food and Drug Administration (FDA) now requires labels on all prescription NSAIDs to include a boxed warning highlighting the potential for increased risk of heart problems, as well as information regarding allergic reactions and internal bleeding.

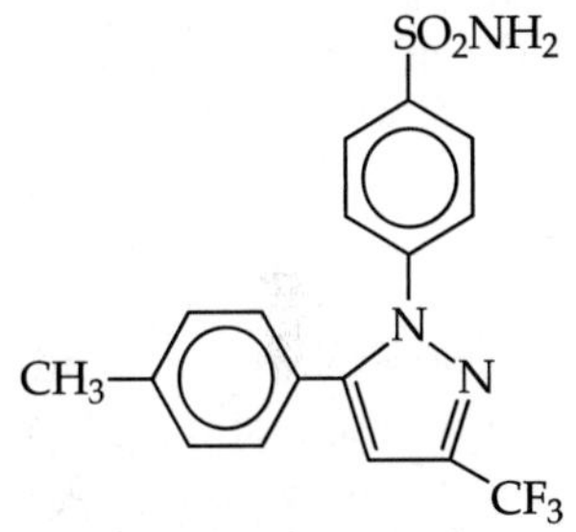

Acetaminophen: A COX-3 Inhibitor

Acetaminophen, perhaps best known by the brand name Tylenol, is the most widely used analgesic in the United States. It provides relief of pain and reduction of fever comparable to that of aspirin. The total world market for acetaminophen (called *paracetamol* in many countries) is about 150 million kg/year. Acetaminophen acts on a third variant of cyclooxygenase enzyme, COX-3. Inhibition of COX-3 may represent a mechanism by which acetaminophen decreases pain and fever. Unlike NSAIDs, it is not an anticoagulant, so people allergic to aspirin or susceptible to bleeding can safely take acetaminophen. However, COX-3 appears to play no role in controlling inflammation, so it is of limited use to people with arthritis. Acetaminophen is often used to relieve the pain that follows minor surgery. Regular acetaminophen tablets contain 325 mg, and extra-strength forms have 500 mg.

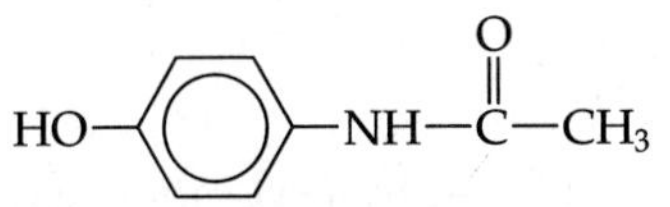

Acetaminophen

Combination Pain Relievers

Many analgesic products are combinations of one or more NSAIDs, acetaminophen, caffeine, antihistamines, and/or other drugs. Acetaminophen, for example, is found in more than 70 combination products for pain relief, allergy symptoms, and cold and flu treatment. These combinations are available under various brand names and as store brands. Familiar brands include Excedrin and Anacin, each of which is available in several different combinations and formulations.

Many other combination pain relievers are available. Most claim to be better than regular aspirin. However, such advantages are marginal at best. For occasional use by most people, plain aspirin may be the cheapest, safest, and most effective product.

1 Why is aspirin not recommended for children?

The use of aspirin in children suffering from viral infections, such as the flu or chickenpox, is associated with *Reye's syndrome*, a potentially fatal illness. For this reason, aspirin products carry a warning that they should not be used to treat children with fevers.

Celecoxib (Celebrex®)

▲ **Figure 18.2** Celecoxib is the only COX-2–inhibiting prescription NSAID that remains on the market.

2 What are the dangers of acetaminophen (Tylenol)?

Although the normal dose of acetaminophen is safe, larger doses are toxic, causing liver damage. This effect is synergistic with alcohol. As little as 4 g (eight extra-strength tablets) in 24 hours, coupled with recent alcohol use, can cause severe liver toxicity. Every year over 33,000 people are hospitalized because of acetaminophen toxicity, and over 150 deaths are recorded. Accidental overdose of acetaminophen is far too common, and it often occurs because people do not realize that acetaminophen is an ingredient in many OTC medicines for treatment of cold symptoms, allergies, insomnia, and other maladies. A careful review of the active ingredients in such medications can save a life!

Chemistry, Allergies, and the Common Cold

Store shelves are stocked with a variety of cold and allergy medicines, which contain antihistamines, cough suppressants, expectorants, bronchodilators, and/or nasal decongestants. None cures the common cold, which is caused by as many as 200 related viruses. In fact, *no* drug provides a cure, and most cold remedies just treat the symptoms.

Many cold medicines contain an **antihistamine**, such as diphenhydramine or cetirizine (Figure 18.3). Antihistamines temporarily relieve the symptoms of allergies: sneezing, itchy eyes, and runny nose.

When an **allergen** (a substance that triggers an allergic reaction) binds to the surfaces of certain cells, it triggers the release of *histamine*, which causes the redness, swelling, and itching associated with allergies. An antihistamine inhibits the release of histamine. Many cough and cold products contain more than one ingredient.

Other OTC cold medicines include *antitussives* (cough suppressants) and *expectorants* (substances that

help bring up mucus from the bronchial passages). Ordinarily, a cough is functional; the respiratory tract uses the coughing mechanism to rid itself of congestion. However, when a cough is dry or interferes with needed rest, temporary cough suppression may be advisable. Only one expectorant, glyceryl guaiacolate (guaifenesin), is rated as safe and effective, and its effectiveness is not well documented.

Nasal *decongestants*, such as naphazoline, phenylephrine, oxymetazoline, and xylometazoline, seem to be safe and effective for occasional use, although repeated use leads to a *rebound effect* in which the nasal passages swell and make congestion seem worse.

How should you treat a common cold? First, you should drink plenty of liquids and get lots of rest. When you are in good physical condition and have a strong immune system, colds seem to strike less often. Frequent hand washing, especially when colds are prevalent, can dramatically reduce transmission of the virus. And many people still swear by chicken soup!

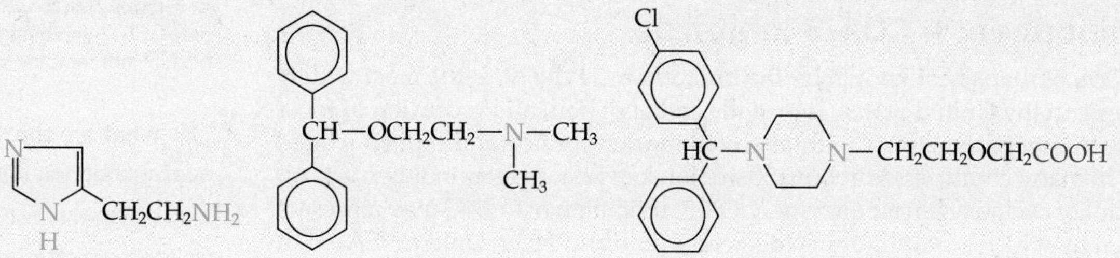

(a) Histamine (b) Diphenhydramine (c) Cetirizine

▲ **Figure 18.3** (a) Histamine and two antihistamines: (b) diphenhydramine (Benadryl) and (c) cetirizine (Zyrtec). Diphenhydramine is also used as a cough suppressant.

Q *To what family of organic compounds do these three substances belong?*

Narcotics

When pain is too great to be relieved by OTC analgesics, stronger drugs are available by prescription. Most such drugs are classified as **narcotics**, drugs that produce narcosis (stupor or general anesthesia) and analgesia (relief of pain). Many drugs produce these effects, but in the United States only those that also *addictive* are legally classified as narcotics.

Several narcotics are products of opium, the dried, resinous juice of the unripe seeds of the opium poppy (*Papaver somniferum*) (Figure 18.4). Opium is a complex mixture of twenty or so alkaloids, plus sugars, resins, waxes, and water. The principal alkaloid, morphine, makes up about 10% of the weight of raw opium. Opium was used in many patent medicines during the nineteenth century. Laudanum, a solution of opium in alcohol, contained about 10 mg of morphine per mL of solution and was widely used during the Victorian era for everything from toothaches to tuberculosis, and even for quieting colicky babies. Paregoric, a much weaker solution that contained only 0.4 mg of morphine per mL, was also widely used for diarrhea until 1970.

Morphine was first isolated in 1805 by Friedrich Sertürner (1783–1841), a German pharmacist. With the invention of the hypodermic syringe in the 1850s, a new method of administration became available. Injection of morphine directly into the bloodstream was more effective for the relief of pain, but this method also seriously escalated the problem of addiction.

During the American Civil War (1861–1865), morphine was used widely to relieve pain due to battle wounds. Soldiers also came to use morphine as a treatment for another common malady of men on the battlefront—dysentery—because constipation is one side effect of morphine use. More than 100,000 soldiers became addicted to morphine while serving in the war. The affliction was so common among veterans that it came to be known as "soldier's disease."

Morphine and other narcotics were placed under control of the federal government by the Harrison Act of 1914. Morphine is still prescribed for the relief of severe pain. It induces lethargy, drowsiness, confusion, euphoria, chronic constipation, and depression of the respiratory system. Morphine becomes addictive when it is administered in amounts greater than the prescribed dose or for a period longer than the prescribed time.

Changes in the morphine molecule alter its physiological properties. Replacement of the phenolic hydroxyl group (—OH) by a methoxy group (—OCH$_3$) produces codeine. Codeine is present in opium at levels of about 0.7% to 2.5%, but it is usually synthesized by methylating the more abundant morphine molecules. Codeine resembles morphine in its action, but it is less potent, has less tendency to induce sleep, and is less addictive. For relief of moderate pain, codeine is often combined with acetaminophen (Tylenol 2 and Tylenol 3) or with aspirin (Empirin Codeine). For cough suppression, codeine is given in liquid preparations that are regulated less stringently than stronger narcotics.

Conversion of both —OH groups of the morphine molecule to acetate ester groups produces heroin. This semisynthetic morphine derivative was first prepared by chemists at the Bayer Company in Germany in 1874. It received little attention until 1890, when it was proposed as an antidote for morphine addiction. Shortly thereafter, Bayer advertised heroin widely as a sedative for coughs, often in the same ads describing aspirin. However, it was soon found that heroin induced addiction even more quickly than morphine and that heroin addiction was harder to cure.

The physiological action of heroin resembles that of morphine. Heroin is less polar than morphine, so it enters the fatty tissues of the brain more rapidly, and it seems to produce a stronger feeling of euphoria than morphine does. Heroin is not legal in the United States, even by prescription. It is, however, used in Britain for pain relief in terminal cancer patients.

Addiction probably has three components: psychological dependence, physical dependence, and tolerance. Psychological dependence is evident in the uncontrollable desire for the drug. Physical dependence is shown by acute withdrawal symptoms such as convulsions. Tolerance for the drug is evidenced by the fact that increasing dosages are required to produce the same degree of narcosis and analgesia over time.

Deaths from heroin and other narcotics are usually attributed to overdoses, but the situation is not always clear. The problem is often a matter of quality control. Street drugs often contain contaminants and can vary considerably in potency.

Much research has gone into developing a drug that would be as effective as morphine for the relief of pain but would not be addictive. Several synthetic narcotics are available. Some common ones are listed in Table 18.1. Structures of some of these and of other narcotics are given in Figure 18.5.

Oxycodone is a semisynthetic opioid related to codeine. It is used for relief of moderate to severe pain. It is available alone and in combination with acetaminophen, ibuprofen, or aspirin. A sustained-release form of oxycodone is called Oxy-Contin®. Abused for its euphoric effects, OxyContin and its generic equivalents are widely available illegally. Hydrocodone, another synthetic narcotic, is always combined with acetaminophen or another medication. Sold as HYCD/APAP (hydrocodone/acetaminophen), this combination was the most prescribed drug in the United States in 2013, with 137 million new and refilled prescriptions. Fentanyl is another particularly potent drug that can be fatal even in small doses. In fact, many law enforcement agencies now recommend that personnel take extreme caution when handling any suspected drug paraphernalia.

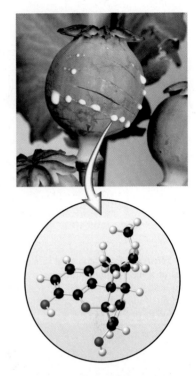

▲ **Figure 18.4** Opium poppy seed pod. The slits on the seed pod exude a resinous juice, which is dried to form opium. About 10% of the weight of raw opium is morphine.

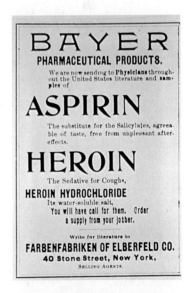

▲ Heroin was regarded as a safe medicine in 1900 and was widely used as a cough suppressant. It was also thought to be a nonaddictive substitute for morphine.

TABLE 18.1 Some Synthetic Narcotics

Substance	Duration of Action (h)	Approximate Oral Dose (mg) Equivalent to 30 mg of Morphine
Meperidine (Demerol®)	2–4	300
Hydrocodone	4–8	20
Fentanyl	1–2[a]	<1
Oxycodone	4–6[b]	20
Hydromorphone (Dilaudid®)	4–5	7.5
Methadone	4–6[c]	Not applicable

[a]Used in transdermal patches for chronic pain.
[b]Often dispensed in a time-release form (OxyContin) that allows 12 h between doses.
[c]Used for treatment of heroin addiction; not usually given for pain relief.

▶ **Figure 18.5** Structural formulas of some synthetic narcotics: (a) meperidine (Demerol), (b) methadone, (c) pentazocine (Talwin), and (d) naloxone.

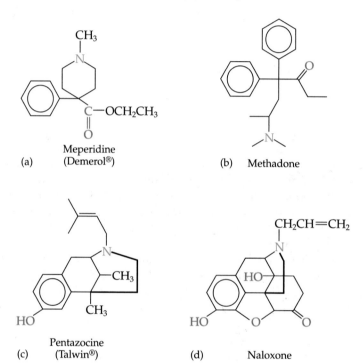

(a) Meperidine (Demerol®)

(b) Methadone

(c) Pentazocine (Talwin®)

(d) Naloxone

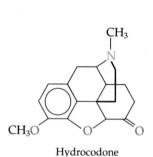

Hydrocodone

Anionic site binds N

Cavity binds carbon atoms

Flat hole binds benzene ring

▲ Diagram of a morphine receptor.

In the United States, the problem of illegal opioid use has grown rapidly. In 2016, there were over 64,000 deaths resulting from drug overdoses. Indeed, roughly two-thirds of deaths of Americans under 50 are now tied to drug overdoses. At least half of these deaths were related to "prescription" drugs, some of which cross the border, but many of which are made readily available through "pill mills." The opioid crisis has been declared a public health emergency, and it carries not only a human toll, but an economic cost (over $500 billion in 2016). Some steps have been taken to limit the use of prescription opioids. In 2014, some drugs were reclassified, and prescriptions for them can no longer be refilled; a new prescription must be issued.

The synthetic narcotic methadone is widely used to treat heroin addiction. Like heroin, methadone is highly addictive. However, when taken orally, it does not induce the sleepy stupor characteristic of heroin intoxication, so a person on methadone maintenance is usually able to hold a productive job. If an addict who has been taking methadone reverts to heroin, the methadone in the individual's system effectively blocks the euphoric rush normally produced by heroin and so reduces the addict's temptation to use heroin.

Morphine Agonists and Antagonists

Chemists have synthesized many morphine analogs, but few have shown significant analgesic activity, and most of those are addictive. Morphine acts by binding to receptors in the brain. A morphine receptor has a flat hole that binds to the benzene ring, a cavity that binds two carbon atoms, and an anionic site that binds the nitrogen atom. A molecule that has morphine-like action is called a morphine **agonist**. A morphine **antagonist** inhibits the action of morphine by blocking morphine receptors. Some molecules have both agonist and antagonist effects. An example is pentazocine (Figure 18.5c), which is less addictive than morphine and is effective for the relief of pain.

Natural Opiates: Endorphins and Enkephalins

Morphine acts by binding to specific receptor sites in the brain. Why should the human brain have receptors for a plant-derived drug such as morphine? Scientists concluded that the body must produce its own substances that fit these receptors. Actually, it produces several morphine-like substances, called **endorphins** (from "endogenous morphines"). Each endorphin is a short peptide chain composed of amino-acid units. Those with five amino-acid units are called *enkephalins*. There are two enkephalins, which differ only in the amino acid at the end of the chain. *Leu*-enkephalin has the sequence Tyr-Gly-Gly-Phe-Leu, and *Met*-enkephalin has the sequence Tyr-Gly-Gly-Phe-Met. Other endorphins have chains of 30 or more amino acids.

Some enkephalins have been synthesized and shown to be potent pain relievers. Their use in medicine is quite limited, however, because they are rapidly broken down in the body by the enzymes that hydrolyze proteins. Researchers have sought to make analogs that are more resistant to hydrolysis and can be employed as morphine substitutes for the relief of pain. Unfortunately, both natural enkephalins and their analogs, such as morphine, seem to be addictive.

It appears that endorphins are released during strenuous exercise and in response to pain. Exercise also directly affects the brain by increasing its blood supply and by producing **neurotrophins**, substances that enhance the growth of brain cells. Exercises that involve complex motions, such as dance movements, lead to an increase in the connections between brain cells. Both our muscles and our brains work better when we exercise regularly.

Some evidence also indicates that acupuncture causes the release of brain "opiates" that relieve pain. The long needles stimulate deep sensory nerves, which release peptides that then block pain signals for a brief time.

Endorphin release has also been used to explain other phenomena once thought to be largely psychological. A soldier who has been wounded in battle may feel no pain until the skirmish is over. His body has secreted its own painkiller.

Pure antagonists such as naloxone, Figure 18.5d, can be used to treat opiate addicts. An addict who has overdosed can be brought back from death's door by an injection of naloxone. Another relatively new drug, buprenorphine, shows promise in the treatment of opioid addition. Methadone must be administered daily on site, but current studies on the efficacy of providing monthly prescriptions for buprenorphine are under way.

SELF-ASSESSMENT Questions

1. Which of the following types of compounds works as a pain reducer?
   a. analgesic   b. anticoagulant
   c. antihistamine   d. antipyretic

2. Which of the following is not an anticoagulant?
   a. acetaminophen   b. aspirin
   c. ibuprofen   d. naproxen

3. Allergens trigger the release of
   a. endorphins   b. histamines
   c. prostaglandins   d. pyrogens

4. A semisynthetic opiate produced by adding acetyl groups to morphine molecules is
   a. heroin   b. MDMA   c. meperidine   d. methadone

5. Which of the following has been used to treat heroin addiction?
   a. endorphins   b. methadone   c. morphine   d. opium

6. Compounds that mimic the effects of opiates and produce a "high" following strenuous exercise are
   a. endorphins   b. histamines
   c. pyrogens   d. thyroxins

7. In order to be classified as a narcotic in the United States, a drug must
   a. be addictive   b. be an analgesic
   c. induce stupor   d. be a general anesthetic

8. Approximately what percent of the dry weight of raw opium is morphine?
   a. 2%   b. 5%   c. 10%   d. 25%

Answers: 1. a; 2. a; 3. b; 4. a; 5. b; 6. a; 7. a; 8. c

18.3 Drugs and Infectious Diseases

Learning Objectives • List the common antibacterial drugs, and describe the action of each. • Name the common categories of antiviral drugs, and describe the action of each.

A century ago, infectious diseases were the main cause of death in the United States (Figure 18.6). At the time of the American Civil War, thousands of wounded soldiers

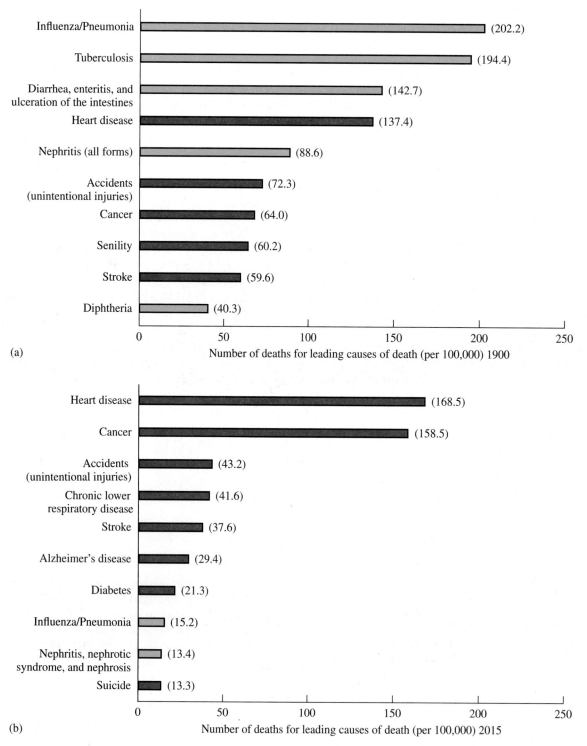

▲ **Figure 18.6** (a) In 1900, five of the ten leading causes of death—responsible for three-fifths of all deaths—in the United States were infectious diseases (blue bars). (b) The leading causes of death in 2015 were so-called lifestyle diseases (red bars): heart disease, stroke, and diabetes, related in part to our diets; cancer and lung diseases, to cigarette smoking; and accidents, often to excessive consumption of alcohol.

required amputations of arms or legs, and nearly 80% of the patients died after coming to a hospital, mostly from infections. Today, infectious disease categories have moved toward the bottom of the list of the top ten leading causes of death. (These ten causes account for more than three-quarters of all deaths in the United States.) Many diseases have been brought under control through the use of *antibacterial drugs*.

Antibacterial Drugs

The first antibacterial drugs were *sulfa drugs*, whose prototype was discovered in 1935 by the German chemist Gerhard Domagk (1895–1964). Sulfa drugs were used extensively during World War II to prevent wound infections. Many soldiers who would have died in earlier wars survived.

Sulfanilamide, the simplest sulfa drug, was one of the first drugs to have its action understood at the molecular level. Its effectiveness is based on a case of mistaken identity. Bacteria need *para*-aminobenzoic acid (PABA) to make folic acid, which is essential for the formation of certain compounds the bacteria require for proper growth. But bacterial enzymes can't tell the difference between sulfanilamide and PABA, because their molecules are so similar.

Sulfanilamide   *para*-Aminobenzoic acid

When sulfanilamide is applied to an infection in large amounts, bacteria incorporate it into pseudo–folic acid molecules that cannot act normally, and the bacteria cease to grow. Of the thousands of sulfanilamide analogs that have been developed and tested, only a few are used today. Two common ones are sulfathiazole and sulfaguanidine. Some sulfa drugs tend to cause kidney damage or other problems.

The next important discovery was that of penicillin, an antibiotic. An **antibiotic** is a soluble substance that is derived from molds or bacteria and inhibits the growth of other microorganisms. Penicillin was first discovered in 1928 but was not tried on humans until 1941. Scottish microbiologist Alexander Fleming (1881–1955), then working at the University of London, first observed the antibacterial action of a mold, *Penicillium notatum*, when one of his bacterial cultures of *Staphylococcus aureus* became contaminated with a blue mold. Fleming astutely noted that bacterial colonies had been destroyed in the vicinity of the mold.

Fleming was able to make crude extracts of the active substance. This material, later called *penicillin*, was further purified and improved by Howard Florey (1898–1968), an Australian, and Ernst Boris Chain (1906–1979), a refugee from Nazi Germany, both working at Oxford University in England. Fleming, Florey, and Chain shared the 1945 Nobel Prize in Physiology or Medicine for their work on penicillin.

Penicillin is not a single substance but a group of related compounds (Figure 18.7). By designing molecules with different structures, chemists can change the properties

Sulfathiazole

Sulfaguanidine

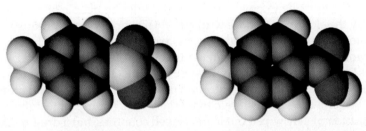

(a) General formula   (b) Penicillin G   (c) Amoxicillin

▲ **Figure 18.7** Penicillins. (a) A general formula, where R is a variable side chain. (b) Penicillin G is generally considered the most effective penicillin but can only be given by injection. (c) Amoxicillin can be administered orally.

3 **Why must I continue taking an antibiotic even after I am feeling better?**

Antibiotics inhibit the growth of bacteria. A prescription will be written for the time it will take to completely kill off all of the bacteria. A large percentage of bacteria will be killed fairly early during treatment, so a person will feel much better. However, some individual bacteria may take longer to kill. It is important to take all of the prescription to make sure that there are no remaining bacteria that can grow and start the infection again. Any bacteria not killed have a greater potential to develop resistance to the drug.

Cephalexin

of the drug, producing penicillins that vary in effectiveness. Some can be taken orally; others must be injected. Bacteria resistant to one penicillin may be killed by another. Amoxicillin has a broad spectrum of activity against many types of microorganisms. It is among the top ten prescribed drugs, with more than 54.8 million prescriptions in 2015, including refills, in the United States. Overall, penicillins account for about half of all antimicrobial prescriptions.

Penicillin works by inhibiting enzymes that the bacteria use to make their cell walls. The bacterial cell walls are made up of *mucoproteins*, polymers in which amino sugars are combined with protein molecules. (An amino sugar contains an $-NH_2$ group in place of an $-OH$ group.) Penicillin prevents cross-linking between these large molecules. This leaves holes in the cell walls, and the bacteria swell and rupture. Cells of higher animals have only external membranes, not mucoprotein walls, and are not affected by penicillin. Thus, penicillin can destroy bacteria without harming human cells. However, some people are allergic to penicillin.

In their early days, antibiotics reduced the number of deaths from blood poisoning (septicemia), pneumonia, and other infectious diseases substantially. Before 1941, a person with a major bacterial infection almost always died. Today, such deaths are rare except for those ill with other conditions that compromise the immune system. Seven decades ago, pneumonia was a dreaded killer of people of all ages. Today, it kills mainly the elderly and those with HIV/AIDS in developed countries.

Antibiotics have been called miracle drugs, but even miracle drugs are not without problems. It wasn't long after antibiotics were first introduced that disease organisms began to develop strains resistant to them. One strain can spread antibiotic resistance to another by sharing genes, and the emergence of more and more resistant strains of bacteria is a serious threat to world health. For example, tuberculosis had been almost completely wiped out in developed countries, but a drug-resistant strain has become rampant in Russia and among HIV/AIDS patients everywhere.

Another example of the development of resistance involves erythromycin, an antibiotic obtained from *Streptomyces erythreus*. As long as it had only limited use, erythromycin could handle all strains of staphylococci. After it was put into extensive use, resistant strains of these bacteria began to appear. Staph infections are now a serious problem in hospitals. People who are admitted to a hospital for treatment sometimes develop serious bacterial infections. Some even die of these staph infections. The Centers for Disease Control and Prevention (CDC) estimated in 2011 that staph infections cause more than 80,000 serious illnesses and about 11,285 deaths each year. Later studies have shown a decline in staph infections, as hospitals have introduced new procedures to prevent infection.

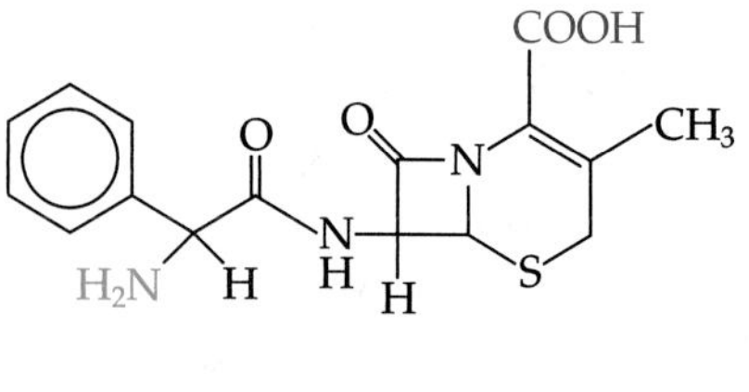

In a race to stay ahead of resistant strains of bacteria, scientists continue to seek new antibiotics. Penicillins have been partially displaced by related compounds called *cephalosporins*, such as cephalexin (Keflex®). Unfortunately, some strains of bacteria are now resistant to cephalosporins.

Another important development in the field of antibiotics was the discovery of a group of compounds called *tetracyclines*. The first of these four-ring compounds, chlortetracycline (Aureomycin®), was isolated by American Benjamin Duggar (1872–1956) in 1948 from *Streptomyces aureofaciens*. Scientists at Pfizer Laboratories isolated oxytetracycline (Terramycin®) from *Streptomyces rimosus* in 1950, and both drugs were later found to be derivatives of tetracycline, a compound now obtained from *Streptomyces viridifaciens*. All three compounds (Figure 18.8) are **broad-spectrum antibiotics**, so called because they are effective against a wide variety of bacteria.

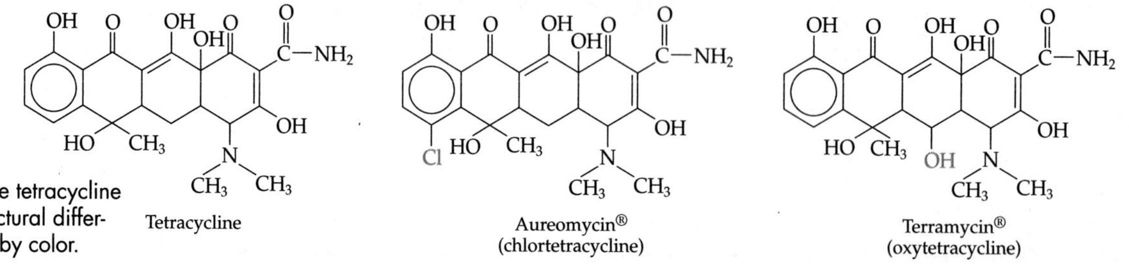

▶ **Figure 18.8** Three tetracycline antibiotics. Subtle structural differences are highlighted by color.

Tetracycline

Aureomycin® (chlortetracycline)

Terramycin® (oxytetracycline)

Tetracyclines bind to bacterial ribosomes, inhibiting bacterial protein synthesis and blocking growth of the bacteria. They do not bind to mammalian ribosomes and do not affect protein synthesis in host cells. Several disease-causing organisms have developed strains resistant to tetracyclines. When given to young children, tetracyclines can cause the discoloration of permanent teeth, even though the teeth may not appear until several years later. Women are told to avoid tetracyclines during pregnancy for the same reason. This probably results from the interaction of tetracyclines with calcium during the period of tooth development. Calcium ions in milk and other foods combine with hydroxyl groups on tetracycline molecules.

Fluoroquinolone antibiotics were first introduced in 1986 and now represent about one-third of the $42 billion global antibiotics market. They act against a broad spectrum of bacteria and seemingly have few side effects. A major use is against bacteria with penicillin resistance. Fluoroquinolones act by inhibiting DNA replication in bacteria through interference with the action of an enzyme called DNA gyrase. Humans do not have this enzyme, so fluoroquinolones do not harm human cells. Because their mechanism of action differs from that of other antibiotics, it was hoped that resistance to fluoroquinolones would be slow to emerge, but resistance has already been observed.

The fluoroquinolones are also of particular interest to chemists and pharmacologists because they have a number of structural features that can be correlated with their activity. This provides a rational and logical approach to designing and synthesizing new drugs in this class. The basic skeletal structure for a fluoroquinolone is

The fluorine atom at position 6 is needed for broad antimicrobial activity when taken orally. Effectiveness also depends on the carboxyl group at position 3 and the carbonyl oxygen at position 4, which are responsible for binding to the bacterial DNA complex. A nitrogen-containing ring structure at position 7 or a methoxy group at position 8 broadens the range of bacteria affected, and a bulky substituent at position 8 results in fewer side effects on the nervous system. Several of these features can be seen in ciprofloxacin (Cipro®).

Ciprofloxacin (Cipro®)

Viruses and Antiviral Drugs

For most of us, antibiotics have eliminated the terror of bacterial infections such as pneumonia and diphtheria. We worry about resistant strains of bacteria, but for most people these problems are not insurmountable. However, viral diseases cannot be cured by antibiotics, and viral infections, from colds and influenza to herpes and **acquired immune deficiency syndrome (AIDS)**, still plague us. Some viral infections—such as poliomyelitis, mumps, measles, and smallpox—can be prevented by vaccination. Influenza vaccines are quite effective against common recurrent strains of flu viruses, but there are many different strains of these viruses, and new ones appear periodically.

DNA Viruses and RNA Viruses

Viruses are composed of nucleic acids and proteins (Figure 18.9). They have an external coat that shows a repetitive pattern of protein molecules. Some coats also include a lipid membrane, and others have sugar–protein combinations called *glycoproteins*. The genetic material of a virus is either DNA or RNA. A *DNA*

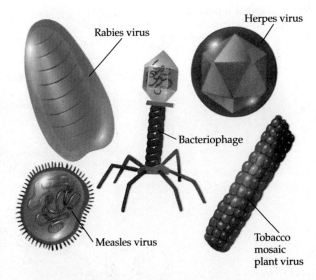

▲ **Figure 18.9** Viruses come in a variety of shapes, which are determined by their protein coats.

virus moves to the location in a host cell where the DNA is replicated and directs the host cell to produce viral proteins. The viral proteins and viral DNA assemble into new viruses that are released by the host cell. These new viruses can then invade other cells and continue the process.

Most *RNA viruses* use their nucleic acids in much the same way. The virus penetrates to where the RNA strands of a host cell are replicated and induces the synthesis of viral proteins; the new RNA strands and viral proteins are then assembled into new viruses. Some RNA viruses, called **retroviruses**, synthesize DNA in the host cell. This process is the opposite of the transcription of DNA into RNA (Section 16.10) that normally occurs in cells. (The term *retro* implies working backward.) The synthesis of DNA from an RNA template is catalyzed by an enzyme called *reverse transcriptase*. The human immunodeficiency virus (HIV) that causes AIDS is a retrovirus. HIV invades and eventually destroys *T cells*, white blood cells that normally help protect the body from infections. Once the T cells are destroyed, a person with AIDS often succumbs to pneumonia or some other infection.

Worldwide, 36.7 million people were living with AIDS in 2016. More than 35 million have already died from AIDS, and 1 million more died in 2016. Some 1.8 million people became infected with HIV in 2016. In the United States, more than 1.1 million persons are living with HIV/AIDS, and about 40,000 new HIV infections occur each year.

Antiviral Drugs

Scientists have developed a variety of drugs that are effective against some viruses, but none of them provides cures. Structures of some common antiviral agents are given in Figure 18.10. Another antiviral agent, acyclovir (Zovirax®), is used to treat chickenpox, shingles, cold sores, and the symptoms of genital herpes.

Antiretroviral drugs prevent the reproduction of retroviruses such as HIV and are used against AIDS. The following are three important kinds of antiretrovirals.

▪ *Nucleoside analogs*, or *nucleoside reverse transcriptase inhibitors (NRTIs)*, substitute an analog for a nucleoside in viral DNA, which cripples the retrovirus and slows down its replication. NRTIs include didanosine (ddI, Videx®), lamivudine (3TC, Epivir®), stavudine (d4T, Zerit®), zalcitabine (ddC, Hivid®), zidovudine (AZT, Retrovir®), and tenofovir (part of the combination drug Atripla®).

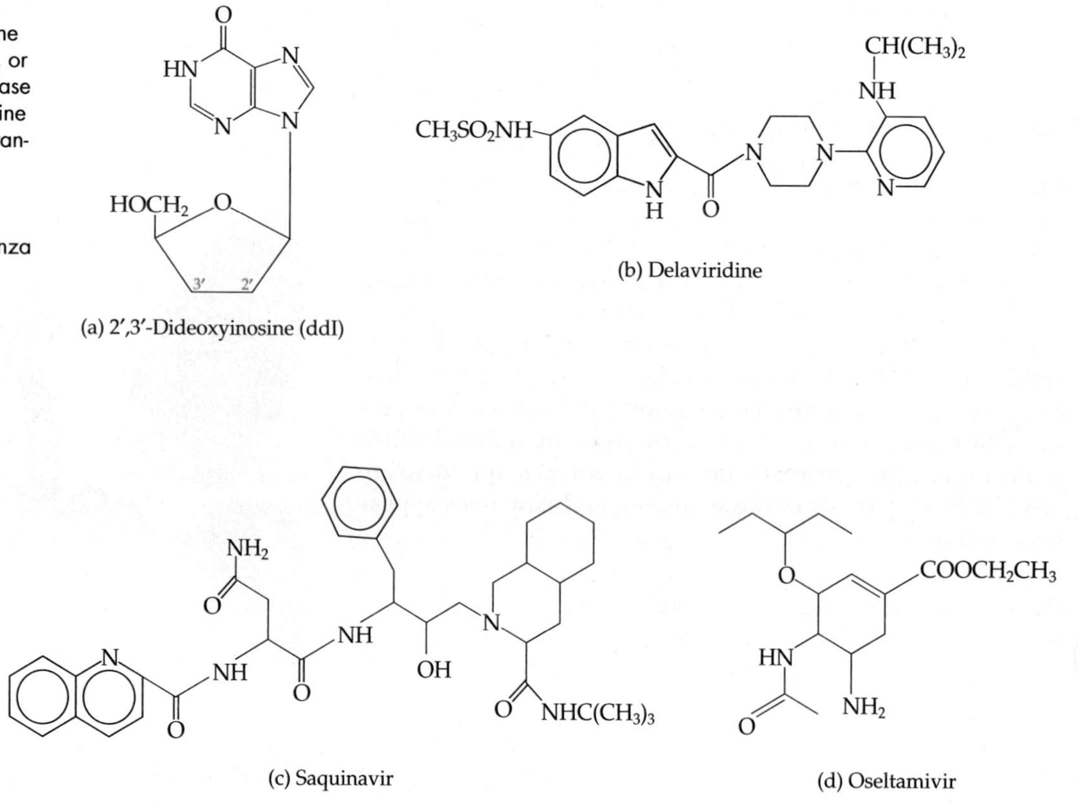

▶ **Figure 18.10** Some antiviral drugs. (a) Didanosine (ddI) is a nucleoside analog, or nucleoside reverse transcriptase inhibitor (NRTI). (b) Delavirdine is a nonnucleoside reverse transcriptase inhibitor (NNRTI). (c) Saquinavir is a protease inhibitor. (d) Oseltamivir (Tamiflu®) can prevent influenza if it is given in early stages.

(a) 2′,3′-Dideoxyinosine (ddI)

(b) Delaviridine

(c) Saquinavir

(d) Oseltamivir

- *Nonnucleoside reverse transcriptase inhibitors (NNRTIs)* stop the reverse transcriptase from working properly to make more of the retrovirus. NNRTIs include delavirdine (Rescriptor®), efavirenz, rilpivirine, and nevirapine (Viramune®).
- *Protease inhibitors* block the protease enzyme so that new copies of the retrovirus can't infect new cells. Protease inhibitors include ritonavir (Norvir®), darunavir, atazanavir, and saquinavir (Invirase®).

Combinations of AIDS drugs seem to be more effective than any one of them alone. For example, Atripla® contains efavirenz, emtricitabine, and tenofovir. It provides a once-daily single-pill treatment that can control HIV for long periods of time.

Basic Research and Drug Development

Scientists had to understand the normal biochemistry of cells before they could develop drugs to treat diseases caused by viruses. Americans Gertrude Elion (1918–1999) and George Hitchings (1905–1998), of Burroughs Wellcome Research Laboratories in North Carolina, and Sir James Black (Scotland, 1924–2010) of Kings College in London did much of the basic biochemical research that led to the development of antiviral drugs and many anticancer drugs (Section 18.4). They determined the shapes of cell membrane receptors and learned how normal cells work. They and other scientists were then able to design drugs to block receptors in infected cells. Elion, Hitchings, and Black shared the 1988 Nobel Prize in Physiology or Medicine for their work.

Drug design in its early days was often a hit-or-miss procedure. Today, scientists use powerful computers to design molecules to fit receptors, making drug design a more precise science.

Prevention of Viral Diseases with Vaccination

For many viral diseases, there are no effective drugs. However, these diseases can often be prevented by vaccination, which causes the body to develop immunity to the virus or microorganism. Examples include poliomyelitis (polio) and smallpox, as well as formerly common childhood diseases such as measles, mumps, and whooping cough.

Polio had been a major public health problem throughout the world, including among its victims President Franklin Roosevelt. In a 1952 epidemic, more than 50,000 victims were reported in the United States, with roughly half of them suffering paralysis. Jonas Salk (American, 1914–1995) developed a vaccine that was introduced in 1955 and administered widely. Albert Sabin (Polish-American, 1906–1993) developed an oral vaccine in the early 1960s. Introduction of these vaccines ended outbreaks of polio in the United States and almost all of the rest of the world.

Smallpox vaccination has been administered for a much longer time, with the best-known vaccine developed by Edward Jenner (English, 1749–1823) in 1796. Smallpox vaccination was discontinued in the United States in 1972, when the tiny statistical risk from the vaccine was judged to be greater than the risk of acquiring the disease.

The MMR (measles, mumps, rubella) vaccine was developed in the 1970s and ordinarily is given to children before they enter kindergarten. Although some of these diseases can be treated after onset, preventing them is much preferred.

Unfortunately, many parents have recently decided not to have their children vaccinated, based on a 1998 study by Andrew Wakefield that purported to link MMR vaccination to autism. Although this work has been found to be fraudulent and Wakefield has been barred from practicing medicine, some well-known public figures have continued to support his work. Uninformed parents, fearful of autism, have elected not to have their children vaccinated. The antivaccination movement has resulted in epidemics in several states. For example, in 2015 and 2016, there were over roughly 6000 confirmed cases of mumps each year. There have also been large increases in incidence of other preventable diseases in recent years, including outbreaks of measles and whooping cough.

The past few years have also seen the emergence of both Ebola and Zika viruses as major threats to public health. Ebola causes hemorrhages and is frequently fatal. Zika causes a high rate of birth defects when a pregnant woman is infected. The 2013–2016 Ebola outbreak in Western Africa spurred work on the development of a

4 **What kinds of drugs work for colds or the flu?**

Colds and the flu are results of viral infection. Antibiotics work against bacteria, but viruses have proved to be much more difficult to treat, in part because viruses mutate quite rapidly. There are some prescription drugs, such as oseltamivir (Tamiflu), which can minimize the effects of the flu when taken within 1–2 days of onset. However, there still is no effective treatment for colds, which can be triggered by any of some 200 rhinoviruses. Treatments to help alleviate the symptoms rather than the cause are widely used.

vaccine. By late 2016 there were vaccines that looked promising in trials, and they were used to try to contain a May 2018 outbreak in the Democratic Republic of the Congo. Similarly, there has been a surge of research in developing a vaccine against the Zika virus, and several candidates have emerged for further testing.

SELF-ASSESSMENT Questions

1. Penicillins act by inhibiting the synthesis of
   a. bacterial cell walls **b.** folic acid
   c. β-lactams **d.** viral RNA

2. A compound that is derived from bacteria or molds and that inhibits the growth of other microorganisms is a(n)
   a. antibiotic **b.** antiviral
   c. anticarcinogen **d.** retroviral

3. Taking antibiotics for a stomach virus will
   a. cure the illness
   b. increase the chance that drug-resistant bacteria will develop
   c. lessen the symptoms
   d. shorten the duration of the illness

4. Viruses consist of a protein coat surrounding a core of
   a. lipids **b.** nucleic acids
   c. polysaccharides **d.** proteins

5. Most RNA viruses replicate in host cells by replicating RNA strands and synthesizing
   a. host-cell lipids **b.** host-cell proteins
   c. viral polysaccharides **d.** viral proteins

6. Retroviruses replicate in host cells by
   a. replicating RNA strands and synthesizing viral proteins
   b. replicating RNA strands and synthesizing host-cell proteins
   c. synthesizing DNA and forming new viruses
   d. synthesizing DNA and replicating host cells

7. What is the best way of dealing with viral diseases such as diphtheria, mumps, and smallpox?
   a. antibacterial drugs **b.** antiviral drugs
   c. tetracyclines **d.** vaccinations

Answers: 1, a; 2, a; 3, b; 4, b; 5, d; 6, c; 7, d

18.4 Chemicals against Cancer

Learning Objective • Describe the action of the common types of anticancer drugs.

Despite decades of "war" on cancer, it remains a dread disease. Each year in the United States, more than 1.7 million new cases are diagnosed, cancer kills over 600,000 people, and we spend over $150 billion in treatment. Scientists have made progress, but much more remains to be done. A major difficulty is that the drugs that kill cancer cells also damage normal cells, especially rapidly growing cells such as those that line the digestive tract or produce hair. Side effects of chemotherapy therefore often include nausea and loss of hair. Thus, a main aim of cancer research is to find a way to kill cancerous tissue without killing too many normal cells.

Treatment with drugs, radiation, and/or surgery yields a high rate of cure for some kinds of cancer. For example, with treatment, early-stage prostate cancer has an overall five-year survival rate of nearly 100%. For some other types, such as lung and bronchus cancer, the overall five-year survival rate is only 18%, mainly because lung cancer is often advanced at the time of diagnosis. The five-year survival rate across all cancers is nearly 70%.

Several categories of anticancer drugs are used widely, and we look at some of them here.

Antimetabolites: Inhibition of Nucleic Acid Synthesis

An **antimetabolite** is a compound that closely resembles a substance essential to normal body metabolism, which allows it to interfere with physiological reactions involving that substance. Rapidly dividing cells, characteristic of cancer, require an abundance of DNA. Cancer antimetabolites block DNA synthesis and therefore block the increase in the number of cancer cells. Because cancer cells are undergoing rapid growth and cell division, they are generally affected to a greater extent than are normal cells. Research over the last two decades has led to refinements in how antimetabolites are administered, increasing their effectiveness.

Gertrude Elion and George Hitchings patented 6-mercaptopurine (6-MP) in 1954. This compound can substitute for adenine in nucleotides, the phosphate–sugar–base units of both DNA and RNA (Section 16.9), and the pseudonucleotide then inhibits the synthesis of nucleotides incorporating adenine and guanine. This slows DNA synthesis and cell division, thus inhibiting the multiplication of cancer cells. Before the use of 6-MP, half of all children with acute leukemia died within a few months. Combined with other medications, 6-MP was able to cure approximately 80% of child leukemia patients.

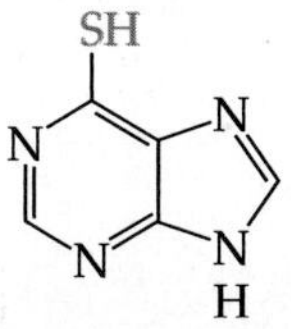

Another prominent antimetabolite is 5-fluorouracil (5-FU), which blocks synthesis of the thymine-containing nucleotide thymidine, an essential component for DNA replication. 5-FU is employed against a variety of cancers, especially those of the breast and the digestive tract.

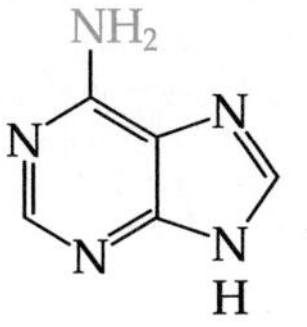

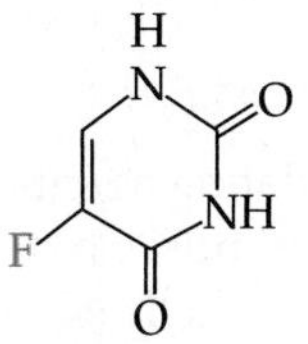

6-Mercaptopurine

Adenine

5-Fluorouracil 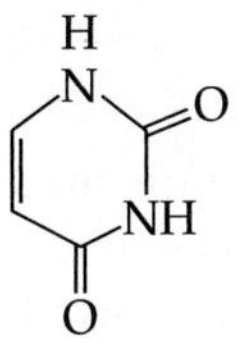   Uracil

The antimetabolite methotrexate acts somewhat differently. It is a folic acid antagonist that interferes with cellular reproduction. Note the similarity between its structure and that of folic acid. Like the pseudo–folic acid formed from sulfanilamide, methotrexate competes successfully with folic acid for an essential enzyme but cannot perform the growth-enhancing function of folic acid. Again, cell division is slowed and cancer growth is retarded. Methotrexate is used frequently against leukemia. It is also used in the treatment of psoriasis and certain inflammatory diseases, such as rheumatoid arthritis.

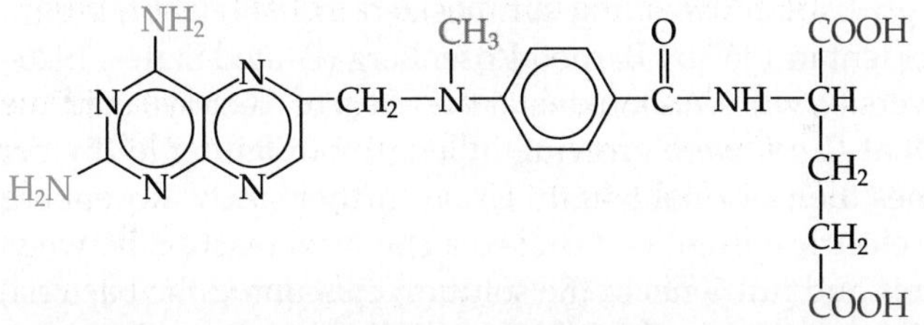

Methotrexate 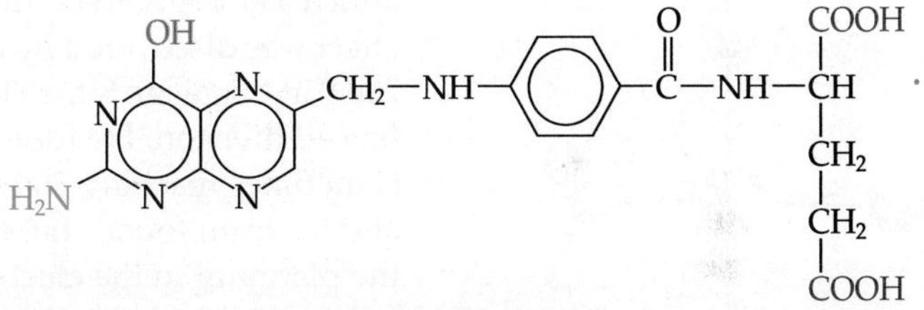   Folic acid

Alkylating Agents: Turning Old Weapons into Anticancer Drugs

Alkylating agents are highly reactive compounds that can transfer alkyl groups to compounds of biological importance. These added alkyl groups then block the usual action of the biological molecules. Some alkylating agents are used against cancer, including nitrogen mustards.

Interestingly enough, these compounds grew out of research for chemical warfare agents. More than thirty chemical agents were employed during World War I, killing 91,000 and wounding 1.2 million (disabling many of them for life).

The use of chemical warfare agents was largely avoided during World War II. However, they have since been used in smaller wars—for example, by Iraq against Iran in the 1980s and by the Iraqi government against Kurdish rebels. Most recently, sarin has been used in Syria. Chemical warfare agents are considered to be one of the *weapons of mass destruction (WMDs)* that can kill huge numbers of people and might be used by a rogue nation or a terrorist group. (The other WMDs are nuclear devices and biological agents such as anthrax and botulism.)

The original mustard gas was a sulfur-containing blister agent used during World War I. Contact with either the liquid or the vapor causes blisters that are painful and slow to heal. It is easily detected, however, by its garlic or horseradish odor. Mustard gas is denoted by the military symbol H.

$$Cl—CH_2CH_2—S—CH_2CH_2—Cl$$
Mustard gas (H)

Nitrogen mustards (symbol HN) were developed around 1935. Though not quite as effective overall as mustard gas, nitrogen mustards produce greater eye damage and don't have an obvious odor. Structurally, nitrogen mustards are chlorinated amines.

$$CH_3CH_2-N\begin{array}{c}CH_2CH_2-Cl\\CH_2CH_2-Cl\end{array} \qquad CH_3-N\begin{array}{c}CH_2CH_2-Cl\\CH_2CH_2-Cl\end{array} \qquad Cl-CH_2CH_2-N\begin{array}{c}CH_2CH_2-Cl\\CH_2CH_2-Cl\end{array}$$

HN_1 HN_2 HN_3

Nitrogen mustards are bifunctional alkylating agents (agents with two functional groups) and act by cross-linking two DNA strands. Cross-linking prevents or hinders replication and thus impedes growth of cancer cells. Knowledge gained through science is neither good nor evil. The same knowledge—in this case, the ability to make nitrogen mustards—can be used either for our benefit or for our destruction.

The nitrogen mustard often used for cancer therapy is a compound called cyclophosphamide (Cytoxan®). It is used to treat Hodgkin's disease, lymphomas, leukemias, and other cancers.

Alkylating agents can cause cancer as well as cure it. For example, the nitrogen mustard HN_2 causes lung, mammary, and liver tumors when injected into mice, yet it can be used with some success in the management of certain human tumors. There is still a lot of mystery—and seeming contradiction—regarding the causes and cures of cancer.

The platinum-containing compound cisplatin $[PtCl_2(NH_3)_2]$ is a prominent anticancer drug. Even though cisplatin was first synthesized in 1844, its biological effect was discovered by accident in 1965 by Barnett Rosenberg (United States, 1926–2009) at Michigan State University, who was looking at the effect of electrical currents on cell division. He found that *E. coli* were growing in length but failing to divide, sometimes reaching 300 times their normal length. Upon further study, Rosenberg and his team found that the electric current had caused a chemical reaction between the platinum in the electrodes and nutrients in the solution containing the bacteria, producing cisplatin. Because the compound inhibited cell division, Rosenberg reasoned that it might be effective as an anticancer drug. It is. Cisplatin, an alkylating agent, binds to DNA and blocks its replication. It is widely used to treat cancers of the ovaries, testes, uterus, head, neck, breast, and lungs, and advanced bladder cancers.

Other Anticancer Agents

Many other anticancer agents are in use today. Alkaloids from vinca plants have been shown to be effective against leukemia and Hodgkin's disease. Paclitaxel (Taxol®), first obtained from the Pacific yew tree, is effective against cancers of the breast, ovary, and cervix.

Imatinib (Gleevec®) is a synthetic drug that is particularly effective against chronic myeloid leukemia. Cell growth is controlled by substances called *growth factors*, which act by attaching to receptor proteins on the surface of certain types of cells. People with chronic myeloid leukemia produce cells with damaged receptors that send grow-and-divide signals to the cells even with no growth factor present. Imatinib inhibits the faulty receptors, preventing them from stimulating the cells to grow and divide; the cancer cells cannot multiply.

Other drugs that are quite effective for treating chronic leukocytic leukemia have come onto the market quite recently. These include ibrutinib (Imbruvica®) and venetoclax, and there are others in clinical trials. Ibrutinib works by inhibiting an enzyme needed for growth of the cancer, and it has the advantage of being an oral medication taken once a day. Annual sales were $1 billion in 2016, and it is estimated to have sales of $5 billion by 2020.

Several antibiotics have been found to kill cancer cells as well as bacteria. Actinomycin, obtained from the molds *Streptomyces antibioticus* and *S. parvus*, is used against Hodgkin's disease and other types of cancer. It is quite effective but

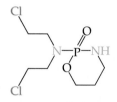

Cyclophosphamide

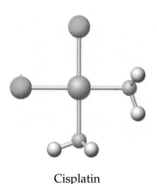

Cisplatin

Imatinib

extremely toxic. Actinomycin acts by binding to the double helix of DNA, thus blocking the replication of RNA on the DNA template. Protein synthesis is inhibited.

Chemotherapy is only part of the treatment for cancer. Surgical removal of tumors and radiation treatment remain major tools. It is unlikely that research will discover a single agent that can cure all cancers. Active research is under way on the mechanisms of carcinogenesis, and a better understanding will lead to better cures. Prevention of cancer is a greater hope. Deaths from smoking-related cancers account for at least 30% of all cancer deaths. Reducing cigarette smoking can significantly reduce cancer deaths (Section 21.6).

SELF-ASSESSMENT Questions

1. Which of the following is an antimetabolite?
   a. 5-fluorouracil   **b.** cyclophosphamide
   c. cisplatin   **d.** nitrogen mustard

2. 6-mercaptopurine mimics the base adenine, forming a pseudonucleotide and thus slowing
   a. cell-mediated immune reactions   **b.** DNA synthesis
   c. RNA transport   **d.** viral replication

3. Methotrexate is a folic-acid antagonist that slows
   a. cell division
   b. cell-mediated immune reactions
   c. transport of nutrients to the cancer cell
   d. viral replication

4. Cisplatin inhibits cell division by
   a. being incorporated in RNA
   b. binding to DNA and blocking its replication
   c. halting viral replication
   d. substituting for cytosine in a nucleotide

5. Cyclophosphamide acts by
   a. being incorporated in RNA
   b. cross-linking two DNA strands and blocking replication
   c. halting viral replication
   d. substituting for cytosine in a nucleotide

Answers: 1, a; 2, b; 3, a; 4, b; 5, b

18.5 Hormones: The Regulators

Learning Objectives • Define the terms *hormone*, *prostaglandin*, and *steroid*, and explain the function of each. • List the three types of sex hormones, and explain how each acts and how birth control drugs work.

Before we discuss the next group of drugs, we need to take a brief look at the human endocrine system and some of the chemical compounds, called *hormones*, that this system manufactures. A **hormone** is a chemical messenger produced in the endocrine glands (Figure 18.11). Hormones are released in one part of the body and send signals for profound physiological changes in other parts of the body. By causing reactions to speed up or slow down, hormones control growth, metabolism, reproduction, and many other functions of body and mind. Some of the important human hormones and their physiological effects are listed in Table 18.2. Still other hormones are formed in tissues of the heart, liver, kidney, gut, and placenta. Most hormones are proteins, peptides, or amino acid derivatives, although hormones produced by the adrenal gland and the gonads are steroids.

Prostaglandins: Hormone Mediators

A *prostaglandin* (Section 18.2) is a hormone-like lipid derived from a fatty acid. Each prostaglandin molecule has 20 carbon atoms, including a 5-carbon ring. Prostaglandins function similarly to hormones in that they act on target cells. However, they differ from hormones in that they (1) act near the site where they are produced, (2) can have different effects in different tissues, and (3) are rapidly metabolized. Prostaglandins are synthesized in the body from arachidonic acid (Figure 18.12). There are several primary prostaglandins, which are widely distributed throughout the body. Extremely small doses of these potent biological chemicals can elicit marked changes. Besides the primary prostaglandins, many others have been identified.

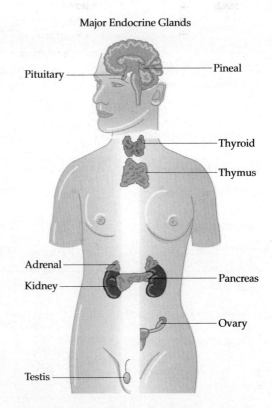

Major Endocrine Glands

Pituitary — Pineal

Thyroid

Thymus

Adrenal —
Kidney — Pancreas

Ovary

Testis —

▲ **Figure 18.11** The approximate locations of the eight major endocrine glands in the human body.

TABLE 18.2 **Some Human Hormones and Their Physiological Effects**

Name	Chemical Nature	Function(s)
Pituitary Hormones		
Vasopressin (antidiuretic hormone)	Nonapeptide	Stimulates contractions of smooth muscle; regulates water retention and blood pressure
Oxytocin	Nonapeptide	Stimulates contraction of the smooth muscle of the uterus; stimulates secretion of milk
Growth hormone (GH) (or somatotropin)	Protein	Stimulates body growth and bone growth
Thyroid-stimulating hormone (TSH)	Protein	Stimulates growth of the thyroid gland and production of thyroid hormones
Adrenocorticotropic hormone (ACTH)	Protein	Stimulates growth of the adrenal cortex and production of cortical hormones
Follicle-stimulating hormone (FSH)	Protein	Stimulates maturation of egg follicles in ovaries of females and of sperm cells in testes of males
Luteinizing hormone (LH)	Protein	Stimulates female ovulation and male secretion of testosterone
Prolactin	Protein	Maintains the production of estrogens and progesterone; stimulates the formation of milk
Thyroid Hormone		
Thyroxine	Amino acid derivative	Increases rate of cellular metabolism
Pancreatic Hormones		
Insulin	Protein	Increases cellular usage of glucose; increases glycogen storage
Glucagon	Protein	Stimulates conversion of liver glycogen to glucose
Adrenal Cortical Hormones		
Cortisol	Steroid	Stimulates conversion of proteins to carbohydrates
Aldosterone	Steroid	Regulates salt metabolism; stimulates kidneys to retain Na^+ and excrete K^+
Adrenal Medullary Hormones		
Epinephrine (adrenaline)	Amino acid derivative	Stimulates a variety of mechanisms to prepare the body for emergency action, including the conversion of glycogen to glucose
Norepinephrine (noradrenaline)	Amino acid derivative	Stimulates the sympathetic nervous system; constricts blood vessels; stimulates other glands
Gonadal Hormones		
Estradiol	Steroid	Stimulates development of female sex characteristics; regulates changes during menstrual cycle
Progesterone	Steroid	Regulates menstrual cycle; maintains pregnancy
Testosterone	Steroid	Stimulates development of and maintains male sex characteristics

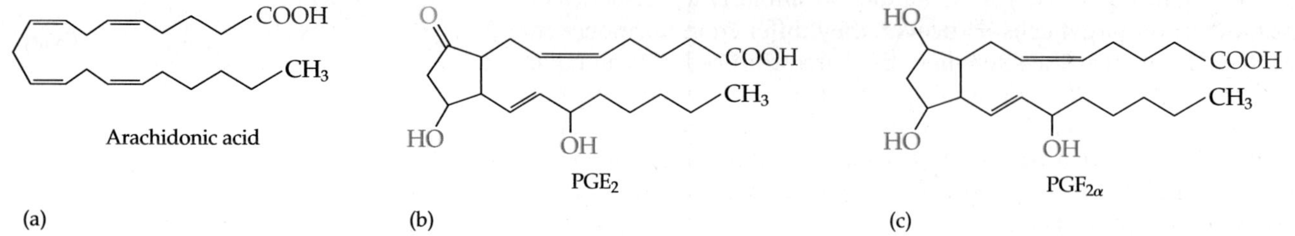

▲ **Figure 18.12** (a) Prostaglandins are derived from arachidonic acid, an unsaturated carboxylic acid with 20 carbon atoms. Two representative prostaglandins are shown here: (b) prostaglandin E_2 (PGE_2) and (c) prostaglandin $F_{2\alpha}$ ($PGF_{2\alpha}$).

Prostaglandins act as mediators of hormone action. They regulate such things as blood pressure, blood clotting, pain sensation, smooth muscle activity, and secretion of substances related to reproduction. A single prostaglandin can have different, or even opposite, effects in different tissues. This range of physiological activity has led to the synthesis of hundreds of prostaglandins and analogs. Prostaglandin E_2 (PGE_2), also known as dinoprostone (Cervidil®), is used to induce labor. PGE_1 can lower blood pressure. A PGE_1 analog called misoprostol is used to help prevent peptic ulcers, a common side effect of large doses of NSAIDs. Other prostaglandins can be used clinically to relieve nasal congestion, to provide relief from asthma, to treat glaucoma, to treat pulmonary hypertension, and to prevent formation of blood clots associated with heart attacks and strokes.

$PGE_{2\alpha}$ is used in cattle to synchronize breeding. It is also used in the artificial insemination of prize cows. Injection of $PGE_{2\alpha}$ along with a hormone into a prize cow will induce the release of many ova, which are then fertilized with sperm from a champion bull. The developing embryos are implanted in less valuable cows, enabling a farmer to get several calves a year from one outstanding cow.

Steroids

A **steroid** is a compound with a characteristic four-ring skeletal structure. Steroids occur widely in living organisms, but not all steroids are hormones (Figure 18.13). Cholesterol, for example, is a steroid component of all animal tissues and is the starting material for the synthesis of all other naturally occurring steroids. About 10% of the brain is cholesterol; it is a vital part of the membranes of nerve cells. Cholesterol is also found in deposits in hardened arteries and is a major component of certain types of gallstones. Cortisol, another natural steroid, is an adrenal hormone (Table 18.2). Prednisone is a synthetic steroid used to reduce inflammation in arthritis sufferers and to treat injuries.

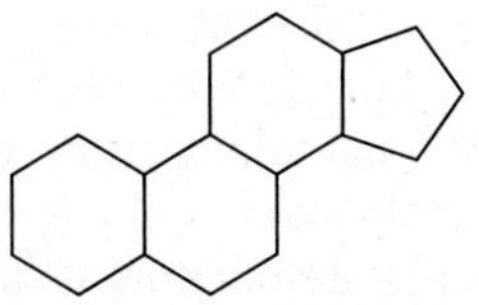

▲ Skeletal structure of steroids.

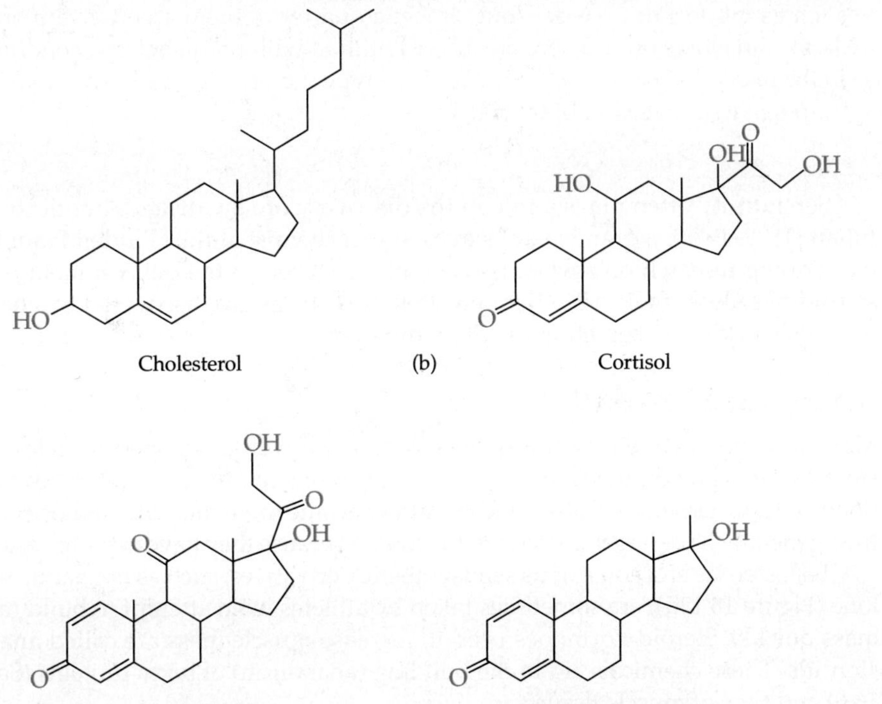

▲ **Figure 18.13** Structural formulas of four steroids. (a) Cholesterol is an essential component of all animal cells. (b) Cortisol is a hormone secreted by the adrenal glands in response to physical or psychological stress. (c) Prednisone is a synthetic anti-inflammatory substance. (d) Methandrostenolone (Dianabol®) is an anabolic steroid. Note that all have the same basic four-ring structure.

Many drugs, both natural and synthetic, are based on steroids. Some are anti-inflammatories used to treat arthritis, bronchial asthma, dermatitis, and eye infections. Some are sex hormones that are active ingredients in birth control pills. Anabolic steroids are often used by athletes and body builders to build muscle and can cause serious damage to health and even result in sterility.

Diagnosis

Diabetes

The pancreas produces the hormone *insulin*, which increases cellular usage of glucose. *Diabetes* arises when the pancreas does not produce enough insulin (type 1) or when the insulin is not properly used by the body (type 2). In both cases, elevated blood glucose levels are the result. Over 90% of Americans diagnosed with diabetes suffer from type 2, which has a strong genetic component as well as a close tie to lifestyle. Poor diet, obesity, and lack of exercise all appear to be contributing factors to type 2. Gestational diabetes is a form of type 2 that occurs in 2 to 10% of pregnant women with no history of the disease. This form generally disappears after delivery but does persist in some women permanently. Paralleling the increase in obesity in the United States, type 2 diabetes has become increasingly common, even among children and adolescents.

People who suffer from type 1 diabetes ordinarily must take insulin, which has to be injected because insulin taken orally is broken down by the body's digestive system before it can reach the bloodstream. Type 2 diabetics may be able to control their disease with careful diet, exercise, and loss of weight. When these measures aren't enough, a biguanide such as metformin (Glucophage®) is often prescribed. Metformin works by reducing the production of glucose in the liver. It is one of the most widely prescribed medications, with over 60 million new and refilled prescriptions dispensed in the United States in 2014. Sulfonylureas, which increase the production of insulin from the pancreas, are also used to treat type 2 diabetes. Sometimes these oral drugs are not enough to maintain glucose levels in individuals with type 2 diabetes, and insulin must be administered. Newer medications have been approved, but are not yet available in generic form, and they often are considerably more expensive. The economic cost of diabetes in the United States was estimated to be $327 billion in 2017.

$$CH_3-N-C-NH-C-NH_2$$

Metformin

Diabetes can cause a number of complications, including blindness, cardiovascular disease, kidney disease, and nerve damage. Thus, diagnosis and control of diabetes are very important. As of 2015, 30.3 million Americans (9.4% of the total population) have the disease—but almost a quarter of them aren't aware of it. There are also 84 million with prediabetes, a condition that often leads to type 2 diabetes within five years if not treated.

Serendipity often plays a role in the discovery of new drugs. American Percy Julian (1899–1975), a grandson of slaves, was a chemist at the Glidden Paint Company doing research on soybeans when his work led to the development of new steroid-based drugs. It often happens that new drugs are discovered by chemists who are working on something entirely different.

▲ Percy Lavon Julian (United States, 1899–1975) was involved in the synthesis of a variety of steroids, including physostigmine, a drug used to treat glaucoma, which can cause blindness.

Anabolic Steroids

Many strenuous athletic performances depend on well-developed muscles. Muscle mass depends on the level of the male hormone *testosterone*. When boys reach puberty, testosterone levels rise, and the boys become more muscular if they exercise. Men generally have larger muscles than women because they have more testosterone.

Testosterone and some of its semisynthetic derivatives, such as methandrostenolone (Figure 18.13d), are sometimes taken by athletes in an attempt to build muscle mass quickly. Steroid hormones used to increase muscle mass are called **anabolic steroids**. These chemicals aid in the building (anabolism) of body proteins (Section 16.6) and thus of muscle tissue.

There are no reputable controlled studies demonstrating the effectiveness of anabolic steroids. They may seem to work—at least for some people—but the side effects are many. In males, side effects include testicular atrophy and loss of function, impotence, acne, liver damage that may lead to cancer, edema (swelling), elevated cholesterol levels, and growth of breasts.

Anabolic steroids act as male hormones (androgens), making women more masculine. They help women build larger muscles, but they also result in balding, extra body hair, a deep voice, and menstrual irregularities.

Because drug use for the enhancement or improvement of athletic performance is illegal, chemists play an important role in sports: screening blood and urine samples for illegal drugs. Using sophisticated instruments, chemists can detect minute amounts of illegal drugs. Drug testing has been used at the Olympic Games since 1968, and it is rapidly becoming standard practice for athletes in professional and even college sports.

The use of illegal drugs not only violates the spirit of fair athletic competition, but it also carries great risks for health and legal problems.

Sex Hormones

Sex hormones are steroids (Figure 18.14). Note that male sex hormones differ only slightly in structure from female sex hormones. In fact, the female hormone progesterone can be converted to the male hormone testosterone by a simple biochemical reaction. The physiological actions of these structurally similar compounds, however, are markedly different.

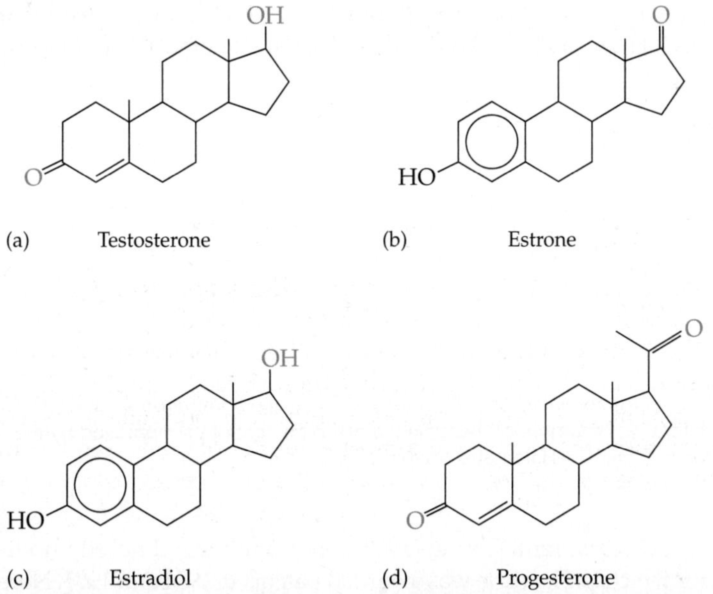

(a) Testosterone   (b) Estrone

(c) Estradiol   (d) Progesterone

▲ **Figure 18.14** Structural formulas of the principal sex hormones. (a) Testosterone is the main male sex hormone (an androgen). (b) Estrone and (c) estradiol are female sex hormones (estrogens). (d) Progesterone, also produced by females, is essential to the maintenance of pregnancy.

An **androgen** is any compound that stimulates the development or controls the maintenance of masculine characteristics. These male sex hormones are secreted by the testes. In males, the pituitary hormones FSH and LH (Table 18.2) are required for the continued production of sperm and of androgens. These hormones are responsible for development of the sex organs and for secondary sexual characteristics, such as voice and hair distribution. The most important androgen is testosterone.

An **estrogen** is a compound that controls female sexual functions, such as the menstrual cycle, breast development, and other secondary sexual characteristics. Two important estrogens are estradiol and estrone, which are produced mainly in the ovaries. Another female sex hormone is progesterone, which prepares the uterus for pregnancy and prevents the further release of eggs from the ovaries during pregnancy. A related type of compound, called a **progestin**, is any steroid hormone that has the effect of progesterone. In females, the pituitary hormones FSH and LH (Table 18.2) are required for the production of ova, progesterone, and estrogens.

Sex hormones—both natural and synthetic—are sometimes used therapeutically. For example, women who have passed menopause used to be given hormones to compensate for those no longer being produced by their ovaries. However, hormone replacement therapy has largely been discontinued because of concerns that the increased health risks it posed outweighed the benefits.

Chemistry and Social Revolution: The Pill

When administered by injection, progesterone serves as an effective birth control drug. It fools the body into acting as if it were already pregnant. The structure of progesterone was determined in 1934 by Adolf Butenandt (Germany, 1903–1995), who received the Nobel Prize in Chemistry in 1939. Other chemists began to try to design a contraceptive that would be effective when taken orally.

In 1938, Hans Inhoffen (Germany, 1906–1992) synthesized the first oral contraceptive, ethisterone, by incorporating an ethynyl group ($-C{\equiv}CH$). (The ethynyl group is derived from ethyne, $HC{\equiv}CH$, also called acetylene.) However, ethisterone had to be taken in large doses to be effective and was not widely used. In 1951, Carl Djerassi (Austria, United States; 1923–2015) synthesized 19-norprogesterone, which is simply progesterone with one of its methyl groups missing. It was 4–8 times as effective as progesterone as a birth control agent, but it had to be given by injection, an undesirable property.

Djerassi then combined the two improvements—removal of a methyl group to make the drug more effective, and addition of an ethynyl group to allow oral administration—to synthesize norethindrone (Norlutin®), which proved effective when taken in small doses (Figure 18.15), and which Djerassi patented in 1956. Working at about the

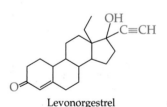

▲ **Figure 18.15** Formulas of some synthetic sex hormones: (a) norethindrone, (b) norethynodrel, and (c) mestranol. In 1960, G. D. Searle's Enovid, which contained 9.85 mg of norethynodrel and 150 mg of mestranol, became the first birth control pill approved by the FDA. Djerassi's norethindrone and Colton's norethynodrel differ only in the position of a double bond.

(a) Norethindrone (Norlutin®) (b) Norethynodrel (c) Mestranol

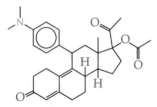

Levonorgestrel

Ulipristal acetate

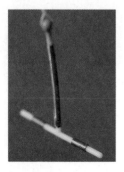

▲ The Copper T 380A IUD (ParaGard).

same time, American Frank Colton (1923–2003) synthesized norethynodrel, another progestin, for which G. D. Searle was awarded patents in 1954 and 1955. Norethynodrel, a progestin, was used in Enovid®, the first birth control pill approved by the FDA (in 1960).

The progestins—norethynodrel, norethindrone, and related compounds—mimic the action of progesterone. Mestranol, a synthetic estrogen, regulates the menstrual cycle, and the progestin establishes a state of false pregnancy. A woman does not ovulate or conceive when she is pregnant or in a state of false pregnancy established by a progestin.

Emergency Contraceptives

Products called *emergency contraceptive pills (ECPs)* can be used to prevent pregnancy after unprotected intercourse. Two common products are Plan B One-Step® and ella®. Plan B One-Step contains only a progestin, levonorgestrel, but in higher doses than regular birth control pills. It is available over the counter without a prescription. Plan B, like progestins in general, acts by tricking the body into thinking it's already pregnant, halting development of the uterine lining and inhibiting ovulation and fertilization. Another choice is ella, a non-hormonal pill available only by prescription that contains ulipristal acetate, a progesterone antagonist. This delays ovulation for up to five days, by which time sperm will have died.

An intrauterine device (IUD), the Copper T 380A IUD (ParaGard®), is also used for emergency contraception. IUD insertion reduces the incidence of pregnancy by 99.9% if done within 3–5 days of unprotected intercourse. Copper(II) ions in the IUD interfere with sperm transport and fertilization. The copper IUD probably causes an inflammation that makes the endometrium unsuitable for implantation.

ECPs act differently from mifepristone (Mifeprex® or RU-486), which causes the uterus to expel an implanted egg and end the pregnancy. Thus, ECPs prevent pregnancy after sexual intercourse, while mifepristone ends an unwanted pregnancy at an early stage.

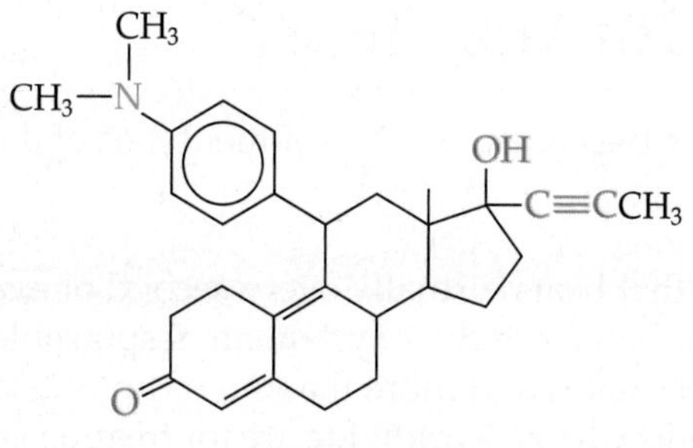

Mifepristone

People who believe that human life begins when the ovum is fertilized by the sperm equate the use of mifepristone with abortion and oppose the use of the drug. Others consider the drug another method of birth control and see its use as being safer than a surgical abortion.

Risks of Taking Birth Control Pills

Because most of the side effects of oral contraceptives are associated with the estrogen component, the amount of estrogen in birth control pills has been greatly reduced over the years, and today's pills contain only a fraction of a milligram of estrogen. "Minipills" are now available that contain only small amounts of progestin and no estrogen at all. While minipills are not quite as effective as the combination pills, they have fewer side effects.

Oral contraceptives are currently used by more than 100 million women world-wide and by about 12 million women in the United States. Most women report no or relatively minor side effects. Some women experience hypertension, acne, or abnormal bleeding. These pills increase the risk of blood clotting in some women, but so does pregnancy. Blood clots can clog arteries and cause death by stroke or heart attack. The death rate associated with birth control pills is about 3 in 100,000, only one-tenth of the death rate associated with childbirth. For smokers, however, the risks are much higher. For women over 40 who smoke 15 cigarettes per day and take an oral contraceptive, the risk of death from stroke or heart attack is 1 in 5000. The FDA advises all women who smoke, especially those over 40, to use some other method of contraception.

A Contraceptive Pill for Males?

Why should females have to bear all the responsibility for contraception? Why not a pill for males? Many men are willing to share the risks and the responsibility of contraception, but condoms and vasectomies are the only effective forms of contraception available to them. There are biological reasons for females to bear the burden. Women are the ones who get pregnant when contraception fails, and in females, contraception has to interfere with only one monthly event: ovulation. Males produce sperm continuously.

In males, the pituitary hormones FSH and LH (Table 18.2) are required for the continued production of sperm and of the male hormone testosterone. Several research groups are working on developing a safe, effective, and reversible contraceptive for males. One approach puts a progestin and testosterone into the same pill. The progestin seems to function much as it does in the women's pill and suppresses the rate of sperm production as well as the quantity of sperm produced. When will such male birth control pills be available to the public? Optimistic projections still estimate a wait of several years.

SELF-ASSESSMENT Questions

1. Which organ system consists of hormone-secreting glands?
   a. endocrine   **b.** lymphatic   **c.** respiratory   **d.** urinary

2. Prostaglandins are synthesized in the body from
   a. arachidonic acid   **b.** ascorbic acid
   c. cholesterol   **d.** progesterone

3. How many contraceptives have been approved by the FDA for use by males?
   a. none   **b.** one   **c.** three   **d.** five

4. The hormone that maintains pregnancy is
   a. estradiol   **b.** estrone   **c.** progesterone   **d.** prolactin

5. The functional group that makes a birth control drug effective orally is
   a. CH_3CH_2- **b.** $CH_2{=}CH_2-$
   c. $HC{\equiv}C-$ **d.** CH_3CO-

18.6 Drugs for the Heart

Learning Objective • Describe the action of four types of drugs used to treat heart disease.

The heart is a muscle that beats virtually every second of every day for as long as we live. Diseases of the cardiovascular system are responsible for over one-quarter of deaths in the United States and more than one-fifth of deaths worldwide. Four of the top ten prescription drugs worldwide are for treating cardiovascular disease. These drugs prolong life and improve its quality.

Major diseases of the heart and blood vessels include

- Ischemic ("lacking oxygen") coronary artery disease
- Heart arrhythmias (abnormal heartbeat)
- Hypertension (high blood pressure)
- Congestive heart failure

Atherosclerosis (buildup of fatty deposits in the lining of the arteries) is the primary cause of coronary artery disease, which in turn causes myocardial infarction (heart attack). Thus, most drug treatments for the heart aim to increase its supply of blood (and oxygen), to normalize its rhythm, to lower blood pressure, or to prevent accumulation of lipid plaque deposits in blood vessels.

Lowering Blood Pressure

Blood pressure units are given in millimeters of mercury (mmHg). Normal blood pressure is defined as lower than 120/80. Recent guidelines from the American Heart Association now label the range between 120/80 to 130/80 as elevated. *Hypertension*, or high blood pressure, is now considered to begin at lower levels, with the range between 130/80 and 140/90 being hypertension stage 1. Levels over 140/90 are now considered to be hypertension stage 2. Hypertension is the most common cardiovascular disease, affecting nearly half of all adults in the United States under the new guidelines. Because it seldom produces symptoms, many people have hypertension without realizing it. There are four major categories of drugs for lowering blood pressure.

- *Diuretics* such as hydrochlorothiazide [HCT] cause the kidneys to excrete more water and thus lower the blood volume.
- *Beta blockers* such as propranolol and metoprolol (Section 18.7) slow the heart rate and reduce the force of the heartbeat.
- *Calcium channel blockers* such as amlodipine are powerful vasodilators, inducing muscles around the blood vessels to relax.
- *Angiotensin-converting enzyme (ACE) inhibitors* such as lisinopril decrease the action of an enzyme that causes blood vessels to contract.

Studies show that diuretics, the oldest, simplest, and cheapest of the blood pressure–lowering drugs, may also be the most effective for treating hypertension in most people. However, both a diuretic and another medication are often prescribed.

Normalizing Heart Rhythm

An *arrhythmia* is an abnormal heartbeat. Some arrhythmias exhibit no symptoms and are discovered only during a physical examination, but others can be life-threatening. The electrical properties of nerves and muscles arise from the flow of ions across cell membranes, and drugs for arrhythmia alter this flow. There are several types of such drugs, with different mechanisms of action, but all of them tend to have a narrow margin between the therapeutic dose and a harmful amount. A change in the heart's rhythm can have serious consequences. *Tachycardia* (too-rapid heartbeat) and abnormal heart rhythms are quite common. *Atrial fibrillation*, a type of irregular heartbeat, is so common that defibrillator devices are now available on many airplanes and in various other public places.

5 **Do larger tablets contain more medicine?**

Tablets contain not only the active ingredient (the actual drug), but they also have a number of other ingredients called excipients. Excipients include binders, often carbohydrate polymers, that give the tablet bulk, especially when the physical amount of the drug is small. They also can act to hold the tablet together so it does not crumble, and can give it color. Often tablets that contain different dosages of the same drug have different colors.

NO—A Messenger Molecule

The simple nitrogen monoxide molecule (NO), notorious as an air pollutant that contributes to smog (Chapter 13), was found in 1986 to act as a *messenger molecule* that carries signals between cells in the body. All previously known messenger molecules are more complex substances, such as norepinephrine and serotonin (Section 18.7), that act by fitting specific receptors in cell membranes. Commonly called nitric oxide, NO is essential to maintaining blood pressure and establishing long-term memory. It also aids in the immune response to foreign invaders in the body and mediates the relaxation phase of intestinal contractions in the digestion of food.

NO is formed in cells from arginine, a nitrogen-rich amino acid (Section 16.6), in an enzyme-catalyzed reaction. NO kills invading microorganisms, probably by deactivating iron-containing enzymes in much the same way that carbon monoxide destroys the oxygen-carrying capacity of hemoglobin (Section 13.4).

Three Americans—Louis Ignarro (1941–), Robert F. Furchgott (1916–2009), and Ferid Murad (1936–)—who discovered the physiological role of NO were awarded the 1998 Nobel Prize in Physiology or Medicine. This award has a fortuitous link to Alfred Nobel (Sweden, 1833–1896), whose invention of dynamite made him wealthy and provided the financial basis of the Nobel prizes. Nitroglycerin, the explosive ingredient of dynamite, relieves the chest pain of heart disease. In his later years, Nobel refused to take nitroglycerin for his heart disease because it causes headaches, and he did not think it would relieve his chest pain. Murad showed that nitroglycerin acts by releasing NO.

NO also dilates the blood vessels that allow blood flow into the penis to cause an erection. Research on this role of NO led to the development of the anti-impotence drug sildenafil (Viagra). One of the physiological effects of sildenafil is the production of small quantities of NO in the bloodstream. Related research has led to drugs for treating shock and a drug for treating high blood nitric oxide pressure in newborn babies.

Treating Coronary Artery Disease

A common symptom of coronary artery disease is chest pain, called *angina pectoris*. This pain is caused by an insufficient supply of oxygen to the heart, usually due to partial blockage of the coronary arteries by lipid-containing plaque (arteriosclerosis). When the blockage is complete, a heart attack occurs, and some of the heart muscle dies. Medical treatment for coronary artery disease usually involves dilation (widening) of the blood vessels to the heart to increase the blood flow and slowing of the heart rate to decrease its workload and its demand for oxygen. Some of the drugs used are the same as those used to lower high blood pressure, such as beta blockers.

In addition, several organic nitro compounds, especially nitroglycerin and amyl nitrite ($CH_3CH_2CH_2CH_2CH_2ONO$), have a long history in the treatment of angina pectoris. These compounds act by releasing nitric oxide (NO), which relaxes the constricted vessels that are reducing the supply of blood and oxygen to the heart.

Although many new drugs are being used to treat heart failure, one that has been used for centuries still plays an important role. The foxglove plant was used by the ancient Egyptians and Romans. The plant contains a mixture of glycosides that yield carbohydrates and steroids on hydrolysis. Hydrolysis of digoxin (one of those glycosides) yields the steroid digitoxigenin, which affects the rhythm and strength of heart muscle contractions. Digoxin is still used to treat patients with heart failure.

$$H_2C-O-NO_2$$
$$HC-O-NO_2$$
$$H_2C-O-NO_2$$

Nitroglycerin

SELF-ASSESSMENT Questions

1. All of the following types of drugs are used to lower blood pressure except
   - **a.** amphetamines
   - **b.** calcium channel blockers
   - **c.** diuretics
   - **d.** angiotensin-converting enzymes

2. Most types of drug treatment for the heart focus on all of the following goals except
   - **a.** regulating heart rate
   - **b.** increasing blood flow to the heart
   - **c.** lowering blood pressure
   - **d.** increasing the deposition of lipid plaque in blood vessels

3. Drugs such as amyl nitrite and nitroglycerin act by releasing a substance that relaxes the smooth muscles in blood vessels. That substance is
   - **a.** an amino acid
   - **b.** digitalis
   - **c.** nitric oxide
   - **d.** a steroid

Answers: 1, a; 2, d; 3, c

18.7 Drugs and the Mind

Learning Objectives • Explain how the brain amines norepinephrine and serotonin affect the mind and how various drugs change their action. • Identify some stimulant drugs, depressant drugs, and hallucinogenic drugs, and describe how they affect the mind.

A **psychotropic drug** is one that affects the human mind. The first drugs used probably were mind-altering substances. Alcoholic beverages, marijuana, opium, coca leaves, peyote, and other plant materials have been used for their psychotropic effects for thousands of years. Generally, only one or two of these were used in any given society. Today, thousands of psychotropic drugs are readily available. Some still come from plants, but many are synthetic.

There is no clear distinction between drugs that affect the mind and those that affect the body. Most drugs probably affect both mind and body. It can still be helpful, however, to distinguish between drugs that act mainly on the body and drugs that primarily affect the mind.

Psychotropic drugs are divided into three classes.

- A **stimulant drug**, such as cocaine or an amphetamine, increases alertness, speeds mental processes, and generally elevates the mood.
- A **depressant drug**, such as alcohol, almost any anesthetic, an opiate, a barbiturate, or any of certain tranquilizers, reduces the level of consciousness and the intensity of reactions to environmental stimuli. In general, depressants dull emotional responses.
- A **hallucinogenic drug**, such as lysergic acid diethylamide (LSD) or mescaline, qualitatively alters the perception of one's surroundings.

We will examine some of the major drugs in each category, but first we need to look at some chemistry of the nervous system.

Chemistry of the Nervous System

The nervous system is made up of about 12 billion **neurons** (nerve cells) with 10^{13} connections between them. The brain has the capacity to handle about 10 trillion bits of information. Nerve cells vary a great deal in shape and size. One type is shown in Figure 18.16. The essential parts of each cell are the cell body, the axon, and the dendrites. We discuss here only the nerves that make up the involuntary (autonomic) nervous system. These nerves carry messages between organs and glands that act involuntarily (such as the heart, the digestive organs, and the lungs) and the brain and spinal column.

The axon on a nerve cell may be up to 60 cm long, but there is no continuous pathway from an organ to the central nervous system. Messages must be transmitted across tiny, fluid-filled gaps, or **synapses** (Figure 18.17). When an electrical signal from a nerve cell in the brain reaches the end of an axon of a presynaptic nerve cell, chemicals called **neurotransmitters** are released from tiny vesicles (small fluid-filled sacs) into the synapse. Receptors on the postsynaptic cell bind the neurotransmitters, causing changes in that receiving cell. After a brief interval, the neurotransmitters are released from the receptor site and carried back to the presynaptic cell by substances called *transporters*.

There are many neurotransmitters. Substances that act as neurotransmitters include amino acids, peptides, and monosubstituted amines (monoamines). For example, glutamate, the carboxylate anion of the amino acid glutamic acid, is the most common neurotransmitter in the brain. More than 50 peptides function as neurotransmitters. Messages are carried to other nerve cells, to muscles, and to endocrine glands (such as the adrenal glands). Each neurotransmitter fits one or more receptor sites on the receptor cell. For example, when nerve impulses trigger the release of glutamate from the presynaptic cell, the glutamate receptors in

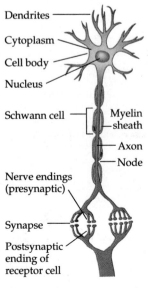

Dendrites
Cytoplasm
Cell body
Nucleus
Schwann cell — Myelin sheath
— Axon
— Node
Nerve endings (presynaptic)
Synapse
Postsynaptic ending of receptor cell

▲ **Figure 18.16** Diagram of a human nerve cell.

the postsynaptic cell bind glutamate and are activated. Neurotransmitters determine to a large degree how you think, feel, and move about. If there are problems with their synthesis or uptake, things can go quite wrong. Many drugs (and some poisons) act by mimicking the action of a neurotransmitter. Others act by blocking the receptor sites and preventing neurotransmitter molecules from acting on them.

Biochemical Theories of Brain Diseases

Like other organs, the brain can suffer from disease. Normal brain activity involves many chemical substances. Problems arise when an imbalance of these substances develops. We all have ups and downs in our lives. These moods probably result from multiple causes, but it is likely that a variety of compounds formed in the brain are involved. Before we consider these ups and downs, however, let's take a look at epinephrine, an amine formed in the adrenal glands and in some central nervous system cells (Figure 18.18).

Epinephrine is secreted by the adrenal glands—and is thus often called *adrenaline*—when a person is under stress or is frightened. A tiny amount of epinephrine causes a large increase in blood pressure, and its flow prepares the body for fight or flight. Because culturally imposed inhibitions prevent fighting or fleeing in most modern situations, the adrenaline-induced supercharge is often not used. This sort of frustration has been implicated in some forms of mental illness.

Biochemical theories of mental illness usually involve chemical imbalances of brain amines, which can be improved with medication. One such brain amine is norepinephrine (NE), a relative of epinephrine. NE is a neurotransmitter formed in the brain (Figure 18.18). When produced in excess, NE causes euphoria. In large excess, NE induces a manic state.

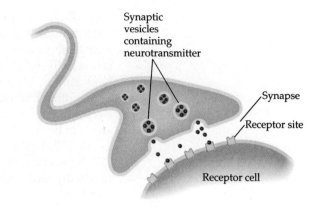

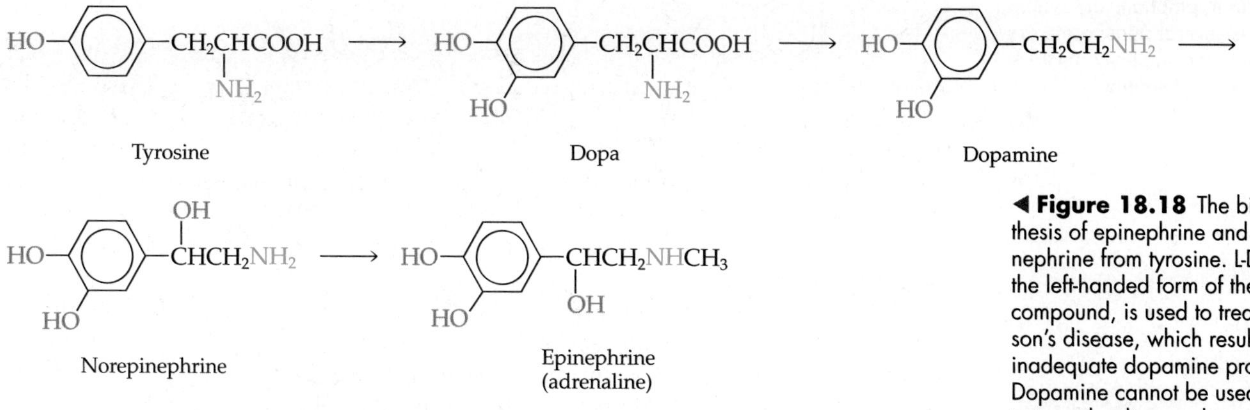

▲ **Figure 18.17** Diagram of a synapse. When an electrical signal reaches the presynaptic nerve ending, neurotransmitter molecules are released from the vesicles. They migrate across the synapse to the postsynaptic nerve cell, where they fit specific receptor sites.

NE and related compounds fall into several general categories. NE *agonists* (drugs that enhance or mimic the action of NE) are stimulants. NE *antagonists* (drugs that block the action of NE) slow down various processes (Section 18.2). For example, beta blockers (Figure 18.19) reduce the stimulant action of epinephrine

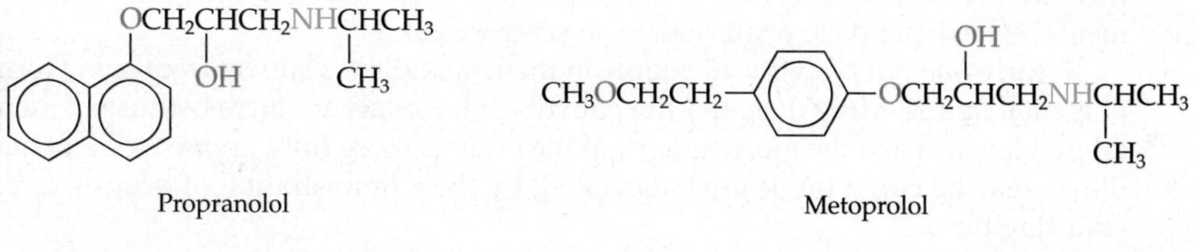

◀ **Figure 18.18** The biosynthesis of epinephrine and norepinephrine from tyrosine. L-Dopa, the left-handed form of the compound, is used to treat Parkinson's disease, which results from inadequate dopamine production. Dopamine cannot be used directly to treat the disease, because it is not absorbed into the brain. Schizophrenia has been attributed to an overabundance of dopamine in nerve cells.

◀ **Figure 18.19** Propranolol and metoprolol are beta blockers, widely used to treat high blood pressure.

and NE on various kinds of cells. Propranolol (Inderal®) is used to treat cardiac arrhythmias, angina, and hypertension by slightly lessening the force of the heartbeat. Unfortunately, it also causes lethargy and depression. Metoprolol (Lopressor®) acts selectively on the cells of the heart. It can be used by hypertensive patients who have asthma, because it does not act on receptors in the bronchi.

Another brain amine is the neurotransmitter *serotonin*. Serotonin is involved in sleep, appetite, memory, learning, sensory perception, mood, sexual behavior, regulation of body temperature, and aggression. Serotonin inhibits the nervous system in ways that calm, soothe, and generate feelings of contentment and satisfaction. The main metabolite of serotonin, 5-hydroxyindoleacetic acid (5-HIAA), is found in unusually low levels in the cerebrospinal fluid of severely depressed patients with a history of suicide attempts; this finding indicates that abnormal metabolism of serotonin plays a role in depression. Some research suggests that a reduced flow of serotonin through the synapses in the frontal lobe of the brain causes depression. Cerebrospinal fluid levels of 5-HIAA are also low in murderers and other violent offenders. However, they are higher than normal in people with obsessive–compulsive disorder, sociopaths, and people with guilt complexes.

Serotonin agonists are used to treat depression, anxiety, and obsessive–compulsive disorder. Serotonin antagonists are used to treat migraine headaches and to relieve the nausea caused by cancer chemotherapy.

Brain Amines and Diet: You Feel What You Eat

Richard Wurtman of the Massachusetts Institute of Technology has demonstrated a relationship between diet and serotonin levels in the brain. As shown in Figure 18.20, serotonin is produced in the body from the amino acid tryptophan. Wurtman found that diets high in carbohydrates lead to high levels of serotonin. High levels of protein lower the serotonin concentration.

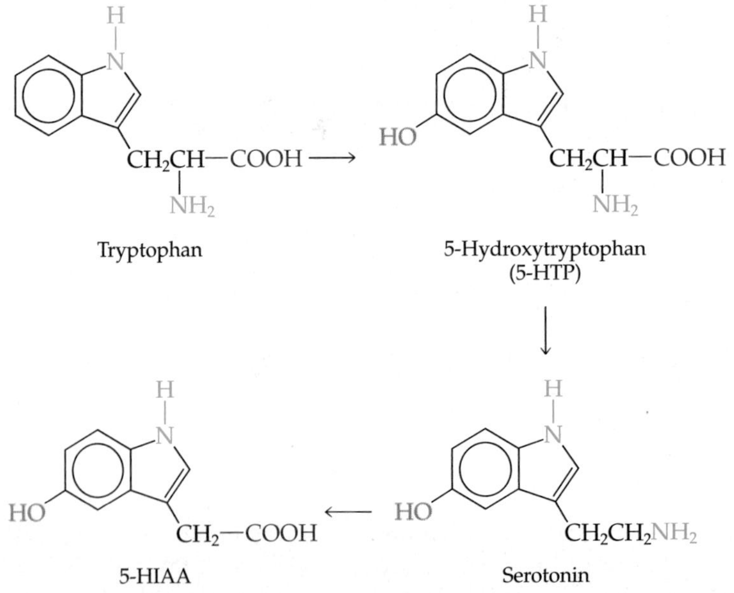

▶ **Figure 18.20** Serotonin is produced in the brain from the amino acid tryptophan. The synthesis involves several steps; some are omitted here for simplicity. 5-HIAA is a metabolite of serotonin.

Norepinephrine is synthesized in the body from the amino acid tyrosine. As noted in Figure 18.18, the synthesis is complex and proceeds through several intermediates. Because tyrosine is also a component of our diets, it may well be that our mental state depends to a fair degree on what we eat.

Nearly one out of every 10 people in the United States suffers from some form of mental illness. More than half the patients in hospitals are there because of mental problems. When the biochemistry of the brain is more fully understood, mental illness may be cured (or at least alleviated) by the administration of drugs—or by adjusting the diet.

Right-Handed and Left-Handed Molecules

Like people, molecules can be either right-handed or left-handed. Ball-and-stick models of two structures for the amino acid alanine $[CH_3CH(NH_2)COOH]$ are shown in Figure 18.21. These structures are **stereoisomers**, isomers having the same structural formula but differing in the arrangement of atoms or groups of atoms in three-dimensional space. Models of the two stereoisomers of alanine are as alike—and as different—as a pair of gloves. If you place one of the models in front of a mirror, the image in the mirror will be identical to the other model. Molecules that are nonsuperimposable (nonidentical) mirror images of each other are stereoisomers of a specific type called **enantiomers** (from the Greek *enantios*, meaning "opposite").

Enantiomers have a **chiral carbon**, a carbon atom that has four different groups attached to it. For example, the central carbon of alanine has four groups $(H, CH_3, COOH,$ and $NH_2)$ joined to it. If a molecule contains one or more chiral carbons, it is likely to exist as two or more stereoisomers. In contrast, propane $(CH_3CH_2CH_3)$, for example, does not contain a chiral carbon and thus does not exist as a pair of stereoisomers.

Stereoisomerism is quite common in organic chemicals of biological importance. The simple sugars (Section 16.2) are all right-handed. While 19 of the 20 amino acids that make up proteins are left-handed, the other, glycine (H_2NCH_2COOH), has no handedness.

A right-hand glove doesn't fit on a left hand, or vice versa. Similarly, enantiomers fit enzymes differently, and they have different effects. Mixtures of compounds that have equal amounts of each enantiomer are called racemates. Many older drugs were sold as racemic mixtures, even though we now know that only one enantiomer is effective and the other either does nothing or, worse yet, is toxic. As an example, consider atorvastatin (Lipitor®), which is used to lower blood cholesterol. Initial research on this drug in 1989 was promising but not outstanding. Later it was found that only the left-handed isomer is active; thus, it is much more effective when separated from its right-handed form. The development of drugs that consist of only one enantiomer is becoming increasingly important in the pharmaceutical industry.

Patent rights for new drugs exist only for a limited period. In recent years, as patents expired on racemic drugs, many pharmaceutical companies have engaged in *chiral switching* and brought forth single-enantiomer drugs, often claiming greater effectiveness or safety. The FDA is now requiring information on each enantiomer in new racemic mixtures.

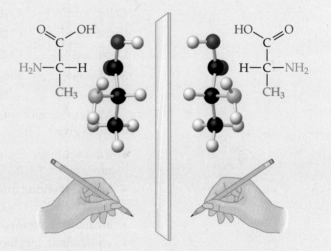

▲ **Figure 18.21** Ball-and-stick models and structural formulas of D- and L-alanine. The molecules are *enantiomers*, nonsuperimposable mirror images of one another. Like right and left hands, enantiomers cannot be superimposed on one another.

Anesthetics

An **anesthetic** is a substance that causes lack of feeling or awareness. A **general anesthetic** acts on the brain to produce unconsciousness and a general insensitivity to pain. A local anesthetic causes loss of feeling in a part of the body.

The first general anesthetic, diethyl ether $(CH_3CH_2OCH_2CH_3)$, was initially used in surgery in the 1840s. Inhalation of ether vapor produces unconsciousness by depressing the activity of the central nervous system. Ether itself is relatively safe because there is a fairly wide gap between the dose that produces an effective level of anesthesia and the lethal dose, but its high flammability and its side effect of nausea are disadvantages.

Nitrous oxide $(N_2O,$ laughing gas$)$ was discovered by Joseph Priestley (England, 1733–1804) in 1772, and its narcotic effect was soon noted. Mixed with oxygen, nitrous oxide is used in modern anesthesia. It acts quickly but is not very potent. Concentrations of 50% or greater must be used for it to be effective, and it must be mixed with oxygen rather than with air to make sure enough oxygen gets into the patient's blood to prevent permanent brain damage.

Chloroform ($CHCl_3$) was introduced as a general anesthetic in 1847. It quickly became popular after Great Britain's Queen Victoria gave birth to her eighth child in 1853 while anesthetized by chloroform. Although chloroform was used widely for years, it is no longer dispensed. It has a narrow safety margin; the effective dose is close to the lethal dose. It also causes liver damage, is a suspected carcinogen, and must be protected from oxygen during storage to prevent the formation of deadly phosgene gas.

Modern inhalant anesthetics include fluorine-containing compounds (Table 18.3). These compounds are nonflammable and relatively safe for patients. Their safety for operating room personnel, however, has been questioned. For example, female operating room workers suffer a higher rate of miscarriages than do women in the general population. Capture of waste anesthetic gases helps to minimize any risk.

TABLE 18.3 **Inhaled Anesthetic Agents**

Generic or Chemical Name	Formula	Commercial Name	Year Introduced	Currently in Use?
Diethyl ether	$CH_3CH_2OCH_2CH_3$	Ether	1842	No
Nitrous oxide	N_2O	Nitrous oxide	1844	Yes
Chloroform	$CHCl_3$	Chloroform	1847	No
Isoflurane	$CHF_2OCHClCF_3$	Forane	1980	Yes
Desflurane	$CHF_2OCHFCF_3$	Suprane	1992	Yes
Sevoflurane	$CH_2FOCH(CF_3)_2$	Ultane	1995	Yes

Modern surgical practice usually makes use of a variety of drugs. Generally, a patient is given

- a tranquilizer such as a benzodiazepine (Valium® or Versed®) to decrease anxiety,
- an intravenous anesthetic such as thiopental to produce unconsciousness quickly,
- a narcotic pain medication such as fentanyl to block pain,
- an inhalant anesthetic to provide insensitivity to pain and to keep the patient unconscious, often combined with oxygen and nitrous oxide to support life, and
- a relaxant such as pancuronium bromide to relax the muscles and make it easier to insert the breathing tube.

General anesthetics reduce nerve transmission at the synapses by inhibiting the excitatory effect of certain neurotransmitters on central nervous system receptors. However, the anesthetics bind only weakly, and it is difficult to determine their exact mode of action.

Local Anesthetics

Local anesthetics are used to render a part of the body insensitive to pain. They block nerve conduction by reducing the permeability of the nerve cell membrane to sodium ions. The patient remains conscious during dental work or minor surgery.

The first local anesthetic to be used successfully was cocaine, a drug first isolated in 1860 from the leaves of the coca plant (Figure 18.22). Its structure was determined in 1898 by Richard Willstätter (Germany, 1872–1942). Scientists have made many attempts to develop synthetic compounds with similar properties. Cocaine is a powerful stimulant whose abuse will also be discussed in this chapter.

Because local anesthetics are weak bases (amines), they are usually used in the form of hydrochloride salts, which are water-soluble. Some local anesthetics (Figure 18.23) are amine–esters (cocaine, procaine, and chloroprocaine), and others

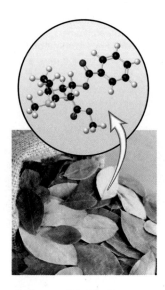

▲ **Figure 18.22** Coca leaves contain the alkaloid cocaine, a local anesthetic and stimulant.

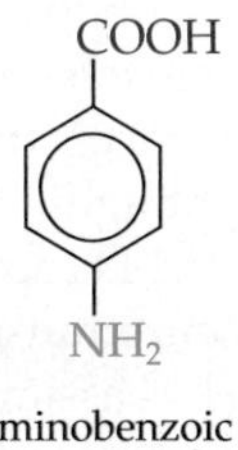

para-Aminobenzoic acid

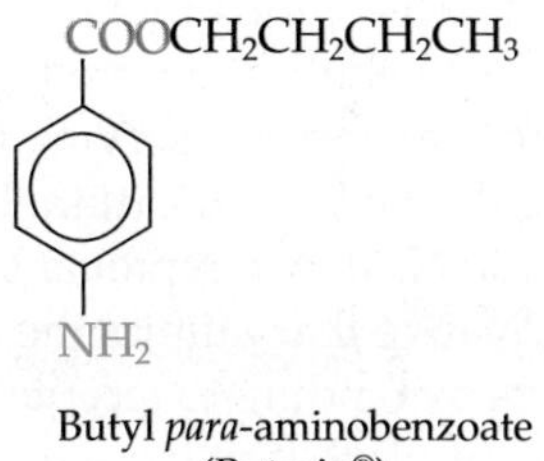

Butyl *para*-aminobenzoate (Butesin®)

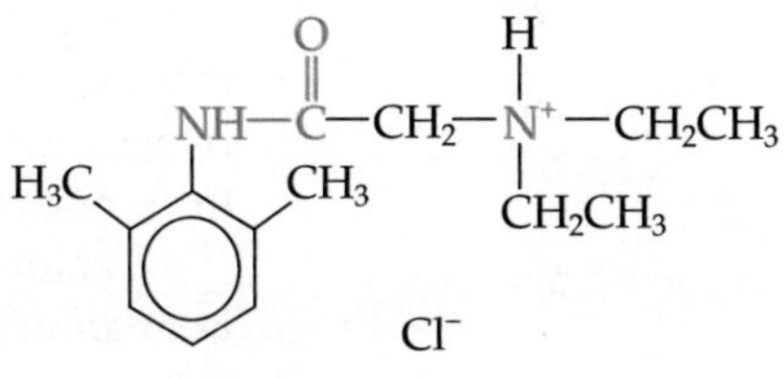

Lidocaine (Xylocaine®)

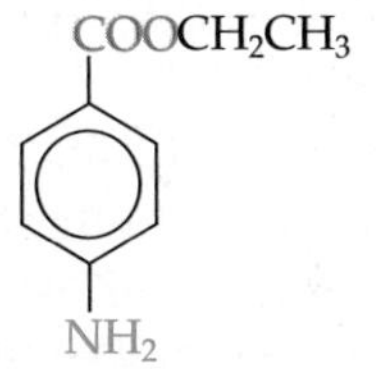

Ethyl *para*-aminobenzoate (Benzocaine®)

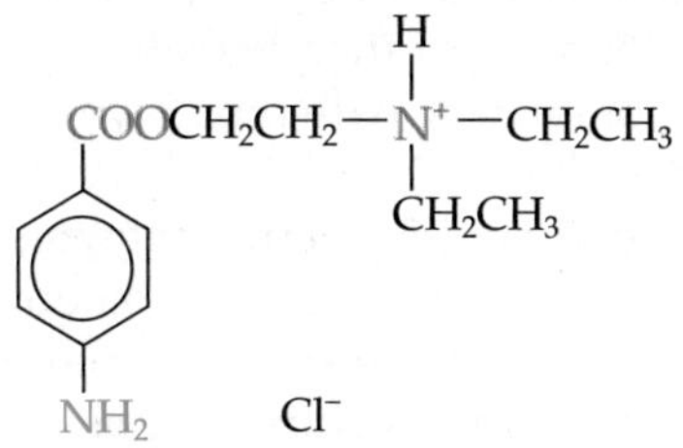

Procaine (Novocaine®)

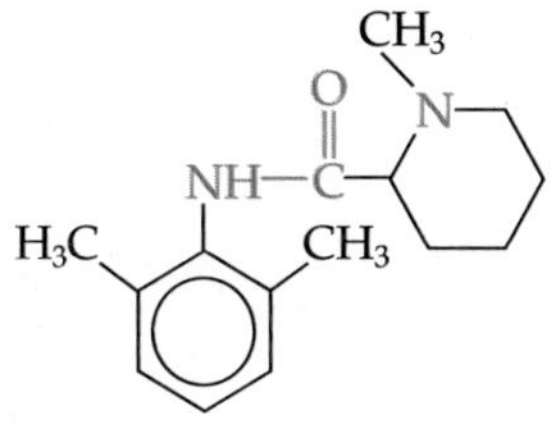

Mepivacaine

▲ **Figure 18.23** Some local anesthetics. They are often used in the form of a hydrochloride or picrate salt, which is more soluble in water than the free base.

Q *Which of these are derived from para-aminobenzoic acid? Which have amide functional groups? Which are esters?*

are amine–amides (lidocaine, mepivacaine, prilocaine, bupivacaine, and etidocaine). The amine–esters cause allergic reactions in some people; the amine–amides do not. The ethyl and butyl esters of *para*-aminobenzoic acid (PABA) are topical anesthetics, usually used in the form of picrate salts in ointments that are applied to the skin to relieve the pain of burns and open wounds.

Introduction of a second nitrogen atom in the alkyl group of the ester produces a more powerful anesthetic (Table 18.4). Perhaps the best known of these is procaine (Novocaine®), first synthesized in 1905 by Alfred Einhorn (Germany, 1865–1917), who had worked with Willstätter on the structure of cocaine. Procaine takes a fairly long time to take effect, wears off very quickly, and causes allergic reactions. It is no longer used in dentistry, having been replaced largely by the amine–amide anesthetics.

TABLE 18.4 Substances Commonly Used as Local Anesthetics

Substance	Duration of Action
Procaine (Novocaine®)	Short
Lidocaine (Xylocaine®)	Medium (30–60 min)
Mepivacaine (Carbocaine®)	Fast (6–10 min)
Bupivacaine (Marcaine®)	Moderate (8–12 min)
Prilocaine (Citanest®)	Medium (30–90 min)
Chloroprocaine (Nesacaine®)	Short (15–30 min)
Cocaine	Medium
Etidocaine (Duranest®)	Long (120–180 min)

Lidocaine was introduced in the 1940s and is still the most widely used local anesthetic. It takes effect quickly. When combined with a small quantity of epinephrine, it acts for an hour or more.

Dissociative Anesthetics: Ketamine and PCP

Ketamine, an intravenous anesthetic, is called a *dissociative anesthetic* because it disconnects one's perceptions from one's sensations. It reduces or blocks signals to the

Ketamine

Phencyclidine
(PCP)

conscious mind from parts of the brain that are associated with the senses. Ketamine induces hallucinations like those reported by people who have had near-death experiences. They seem to remember observing their rescuers from a vantage point above the scene or moving through a dark tunnel toward a bright light.

Ketamine acts as an NMDA-receptor antagonist. N-methyl-D-aspartate, or NMDA, is an amino acid derivative that mimics the action of the neurotransmitter glutamate. Because ketamine acts by binding to receptors in the body, it is assumed that the body produces its own chemicals that fit these receptors. These compounds may be synthesized or released only in extreme circumstances—such as in near-death experiences.

Ketamine is used widely in veterinary medicine. Because it suppresses breathing much less than most other available anesthetics, ketamine is sometimes used as an anesthetic for humans with unknown medical histories and for children and persons with poor health.

Phencyclidine, commonly known as PCP, is closely related to ketamine. PCP was formerly used as an animal tranquilizer and for a brief time as a general anesthetic for humans. It is now fairly common on the illegal drug scene. PCP, which is soluble in fat and has no appreciable water solubility, is stored in fatty tissue and released when the fat is metabolized, accounting for the "flashbacks" commonly experienced by users. Many users experience bad "trips" with PCP, and about 1 in 1000 develops a severe form of schizophrenia. High doses cause illusions and hallucinations and can also cause seizures, coma, and death. However, death of PCP users more often results from an accident or suicide during a "trip."

Depressant Drugs

We will start our exploration of depressant drugs by reconsidering ethyl alcohol, discussed in some detail in Chapter 9, because it is by far the most used and abused depressant drug in the world.

In nature, sugars in fruit are often fermented into alcohol by airborne yeasts, but purposeful production of ethanol probably did not begin until people began to farm. As early as 3700 B.C.E., the Egyptians fermented fruit to make wine and the Babylonians made beer from barley. People have been fermenting fruits and grains ever since, but relatively pure ethanol was first produced by Muslim alchemists who invented distillation in the eighth or ninth century.

People often think they are stimulated ("get high") when they drink, but ethanol is actually a depressant and slows down both physical and mental activity. Nearly two-thirds of adults in the United States drink alcohol, and nearly one-third have five or more drinks on at least one day each year. There are at least 15 million alcoholics in the United States, and alcoholism costs over $250 billion a year. Approximately 30% of current drinkers in the United States drink to excess. According to the CDC, excessive drinking of alcohol kills about 88,000 people each year in the United States, and underage drinking is involved in about 5000 deaths of people under 21 each year. For 15- to 24-year-olds, the three leading causes of death are automobile accidents, homicides, and suicides, and alcohol is a leading factor in all three.

For some, there is a small positive side to drinking alcohol. Longevity studies indicate that those who use alcohol moderately (no more than a drink or two a day) live longer than nondrinkers. This may result from alcohol's relaxing effect. Heavier drinking, however, can cause many health problems and can even shorten the life span by 10–12 years.

Alcoholic beverages are generally high in calories. Pure ethanol furnishes about 7 kcal/g. Ethanol is metabolized by oxidation to acetaldehyde at a rate of about 1 oz (about 30 mL) per hour, and the buildup of ethanol concentration in the blood caused by drinking at a rate faster than this results in intoxication.

Oxidation of ethanol uses up a substance called NAD^+, the oxidized form of nicotinamide adenine dinucleotide; the reduced form is represented as NADH, where H is a hydrogen atom:

$$CH_3CH_2OH + NAD^+ \longrightarrow CH_3CHO + NADH + H^+$$

Ethanol   Acetaldehyde

Because NAD$^+$ is usually employed in the oxidation of fats, drinking too much alcohol causes fat to be deposited—mainly in the abdomen—rather than metabolized. This results in the familiar "beer belly" of many heavy drinkers.

Just how ethanol intoxicates is still somewhat of a mystery. Researchers have found that ethanol disrupts receptors for two neurotransmitters: *gamma*-aminobutyric acid (GABA), which inhibits impulsiveness, and glutamate, which excites certain nerve cells. Long-term drinking changes the balance of GABA and glutamate, disrupting the brain's control over muscle activity and causing the drinker to stagger and fall. Ethanol also raises dopamine levels in the brain; dopamine is associated with the pleasurable aspects of alcohol and other drugs. One recent explanation for disruption of the receptors is that ethanol can replace some water molecules involved in maintaining a protein's three-dimensional structure, thus affecting the receptor's function.

Excessive drinking over time changes the levels of some of the brain chemicals, causing a craving for ethanol in order to bring back good feelings or to avoid bad feelings. Studies on twins raised separately show that there is undoubtedly a genetic component to alcoholism. There are also emotional, psychological, social, and cultural factors. We still have much to learn about this ancient drug.

Barbiturates

A family of depressants, the barbiturates display a wide range of properties. They can be employed to produce mild sedation, deep sleep, or even death. Barbiturates are cyclic amides that act on GABA receptors. The mechanism of their action is similar in some respects to that of alcohol. More than 2500 barbiturates have been synthesized over the years, but only a few have found widespread use in medicine (Figure 18.24).

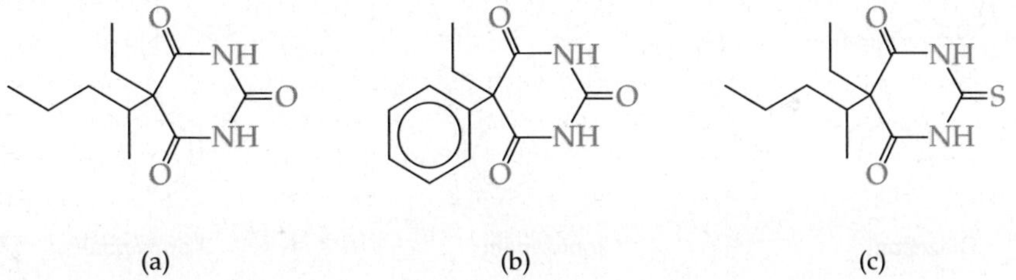

(a)   (b)   (c)

▲ **Figure 18.24** Formulas of some barbiturate drugs: (a) pentobarbital (Nembutal®), (b) phenobarbital (Luminal®), and (c) thiopental (Pentothal®). These drugs, derived from barbituric acid, are often used in the form of their sodium salts; for example, thiopental is used as sodium pentothal.

- Pentobarbital is employed as a short-acting hypnotic drug. Before the discovery of modern tranquilizers, it was used widely to calm anxiety.
- Phenobarbital is a long-acting drug. It is employed as an anticonvulsant for people suffering from seizure disorders such as epilepsy.
- Thiopental, which differs from pentobarbital only in having a sulfur atom on the ring in place of one of the oxygen atoms, is used as an anesthetic.

Barbiturates were once used in small doses (a few milligrams) as sedatives. In larger doses (about 100 mg), barbiturates induce sleep. They were widely prescribed as sleeping pills but have now been replaced by drugs such as benzodiazepines, which have less potential for abuse and are less likely to be lethal.

Barbiturates are especially dangerous when ingested along with ethyl alcohol. This combination produces an effect perhaps 10 times greater than the sum of the effects of the two depressants. In the past, many people died by accident or through suicide by ingesting this combination.

The phenomenon, in which two chemicals bring about an effect greater than that from each individual chemical, is called a **synergistic effect** (Section 12.2). Synergistic effects can be deadly, and they are not limited to alcohol–barbiturate combinations. Two drugs should never be taken at the same time without competent medical supervision.

Barbiturates, like ethanol, are intoxicating, and they are strongly addictive. Habitual use leads to the development of a tolerance, meaning that ever-larger doses are required to produce the same degree of intoxication. The side effects of barbiturates are similar to those of alcohol: hangovers, drowsiness, dizziness, and headaches. Withdrawal symptoms are often severe, including convulsions and delirium, and withdrawal can cause death.

Antianxiety Agents

The hectic pace of life in the modern world causes some people to seek rest and relaxation in chemicals. Many turn to ethyl alcohol. The drink before dinner—to "unwind" from the tensions of the day—is a part of the way of life for many people. This search for relief from stress and anxiety has made antidepressant drugs highly popular. According to the CDC, antidepressants are taken by one in eight Americans over the age of 12.

One class of antianxiety drugs (anxiolytics) is the *benzodiazepines*, compounds that feature seven-member heterocyclic rings (Figure 18.25). Three common ones are diazepam (Valium), a classic antianxiety agent; clonazepam, an anticonvulsant as well as an antianxiety drug; and lorazepam, used to treat insomnia. Antianxiety agents—sometimes called *minor tranquilizers*—make people feel better simply by making them feel dull and insensitive, but they do not solve any of the underlying problems that cause anxiety. They are thought to act on GABA receptors, dampening neuron activity in parts of the brain associated with cognitive functioning.

Diazepam Clonazepam Lorazepam

▲ **Figure 18.25** Three benzodiazepines: diazepam (Valium®), clonazepam (Klonopin®), and lorazepam (Ativan®).

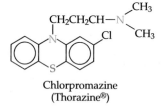

Chlorpromazine
(Thorazine®)

The first antipsychotic drugs—sometimes referred to as *major tranquilizers*—were compounds called *phenothiazines*. In 1952, chlorpromazine (Thorazine®) was administered as a tranquilizer to patients with psychosis in the United States. The drug had been tested in France as an antihistamine, and medical workers there had noted that it calmed patients with mental health disorders being treated for allergies. Found to be helpful in controlling the symptoms of schizophrenia, chlorpromazine revolutionized therapy for mental illness. Chlorpromazine is one of several phenothiazines used as antipsychotics.

Phenothiazines act in part as dopamine antagonists. They block postsynaptic receptors for dopamine, a neurotransmitter important in the control of detailed motion (such as grasping small objects), in memory and emotions, and in exciting the cells of the brain. Some researchers think patients with schizophrenia produce too much dopamine, whereas others think that they have too many dopamine receptors. In either case, blocking the action of dopamine relieves the symptoms of schizophrenia.

A second generation of antipsychotics, called *atypical antipsychotics*, includes aripiprazole (Abilify®), risperidone (Risperdal®), clozapine (Clozaril®), and olanzapine (Zyprexa®). These drugs are thought to act on a type of serotonin receptor that loosens the binding of dopamine receptors, allowing more dopamine to reach the neurons. Aripiprazole and risperidone are used to treat schizophrenia, acute manic episodes of bipolar disorder, and depression. Clozapine, the first of the atypical antipsychotics to be developed, has actually been shown to be the most effective drug for treating schizophrenia, but its severe side effects—such as destruction of white blood cells—cause it to be used only if other antipsychotic drugs have failed.

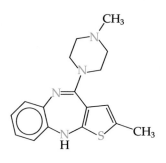

Olanzapine (Zyprexa®)

Olanzapine is used to treat psychotic disorders, such as schizophrenia and acute manic episodes, and to stabilize bipolar disorder.

Antipsychotic drugs have served to reduce greatly the number of patients confined to mental hospitals. They do not cure schizophrenia, but they control its symptoms well enough that 95% of all people with schizophrenia no longer need hospitalization. Patients who stop taking these medications relapse.

The oldest class of antidepressant drugs, the tricyclic (three-ring) antidepressants, such as amitriptyline, block the reabsorption ("reuptake") of neurotransmitters such as norepinephrine and serotonin into the presynaptic neuron. These antidepressants have serious side effects and have been largely replaced by newer drugs.

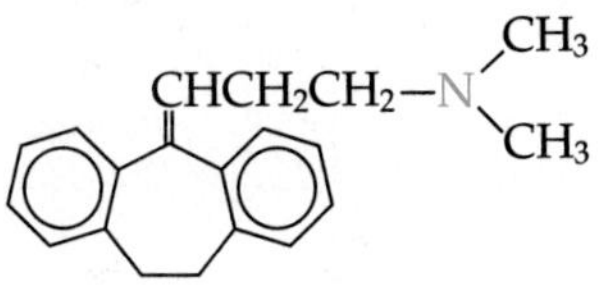

Antidepressants commonly prescribed today include drugs called *selective serotonin-reuptake inhibitors* (SSRIs). Three major ones are shown in Figure 18.26. Doctors prescribe these drugs to help people cope with a wide variety of psychological disorders, including anxiety syndromes such as panic disorder, obsessive–compulsive disorders (including gambling problems and overeating), and premenstrual syndrome (PMS). The drugs enhance the effect of serotonin, blocking its reuptake by nerve cells. They seem to be safer than tricyclic antidepressants and more easily tolerated.

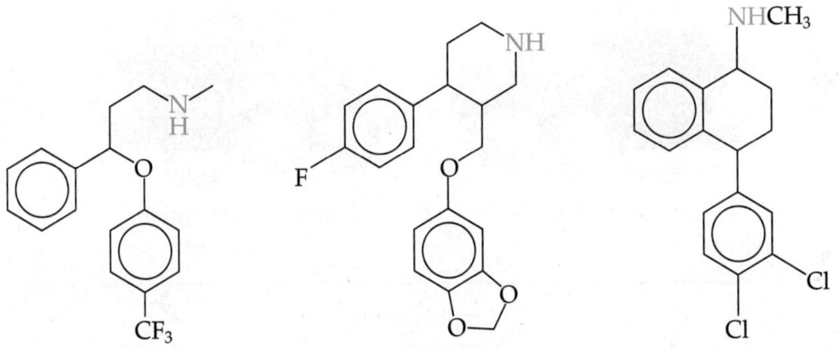

Amitriptyline

◀ **Figure 18.26** Three selective serotonin reuptake inhibitors (SSRIs): fluoxetine (Prozac®), paroxetine (Paxil®), and sertraline (Zoloft®).

Fluoxetine   Paroxetine   Sertraline

Stimulant Drugs

Among the more widely known stimulant drugs are a variety of synthetic amines related to β-phenylethylamine (Figure 18.27). These drugs, called **amphetamines**, are similar in structure to epinephrine and norepinephrine (Figure 18.18). They seem to act by increasing levels of norepinephrine, serotonin, and dopamine in the brain.

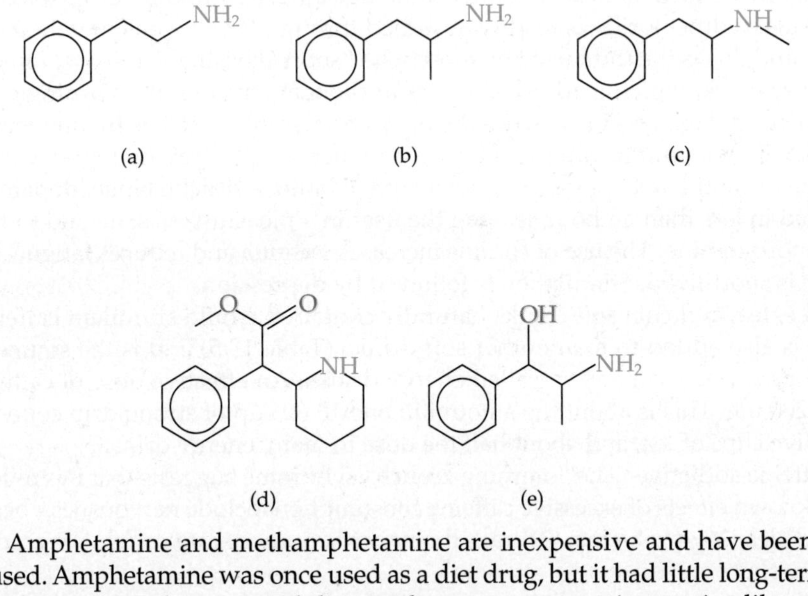

(a)   (b)   (c)

◀ **Figure 18.27** (a) β-phenylethylamine and related compounds, (b) amphetamine, (c) methamphetamine, (d) methylphenidate, and (e) phenylpropanolamine.

(d)   (e)

Amphetamine and methamphetamine are inexpensive and have been widely abused. Amphetamine was once used as a diet drug, but it had little long-term effect. Amphetamine induces excitability, restlessness, tremors, insomnia, dilated pupils, increased pulse rate and blood pressure, hallucinations, and psychoses.

Methamphetamine has a much more pronounced psychological effect than amphetamine has. Methamphetamine is made readily from an antihistamine and household chemicals. Illegal "meth labs" have operated in all parts of the United States. Like other amine drugs, amphetamines are often distributed as hydrochloride salts. The free-base form (the basic amine form that has not been neutralized by an acid) of methamphetamine is smoked because it vaporizes more readily than the salts. Long-term effects of methamphetamine use include severe psychotic problems, loss of memory, impotence, and serious dental problems.

Either methylphenidate (Ritalin®) or a mixture of amphetamines is used to treat attention-deficit/hyperactivity disorder (ADHD) in children. Although they are stimulants, these drugs seem to calm children who otherwise can't sit still. This use has received criticism that it "leads to drug abuse" and "solves the teacher's problem, not the kid's."

Like many other drugs, amphetamine exists as mirror-image isomers (enantiomers). Benzedrine® is a mixture of the two isomers in equal amounts. The dextro (right-handed) isomer is a stronger stimulant than the levo (left-handed) isomer. The pure dextro isomer is sold under the trade name Dexedrine®.

▶ Benzedrine is a mixture of Dexedrine and its enantiomer, levoamphetamine. As is often the case, the two isomers differ greatly in their biological activity.

Levoamphetamine

Dextroamphetamine
(Dexedrine®)

Amphetamine
(Benzedrine®)

Cocaine, Caffeine, and Nicotine

Three plant alkaloids have long been used as stimulants. Cocaine, first used as a local anesthetic, is also a powerful stimulant. The drug is obtained from the leaves of a shrub that grows almost exclusively on the eastern slopes of the Andes Mountains. Many of the Indigenous people living in and around the area of cultivation chew coca leaves—mixed with lime and ashes—to experience the stimulant effect. Cocaine used to arrive in the United States as the salt cocaine hydrochloride, but now much of it comes in the form of broken lumps of the free base, a form called *crack cocaine*.

Cocaine hydrochloride is readily absorbed through the mucous membrane of the nose, and this is the form used by those who "snort" cocaine. Those who smoke cocaine use crack, which readily vaporizes at the temperature of a burning cigarette. When smoked, cocaine reaches the brain in 15 seconds. It acts by preventing the reuptake of dopamine after it is released by nerve cells, leaving high levels of dopamine to stimulate the pleasure centers of the brain. After the binge, dopamine is depleted in less than an hour, leaving the user in a pleasureless state and (often) craving more cocaine. The use of cocaine increases stamina and reduces fatigue, but the effect is short-lived. Stimulation is followed by depression.

Coffee, tea, and cola soft drinks naturally contain the mild stimulant caffeine. Caffeine is also added to many other soft drinks (Table 18.5) and is the source of the "lift" provided by many so-called energy drinks. An effective dose of caffeine is about 200 mg. This is about the amount in one 10 oz cup of strong drip coffee or three to five cups of tea, and about half the dose in many energy drinks.

Is caffeine addictive? The "morning grouch" syndrome suggests that it is mildly so. Well-known effects of excessive caffeine consumption include nervousness, headache, and sleeping problems. Overall, the hazards of moderate caffeine ingestion seem to be slight.

Another common stimulant is nicotine. This drug is taken by smoking or chewing tobacco. Nicotine is highly toxic to animals and has been used in agriculture as a

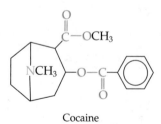

Cocaine

Caffeine

contact insecticide. It is especially deadly when injected. The lethal dose for a human is estimated to be about 50 mg. Nicotine seems to have a rather transient effect as a stimulant, with the initial response followed by depression. Most smokers keep a near-constant level of nicotine in their bloodstreams by indulging often.

Nicotine is powerfully addictive. Consider the 1972 memorandum from a Philip Morris scientist who noted that "no one has ever become a cigarette smoker by smoking cigarettes without nicotine." He suggested that the company "think of the cigarette as a dispenser for a dose unit of nicotine."

Hallucinogenic Drugs

Hallucinogenic drugs are consciousness-altering substances that induce changes in sensory perception, qualitatively altering the way users perceive things. Common hallucinogenic drugs include natural products from psilocybin mushrooms and peyote cactus, synthetic chemicals such as MDMA ("Ecstasy"), dimethyltryptamine (DMT), PCP, ketamine, and semisynthetic drugs such as LSD and mescaline. Marijuana is sometimes considered a mild hallucinogen.

Lysergic acid diethylamide (LSD), a semisynthetic drug, is a powerful hallucinogen. Its physiological properties were discovered accidentally by Swiss chemist Albert Hofmann (1906–2008) in 1943 when he unintentionally ingested some LSD. He later took 250 mg, which he considered a small dose, to verify that LSD had caused the symptoms he had experienced. Hofmann had a rough time for the next few hours, experiencing such symptoms as visual disturbances and schizophrenic behavior.

LSD can create a feeling of lack of self-control and sometimes of extreme terror. The exact mechanism by which LSD exerts its effects is still unknown, but it is thought to act on dopamine receptors and to excite by increasing the release of glutamate.

Lysergic acid is obtained from ergot, a fungus that grows on rye. This carboxylic acid is converted to its diethylamide derivative.

The potency of LSD is indicated by the small amount required for a person to experience its extraordinary effects. The usual dose is probably about 10–100 μg (no wonder Hofmann had a bad time after taking 250 mg). To give you an idea of how small this is, a portion of LSD the size of an aspirin tablet could provide up to 30,000 doses.

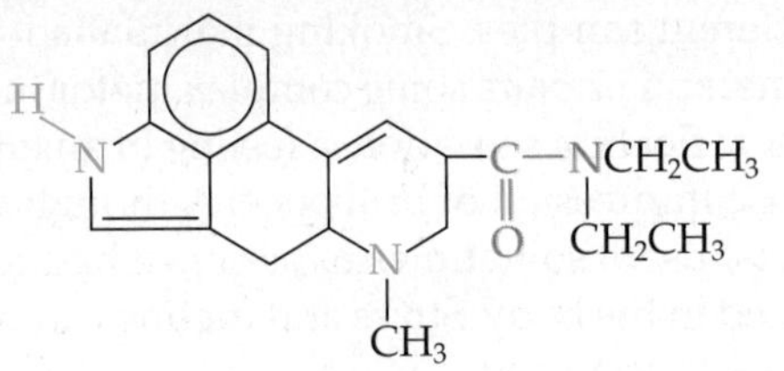

Lysergic acid diethylamide
(LSD)

Many kinds of plants or fungi produce hallucinogenic compounds. The following are two examples.

- Psilocybin, a hallucinogenic alkaloid of the tryptamine family, is found in several species of mushrooms, including *Psilocybe cubensis*. Its effects resemble those of LSD but are not as long-lasting.
- Mescaline (3,4,5-trimethoxyphenylethylamine) is a hallucinogenic drug related to phenylethylamine (see Figure 18.27). It is found in the peyote cactus (and other species), which is used in the religious ceremonies of some Native American tribes. Mescaline is also made synthetically. The effects last up to 12 hours.

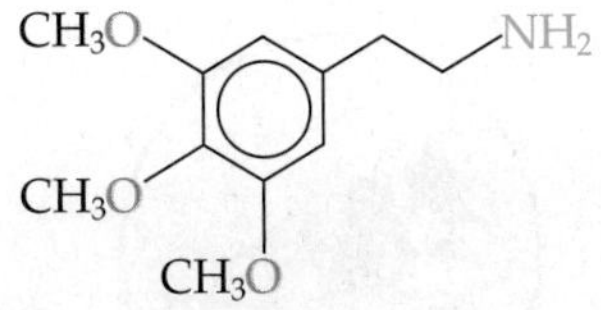

Marijuana

The plant *Cannabis sativa*, commonly known as marijuana, has long been useful. The plant stems yield tough fibers (hemp) for making ropes. *Cannabis* has been used as a drug in tribal religious rituals and also has a long history as a medicine, particularly in India. In the United States, marijuana is the most commonly used illegal drug.

TABLE 18.5 Caffeine in Soft Drinks

Brand	Caffeine[a]
5-Hour Energy (2 oz)	138
Red Bull (8.2 oz)	80
Sun Drop	63
Mountain Dew	55
Mello Yello	53
Diet Coke	46
Dr Pepper	41
Diet Sunkist Orange	41
Pepsi-Cola	38
Diet Pepsi	36

[a]Milligrams per 12 oz serving
Sources: National Soft Drink Association; FDA.

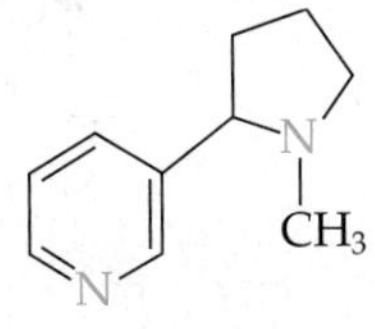

Nicotine

Mescaline

Club Drugs: Raves and Rapes

Club drugs are substances used by teenagers and young adults at all-night dance parties called "raves." These drugs include 3,4-methylenedioxymethamphetamine (MDMA, or Ecstasy), *gamma*-hydroxybutyrate ($HOCH_2CH_2CH_2COOH$; GHB), flunitrazepam (Rohypnol®), ketamine, methamphetamine, and LSD. Use of such drugs can cause serious health problems, especially in combination with alcohol.

Especially troubling are substances known as *date rape drugs*: Rohypnol, ketamine, and GHB have been used to facilitate sexual assault because they render the victim incapable of resisting. Events that happen while a person is under the influence of these drugs often cannot be remembered. The GHB analogs 1,4-butanediol ($HOCH_2CH_2CH_2CH_2OH$) and *gamma*-butyrolactone (GBL) have similar effects.

Flunitrazepam (Rohypnol®)

MDMA (Ecstasy)

gamma-Butyrolactone (GBL)

Tetrahydrocannabinol
(THC)

The term **marijuana** refers to a mixture made by gathering the leaves and flower buds of the plant (Figure 18.28), which are generally dried and smoked. There are several active cannabinoids in marijuana. The main one is *tetrahydrocannabinol (THC)*.

Marijuana plants vary considerably in THC potency, depending on their genetic variety. Wild plants native to the United States have a low THC content, usually about 0.1%, but some marijuana sold in North America has a THC content approaching 6%.

The effects of marijuana are difficult to measure, partly because of the variable amount of THC in different samples. Smoking marijuana increases the pulse rate, distorts the sense of time, and impairs some complex motor functions. Other possible effects include a euphoric floating sensation, a feeling of anxiety, a heightened enjoyment of food, and a false impression of brilliance. Although studies have shown no mind-expanding effects, users sometimes experience hallucinations. Because it is fat-soluble, THC is stored in the body. Stress and dieting can produce a positive THC test long after the drug was last used.

The long-term effects of marijuana, such as causing psychotic symptoms that persist beyond temporary intoxication, are unclear. There is some evidence of increased risk of psychotic problems in people who have used marijuana, with the greatest increase (50% to 200%) observed in people who used the drug most frequently. People who use marijuana heavily often appear to be lazy, passive, and mentally sluggish, but it is difficult to prove that marijuana is the cause. Even if it is, the damage is less extensive than that caused by heavy use of alcohol.

When a person smokes marijuana, THC rapidly passes from the lungs into the bloodstream and is carried to the brain, where it binds to specific sites on nerve cells called *cannabinoid receptors*. Some regions of the brain have many such receptors; other areas have few or none. The parts of the brain that influence pleasure, memory, thought, concentration, judgment, and sensory and time perception, and that regulate movement, are rich in these receptors. As THC enters the brain, it activates the brain's reward system in the same way that food and drink do. Like nearly all drugs of abuse, THC causes a euphoric feeling by stimulating the release of dopamine.

The location of receptors in movement control centers in the brain explains the loss of coordination seen in those intoxicated by the drug. The presence of receptors in memory and cognition areas of the brain explains why marijuana users do poorly

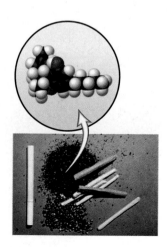

▲ **Figure 18.28** Prepared marijuana.

on tests. That few receptors are present in the brainstem, where breathing and heartbeat are controlled, is probably why it is hard to get a lethal dose of pure marijuana.

The brain also produces *anandamide*, a THC-like substance that binds to cannabinoid receptors in the brain. This substance is derived from arachidonic acid, which is the precursor for prostaglandin synthesis.

Anandamide

Marijuana has some legitimate medical uses. It reduces eye pressure in people who have glaucoma. If not treated, this increasing pressure eventually causes blindness. Marijuana also relieves the nausea that afflicts cancer patients undergoing radiation treatment and chemotherapy and may be useful for alleviating pain. The FDA approved the use of cannabidiol (a constituent of cannabis) for treatment of rare childhood epilepsies in mid-2018.

The legal status of marijuana is in a state of flux. As of early 2018, the United States government still classified it as illegal for any purpose. State laws differ greatly and often conflict with federal laws. It is legal for recreational use in nine states and for medical use in 31 states, although there are many variations in the type of marijuana and the medical uses covered among states. In some cases, the sale of medical marijuana is restricted to pill form.

WHY IT MATTERS

University of California, Irvine researchers Daniele Piomelli, Nicholas DiPatrizio, and colleagues found that when rats tasted fatty foods, such as potato chips or French fries, a signal was generated that stimulated the production of anandamide and other natural compounds similar to THC. These compounds trigger the desire for more fatty foods, making it hard to resist eating more chips or fries.

SELF-ASSESSMENT Questions

1. Depression may result from abnormal metabolism of
   a. acetylcholine   **b.** GABA
   c. nitric oxide   **d.** serotonin

2. A hazard of the use of nitrous oxide as an anesthetic is that it
   a. can cause brain damage if not combined with O_2
   b. has too long a recovery time
   c. induces anesthesia that is too deep
   d. is flammable

3. Which of the following is the most used and abused depressant drug in the world?
   a. amphetamines   **b.** cocaine
   c. ethyl alcohol   **d.** marijuana

4. Which of the following classes of drugs is commonly used to treat anxiety?
   a. amphetamines   **b.** benzodiazepines
   c. opiates   **d.** steroids

5. Dextroamphetamine is composed of molecules that
   a. differ from amphetamine by a —CH_3 group
   b. are a mixture of left- and right-handed isomers
   c. are all the left-handed isomer
   d. are all the right-handed isomer

6. LSD acts on the receptors for the neurotransmitter
   a. acetylcholine   **b.** dopamine
   c. GABA   **d.** serotonin

7. How do antagonists and agonists for a drug work?
   a. Agonists mimic the action of the drug, but antagonists block the action of the drug.
   b. Both agonists and antagonists mimic the action of the drug.
   c. Both agonists and antagonists block the action of the drug.
   d. Antagonists mimic the action of the drug, but agonists block the action of the drug.

Answers: 1, d; 2, a; 3, c; 4, b; 5, d; 6, b; 7, a

18.8 Drugs and Society

Learning Objectives • Differentiate between drug abuse and drug misuse.
• Describe how a new drug is developed and brought to market.

Illegal drugs cause enormous problems. The market is covert, with sellers failing to report their income or pay taxes on it, making the illicit drug business enormously lucrative. Annual revenues for drug dealers amount to more than $500 billion, about half of which comes from the United States. Worldwide, people spend more money on illegal drugs than on food. The large amounts of money involved make illegal drugs a dangerous business in which crimes are frequent, murders are common, and corrupt politicians abound. **Drug abuse** (using drugs for their intoxicating effects) is a serious problem not only for the abusers but for society in general.

The illegal drug user is the biggest loser. Street drugs cost money, and most are addictive. Many addicted users steal to pay for drugs and eventually end up in jail. One study found that 16% of convicted jail inmates admitted that they committed their offense to get money for drugs. In the workplace, 50% to 80% of all accidents and personal injuries are drug related. Drug users are absent from work over twice as often as nonusers, and they are five times as likely to file claims for worker's compensation. They are a drain on their employers' income, and they often have trouble keeping a job.

Illegal drugs are not always what their sellers claim they are. Buyers simply have to trust the information sellers give them about the identity and quality of the products they buy. In fact, crime labs have generally found that nearly two-thirds of all drugs (other than marijuana) brought in for analysis are something other than what the dealers said they were.

There can be difficulties even with legal drugs that have been approved and tested. According to the CDC, drug overdoses killed more than 64,000 people in 2016. Contrary to popular belief, most were not caused by cocaine or heroin, but by OxyContin and other painkillers diverted into the illegal drug trade. Nearly 30 million Americans admitted to using prescription drugs for nonmedical reasons.

Other problems with legal drugs range from faulty prescriptions written by physicians to pharmacies' errors in filling prescriptions to patients' mistakes in taking medications. Some drugs have undesirable side effects that are worse than the condition being treated. Some drugs (both prescription and over-the-counter) have such similar names that they are easily confused. **Drug misuse**—for example, using penicillin, which has no effect on viruses, to treat a viral infection—is all too common. Such overuse of penicillin and other antibiotics has led to strains of bacteria that are now resistant to antibiotics.

Design and Approval of New Drugs

The process of developing and bringing a new drug to market is both costly and lengthy. As noted earlier, there are several approaches to designing and synthesizing new drugs. One is to modify an already existing drug with the goal of improving its effectiveness or its ability to reach the target organ or molecules. Initially, most modifications were somewhat haphazard, but more recently drug design has incorporated knowing what structural features are involved in the mechanism of the drug's action, as is the case with the fluoroquinolones.

Another relatively new approach is combinatorial chemistry, in which several starting chemicals are reacted in all possible ways to give a large number of compounds. These products are tested for their biological activity, and then the process of testing the more promising compounds for safety and effectiveness begins, in anticipation of receiving FDA approval.

Preliminary testing of a drug is not carried out on humans but is instead conducted with other systems, often in bacteria, tissue culture, or animal models. The drug is examined for a number of attributes: how effective it is in treating the malady, its ability to reach the site of action, and of course its safety. If initial tests in laboratory or animal models are positive, then clinical trials will be undertaken to determine the drug's effects in humans. Phase 1 trials involve a small number of healthy volunteers to determine whether there are side effects and to optimize the dosage and means of administration. Phase 2 trials include a somewhat larger number of people and focus on whether the drug is effective against the disease. If a drug appears to be effective, Phase 3 trials will be conducted on a larger group. In Phase 2 and 3 trials, the new drug will be compared with placebo or other treatments for the disease. A safe and effective drug that receives FDA approval will be monitored for safety and side effects even after its release to the market.

A common way to evaluate a new drug is to administer it to one group of patients and to give a placebo to a similar "control" group. A placebo is an inactive substance given in the form of medication to a patient. In a double-blind study,

6 **What is the difference between a brand-name drug and a generic drug?**

When a new drug is developed, the manufacturer has patent rights to produce it for a given period of time. This enables the pharmaceutical company to recoup the high costs of developing and testing the drug. Once patent rights expire, other companies can market a generic version of the drug, but they must show that the same amount of the active ingredient is absorbed into the bloodstream and that it has the same activity as the brand-name drug. There may be differences in inactive ingredients. Generic drugs cost much less than brand-name drugs do.

which is the type most accepted by the medical community, neither the patients nor the doctors know who is receiving the real drug and who is receiving the placebo.

An interesting phenomenon associated with the testing of drugs is the **placebo effect**. Sometimes people who think they are receiving a certain drug expect positive results and may actually experience such results, even though they have not been given the actual drug. The psychological effect of a placebo can be very powerful. The placebo effect confers health benefits from a treatment that should have no effect.

There are also a number of diseases that are quite rare and that affect relatively small numbers of individuals. (Many of these involve genetic defects.) As a result, they have been classified as orphan diseases, and various ways of encouraging development of **orphan drugs** to treat them have been pursued. Governments sometimes intervene to encourage research on orphan drugs, offering tax incentives, enhancing patent protection, or even developing a government-run entity to pursue research in this area. Such drugs still have to meet safety and effectiveness standards, but variations in testing procedure have included smaller test groups for clinical trials on very rare diseases. Before passage of the Orphan Drug Act in 1983, fewer than 40 drugs had been approved for treatment of orphan diseases in the United States. In the three decades following its passage, more than 600 drugs received marketing approval, and many others are being developed or tested.

What Does the Future Hold?

There are a couple of trends and new approaches that seem to be surfacing in the development of new drugs. One is the testing of drugs that are currently approved for one use to see if they are effective in treating other conditions. One example is the recent response to the Zika virus. While much effort has gone into the development of vaccines, some laboratories have been testing known drugs to see if they can be used against the virus. An advantage of this approach is that the safety of approved drugs has already been established, so some clinical trials need not be repeated.

In addition to the public's increasingly widespread reliance on generic drugs, biosimilar products are becoming more common. **Biosimilars** are biologic products that are used to diagnose, prevent, cure, or treat diseases. Generally, they are large complex molecules, and they must demonstrate a highly similar effect that has no clinically meaningful differences from an existing FDA-approved reference product in terms of safety and effectiveness.

Immunotherapy approaches use a person's own immune system to fight disease more effectively. These can involve stimulating the person's immune system generally or by treating a person with immune system components, such as monoclonal antibodies or other immune system proteins.

Gene therapy involves introducing a healthy gene into an individual's cells to replace a defective gene. Although there have been some attempts to utilize this approach, this type of treatment is in its infancy. It is still highly experimental, risky, and very expensive, and it raises a number of ethical issues.

Many drugs have supplied enormous benefits to society, including longer, healthier lives and relief from pain, but sometimes drugs can create problems, such as addiction, dangerous side effects, and even death. We must use them wisely.

SELF-ASSESSMENT Questions

1. Using an antibiotic to treat a cold is an example of
   a. appropriate therapy **b.** drug abuse
   c. drug misuse **d.** a placebo effect

2. Using the narcotic OxyContin for its intoxicating effect is an example of
   a. appropriate recreation **b.** drug abuse
   c. drug misuse **d.** legal intoxication

3. A placebo is
   a. an alcohol solution of an active drug
   b. an inactive substance that looks like real medication
   c. a PCB
   d. a platinum/cesium/boron drug

4. In a double-blind study, some patients are given the drug being tested and others are given
   a. acupuncture **b.** a homeopathic formulation
   c. morphine **d.** a placebo

5. An orphan drug is
   a. derived from an unknown source
   b. used to treat diseases for which vaccines exist
   c. a drug whose use has been abandoned
   d. used against rare diseases

Answers: 1, c; 2, b; 3, b; 4, d; 5, d

GREEN CHEMISTRY Green Pharmaceutical Production
Joseph M. Fortunak

Learning Objectives • Identify green chemistry principles that can improve the E-factor of chemical syntheses.
• Explain how prodrugs can be important for drug delivery and green chemistry.

This chapter discusses principles for finding new medicines. First, we investigate the pathway of a disease, select a molecular target for treatment, and then identify drugs that affect the molecular target and provide benefit. Because drugs must be selective and safe, newer drugs tend to be structurally complex and require lengthy syntheses. Companies spend about twice the amount of money on manufacture as on new drug research, so there is a powerful incentive to implement green chemistry to reduce waste and cost. Green chemistry essays in Chapters 4 and 9 discuss organic synthesis and drug manufacture. Now that you have learned how drugs target things like bacterial infections, cancer, and pain, we'll discuss how green chemistry principles apply.

In 1992, Roger Sheldon at Delft University of Technology, Netherlands, coined the term *E-factor* (environmental factor) for chemical production (see Chapter 8 essay). The E-factor is the mass of generated waste for each kilogram of manufactured product. Drug synthesis often uses large amounts of solvents and complex reagents. Drug manufacturers commonly generate more than 100 kg of waste for every kilogram of drug—an E-factor greater than 100. Syntheses in water, enzymatic processes, catalysts that mimic biological processes, and processes without hazardous reagents are increasingly used to make drug production more efficient and green (Principles 5 and 9).

"The Holý Trinity" is a pun on the name of Dr. Antonín Holý (1936–2012, Prague) and his collaborations with Drs. John C. Martin (Gilead Sciences, U.S.A.) and Erik De Clercq (Rega Institute for Medical Research, Belgium). These scientists pioneered a new class of drugs to treat viral infections including tenofovir disoproxil fumarate (TDF) and tenofovir alafenamide fumarate (TAF). TDF is heavily used worldwide to treat HIV/AIDS. In 2013, over 6 million people, mainly in low- and middle-income countries (LMICs) where most HIV/AIDS infections occur, took TDF.

The synthesis and use of TDF and TAF demonstrate several principles (1–5 and 8–9) of green chemistry. The HIV reverse transcriptase enzyme mistakenly incorporates tenofovir as a raw material for synthesizing viral DNA. When tenofovir is incorporated into viral DNA particles, viral replication stops.

Tenofovir is synthesized from adenine and (R)-propylene carbonate (RPC) as shown.

Tenofovir is the active drug, but tenofovir has a hard time crossing lipophilic (fat-loving) cell membranes to enter HIV-infected cells because it is very hydrophilic (water-loving). Because of this, tenofovir is delivered in a lipophilic "prodrug" form known as TDF. A prodrug is a masked or protected form of a drug that enhances one or more

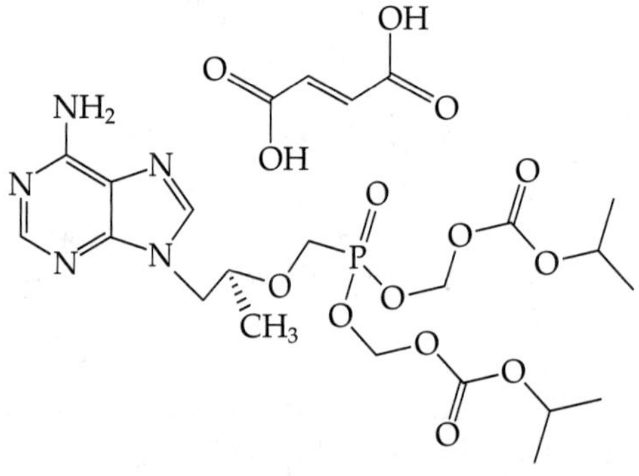

Tenofovir diisopoxil fumarate (TDF)

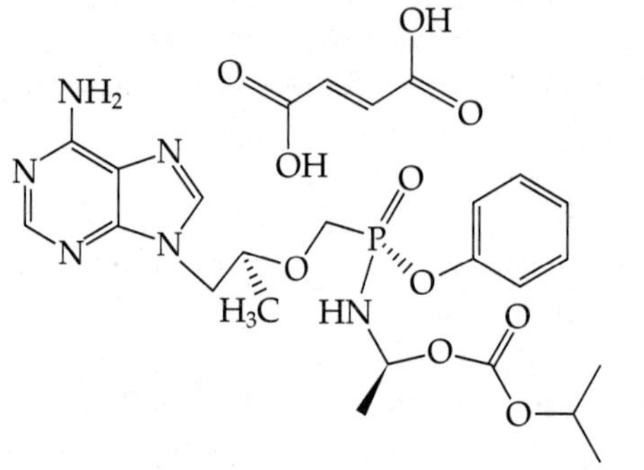

Tenofovir alafenamide fumarate (TAF)

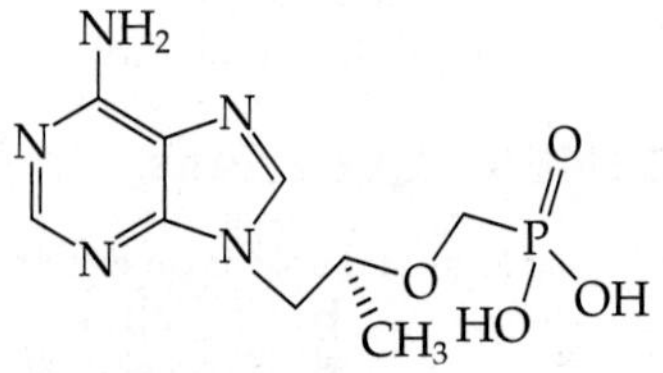

Tenofovir

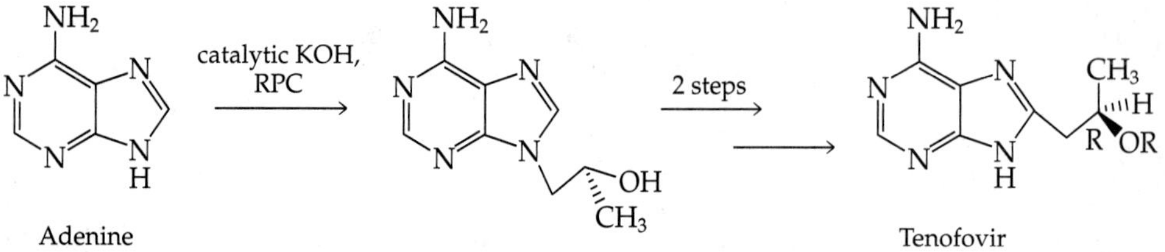

Adenine

Tenofovir

of the properties of absorption, distribution, metabolism, and excretion (ADME). TDF is much better distributed into cells than tenofovir because TDF is lipophilic. After TDF is hydrolyzed to tenofovir inside infected cells, it is in the proper form to inhibit viral replication.

Carbonate derivative of hydroxyacetone → Intermediate (not isolated) → (R)-propylene carbonate (RPC)

(aqueous buffer, *E. coli* containing ADH and GDH, glucose)

TDF is a very good drug at a daily dose of 300 mg. But TDF is still cleaved rapidly into tenofovir in blood plasma. Tenofovir in blood plasma is rapidly excreted, so the proportion of TDF that enters infected cells is low. TAF incorporates a more selective prodrug approach. The prodrug functions of TAF are only removed by intracellular enzymes *after* it enters a cell. Because of this, a 10 mg daily dose of TAF is as effective as a 300 mg daily dose of TDF. The lower level of drug circulating in blood plasma may also reduce the toxic side effects of TAF versus TDF.

If 6 million people per year take 300 mg of TDF per day, the annual volume demand for TDF is 657 tons. If 6 million people per year take 10 mg of TAF per day, the demand is only 21.9 tons. Switching from TDF to TAF, therefore, reduces waste generation and is very green. This is also potentially very important for reducing cost and making tenofovir more available to low- and middle-income markets.

In another example of green chemistry, the compound RPC for synthesizing tenofovir is produced in aqueous media from the very inexpensive carbonate derivative of hydroxyacetone (shown) using *E. coli* bacteria and the enzymes alcohol dehydrogenase and glucose dehydrogenase (ADH, GDH). This synthesis of RPC is atom-economical and has an E-factor of about 6. Previous preparations of RPC had an E-factor as high as 45 and used more hazardous reagents and solvents.

Green chemistry has become more important as drug molecules have become increasingly complex and more difficult to produce. The expense and high E-factors common to drug manufacturing provide incentives to develop green chemistry approaches. Reducing drug manufacturing expense also improves worldwide access to essential medicines for diseases such as HIV/AIDS, malaria, and tuberculosis. The number of people in low-and-middle-income countries taking HIV/AIDS medicines has increased from 235,000 in 2003 to over 11 million at the end of 2013. The cost of treating each patient over that timeframe, however, has dropped by about 80%, and Green Chemistry has been a significant contributor to this effort.

⬤ Summary

Section 18.1—A **drug** is a substance that relieves pain, treats illness, or improves one's health or well-being. A drug may be obtained from natural sources, or it may be semisynthetic or synthetic. **Chemotherapy** is the use of chemicals to treat disease.

Section 18.2—Aspirin is a **nonsteroidal anti-inflammatory drug (NSAID)**. It is an **analgesic** (pain reliever), **antipyretic** (fever reducer), **anti-inflammatory** (inflammation reducer) and **anticoagulant** (blood clotting inhibitor). NSAIDs inhibit the production of **prostaglandins**, hormone-like lipids that can send pain messages to the brain. Ibuprofen and naproxen are NSAIDs. Acetaminophen is an analgesic and antipyretic but does not reduce inflammation. Cold medicines often contain an antihistamine to relieve symptoms caused by **allergens**, which trigger the release of histamine. No treatment cures a cold.

A **narcotic** relieves pain and produces stupor or anesthesia but is addictive. Opium contains morphine, a natural narcotic. Codeine and heroin are made from morphine. Codeine is often combined with acetaminophen for pain relief and is used to suppress coughs. Heroin is a much more potent and highly addictive derivative of morphine. A morphine **agonist** has morphine-like action; a morphine **antagonist** blocks morphine receptors. The body produces morphine-like substances called **endorphins** in response to extreme pain. Exercise increases production of **neurotrophins**, substances that enhance the growth of brain cells.

Section 18.3—Sulfa drugs were the first antibacterial drugs; they mimic compounds essential for bacterial growth. An **antibiotic** is a soluble substance that inhibits the growth of other microorganisms and is derived from mold or bacteria.

Penicillins prevent bacteria from forming mucoprotein cell walls. Tetracyclines and fluoroquinolones are **broad-spectrum antibiotics** and are effective against a wide variety of bacteria. Antibiotic-resistant bacteria often develop after an antibiotic has been in use for a while.

Antibiotics cannot cure viral diseases. Some viral diseases, such as measles, mumps, and smallpox, can be prevented by vaccination, but vaccines are not available for all viral diseases. Herpes and **acquired immune deficiency syndrome (AIDS)** are viral infections. Viruses are composed of nucleic acids—either DNA or RNA—and proteins. A **retrovirus** such as the AIDS virus is an RNA virus that synthesizes DNA in the host cell. Antiviral drugs are used to treat some viral diseases.

Section 18.4—Many anticancer drugs are **antimetabolites**, which inhibit DNA synthesis and thus slow the growth of cancer cells. Alkylating agents such as nitrogen mustards transfer alkyl groups to biological molecules, blocking their usual action.

Section 18.5—A **hormone** is a chemical messenger that is produced in the endocrine glands that signals physiological changes in other parts of the body. Prostaglandins mediate hormone action. Some hormones are **steroids**, compounds with a characteristic four-ring skeletal structure. **Anabolic steroids** are used to increase muscle mass, but have negative side effects. Cholesterol and cortisol are steroids, as are sex hormones. An **androgen** controls masculine characteristics. An **estrogen** controls female sexual functions. A **progestin** is a steroid hormone that has the effect of progesterone. Birth control pills contain a synthetic estrogen or a progestin, and prevent ovulation and conception.

Section 18.6—Coronary artery disease, arrhythmias, hypertension, and congestive heart failure are diseases of the cardiovascular system. **Diuretics**, beta blockers, calcium channel blockers, and ACE inhibitors are used for hypertension. Arrhythmia drugs alter ion flow across cell membranes. Coronary artery drugs include organic nitrate compounds.

Section 18.7—**Psychotropic drugs** affect the mind. They can be divided into three classes: **stimulant drugs** such as cocaine and amphetamines; **depressant drugs** such as alcohol, anesthetics, opiates, barbiturates, and tranquilizers; and **hallucinogenic drugs** such as LSD.

Neurons (nerve cells) have tiny gaps called **synapses** between them. **Neurotransmitters** are released into the synapses during nerve transmissions and cause changes in the receiving cells. Mental illness is thought to involve brain amines such as norepinephrine (NE) and serotonin. NE agonists enhance or mimic the action of NE, while NE antagonists block or slow its action. Many compounds exist as **stereoisomers**, which have the same formula but different arrangements of atoms. Stereoisomers that are not superimposable on their mirror images are called **enantiomers** and have one or more **chiral carbons** (carbon atoms with four different groups attached).

Anesthetics are substances that causes lack of feeling or awareness. A **general anesthetic** produces unconsciousness and general insensitivity to pain. A local anesthetic causes loss of feeling in a part of the body.

Alcohol is the most common depressant used worldwide. Barbiturates like pentobarbital and thiopental are used as sedatives and anesthetics. Ingesting alcohol and a barbiturate together produces a **synergistic effect**, with the action of one enhancing action of the other. Phenothiazines, a class of tranquilizer, act as dopamine antagonists and relieve symptoms of schizophrenia. SSRIs are commonly prescribed antidepressants that enhance the effect of serotonin.

Amphetamines are stimulants, similar in structure to epinephrine and NE. Amphetamine exists as enantiomers, one of which is much more potent than the other. Methamphetamine is even more potent than amphetamine. Caffeine and nicotine are widely used legal stimulants.

Hallucinogenic drugs induce changes in sensory perception. The best-known hallucinogenic drugs are LSD and **marijuana**. Psilocybin from certain mushrooms and mescaline from peyote cactus are natural hallucinogens.

Section 18.8—**Drug abuse**, or using drugs for their intoxicating effects, is a serious problem for abusers and society. **Drug misuse**, in which legal drugs are used incorrectly or by mistake, is also a problem. New drugs must be tested properly to avoid the **placebo effect**, in which an inactive substance, given as medication, produces results in patients for psychological reasons. Drugs must go through a rigorous vetting process before gaining FDA approval. The steps include initial testing in laboratory and animal models and several stages of clinical trials with human subjects. **Orphan drugs** treat diseases that affect a small number of individuals. Use of **biosimilars** is increasing.

GREEN CHEMISTRY—Modern drugs have complex structures and are often synthesized with poor efficiency (high E-factors). Green chemistry reduces the expense of drug manufacturing and increases sustainability by reducing waste generation. Sustainability and improved efficiency also can increase access to medicines globally, providing these important contributions to low- and middle-income populations as well as industrialized countries.

Learning Objectives	Associated Problems
• Classify common drugs as natural, semisynthetic, or synthetic. (18.1)	1, 8, 11, 12
• Define *chemotherapy*, and explain its origin. (18.1)	13, 14
• List common over-the-counter analgesics, antipyretics, and anti-inflammatory drugs, and describe how each works. (18.2)	2, 15–18, 20, 60, 65, 71, 79, 81
• Name several common narcotics, describe how each functions, and state its potential for addiction. (18.2)	9, 19, 21, 22, 68, 73
• List the common antibacterial drugs, and describe the action of each. (18.3)	3, 23–25, 59, 61, 69, 80, 82, 85
• Name the common categories of antiviral drugs, and describe the action of each. (18.3)	26–28, 62
• Describe the action of the common types of anticancer drugs. (18.4)	29–34, 86
• Define the terms *hormone*, *prostaglandin*, and *steroid*, and explain the function of each. (18.5)	4, 5, 35, 38, 63, 78, 83, 84
• List the three types of sex hormones, and explain how each acts and how birth control drugs work. (18.5)	6, 36, 37, 39, 40, 77
• Describe the action of four types of drugs used to treat heart disease. (18.6)	41–46
• Explain how the brain amines norepinephrine and serotonin affect the mind and how various drugs change their action. (18.7)	48, 50, 54
• Identify some stimulant drugs, depressant drugs, and hallucinogenic drugs, and describe how they affect the mind. (18.7)	7, 47, 49, 51–53, 64, 66, 67, 72, 74, 75
• Differentiate between drug abuse and drug misuse. (18.8)	56, 58
• Describe how a new drug is developed and brought to market. (18.8)	10, 55, 57, 70, 76
• Identify green chemistry principles that can improve the E-factor of chemical syntheses.	87, 88

Conceptual Questions

1. What is the difference between a natural drug, a semisynthetic drug, and a synthetic drug?

2. What is the difference between an antihistamine, an antitussive, and an expectorant?

3. What is an antibiotic?

4. What is a hormone?

5. What are prostaglandins? How do prostaglandins differ from hormones?

6. How do birth control pills work?

7. What is the most widely used depressant drug in the world?

8. Classify the following drugs as natural, semisynthetic, or synthetic: heroin, morphine, penicillin, and cisplatin.

9. What is a narcotic?

10. What is the placebo effect?

 Problems

Drugs from Nature and the Laboratory

11. Give an example of a natural drug, a semisynthetic drug, and a synthetic drug.

12. Classify the following drugs as natural, semisynthetic, or synthetic: codeine, fluoroquinolone, quinine, vinblastine, Taxol, and 6-mercaptopurine.

13. What is chemotherapy?

14. Give four possible reasons that people use drugs.

Pain Relievers

15. What is the chemical name for aspirin?

16. Why is acetaminophen used instead of aspirin for the relief of pain associated with surgical procedures?

17. What is a COX-2 inhibitor? How does a COX-2 inhibitor work?

18. What does each of the following types of drugs do?
    a. analgesic   **b.** antipyretic   **c.** anti-inflammatory

19. Which of the following was originally marketed as a cough suppressant?
    a. acetaminophen   **b.** aspirin
    c. heroin   **d.** methadone

20. Which of the following is *not* considered to be an NSAID?
    a. acetaminophen   **b.** aspirin
    c. ibuprofen   **d.** naproxen

21. How does codeine differ from morphine in its structure?

22. How are endorphins related to **(a)** the anesthetic effect of acupuncture and **(b)** the absence of pain in a wounded soldier?

Antibacterial and Antiviral Drugs

23. Tetracyclines are broad-spectrum antibiotics. What does this mean?

24. What are penicillins? How do they kill bacteria?

25. Quinolone antibiotics are
    a. antimetabolites   **b.** broad-spectrum
    c. *beta*-lactams   **d.** very specific narrow-spectrum

26. Do antiviral drugs cure viral diseases? What is the best way to deal with viral diseases?

27. What are the three classes of antiretroviral drugs? How does each work?

28. Which of the following is an antiviral drug?
    a. acyclovir   **b.** ciprofloxacin
    c. penicillin   **d.** sulfanilamide

Anticancer Drugs

29. All of the following are antimetabolites except
    a. 6-mercaptopurine   **b.** cyclophosphamide
    c. 5-fluorouracil   **d.** methotrexate

30. List two major classes of anticancer drugs.

31. How do antimetabolites and alkylating agents work?

32. How does cisplatin work?

33. What compound does the following formula represent? To which class of anticancer drugs does it belong?

34. What compound does the following formula represent? To which class of anticancer drugs does it belong?

Hormones

35. Give three differences between hormones and prostaglandins.

36. Carl Djerassi altered the progesterone structure by removing a methyl group and adding an ethynyl group when he synthesized norethindrone. How did these changes make norethindrone a better drug?

37. Identify the three functional groups in the progesterone molecule (Figure 18.14).

38. What are anabolic steroids used for? What risks are associated with using them?

39. What are emergency contraceptives? How do they work?

40. Which of the following is *not* a steroid?
    a. estradiol
    b. enkephalin
    c. progesterone
    d. testosterone

Drugs for the Heart

41. Diuretics work to lower blood pressure by
  a. causing the kidneys to excrete more water and thus lower the blood volume
  b. slowing the heart rate
  c. acting as vasodilators, causing the muscles around blood vessels to relax
  d. inhibiting an enzyme that causes blood vessels to contract

42. Beta blockers work to lower blood pressure by
  a. causing the kidneys to excrete more water and thus lower the blood volume
  b. slowing the heart rate and lessening the force of the heartbeat
  c. acting as vasodilators, causing the muscles around blood vessels to relax
  d. inhibiting an enzyme that causes blood vessels to contract

43. Which of the following small molecules acts as a messenger molecule that allows the dilation of blood vessels, thereby relieving chest pain and alleviating the symptoms of erectile dysfunction?
  a. carbon dioxide, CO_2 **b.** ammonia, NH_3
  c. nitric oxide, NO   **d.** sulfur dioxide, SO_2

44. Which of the following is often treated with organic nitro compounds?
  a. angina pectoris   **b.** erectile dysfunction
  c. heart arrhythmias   **d.** hypertension

45. Which of the following is *not* normally a goal of treatment with heart drugs?
  a. blocking the effect of progesterone on the heart
  b. lowering blood pressure
  c. normalizing heart rhythm
  d. preventing buildup of lipid plaque in blood vessels

46. Which of the following is probably the most effective and cheapest type of drug for the treatment of hypertension?
  a. angiotensin-converting enzyme (ACE) inhibitors
  b. beta blockers
  c. calcium channel blockers
  d. diuretics

Drugs and the Mind

47. What is the difference between a psychotropic drug and a hallucinogenic drug?

48. What are synapses, and what is their role in the nervous system?

49. Which of the following is *not* considered to be a class of psychotropic drugs?
  a. antimetabolites   **b.** depressants
  c. hallucinogenics   **d.** stimulants

50. Beta blockers work by
  a. decreasing the stimulant action of serotonin
  b. limiting the action of a drug so that only one enantiomer is active
  c. acting as an anesthetic
  d. decreasing the stimulant action of epinephrine and norepinephrine

51. Which of the following has been used as a local anesthetic?
  a. ketamine   **b.** lidocaine
  c. nitrous oxide   **d.** thiopental

52. Which of the following is a dissociative anesthetic?
  a. chloroform   **b.** halothane
  c. ketamine   **d.** mescaline

53. For each of the following anesthetics, identify a disadvantage other than flammability that is associated with its use.
  a. nitrous oxide   **b.** halothane   **c.** diethyl ether

54. What is an agonist? What is an antagonist?

Drugs and Society

55. For a drug to receive FDA approval for further testing, all of the following requirements must be satisfied except
  a. safety for human use
  b. low cost
  c. effectiveness
  d. ability to reach the site of action

56. Which of the following statements about illegal drug use is *not* true?
  a. Buyers have no good way of knowing what the actual content or concentration of the drug is.
  b. Many users steal to obtain money to pay for drugs.
  c. Illegal drug use affects only the users.
  d. Many accidents and personal injuries are drug-related.

57. Tests on a new investigative drug include using a double-blind test procedure. In a double-blind study
  a. all test subjects receive the drug
  b. some test subjects receive the drug and others receive a placebo
  c. the doctor administering the drug knows who is receiving the drug or the placebo
  d. patients know whether they are receiving the drug

58. What is the difference between drug abuse and drug misuse?

Expand Your Skills

59. Review the information on structure and function for fluoroquinolone drugs. Draw the structure of a fluoroquinolone drug, which is similar to ciprofloxacin in action but does not have the large nitrogen-containing ring at position 7.

60. Why is acetaminophen not classified as an NSAID?

61. A 140 lb. patient was treated with the antibiotic ampicillin for a urinary tract infection. The daily dose was 40 mg/kg divided into three doses per day. **(a)** What was the daily dose, in grams, for the patient? **(b)** Each ampicillin capsule contains 500 mg of active ingredient. How many capsules did the patient have to take every 8 hours?

62. A retrovirus
  a. breaks down RNA into nucleotides
  b. breaks down DNA into nucleotides
  c. synthesizes RNA from a DNA template
  d. synthesizes DNA from an RNA template

63. What structural feature is shared by all steroids?

64. Some psychotropic drugs can cause flashbacks long after the drug was originally used. Identify two such drugs and explain why they can cause flashbacks.

65. Diphenhydramine (Benadryl) has the chemical formula $C_{17}H_{21}NO$. Is Benadryl water-soluble or fat-soluble? Explain.

66. What structural feature is common to all barbiturate molecules? How is this structure modified to change the properties of individual barbiturate drugs?

67. Alcohol can interact with other drugs and greatly increase their potency, even resulting in death. Identify two such drugs.

68. The LD_{50} of a substance is the dose (in mg substance per kg body mass) that would be lethal to 50% of a population of test animals. If drugs A and B have LD_{50} values of 750 mg/kg and 45 mg/kg, respectively, which is more toxic? Explain.

69. Which of the following drugs will cause staining of teeth when given to small children?
  a. cyclosporins   **b.** penicillin
  c. sulfa drugs   **d.** tetracycline

70. What is an orphan drug?

71. What do antitussive drugs, decongestants, and expectorants do?

72. If the minimum lethal dose (MLD) of amphetamine is 5 mg/kg, what is the MLD for a 70 kg person? Can toxicity studies on animals always be extrapolated to humans?

73. As noted in Section 18.2, replacement of the phenolic —OH group of the morphine molecule by a methoxy (—OCH₃) group produces codeine. Draw the structure of the codeine molecule.

74. Some states have recently legalized the use of marijuana for medicinal purposes. Identify some medical uses of marijuana.

75. What is a synergistic effect for drugs?

76. How do generic drugs differ from name-brand drugs?

77. Rank the following from lowest to highest risk of death:
  a. pregnancy and childbirth
  b. taking birth control pills
  c. taking birth control pills while continuing to smoke

78. Which of the following compounds would *not* be classified as a steroid?
  a. cholesterol   **b.** prednisone
  c. progesterone   **d.** prostaglandin E_2

79. How do NSAIDs work to counter the effects of prostaglandins in inflammation?

80. How would removal of the *para*-hydroxyl group from the molecule represented below affect its solubility in **(a)** water and **(b)** fat?

81. Which NSAID does the following formula represent? To what aromatic hydrocarbon is it related? Name the two oxygen-containing functional groups.

82. Chloramphenicol is a broad-spectrum antibiotic that acts by inhibiting bacterial protein synthesis. How would conversion of chloramphenicol ($R = -H$) to **(a)** its palmitate ester [$R = -CO(CH_2)_{14}CH_3$] and **(b)** its sodium succinate ($R = -COCH_2CH_2COONa$) derivative affect its solubility? Explain.

83. What type of drug does the following formula represent? Name the three types of oxygen-containing functional groups present.

84. What type of drug is megestrol acetate (shown below), which is used for the treatment of some cancers?

  a. phencyclidine   **b.** prostaglandin
  c. steroid   **d.** tetracycline

85. Which statement is true regarding antibiotics?
  a. Antibiotics should be used to treat bacterial infections, because they are active against bacteria, but not against viruses.
  b. Antibiotics should be used to treat both bacterial and viral infections, because they are active against both bacteria and viruses.
  c. Antibiotics should be used to treat viral infections, because they are active against viruses but not against bacteria.
  d. Antibiotics should not be used to treat either bacterial or viral infections, because they are not active against either bacteria or viruses.

86. Which of the following is true of anticancer drugs?
  a. They kill only cancer cells, because cancer cells grow more rapidly than normal cells.
  b. They kill both cancer cells and normal cells, because they are toxic to both.
  c. They often have nasty side effects, because the cells growing in the intestines grow more slowly than most normal cells do.
  d. They kill both cancer cells and normal cells, but **have** a greater effect on normal cells, because normal cells grow more slowly than cancer cells do.

87. If the E-factor for synthesizing TDF is about 70 and the E-factor for synthesizing TAF is about 85, what is the difference in waste generated per year during drug manufacturing of TAF versus TDF for 6 million patients?

88. What are the green chemistry principles exercised in the enzymatic synthesis of (R)-propylene carbonate (RPC) versus a chemical synthesis in organic solvents?

 Critical Thinking Exercises

Apply knowledge that you have gained in this chapter and one or more of the FLaReS principles (Chapter 1) to evaluate the following statements or claims.

18.1. A television advertisement encourages consumers to "talk to their doctor" about using a new drug to combat their depression. The ad ends with a woman happily playing with her children in an outdoor setting.

18.2. A nurse claims to be able to help patients by "therapeutic touch," a technique that she says allows her to sense a patient's energy fields by moving her hands above the patient's body.

18.3. A television advertisement shows an actor taking a medicine that inhibits the production of stomach acid, then eating a large meal of rich, spicy food.

18.4. A television advertisement shows an actor taking an antihistamine and then joyously walking through a grassy, flower-filled meadow without experiencing the usual allergy symptoms.

18.5. Many people believe that the MMR vaccine is responsible for autism and cite a study on 12 children by Andrew Wakefield. However, autism is generally diagnosed at approximately the age that children usually receive the vaccine, and the study did not compare autism rates of vaccinated and unvaccinated children.

18.6. A website claims that red wine is good for your heart.

 Collaborative Group Projects

Prepare a PowerPoint, poster, or other presentation (as directed by your instructor) for presentation to the class. Projects 1 and 2 are best done by a group of four students.

1. Prepare a brief report, including a list of the ingredients and their amounts, on a combination product such as Advil® Cold & Sinus, Aleve® Cold & Sinus, Dimetapp® Cold & Flu, Theraflu®, Excedrin PM®, Alka-Seltzer® Plus, DayQuil®, or NyQuil®. Give the chemical structure, medical use, toxicity and side effects for each substance in the product.

2. Four major drug categories for lowering blood pressure are listed in Section 18.6. Give the chemical structure, toxicity, and side effects for a drug in each of the categories listed.

3. Do a cost analysis of five brands of plain aspirin, calculating the cost per gram of each. Compare the cost per gram of an extra-strength aspirin formulation with that of plain aspirin.

4. Search the Internet to learn about drug trials, and answer the following questions. **(a)** Why might a person who is gravely ill not want to participate in a placebo-controlled drug study? **(b)** What are the advantages and disadvantages of participating in such a drug study? **(c)** Why is it important that studies use placebos as well as the test

drug? **(d)** Sometimes the control group receives the standard treatment for the disease instead of a placebo. Why does this happen?

5. Search the Internet for information on new drugs for treating one of the following diseases.
   a. pneumonia   **b.** arthritis
   c. cancer   **d.** diabetes

6. Use the Internet or *The Merck Index* or a similar reference work to look up the toxicities of cocaine, procaine, lidocaine, mepivacaine, and bupivacaine. Is it always accurate to compare toxicities in different animals and extrapolate those animal toxicities to humans? Does the method of administration have an effect on the observed toxicity?

7. Prepare a brief report on a neurotransmitter. Describe its function and site of action.

8. Prepare a brief report on thalidomide—its structure, original intended use, problems that arose, and proposed new uses. Identify the chiral carbon atom(s) in thalidomide that were the source of the problems.

9. Prepare a brief report on an antihistamine, one of the narcotics listed in Table 18.1, an antibiotic, an anticancer drug, or an antidepressant. Include the chemical structure, medical use, toxicity and side effects for the drug you selected.

LET'S EXPERIMENT! Heal My Heartburn

Materials Needed

- Three different kinds of commercial chewable antacids (Hint: Use white tablets, not colored tablets, and look for brands with different active ingredients.)
- Four small clear plastic cups
- Mortar and pestle, rolling pin, or kitchen mallet (to crush the pills)

- Red cabbage leaves (to prepare indicator)
- Cabbage juice indicator (See preparation instructions below.)
- Vinegar
- Measuring spoons (a tablespoon and a teaspoon)

Why do doctors recommend particular medicines over others? How do over-the-counter medicines differ in effectiveness? In this experiment, you will analyze three different brands of chewable antacids to determine which active ingredient is the most effective.

Medicines and drugs are seldom given to patients in pure form. Typically, the active ingredient is combined with other (inactive) ingredients so that the drug can be delivered in a pill form or intravenously, for example. The inactive ingredients help to control the timing and method of delivery of a particular dose of the drug. The active ingredient is the molecule that functions biologically to provide a cure or remedy an adverse symptom. Usually, several different active ingredients can be used to cure an ailment.

For this experiment, we will be using commercial chewable antacid tablets to analyze the effectiveness of the different active ingredients. Antacids are taken to help relieve the symptoms of heartburn, and almost all of the active ingredients function by neutralizing the excess acid in the stomach with weak bases. The most common bases used are magnesium, aluminum, and calcium hydroxides and carbonates.

First, gather three different brands of antacids that have three different active ingredients. Many antacids will have more than one active ingredient and will be combinations of different hydroxides and carbonates.

Prepare four small clear plastic cups by labeling three of them with the different names of the antacids and labeling the fourth as "control." Prepare the control solution by adding 1 tablespoon of vinegar and 1 teaspoon of cabbage juice indicator to the plastic cup. You can use this control solution to compare the colors of the reactions.

Next, take two tablets of the first brand of antacid and crush them using a mortar and pestle, a kitchen mallet, or another kitchen device. Add the crushed tablets to the appropriately labeled plastic cup. Do the same for the other two brands of antacids.

Then, measure and add 1 tablespoon of vinegar, followed immediately by 1 teaspoon of the cabbage juice indicator, to each plastic cup. Note the time that you added the vinegar and indicator. Gently swirl each mixture and observe the reactions for 5–10 minutes. What do you notice about the color of each solution? What

does the color change indicate? According to your results, rank the antacid brands in order of most effective to least effective.

How to make cabbage juice indicator: To prepare the cabbage juice indicator, first peel off six leaves from a head of red cabbage and chop them into fine pieces. Bring 8 cups of water to boil and add the chopped cabbage leaves to the boiling water. Lower the heat and allow the cabbage mixture to simmer for 30 minutes. Pour the mixture through a strainer into a large container and let the juice cool to room temperature. This juice will be your indicator solution.

Typical Results

Questions

1. Write out the chemical reactions for each brand of the active ingredients. Remember to balance your equations.
2. Do you think that the conditions of this experiment mimic those of the stomach? What differences might exist between the conditions of this experiment and those of the stomach (and the conditions associated with heartburn)?
3. What other factors associated with the antacid tablets might affect this experiment? For example, do the different brands have the same inactive ingredients?

19

Chemistry Down on the Farm

. . . and in the Garden and on the Lawn

Modern agriculture produces abundant food for much of Earth's growing population, but it has led to many problems, both economic and environmental. Many people are turning to sustainable practices.

Have You Ever Wondered?

1 What is anhydrous ammonia? **2** Why can't I buy a lawn fertilizer with phosphorus in my city? **3** Are pesticides causing honeybees to die off? **4** Why does my "weed-killer" not kill my grass, too? **5** Are "organic" fruits and vegetables healthier for us? **6** How can there be a food crisis if I see plenty of corn growing in the fields?

The earliest humans obtained their food by hunting and gathering. The dawn of the agricultural revolution about 10,000 years ago gradually led to more organized ways to obtain food for an increasing population of people settling into villages and cities. The industrial revolution and scientific advancements of the last few hundred years sparked the development of modern agriculture, with all of its benefits and drawbacks. The most dramatic changes in agricultural chemistry have happened over the last century. In this chapter, we examine chemical aspects of past, present, and potential future agricultural practice.

Our food comes ultimately from green plants. All organisms can transform one type of food into another, but green plants can use the energy from sunlight and the chlorophyll in the plant cells to convert carbon dioxide and water to the sugars that directly or indirectly fuel all living things.

$$6\ CO_2 + 6\ H_2O \xrightarrow{\text{sunlight}} C_6H_{12}O_6 + 6\ O_2$$

This simplified reaction also replenishes the oxygen in the atmosphere and removes carbon dioxide (Section 13.2). Using other nutrients, particularly compounds of nitrogen and phosphorus, plants can convert the sugars produced by photosynthesis to proteins, fats, and other chemicals that we use as food.

The main chemical elements in plants—carbon, hydrogen, and oxygen—are found in cellulose, the primary compound in their stems, leaves, and roots; those elements come from air and water. Other plant nutrients are taken from the soil, and energy is supplied by the sun. In early agricultural societies, people grew plants for food and obtained energy from the food. Much of this energy was reinvested in the production of food, although a portion went into making clothing and building shelter. Figure 19.1 shows a simplified diagram of the energy flow in such a society. Our modern society is much more complicated than indicated here. Some plants, for example, might be fed to animals, and some human energy would then be obtained from animal flesh or animal products such as milk and eggs.

In early societies, nearly all the energy came from renewable resources. One unit of human work energy, supplemented liberally by energy from the Sun, might produce ten units of food energy. Some human energy could be used to grow crops for cloth to make clothing and to provide shelter. It also might be used in games or cultural activities.

The flow of nutrients is also rather simple. Unused portions of plants as well as human and animal wastes are returned to the soil, forming "humus," organic material formed by the decomposition of leaves and other plant parts broken down by microorganisms. This material provides nutrients for the growth of new plants (Figure 19.2). Properly practiced, this kind of agriculture could be continued

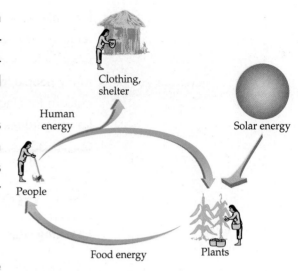

▲ **Figure 19.1** Energy flow in an early agricultural society.

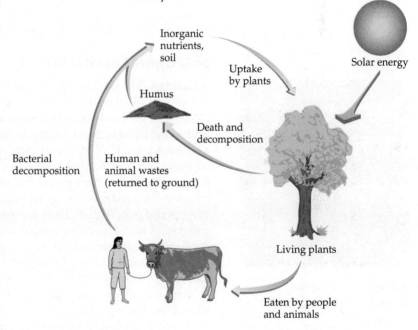

▲ **Figure 19.2** Flow of nutrients in a simple system.

▲ Soil consists of rock broken down by weathering, plus various types of decaying plant and animal matter. To build even a few millimeters of soil from rock takes many years. The United States loses 5 billion tons of soil each year to erosion. Developing countries often suffer far greater soil losses as more land is cleared for farming in an attempt to feed an increasing population.

for centuries without seriously depleting the soil, but farming at this level supports relatively few people. By making use of fertilizers, pesticides, and energy from fossil fuels, modern high-production farming has greatly expanded crop yields. Many people living in urban areas use materials and methods similar to those of modern farming to maintain green lawns and sustain productive vegetable and flower gardens.

19.1 Growing Food with Fertilizers

Learning Objectives • List the three primary plant nutrients as well as several secondary nutrients and micronutrients, and describe the function of each.
• Differentiate between organic fertilizers and conventional fertilizers.

To replace nutrients lost from the soil and to increase crop production, modern farmers use a variety of chemical fertilizers. The three **primary plant nutrients** are nitrogen, phosphorus, and potassium. Let's consider nitrogen first.

Nitrogen Fertilizers

A German chemist, Justus von Liebig (1803–1873), discovered in 1843 that nitrogen is a critical plant nutrient. Nitrogen, although present in air in the elemental form (N_2), is not generally available to plants. Every flash of lightning forms some nitric acid in the air, which is brought to the ground by rain. Lightning contributes about a billion tons of fixed nitrogen each year to soil, but lightning is random rather than frequent and dependable. Some bacteria are able to *fix* nitrogen—that is, to convert it to water-soluble nitrogen-containing compounds. Colonies of bacteria that can perform this vital function grow in nodules on the roots of legumes—plants such as clovers and peas (Figure 19.3). Farmers are able to restore fertility to the soil by crop rotation. A nitrogen-fixing crop (such as clover) can be alternated with a nitrogen-consuming crop (such as corn).

▲ **Figure 19.3** Bacteria in nodules on the roots of legumes fix atmospheric nitrogen by converting it to soluble compounds that plants can use as nutrients.

Plants usually take up nitrogen in the form of nitrate ions, NO_3^-, or ammonium ions, NH_4^+. Plants combine these ions with carbon compounds from photosynthesis to form the amino acids that make up proteins, compounds essential to all life processes. For thousands of years, farmers were dependent on manure as a soil fertilizer. Even today, farmers in underdeveloped nations as well as some small farmers in developed nations still add manure to their fields. We now know that manure does contain nitrates, but it is also full of *E. coli* bacteria.

After the discovery of deposits of sodium nitrate (called *Chile saltpeter*) in the deserts of northern Chile, this substance ($NaNO_3$) was widely used as a source of nitrogen as soil fertilizer. By applying fertilizers, farmers don't have to alternate crops to restore soil fertility. Why raise a corn crop every other year if you can raise one every year, just by adding fertilizer?

A rapid rise in population growth during the late nineteenth and early twentieth centuries put increasing pressure on the available food supply, which led to an increasing demand for nitrogen fertilizers. The first real breakthrough in nitrogen fertilizer manufacturing came in Germany on the eve of World War I. An earlier process developed by Fritz Haber (Germany, 1868–1934), transformed by Carl Bosch (Germany, 1874–1940) into a large-scale method using a catalyst and high pressure, made possible the combination of nitrogen and hydrogen to make ammonia.

$$3\,H_2 + N_2 \longrightarrow 2\,NH_3$$

By 1913, one nitrogen-fixation plant was in production, and several more were under construction. The Germans were interested in ammonium nitrate mainly as an explosive, but it turned out to be a valuable nitrogen fertilizer as well. The Germans were able to make ammonium nitrate (NH_4NO_3), by oxidizing some ammonia to nitric acid.

$$NH_3 + 2\,O_2 \longrightarrow HNO_3 + H_2O$$

The nitric acid was then reacted with more ammonia to produce ammonium nitrate.

$$HNO_3 + NH_3 \longrightarrow NH_4NO_3$$

A gas at room temperature, ammonia is easily compressed into a liquid that can be stored and transported in tanks. It is applied directly to the soil as fertilizer (Figure 19.4). Be aware that this form of ammonia [$NH_3(l)$] is not the same as household ammonia [$NH_3(aq)$], which is a water solution that usually contains detergents and other ingredients as well as NH_3.

Some solids made from ammonia for application as fertilizers are listed in Table 19.1.

▲ **Figure 19.4** A farmer applies anhydrous ammonia, a source of the plant nutrient nitrogen, to his fields.

TABLE 19.1 **Various Nitrogen Fertilizers Made from Ammonia**

Reacted with:	Product Fertilizer	Formula
None	Anhydrous ammonia	NH_3
Carbon dioxide	Urea	NH_2CONH_2
Sulfuric acid	Ammonium sulfate	$(NH_4)_2SO_4$
Nitric acid	Ammonium nitrate	NH_4NO_3
Phosphoric acid	Ammonium hydrogen phosphate	$(NH_4)_2HPO_4$

The atmosphere has vast amounts of nitrogen, but current industrial methods of fixing it require a lot of energy and use nonrenewable resources. Genes for nitrogen fixation have been transferred from bacteria into nonlegume plants. Perhaps someday soon we will see the development of transgenic corn and cotton capable of producing their own nitrogen fertilizer the way clovers and peas do.

The use of fertilizers containing nitrogen has negative impacts on those living on farms. Water from wells on farms will likely be high in nitrates because of fertilizer runoff and runoff of manure from livestock. The use of well water to mix powdered baby formula can be disastrous to infants. (See Section 14.3.)

Phosphorus Fertilizers

The limiting factor in plant growth is often the availability of phosphorus, usually in the form of phosphate ions, $PO_4{}^{3-}$. Plants incorporate phosphates into DNA and RNA (Chapter 16) and other compounds essential to plant growth. Phosphates have probably been used as fertilizers since ancient times because those ions are found in bone, guano (bird droppings), and fish meal. However, phosphates as such were not recognized as plant nutrients until about 1800. Following this discovery, the great battlefields of Europe were dug up and bones were shipped to chemical plants for processing into fertilizer.

Animal bones are rich in phosphorus, but most phosphates are poorly soluble in water, so the phosphorus is not readily available to plants. In the 1840s, chemists learned how to treat animal bones with sulfuric acid to convert them to a mixture of calcium sulfate and calcium dihydrogen phosphate, which is sometimes nicknamed

1 What is anhydrous ammonia?

"Anhydrous" means "without water." The ammonia solution used in laboratories and household cleaning products is ammonia gas dissolved in water. Agricultural vehicles transporting "anhydrous ammonia" contain ammonia that has been compressed until it liquefies. Anhydrous ammonia is an effective nitrogen-containing fertilizer, partly because strong hydrogen bonds form between ammonia and water in the soil.

2 **Why can't I buy a lawn fertilizer with phosphorus in my city?**

Some large cities, such as Minneapolis, Minnesota, banned the use of lawn fertilizers containing phosphorus in the early 2000s because the soil there had enough phosphorus, and because excess fertilizer runoff had caused rapid algae growth in the cities' lakes, which was not only an aesthetic but an economic problem.

superphosphate. By the 1860s, the same treatment was being applied to phosphate rock to produce thousands of tons a year. The crucial reaction for forming super-phosphate is

$$Ca_3(PO_4)_2 + 2\,H_2SO_4 \longrightarrow \underbrace{Ca(H_2PO_4)_2 + 2\,CaSO_4}$$

Phosphate rock   Superphosphate
or bone (insoluble)   (more soluble)

Modern phosphate fertilizers are often produced by treating phosphate rock with phosphoric acid to make water-soluble calcium dihydrogen phosphate.

$$Ca_3(PO_4)_2 + 4\,H_3PO_4 \longrightarrow 3\,Ca(H_2PO_4)_2$$

More commonly used today is ammonium monohydrogen phosphate [$(NH_4)_2HPO_4$], which supplies both nitrogen and phosphorus.

About 90% of all phosphates produced are used in agriculture. Until 2006, the United States was the leading producer of phosphate fertilizer, from rich phosphate deposits in Florida, North Carolina, Utah, and Montana. In 2006 China took over the lead, with Morocco second in production. Our use of phosphates unfortunately scatters them irretrievably throughout the environment.

A major problem with fertilizer runoff is the increased amount of phosphates that enters lakes, ponds, and rivers, as discussed in Chapter 14. Since phosphates are plant nutrients, it makes sense that excess phosphates would cause rapid growth of blue-green algae that use up dissolved oxygen. This eutrophication process (recall Section 14.3) has become a major concern to environmentalists and the general public, and its economic impact is sizable. Fertilizer runoff from farms in the north-central part of the United States has ended up in the Mississippi River and ultimately in the Gulf of Mexico. In August 2017, the National Oceanic and Atmospheric Administration (NOAA) measured a dead zone in the Gulf of Mexico near the Mississippi Delta to be 8776 square miles (the size of New Jersey). Water there is so low in oxygen that fish and marine life cannot survive. This dead zone is having a deleterious effect on the fishing industry in Louisiana.

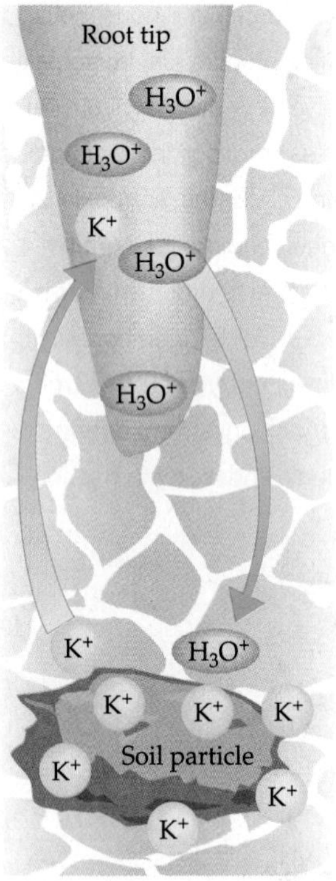

Potassium Fertilizers

The third major element necessary for plant growth is potassium. Plants use it in the form of the simple ion K^+. Generally, potassium is abundant, and there are no problems with solubility. Potassium ions, along with Na^+ ions, are essential to the fluid balance of cells. They also seem to be involved in the formation and transport of carbohydrates and may be necessary for the assembly of proteins from amino acids. Uptake of potassium ions from the soil leaves the soil acidic. Each time a positive potassium ion enters the root tip, a positive hydronium ion must leave for the plant to maintain electrical neutrality (Figure 19.5).

The usual chemical form of potassium in commercial fertilizers is potassium chloride (KCl). Vast deposits of this salt occur in and around Stassfurt, Germany, and for years this source supplied nearly all the world's potassium fertilizer. With the coming of World War I, the United States sought supplies within its own borders. Deposits at Searles Lake, California, and Carlsbad, New Mexico, now supply most U.S. needs. Canada has vast deposits in Saskatchewan and Alberta. Beds of potassium chloride up to 200 m thick lie about 1.5 km below the Canadian prairies. Although the reserves are large, potassium salts are a nonrenewable resource that we should use wisely.

Other Essential Elements

In addition to the three major nutrients (nitrogen, phosphorus, and potassium), other elements are necessary for proper plant growth. Three **secondary plant nutrients**—magnesium, calcium, and sulfur—are needed in moderate amounts. Calcium, in the form of lime (calcium oxide), is used to neutralize acidic soils.

$$CaO(s) + 2\,H^+(aq) \longrightarrow Ca^{2+}(aq) + H_2O(l)$$

▲ **Figure 19.5** When a root tip takes up K^+ ions from the soil, H_3O^+ ions are transferred to the soil. The uptake of potassium ions therefore tends to make soil acidic.

Calcium ions are also necessary plant nutrients. Magnesium ions (Mg^{2+}) are incorporated into chlorophyll molecules and therefore are necessary for photosynthesis. Sulfur is a constituent of several amino acids and is necessary for protein synthesis.

Plants need eight other elements, called **micronutrients**, in small amounts. These elements are summarized in Table 19.2. Many soils contain these trace elements in sufficient quantities. Some soils are deficient in one or more, and their productivity can be markedly increased by adding small amounts of the needed elements.

TABLE 19.2 **Eight Micronutrients Necessary for Proper Plant Growth**

Element	Form Used by Plants	Function	Symptoms of Deficiency
Boron	H_3BO_3	Required for protein synthesis; essential for reproduction and for carbohydrate metabolism	Death of growing points of stems; poor growth of roots; poor flower and seed production
Chlorine	Cl^-	Increases water content of plant tissue; involved in carbohydrate metabolism	Shriveling
Copper	Cu^{2+}	Constituent of enzymes; essential for reproduction and for chlorophyll production	Twig dieback; yellowing of newer leaves
Iron	Fe^{2+}	Constituent of enzymes; essential for chlorophyll production	Yellowing of leaves, particularly between veins
Manganese	Mn^{2+}	Essential for redox reactions and for the transformation of carbohydrates	Yellowing of leaves; brown streaks of dead tissue
Molybdenum	MoO_4^{2-}	Essential to nitrogen fixation by legumes and to reduction of nitrates for protein synthesis	Stunting; pale-green or yellow leaves
Nickel	Ni^{2+}	Required for iron absorption; constituent of enzymes	Failure of seeds to germinate
Zinc	Zn^{2+}	Essential for early growth and maturing	Stunting; reduced seed and grain yields

Plants may also require other elements, including sodium, silicon, vanadium, chromium, selenium, cobalt, fluorine, and arsenic. Most of these elements are present in soil, but it is not known yet if they are necessary for plant growth.

Fertilizers: A Mixed Bag

Farmers and gardeners often buy *complete fertilizers*, which—despite the name—usually contain only the three main nutrients. The three numbers—for example, 18-18-21—on fertilizer bags (Figure 19.6) indicate the proportions of nitrogen, phosphorus, and potassium (NPK). The first number represents the percentage of nitrogen (N); the second, the percentage of phosphorus (calculated as P_2O_5); and the third, the percentage of potassium (calculated as K_2O). So, 18-18-21 means that a fertilizer contains the equivalent of 18% N, 18% P_2O_5, and 21% K_2O; the rest is inert material. The use of fertilizers by farmers has significantly increased crop yields (Figure 19.7).

Organic Fertilizers

Most of the fertilizers we have discussed so far are "inorganic" in the chemical sense, because they are not carbon-based compounds. They are simple molecules, such as the NH_3 of anhydrous ammonia, or ionic compounds such as NH_4NO_3, $Ca(H_2PO_4)_2$, and KCl.

Recall from Chapter 9 that the word *organic* has several meanings. Urea is organic in the chemical sense: It is a carbon-based compound with the structural formula NH_2CONH_2. People engaged in organic farming and gardening use the word *organic* not to refer to carbon-containing compounds but to imply the use of natural plant and animal products—such as cottonseed meal, blood meal, fish emulsion or fish meal, manure, compost, and sewage sludge—as fertilizers. Rock phosphate is

▲ **Figure 19.6** The numbers on this container of fertilizer indicate that the fertilizer is 18% N, 18% P_2O_5, and 21% K_2O. Actually, there is no K_2O or P_2O_5 in fertilizer. Those formulas are used merely as a basis for calculation. The actual form of potassium in fertilizers is nearly always KCl, although any potassium salt—such as KNO_3—can furnish the needed K^+ ions. Phosphorus is supplied as one of several salts with soluble phosphates.

▲ **Figure 19.7** In 1912, American farmers produced an average of 26 bushels of corn per acre. In 2017, the yield per acre was 177 bushels. This seven-fold increase is mainly due to the increased use of fertilizers.

WHY IT MATTERS

Some soil additives are used for purposes other than fertilization. Calcium hydroxide [Ca(OH)$_2$, slaked lime] brings the acidity of soil to a better range for most plants. Vermiculite (above) is made from mica, in a process similar to that used in making puffed rice cereal. It lightens the soil and increases its ability to hold moisture.

sold as an "organic" fertilizer even though it is clearly inorganic in the chemical sense. Organic fertilizers generally must be broken down by soil organisms to release the nutrients that the plants use. Whether from cow manure or a chemical factory, nitrogen is taken up by plants as NO_3^- and NH_4^+ ions. Your garden plants don't care whether the nitrogen they need came from a compost pile or a chemical factory. Nutrient release from organic fertilizers generally takes some time. Usually, slow release over a long growing period is desirable, but organic materials may not release sufficient nutrients when needed by the plants.

Organic materials also have the advantage that they provide *humus*, a dark-colored substance consisting of partially decayed plant or animal matter. Humus improves the texture of soils, thus improving water retention, reducing the loss of nutrients through leaching, and helping to resist erosion. It also provides a better environment for beneficial soil organisms such as earthworms.

If sold as fertilizers, organic products have an NPK ratio on the package label, just as synthetic fertilizers do. However, some organic materials, such as composted manure and sewage sludge, are sold as soil "conditioners." These products have no claim as to nutrient content on the package, although they do furnish some nutrients.

Fresh human manure should not be used to fertilize fruits and vegetables, to avoid the possibility of transmitting human pathogens, such as *E. coli*, a common fecal bacterium. The variant *E. coli* O157:H7 causes a food-borne illness that has increasingly been linked to contaminated vegetables. In Germany in 2011, bean sprouts contaminated with *E. coli* killed 22 people and infected over 2400 more. The illness can cause severe problems in susceptible people, particularly children and the elderly. Pig, dog, and cat feces should also not be used as fertilizers, as they can carry internal parasitic worms from those animals to humans.

Organic fertilizers are often inexpensive or even free, such as animal manure. In contrast, manufactured nitrogen fertilizers are more expensive, and are usually made using natural gas and require a high energy input. Intelligent use of organic fertilizers can save money and be environmentally friendly. Unwise use can be expensive or even disastrous.

 EXAMPLE 19.1 Fertilizer Production Reactions

Write an equation for a reaction showing the formation of ammonium sulfate (Table 19.1) from ammonia. What other compound is needed?

Solution

We have seen that ammonium nitrate is made by reacting ammonia with nitric acid. By analogy, ammonium sulfate should therefore be formed by reacting ammonia with sulfuric acid. We write the equation for the reaction as

$$NH_3 + H_2SO_4 \longrightarrow (NH_4)_2SO_4 \text{ (not balanced)}$$

Balancing the equation requires the coefficient 2 for NH_3.

$$2\,NH_3 + H_2SO_4 \longrightarrow (NH_4)_2SO_4$$

> **Exercise 19.1A**
Write an equation to show the formation of ammonium monohydrogen phosphate (Table 19.1) from ammonia. What other compound is needed?

> **Exercise 19.1B**
The micronutrient zinc is often applied in the form of zinc sulfate. Write the equation for the formation of zinc sulfate from zinc oxide. What other compound is needed?

SELF-ASSESSMENT Questions

1. The three primary plant nutrients are nitrogen,
   a. phosphorus, and kaolin
   b. potassium, and plutonium
   c. potash, and phosphorus
   d. potassium, and phosphorus

2. Nitrogen is taken up by plants in the form of
   a. N_2 **b.** NO_2 **c.** NO_2^- **d.** NO_3^-

3. The third number in 24-5-10 on a fertilizer label represents 10%
   a. active ingredients   **b.** inert ingredients
   c. K_2O **d.** P_2O_5

4. Which of the following compounds furnishes two primary plant nutrients?
   a. NH_4NO_3 **b.** $(NH_4)_2HPO_4$
   c. $(NH_4)_2SO_4$ **d.** NH_2CONH_2

5. To make phosphates more available to plants, phosphate rock is treated with
   a. CO_2 **b.** H_2O **c.** H_2SO_4 **d.** NH_3

6. Potassium uptake in plants leaves the soil
   a. more acidic
   b. more basic
   c. neutral
   d. less contaminated with toxins

7. The secondary plant nutrients are
   a. C, H, and O   **b.** Ca, Mg, and S
   c. N, P, and K   **d.** Zn, Cu, and Fe

8. The plant nutrient essential to the function of chlorophyll is
   a. Ca^{2+} **b.** Co^{2+} **c.** Fe^{2+} **d.** Mg^{2+}

9. What is the percent nitrogen in a bag of 8-0-24 fertilizer?
   a. 8%   **b.** 0%   **c.** 24%   **d.** 32%

10. The main difference between organic fertilizers and synthetic fertilizers is that organic fertilizers
    a. are more soluble   **b.** are risk-free
    c. provide humus   **d.** smell better

11. What substance is added to make soil less acidic AND provide calcium for plants?
    a. 20-10-5 fertilizer   **b.** 5-10-5 fertilizer
    c. lime   **d.** vermiculite

12. Iron is added to some fertilizers to
    a. accelerate growth   **b.** adjust the soil pH
    c. produce larger flowers   **d.** prevent yellowing of leaves

13. About how many elements are known to be important to a plant's growth and survival?
    a. 3   **b.** 6   **c.** 14   **d.** 26

Answers: 1, d; 2, d; 3, c; 4, b; 5, c; 6, a; 7, b; 8, d; 9, a; 10, c; 11, c; 12, d; 13, c

19.2 The War against Pests

Learning Objectives • Name and describe the action of the main kinds of pesticides. • List several biological pest controls, and explain how they work.

Insects are critical to life on Earth. Honeybees, butterflies, and other insects pollinate flowering plants, including trees that provide shade, oxygen production, fruits, and wood. Thousands of species of birds and many amphibians are insectivores. Insects are extremely important for the survival of nearly every ecosystem on our planet, and they are critical for fruit and flower production and maintenance of many animal species.

However, people have always been plagued by insect pests. Three of the ten plagues of Egypt (described in the book of *Exodus*) were insect plagues—lice, flies, and locusts. The decline of Roman civilization has been attributed in part to malaria, a disease that is carried by mosquitoes and that destroys vigor and vitality when it does not kill. Bubonic plague, carried by fleas from rats to humans, swept through the Western world repeatedly during the Middle Ages. One such plague during the 1660s is estimated to have killed 25 million people, 25% of the population of Europe at that time. The first attempt (by the French during the 1880s) to dig a canal across Panama was defeated by outbreaks of yellow fever and malaria. During the Dust Bowl era in the central United States in the 1930s, swarms of locusts plagued Kansas and Oklahoma farmers who were already suffering from drought, massive black clouds of lost topsoil and dust, and nearly complete crop failure, resulting in economic disaster.

Only a few species—of the millions of insect species that exist—are harmful. Yet we have to exercise some control of insects that destroy specific crops or cause human diseases. The use of modern chemical **pesticides** (substances that kill organisms that we consider pests) may be all that stands between us and some of these plagues. Note that the term *pesticides* includes insecticides specifically targeted to insects and rodenticides specifically targeted to rats, mice, and other rodents. Pesticides prevent the loss of a major portion of our food supply that would otherwise

WHY IT MATTERS

Locust plagues have devastated crops throughout history. Desert locusts (*Schistocerca gregaria*) are usually solitary insects, but under certain environmental conditions, they change to a swarm-forming "gregarious" phase. A neurochemical mechanism underlies this change, which also includes a change in body color from green to yellow. The key is the neurotransmitter serotonin, which in humans modulates anger, body temperature, mood, sleep, appetite, and metabolism. In locusts, serotonin is synthesized in response to increased numbers and increased social contacts, and perhaps other factors.

be eaten by insects and other pests. Pesticides are also used extensively in urban areas to control annoying insects such as backyard mosquitoes, box elder bugs, or ants. Mosquito and insect repellants do not kill those insects; they only keep them away.

Up until the last few decades, people tried to control insect pests by draining swamps, pouring oil on ponds to kill mosquito larvae, and applying arsenic compounds to areas where rodents were a problem. Lead arsenate $[Pb_3(AsO_4)_2]$ is a particularly effective poison because of both the lead and the arsenic in it. Yet it was banned in the United States about 25 years ago, because it can kill other wildlife and pets instead of the rats or other rodents that were targeted by its use.

A few pesticides, such as pyrethrum (used in mosquito control) and nicotine sulfate (Black Leaf 40), have been obtained from plant matter. To protect themselves from being eaten, some plants produce their own pesticides, which often make up 5% to 10% of their dry weight. Nicotine protects tobacco plants. Pyrethrins are also plant-produced pesticides.

The United States produces about 500 million kg of pesticides annually. About a quarter of this total is used in houses, yards, parks, and golf courses. More than 1000 active ingredients are found in over 16,000 pesticide products marketed in the United States. Most poisons are indiscriminate. They kill all insects, not just those we consider pests. Many of those poisons are also toxic to humans and other animals. Some say we should call such poisons *biocides* (because they kill living things) rather than **insecticides** (substances that kill insects).

Table 19.3 lists the acute toxicities (from short-term exposures) of some insecticides. Effects of long-term exposure are more difficult to measure. Toxic substances in general are discussed in Chapter 21. Table 19.3 lists the amount in milligrams of specific pesticides that will kill 50% of a test population of rats when ingested orally. Ingestion by humans can occur if pesticides are accidentally left on the hands or skin and are not effectively washed off. Then any home gardener, farmer, or lawn-maintenance worker could ingest that pesticide when handling food or a beverage. Accidents have occurred when children or pets inadvertently found a container of pesticide in a garage or garden shed and ate or drank it. Pesticides such as lindane, found in some head lice shampoos, are not meant to be ingested, but accidental swallowing can occur.

TABLE 19.3 **Approximate Acute Toxicity of Insecticidal Preparations Administered Orally to Rats**

Pesticide	LD$_{50}$[a]	Category of Acute Toxicity	Selected Uses
Pyrethrins[b]	200–2600	Slightly toxic	Food crops; pest control (flea, tick, fly)
Malathion	1000 (1375)	Slightly toxic	Crops (alfalfa, lettuce)
Lead arsenate	825	Slightly toxic	Banned in 1988
Diazinon	285 (250)	Moderately toxic	Crops (lettuce, almonds, broccoli)
Carbaryl	250	Moderately toxic	Crops (apples, pecans, grapes, citrus fruits)
Nicotine[c]	230	Moderately toxic	Restricted to use by certified applicators
DDT[d]	118 (113)	Moderately toxic	Use banned in United States; malaria-carrying mosquito control
Lindane	91 (88)	Moderately toxic	Crops (cotton); head lice shampoos
Methyl parathion	14 (24)	Highly toxic	Crops (cotton, rice, fruit trees)
Parathion	3.6 (13)	Highly toxic	Crops (alfalfa, barley, corn, cotton)
Carbofuran[e]	2	Highly toxic	Field crops; some fruits and vegetables
Aldicarb	1	Highly toxic	Crops (cotton)

[a]Dose in milligrams per kilogram of body weight that will kill 50% of a test population. Values in parentheses are for male rats. (LD$_{50}$ is discussed in more detail in Chapter 21.) [b]Active ingredients of pyrethrum; LD$_{50}$ depends on variability of concentration of active components in the formulation. [c]Oral in mice. Nicotine is much more toxic by injection. [d]Estimated LD50 for humans is 500 mg/kg. [e]In mice.

Sources: O'Neill, Maryadele J., Patricia E. Heckelman, Cherie B. Koch, Kristin J. Roman, and Catherine M. Kenny, eds. 2006. *The Merck Index*, 14th ed. (Whitehouse Station, NJ: Merck & Co.; and the Extension Toxicology Network.

DDT: The Dream Insecticide—or Nightmare?

Shortly before World War II, Swiss scientist Paul Müller (1899–1965) found that DDT (dichlorodiphenyltrichloroethane; see insert on Figure 19.8), a chlorinated hydrocarbon with numerous chlorine atoms in its structure, is a potent insecticide. DDT was soon used effectively against grapevine pests and against a particularly severe potato beetle infestation. DDT is easily synthesized from cheap, readily available chemicals. Like other chlorinated hydrocarbons, it is almost insoluble in water.

When World War II began, supplies of pyrethrum, a major insecticide of the time, were cut off by the Japanese occupation of Southeast Asia and the Dutch East Indies (now Indonesia). Lead, arsenic, and copper that were used in insecticides were diverted for armaments and other military purposes. The Allies, desperately needing an insecticide to protect soldiers from disease-bearing lice, ticks, and mosquitoes, obtained a small quantity of DDT and quickly tested it. Combined with talcum powder, DDT was an effective delousing powder. Clothing was impregnated with DDT, and it seemed to have no harmful effects even on those exposed to large doses. Allied soldiers were nearly free of lice, but German troops were heavily infested and many were sick with typhus, a disease carried by lice. (In earlier wars, more soldiers probably died from typhus than from bullets.)

A cheap insecticide that is effective against a variety of insect pests, DDT came into widespread use after the war, mainly applied by "crop-dusters," small planes spraying large areas. Although invaluable to farmers in the production of food, chlorinated hydrocarbons won their most dramatic victories in the field of public health. According to the World Health Organization (WHO), approximately 25 million lives have been saved and hundreds of millions of illnesses prevented by the use of DDT and related pesticides, particularly in tropical areas such as Central and South America and in Africa.

DDT seemed to be a dream come true. It would at last free the world from insect-borne diseases. It would protect crops from the ravages of insects and thus increase food production. In recognition of his discovery, Müller was awarded the Nobel Prize in Physiology and Medicine in 1948.

Even before Müller received his prize, however, there were warning signs that all was not well. Houseflies resistant to DDT were reported as early as 1946, and DDT's toxicity to fish was documented by 1947. Such early evidence was largely ignored, and it was also assumed that the toxicity would disappear soon after the chemical was discharged into the environment. By 1962, the year Rachel Carson's book *Silent Spring* (Section 1.1) appeared, U.S. production of DDT had reached 76 million kg/year.

Chlorinated hydrocarbons generally are fairly stable, which was a major advantage of DDT. Sprayed on a crop, DDT stayed there and killed insects for weeks or even months. This *pesticide persistence* also turned out to be a major disadvantage: The substance does not break down readily in the environment, and persists at least a decade after application.

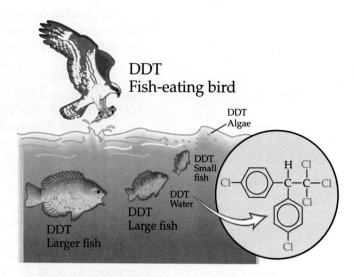

▲ **Figure 19.8** Biomagnification: concentration of DDT going up a food chain. Starting with algae that incorporate DDT in ponds from contaminated groundwater, and up through the food chain to ospreys, fish-eating birds, the concentrations of pesticide increase. The larger the fish, the more DDT in the fatty membranes of cells. The wedge-shaped DDT molecules open channels in the membranes, causing them to leak.

Biological Magnification: Concentration in Fatty Tissues

DDT's lack of reactivity to oxygen, water, and components of the soil led to its buildup in the environment, where it threatened fish, birds, and other wildlife. Although not very toxic to humans and other warm-blooded creatures, DDT is highly toxic to cold-blooded organisms, which include insects, of course, and fish. Chlorinated hydrocarbons are fat-soluble, so chlorinated hydrocarbons tend to become concentrated in animals higher on the food chain.

This *biological magnification* or *biomagnification* was graphically demonstrated in California in 1957. Clear Lake, about 100 miles north of San Francisco, was sprayed with DDT in an effort to control gnats. After spraying, the water contained only 0.02 ppm of DDT, but the microscopic plant and animal life contained 5 ppm—250 times as much. Fish feeding on these microorganisms contained up to 2000 ppm. Grebes, the diving birds that ate the fish, died by the hundreds (Figure 19.8). Even a few parts per billion of DDT interfere with the growth of plankton and the reproduction of crustaceans such as shrimp.

DDT interferes with calcium metabolism essential to the formation of healthy bones and teeth. Although the extent of harm to humans has never been conclusively demonstrated, the disruption of calcium metabolism in birds was disastrous for some species. The bald eagle, peregrine falcon, brown pelican, and other species became endangered because the shells of their eggs, composed mainly of calcium compounds, were thin and poorly formed, and therefore many eggs cracked and broke before the young birds could develop. Those bird species have fortunately made dramatic recoveries since DDT was banned in the United States and other industrialized countries in the 1970s.

▲ **Figure 19.9** A mosquito netting treated with permethrin is generally hung from the ceiling of a dwelling in Africa and tucked under and all around the sleeping mat overnight, when mosquitoes carrying the *Plasmodium* parasite that causes malaria are most active. This practice has been an important factor in the decrease in malaria cases in Asia and Africa, where the majority of malaria cases and deaths have occurred.

DDT is still used in some countries where malaria and typhus are major health problems. Malaria is caused by a *Plasmodium* parasite carried by mosquitoes, and infects people through mosquito bites. In 2006, malaria infected roughly 247 million people and killed an estimated 881,000. About 90% of the deaths occurred in sub-Saharan Africa, and 70% the victims were children under 5. DDT applied mainly on the walls of homes has helped decrease those fatalities. Unfortunately, exposure to DDT likely causes health problems in humans, and so even in Africa, it is now being used as a last resort. In the last few years, extensive use of netting in sleeping areas (often called "bednets") treated with the insecticide permethrin (see Figure 19.9), together with significantly improved early diagnosis and access to life-saving medications, has decreased the number of fatalities due to malaria to about 440,000 in 2017, cutting those fatalities by about half in a little over a decade.

Today, in developed countries, DDT is known mostly for its harmful environmental effects. In 2000, more than 120 countries signed a treaty to phase out persistent organic pollutants (POPs), a group of chemicals—many of them chlorinated hydrocarbons—that includes DDT. This treaty has the goal of reducing and ultimately eliminating the use of DDT, but individual countries may continue to use it for controlling malaria. As bad as DDT may sound, it may have saved the lives of more people than any other chemical substance.

Development of Insect Resistance

In the six decades or so in which pesticides have been widely used, hundreds of insect species have become resistant to one or more pesticides. Insects develop resistance to a pesticide by natural selection. A genetic mutation for resistance can spread quickly: For example, if a farmer uses a new pesticide on a field with 10.0 billion pest insects, perhaps 9.9 billion are killed. Resistant insects survive and pass the gene for resistance to the next generation. The resistant insect group becomes a larger percentage of the whole. Through repetition of this process over several generations, the resistant insects become the main form and the pesticide is no longer effective.

Resistance has resulted in increases in crop losses and in insect-transmitted diseases. Some insect species such as the cotton bollworm, Colorado potato beetle, malaria-transmitting Anopheles mosquitoes, and German cockroach exhibit resistance to all commercially available pest control agents. But what would crop losses have been without any insecticides? Probably much higher.

Organophosphorus Compounds

Bans and restrictions on chlorinated hydrocarbon insecticides have led to increased use of *organophosphorus* compounds (containing carbon and phosphorus) such as malathion, diazinon, and parathion (Figure 19.10). More than two dozen of these insecticides are available commercially. They have been extensively studied and evaluated for effectiveness against insects and for toxicity to people, laboratory animals, and farm animals.

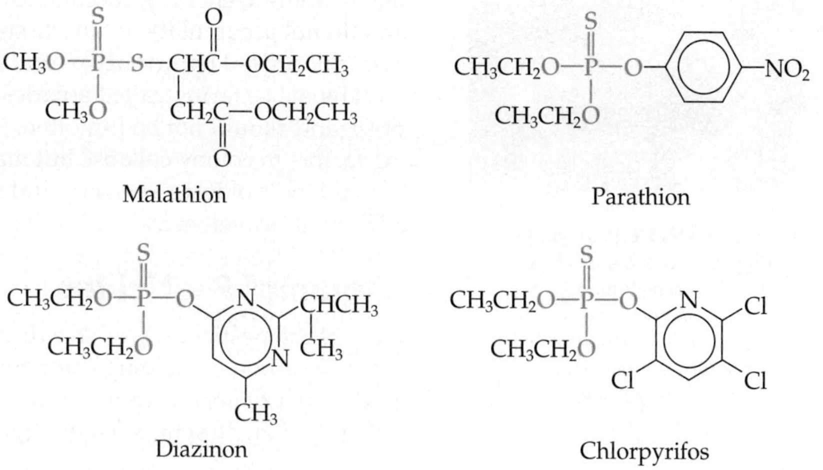

Organophosphorus compounds are nerve poisons (Chapter 21); they interfere with the conduction of nerve signals in insects. At high levels of exposure, they do the same in humans. Most are more toxic to mammals than are chlorinated hydrocarbons (Table 19.3). Malathion is a notable exception; it is less toxic than DDT. Like chlorinated hydrocarbons, organophosphorus pesticides concentrate in fatty tissues. However, the phosphorus compounds are less persistent in the environment. They break down in days or weeks, whereas chlorinated hydrocarbons often persist for years.

An organophosphate insecticide called chlorpyrifos has been used extensively around the world, applied by spraying a concentrated solution. The airborne insecticide can then be inhaled by farm workers or others living in rural areas. Medical studies showed significant neurodetrimental effects on fetuses, infants, and young children at higher levels of exposure, so chlorpyrifos was banned from use in the U.S. in October of 2015. However the EPA rescinded the ban on its use in November of 2016, a controversial decision.

Malathion

Parathion

Diazinon

Chlorpyrifos

▲ **Figure 19.10** Four organic phosphorus compounds used as insecticides.

Carbamates: Targeting the Pest

Another widely used family of insecticides is the carbamates. Examples are carbaryl (Sevin®), carbofuran (Furadan®), and aldicarb (Temik®) (Figure 19.11). Like the phosphorus compounds, carbamates are nerve poisons, but they act over a shorter span of time; they work by inactivating an enzyme called acetylcholinesterase. Most carbamates are *narrow-spectrum insecticides* directed specifically at one, or a few, insect pests. In contrast, chlorinated hydrocarbons and organophosphorus compounds kill many kinds of insects and are called *broad-spectrum insecticides*. Carbaryl has a low toxicity to mammals, but carbofuran and aldicarb have toxicities similar to that of parathion. Carbaryl is particularly effective against wasps but unfortunately is also quite toxic to honeybees. Honeybees are important pollinators, and their declining numbers could affect some crop yields by 30% or more in years to come.

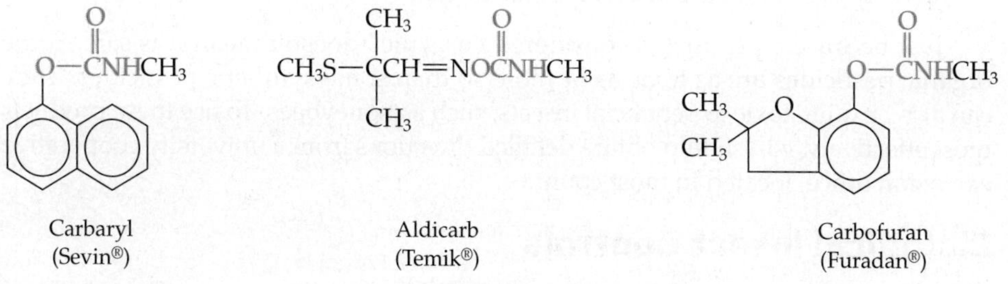

Carbaryl (Sevin®)

Aldicarb (Temik®)

Carbofuran (Furadan®)

◀ **Figure 19.11** Three carbamate insecticides.

Carbamates are characterized by the presence of the carbamate group:

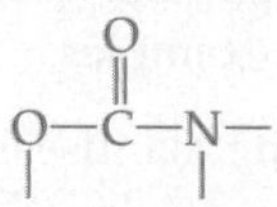

▲ **Figure 19.12** Honeybees residing in a beekeeper's hive produce humanity's oldest sweetener, honey, but are susceptible to infestation by mites, destruction by extreme weather, and exposure to pesticides.

3 Are pesticides causing honeybees to die off?

Bees not only fly significant distances from hives, but are often transported to new fields by beekeepers, and thus are exposed to a wide variety of agricultural pesticides. According to the U.S. Department of Agriculture, many studies have shown that pesticides are a contributing factor to *colony collapse disorder*, which first appeared in 2006, but viruses and mites have been found in higher concentrations in hives experiencing significant loss as well, indicating more than one contributing factor. One of the suspect pesticides is the one used to kill varroa mites in hives.

▲ Many pesticides are used in homes against insect infestations and on gardens and lawns to prevent plant damage and to kill weeds. But less toxic alternatives are available. For example, diatomaceous earth is effective against snails and slugs.

Many variations can be obtained by changing the groups attached to the O and the N atoms. Generally, carbamates break down fairly rapidly in the environment and do not accumulate in fatty tissue, which gives these insecticides an advantage over chlorinated hydrocarbons and organic phosphorus compounds.

Honeybee farms, called apiaries, are an important part of the agriculture enterprise, and should not be forgotten. (See Figure 19.12.) Pesticides may be contributing factors to colony collapse but many other factors must be considered. Honey is humankind's oldest sweetener, and is becoming more popular recently because it is a "natural" sweetener.

"Organic" Pesticides

Most of the pesticides we have discussed so far are organic in the chemical sense of being carbon-containing compounds. As mentioned earlier, organic farmers and gardeners use *organic* to imply the use of materials derived from natural sources rather than made synthetically. Organic pesticides include the following.

- Insecticidal soap acts by breaking down the insect's cuticle (outer covering). The cuticle is mainly chitin, a polysaccharide. The soap only acts while wet and in direct contact with the insect. It is not toxic to animals, and it leaves no harmful residue. This method is usually used by gardeners, who must spray the diluted soap directly on the insects, then "wash" the plants with water two to three hours after applying. Repetition may be needed four to seven days later. It is impractical for commercial farms.
- Pyrethrins are obtained from the perennial plant pyrethrum (*Chrysanthemum cinerariaefolium*). Pyrethrins are low in toxicity to mammals because they are quickly broken down into inactive forms by mammalian liver enzymes. Some are moderately toxic to mammals and quite toxic to fish and tadpoles. Insects, which have no liver, are much more susceptible to pyrethrins.
- Rotenone is a natural product obtained from the roots and stems of several tropical and subtropical plant species. It is slightly toxic to mammals, but extremely toxic to insects and fish.
- Boric acid (H_3BO_3), usually applied as a bait, dust, or powder, is mainly used for control of ants and cockroaches. It acts as a stomach poison and as a desiccant by abrading the wax from the insect's cuticle. Boric acid is moderately toxic to humans.
- Diatomaceous earth is the fossilized remains of diatoms, a type of algae whose shells consist largely of silica. The shells are ground to a fine powder with sharp-edged particles that cut the insects' cuticles and absorb oils, causing the insects to dehydrate and die. If ingested, the particles cause the same sort of damage to the digestive system. Diatomaceous earth is harmless to people.
- Ryania is a botanical insecticide obtained from *Ryania speciosa*, a tropical plant. The active principle is an alkaloid called *ryanodine*. Ryania is highly toxic to the caterpillars of fruit moths, codling moths, and corn earworms. It has a very low toxicity to mammals but is very toxic to fish.

Just because a product is considered "organic" doesn't mean it is safe. Some organic pesticides are as toxic as or more so than some synthetic pesticides. They can also be quite toxic to beneficial insects, such as honeybees. To use these products most effectively, you should obtain detailed directions from a university cooperative extension office, located in most counties.

Biological Insect Controls

The use of natural enemies is another method for controlling pests. These biological controls include predatory insects, mites, and mollusks; parasitic insects; and microbial controls, which are insect pathogens such as bacteria, viruses, fungi, and nematodes. The following are a few examples.

- Predatory insects such as praying mantises and ladybugs will destroy garden pests and are sold commercially.

- Parasitic insects such as *Trichogramma* wasps are used to control moth larvae such as tomato hornworm, corn earworm, cabbage looper, codling moth, cutworm, and armyworm.
- *Bacillus thuringiensis (Bt)*, sold under trade names such as DiPel®, is widely used by home gardeners against cabbage loopers, hornworms, and other moth caterpillars. Bt is also used against Colorado potato beetles, mosquito larvae, black flies, European corn borers, grape leaf rollers, and gypsy moths.
- Naturally occurring nuclear polyhedrosis virus (NPV) is present at low levels in many insect populations. These viruses can be grown in culture and applied to crops. NPV is highly selective, affecting only *Heliothis* caterpillars. Natural viral pesticides appear to be harmless to humans, wildlife, and beneficial insects. They are completely biodegradable but are generally quite expensive to produce.

Human Risks

The U.S. Environmental Protection Agency estimates that physicians diagnose 10,000–20,000 pesticide poisonings each year among agricultural workers. Many agricultural poisoning cases are never reported, and many other incidents affect people who have used pesticides in their home, lawn, or garden. The WHO estimates that pesticide poisonings kill some 300,000 agricultural workers yearly worldwide and adversely affect the health of 2 million to 5 million more. Many homeowners routinely hire lawncare companies to spray weed-killers and fertilizers in order to create the perfectly green lawn, and use insect sprays to kill mosquitoes and other flying insects in order to enjoy a pleasant backyard get-together. But there are potential adverse effects from inhaling airborne sprays; even homeowners would be wise to read labels carefully before use, and consider avoiding such products as much as possible.

Genetically Engineered Insect Resistance

A highly successful biological approach is the breeding of insect- and fungus-resistant plants. This field of biotechnology, sometimes called genetic engineering, is a very active field of current research worldwide. Researchers in Thailand, China, and other parts of Asia are working on rice; those in the Western world are focusing on corn, soybeans, and wheat. Yields of corn (maize), wheat, rice, and other grains have already been increased substantially in this manner. The gene for the toxin produced by Bt (*Bacillus thuringiensis*) has been inserted into cotton, corn, potato, and other plants. Genetically modified (GM) cotton plants are protected from damage by cotton bollworms. Corn has been genetically modified to resist corn root worms and corn borers.

Statistics show that the use of insecticides and pesticides has decreased significantly due mainly to the introduction of genetically modified plants. In the United States, insecticide use fell from 25% of corn crops treated in 2005 to 9% of cornfields treated in 2010. Unfortunately, by 2013, corn root worms increased again, so pesticide use for corn is on the upswing.

No scientific development is without controversy. Some people fear that the introduced genes will spread to wild relatives of the cultivated species, leading to the development of "super weeds." People also fear that proteins produced in GM plants will cause allergic reactions in susceptible individuals. For example, introduction of a gene for a peanut protein into another food plant could cause a dangerous reaction in people allergic to peanuts. Plant breeders produced a potato that is insect-resistant, but it had to be taken off the market because it is also toxic to people. Public fears about foods from genetically modified plants have slowed developments, particularly in Europe. A main argument against GM plants is that more herbicides need to be applied. The question remains: Is the fact that more herbicides are now required than even ten years ago the result of GM technology, or is it the result of weeds developing a resistance to herbicides that have been used for some time, just as insects develop a resistance to insecticides over time?

▲ **Figure 19.13** A male gypsy moth uses its large antennae to detect pheromones from a female.

Sterile Insect Technique

A *sterile insect technique (SIT)* is a method of insect control that involves rearing large numbers of males, sterilizing them with radiation or chemicals or by cross-breeding, and then releasing them in areas of infestation. These sterile males, which far outnumber the local fertile males, mate with wild females. If a female insect mates with a sterile male, no offspring are produced, thus reducing the reproductive potential of the insect population and controlling its growth.

Successful SIT programs have been conducted against screwworm, a pest that seriously affects cattle in the southern United States, Mexico, and Central America, as well as in Libya. SIT programs have also been employed against the Mediterranean fruit fly (medfly) in Latin America and against the codling moth in Canada. The great expense and limited applicability of SIT probably mean that it will not become a major method of insect control.

Pheromones: The Sex Trap

Substances called *pheromones* are increasingly important in insect control. A **pheromone** is a chemical that is secreted externally by an insect to mark a trail, send an alarm, or attract a mate. Insect **sex attractants** are usually secreted by females to attract males (Figure 19.13).

Chemical research has identified pheromones of hundreds of insect species. Most are blends of two or more chemicals that must be present in exactly the right proportions to be biologically active. Chemists can synthesize these compounds, and workers can use them to lure male insects into traps, allowing a determination of which pests are present and the level of infestation. The workers can then undertake measures, including the use of conventional pesticides, to minimize damage to the crops.

If the attractant is exceptionally powerful and the insect population level is very low, a pheromone trap technique called "attract and kill" may achieve sufficient control. Otherwise, the attractant can be used in quantities sufficient to confuse and disorient males, who detect a female in every direction but can't find one to mate with. Mating disruption has been used successfully in controlling some insect pests. Many grape growers in Germany and Switzerland use this technique, allowing them to produce wine without using conventional insecticides.

Some sex attractants have relatively simple structures. The sex attractant for the codling moth, which infests apples, is an alcohol with a straight chain of 12 carbon atoms and two double bonds.

However, most sex attractants have complicated structures, and all are secreted in extremely tiny amounts. This can make research on these attractants difficult and tedious. For example, a team of U.S. Department of Agriculture researchers had to use the tips of the abdomens of 87,000 female gypsy moths to isolate a minute amount of a powerful sex attractant.

Yet pheromones are effective at extremely low levels. A male silkworm moth can detect as few as 40 molecules per second. If a female silkworm moth releases as little as 0.01 mg, she can attract every male within 1 km. Gypsy moth larvae have defoliated great areas of forest, mainly in the northeastern United States, and have now spread over much of the country. The gypsy moth pheromone has been used mainly in traps to monitor insect populations.

Pheromones are usually too expensive to play a huge role in insect control, though the use of recombinant DNA methods (Section 16.11) to synthesize them holds promise for future work. For now, the cost is high, and research is painstaking and time-consuming. Workers must be careful not to get the attractants on their clothes. Who wants to be attacked on a warm summer night by a million sex-crazed gypsy moths?

Juvenile Hormones

Some insects can be controlled by the use of *juvenile hormones*. Hormones are the chemical messengers that control many life functions in plants and animals, and minute quantities produce profound physiological changes. In the insect world, a **juvenile hormone** controls the rate of development of the young. Normally, production of the hormone is shut off at the appropriate time to allow proper maturation to the adult stage.

Chemists have been able to isolate insect juvenile hormones and determine their structures. With knowledge of the structure, they can synthesize a hormone or one of its analogs. The application of juvenile hormones to ponds where mosquitoes breed keeps the mosquitoes in the harmless pre-adult stage. Because only adult insects can reproduce, juvenile hormones appear to be a nearly perfect method of mosquito control.

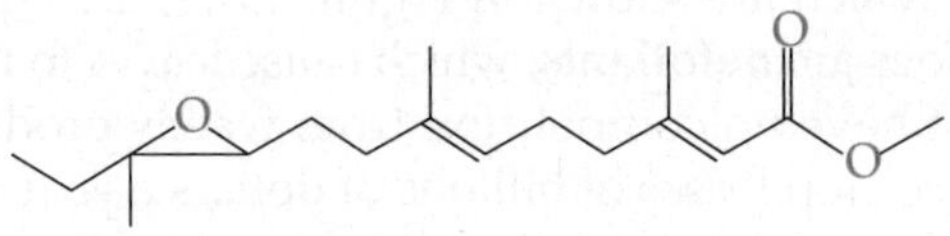

A natural juvenile hormone

Methoprene, a juvenile hormone analog, is approved by the EPA for use against mosquitoes and fleas.

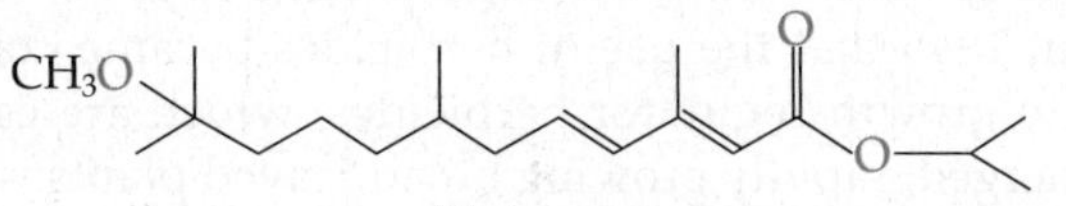

Methoprene

The synthesis of juvenile hormones is difficult and expensive, and they can only be used against insects that are pests at the adult stage. Little would be gained by keeping a moth or a butterfly in the caterpillar stage for a longer period of time. Caterpillars have voracious appetites and do a lot of damage to crops.

SELF-ASSESSMENT Questions

1. A pesticide is a product that kills
   a. insects and weeds only
   b. insects or other animals that are considered to be pests
   c. ants or mosquitoes only
   d. ticks and fleas but not mice

2. The use of genetically modified crops has led the use of pesticides on farms to
   a. decrease on many crops
   b. go up dramatically
   c. go up on some crops
   d. remain fundamentally unchanged

3. If pesticides get into a lake, which organisms will show the highest level in their fat tissues?
   a. fish
   b. microscopic animals
   c. microscopic plants
   d. great blue heron (fish eater)

4. Spraying of pesticides on crops
   a. affects only the sprayed crops
   b. has no effect on the food chain
   c. is necessary if farmers are to grow crops
   d. may result in pesticides ending up in rivers and lakes

5. Compared with DDT, organic phosphorus compounds are generally
   a. less persistent and less toxic
   b. less persistent and more toxic
   c. more persistent and less toxic
   d. more persistent and more toxic

6. "Organic" pesticides include all the following except
   a. carbamates   b. pyrethrins
   c. rotenone   d. insecticidal soaps

7. The sterile insect technique method of insect control works best when
   a. there is a large target insect population
   b. the females mate repeatedly
   c. the males of the pest species can be reared en masse
   d. the males mate with related species

8. A substance produced by an organism to influence the behavior of another member of the same species is a(n)
   a. agonist   b. inducer   c. marker   d. pheromone

9. The use of ladybugs and praying mantises to control insect pests is an example of
   a. biocidal control   b. biological control
   c. abiotic control   d. exploitation of insect pests

Answers: 1. b; 2. a; 3. d; 4. d; 5. b; 6. a; 7. c; 8. d; 9. b

19.3 Herbicides and Defoliants

Learning Objective • Name and describe the action of the major herbicides and defoliants.

▲ Dandelions, introduced into North America from Europe because of their attractive flowers and their uses as a salad vegetable and a folk medicine, are now generally considered a lawn weed. About half of all homeowners treat their lawns with herbicides such as 2,4-D (Weed-B-Gon) to get rid of dandelions, crabgrass, common plantain, and other lawn weeds. More than 10 million pounds of 2,4-D are used on lawns in the United States each year.

Ralph Waldo Emerson once wrote that "a weed is a plant whose virtues have not yet been discovered." Unfortunately, most of the plants defined as weeds in the nineteenth century are still defined as weeds today, and they appear to have few virtues. Colloquially, a weed is a rapidly growing plant that is hard to get rid of and that tends to choke out plants that we seek to grow. A dictionary might define a weed as a wild plant growing where it is not wanted and in competition with cultivated plants.

The United States produces about 225 million kg of herbicides annually. **Herbicides**, four of which are shown in Figure 19.14, are chemicals used to kill weeds. Some herbicides are **defoliants**, which cause leaves to fall off plants.

Crop plants that have no competition from weeds produce more abundant harvests. Weeds cause crop losses of billions of dollars a year in the United States. Removing weeds by hand or hoe is tedious, backbreaking work, so chemical herbicides are used to kill unwanted plants in large-scale agricultural operations. Many small-scale gardeners prefer to dig up or pull out weeds rather than use chemical herbicides.

Early herbicides included solutions of copper salts, sulfuric acid, and sodium chlorate ($NaClO_3$). It wasn't until the introduction of 2,4-D (2,4-dichlorophenoxyacetic acid) in 1945 that the use of herbicides became common. 2,4-D and its derivatives are growth-regulator herbicides, which are especially effective against newly emerged, rapidly growing broad-leaved plants, since they work by *defoliation*, which causes leaves to fall off plants. One derivative of 2,4-D is 2,4,5-T (2,4,5-trichlorophenoxyacetic acid), which is especially effective against woody plants and works by defoliation.

Combined in a formulation called **Agent Orange**, 2,4-D and 2,4,5-T were used extensively during the Vietnam War to remove enemy cover and to destroy crops that maintained enemy armies. Agent Orange caused vast ecological damage, and is suspected of causing birth defects in children born to both American soldiers and Vietnamese exposed to the herbicides. Laboratory studies show that 2,4-D and 2,4,5-T, when pure, do not cause abnormalities in fetuses of laboratory animals. Extensive birth defects are caused, however, by **dioxins**, contaminants of 2,4,5-T frequently found in the herbicide mixture. The most toxic dioxin is 2,3,7,8-tetrachlorodibenzo-*para*-dioxin, abbreviated 2,3,7,8-TCDD, or just TCDD. Continuing concern about dioxin contamination led the EPA to ban 2,4,5-T in 1985.

2,4-Dichlorophenoxyacetic acid (2,4-D)

2,4,5-Trichlorophenoxyacetic acid (2,4,5-T)

Atrazine

Glyphosate

▲ **Figure 19.14** Four common herbicides: 2,4-dichlorophenoxyacetic acid (2,4-D), 2,4,5-trichlorophenoxyacetic acid (2,4,5-T), atrazine, and glyphosate.

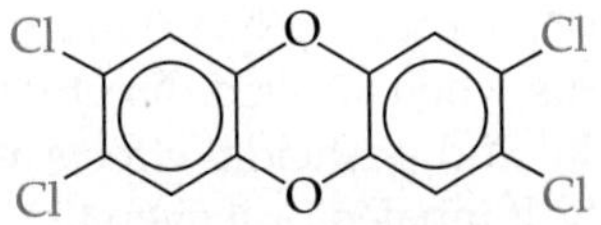

2,3,7,8-Tetrachlorodibenzo-*para*-dioxin
(a dioxin)

Almost all "selective" herbicides (that kill only certain types of plants, not every plant with which they come in contact) contain 2,4-D or a related compound abbreviated as MCPA (2-methyl, 4-chlorophenoxyacetic acid). Nonselective herbicides will kill all plants.

Atrazine and Glyphosate

The most widely used nonselective herbicides in the United States are atrazine and glyphosate. Atrazine binds to a protein in the chloroplasts of plant cells, shutting off the electron-transfer reactions of photosynthesis. Atrazine is often used on corn. Corn plants deactivate atrazine by removing the chlorine atom from the atrazine molecule. Weeds cannot deactivate the compound, and are killed. Some studies indicate that atrazine interferes with the hormone system, disrupting estrogen function. Atrazine has been linked to sexual abnormalities in frogs. Atrazine is the second most commonly used herbicide in the United States in agriculture, although its use has been banned in the European Union.

Glyphosate—sold under several trade names, including Roundup®, is the most commonly used herbicide in the United States. It is a derivative of the amino acid glycine and is a nonselective herbicide that kills all vegetation by inhibiting the function of a certain plant enzyme. Glyphosate can be used to treat the soil *before* a crop is planted, and will effectively kill weeds before they emerge through the soil. When used in this way, glyphosate is called a **pre-emergent herbicide**. Fortunately, it is metabolized by bacteria in the soil, so that other plants can be sown or transplanted into treated areas shortly after spraying the area with glyphosate. Glyphosate can also be applied to plants that have already emerged from the soil, but it will kill almost every plant it contacts.

Genetically modified soybeans, corn, alfalfa, and other plants that are resistant to glyphosate have been developed. When glyphosate is applied to a field, the weeds are killed but the crop plants remain unharmed. However, widespread use of glyphosate has resulted in many glyphosate-resistant weeds.

Paraquat: A Preemergent Herbicide—and More

Paraquat, an ionic compound that is toxic to most plants but is rapidly broken down in the soil, is a notable example of a preemergent herbicide. Paraquat inhibits photosynthesis by accepting the electrons that otherwise would reduce carbon dioxide.

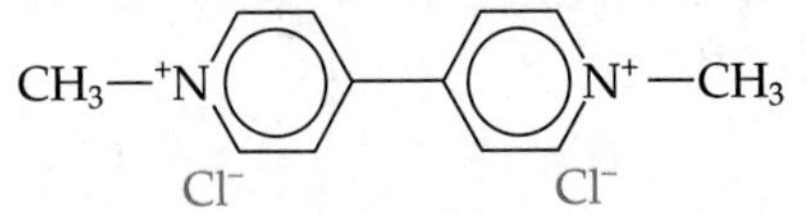

Paraquat

Paraquat can also be used to destroy the leaves of almost any plant. Sugarcane, an important crop used for the production of white sugar, brown sugar, molasses, and rum, has large leaves with sharp edges on top of the sugar-containing stems. The leaves are traditionally burned before the sugarcane is harvested. The use of paraquat, applied just before the sugarcane is harvested to quickly destroy the leaves, is becoming popular in sugar-producing countries. The leaves are destroyed, but not the stems, so harvesting is then easier.

4 Why does my "weed-killer" not kill my grass, too?

Most garden and lawn weeds are classified as "broad-leaf plants," including dandelions, henbit, and plantain, to name a few. Lawn grass is not classified as a broad-leaf plant, and so is unaffected by the application of a selective herbicide, such as Weed-B-Gon. Read the labels carefully. If glyphosate is an ingredient, it will kill your entire lawn if applied to all the grass, because glyphosate is a nonselective herbicide.

Paraquat is one of the most potent toxins used in agriculture. It not only has astonishingly rapid herbicidal effects when applied directly to plants, killing them within minutes, but it has insecticidal and pesticidal effects as well. It is also a dangerous herbicide for workers to apply. If inhaled or if even tiny amounts somehow enter the mouth, paraquat quickly affects the lungs, liver, and kidneys, and has the potential to kill. Farm workers exposed to crops that have had paraquat applied, either as a pre-emergent herbicide or postemergent, have a greater probability to develop Parkinson's disease or other health problems. Paraquat was banned in China in 2012.

Organic Herbicides

Weeds can be killed by methods other than synthetic herbicides. The manual labor of hoeing and other cultivation practices such as mulching have been used for millennia. These practices are readily applied to small areas such as lawns and gardens. There are also herbicides made from natural ingredients. The following are some examples.

- Oil of cloves (in which eugenol is the active ingredient) helps control young broadleaf weeds.
- Acetic acid, often combined with citric acid, helps kill young grasses. (Some preparations contain both oil of cloves and acetic acid, making them useful against a broad assortment of weeds.)
- Soap-based herbicides cause clogging of plants' *stomata*, the respiratory organs of the leaves.
- Gluten meal, a by-product of the wet-milling process for corn, acts as a preemergent herbicide by inhibiting the root formation of germinating plant seeds.

These types of organic herbicides do not work as rapidly or as well as synthetic herbicides. For instance, corn gluten meal often requires four years of repeated applications to achieve good weed control. And some organic herbicides are effective only at fairly high concentrations, making them quite expensive for widespread use.

Genetically Engineered Herbicide Resistance

Because weeds tend to choke out planted crops, herbicides often are applied to crops and lawns. However, some of those herbicides, as we have discussed, are so potent that they may negatively affect the very crops and plants we wish to grow. Biotechnology research has developed genetically modified organisms (GMOs) that are herbicide-tolerant. Herbicide-resistant corn and soybeans have been developed so far. The use of glyphosate now appears to be increasing, probably due to the natural process of plants evolving further herbicide resistance, much like the development of insect resistance in plants. Those who argue against GMO crops will cite the increase in use of herbicides, yet often fail to recognize the general decrease in use of insecticides with such crops.

SELF-ASSESSMENT Questions

1. Agent Orange, a mixture of 2,4-D and 2,4,5-T used as a defoliant by the U.S. military during the Vietnam War, was harmful to humans mainly due to
   a. 2,4-D　　　　　　　**b.** 2,4,5-T
   c. an impurity in 2,4-D　**d.** dioxin impurity in 2,4,5-T

2. One of the most potent toxins used as an herbicide is
   a. paraquat　**b.** atrazine
   c. 2,4-D　　　**d.** eugenol

3. The herbicide atrazine acts by inhibiting
   a. amino-acid synthesis　**b.** cell division
   c. lipid synthesis　　　　**d.** photosynthesis

4. Glyphosate
   a. is a selective herbicide
   b. is quite long-lasting in the soil
   c. does not kill crop plants
   d. can be used as a preemergent herbicide

5. The most common compound found in selective herbicides is
   a. atrazine　　**b.** 2,4-D
   c. glyphosate　**d.** paraquat

Answers: 1, d; 2, a; 3, d; 4, d; 5, b

19.4 Sustainable Agriculture

Learning Objective ▪ Describe sustainable agriculture and organic farming.

Agriculture changed dramatically in the twentieth century. Mechanization and increased use of pesticides, synthetic fertilizers, and fossil fuel energy led to vastly increased production. With reduced labor demands, fewer and fewer farmers could produce more and more food and fiber.

These changes also brought problems that have led to calls for alternatives to conventional farming. Most prominent among these alternatives is **sustainable agriculture**, which is the ability of a farm to produce food and fiber indefinitely, without causing irreparable damage to the ecosystem. Sustainable agriculture has three main goals: a healthy environment, farm profitability, and social and economic equity. Much of sustainable agriculture is similar to the practices that organic farmers and gardeners have followed for generations.

Modern agriculture is energy-intensive. Nonrenewable petroleum energy is required for the production of fertilizers, pesticides, and farm machinery. Energy is also required to run the machinery needed to till, harvest, dry, and transport the crops; to process and package the food; and to preserve and prepare the food in consumers' households. A breakdown of the energy used in agriculture and food production is presented in Figure 19.15.

A 21-year Swiss study compared conventional plots that used synthetic fertilizers and synthetic pesticides with organic plots that were fertilized with manure and treated only occasionally with a copper fungicide. Crops of potatoes, winter wheat, grass clover, barley, and beets were grown under otherwise identical conditions. The organic method appeared to be more efficient, with yields only 20% less, even though the nutrient input had been reduced by 50%. The organic method was said to use 20% to 56% less energy than the conventional approach when the energy required for production of fertilizers and pesticides was taken into account.

Organic farming is carried out without the use of synthetic fertilizers or pesticides. As was mentioned earlier, organic farmers (and gardeners) use manure from farm animals for fertilizer, and they rotate other crops with legumes to restore nitrogen to the soil. They control insects by planting a variety of crops, alternating the use of fields. (A corn pest has a hard time surviving during the year that its home field is planted in soybeans.) Organic farming is also less energy-intensive. According to a study by the Center for the Biology of Natural Systems at Washington University, comparable conventional farms used 2.3 times as much energy as did organic farms. Production on organic farms was 10% lower, but costs were comparably lower. Organic farms require 12% more labor than conventional ones. Human labor is a renewable resource, though, whereas petroleum is not. Compared with conventional methods, organic farming uses less energy and leads to healthier soils.

In addition to organic practices, sustainable agriculture involves buying local products and using local services when possible, thus avoiding the cost of transportation while getting fresher food and strengthening the economy of the local community. Transportation's share of food costs soars as fuel prices rise. Half the cost of broccoli in New York City may well be freight costs.

Sustainable agriculture promotes independent farmers and ranchers producing good food and making a good living while protecting the environment. Conventional agriculture can result in severe soil erosion and is the source of considerable water pollution. No doubt we should practice organic farming to the limit of our ability to do so. But we should not delude ourselves. Abrupt banning of synthetic fertilizers and pesticides would likely lead to a drastic drop in food production.

As far as human energy is concerned, U.S. agriculture is enormously efficient. Each farm worker produces enough food for about 80 people. But this productivity is based on fossil fuels. About ten units of petroleum energy are required to produce one unit of food energy. In terms of production per hectare, modern

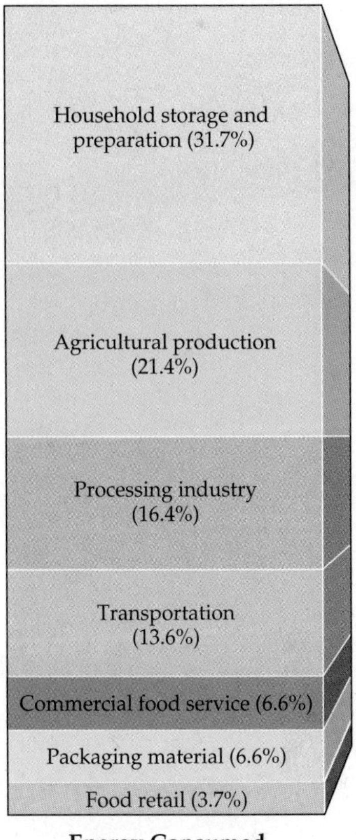

Household storage and preparation (31.7%)

Agricultural production (21.4%)

Processing industry (16.4%)

Transportation (13.6%)

Commercial food service (6.6%)

Packaging material (6.6%)

Food retail (3.7%)

Energy Consumed

▲ **Figure 19.15** Energy use in modern agriculture and food production. (Data from the Center for Sustainable Systems, School for Environment and Sustainability, University of Michigan.)

5 Are organic fruits and vegetables healthier for us?

Organically grown fruits and vegetables do not contain the pesticides and fertilizer residues found in foods grown by conventional methods, but they can contain natural toxins or contaminants such as *E. coli* from organic fertilizer. It is worth noting that most applied pesticides are concentrated on the surfaces of fruits or vegetables. Organic bananas, oranges, and grapefruit may be no better than the conventionally grown fruits because their peels are not eaten.

farming is marvelously efficient. But in terms of the energy used in relation to the energy produced, it is remarkably inefficient. It should be noted, however, that in energy-efficient early agricultural societies, nearly all human energy went into food production. In modern societies, only about 10% of human energy is devoted to producing food. The other 90% is used to provide the materials and services that are so much a part of our civilization. We should try to make our food production more energy-efficient, but it is unlikely that we will want to return to an outmoded way of life.

SELF-ASSESSMENT Questions

1. Farming that meets human needs without degrading the environment or depleting resources is called
   a. corporate farming
   b. industrial farming
   c. local farming
   d. sustainable agriculture

2. Organic farming differs from conventional agriculture in that organic farmers use no
   a. animal manure
   b. crop rotation
   c. legumes
   d. synthetic pesticides

3. Comparing modern agriculture with organic farming, which of the following is *not* true?
   a. Modern farming uses far less human energy and much more mechanical energy.
   b. Each modern farm worker provides food for about 80 people, many more than organic farmers do.
   c. Organic farmers use far more petroleum-based energy overall than do farmers who use commercially available fertilizers and herbicides.
   d. Organic farmers use fewer inorganic fertilizers and more carbon-based fertilizers.

Answers: 1, d; 2, d; 3, c

19.5 Looking to the Future: Feeding a Growing, Hungry World

Learning Objective • Assess the challenges of feeding an increasing human population.

In 1830, Thomas Robert Malthus, an English clergyman and political economist, made the prediction that the world's population would increase faster than the world's food supply. Unless the birthrate was controlled, he said, poverty and war would have to serve as restrictions on the increase in population. That was a startling statement!

Malthus's prediction was based on simple mathematics: Population grows geometrically, while the food supply increases arithmetically. In **arithmetic growth**, a constant amount is added during each growth period. As an example, consider a child who starts to save money in a piggy bank. The first week, the child puts in 25¢. Each week thereafter, she adds 25¢. The growth of her savings is *arithmetic*, and it increases by a constant amount (25¢) each week. At the end of five weeks she will have $1.25 saved.

In **geometric growth**, the increment increases in size for each growth period. Again, let's use a child's bank as an example. The first week she puts in 25¢. The second week she puts in 50¢ to double the amount deposited, for a total of 75¢ saved. After five weeks, she has $20.25 deposited. Before long, she will have to start robbing banks to keep up her geometrically growing deposits! Population growth rates obviously are not doubled weekly, but an overview of population growth shows an approximately geometric increase. (See Table 19.4.)

TABLE 19.4 World Population Milestones

World Population	Year	Time to Add 1Billion
1 billion people	1804	
2 billion	1927	123 years
3 billion	1960	33 years
4 billion	1974	14 years
5 billion	1987	13 years
6 billion	1999	12 years
7 billion	2011	12 years
8 billion	?	

For a world population growing geometrically, we can estimate the **doubling time** using the **rule of 72**. Simply divide 72 by the percentage of annual growth.

EXAMPLE 19.2 Population Doubling Time

Earth's population was 7.0 billion in 2011 and growing by 1.092% per year. If it continues to grow at this rate, how long will it (theoretically) have taken to double to 14 billion?

Solution

$$\frac{72}{1.092\%} = 66 \text{ years from 2011, or in 2077}$$

Be cautious. The world's population growth rate is not steady at 1.092% per year!

> **Exercise 19.2A**
Consult Table 19.4. What was the average population growth rate from 1927 to 1974, when the world population doubled from 2 billion to 4 billion?

> **Exercise 19.2B**
What was the average population growth rate from 1804 to 1927, when the world population doubled from 1 billion to 2 billion? Compare your answer Exercise 19.2A.

The calculations using the rule of 72 are only speculative. World population growth, individual country growth, and even small-region growth rates vary significantly. In 2011, for example, the U.S. growth rate was 0.963% per year. For many years, China enforced a rule of one child per family in urban areas, but as of January 1, 2016, two children per family are allowed. India's growth rate is such that it may within six or seven years surpass China as the most populous country on Earth. Some areas of Africa are growing at a rapid rate, which unfortunately often mirrors the poverty rate. It is simply more difficult to feed larger families with diminishing economic resources. So the rule of 72 is, at best, an approximation. Many factors will affect the growth rate. An epidemic or other catastrophe might also change the death rate.

Earth's population has nonetheless grown enormously since the time of Malthus, reaching 7.6 billion in December 2017. Though modern farming often produces surpluses in developed countries, famine has brought suffering and death to millions in war-ravaged areas around the world. As an example, in 2008 more than 2 million people faced death from starvation and disease in the Darfur region of Sudan. Since then, the southern part of the country has become independent as South Sudan, which has continued to suffer political turmoil and violence, and malnourishment still is widespread. Similar examples could be cited from Nigeria, Syria, Iraq, Afghanistan, and many other countries involved in wars.

Some developing nations have made great progress in food production, with many of them becoming self-sufficient. Despite surpluses in some parts of the world, an estimated 820 million people are seriously malnourished, including people in the United States. Most of the 820 million are in Asia or Africa (Figure 19.16). Half are children under 5 years old who will carry the physical and mental scars of this deprivation for the rest of their lives. Undernutrition puts children at risk of dying of common infections, and results in stunted growth. Scientific developments, such as those outlined in this chapter, have brought abundance to many, but millions still go hungry.

GREEN CHEMISTRY Safer Pesticides through Biomimicry and Green Chemistry

Amy S. Cannon, *Beyond Benign*

Principles 3–6, 9–10

Learning Objectives • Identify the benefits of pesticides to humans and the challenges in their use. • Give examples of how pesticides can be designed to reduce hazard to humans and the environment.

The practice of agriculture has been largely responsible for the growth of human civilization, allowing humans to settle in societies and raise food to sustain the community. Population growth brought the challenge of producing enough food, in a sustainable manner, to feed the world's populations. This challenge today is one of the toughest that humans have faced in history. The solution involves global issues such as clean water, population growth, land use, energy use, genetic modification, and pesticide and fertilizer use.

From a scientific perspective, the chemistry we use on the farm and in the garden can have a drastic impact on the output of our food production and the sustainability of agricultural practices. Pesticide use in agriculture has led to a drastic reduction in crop failure and increased crop yields, but pesticides often are toxic toward other than the target organisms. For example, the use of neonicotinoids, a class of neuro-active pesticides, has been found to contribute to honeybee colony collapse disorder and the decline of honeybee populations worldwide. The European Commission issued a ban on three neonicotinoids that went into effect in December 2013 to further protect honeybee populations. How do we address the issue of creating pesticides that target pest organisms and not beneficial ones, such as honeybees?

Green chemists are addressing the need for more highly targeted pesticides. Green chemistry principle 4 encourages the design of safer, more benign products and processes without compromising on efficacy or cost. This is perhaps the most important application of green chemistry in the production and use of pesticides. Generally, a safer pesticide means higher selectivity toward the pest organism and fewer effects on the surrounding environment and beneficial organisms.

One place to look for inspiration when making safer pesticides is nature. There are many examples of organisms in nature that naturally and selectively ward off pests. A recent example is that of a pesticide based on spider venom. Years ago, Glenn King, from the University of Queensland in Australia, identified a peptide from the venom of the Blue Mountains funnel-web spider. The venom of the spider was found to be toxic to insects but nontoxic to mammals. This selectivity inspired additional researchers to investigate this unique venom. A recent publication reports the development of a biopesticide that is inspired by the spider-venom peptide. The biopesticide is found to target pest organisms but has no adverse effect on honeybees.

In creating more targeted pesticides, scientists focus on structure–function relationships to find specific modes of action that are unique to the target pest. New research is tapping into naturally inspired, or biomimetic, methods for repelling or mitigating pests, such as the biopesticide inspired by spider venom. These pesticides represent a growing trend in the field of green chemistry: that of looking to nature for inspiration. Biomimicry, derived from the two terms *bios* (life) and *mimic* (to copy or emulate), is the imitation of the models, systems, and elements of nature for the purpose of solving complex human problems. By looking to nature, we can find inspiration for solving the challenges our society faces, such as pesticides and crop protection.

Through green chemistry we can make greener pesticides and also reduce the environmental impact of processes for making the pesticides. We can design less hazardous chemical synthesis (Principle 3), use catalytic reagents (Principle 9), improve the energy efficiency of processes (Principle 6), rely on safer solvents (Principle 5), and design the pesticides to be biodegradable (Principle 10). Biomimicry and the application of the green chemistry may be the key to providing food for our people and protecting our environment at the same time.

▲ Australian funnel web spider *Hadronyche versuta*.

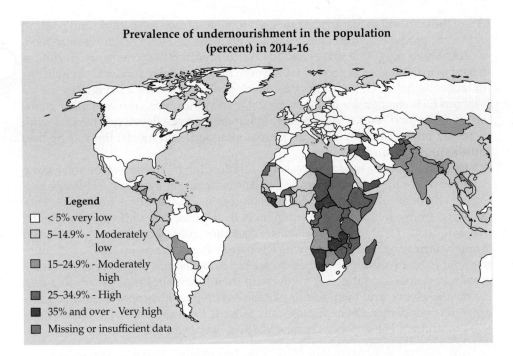

Prevalence of undernourishment in the population (percent) in 2014-16

Legend
- □ < 5% very low
- □ 5–14.9% - Moderately low
- ▨ 15–24.9% - Moderately high
- ▨ 25–34.9% - High
- ■ 35% and over - Very high
- ▨ Missing or insufficient data

▲ **Figure 19.16** In developed countries, science has by and large thwarted the Malthusian prediction of hunger, though many people still have insufficient food intake or vitamin-deficient diets. Starvation is still a fact of life for many people in developing countries, particularly in Asia and in Africa (map from the Food and Agriculture Organization of the United Nations, 2015).

Can We Feed a Hungry World?

The United Nations Population Division projects that Earth's population will reach 9.8 billion in 2050. It could be sooner. It could be later. Can we feed all those people?

An alliance between science and agriculture has increased food supplies beyond the imagination of people of a few generations ago, as a result of irrigation, synthetic fertilizers, pesticides, and improved genetic varieties of plants and animals. Food production can be further increased through genetic engineering. Scientists are designing plants that produce more food, that are resistant to disease and insect pests, and that grow well in hostile environments. They can design animals that grow larger and produce more meat or milk. Such efforts may someday provide us with food in unimaginable abundance. But the hungry are with us now, and 75 million more people come to dinner each year. One challenge is that many people oppose the genetic modification of food crops, which may hamper the use of this approach.

Even quadrupling current food production would meet our needs for only a few decades. Virtually all of the world's available arable land is now under cultivation, about 38.6% of ice-free land available on our planet. About 46.5% of land on Earth is desert, high mountains, tundra, or deep forest, and so not really amenable to expansion of agriculture. About 14.9% of land has been human-modified, as cities and towns, highways, mines, and logged areas. We lose farmland every day to housing, roads, erosion, encroaching deserts, and the increasing salt content of irrigated soils. Corn growers in the midwestern area of the United States are placing plants much closer together than they did just a decade ago, to save space. New methodologies for fish hatcheries and more efficient, space-saving chicken egg production facilities will help.

Regional climate swings, such as severe droughts in parts of Africa or monsoons in Asia, affect the ability of nations to grow food. Distribution of food to famine-ravaged areas is not a simple task, due to political and transportation challenges. It does appear that keeping food production ahead of the rate of population growth, particularly in developing countries, will be increasingly difficult.

An interesting aspect of the world's food crisis is the percent of "food" produced that is used for human food and the percent that is used to feed livestock or for other uses. In developing nations, the vast majority of food grown is used for human food, including rice and quinoa. In developed nations, for example in the U.S. Midwest,

6 How can there be a food crisis if I see plenty of corn growing in the fields?

Surprisingly, the vast majority of that corn is used to feed farm animals, including chickens and beef cattle—not to make cornflakes and corn meal for people. Much is used to make ethanol, most likely 10% of the fuel for your car. Just because you see grocery stores filled with a vast variety of foodstuffs, and see grains growing in fields does not mean that no one is hungry. Widespread poverty, poor transportation systems, poor soils, and political factors must be considered in discussing the worldwide food crisis.

83% of the corn and soybean crops are used for animal food (beef, pork, and poultry) and for the production of ethanol, a gasoline additive, and only 17% for human food.

It has been suggested that one solution to the food crisis is changing our choices in food to a diet higher in grains and lower in meat and dairy products. A challenge for that solution is that it appears that developed nations have a larger appetite for meat products compared with developing nations, where rice, corn, and other grains are the main food staples. As developing nations become developed nations, with increased prosperity may come an increased appetite for meats. But raising beef cattle requires pasture and corn.

Despite a growing movement toward sustainable agriculture, we are still greatly dependent on a high-energy form of agriculture that uses synthetic fertilizers, pesticides, and herbicides, and that depends on machinery that burns fossil fuels. Ultimately, the only solution to the problem is to stabilize population. Some predict that will happen within a few decades. Population control could be accomplished through decreasing the birthrate, which is happening in almost all developed countries and in many developing ones. However, some nations still have populations growing at explosive rates that far outstrip their food supplies. Catastrophic war, famine, pestilence, and the poisoning of the environment by the wastes of an ever-expanding population all could drastically increase the death rate, which we all hope will not happen. The ghost of Thomas Malthus haunts us yet.

SELF-ASSESSMENT Questions

1. Factors that will have to be considered in addressing the world's food crisis include all the following except
   a. the production levels of corn and soybeans, irrespective of their food and non-food use
   b. variations in population increases
   c. loss of farmland to urban sprawl
   d. efficiency of crop production

2. The rapid population growth that occurred in the late nineteenth and early twentieth centuries was due in large part to
   a. conservation of natural resources
   b. global peace
   c. use of scientific agricultural practices
   d. equitable distribution of wealth

3. Uganda's estimated population in 2013 was 35.9 million, and it is growing at a rate of 3.32% per year. If that growth rate continues, the population will have doubled in about
   a. 10 y
   b. 20 y
   c. 40 y
   d. 80 y

4. By 2025, what country is estimated to have the highest population?
   a. China
   b. India
   c. USA
   d. Nigeria

Answers: 1, a; 2, c; 3, b; 4, b

Summary

Section 19.1—The three **primary plant nutrients** are nitrogen, phosphorus, and potassium. Atmospheric nitrogen is fixed naturally by soil bacteria and by lightning. Artificial fixation of nitrogen produces ammonia and nitrates that can be added to the soil. Phosphorus can be made available by treating phosphate rock. Potassium, as potassium chloride, is usually mined. The **secondary plant nutrients** are calcium, magnesium, and sulfur. Eight other **micronutrients** are needed in small amounts. The three numbers on the packaging for commercial fertilizers indicate the percentages of N, P_2O_5, and K_2O in the fertilizer. Organic fertilizers include composted manure and sewage sludge.

Section 19.2—Pests can cause plagues and ruin food supplies. **Pesticides** are used to kill organisms that we consider harmful, including many insects. Most **insecticides** kill all insects they come in contact with, not just the harmful ones. The insecticide DDT saved millions of lives during and after World War II, yet its persistence in the environment caused problems. Widely banned, DDT continues to be used in some developing countries for malaria control. Organic phosphorus compounds are nerve poisons. Although they accumulate in fatty tissues, they are less persistent in the environment than are chlorinated compounds like DDT. Carbamates are mostly narrow-spectrum insecticides, harmful to just a few insect pests. They neither accumulate in fatty tissue nor persist in the environment. Organic pesticides include pyrethrins, permethrins, rotenone, boric acid, and diatomaceous earth. Some pests can be controlled biologically, using praying mantises, ladybugs, or pest-specific bacteria and viruses. Some insects can be controlled by SIT, or sterilization and release of males. A **pheromone** is a chemical secreted by an insect to mark a trail, send an alarm, or—in the case of a **sex attractant**—attract a mate. **Juvenile hormones** can be used to keep insect pests in an immature stage, in which they cannot reproduce. Pheromones and juvenile hormones have proved useful to some extent in monitoring and control of insect pests but are difficult to produce and expensive to use.

Section 19.3—**Herbicides** kill weeds, and **defoliants** cause leaves to fall off plants. The herbicide 2,4-D and the defoliant 2,4,5-T were combined in **Agent Orange**, used to remove ground cover in Vietnam. Highly toxic **dioxins** that were contaminants in this mixture caused birth defects, resulting in a ban on 2,4,5-T. Atrazine and glyphosate are the most widely used herbicides in the United States. Paraquat is a **preemergent herbicide** that kills weed plants before crop seedlings emerge, but it also can be used to quickly kill emergent weeds. Organic herbicides include oil of cloves, acetic acid, and gluten meal.

Section 19.4—**Sustainable agriculture** is the ability of a farm to produce food and fiber indefinitely, without causing irreparable damage to the ecosystem. **Organic farming** avoids the use of synthetic fertilizers and pesticides. It is less energy-intensive and less productive than conventional farming but costs less and uses renewable resources.

Section 19.5—According to Malthus, food production shows a steady increase called **arithmetic growth**. Because population increase historically has occurred as **geometric growth**, with the rate of increase itself increasing, food production cannot keep up with population. The **rule of 72** predicts the **doubling time** for populations, but cultural and political conditions and changes make such predictions speculative. Additional scientific and technical advances may help increase food production or control population in years to come, but Malthus's prediction still hangs over us.

GREEN CHEMISTRY—Through inspiration from nature and the application of green chemistry principles in the design of new pesticides, scientists can create effective pest control products that target only the pest and not the surrounding environment or beneficial organisms. These new pesticides can be designed using more benign reagents and processes that require less energy. The pesticides can also be designed so that they biodegrade, leaving behind no harmful residues.

Learning Objectives	Associated Problems
• List the three primary plant nutrients as well as several secondary nutrients and micronutrients, and describe the function of each. (19.1)	3, 4, 15–28, 53–56
• Differentiate between organic fertilizers and conventional fertilizers. (19.1)	12
• Name and describe the action of the main kinds of pesticides. (19.2)	10, 11, 13, 14, 29–32
• List several biological pest controls, and explain how they work. (19.2)	33–35, 57
• Name and describe the action of the major herbicides and defoliants. (19.3)	5, 36, 37
• Describe sustainable agriculture and organic farming. (19.4)	2, 6, 51, 52, 63
• Assess the challenges of feeding an increasing human population. (19.5)	7–9, 43–45
• Identify the benefits of pesticides to humans and the challenges in their use.	11, 64, 65

 Conceptual Questions

1. From where does most of the matter of a growing plant come?

2. List the three main goals of sustainable agriculture.

3. What metal ion is needed for chlorophyll to function in a plant?

4. What is the function of chlorophyll in green plants?

5. What does the phrase "preemergent herbicide" mean? Give an example.

6. Compare "organic" farming with conventional farming. Consider energy requirements, labor, profitability, and crop yields.

7. What did Thomas Malthus predict in his famous 1830 statement?

8. What are the primary challenges in feeding an increasing human population?

9. Why can't we simply make more land into farms to grow more food?

10. Describe how chlorinated hydrocarbons become increasingly concentrated in a food chain.

11. Give two potential problems inherent in using pesticides on food crops.

12. What is the primary reason why human and animal waste should not be used as fertilizer, despite the fact that this practice has been used for thousands of years?

13. What is the difference between a pesticide and an insecticide?

14. What is the difference between a pesticide and an herbicide?

 Problems

Fertilizers

15. List the three chemical elements that form the structural parts of plants. What is the name of the compound that is the primary substance in the structure of plants?

16. What are the three elements—other than those used to make roots, stems, and leaves—that are needed by plants?

17. List several ways that nitrogen can be fixed.

18. How is ammonia made?

19. What is the difference between an ammonia solution and anhydrous ammonia?

20. Why do plants need nitrogen?

21. What is the source of phosphate fertilizers?

22. What is the role of phosphorus in plant nutrition?

23. How can phosphate rock be changed into a soluble form?

24. Why does soil become acidic when potassium ions are absorbed by plants?

25. What is the common name for CaO, chemically known as calcium oxide?

26. When would it be advantageous to add CaO to your garden or farm soil? Why?

27. Other than N, P, and K, name at least five other ions that are necessary for plant growth.

28. What is eutrophication? What ions are primarily responsible for this problem in ponds and lakes?

Pesticides and Herbicides

29. List the advantages and disadvantages of DDT as an insecticide.

30. Why is DDT especially harmful to birds?

31. What are some situations when DDT might be advantageous?

32. Define and give an example of (a) a narrow-spectrum insecticide and (b) a broad-spectrum insecticide.

33. How are pheromones used in insect control? Give an example.

34. How are juvenile hormones used against insect pests?

35. Describe SIT as a method for controlling insects. What does SIT stand for? Why is it not more widely used?

36. List the two most commonly used compounds in weed-killers, and differentiate between them in terms of the types of plants that are killed by each.

37. What compounds were used in the Vietnam War, and what detrimental effects did this have on humans?

Toxicity

38. How do we quantitatively describe the toxicity of a substance? (That is, in what units?)

39. How much DDT would it take to kill a person weighing 80 kg if the lethal dose is 0.50 g per kilogram of body weight?

40. Refer to the toxicity of carbofuran in Table 19.3. If the toxicity in mice approximates that in humans, what is the lethal dose in a 15 kg child?

Growth, Population Changes, and the Food Crisis

41. You start a savings account, planning to add $150 each month. How many months will it take you to save at least $1000? Do your savings grow arithmetically or geometrically?

42. You decide to raise gerbils. Starting with two, you have four at the end of 6 weeks, eight at the end of 12 weeks, and so on. The gerbil population doubles every 6 weeks. How many gerbils will you have at the end of 24 weeks? Is the gerbil population growing arithmetically or geometrically? Note that the gerbil litters range from two to five, and the gestation period is only about 3 1/2 weeks.

43. The population of India was estimated to be 1.324 billion in 2016 and was growing at the rate of 1.35% a year. At this rate, how many years will it take for the population to double to 2.65 billion? What factors are likely to change this rate of growth?

44. The population of China was estimated to be 1.379 billion in 2016 and was growing at an annual rate of 0.5%. At this rate, how many years will it take for the population to double to 2.758 billion? What factors are likely to change this rate of growth?

45. If there are about 7.7 billion acres of ice-free land on Earth, how many billion acres of land are currently used for agriculture? (See Section 19.5 for more information.)

46. How many billion acres of land on Earth are currently used for urban areas, roads, and mines? (See Section 19.5 for further information.)

47. If the percent of corn produced that was used to make ethanol was 7% in 2000, and about 40% in 2014, what percent increase is that? What are some impacts of that trend?

Expand Your Skills

48. Give the balanced overall chemical equation for photosynthesis.

49. Give the chemical equation for the synthesis of ammonia.

50. Give the chemical equation for the synthesis of ammonium sulfate.

51. Indicate whether each of the following pesticides is organic (a) in the chemical sense only, (b) in the sense used for organic farming only, (c) in both senses, or (d) in no way at all.
    1. boric acid
    2. diatomaceous earth
    3. lead arsenate
    4. malathion
    5. rotenone

52. Indicate whether each of the following herbicides is organic **(a)** in the chemical sense only, **(b)** in the sense used for organic farming only, **(c)** in both senses, or **(d)** in no way at all.
1. acetic acid
2. atrazine
3. 2,4-D
4. gluten meal
5. glyphosate

53. The compound potassium nitrate may be used as a fertilizer. Write the correct formula for potassium nitrate, and list what plant nutrients that compound would supply.

54. The compound ammonium monohydrogen phosphate is frequently used as a fertilizer. **(a)** Write the chemical formula for that compound. **(b)** What two plant nutrients does it supply?

55. Write the equation for the formation of ammonium monohydrogen phosphate from ammonia and the appropriate acid.

56. The micronutrient copper is often applied in the form of copper(II) sulfate. Write the equation for the formation of copper(II) sulfate from copper(II) oxide and the appropriate acid.

57. Examine the chemical formula for methoprene in the section discussing juvenile hormones. What are the functional groups found in that molecule?

58. What is lime, and what is its chemical formula? Why is it commonly added to gardens and farm soil?

59. Oil of cloves, used as an organic herbicide, has eugenol as its principal active component. Name the functional groups in the eugenol molecule.

Eugenol

Critical Thinking Exercises

Apply knowledge that you have gained in this chapter and one or more of the FLaReS principles (Chapter 1) to evaluate the following statements or claims.

19.1. An advertisement claims that a certain cereal made from organically grown grains provides a "life energy" that cannot be obtained from ordinary processed cereals.

19.2. A newspaper claims that certain genetically modified crops may become weeds if not properly controlled.

Collaborative Group Projects

Prepare a PowerPoint, poster, or other presentation (as directed by your instructor) to share with the class.

1. Go to a garden supply center and find several fertilizers for use with houseplants or garden flowers. Compare the percents of N, P, and K for different targeted plants, such as roses, or indoor plants.

2. Examine the label on a package of garden pesticide. List its trade name and active ingredients.

60. Potassium bicarbonate, usually combined with horticultural oil (a highly refined petroleum product), is often sold as an organic pesticide. To make potassium bicarbonate, we start with potassium hydroxide and then react it with carbon dioxide to form potassium carbonate and water. The potassium carbonate is reacted with more carbon dioxide and water to form the potassium bicarbonate. Write balanced equations for the two reactions. Is potassium bicarbonate an organic pesticide? Explain.

61. The herbicide dicamba (Banvel), widely used to control broadleaf weeds, has the systematic name 2-methoxy-3,6-dichlorobenzoic acid. The methoxy group is —OCH_3. Draw the structure of dicamba.

62. The average woman in Uganda has 5.8 children. The population of Uganda—which, by some estimates, is growing faster than that of any other country in the world—was estimated to be 42.9 million in 2017, compared with 44.3 million in 2018. **(a)** What is the annual rate of increase in population during that yearlong span? **(b)** At this rate, how many years will it take for the population to double to 88.6 million? **(c)** What factors are likely to change this rate of growth?

63. Why is the use of insecticidal soaps not practical for large-scale agriculture?

64. Which of the following are benefits of using pesticides on food crops? Choose all that apply.
a. increased yield of food production
b. reduced crop failure
c. identification of new pests
d. improved taste of foods

65. Consider the statement "The development of more selective pesticides is advantageous to farmers, because these pesticides would easily degrade quickly in the environment." Choose one of these:
a. Both the statement and reason are correct.
b. The statement is correct, but the reason is not.
c. The statement is incorrect, but the reason is correct.
d. Neither the statement nor the reason is correct.

19.3. A website states that nitrate ions from synthetic fertilizers such as ammonium nitrate are less effective as plant nutrients than nitrate ions from the breakdown of organic fertilizers by soil bacteria, and thus nonorganic foods are less nutritious.

19.4. A newspaper article suggests, "Try out tobacco 'sun tea' as a nontoxic pesticide." In the article is this claim: "This special tea can help you eliminate most garden pests without any unwanted toxicity."

3. Compare labels for several brands of commercially available spray insecticides targeting the same type of insect. Look for the active ingredients, note their concentrations, and note safety precautions on the label for its use. Find commonalities and differences. Focus on one type of targeted pest:
a. ant killers
b. box elder bugs
c. wasps or hornets
d. mosquitoes or flies

LET'S EXPERIMENT! Wash Away the Weeds

Materials Needed

- 6 cups of vinegar
- 2 tablespoons of dish detergent
- $^1/_8$ cup of table salt
- Measuring cup

- Measuring spoons
- 3 spray bottles (each holding a quart of liquid)
- 4 pots of plants (weeds or inexpensive annuals)
- Permanent marker

(Note: Double vinegar, detergent, and salt if you perform both the indoor and outdoor experiments.)

How do you get rid of unwanted weeds in your garden? Are there any homemade recipes that are just as effective but less expensive and less hazardous than commercially manufactured herbicides?

Get outside! This experiment works best if you can get outdoors and test this homemade herbicide on weeds in your yard. (Your friend or neighbor's yard would work, too.) In this experiment, you will be comparing three different homemade herbicide recipes. Begin by making your weed-killers. Measure out the ingredients and label the spray bottles.

- Weed-killer 1: 2 cups vinegar
- Weed-killer 2: 2 cups vinegar + 1 tbsp liquid dish soap
- Weed-killer 3: 2 cups vinegar + 1 tbsp liquid dish soap + $^1/_8$ c salt

Outdoors Experiment with Recipes 1 and 2: (3 days)

The homemade herbicides you will be experimenting with are not selective; they can potentially harm plants you wish to keep. If you are able to do this experiment outside, be careful to spray the weed-killer directly onto the weeds so there is no floating mist. The objective is to thoroughly coat the visible foliage. Only test weed-killers 1 and 2 outside. (The salt in weed killer 3 could also contaminate ground water.) It's best to do this in a walkway or area where you do not want plants at all.

Day 1 (Outdoors):

▲ Outdoors results.

Indoors Experiment with All Three Recipes:

You will want to prepare four potted plants using either weeds or inexpensive leafy annuals from a gardening store. One will be your control and the others will be labeled weed-killers 1–3.

Place your plants in conditions that are the same for all four plants. A sunny windowsill works well. For the next four days, consistently spray the weed-killers over the weeds in the corresponding pots, saturating the leaves. Spray at the same time each day, and monitor the conditions of the weeds. What happened to the weeds? Are there any color changes in the weeds? Record your results.

Day 1 (Indoors): | Day 2 (Indoors):

Day 3 (Indoors): | Day 4 (Indoors):

▲ Indoors results.

Questions

1. Which weed-killer was the most effective indoors? Why?
2. Which weed-killer was most effective outdoors? Why?
3. Which weed-killer was the least effective indoors? Why?
4. Which weed-killer was least effective outdoors? Why?
5. Will you consider making your own herbicide rather than buying a synthetic weed-killer?

Household Chemicals

20

The household chemicals sold in greatest volume are cleaning products— soaps, detergents, and various special-purpose and all-purpose cleaners. In this chapter, we examine a variety of familiar products and discuss how they work.

Have You Ever Wondered?

1 How does soap remove grease from my shirt and hands?

2 Why can't hand dishwashing liquid be used in a dishwasher?

3 Does putting more fabric softener in the washer make fabrics even softer? **4** Should I worry about lead-based paint if I purchase a classic home built in the 1950s? **5** Does a thicker shampoo work better to clean my hair? **6** Why must I leave a hair coloring preparation on my hair for the full time suggested on the bottle if I want to become a blonde?

"CHEMICALS" IN YOUR HOME   One theme of this textbook has been how advancements in chemistry have extensively improved our lives in the twenty-first century, extended our lifespan, improved our health and quality of life, and provided myriad new materials for useful products for us to enjoy. Perhaps half a million chemical products are available for use in the typical home, including soaps and detergents, toothpastes, shampoos, perfumes, lotions, shaving creams, deodorants, waxes,

▲ **Figure 20.1** A modern home is stocked with a variety of chemical products.

paints, paint removers, bleaches, insecticides, spot removers, solvents, disinfectants, cosmetics, hair sprays, and hair dyes (Figure 20.1).

We discussed the chemical composition of food and the chemicals that are used in producing it in Chapter 17. Some agricultural chemicals are also used around the home, especially in the yard and garden (Chapter 19). We discussed polymers used in clothing, home furnishings, structural materials, toys, and storage containers in Chapter 10 and the fuels burned in our furnaces and automobiles in Chapter 15. We discussed the chemistry of our bodies in Chapter 16 and pharmaceutical products present in our medicine cabinets in Chapter 18. In this chapter, we will look at other chemicals around the house—in cleaning products, personal care supplies, and paint, to name just a few. We will be discussing precautions to be followed with some of them. Frequently the labels on bottles are not read at all, and the *misuse* of household chemicals can sometimes end in tragedy. Let's begin with cleaning agents, which make up the largest volume of household chemicals.

20.1 Cleaning with Soap

Learning Objectives • Describe the structure of soap, and explain how it is made and how it works to remove greasy dirt. • Explain the advantages and disadvantages of soap. • Explain how water softeners work, and what substances are used as water softeners.

The use of soap and the attitudes of various cultures on cleanliness have a fascinating history. In developing societies, clothes were cleaned by beating them with rocks in the nearest stream. Sometimes plants, such as the soapworts of Europe or the soapberries of tropical America, were used as cleansing agents. The leaves of soapworts and soapberries contain **saponins**, chemical compounds that produce a soapy lather.

Ashes of burned plants contain potassium carbonate and sodium carbonate. The carbonate ion present in both these compounds reacts with water to form an alkaline solution that can attack grease and oil. The word "alkali" comes from an Arabic word meaning "the ashes." The Babylonians used these alkaline plant ashes to make cleansing agents at least 4000 years ago, by heating those ashes with animal fats to make simple soaps.

Personal Cleanliness

The Romans, with their great public baths, probably did not use any sort of soap. They covered their bodies with oil, worked up a sweat in a steam bath, and then rubbed off the oil. A dip in a pool of fresh water completed the cleansing.

During the Middle Ages, bodily cleanliness was prized in some cultures but not in others. For example, twelfth-century Paris, with a population of about 100,000, had many public bathhouses. In contrast, the Renaissance, which was the revival of learning and art that lasted from the fourteenth to the seventeenth century, was surprisingly not noted for cleanliness. Queen Elizabeth I of England (1533–1603) supposedly bathed once a month, a habit that caused many to think her overly fastidious. Washing the hands and face using a bowl of water on a daily basis was common for the nobility. But remember that in those days, people did not have faucets with clean, hot water in their homes, nor was soap or shampoo easily available. The lower classes in Elizabethan England probably bathed only a few times a year, using a large wooden tub next to a fireplace to warm water brought in from a well or river. A common remedy for unpleasant body odor back then, at least among the upper classes, was the liberal use of perfume.

And today? Perhaps we have gone too far in the other direction. With soap, detergent, body wash, shampoo, conditioner, deodorant, antiperspirant, aftershave, cologne, and perfume, we may add more during and after a shower than we remove in the shower.

Although soap has been known for hundreds of years, it was first used mainly for medicinal purposes. Even today, particularly in Asia and Africa, soaps made from plants such as the neem tree (*Azadirachta indica*) are used for skin treatments. In India, 80% of neem oil is used to make soap that treats skin rashes, psoriasis, eczema, acne, and chicken pox.

The discovery that disease-causing microorganisms cause infections and illness increased interest in cleanliness and improved public health practices during the eighteenth century. By the middle of the nineteenth century, soap was in common use as a cleansing agent.

Soap-Making: Fat Plus Lye Forms a Soap Plus Glycerol

The first written record of soap-making is found in the writings of Pliny the Elder (23–79 C.E.), a Roman scholar who described the Phoenicians' synthesis of soap using goat tallow and ashes. It may seem strange, but soap can actually be made from fats! American pioneers in the nineteenth century, nearly two millennia later, made soap in a process similar to that used by ancient people. They filled a large pot with wood ashes and allowed water to trickle down through the ashes, dissolving the basic compounds in the ashes. Using a pot of ash gave rise to the term *potash* for the mixture of these compounds. After filtering off the wet ashes by pouring the contents of the pot through a large piece of cloth, the potash solution was added to animal fat in a huge iron kettle. That mixture was cooked over a wood fire for several hours. The soap rose to the surface and, on cooling, solidified. A by-product of the reaction was glycerol, which remained as a liquid on the bottom of the pot. Both the glycerol and the soap often contained unreacted alkali, which was a strong irritant on the skin, since its pH was often over ten. The harshness of Grandma's lye soap is not just a myth.

The general reaction for making soap is

$$\text{a fat + a base} \longrightarrow \text{soap + glycerol}$$

Sodium hydroxide is the best base to use in making soap. The glycerol by-product (sometimes called glycerin on labels of commercial products) as shown in Figure 20.2 is a useful, oily liquid found in many hand creams and lotions. (See Section 20.6.) Note that three soap molecules are formed from one fat molecule and three NaOH formula units. Soap molecules vary in the length of the carbon chains depending on the source of the fat. The soap molecule in Figure 20.2 is called sodium stearate, and is made from beef fat.

Look carefully at the structure of the soap molecule in Figure 20.2 to see that a **soap** is a salt of a long-chain carboxylic acid (Chapter 9). Another example of a soap is sodium palmitate, made from palm oil, with the formula $CH_3CH_2CH_2CH_2CH_2CH_2CH_2CH_2CH_2CH_2CH_2CH_2CH_2CH_2CH_2COO^- —Na^+$. To save space, the CH_2 groups can be lumped together within parentheses to give a shorter formula, $CH_3(CH_2)_{14}COO^-Na^+$.

In modern commercial soap-making, molecules of fats and oils are often split up into fatty acids and glycerol with superheated steam. The fatty acids are then neutralized to make soap. Hand soaps usually contain additives such as dyes, perfumes, creams, oils, or even compounds with deodorizing properties. Scouring soaps contain abrasives such as silica and pumice. Some soaps have air blown in before they solidify to lower their density so that they float.

Potassium soaps are softer than sodium soaps and produce a finer lather. They are used alone, or in combination with sodium soaps, in some liquid soaps and shaving creams. However, beware that most liquids for hand cleaning contain synthetic **detergents**, despite the labels, which may erroneously read "liquid hand soap."

▲ Queen Isabella of Castile (1451–1504), who supported the voyage of Columbus to the New World in 1492, is reported to have bathed only twice in her life.

1 **How does soap remove grease from my shirt and hands?**

Soap molecules have a long hydrocarbon end that is nonpolar so it attracts nonpolar molecules, such as grease. The other end of the molecule is ionic (usually negatively charged), so it is attracted to the positive "poles" of polar water molecules. Large, spherical masses, called *micelles*, consisting of many soap molecules surrounding a nonpolar one, are pulled down the drain by the myriad water molecules.

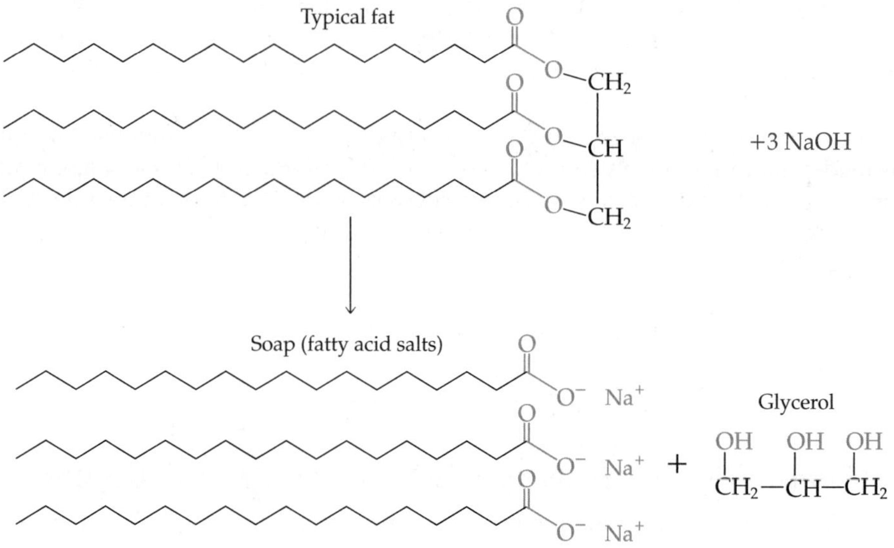

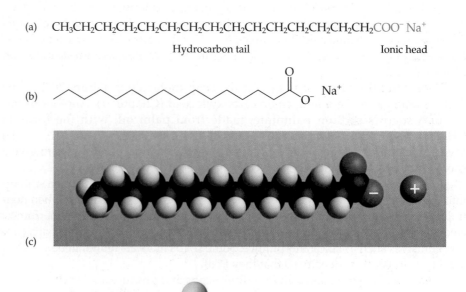

Typical fat

+3 NaOH

Soap (fatty acid salts)

O⁻ Na⁺

O⁻ Na⁺   Glycerol

$$OH \quad OH \quad OH$$
$$CH_2-CH-CH_2$$

O⁻ Na⁺

▲ **Figure 20.2** Soap can be made by reacting animal fat or vegetable oil with sodium hydroxide. Vegetable oils, with unsaturated carbon chains, generally produce softer soaps. Coconut oil, with shorter carbon chains, yields soap that is more soluble in water.

How Soap Works

Dirt and grime usually adhere to skin, clothing, and other surfaces because they are combined with greases and oils—body oils, cooking fats, lubricating greases, and other similar substances—that act a little like sticky glues. Because oils are not miscible with water, washing with water alone does little good.

A soap molecule has a dual nature, which is the clue as to how a soap actually works to remove grease and dirt. One end of the molecule is ionic and, therefore, **hydrophilic** (water-attracting). The rest of the molecule is a hydrocarbon chain, and therefore nonpolar and **hydrophobic** (water-repelling). Figure 20.3 shows a typical soap. The hydrophilic "head" dissolves in water, while the hydrophobic "tail" dissolves in nonpolar substances, such as oils.

(a) $CH_3CH_2CH_2CH_2CH_2CH_2CH_2CH_2CH_2CH_2CH_2CH_2CH_2CH_2CH_2COO^- \ Na^+$

Hydrocarbon tail   Ionic head

(b)

Na⁺
O⁻

(c)

−   +

(d)

−

▲ **Figure 20.3** Representations of sodium palmitate, a soap. (a) Condensed formula. (b) Structural formula. (c) Space-filling model. (d) Schematic representation of the palmitate anion.

The cleansing action of soap is illustrated in Figure 20.4. A spherical collection of molecules like that shown in the figure is called a **micelle**. The hydrocarbon tails stick in the oil, with their ionic heads remaining in the aqueous phase. The formation of micelles breaks the oil into extremely tiny droplets and disperses them throughout the solution. The droplets don't coalesce, because of the repulsion of the charged groups (the carboxyl anions) on their surfaces. The oil and water form an **emulsion**, with soap acting as the **emulsifying agent**. Because the oil no longer is "glued" to hair, skin, or any other "dirty" surface, the dirt can be removed easily, and washed down the drain with the rinse water. Any substance, including soap, that stabilizes the suspension of nonpolar substances in water—such as oil and grease—is called a **surface-active agent** (or **surfactant**).

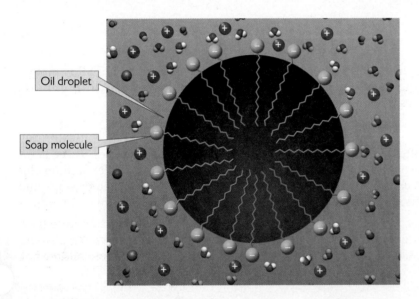

Oil droplet

Soap molecule

◀ **Figure 20.4** The cleansing action of soap is visualized in this diagram of a soap micelle. A tiny oil droplet is suspended in water because the hydrophobic hydrocarbon tails of soap molecules are immersed in the oil, while their hydrophilic ionic heads extend into the water. Attraction between the water and the ionic ends of the soap molecules carries the oil droplet into the water and down the drain.

Disadvantages and Advantages of Soap

The major disadvantage of soap is that it doesn't work well in hard water. Hard water contains calcium, magnesium, and/or iron ions. Soap anions react with these metal ions to form sticky, insoluble curds (Figure 20.5). As an example, when calcium ions are the reason for the water being hard and the soap is sodium tetradecanoate, the reaction is

$$2\ CH_3(CH_2)_{12}COO^-\ Na^+(aq) + Ca^{2+}(aq) \longrightarrow$$
$$[CH_3(CH_2)_{12}COO]_2Ca(s) + 2\ Na^+\ (aq)$$

The calcium tetradecanoate is an insoluble, sticky deposit that may be seen as the familiar ring around the bathtub. Essentially all soaps with varying lengths of the hydrocarbon tail will have the same result. More than seven decades ago, freshly washed hair was left sticky, and soap-laundered clothing took on a "tattletale gray," because soaps were the only products available at the time for cleaning. Since then, soap has been largely replaced by synthetic detergents (Section 20.2), which do not form those sticky, insoluble substances in hard water.

Soap does have some advantages. It is an excellent cleanser in soft water, it is relatively nontoxic, it is derived from renewable resources (animal fats and vegetable oils), and it is biodegradable.

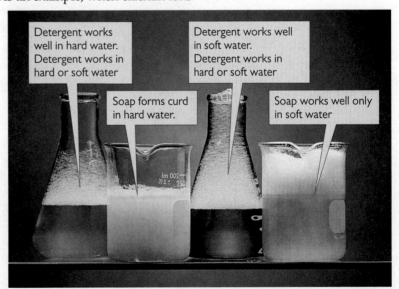

Detergent works well in hard water. Detergent works in hard or soft water

Detergent works well in soft water. Detergent works in hard or soft water

Soap forms curd in hard water.

Soap works well only in soft water

▲ **Figure 20.5** The sudsing quality of hard water versus soft water. *From left:* detergent in hard water, soap in hard water, detergent in soft water, and soap in soft water. Note that the sudsing of the detergent is about the same in both hard and soft water. Note also that there is little sudsing and that insoluble material is formed when soap is used in hard water.

Water Softeners

To alleviate the problem of sticky precipitates in bathtubs and to improve the sudsing action of soaps, a variety of water-softening agents and devices may be used. Water softeners are used to remove ions of calcium, magnesium, and iron from water. An effective water softener is washing soda, sodium carbonate ($Na_2CO_3 \cdot 10\,H_2O$). Carbonate ions react with water molecules to raise the pH and thus prevent the precipitation of fatty acids.

$$CO_3^{2-}(aq) + H_2O(l) \longrightarrow HCO_3^-(aq) + OH^-(aq)$$

The carbonate ions also react with the ions that cause hard water and remove them as insoluble salts.

$$Mg^{2+}(aq) + CO_3^{2-}(aq) \longrightarrow MgCO_3(s)$$
$$Ca^{2+}(aq) + CO_3^{2-}(aq) \longrightarrow CaCO_3(s)$$

Sodium phosphate, Na_3PO_4, commonly called *trisodium phosphate*, or *TSP*, is another water-softening agent. The phosphate ions form precipitates with hard water ions such as magnesium ions, shown below.

$$2\,PO_4^{3-}(aq) + 3\,Mg^{2+}(aq) \longrightarrow Mg_3(PO_4)_2(s)$$

Water-softening tanks are widely used in homes and businesses. These tanks contain an insoluble polymeric resin that usually consists of complex sodium aluminosilicates called *zeolites*. The resin attracts and holds calcium, magnesium, and iron ions to its surface, releasing sodium ions into the water and thus softening it (Figure 20.6)

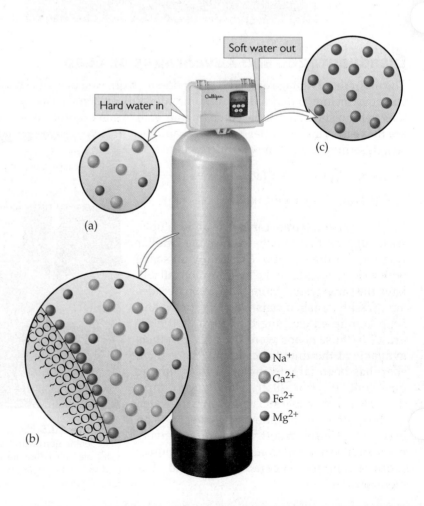

▶ **Figure 20.6** Water softeners work by *ion exchange*. Hard water going into the softener contains doubly charged cations—Ca^{2+}, Mg^{2+}, and Fe^{2+}. Those ions attach to the ion-exchange resin, and Na^+ (violet) ions are released. The water leaving the softener contains mostly Na^+ ions.

This process is called *ion exchange:*

$$Ca^{2+}(aq) + Na_2Al_2Si_2O_8(s) \longrightarrow 2\ Na^+(aq) + CaAl_2Si_2O_8(s)$$

After a period of use, the resin becomes saturated with hard-water ions and must be regenerated by adding salt (NaCl) to replace its sodium ions. Note that water filters such as Brita's® that attach to a faucet usually contain activated carbon, which absorbs dissolved organic substances that give water an off taste; most do not remove hard-water ions.

SELF-ASSESSMENT Questions

1. Which of the following is NOT a soap?
   a. sodium stearate   **b.** sodium palmitate
   c. sodium myristate   **d.** sodium phosphate

2. Which of the following is a soap formula?
   a. $CH_3(CH_2)_{12}CH_2OH$ **b.** $CH_3(CH_2)_{12}COOH$
   c. $CH_3(CH_2)_{14}COONa$ **d.** $CH_3(CH_2)_{10}SO_3Na$

3. Which of the following does NOT function as a water softener?
   a. sodium carbonate   **b.** sodium chloride
   c. sodium phosphate   **d.** sodium zeolite

4. The hydrocarbon part of a soap molecule
   a. is attracted to nonpolar oil
   b. is repelled by nonpolar oil
   c. is attracted to polar water
   d. is repelled by polar oil

5. The ionic end of a soap molecule is
   a. hydrophobic   **b.** hydrophilic
   c. polar covalent   **d.** acidic

6. Soap removes oily dirt by forming
   a. an acidic solution   **b.** curds
   c. fat globules   **d.** micelles

7. In the cleansing action of soap, the hydrocarbon tail
   a. intermingles with the oily dirt, while the ionic head interacts with the water
   b. intermingles with the water, while the ionic head interacts with the oily dirt
   c. remains inert and takes no part in the process
   d. separates from the ionic head and dissolves in the oily dirt

8. Which of the following ions does not contribute to making water hard?
   a. Ca^{2+} **b.** Fe^{2+} **c.** Mg^{2+} **d.** Na^+

9. In hard water, metal ions combine with soap anions, and the soap
   a. cleans about as well as in soft water
   b. dissolves completely but doesn't clean
   c. forms suds but doesn't clean
   d. precipitates and doesn't clean

Answers: 1. d; 2. c; 3. b; 4. a; 5. b; 6. d; 7. a; 8. d; 9. d

20.2 Synthetic Detergents

Learning Objectives • List the structures, advantages, and disadvantages of synthetic detergents. • Classify surfactants as amphoteric, anionic, cationic, and nonionic, and describe how they are used in various detergent formulations.

A technological approach to the problems associated with soap was the development of new synthetic detergents. Molecules of synthetic detergents are similar to soap molecules because they have an ionic head and a long hydrocarbon tail, so they have the same cleaning action but are different enough to resist the effect of ions in hard water. The raw materials for soap manufacture were scarce and expensive during and immediately after World War II. The synthetic detergent industry developed rapidly in the postwar period.

ABS Detergents: Nonbiodegradable

Within a few years of the war's end, cheap synthetic detergents, made from petroleum products, were widely available. Alkylbenzenesulfonate (ABS) detergents were made from propylene ($CH_2{=}CHCH_3$), benzene (C_6H_6), and sulfuric acid (H_2SO_4). The resulting sulfonic acid (RSO_3H) was neutralized with a base, usually sodium carbonate, to yield the final product.

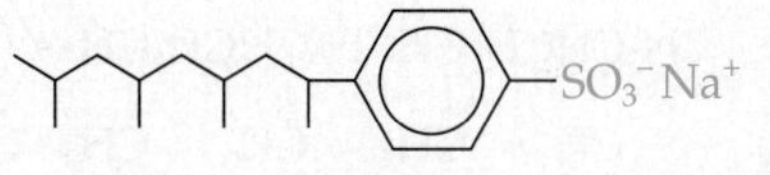

Note that the detergent molecule is much like a soap molecule. It, too, has a long hydrocarbon tail and an ionic head, but in this case, it is $-SO_3^-$, not $-CO_2^-$. Micelles still form around nonpolar grease droplets, and the ionic ends are attracted

to polar water molecules (Figure 20.4). The significant difference is that the sulfonate detergents would not form insoluble precipitates in hard water. As Figure 20.5 shows, the cleansing action of a synthetic detergent is little affected by hard water. Detergents work well in acidic water as well.

Sales of ABS detergents soared. For a decade or more, nearly everyone was happy, but suds began to accumulate in sewage treatment plants. Foam piled high in rivers, and in some areas, a head of foam even appeared on drinking water (Figure 20.7). The new problem was that the branched-chain structure of ABS molecules (look carefully at the structure) was not readily broken down by the microorganisms in sewage treatment plants, and the groundwater supply was threatened. Public outcries caused laws to be passed and industries to change their processes. Biodegradable detergents were quickly put on the market, and nonbiodegradable detergents were banned.

▶ **Figure 20.7** ABS detergents degrade very slowly in nature. (a) Their use led to foam on rivers such as that on the Bogotá River in Colombia. (b) They also caused contaminated water from a ground-water source to foam as it came from the tap.

(a)
(b)

LAS Detergents: Biodegradable

Biodegradable detergents called *linear alkyl sulfonates (LAS's)* have a linear chain of carbon atoms, rather than the branched chains of the ABS detergents.

$$\text{—SO}_3^-\text{Na}^+$$

Microorganisms can break down LAS molecules by producing enzymes that degrade the molecule two (and only two) carbon atoms at a time (Figure 20.8). The branched chain of ABS molecules blocked this enzyme action, preventing their degradation. In this way technology solved the problem of foaming rivers.

Surfactants are used in thousands of products, not just in cleaning products. A small amount of lecithin, an edible surfactant, is often added to chocolate in the manufacturing process. The lecithin helps the chocolate disperse on the tongue more easily. The chocolate then tastes "more chocolatey."

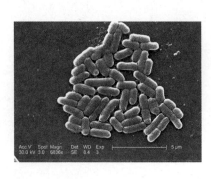
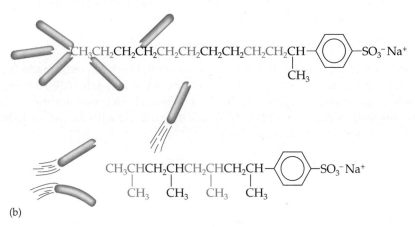

(a)
(b)

▲ **Figure 20.8** Microorganisms such as *Escherichia coli*, shown (a) magnified 42,500 times by scanning electron microscopy, are able to degrade LAS detergents. They can readily metabolize these detergents (b), removing two carbon atoms at a time from each molecule. It takes them much longer to break down the branched chains of the ABS molecules.

Classification of Surfactants Based on Charge

Surfactants are generally classified according to the ionic charge, if any, at one end of the long molecule. Soaps and LAS detergents are all **anionic surfactants**; they have a negative charge on one end of the molecule. Other common types of anionic surfactants are the alkyl sulfates, such as sodium dodecyl sulfate (often called sodium lauryl sulfate on commercial product labels), and the alcohol ethoxysulfates (AES's). An ionic detergent is always associated with a small ion of opposite charge, called a *counter ion*. The counter ion for most anionic surfactants is Na^+.

$$CH_3(CH_2)_nOSO_3^-Na^+ \qquad CH_3(CH_2)_mO(CH_2CH_2O)_nSO_3^-Na^+$$

<table>
<tr><td>An alkyl sulfate</td><td>An alcohol ethoxysulfate</td></tr>
<tr><td>$n = 11$ to 13</td><td>$m = 6$ to 13; $n = 7$ to 13</td></tr>
</table>

Anionic surfactants are used in laundry and hand dishwashing detergents, household cleaners, and personal cleansing products, such as shampoo. They have excellent cleaning properties and generally make a lot of suds.

As the name indicates, **nonionic surfactants** have no electrical charge. They are low-sudsing and not affected by water hardness. They work well on most soils and are typically used in laundry and automatic dishwasher formulations. The most widely used nonionic surfactants are the *alcohol ethoxylates*.

$$CH_3(CH_2)_mO(CH_2CH_2O)_nH$$

An alcohol ethoxylate

$m = 6$ to 13; $n = 7$ to 13

Cationic surfactants have a positive charge on the active part. These surfactants are not particularly good detergents, but they have a germicidal action. They are therefore used as cleansers and disinfectants in the food and dairy industries and as the sanitizing ingredient in some household cleaners. The most common cationic surfactants are *quaternary ammonium salts* (or quats), so called because they have four hydrocarbon groups attached to a nitrogen atom that bears a positive charge.

$$CH_3(CH_2)_nCH_2N^+(CH_3)_3Cl^-$$

A quaternary ammonium salt

$n = 10$ to 16

Sometimes, cationic surfactants are used along with nonionic surfactants. Cationic surfactants are seldom used with anionic ones, because the ions of opposite charge tend to clump together and precipitate from solution, destroying the detergent action of both.

Amphoteric surfactants carry both a positive and a negative charge. They react with both acids and bases and are noted for their mildness, sudsing, and stability. They are used in personal cleansing products such as shampoos for babies and in household cleaning products such as liquid handwashing formulations. Typical amphoteric surfactants are compounds called *betaines*.

$$CH_3(CH_2)_nCH_2NH_2^+CH_2COO^-$$

A betaine

$n = 10$ to 16

 CONCEPTUAL Example 20.1 Classification of Surfactants

Classify each of the following surfactants as anionic, cationic, nonionic, or amphoteric.
  a. $CH_3(CH_2)_{14}CH_2N^+(CH_3)_3\ Cl^-$
  b. $CH_3(CH_2)_{13}O(CH_2CH_2O)_7SO_3^-Na^+$
  c. $CH_3(CH_2)_{12}CH_2NH_2^+CH_2COO^-$
  d. $CH_3(CH_2)_{12}CON(CH_2CH_2OH)_2$

Solution

  a. The N atom of the active part bears a positive charge, making this a cationic surfactant. (The counter ion is Cl^-.)

WHY IT MATTERS

Using more than the recommended amount of dishwasher detergent generally does not result in cleaner dishes. The strong agitation of water on the surfaces of the dishes is the main mechanism of cleaning. Using more detergent leads to faster etching of the glass in the dishwasher.

▲ **Figure 20.9** Zeolites used in detergent formulations can trap positive ions such as Ca^{2+} and Mg^{2+} in their cagelike structures and hold the ions in suspension in the wash water.

b. There is a negative charge on the $-SO_3$ group of the active part, making this an anionic surfactant. (The counter ion is Na^+.)

c. This substance has both a positive charge (on the N atom) and a negative charge (on the $-COO^-$ group), making this an amphoteric surfactant.

d. This substance has neither a positive charge nor a negative charge, so it is a nonionic surfactant.

> **EXERCISE 20.1A**

Classify each of the following surfactants as anionic, cationic, nonionic, or amphoteric.

a. $CH_3(CH_2)_8-C_6H_4-O(CH_2CH_2O)_9H$

b. $CH_3(CH_2)_{12}CH_2N^+(CH_3)_3Cl^-$

> **EXERCISE 20.1B**

Classify each of the following surfactants as anionic, cationic, nonionic or amphoteric.

a. $CH_3(CH_2)_9-C_6H_4-SO_3^-\ Na^+$

b. $CH_3(CH_2)_{14}COO^-\ K^+$

Laundry Detergent Formulations

Detergent products used in homes and commercial laundries usually contain a variety of ingredients. The main component is a surfactant, but other additives include builders, brighteners, dyes, fabric softeners, enzymes, and other substances that lessen the re-deposition of dirt.

Any substance added to a surfactant to increase its detergency is called a **builder**. Most builders are water softeners that reduce water hardness. Sodium citrate ($Na_3C_6H_5O_7$) and complex phosphates such as sodium tripolyphosphate ($Na_5P_3O_{10}$) and sodium hexametaphosphate [$(NaPO_3)_6$] tie up Ca^{2+} and Mg^{2+} ions in soluble complexes, a process called *sequestration*. Sometimes, mixtures of builders enhance cleaning more than would either component alone. A major problem with builders is the eutrophication of lakes by phosphates (Chapter 14 and Chapter 19). Several state and local governments have banned the sale of detergents containing phosphates.

Zeolites (see Section 20.1 regarding water softeners) also may be used as builders, but in this case the calcium and magnesium ions from hard water are held in suspension by the zeolites rather than being precipitated (Figure 20.9).

Almost all detergent formulations include fluorescent dyes called **optical brighteners**. After white clothing is worn quite a few times, it often looks yellowish and drab. Optical brighteners absorb ultraviolet rays in sunlight and reemit that energy primarily as blue light. This blue light both camouflages the yellowish color and increases the amount of visible light reaching the eye, giving the garment the appearance of being "whiter than white." Brighteners are also used in cosmetics, paper, soap, plastics, and other products. Optical brighteners may cause skin rashes in some people. Their effect on lakes and streams is largely unknown, despite the large amounts entering our waterways. The only benefit of these compounds is visual.

Early laundry detergent preparations were powders, but liquid laundry detergent formulations now command most of the market. Laundry detergent formulations two or three times more concentrated than in the past are often called "ultra" detergents. Liquid products may also have a solvent such as ethanol, isopropyl alcohol, or propylene glycol to help prevent separation of the other components and to help dissolve greasy soils. Some liquid laundry detergents use sodium citrate, sodium silicate, sodium carbonate, or zeolites as builders, or they may include bleaches and colorants.

Some formulations of liquid laundry detergents are *unbuilt*—high in surfactants but containing no builders. LAS surfactants are the cheapest surfactants used in liquid laundry detergents. In unbuilt formulations, they usually are in the form of the sodium salt or the triethanolamine [$N(CH_2CH_2OH)_3$] salt.

$$CH_3(CH_2)_9CH(CH_3)-C_6H_4-SO_3^-\ HN^+(CH_2CH_2OH)_3$$
triethanolamine salt of an LAS

In built varieties, LAS surfactants are often present as the potassium salt.

Other popular surfactants for liquid formulations are the AES compounds, that have the unusual property of being more soluble in cold water than in hot, making them particularly suitable for cold-water laundering.

Some detergent formulations include *lipases* (enzymes that work on lipids) to help remove fats and oils, *amylases* (enzymes that work on starches) to help remove starchy stains such as those from pasta, and *proteases* (enzymes that work on proteins) to aid in the removal of protein stains such as blood.

Laundry detergent "pods" are small packets of concentrated laundry detergent wrapped in water-soluble polyvinylalcohol (PVA) that were first made available in 2013. Tragically, the fad of teenagers chewing and eating them, popularized through social media, causes very serious physiological results, including severe damage to the esophagus, severe breathing problems, kidney and blood pressure problems, unconsciousness, and in several cases, death. Over 10,000 cases of poisonings reported to Poison Control centers and at least 8 deaths have been attributed to consumption of detergent "pods."

Dishwashing Detergents

Although colloquially called "dish soap," liquid cleaning formulations for washing dishes by hand generally contain at least one LAS detergent as the main active ingredient. Some use nonionic surfactants such as cocamide DEA, an amide made from a fatty acid and diethanolamine [$HN(CH_2CH_2OH)_2$]. Dishwashing liquids may also contain enzymes to help remove greases and protein stains, fragrances, preservatives, solvents, bleaches, and colorants.

$$CH_3(CH_2)_nCON(CH_2CH_2OH)_2$$
Cocamide DEA
$$n = 8 \text{ to } 12$$

Detergents for automatic dishwashers are quite different from dishwashing liquids. They are strongly alkaline and should never be used for hand dishwashing. They contain sodium tripolyphosphate ($Na_5P_3O_{10}$), sodium carbonate, sodium metasilicate (Na_2SiO_3), sodium sulfate, a chlorine or oxygen bleach, and only a small amount of surfactant, usually a nonionic type. Some contain sodium hydroxide, a strong base that attacks skin. They depend mainly on their strong alkalis, heat, and the vigorous agitation of the machine for cleaning. Dishes that are washed repeatedly in an automatic dishwasher may be permanently etched by these strong alkalis, which is why delicate crystal stemware is often washed by hand.

2 Why can't hand dishwashing liquid be used in a dishwasher?

The surfactants in hand dishwashing liquids generally form lots of suds. In a dishwasher, those suds would fill the chamber and inhibit the sprays of water that do much of the cleaning. Removal of suds from glassware and plastic or ceramic plates would require much more water in the rinse cycles. Use a product specifically designed for an automatic dishwasher.

SELF-ASSESSMENT Questions

1. The critical part(s) of a molecule used to mimic the function of a soap is/are
   a. a benzene ring in a long hydrocarbon chain and a nonpolar end
   b. a long hydrocarbon chain with —OH at one end
   c. a long hydrocarbon chain with an anionic end
   d. a long hydrocarbon chain with a hydrophilic end

2. Which of the following are not biodegradable?
   a. LAS detergents
   b. ABS detergents
   c. sodium lauryl sulfate
   d. quaternary ammonium salts

3. Which of the following is an LAS detergent?
   a. $CH_3(CH_2)_{10}CH_2N^+(CH_3)_3Cl^-$
   b. $CH_3(CH_2)_{11}OSO_3^-Na^+$
   c. $CH_3(CH_2)_9CH(CH_3)C_6H_4SO_3^-Na^+$
   d. $CH_3(CH_2)_8O(OCH_2CH_2)_9H$

4. Hard-water ions react with carbonate ions to form products such as
   a. $CaCO_3(s)$
   b. $Ca(OH)_2(s)$
   c. $Na_2CO_3(s)$
   d. soap curds

5. When it softens water, an ion-exchange resin retains
   a. magnesium ions
   b. chloride ions
   c. hydroxide ions
   d. sodium ions

6. An optical brightener converts ultraviolet light to
   a. black light
   b. blue light
   c. heat
   d. yellow light

7. What kind of surfactant is cocamide DEA, $CH_3(CH_2)_{10}CON(CH_2CH_2OH)_2$, which is used in hand dishwashing detergents?
   a. amphoteric
   b. anionic
   c. cationic
   d. nonionic

8. Which of the following is the most likely to irritate the skin?
   a. detergent for automatic dishwashers
   b. liquid handwashing soap
   c. liquid dishwashing detergent
   d. powdered laundry detergent

9. Which of the following cleaning agents is likely to have the highest pH?
   a. a liquid dish detergent for hand washing
   b. a liquid laundry detergent
   c. an automatic dishwasher detergent
   d. an LAS detergent

Answers: 1, d; 2, b; 3, b; 4, a; 5, a; 6, b; 7, d; 8, a; 9, c

20.3 Laundry Auxiliaries: Softeners and Bleaches

Learning Objectives • Identify a fabric softener from its structure, and describe its action. • Name two types of laundry bleaches, and describe how they work.

Section 20.2 discussed surfactants that are quaternary ammonium salts with one long hydrocarbon chain and three small, usually methyl groups attached to the nitrogen atom. An example of is octadecyltrimethylammonium chloride. The chloride counter ion is not shown.

$$CH_3CH_2CH_2CH_2CH_2CH_2CH_2CH_2CH_2CH_2CH_2CH_2CH_2CH_2CH_2CH_2CH_2CH_2 \overset{\overset{CH_3}{|}}{\underset{\underset{CH_3}{|}}{N^+}} - CH_3$$

3 **Does putting more fabric softener in the washer make the fabrics even softer?**

A very small amount of fabric softener gives a soft feeling to the laundry. Too much produces a coating that can make the laundry feel damp or waxy.

Fabric Softeners

Another kind of quaternary salt with *two* long carbon chains and two smaller groups on a nitrogen atom is used as a fabric softener. An example is dioctadecyldimethylammonium chloride. This compound is strongly adsorbed by the fabric, forming a film one molecule thick on the surface. The long hydrocarbon chains lubricate the fibers, imparting increased flexibility and softness to the fabric.

$$CH_3CH_2CH_2CH_2CH_2CH_2CH_2CH_2CH_2CH_2CH_2CH_2CH_2CH_2CH_2CH_2CH_2CH_2 - \overset{\overset{CH_3}{|}}{\underset{\underset{CH_3CH_2CH_2CH_2CH_2CH_2CH_2CH_2CH_2CH_2CH_2CH_2CH_2CH_2CH_2CH_2CH_2CH_2}{|}}{N^+}} - CH_3$$

Laundry Bleaches: Whiter Whites

Bleaches are oxidizing agents (Chapter 8) that remove colored stains from fabrics. Two main types of laundry bleaches are chlorine bleaches and oxygen bleaches. Most familiar liquid chlorine laundry bleaches (such as Clorox® and Purex®) are dilute solutions (usually about 6%) of sodium hypochlorite (NaOCl) that differ mainly in price. Hypochlorite bleaches release chlorine rapidly, and high concentrations of chlorine can damage fabrics. These bleaches work best on cotton clothing or towels. They do not work well on polyester fabrics, often causing yellowing rather than the desired whitening.

Other chlorine-based bleaches are available as a solid powder that releases chlorine slowly in water to minimize damage to fabrics. Trichloroisocyanuric acid, also known as trichlor or symclosene, is an example of a cyanurate-type bleach.

Oxygen bleaches usually contain sodium percarbonate ($2 Na_2CO_3 \cdot 3 H_2O_2$) or sodium perborate ($NaBO_2 \cdot H_2O_2$). As indicated by the formulas, these compounds are complexes of Na_2CO_3 or $NaBO_2$ and hydrogen peroxide (H_2O_2). In hot water, the hydrogen peroxide is liberated and acts as a strong oxidizing agent.

Borates are somewhat toxic. *Perborate bleaches* are less active than chlorine bleaches and require higher temperatures, higher alkalinity, and higher concentrations to do an equivalent job. They are used mainly for bleaching white polyester–cotton fabrics, which last much longer with oxygen bleaching than with chlorine bleaching. Less toxic than the perborate bleaches, *percarbonate bleaches* are rapidly becoming the predominant type of oxygen bleach. Properly used, oxygen bleaches make fabrics whiter than do chlorine bleaches.

Bleaches work by acting on certain light-absorbing chemical groups, called *chromophores*, that cause a substance to be colored; most have a number of

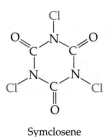

Symclosene

carbon-to-carbon double bonds that are attacked by the bleach. The oxidation of a chromophore can be written as follows:

$$-CH=CH-\overset{\overset{\displaystyle O}{\|}}{C}-CH_2- \xrightarrow{[O]} -\overset{\overset{\displaystyle O}{\|}}{C}-OH + HO-\overset{\overset{\displaystyle O}{\|}}{C}-\overset{\overset{\displaystyle O}{\|}}{C}-CH_2-$$

Chromophore   Fragments
(colored)   (colorless)

Thus, the bleach does not actually remove the stain, but merely changes it to a colorless form. Sometimes not all of the stain is oxidized, and a very faint ghost of the original stain can be seen.

SELF-ASSESSMENT Questions

1. Which of the following could serve as a fabric softener?
   a. $CH_3(CH_2)_{10}CON(CH_2CH_2OH)_2$
   b. $CH_3(CH_2)_{10}CH_2N^+(CH_3)_3Cl^-$
   c. $CH_3(CH_2)_{16}CH_2N^+(CH_3)_3Cl^-$
   d. $CH_3(CH_2)_{16}CH_2N^+(CH_3)_2CH_2(CH_2)_{16}CH_3Cl^-$

2. To what class of compounds do most fabric softeners belong?
   a. amphoteric detergents
   b. anionic detergents
   c. nonionic detergents
   d. quaternary ammonium salts

3. Oxygen bleaches
   a. release hydrogen peroxide when placed in hot water
   b. release oxygen molecules to act as an oxidizing agent
   c. include borates and hypochlorites
   d. are mild, nontoxic chemicals

4. The main component (other than water) of a chlorine-based household bleach is
   a. NaOH   b. Cl_2 c. NH_4Cl d. NaOCl

5. Bleaches act by
   a. enhancing surfactant action
   b. opening active sites
   c. sequestering hard-water ions
   d. oxidizing chromophores

6. Which is not a disadvantage of powdered perborate bleaches?
   a. They work best in concentrated solutions only.
   b. They don't work well with cotton fabrics.
   c. They are less active than chlorine bleaches.
   d. They require higher temperatures.

Answers: 1, d; 2, d; 3, b; 4, a; 5, d; 6, b

20.4 All-Purpose and Special-Purpose Cleaning Products

Learning Objective • List the major ingredient(s) (and their purposes) of some all-purpose cleaning products and some special-purpose cleaning products.

All-Purpose Cleaners

Various all-purpose cleaning products are available for use on walls, floors, countertops, appliances, and other tough, durable surfaces. Those for use in water solution may contain surfactants, sodium carbonate, ammonia, solvent-type grease cutters, disinfectants, bleaches, deodorants, and other ingredients. Some are great for certain jobs but not as good for others. They damage some surfaces but work especially well on others. Most important, they may be harmful when used improperly. Reading the labels is important.

Household ammonia solutions, straight from the bottle, are effective for loosening baked-on grease or burned-on food. Diluted with water, these solutions clean mirrors, windows, and other glass surfaces well. Mixed with detergent, ammonia rapidly removes wax from vinyl floor coverings. Ammonia vapors are highly irritating, so these products should not be used in a closed room. Ammonia should not be used on asphalt tile, wood surfaces, or aluminum because it may stain, pit, or erode these materials.

Baking soda (sodium bicarbonate, $NaHCO_3$) straight from the box is a mild abrasive cleanser. It absorbs food odors to some extent, making it good for cleaning the inside of a refrigerator.

Vinegar (acetic acid) cuts grease film. Vinegar should not be used on marble, though, because it reacts with and dissolves the marble, which is primarily calcium carbonate, pitting the surface.

$$\underset{\text{Marble}}{CaCO_3(s)} + \underset{\text{Vinegar}}{2\ CH_3COOH(aq)} \longrightarrow Ca^{2+}(aq) + 2\ CH_3COO^-(aq) + CO_2(g) + H_2O(l)$$

The physical and chemical processes that clean a surface take some time. An often-overlooked technique in cleaning is allowing the surface to soak in plain water or dilute soap solution. Diluted soap solution is an amazingly effective, cheap, and nonpolluting cleaner when it is simply sprayed on a surface and allowed to stand for a few minutes.

▲ Vinegar reacts with marble, pitting the surface. Carbon dioxide gas is given off, forming bubbles.

Hazards of Mixing Cleaners

Mixing bleach with other household chemicals can be quite dangerous. For example, mixing a hypochlorite bleach with hydrochloric acid produces poisonous chlorine gas.

$$2\,HCl(aq) + ClO^-(aq) \longrightarrow Cl^-(aq) + H_2O(l) + Cl_2(g)$$

Mixing bleach with toilet bowl cleaners that contain HCl is especially dangerous. Most bathrooms are small and poorly ventilated, so generating chlorine in such a limited space is hazardous. Chlorine can do enormous damage to the throat and the entire respiratory tract. If the concentration of chlorine in the air is high enough, it can kill.

Mixing bleach with ammonia is also extremely hazardous. Two of the gases produced are chloramine (NH_2Cl) and hydrazine (NH_2NH_2), both of which are quite toxic. *Never* mix bleach with other chemicals without specific directions to do so. In fact, it is a good rule to *never* mix any chemicals unless you know exactly what you are doing.

Special-Purpose Cleaners

Many highly specialized cleaning products are on the market. For metals, there are chrome cleaners, brass cleaners, copper cleaners, and silver cleaners. There are lime removers, rust removers, and grease removers, and cleaners specifically for wood surfaces, for vinyl floor coverings, and for ceramic tile. Let's look at just a few of the special-purpose cleaners found in almost any home.

Toilet bowl cleaners are acids that dissolve the "lime" buildup that forms in toilet bowls. The residue is mainly calcium carbonate ($CaCO_3$), deposited from hard water and often discolored by iron compounds and fungal growth. Calcium carbonate is readily dissolved by acid. The solid crystalline cleaners usually contain sodium bisulfate ($NaHSO_4$), and the liquid cleaners include hydrochloric acid (HCl), citric acid, or some other acidic material.

If rust is the main discoloration, a product specifically meant for rust removal may be the best. It will usually contain oxalic acid, $H_2C_2O_4$, or oxalate $(C_2O_4)^{2-}$ ions to make a complex of the iron ions.

Most scouring powder cleansers contain an abrasive such as sodium carbonate, calcium carbonate, or silica (SiO_2) that scrapes soil from hard surfaces, and a surfactant to dissolve grease. Some include a powdered bleach. Scouring powders are mainly intended for removing stains from porcelain tubs and sinks. Such abrasive cleansers may scratch the finish on appliances, countertops, and metal utensils. They may even scratch the surfaces of sinks, toilet bowls, and bathtubs. Dirt gets into the scratches and makes future cleaning even more difficult.

Glass cleaners are volatile liquids that evaporate without leaving a residue. A common glass cleaner is simply isopropyl alcohol (rubbing alcohol) diluted with water. Most commercial glass cleaners contain ammonia or vinegar for greater cleaning power.

When a kitchen drain becomes clogged, it is usually because the pipe has been blocked by grease. Drain cleaners often contain sodium hydroxide, either in the solid form or as a concentrated liquid solution. Solid sodium hydroxide dissolves in the water in the pipe, which is an exothermic process; the heat melts much of the solid or semi-solid grease. The sodium hydroxide then reacts with some of the grease, converting it to soap, which is easier to remove than the grease. Some products also contain bits of aluminum metal that react with the sodium hydroxide solution to form hydrogen gas, which bubbles out of the clogged area of the drain, creating a stirring action.

H₂C—COOH
|
HO—C—COOH
|
H₂C—COOH

Citric acid

You can make your own glass cleaner. Put about 0.8 L of cold water into a bucket. Carefully add 30 mL of sudsy ammonia solution and 120 mL of isopropyl (rubbing) alcohol. Mix well and pour the solution into clean spray bottles. Add a drop of blue food coloring if you want your cleaner to look like a popular name-brand window cleaner. (Adapted from *How to Clean and Care for Practically Anything*, Yonkers, NY: Consumer Reports Special Publications, 2002.)

Bathtub or shower drains are often clogged with hair, so the sodium hydroxide in the drain cleaner, like all bases, attacks the protein-based hair and breaks it apart. Many liquid drain cleaners contain bleach (sodium hypochlorite, NaOCl) as well. The best drain cleaner is prevention: Don't pour grease down the kitchen sink, and try to keep hair out of the bathtub drain. If a drain does get plugged, a mechanical device such as a plunger or plumber's snake is often a better choice than a chemical drain cleaner.

The active ingredient in most oven cleaners is sodium hydroxide. Several popular products are dispensed as an aerosol foam. The greasy deposits on oven walls are converted to soaps when they react with the sodium hydroxide. The resulting mixture can then be washed off with a wet sponge. Wear rubber gloves, because sodium hydroxide is extremely caustic and damaging to the skin. Wearing safety goggles to protect the eyes from spray is also highly recommended, since strong bases can cause blindness.

A hand sanitizer is useful when there is no water available for hand washing. Most hand sanitizers contain a high concentration of alcohol, which kills bacteria and evaporates rapidly during use, leaving the hands clean and dry. However, alcohol is flammable, so care should be taken around flames. A surfactant, often a quaternary ammonium salt, provides additional germicidal properties, and a thickening agent makes the sanitizer easier to spread over the hands.

The Anti-bacterial Agent Triclosan

Triclosan is a germicide that was used in surgical scrub formulations from the 1970s until fairly recently.

Triclosan

The *Staphylococcus aureus* bacterium (causing "staph" infections) was generally controlled by the use of such products, when the proper procedure for hand cleaning by surgeons was strictly followed. Over the decades since then, many manufacturers of bar soaps, liquid soaps, shampoos, and toothpastes added triclosan to their formulations and labeled them "antibacterial" or "antimicrobial." By the year 2000 about 75% of liquid soaps and at least 25% of bar soaps contained triclosan. The popularity of hand sanitizers and liquid soaps soared, and by 2014 over 2000 consumer products contained the "antibacterial agent" triclosan, including products to destroy mildew and mold. However, extensive scientific studies led the U.S. Food and Drug Administration (FDA), which has been studying this substance since 1974, to conclude that triclosan is only mildly effective in toothpaste and mouthwash formulations in decreasing plaque, but no more effective in hand soaps than is ordinary soap. In September 2017, the FDA prohibited the inclusion of triclosan in "consumer antiseptic washes" due to lack of efficacy in those products, as well as *possible* other detrimental health effects, including allergy sensitization, microbial resistance, endocrine disruption, and skin rashes. It seems obvious that triclosan would be found in the waste water stream, since the hands are washed with water that goes down the drain, and studies have shown this to be true. The FDA and U.S. Environmental Protection Agency (EPA) now state that "antibacterial agents [including triclosan] to be used with water are not generally recognized as safe." Surgical scrub formulations currently contain chlorhexidine gluconate, hexachlorophene or other antibacterial substances.

Do-it-yourselfers and garage mechanics often use a waterless hand cleaner to remove grease and grime from their hands. Like hand sanitizers, these hand cleaners do not need water to work. In fact, they work best if applied without water at first. Most waterless hand cleaners contain lanolin (Section 20.5), plus lemon or orange oil and a surfactant that thickens the mixture into a creamy emulsion. Pumice (powdered volcanic ash) is often added to help loosen the dirt from the creases and folds of the skin. The oil acts as a solvent for the grease on the hands, and the surfactant keeps the oil and grease suspended so that it can be wiped off with a towel.

"Green" and "Natural" Cleaners

Many cleaning products marketed for home use are claimed to be "natural" or "green" and thus "safer to use" than traditional products. Consumer choices are hampered by lack of a standard definition or federal oversight of such cleaners. Manufacturers do not have to list all the ingredients, because doing so could disclose trade secrets. There are 84,000 chemical compounds in the EPA's inventory, and the EPA requires manufacturers to provide only a caution of toxicity. Products claiming to be green can contain harmful substances. For example, a solvent made with a natural compound called *limonene*, derived from orange peels, is combustible and can irritate the skin and eyes.

In general, green products have more ingredients derived from plants. If scented, they usually employ a natural scent such as citrus or lavender. Green substances derived from corn or soy are common. One useful example is the ester ethyl lactate ($CH_3CHOHCOOCH_2CH_3$); both the alcohol (ethanol) and the carboxylic acid (lactic acid) from which the ester is made can be produced from corn. Ethyl lactate dissolves many different substances and has a low volatility, thus contributing little to air pollution. It also degrades readily in the environment.

SELF-ASSESSMENT Questions

1. The most dangerous mixture of cleaning agents, because of the release of a toxic gas, is
   a. rubbing alcohol and ammonia
   b. rubbing alcohol and vinegar
   c. acidic toilet cleaner and bleach
   d. bleach and rubbing alcohol

2. Household ammonia can be used safely for all of the following *except*
   a. cleaning aluminum
   b. loosening baked-on grease
   c. polishing mirrors
   d. wax removal

3. Vinegar can be used safely for all of the following *except*
   a. cutting grease
   b. dissolving mineral deposits
   c. polishing marble surfaces
   d. removing soap scum

4. Baking soda (sodium bicarbonate) can be used safely for all of the following *except*
   a. deodorizing a carpet
   b. deodorizing a refrigerator
   c. neutralizing acids
   d. neutralizing bleach

5. If you mix ammonia solution with vinegar, the products are
   a. a better toilet cleaner
   b. a salt solution
   c. a good solution for cleaning metals
   d. a surfactant effective on lime deposits

6. Mixing hypochlorite bleach and ammonia produces
   a. ammonium hypochlorite
   b. a more effective cleaner
   c. a surfactant
   d. toxic gases

7. Lime ($CaCO_3$) deposits in toilet bowls can be removed by treatment with
   a. NaOCl
   b. NaCl
   c. HCl
   d. NaOH

8. The most effective way to keep drains open is to
   a. avoid letting grease and hair get in them
   b. convert grease to soap before putting it down a drain
   c. treat them once a week with NaOCl
   d. treat them twice a month with NaOH

Answers: 1, c; 2, a; 3, c; 4, d; 5, b; 6, d; 7, c; 8, a

20.5 Solvents, Paints, and Waxes

Learning Objectives • Identify compounds used in solvents and paints, and explain their purposes. • Identify waxes by their structure.

Of the thousands of chemicals available for use in the home, three common categories are *solvents*, *paints*, and *waxes*.

Solvents

Solvents are used in the home to remove paint, varnish, adhesives, waxes, and other materials, from surfaces or even from skin, if paint is spilled on skin. "Turpentine" is a mixture of liquids obtained from pine trees. It has been largely replaced by petroleum solvents, called petroleum distillates or "mineral spirits" on labels in the United States and Canada. These solvents are mixtures of hydrocarbons, mainly with 7 to 14 carbons, that sometimes are added to some all-purpose cleansers as grease cutters (Figure 20.10). They dissolve grease and oil-based paints readily but, like gasoline, are highly flammable and deadly when swallowed. The lungs become saturated with hydrocarbon vapors, fill with fluid, and fail to function. Therefore, such solvents should be used only with adequate ventilation and never around a flame. Be sure to read—and heed—all precautions before you use any solvent. Gasoline should not be used for cleaning. It is too hazardous in too many ways.

A troubling problem connected with household solvents is the practice of inhaling fumes to get "high." The popularity of solvent sniffing seems to be related mainly to peer pressure, especially among young teenagers. Long-term sniffing can cause permanent damage to vital organs, especially the lungs. There can be irreversible brain damage, and some individuals have died of heart failure while sniffing solvents.

Paints

Paint is a broad term used to cover a wide variety of products—lacquers, enamels, varnishes, oil-base coatings, and a number of different acrylic- or latex-base finishes. Any or all of these materials can be found in any home. A paint contains three basic ingredients: a pigment, a binder, and a solvent. A pigment is a solid substance that is insoluble in water and that generally provides opaqueness; it may be white or colored. Binders hold the pigment to the surface to which it is applied, and the solvent helps create an appropriate consistency for ease of application.

The history of paint dates back to the beginnings of human civilization. In 2011, grinding tools and some ground red ochre pigment were discovered in abalone shells in Blombos Cave in South Africa and dated to 100,000 B.C.E. Cave paintings have been found in Europe and Africa that depict animals and cultural activities of prehistoric peoples. Since the Middle Ages, artists have used crushed minerals as pigments mixed with natural plant oils to create paintings. Until fairly recently, relatively few pigments were available as colorants. Many were oxides or hydroxides of iron, which were called "ochres" or "umbers," including yellow and red ochre, and burnt and raw umber. Green pigments were made of chromium oxides or copper compounds. Blues were more difficult to find; ground lapis lazuli (a mineral) was the best but rarest source of blue pigment. The red pigment *alizarin* was obtained from the roots of a plant called madder or from crushed cochineal insects. Even today, most of these colored pigments are available in art stores.

For purposes of painting walls or other surfaces, "white lead" [basic lead carbonate, $2 \, PbCO_3 \cdot Pb(OH)_2$] was the white pigment of choice for most paints until the 1970s. Its use was banned in 1977, and it has mainly been replaced by titanium dioxide (TiO_2). Titanium dioxide is a brilliant white nontoxic pigment with great stability and excellent hiding power. All ordinary paints are pigmented with titanium dioxide. For colored paint, small amounts of manufactured colored pigments or dyes are added to the white base mixture to create an almost unlimited number of tints.

The binder, or film former, is the substance that binds the pigment particles together and holds them on the painted surface. The most ancient binder consisted of egg yolks mixed with water. In oil paints, the binder is usually tung oil or linseed oil; cleanup requires mineral spirits or turpentine. In water-based paints, the binder is a latex-based or acrylic polymer. Most interior paints have polyvinyl acetate as the binder, and spills can be washed off with water. Artists' watercolor binders usually contain gum arabic, obtained from acacia trees. Exterior water-based paints use viscous acrylic polymers, called *resins*, as binders. Acrylic latex paints are much more resistant to rain and sunlight.

The solvent in paint keeps the paint fluid until it is applied to a surface. The solvent might be an alcohol, a hydrocarbon, an ester, or a ketone (or some mixture thereof), or water.

Paint may also contain additives such as a drier (or activator) to make the paint dry faster, a fungicide to resist mold growth, a thickener to increase the paint's viscosity, an anti-skinning agent to keep the paint from forming a skin inside the can, and a surfactant to stabilize the mixture and keep the pigment particles separated.

Paint, whether it is enamel, oil-based, or latex, hardens in different ways. Oil-based paints commonly use linseed oil that oxidizes to form a hard surface. Lacquer contains a polymer in a solvent, and it hardens by evaporation of the solvent.

▲ **Figure 20.10** Cleansers that contain petroleum distillates have labels warning of their combustibility and of the hazard of swallowing them.

4 Should I worry about lead-based paint if I purchase a classic home built in the 1950s?

Yes and no. Flaking paint in older homes can be inadvertently inhaled or ingested by small children or even adults, with serious health results. However, federal laws require an analysis of the paint in homes built before 1978 to determine if any lead-based paint was used. When you purchase a home, you should receive an analysis of the paint used in the home to be sure it is lead-free. If the exterior is found to contain lead-based paint, then special precautions must be taken when either sanding the siding for repainting or even removing the old siding in order to replace it with steel or vinyl siding. Workers must wear protective clothing and face masks to be sure that no airborne particles are inhaled.

Waxes

Chemically, a **wax** is an ester of a long-chain organic acid (fatty acid) with a long-chain alcohol. Waxes are produced by plants and animals mainly as protective coatings. These compounds are not to be confused with paraffin wax, which is made up of hydrocarbons and used for birthday candles.

Three typical waxes are shown in Figure 20.11. Beeswax is the material with which bees build honeycombs and is used in such household products as candles and shoe polish. Carnauba wax is a coating that forms on the leaves of certain palm trees found in Brazil. It is a mixture of esters similar to those in beeswax and is used in making automobile wax, floor wax, and furniture polish, as well as lip balms such as Chapstick™ and eye make-up. Spermaceti wax, which is extracted from the head of a sperm whale, is largely cetyl palmitate. Once widely used in making cosmetics and other products, spermaceti wax is now in short supply because the sperm whale was hunted almost to extinction and is now protected by international treaties.

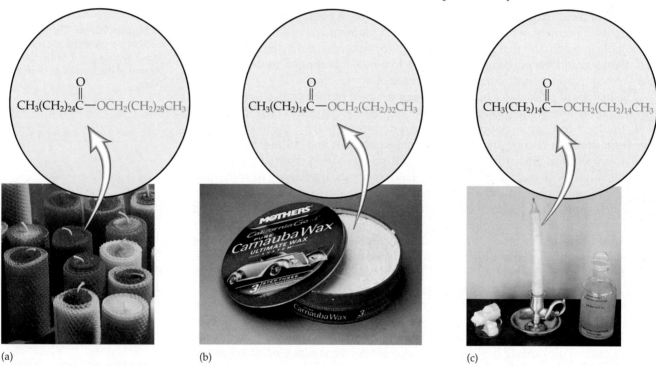

(a) (b) (c)

▲ **Figure 20.11** The molecular views show the esters that are typical components of (a) beeswax, (b) carnauba wax, and (c) spermaceti.

Lanolin, the grease in sheep's wool, is also a wax. Because it forms stable emulsions with water, it is used in various skin creams and lotions. Many natural waxes have been replaced by synthetic polymers such as silicones (Section 10.5), which are often cheaper and more effective.

SELF-ASSESSMENT Questions

1. Which is *not* a property of organic solvents in household and commercial products?
   a. They are generally good for removing varnishes and adhesives.
   b. They are usually flammable.
   c. They are volatile, and the vapors affect breathing and function of the heart and brain when inhaled.
   d. They have a rather low toxicity if ingested.

2. The three basic components of a paint are a pigment, a solvent, and a
   a. builder   b. preservative   c. binder   d. wax

3. Which of the following is a wax?
   a. $CH_3(CH_2)_{14}COOCH_3$
   b. $CH_3(CH_2)_{28}CH_2OH$
   c. $CH_3(CH_2)_{28}COOCH_3$
   d. $CH_3(CH_2)_{14}COOCH_2(CH_2)_{28}CH_3$

4. Waxes are
   a. long-chain fatty acids   b. long-chain alcohols
   c. esters with two long chains   d. proteins

5. Modern paint pigments
   a. use PbS for black
   b. use copper compounds for blues
   c. substitute TiO_2 for $PbCO_3$
   d. contain no inorganic components

20.6 Cosmetics: Personal-Care Chemicals

Learning Objectives • Describe the chemical nature of skin, hair, and teeth.
• Identify the principal ingredients in various cosmetic products and their purposes.

Ages ago, people used materials from nature for cleansing, beautifying, and otherwise altering their appearance. Evidence indicates that at least 6000 years ago, Egyptians used powdered antimony and the green copper ore *malachite* as eye shadow. Egyptian pharaohs used perfumed hair oils as far back as 3500 B.C.E. Galen, a Greek physician of the second century C.E., is said to have invented cold cream. Fashion-conscious gentlemen of seventeenth-century Europe used cosmetics lavishly, often to cover the fact that they seldom bathed. Ladies of eighteenth-century Europe as well as Japanese women whitened their faces with lead carbonate ($PbCO_3$), and many died from lead poisoning.

The use of cosmetics has a long and interesting history, but past usage doesn't even come close to the amounts and varieties of cosmetics used by people in the modern industrial world. Each year, we spend billions of dollars on everything from hair sprays to nail polishes, from mouthwashes to foot powders. Combined sales of the world's 100 largest cosmetics companies are more than $190 billion annually, and increasing at a rate of at least 5% per year.

What is a cosmetic? The U.S. Food, Drug, and Cosmetic Act of 1938 defined **cosmetics** as "articles intended to be rubbed, poured, sprinkled or sprayed on, introduced into, or otherwise applied to the human body or any part thereof, for cleansing, beautifying, promoting attractiveness or altering the appearance." Soap, though obviously used for cleansing, is specifically excluded from restrictions in the law. Also excluded are substances that affect the body's structure or functions. Antiperspirants, products that reduce perspiration, are legally classified as drugs, as are antidandruff shampoos.

The main difference between drugs and cosmetics is that drugs must be proven safe and effective before they are marketed, while cosmetics generally do not have to be tested before being marketed. The Food, Drug, and Cosmetic Act states that if a product has drug properties, it must be approved as a drug. Nevertheless, the cosmetics industry uses the term *cosmeceutical* to refer to cosmetic products that allegedly have medicinal or drug-like benefit.

More than 8000 different chemical compounds are used in cosmetics, sold in tens of thousands of different combinations. Most brands of a given type of cosmetic contain the same (or quite similar) active ingredients. Thus, advertising is usually geared toward selling a name, a container, or a fragrance rather than the product itself.

Skin Creams and Lotions

Skin is the body's largest organ, with an average area in adults of about 18 ft². It encloses the body, forming a barrier to keep harmful substances out and moisture and nutrients in (Figure 20.12). The outer layer of skin, the *epidermis*, is divided into two parts: dead cells on the outside (the corneal layer) and living cells on the inside, continually replacing corneal cells, which are then sloughed off.

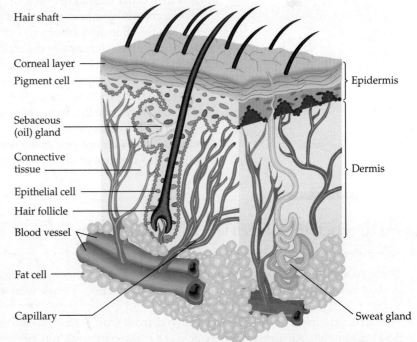

Hair shaft

Corneal layer

Pigment cell

Sebaceous (oil) gland

Connective tissue

Epithelial cell

Hair follicle

Blood vessel

Fat cell

Capillary

Epidermis

Dermis

Sweat gland

▲ **Figure 20.12** Cross section of skin. Skin has two main layers: Each square inch of the dermis contains about 1 m of blood vessels, 4 m of nerves, 100 sweat glands, several oil glands, more than 3 million cells, and the active part of hair follicles. The epidermis, lying over the dermis, consists of a thin layer of cells that divide continuously. These cells move upward to the corneal layer, dying as they do so. Cosmetics generally affect only the outer corneal layer.

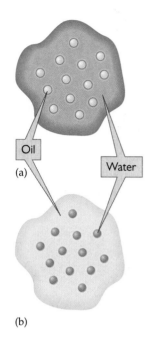

▲ Figure 20.13 (a) A lotion is an emulsion of oil in water. (b) A cream is an emulsion of water in oil. A lotion feels cool because evaporating water removes heat from the skin. A cream feels greasy.

The corneal layer is composed mainly of a tough, fibrous protein called **keratin**. Keratin has a moisture content of about 10%. Below 10% moisture, human skin is dry and flaky. Above 10%, conditions are ideal for the growth of harmful microorganisms. Skin is protected from loss of moisture by **sebum**, an oily secretion of the sebaceous glands. Exposure to sun and wind can leave the skin dry and scaly, and too frequent washing removes natural skin oils.

Cosmetics are applied to the dead cells of the corneal layer. Most of the preparations applied to the skin are in the form of lotions or creams. A **lotion** is an emulsion of tiny oil droplets dispersed in water. A **cream** is the opposite, having tiny water droplets dispersed in oil (Figure 20.13), and generally has a much higher viscosity than a lotion. The essential ingredients of creams and lotions are water, an oily substance that forms a protective film as well as functioning as a softening agent, and an emulsifier that binds the polar water to the nonpolar oily component, usually glyceryl stearate. The emulsifier helps transport the water to the lower layers of the skin, where it is needed, which may be the lotion or cream's primary function. Products claiming to be "moisturizers" must contain all three components.

Among the substances typically used as the oily component are glycerol and mixtures of alkanes from crude oil, including mineral oil and petroleum jelly. Others are natural fats and oils, such as lanolin obtained from sheep's wool, *aloe vera* gel, olive oil, and palm oil. Petroleum jelly (such as Vaseline®) or mineral oil (baby oil) soften the skin but do not restore moisture to the lower layers of skin. It may seem strange that gasoline, a mixture of alkanes, dries out the skin but that higher alkanes in mineral oil and petroleum jelly soften it. Keep in mind, however, that gasoline is a thin, free-flowing liquid. It dissolves natural skin oils and carries them away. Higher alkanes are viscous, staying right on the skin and serving as emollients.

Some cosmetics contain *humectants* to curb the loss of water from the skin. These humectants include glycerol, lactic acid, and urea. Glycerol, with its three hydroxyl groups, holds water by hydrogen bonding. Urea binds water below the surface of the epidermis, right down to the skin-building cells. Urea and lactic acid also act as *exfoliants*, causing the dead cells to fall off more rapidly, leaving new, fresher-looking skin exposed. Perfumes and fragrances provide no real function in lotions and creams, other than to help sell the product.

Some creams have been formulated with vitamins, antibacterials, and other ingredients. Vitamin E in skin lotions and creams has been demonstrated to accelerate the healing of wounds, scrapes, and surgical incisions, as it is an antioxidant and anti-inflammatory. Vitamin C applied to the skin may not be nearly as useful to the body in a cream as when it's ingested via fruits and vegetables, but many manufacturers claim its benefits in skin products. Creams for diaper rash in babies usually contain zinc oxide, an inorganic substance that helps the body heal the rash; such products can be used for healing skin scrapes or minor skin wounds in adults as well. A variety of antibacterial creams contain bacitracin zinc as the active ingredient, and anti-itch formulations may contain hydrocortisone or diphenhydramine hydrochloride. (See Chapter 18.)

Anti-aging Creams and Lotions

Some cosmetics are claimed to reduce wrinkles, spots, and other signs of aging skin, and consumers spend billions of dollars each year for such products. Many contain alpha hydroxy acids (AHAs), derived from fruit and milk sugars. An alpha hydroxy acid is a carboxylic acid that has a hydroxyl group on the carbon atom next to the carboxyl carbon. Glycolic acid and lactic acid are typical examples. AHAs and related substances act as exfoliants. They seem relatively safe, but they may cause redness, swelling, burning, blistering, bleeding, rash, itching, and skin discoloration in some people. People who use AHA-containing products have greater sensitivity to sun, perhaps putting them at greater risk of skin cancer. The best treatment for wrinkling of the skin is prevention, by avoiding excessive exposure to the sun and by not smoking cigarettes. Cigarette smoking leads to premature aging of the skin. Nicotine causes constriction of the tiny blood vessels that feed the skin. Repeated constriction over the years causes the skin to lose its elasticity and become wrinkled.

Shaving Creams

Shaving creams consist of a soap or soap-like compound such as stearic acid, glycerol, water, plus fragrances, preservatives, and possibly a foaming agent such as cocamide. Stearic acid, a long-chain saturated fatty acid (see Chapter 17), acts like a soap, forming a smooth, slippery surface on the skin and hair. Glycerol is attracted to water by hydrogen bonding and acts as a humectant. It hydrates the beard, easing hair removal. It also increases the viscosity of the product.

Sunscreens

Ultraviolet rays in sunlight turn light skin darker by triggering production of **melanins**. The dark melanin protects the deeper layers of the skin from damage. Excessive exposure to ultraviolet radiation causes premature aging of the skin and may lead to skin cancer, especially squamous cell carcinomas. Excessive tanning has resulted in an epidemic of skin cancers, including melanomas. Shorter-wavelength ultraviolet rays (UV-B) are more energetic and are especially harmful.

Sunscreens offer either physical or chemical protection from UV rays. Physical sunscreens, or sunblocks, are opaque films that prevent UV rays from reaching the skin, thus protecting against both UV-A and UV-B. Most contain white zinc oxide (ZnO) or titanium dioxide TiO_2 pigments to reflect the sun's rays. The most recent formulations contain nanoparticles of ZnO or TiO_2, and so do not appear to be as white as the earlier formulations, and thus are more acceptable to most users of sunscreen. Chemical sunscreens absorb UV rays before they can cause any damage. Typical ingredients include avobenzone, oxybenzone, homosalate, octyl methoxycinnamate, and octinoxate, which protect against UV-A. Recent studies have shown that most of those UV-absorbents may possibly cause hormonal changes in the body, if excessive amounts are rubbed into the skin. The problem, of course, is that better protection from UV rays is afforded with larger amounts applied.

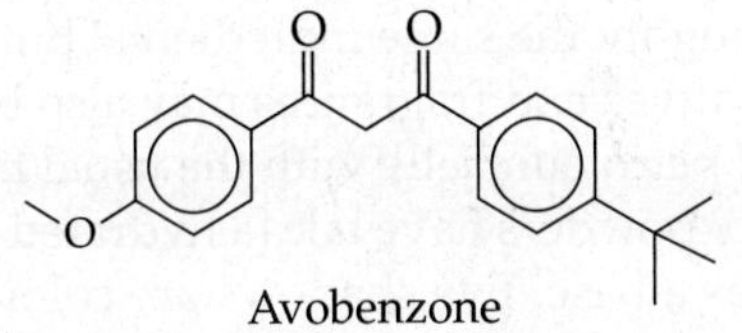

Avobenzone

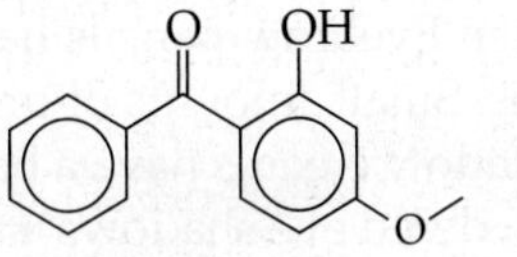

Oxybenzone

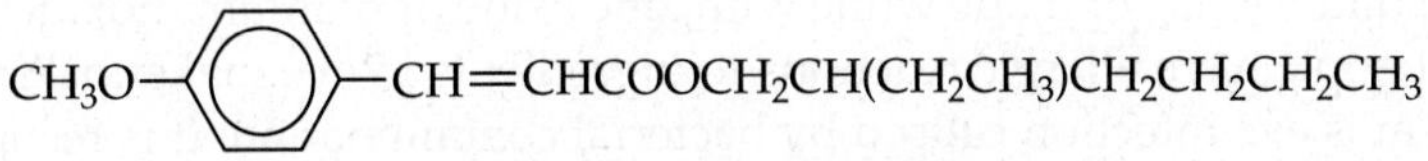

$$CH_3O-\bigcirc-CH=CHCOOCH_2CH(CH_2CH_3)CH_2CH_2CH_2CH_3$$

Octyl methoxycinnamate (OMC)
(2-ethylhexyl-4'-methoxycinnamate)

The **sun protection factor (SPF)** is a rating that indicates the effectiveness of a product's UV absorption in terms of length of time. SPF values range from 2 to 45 or more. An SPF of 15, for example, indicates that the product should allow you to stay in the sun 15 times longer without burning than you could with unprotected skin. Consumers should be aware, however, that the required amount applied for such protection is usually much greater than what people apply and that the creams can easily be rubbed off. Reapplying often is recommended no matter what the SPF rating. Sunscreens with quite high SPF ratings may give consumers false hope that they will not get sunburn. Some studies show there is relatively little difference in protection between an SPF 30 and an SPF 45 product. Skiers and winter outdoors enthusiasts may be well advised to use sunscreen even in the winter because the sun's rays readily bounce off white snow.

Lipsticks and Lip Balms

Lipsticks and lip balms are quite similar to skin creams in composition and function. They are made of an oil and a wax, with a higher proportion of wax to provide solidity. Because the lips have little in the way of protective oils, they often dry out,

Tetrabromofluorescein

leading to chapped lips. Lipsticks and lip balms prevent moisture loss as well as soften and brighten the lips. The oil in these products is frequently castor oil, sesame oil, or mineral oil. Waxes often employed are beeswax, carnauba wax, and candelilla wax, obtained from a shrub in the Euphorbia family, found in northern Mexico and adjoining areas of the southern United States. Dyes and pigments provide color to lipsticks. Perfumes are added to cover up the unpleasant fatty odor of the oil, and antioxidants are used to retard rancidity. Bromo acid dyes such as tetrabromofluorescein, a bluish-red compound, are responsible for the color of many lipsticks.

Eye Makeup

A thick, black, slightly oily mixture called *kohl* has been used around the eyes for thousands of years. Egyptian queens and noblewomen used such mixtures to enhance the look of the eyes. Even today, kohl is widely used to decorate the face and eyes in southern Asia, especially in India, the Middle East, western Africa, and the Horn of Africa, where it is used in social and religious ceremonies, particularly Islamic festivals; for traditional dances; and for medicinal reasons. Many believe that kohl protects eyesight and has a cooling and healing effect on eyes. Certainly, black substances absorb much of the sun's radiation, including harmful UV light. Even twenty-first-century American football players who apply streaks of eyeblack under the eyes on game days take advantage of this property.

So what is kohl? In the Western world, carbon is the main ingredient, but traditional kohl in other cultures has consisted of finely ground lead(II) sulfide (galena) or, less commonly, antimony sulfide. Naturally, there is a concern about the use of lead and antimony compounds on the skin and near the eyes, especially with children. In the United States, it is illegal to import or sell foreign-made kohl; it is not on the list of color additives approved by the FDA.

Commercially available mascara and eyeliners have a base of soap, oils, fats, and waxes. Mascara is colored brown by iron(III) oxide pigments, and black by carbon (lampblack). A typical composition is 40% wax, 50% soap, 5% lanolin, and 5% coloring matter. Eyebrow pencils have roughly the same ingredients, but in different proportions. Small amounts of preservatives and fragrances may also be added.

Eye-shadow creams have a base of petroleum jelly with the usual fats, oils, and waxes. Powdered eye shadows and face powders have talc [a hydrated magnesium silicate, $Mg_3Si_4O_{10}(OH)_2$] or corn silk as a base. Eye shadows are colored green by chromium(III) oxide (Cr_2O_3) or blue by ultramarine (a silicate mineral that contains some polysulfide ions), or made white with zinc oxide or titanium dioxide pigments.

Some people have allergic reactions to ingredients in eye makeup. A more serious problem is eye infection caused by bacterial contamination. It is recommended that eye makeup be discarded after 3 months. Mascara is of special concern because of the wand applicator, which can easily scratch the cornea.

As in other industries, cosmetic companies are moving toward greener products. Sales of natural cosmetics topped $14 billion in the United States in 2017, twice as much as in 2007. Made from renewable plant, animal, and mineral products, these cosmetics may be better for the environment, but there is little evidence that they work better than traditional cosmetics. In fact, natural components may be more likely to cause allergic reactions than the synthetic ingredients. Natural is not necessarily better: Think of poison ivy as an extreme example.

Deodorants and Antiperspirants

Many commercially available products are called "deodorants" by most people, who think that "antiperspirant" is a more sophisticated term for the same thing. In reality, a **deodorant** and an **antiperspirant** have different functions. Most commercial products contain both of those components, plus ingredients to create the correct consistency, plus perfumes.

A chemical deodorant is a germicide to kill odor-causing bacteria. Bacteria act on perspiration residue and sebum (natural body oil), in the warm, moist environment of armpits, to produce malodorous short-chain fatty acids such as trans-3-methyl-2-hexenoic acid. The germicide is usually an alcohol such as stearyl alcohol, or sodium stearate. (Triclosan is no longer used; see Feature box on page 681.) Powdered deodorants for the feet also contain germicides.

Antiperspirants retard perspiration, and are classified as drugs by the FDA. Active antiperspirant ingredients include complexes of chlorides and hydroxides of aluminum and zirconium. Either aluminum chlorohydrate [$Al_2(OH)_5Cl \cdot 2 H_2O$] or an Al-Zr complex (called aluminum zirconium trichlorohydrex gly) is found in antiperspirants. Aluminum salts function as mild **astringents**, which constrict the openings of the sweat glands, and react with keratin fibrils in the sweat ducts, forming gel-like plugs. In those two ways they restrict the amount of perspiration that can escape.

Commercial products for underarm use can be formulated into creams or lotions, or they can be dissolved in alcohol and applied as sprays. The roll-on mechanism was invented in the 1960s, but deodorant "sticks" have now taken over the market share. Products targeted at men typically contain a higher percentage of both major active components compared with products meant for women. Sweating is a natural, healthy body process, and to stop it routinely may not be wise. Regular bathing and changing of clothes make antiperspirants unnecessary for most people, but modern culture seems to require its use.

Toothpaste: Soap with Grit and Flavor

Centuries ago, people used small sticks, crushed charcoal, bird feathers, or porcupine quills to clean the teeth. People still buy toothpicks for removing food particles stuck between teeth, though dental floss is by far the norm in twenty-first-century culture. Tooth powders were invented in the nineteenth century and used commonly until at least the mid-twentieth century. Toothpastes and gels are a more recent development, and now, one may choose among a vast variety of flavors and additives.

The only essential components of toothpaste are a detergent and an abrasive. Abrasives may account for at least 50% of a typical toothpaste formulation. The ideal abrasive should be hard enough to clean teeth but not hard enough to damage tooth enamel. Most toothpastes do not feel gritty when rubbed between fingers or in the mouth, because the particles are so small. Abrasives frequently used in toothpaste are listed in Table 20.1.

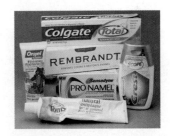

▲ Toothpastes are available under many brand names and in various formulations. Some contain baking soda (sodium bicarbonate) and/or hydrogen peroxide as ingredients, but there is only marginal evidence of their effectiveness. The only essential ingredients in a toothpaste are a detergent and an abrasive.

TABLE 20.1 Abrasives Commonly Used in Toothpastes

Name	Chemical Formula
Precipitated calcium carbonate	$CaCO_3$
Insoluble sodium metaphosphate	$(NaPO_3)_n$
Calcium hydrogen phosphate	$CaHPO_4$
Tricalcium phosphate	$Ca_3(PO_4)_2$
Calcium pyrophosphate	$Ca_2P_2O_7$
Hydrated alumina	$Al_2O_3 \cdot nH_2O$
Hydrated silica	$SiO_2 \cdot nH_2O$

The typical detergent found in toothpaste is sodium dodecyl sulfate [sodium lauryl sulfate, $CH_3(CH_2)_{11}OSO_3^-Na^+$], which is surprisingly the same detergent found in most hair shampoos (see "Hairy Chemistry" below). Toothpastes have a variety of flavors, colors, and sweeteners such as sorbitol, glycerol, or saccharin (Section 17.5). The classic flavors include spearmint, wintergreen, and peppermint, but more recent additions to the flavor options include ginger, cinnamon, bubble gum (especially for children), lemon, orange, and vanilla. Some toothpastes are colored or even striped with food dyes, or contain fine mica to make it sparkly. Thickeners such as cellulose gum and polyethylene glycols, and preservatives such as methyl *para*-hydroxybenzoate, are important in toothpaste formulations. Titanium dioxide often accounts for a significant percentage of a toothpaste formulation, functioning mainly as a white pigment, but also as a mild abrasive.

Methyl *para*-hydroxybenzoate
(Methylparaben)

You can make your own tooth-paste. Stir together 180 mL (about $^2/_3$ cup) of sodium bicar-bonate (baking soda) and 60 mL (about $^1/_4$ cup) of sodium chloride (table salt). To this dry mixture, add 15 mL (about 1 tablespoon) of glycerol. Stir in just enough water to make a thick paste. Add a few drops of peppermint flavor-ing or one drop of peppermint extract to improve the taste. Store in a closed container.

Fluorine compounds are added to most modern toothpastes, even the least expensive generic products, to reduce tooth decay. Sodium fluoride, NaF, and sodium monofluorophosphate, Na_2PO_3F, are the most common fluorine-containing compounds used, though tin(II) fluoride (stannous fluoride) is still found in some products. The enamel of teeth is composed mainly of hydroxyapatite [$Ca_5(PO_4)_3OH$]. Fluoride from toothpaste (or fluorinated drinking water) converts part of the enamel to fluoroapatite [$Ca_5(PO_4)_3F$], which is stronger and more resistant to decay than hydroxyapatite.

Tooth decay is caused mainly by bacteria that convert sugars to *plaque*, a biofilm that is composed of thousands of bacteria, small particles, proteins, and mucus that is left on the teeth when they are not cleaned adequately. These bacteria convert sugars to acids such as lactic acid. Acids dissolve tooth enamel. If not removed, plaque can harden into *calculus*, also known as *tartar*. Brushing and flossing remove plaque and thus prevent decay. Decay can be minimized by eating sugars only at meals rather than in snacks and by brushing immediately after eating. Acids such as lactic acid in beer and phosphoric acid (H_3PO_4) in cola drinks may also erode the tooth enamel of people who consume these beverages in large amounts.

Specialized toothpastes for highly sensitive teeth may contain potassium nitrate or strontium chloride. Tartar-control products and whitening toothpastes contain higher percentages of abrasives or hydrogen peroxide. Toothpaste formulations with baking soda ($NaHCO_3$) and hydrogen peroxide (H_2O_2) as ingredients are purported to prevent gum disease, the main cause of tooth loss in adults, but there seems to be little evidence that they do.

Popular teeth whiteners often include hydrogen peroxide as a bleaching agent. The peroxide works on colored compounds in teeth in much the same way that bleaches work on stained clothing. However, the action of the H_2O_2 is slow. In tooth-paste, the peroxide is not in contact with the teeth long enough to have a significant effect. Most of the successful teeth whiteners require application for 30 minutes or more. Dentists' whitening preparations often use higher concentrations of peroxide along with ultraviolet light, which speeds the reaction. Caution is advised concern-ing teeth-whitening preparations. If used too often, they can cause teeth to become brittle.

Perfumes, Colognes, and Aftershaves

Perfumes are among the most ancient and widely used cosmetics. Originally, all per-fumes were extracted from natural sources such as fragrant plants. Their chemistry is exceedingly complex, but chemists have identified many of the components of perfumes and synthesized them in the laboratory, allowing the production of many synthetic perfumes. The best perfumes are probably still made from natural materi-als, because chemists have so far been unable to identify all the many important, but sometimes minor, ingredients.

A good perfume may have a hundred or more constituents. The components are divided into three categories, called *notes*, based on differences in volatility. The most volatile fraction (the one that vaporizes most readily) is called the **top note**. This frac-tion, made up of relatively small molecules, is responsible for the fragrance when a perfume is first applied. Common top notes include citrus and ginger scents and chemical compounds such as phenylacetaldehyde (Table 20.2), which has a lilac or hyacinth odor. Note that the fragrance molecules in Table 20.2 are fat-soluble. They dissolve in skin oils and can be retained for hours.

The **middle note** (*or heart note*) is intermediate in volatility. This fraction is responsible for the lingering aroma after most of the top-note compounds have vaporized. Typical middle notes include lavender and rose scents and chemical compounds such as 2-phenylethanol, which has the aroma of roses. The **end note** (*or base note*) is the low-volatility fraction and is made up of compounds with large molecules, often with musky odors. Essential oils—volatile oily compounds from

TABLE 20.2 Compounds with Flowery and Fruity Odors Used in Perfumes

Name	Structure[a]	Odor	Molecular Formula
Citral	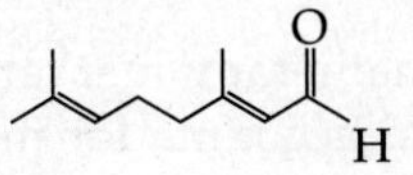	Lemon	$C_{10}H_{16}O$
Irone	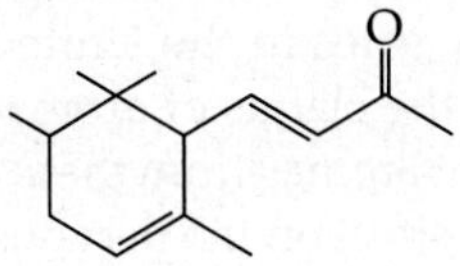	Orris root, iris	$C_{14}H_{22}O$
Jasmone	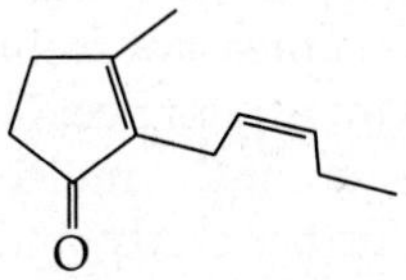	Jasmine	$C_{11}H_{16}O$
Phenylacetaldehyde	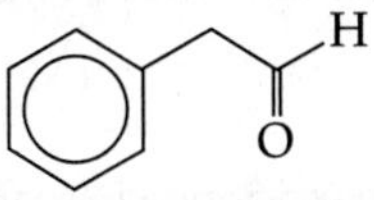	Lilac, hyacinth	C_8H_8O
2-Phenylethanol	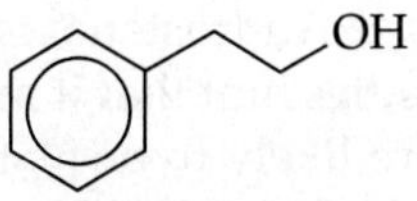	Rose	$C_8H_{10}O$

[a]Note that each compound has only one oxygen atom and ten or more carbon atoms.

plants with pleasant aromas—such as those from sandalwood and patchouli are typically used to supply base notes.

Odors vary with dilution. A concentrated solution of a compound may have an unpleasant odor, yet a dilute solution of it will have a pleasant aroma. Such compounds include musks (Table 20.3), which are often added to perfumes to moderate the odors of flowery or fruity top and middle notes. The civet, from which the compound civetone is obtained, is a skunklike animal. Its secretion, like that of the skunk, is a defensive weapon. The secretion from musk deer is probably a pheromone that serves as a sex attractant for the deer. Synthetic civetone and muscone have been produced, but many upscale perfume makers still prefer natural civetone. Almost all muscone used in perfumes today is synthetic.

TABLE 20.3 Unpleasant Compounds Used to Fix Delicate Odors in Perfumes

Compound	Structure	Natural Source	Molecular Formula	Molecular Mass (u)
Civetone	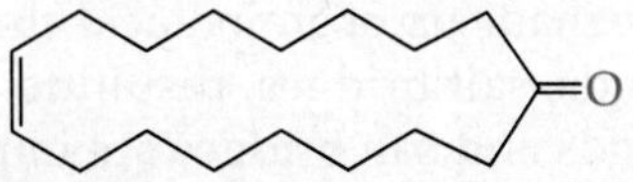	Civet	$C_{17}H_{30}O$	250.42
Muscone	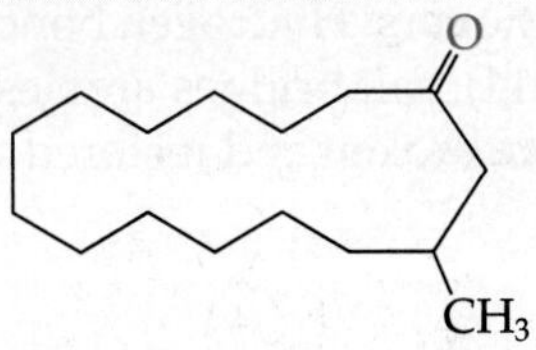	Musk deer	$C_{16}H_{30}O$	238.41
Indole	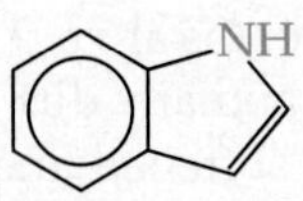	Feces	C_8H_7N	117.15

Aromatherapy

Aromatherapy is the practice of using essential oils as a purported way to promote health and well-being. Certainly olfactory sensations are very important in everyday life. Aromatherapy includes the use of soap, bath additives, shower gels, candles, room fragrances, oil burners, and cosmetic products containing essential oils.

Commonly used essential oils are those from marigold, lemon, orange, grapefruit, geranium, lavender, jasmine, and bergamot. The essential oils are diluted by carrier oils, which are vegetable oils such as sweet almond oil, apricot kernel oil, and grapeseed oil, before being applied to the skin.

The U.S. aromatherapy market is valued at $800 million a year and accounts for most of worldwide sales. Home fragrance products account for nearly half the aromatherapy sales in the United States. It is difficult to evaluate the claims of aromatherapy's proponents. The pleasant aromas from these oils are supposed to have a positive effect on the nervous system or, minimally, have a positive psychological effect on people. It may be as simple as a pleasant feeling caused when a scent recalls a pleasant experience. The chemical effects on the nervous system of the molecules found in the essential oils need further study, and the sensitivity of the human nose is more complex than we may realize.

α-Androstenol

Menthol

Are there sex attractants for humans? The evidence is scanty, but some perfume makers add α-androstenol to their products. α-Androstenol is a steroid that occurs naturally in human hair and urine. Some studies hint that it may act as a sex attractant for human females, but this seems more likely to be just an advertising ploy. Even if there are human sex attractants, it is doubtful that they have much influence on human behavior, because we probably have overriding cultural constraints. Anyway, we seem to prefer the sex attractant of the musk deer.

A perfume usually consists of 10% to 25% fragrant compounds and fixatives dissolved in ethyl alcohol. **Colognes** are perfumes diluted with ethyl alcohol or with an alcohol–water mixture. They are only about one-tenth as strong as perfumes, usually containing only 1% to 2% perfume essence.

Aftershave lotions are similar to colognes. Most are about 50% to 70% ethanol, the remainder being water, perfume, and food coloring. Some have menthol or eucalyptol added for a cooling effect on the skin. Others contain an emollient to soothe chapped skin.

Perfumes are the source of many allergic reactions associated with cosmetics and other consumer products. **Hypoallergenic cosmetics** are those that purport to cause fewer allergic reactions than regular products cause. The term has no legal meaning, but most of these cosmetics do not contain perfume.

Hairy Chemistry

Like skin, hair is composed mainly of the fibrous protein keratin. The keratin of hair has five or six times as many disulfide linkages as the keratin of skin. Recall (Section 16.6) that protein molecules are made up of amino-acid chains held together by four types of forces: hydrogen bonds, salt bridges, disulfide linkages, and dispersion forces. Of these, hydrogen bonds and salt bridges are important to understanding the actions of shampoos and conditioners. Hydrogen bonds between protein chains are disrupted by water (Figure 20.14); salt bridges are destroyed by changes in pH (Figure 20.15). Disulfide linkages are broken and restored when hair is permanently waved or straightened.

Shampoo

▲ Shampoos are available in many forms and under many brand names. The only essential ingredient in any shampoo is a detergent.

The visible portion of hair is dead. Only the root is alive. The hair shaft is lubricated by sebum. Washing the hair removes this oil and any dirt adhering to it. When hair is washed, the keratin absorbs water and is softened and made more elastic. The water disrupts hydrogen bonds and some of the salt bridges. Acids and bases are particularly disruptive to salt bridges, making control of pH important in hair care.

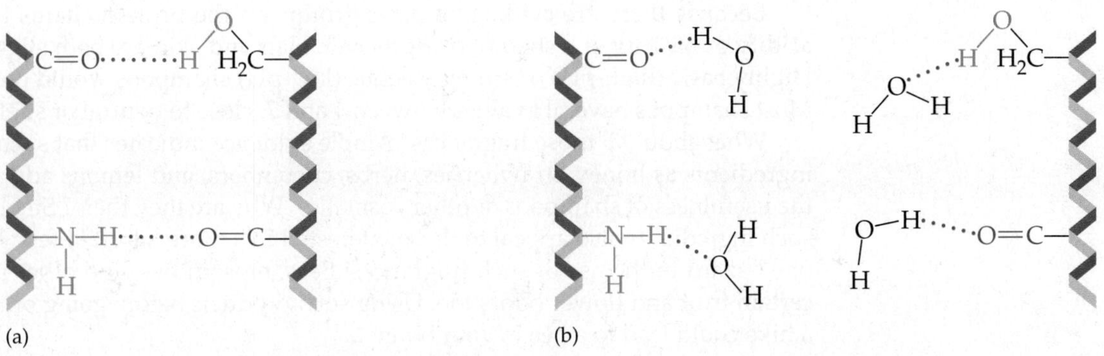

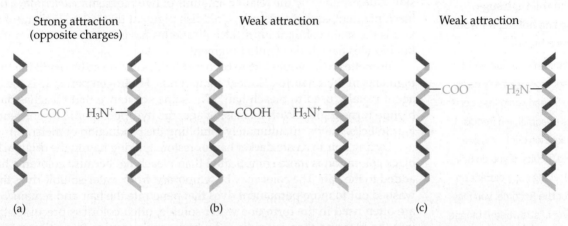

▲ **Figure 20.14** (a) In a strand of hair, adjacent protein chains are held together by hydrogen bonds that link the carbonyl group of one chain to the amide group of another. (b) When hair is wet, these groups can hydrogen-bond with water rather than with each other, disrupting the hydrogen bonds between chains.

Strong attraction (opposite charges)   Weak attraction   Weak attraction

—COO⁻ H₃N⁺—   —COOH H₃N⁺—   —COO⁻ H₂N—

(a)   (b)   (c)

▲ **Figure 20.15** (a) Strong electrostatic forces between an ionized carboxyl group on one protein chain and an ionized amino function on another hold the chains together. (b) A change in pH can disrupt this salt bridge. If the pH drops, the carboxyl group takes on a proton and loses its charge. (c) If the pH rises, a proton is removed from the amino group and it is left uncharged. In either case, the forces between the protein chains are weakened considerably.

The number of salt bridges is maximized at a pH of 4.1. Before World War II, the cleansing agent in shampoos was a soap, which worked well in soft water but left a dulling film on hair in hard water. People often removed the film by using a rinse containing vinegar or lemon juice. Such rinses are not needed with today's products.

Modern shampoos use a synthetic detergent as a cleansing agent, which is the only essential ingredient, usually 20% to 25% of the formulation. In adult shampoos, the detergent is often an anionic type such as sodium dodecyl sulfate (usually written as sodium lauryl sulfate on product labels), ammonium lauryl sulfate, or sodium laureth sulfate, or a mixture of detergents. In shampoos used for babies and children, the detergent is often an amphoteric surfactant (Section 20.2) that is less irritating to the eyes.

Why is there such a significant difference in cost between the least expensive and the most expensive shampoos? Most of the difference is traced to additives. You can buy shampoos that are fruit- or herb-scented, protein- or vitamin-enriched, or made especially for oily, dry, or color-treated hair. Shampoos for oily, normal, and dry hair seem to differ mainly in the concentration of the detergent. Shampoo for oily hair is more concentrated; shampoo for dry hair is more dilute. Vitamins in shampoos serve no purpose. Shampoos meant for color-treated hair contain sunscreen, since UV radiation destroys the molecules used to darken gray or blonde hair. Whether the sunscreen stays on the hair after the hair is rinsed is questionable. Because hair is protein, a protein-enriched shampoo does give it more body. The protein (usually keratin or collagen) coats the hair and literally glues split ends together. Protein is often added to conditioners, too. Conditioners are mainly long-chain alcohols or long-chain quaternary ammonium salts, which are similar to the compounds used in fabric softeners (Section 20.3) and work in much the same way to coat hair fibers.

5 Does a thicker shampoo work better to clean my hair?

Shampoo is thickened with fine silica or a polymer. A thicker shampoo simply has more of the thickening agent added, rather than more detergent. Consumers seem to want thicker shampoo for greater convenience in the shower: A thinner shampoo would dribble through the fingers.

Because there are acidic and basic groups on the protein chains in hair, the acidity or basicity of a shampoo affects hair. Hair and skin are both slightly acidic. Highly basic (high-pH) or strongly acidic (low-pH) shampoos would damage hair. Most shampoos have pH values between 4 and 7, close to neutral or slightly acidic.

What about all those fragrances? Ample evidence indicates that such "natural" ingredients as honey, strawberries, herbs, cucumbers, and lemons add nothing to the usefulness of shampoos or other cosmetics. Why are they there? Smells sell, and such ingredients also appeal to those interested in a more natural lifestyle. There is one hazard in the use of such fragrances: Bees, mosquitoes, and other insects like certain fruit and flower odors too. Using such products before going on a picnic or a hike could lead to a bee in your bonnet.

Hair Coloring

Melanin is a broad term that refers to colorants in skin and hair. The color of hair and skin is determined by the relative amounts of two pigments: *eumelanin*, a brownish-black pigment, and *pheomelanin*, a reddish pigment that colors the hair of redheads and is the main colorant in freckles. Brunettes have mostly eumelanin in the hair, but blondes have little of either pigment.

Brunettes who would like to become blondes can do so by oxidizing the colored pigments in their hair to colorless compounds. Hydrogen peroxide is the oxidizing agent usually used to bleach hair. (The same substance that bleaches darker hair blonde turns it gray. With increasing age, the hydrogen peroxide concentration in hair follicles builds up, ultimately inhibiting the production of melanin.)

Dyeing hair to create darker hair or restore graying hair to the original brown or black coloration is more complicated than bleaching, because colorants have to be added to the hair. The color may be temporary, from water-soluble dyes that can be washed out to more permanent dyes that penetrate the hair and remain. Such dyes are often used in the form of a water-soluble, often colorless precursor that soaks into the hair and then is oxidized by hydrogen peroxide to a colored compound. Permanent dyes affect only the dead outer portion of the hair shaft. New hair, as it grows from the scalp, has its natural color.

Permanent dyes are often derivatives of an aromatic amine called *para*-phenylenediamine (Figure 20.16). Variations in color can be obtained by placing substituents on this molecule. The *para*-phenylenediamine molecule produces a black color, and the derivative *para*-aminodiphenylaminesulfonic acid is used in blonde formulations. Intermediate colors can be obtained by the use of other derivatives. One derivative, *para*-methoxy-*meta*-phenylenediamine (MMPD), has been shown to be carcinogenic when fed to rats and mice, but the hazard to those who use it as hair dye is not yet known. It is interesting to note that one substitute for MMPD was its homolog, *para*-ethoxy-*meta*-phenylenediamine (EMPD). Screening revealed that EMPD causes mutations in bacteria. Such mutagens are often also carcinogens.

6 **Why must I leave a hair coloring preparation on my hair for the full time suggested on the bottle, if I want to become a blonde?**

The eumelanin in your hair, which is black or dark brown, is fairly easily oxidized to a nearly colorless compound. Any pheomelanin (reddish) that may exist in your hair is oxidized with difficulty. If you do not leave a bleaching hair colorant on your hair for the full time, you may end up being a "strawberry" blonde with definite reddish tints instead of the pale blonde you intended.

para-Phenylenediamine
(a)

para-Aminodiphenylaminesulfonic acid
(b)

MMPD
(c)

EMPD
(d)

▲ **Figure 20.16** (a) Most hair dyes are derivatives of *para*-phenylenediamine. These include (b) *para*-aminodiphenylaminesulfonic acid, (c) *para*-methoxy-*meta*-phenylenediamine (MMPD), and (d) *para*-ethoxy-*meta*-phenylenediamine (EMPD).

Henna is a natural orange-brown dye obtained from the *Lawsonia inermis* plant, and has been used since antiquity for coloring hair and decorating the skin and fingernails in detailed designs in the Indian subcontinent, the Middle East, and the Horn of Africa.

Hair treatments that develop color gradually, such as Grecian Formula, use rather simple chemistry. A solution containing colorless lead acetate [Pb(CH₃COO)₂] is rubbed on the hair. As it penetrates the hair shaft, Pb^{2+} ions react with sulfur atoms in the hair to form black lead(II) sulfide (PbS). Repeated applications produce darker colors as more lead sulfide is formed. The safety of using lead compounds in contact with skin has been questioned.

Permanent Waving: Chemistry to Curl Your Hair

The chemistry of curly hair is interesting. Hair is protein, and adjacent protein chains are held together by disulfide linkages. Permanent wave lotion contains a reducing agent such as thioglycolic acid (HSCH₂COOH), which ruptures the disulfide linkages (Figure 20.17), allowing the protein chains to be pulled apart as the hair is held in a curled position on rollers. The hair is then treated with a mild oxidizing agent such as hydrogen peroxide. Disulfide linkages are formed in new positions to give shape to the hair.

The same chemical process can be used to straighten naturally curly hair. The change in curliness depends only on how the hair is arranged after the disulfide bonds have been reduced and before the linkages have been restored.

Hair Sprays

Hair can be held in place by **resins**, solid or semisolid organic materials that form a sticky film on the hair. Common resins used on hair are polyvinylpyrrolidone (PVP) and its copolymers.

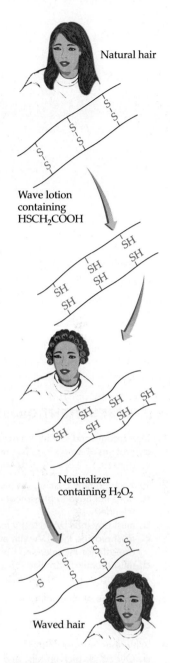

▲ **Figure 20.17** Permanent waving of hair is accomplished by breaking disulfide bonds between protein chains and then re-forming them in new positions.

The resin is dissolved in a solvent, and the mixture is sprayed on the hair, where the solvent evaporates. The propellants in hair sprays are usually volatile hydrocarbons, such as butane, which are flammable.

Holding resins are also available as mousses. A **mousse** is simply a foam or froth. The active ingredients, like those of hair sprays, are resins such as PVP. Coloring agents and conditioners are also available as mousses.

Hair Removers

Chemicals that remove unwanted hair are called **depilatories**. Most of them contain a soluble sulfur compound, such as sodium sulfide or calcium thioglycolate [Ca(OOCCH₂SH)₂], formulated into a cream or lotion. These strongly basic mixtures destroy some of the peptide bonds in the hair so that it can be washed off. Remember that skin is made of protein, too, so any chemical that attacks the hair can also damage the skin.

Hair Restorers

For years, men have been searching for a way to make hair grow on bald spots. Women suffer from baldness, too, but seem to consider wigs more acceptable than men do.

Minoxidil was first introduced as a drug for treating high blood pressure. It acts by dilating the blood vessels. When people who were taking the drug started growing hair on various parts of their bodies, it was applied to the scalps of people who were becoming bald. Minoxidil can produce a growth of fine hair anyplace on the skin where there are hair follicles. It is sold under the trade name Rogaine®, although generic versions are now available. To be effective, it must be used continuously, and the cost can be several hundred dollars a year or more.

The Well-Informed Consumer

This chapter cannot even attempt to tell you everything about the many chemical products you have in your home. Books have been written about some of them, and there are many volumes in the library that can help you if you want to know more. There is also much information available on the Internet, but much of it is little more than advertising hype; read it with caution! We hope that the knowledge you have gained here and the information you read on product labels will make you a better-informed consumer, and help prevent accidents.

If you are unhappy with a product you have bought, tell the company that made it. The cosmetic industry gets thousands of complaints from customers every year. You can also complain to the FDA. Many reports are presented each year to the FDA regarding adverse effects of cosmetic products. The FDA cannot ban a cosmetic without proof of harm, and testing a product for carcinogenicity (Chapter 21) is an involved process, takes several years, and costs a minimum of half a million dollars.

Most cosmetics are made from inexpensive ingredients, yet many highly advertised cosmetics have extremely high price tags. Are they worth it? When you are faced with many different brands of a product at the supermarket or drugstore, remember that the most expensive one is not necessarily better than the others. This is a judgment that lies beyond the realm of chemistry. The $100 product inside that fancy little bottle with the famous label may have cost the manufacturer less than $5 to produce. On the other hand, if the product makes you look better or feel better, perhaps the price is not important. Only you can decide.

SELF-ASSESSMENT Questions

1. Skin creams and lotions may contain all the following except
   a. proteins   b. mineral oil
   c. glycerin   d. an emulsifying agent

2. To market a new cosmetic, a manufacturer
   a. does not have to prove that it is safe and effective for its intended use
   b. must pay the FDA to test it for safety
   c. must provide the FDA with evidence that the ingredients are safe for the intended use
   d. must submit an application to the FDA showing that the product is not a cosmeceutical

3. Which type of ultraviolet radiation is most responsible for sunburn?
   a. UV-A   b. UV-B   c. UV-C   d. UV-X

4. Why does hair get "dirty?"
   a. Our hands pick up oils, and they get into the hair by touching the hair.
   b. Dirt and dust are naturally in the air and settle on the hair.
   c. Hair, being a protein, attracts greasy dirt.
   d. Sebaceous glands near the hair roots secrete oils that pick up airborne dirt.

5. Sucrose contributes to tooth decay by
   a. being converted by bacteria into acid that erodes tooth enamel
   b. directly attacking tooth enamel
   c. mechanically abrading tooth enamel
   d. triggering the release of stomach acid

6. Which of the fractions of a perfume is made up of the smallest molecules?
   a. end note   b. middle note
   c. base note   d. top note

7. Aftershave lotions contain all the following except
   a. ethanol   b. menthol   c. water   d. protein

8. If a cosmetic is labeled as hypoallergenic, it
   a. is all-natural
   b. is dermatologist-tested
   c. probably has no perfume in it
   d. still may contain substances that can cause allergic reactions

9. When hair is wetted, what type of force is disrupted?
   a. hydrogen bonds   b. disulfide linkages
   c. salt bridges   d. peptide bonds

10. Kohl is
    a. simply crushed coal
    b. used in face decorating in many cultures
    c. a good reflector of UV and visible radiation from the sun
    d. effective at protecting the eyes from disease

11. The essential ingredient in a shampoo is
    a. a conditioner   b. a fragrance
    c. a detergent   d. protein

12. What chemical forces are disrupted during home permanents?
    a. hydrogen bonds   b. disulfide linkages
    c. salt bridges   d. peptide bonds

13. People who have little melanin in the skin
    a. are more susceptible to sunburn and skin cancer
    b. usually are naturally dark-haired
    c. have hair that turns gray early in life
    d. seldom have to use sunscreen

Answers: 1, a; 2, a; 3, b; 4, d; 5, a; 6, d; 7, d; 8, c; 9, a; 10, b; 11, c; 12, b; 13, a

GREEN CHEMISTRY Practicing Green Chemistry at Home
Marty Mulvihill, University of California-Berkeley

Learning Objective • Identify how green chemistry principles can be applied when selecting greener household products.

The Twelve Principles of Green Chemistry help chemists design and use chemicals safely and more efficiently. These principles can also guide the use of chemical products in our own homes. While we don't often get the opportunity to design the products found in our homes, we do control which ones we buy and how we use them. You will see how we can become greener consumers by combining these principles with an understanding of the chemistry found in this chapter. The guidelines below are an example of the way that the twelve principles of green chemistry can be used to inspire greener consumer habits.

1. **Buy and use only what you need.** Principle 1 helps a chemist avoid unnecessary waste. You can avoid waste at home by only buying and using the products that you need.
2. **Choose safer products.** Principles 3, 4, and 5 all remind chemists to choose the safest ingredients possible that still provide the desired function. The same is true at home. Consult expert resources to determine which products are the safest.

3. **Choose products that minimize resources depletion.** Principle 7 encourages the chemist to use renewable feedstock chemicals and Principle 6 promotes energy efficiency. We can minimize depletion of resources by choosing products that are energy efficient or made from renewable resources.
4. **Choose reusable, recyclable, or readily degradable materials.** Principle 10 tells the chemist to design chemicals that degrade in the environment. Look for products that readily degrade in the environment. Buy items in packaging that can be reused or recycled.
5. **Stay informed.** Principle 11 encourages chemists to monitor reactions in real time so that they can avoid hazards and waste. As consumers, we can monitor new developments in product safety and regulations so that we can choose the best products.
6. **Prevent accidents.** Principle 12 reminds chemists that they should strive to prevent accidents by using inherently safer chemistry. Accidents are prevented by reading product labels, using proper protective gear, maximizing ventilation, and choosing the safest products.

Summary

Section 20.1—Some plants contain **saponins**, which produce lather. Plant ashes contain K_2CO_3 and Na_2CO_3, which are alkaline because they react with water to form hydroxide ions. A **soap** is a salt of a long-chain carboxylic acid, traditionally made from fat and lye (NaOH), with glycerol as a by-product. A soap molecule has a **hydrophobic** hydrocarbon tail (soluble in oil or grease) and a **hydrophilic** ionic head (soluble in water), usually COO^-. Soap molecules disperse oil or grease in water by forming **micelles**, tiny spherical oil droplets in which the hydrocarbon tails of soap molecules are embedded. Agents that stabilize suspensions in this way are called **surface-active agents** (or **surfactants**). Soaps do not work well in hard water (water containing Ca^{2+}, Fe^{2+}, or Mg^{2+} ions) because they form insoluble salts. Water softeners contain an ion exchange resin that removes hard water ions and replaces them with Na^+ ions. Other water-softening agents can precipitate hard water ions as phosphates or carbonates.

Section 20.2—A synthetic **detergent** usually has a long hydrocarbon tail and a polar or ionic head, usually with SO_3^- in the structure. ABS detergents have branched-chain structures and are not easily biodegraded. LAS detergents have a linear chain of carbon atoms that can be broken down by microorganisms. Soaps, ABS detergents, and LAS detergents are **anionic surfactants** that have a negative charge on the active part. A **nonionic surfactant** has no charge. A **cationic surfactant** such as a quaternary ammonium salt (quat) has a positive charge. An **amphoteric surfactant** carries both a positive and a negative charge. Detergent formulations often include **builders** such as

complex phosphates to increase detergency. **Optical brighteners**, which are fluorescent dyes, are often added to detergents to make whites appear "whiter." Bleaches, fragrances, and fabric softeners may be in laundry detergents. Liquid dishwashing detergents contain mainly LAS surfactants as the active ingredients. Detergents for automatic dishwashers are strongly alkaline and should not be used for hand dishwashing.

Section 20.3—Cationic surfactants include quaternary ammonium salts, and those with two long hydrocarbon chains are used as fabric softeners. **Bleaches** are oxidizing agents. Hypochlorite bleaches release chlorine. Oxygen bleaches usually contain perborates or percarbonates. Bleaches act by oxidizing colored chromophores to colorless fragments.

Section 20.4—All-purpose cleaners may contain surfactants, ammonia, and disinfectants. Reading the label is imperative, so that a cleaner may be used properly and safely. Household ammonia, sodium bicarbonate (baking soda), and vinegar are good cleaners for many purposes. Special-purpose cleaners include toilet bowl cleaners that contain an acid, scouring powders that contain an abrasive, and glass cleaners that contain ammonia. Drain cleaners contain NaOH and sometimes bleach. The active ingredient in oven cleaners is NaOH. Bleach should never be combined with any other cleaning product.

Section 20.5—Solvents are used to remove paint and other materials. Many organic solvents are volatile and flammable. **Paint** contains a pigment, a binder, and a solvent. Titanium dioxide is the most common pigment. Water-based paints contain a polymer as the binder; oil-based paints use tung oil or linseed oil. A **wax** is an ester of a fatty acid with a long-chain alcohol. Beeswax and carnauba wax are commonly used waxes. **Lanolin**, from sheep's wool, is also a wax, used in skin creams and lotions.

Section 20.6—A **cosmetic** is something applied to the body for cleansing or promoting attractiveness, but cosmetics do not have to be proven safe and effective like drugs do. Skin has an outer layer made of a tough, fibrous protein called **keratin**. **Sebum**, an oily secretion, protects skin from loss of moisture. A **lotion** is an emulsion of oil droplets in water; a **cream** is an emulsion of water droplets in oil. Each of these can form a protective physical barrier over the skin to hold in moisture. Creams and lotions also soften skin; such skin softeners are called **emollients**. Humectants—such as glycerol hydrogen—bond to water and hold it to the skin. Ultraviolet rays trigger the production of the pigment **melanin**, which darkens the skin. **Sunscreens** block or absorb ultraviolet (UV) radiation, of which UV-B rays are more harmful than UV-A rays. The **sun protection factor (SPF)** indicates how effective a sunscreen's UV-absorbing substances are. Eye makeup and lipstick are mainly blends of oil, waxes, and pigments. Some traditional makeup components are dangerous. **Deodorants** mask body odor and kill odor-causing bacteria. **Antiperspirants** retard perspiration, acting as **astringents**, constricting the openings of the sweat glands. Toothpaste contains a detergent and an abrasive, as well as a thickener, flavoring, and often an added fluoride compound. The fragrance of a **perfume** is due to three fractions: the **top note** (the most volatile), the **middle note** (of intermediate volatility), and the **end note** (of lowest volatility). **Colognes** are diluted perfumes. **Hypoallergenic cosmetics** purport to cause fewer allergic reactions than do regular products, but the term has no legal meaning; they usually have no perfumes. Shampoos are synthetic LAS detergents blended with water, thickening agents, and fragrances. Dark hair contains the pigment eumelanin, red hair contains pheomelanin, and blonde hair contains very little of either pigment. Many hair dyes are based on *para*-phenylenediamine. Hair is bleached with hydrogen peroxide. Permanent waving or straightening of the hair uses a reducing agent to break the disulfide linkages between protein chains and then an oxidizing neutralizer to re-form the linkages in different positions. Hair sprays contain **resins**, organic materials that form a sticky film on the hair. Such resins are also available in foam form as **mousses**. Chemicals that remove hair are called **depilatories**, and most such chemicals can also damage the skin.

GREEN CHEMISTRY—The principles of green chemistry can be interpreted to help us make greener choices. Understanding chemical properties can lead to the selection of greener products.

Learning Objectives

Learning Objectives	Associated Problems
• Describe the structure of soap, and explain how it is made and how it works to remove greasy dirt. (20.1)	1, 2, 17–23, 89, 90
• Explain the advantages and disadvantages of soap. (20.1)	3, 6
• Explain how water softeners work, and what substances are used as water softeners. (20.1)	5, 13, 24–28, 30, 31
• List the structures, advantages, and disadvantages of synthetic detergents. (20.2)	4, 29, 38–42, 91
• Classify surfactants as amphoteric, anionic, cationic, and nonionic, and describe how they are used in various detergent formulations. (20.2)	7, 8, 33–35, 44, 45, 93, 97
• Identify a fabric softener from its structure, and describe its action. (20.3)	46–48
• Name two types of laundry bleaches, and describe how they work. (20.3)	49, 50
• List the major ingredient(s) (and their purposes) of some all-purpose cleaning products and some special-purpose cleaning products. (20.4)	9–12, 51–57, 98, 99, 102, 103, 106, 108
• Identify compounds used in solvents and paints, and explain their purposes. (20.5)	58–60
• Identify waxes by their structure. (20.5)	61, 101, 108
• Describe the chemical nature of skin, hair and teeth. (20.6)	12, 65, 76, 77, 104, 105
• Identify the principal ingredients in various cosmetic products and their purposes. (20.6)	14–16, 36, 37, 62–75, 78–88, 94, 103
• Identify how green chemistry principles can be applied when selecting greener household products.	107, 108

Conceptual Questions

1. Explain in detail how soap molecules remove greasy dirt from a surface, in terms of the shape and nature of soap molecules, polarity, and intermolecular interactions.

2. What is the function of water in the cleaning process, when the cleaning agent is classified as a soap? Be sure to include intermolecular interactions in your answer.

3. What are some advantages and disadvantages of soaps?

4. What are some advantages and disadvantages of synthetic detergents?

5. Explain why and how soap does not work well in "hard" water. Be sure to include the ions present in water to make it "hard."

6. What is a micelle?

7. What is the structural difference between a soap and an anionic detergent?

8. Cationic surfactants are often used with nonionic surfactants but are seldom used with anionic surfactants. Why?

9. What ingredients are used in liquid dishwashing detergents (for hand dishwashing)? How do detergents for automatic dishwashers differ from those used for hand dishwashing?

10. List some uses of diluted household ammonia. What safety precautions should be followed when using ammonia?

11. What are some properties of ammonia solution that make it useful? What are some drawbacks?

12. List two properties of baking soda that make it useful as a cleaning agent.

13. How does a water-softening tank work?

14. What is the legal definition of a cosmetic? Is soap a cosmetic?

15. Is a deodorant classified as a cosmetic or as a drug? Why?

16. Is an antiperspirant classified as a cosmetic or a drug? Why?

Problems

Soaps

17. How does a potassium soap differ from a sodium soap? Write the formulas for sodium dodecyl sulfate and potassium dodecyl sulfate.

18. What ingredient is added to scouring soaps? What property of that additive makes it useful for scouring?

19. Write a generalized word equation for the formation of soap from lye and a fat.

20. Write a word equation that describes the reaction of tripalmitin with lye (sodium hydroxide).

21. Give the structural formula for each of the following.
    a. sodium stearate   **b.** potassium palmitate

22. Give the structural formula for each of the following.
    a. potassium stearate   **b.** sodium palmitate

23. Give the formulas of the products formed in the following reaction.

$$
\begin{array}{l}
\text{CH}_2\text{O}-\overset{\overset{\text{O}}{\|}}{\text{C}}(\text{CH}_2)_{10}\text{CH}_3 \\[4pt]
\text{CH}-\text{O}-\overset{\overset{\text{O}}{\|}}{\text{C}}(\text{CH}_2)_{10}\text{CH}_3 \quad + \ 3\,\text{NaOH} \longrightarrow \\[4pt]
\text{CH}_2\text{O}-\overset{\overset{\text{O}}{\|}}{\text{C}}(\text{CH}_2)_{10}\text{CH}_3
\end{array}
$$

Water Softeners

24. What ions make water "hard?"

25. Write the equation that shows how carbonate ions react with magnesium ions in hard water to form an insoluble precipitate.

26. Write the equation that shows how phosphate ions react with calcium ions in hard water to form an insoluble precipitate.

27. Write a word equation that shows how a water softener using sodium zeolite removes calcium ions from the water.

28. Describe how trisodium phosphate softens water. Write a word equation that shows how that reaction occurs.

Builders

29. What is a (detergent) builder?

30. How does sodium tripolyphosphate aid the cleansing action of a detergent?

31. How do zeolites aid the cleansing action of a detergent?

32. What advantage do zeolites have over carbonates as builders?

Detergents: Structures and Properties

33. Classify each of the following surfactants as anionic, cationic, nonionic, or amphoteric.
    a. $\text{CH}_3(\text{CH}_2)_9\text{CH}_2\text{CH}_2\text{OSO}_3{}^-\ \text{Na}^+$
    b. $\text{CH}_3-\text{C}_6\text{H}_4-(\text{CH})_{10}\text{N}^+(\text{CH}_3)_3\text{Cl}^-$

34. What is an amphoteric surfactant? In what kind of product would one be found?

35. What is a nonionic surfactant? In what kind of product would one be found?

36. What is the most common type of detergent found in shampoos and toothpaste?

37. Name the most common detergent found in shampoos and toothpaste. Give the "common" name found on product labels and the IUPAC name.

38. What is the structural difference between an ABS and an LAS detergent? What are the similarities?

Problems 39 through 43 refer to compounds I, II, and III below; C_6H_4 represents a 1, 4-disubstituted benzene ring.

    I. $\text{CH}_3(\text{CH}_2)_9\text{CH}(\text{CH}_3)-\text{C}_6\text{H}_4-\text{SO}_3{}^-\text{Na}^+$
    II. $\text{CH}_3(\text{CH}_2)_{11}\text{OSO}_3{}^-\text{Na}^+$
    III. $\text{CH}_3(\text{CH}_2)_{15}\text{N}^+(\text{CH}_3)_3\text{Br}^-$

39. Which is a linear alkyl sulfonate (LAS)?

40. Which is an alkyl sulfate?

41. Which is/are not biodegradable?

42. Which is a quaternary ammonium salt?

43. Which is/are anionic surfactants? Which is a cationic surfactant?

44. What sorts of surfaces should not be cleaned with ammonia? Why?

45. What sorts of surfaces should not be cleaned with vinegar? Why?

Fabric Softeners

46. Give an example of a compound that acts as a fabric softener.

47. How do fabric softeners work?

48. Why should you not add more fabric softener sheets to the dryer or more liquid fabric softener to the wash water?

Bleaches

49. Give the names of two important compounds found in powdered bleaches.

50. What is the name and chemical formula of the compound found in all liquid bleaches?

51. How do perborate bleaches work?

52. List some advantages and disadvantages of perborate bleaches.

53. How do chlorine bleaches work?

54. List some advantages and disadvantages of chlorine bleaches.

Solvents and Paint

55. What are the main hazards of using solvents in the home?

56. List two hazards of cleansers that contain petroleum distillates.

57. Should gasoline be used as a cleaning solvent? Why or why not?

58. What are the name and chemical formula for the white base pigment found in most paints?

59. What is the main difference between latex-based paints and oil-based paints in terms of the solvent needed for clean-up? Be sure to name solvents that would be appropriate for each.

60. What is the purpose of a binder in a paint? Name some modern and ancient substances used in paint.

61. Describe the basic structure of a wax.

Skin Treatments and Skin Products

62. What is an emollient? Name three substances commonly used as emollients.

63. List the three main components of a skin lotion.

64. Why is an emulsifying agent such as glyceryl stearate important in a skin lotion or cream?

65. What is the primary reason why skin becomes rough and flaky?

66. What is one dangerous ingredient in kohl, used in many cultures for religious, social, and supposedly medicinal purposes?

67. What is the difference between a sunscreen and a sunblock?

68. What substances are commonly used in sunblocks?

69. What are the names of at least two substances used as sunscreens?

Lipsticks and Lip Balms

70. What are the main ingredients in lip balms and lipsticks? Provide the specific name of one of those ingredients.

71. What materials are used to color lipsticks?

Toothpastes

72. What is the principal cause of tooth decay? What are the best ways to prevent tooth decay?

73. What is the chemical formula for sodium monofluorophosphate? What is its purpose in toothpaste?

74. What is the purpose of (a) glycerol, (b) sodium fluoride, and (c) sodium dodecyl sulfate in toothpaste?

75. What is the purpose of (a) potassium nitrate, (b) hydrogen peroxide, and (c) hydrated silica in toothpaste?

76. What is the chemical formula for tooth enamel? Write out its chemical name.

77. How do fluorides strengthen tooth enamel? What is the name and chemical formula for the substance formed when tooth enamel is treated with fluoride?

Perfumes and Colognes

78. What is a perfume? What is a cologne?

79. What is a musk? Why are musks added to perfumes?

80. What is the function of menthol in some aftershave lotions?

81. Give the name of a compound that functions as an antiperspirant and tell how it functions.

82. Give the name of a compound that functions as a deodorant and tell how it functions.

Hair and Hair Care

83. What is the difference between a temporary and a permanent hair dye? Why is a permanent dye not really permanent?

84. What are resins? How are resins used in hair care?

85. Briefly describe the two processes involved in curling straight or straightening curly hair, with respect to the disulfide linkages in the hair.

86. What is the most important chemical component of hair shampoo? Write the name and chemical formula for a commonly used compound.

87. Describe the difference between hair coloring that allows one to go from dark- to light-haired and one used to change gray hair to a darker color.

88. What is the difference between a temporary, semipermanent, and permanent hair dye in terms of how they work?

Expand Your Skills

89. Match the formula in the left column with the description or use in the right column.

   1. CH_3COOH a. ingredient in vinegar
   2. $Na_2Al_2Si_2O_8$ b. mild abrasive cleaner
   3. $NaHCO_3$ c. used as bleach
   4. $NaOCl$ d. used to make soap
   5. $NaOH$ e. water softener
   6. Na_3PO_4 f. zeolite

Problems 90 through 93 refer to structures I through IV below.

I. $CH_3(CH_2)_{15}N^+H_2CH_2COO^-$
II. $CH_3(CH_2)_{10}COO^-$
III. $CH_3(CH_2)_9CH(CH_3)-C_6H_4-SO_3^- Na^+$
IV. $CH_3(CH_2)_{11}-C_6H_4-N^+(CH_3)_3Cl^-$

90. Which are soaps?

91. Which are synthetic detergents?

92. Which is a principal ingredient in baby shampoo?

93. Which is an amphoteric surfactant?

94. Write the structure for isopropyl palmitate, an ingredient in skin creams.

95. The hardness of water is reported as milligrams of calcium carbonate per liter of water. Calculate the quantity of sodium carbonate needed to soften 10.0 L of water with a hardness of 295 mg/L. (*Hint*: How much Ca^{2+} is present in the water?)

96. Phenylacetaldehyde and 2-phenylethanol have similar molecular masses, but phenylacetaldehyde is a top-note component and 2-phenylethanol is a middle-note component of perfumes. Referring to Table 20.2, explain why this is so.

97. When you search the Internet for information about the composition of a particular industrial detergent mixture, you find that it contains quats. What is the full name of this type of detergent (not the nickname)? What type of surfactant are these ingredients?

98. Sodium hydroxide performs several functions in a drain cleaner. Describe two of these functions.

99. Optical brighteners make fabric appear to reflect more light (energy) than strikes the fabric. Do optical brighteners violate the law of conservation of energy? Explain.

100. Referring to Figure 20.11, describe how the esters of carnauba wax and spermaceti are similar and how they are different.

101. Review Figure 20.11a and Section 9.8. Draw the condensed structure of **(a)** the carboxylic acid and **(b)** the alcohol that are formed when the ester in beeswax undergoes hydrolysis.

Critical Thinking Exercises

Apply knowledge that you have gained in this chapter and one or more of the FLaReS principles (Chapter 1) to evaluate the following advertising claims.

20.1 A laundry detergent "leaves clothes 50% brighter, 50% whiter, [and is] 100% new and improved."

20.2 A product is claimed to be an odor-removing spray.

20.3 A brand of toothpaste claims to be "all-natural" and contains silica and SLS [sodium lauryl sulfate].

20.4 A cosmetics advertisement states "Most mineral-based cosmetics have synthetic preservatives such as parabens, dyes, and fillers which may be harmful. Our brand uses bismuth oxychloride, a natural ingredient."

20.5 A new skin cream "reduces wrinkles and prevents or reverses damage caused by aging and sun exposure."

Collaborative Group Projects

Prepare a PowerPoint, poster, or other presentation (as directed by your instructor) for presentation to the class.

1. Research the kinds of pigments used by Renaissance artists and the Dutch Masters, how they were obtained and prepared, and some of the dangers of those pigments.

2. Research the use of henna and its importance in South Asian and other cultures.

3. Compare facial makeup in three different cultures and/or historical periods, including the social significance and the chemical components.

102. Write equations for the dissolution of the calcium carbonate buildup in a toilet bowl by each of the following toilet bowl cleaners.
    a. hydrochloric acid   b. sodium bisulfate
    c. citric acid ($HC_6H_7O_7$)

103. If a sunscreen label states "SPF 30," and you have skin that will burn in about 15 minutes when exposed to direct, bright sun, how long could you theoretically stay in the sun before reapplying the cream? What are some reasons you should reapply sooner than your calculation?

104. Into what category of organic compounds do the primary components of skin primarily fall?

105. Into what category of organic compounds does the primary component of hair primarily fall?

106. Which of the following molecules would not be considered a volatile organic compound?
    a. hexane   b. ether   c. formaldehyde
    d. water   e. acetone

107. Calculate how much energy we can save by using cold (unheated tap water) compared to hot water (65 °C) per load of laundry. A typical modern high-efficiency washer uses between 50–100 L of water per load (older top loading washers used as much as 170 L). Complete your calculation assuming that tap water is 10 °C and that your washer uses 75 L of water.

108. Identify the purpose for each of the following ingredients found in household products.
    a. NaOCl   b. $CH_3(CH_2)_{14}COONa$
    c. $(Na_5P_3O_{10})$ d. $CH_3(CH_2)_{14}COOCH_2(CH_2)_{28}CH_3$

20.6 A vitamin in a skin cream "nourishes the skin."

20.7 "Products that contain artificial ingredients do not provide true aromatherapy benefits."

20.8 "Laundry balls contain no chemicals and yet are reusable indefinitely in the washing machine to clean, deodorize, sterilize, bleach, and soften clothes."

20.9 An advertisement claims a deodorant to be "pH-balanced for a woman's tender skin"; its proof is showing that a wet piece of litmus paper placed on her wrist does not change color.

20.10 A YouTube video urges you to have someone make a video of you chewing and swallowing a detergent pod; after all, it must be harmless if you wash clothes with it.

4. Go to a local grocery store or discount store, find ten different variations of one or more of the following personal care products, and compare the ingredients, finding the commonalities and the unique ingredients. Compare the cost per ounce.
    a. toothpaste or tooth gel
    b. shampoos
    c. deodorants/antiperspirants
    d. hand creams and lotions

LET'S EXPERIMENT! Happy Hands

Materials Needed:

- ²/₃ cup rubbing alcohol (usually 70% isopropyl alcohol)
- ¹/₃ cup plain aloe vera gel (the fewer additives, the better)
- 8–10 drops essential oil (such as lavender, thyme, clove, cinnamon leaf, peppermint, chamomile, or a few drops of orange or vanilla extract), optional
- Liquid soap
- Mixing bowl

- Spatula/wooden spoon
- Funnel
- Plastic bottle with pump or squirt top
- Glo Germ Mini Kit (to visualize bacteria)
- UV pen light
- Friend

How clean are your hands? Do hand sanitizers kill more bacteria than washing with soap and water?

It is important to wash your hands several times a day to prevent the spread of disease and germs. For this experiment, you will make your own hand sanitizer and compare it with a liquid soap, for antibacterial efficacy. Neither contains triclosan, an antibacterial chemical recently banned by the FDA and EPA.

Part 1: Homemade Hand Sanitizer

Start by acquiring your supplies and making homemade hand sanitizer. You will pour ²/₃ cup of rubbing alcohol (~70 % isopropyl) into a small mixing bowl.

Next, you will measure out ¹/₃ cup of aloe vera. To lessen the alcohol scent, you have the option of adding 8–10 drops of an essential oil or fruit extract. Mix the ingredients using a spatula or wooden spoon. Pour your solution into a container with a pump or squirt top using a funnel to prevent spills (Figure 1).

Figure 1

Part 2: Hand-Cleaning Effectiveness Experiment

Teaming up with a partner for this experiment will work best. Start by applying a nickel-sized amount of the Glo Germ gel on the palms of your hands. Rub it on your hands as you would lotion, completely covering your hands, your fingernails, and between your fingers.

Once the Glo Germ gel has been applied, have your partner shine the UV pen light on your hands. Doing this in a darkened room or closet makes it easier to see. What do you see? Is there any evidence of germs? (See Figure 2.)

Figure 2

Wash your hands with soap and water well for 30 seconds (Figure 3). Now return to the darkened room to shine the UV pen light on your hands again as in Figure 2.

Figure 3

What do you see? Was there a change in the number or size of spots highlighted under the UV light?

Switch roles and have your partner apply the Glo Germ gel. Follow the steps as you did above, but this time have your partner wash their hands using the hand sanitizer instead of soap and water. Did you notice a difference?

Questions

1. Whose hands appeared to have less area with glowing germs after using two different methods of hand cleaning?
2. Not all soaps and hand sanitizers are created equal. What other variables could you include to expand this experiment?
3. How are the mechanisms of soap and hand sanitizer different?

Poisons

Have You Ever Wondered?

1 How can something be both necessary to life and a poison at the same time? **2** Why is chocolate so poisonous to dogs? **3** Why do scientists use such high doses when testing possible cancer-causing substances? **4** Why can't we dispose of wastes like PCBs, lead, and dioxins by burning them?

We tend not to think of *poisons* and *natural* in the same context. But when it comes to poisons, nature has outperformed human beings for millennia. Even today, the most toxic substances come from nature, not scientific laboratories. The bulbs and main bodies of the daffodil and other members of the amaryllis family are quite poisonous. All of the *Narcissus* species contain lycorine, an alkaloid poison that causes vomiting, diarrhea, convulsions, and a touch dermatitis called "daffodil itch." The bulb's similarity to a common onion has led to incidents of accidental poisoning.

TOXICOLOGY: WHAT MAKES A POISON? What is a poison? Perhaps a better question is: How much is a poison? A substance may be harmless in one amount but injurious—or even deadly—in another. It was probably Paracelsus who, around 1500 C.E., first suggested that "the dose makes the poison." Even a common substance such as table salt or sugar can be poisonous if eaten in abnormally large amounts.

A **poison** is a substance that causes injury, illness, or death of a living organism. We usually think of a substance as poisonous if it can kill or injure at a low dosage. The toxic dose need not be the same for everyone. An amount of salt that is safe for an adult could be deadly for a tiny infant. An amount of sugar that is all right for a healthy teenager might be quite harmful to her diabetic father.

Of course, some compounds are much more toxic than others. It would take a massive dose of salt to kill a healthy adult, whereas a few nanograms of botulin could be fatal. How the substance is administered also matters. Nicotine given intravenously is more than 50 times as toxic as when taken orally. Pure fresh water that is delightful to drink can be deadly if imbibed in large quantity. Recall from Chapter 17 that vitamin A is essential to health but that it can build to toxic levels in the body's fatty tissues.

In this chapter we take a look at poisons—both natural and human-made. We will learn what those poisons do to the body and how various poisonings are treated. **Toxicology** is the study of the effects of poisons, their detection and identification, and their antidotes. This chapter covers only a few of the many toxic substances that are known, giving priority to those that are more likely to be encountered in everyday life.

1 **How can something be both necessary to life and a poison at the same time?**

There are many substances that, at low doses, carry out a needed function in the body. At high doses, the same substances cause a detrimental biochemical reaction. For example, iron ions are needed to make red blood cells. But an excess of iron can be absorbed by other cells, causing cellular death. Iron can also interfere with liver function.

21.1 Natural Poisons

Learning Objective • Name some natural poisons and their sources.

When Socrates was accused of corrupting the youth of Athens in 399 B.C.E., he was given the choice of exile or death. He chose death by drinking a cup of hemlock. A **toxin** is a poison that is naturally produced by a plant or animal. Toxins were well known in the ancient world. Poisoning was an official means of execution in many ancient societies. Hemlock was the official Athenian poison. Snake and insect venoms as well as other plant alkaloids have also been widely used throughout the world. Today, many natural poisons such as curare are still used by certain Indigenous peoples in South America to poison their arrows when hunting wild game.

Within the past 150 years, we have learned much more about these natural poisons. The hemlock taken by Socrates was probably prepared from the unripe fruit of *Conium maculatum* (poison hemlock), dried and brewed into a tea (Figure 21.1). That fruit contains coniine, which causes, nausea, weakness, paralysis, and—as in the case of Socrates—death.

► **Figure 21.1** Poison hemlock (*Conium maculatum*), the plant whose unripe fruit was probably the source of the hemlock taken by Socrates. The main alkaloid in poison hemlock is coniine.

Alkaloids are heterocyclic amines (Section 9.9) that occur naturally in plants. In addition to coniine and the toxins in curare, alkaloids include caffeine, nicotine, morphine, and cocaine (Section 18.7). One of the most poisonous alkaloids is strychnine, which occurs in the seeds of several varieties of trees in the species *Strychnos*. A dose of only 30 mg can be fatal.

Toxins are produced by animals—snakes, lizards, frogs, and others—as well as by plants. Many insects and microorganisms also produce poisons. For example, each year about 100,000 Americans get *E. coli* infections, and approximately 1% of them die. Most *E. coli* bacteria are harmless, but some are particularly virulent and some produce verotoxin. Perhaps the most poisonous bacterial toxin of all is *botulin*, the nerve poison that causes botulism (Section 21.4).

Poisonous Plants in the Garden and Home

Because many natural poisons are alkaloids that occur in plants, it is not surprising that poisons are found in gardens and on farms and ranches. In addition to the toxic pesticides that might be found on a shelf (Chapter 19), some of the plants themselves are toxic. For example, oleander (*Nerium oleander*), a beautiful shrub, contains several types of poison, including the potent cardiac glycosides *oleandrin* and *neriine*. (Cardiac glycosides are compounds with a steroid part and a sugar part that increase the force of the heart's contraction.) Oleander's poisons are so strong that a person can be poisoned just by eating the honey made by bees that have fed on oleander nectar.

Irises, azaleas, and hydrangeas all are poisonous. So are holly berries, wisteria seeds, and the leaves and berries of privet hedges. Indoor plants can also be poisonous. Philodendron, one of the most popular, is quite toxic. The rosary pea (*Abrus precatorius*) has beautiful seeds that are often used to make jewelry. However, the seeds contain *abrin*. Abrin is composed of two peptide chains that act together to inhibit protein synthesis by shutting down ribosomes. The amount of poison in one pea seed, about 3 μg, is enough to kill a person. Fortunately, the seeds present little danger when swallowed whole. However, if the coating is broken, the toxin is released.

Some foods also contain poisons. For example, rhubarb leaves contain toxic oxalic acid. Bruised celery produces *psoralens*, and molds growing on stored peanuts and grain produce *aflatoxins*. These compounds are powerful mutagens and carcinogens (Section 21.6). People in Japan relish a variety of puffer fish that contains deadly poison in its ovaries and liver. Several people die each year from improperly prepared puffers.

▲ Oleander blossoms are beautiful—but all parts of the plant are poisonous.

SELF-ASSESSMENT Questions

1. The study of poisons is called
   a. oncology
   b. pharmacology
   c. proctology
   d. toxicology

2. A poison is
   a. any alkaloid
   b. any chemical
   c. best defined by dose
   d. a synthetic chemical

3. Which of the following is not an alkaloid?
   a. caffeine   b. botulin
   c. nicotine   d. cocaine

4. The active ingredient in the poison that Socrates took was
   a. arsenic   b. botulin
   c. coniine   d. ricin

Answers: 1, d; 2, c; 3, b; 4, c

▲ Rosary peas are often used to make necklaces and have been used for rosaries.

21.2 Poisons and How They Act

Learning Objectives • Distinguish among corrosive poisons, metabolic poisons, and heavy metal poisons, and explain how each acts. • Identify antidotes for some common metabolic poisons and heavy metal poisons.

Poisons act in many ways. We examine the action of a few prominent poisons here. Recall from Chapter 7 that strong acids and strong bases have corrosive effects on human tissue. These substances indiscriminately destroy living cells. Corrosive chemicals in lesser concentrations can exhibit more subtle effects.

Strong Acids and Bases as Poisons

Both acids and bases catalyze the hydrolysis of amides, including proteins (polyamides).

Intact protein molecule   + H_2O $\xrightarrow{H^+ \text{ or } OH^-}$ Fragments

Shape is vital to the function of a protein, and the hydrolysis products cannot carry out the functions of the original protein. For example, if the protein is an enzyme, it is inactivated by hydrolysis. In cases of severe exposure, fragmentation continues until the tissue is completely destroyed.

Acids in the lungs are particularly destructive. In Chapter 13, we saw how sulfuric acid is formed when sulfur-containing coal is burned. Acids are also formed when certain plastics and other wastes are burned. These acid air pollutants cause the breakdown of lung tissue.

Oxidizing Agents as Poisons

Other air pollutants also damage living cells. Ozone, peroxyacetyl nitrate (PAN), and the other oxidizing components of photochemical smog probably do their main damage by inactivating enzymes. The active sites of enzymes often include the sulfur-containing amino acids cysteine and methionine. Cysteine is readily oxidized by ozone to cysteic acid.

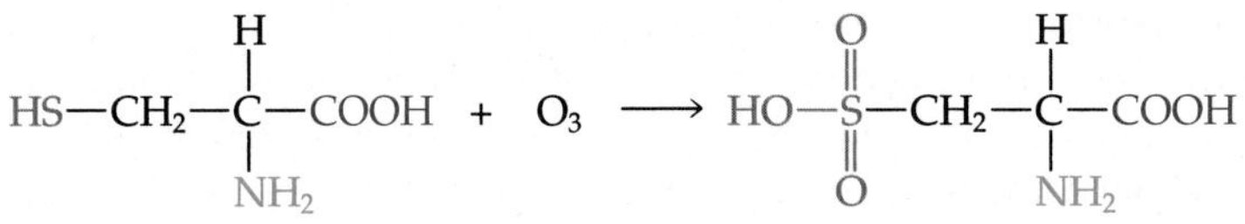

$$HS-CH_2-\underset{\underset{NH_2}{|}}{\overset{\overset{H}{|}}{C}}-COOH \ + \ O_3 \ \longrightarrow \ HO-\underset{\underset{O}{\|}}{\overset{\overset{O}{\|}}{S}}-CH_2-\underset{\underset{NH_2}{|}}{\overset{\overset{H}{|}}{C}}-COOH$$

Cysteine   Cysteic acid

Methionine is oxidized to methionine sulfoxide.

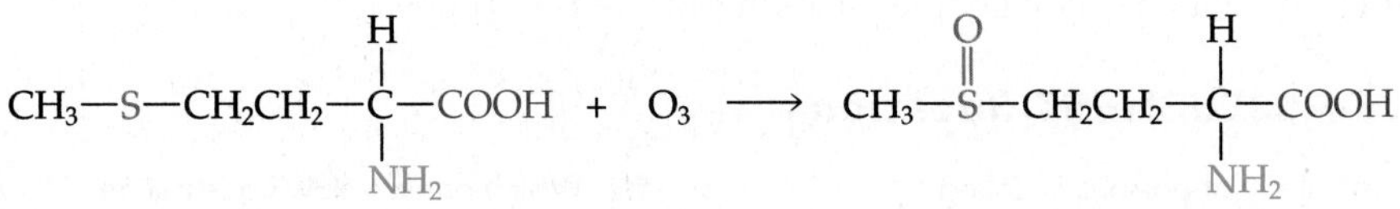

$$CH_3-S-CH_2CH_2-\underset{\underset{NH_2}{|}}{\overset{\overset{H}{|}}{C}}-COOH \ + \ O_3 \ \longrightarrow \ CH_3-\overset{\overset{O}{\|}}{S}-CH_2CH_2-\underset{\underset{NH_2}{|}}{\overset{\overset{H}{|}}{C}}-COOH$$

Methionine   Methionine sulfoxide

Tryptophan also reacts with ozone. This amino acid does not contain sulfur, but it undergoes a ring-opening oxidation at the double bond.

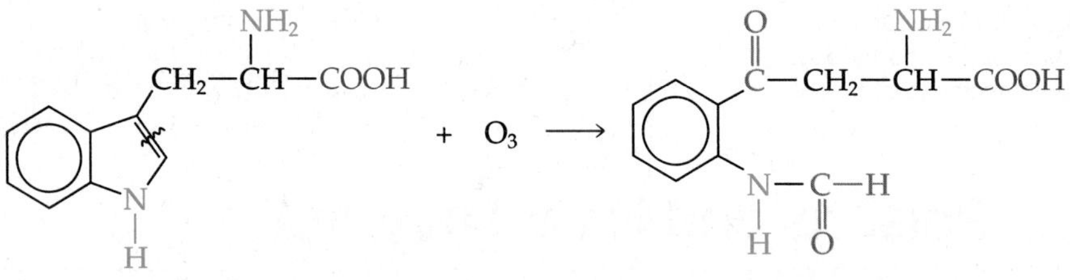

Tryptophan   Oxidation product

Oxidizing agents can also break bonds in many other chemical substances in a cell. Powerful agents such as ozone are more likely to make an indiscriminate attack than to react in a highly specific way.

Metabolic Poisons

Certain chemical substances prevent cellular oxidation of metabolites by blocking the transport of oxygen in the bloodstream or by interfering with oxidative processes in the cells. These chemicals act on the iron atoms in complex protein molecules. Probably the best known of these metabolic poisons is carbon monoxide. Recall (Section 13.4) that CO blocks the transport of oxygen by binding tightly to the iron atom in hemoglobin. Hemoglobin is bright red and is responsible for the red color of blood. Hemoglobin contains Fe^{2+}. During cooking, red meat turns brown because hemoglobin is oxidized to *methemoglobin*. Methemoglobin, which contains Fe^{3+}, is brown. Dried bloodstains turn brown for the same reason.

Nitrate ions, found in dangerous amounts in the groundwater in some agricultural areas (Section 14.3), also diminish the ability of hemoglobin to carry oxygen. Microorganisms in the digestive tract reduce nitrates to nitrites. We can write the process as a reduction half-reaction (Section 8.3).

$$2\,H^+(aq) \;+\; \underset{\text{Nitrate ion}}{NO_3^-(aq)} \;+\; 2\,e^- \;\longrightarrow\; \underset{\text{Nitrite ion}}{NO_2^-(aq)} \;+\; H_2O$$

The nitrite ions oxidize the Fe^{2+} in hemoglobin to Fe^{3+}, forming methemoglobin, which is incapable of carrying oxygen. The resulting oxygen-deficiency disease is called *methemoglobinemia*, known when it affects infants as "blue-baby syndrome." A compound called *methylene blue* is capable of reducing the methemoglobin back to hemoglobin.

Cyanides, compounds that contain a $C\equiv N$ group, are among the most notorious poisons in both fact and fiction. They include salts such as sodium cyanide (NaCN) and the covalent compound hydrogen cyanide ($H-C\equiv N$). Cyanides act quickly and are powerful. It takes only about 50 mg of HCN or 200–300 mg of a cyanide salt to kill. Cyanide exposure occurs frequently in victims of smoke inhalation from house fires. Experts use HCN gas to exterminate pests in ships, warehouses, and railway cars, and on citrus and other fruit trees. Sodium cyanide (NaCN) is employed to extract gold and silver from ores and in electroplating baths.

Worldwide, about 1.1 billion kg of hydrogen cyanide is produced annually. About 150 million kg is used to produce sodium cyanide for gold processing. The rest is used industrially in the production of plastics, adhesives, fire retardants, cosmetics, pharmaceuticals, and other materials.

Deaths that occur in house fires are often actually caused by toxic gases. Smoldering fires produce carbon monoxide, but hydrogen cyanide formed when plastics and fabrics burn is also important (Chapter 10). Some fire department rescue crews carry spring-loaded hypodermic syringes filled with sodium thiosulfate solution for emergency treatment of victims of smoke inhalation. In the Nazi death camps during World War II, gases were used to murder 15 million people, including 6 million Jews. Treblinka, Belzec, and Sobibor used CO from internal combustion engines in closed chambers. Auschwitz/Birkenau, Stutthof, and other camps used HCN in the form of *Zyklon-B* (HCN absorbed in a carrier, typically wood pulp or diatomaceous earth).

Cyanides act by blocking the oxidation of glucose inside cells; cyanide ions form a stable complex with iron(III) ions in oxidative enzymes called *cytochrome oxidases*. These enzymes normally act by providing electrons for the reduction of oxygen in the cell. Cyanide blocks this action and brings an abrupt end to cellular respiration, causing death in minutes. Cyanides are sometimes used to commit suicide, particularly by health care and laboratory workers.

Any antidote for cyanide poisoning must be administered quickly. Providing 100% oxygen to support respiration can sometimes help. Sodium nitrite is often given intravenously to oxidize the iron atoms in the enzymes back to the inactive Fe^{3+} form. The resultant methemoglobin strongly binds the cyanide anion, which frees the cytochrome oxidase. As before, the methemoglobin must be removed. Sodium thiosulfate ($Na_2S_2O_3$) is then used if time permits. The thiosulfate ion transfers a sulfur atom to the cyanide ion, converting it to the relatively innocuous thiocyanate ion.

$$\underset{\text{Cyanide ion}}{CN^-(aq)} \;+\; \underset{\text{Thiosulfate ion}}{S_2O_3^{2-}(aq)} \;\longrightarrow\; \underset{\text{Thiocyanate ion}}{SCN^-(aq)} \;+\; \underset{\text{Sulfite ion}}{SO_3^{2-}(aq)}$$

Unfortunately, few victims of cyanide poisoning survive long enough to be treated.

Make Your Own Poison: Fluoroacetic Acid

The body generally acts to detoxify poisons (Section 21.5), but it can also convert an essentially harmless chemical to a deadly poison, such as fluoroacetic acid (FCH_2COOH). Our cells use acetic acid to produce citric acid, which is then broken

$$\begin{array}{c} CH_2-COOH \\ | \\ HO-C-COOH \\ | \\ CH_2-COOH \end{array}$$

Citric acid

$$\begin{array}{c} F-CH-COOH \\ | \\ HO-C-COOH \\ | \\ CH_2-COOH \end{array}$$

Fluorocitric acid

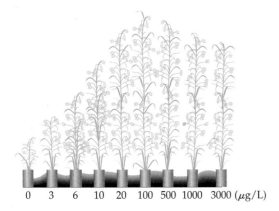

▲ **Figure 21.2** The effect of copper ions on the height of oat seedlings. From left to right, the concentrations of Cu^{2+} are 0, 3, 6, 10, 20, 100, 500, 1000, and 3000 $\mu g/L$. The plants on the left show varying degrees of deficiency; those on the right show copper ion toxicity. The optimum level of Cu^{2+} for oat seedlings is therefore about 100 $\mu g/L$.

Active enzyme

SH

SH

Active site

+

Hg^{2+}

↓

S
Hg
S

Inactive enzyme

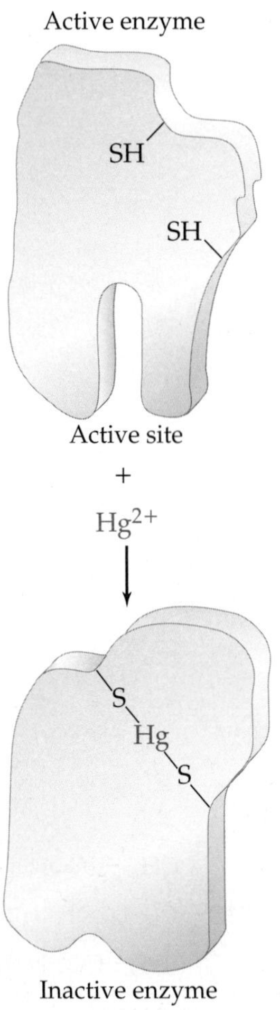

▲ **Figure 21.3** Mercury poisoning. Enzymes catalyze reactions by binding a reactant molecule at the active site (Section 16.7). Mercury ions react with sulfhydryl groups to change the shape of the enzyme and destroy the active site.

down to release energy in a series of steps called the citric acid cycle. When fluoroacetic acid is ingested, it is incorporated into fluorocitric acid. The latter effectively blocks the citric acid cycle by tying up the enzyme that acts on citric acid. Thus, the energy-producing mechanism of the cell is shut off, and death comes quickly.

Sodium fluoroacetate (FCH_2COONa), a salt of fluoroacetic acid and also known as *Compound 1080*, is used to poison rodents and predatory animals. It is not selective, making it dangerous to humans, pets, and other animals. Fluoroacetic acid occurs in nature in an extremely poisonous South African plant called *gifblaar*.

Heavy Metal Poisons

Metals with densities at least five times that of water are called *heavy metals*. Many heavy metals are toxic, but lead, mercury, and cadmium are especially notable. Of course, metals need not be dense to be toxic. One especially toxic metal is beryllium, a very light metal. All beryllium compounds are poisonous and can cause a painful and usually fatal disease known as *berylliosis*.

People have long used a variety of metals in industry and agriculture and around the home. Most metals and their compounds show some toxicity when ingested in large amounts. Even essential mineral nutrients can be toxic when taken in excessive amounts. In many cases, too little of a metal ion (a deficiency) can be as dangerous as too much (toxicity). This effect is illustrated in Figure 21.2.

Some metal ions are critical to human health—in the right amounts. The average adult requires 10–18 mg of iron every day. When less is taken in, the person suffers from anemia. Yet an overdose can cause vomiting, diarrhea, shock, coma, and even death. As few as 10–15 tablets, each containing 325 mg of $FeSO_4$ (65 mg Fe per tablet), have been fatal to children. Large excesses of Fe^{2+} can damage the intestinal lining. Iron poisoning occurs when the concentration of circulating Fe^{2+} is more than the iron-binding proteins in the blood can take up.

Other heavy metals exert their action primarily by inactivating enzymes. In simple laboratory chemical reactions, heavy metal ions react with hydrogen sulfide to form insoluble sulfides.

$$Pb^{2+}(aq) + H_2S(g) \longrightarrow PbS(s) + 2 H^+(aq)$$

$$Hg^{2+}(aq) + H_2S(g) \longrightarrow HgS(s) + 2 H^+(aq)$$

Most enzymes have amino acids with sulfhydryl (—SH) groups. These groups bind with heavy metal ions in the same way that H_2S does, rendering the enzymes inactive (Figure 21.3).

Mercury (Hg) is a most unusual metal. It is the only common metal that is a liquid at room temperature. People have long been fascinated by this bright, silvery, dense liquid, once known as *quicksilver*. Children sometimes played with the mercury from a broken thermometer. Household fever thermometers with mercury have been replaced by other types of thermometers, although some households may still have old ones with mercury.

Mercury metal has many uses. It has been used in barometers. Dentists use it to make amalgams for filling teeth. It is used as a liquid metal electrical contact in switches. These uses are now restricted in many areas.

In the United States, about 140,000 kg of mercury is released each year into the air, about 40% from coal-fired power plants. (Although mercury occurs in only trace amounts in coal, the United States burns around 700 million metric tons of coal a year.) Other sources of mercury are waste incinerators and factories that produce chlorine and sodium hydroxide (chlor-alkali plants). The mercury returns to the land and water.

Mercury presents a hazard to those who work with it because its vapor is toxic. Mercury vapor is quite hazardous when inhaled, particularly when exposure takes place over a long period of time. An open container of mercury or a spill on the floor can release enough mercury vapor into the air to exceed the established maximum safe level by a factor of 200. Workers who were involved in the production of felt hats had long-term exposure to mercury vapor and developed neurological symptoms that became known as *mad hatter disease*.

Hazardous chronic exposure usually occurs with regular occupational use of the metal, such as in mining or extraction. The body converts the inhaled mercury to Hg^{2+} ions. Because mercury is a cumulative poison (it takes the body about 70 days to rid itself of *half* of a given dose), chronic poisoning is a threat to those continually exposed.

In aquatic systems, mercury vapor is converted by anaerobic organisms to highly toxic methylmercury ions (CH_3Hg^+). Fish tend to concentrate CH_3Hg^+ in their bodies. Federal and state environmental agencies have warned people to limit consumption of certain species of fish—particularly swordfish and shark—or to avoid them entirely.

Fortunately, there are antidotes for mercury poisoning. While searching for an antidote for the arsenic-containing war gas lewisite (see box on Arsenic Poisoning), British scientists came up with a compound for treating heavy metal poisoning as well. The compound, a derivative of glycerol, is called *British antilewisite (BAL)*. It acts by **chelating** (from the Greek *chela*, meaning "claw") Hg^{2+} ions, surrounding the ions so that they are tied up and cannot attack vital enzymes.

CH₂—CH—CH₂
OH   SH   SH
BAL

Mercury atom chelated by two BAL molecules

The bad news is that the effects of mercury poisoning may not show up for several weeks. By the time the symptoms—loss of equilibrium, sight, feeling, and hearing—are recognizable, extensive damage has already been done to the brain and nervous system. Such damage is largely irreversible. BAL and similar antidotes are effective only when people know that they have been poisoned and seek treatment right away.

All compounds of mercury, except those that are virtually insoluble in water, are poisonous no matter how they are administered. However, metallic mercury does not seem to be very toxic when ingested (swallowed). Most of it passes through the system unchanged. Indeed, there are numerous reports of mercury being given orally in the eighteenth and nineteenth centuries as a remedy for obstruction of the bowels. Benjamin Franklin was treated with mercury late in his life. Doses varied from a few ounces to a pound or more!

Lead is widespread in the environment, reflecting the many uses we have for this soft, dense, corrosion-resistant metal and its compounds. Lead (as Pb^{2+} ions) is present in many foods, generally in concentrations of less than 0.3 ppm. Lead ions also get into our drinking water (up to 0.1 ppm) from lead-sealed pipes. Lead compounds were once widely used in house paints, and tetraethyl lead was used in most gasolines. Since those two uses were banned, exposure to lead has decreased dramatically, from about 15 μg/dL in the 1970s to less than 2 μg/dL today. The CDC sets a blood lead level above 10 μg/dL as an unacceptable health risk, but adverse health effects can occur at lower concentrations. Effects include lower intelligence scores and poor school performance.

A major health crisis occurred in Flint, Michigan, after the source for the city's water supply was changed in 2014 to save money. This resulted in the exposure of thousands of Flint children to high lead levels, and the number of children in Flint with elevated lead blood levels doubled to nearly 5% (Section 14.5). Many of our cities have aging infrastructure, including water pipes.

Lead and its compounds are quite toxic. Metallic lead is generally converted to Pb^{2+} in the body. Lead can damage the brain, liver, and kidneys. Extreme cases can be fatal.

Lead poisoning is especially harmful to children. Some children develop a craving that causes them to eat unusual things, and children with this syndrome (called *pica*) eat chips of peeling lead-based paints. They also ingest lead from food and

▲ In 2007, Fisher-Price voluntarily recalled almost a million children's toys because the paint may have contained excessive levels of lead. Less extensive children's toy recalls in 2011 affected particular varieties of memory-testing cards, toy garden rakes, sketchbooks, and ESI-R screening materials.

even from toys coated with lead-containing paint. The elimination of tetraethyl lead as a gasoline additive has lowered children's exposure to lead compounds from the streets, where they were deposited by automobile exhausts. Large amounts of Pb^{2+} in a child's blood can cause cognitive and behavioral problems, anemia, hearing loss, developmental delays, and other physical and mental problems.

Arsenic Poisoning

Arsenic is not a metal, but it has some metallic properties. In commercial poisons, arsenic is usually found as arsenate (AsO_4^{3-}) or arsenite (AsO_3^{3-}) ions. Like heavy metal ions, these ions render enzymes inactive by tying up sulfhydryl groups.

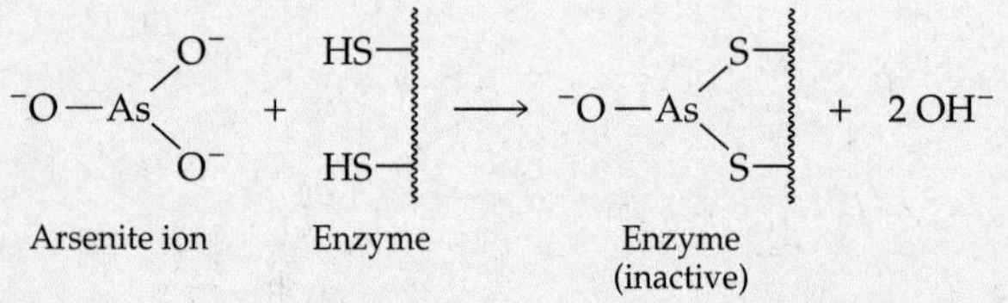

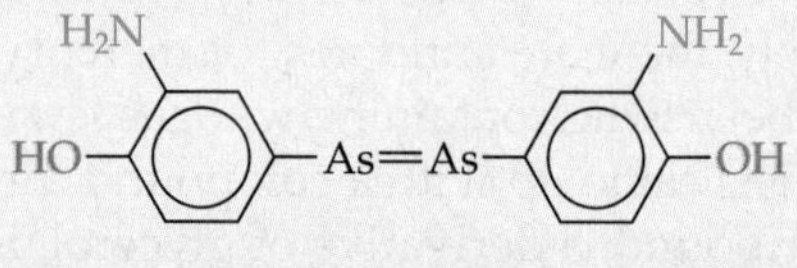

Arsphenamine

Another arsenic compound, *lewisite*, first synthesized by (and named for) W. Lee Lewis (United States, 1878–1943), was developed as a blister agent for use in chemical warfare. The United States started large-scale production of lewisite in 1918, but fortunately World War I came to an end before this gas could be employed.

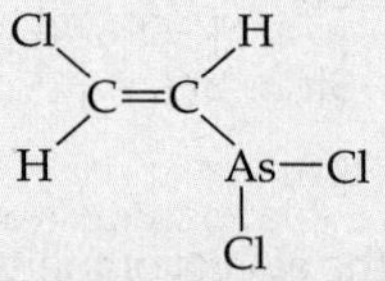

Arsenite ion   Enzyme   Enzyme (inactive)

There are numerous organic compounds containing arsenic. One such compound, *arsphenamine*, was the first antibacterial agent and was once used widely in the treatment of syphilis.

Lewisite

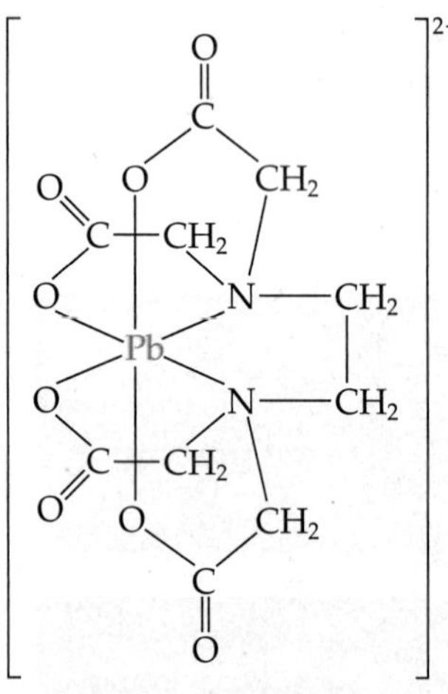

Lead–EDTA complex

Adults can excrete about 2 mg of lead per day. Most people take in less than that from air, food, and water, and thus generally do not accumulate toxic levels. If intake exceeds excretion, however, lead builds up in the body and chronic irreversible lead poisoning can result.

Lead poisoning is usually treated with a combination of BAL and another chelating agent called *ethylenediaminetetraacetic acid (EDTA)*.

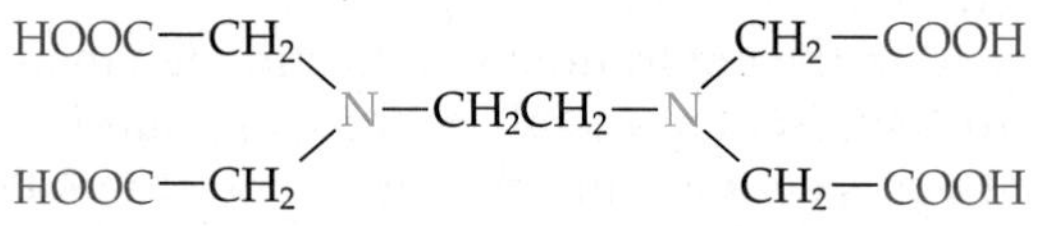

EDTA

The calcium salt of EDTA is administered intravenously. In the body, calcium ions are displaced by lead ions, which bind to EDTA more tightly.

$$CaEDTA^{2-} + Pb^{2+} \longrightarrow PbEDTA^{2-} + Ca^{2+}$$

The lead–EDTA complex is then excreted.

As with mercury poisoning, the neurological damage caused by lead compounds is irreversible. Treatment must begin early to be effective.

Cadmium is another useful but troublesome heavy metal. It is used in alloys, in the electronics industry, in nickel–cadmium rechargeable batteries, and in many other applications. Cadmium (as Cd^{2+} ions) is quite toxic. Cadmium poisoning leads to loss of calcium ions $\left(Ca^{2+}\right)$ from the bones, leaving them brittle and easily broken. It also causes severe abdominal pain, vomiting, diarrhea, and a choking sensation.

The most notable cases of cadmium poisoning occurred along the upper Jinzū River in Japan, where cadmium ions entered the water in milling wastes from a mine. Downstream, farm families used the water to irrigate their rice fields and for drinking, cooking, and other household purposes. Many of these people soon began to suffer from a strange, painful malady that became known as *itai-itai*, the "it hurts, it hurts" disease. Over 200 people died, and thousands were disabled.

SELF-ASSESSMENT Questions

1. Which of the following is *not* a corrosive poison?
   a. HCN   **b.** HNO_3 **c.** O_3 **d.** NaOH

2. Nitrites poison by
   a. blocking the action of cytochrome oxidase
   b. blocking the action of acetylcholinesterase
   c. displacing O_2 from oxyhemoglobin
   d. oxidizing Fe^{2+} in hemoglobin

3. Cyanides block normal respiration by acting as a(n)
   a. coenzyme
   b. enzyme
   c. enzyme activator
   d. enzyme inhibitor

4. The most important toxic heavy metals are
   a. gold, platinum, and silver
   b. iron, calcium, and mercury
   c. lead, silver, and gold
   d. lead, mercury, and cadmium

5. Which of the following is most likely to be deadly to a young child? An overdose of a multivitamin that contains
   a. ascorbic acid   **b.** biotin
   c. iron   **d.** lycopene

6. A chelating agent will bind to
   a. the —SH groups in methionine   **b.** the iron in hemoglobin
   c. lead ions   **d.** a strong acid

Answers: 1, a; 2, d; 3, d; 4, d; 5, c; 6, c

21.3 More Chemistry of the Nervous System

Learning Objective • Define *neurotransmitter*, and describe how various substances interfere with the action of neurotransmitters.

Some of the most toxic substances known act on the nervous system. Signals are shuttled across synapses between nerve cells by chemical messengers called *neurotransmitters* (Section 18.7). Neurotoxins can disrupt the action of a neurotransmitter in several ways: by interfering with its synthesis or transport, by occupying its receptor site, or by blocking its degradation.

One such chemical messenger is *acetylcholine (ACh)*. It activates the postsynaptic nerve cell by fitting into a specific receptor and thus changing the permeability of the cell membrane to certain ions. Once ACh has carried the impulse across the synapse, it is rapidly hydrolyzed to acetic acid and relatively inactive choline in a reaction catalyzed by an enzyme, acetylcholinesterase.

$$CH_3\overset{\text{O}}{\overset{\|}{C}}OCH_2CH_2-\overset{CH_3}{\underset{CH_3}{\overset{|}{N^+}}}-CH_3 + H_2O \xrightarrow{\text{acetylcholinesterase}} CH_3\overset{\text{O}}{\overset{\|}{C}}OH + HOCH_2CH_2-\overset{CH_3}{\underset{CH_3}{\overset{|}{N^+}}}-CH_3$$

$\qquad\qquad$ Acetylcholine $\qquad\qquad\qquad\qquad\qquad\qquad\qquad\qquad\qquad$ Acetic acid $\qquad\qquad$ Choline

The postsynaptic cell releases the hydrolysis products and is then ready to receive further impulses. Other enzymes, such as acetylase, convert the acetic acid and choline back to acetylcholine, completing the cycle (Figure 21.4).

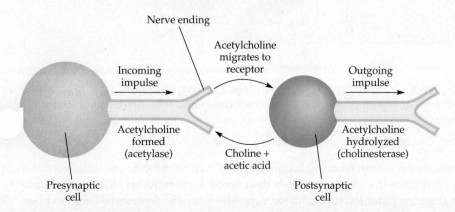

◀ **Figure 21.4** The acetylcholine cycle.

People with Alzheimer's disease are deficient in the enzyme acetylase. They produce too little acetylcholine for proper brain function. Drugs to treat the disease, such as donepezil (Aricept®), inhibit the action of acetylcholinesterase, thus helping to prevent the decline of ACh levels. These drugs are only moderately effective.

Nerve Poisons and the Acetylcholine Cycle

Various substances can disrupt the acetylcholine cycle at one of three different points, as illustrated by the following examples.

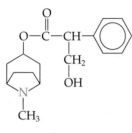

Atropine

- Botulin, the deadly toxin produced by *Clostridium botulinum* (an anaerobic bacterium) found in improperly processed canned food, has a powerful effect on ACh; it blocks the synthesis of ACh. When no messengers are formed, no messages are carried. Paralysis sets in and death occurs, usually from respiratory failure.
- Curare, atropine, and some local anesthetics act by blocking ACh receptor sites. In this case, the message is sent but not received. Local anesthetics use this mechanism to provide pain relief in a limited area, but these drugs, too, can be lethal in sufficient quantity.
- Anticholinesterase poisons act by inhibiting the enzyme acetylcholinesterase.

Organophosphorus Compounds as Insecticides and Weapons of War

Organic phosphorus insecticides (Section 19.2) are well-known nerve poisons. The phosphorus–oxygen linkage is thought to bond tightly to acetylcholinesterase, blocking the breakdown of acetylcholine (Figure 21.5). Acetylcholine therefore builds up, causing postsynaptic nerve cells to fire repeatedly. This overstimulates the muscles, glands, and organs. The heart beats wildly and irregularly, and the victim goes into convulsions and dies quickly.

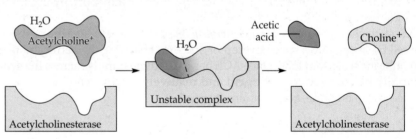

(a) Normal catalysis of acetylcholine

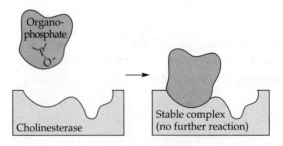

(b) Catalysis inhibited by organophosphate toxin

▲ **Figure 21.5** (a) Acetylcholinesterase catalyzes the hydrolysis of acetylcholine to acetic acid and choline. (b) An organophosphate ties up acetylcholinesterase, preventing it from breaking down acetylcholine.

Q *What functional group is split in the hydrolysis of acetylcholine?*

While doing research on organophosphorus compounds for use as insecticides during World War II, German scientists discovered some extremely toxic compounds with frightening potential for use in warfare. The Russians captured a German plant that manufactured a compound called *tabun* (referred to as *agent GA* by the U.S. Army), a compound with a fruity odor. The Soviets dismantled the factory and moved it to Russia. The U.S. Army, though, captured some of the stock.

The United States has developed other nerve poisons (Figure 21.6). One, called *sarin* (*agent GB*), is four times as toxic as tabun and was used in the 1995 attack on the Tokyo subway. It also has the "advantage" of being odorless. Another organophosphorus nerve poison is *soman* (or *agent GD*). It is moderately persistent, whereas tabun and sarin are generally nonpersistent. Yet another nerve poison is referred to as *VX*. It is only slightly more toxic than sarin, but it is more persistent in the environment. Note the variation in structure among these compounds. The approach to developing chemical warfare agents is much the same as that for developing drugs—find one that works, then synthesize and test structural variations.

During the 1970s and 1980s, another series of nerve agents was developed by Russian scientists, some of which are reputed to be several times more toxic than VX. One of these *Novichok* agents was implicated in the March 2018 poisonings of Sergei and Yulia Skripal in England.

These nerve poisons are among the most toxic synthetic chemicals known (although still not nearly as toxic as the natural toxin botulin). They are inhaled or absorbed through the skin, causing a complete loss of muscular coordination and subsequent death by cessation of breathing through paralysis, as described previously. The usual antidote consists of atropine injection—which blocks the acetylcholine receptor sites, preventing the paralysis by the nerve agent—and artificial respiration. Care must be taken, because atropine in too large a dose is itself a poison. Without an antidote, death usually occurs in 2–10 minutes, depending on the dose absorbed.

The phosphorus-based insecticides malathion and parathion (Section 19.2) are similar to the nerve poisons used as chemical warfare agents but are far less toxic. Nonetheless, they and other insecticides like them should be used with great caution.

Nerve poisons have helped us gain an understanding of the chemistry of the nervous system. This knowledge enables scientists to design antidotes for nerve poisons, to better understand diseases of the nervous system, and to develop new drugs to treat pain and diseases such as Alzheimer's disease and Parkinson's disease.

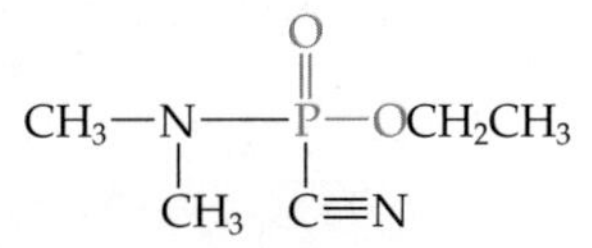

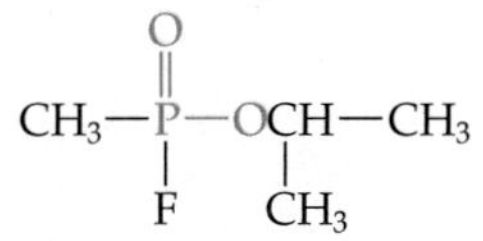

▲ **Figure 21.6** Four organophosphorus compounds used as nerve poisons in chemical warfare. Compare these structures with those of the insecticides malathion and parathion (Figure 19.10).

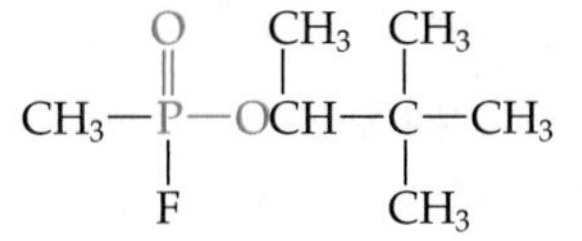

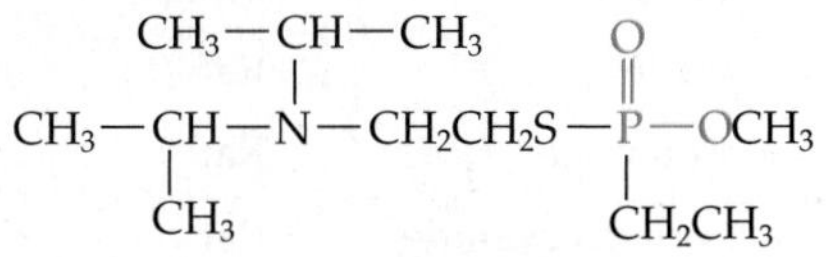

SELF-ASSESSMENT Questions

1. Receptor sites for the neurotransmitter acetylcholine (ACh) are blocked by
   a. atropine
   b. choline
   c. opiates
   d. naloxone

2. Acetylcholinesterase is a(n)
   a. enzyme that breaks down acetylcholine
   b. hormone
   c. neurotransmitter
   d. stimulant drug

3. Curare blocks the acetylcholine receptors on muscle tissue, resulting in
   a. contraction of the muscle fiber
   b. excessive contractions and convulsions
   c. inability of the muscle to respond to motor nerve stimulus
   d. increased stimulation of the muscle fiber

4. Which of the following nerve poisons is much less toxic than the others?
   a. malathion
   b. sarin
   c. soman
   d. tabun

5. The most hazardous nerve poisons are
   a. alkaloids such as cocaine
   b. chlorinated hydrocarbons such as DDT
   c. hydrocarbons such as gasoline
   d. organophosphorus compounds

Answers: 1, a; 2, a; 3, c; 4, a; 5, d

21.4 The Lethal Dose

Learning Objectives • Explain the concept of an LD_{50} value, and list some of its limitations as a measure of toxicity. • Calculate lethal doses from LD_{50} values and body weights.

Some substances are much more poisonous than others. To quantify toxicity, scientists use the **LD_{50}** (lethal dose for 50%) value, which specifies the dosage that kills 50% of a population of test animals. This statistic is used because animals vary in strength and in their susceptibility to toxins. The dose that kills 50% is the average lethal dose.

Usually, LD_{50} values are given in terms of mass of poison per unit of body weight of the test animal (for example, milligrams of poison per kilogram of animal).

Tested substances are most often given orally. (Values for intravenous administration can differ greatly from those for oral ingestion.) Although LD_{50} values are useful in comparing the relative toxicities of various substances, the values for test animals are not necessarily the same values that would be measured for humans. Clearly, it is unethical to test such toxins on human subjects, though data can be gathered through the testing of toxic substances on cultured human cells.

Table 21.1 lists LD_{50} values for a variety of substances. The larger the LD_{50} value, the less toxic the substance. In general, substances with oral LD_{50} values greater than 15,000 mg/kg are considered nearly nontoxic. Substances with LD_{50} values from 7500 to 15,000 mg/kg are considered to be of low toxicity; those with values from 500 to 7500 mg/kg, moderately toxic; and those with values less than 500 mg/kg, highly toxic.

TABLE 21.1 **Approximate LD$_{50}$ Values for Selected Substances**

Substance	Test Animal	Route	LD$_{50}$ (mg/kg)	Use or Occurrence
Sucrose	Rat	Oral	29,700	Table sugar
Sodium chloride	Rat	Oral	3000	Table salt
Ethyl alcohol	Rat	Oral	2080	Beverage alcohol
Aspirin	Rat	Oral	1500	Analgesic
Acetaminophen	Mouse	Oral	340	Analgesic
Nicotine	Mouse	Oral (intravenous)	230 (0.3)	Tobacco component; insecticide
Caffeine	Rat	Oral	192	Coffee and cola ingredient
Rotenone	Rat	Oral (intravenous)	132 (6)	Insecticide
Sodium cyanide	Rat	Oral	15	Gold extraction
Arsenic trioxide	Rat	Oral	15	Manufacture of insecticides and drugs
Ethylene glycol	Rat	Oral	6.86	Antifreeze
Ketamine	Rat	Oral	0.229	Veterinary anesthetic
Sarin	Human	Skin contact	0.01	Chemical warfare agent
Agent VX	Human	Skin contact	0.01	Chemical warfare agent
Tetanus toxin	Human	Puncture wound	0.0000025	In soil
Ricin	Mouse	Oral	0.000005	Castor beans
Botulin toxin	Human	Inhalation	0.000003	Possible biological warfare agent
		Oral	0.0000002	Food poisoning

The Botox Enigma

Botulin is the most toxic substance known, with a median lethal dose of only about 0.2 ng/kg of body weight. (Recall that 1 nanogram is one billionth of a gram.) Yet botulin, in a commercial formulation called Botox®, is used medically to treat intractable muscle spasms and is widely used in cosmetic surgery to remove wrinkles. When a muscle contracts over a long period of time without relaxing, it can cause severe pain. Tiny amounts of Botox can be injected into the muscle, where it binds to receptors on nerve endings. This stops the release of acetylcholine, and the signal for the muscle to contract is blocked. The muscle fibers that were pulling too hard are paralyzed, providing relief from the spasm.

Botox is also used to treat afflictions such as uncontrollable blinking, crossed eyes, and Parkinson's disease. In the case of crossed eyes, for example, the optic muscles cause the eyeballs to be drawn to focus inward. Treatment with botulin prevents the muscles from doing so, and repeated treatments correct the problem.

Botox curbs migraine headache pain in some people, probably by blocking nerves that carry pain messages to the brain and by relaxing muscles, making them less sensitive to pain. It is also used to treat excessive sweating, most likely by blocking the release of acetylcholine, the chemical that stimulates the sweat glands. Botox injections are used to treat bladder problems in which the patients are suffering from involuntary

Botox is a purified and extremely dilute preparation of the botulin toxin that causes the food poisoning known as *botulism*. When Botox is injected into specific muscles, it blocks the signals that cause muscles to contract. Because the muscles can't constrict, the skin levels out and can appear less wrinkled.

contractions of the bladder muscles. These contractions can cause incontinence or make it impossible to completely empty the bladder. Botox blocks the contractions.

In cosmetic surgery, small, carefully delivered amounts of Botox inactivate small facial muscles, markedly reducing "surprise wrinkles" in the forehead, frown lines between the eyebrows, and crow's feet at the corners of the eyes. These Botox treatments wear off in a few months.

SELF-ASSESSMENT Questions

1. Toxicity is often reported in terms of an LD_{50} value, which is the
   a. dose that kills half of a tested population
   b. lowest dose that kills an average 50-year-old person
   c. lowest dose likely to kill a 50-month-old child
   d. portion of the tested population killed by a 50 mg dose

2. According to Table 21.1, which of the following substances is *least* toxic?
   a. antifreeze   **b.** nicotine
   c. ketamine   **d.** sodium cyanide

3. The LD_{50} value for a compound is 4,000 mg/kg body weight. This compound
   a. is nearly nontoxic
   b. has low toxicity
   c. has moderate toxicity
   d. is highly toxic

Answers: 1, a; 2, b; 3, c

21.5 The Liver as a Detox Facility

Learning Objective • Describe how the liver is able to detoxify some substances.

The human body can handle moderate amounts of some poisons. The liver is able to detoxify some compounds by oxidation or reduction, or by coupling them with amino acids or other normal body chemicals.

Perhaps the most common route is oxidation. Ethanol (Section 9.6) is detoxified by oxidation to acetaldehyde, which in turn is oxidized to acetic acid and then to carbon dioxide and water.

$$CH_3CH_2OH \longrightarrow CH_3CHO \longrightarrow CH_3COOH \longrightarrow CO_2 + H_2O$$

A traditional antidote for methanol or ethylene glycol poisoning is ethanol, administered intravenously. This "loads up" the liver enzymes with ethanol, blocking oxidation of methanol until the compound can be excreted. A safer antidote, the drug fomepizole, inhibits the liver enzyme that catalyzes the oxidation of alcohols.

Highly toxic nicotine from tobacco is detoxified by oxidation to cotinine.

Nicotine   Cotinine

Cotinine is less toxic than nicotine, and the added oxygen atom makes it more water-soluble, and thus more readily excreted in the urine, than nicotine.

The liver has a system of iron-containing enzymes called *P-450* enzymes. The iron atoms are able to transfer oxygen to compounds that are otherwise highly resistant to oxidation. The P-450 enzymes catalyze the oxidation of fat-soluble substances (which are otherwise likely to be retained in the body) into water-soluble ones that are readily excreted. These enzymes also conjugate compounds with amino acids. For example, toluene is virtually insoluble in water. P-450 enzymes oxidize toluene to the more soluble benzoic acid, and then couple the benzoic acid with the amino acid glycine to form hippuric acid, which is even more soluble and is readily excreted.

Toluene   Benzoic acid   Hippuric acid

2 Why is chocolate so poisonous to dogs?

Chocolate contains several compounds related to caffeine, including the alkaloid *theobromine*. Dogs metabolize theobromine much more slowly than do humans, so its toxic effects of hyperactivity, high blood pressure, rapid heart rate, and seizures last longer. This can lead to respiratory failure and cardiac arrest.

Theobromine

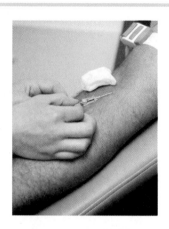

WHY IT MATTERS

Before prescribing certain drugs such as statins, doctors will often perform a blood test for liver enzymes. The liver must function properly for these drugs to work; if the liver is damaged, some of these drugs can worsen the damage. "Leakage" of liver enzymes into the bloodstream indicates a problem that must be rectified before the drug is administered.

Note that liver enzymes simply oxidize, reduce, or conjugate. The end product is not always less toxic. For example, as was noted in Section 9.6, methanol is oxidized to more toxic formaldehyde, which then reacts with proteins in the cells to cause blindness, convulsions, respiratory failure, and death.

The liver enzymes that oxidize alcohols also deactivate the male hormone testosterone. Buildup of these enzymes in a person with chronic alcoholism leads to a more rapid destruction of testosterone. This is the mechanism underlying alcoholic impotence, one of the well-known characteristics of the disease.

Benzene, because of its general inertness in the body, does not react until it reaches the liver. There it is slowly oxidized to an epoxide.

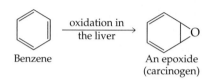

Benzene   oxidation in the liver   An epoxide (carcinogen)

The epoxide is a highly reactive molecule that can attack certain key proteins. The damage done by this epoxide sometimes results in leukemia.

Carbon tetrachloride (CCl_4) is also quite inert in the body. But when it reaches the liver, it is converted to the reactive trichloromethyl free radical ($Cl_3C\cdot$), which in turn attacks the unsaturated fatty acids in the body. This action can trigger cancer.

Getting Rid of "Toxins"

Two thousand years ago, Greek physicians applied the theory of the four *humors*—blood, phlegm, black bile, and yellow bile—to diagnose illnesses. They held that people in good health had a balance of these fluids, and that too much or too little of one or more of them caused disease. A person who is positive and happy is still said to be "in a good humor." For a patient short of a humor, treatment included bed rest, a change in diet, or medicines. For those with too much of a humor, treatment included sweating, purging, and bloodletting. Today, some people believe that sweating gets rid of "toxins." And there is a rather large business built on the idea that laxatives and other "colon cleansers" somehow purify the body. Erroneous beliefs often persist for centuries.

▲ In the Middle Ages, bloodletting was often the treatment for inflammation from infection.

SELF-ASSESSMENT Questions

1. Which of the following statements regarding activity in the human body is *not* true?
   a. Benzene is inert and thus nontoxic.
   b. Mercury vapor is converted to Hg^{2+}.
   c. Some nontoxic substances can be converted to toxic forms.
   d. Some poisons can be converted to nontoxic forms.

2. What substance is oxidized to formaldehyde in the liver?
   a. benzene   b. methanol
   c. nicotine   d. toluene

3. Nicotine is oxidized to cotinine in the liver. Why is cotinine less toxic than nicotine?
   a. Cotinine is reduced in the liver, so it is no longer active.
   b. Cotinine has an added oxygen, so it is more soluble in the urine.
   c. Cotinine contains an epoxide, so it is more reactive.
   d. Cotinine removes fatty acids from the bloodstream.

Answers: 1, a; 2, b; 3, b

21.6 Carcinogens and Teratogens

Learning Objectives • Describe how cancers develop. • Name and describe three ways to test for carcinogens.

A **carcinogen** is something that causes the growth of tumors. A *tumor* is an abnormal growth of tissue and can be either benign or malignant. *Benign tumors* do not invade neighboring tissues. They often regress spontaneously, and they are characterized by slow growth. *Malignant tumors*, often called *cancers*, can invade and destroy neighboring tissues. Their growth can be slow or rapid, but the growth is unrestricted.

Cancer is not a single disease. Rather, the term is a catchall for about two hundred different afflictions, many of them not even closely related to each other. We worry about carcinogens because half of the men and one-third of the women in the United States will develop cancer at some point in their lives.

What Causes Cancer?

Most people seem to believe that synthetic chemicals are a major cause of cancer, but the facts indicate otherwise. Like deaths in general (Section 18.3), most cancers are caused by lifestyle factors. Nearly two-thirds of all cancer deaths in the United States are linked to diet, tobacco, or a lack of exercise (Figure 21.7).

The U.S. Occupational Safety and Health Administration (OSHA) has a National Toxicology Program (NTP) that compiles ongoing lists of carcinogens in two categories. The 2016 list titled "Known Human Carcinogens" has 62 entries. For these, the NTP thinks there is sufficient evidence of carcinogenicity from studies in humans to establish a causal relationship between exposure to the agent, substance, or mixture and human cancer. Another 186 entries are included in the list titled "Reasonably Anticipated to Be Human Carcinogens." These suspected carcinogens often are structurally related to substances that are known to be carcinogens, or there is only limited evidence of their carcinogenicity from studies in humans or in experimental animals.

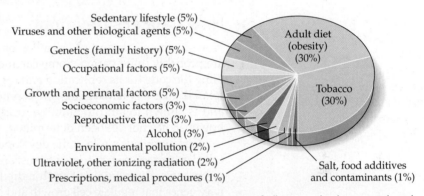

▲ **Figure 21.7** Ranking cancer risks. About 30% of all cancer deaths are attributed to tobacco and another 30% or so to a diet high in fat and calories and low in fruits and vegetables. Environmental pollution and food additives rank high in the public's perception of cancer causes but are only relatively minor contributors to cancer deaths. (Based on data from Harvard University School of Public Health.)

Some carcinogens, such as sunlight, radon, and safrole in sassafras, occur naturally. Some scientists estimate that 99.99% of all carcinogens that we ingest are natural ones. Plants produce compounds to protect themselves from fungi, insects, and higher animals, including humans. Some carcinogenic compounds are found in mushrooms, basil, celery, figs, mustard, pepper, fennel, parsnips, and citrus oils—almost every place that a curious chemist looks. Carcinogens are also produced during cooking and as products of normal metabolism. Because carcinogens are so widespread, we must have some way of protecting ourselves from them.

Cigarette Smoking, E-Cigarettes, and Cancer

The association between cigarette smoking and cancer has been known for decades, but the precise mechanism was not determined until 1996. The research scientists who found the link focused on a tumor-suppressor gene called *P53*, a gene that is mutated in about 60% of all lung cancers, and on a metabolite of benzpyrene, a carcinogen found in tobacco smoke.

In the body, benzpyrene is oxidized to an active carcinogen (an epoxide) that binds to specific nucleotides in the gene—sites called "hot spots"—where mutations frequently occur. It seems quite likely that the benzpyrene metabolite causes many of the mutations.

In 2007, electronic cigarettes (or *e-cigarettes*) were introduced into the United States from China. These devices are inhalers that vaporize nicotine solutions. Some evidence exists that teenagers who use these devices may eventually become smokers. More than 400,000 children already become regular smokers each year, and almost a third will eventually die from smoking. Many of these e-cigarette devices use flavored nicotine solutions that taste like candy or fruit. Young children may consume the refill solution, which contain potentially lethal levels of nicotine as well as other harmful chemicals, including diethylene glycol, heavy metals, volatile organic chemicals, and ultrafine particulates, which can get into the lungs or interfere with brain development.

Cigarette smoking is the chief preventable cause of premature death in the United States. One in five deaths in the United States is smoking-related. According to the CDC, cigarette smoking is associated with death from 24 different medical issues of which include heart disease, strokes, various cancers, chronic obstructive pulmonary disease (COPD), emphysema, and pneumonia. Smoking causes diseases of the gums and mouth, chronic hoarseness, vocal cord polyps, and premature aging and wrinkling of the skin. Lung cancer is not the only malignancy caused by cigarette smoking. Cancers of the pancreas, bladder, breast, kidney, mouth and throat, stomach, larynx, and cervix are also associated with smoking. In fact, smoking kills more people than AIDS, alcohol, illegal drugs, car accidents, murders, and suicides combined. On average, smokers die 10 years earlier than nonsmokers do.

How Cancers Develop

How do chemicals and physical factors cause cancer? Their mechanisms of action are probably quite varied. Some carcinogens chemically modify DNA, thus scrambling the code for its replication and for the synthesis of proteins. For example, aflatoxin B, a very potent carcinogen, is produced by molds on peanuts and other foods and is known to bind to guanine residues in DNA. Just how this initiates cancer, however, has not yet been determined.

Genetics plays a role in the development of many forms of cancer. Certain genes, called *oncogenes*, seem to trigger or sustain the processes that convert normal cells to cancerous ones. There are about one hundred known oncogenes, which develop from ordinary genes that regulate cell growth and cell division. Chemical carcinogens, radiation, or perhaps some viruses can activate oncogenes. It seems that more than one oncogene must be turned on, perhaps at different stages of the process, before a cancer develops. There are also *suppressor genes* that ordinarily prevent the development of cancers. These genes must be inactivated before a cancer develops. Suppressor gene inactivation can occur through mutation, alteration, or loss. In all, several mutations may be required in a cell before it turns cancerous. A hereditary form of breast cancer that usually develops before age 50 results from a mutation in a tumor-suppressor gene called *BRCA1*. Women who have a mutant form of *BRCA1* have an 80% risk of developing breast cancer, compared with most women in the United States, who have about a 10% risk.

Chemical Carcinogens

A variety of widely different chemical compounds are carcinogenic. We will concentrate on only a few major classes here.

Polycyclic aromatic hydrocarbons (PAHs), of which 3,4-benzpyrene (Section 13.7) is perhaps the best known, are among the more notorious carcinogens. These hydrocarbons are formed during the incomplete burning of nearly any organic material. They have been found in charcoal-grilled meats, cigarette smoke, automobile exhausts, coffee, burnt sugar, and many other materials. Not all polycyclic aromatic hydrocarbons are carcinogenic. However, there are strong correlations between carcinogenicity and certain molecular sizes and shapes.

Aromatic amines make up another important class of carcinogens. Two prominent ones are β-naphthylamine and benzidine. Once widely used in the dye industry, these compounds were responsible for a high incidence of bladder cancer among workers whose jobs brought them into prolonged contact with them.

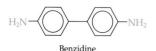

Benzidine

Not all carcinogens are aromatic compounds. Two prominent aliphatic (nonaromatic) ones are dimethylnitrosamine (Section 17.5) and vinyl chloride (Section 10.3). Other nonaromatic carcinogens include three- and four-membered heterocyclic rings containing nitrogen or oxygen, such as epoxides and derivatives of ethyleneimine, as well as cyclic esters called *lactones* (Figure 21.8).

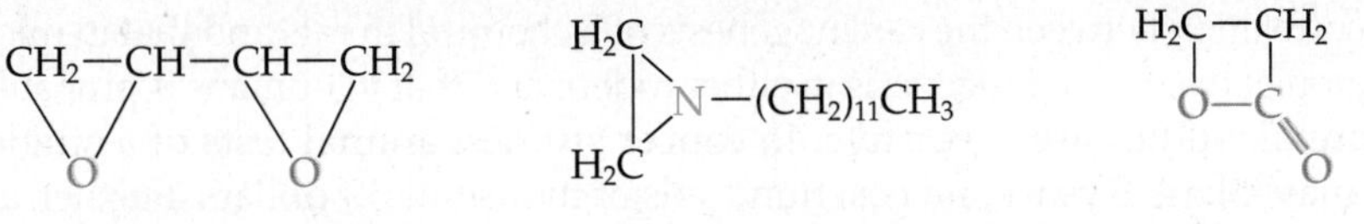

$$CH_2\text{—}CH\text{—}CH\text{—}CH_2$$

Bis(epoxy)butane

$$\begin{array}{c} H_2C \\ | \\ H_2C \end{array} N\text{—}(CH_2)_{11}CH_3$$

N-Laurylethyleneimine

$$\begin{array}{c} H_2C\text{—}CH_2 \\ | \quad | \\ O\text{—}C \\ \quad \| \\ \quad O \end{array}$$

β-Propiolactone

▲ **Figure 21.8** Three small-ring heterocyclic carcinogens.

Q *To what family of compounds does each compound belong?*

This list of carcinogens is far from all-inclusive. Its purpose is to give you an idea of some of the kinds of compounds that have tumor-inducing properties.

Anticarcinogens

If the food we eat has many natural carcinogens, why don't we all get cancer? Probably because other substances in our food act as **anticarcinogens**. Antioxidant vitamins are believed to protect against some forms of cancer, and the food additive butylated hydroxytoluene (BHT) may give protection against stomach cancer. Certain vitamins have also been shown to have anticarcinogenic effects.

Antioxidant vitamins include vitamin C, vitamin E, and β-carotene, a precursor of vitamin A. A diet rich in cruciferous vegetables (cabbage, broccoli, Brussels sprouts, kale, and cauliflower) has been shown to reduce the incidence of cancer both in animals and in human population groups. The evidence for the anticarcinogenicity of vitamins taken as supplements is less conclusive and, in the case of β-carotene, even contradictory. Yellow, red, or blue plant pigments called *flavonoids* are thought to be anticarcinogens. Good sources of flavonoids are citrus fruits, berries, red onions, green tea, and dark chocolate.

There are probably many other anticarcinogens in our food that have not yet been identified.

Three Ways to Test for Carcinogens

How do we know that a chemical causes cancer? Obviously, we can't experiment on humans to see what happens. That leaves us with no way to prove absolutely whether a chemical causes cancer in humans. Three ways to gain evidence about a substance are bacterial screening for mutagenesis, animal tests, and epidemiological studies.

The quickest and cheapest way to find out whether a substance may be carcinogenic is to use a screening test such as that developed by Bruce N. Ames of the University of California at Berkeley. The **Ames test** is a simple laboratory procedure that can be carried out in a petri dish (Figure 21.9). It assumes that most carcinogens are also mutagens that alter genes in some way. This usually seems to be the case. About 90% of the chemicals that appear on a list of either mutagens or carcinogens are found on the other list as well.

The Ames test uses a special strain of *Salmonella* bacteria that have been modified so that they cannot produce histidine, which must be added for growth . The bacteria are placed in an agar medium containing all nutrients except histidine. Incubating the mixture in the presence of a mutagenic chemical causes the bacteria to mutate so that they no longer require external histidine and can grow like normal bacteria. Cancer is an uncontrolled growth of cells, so growth of bacterial colonies in a Petri dish means that the chemical added was a mutagen and, probably, also a carcinogen.

Chemicals suspected of being carcinogens can be tested on animals. Tests involving low dosages and millions of rats would cost too much, so tests are usually done by using large doses and a few dozen rats. An equal number of rats serve as controls; these are exposed to the same diet and environment but not given the suspected

Mutant induced revertants

Spontaneous revertants

▲ **Figure 21.9** The Ames test. Both Petri dishes contain a modified strain of *Salmonella* bacteria. A chemical added to the upper Petri dish has caused the *Salmonella* to revert back to their unmodified form, indicating that mutations have occurred and the chemical is a mutagen. The bottom dish is a control that shows only a few spontaneous mutations and little reversion.

3 Why do scientists use such high doses when testing possible cancer-causing substances?

The validity of some animal tests is sometimes questioned in the media. However, it has been observed repeatedly that a substance that is not a carcinogen will not cause cancer even in high doses. The high doses often used in animal testing simply make positive results appear sooner.

Thalidomide

▲ Structure of thalidomide.

▲ Frances O. Kelsey (left front) of the FDA receives a Distinguished Federal Civilian Service Award from President John F. Kennedy for her refusal to approve thalidomide for use in the United States.

carcinogen. A higher incidence of cancer in the experimental animals than in the controls indicates that the compound is carcinogenic.

Animal tests are not conclusive. Humans are not usually exposed to comparably high doses, and there may be a threshold below which a compound is not carcinogenic. Furthermore, human metabolism is somewhat different from that of the test animals. A substance might be carcinogenic in rats but not in humans (or vice versa). There is only a 70% correlation between the carcinogenesis of a chemical in rats and that in mice. The correlation between carcinogenesis in either rodent and that in humans is probably less.

Animal studies are expensive. In cancer studies, animal tests of a single substance may take 4–8 years and cost hundreds of thousands of dollars. Further, animal testing is controversial. Supporters and opponents argue over both the ethics and the value of the tests.

Epidemiological studies use case series studies to provide evidence that a substance causes cancer in humans. A population that has a higher-than-normal rate for a particular kind of cancer is studied for common factors in the individuals' backgrounds. Studies of this sort showed that cigarette smoking causes lung cancer, that vinyl chloride causes a rare form of liver cancer, and that asbestos causes cancer of the lining of the pleural cavity (the body cavity containing the lungs). These studies sometimes require sophisticated mathematical analyses, and there is always the chance that some other (unknown) factor is involved in the carcinogenesis.

More recently, epidemiological studies were carried out to determine whether electromagnetic fields, such as those surrounding high-voltage power lines, certain electric appliances, and cellular phones, could cause cancer. The results so far indicate that if there is such an effect, it is exceedingly small.

Physical and Biochemical Abnormalities: Teratogens

A **teratogen** is a substance that causes physical and biochemical abnormalities. Perhaps the most notorious teratogen is the tranquilizer *thalidomide*. Over fifty years ago, thalidomide was considered so safe, based on laboratory studies, that it was often prescribed for pregnant women. In Germany, it was even available without a prescription. It took several years for the human population to provide evidence that laboratory animals had not provided.

The drug had a disastrous effect on developing human embryos. About 12,000 women who had taken the drug during the first 12 weeks of pregnancy had babies who suffered from *phocomelia*, a condition characterized by shortened or absent arms and legs and other physical defects. The drug was used widely in Germany and Great Britain, and these two countries bore the brunt of the tragedy. The United States escaped relatively unscathed because Frances O. Kelsey of the U.S. Food & Drug Administration (FDA) believed that there was evidence to doubt the drug's safety and therefore did not approve it for use in the United States. Kelsey received the Distinguished Federal Civilian Service Award from President John F. Kennedy in 1962.

Thalidomide is now being used for treating certain conditions, with strict restrictions in place to prevent its use by pregnant women. It has a unique anti-inflammatory action against a debilitating skin condition that often occurs with leprosy. Thalidomide is also being studied for use in the treatment of *wasting*, the severe weight-loss condition that often accompanies AIDS. Scientists are also studying thalidomide derivatives, hoping to enhance the beneficial effects while reducing the harmful properties.

Other chemicals that act as teratogens include isotretinoin (Accutane), a prescription medication approved for use in treating severe acne. When taken by women during the first trimester of pregnancy, it can cause multiple major malformations. Educational materials provided to physicians and to patients have prevented all but a few tragedies associated with the use of isotretinoin.

Isotretinoin
(13-*cis*-retinoic acid)

By far the most hazardous teratogen is ethyl alcohol, which causes fetal alcohol syndrome and can result in cognitive and physical disabilities in babies.

SELF-ASSESSMENT Questions

1. Some genes that are involved in cancer development when they are switched on are called
   a. DNA repair genes   **b.** oncogenes
   c. P53   **d.** suppressor genes

2. Some genes are involved in preventing the development of cancers. These genes, when switched off, allow cancerous cells to grow. They are called
   a. DNA repair genes   **b.** oncogenes
   c. BRCA1 genes   **d.** suppressor genes

3. Many polycyclic aromatic hydrocarbons (PAHs) are known carcinogens. Sources of PAHs include
   a. artificial flavorings **b.** automobile exhausts
   c. pesticides   **d.** plastics

4. A prominent aliphatic carcinogen is
   a. arachidonic acid   **b.** butane
   c. valine   **d.** vinyl chloride

5. The fastest way to determine whether a substance is likely to be a carcinogen is to use
   a. the Ames test   **b.** animal tests
   c. epidemiological studies **d.** metabolic testing

6. The Ames test is a(n)
   a. evaluation of new compounds for hepatotoxicity
   b. immunological test for melanoma
   c. standardized final exam for introductory chemistry courses
   d. test for mutagenicity used as a screen for potential carcinogens

7. The teratogen that caused phocomelia—missing or shortened limbs in newborn babies—is
   a. ethanol   **b.** isotretinoin
   c. malathion   **d.** thalidomide

21.7 Hazardous Wastes

Learning Objectives • Define *hazardous waste*, and list and give an example of the four types of such waste.

The public has become increasingly concerned in recent years about hazardous wastes in the environment. Problems created by chemical dumps made household names out of Love Canal in New York and Valley of the Drums in Kentucky. Serious problems do exist, and they can be worsened by natural events, such as Hurricane Harvey, which hit the Houston, Texas, region in 2017 and triggered massive flooding and power outages. Hazardous wastes can cause fires or explosions. They can pollute the air, and they can contaminate our food and water. Occasionally, they poison by direct contact. As long as we want the products that industries produce, however, we will have to deal with the problems of hazardous wastes (Table 21.2).

The first step in dealing with any problem is to understand the problem. A **hazardous waste** is one that can cause or contribute to death or illness or that threatens human health or the environment when improperly managed. For convenience, hazardous wastes are divided into four types: *reactive, flammable, toxic,* and *corrosive.*

TABLE 21.2 Industrial Products and Hazardous Waste By-Products

Product	Associated Waste
Plastics	Organic chlorine compounds
Pesticides	Organic chlorine compounds; organophosphate compounds
Medicines	Organic solvents and residues; heavy metals (for example, mercury and zinc)
Paints	Heavy metals; pigments; solvents; organic residues
Oil, gasoline	Oil; phenols and other organic compounds; heavy metals; ammonium salts; acids; strong bases
Metals	Heavy metals; fluorides; cyanides; acid and alkaline cleaners; solvents; pigments; abrasives; plating salts; oils; phenols
Leather	Heavy metals; organic solvents
Textiles	Heavy metals; dyes; organic chlorine compounds; solvents

A **reactive waste** tends to react spontaneously or to react vigorously with air or water. Such wastes can generate toxic gases, such as hydrogen cyanide (HCN) or hydrogen sulfide (H_2S), or explode when exposed to shock or heat. Explosives such as trinitrotoluene (TNT) and nitroglycerin obviously are reactive wastes. Examples of household wastes that can explode include improperly handled propane tanks and aerosol cans. Another example of a reactive waste is sodium metal. It reacts with water to form hydrogen gas, which can then explode when ignited in air.

$$2\,Na(s) + 2\,H_2O(l) \longrightarrow 2\,NaOH(aq) + H_2(g)$$
$$2\,H_2(g) + O_2(g) \longrightarrow 2\,H_2O(g)$$

Reactive wastes can usually be deactivated before disposal. Propane tanks and aerosol cans can be thoroughly emptied and properly discarded. Sodium can be treated with isopropyl alcohol, with which it reacts slowly, rather than being dumped without treatment.

A **flammable waste** is one that burns readily on ignition, presenting a fire hazard. An example is hexane, a hydrocarbon solvent. Hexane (presumably dumped accidentally by Ralston-Purina) was ignited in the sewers of Louisville, Kentucky, in 1981, causing explosions that blew up several blocks of streets.

$$2\,C_6H_{14}(l) + 19\,O_2(g) \longrightarrow 12\,CO_2(g) + 14\,H_2O(g)$$

Examples of flammable household wastes include gasoline, lighter fluid, many solvents, and nail polish.

A **toxic waste** contains or releases toxic substances in quantities sufficient to pose a hazard to human health or to the environment. Most of the toxic substances discussed in this chapter would qualify as toxic wastes if they were improperly dumped in the environment. Some toxic wastes can be incinerated safely. For example, polychlorinated biphenyls (PCBs; Section 10.7) are burned at high temperatures to form carbon dioxide, water, and hydrogen chloride. If the hydrogen chloride is removed by scrubbing or is safely diluted and dispersed, this is a satisfactory way to dispose of PCBs. Some toxic wastes cannot be incinerated, however, and must be contained and monitored for years. Examples of household toxic wastes include paint, pesticides, motor oil, medicines, and cleansers.

A **corrosive waste** is one that requires a special container because it corrodes conventional container materials. Acids cannot be stored in steel drums, because they react with and dissolve the iron.

$$Fe(s) + 2\,H^+(aq) \longrightarrow Fe^{2+}(aq) + H_2(aq)$$

Acid wastes can be neutralized (Chapter 7) before disposal, and lime (CaO) is an inexpensive base often used for this neutralization.

$$2\,H^+(aq) + CaO(s) \longrightarrow Ca^{2+}(aq) + H_2O(l)$$

Examples of household corrosive wastes include battery acid, drain cleaners, and oven cleaners.

The best way to handle hazardous wastes is to not produce them in the first place. This is a major focus of green chemistry. Many industries have modified manufacturing processes to prevent wastes, and some wastes can be reprocessed to recover energy or materials. Hydrocarbon solvents such as hexane can be purified and reused or burned as fuels. Often the waste from one industry can become raw material for another industry. For example, waste nitric acid from the metals industry can be converted to fertilizer.

If a hazardous waste cannot be used or incinerated or treated to render it less hazardous, it must be stored in a secure landfill. Unfortunately, landfills often leak, contaminating the groundwater. We clean up one toxic waste dump and move the materials to another, playing a rather macabre shell game. The best current technology for treating organic wastes, including chlorinated compounds, is incineration (Figure 21.10). At 1260 °C, greater than 99.9999% destruction is achieved. A major issue with incineration is finding a place to build the incinerator. No one wants a hazardous waste incinerator nearby.

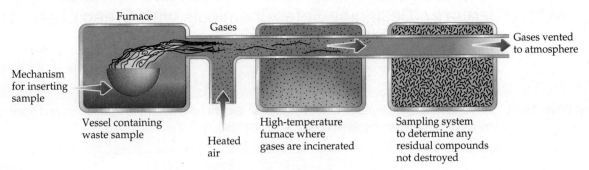

▲ **Figure 21.10** A schematic diagram of an incinerator for hazardous wastes.

Q *PCBs are burned in air at high temperatures, forming carbon dioxide, water, and hydrogen chloride. Write the equation for the incineration of the PCB component $C_6H_2Cl_3 — C_6H_3Cl_2$.*

Biodegradation is another method that shows great promise. Scientists have identified microorganisms that degrade hydrocarbons such as those in gasoline. Other bacteria, when provided with proper nutrients, can degrade chlorinated hydrocarbons. Certain bacteria in the soil can even break down TNT and nitroglycerin. Through genetic engineering, scientists are developing new strains of bacteria that can decompose a great variety of wastes.

At What Price Poisons?

We use so many poisons in and around our homes and workplaces that accidents are bound to happen. Poison control centers have been established in many cities to help physicians deal with emergency poisonings. Are insecticides, medicines, cleansers, and other chemicals worth the price we pay in accidental poisonings? That is for you to decide. Generally, it is misuse of these chemicals that leads to tragedy.

Perhaps it is easy to be negative about chemists and chemistry when you think of such horrors as nerve gases, carcinogens, and teratogens. But keep in mind that many chemicals are of enormous benefit to us and that they can be used safely, despite the hazardous nature of toxic chemicals. The plastics industry was able to control vinyl chloride emissions once the hazard was known. We are still able to use valuable vinyl plastics, even though the vinyl chloride from which they are made is a carcinogen.

We have to decide whether the benefits we gain from hazardous substances are worth the risks we assume by using them. Many issues involving toxic chemicals are emotional. Most of the decisions regarding them are political. Yet possible solutions to such problems lie mainly in the field of chemistry.

We hope that the chemistry you have learned here will help you make wise decisions. Most of all, we hope that you continue to learn more about chemistry throughout your life, because chemistry affects nearly everything you do. We wish you success and happiness, and may the joy of learning go with you always.

4 **Why can't we dispose of wastes like PCBs, lead, and dioxins by burning them?**

PCBs, nerve gases, and other organic materials usually can be disposed of by incineration (Figure 21.10) because they are oxidized to simple compounds like CO_2 and H_2O. However, toxins such as lead, mercury, and arsenic are elements. If a waste containing lead compounds is incinerated, the lead will still be present in some form, such as PbO or elemental Pb, after incineration.

SELF-ASSESSMENT Questions

For items 1–4, match the waste described or named in the left column with the category in the right column.

1. tends to react vigorously with water
2. common strong acids and bases
3. discarded unused pesticides
4. charcoal lighter fluid

a. corrosive
b. flammable
c. reactive
d. toxic

5. The best way to deal with hazardous wastes is to
   a. burn them in an incinerator
   b. bury them in a landfill
   c. compost them
   d. minimize or eliminate their use

Answers: 1, c; 2, a; 3, d; 4, b; 5, d

GREEN CHEMISTRY Designing Safer Chemicals with Green Chemistry
Richard Williams

Principles 3, 4, 10

Learning Objectives • Explain how the structure of a chemical can lead to both hazardous properties and commercial effectiveness. • Trace the process for designing safer chemical structures.

The chemical industry makes a large number of products that improve the quality of life for humans. However, uncertainty about the safety of many human-made chemicals has proved to be a concern. A good question is, "How can chemical developers identify and respond to hazards early in the research and development process?" Green chemists can address this challenge by understanding what makes a chemical toxic and using this information to design products that are both effective and safe.

The public and the regulators they influence are generally becoming much more concerned about risk, as shown by the European Union's Registration, Evaluation, Authorisation, and Restriction of Chemicals (REACH) legislation and proposed changes to the Toxic Substances Control Act in the United States. REACH requires companies to submit human health and environmental safety test data and an assessment of risks before manufacturing or importing chemicals. Although these are a step in the right direction, most chemical safety testing is typically done near the end of the research-and-development process. At that point, the product testing might have consumed many years and millions of dollars, and changes to a chemical product can be hard to make at that time.

Green chemistry offers the opportunity to optimize the commercial effectiveness of a chemical during R&D while minimizing the hazard and risk (Principle 4). Risk is a function of both a molecule's inherent hazard and exposure—contact between a chemical and an organism. As you learned in Section 21.2, hazard, the capability to cause harm, is an inherent characteristic of a molecule. Likewise, the commercial function is dependent on the molecule's content and three-dimensional arrangement of atoms. Hazards may be either physical (such as combustion or explosion) or toxicological (such as acute or chronic poisoning, carcinogens, reproductive toxins, or irritants). Risk often is reduced by the use of protective equipment that prevents exposure, but green chemistry provides a better idea.

Designing safer chemicals relies on understanding what makes a molecule hazardous. For example, halogen atoms are often used to make molecules stable, but that stability also prevents biodegradation and can increase the concentration of these materials in the environment. Polychlorinated biphenyls (PCBs) have become a health hazard because they build up in fish that live in polluted waters. The dose to a human who eats the fish can therefore be much higher than is healthy (Section 21.4). Design of molecules to include fewer chlorine atoms on aromatic rings can promote degradation (Principle 10) and eliminate significant exposure and risk. Current innovative design strategies specifically target unique physiological features of the organisms of concern using chemicals with only the degree of stability needed to achieve the commercial function.

Understanding how molecular characteristics such as the presence or absence of toxicity are controlled by chemical structure is critical. Stating that the nonselective contact herbicide paraquat is toxic to mammals provides little insight to a design chemist. But by knowing that paraquat accumulates and causes damage to the lungs because of its charge, shape, and number of carbon atoms between the charged nitrogen atoms, chemists were able to design diquat. Diquat is equally effective as a product but much less toxic because its structure lowers its ability to get into lung cells and reduces cellular damage (Principle 3).

$$CH_3-{}^+N \qquad N^+-CH_3 \quad 2\,Cl^-$$
Paraquat

$$2\,Br^-$$
$$CH_2-CH_2$$
Diquat

Scientific advances are being sought to more effectively identify hazards based on structure during chemical design because traditional current safety testing methods are slow and expensive, rely on the use of lots of test animals, and do not provide information about how and why molecules are toxic. What we do have at this point are approaches including rules of thumb that link structural features to hazards, computational methods, databases of known hazards, models, and short-term assays. Green chemistry supports these ways of helping chemists, who make thousands of new chemicals annually, to design safer molecules. These approaches can save money and time if they are introduced early in the design process.

For the future, researchers are trying to develop chemical design tools that can quickly screen large groups of structurally similar or diverse chemicals and build the knowledge base to support the design of safer and commercially successful chemicals. The innovation challenges are significant, but incorporating safety early in the chemical design process holds great promise.

 Summary

A **poison** is a substance that causes injury, illness, or death. **Toxicology** is the branch of pharmacology that deals with poisons.

Section 21.1—A **toxin** is a poison that is naturally produced by a plant or animal. Many natural poisons exist, including curare, strychnine, and botulin. Many plants, including hemlock, holly berries, iris, azalea, and hydrangea, are toxic.

Section 21.2—Poisons act in different ways. Strong acids and strong bases are toxic because they are corrosive. They cause hydrolysis of proteins. Strong oxidizing agents such as ozone and PAN can cause damage largely by deactivating enzymes. Carbon monoxide and nitrite ions interfere with the transport of oxygen by the blood. Cyanide shuts down cell respiration. Several substances, including fluoroacetic acid, are not highly toxic themselves, but in the body, they form substances that are toxic. Heavy metal poisons such as lead and mercury inactivate enzymes by tying up their —SH groups. Cadmium leads to loss of calcium ions from bones. Heavy metal poisoning can be treated with **chelating** agents that tie up the metal ions so that they can be excreted.

Section 21.3—Nerve poisons, such as organophosphorus compounds, interfere with the acetylcholine cycle. Nerve poisons are among the most toxic synthetic chemicals known.

Section 21.4—An **LD_{50}** value is the dosage that kills 50% of a population of test animals. Many substances not ordinarily thought of as poisons do have some toxicity, indicated by a high LD_{50} value. Botulism toxin has the lowest LD_{50} value and is the most toxic substance known.

Section 21.5—The liver can detoxify substances by oxidizing or reducing them or by coupling them with other chemicals normally found in the body. The liver's P-450 enzymes carry out many such detoxifying reactions. Sometimes liver enzymes convert a less toxic substance, such as benzene or CCl_4, into one that is more toxic.

Section 21.6—A **carcinogen** causes the growth of tumors. A benign tumor grows slowly and does not invade neighboring tissues. A malignant tumor, or cancer, can invade and destroy neighboring tissues. Some carcinogens are natural, but most cancer deaths are linked to tobacco, diet, and lack of exercise. Oncogenes appear to trigger processes that convert normal cells to cancer, and suppressor genes act to prevent the development of cancers. Some substances in our food, including antioxidant vitamins, act as **anticarcinogens**.

The **Ames test** screens substances for mutagenicity and possible carcinogenicity by observing bacterial growth. Animal testing is often necessary but is not conclusive. Epidemiological studies appear to be most useful for identifying carcinogens.

A **teratogen** is a substance that causes cognitive and physical disabilities in newborns.

Section 21.7—**Hazardous waste** can be classified as **reactive waste**, which tends to react spontaneously or to react vigorously with air or water; **flammable waste**, which burns readily on ignition; **toxic waste**, which contains or releases toxic substances; and **corrosive waste**, which destroys conventional container materials. The best way to handle hazardous wastes is not to produce them in the first place. Wastes may be stored in a landfill or incinerated, and biodegradation shows promise for disposing of some wastes.

GREEN CHEMISTRY—Advances in chemistry can enable the identification and control or elimination of the hazards and risks that a new chemical might contain. Designing to eliminate non-preferred features can be pursued early in the development of new chemicals to achieve commercial effectiveness and maximize safety.

Learning Objectives

Learning Objectives	Associated Problems
• Name some natural poisons and their sources. (21.1)	9, 10, 63
• Distinguish among corrosive poisons, metabolic poisons, and heavy metal poisons, and explain how each type acts. (21.2)	4, 5, 11–14, 18, 50, 52, 57
• Identify antidotes for some common metabolic poisons and heavy metal poisons. (21.2)	15–17
• Define *neurotransmitter*, and describe how various substances interfere with the action of neurotransmitters. (21.3)	19–22, 47, 53
• Explain the concept of an LD_{50}, and list some of its limitations as a measure of toxicity. (21.4)	1–3, 25, 26, 55, 64
• Calculate lethal doses from LD_{50} values and body weights. (21.4)	23, 24, 54, 60
• Describe how the liver is able to detoxify some substances. (21.5)	27–32, 58, 59
• Describe how cancers develop. (21.6)	6–8, 37, 39, 40, 48, 49, 51, 56, 62
• Name and describe three ways to test for carcinogens. (21.6)	33–36, 38, 61
• Define *hazardous waste*, and list and give an example of the four types of such waste. (21.7)	41–46
• Explain how the structure of a chemical can lead to both hazardous properties and commercial effectiveness.	65
• Trace the process for designing safer chemical structures.	66, 67

Conceptual Questions

1. Is ethyl alcohol (drinking alcohol) poisonous? Explain.
2. How can a substance be more harmful to one person than to another? Give an example.
3. How does the toxicity of a substance depend on the route of administration?
4. How does fluoroacetic acid exert its toxic effect?
5. List some sources of mercury poisoning and of lead poisoning.
6. What are the two leading causes of cancer?
7. Name several natural carcinogens.
8. How does cigarette smoking cause cancer?

 Problems

Natural Poisons

9. Which of the following would *not* be considered to be a natural poison?
   a. curare
   b. DDT
   c. nicotine
   d. strychnine
10. What is a toxin?

Poisons and How They Act

11. Describe how hydrolysis affects the function of proteins.
12. What is a metabolic poison? List two such poisons.
13. What is methemoglobin? How does methylene blue affect it?
14. How do nitrates exert their toxic effect?
15. How does sodium thiosulfate act as an antidote for cyanide poisoning?
16. How does BAL act as an antidote for mercury poisoning?
17. How does EDTA act as an antidote for lead poisoning?
18. How does cadmium (as Cd^{2+}) exert its toxic effect? What is *itai-itai*?

More Chemistry of the Nervous System

19. What are the acetylcholine-blocking poisons? How do they act on the acetylcholine cycle?
20. Describe some uses of botulin in medicine.
21. All of the following may result from exposure to nerve poisons except
    a. absorption through the skin
    b. activation of acetylcholinesterase
    c. loss of muscle coordination
    d. paralysis that causes breathing to stop
22. A common source of botulism poisoning is
    a. contaminated fresh fruits and vegetables
    b. exposure to heavy metals
    c. improperly canned foods
    d. use of local anesthetics

Lethal Dose

23. The LD_{50} value for methyl isocyanate (orally in rats), the substance that caused the Bhopal tragedy, is 140 mg/kg. Estimate the lethal dose for a 125 lb human.
24. The LD_{50} value for carbaryl (Sevin), given orally to rats, is 307 mg/kg. If this value could be extrapolated to humans, what would be the lethal dose for an 18 kg child?

25. For rats, the LD_{50} value for ethyl alcohol is 2080 mg/kg body weight and an LD_{50} value for ethylene glycol is 6.86 mg/kg body weight. Which is more toxic? Explain.
26. An LD_{50} value of greater than 15,000 mg/kg would be considered to have
    a. high toxicity
    b. low toxicity
    c. moderate toxicity
    d. negligible toxicity

The Liver as a Detox Facility

27. How does the liver detoxify nicotine?
28. What is the P-450 system? What is its function?
29. Do the P-450 enzymes always detoxify foreign substances? Explain.
30. List two steps in the detoxification of ingested toluene. What is the effect of these steps?
31. Why does chronic alcoholism also lead to decreases in testosterone?
32. Why is fomepizole a good antidote for methanol poisoning?

Carcinogens and Teratogens

33. A mutagen is a generic term for a mutation-causing agent. Some mutagens are also carcinogens, meaning that they cause cancer. How would a scientist determine if a substance is a mutagen?
34. What are teratogens? Why can teratogens be particularly difficult to detect?
35. List some of the limitations involved in testing compounds for carcinogenicity by using laboratory animals.
36. What is an epidemiological study? Can such a study prove absolutely that a substance causes cancer?
37. What is the *P35* gene?
38. Describe the Ames test for mutagenicity. What are its limitations as a screening test for carcinogens?
39. Which of the following teratogens is responsible for the greatest number of cognitive and physical disabilities in newborns?
    a. aflatoxin
    b. ethyl alcohol
    c. isotretinoin
    d. thalidomide
40. Approximately what percentage of cancers is caused by lifestyle factors, such as smoking, diet, or lack of exercise?
    a. 15%
    b. 35%
    c. 65%
    d. 95%

Hazardous Wastes

41. What is the best way to deal with hazardous waste?
    a. burn it   b. bury it
    c. neutralize it   d. not generate it

42. Which of the following are considered to be types of hazardous wastes?
    a. biodegradable, corrosive, flammable, and toxic
    b. reactive, biodegradable, corrosive, and flammable
    c. toxic, reactive, corrosive, and flammable
    d. flammable, neutralizable, corrosive, and toxic

43. Define and give an example of a toxic waste.

44. Which of the following are corrosive wastes?
    a. sodium hydroxide, which is used as a drain cleaner
    b. Sr metal, which reacts with water to form flammable H_2 gas
    c. trinitrotoluene, an explosive
    d. phosphoric acid, which dissolves steel drums

45. What is a corrosive waste?

46. What is a hazardous waste?

 ## Expand Your Skills

47. All of the following are ways in which the acetylcholine cycle can be interrupted except
    a. adding hippuric acid
    b. blocking acetylcholine receptor sites
    c. blocking synthesis of acetylcholine
    d. inhibition of acetylcholinesterase

48. How does benzene act to cause cancers?

49. How does carbon tetrachloride (CCl_4) act to cause cancer?

50. All of the following amino acids can be oxidized by ozone except
    a. cysteine   b. glycine
    c. methionine   d. tryptophan

51. What is the difference between an oncogene and a suppressor gene?

52. Which of the following act(s) by hydrolyzing proteins?
    a. acids and bases   b. cadmium
    c. carbon monoxide   d. sodium cyanide

53. Which of the following is a neurotransmitter that is released into synapses?
    a. acetylcholine   b. arsphenamine
    c. continine   d. hippuric acid

54. Absinthe, a wormwood-flavored liqueur that was highly popular in France from about 1880 to 1914, contains thujone, a neurotoxin with an LD_{50} value (orally in mice) of 87.5 mg/kg. Absinthe was the subject of famous artworks by Degas, van Gogh, and others. Banned in 1912 in the United States, the liqueur became legal again in 2007, with levels in the commercially distilled product limited to 10 mg/L or less. (a) Determine the molecular formula of thujone from the structure below. (b) Estimate the likely lethal dose for a 55 kg person. (Assume that the LD_{50} value s for mice and humans are the same.) (c) What is the maximum mass of thujone that a 0.50 L bottle of absinthe can contain?

55. Phosphorus comes in two forms, or *allotropes*. Red phosphorus is used in manufacture of matches and is only mildly toxic. White phosphorus, a waxy yellow solid, is highly toxic, with an orally ingested lethal dose of just 70 mg for a 55 kg person. (a) What is the LD_{50} (in milligrams per kilogram) for white phosphorus? (b) Where does white phosphorus fall in Table 21.1?

56. A British study more than 50 years ago involved 1357 people with lung cancer. Of these, 1350 were smokers and 7 were nonsmokers. A control group without cancer, matched in age and other characteristics, consisted of 1296 smokers and 61 nonsmokers. (a) Calculate the percentage of cancer sufferers who smoked and the percentage of controls who smoked. What can you infer from these proportions? (b) Calculate the odds of smoking for the cancer sufferers and the odds of smoking for the controls. (c) Calculate the odds ratio. Interpret this result.

57. Which of the nutrients listed in Table 17.2 are heavy metals?

58. Justify the following statement: "Benzene itself is not toxic; the body makes it toxic."

59. Why does the added oxygen atom in cotinine make it more soluble in water? Why is this important?

60. The LD_{50} value for ethyl alcohol given orally to rats is 10.3 g/kg. (a) What volume in milliliters of 100 proof vodka (50% v/v ethanol) would be likely to kill a typical 395 g rat? (b) Assuming that the rat's mass is 67% water, what would its blood alcohol level be, in milligrams per milliliter? (Density of 100 proof vodka is 0.92 g/mL and the density of pure ethanol is 0.79 g/mL.)

61. What is the current "gold standard" for chemical toxicity testing?
    a. cell culture testing   b. rodent testing
    c. human testing   d. nanotechnology testing

62. Which of the following statements about mutagens and carcinogens is true?
    a. A mutagen is always a carcinogen, because a mutagen alters DNA.
    b. A mutagen may be a carcinogen, because a mutagen can alter DNA.
    c. A mutagen will not be a carcinogen, because a carcinogen cannot alter DNA.
    d. A mutagen will not be a carcinogen, because mutagens cause cognitive and physical disabilities in newborns and not cancer.

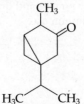

Thujone

63. Which of the following statements about poisons and toxins is true?
    a. Toxins are poisons, because they are not made by natural processes.
    b. Toxins are poisons, because they are made naturally.
    c. A poison may be a toxin, because it can cause injury, illness, or death to an organism.
    d. A poison may be a toxin, because toxins are made by synthetic processes.

64. The LD_{50} value for a substance is 400 mg/kg when it is administered to rats orally. The LD_{50} value for humans would be
    a. less than 400 mg/kg, because humans are bigger than rats
    b. 400 mg/kg, because the LD_{50} value is the same for all species
    c. greater than 400 mg/kg, because humans are bigger than rats
    d. unknown, because different species metabolize substances differently

65. Explain why it would be greener to replace paraquat with diquat as an herbicide.

66. How can green chemistry lead to the design of safer chemicals?
    a. Testing chemicals at the end of the R&D process will make sure they are safe.
    b. By linking specific structural features to hazards, chemists can design those features out of new chemicals.
    c. Risk can be reduced by the use of protective equipment.
    d. Chemists can make molecules less biodegradable so that they are not introduced to the environment.

67. Which of these is an example of how green chemistry can contribute to the design of safer chemicals?
    a. A consumer chooses a safer product that does not work as well as the original version of the product.
    b. A chemist identifies structural features of a chemical that are responsible for hazardous properties and replaces those portions of the molecule with safer alternatives.
    c. Legislators pass a law that bans the use of any chemical that contains chlorine.
    d. A chemist suggests that protective equipment be worn during the use of a toxic chemical.

 ## Critical Thinking Exercises

Apply knowledge that you have gained in this chapter and one or more of the FLaReS principles (Chapter 1) to evaluate the following statements or claims.

21.1. An online advertisement implies that echinacea is nontoxic because it is natural.

21.2. A book claims that all aromatic compounds are carcinogens and should be banned.

21.3. A natural-foods enthusiast contends that no food that contains a carcinogen should be allowed on the market.

21.4. An advertisement promotes "Organic Natural Chemical-Free Mattresses." The company claims its products contain "no toxins, formaldehyde, VOCs, pesticides, or synthetic or chemical foams."

21.5. An advertisement implies that a supplement containing β-carotene should be taken in megadoses because it is an antioxidant and can therefore prevent cancer.

 ## Collaborative Group Projects

Prepare a PowerPoint, poster, or other presentation (as directed by your instructor) to be shared with the class.

1. Choose one of the following arguments, and be prepared to support either side (for or against).
   a. Carcinogens that occur naturally in foods should be subjected to the same tests used to evaluate synthetic pesticides.
   b. Substances should be tested for toxicity or carcinogenicity on laboratory animals.
   c. Nerve gases are less humane than bullets in warfare.

2. Search the Internet for information on brownfields. What are they? What is being done about them? Are there any near you?

3. Many tooth fillings consist of amalgams containing mercury, and the safety of these amalgams is a recurring controversy. Using your favorite search methods, find some websites dealing with this issue and write a brief analysis of the opposing points of view.

4. Search the Internet for information on the effects of cadmium in the environment and do a risk–benefit analysis (Chapter 1) of cadmium use.

LETS EXPERIMENT! Salty Seeds

Materials Needed

- Table salt (NaCl)
- 6 small plastic bags
- 5 small clear plastic cups
- Kitchen scale
- Dropper

- Teaspoon
- Lettuce seeds
- Paper towels
- Measuring cup
- Distilled water

Who doesn't like salt on popcorn, French fries, or buttered corn on the cob? Table salt is safe for humans to consume in small amounts, but what about plants? How sensitive are certain plants to saltwater? How much salt is toxic to lettuce seeds?

We all have sodium in our diets. But where else do we use sodium chloride? It is a common chemical in many industrial processes. Sodium chloride (rock salt) is a common de-icer, but more and more places are using alternatives, since it damages concrete, contaminates soil, and harms plants and animals. One goal of chemists is to design materials and processes that are inherently safer for human health and the environment. It is important that chemists be trained to consider the effects of the materials they use on human health and the environment. Understanding ecotoxicity is essential to designing safer chemicals and processes.

This is an ecotoxicity experiment that uses different concentrations of table salt (sodium chloride, NaCl) on samples of lettuce seed. In this experiment, you will determine at what concentrations NaCl is toxic to lettuce seeds.

Start by making a 10% by weight NaCl stock solution. Weigh out 10 g of NaCl and fill the measuring cup with water just short of the $1/2$ cup line.

Next you will prepare serial dilutions for which each solution is 1/10 the concentration of the previous solution. Label five cups with 10%, 1%, 0.1%, 0.01%, and 0.001% NaCl, respectively.

Take $1/4$ teaspoon of the 10% solution and dilute with nine $1/4$ teaspoon portions of distilled water to make the 1% solution. Then take $1/4$ teaspoon of the 1% solution with nine $1/4$ teaspoon portions of distilled water to make the next, and so on.

Label six plastic bags control, 10%, 1%, 0.1%, 0.01%, and 0.001% NaCl, respectively. Place a folded paper towel in each plastic bag.

Pour 1 teaspoon of distilled water into the control bag and into each of the other bags.

Carefully add six lettuce seeds to each of the bags, spacing the seeds evenly on the wet paper towel inside the bag. Figure 1 shows each bag aligned with the cup of the solution it contains.

◀ **Figure 1**

Carefully trickle the different concentration NaCl solutions into the corresponding bags. Check to make sure the seeds are still on the paper towel after adding the solutions. Seal the bags to retain moisture and place in a well-lit location but out of direct sunlight. Figure 2 shows the control.

▲ **Figure 2**

Check the seeds each day. Record how many seeds have germinated in each bag. How many seeds sprouted in each bag? Wait five days to allow the seeds to germinate.

After germination is complete, use the figure below as a model to measure growth of the radical. (Length is indicated in Figure 3.)

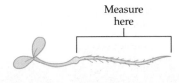

▲ **Figure 3**

Questions

1. Did the seeds in the control bag sprout faster than the others?
2. Which concentration is sufficiently toxic that the seeds do not germinate at all?
3. Can you determine which concentration impedes seed germination?
4. What do your results tell you about the effects of salt toxicity on lettuce seeds?

Appendix
Review of Measurement and Mathematics

Accurate measurements are essential to science. As noted in Chapter 1, measurements can be made in a variety of units, but most scientists use the International System of Units (SI). The data they record sometimes has to be converted from one kind of unit to another and otherwise manipulated mathematically. This appendix extends the discussion of metric measurement and reviews some of the mathematics that you may find useful in this course or other endeavors.

A.1 The International System of Units

The standard SI unit of length is the *meter*. This distance was once defined as 0.0000001 of Earth's quadrant—that is, of the distance from the North Pole to the equator measured along a meridian. The quadrant proved difficult to measure accurately, and today the meter is defined precisely as the distance light travels in a vacuum during 1/299,792,458 of a second.

The SI unit of mass is the *kilogram* (1 kg = 1000 g). It is based on a standard platinum–iridium cylinder kept at the International Bureau of Weights and Measures. The *gram* is a more convenient unit for many chemical operations.

The derived SI unit of volume is the *cubic meter*. The units more frequently employed in chemistry, however, are the *liter* (1 L = 0.001 m³) and the *milliliter* (1 mL = 0.001 L). Other SI units of length, mass, and volume are derived from these basic units. Table A.1 lists some metric units of length, mass, and volume and illustrates the use of prefixes.

TABLE A.1 Some Metric Units of Length, Mass, and Volume

Length	
1 kilometer (km)	= 1000 meters (m)
1 meter (m)	= 100 centimeters (cm)
1 centimeter (cm)	= 10 millimeters (mm)
1 millimeter (mm)	= 1000 micrometers (μm)
1 micrometer (μm)	= 1000 nanometers (nm)
Mass	
1 kilogram (kg)	= 1000 grams (g)
1 gram (g)	= 1000 milligrams (mg)
1 milligram (mg)	= 1000 micrograms (μg)
Volume	
1 liter (L)	= 1000 milliliters (mL)
1 milliliter (mL)	= 1000 microliters (μL)
1 milliliter (mL)	= 1 cubic centimeter (cm³)

A.2 Exponential (Scientific) Notation

Scientists often use numbers that are inconceivably large or small. For example, light travels at about 300,000,000 m/s. There are 602,200,000,000,000,000,000,000 carbon atoms in 12.01 g of carbon. On the small side, the diameter of an atom is about 0.0000000001 m, and the diameter of an atomic nucleus is about 0.000000000000001 m.

Because it is difficult to keep track of the zeros in such numbers, scientists find it convenient to express them in exponential notation.

A number is in *exponential notation* when it is written as the product of a coefficient and a power of 10. (Such a number is called *scientific notation* when the coefficient has a value between 1 and 10). Two examples are

$$4.18 \times 10^3 \quad \text{and} \quad 6.57 \times 10^{-4}$$

Expressing numbers in exponential form generally serves two purposes.

1. We can write very large or very small numbers in a minimum of space and with a reduced chance of typographical error.
2. We can convey explicit information about the precision of measurements: The number of significant figures (Section A.4) in a measured quantity is stated unambiguously.

In the expression 10^n, n is the exponent of 10, and the number 10 is said to be raised to the *n*th power. If n is a *positive quantity*, 10^n has a value *greater than 1*. If n is a *negative quantity*, 10^n has a value *less than 1*. We are particularly interested in cases where n is an integer. For example,

<table>
<tr><td align="center">*Positive Powers of 10*</td><td align="center">*Negative Powers of 10*</td></tr>
<tr><td>

$10^0 = 1$
$10^1 = 10$
$10^2 = 10 \times 10 = 100$
$10^3 = 10 \times 10 \times 10 = 1000$
and so on

</td><td>

$10^0 = 1$
$10^{-1} = 1/10 = 0.1$
$10^{-2} = 1/(10 \times 10) = 0.01$
$10^{-3} = 1/(10 \times 10 \times 10) = 0.001$
and so on

</td></tr>
<tr><td>

The power of 10 determines the number of zeros that follow the digit 1

</td><td>

The power of 10 determines the number of places to the right of the decimal point where the digit 1 appears

</td></tr>
</table>

We express 612,000 in scientific notation as follows:

$$612,000 = 6.12 \times 100,000 = 6.12 \times 10^5$$

We express 0.000505 in scientific notation as

$$0.000505 = 5.05 \times 0.0001 = 5.05 \times 10^{-4}$$

We can use a more direct approach to converting numbers to scientific notation.

- Count the number of places a decimal point must be moved to produce a coefficient having a value between 1 and 10.
- The number of places counted then becomes the power of 10.
- The power of 10 is *positive* if the decimal point is moved to the *left*.

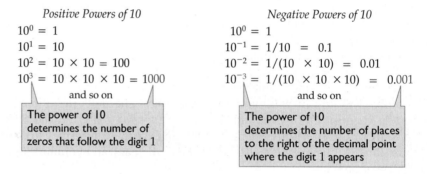

$$6\,1\,2\,0\,0\,0 \quad = \quad 6.12 \times 10^5$$

Move the decimal point (here understood) five places to the *left*

The exponent is (positive) 5

- The power of 10 is *negative* if the decimal point is moved to the *right*.

$$0.0\,0\,0\,5\,0\,5 \quad = \quad 5.05 \times 10^{-4}$$

Move the decimal point *four places to the right*

The exponent is (negative) 4

To convert a number from exponential form to the conventional form, move the decimal point in the opposite direction.

$$3.75 \times 10^6 = 3.750000$$

| The exponent is 6 | Move the decimal point *six* places to the *right* |

$$7.91 \times 10^{-7} = 0.0000007.91$$

| The exponent is −7 | Move the decimal point *seven* places to the *left* |

It is easy to handle exponential numbers on most calculators. A typical procedure is to enter the number, hit the key [EE] or [EXP], and then enter the exponent of 10. The keystrokes required for the number 2.85×10^7 are [2] [.] [8] [5] [EE] [7].

For the number 1.67×10^{-5}, the keystrokes are [1] [.] [6] [7] [EXP] [5] [±]; the last keystroke changes the sign of the exponent. Many calculators can be set to convert all numbers and calculated results to the exponential form, regardless of the form in which the numbers are entered. Most scientific and graphing calculators can also be set to display a fixed number of digits in the results.

Addition and Subtraction

To add or subtract numbers in exponential notation using a scientific or graphing calculator, simply enter the numbers as usual and perform the desired operations. However, to add or subtract numbers *by hand* in exponential notation, it is necessary to express each quantity as the *same power of 10*. In calculations, this approach treats the power of 10 in the same way as a unit—it is simply "carried along." For example, we express each quantity in the following calculation with the power 10^{-3}.

$$(3.22 \times 10^{-3}) + (7.3 \times 10^{-4}) - (4.8 \times 10^{-4}) =$$
$$(3.22 \times 10^{-3}) + (0.73 \times 10^{-3}) - (0.48 \times 10^{-3}) = (3.22 + 0.73 - 0.48) \times 10^{-3}$$
$$= 3.47 \times 10^{-3}$$

Multiplication and Division

To multiply numbers expressed in exponential form by hand, *multiply* all coefficients to obtain the coefficient of the result and *add* all exponents to obtain the power of 10 in the result. Generally, most calculators perform these operations automatically, and no intermediate results need to be recorded.

To divide two numbers in exponential form, *divide* the coefficients to obtain the coefficient of the result and *subtract* the exponent in the denominator from the exponent in the numerator to obtain the power of 10. In the example below, multiplication and division are combined. First, the rule for multiplication is applied to the numerator and to the denominator, and then the rule for division is used.

Rewrite in exponential form

$$\frac{0.015 \times 0.0088 \times 822}{0.092 \times 0.48} = \frac{(1.5 \times 10^{-2})(8.8 \times 10^{-3})(8.22 \times 10^2)}{(9.2 \times 10^{-2})(4.8 \times 10^{-1})}$$

Apply the rule for multiplication to the numerator and denominator

Apply the rule for division

$$= \frac{1.1 \times 10^{-1}}{4.4 \times 10^{-2}} = 0.25 \times 10^{-1-(-2)} = 0.25 \times 10^1 = 2.5$$

Raising a Number to a Power and Extracting the Root of an Exponential Number

To raise an exponential number to a given power, raise the coefficient to that power and multiply the exponent by that power. For example, we can cube a number (that is, raise it to the *third* power) in the following manner.

Rewrite in exponential form	Cube the coefficient	Multiply the exponent by 3

$$(0.0066)^3 = (6.6 \times 10^{-3})^3 = (6.6)^3 \times 10^{(-3 \times 3)}$$

$$= (2.9 \times 10^2) \times 10^{-9} = 2.9 \times 10^{-7}$$

To extract the root of an exponential number, we raise the number to a fractional power—one-half power for a square root, one-third power for a cube root, and so on. Most calculators have keys designed for extracting square roots and cube roots. Thus, to extract the square root of 1.57×10^5, enter the number 1.57×10^5 into a calculator, and use the $[\sqrt{}]$ key.

$$\sqrt{1.57 \times 10^{-5}} = 3.96 \times 10^{-3}$$

Some calculators allow you to extract roots by keying the root in as a fractional exponent. For example, we can take a cube root as follows.

$$(2.75 \times 10^{-9})^{1/3} = 1.40 \times 10^{-3}$$

SELF-ASSESSMENT Questions

For items 1–4, express the number in scientific notation, and then select the correct answer.

1. 35,400
   a. 3.54×10^{-4} b. 35.4×10^{-3}
   c. 3.54×10^4 d. 35.4×10^4

2. 43,600,000
   a. 436×10^4 b. 43.6×10^{-6}
   c. 4.36×10^7 d. 4.36×10^{-7}

3. 0.000000000000312
   a. 3.12×10^{-15} b. 3.12×10^{-13}
   c. 3.12×10^{-12} d. 3.12×10^{13}

4. 0.000438
   a. 4.38×10^{-3} b. 4.38×10^{-4}
   c. 4.38×10^4 d. 4.38×10^6

For items 5–6, express the number in decimal (ordinary) notation, and then select the correct answer.

5. 7.1×10^4
   a. 0.00071 b. 0.000071
   c. 71,000 d. 710,000

6. 2.83×10^{-4}
   a. 0.000283 b. 0.00283
   c. 2830 d. 28,300

For items 7–14, perform the indicated operation, and then select the correct answer.

7. $(3.0 \times 10^4) \times (2.1 \times 10^5) =$?
   a. 5.1×10^9 b. 6.3×10^1
   c. 6.3×10^9 d. 6.3×10^{20}

8. $(6.27 \times 10^{-5}) \times (4.12 \times 10^{-12}) =$?
   a. 2.58×10^{-17} b. 2.58×10^{-16}
   c. 2.58×10^{-59} d. 2.58×10^{59}

9. $\dfrac{4.33 \times 10^{-7}}{7.61 \times 10^{22}} =$?
   a. 5.69×10^{-30} b. 5.69×10^{-15}
   c. 5.69×10^{15} d. 5.69×10^{30}

10. $\dfrac{9.37 \times 10^9}{3.79 \times 10^{27}} =$?
    a. 2.47×10^{-36} b. 2.47×10^{-35}
    c. 2.47×10^{-18} d. 2.47×10^{36}

11. $\dfrac{2.21 \times 10^5}{9.80 \times 10^{-7}} =$?
    a. 2.26×10^{-12} b. 2.26×10^{-11}
    c. 2.26×10^{11} d. 4.43×10^{12}

12. $\dfrac{4.60 \times 10^{-12}}{2.17 \times 10^3} =$?
    a. 2.12×10^{-15} b. 2.12×10^{-14}
    c. 2.12×10^{14} d. 2.12×10^{15}

13. $(2.19 \times 10^{-6})^2 =$?
    a. 4.80×10^{-12} b. 4.80×10^{-6}
    c. 4.80×10^6 d. 4.80×10^{12}

14. $\sqrt{1.74 \times 10^{-7}} =$?
    a. 1.32×10^{-7} b. 4.17×10^{-7}
    c. 4.17×10^{-4} d. 1.32×10^{-4}

Answers: 1, c; 2, c; 3, b; 4, b; 5, c; 6, a; 7, c; 8, b; 9, a; 10, c; 11, c; 12, a; 13, a; 14, c

A.3 Unit Conversions

Chapter 1 discusses conversions within the metric system. Because almost all countries except the United States use metric measurements, we sometimes need to convert between common and metric units. Suppose that an American applies for a job in another country and the job application asks for her height in centimeters. She knows that her height is 66.0 in. An actual measured quantity is the same, no matter the unit we use to express it. The woman's height is the same, whether we express it in inches, feet, centimeters, or millimeters. We can use equivalencies (given here to three significant figures; Section A.4) such as those in Table A.2 to derive conversion factors.

In mathematics, multiplying a quantity by 1 does not change its value in any fundamental way. We can therefore use a factor equivalent to 1 to convert between inches and centimeters. We find our factor in the definition of the inch in Table A.2.

$$1 \text{ in} = 2.54 \text{ cm}$$

TABLE A.2 Some Conversions between Common and Metric Units

Length
1 mile (mi) = 1.61 kilometers (km)
1 yard (yd) = 0.914 meter (m)
1 inch (in) = 2.54 centimeters (cm)[a]

Mass
1 pound (lb) = 454 grams (g)
1 ounce (oz) = 28.4 grams (g)
1 pound (lb) = 0.454 kilogram (kg)

Volume
1 U.S. quart (qt) = 0.946 liter (L)
1 U.S. pint (pt) = 0.473 liter (L)
1 fluid ounce (fl oz) = 29.6 milliliters (mL)
1 gallon (gal) = 3.78 liters (L)

[a] This conversion is exact. The other conversions are available with more significant digits if needed; see any edition of the *CRC Handbook of Chemistry and Physics*.

Notice that if we divide both sides of this equation by 1 in, we obtain a ratio of quantities that is equal to 1.

$$1 = \frac{1 \text{ in}}{1 \text{ in}} = \frac{2.54 \text{ cm}}{1 \text{ in}}$$

If we divide both sides of the equation by 2.54 cm, we also obtain a ratio of quantities that is equal to 1.

$$\frac{1 \text{ in}}{2.54 \text{ cm}} = \frac{2.54 \text{ cm}}{2.54 \text{ cm}} = 1$$

The two ratios, one shown in red and the other in blue, are conversion factors. A *conversion factor* is a ratio of terms, equivalent to the number 1, used to change the unit in which a quantity is expressed. We call this process the *unit conversion method* of problem solving. Because we set up calculations by examining the *dimensions* associated with the given quantity, those associated with the desired result, and those needed in the conversion factors, the process is sometimes called *dimensional analysis*.

What happens when we multiply a known quantity by a conversion factor? The original unit cancels out and is replaced by the desired unit. Thus, our general approach to using conversion factors is

Desired quantity and unit = given quantity and unit × conversion factors

Now let's return to the question about the woman's height, a measured quantity of 66.0 in. To get an answer in centimeters, we must use the appropriate conversion factor (in blue). Note that the desired unit (cm) is in the numerator and the unit to be replaced (in.) is in the denominator. Thus, we can cancel the unit in so that only the unit cm remains.

$$66.0 \ \text{in} \times \frac{2.54 \ \text{cm}}{1 \ \text{in}} = 168 \ \text{cm} \ \text{(rounded off from 167.64)}$$

You can see why the other conversion factor (in red) won't work. No units cancel. Instead, we get a nonsensical unit.

$$66.0 \ \text{in} \times \frac{1 \ \text{in}}{2.54 \ \text{cm}} = \frac{26.0 \ \text{in}^2}{\text{cm}}$$

Following are some examples and exercises to give you some practice in this method of problem solving. A variety of conversion calculators are available on the Internet. Use the key words *conversion calculator*.

EXAMPLE A.1 Unit Conversions

   a. Convert 0.742 kg to grams.
   b. Convert 0.615 lb to ounces.
   c. Convert 135 lb to kilograms.

Solution

   a. This is a conversion from one metric unit to another. We simply use our knowledge of prefixes to convert from kilograms to grams.

We start here	This converts kg to g	Our answer: the number and the unit

   $$0.742 \ \text{kg} \times \frac{1000 \ \text{g}}{1 \ \text{kg}} = 742 \ \text{g}$$

   b. This is a conversion from one common unit to another. Here we use the fact that 1 lb = 16 oz to convert from pounds to ounces. Then we proceed as in part (a), arranging the conversion factor to cancel the unit lb.

   $$0.615 \ \text{lb} \times \frac{16 \ \text{oz}}{1 \ \text{lb}} = 9.84 \ \text{oz}$$

   c. This is a conversion from a common unit to a metric unit. We need data from Table A.2 to convert from pounds to kilograms. Then we proceed as in part (b), arranging the conversion factor to cancel the unit lb.

   $$135 \ \text{lb} \times \frac{0.454 \ \text{kg}}{1 \ \text{lb}} = 61.3 \ \text{kg}$$

(Our answers have the proper number of significant figures. If you do not understand significant figures and wish to do so, they are discussed in Section A.4. This text uses three significant figures for most calculations.)

> **Exercise A.1**
   a. Convert 16.3 mg to grams.
   b. Convert 24.5 oz to pounds.
   c. Convert 12.5 fl oz to milliliters.

Quite often, to get the desired unit, we must use more than one conversion factor. We can do so by arranging all the necessary conversion factors in a single setup that yields the final answer in the desired unit.

EXAMPLE A.2 **Unit Conversions**

What is the length, in millimeters, of a 3.25-ft piece of tubing?

Solution

No relationship between feet and millimeters is given in Table A.2, so we need more than one conversion factor. We can think of the problem as a series of three conversions.

1. Use the fact that 1 ft = 12 in to convert from feet to inches.
2. Use data from Table A.2 to convert from inches to centimeters.
3. Use our knowledge of prefixes to convert from centimeters to millimeters.

We could solve this problem in three distinct steps by making one conversion in each step, but it is just as easy to combine three conversion factors into a single setup. Then we proceed as indicated below.

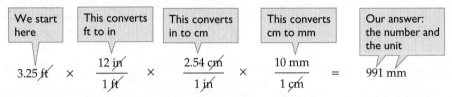

We start here | This converts ft to in | This converts in to cm | This converts cm to mm | Our answer: the number and the unit

$$3.25 \text{ ft} \times \frac{12 \text{ in}}{1 \text{ ft}} \times \frac{2.54 \text{ cm}}{1 \text{ in}} \times \frac{10 \text{ mm}}{1 \text{ cm}} = 991 \text{ mm}$$

❯ **Exercise A.2**

Carry out the following conversions.
 a. 729.9 ft to kilometers
 b. 1.17 gal to fluid ounces

Sometimes we need to convert two or more units in the measured quantity. We can do this by arranging all the necessary conversion factors in a single setup so that the starting units cancel and the conversions yield the final answer in the desired units.

 EXAMPLE A.3 **Unit Conversions**

A saline solution has 1.00 lb of salt in 1.00 gal of solution. Calculate the concentration in grams per liter of solution.

Solution

First, let's identify the measured quantities. They can be expressed in the form of a ratio of mass of salt in pounds to a volume in gallons.

$$\frac{1.00 \text{ lb (salt)}}{1.00 \text{ gal (solution)}}$$

This ratio must be converted to one expressed in grams per liter. We must convert from pounds to grams in the numerator and from gallons to liters in the denominator. The following set of equivalent values from Table A.2 can be used to formulate conversion factors.

$$1 \text{ lb} = 454 \text{ g} \qquad 1 \text{ gal} = 3.78 \text{ L}$$

We start here | To convert lb to g in numerator | To convert gal to L in denominator | Our answer: the number and the unit

$$\frac{1.00 \text{ lb}}{1.00 \text{ gal}} \times \frac{454 \text{ g}}{1 \text{ lb}} \times \frac{1 \text{ gal}}{3.78 \text{ L}} = 120 \text{ g/L}$$

Note that we could also do the conversions in the numerator and denominator separately and then divide the numerator by the denominator.

$$\text{Numerator:} \quad 1.00 \ \cancel{lb} \times \frac{454 \ g}{1 \ \cancel{lb}} = 454 \ g$$

$$\text{Denominator:} \quad 1.00 \ \cancel{gal} \times \frac{3.78 \ L}{1 \ \cancel{gal}} = 3.78 \ L$$

$$\text{Division:} \quad \frac{454 \ g}{3.78 \ L} = 120 \ g/L$$

Both methods give the same answer to three significant figures.

> **Exercise A.3**
Carry out the following conversions.
  a. 88.0 km/h to meters per second
  b. 4.07 g/L to ounces per quart

SELF-ASSESSMENT Questions

For items 1–5, perform the indicated conversion, and then select the correct answer. You may need data from Table A.2 for some of the items.

1. 645 μs to seconds
  a. 6.45×10^{-6} s   **b.** 6.45×10^{-4} s
  c. 6.45×10^{-2} s   **d.** 6.45×10^{8} s

2. 1445 oz to grams
  a. 50.9 g   **b.** 90.3 g
  c. 144.5 g   **d.** 4.10×10^{4} g

3. 1.24×10^5 mm to feet
  a. 10.3 ft   **b.** 103 ft
  c. 407 ft   **d.** 4.88×10^3 ft

4. 12.0 fl oz to milliliters
  a. 0.405 mL   **b.** 240 mL
  c. 355 mL   **d.** 3.55×10^5 mL

5. 343 m/s to miles per hour
  a. 0.0592 mi/h   **b.** 12.8 mi/h
  c. 767 mi/h   **d.** 1220 mi/h

Answers: 1, b; 2, d; 3, c; 4, c; 5, c

A.4 Precision, Accuracy, and Significant Figures

Counting can give exact numbers. For example, we can count exactly 24 students in a room. Measurements, on the other hand, are subject to error. One source of error is the measuring instruments themselves. For example, an incorrectly calibrated thermometer may consistently yield a result that is 0.2 °C too low. Other errors may result from the experimenter's lack of skill or care in using measuring instruments. But even the most careful measurement will have some uncertainty associated with it. For example, a meter stick can be used to measure to the nearest millimeter or so. But there is no way for a meter stick to give a measurement that is correct to the nearest 0.001 mm.

Precision and Accuracy

Suppose that five students were asked to measure a person's height using a meter stick marked off in millimeters. The five measurements are recorded in Table A.3. The *precision* of a set of measurements refers to how closely individual measurements agree with one another. We say that the precision is good if each of the measurements is close to the average or poor if there is a wide deviation from the average value.

TABLE A.3 Five Measurements of a Person's Height

Student	Height (m)
1	1.827
2	1.824
3	1.826
4	1.828
5	1.829
Average	1.827

How would you describe the precision of the data in Table A.3? Examine the individual data, note the average value, and determine how much the individual data differ from the average. Because the maximum deviation from the average value is 0.003 m, the precision is quite good.

The *accuracy* of a set of measurements refers to the closeness of the average of the set to the "correct," or most probable, value. Measurements of high precision are more likely to be accurate than are those of poor precision, but even highly precise measurements are sometimes inaccurate. For example, what if the meter sticks used to obtain the data in Table A.3 were actually 1005 mm long but still had 1000-mm markings? The accuracy of the measurements would be rather poor, even though the precision would remain good.

Sampling Errors

No matter how accurate an analysis, it will not mean much unless it is performed on valid, representative samples. Consider determining the level of glucose in the blood of a patient. High glucose levels are associated with diabetes, a debilitating disease. Results vary, depending on several factors such as the time of day and what and when the person last ate. Glucose levels are much higher soon after a meal high in sugars. They also depend on other factors, such as stress. Medical doctors usually take blood for analysis after a night of fasting. These fasting glucose levels tend to be more reliable, but physicians often repeat the analysis when high or otherwise suspicious results are obtained. Therefore, there is no one true value for the level of glucose in a person's blood. Repeated samplings provide an average level. Similar results from several measurements give much more confidence in the findings. For example, a diagnosis of diabetes is usually based on two consecutive fasting blood glucose levels above 120 mg/dL.

Significant Figures

Look again at Table A.3. Notice that the five measurements of height agree in the first three digits (1.82); they differ only in the fourth digit. We say that the fourth digit is uncertain. All digits known with certainty, plus the first uncertain one, are called *significant figures*. The precision of a measurement is reflected in the number of significant figures—the more significant figures, the more precise the measurement. The measurements in Table A.3 have four significant figures. In other words, we are quite sure that the person's height is between 1.82 m and 1.83 m. Our best estimate of the average value, including the uncertain digit, is 1.827 m.

The number 1.827 has four digits; we say it has four significant figures. In any properly reported measurement, all nonzero digits are significant. Zeros, however, may or may not be significant because they can be used in two ways: as a part of the measured value or to position a decimal point.

- Zeros between two other significant digits are significant. *Examples*: 4807 (four significant figures); 70.004 (five).
- The lone zero preceding a decimal point is there for readability; it is not significant. *Example*: 0.352 (three significant figures).

(a) Low accuracy   (b) Low accuracy
Low precision   High precision

(c) High accuracy   (d) High accuracy
Low precision   High precision

▲ Comparing precision and accuracy: a dartboard analogy. (a) The darts are both scattered (low precision) and off-center (low accuracy). (b) The darts are in a tight cluster (high precision) but still off-center (low accuracy). (c) The darts are somewhat scattered (low precision) but evenly distributed about the center (high accuracy). (d) The darts are in a tight cluster (high precision) and well centered (high accuracy).

- Zeros that precede the first nonzero digit are also not significant. *Examples*: 0.000819 (three significant figures); 0.03307 (four).
- Zeros at the end of a number are significant if they are to the right of the decimal point. *Examples*: 0.2000 (four significant figures); 0.050120 (five).

We can summarize these four situations with a general rule: When we read a number from left to right, all the digits starting with the first nonzero digit are significant. Numbers without a decimal point that end in zeros are a special case, however.

- Zeros at the end of a number may or may not be significant if the number is written without a decimal point. *Example*: 700. We do not know whether the number 700 was measured to the nearest unit, ten, or hundred. To avoid this confusion, we can use exponential notation (Section A.2). In exponential notation, 700 is recorded as 7×10^2 or 7.0×10^2 or 7.00×10^2 to indicate one, two, or three significant figures, respectively. The only significant digits are those in the coefficient, not in the power of 10.

We use significant figures only with measurements—quantities subject to error. The concept does not apply to a quantity that is

1. inherently an integer, such as 3 sides to a triangle or 12 items in a dozen;
2. inherently a fraction, such as the radius of a circle equals $\frac{1}{2}$ of the diameter;
3. obtained by an accurate count, such as 18 students in a class; or
4. a defined quantity, such as 1 km = 1000 m.

In these contexts, the numbers 3, 12, $\frac{1}{2}$, 18, and 1000 can have as many significant figures as we want. More properly, we say that each is an *exact value*.

 EXAMPLE A.4   Significant Figures

In everyday life, common units are often used with metric units in parentheses (or vice versa), but significant figures are not always considered. Which of the following has followed proper significant figure usage?

a. A Chinese high-speed train broke a world record in 2010 for fastest unmodified commercial train, reaching speeds of up to 481.1 kph (298.9 mph), state media reported.
b. The Earth warmed by about 0.7° Celsius (1.26° Fahrenheit) over the past century

Solution

a. The quantity 481.1 kph has four significant figures, as does 298.9 mph; significant figure rules were followed.
b. The quantity 0.7° has only one significant figure, but 1.26° Fahrenheit has three; significant figure rules were not followed.

› Exercise A.4

Which of the following has followed proper significant figure usage?

a. A set of exercise room weights are rated at 45 lb (20.4 kg).
b. Fossils of an ancient human-like creature, named *Ardipithecus ramidus*, was found in Ethiopia in 1992 and shown to be 4,400,000 years old. In 2009 researchers demonstrated that *A. ramidus* was likely to be a direct human ancestor. Because it took 17 years to assess the significance of the fossils, a Twitter note said that made the fossil 4,400,017 years old.

Significant Figures in Calculations: Multiplication and Division

If we measure a sheet of notepaper and find it to be 14.5 cm wide and 21.7 cm long, we can find the area of the paper by multiplying the two quantities. A calculator gives the answer as 314.65. Can we conclude that the area is 314.65 cm²? That is, can we know the area to the nearest hundredth of a square centimeter when we know the width and length only to the nearest tenth of a centimeter? It just doesn't seem reasonable—and it isn't. A calculated quantity can be no more precise than the data used in the calculation, and the reported result should reflect this fact.

A strict application of this principle involves a fairly complicated statistical analysis that we will not attempt here, but we can do a fairly good job by using a practical rule involving significant figures:

In multiplication and division, the reported result should have no more significant figures than the factor with the fewest significant figures.

In other words, a calculation is only as precise as the least precise measurement that enters into the calculation.

To obtain a numerical answer with the proper number of significant figures often requires that we round off numbers. In rounding, we drop all digits that are not significant and, if necessary, adjust the last reported digit. We use the following rules in rounding.

- If the leftmost digit to be dropped is *4 or less*, drop it and all following digits. *Example*: If we need four significant figures, 69.744 rounds to 69.74, or if we need three significant figures, to 69.7.

- If the leftmost digit to be dropped is *5 or greater*, increase the final retained digit by 1. *Example*: 538.76 rounds to 538.8 if we need four significant figures. Similarly, 74.397 rounds to 74.40 if we need four significant figures or to 74.4 if we need three.

 EXAMPLE A.5   Unit Conversions

What is the area, in square centimeters, of a rectangular gauze bandage that is 2.54 cm wide and 12.42 cm long? Use the correct number of significant figures in the answer.

Solution

The area of a rectangle is the product of its length and width. In the result, we can show only as many significant figures as there are in the least precisely stated dimension, the width, which has three significant figures.

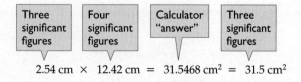

Three significant figures	Four significant figures	Calculator "answer"	Three significant figures

2.54 cm × 12.42 cm = 31.5468 cm² = 31.5 cm²

We use the rules for rounding off numbers as the basis for dropping the digits 468.

▸ Exercise A.5

Calculate the volume, in cubic meters, of a rectangular block of foamed plastic that is 1.827 m long, 1.04 m wide, and 0.064 m thick. Use the correct number of significant figures.

EXAMPLE A.6 Significant Figures

For a laboratory experiment, a teacher wants to divide all of a 226.8-g sample of glucose equally among the 18 members of her class. How many grams of glucose should each student receive?

Solution

The number 18 is a counted number, that is, an exact number that is not subject to significant figure rules. The answer should carry four significant figures, the same as in 226.8 g.

$$\frac{226.8 \text{ g}}{18 \text{ students}} = 12.60 \text{ g/student}$$

In this calculation, a calculator displays the result as 12.6. We add the digit 0 to emphasize that the result is precise to four significant figures.

❭ Exercise A.6

A dozen eggs has a mass of 681 g. What is the average mass of one of the eggs, expressed with the appropriate number of significant figures?

Significant Figures in Calculations: Addition and Subtraction

In addition or subtraction, we are concerned not with the number of significant figures but with the number of digits to the right of the decimal point. When we add or subtract quantities with varying numbers of digits to the right of the decimal point, we need to note the one with the fewest such digits. The result should contain the same number of digits to the right of its decimal point. For example, if you are adding several masses and one of them is measured only to the nearest gram, the total mass cannot be stated to the nearest milligram no matter how precise the other measurements are.

We apply this idea in Example A.7. Note that in a calculation involving several steps, we need round off only the final result.

 EXAMPLE A.7 Significant Figures

Perform the following calculation and round off the answer to the correct number of significant figures.

$$2.146 \text{ g} + 72.1 \text{ g} - 9.1434 \text{ g}$$

Solution

In this calculation, we add two numbers and subtract a third from the sum of the first two.

2.146 g ◄ three decimal places
+ 72.1 g ◄ one decimal place
74.246 g

− 9.1434 g ◄ four decimal places
65.1026 g = 65.1 g ◄ one decimal place

Note that we do not round off the intermediate result (74.246). When using a calculator, you generally don't need to write down an intermediate result.

⟩ Exercise A.7

Perform the indicated operations and give answers with the proper number of significant figures. Note that in addition and subtraction all terms must be expressed in the same unit.

a. 48.2 m + 3.82 m + 48.4394 m

b. 15.436 L + 5.3 L − 6.24 L − 8.177 L

c. (51.5 m + 2.67 m) × (33.42 m − 0.124 m)

d. $\dfrac{125.1 \text{ g} - 1.22 \text{ g}}{52.5 \text{ mL} + 0.63 \text{ mL}}$

SELF-ASSESSMENT Questions

1. A measurement has good precision if it
   a. is close to an accepted standard
   b. is close to similar measurements
   c. has few significant figures
   d. is the only value determined

2. Following are data for the length of a pencil as measured by students X, Y, and Z. The length is known to be 11.54 cm.

	Trial 1	Trial 2	Trial 3
X	11.8	11.1	11.5
Y	11.7	11.2	11.6
Z	11.3	11.4	11.5

Which of the following best characterizes the data?
   a. Y has the most precise data, X the most accurate.
   b. Y has the most precise data, Z the most accurate.
   c. Z has the most precise data, X the most accurate.
   d. Z has the most precise data, Y the most accurate.

3. 725.8055 rounded to three significant figures is
   a. 725.806   b. 725.81
   c. 725.8   d. 726

For items 4–6, determine the number of significant figures, and then select the correct answer.

4. 0.0000073
   a. 2   b. 5   c. 6   d. 8

5. 0.04000
   a. 2   b. 4   c. 5   d. 6

6. 80.0040
   a. 2   b. 4   c. 5   d. 6

For items 7–16, perform the indicated calculation, and then select the correct answer, with the proper number of significant figures.

7. 7.20 + 3.013 + 0.04327 = ?
   a. 10.26   b. 10.256
   c. 10.2563   d. 10.25627

8. 4.702 − 0.4123 = ?
   a. 4.2897   b. 4.29
   c. 4.290   d. 4.30

9. 10.03 + 4.555 = ?
   a. 14   b. 14.59
   c. 14.6   d. 15

10. 15.3 − 4.001 = ?
    a. 11   b. 11.2
    c. 11.299   d. 11.3

11. 4.602 ÷ 0.0240 = ?
    a. 191   b. 191.75
    c. 191.8   d. 192

12. 40.625 × 0.0028 = ?
    a. 0.11   b. 0.11375   c. 0.1138   d. 0.114

13. 8.64 ÷ 0.1216 = ?
    a. 71.1   b. 71.05   c. 71.0526   d. 71.053

14. (10.30)(0.186) ÷ 0.085 = ?
    a. 23   b. 22.5   c. 22.5388   d. 2.254

15. (42.5 + 0.459) ÷ 28.45 = ?
    a. 1.510   b. 1.5103   c. 1.51   d. 1.510

16. (7.06 ÷ 0.084) − (29.6 × 0.023) = ?
    a. 83   b. 83.3   c. 83.4   d. 83.369

Answers: 1, b; 2, d; 3, d; 4, a; 5, b; 6, d; 7, a; 8, c; 9, b; 10, d; 11, d; 12, d; 13, a; 14, a; 15, c; 16, a

A.5 Calculations Involving Temperature

As noted in Chapter 1, we sometimes need to make conversions from one temperature scale to another. On the Fahrenheit scale, the freezing point of water is 32 °F and the boiling point is 212 °F; on the Celsius scale, the freezing point of water is 0 °C and the boiling point is 100 °C. A 100° temperature interval (100 − 0) on the Celsius scale therefore equals a 180° interval (212 − 32) on the Fahrenheit scale. From these facts, we can derive two equations that relate temperatures on the two scales. One of these requires multiplying the degrees of Celsius temperature by the factor 1.8 (that is, 180/100) to obtain the degrees of Fahrenheit temperature, followed by adding 32 to account for the fact that 0 °C = 32 °F.

$$°F = (1.8 \times °C) + 32$$

In the other equation, we subtract 32 from the Fahrenheit temperature to get the number of degrees Fahrenheit above the freezing point of water. Then this quantity is divided by 1.8.

$$°C = \frac{°F - 32}{1.8}$$

The SI unit of temperature is the kelvin (Chapters 1 and 6). Recall that to convert from °C to K, we simply add 273.15. And to convert from K to °C, we subtract 273.15.

$$K = °C + 273.15 \quad \text{and} \quad °C = K - 273.15$$

Example A.8 illustrates a practical situation where conversion between Celsius and Fahrenheit temperatures is necessary.

In everyday life, the following ditty will often suffice to assess the meaning of a Celsius temperature.
THE CELSIUS SCALE Thirty is hot, Twenty is pleasing, Ten is quite cool, And zero is freezing.

 EXAMPLE A.8   Temperature Conversion

At home, you keep your thermostat set at 68 °F. When traveling, you have a room with a thermostat that uses the Celsius scale. What Celsius temperature will give you the same temperature as at home?

Solution

$$°C = \frac{°F - 32}{1.8} = \frac{68 - 32}{1.8} = 20 \, °C$$

〉 Exercise A.8
  a. Convert 85.0 °C to degrees Fahrenheit.
  b. Convert −12.2 °C to degrees Fahrenheit.
  c. Convert 355 °F to degrees Celsius.
  d. Convert −20.8 °F to degrees Celsius.

SELF-ASSESSMENT Questions

For items 1–10, perform the indicated conversion, and then select the correct answer.

1. 25 °C to kelvins
   a. −298 K   **b.** −248 K   **c.** 248 K   **d.** 298 K

2. 373 K to degrees Celsius
   a. −73 °C   **b.** 0 °C   **c.** 73 °C   **d.** 100 °C

3. 301 K to degrees Celsius
   a. −72 °C   **b.** 28 °C   **c.** 128 °C   **d.** 374 °C

4. 473 °C to kelvins
   a. 100 K   **b.** 200 K   **c.** 300 K   **d.** 746 K

5. 37.0 °C to degrees Fahrenheit
   a. −20.6 °F   **b.** −38.3 °F   **c.** 69.0 °F   **d.** 98.6 °F

6. 5.50 °F to degrees Celsius
   a. −20.8°C   **b.** −14.7 °C   **c.** 3.06 °C   **d.** 6.31 °C

7. 273 °C to degrees Fahrenheit
   a. 152 °F   **b.** 491 °F   **c.** 523 °F   **d.** 549 °F

8. 98.2 °F to degrees Celsius
   a. 36.8 °C   **b.** 54.6 °C
   c. 72.3 °C   **d.** 177 °C

9. 2175 °C to degrees Fahrenheit
   a. 1190 °F   **b.** 1208 °F
   c. 1226 °F   **d.** 3947 °F

10. 25.0 °F to degrees Celsius
   a. −31.6 °C   **b.** −13.9 °C
   c. −3.89 °C   **d.** 42.8 °C

11. The temperatures, 0 K, 0 °C, and 0 °F, arranged from coldest to hottest, are
   a. 0 °C, 0 °F, 0 K   **b.** 0 K, 0 °C, 0 °F
   c. 0 °F, 0 °C, 0 K   **d.** 0 K, 0 °F, 0 °C

Answers: 1. d; 2. d; 3. b; 4. d; 5. d; 6. b; 7. c; 8. c; 9. d; 10. c; 11. d

Glossary

absolute zero The lowest possible temperature, 0 K, -273.15 °C, or -459.7 °F.

absorb Gather a substance on a surface in a condensed layer.

acid A substance that, when added to water, produces an excess of hydrogen ions; a proton donor.

acid–base indicator A substance that is one color in acid and another color in base.

acidic anhydride A substance, such as a nonmetal oxide, that reacts with water to form an acid.

acid rain Precipitation having a pH less than 5.6.

acquired immune deficiency syndrome (AIDS) A disease caused by a retrovirus (human immunodeficiency virus) that weakens the immune system.

activated charcoal (activated carbon) Charcoal that has been treated with hot gases to greatly increase its surface area. In tertiary water treatment water is filtered through powdered activated charcoal to adsorb organic compounds.

activated sludge method Secondary sewage treatment technique in which sewage is aerated and some sludge is recycled.

activation energy The minimum quantity of energy that must be available before a chemical reaction can take place.

active site The region on an enzyme or a catalyst where a reaction occurs.

addition polymerization (chain-reaction polymerization) A polymerization reaction in which all the atoms of the monomer molecules are included in the polymer.

addition reaction A reaction in which the single product contains all the atoms of two reactant molecules.

adipose tissue Connective tissue where fat is stored.

advanced treatment (tertiary treatment) Sewage treatment designed to remove phosphates, nitrates, other soluble impurities, metal compounds, and other contaminants remaining after secondary treatment.

aerobic exercise Physical activity in which muscle contractions occur in the presence of oxygen.

aerobic oxidation An oxidation process occurring in the presence of oxygen.

aerosol Solid or liquid particles of 1 mm diameter or less, dispersed in air.

aflatoxins Toxins produced by molds growing on stored peanuts and grains.

Agent Orange A mixture of 2,4-D and 2,4,5-T used extensively as a defoliant in the Vietnam War; transported in barrels with orange bands.

agonist A molecule that fits and activates a specific receptor.

AIDS See **acquired immune deficiency syndrome.**

alchemy A mixture of chemistry and magic practiced in Europe during the Middle Ages (roughly 500 to 1700 c.e.).

alcohol (ROH) An organic compound composed of an alkyl group and a hydroxyl group.

aldehyde (RCHO) An organic compound with a carbonyl group that has a hydrogen atom attached to the carbonyl carbon.

aldose A monosaccharide with an aldehyde functional group.

aliphatic compound A nonaromatic substance. *See also* **aromatic compound.**

alkali metal A metal in group 1A in the customary U.S. arrangement of the periodic table or in group 1 of the IUPAC-recommended table.

alkaline earth metal An element in group 2A in the customary U.S. arrangement of the periodic table or in group 2 of the IUPAC-recommended table.

alkaloid A physiologically active nitrogen-containing organic compound that occurs naturally in a plant.

alkalosis A physiological condition in which the pH of the blood is too high.

alkane A hydrocarbon with only single bonds; a saturated hydrocarbon.

alkene A hydrocarbon containing one or more double bonds.

alkyl group (—R) The group of atoms that results when a hydrogen atom is removed from an alkane.

alkyne A hydrocarbon containing one or more triple bonds.

allergen A substance that triggers an allergic reaction.

allotropes Different forms of the same element in the same physical state.

alloy A homogeneous mixture of two or more elements, at least one of which is a metal; an alloy has metallic properties.

alpha decay Emission of an alpha particle ($^4_2$He) by a radioactive nucleus.

alpha (α) helix A secondary structure of a protein molecule in which the chains coil around one another in a spiral arrangement.

alpha (α) particle A particle of radiation consisting of two protons and two neutrons; a helium nucleus.

Ames test A laboratory test that screens for mutagens, which are usually also carcinogens.

amide group (—CON—) A functional group in which a carbon is joined to an oxygen atom by a double bond and to a nitrogen atom by a single bond.

amine A nitrogen compound derived from ammonia by replacing one or more hydrogen atoms with alkyl or aromatic group(s).

amino acid An organic compound that contains both an amino group and a carboxyl group; amino acids combine to produce proteins.

amino group (—NH₂) A functional group comprised of a nitrogen atom bonded to two hydrogen atoms.

amphetamines Stimulant drugs that are similar in structure to epinephrine and norepinephrine.

amphiprotic The ability of a substance, such as water, to either accept or donate a proton (H^+ ion).

amphoteric surfactant A surfactant that carries both a positive and negative charge.

anabolic steroid A drug that aids in the building (anabolism) of body proteins and thus of muscle tissue.

anabolism The building up of molecules through metabolic processes.

anaerobic decay The process of anaerobic (not using oxygen) bacteria in bodies of water depleted of oxygen, reducing organic matter to methane and hydrogen sulfide.

anaerobic exercise Physical activity that takes place without sufficient oxygen.

analgesic A substance that provides pain relief.

androgen A male sex hormone.

anesthetic A substance that causes loss of feeling or awareness.

anion A negatively charged ion.

anionic surfactant A surfactant having a negative charge on the water-soluble end of the molecule.

anode A positive electrode at which oxidation occurs.

antagonist A molecule that prevents the action of an agonist by blocking its receptor.

antibiotic A soluble substance, produced by a mold or bacterium, which inhibits growth of other microorganisms.

anticarcinogen A substance that inhibits the development of cancer.

anticholinergic A drug that acts on nerves using acetylcholine as a neurotransmitter.

anticoagulant A substance that inhibits the clotting of blood.

anticodon The sequence of three adjacent nucleotides in a tRNA molecule that is complementary to a codon on mRNA.

antihistamine A substance that relieves the symptoms caused by allergens: sneezing, itchy eyes, and runny nose.

anti-inflammatory A substance that inhibits inflammation.

antimetabolite A compound that inhibits the synthesis of DNA and thus slows the growth of cancer cells.

antioxidant A reducing agent that retards damaging oxidation reactions in living cells or reacts with free radicals to prevent rancidity in foods.

antiperspirant A substance or mixture that retards perspiration by constricting the openings of sweat glands.

antipyretic A fever-reducing substance.

apoenzyme The pure protein part of an enzyme.

applied research An investigation aimed at creating a useful product or solving a particular problem.

aqueous solution A solution in which the solvent is water.

arithmetic growth A process in which a constant quantity is added during each period of time.

aromatic compound A compound that has a ring structure and properties like those of benzene.

asbestos A fibrous silicate mineral composed of chains of SiO_4^{4-} tetrahedra.

astringent A substance that constricts the openings of sweat glands or skin pores.

atmosphere The thin blanket of air surrounding Earth.

atmosphere (atm) A unit of pressure equal to 760 mmHg.

atom The smallest characteristic particle of an element.

atomic mass unit (amu) The unit of relative atomic masses, equal to $\frac{1}{12}$ the mass of a carbon-12 atom $(1.661 \times 10^{-24}\,g)$.

atomic number (Z) The number of protons in the nucleus of an atom.

atomic theory A model that explains the law of multiple proportions and the law of definite proportions by stating that all elements are composed of atoms.

Avogadro's law States that at a fixed temperature and pressure, the volume of a gas is directly proportional to the amount (number of moles) of gas. This means in turn that equal volumes of gases, regardless of their compositions, contain equal numbers of molecules when measured at a given temperature and pressure.

Avogadro's number The number of atoms (6.022×10^{23}) in exactly 12 g of pure carbon-12.

background radiation Constantly occurring radiation from cosmic rays and from natural radioactive isotopes in air, water, soil, and rocks.

base A substance that, when added to water, produces an excess of hydroxide ions; a proton acceptor.

base triplet The sequence of three bases on a tRNA molecule that determine which amino acid it can carry.

basic anhydride A substance, such as a metal oxide, that reacts with water to form a base.

basic research The search for knowledge for its own sake.

battery A series of two or more connected electrochemical cells.

becquerel (Bq) A measure of the rate of radioactive decay; 1 becquerel $(Bq) = 1$ disintegration per second.

benefit Anything that promotes well-being or has a positive effect.

beta decay Emission of a beta particle by a radioactive nucleus.

beta (β) particle An electron emitted by a radioactive nucleus.

beta (β) pleated sheet A secondary protein structure in which arrays of chains form a zigzag sheet.

binary ionic compound A compound consisting of cations of a metal and anions of a nonmetal.

binding energy The energy that holds the nucleons together in an atom's nucleus.

biochemical oxygen demand (BOD) The quantity of oxygen required by microorganisms to remove organic matter from water.

biochemistry The study of the chemistry of living things and life processes.

biocide A chemical substance that will kill any living organism, including both plants and animals.

biomass Dry plant material used as fuel.

biosimilar A biologic product that is used to diagnose, prevent, cure, or treat disease, and that produces effects essentially identical to (no clinically meaningful differences from) an existing FDA-approved reference product.

bitumen A hydrocarbon mixture obtained from tar sands by heating.

bleach An oxidizing agent that removes unwanted color from fabric or other material.

blood sugar Glucose, a simple sugar that is circulated in the bloodstream and used directly by cells for energy.

body mass index (BMI) A commonly used measure of body fat, defined as weight (in kilograms) divided by the square of the height (in meters).

boiling point The temperature at which the pressure of the vapor escaping from a liquid equals the outside pressure.

bond See **chemical bond**.

bonding pair A pair of electrons shared by two atoms, forming a chemical bond.

Boyle's law States that for a given mass of gas at constant temperature, the volume varies inversely with the pressure.

breeder reactor A nuclear reactor that converts nonfissionable isotopes to nuclear fuel; most commonly, uranium-238 is converted to plutonium-239 by neutron bombardment.

broad-spectrum antibiotic An antibiotic that is effective against a wide variety of microorganisms.

bronze An alloy containing copper and tin as the main constituents.

buffer solution A mixture of a weak acid and its conjugate base or a weak base and its conjugate acid that maintains a nearly constant pH when a small amount of strong acid or strong base is added.

builder Any substance (often a complex phosphate) added to a surfactant to increase its detergency.

calorie (cal) The amount of heat required to raise the temperature of 1 g of water by 1°C.

carbohydrate A compound consisting of carbon, hydrogen, and oxygen, with hydrogen and oxygen in a 2:1 ratio; a starch or sugar.

carbon-14 dating A radioisotopic technique for determining the age of artifacts, based on the half-life of carbon-14.

carbon sequestration Removal of CO_2 from the atmosphere by collecting and transporting CO_2 from large emission sources, such as power plants and factories, and storing it in underground reservoirs, or by converting it to solid form such as carbonates.

carbonyl group (C=O) A functional group consisting of a carbon atom joined to an oxygen atom by a double bond.

carboxyl group (—COOH) A carbon atom with a double bond to one oxygen atom and a single bond to a second oxygen atom, which in turn is bonded to a hydrogen atom; the functional group of carboxylic acids.

carboxylic acid (RCOOH) An organic compound that contains the carboxyl functional group.

carcinogen A substance or physical entity that causes the growth of tumors.

catabolism Any metabolic process in which complex compounds are broken down into simpler substances.

catalyst A substance that increases the rate of a chemical reaction without itself being used up.

catalytic converter A device that uses catalysts that oxidize carbon monoxide and hydrocarbons to carbon dioxide and that reduce nitrogen oxides to nitrogen gas.

catalytic reforming A process that converts straight-chain alkanes to aromatic hydrocarbons.

cathode A negative electrode at which reduction occurs.

cathode ray A stream of high-speed electrons emitted from a cathode in an evacuated tube.

cation A positively charged ion.

cationic surfactant A surfactant having a positive charge on the water-soluble end of the molecule.

cell Structural unit of all living things.

cell membrane The double layer of lipid molecules enclosing a cell, through which the cell absorbs nutrients and eliminates wastes.

celluloid Cellulose nitrate, a synthetic material derived from natural cellulose by treating it with nitric acid.

cellulose A polymer comprised of glucose units joined by beta linkages.

Celsius (°C) scale A temperature scale on which water freezes at 0° and boils at 100°.

cement A complex mixture of calcium and aluminum silicates plus calcium sulfate, mixed with water, sand, and gravel to make concrete.

ceramic A hard, solid product made by heating clay with other materials to fuse them.

chain reaction A self-sustaining reaction in which one or more products of one event cause one or more new events.

chain-reaction polymerization See **addition polymerization**.

charcoal filtration Filtration of water through charcoal to adsorb organic compounds.

Charles's law States that for a given mass of gas at constant pressure, the volume varies directly with the absolute temperature.

chelating Tying up metal ions by surrounding them.

chemical bond The force of attraction that holds atoms or ions together in compounds.

chemical change A chemical reaction; a change in chemical composition.

chemical equation A shorthand representation of a chemical change that uses symbols and formulas instead of words.

chemical property A characteristic of a substance that, when observed, causes new types of matter with different compositions to be formed.

chemical symbol An abbreviation, consisting of one or two letters, that stands for an element.

chemistry The study of the behavior of matter and of how it interacts with other matter and with energy.

chemotherapy The use of chemicals to control or cure diseases.

chiral carbon A carbon atom that has four different groups attached to it.

chromosome Structure made of DNA and protein that includes the basic units of heredity, the genes.

coal A solid fossil fuel that is rich in carbon.

codon Set of three units on mRNA that codes for the specific amino acid to be inserted into a protein during transcription.

coenzyme An organic molecule (often a vitamin) that combines with an apoenzyme to make a complete, functioning enzyme.

cofactor An ion or molecule that combines with an apoenzyme to make a complete, functioning enzyme.

cologne A diluted perfume.

combined gas law The single relationship that incorporates the simple gas laws.

compound A substance made up of two or more elements combined in a fixed ratio.

concentrated solution A solution that has a relatively large amount of solute per unit volume of solution.

concrete A building material made from cement, sand, gravel, and water.

condensation The reverse of vaporization; a change from the gaseous state to the liquid state.

condensation polymerization (step-wise polymerization) A polymerization reaction in which not all the atoms in the starting monomers are incorporated in the polymer because water (or other small) molecules are formed as by-products.

condensed structural formula A chemical formula for an organic compound that omits the bonds joining hydrogens to carbons.

conjugate acid–base pair Two molecules or ions that differ by one proton (H^+ ion).

constitutional isomers Compounds having the same molecular formula but different structural formulas.

copolymer A polymer formed by the combination of two (or more) different monomers.

core electrons Electrons in any shell of an atom except the outermost shell.

corrosive waste A hazardous waste that requires a special container because it destroys conventional container materials.

cosmetics Substances defined in the 1938 U.S. Food, Drug, and Cosmetic Act as "articles intended to be rubbed, poured, sprinkled or sprayed on, or introduced into or otherwise applied to the human body or any part thereof, for cleaning, beautifying, promoting attractiveness or altering the appearance."

cosmic rays Extremely high-energy radiation from outer space.

covalent bond A bond formed when two atoms share one or more pairs of electrons.

cream An emulsion of tiny water droplets in oil.

CRISPR/cas9 A gene-editing system that results in very specific alterations to DNA, including the replacement of a single base in a gene.

critical mass The minimum amount of fissionable material required to achieve a self-sustaining chain reaction.

critical thinking The process of gathering and assessing facts, using logic to reach a conclusion, and then evaluating the conclusion.

cross-linking Connecting long chains of molecules through covalently bonded atoms or small groups of atoms.

crystal A solid, regular array of ions, atoms, or molecules.

cyclic hydrocarbon A ring-containing hydrocarbon.

daughter isotope An isotope formed by the radioactive decay of another isotope.

dead zone A region, usually in a large body of water, that does not support fish or other aquatic life because of excess algae and phosphates in the water.

defoliant A substance that causes premature dropping of leaves by plants.

denature To destroy the properties of a protein through heat, strong acids, surfactants, etc.

density The amount of mass per unit volume.

deodorant A product that contains perfume to mask body odor; some have a germicide such as alcohol, to kill odor-causing bacteria.

deoxyribonucleic acid (DNA) The type of nucleic acid found primarily in the nuclei of cells; contains the sugar deoxyribose.

depilatory Compound that removes hair by chemical action.

deposition The direct formation of a solid from a gas without passing through the liquid state; the reverse of sublimation.

depressant drug A drug that slows both physical and mental activity.

detergent A synthetic compound similar to a soap, but having a different hydrophilic end group.

deuterium An isotope of hydrogen with a proton and a neutron in the nucleus (mass of 2 amu).

dextro isomer A "right-handed" isomer.

dietary mineral An inorganic substance required in the diet for proper health and well-being.

Dietary Reference Intakes (DRIs) A set of reference values for nutrients established by the U.S. Academy of Sciences for planning and assessing diets.

dilute solution A solution that has a relatively small amount of solute per unit volume of solution.

dioxins Highly toxic chlorinated cyclic compounds that can cause extensive birth defects, the most infamous of which was a contaminant in "Agent Orange" used during the Vietnam War; can be formed by burning wastes containing chlorinated hydrocarbons.

dipole A molecule that is polar.

dipole–dipole forces The attractive forces that exist among polar covalent molecules.

disaccharide A sugar that on hydrolysis yields two monosaccharide molecules per molecule of disaccharide.

dispersion forces The momentary, usually weak attractive forces between molecules resulting from electron motions that create short-lived dipoles.

dissociative anesthetic A substance that causes gross personality disorders, including hallucinations similar to those in near-death experiences.

dissolved oxygen Oxygen dissolved in water; provides a measure of the water's ability to support fish and other aquatic life.

disulfide linkage A covalent linkage between cysteine units through two sulfur atoms.

diuretic A substance that increases the output of urine.

double bond Two pairs of electrons shared between two atoms.

doubling time The time it takes a population to double in size.

drug A substance that affects the functioning of living things; used to relieve pain, to treat illness, or to improve health or well-being.

drug abuse The use of a drug for its intoxicating effect.

drug misuse The use of a drug in a manner other than its intended use.

duet rule States that hydrogen and helium may have at most two electrons in their outermost shells.

elastomer A polymeric material that returns to its original shape after being stretched.

electrochemical cell A device that produces electricity by means of a chemical reaction.

electrode A carbon rod or a metal strip inserted into an electrochemical cell, at which oxidation or reduction occurs.

electrolysis The process of using electricity to cause chemical change.

electrolyte A compound that conducts an electric current in water solution.

electron The subatomic particle that bears a unit of negative charge.

electron capture (EC) A type of radioactive decay in which a nucleus absorbs an electron from the first or second shell of the atom.

electron configuration The arrangement of an atom's electrons in its energy levels.

electron-dot symbol See **Lewis symbol.**

electronegativity The attraction of an atom in a molecule for a bonding pair of electrons.

electrostatic precipitator A device that removes particulate matter from smokestack gases by creating an electric charge on the particles, which are then removed by attraction to a surface of opposite charge.

element A substance composed of atoms that have the same number of protons.

emollient An oil or grease used as a skin softener.

emulsion A suspension of submicroscopic particles of fat or oil in water.

enantiomers Isomers that are not superimposable on their mirror image.

end note The fraction of a perfume that has the lowest volatility, and remains for some time after application; composed of large molecules.

endorphins Naturally occurring peptides that bond to the same receptor sites as morphine.

endothermic Describes a process that requires heat to occur, taking that energy from the surroundings.

energy The ability to do work; the ability to change matter either physically or chemically.

energy levels (shells) The specific, quantized values of energy that an electron can have in an atom.

enrichment (of food) Replacement of B vitamins and addition of iron to flour.

enrichment (of an isotope) The process by which the proportion of one isotope of an element is increased relative to those of the others.

enthalpy change The energy needed to break all the chemical bonds in the reactants minus the energy released by forming all the bonds in the products.

enthalpy of reaction (heat of reaction) The amount of energy released (a negative number, in an exothermic reaction) or absorbed (a positive number, in an endothermic reaction) in a reaction with stoichiometric amounts of each reactant.

entropy A measure of the dispersal of energy among the possible energy states of a system.

enzyme A biological catalyst produced by cells.

essential amino acid An amino acid that is not produced in the body and must be included in the diet.

ester (RCOOR′) A compound derived from a carboxylic acid and an alcohol; the —OH of the acid is replaced by an —OR group.

estrogen A female sex hormone.

ether (ROR′) A molecule with two hydrocarbon groups attached to the same oxygen atom.

eutrophication The process by which growth of algae in a body of water is greatly accelerated, usually by phosphates, resulting in depleted oxygen and destruction of that body of water for fish and recreation.

excited state A state in which an atom has at least one electron that is not in its lowest energy level.

exothermic Describes a process that releases heat to the surroundings.

exponential notation A number written as the product of a coefficient and a power of ten. *See also* **scientific notation**.

falsifiability The capacity for a statement or claim to be proven wrong.

fast-twitch fibers The stronger, larger muscle fibers that are suited for short bursts of vigorous exercise.

fat An ester formed by the reaction of glycerol with three fatty-acid units; a triglyceride or triacylglycerol.

fat depots Storage places for fats in the body.

fatty acid A carboxylic acid that contains 4 to 20 or more carbon atoms in a chain.

first law of thermodynamics (law of conservation of energy) States that energy cannot be created or destroyed, only transformed from one form to another.

flammable waste A hazardous waste that burns readily on ignition, presenting a fire hazard.

FLaReS An acronym representing four rules used to test a claim: falsifiability, logic, replicability, and sufficiency.

flocculent A substance that causes particles to clump together and settle out.

food additive Any substance other than a basic foodstuff that is added to food to aid nutrition, enhance color or flavor, or provide texture.

formula A representation of a chemical substance in which the component chemical elements are represented by their symbols.

formula mass The sum of the masses of the atoms represented in the formula of a substance, expressed in atomic mass units (amu).

fossil fuels Natural fuels, especially coal, petroleum, and natural gas, derived from once-living plants and animals.

fracking See **hydraulic fracturing**.

free radical A highly reactive chemical species that contains an unpaired electron.

freezing The reverse of melting; changing from the liquid to the solid state.

fuel A substance that burns readily with the release of significant energy, ideally in a controllable manner.

fuel cell An electrochemical cell that produces electricity directly from continuously supplied fuel and oxygen.

functional group An atom or group of atoms that confers characteristic properties to a family of organic compounds.

fundamental particle An electron, proton, or neutron.

gamma decay Emission of a gamma ray ($_0^0\gamma$) by a radioactive nucleus.

gamma (γ) rays Electromagnetic radiation that is emitted by radioactive nuclei; have higher energy and are more penetrating than X-rays.

gas The state of matter in which the substance takes both the shape and volume of a container that it occupies.

gasoline A liquid mixture of (mostly) hydrocarbons, mainly alkanes with 5–12 carbon atoms, plus anti-knock and detergent additives, for use as an automotive fuel.

Gay-Lussac's law States that the pressure of a fixed amount of gas at fixed volume is directly proportional to its temperature in kelvins.

gene The segment of a DNA molecule that contains the information necessary to produce a protein; the smallest unit of hereditary information.

gene therapy Treatment that introduces a healthy gene into an individual's cells to replace a defective gene.

general anesthetic A depressant that acts on the brain to produce unconsciousness and insensitivity to pain.

geometric growth A process in which the rate of growth itself increases during each period of time.

geometric isomers Molecules having the same molecular formula that are locked into their spatial positions with respect to one another due to a double bond or a ring structure.

geothermal energy Energy derived from the heat of Earth's interior.

glass A noncrystalline solid material made by melting sand with sodium carbonate, calcium carbonate, and other metal oxides.

glass transition temperature (T_g) The temperature above which a polymer is rubbery and tough and below which the polymer is brittle.

global warming An increase in Earth's average temperature.

globular protein A protein whose molecules are folded into compact spherical or ovoid shapes.

glycogen A polymer of glucose with alpha linkages and branched chains; stored in the liver and muscles.

GRAS list A list, established by the U.S. Congress in 1958, of food additives generally recognized as safe.

green chemistry An approach that uses materials and processes that are intended to prevent or reduce pollution at its source.

greenhouse effect The retention of the sun's heat energy by Earth's atmosphere as a result of excess carbon dioxide and other gases in the atmosphere.

ground state The state of an atom in which all of its electrons are in the lowest possible energy levels.

group The elements in a column in the periodic table; a family of elements.

half-life The length of time required for one-half of the radioactive nuclei in a sample to decay.

hallucinogenic drug A drug that produces visions and sensations that are not part of reality.

halogen An element in group 7A in the customary U.S. arrangement of the periodic table or in group 17 of the IUPAC-recommended table.

hard water Water containing excessive concentrations of ions of calcium, magnesium, and/or iron.

hazardous waste A waste that, when improperly managed, can cause or contribute to death or illness, or threaten human health or the environment.

heat Energy transfer that occurs as a result of a temperature difference.

heat of vaporization The amount of heat needed (in kJ/mol or kcal/mol) to convert one mole of a substance from liquid to gas.

heat stroke A failure of the body's heat-regulatory system; unless the victim is treated promptly, the rapid rise in body temperature will cause brain damage or death.

herbicide Any chemical species that will kill weeds; some herbicides are capable of killing any plant they contact.

heterocyclic compound A cyclic compound in which one or more atoms in the ring are not carbon.

heterogeneous mixture A mixture in which different parts of the mixture have different compositions.

homogeneous mixture A mixture that is completely uniform; all parts of the mixture have the same composition.

homologous series A series of compounds whose adjacent members differ by a fixed unit of structure.

hormone A chemical messenger secreted into the blood by an endocrine gland.

humectant A moistening agent added to food.

hydraulic fracturing (fracking) A process for obtaining oil or natural gas, whereby deep holes are drilled into geological formations, and high-pressure fluids (usually water mixed with fine sand) are pumped in to form cracks in the rocks through which the desired fluid may flow.

hydrocarbon An organic compound that contains only carbon and hydrogen.

hydrogen bomb A bomb based on the nuclear fusion of isotopes of hydrogen.

hydrogen bond A type of intermolecular force in which a hydrogen atom covalently bonded in one molecule is attracted to a nonmetal atom in a neighboring molecule; both the atom to which the hydrogen atom is bonded and the one to which it is attracted are small, highly electronegative atoms, usually N, O, or F.

hydrolysis The reaction of a substance with water; literally, a splitting by water.

hydrophilic "Water-loving"; attracted to polar solvents such as water.

hydrophobic "Water-fearing"; not attracted significantly to water; attracted to non-polar solvents, such as oils and grease.

hypoallergenic cosmetic A cosmetic claimed to reduce or eliminate allergic reactions.

hypothesis A tentative explanation of observations that can be tested by experiment.

ideal gas law States that the volume of a gas is proportional to the amount of gas and its Kelvin temperature and inversely proportional to its pressure.

immunotherapy Therapeutic approaches that stimulate a person's immune system to become more effective in fighting disease.

induced radioactivity Radioactivity caused by formation of a radioactive isotope from bombardment of a stable isotope with elemental particles.

industrial smog (sulfurous smog) Polluted air associated with industrial activities, characterized by sulfur oxides and particulate matter.

inorganic chemistry The study of the compounds of all elements other than carbon.

insecticide A substance that will kill insects, particularly those that destroy crops.

intermolecular forces Attractive forces between two or more molecules.

intramolecular forces Attractive forces between atoms in a molecule; bonds.

iodine number The mass in grams of iodine that reacts with 100 g of a fat or oil; an indication of the degree of unsaturation.

ion A charged atom or group of atoms.

ionic bond The chemical bond that results when electrons are transferred from a metal to a nonmetal; the electrostatic attraction between ions of opposite charge.

ionizing radiation Radiation that produces ions as it passes through matter.

isoelectronic Having the same electron configuration.

isomers Compounds that have the same molecular formula but different structures.

isotopes Atoms that have the same number of protons but different numbers of neutrons.

joule (J) The SI unit of energy ($1 J = 0.239 cal$).

juvenile hormone A chemical messenger that controls the rate of development of the young of any species; can be used to prevent insects from maturing.

kelvin (K) The SI unit of temperature; zero on the Kelvin scale is absolute zero.

keratin The tough, fibrous protein that composes most of the outermost layer of the epidermis.

kerogen The complex material found in oil shale; has an approximate composition of $(C_6H_8O)_n$, where n is a large number.

ketone (RCOR') An organic compound with a carbonyl group between two carbon atoms.

ketose A monosaccharide with a ketone functional group.

kilocalorie (kcal) A unit of energy equal to 1000 cal; one food calorie.

kilogram (kg) The SI unit of mass, a quantity equal to about 2.2 lb.

kinetic energy The energy of motion.

kinetic–molecular theory An explanation of the behavior of gases based on the motion and energy of particles.

kinetics The study of reaction rates and factors that affect those rates: temperature, solute concentration, presence of a catalyst, etc.

lanolin An oily substance obtained from sheep's wool or skin.

law of combining volumes States that the volumes of gaseous reactants and products are in a small whole-number ratio when all measurements are made at the same temperature and pressure.

law of conservation of energy See **first law of thermodynamics**.

law of conservation of mass States that matter is neither created nor destroyed during a chemical change.

law of definite proportions (law of constant composition) States that a compound always contains the same elements in exactly the same proportions by mass.

law of multiple proportions States that elements may combine in different proportions to form more than one compound—for example, CO and CO_2.

LD₅₀ The dosage that is lethal to 50% of a population of test animals.

levo isomer A "left-handed" isomer.

Lewis formula (Lewis structure) A structural formula of a molecule or polyatomic ion that shows the arrangement of atoms, bonds, and lone pairs.

Lewis symbol A symbol consisting of the element's symbol surrounded by dots representing the atom's valence electrons; also referred to as an *electron-dot symbol.*

limiting reactant The reactant that is used up first in a reaction, after which the reaction ceases no matter how much remains of the other reactants.

line-angle formula A representation of a molecule in which the corners and ends of lines are understood to be carbon atoms and each carbon atom is understood to be attached to enough hydrogen atoms to give it four bonds.

line spectrum The pattern of lines of different wavelengths, emitted by an element with atoms in excited states.

lipid A substance from animal or plant cells that is soluble in solvents of low polarity and insoluble in water.

lipoprotein A protein combined with a lipid, such as a triglyceride or cholesterol.

liquid The state of matter in which the substance assumes the shape of its container, flows readily, and maintains a fairly constant volume.

liter (L) A unit of volume equal to a cubic decimeter.

local anesthetic A substance that renders part of the body insensitive to pain while leaving the patient conscious.

logical Characterized by sound, rational reasoning.

lone pair (nonbonding pair) A pair of unshared electrons in the valence shell of an atom.

lotion An emulsion of tiny oil droplets dispersed in water; often contains other ingredients such as perfumes and colorants.

main group elements The elements in the A groups of the customary U.S. arrangement of the periodic table, or in groups 1, 2, and 13–18 of the IUPAC-recommended table.

marijuana A hallucinogenic drug consisting of the leaves, flowers, seeds, and small stems of the *Cannabis* plant.

mass A measure of the inertia of an object; a measure of its amount of matter.

mass–energy equation Einstein's equation $E = mc^2$, in which E is energy, m is mass, and c is the speed of light.

mass number (A; nucleon number) The sum of the numbers of protons and neutrons in the nucleus of an atom.

matter The stuff of which all materials are made; anything that has mass and occupies space.

melanin A brownish-black pigment that determines the color of skin and hair.

melting point The temperature at which a substance changes from the solid to the liquid state.

messenger RNA (mRNA) The type of RNA that contains the codons for a protein; travels from the nucleus of the cell to a ribosome.

metabolism The set of coordinated chemical reactions that keep the cells of an organism alive.

metalloid An element with properties intermediate between those of metals and those of nonmetals.

metals The elements that are to the left of the heavy, stepped line in the periodic table, having properties of malleability, ductility, electrical conductivity, and metallic luster.

meter (m) The SI unit of length, slightly longer than a yard.

mica A mineral composed of SiO_4^{4-} tetrahedra in a two-dimensional, sheet-like array.

micelle A tiny spherical oil droplet into which the hydrocarbon tails of surfactant molecules are embedded, with their hydrophilic heads on the outer surface.

micronutrient A substance needed by a plant or animal in only very low concentrations, including Se, Cu, B, Cl, Fe, Mn, Ni, Mo, and Zn.

middle note The fraction of a perfume that is intermediate in volatility and is responsible for the lingering aroma after most top-note compounds have vaporized.

mineral (dietary) An inorganic substance required in the diet for good health.

mineral (geological) A naturally occurring inorganic solid with a definite composition.

mixture Matter with a variable composition.

moisturizer A substance that acts to retain moisture in the skin by forming a protective physical barrier.

molarity (M) The concentration of a solution in moles of solute per liter of solution.

molar mass The mass of one mole of a substance expressed in units of grams/mole.

molar volume The volume occupied by 1 mol of a substance (usually a gas) under specified conditions.

mole (mol) The amount of a substance that contains 6.022×10^{23} elementary units (atoms, molecules, or formula units) of the substance.

molecular mass The mass of a molecule of a substance; the sum of the atomic masses as indicated by the molecular formula, usually expressed in atomic mass units (amu).

molecule An electrically neutral unit of two or more atoms joined by covalent bonds; the smallest fundamental unit of a molecular substance.

monomer A substance of relatively low molecular mass; monomer molecules are the building blocks of polymers.

monosaccharide A carbohydrate that cannot be hydrolyzed into simpler sugars.

mousse A foamy hair care product composed of resins and used to hold hair in place.

mutagen Any entity that causes changes in genes without destroying the genetic material.

narcotic A depressant, analgesic drug that induces narcosis (sleep).

natural gas A mixture of gases, mainly methane, found in underground deposits.

natural philosophy Philosophical speculation about nature, a precursor to the modern study of science.

neuron A nerve cell.

neurotransmitter A chemical that carries an impulse across a synapse from one nerve cell to the next.

neurotrophin A substance produced during exercise that promotes the growth of brain cells.

neutralization The combination of H^+ and OH^- to form water, or the reaction of an acid and a base to produce a salt and (usually) water.

neutron A nuclear particle with a mass of approximately 1 amu and no electric charge.

nitrogen cycle The various processes by which nitrogen is cycled among the atmosphere, soil, water, and living organisms.

nitrogen fixation A process that combines nitrogen with one or more other elements.

noble gases Generally unreactive elements that appear in group 8A of the customary U.S. arrangement of the periodic table or in group 18 of the IUPAC-recommended table.

nonbonding pair See **lone pair**.

nonionic surfactant A surfactant having no charge on the water-soluble end of the molecule..

nonmetals The elements to the right of the heavy, stepped line in the periodic table.

nonpolar covalent bond A covalent bond in which there is equal sharing of the bonding pair of electrons.

nonsteroidal anti-inflammatory drug (NSAID) An anti-inflammatory drug, such as aspirin or ibuprofen, that is milder than the more potent steroidal anti-inflammatory drugs, such as cortisone and prednisone.

nuclear fission The splitting of an atomic nucleus into two smaller ones.

nuclear fusion The combination of two small atomic nuclei to produce one larger nucleus.

nuclear reactor A power plant that produces electricity using nuclear fission reactions.

nucleic acid A nucleotide polymer; DNA or RNA.

nucleon A proton or a neutron.

nucleon number See **mass number**.

nucleotide A combination of an amine base, a sugar unit, and a phosphate unit; the monomer unit of nucleic acids.

nucleus The tiny core of an atom, composed of protons and neutrons and containing all the positive charge and most of the mass of the atom.

octane rating Comparison of the anti-knock quality of a gasoline with that of pure isooctane, which has a rating of 100.

octet rule States that atoms tend to have eight electrons in their outermost shell.

oil (food) A substance, formed from glycerol and fatty acids, which is liquid at room temperature.

oil shale Fossil rock from which oil can be obtained at high cost.

optical brightener A fluorescent dye, added to detergent formulations, that absorbs invisible UV radiation, and re-emits it in the blue portion of the visible spectrum, resulting in the yellowed fabric's appearing brighter or whiter.

orbital A volume of space in an atom that is occupied by one or two electrons.

organic chemistry The study of the compounds of carbon.

organic farming Farming without using synthetic fertilizers or pesticides.

orphan drugs Medicinal products intended for diagnosis, prevention, or treatment of rare serious or life-threatening diseases.

oxidation An increase in oxidation number; combination of an element or compound with oxygen; loss of hydrogen; loss of electrons.

oxidation number A hypothetical number signifying the charge that an atom would have in a molecule or polyatomic ion if all bonds in the molecule/ion were ionic.

oxidizing agent A substance that causes oxidation and is itself reduced.

oxygen cycle The various processes by which oxygen is cycled among the atmosphere, soil, water, and living organisms.

oxygen debt The demand for oxygen in muscle cells that builds up during anaerobic exercise.

ozone layer The layer of the stratosphere that contains ozone and shields living creatures on Earth from the sun's ultraviolet radiation.

paint A liquid mixture of a pigment, binder, and solvent that forms a stable, smooth protective surface when dry.

particulate matter (PM) An air pollutant composed of solid and liquid particles whose size is greater than that of a molecule.

peptide bond The amide linkage that joins amino acids in chains of peptides, polypeptides, and proteins.

percent by mass The concentration of a solution, expressed as (mass of solute/mass of solution) × 100%.

percent by volume The concentration of a solution, expressed as (volume of solute/volume of solution) × 100%.

perfume A fragrant mixture of plant extracts and other chemicals dissolved in alcohol.

period A horizontal row of the periodic table.

periodic table A systematic arrangement of the elements in columns and rows; elements in a given column have similar properties.

pesticide A broad term referring to any chemical substance that will kill weeds, insects, rodents, spiders, or anything considered to be a pest.

petroleum (crude oil) A thick liquid mixture of (mostly) hydrocarbons with various impurities including sulfur, occurring in various geologic deposits.

pH The negative logarithm of the hydronium ion concentration, which indicates the degree of acidity or basicity of a solution.

pharmacology The study of the response of living organisms to drugs.

phenol A compound with an —OH group attached to a benzene ring.

pheromone A chemical secreted by an insect or other organism to mark a trail, send an alarm, or attract a mate.

photochemical smog Smog created by the action of sunlight on hydrocarbons and nitrogen oxides, which come mainly from automobile exhaust.

photon A unit particle of radiant energy, which moves at the speed of light.

photosynthesis The chemical process used by green plants to convert solar energy into chemical energy by reducing carbon dioxide and producing glucose and oxygen.

photovoltaic cell (solar cell) A cell that uses semiconductors to convert sunlight directly to electrical energy.

physical change A change in physical appearance of matter without changing its chemical identity or composition.

physical property A quality of a substance that can be demonstrated without changing the composition of the substance.

phytoremediation A method of water and soil treatment that uses plants having an affinity for metal ions and other contaminants to remove those contaminants.

placebo effect The phenomenon in which an inactive substance produces results in recipients for psychological reasons.

plasma A state of matter similar to a gas but composed of isolated electrons and nuclei rather than discrete whole atoms or molecules.

plasticizer A substance added to some plastics to make them more flexible and easier to work with.

poison A substance that causes injury, illness, or death of a living organism.

polar covalent bond A covalent bond in which the bonding pair of electrons is shared unequally by the two atoms, giving each atom a partial positive or negative charge.

polar molecule A molecule that has a separation between centers of positive and negative charge; a dipole.

pollutant A chemical that causes undesirable effects by being in the wrong place and/or in the wrong concentration.

polyamide A polymer that has monomer units joined by amide linkages.

polyatomic ion A charged particle consisting of two or more covalently bonded atoms.

polyester A polymer that has monomer units joined by ester linkages.

polymer A molecule with a large molecular mass that is formed of repeating smaller units (monomers).

polymerase chain reaction (PCR) A process that reproduces many copies of a DNA fragment.

polymerization A process by which monomers are made into polymers.

polypeptide A polymer of amino acids, usually of lower molecular mass than a protein.

polysaccharide A carbohydrate, such as starch or cellulose, that consists of many monosaccharide units linked together.

polyunsaturated fat A fat containing fatty-acid units that have two or more carbon-to-carbon double bonds.

positron (β^+ or $_{+1}^{0}e$) A positively charged particle with the mass of an electron.

potential energy Energy due to position or composition.

preemergent herbicide A herbicide that is rapidly broken down in the soil and can therefore be used to kill weed plants before crop seedlings emerge.

primary plant nutrients Nitrogen, phosphorus, and potassium as the main elements needed by plants.

primary sewage treatment Treatment of wastewater in a holding pond to allow some of the sewage solids to settle out as sludge.

primary structure The amino-acid sequence in a protein.

product A substance produced by a chemical reaction; product formulas follow the arrow in a chemical equation.

progestin A steroid hormone that mimics the action of progesterone.

prostaglandin One of several hormone-like lipids that are derived from the fatty acid arachidonic acid; involved in increased blood pressure, the contractions of smooth muscle, and other physiological processes.

protein An amino acid polymer with a molecular weight exceeding about 10,000 amu.

proton (H^+) The hydrogen ion in acid–base chemistry.

proton (nuclear) The unit of positive charge in the nucleus of an atom.

psychotropic drugs Drugs that affect the mind.

purine A heterocyclic amine base with two fused rings, found in nucleic acids.

pyrimidine A heterocyclic amine base with one ring, found in nucleic acids.

quantum A discrete unit of energy; one photon.

quartz A very hard mineral composed of SiO_4^{4-} tetrahedra arranged in a three-dimensional array.

quaternary structure An arrangement of protein subunits in a particular pattern.

radiation therapy Use of radioisotopes to destroy cancer cells.

radioactive decay The disintegration of an unstable atomic nucleus with spontaneous emission of radiation.

radioactive fallout Radioactive debris produced by explosion of a nuclear bomb.

radioactivity The spontaneous emission of particles (for example, alpha or beta) or rays (gamma) from unstable atomic nuclei.

radioisotopes Atoms or ions with radioactive nuclei.

reactant A starting material in a chemical change; reactant formulas precede the arrow in a chemical equation.

reactive wastes Hazardous wastes that tend to react spontaneously or to react vigorously with air or water.

Recommended Dietary Allowance (RDA) The recommended level of a nutrient necessary for a balanced diet.

recycle To make something new from something previously used.

reduce To make something smaller in size, quantity, or number.

reducing agent A substance that causes reduction and is itself oxidized.

reduction A decrease in oxidation number; a gain of electrons; a loss of oxygen; a gain of hydrogen.

replicability The capacity to be reproduced.

replication Copying or duplication; the process by which DNA reproduces itself.

resin A polymeric organic material, usually a sticky solid, semisolid, or thick liquid.

restorative drug A drug used to relieve the pain and reduce the inflammation resulting from overuse of muscles.

retrovirus An RNA virus that synthesizes DNA in a host cell.

reuse To use something again, especially in a different way or after reclaiming or reprocessing it.

reverse osmosis A method of pressure filtration through a semipermeable membrane; water is forced to flow from a region of high solute concentration to a region of low solute concentration.

ribonucleic acid (RNA) The form of nucleic acid found mainly in the cytoplasm but also present in all other parts of the cell; contains the sugar ribose.

risk Any hazard that leads to loss or injury.

risk-benefit analysis A technique for estimating a desirability quotient by dividing the benefits by the risks.

rule of 72 A mathematical formula that gives the doubling time for a population growing geometrically; 72 divided by the annual rate of growth expressed as a percentage equals the doubling time.

salt An ionic compound produced by the reaction of an acid with a base.

saponin A natural compound in some plants that produces a soapy lather.

saturated fat A triglyceride composed of a large proportion of saturated fatty acids esterified with glycerol.

saturated hydrocarbon An alkane; a compound of carbon and hydrogen with only single bonds.

science A search for understanding of and explanations for natural phenomena through careful observation and experimentation; an accumulation of knowledge about nature and the physical world.

scientific law A summary of experimental data; often expressed in the form of a mathematical equation.

scientific model A representation of an invisible process that uses tangible items or pictures.

scientific notation A form of exponential notation wherein the coefficient has a value between one and ten.

scientific theory The best current explanation for a phenomenon, based on experimentation; may be revised if new data warrant.

sebum An oily secretion that protects the skin from moisture loss.

second The SI base unit of time.

second law of thermodynamics States that the entropy of the universe increases in any spontaneous process.

secondary plant nutrients Elements needed by plants but in relatively low concentration, including magnesium, calcium, and sulfur.

secondary sewage treatment Passing effluent from primary sewage treatment through gravel and sand filters, and aerating the water.

secondary structure The arrangement of a protein's polypeptide chains—for example, helix or pleated sheet.

sex attractant A pheromone secreted by an insect to attract a mate.

shell (principal shell) One of the general quantized energy levels that an electron can occupy in an atom; the first shell can contain up to 2 electrons, the second up to 8 electrons, the third up to 18 electrons, and so on.

significant figures Those measured digits that are known with certainty plus one uncertain digit.

silicone A polymer whose chains consists of alternating silicon and oxygen atoms.

single bond A pair of electrons shared between two atoms.

SI units (International System of Units) A measuring system that is used by scientists worldwide and has seven base units (quantities) with their multiples and submultiples.

skeletal structure Representation of the carbon skeleton of a molecule, often shown as a geometric figure or shape.

slag A relatively low-melting product of the reaction of limestone with silicate impurities in iron ore.

slow-twitch fibers Muscle fibers suited for steady exercise of long duration.

smog A combination of smoke and other pollutants, forming a visible haze.

soap A salt (usually a sodium salt) of a long-chain carboxylic acid.

solar cell A device used for converting sunlight to electricity; a photoelectric cell.

solid A state of matter in which the substance has a definite shape and volume.

solute The substance that is dissolved in another substance (solvent) to form a solution; usually present in a smaller amount than the solvent.

solution A homogeneous mixture of two or more substances.

solvent The substance that dissolves another substance (solute) to form a solution; determines the physical state of the solution and is usually present in a larger amount than the solute.

specific heat capacity (specific heat) The amount of heat required to raise the temperature of 1 g of a substance by 1 °C.

standard temperature and pressure (STP) Conditions of 0 °C and 1 atm pressure.

starch A polymer of glucose units joined by alpha linkages; a complex carbohydrate.

starvation The voluntary or involuntary withholding of nutrition from the body.

steel An alloy of iron and carbon, usually also containing other metals such as manganese, nickel, or chromium.

step-wise polymerization See **condensation polymerization**.

stereoisomers Isomers having the same formula but differing in the arrangement of atoms or groups of atoms in three-dimensional space.

steroid A molecule that has a four-ring skeletal structure, with one cyclopentane and three cyclohexane fused rings.

stimulant drug A drug that increases alertness, speeds up mental processes, and generally elevates the mood.

stoichiometric factor A conversion factor that relates the numbers of moles of two substances through their coefficients in a chemical equation.

stoichiometry The quantitative relationship between reactants and/or products in a chemical reaction.

strong acid An acid that ionizes completely in water; a potent proton donor.

strong base A base that dissociates completely in water; a potent proton acceptor.

structural formula A chemical formula that shows how the atoms of a molecule are arranged, to which other atom(s) they are bonded, and the kinds of bonds.

sublevel See **subshell**.

sublimation Conversion of a solid directly to the gaseous state without going through the liquid state.

subshell (sublevel) A set of orbitals in an atom that are in the same shell and have the same energy.

substance A sample of matter that always has the same composition, no matter how it is made or where it is found; an element or compound.

substituent An atom or group of atoms substituted for a hydrogen atom on a hydrocarbon.

substrate The substance that attaches to the active site of an enzyme and is then acted upon.

sufficiency The quality of being adequate for a particular purpose.

sulfurous smog See **industrial smog**.

sun protection factor (SPF) The ability of a sunscreen to absorb UV radiation, theoretically equal to the relative increase in time that a person can be in the sun before "burning."

sunscreen A substance or mixture that blocks, reflects, or absorbs ultraviolet (UV) radiation.

supercritical fluid An intermediate state having properties of both gases and liquids.

surface-active agent (surfactant) Any substance that can stabilize the suspension of a nonpolar substance (such as oil) in water.

surroundings Everything that is not part of the system being observed in a thermochemical study.

sustainable agriculture Farming practices that produce food and fiber indefinitely, without causing irreparable damage to the ecosystem.

sustainable chemistry An approach designed to meet the needs of the present generation without compromising the needs of future generations.

synapse A tiny gap between nerve cells.

synergistic effect An effect greater than the sum of the effects expected from two or more interacting processes.

synthesis gas A mixture of carbon monoxide and hydrogen gas used for organic synthesis.

system The part of the universe under consideration in a thermochemical study.

tar sands Sands that contain bitumen, a thick hydrocarbon material.

technology The practical application of knowledge by which humans modify the materials of nature to better satisfy their needs and wants.

temperature A measure of heat intensity, or how energetic the particles of a sample are.

temperature inversion A warm layer of air above a cool, stagnant lower layer.

teratogen A substance that causes birth defects when introduced into the body of a pregnant female.

tertiary structure The folding pattern of a protein.

tetracyclines Antibacterial drugs with four fused rings.

theory See **scientific theory**.

thermochemistry The study of energy changes that occur during chemical reactions.

thermonuclear reactions Nuclear fusion reactions that require extremely high temperatures and pressures.

thermoplastic polymer A kind of polymer that can be heated and reshaped.

thermosetting polymer A kind of polymer that cannot be softened and remolded.

thermosetting resins Polymeric materials that harden irreversibly.

thiol (R—SH) An organic compound composed of an alkyl group and a thiol (—SH) group.

top note The fraction of a perfume that vaporizes most quickly, and is responsible for the fragrance when the perfume is first applied; composed of relatively small molecules.

toxicology The branch of pharmacology that deals with the effects of poisons on the body, their identification and detection, and remedies for them.

toxic waste A waste that contains or releases poisonous substances in amounts large enough to threaten human health or the environment.

toxin A poisonous substance produced by a living organism.

tracers Radioisotopes used to trace the movement of substances or locate the sites of activity in physical, chemical, and biological systems.

transcription The process by which a segment of DNA transfers its information to a messenger RNA (mRNA) molecule during protein synthesis.

transfer RNA (tRNA) A small molecule that contains anticodon nucleotides; the RNA molecule that bonds to and carries an amino acid.

transition elements Metallic elements in the B groups of the customary U.S. arrangement of the periodic table or in groups 3–12 of the IUPAC-recommended table.

translation The process by which the information contained in the codon of an mRNA molecule is converted to a protein structure.

transmutation The conversion of one element into another.

triglyceride An ester of glycerol with three fatty-acid units; also called a triacylglycerol.

triple bond The sharing of three pairs of electrons between two atoms.

tritium (hydrogen-3) A radioactive isotope of hydrogen with two neutrons and one proton in the nucleus.

unsaturated hydrocarbon An alkene, alkyne, or aromatic hydrocarbon; a hydrocarbon containing one or more double or triple bonds or aromatic rings.

valence electrons Electrons in the outermost shell of an atom.

valence shell electron pair repulsion theory (VSEPR theory) A theory of chemical bonding useful in determining the shapes of molecules; it states that valence shell electron pairs around a central atom locate themselves as far apart as possible.

vaporization The process by which a substance changes from the liquid to the gaseous (vapor) state.

variable A factor that changes and potentially affects results during an experiment.

vitamin An organic compound that is required in the diet to protect against some diseases.

volatile organic compounds (VOCs) Organic compounds that cause pollution because they vaporize readily.

VSEPR theory See **valence shell electron pair repulsion theory**.

vulcanization The process of making naturally soft rubber harder by reacting it with sulfur.

wax An ester of a long-chain fatty acid and a long-chain alcohol, or a solid mixture of higher-molecular-mass hydrocarbons from petroleum.

weak acid An acid that ionizes only slightly in water; a poor proton donor.

weak base A base that ionizes only slightly in water; a poor proton acceptor.

weight A measure of the force of attraction between Earth (or another planet) and an object.

wet scrubber A pollution-control device that uses water to remove pollutants from smokestack gases.

X-rays Radiation similar to visible light but of much higher energy and that is much more penetrating.

zeolite A naturally occurring mineral that has ion-exchange properties. Zeolites are often used in water softener equipment, substituting sodium ions for hard-water (Mg^{2+}, Ca^{2+}, Fe^{2+}) ions.

zwitterion A molecule that contains both a positive charge and a negative charge; a dipolar ion.

Brief Answers to Selected Problems

Answers are provided for *all in-chapter exercises*. Brief answers are given for *odd-numbered Review Questions*; more complete answers can be obtained by reviewing the text. Answers are provided for *all odd-numbered Problems and Additional Problems*.

Note: For numerical problems, your answer may differ slightly from ours because of rounding and the use of significant figures. (See Appendix.)

Chapter 1

1.1 **A. a.** DQ would probably be small.
    b. DQ would probably be large.
  B. a. DQ would be uncertain.
    b. DQ would probably be large.
1.2 **A. a.** 1.00 kg   **b.** 179 lb
1.3 **A.** physical: b; chemical: a, c
  B. physical: a, c; chemical: b
1.4 **A.** elements: Hf, No, Fm; compounds: CuO, NO, HF   **B.** 8
1.5 **A. a.** 7.24 kg   **b.** 4.29 μm   **c.** 7.91 ms
  B. a. 3.8×10^{-9} s   **b.** 7.54×10^{-3} m **c.** 2.9×10^3 A
1.6 **A. a.** 7.45×10^{-9} m   **b.** 5.25×10^{-6} s **c.** 1.415×10^3 m
    d. 2.06×10^{-3} m   **e.** 6.19×10^3 m
  B. a. 5.7×10^4 m   **b.** 1.1×10^{-2} A
1.7 **A.** 500 cm^2 **B.** 1600 cm^3
1.8 **A. a.** 755 mm   **b.** 0.2056 L   **c.** 206,000 μg
  B. a. 409,000 mg   **b.** 2.45×10^8 ns
1.9 **A.** sink; float   **B.** 0.88 g/mL (oil floats on water)
1.10 **A.** 1.11 g/mL   **B.** magnesium
1.11 **A.** 351 K   **B.** 77 K
1.12 **A.** 1799 kcal   **B.** 34 kcal

1. Science is testable, reproducible, explanatory, predictive, and tentative. Testability best distinguishes science.
3. These problems usually have too many variables to be treated by the scientific method.
5. Risk–benefit analysis compares benefits of an action with risks of that action.
7. DQ, the desirability quotient, is benefits divided by risks. A large DQ means that risks are minimal compared to benefits. Often it is hard to quantify risks and benefits.
9. The SI-derived unit for volume is liter (L), a relatively large quantity. Milliliters (mL) and microliters (μL) are more often used.
11. **a.** applied   **b.** applied   **c.** basic
13. Society's benefits are very great; risk to selected individuals is also great but is managed by restricting access via prescription. The DQ is high.
15. **a.** high
   b. high (but lower than in scenario a); protective equipment
17. **a.** The DQ is low, especially if the problem's origin is unknown. The risk of development of MRSA likely outweighs the possible benefits of the treatment. The flu is a viral problem that does not respond to antibiotics; the DQ is even lower than in (a).
19. 100 g; 2 kg
21. Both a and b are reasonable; c, 100 kg; d, 10 kg
23. 250 mL
25. 1.46×10^9 km^3
27. Yes; the 26.3-mm i.d. tube is larger than 25.4 mm (1 inch).
29. **a.** physical   **b.** chemical
   c. chemical   **d.** physical
31. **a.** physical   **b.** chemical   **c.** physical
33. **a.** mixture   **b.** substance
   c. substance   **d.** substance

35. **a.** heterogeneous   **b.** homogeneous
   c. heterogeneous   **d.** homogeneous
37. Substance; properties do not vary
39. **a.** compound   **b.** element
   c. element   **d.** element
41. **a.** carbon   **b.** magnesium
   c. helium   **d.** nitrogen
43. f
45. **a.** 4.4 ns   **b.** 8.5 cg   **c.** 3.38 Mm
47. **a.** 55.2 L   **b.** 0.325 g   **c.** 0.27 m
   d. 2.7 cm   **e.** 0.078 ms
49. **a.** cm   **b.** kg   **c.** dL
51. **a.** 1.4×10^6 mg   **b.** 1.4×10^3 g   **c.** 1.4×10^{12} ng
53. d
55. **a.** 222 g (Lucara), 621 g (Cullinan)   **b.** 0.49 lb., 1.37 lb.
57. 0.918 g/mL
59. 18.9 g
61. **a.** 344 mL   **b.** 495 cm^3
63.

Hexane
Water
Mercury

65. Yes (Total load is 145 lb.)
67. 5.40 kg
69. $-196\,°C$
71. 38.5 kcal
73. a, b, and c are hypotheses.
75. 52.6 min.
77. b
79. 2.09 g hardener
81. (1) d; (2) e; (3) b; (4) a; (5) d
83. 6.0×10^2 g
85. Cabbage < potatoes < sugar (heaviest)
87. 5.23 g/cm^3
89. 6.73×10^{-6} cm
91. 0.908 mt
93. **a.** 1.3 g/cm^3 **b.** 5.44 g/cm^3 **c.** 0.688 g/cm^3
95. The design of chemical products and processes that reduce or eliminate the generation and use of hazardous substances.
97. a

Let's Experiment!

1. They all have slightly different densities.
2. The liquids at the bottom of the glass have a higher density than those at the top. Those liquids at the top have the lowest density of the group.
3. Yes, several of the liquids will mix together; water and alcohol are miscible, meaning they will mix together. Vegetable oil and syrup will separate, as the oil is immiscible with the water-based syrup.
4. The water and alcohol will not separate back out but the vegetable oil and syrup will separate back out, as they are oil-based and immiscible.
5. Density (mass/volume) allows you to convert between the volume of a substance and its mass (or weight). Density becomes important anytime you want to build something where weight and distribution of weight are critical. For example, a boat made of aluminum foil can float on water; crumple the same piece of foil and it will sink.

Chapter 2

2.1 **A.** 3 atoms of hydrogen to 1 atom of phosphorus
   B. 63.0 g nitrogen
2.2 **A.** 112 g CaO
   B. 87.9 g CO_2
2.3 **A.** average Cl+I is 81.2; actual Br is 79.9 amu.
   B. He/Ne/Ar
 1. **a.** Atomic: Matter is made of discrete particles. Continuous: Matter is infinitely divisible.
   b. Greek: four elements, no atoms. Modern: Each element has its own kind of atom.
 3. Discrete: people, calculators, M&M candies. Continuous: Cloth and milk chocolate give the impression that they can be infinitely divided.
 5. As all hydrogen atoms weigh the same, and all oxygen atoms weigh the same, a substance containing 2 H and 1 O will always have the same mass ratio of H to O.
 7. Law of definite proportions
 9. Law of multiple proportions
11. Box C has 15 oxygen atoms; the initial mixture has only 14.
13. By placing elements having similar properties in columns, there occasionally would be empty spaces between elements. Mendeleev interpolated between the elements above and below an empty space, and on either side of the space, to predict properties of the undiscovered element.
15. **a.** If the container is completely sealed then its weight after one or two weeks will be the same as it weighed at the start of the experiment, because all the life processes involving reactions conserve the atoms involved. Some of those atoms will appear in the form of gases, some as solids, and others as liquids, but all will be retained in the container.
   b. No; if the mouse were in a wire container, any gaseous products of its metabolism would escape. The weight after one or two weeks would be less than the original weight.
17. c. The effervescence is gas, which is matter and has mass, leaving the water.
19. 36.6 g O_2
21. The element ratio means that 88.20 g of propane contains 72.06 g C. From the element ratio of CO_2, 72.06 g C can produce at most 264 g CO_2.
23. b
25. **a.** 54.8 g of hydrogen **b.** 434 g of oxygen
27. 3.8 kg C
29. d
31. Dalton's theory says that chemical reactions involve rearrangement of atoms, so the carbon atoms in the diamond would have combined with oxygen atoms to form a new compound. The hydrogen and oxygen atoms in water would be separated but not destroyed by electrolysis.
33. a
35. c
37. SnO_2
39. Yes. Fe:O is 3.48:1 for wustite, 2.33:1 for hematite, 2.61 for magnetite. 3.48/2.61 = 1.333 or 4:3; 3.48/2.33 = 1.49 or 3:2.
41. 0.050 mol C; 3.0×10^{22} C atoms
43. 60.00 g carbon-12; 60.05 g C
45. Yes, the ratio of carbon to hydrogen in the three samples is the same.
47. 6 F atoms for every U atom
49. 2.61 g SO_2
51. Divide 0.5836 g by 0.4375 to get 1.334. This gives a ratio of N to O of 1.334 to 1. Multiplying both by 3 gives 4.002 to 3, or 4 to 3.
53. Law of conservation of mass. All the atoms were retained in the balloon, so their mass did not change.
55. **a.** mass of Se = (32 + 125)/2 = 78.6
   b. mass of Se = (32.06 + 127.60)/2 = 79.83
57. **c.** and **e.** are hazardous; **b.** is rare; others are neither.

59. Mercury is an element and cannot be destroyed. Accumulation in landfills could make the region's groundwater and soil high in mercury.

Let's Experiment!
 1. Both reactions will "fizz" and a gas will bubble out of the mixture. The weight should go down for the mixture that was NOT in the bag due to the loss of carbon dioxide. The mixture that was in the bag should remain the same weight.
 2. The weight should not have changed significantly, since the carbon dioxide should be captured within the sealed bag.
 3. It should have supported the law of conservation of matter. In the Alka-Seltzer reaction, the matter within the tablet does not disappear, but some of it is transformed in to a gas, which escapes. If we can trap that gas, as we did with the bag, then we can see that the mass is conserved.

Chapter 3

3.1 **A.** 20 neutrons   **B.** mass number 80; bromine-80
3.2 **A.** $^{90}_{37}X$ and $^{88}_{37}X$; $^{88}_{38}X$ and $^{93}_{38}X$ **B.** three; Br, Rb, and Sr
3.3 **A.** 32 electrons   **B.** 3
3.4 **A. a.** (B) 2, 2,1
       b. (Al) 2, 8, 3; both have the same number of outer electrons.
   B. a. (N) 2, 5
       b. (S) 2, 8, 6. No, different columns have different numbers of outer electrons.
3.5 **A. a.** (F) $1s^2 2s^2 2p^5$
       b. (Cl) $1s^2 2s^2 2p^6 3s^2 3p^5$
   B. a. (Ti) $1s^2 2s^2 2p^6 3s^2 3p^6 4s^2 3d^2$ (See Figure 3.2.)
       b. (Sn) $1s^2 2s^2 2p^6 3s^2 3p^6 4s^2 3d^{10} 4p^6 5s^2 4d^{10} 5p^2$
3.6 **A. a.** (Cs) $6s^1$ **b.** (Sb) $5s^2 4d^{10} 5p^3$ **c.** (Si) $3s^2 3p^2$
   B. a. (Ga) $4s^2 4p^1$ **b.** (In) $5s^2 5p^1$
 1. Electrons
 3. Goldstein examined particles traveling in the opposite direction from the particles Thomson examined. Goldstein showed that matter contained positively charged particles.
 5. Rutherford decided that almost all the mass of an atom was in the center, and that "core" was tiny and positively charged.
 7. Bohr concluded that electrons did not orbit the nucleus randomly, but orbited at fixed distances and energies.
 9. "Quantum" means a fixed, definite amount of energy.
11. In the first step the atom absorbs a fixed amount (quantum) of energy to send an electron into a higher energy level. In the second step that electron drops back to a lower energy level, giving off the absorbed energy as a photon of light.
13. 19 electrons
15. **a.** 5   **b.** 16   **c.** 29
17. **a.** 5   **b.** 16   **c.** 29
19. Your sketch should show 16 protons and 17 neutrons in the nucleus; also 2 electrons in the first main shell, 8 in the second main shell outside the first, and 6 in the third main shell outside the second.
21. Os, Osmium, 190.23 amu
23. **a.** no   **b.** no   **c.** yes
   d. chlorine   **e.** argon
25. Atom A and Atom B
27. a
29. $^{37}_{17}Cl$, chlorine
31.

Element	Mass Number	Number of Protons	Number of Neutrons
Nickel	60	28	32
Palladium	108	46	62
Nitrogen	14	7	7
Iodine	127	53	74

33. Four
35. 28

37. +3

39. second

41. a. 8   **b.** 14   **c.** 28.

45. Li: 2,1   Be: 2,2   B: 2,3   C: 2,4   N: 2,5   O: 2,6   F: 2,7   Ne: 2,8

$1s^2\, 2s^1$ $1s^2\, 2s^2$ $1s^2\, 2s^2\, 2p^1$ $1s^2\, 2s^2\, 2p^2$ $1s^2\, 2s^2\, 2p^3$ $1s^2\, 2s^2\, 2p^4$ $1s^2\, 2s^2\, 2p^5$ $1s^2\, 2s^2\, 2p^6$

Na: 2,8,1   Mg: 2,8,2   Al: 2,8,3   Si: 2,8,4   P: 2,8,5   S: 2,8,6   Cl: 2,8,7   Ar: 2,8,8

$1s^2\, 2s^2\, 2p^6\, 3s^1$ $1s^2\, 2s^2\, 2p^6\, 3s^2$ $1s^2\, 2s^2\, 2p^6\, 3s^2\, 3p^1$ $1s^2\, 2s^2\, 2p^6\, 3s^2\, 3p^2$ $1s^2\, 2s^2\, 2p^6\, 3s^2\, 3p^3$ $1s^2\, 2s^2\, 2p^6\, 3s^2\, 3p^4$ $1s^2\, 2s^2\, 2p^6\, 3s^2\, 3p^5$ $1s^2\, 2s^2\, 2p^6\, 3s^2\, 3p^6$

47. Similar: Both have outermost shells $s^2\, p^2$. Different: Outermost shell is 3 for Si, 4 for Ge.

     a. Kr   **b.** Na   **c.** Br   **d.** Sc

49. a, b, and c are metals; d is a nonmetal

51. Cs and K

53. Fe and Mo

55. (a) 15;   **(b)** phosphorus;   **(c)** 15;   **(d)** 6;   **(e)** 0

57. (a) Ne;   **(b)** No, it still has 12 protons.

59. (a) Fe: $1s^2\, 2s^2\, 2p^6\, 3s^2\, 3p^6\, 4s^2\, 3d^6$

     (b) Sn: $1s^2\, 2s^2\, 2p^6\, 3s^2\, 3p^6\, 4s^2\, 3d^{10}\, 4p^6\, 5s^2\, 4d^{10}\, 5p^2$

61. A is zinc (Zn), B is gallium (Ga).

63. Q is calcium (Ca), R is strontium (Sr).

65. c

67. Refutes. If the oxygen were coming from the CO_2 it would not have any O-18.

Let's Experiment!

1. Answers vary. There should be a large list of colors when viewed with the diffraction film.

2. There are different metals in the different colored candles, which give the flames their colors.

3. When the atoms are heated, electrons gain energy and transition to higher energy levels. When the energy is released as photons, the colors are observed.

4. There are several potential sources of contamination, such as the wick, the wax, dyes added to the wax, etc. Also, looking through the diffraction grating film is subjective.

5. Typical results blended with information from the ColorFlame website regarding candle ingredients.

Candle	Flame color (to naked eye)	Diffraction emissions	Element identified
Red	Red	Red/yellow/green (possibly indigo)	Lithium strontium calcium
Purple	Yellow and violet	Orange/green/indigo	Potassium
Green	White with traces of indigo and/or yellow	Red/yellow/green/indigo	Phosphorus copper zinc
Blue	Green/indigo	Yellow/green/indigo	Copper selenium
Orange	Yellow/orange (intense)	Orange	Sodium

Chapter 4

4.1 A. a. :Kr: **b.** ·Ba·   **c.** :I·   **d.** ·N·   **e.** K·   **f.** :I·

     B. a. $[: N :]^{3-}$ **b.** $[Mg]^{2+}$ **c.** $[K]^{+}$ **d.** $[: Br :]^{-}$

4.2 A. Li· + :F· ⟶ Li⁺ + :F:⁻

     B. Cs· + :F· ⟶ Cs⁺ + :F:⁻

4.3 A. ·Ca· + :O· ⟶ $[Ca]^{2+}\ [: O :]^{2-}$

     B. 2 ·Al· + 3 :O· ⟶ 2 Al³⁺ + 3 :O:²⁻

4.4 A. a. CaF_2 **b.** Li_2O

     B. $FeCl_2$ and $FeCl_3$

4.5 A. a. K_2O **b.** Ca_3N_2

     B. a. Mg_3P_2 **b.** AlP

4.6 A. a. calcium fluoride   **b.** copper(II) bromide

     B. a. lithium sulfide   **b.** iron(III) sulfide

4.8 A. a. :Br· + ·Br: ⟶ :Br:Br:

     b. :F· + ·Cl: ⟶ :F:Cl:

     B. a. :I· + ·I: ⟶ :I:I:

     b. H· + ·I: ⟶ H:I:

4.9 A. a. polar covalent   **b.** ionic

     c. nonpolar covalent   **d.** polar covalent

     B. a. polar covalent   **b.** polar covalent

     c. polar covalent   **d.** nonpolar covalent

     C. b < a < d < c

4.10 A. a. bromine trifluoride   **b.** bromine pentafluoride

     c. dinitrogen monoxide   **d.** nitrogen dioxide

     B. a. carbon monoxide   **b.** sulfur hexafluoride

     c. sulfur trioxide   **d.** dichlorine pentoxide

4.11 A. a. CO_2 **b.** Cl_2O_7

     B. a. NI_3 **b.** S_2F_{10}

4.12 a. $CaCO_3$ **b.** K_3PO_4 **c.** $(NH_4)_2S$

4.13 A. a. magnesium carbonate   **b.** calcium phosphate

     c. ammonium acetate

     B. a. ammonium sulfate

     b. potassium dihydrogen phosphate

     c. copper(II) dichromate

4.14 A. a. H—S: **b.** :F—O—F:
             |
             H

     B. a. H—C—Cl: **b.** $[: O — C — O :]^{2-}$ with ‖ O:

4.15 A. a. trigonal pyramidal   **b.** tetrahedral

     B. a. P in PH_3 **b.** O in H_2O

1. Sodium metal is quite reactive; sodium ions (as in NaCl) are quite unreactive.

3. Chlorine atoms and Cl_2 molecules have the same number of protons as electrons, while Cl⁻ ions have one more electron than protons. A chlorine molecule has twice as many protons and twice as many electrons as a Cl atom. Chlorine atoms and Cl_2 molecules are very reactive; Cl⁻ ions are not reactive; they are strongly attracted to positive ions such as Na⁺, with which they form ionic bonds.

5. **a.** 1   **b.** 2   **c.** 3   **d.** none

7. Covalent bond: electrons shared. Ionic bond: electrons transferred.

9. **a.** ·Ca·   **b.** :S: **c.** ·Si·

43. a. possible excited   **b.** incorrect

     c. possible excited   **d.** ground

11. a. Na·, Na$^+$ **b.** :Ċl·, $\left[:\ddot{Cl}:\right]^-$

13. a. K$^+$ $\left[:\ddot{Br}:\right]^-$ **b.** 2 Na$^+$ $\left[:\ddot{O}:\right]^{2-}$

  c. Mg^{2+} 2$\left[:\ddot{F}:\right]^-$ **d.** Al^{3+} 3$\left[:\ddot{Cl}:\right]^-$

15. a. T **b.** F **c.** T **d.** T
17. a. S^{2-} **b.** potassium ion **c.** bromide ion
  d. F$^-$ **e.** calcium ion **f.** Fe^{3+}
19. a. cobalt(III) ion **b.** cobalt(VI) ion
21. a. Mo^{2+} **b.** Mo^{4+} **c.** Mo^{6+}
23. a with d; c with h; e with b; g with f
25. Cr$_2$O$_3$, chromium(III) oxide; and CrO$_3$, chromium(VI) oxide
27. a with d; c with h; e with b; g with f
29. a. silver nitrate **b.** Na$_2$CrO$_4$
  c. ammonium sulfite **d.** Sr(HCO$_3$)$_2$
  e. aluminum permanganate **f.** Cu$_3$(PO$_4$)$_2$

31. H:Ï:

33. 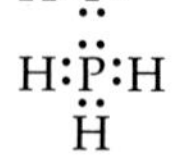 H:P̈:H
    H

35. 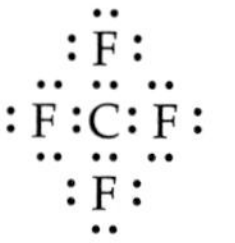 :F̈:C̈:F̈:
    :F̈:

37. One electron, because H has only one valence electron and forms a single covalent bond.
39. a. N$_2$O$_4$ **b.** BrCl$_3$
  c. oxygen difluoride **d.** NI$_3$
  e. carbon tetraiodide **f.** dinitrogen trioxide

41. a. 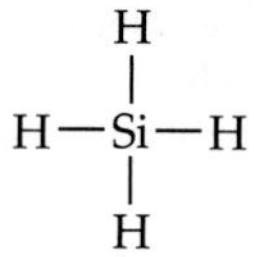 H—Si—H (with H above and below) **b.** 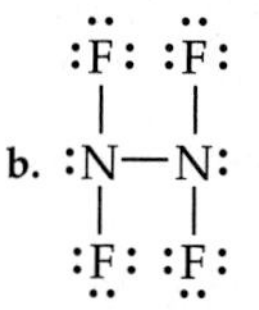 :N—N: (with :F̈: :F̈: above and below)

  c. H—C—N: (with H H, H H) **d.** H—C—H (with :O: double bond)

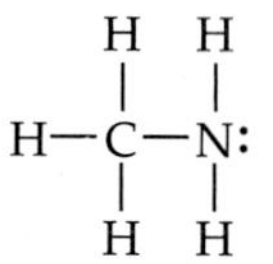

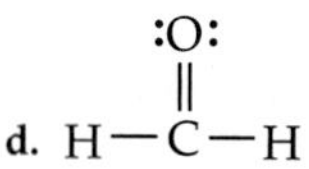

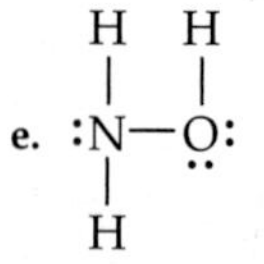

  e. :N—O: (with H) **f.** H—Ö—P—Ö—H (with :O—H)

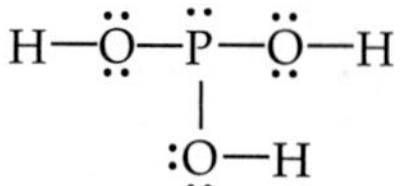

43. a. 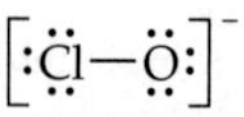 $\left[:\ddot{Cl}—\ddot{O}:\right]^-$ **b.** 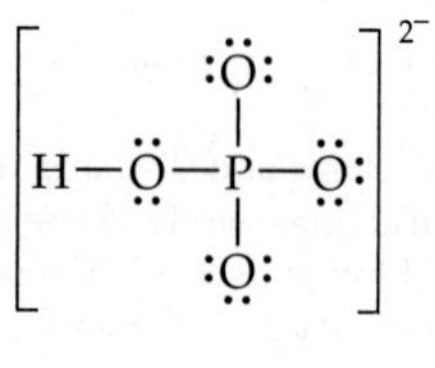 $\left[H—\ddot{O}—P—\ddot{O}:\right]^{2-}$ (with :O: above and :O: below)

  c. 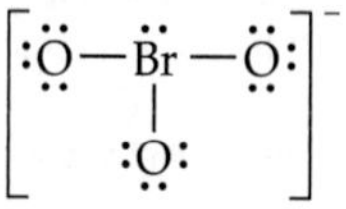 $\left[:\ddot{O}—\ddot{Br}—\ddot{O}:\right]^-$ (with :O: below)

45. between 0.5 and 2.0
47. a. polar **b.** polar **c.** polar

49. 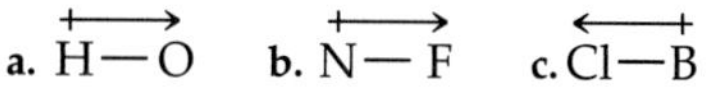 a. H—O (arrow) **b.** N—F (arrow) **c.** Cl—B (arrow)
51. a. Si$^{\delta+}$—O$^{\delta-}$ **b.** F—F **c.** F$^{\delta-}$—N$^{\delta+}$
53. a. ionic **b.** nonpolar covalent
  c. ionic **d.** nonpolar covalent
55. b < d < a < c
57. a with d; c with b; e with f; g with h
59. a. tetrahedral **b.** tetrahedral
  c. bent **d.** bent
61. Nonpolar; the two bond polarities are the same, and in a linear molecule, they will cancel.
63. a. nonpolar bonds; 109°; nonpolar molecule
  b. polar bonds; 120°; polar molecule
  c. polar bonds; 109°; polar molecule
  d. polar bonds; 109°; nonpolar molecule
65. It has two polar bonds but is not linear, so the bond polarities add.
67. CH$_2$Cl$_2$ is polar; the C—H bonds and the C—Cl bonds have different polarities and do not cancel one another as they would in either CH$_4$ or CCl$_4$.
69. a and c
71. a. 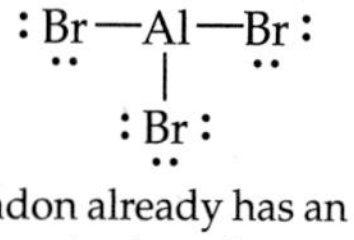 :Br—Al—Br: (with :Br: below) **b.** H—Be—H **c.** 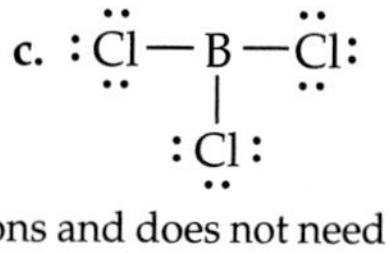 :Cl—B—Cl: (with :Cl: below)
73. Radon already has an octet of valence electrons and does not need more (or fewer).
75. AlP and Mg$_3$P$_2$
77.

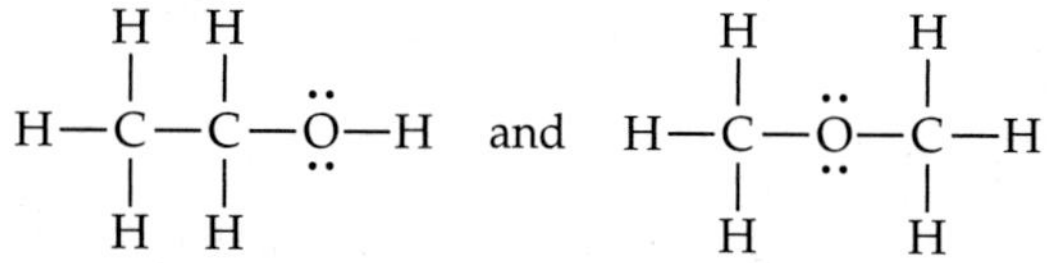

H—C—C—Ö—H (with H H top and H H bottom) and H—C—Ö—C—H (with H H top and H H bottom)

79. a. X is group 7a, Y is group 6a, Z is group 5a.
  b. 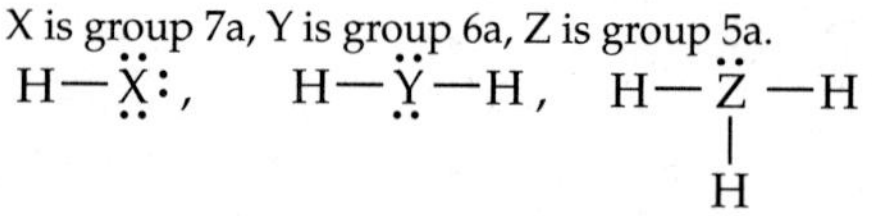 H—Ẍ:, H—Ÿ—H, H—Z̈—H (with H below)
  c. Na$^+$ + $\left[:\ddot{X}:\right]^-$, 2Na$^+$ + $\left[:\ddot{Y}:\right]^{2-}$
81. Potassium iodide is an ionic compound and does not consist of molecules; it consists of K$^+$ ions and I$^-$ ions arranged in rows, ranks, and columns.
83. [He:H]$^+$; He: + H$^+$ → [He:H]$^+$
85. Carbon exists as C atoms, not molecules. Also, the formula for methane is CH$_4$, not CH$_6$. Finally, methane consists of atoms bonded to atoms. The correct statement would end "—just a single carbon atom attached to four hydrogen atoms."
87. Copper(I): 1s^2 2s^2 2p^6 3s^2 3p^6 4s^1 3d^9 (Instructor: In advanced classes the student would find that copper is one of the exceptions to the filling order. The last two subshells in Cu are really 4s^1 3d^{10}, so Cu(I) ion ends with 3d^{10}.) Vanadium(II): 1s^2 2s^2 2p^6 3s^2 3p^6 3d^3
89. The angle 120° is as far apart as three bonds can be. If a bond is moved so that its angle with another bond is increased, one of the other bonds will have an angle less than 120°.
91. Column 4A elements tend not to form ions at all, instead forming covalent bonds, so **d** is the best response.
93. New drug molecules are designed to resemble biological molecules, and to bind to enzymes or receptors in the body.
95. Use lower energy; use less (or no) solvent; use milder conditions.

Let's Experiment!
1. SO$_2$ has a lone pair of electrons on the central sulfur atom, which pushes the oxygen molecules away and gives it a 120° angle. Water has two pairs of lone electrons on the central oxygen atoms, which push away the hydrogen atoms and give the molecule a tetrahedral angle of 109.5°.

Chapter 5

5.1 **A.** 3 H$_2$ + N$_2$ ⟶ 2 NH$_3$
  B. 2 Fe$_2$O$_3$ + 3 C ⟶ 3 CO$_2$ + 4 Fe

5.2 **A.** 1.66 L
 B. propane; 3.20 L (4.00 L oxygen will consume 0.80 L propane.)
5.3 **A. a.** 65.0 amu   **b.** 76.1 amu
 B. a. 60.1 amu   **d.** 234.1 amu
5.4 **A.** 41.1% N   **B.** Ammonium nitrate
5.5 **A. a.** 0.874 g C   **b.** 1000 g H_2O
      c. 17.0 g $Ca(HSO_4)_2$
 B. a. 0.0664 mol Fe   **b.** 2.29 mol C_5H_{12}
      c. 0.000674 mol $Mg(NO_3)_2$
5.6 **A.** *Molecular:* 2 molecules H_2S react with 3 molecules O_2 to form 2 molecules SO_2 and 2 molecules H_2O; *Molar:* 2 mol H_2S react with 3 mol O_2 to form 2 mol SO_2 and 2 mol H_2O; *Mass:* 68.2 g H_2S react with 96.0 g O_2 to form 128.1 g SO_2 and 36.0 g H_2O.
 B. a. Atoms are conserved.
      b. Molecules are not conserved.
      c. Moles are not conserved.
      d. Mass is conserved.
5.7 **A. a.** 3.750 mol CO_2 **b.** 0.0354 mol CO_2
 B. Propane, 7.0 mol
5.8 **A.** 3.44 g O_2
 B. a. 974 g CO_2 **b.** 1083 g CO_2
5.9 **A.** 0.00672 M   **B.** 1.10 M
5.10 **A. a.** 0.0909 M   **b.** 0.0168 M
 B. 0.00881 M
5.11 **A. a.** 337 g KOH   **b.** 11.2 g KOH
 B. 9.55 kg NaOH
5.12 **A.** 24.7 mL
 B. 0.000630 g
5.13 **A.** 89.3% ethanol   **B.** 30.7% toluene
5.14 **A.** 2.06% H_2O_2 **B.** 49.4% NaOH
5.15 **A.** Dissolve 11.3 g glucose in enough water to make 250 g of solution.
 B. Dissolve 11.1 g NaCl in enough water to make 1.25 kg of solution.
1. **a.** the smallest repeating unit of a substance
 b. the sum of the atomic masses of a chemical compound
 c. the amount of a substance that contains 6.022×10^{23} formula units of the substance
 d. the number of atoms (6.022×10^{23}) in exactly 12 g of pure carbon-12
 e. the formula mass of a substance expressed in grams; mass of one mole
 f. the volume occupied by 1 mol of a gas under specified conditions
3. A handful. A pinch would represent roughly a gram of a substance (meaning 1 g/mol), a bucketful would be several kilograms (>1000 g/mol), and a truckload would be a lot more than that!
5. **a.** A solution is a homogeneous mixture of two or more substances.
 b. The solvent is the substance that disperses the solvent. It is usually present in the greatest quantity in a solution.
 c. A solute is a substance dissolved in a solvent to form a solution.
 d. An aqueous solution is a solution in which water is the solvent.
7. **a.** 12   **b.** 3   **c.** 6
9. 6 N, 3 P, 27 H, 12 O
11. **a.** 2 molecules of H_2O_2 produce 2 molecules of H_2O and 1 molecule of O_2.
 b. 2 mol of H_2O_2 produce 2 mol of H_2O and 1 mol of O_2.
 c. 68 g of H_2O_2 produce 36 g of H_2O and 32 g of O_2.
13. **a.** $4\,Li + O_2 \rightarrow 2\,Li_2O$
 b. $3\,Mg + Co_2O_3 \rightarrow 2\,Co + 3\,MgO$
 c. $Zr + 2\,H_2S \rightarrow ZrS_2 + 2\,H_2$
15. **a.** $N_2 + 2\,O_2 \rightarrow N_2O_4$ **b.** $9\,C + 2\,O_3 \rightarrow 3\,C_3O_2$
 c. $UO_3 + 6\,HF \rightarrow UF_6 + 3\,H_2O$
17. **a.** They contain equal numbers of molecules.
 b. A molecule of Cl_2 has the greatest mass, 70.9 amu, so the chlorine balloon will be heaviest.
19. **a.** 103 L CO_2 **b.** 7.30 mL

21. 5:1
23. **a.** 6.02×10^{23} molecules of P_4 **b.** 2.41×10^{24} P atoms.
25. c
27. **a.** 159.6 g/mol   **b.** 286.6 g/mol
 c. 368.2 g/mol   **d.** 74.1 g/mol
29. **a.** 535 g $AgNO_3$ **b.** 10.0 g $CaCl_2$
 c. 404.4 g H_2S
31. **a.** 0.228 mol Sb_2S_3 **b.** 2.23 mol MoO_3
 c. 7.44 mol $AlPO_4$
33. **a.** 17.1%   **b.** 26.2%
35. **a.** 2.54 mol CO_2 **b.** 11.4 mol O_2
37. **a.** 1400 g NH_3 **b.** 200 g H_2
39. **a.** 1.09 M   **b.** 0.602 M
41. **a.** 23.7 g HCl   **b.** 12.7 g K_2CrO_4
43. **a.** 0.333 L   **b.** 0.64 L
45. **a.** 3.68 %   **b.** 50.5 %
47. Dissolve 310 g of NaCl in 3440 g of water.
49. Add water to 40.0 mL of acetic acid to bring the total volume of the solution to 2.00 L.
51. The second equation is not correct, because the total charge on each side is different.
53. **a.** $3\,FeO(s) + 2\,Al(s) \rightarrow 3\,Fe(l) + Al_2O_3(s)$
 b. $K_2S(aq) + Cu(NO_3)_2(aq) \rightarrow 2\,KNO_3(aq) + CuS(s)$
55. **a.** yes   **b.** 0.50 mol C_2H_2
57. 1.4×10^5 g CaO
59. 0.418 mol H_2O_2
61. **a.** 94.9% ethanol   **b.** 0.258% acetone
63. **a.** 357 yg U   **b.** 722 U atoms
65. 8.25 mL
67. The number of cups of water in all the oceans is 6.10×10^{21} and the number of H_2O molecules in a cup of water is 7.89×10^{24}, so the statement is valid.
69. 8.57×10^7 g or 8.57×10^4 kg $NaHCO_3$
71. 111 mol H_2O
73. $975
75. **a.** $C_6H_{12}O_6 \longrightarrow 2\,C_2H_6O + 2\,CO_2;\ C_2H_4 + H_2O \longrightarrow C_2H_6O$
 b. (1) 51.1%; (2) 100%
 c. (1) CO_2 is a byproduct ("waste"); (2) no byproduct
 d. (1) sustainable; (2) not sustainable
 e. Reaction (1) does produce a byproduct, but is sustainable and does not consume valuable, non-renewable petroleum.
77. **a.** 51.3%   **b.** 52.6%   **c.** 81.5%

Let's Experiment!
1. Atoms are conserved, molecules are not.
2. While carbon dioxide is essential to life on Earth, too much carbon dioxide is contributing to global climate change. The burning of fossil fuels has made a considerable impact on the environment and will continue to cause environmental harms if we are not able to reduce our carbon footprint.
3. The law of the conservation of matter states that matter is neither created nor destroyed during the course of a chemical reaction. Although they may be combined differently, the number and mass of the atoms in the reactants must equal the number and mass of the atoms in the product. Balancing an equation assures that the equation represents reality.

Chapter 6

6.1 **A. a.** dispersion   **b.** dipole–dipole
 B. Hydrogen bonding
6.2 **A.** 30 mL   **B.** 400 mmHg
6.3 **A.** 1800 mL   **B.** 503 mL
6.4 **A.** 21.5 L   **B.** −259.7 °C
6.5 **A.** 112 psi   **B.** −95 °C
6.6 **A.** 5.86 g/L   **B.** 1.29 g/L, 4.5 times less than Xe
6.7 **A.** 922 mL   **B.** 426 mmHg
6.8 **A. a.** 0.0526 atm   **b.** 9.99 L
 B. 112 L
1. Both solids and liquids are difficult to compress (fixed volume) because the particles are close together. Liquids flow because their

particles are free to move; solids have fixed shapes because their particles have fixed positions.

3. Solid to liquid: melting. Liquid to gas: vaporization. Gas to liquid: condensation. Liquid to solid: freezing.

5. Ionic interactions (bonds), NaCl. Dipole–dipole forces, $H_2C{=}O$. Hydrogen bonding, H_2O. Dispersion forces, CCl_4.

7. Volume is inversely proportional to pressure and directly proportional to the absolute temperature.

9. NaBr

11. F_2 and CF_4

13. b < a < c

15. Water: b and d. Carbon tetrachloride: a and c.

17. 42.9 m^3

19. 1.7×10^5 L

21. 3.76 L

23. 226 mL

25. 1.07 L

27. 160 °C

29. 0.31 atm

31. Yes (Final pressure would have been 5.3 atm.)

33. **a.** 22.4 L **b.** 182 L **c.** 4.41 L

35. 5.86 g/L

37. **a.** 47.4 g/mol **b.** 66.5 g/mol

39. The ideal gas law is $PV = nRT$. If n is constant:
    a. decrease **b.** decrease **c.** increase

41. **a.** Temperature will decrease. **b.** Pressure will decrease.

43. **a.** 0.172 L **b.** 1.04 atm

45. 0.0208 mol Kr

47. 0.00415 moles

49. 14.5 L

51. See list on page 179.

53. 1.24 g/mol

55. 3.0 g He

57. 0.394 L

59. 44.0 g/mol

61. 226 kg

63. **a.** 380 mmHg **b.** 38 mmHg **c.** 7.6 mmHg

65. b, c, and d

67. 1.16 L

69. e

71. Perchloroethylene in the dry cleaning industry, and methylene chloride for the decaffeination of coffee

Let's Experiment!

3. Heat makes the air molecules inside the balloon move faster and exert greater force on the balloon walls, making it grow. Cooling the balloon slows the molecules and reduces the force exerted on the walls; the balloon shrinks.

4. Volume is directly related to temperature.

5. A rise in pressure if heated, a decrease in pressure if cooled

Chapter 7

7.1 **A. a.** $HI(aq) \longrightarrow H^+(aq) + I^-(aq)$
 b. $Ca(OH)_2(s) \xrightarrow{H_2O} Ca^{2+}(aq) + 2\,OH^-(aq)$
 B. $CH_3COOH(aq) \xrightarrow{H_2O} H^+(aq) + CH_3COO^-(aq)$

7.2 **A.** $HBr(aq) + H_2O \longrightarrow H_3O^+(aq) + Br^-(aq)$
 B. $HClO_4 + CH_3OH \longrightarrow CH_3OH_2^+ + ClO_4^-$

7.3 **A.** H_2SeO_3 **B.** HNO_3

7.4 **A.** $Sr(OH)_2$ **B.** KOH

7.5 **A.** $NaOH(aq) + HC_2H_3O_2(aq) \longrightarrow NaC_2H_3O_2(aq) + H_2O$
 B. $2\,HCl(aq) + Mg(OH)_2(aq) \longrightarrow MgCl_2(aq) + 2\,H_2O$

7.6 **A.** pH = 9 **B.** pH = 2

7.7 **A.** 1×10^{-2} M (0.01 M) **B.** 1×10^{-3} M (0.001 M)

7.8 **A.** c **B.** pH = 2.4

7.9 **A.** Beaker III: HNO_3; Beaker II: KOH
 B. five OH^-, six NO_3^-

7.10 **A. a.** HSO_4^- **b.** H_2CO_3
 B. a. CN^- **b.** NH_2^-

1. **a.** An acid is a substance (or ion) that produces H^+ ions in water (Arrhenius definition); HCl.

b. A base is a substance (or ion) that produces OH^- ions in a solution (Arrhenius definition); KOH.

c. A salt is the ionic compound formed from an acid–base reaction; K_2SO_4 (from H_2SO_4 and KOH).

3. Litmus would turn violet (or there would be no effect). No effect on iron or zinc.

5. Most hydrogen atoms contain no neutrons, just a proton in the nucleus. In acid–base chemistry, a proton is a hydrogen ion, H^+ —a hydrogen atom that has lost its electron. A nuclear proton is the positively charged particle at the center of any atom.

7. Neutralization occurs when acid is added to a base, or a base is added to an acid, until the solution is neutral: same amount (concentration) of H^+ as OH^-.

9. Strong bases such as NaOH ionize completely in water to produce OH^-. Weak bases react only slightly with water to produce OH^- ions, and they do so by accepting H^+ from water.

11. Strong acids and strong bases are corrosive and can cause burns if not washed off quickly. Strong bases initially feel slippery to the skin.

13. **a.** $HBr \longrightarrow H^+(aq) + Br^-(aq)$
 b. $CsOH \longrightarrow Cs^+(aq) + OH^-(aq)$

15. HClO, hypochlorous acid

17. **a.** base **b.** acid **c.** base

19. **a.** $HCOOH(aq) + H_2O(l) \longrightarrow H_3O^+(aq) + HCOO^-(aq)$
 b. $C_5H_5N(aq) + H_2O(l) \longrightarrow C_5H_5NH^+(aq) + OH^-(aq)$

21. $NH_3(aq) + H_2O \longrightarrow NH_4^+(aq) + OH^-(aq)$

23. **a.** hydrochloric acid [HCl(g) is hydrogen chloride.]
 b. $Sr(OH)_2$
 c. potassium hydroxide **d.** H_3BO_3

25. **a.** phosphoric acid, acid **b.** cesium hydroxide, base
 c. carbonic acid, acid

27. **a.** HNO_2 **b.** H_3PO_3

29. **a.** H_2SeO_3, acid **b.** $Sr(OH)_2$, base

31. Strong acid

33. Weak base

35. **a.** strong base **b.** strong acid
 c. weak acid **d.** salt

37. 0.10 M $HClO_4$ (strong acid) > 0.10 M HClO (weak acid) > 0.20 M NH_3 (weak base)

39. **a.** $AgOH + HCl \longrightarrow AgCl + H_2O$
 b. $RbOH + HNO_3 \longrightarrow RbNO_3 + H_2O$

41. **a.** $H_2SO_3(aq) + Mg(OH)_2(aq) \longrightarrow 2\,H_2O(aq) + MgSO_3(aq)$

43. **a.** acidic **b.** basic **c.** acidic **d.** basic

45. 8.0

47. 11.0

49. 1.0×10^{-6} M H^+

51. The pH is between 5 and 6.

53. NH_3OH^+ is the conjugate acid of NH_2OH; Cl^- is the conjugate base of HCl.

55. $Al(OH)_3(s) + 3\,HCl(aq) \longrightarrow AlCl_3(aq) + 3\,H_2O$
 $Mg(OH)_2(s) + 2\,HCl(aq) \longrightarrow MgCl_2(aq) + 2\,H_2O$

57. Bases are reported to taste bitter, so soaps are likely to be basic (pH greater than 7.0).

59. c

61. $CaCO_3(s) + 2\,HCl(aq) \longrightarrow CaCl_2(aq) + H_2CO_3(aq)$ [instead of H_2CO_3, $H_2O + CO_2$ is ok]

63. $HS^- + H_2O \longrightarrow H_2S + OH^-$

65. $NaOH > NH_3 > NaCl > HC_2H_3O_2 > HNO_3$

67. HPO_4^{2-}(acid) $ + H_2O \rightarrow H_3O^+ + PO_4^{3-}$;
 HPO_4^{2-}(base) $ + H_2O \rightarrow H_2PO_4^- + OH^-$

69. 3 regular; 4 regular

71. $NH_3 + NH_3 \rightarrow NH_4^+ + NH_2^-$

73. Reduce the pH.

75. 1000 L

77. Montmorillonite clay

Let's Experiment!

2. Lemon juice, vinegar, orange juice, seltzer water

3. Ammonia, borax, baking soda, milk of magnesia, shampoo

4. Add a few drops of one of the basic solutions, such as ammonia, baking soda solution, or milk of magnesia.

5. Possibly the pH of the water could impact your results. It is also possible that the substances could have unknown additives, which could also influence your results.

Chapter 8

8.1 **A. a.** oxidation   **b.** oxidation   **c.** reduction
    B. a. oxidation   **b.** reduction   **c.** oxidation

8.2 **A. a.** oxidation   **b.** oxidation
    B. a. reduction   **b.** oxidation

8.3 **A. a.** oxidation   **b.** reduction   **c.** oxidation
    B. a. reduction   **b.** oxidation   **c.** reduction

8.4 **A. a.** +2   **b.** −3   **c.** 0
    B. a. 0   **b.** −1   **c.** +5   **d.** +3

8.5 **A. a.** oxidation   **b.** reduction   **c.** oxidation
    B. a. reduction   **b.** oxidation   **c.** reduction

8.6 **A. a.** F_2, oxidizing agent; Mg, reducing agent
    b. Fe_3O_4, oxidizing agent; H_2, reducing agent
    c. Ce^{4+}, oxidizing agent; Ni^{2+}, reducing agent
    B. a. O_2, oxidizing agent; Se, reducing agent
    b. Br_2, oxidizing agent; K, reducing agent
    c. O_2, oxidizing agent; Fe, reducing agent

8.7 **A.** oxidation: $Co \longrightarrow Co^{3+} + 3\,e^-$; reduction: $S + 2\,e^- \longrightarrow S^{2-}$
    B. oxidation: $Mg \longrightarrow Mg^{2+} + 2\,e^-$;
    reduction: $Zn^{2+} + 2\,e^- \longrightarrow Zn$

8.8 **A.** $Zn + 2\,H^+ \longrightarrow Zn^{2+} + H_2$
    B. $Pb + 2\,Ag(NH_3)_2^+ \longrightarrow Pb^{2+} + 2\,Ag + 4\,NH_3$

8.9 **A. a.** $2\,Zn + O_2 \longrightarrow 2\,ZnO$ **b.** $C + O_2 \longrightarrow CO_2$
    B. a. $4\,Al + 3\,O_2 \longrightarrow 2\,Al_2O_3$ **b.** $4\,Li + O_2 \longrightarrow 2\,Li_2O$

1. The oxidation number is increased during oxidation and decreased in reduction.

3. Oxidation occurs at the anode, reduction at the cathode. Oxidation involves loss of electrons, so the "lost" electrons flow from anode to cathode, where they combine with the reduced substance.

5. The porous plate between two electrode compartments allows ions to pass through, thereby completing the electrical circuit and keeping the solutions electrically neutral.

7. Zinc, which is the anode and is oxidized to Zn^{2+} in the cell reaction

9. It is recharged by applying voltage to reverse the discharging reaction. The recharging reaction is:
$$2\,PbSO_4 + 2\,H_2O \longrightarrow Pb + PbO_2 + 2\,H_2SO_4.$$

11. Lithium has a very low density and can provide higher voltages than other metals.

13. Ag is oxidized by H_2S to produce black Ag_2S. Aluminum metal can be used to reduce Ag^+ back to Ag, so that there is no loss of silver.

15. In photosynthesis, plants store energy from the sun in the form of chemical bonds in reduced forms of carbon. We consume foods containing these compounds and reverse the photosynthetic reaction by oxidizing the carbon atoms and releasing the stored energy. That energy fuels the body's processes, including your heartbeat.

17. b

19. **a.** oxidation, 0 to +4   **b.** oxidation, 0 to +2
    c. oxidation, 0 to +4   **d.** oxidation, −1 to 0
    e. reduction, 0 to −4

21. **a.** $K \longrightarrow K^+ + e^-$ **b.** $Mg \longrightarrow Mg^{2+} + 2\,e^-$
    c. $Fe \longrightarrow Fe^{2+} + 2\,e^-$, $Fe \longrightarrow Fe^{3+} + 3\,e^-$

23. **a.** Cu is oxidized; O_2 is reduced.
    b. Mg is oxidized; H^+ is reduced.
    c. The O in H_2O_2 is oxidized; Br_2 is reduced.
    d. S in $S_2O_3^{2-}$ is oxidized; O_2 is reduced.

25. **a.** H_2, reducing agent; Cu^{2+}, oxidizing agent
    b. C in HCOH, reducing agent; O_2, oxidizing agent
    c. Mn, reducing agent; Al in Al_2O_3, oxidizing agent
    d. Cr^{2+}, reducing agent; Sn^{4+}, oxidizing agent

27. Copper metal

29. MnO_2 (Mn^{4+}), oxidizing agent; Zn, reducing agent

31. Anode, reducing agent; cathode, oxidizing agent

33. b

35. $Mg^{2+} + 2\,e^- \longrightarrow Mg$; $2\,K + Mg^{2+} \longrightarrow Mg + 2\,K^+$

37. **a.** $Mo^{6+} + 6\,e^- \longrightarrow Mo$, reduction; $H_2 \longrightarrow 2\,H^+ + 2\,e^-$, oxidation
    a. $Zr^{4+} + 2\,e^- \longrightarrow Zr^{2+}$, reduction; $Cd \longrightarrow Cd^{2+} + 2\,e^-$, oxidation

39. **a.** reduction; oxidation; $Cu^{2+} + Fe \longrightarrow Cu + Fe^{2+}$
    b. oxidation; reduction; $2\,H_2O_2 + 4\,Fe^{3+} \longrightarrow 2\,O_2 + 4\,H^+ + 4\,Fe^{2+}$
    c. reduction; oxidation; $WO_3 + 3\,C_2H_6O \longrightarrow W + 3\,H_2O + 3\,C_2H_4O$

41. **a.** oxidation; reduction; $Pb^{4+} + Ti \longrightarrow Ti^{4+} + Pb$
    b. oxidation; reduction; $S + O_2 \longrightarrow S^{4+} + 2\,O^{2-}$ (or SO_2)

43. **a.** A C atom in CH_3CHO is oxidized; an O atom in H_2O_2 is reduced.
    b. A C atom is oxidized; Mn^{7+} is reduced.

45. Reduced in both

47. Reduced because of an increase in H atoms

49. I^- is oxidized; Cl_2 is reduced

51. Reduced; gain of H and loss of O atoms

53. Oxidized; loss of H atoms

55. Fe^{3+} is reduced from to Fe^{2+}; so Fe^{3+} is the oxidizing agent. C is reduced from the 0 to the +4 state.

57. **a.** $N_2 + 2\,O_2 \longrightarrow 2\,NO_2$ (other possible products; N_2O, NO, N_2O_4)
    b. $CS_2 + 3\,O_2 \longrightarrow CO_2 + 2\,SO_2$
    c. $C_5H_{12} + 8\,O_2 \longrightarrow 5\,CO_2 + 6\,H_2O$

59. **a.** H is −1.   **b.** O is −1.
    c. O is −1/2, a fraction. There is no such thing as half an electron. (Actually, when Cs reacts with O_2, the oxygen molecule loses one electron to form O_2^-, the superoxide ion.)

61. $F > Cl > Br > I$

63. The light grease or oil coating prevents oxygen from reaching the metal, minimizing oxidation.

65. $2\,CO_2 + 2\,e^- \longrightarrow C_2O_4^{2-}$; CO_2 is reduced.

67. 7.5 L of air per minute; about 2000 L O_2/day

69. **a.** oxidation
    b. $6\,CO_2 + 24\,H^+ + 24\,e^- \longrightarrow 6\,H_2O + C_6H_{12}O_6$

71. +4; 0

73. Metals can act only as reducing agents because they do not form anions; in reaction, a metal goes from the 0 state to a positive state. Nonmetals can act as either oxidizing or reducing agents because they can form either positive or negative oxidation numbers. (See carbon in Problem 19.)

75. $H_2 + Cl_2 \longrightarrow 2\,HCl$; Cl_2 is the oxidizing agent.

77. $2\,Al + 6\,H_2O \longrightarrow 2\,Al(OH)_3 + 3\,H_2$; Al is the reducing agent and H_2O is the oxidizing agent.

79. H_2O_2 acts as both oxidizing and reducing agent. The oxidation number of oxygen in H_2O_2 is −1. One oxygen atom forms water (ox. no. −2). The other oxygen atom forms O_2 (ox. no. zero).

81. **a.** Reduction: $NAD^+ + H^+ + 2\,e^- \longrightarrow NADH$; oxidation: $CH_3CH_2OH \longrightarrow CH_3CHO + 2\,H^+ + 2\,e^-$. Overall: $NAD^+ + CH_3CH_2OH \longrightarrow CH_3CHO + H^+ + NADH$.
    b. Reduction:$CH_3COCOOH + 2\,H^+ + 2\,e^- \rightarrow CH_3CHOHCOOH$; oxidation: $NADH \longrightarrow NAD^+ + H^+ + 2\,e^-$. Overall: $CH_3COCOOH + NADH + H^+ \longrightarrow CH_3CHOHCOOH + NAD^+$.

83. **a.** propylene + Cl_2 + $H_2O \longrightarrow$ propylene chlorohydrin + HCl; 2 propylene chlorohydrin + $Ca(OH)_2 \longrightarrow$ 2 propylene oxide + $2\,H_2O$ + $CaCl_2$; propylene + H_2O_2 (with titanium silicalite) $\longrightarrow$ propylene oxide + H_2O
    b. AE of traditional process: 25%. AE of alternative process: 55%.
    c. The alternative process is superior in both respects. Higher AE, and the only waste is water.

85. Many oxidation catalysts have organic molecules built into their structures. These organic components may undergo chemical oxidation, causing the catalyst to lose effectiveness.

Let's Experiment!

1. Zinc; it loses electrons more easily than copper.
2. The lemon should produce close to 1 volt and about 2 milliamps.
3. The kiwi should produce just over 1 volt and about 2 milliamps.
4. The kiwi will most likely produce higher voltage.
5. The LED may not have lit with a single "cell" because the voltage may have been too low, but with two cells it probably lit the LED. (Depends on the type of LED used.)

Chapter 9

9.1 **A. a.** Molecular: C_6H_{14}; complete structural:

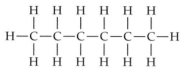

condensed structural: $CH_3CH_2CH_2CH_2CH_2CH_3$

b. Molecular: C_8H_{18}; complete structural:

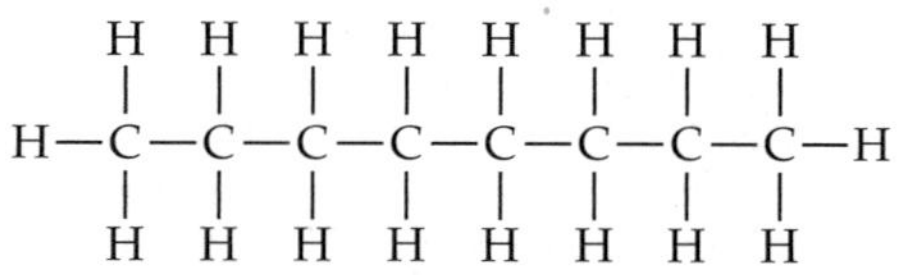

complete structural (H—C—C—C—C—C—C—C—C—H with all H)

condensed structural: $CH_3CH_2CH_2CH_2CH_2CH_2CH_2CH_3$

B. a. molecular: C_5H_{12}; complete structural:

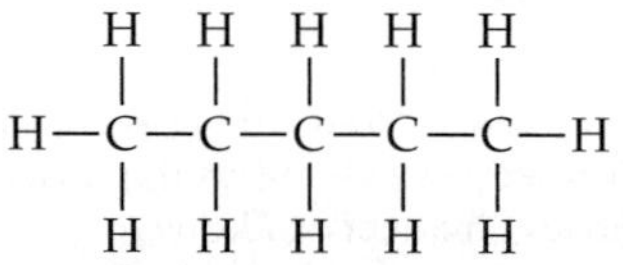

(H—C—C—C—C—C—H with all H)

condensed structural: $CH_3CH_2CH_2CH_2CH_3$

b. molecular: $C_{10}H_{22}$; complete structural:

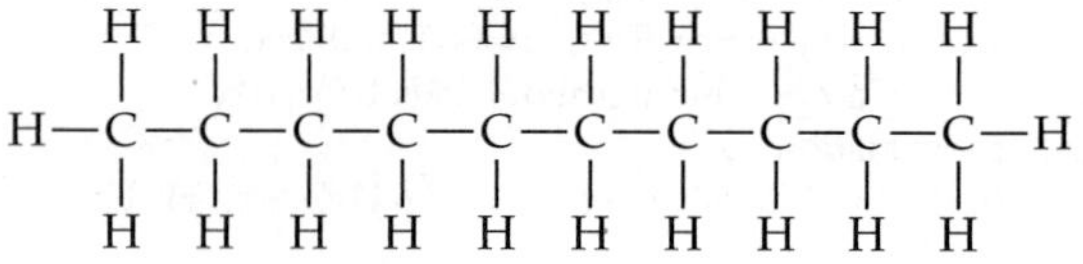

(H—C—C—C—C—C—C—C—C—C—C—H with all H)

condensed structural:
$CH_3CH_2CH_2CH_2CH_2CH_2CH_2CH_2CH_2CH_3$

9.2 **A.**
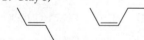

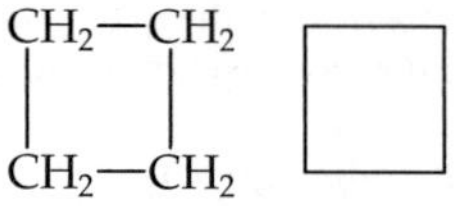

B. When the C—C bond is formed to make a ring, there are two fewer C—H bonds and so two fewer H atoms.

9.3 **A. a.** C_6H_{12}; $CH_3CH=CHCH_2CH_2CH_3$
 b. C_7H_{12}; $CH_3CH_2C≡CCH_2CH_2CH_3$
 B. a. C_9H_{18}; $CH_3CH_2CH_2CH=CHCH_2CH_2CH_2CH_3$
 b. C_5H_8; $CH_3C≡CCH_2CH_3$

9.4 **A.** No; there are two H atoms on one of the C=C atoms, which fails Rule 2.

B. Only b;

9.5 **A. a.** $CHCl_3$ **b.** CCl_4
 B. a. CF_4 **b.** CH_2F_2

9.6 **A.**

(H—C—C—C—C—O—H with all H)
 a.

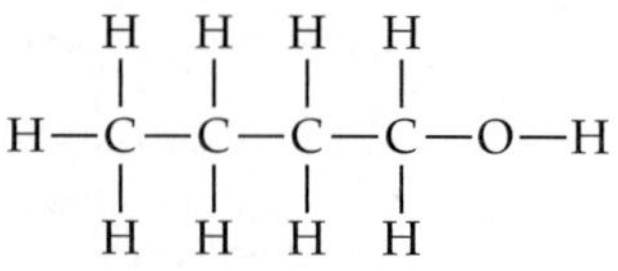

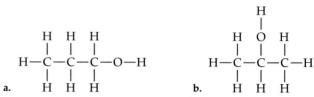
 b.

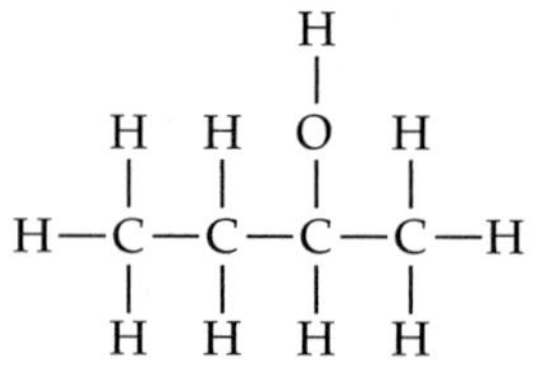

B.

(H—C—C—C—O—H with OH at top on third carbon)
 a.

(H—C—C—C—H with O—H groups)
 b.

9.7 **A. a.** alcohol   **b.** ether   **c.** ether
 B. a. phenol   **b.** ether   **c.** thiol

9.8 **A. a.** $CH_3OCH_2CH_2CH_2CH_3$
 b. $CH_3CH_2OC(CH_3)_3$
 B. a. $CH_3OCH_2CH_2CH_2CH_2CH_3$
 b. $CH_3CH_2CH_2OCH_2CH_2CH_2CH_2CH_2CH_3$

9.9 **A. a.** ketone   **b.** aldehyde
 B. a. aldehyde   **b.** ketone

9.10 **A.**

(H—C—C—C—C—OH with =O, a structure)
 a.

(H—C—C(=O) structure)
 b.

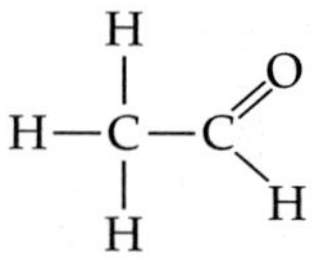

(H—C—C—C—C—C—H with O=)
 c.

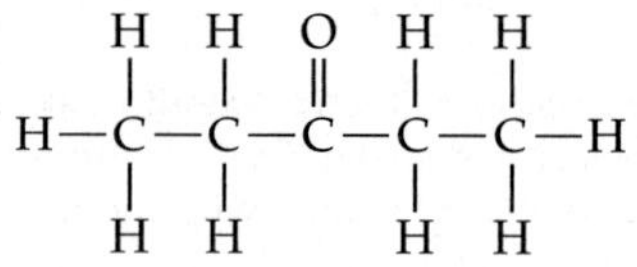

B.

(H—C—C—C—C—C—C—C—C—H with O=)
 a.

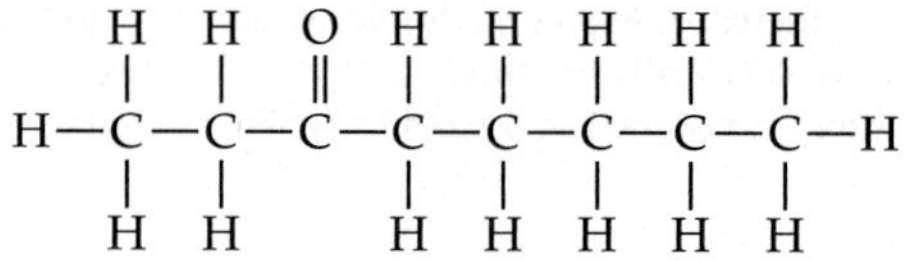

(H—C—C—C—C—C—C—C—H with O=)
 b.

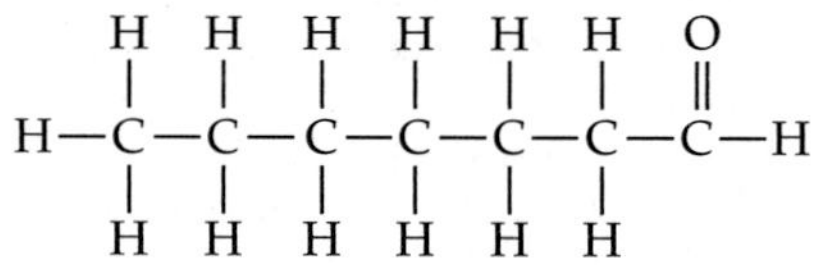

(H—C—C—C—C—C—C—OH with O=)
 c.

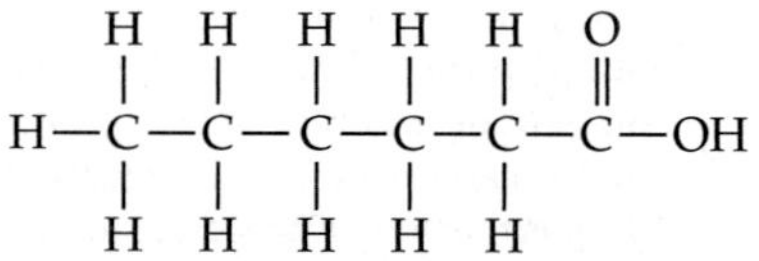

9.11 **A. a.**

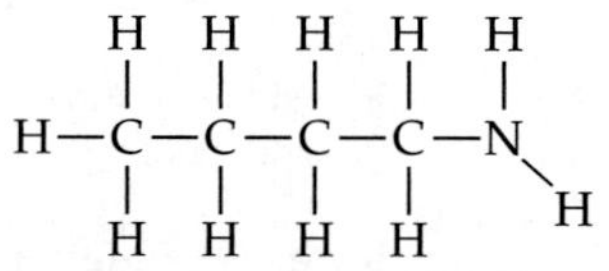

(H—C—C—C—C—N with H)

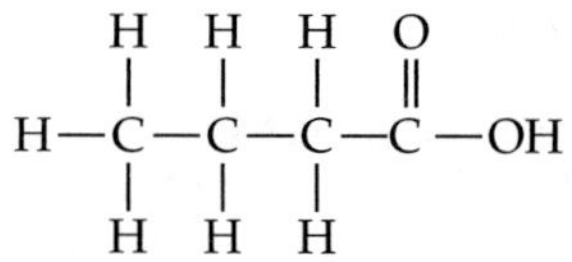

 b.

(H—C—C—N—C—C—H)

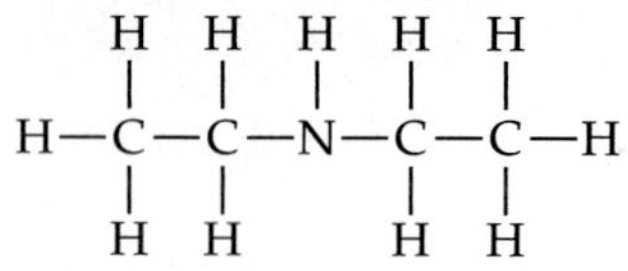

B.

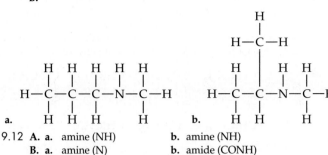

a. (structure) b. (structure)

9.12 A. a. amine (NH)   **b.** amine (NH)
    B. a. amine (N)   **b.** amide (CONH)

9.13 A. a   **B.** b

1. Organic: Bonding is usually covalent; low melting and boiling points; few are water-soluble. Inorganic: Bonding is often ionic; relatively high melting and boiling points; many are water-soluble.
3. Carbon atoms can bond to form long chains; can form multiple bonds; can form rings.
5. Isomers have the same chemical formula but different structures. Constitutional isomers have different structural formulas; geometric isomers have different arrangements of groups on a double bond.
7. It represents six delocalized electrons that give benzene different properties from other unsaturated compounds.
9. 1–4 C, gases; 5–16 C, liquids; more than 16–17 C, solids
11. **a.** ethanol   **b.** 2-propanol   **c.** methanol
13. Historical use: anesthetic; modern use: solvent
15. **a.** alcohol   **b.** ketone   **c.** ester
    d. ether   **e.** carboxylic acid   **f.** aldehyde
    g. thiol   **h.** amide   **i.** amine
17. a and d are organic.
19. c
21. **a.** 6   **b.** 9   **c.** 5   **d.** 5
23. **a.** butane   **b.** 1-butene   **c.** acetylene (ethyne)
25. **a.** C_5H_{12}; $CH_3CH_2CH_2CH_2CH_3$
    b. $C_{10}H_{22}$; $CH_3CH_2CH_2CH_2CH_2CH_2CH_2CH_2CH_2CH_3$
27. **a.** unsaturated, alkyne   **b.** saturated, alkane
29. **a.** isomers   **b.** same compound
31. c
33. d
35. a
37. d
39. b
41. **a.** methyl   **b.** *sec*-butyl
43. A functional group bonds to a propyl group at an end carbon; to an isopropyl group on the center carbon. Both groups have the chemical formula C_3H_7.
45. **a.** ethanol   **b.** propanol   **c.** CH_3OH
    d. $CH_3\overset{\text{OH}}{\underset{|}{C}}HCH_2CH_2CH_2CH_2CH_3$
47. **a.** (structure: o-cresol)   **b.** (structure: 4-iodophenol)
49. **a.** $CH_3CH_2CH_2OCH_2CH_2CH_3$
    b. $CH_3CH_2CH_2CH_2OCH_2CH_3$
51. d
53. **a.** $CH_3CH_2\overset{O}{\overset{\|}{C}}CH_3$ **b.** $H\overset{O}{\overset{\|}{C}}H$
    c. butanal   **d.** 3-heptanone
55. a
57. **a.** propanoic acid   **b.** pentanoic acid
    c. HCOOH   **d.** (structure: benzoic acid)—COOH

59. **a.** $CH_3\overset{O}{\overset{\|}{C}}OCH_3$ **b.** $CH_3CH_2O\overset{O}{\overset{\|}{C}}CH_2CH_2CH_3$
    c. ethyl propanoate (propionate)
61. **a.** CH_3NH_2 **b.** $CH_3\overset{CH_3}{\underset{|}{C}}HNH_2$
    c. diethylamine   **d.** aniline
63. a
65. **a.** $CH_3CH_2C{\equiv}CCH_3 + 2 H_2 \longrightarrow CH_3CH_2CH_2CH_2CH_3$
    b. $CH_3CH{=}C(CH_3)CH_2CH_2CH_2CH_3 + H_2 \longrightarrow$
    $CH_3CH_2CH(CH_3)CH_2CH_2CH_2CH_3$
67. **a.** $CH_3CH_2CH_2CH_2OH$ **b.** $CH_3CH_2CH(OH)CH_2CH_3$
    c. $CH_3CH(OH)CH(CH_3)_2$ **d.** $C_6H_5CH_2OH$
69. $2\,CH_3OH + 3\,O_2 \rightarrow 2\,CO_2 + 4\,H_2O$; 1.06 kg
71. **a.** dipole–dipole (and dispersion) forces
    b. Hydrogen bonding (and dispersion) forces. Hydrogen bonding is much stronger than most dipole–dipole forces, so the ethanol molecules stay close together (liquid) at a higher temperature than dimethyl ether.
73. **a.** hexane   **b.** pentane   **c.** heptane
    d. Octane. A straight-chain compound has a higher boiling point than its branched isomer.
75. d
77. b
79. one solvent instead of four; less-hazardous solvent; reduced solvent use
81. lower energy consumption because microwave heating is more efficient than a hotplate or other heating method

Let's Experiment!

2. Because they are solid at room temperature, we would expect that animal fats would have a higher amount of saturated fat and therefore a lower degree of unsaturation than oils.
3. *Trans* double bonds have structures similar to saturated fats; the two can pack closely together, rendering them as solids. *Cis* double bonds have a kink in the fatty acid chain, and the fat will be more fluid due to the inability to stack.

Chapter 10

10.1 **A.** $CH_2{=}CH{-}CN$ **B.** $CH_2{=}CH{-}Br$

10.2 **A.** $\left[\!\!-CH_2CH-CH_2CH-CH_2CH-CH_2CH-\!\!\right]$
              $|\quad\quad|\quad\quad|\quad\quad|$
              $F\quad\quad F\quad\quad F\quad\quad F$

    B. $+CH_2CH(OCOCH_3)CH_2CH(OCOCH_3)CH_2CH(OCOCH_3)$
    $CH_2CH(OCOCH_3)+$

10.3 **A. a.** $+OCH_2COOCH_2COOCH_2COOCH_2CO+$
    b. $+OCH_2C({=}O)+_n$
    B.
a. $+NH(CH_2)_3CONH(CH_2)_3CONH(CH_2)_3CONH(CH_2)_3CO+$

b. $\left[NH(CH_2)_3CO\right]_n$

1. In polyethylene, each carbon atom has two hydrogen atoms attached. PVC's monomer is vinyl chloride, $H_2C{=}CHCl$; in PVC, every other carbon has one hydrogen and one chlorine atom. Uses: toys, plastic bags, trash cans, etc.
3. Polymerization in which all monomer atoms are incorporated in the polymer; a double bond
5. Styrene
7. A semi-synthetic material made from natural cellulose that has been treated with nitric acid
9. Synthetic fibers are cheaper and have a wider range of properties.
11. For many applications, plastics must be separated by type, melted, and formed into new objects.
13. A polymer consists of many small repeating units bonded together; the units are called monomers.

15. LDPE has highly branched molecules—a loosely packed, amorphous structure that makes it soft and flexible.
17. HDPE has mostly linear molecules that can pack closely together, while LDPE molecules have many side chains. HDPE (e.g., milk jugs) is more rigid and retains its shape at higher temperatures than LDPE (e.g., sandwich bags), which is soft and flexible.
19. **a.** $H_2C\!=\!CH_2$　　**b.** $H_2C\!=\!CHC_6H_5$
21. **a.**

b. $\dashleftarrow CH_2CF_2CH_2CF_2CH_2CF_2CH_2CF_2\dashrightarrow$
23. **a.** $\dashleftarrow CH_2CH(CH_2CH_2CH_3)CH_2CH(CH_2CH_2CH_3)CH_2CH$
$(CH_2CH_2CH_3)\dashrightarrow$

b.

25.
27. The molecules in rubber are coiled. When stretched, those coiled molecules are straightened. When released, the molecules coil again.
29. SBR is made by copolymerization of styrene and butadiene.
31. $\dashleftarrow CO(CH_2)_6CONH(CH_2)_8NHCO(CH_2)_6CONH(CH_2)_8NH\dashrightarrow$
33. $\dashleftarrow OCH_2COOCH_2COOCH_2COOCH_2CO\dashrightarrow$
35. atom < element < compound < polymer
37. The temperature at which the properties of the polymer change from hard, stiff, and brittle to rubbery and tough; rubbery materials such as automobile tires should have a low T_g; glass substitutes, a high T_g.
39. A monomer has a much lower melting point than its polymer; intermolecular forces are stronger in a polymer than in a monomer; many monomers can boil (liquid → gas), while polymers merely decompose when heated sufficiently.
41. A plasticizer's molecules act like molecular ball bearings, separating the polymer molecules and allowing them to slide over one another. This lowers the glass transition temperature of the polymer.
43. Phthalates
45. d
47. a = monomer, b = repeat unit, c = polymer; addition polymerization
49. $CH_2\!=\!CCl_2$ and $CH_2\!=\!CHCl$
51.

53. $\{-CH_2-C(CH_3)_2-CH_2-C(CH_3)_2-CH_2-C(CH_3)_2$
$-CH_2-C(CH_3)_2-\}$
55.

57. The CH_3 group is not part of the straight chain; it is attached to the CH group, then the next CH_2 is attached to the CH group, and so on.
59. They are made of elastomers; the golf ball.

61. Because 1-butanol has an OH group on one end of the molecule only, the reaction cannot continue after the first ester forms on both ends of the terephthalic acid. The final product will be limited to a trimer of butanol–terephthalic acid–butanol or $CH_3(CH_2)_3OC(O)C_6H_5C(O)O(CH_2)_3CH_3$. A polymer cannot form.
63.

65.

67. b, c, and d
69. In LDPE the side chains are not bonded to other main chains, so they interfere with the ability of the chains to align closely and reduce the polymer's strength. In vulcanized rubber, the sulfur chains connect two different chains, holding them together and adding strength and elasticity to the rubber.
71. b
73. a, b, d, and e
75. a and c

Let's Experiment!

1. The ball will bounce. The height varies, roughly 12–36 inches.
2. A chemical reaction; the polymer has properties that are different from the original starting materials.
3. Depending on how much of each ingredient that you mix together, you can make something that is "goopy," slimy, or stretchy. For instance, if you add more corn starch, you will be able to bend and stretch the mix. Add less borax and you will get a "goopy" mixture. To make a slimy substance, add more glue.
4. No, the ball is not biodegradable. PVA is made from carbon-based monomers that are generally derived from petroleum.

Chapter 11

11.1 **A. a.** $^{214}_{83}Bi \rightarrow {}^{210}_{81}Tl + {}^4_2He$　　**b.** $^{94}_{41}Nb \rightarrow {}^{94}_{42}Mo + {}^0_{-1}e$
　　c. $^{68}_{31}Ga \rightarrow {}^{68}_{30}Zn + {}^0_{+1}e$　　**d.** $^{37}_{18}Ar + {}^0_{-1}e \rightarrow {}^{37}_{17}Cl$
　　B. a. $^{42}_{18}Ar \rightarrow {}^{42}_{19}K + {}^0_{-1}e$　　**b.** $^{228}_{90}Th \rightarrow {}^{224}_{88}Ra + {}^4_2He$
　　c. $^{44}_{22}Ti + {}^0_{-1}e \rightarrow {}^{44}_{21}Sc$　　**d.** $^{26}_{13}Al \rightarrow {}^{26}_{12}Mg + {}^0_{+1}e$
11.2 **A.** 6.25%　　**B.** 500 Bq; 4 Bq
11.3 **a.** yes (3 Bq left)
　　b. no (4.0×10^{10} Bq remain; over 1 curie, which is a huge amount of radioactive material)
11.4 **A.** 11,460 y　　**B.** 49 y
11.5 **A.** a neutron (^1_0n)　　**B.** Thorium-232 ($^{232}_{90}Th$)
11.6 **A.** Silicon-30 ($^{30}_{14}Si$)　　**B.** Nickel-64 ($^{64}_{28}Ni$)
11.7 **A.** Higher; the gas can be absorbed by inhalation and decay in the body.
　　B. If the person drinks the water. Radioactive decay occurs within the body.

1. d and e
3. **i.** b　　**ii.** a　　**iii.** c
5. **a.** Nucleon number decreases by 4, atomic number decreases by 2.
　b. Both nucleon and atomic numbers are unchanged.
　c. Nucleon number and atomic number both decrease by 1.
7. False; in the second half-life, half of the radioactive material at the start of that second half-life (50%) decays, leaving 25% of the original radioactive sample.
9. 0.625 μg of F-20
11. **a.** alpha particles　　**b.** gamma rays
13. The missing mass or mass defect is realized as energy that holds the nucleus together.
15. Although they have poor penetrating ability, alpha particles have a very high ability to ionize tissue.
17. c
19. d

21. **a.** $^{250}_{98}Cf \longrightarrow ^{4}_{2}He + ^{246}_{96}Cm$ **b.** $^{210}_{83}Bi \longrightarrow ^{210}_{84}Po + ^{0}_{-1}e$
     c. $^{117}_{53}I \longrightarrow ^{0}_{+1}e + ^{117}_{52}Te$

23. **a.** $^{179}_{79}Au \longrightarrow ^{175}_{77}Ir + ^{4}_{2}He$ **b.** $^{12}_{6}C + ^{2}_{1}H \longrightarrow ^{13}_{6}C + ^{1}_{1}H$
     c. $^{154}_{62}Sm + ^{1}_{0}n \longrightarrow 2\,^{1}_{0}n + ^{153}_{62}Sm$

25. $^{235}_{93}Np$

27. 100,000 years

29. b (5.00 s is over 4000 half-lives.)

31. 156 hours

33. 25 days

35. $^{48}_{20}Ca + ^{247}_{97}Bk \longrightarrow ^{293}_{117}X + 2\,^{1}_{0}n$; two neutrons were released.
     $^{48}_{20}Ca + ^{247}_{97}Bk \longrightarrow ^{294}_{117}X + ^{1}_{0}n$; one neutron was released.

37. **a.** $^{64}_{28}Ni + ^{238}_{92}U \longrightarrow ^{302}_{120}X$ **b.** $^{58}_{26}Fe + ^{244}_{94}Pu \longrightarrow ^{302}_{120}X$
     c. $^{54}_{24}Cr + ^{248}_{96}Cm \longrightarrow ^{302}_{120}X$

39. Alpha particle

41. d

43. c

45. a

47. a

49. Nuclear fission releases energy because the lighter product nuclei are more stable—more binding energy per nucleon; closer to Fe-56—than the very heavy nuclei being split. Nuclear fusion releases energy because the heavier product nuclei are more stable than the light reactant nuclei. (See Figure 11.8.)

51. They were separated by gaseous diffusion. The lighter $^{235}UF_6$ molecules diffuse through porous membranes very slightly faster than do heavier $^{235}UF_6$ molecules. After thousands of diffusion steps, the final product is much richer in $^{235}UF_6$.

53. The U-235 bomb created a critical mass by firing a hollow sub-critical "bullet" of U-235 into a sub-critical target of U-235. The Pu-239 bomb used explosives to crush a hollow sphere of Pu-239 together into a single critical mass.

55. The original U-235 that undergoes fission forms two smaller nuclei, both of which are also radioactive.

57. A power plant uses fuel that is about 3% fissionable U-235, while a bomb requires 90% (approx.) pure U-235. A power plant cannot explode (though it can melt down), because of the low concentration of U-235. There are many other differences.

59. b

61. Lead-206

63. $^{223}_{88}Ra \longrightarrow ^{219}_{86}Rn + ^{4}_{2}He$
     $^{223}_{88}Ra \longrightarrow ^{209}_{82}Pb + ^{14}_{6}C$

65. The neutron

67. Between 7 and 8 half-lives (1.6% at seven half-lives, 0.8% at eight)

69. 2 AM Wednesday (38 hours)

71. **a.** 87.5% has decayed.
     b. After 3.6 billion years, somewhat more than half of the U-238 remains, which should not be hard to detect, while only a tiny unmeasurable trace of C-14 (half-life 5700 years) would remain. Also, carbon is found mostly in organic material. Ain't much life on the moon . . .

73. b

75. The amount of radioactive isotope initially; half-life (shorter half-lives more dangerous); type and energy of radiation emitted (other factors as well)

Let's Experiment!

1. The answer should be around 1000 years. (May vary.)

2. Between 3000–4000 years old

3. The curve should look roughly the same, but the amount in each column will be higher. Also, since you are starting with fewer candies, there may be more variation from the predicted shape. (With 1000 candies, getting 510 "decayed" instead of exactly 500 won't change the shape much; with 100 candies, getting 60 decayed instead of 50 would have a greater effect on the shape.)

4. The piece of candy might last one half-life or several—it is random.

Chapter 12

1. The lithosphere, atmosphere, and hydrosphere. These three constitute an extremely small fraction of the mass of Earth.

3. Boron oxide for heat-resistance, lead oxide to make lenses for optical instruments, cobalt compounds to make glass blue. Other answers possible.

5. Limestone is formed from crushed mollusk shells or coral reefs at the edges of oceans. Much of the Earth was covered in water at some point, so limestone is expected in many areas on Earth.

7. Copper can be found in the native (metallic) state, and tin is easier to extract from ores than is iron.

9. Most aluminum is thinly distributed in clays rather than localized (in bauxite).

11. Aluminum, as it can be extracted only with electrolysis, which requires a great deal of electricity

13. A cell dies from dehydration. Water passes from the cell through the cell wall to the salt solution on its surface, to "equalize" the concentrations on either side of the cell membrane.

15. To preserve meat and fish, and to apply to wounds to kill microorganisms that would cause infections

17. Storing wastes in sanitary landfills; incinerating them; recycling them

19. Used metals are already in the metallic form in which they will be used again, so they merely need to be melted and possibly purified slightly. The main limitation is keeping the metals out of landfills or dispersed through the environment.

21. d

23. Silicates (quartz), carbonates (calcite), oxides (hematite), and sulfides (galena)

25. Si (15.9%) H (15.1%) Na (1.8%) Fe (1.5%) = Ca (1.5%) Mg (1.4%) K (1.0%)

27. Soil, fossil fuels, animals, plants

29. SiO_2; silicate tetrahedra are arranged in a 3-D array.

31. SiO_4 tetrahedra in quartz are arranged in a 3-D array much like carbon atoms in diamond, which gives quartz its hardness and high melting point.

33. Micas are composed of SiO_4 tetrahedra arranged in two-dimensional sheet-like arrays. Micas are easily cleaved into thin, transparent sheets. The sheets are linked by bonds between O atoms and cations, mainly Al^{3+}.

35. c

37. Oxygen has both the highest atom percentage and the highest mass percentage. The main ingredient in glass is sand, SiO_2, and two oxygen atoms weigh more than one silicon atom.

39. As molten glass is cooled, the silicate tetrahedra join in an irregular 3-D arrangement, so that the bonds now differ in strength. When heated, the weaker bonds break first, so the glass softens over a range of temperatures.

41. Cement is a complex mixture of calcium and aluminum silicates, made by heating limestone and clay. Concrete is a mixture of cement, sand, gravel, and water that cures to a solid.

43. In warm oceans, often near shore, where coral or mollusks are abundant

45. Statuary; building materials for structures such as the Taj Mahal, churches, and staircases; kitchen countertops

47. **a.** $BaCO_3$ **b.** K_2CO_3

49. $H_2SO_4 + CaCO_3 \rightarrow CaSO_4 + H_2O + CO_2$

51. Aluminum from bauxite; copper from chalcocite; iron from hematite

53. **a.** Copper changes from +1 to 0.
     b. Iron changes from +3 to 0.

55. **a.** 4 electrons
     b. 6 electrons

57. Reduction; in the following, erbium is reduced from +3 to 0. $2\,ErF_3 + 3\,Mg \rightarrow 2\,Er + 3\,MgF_2$.

59. Copper is soft; tin hardens it.

61. $V_2O_5 + 5\,Ca \rightarrow 5\,CaO + 2\,V$
     a. V in V_2O_5 is reduced.   **b.** Ca is the reducing agent.
     c. Ca is oxidized.   **d.** V in V_2O_5 is the oxidizing agent.

63. $2\,EuCl_3 \rightarrow 2\,Eu + 3\,Cl_2$
     a. Eu in $EuCl_3$ is reduced.   **b.** Cl^- is the reducing agent.
     c. Cl^- is oxidized.   **d.** Eu in $EuCl_3$ is the oxidizing agent.

65. A salt is the ionic product of an acid–base reaction. Ammonium nitrate, NH_4NO_3, is a salt formed from nitric acid, HNO_3, and ammonia, NH_3.
67. Diamond, emerald, ruby, and sapphire
69. Ruby and sapphire are both aluminum oxide, Al_2O_3. Rubies contain traces of chromium(III) ions; sapphires, traces of titanium(IV) and Fe(II) ions.
71. 1.2×10^9 kWh
73. 38.5 g $InCl_3$
75. The three Rs stand for reduce, reuse, and recycle.
77. Answers will vary, will require internet search.
    a. Platinum metal is used as a catalyst.
    b. Platinum is chemically inert. Nickel and other metals might be substituted.
79. Most solvents are organic and do not mix with water, making reactions in aqueous solution difficult. They evaporate readily, contributing to VOC pollution. Most are considered hazardous waste.

Let's Experiment!

1. The lemon juice and vinegar should have bubbled or fizzed on the limestone, calcite, and chalk, which all contain calcium carbonate. The lemon juice is more acidic, with a pH of 2.3, and vinegar usually has a pH closer to 3, which means you could get more fizzing with the lemon juice.
2. Only the rocks with calcium carbonate, such as limestone or chalk, will react to form CO_2 gas bubbles.
3. Acid rain produces the same effect on limestone and marble statues as does lemon juice or vinegar. The limestone/marble gradually dissolves.
4. $CaCO_3 + C_2H_4O_2 \longrightarrow Ca(C_2H_3O_2)_2 + CO_2 + H_2O$

Chapter 13

13.1 **A.** 9.3×10^4 kg CO
   B. The mass of CO_2 produced will depend on the mass of carbon in the fuel. Each C atom in CH_3OH is accompanied by 4 H atoms and 1 O atom, while each C atom in CH_4 is bonded only to 4 H atoms. Therefore, CH_3OH contains less carbon per kg and will produce less CO_2.
13.2 **A.** 3790 g H_2O **B.** 4940 kg CO_2
1. Refrigerant; molding of plastic foams
3. Solid mineral matter left behind after burning coal is bottom ash. Fly ash goes out the smokestack.
5. A greenhouse gas allows visible light to pass through to the surface, but absorbs the heat radiated by the Earth, raising Earth's temperature; CO_2, CH_4, H_2O.
7. **a.** Ground-level ozone is a powerful oxidizing agent and is highly reactive. It is a severe respiratory irritant and especially harmful to children. Ozone causes rubber to harden and crack, resulting in damage to automobile tires, and also causes crop damage.
   b. Depletion of stratospheric ozone lessens the ability of the atmosphere to absorb ultraviolet light from the sun. High-energy UV light causes cellular damage, destroys plant matter, and causes cancer.
9. Photochemical smog occurs when unburned hydrocarbons and nitrogen oxides, in the presence of sunlight, undergo a complex series of reactions to produce a brown haze. This type of smog is often associated with warm, sunny climates. Polluted air associated with industrial activities is often called industrial smog. It is characterized by the presence of smoke, fog, sulfur dioxide, and particulate matter such as ash and soot. Most industrial smog occurs as a result of burning coal. The weather conditions associated with industrial smog are cool/cold temperatures, high humidity, and often fog.
11. In nitrogen fixation, atmospheric nitrogen is converted by bacteria to nitrates and ammonia, making them available to plants. Nitrogen fixation is important because animals and most plants cannot use atmospheric nitrogen. Artificial nitrogen fixation has resulted in a great increase in food production.
13. Thermosphere
15. **a.** $4 Fe + 3 O_2 \longrightarrow 2 Fe_2O_3$ **b.** $4 Cr + 3 O_2 \longrightarrow 2 Cr_2O_3$

17. Cold, damp, still air
19. $2 SO_2 + O_2 \longrightarrow 2 SO_3$
21. $SO_2 + 2 H_2S \longrightarrow 3 S + 2 H_2O$
23. $3 NO_2 + H_2O \longrightarrow 2 HNO_3 + NO$
25. At high temperatures such as are found in an automotive engine. $N_2 + O_2 \rightarrow 2 NO$.
27. Peroxyacetyl nitrate ($CH_3COOONO_2$) is formed from hydrocarbons, oxygen, and nitrogen dioxide; it makes breathing difficult and causes eye irritation.
29. **a.** $2 C_8H_{18} + 25 O_2 \longrightarrow 16 CO_2 + 18 H_2O$
   b. 1.3×10^{10} kg CO_2
31. Only NO_2 and NO have odd numbers of electrons.
33. 1.2 g CO
35. The decrease in CO emission from 1992 to 2009 may be caused by older cars (with no, or less effective, catalytic converters) being removed from use. By 2009 most of those older cars were off the road, so CO emissions reached a plateau.
37. Allotropes
39. $3 CH_4 + 4 O_3 \longrightarrow 3 CO_2 + 6 H_2O$
41. Another free radical; odd number + odd number = even number.
43. H_2SO_4 and HNO_3
45. $6 HNO_3 + 2 Fe \longrightarrow 2 Fe(NO_3)_3 + 3 H_2$
47. Carbon monoxide from poorly ventilated heaters indoors, from automobiles outdoors
49. The crawl space allows gaseous radon to dissipate through vents; the concrete slab traps it inside.
51. A gas; it can be inhaled, forming radioactive decay products and giving off radiation *inside* the body.
53. The warming of Earth caused by gases absorbing infrared radiation given off from Earth
55. The CO_2 exhaled by humans is consumed by plants in photosynthesis, forming glucose and oxygen gas. Humans (and animals) fuel their existence by eating food (plants and animals), generating CO_2 again. The result is a cycle with no significant contribution to atmospheric CO_2 levels.
57. Over 33% (from 300 ppm to over 400 ppm CO_2)
59. Human activity inevitably produces wastes, and no process is 100% efficient.
61. 6.9 mg/day
63. The amount of water vapor that the atmosphere will hold depends largely on temperature. If more water vapor is formed than the atmosphere can hold at that temperature (from combustion of fossil fuels), the excess vapor will condense and form clouds (reflect sunlight) or precipitation. An increase in surface temperature, however, would increase the amount of water vapor the atmosphere can hold, increasing the greenhouse effect.
65. There would be about two air molecules from the Buddha's breath in a breath you take today.
67. $3 NO_2 + H_2O \longrightarrow 2 HNO_3 + NO; 2 NO + O_2 \longrightarrow 2 NO_2$
69. The message to U.S. citizens could be that, individually, they are contributing 40% more to the atmospheric CO_2 concentration than are the individuals in China.
71. **a.** 5.1×10^{18} kJ
   b. 5.0×10^{17} kJ, about 10% of the energy needed to cause a 1° temperature increase
   c. 2×10^{17} kJ, only about 4% of the energy needed to cause a 1° temperature increase
73. Most solvents are organic and do not mix with water, making reactions in aqueous solution difficult. They evaporate readily, contributing to VOC pollution. Most are considered hazardous waste.

Let's Experiment!

1. The color change depends on the SPF rating, with higher-SPF sunscreen slowing the color change.
2. Answers will vary.
3. The results should show that the SPF sunscreen is no longer blocking the UV rays, or is no longer as effective. Recently the FDA passed regulation that manufacturers can no longer make claims that sunscreens are "waterproof" or "sweatproof."

Chapter 14

14.1 **A.** 8.00×10^4 cal; 80.0 kcal; 334 kJ   **B.** Copper

1. In liquid water, the molecules are randomly arranged with little space between them. In ice, hydrogen bonding holds water molecules in a rigid hexagonal arrangement. Each oxygen atom is covalently bonded to two hydrogen atoms and hydrogen-bonded to a hydrogen atom on each of two other water molecules. There is only empty space in the center of each hexagon, making the solid structure less dense than liquid water.

3. Crude oil consists mainly of nonpolar molecules with weak intermolecular forces. Water is polar, with strong hydrogen-bonding forces. The weak attraction of oil molecules for water molecules is insufficient to overcome the much stronger attraction of water molecules for one another, so the oil cannot mix with the water.

5. Water has a high specific heat capacity, so during the day the oceans absorb a great deal of heat from the sun, without raising the water temperature very much. At night the water cools, releasing that energy as heat into the air and nearby land. Rocks, plants, and buildings do not have a very high heat capacity, so inland areas experience greater air temperature changes.

7. Water dissolves many ionic salts. As rivers flow over rocks and soil, these salts are dissolved (many of them slowly because of low solubility). The ions ultimately reach the oceans, where they remain when water evaporates. Over eons, the oceans have become salty. This process continues even now.

9. Plant fertilizers deliver phosphates, nitrates, and sodium and potassium ions to waterways.

11. Cholera and typhoid. Most developed countries chlorinate their water or use ozone, to kill waterborne bacteria.

13. If oxygen is required for the breakdown of organic matter in a body of water, the amount of dissolved oxygen in the water decreases. Biochemical oxygen demand expresses the amount of oxygen needed for breakdown. Eutrophication: Nitrates and phosphates from human waste or other sources act as nutrients for algae; as the algae die, they become additional organic matter, further increasing the BOD.

15. Rock often contains limestone, $CaCO_3$. As it neutralizes the acid, calcium ions are liberated, causing hard water.

17. Chlorinated hydrocarbons such as $HCCl_3$ and CH_3Cl are polar molecules, attracted to polar water by dipole–dipole interactions. In water underground, these compounds do not evaporate readily. Very low concentrations can remain dissolved in water for a long time.

19. 2.48×10^4 calories

21. The specific heat capacity of iron is about 1/10 that of water, so it takes about 1/10 the amount of heat to warm the same amount of water by the same temperature change.

23. Vaporization of sweat helps to maintain body temperature. The high heat of vaporization of water means that we will not lose much water in the process.

25. Dissolved CO_2 from the atmosphere, and NOx and SOx from pollution, form acids in rainwater, lowering the pH.

27. Most water is ocean (salt) water and is not potable. Much of the fresh water that does exist is frozen in the polar ice caps.

29. Aerobic bacteria require O_2 in order to break down nutrients. Anaerobic bacteria can break down nutrients in the absence of O_2.

31. b

33. No, it contains only 0.44 g/gallon.

35. b

37. 0.7–1.0 ppm fluoride

39. Yes, by UV irradiation. Filtered, clear water can be exposed to direct sunlight for at least 6 hours to kill most microorganisms.

41. About 2 L

43. Bottled water is safe and convenient but very expensive relative to tap water. It may not be as safe as municipal water, because although it is considered a food and therefore regulated by the EPA, it is not tested as rigorously as municipal water supplies. Perhaps the biggest drawback of bottled water is the need to manufacture the bottles, a process that requires both petroleum supplies and a great deal of energy. Recycling these bottles would help lessen the disadvantages of this product.

45. Sewage is allowed to settle in a pond, removing about half of the suspended solids and a third of the organic matter.

47. Nitrates and phosphates are not removed by secondary wastewater treatment.

49. **a.** tertiary   **b.** primary   **c.** secondary

51. $2\ HNO_3(aq) + CaCO_3(s) \longrightarrow Ca(NO_3)_2(aq) + H_2O(l) + CO_2(g)$

53. $Ca_5(PO_4)_3OH(s) + F^-(aq) \longrightarrow Ca_5(PO_4)_3F(s) + OH^-(aq)$

55. 9 ppb benzene

57. 110 ppm $Ba(NO_3)_2$

59. No, the levels (0.009 mg Cu/L, 8 mg NO_3^-) are below the standards.

61. Heptachlor; 1 mg/L is 2500 times the maximum allowed.

63. a, b, and c are incorrect.

65. b

67. Cl_2 is reduced to Cl^- and is the oxidizing agent. SO_2 is the reducing agent.

69. Dispersion forces; Rn is nonpolar.

71. **a.** 9.5×10^3 kcal   **b.** 715 g CH_4
     c. 35.4 ft^3 **d.** $0.26

73. a

75. Polar substances tend to dissolve other polar substances. Because less substance X remained dissolved in the water, Y must be more polar than X.

Let's Experiment!

1. The color and taste both fade as the solution is diluted, but you can still see a faint color even though you can't taste the sugar.

2. There should be virtually no color or taste to the 6th cup. With a few more dilutions you would get to a point where there would be no color or taste.

3. Dilution is not the solution to pollution. Just because the water is not colored or sweet does not mean that the contamination is entirely gone. Some chemicals can still have detrimental effects at very low concentrations.

4. The pollution may not remain in the water, but it can migrate into soil, where it accumulates in plants, or (liquids) evaporate into the atmosphere.

Chapter 15

15.1 **A.** 39,000 J
     B. 46,800 J fluorescent bulb; 216,000 J incandescent bulb; 78.3% savings
     C. $1.24 savings

15.2 **A.** 183 kJ   **B.** 290.2 kJ

1. Temperature is related to the average kinetic energy of moving molecules. Heat is the total amount of that kinetic energy transferred from reactants to products, or vice versa, in a chemical reaction.

3. Increasing temperature increases the speed of moving molecules, so that collisions that lead to reactions occur more often.

5. Pure oxygen has about five times as many O_2 molecules as air, so the effective (reaction-producing) collisions between the wood and the O_2 molecules occur at a much greater rate.

7. The material used in a reactor is about 3% U-235 (actual fissionable "fuel"). A nuclear bomb requires over 90% pure U-235 or Pu-239 in order to work.

9. Joules and calories. 4.18 J = 1 cal.

11. 5.4×10^3 kJ

13. Water at the top of the waterfall has a maximum potential energy. As the water falls, its potential energy decreases, and its kinetic energy increases by the same amount.

15. Wood

17. hydrocarbon + oxygen gas → carbon dioxide + water

19. $C_3H_8 + 5\ O_2 \rightarrow 3\ CO_2 + 4\ H_2O$

21. $2\ C_3H_8 + 7\ O_2 \rightarrow 6\ CO + 8\ H_2O$

23. $2\ C_8H_{18} + 25\ O_2 \rightarrow 16\ CO_2 + 18\ H_2O$

25. **a.**
$$\overset{\displaystyle H}{\underset{\displaystyle H}{H-\!\!\overset{|}{\underset{|}{C}}\!\!-H}} \qquad :\!\ddot{O}\!=\!\ddot{O}: \qquad :\!\ddot{O}\!=\!C\!=\!\ddot{O}: \qquad H-\!\ddot{\underset{..}{O}}\!-H$$

b. 1652 kJ   **c.** 998 kJ
d. 1596 kJ   **e.** 1868 kJ
f. −816 kJ; the reaction is exothermic, heat is released. (Answer is not the same as seen in problem 27, p. 496, because many bond energies are approximate values.)

27. 1.97×10^4 kJ
29. 4.43 kJ
31. 1.20×10^3 kJ
33. Energy is conserved; the amount of energy in the universe is a constant.
35. Entropy is a measure of the degree of distribution of energy in a system. Entropy increases when a fossil fuel is burned because a few larger molecules (the fuel) are converted into a larger number of smaller molecules; energy is more widely distributed.
37. Nitrogen in the atmosphere
39. Advantages: plentiful; can be converted to more-convenient liquid and gaseous fuels. Disadvantages: produces sulfur oxides that lead to acid rain, and fly ash in the air; traces of mercury in the coal become airborne; hazardous to obtain; inconvenient to use.
41. 311 years
43. 57 years
45. Petroleum is thought to originate from decayed animal matter; coal, from decayed plant matter.
47. In fractional distillation, crude oil is vaporized and the vapor rises in a tall column. As molar mass of the substances increases, their boiling points increase, so that the low-boiling material rises highest, with higher-boiling material rising lower in the column. This effects a separation.
49. Paving material is mostly rock, with asphalt (left over from crude oil distillation) as a binder. Because we use such a large amount of crude oil, a large amount of asphalt is available quite cheaply.
51. 20.0%
53. $^{238}_{92}U + ^{1}_{0}n \rightarrow ^{239}_{92}U \rightarrow ^{239}_{93}Np + ^{0}_{-1}e$; $^{239}_{93}Np \rightarrow ^{0}_{-1}e + ^{239}_{94}Pu$. A U-238 absorbs a neutron to become U-239. U-239 decays fairly quickly by beta emission, forming Np-239. Np-239 also decays quickly by beta emission, producing Pu-239.
55. A breeder reactor does not make the same fuel that it uses. Breeder fuel is a mixture of about 3% U-235 (the actual fuel) and 97% non-usable U-238. The excess neutrons given off during fission of U-235 are absorbed by U-238 atoms, producing U-239, which decays to (useable fuel) Pu-239. Whereas more neutrons are absorbed by the abundant U-238 than the U-235, more fuel is produced than is used. This does not violate the law of conservation of energy.
57. A fusion reactor can use hydrogen-2 (deuterium), readily available from ocean water, as fuel. That fuel is not radioactive; neither is the main product, helium. Plasma is a hot, gaseous mixture of positive nuclei and electrons.
59. $^{232}_{90}Th + ^{1}_{0}n \rightarrow ^{233}_{90}Th \rightarrow ^{233}_{91}Pa + ^{0}_{-1}e$; $^{233}_{91}Pa \rightarrow ^{0}_{-1}e + ^{233}_{92}U$
61. 2.2×10^7 mol CH_4
63. 4.6×10^7 mol C
65. A photovoltaic cell is a semiconductor device that converts solar energy directly into electricity.
67. 13 m^2
69. Advantages: very low operating cost (maintenance); wind is free; initial cost is recovered in a few years. Disadvantages: very high up-front costs; not all areas have enough wind; wind does not blow constantly; they kill thousands of birds each year.
71. Fuel cells do not go dead as long as reactants are supplied; the byproduct is not hazardous waste. ($H_2 - O_2$ cells produce water.)
73. Advantages: very low operating cost (maintenance); constant supply of electricity; no significant wastes. Disadvantages: high up-front cost; most good sites have already been taken; reservoirs silt up; dams block fish migration.
75. Advantages: no emissions; can be plugged into the power grid. Disadvantages: more expensive than gasoline cars; lithium batteries may need to be replaced periodically; short cruising range; few recharging stations; recharging is time-consuming.
77. **a.** $C + 2 H_2 \rightarrow CH_4$ **b.** $C + H_2O \rightarrow CO + H_2$

79. Endothermic; in dissolving, the ammonium nitrate absorbs heat from the water, decreasing the temperature.
81. b
83. **a.** Endothermic; the water is absorbing heat from the stove to boil it.
b. Endothermic; heat (of vaporization) must be absorbed by water in order for it to be changed to steam.
85. **a.** 102 W
b. (102 W)(1 hp/745 W) = 0.14 horsepower = 1 humanpower
87. 551 kW
89. Hydrogen: 1.33×10^5 kJ/kg; isobutene, 4.62×10^4 kJ/kg
91. 6.91 metric tons of CH_4
93. **a.** $S + O_2 \rightarrow SO_2$ **b.** 5000 t
95. **a.** CH_4, which is 75% carbon by mass, will produce the least CO_2. C_4H_{10} is 83% carbon, and carbon is obviously 100% carbon.
b. Methane, CH_4. For 1 kJ of heat, carbon produces 0.11 g CO_2, CH_4 produces 0.055 g, and C_4H_{10} produces 0.061 g.
97. The need for gasoline as a vehicle fuel is very great.

Let's Experiment!
1. The dissolving of the candy was a physical change; the reaction of the hydrogen peroxide was a chemical change.
2. Water and the candy; the mixture cooled, and heat was absorbed into the candy to dissolve it.
3. Yeast and hydrogen peroxide; heat was given off as a result of the reaction.
4. Oxygen; the glowing splint (should have) burst into flame.
5. No; that was a physical change, no oxygen was produced.

Chapter 16

16.1 **A.** Mannose differs from galactose in configuration about C-2 and C-4.
B. Mannose is an aldose; fructose is a ketose.
16.2 **A. a.** H-P-N-A
b. Histidylprolylasparagylalanine
B. a. Thr-Gly-Ala-Ala-Leu
b. T-G-A-A-L
16.3 **A.** 3   **B.** 24
16.4 **A.** Sugar: ribose; base: uracil; RNA
B. Neither; uracil occurs only in RNA, deoxyribose occurs only in DNA.
1. Photosynthesis converts solar energy to carbohydrates, which provide energy to plants. Plants provide energy to animals that eat either plants or other animals.
3. In every cell; muscles, skin, hair, nails
5. Proteins are polyamides.
7. Hydrogen bonding
9. DNA is a double helix; RNA is a single helix with some loops.
11. 1: isolation and amplification; 2: gene is spliced into a plasmid; 3: plasmid is inserted into a host cell; 4: plasmid replicates, making copies of itself
13. c
15. c
17. a, c
19. **a.** polysaccharide;   **b.** polysaccharide;
c. monosaccharide;   **d.** disaccharide
21. Aldehyde and alcohol (hydroxyl) functional groups
23. a
25. Provide energy
27. Most oils are more unsaturated—more C-to-C double bonds—than are fats. At room temperature, most fats are solids, while oils are liquid.
29. **a.** saturated   **b.** polyunsaturated
c. monounsaturated   **d.** saturated
31. Corn oil (right); it is unsaturated, while lard is saturated.
33. A *trans* fat has the two parts of the carbon chain in the *trans* position of a double bond. They are made by adding hydrogen to polyunsaturated oils.
35. d
37. d
39. d

41. **a.** H₂NCH₂(C=O)NHCH[(CH₂)₃NHC(NH₂)=NH]COOH
    b. H₂NCHCH₃(C=O)NHCH(CH₂SH)COOH
43. Lysine and methionine
43. c
45. Hydrogen bonds, ionic bonds, disulfide linkages, and dispersion forces
47. c
49. b
51. a
53. b
55. One of the essential amino acids
57. a
59. **a.** DNA   **b.** RNA   **c.** RNA
61. **a.** guanine   **b.** thymine
    c. cytosine   **d.** adenine
63. Each strand becomes one-half of a new DNA molecule.
65. b
67. b
69. **a.** mRNA   **b.** tRNA
71. c
73. d
75. c
77. d
79. b
81. **a.** Ala-Ser-Cys-Phe-Gly-Gly   **b.** A-S-C-F-G-G
83. **a.** purine   **b.** RNA
85. c
87. d
89. b
91. b

Let's Experiment!
1. Yes, you are able to see the clear strands binding together.
2. The consistency makes it a challenge to pick the clumps up, but you can do it.
3. Since DNA is unique, crime scene investigators will collect samples from crime scenes. When they find a DNA sample that doesn't match the victim, they compare it with samples from possible suspects. If a suspect has left DNA at the crime scene, the suspect is definitely tied to the crime.
4. The Watson-Crick structure of DNA was so important because it helped answer such questions as how cells are able to divide and go on functioning, how genetic data is passed on to new generations, and how proteins are built to required specifications. Many major advances have been possible thanks to this discovery.

Chapter 17

17.1 **A.** About 45% fat
     B. 75% of the maximum recommended fat intake per day; only 150 g more; difficult
17.2 **A.** 4.8 lb   **B.** 7.4 lb
17.3 **A.** 70 miles   **B.** 2700 kcal
17.4 **A.** 16.3   **B.** 174 lb
1. Fat-soluble; an excess is stored and accumulated; excess water-soluble vitamins are excreted.
3. More food energy (calories); increased carbohydrate intake, primarily starches
5. Vitamin E; vitamin C
7. Density; bioelectric impedance analysis. Accurate density measurement requires a large tank of water. Impedance depends on bone density, body water content, etc.
9. Rapid weight gain; leads to heart disease, strokes, type 2 diabetes.
11. FDA; company must demonstrate that the additive is safe.
13. Carbohydrates and fats contain only carbon, hydrogen, and oxygen. Proteins contain those elements plus nitrogen and sulfur.
15. Athletes generally require more food energy, preferably as starches. They also need more water and electrolytes.
17. An adequate protein provides all the essential amino acids in sufficient quantities; meat, eggs, milk.
19. **a.** iodide ions   **b.** iron(II) ions
    c. calcium ions   **d.** sodium ions

21. c
23. It promotes absorption of calcium and phosphorus for bone structure. No; an excess can lead to undesirable calcium deposits.
25. Riboflavin and cyanocobalamin; they are water-soluble, and cannot be stored in the body.
27. Ascorbic acid = vitamin C;   calciferol = vitamin D; cyanocobalamin = vitamin B₁₂;   retinol = vitamin A; tocopherol = vitamin E.
29. A diuretic promotes urine formation and water loss. ADH prevents excessive water loss.
31. b
33. d
35. **a.** Retards spoilage.   **b.** colorant for visual appeal
    c. sweetener
37. b
39. a
41. d
43. **a.** Leptin tells the hypothalamus how much fat is in the body.
    b. Ghrelin is an appetite stimulant produced by the stomach.
45. Low-carbohydrate diets deplete glycogen and water quickly; as soon as carbohydrates are consumed, the glycogen is restored and the weight returns. Total fasting causes metabolism of protein, not of fat. When the weight is regained, it is regained as fat.
47. 2 hr
49. About 20 km
51. 26.8
53. **a.** anaerobic   **b.** aerobic
55. Aerobic oxidation requires oxygen; muscle tissue with high levels of myoglobin provides that oxygen.
57. Actomyosin catalyzes hydrolysis of ATP to provide energy for muscle contraction. High catalytic activity means energy is parceled out rapidly—for brief, intense activity.
59. Yes; it contains cereal grains (wheat, barley) and beans.
61. Beta-carotene is a precursor to vitamin A and can be stored in fat as needed.
63. **a.** 20 mL   **b.** 1200 mL
65. 10 grams saturated fat
67. 0.82 g/mL; fat
69. Three OH groups in sucrose have been replaced by chlorine atoms in sucralose.
71. No; build muscles by exercising.
73. 4.5 L of oxygen
75. c
77. **a.** 3   **b.** 1   **c.** 2
79. **a.** Waste can be turned into fuel by anaerobic digestion.
    b. Some chemicals can be extracted from FSCW for alternative uses (limonene from orange peel).
    c. Making use of FSCW means that the waste does not go into a landfill (expensive).
    d. Some FSCW contains higher-value chemicals that can be extracted, creating a business opportunity.

Let's Experiment!
1. No
2. They dissolve in stomach acid to provide iron ions that the body can use.
3. Beef, liver, chicken, ham, sardines, many others

Chapter 18

1. Natural drugs are usually extracted from plant or animal sources by physical processes. Semi-synthetic drugs are made by modifying natural drugs to improve their therapeutic properties. Synthetic drugs are synthesized in laboratories.
3. A drug that kills or slows the growth of bacteria; originally limited to formulations derived from living organisms
5. Mediators of hormone action; prostaglandins act near where they are produced; hormones act throughout the body.
7. Ethyl alcohol
9. A drug that produces stupor and relief of pain
11. Natural: morphine. Semisynthetic: heroin. Synthetic: hydrocodone.

13. Chemotherapy is the use of drugs to treat cancer.
15. Acetylsalicylic acid
17. A COX-2 inhibitor is a non-steroidal anti-inflammatory drug. It reduces the catalytic effect of the COX-2 enzyme, which is found in tissues where inflammation occurs.
19. c
21. The phenolic hydroxyl group (—OH) in morphine is replaced by a methoxy group (—OCH₃) in codeine.
23. Tetracyclines are effective against a wide variety of bacteria.
25. b
27. Nucleoside analogs substitute an analog for a nucleoside in viral DNA, crippling the retrovirus. Nonnucleoside reverse transcriptase inhibitors stop the reverse transcriptase from making more of the virus. Protease inhibitors block the protease enzyme so new cells cannot be infected.
29. b
31. An antimetabolite closely resembles a metabolically essential substance, and interferes with reactions involving that substance. An alkylating agent transfers alkyl groups to biologically important compounds, blocking the usual action of those compounds.
33. 6-Mercaptopurine; it is an antimetabolite.
35. Prostaglandins act near the site where they are produced; can have different effects in different tissues; are rapidly metabolized. Hormones are produced in the endocrine glands; have the same effect in different locations; are slowly metabolized.
37. Two ketones and one alkene
39. Emergency contraceptives work after intercourse. Plan B makes the body think it's already pregnant. RU-486 causes the uterus to expel an implanted egg.
41. a
43. c
45. a
47. A psychotropic drug affects the human mind; a hallucinogenic drug is one class of psychotropic drug that alters one's perception of one's surroundings.
49. a
51. b
53. a. N₂O is not very potent.
    b. Safety for operating room personnel is questioned.
    c. Causes nausea.
55. b
57. b
59.

61. a. 2.5 g
    b. 1.7 capsule/dose (reality: 2 capsules twice and 1 once per day)
63. Three six-membered rings and one five-membered ring, fused ("three rooms and a bath")
65. Fat-soluble. With only one O and one N atom among 17 C and 21 H, most of the structure is nonpolar, like fat.
67. Barbiturates and acetaminophen
69. d
71. Antitussives reduce coughing; decongestants reduce swelling of nasal passages; expectorants loosen mucus so it can be expelled.
73.

75. One drug magnifies the effect of another.
77. b < a < c
79. NSAIDs inhibit enzymes (cyclooxygenases, COX) that catalyze the formation of prostaglandins.
81. Related to benzene; ketone and carboxylic acid
83. PGE2; ketone, alcohol, carboxylic acid
85. a
87. TAF produces about 4% of the waste produced by TDF.

Let's Experiment!
1. Brand 1:
$$Al(OH)_3(s) + 3\ CH_3COOH(aq) \longrightarrow Al(OOCCH_3)_3(aq) + 3\ H_2O(l)$$
$$MgCO_3(s) + 2\ CH_3COOH(aq) \longrightarrow Mg(OOCCH_3)_2(aq) + H_2O(l) + CO_2(g)$$
   Brand 2:
$$CaCO_3(s) + 2\ CH_3COOH(aq) \longrightarrow Ca(OOCCH_3)_2(aq) + H_2O(l) + CO_2(g)$$
$$Mg(OH)_2(s) + 2\ CH_3COOH(aq) \longrightarrow Mg(OOCCH_3)_2(aq) + 2\ H_2O(l)$$
   Brand 3:
$$CaCO_3(s) + 2\ CH_3COOH(aq) \longrightarrow Ca(OOCCH_3)_2(aq) + H_2O(l) + CO_2(g)$$
2. We used vinegar (acetic acid) for this experiment. The excess gas in the stomach is typically hydrochloric acid. This is one difference. The contents of the stomach are another variable—other things can work against the active ingredient, or enhance it.
3. The tablets are made differently, with different fillers and inactive ingredients, and so might dissolve at different rates.

Chapter 19
19.1 A. $2\ NH_3 + H_3PO_4 \longrightarrow (NH_4)_2HPO_4$; phosphoric acid
     B. $ZnO + H_2SO_4 \longrightarrow ZnSO_4 + H_2O$; sulfuric acid
19.2 A. 1.53%
     B. 0.59%; about 1/3 as fast as part A
1. Carbon dioxide and water through photosynthesis
3. Magnesium ion, Mg^{2+}
5. An herbicide that kills weeds before the crop is planted; glyphosate
7. The population would increase faster than the food supply.
9. Not all land is suitable for growing foodstuffs.
11. Pests develop resistance; the pesticide remains in the food.
13. *Pesticide* is a general term for a substance that kills organisms we consider to be pests, while an insecticide specifically kills insect pests.
15. Carbon, hydrogen, oxygen; cellulose
17. By lightning; by bacteria on legumes; artificially by the Haber process
19. An ammonia solution is NH_3 gas dissolved in water. Anhydrous ammonia is NH_3 gas that has been compressed until it turns into a liquid.
21. Bones, bird droppings, fish meal
23. By treatment with sulfuric acid
25. Lime
27. Mg^{2+}, Ca^{2+}, Cl^-, Fe^{2+}, Cu^{2+}, and others
29. DDT is effective against many insects, but it does not break down, and it accumulates in foodstuffs.
31. In places where insect-carried diseases (malaria, typhus) are a major problem
33. Pheromones are used to attract insects or mark a trail. Sex-attractant pheromones can lure male insects into traps or confuse them, making them unable to find a mate.
35. Sterile insect technique involves sterilizing and then releasing male pests in areas of infestation; a female mating with a sterile male produces no offspring. It is expensive and has limited application.
37. 2,4-dichlorophenoxyacetic acid (2,4-D) and 2,4,5-trichlorophenoxyacetic acid (2,4,5-T), which were contaminated with dioxins that cause birth defects
39. 40 g
41. 20 months; arithmetically
43. 53 years; available living space, food supplies, water availability, other factors

45. 3.0 billion (38.6%)

47. About 570%; if the trend continues, a shortage of corn used for foodstuffs and cattle feed

49. $N_2 + 3 H_2 \rightarrow NH_3$

51. **a.** d   **b.** b   **c.** d   **d.** a   **e.** c

53. KNO_3; supplies potassium and nitrogen.

55. $2 NH_3 + H_3PO_4 \rightarrow (NH_4)_2HPO_4$

57. Ether, ester, two alkene groups

59. Ether, phenol, alkene, aromatic ring

61.

63. It requires spraying with soap, then rinsing with water 2–3 hours later; time constraints.

65. b

Let's Experiment!

1. The weed killer with the dish detergent and the salt is probably the most effective. The surfactant helps spread the solution over the leaves, and the acidity of the vinegar (5% acetic acid) contributes to the suffocation of the plant. Salt contaminates the soil and causes reverse osmosis, starving the plants of the nutrients from the soil. Strong concentrations of salt and vinegar can cause lasting damage to the ground if not applied carefully.

2. The weed killer with the dish detergent and the vinegar is more effective than just vinegar. The surfactant helps spread the solution over the leaves, and the acidity of the vinegar (5% acetic acid) contributes to the suffocation of the plant.

3. Vinegar most likely will be the least effective of the three weed killers, but depending on the weed, it can certainly still work; it just may take longer than the 3-day experiment.

4. Vinegar alone is most likely the less effective of the two weed killers, but depending on the weed, it can certainly still work; it just may take longer than the 3-day experiment.

Chapter 20

20.1 A. a. nonionic   **b.** cationic
B. a. anionic   **b.** anionic

1. Nonpolar molecules attract nonpolar molecules (dispersion forces). The nonpolar hydrocarbon ends of soap molecules are attracted to nonpolar "greasy dirt" molecules, forming a spherical micelle, with the ionic ends on the surface. Those ionic ends are attracted to the (very polar) water surrounding them, and are rinsed down the drain.

3. Advantages: nontoxic, biodegradable, derived from renewable resources. Disadvantages: does not work in hard water, leaves insoluble calcium salts behind.

5. Calcium, magnesium, or iron ions react with a soap, such as sodium palmitate, forming an insoluble precipitate of calcium palmitate (for example). Soap only works if it is in solution.

7. The ionic end of a soap molecule has a $-COO^-$, and the ionic end of an anionic detergent molecule typically has an $-SO_3^-$ end.

9. At least one LAS detergent; may also contain non-ionic surfactants, enzymes, fragrances, other ingredients. Dishwasher detergents are strongly alkaline, containing sodium tripolyphosphate, sodium carbonate, sodium metasilicate, and a small amount of surfactant. (People who have never used a dishwasher may learn very quickly why you don't use hand dishwashing soap in them!)

11. As a weak base, ammonia is effective on grease and burned-on food, glass, and other hard surfaces. Vapors are irritating, and ammonia will stain, pit, or erode asphalt tile, wood, or aluminum.

13. In a process called ion exchange, hard-water ions (calcium, magnesium, and iron ions) are attracted to the zeolite anions, and replace the sodium ions in the zeolite, releasing those sodium ions into the water.

15. Cosmetic; it destroys bacteria that cause chemicals in perspiration to form malodorous compounds.

17. $CH_3(CH_2)_{10}CH_2OSO_3^-\ Na^+$; $CH_3(CH_2)_{10}CH_2OSO_3^-\ K^+$

19. Lye (NaOH) + a fat $\rightarrow$ sodium soap + glycerol

21. **a.** $CH_3(CH_2)_{15}COO^-\ Na^+$ **b.** $CH_3(CH_2)_{14}COO^-\ K^+$

23. $3\ CH_3(CH_2)_{10}COO^-\ Na^+ + CH_2(OH)CH(OH)CH_2OH$

25. $Mg^{2+} + CO_3^{2-} \rightarrow MgCO_3$

27. 2 sodium zeolite + calcium ion $\rightarrow$ calcium zeolite + 2 sodium ions

29. A substance added to a substrate to increase its detergency

31. Calcium and magnesium ions are held in suspension by the zeolite rather than being precipitated as "soap scum."

33. **a.** anionic   **b.** cationic

35. Nonionic surfactants carry no charge; used in laundry and automatic dishwasher detergents.

37. Common name: sodium lauryl sulfate. IUPAC name: sodium dodecyl sulfate.

39. I

41. III

43. I and II are anionic, III is cationic.

45. Marble, which is calcium carbonate and a base. Vinegar will react with it, pitting the surface.

47. They leave a thin film on the fabric that lubricates the fibers, increasing flexibility and softness.

49. Sodium percarbonate, sodium perborate

51. In hot water, perborate bleaches liberate H_2O_2, a strong oxidizing agent.

53. They contain NaOCl, which releases chlorine in hot water, which oxidizes color-producing molecules.

55. Flammability and toxicity

57. No, it is far too flammable.

59. Latex paints clean up with soap and water. Oil-based paints require an organic solvent, such as mineral spirits or turpentine.

61. A wax is an ester of a long-chain carboxylic acid and a long-chain alcohol.

63. Water; an oily substance that protects and softens skin; an emulsifier to bind the two

65. Loss of moisture

67. A sunscreen absorbs harmful UV rays; a sunblock reflects those rays.

69. Avobenzone; oxybenzone; octyl methoxycinnamate

71. Dyes such as bromo acid dyes, and pigments such as iron oxides

73. Na_2PO_3F; it provides fluoride ions to strengthen enamel.

75. **a.** Reduce tooth sensitivity.   **b.** whitener (bleach)
c. abrasive to help clean teeth

77. Fluorides react with the hydroxyapatite of tooth enamel, converting it to fluoroapatite $[Ca_5(PO_4)_3F]$, which is stronger and more resistant to decay.

79. A musk is used as a bottom note (heavy, low volatility) and moderates the odors of flowery or fruity top and middle notes.

81. Aluminum chlorohydrate constricts the openings of the sweat glands.

83. Temporary dyes are water-soluble and can be washed out; permanent dyes last until the hair is cut off or falls out.

85. A reducing agent is used to break the disulfide linkages in hair. The hair is then wrapped around a rod. A mild oxidizing agent is then applied to re-establish disulfide linkages to reconfigure the hair in the shape of the rod.

87. Lightening hair color involves oxidizing the colored pigment to lighten it. Changing gray hair to a darker color may use lead acetate, which reacts with sulfur atoms in the hair to form black lead(II) sulfide.

89. 1 = a; 2 = f; 3 = b; 4 = c; 5 = d; 6 = e

91. I, III, and IV

93. I

95. 3.12 g Na_2CO_3

97. Quaternary ammonium salts; they are cationic surfactants.

99. No; they absorb energy from invisible UV light and emit that energy as visible light.

101. **a.** $CH_3(CH_2)_{24}COOH$ **b.** $CH_3(CH_2)_{28}CH_2OH$

103. 7.5 hours; sweat, swimming, clothing is likely to remove some of the sunscreen in that time.
105. Proteins (keratin)
107. 1.7×10^4 kJ

Let's Experiment!
1. Soap and water works best for cleaning hands.
2. Time of washing, amount of soap/sanitizer, amount of "dirt," and other possibilities
3. Soap is a surfactant that converts oily dirt to tiny charged micelles that are easily washed away with water. Hand sanitizers use an alcohol that actively kills bacteria but is not as effective at removing dirt and grease.

Chapter 21

1. Taken in small amounts, no. Larger amounts can cause liver problems.
3. The route of administration determines the speed of the poison's action, the rate at which it is detoxified, etc.
5. Mercury: occupational exposure; eating contaminated fish, etc. Lead: leaded paint chips; soldered pipes (water) and cans (food).
7. Sunlight, radon, safrole, PAHs
9. b
11. It disrupts the protein's ability to function.
13. It is the product of oxidation of the Fe(II) in hemoglobin to Fe(III); methylene blue reverses the reaction.
15. It oxidizes the cyanide ion to thiocyanate ion.
17. It tightly binds to lead, forming a complex that can then be excreted in the urine.
19. Nerve poisons such as botulin, curare, tabun, and sarin. They tie up acetylcholinesterase, preventing breakdown of acetylcholine.
21. b
23. About 8.0 g
25. Ethylene glycol; it requires less of it to cause death.
27. Oxidizes it to less toxic cotinine.
29. No, they just catalyze oxidation. Sometimes they convert a nontoxic substance to a toxin.
31. In a chronic alcoholic, buildup of enzymes leads to a more rapid destruction of testosterone.
33. The Ames test
35. Humans are not exposed to comparable doses; human metabolism is different from test-animal metabolism.
37. The P35 gene
39. b
41. d
43. A toxic waste contains or releases toxins that pose a hazard to human health or the environment; paint, pesticides, motor oil, medicines, and cleansers.

45. A corrosive waste corrodes or destroys conventional container materials.
47. a
49. The liver converts CCl_4 to the reactive trichloromethyl free radical that can trigger cancer.
51. An oncogene seems to trigger or sustain the conversion of normal cells to cancerous ones. A suppressor gene ordinarily prevents the development of cancers.
53. a
55. a. 1.3 mg/kg
    b. between ethylene glycol and ketamine
57. The first 12 trace elements: iron, copper, zinc, manganese, etc.
59. It forms a polar ketone group that can hydrogen-bond to water, making it more easily excreted.
61. c
63. c
65. Diquat is equally effective but is much less toxic.
67. b

Let's Experiment!
1. Yes, the seeds in the control should have the fastest growth and longest roots.
2. There was no growth at the 10% NaCl concentration.
3. There is a clear difference in growth between the 1% and 0.01% NaCl concentrations.
4. The evidence is that 10% NaCl concentration is toxic to lettuce seeds—no growth. There is more growth at 1% NaCl than at 0.1% NaCl. The expectation would be that there would be no growth with the higher concentrations and more growth with the lower concentrations. The growth for 0.01% and 0.001% is similar to that of the control, so very low concentrations of NaCl do not impede growth.

Appendix

A.1 a. 0.0163 g   b. 1.53 lb   c. 370 ml
A.2 a. 0.2224 km   b. 150 fl oz
A.3 a. 24.4 m/s   b. 0.136 oz/qt
A.4 a. no   b. no
A.5 0.12 m^3
A.6 56.8 g
A.7 a. 100.5 m   b. 6.3 L
     c. 1800 m^2 ($1.80 \times 10^3 \text{ m}^2$)   d. 2.33 g/mL
A.8 a. 185 °F   b. 10.0 °F
     c. 179 °C   d. −29.3 °C

Credits

Photo Credits

Cover The coastline at Big Sur in California: Mint Images/Getty Images **Chapter 1** Opener: Steve Proehl/Getty Images; Page 1 (top): Iaremenko Sergii/Shutterstock; Page 1 (bottom): Zoonar GmbH/Alamy Stock Photo; Page 2: Shanghai Daily/AP Images; Page 3 (top): Universal Art Archive/Alamy Stock Photo; Page 3 (bottom): Bonninstudio/Alamy Stock Photo; Page 5: icollection/Alamy Stock Photo; Page 7: iofoto/Shutterstock; Page 10: Bettmann/Getty Images; Page 11: Karen Tam/AP Images; Figure 1.4: NASA; Figure 1.5 (left): Westend61/Getty Images; Figure 1.5 (right): Tom Boschler/Pearson Education, Inc.; Figure 1.7 (left to right): United States Mint, photomatz/Shutterstock, Grzegorz Raniewicz/Shutterstock, HDesert/Alamy Stock Photo; Page 17: Jim West/Alamy Stock Photo; Figure 1.8: Richard Megna/Fundamental Photographs; Figure 1.9: Eric Schrader/Pearson Education, Inc.; Figure 1.10 (left): Niderlander/Shutterstock; Figure 1.10 (right): emf images/Fotolia; Page 31: Elevance Renewable Science, Inc.; Page 35: McCreary, Terry; Page 36 (left): Buquet Christophe/Shutterstock; Page 36 (center): Reimschuessel/Newscom; Page 36 (right): World History Archive/Newscom; Page 38 (left): Yana Kabangu/Shutterstock; Page 38 (right): David A. Aguilar; Page 40: Wm. Baker/GhostWorx Images/Alamy Stock Photo. **Chapter 2** Opener: maximimages.com/Alamy Stock Photo; Page 41 (top): MS Mikel/Shutterstock; Page 41 (bottom): Wilson Ho; Figure 2.1: John Muggenborg/Alamy Stock Photo; Figure 2.3: Konradlew/Getty Images; Figure 2.3 (inset): Susumu Nishinaga/Science Source; Page 44: DEA/E. LESSING/Getty Images; Figure 2.6a: Manamana/Shutterstock; Figure 2.6b: Katharina Wittfeld/Shutterstock; Figure 2.6c: Richard Megna/Fundamental Photographs; Page 47: World History/TopFoto/The Image Works; Page 49: Matt Meadows/Getty Images; Page 55: Science Source; Page 57 (top): IBM Research/Science Source; Page 57 (left to right): I. Pilon/Shutterstock, Scanrail/Fotolia, Patryk Kosmider/Fotolia, tagstiles.com/Fotolia; Page 58 (left to right): I. Pilon/Shutterstock, Rawpixel.com/Shutterstock, LesPalenik/Shutterstock, Skaljac/Shutterstock; Page 59: Libby Welch/Alamy Stock Photo; Page 63: Richard Megna/Fundamental Photographs. **Chapter 3** Opener: Jeff Hunter/Getty Images; Page 65 (top): Goritza/Shutterstock; Page 65 (bottom): Friedberg/Fotolia; Page 66: GL Archive/Alamy Stock Photo; Figure 3.2b: Richard Megna/Fundamental Photographs; Page 70 (top): Otto Glasser/National Library of Medicine; Page 70 (bottom): INTERFOTO/Alamy Stock Photo; Page 72: ZUMA Press, Inc./Alamy Stock Photo; Figure 3.9 (left, right): Richard Megna/Fundamental Photographs; Figure 3.9 (center): SPL/Science Source; Figure 3.10: ktsdesign/Shutterstock; Figure 3.11: Wabash Instrument Corp/Fundamental Photographs; Page 79: Martí sans/Alamy Stock Photo; Page 81: Elenathewise/Fotolia; Page 84: Rick Pickford; Page 89 (top): Richard Megna/Fundamental Photographs; Page 89 (center): Africa Studio/Shutterstock; Page 89 (bottom): Gary Retherford/Science Source; Page 96: Andrey_Kuzmin/Shutterstock. **Chapter 4** Opener: HERA FOOD/Alamy Stock Photo; Page 97 Fotocrisis/Shutterstock; Page 98: Andreas Argirakis/Alamy Stock Photo; Page 100: Science History Images/Alamy Stock Photo; Figure 4.1: Richard Megna/Fundamental Photographs; Figure 4.2c: Charles D Winters/Science Source; Figure 4.3: Richard Megna/Fundamental Photographs; Page 105: Aleksandar Varbenov/Alamy Stock Photo; Page 124: Scott Abbott; Page 139: Marilyn Duerst. **Chapter 5** Opener: Dmitri Ma/Shutterstock; Page 140 (top): Casther/Shutterstock; Page 140 (bottom): Elizaveta Galitckaia/Shutterstock; Page 144: Norenko Andrey/Shutterstock; Page 146: Pictorial Press Ltd/Alamy Stock Photo; Page 149: Winai Tepsuttinun/Shutterstock; Page 153: Jim Gibson/Alamy Stock Photo; Page 157: Eric Schrader/Pearson; Page 158: Alistair Scott/Alamy Stock Photo; Page 159, 160: Nathan Eldridge/Pearson Education, Inc.; Page 168: JIANG HONGYAN/Shutterstock, Fitria Ramli/Shutterstock. **Chapter 6** Opener: posteriori/Shutterstock; Page 169 (top): CK Foto/Shutterstock; Page 169 (bottom): Jonathan Daniel/Allsport/Getty Images; Page 170: Hercules Milas/Alamy Stock Photo; Page 171: fullempty/Alamy Stock Photo; Page 172: kviktor/Shutterstock; Page 172: holly.w/Stockimo/Alamy Stock Photo; Figure 6.4: Richard Megna/Fundamental Photographs; Figure 6.11a: Tony Freeman/PhotoEdit; Figure 6.11b: George Mattei/Science Source; Figure 6.11c: Liv friis-larsen/Shutterstock; Figure 6.17: Richard Megna/Fundamental Photographs; Figure 6.18: Scanrail1/Shutterstock; Page 192: PJF Military Collection/Alamy Stock Photo; Page 195: amalia19/Shutterstock. **Chapter 7** Opener: SolStock/Getty Images; Page 196 (top): grey_and/Shutterstock; Page 196 (bottom): Zigzag Mountain Art/Shutterstock; Figure 7.1: Eric Schrader/Pearson Education, Inc.; Page 198 (top): Tyler Boyes/Shutterstock; Page 198 (center): Denise Kappa/Shutterstock; Page 198 (bottom): Richard Megna/Fundamental Photographs; Figure 7.2: Eric Schrader/Pearson Education, Inc.; Page 199: Science and Society/SuperStock; Page 200: Andriy Popov/123RF; Page 204: Blue-Horse_pl/Shutterstock; Figure 7.7: Richard Megna/Fundamental Photographs; Page 210: Richard Megna/Fundamental Photographs; Figure 7.9: Creative Digital Visions/Pearson Education, Inc., Inc.; Page 215 (top): Terry Putman/Shutterstock; Page 215 (bottom): Michael P Gadomski/Science Source; Page 223: Farsad Behzad Ghafarian/Shutterstock. **Chapter 8** Opener: Scharfsinn/Shutterstock; Page 224 (top): PHENPHAYOM/Shutterstock; Page 224 (bottom): Car Culture/Getty Images; Figure 8.1: Paul Silverman/Fundamental Photographs; Page 231: Kara/Fotolia; Figure 8.4:

Peticolas/Megna/Fundamental Photographs; Page 235: Pelikh Alexey/Shutterstock; Page 237 (top): Spencer Grant/PhotoEdit; Page 237 (bottom): Marianne de Jong/Shutterstock; Page 239: krishnacreations/Fotolia; Page 241: Walter Drake; Page 242: sirtravelalot/Shutterstock; Page 243: Spacex/Newscom; Page 245 (top): Richard Megna/Fundamental Photographs; Page 245 (bottom): Eric Schrader/Pearson Education, Inc.; Page 246: Paul Reid/Shutterstock; Figure 8.9: Eric Schrader/Pearson Education, Inc.; Figure 8.10: L Lauzuma/Shutterstock; Page 253 (left): sciencephotos/Alamy Stock Photo; Page 253 (right): photowind/Shutterstock; Page 254: Sherri R. Camp/Shutterstock; Page 255: Jacob Kearns/Shutterstock. **Chapter 9** Opener: Mr.Weerayut Chaiwanna/Shutterstock; Page 259: PattayaPhotography/Shutterstock; Page 264: Vladimir Breytberg/Shutterstock; Figure 9.6b: Richard Megna/Fundamental Photographs; Page 268: rommma/123RF; Page 269: JPC-PROD/shutterstock; Figure 9.9: Clive Freeman/The Royal Institution/Science Source; Page 274, 278, 279: Eric Schrader/Pearson Education, Inc. Inc.; Page 280: Southern Illinois University/Science Source; Page 286 (top): Henrik Larsson/Shutterstock; Page 286 (bottom): Eric Schrader/Pearson Education, Inc. **Chapter 10** Opener: Jelle vd Wolf/Shutterstock; Page 303 (top): Good Shop Background/Shutterstock; Page 303 (bottom): BALDUCCI/SINTESI/SIPA/Newscom; Page 304: Brian Nolan/Shutterstock; Figure 10.1b, 10.2: Richard Megna/Fundamental Photographs; Page 310 (top): Dennis MacDonald/PhotoEdit; Page 310 (bottom left): Koksharov Dmitry/Shutterstock; Page 310 (bottom center): James Edward Bates/MCT/Newscom; Page 310 (bottom right): Inara Prusakova/Shutterstock; Page 311: Michele Cozzolino/Shutterstock; Page 312: Global Warming Images/Alamy Stock Photo; Page 313: Freerk Brouwer/Shutterstock; Page 315: Lenscap/Alamy Stock Photo; Page 316 (top): Maridav/Shutterstock; Page 316 (bottom): IMAGENFX/Shutterstock; Page 318: glenda/Shutterstock; Page 320 (bottom): Mar Photographics/Alamy Stock Photo; Page 321: Crisferra/Shutterstock; Page 323: Amoco Fabrics & Fibers/Propex; Page 324: Tommy Trenchard/Alamy Stock Photo; Page 325: Beth Hall/Alamy Stock Photo; Page 334: Daftly Domestic. **Chapter 11** Opener: Media for Medical SARL/Alamy Stock Photo; Page 335 (top): Triff/Shutterstock; Page 335 (bottom): SergeyIT/Shutterstock; Figure 11.2: Nora D. Volkow/National Institute of Health; Page 347: Tim Graham/Alamy Stock Photo; Page 348: Library of Congress, Washington, D.C (Photoduplication); Page 349 (top): Omikron/Science Source; Page 349 (bottom): Khlungcenter/Shutterstock; Page 351: Simon Fraser/Science Source; Page 355: Science Source/Science Source; Page 357 (top): Pictorial Press Ltd/Alamy Stock Photo; Page 357 (bottom): PF-(bygone1)/Alamy Stock Photo; Page 359: Argonne National Laboratory/Science Source; Page 360: Lawrence Berkeley National Laboratory; Figure 11.12: United States Air Force; Figure 11.13: JIJI PRESS/AFP/Getty Images; Figure 11.14: DigitalGlobe/ScapeWare3d Contributor/Getty Images; Page 366: Steve Allen/Shutterstock; Page 373: ACORN 1/Alamy Stock Photo. **Chapter 12** Opener: Marilyn Duerst; Page 374 (top): Ruslan Ivantsov/Shutterstock; Page 374 (bottom): Steve Hamblin/Alamy Stock Photo; Page 377: Galyna Andrushko/Shutterstock; Figure 12.3 (left): Madlen/Shutterstock; Figure 12.3 (right): Gary Cook/Alamy Stock Photo; Figure 12.4: Marilyn Duerst; Figure 12.5: S_E/Fotolia; Figure 12.6: H.Catherine/W. Skinner; Figure 12.7: Benoit Daoust/Shutterstock; Page 381: nikkytok/Shutterstock; Page 382 (top): Lee Prince/Shutterstock; Page 382 (bottom): tawat thanumtieng/Shutterstock; Page 383: BIOPHOTO ASSOCIATES/Getty Images; Page 384 (left): Haider Azim/Alamy Stock Photo; Page 384 (right): maxstock/Alamy Stock Photo; Page 386: Puwalski/Shutterstock; Page 388: Joan Albert Lluch/Fotolia; Page 389: Byjeng/Shutterstock; Page 390: Arturo Limon/Shutterstock; Page 391: John Cancalosi/Alamy Stock Photo; Page 392 (left): vvoe/Shutterstock; Page 392 (right): Arne Dedert/Newscom; Page 397 (left): Stephen St. John/Getty Images; Page 397 (right): Turtle Rock Scientific/Science Source. **Chapter 13** Opener: Paopano/Shutterstock; Page 398: trancedrumer/Shutterstock; Page 400: IrinaK/Shutterstock; Page 403: Jacques Descloitres/MODIS Rapid Response Team/NASA/GSFC; Page 404 (top): Sergei Butorin/Shutterstock; Page 404 (bottom): Dr. Ray Clark FRPS & Mervyn de Calcina-Goff FRPS/Science Source; Page 405: Carl Mydans/The LIFE Picture Collection/Getty Images; Page 408: Jose Gil/Shutterstock; Page 410: BrazilPhotos.com/Alamy Stock Photo; Page 413: Monica Schroeder/Science Source; Page 416: NOAA; Figure 13.11: NASA Ozone Watch; Page 420: U.S. Geological Survey/Science Source. **Chapter 14** Opener: Nito/Shutterstock; Page 431 (top): Gaak/Shutterstock; Page 431 (bottom): Mcarter/Shutterstock; Figure 14.1b: Paul D. Van Hoy II/Getty Images; Page 433: Kokoulina/Shutterstock; Page 434: Oleksandr Kalinichenko/Shutterstock; Page 441: Top Photo Corporation/Shutterstock; Page 442: US Environmental Protection Agency (EPA); Page 443: Design Pics Inc/Alamy Stock Photo; Page 444: PETER PARKS/AFP/Getty Images; Page 448 (top): Sven Torfinn/Panos Pictures; Page 448 (bottom): National Institute of Dental Research. **Chapter 15** Opener: Casey Reed/NASA; Page 459 (top): seeyou/Shutterstock; Page 459 (botttom): videowokart/123RF; Page 462: U.S. Consumer Product Safety Commission; Figure 15.1: Greenshoots Communications/Alamy Stock Photo; Figure 15.2: Richard Megna/Fundamental Photographs; Page 466: Mark Phillips/Alamy Stock Photo; Page 467: PÃ©ter Gudella/123RF; Figure 15.3: D and D Photo Sudbury/Shutterstock; Figure 15.4: Richard Megna/Fundamental Photographs; Page

469: Carolyn Franks/123RF; Page 474: U.S. Department of the Interior Museum; Page 475 (top): BanksPhotos/Getty Images; Page 475 (bottom): John Kaprielian/Getty Images; Figure 15.12: Larry Lee Photography/Getty Images; Figure 15.14: UK Atomic Energy Authority/EUROfusion; Figure 15.15b: Beautiful landscape/Shutterstock; Page 488 (top): zgsxycll/123RF; Page 488 (bottom): Martin Bond/Science Source; Page 489: Frank Fennema/123RF; Figure 15.18: Steve Allen/Science Source; Page 500 (left): Gift of Curiosity; Page 500 (right): Kathy L. Ceceri. **Chapter 16** Opener: TCreativeMedia/Shutterstock; Page 501 (top): NaniP/Shutterstock; Page 501 (bottom): Ed Walls Photography/Shutterstock; Page 505: Eric Schrader/Pearson Education, Inc.; Figure 16.7a: Biophoto Associates/Science Source; Figure 16.7b: CNRI/Science Source; Figure 16.9: Richard Megna/Fundamental Photographs; Page 508: Alex Treadway/Alamy Stock Photo; Figure 16.10: Yakov Oskanov/123RF; Figure 16.11 (left): Marilyn Duerst; Figure 16.11 (right): Baiba Opule/123RF; Page 509: Marianna Day Massey/Newscom; Page 511: Nicolaas Weber/Shutterstock; Figure 16.14: Eric Schrader/Pearson Education, Inc.; Figure 16.16a: Martin M Rotker/Getty Images; Figure 16.16b: Sebastian Kaulitzki/Shutterstock; Page 514, 515: Eric Schrader/Pearson Education, Inc.; Figure 16.18a: Richard Megna/Fundamental Photographs; Page 526: Swapan/Fotolia; Page 530: A. Barrington Brown/Science Source; Page 532: Andrew Syred/Science Source; Page 536: Tono Balaguer/Shutterstock; Page 540: UCLA Henry Samueli School of Engineering and Applied Science; Page 543: Eric Schrader/Pearson Education, Inc.; Page 547: Beret Marie Olsen. **Chapter 17** Opener: Aleksandr Markin/Shutterstock; Page 548 (top): Ivonne Wierink/Shutterstock; Page 548 (bottom): Rido/Shutterstock; Figure 17.1: USDA; Figure 17.2: chatuphot/Shutterstock; Figure 17.3a: Clinical Photography, Central Manchester University Hospitals NHS Foundation Trust, UK/Science Source; Figure 17.3b: DoubleVision/Science Source; Page 557: Corbin O'Grady Studio/Science Source; Figure 17.4: Christopher Edwin Nuzzaco/Shutterstock; Page 561: Charles Taylor/Shutterstock; Page 564: Gvictoria/Shutterstock; Figure 17.11: Andy Crump, TDR, World Health Organization/Science Source; Figure 17.12: David Madison/Getty Images; Page 575: Python Pictures/Everett Collection; Page 576: ROBYN BECK/AFP/Getty Images; Figure 17.14: ESB Professional/Shutterstock; Figure 17.15: Kathy Willens/ASSOCIATED PRESS; Figure 17.16: Jose Luis Calvo/Shutterstock; Figure 17.17: Greg M. Cooper/ASSOCIATED PRESS; Figure 17.18: ammentorp/123RF; Page 580 (left): granata68/Shutterstock; Page 580 (right): Egor Rodynchenko/Shutterstock; Page 580 (bottom left): Green Leaf/Shutterstock; Page 580 (bottom center): Maksim Toome/Shutterstock; Page 580 (bottom right): Curly Pat/Shutterstock. **Chapter 18** Opener: unpict/Fotolia; Page 587 (top): Diana Taliun/Shutterstock; Page 588 (top): mce12/Getty Images; Page 588 (center): Jeff Rotman/Science Source; Page 588 (bottom): Vasiliy Vishnevskiy/123RF; Page 590: Mike Kemp/RubberBall/Alamy Stock Photo; Figure 18.4: Nigel Cattlin/Alamy Stock Photo; Page 593: Library of Congress; Page 608: Walter Oleksy/Alamy Stock Photo; Page 610: Mark Harmel/Alamy Stock Photo; Figure 18.22: Ildi Papp/Shutterstock; Figure 18.28: U.S. Drug Enforcement Administration; Page 627: Monkey Business Images/Shutterstock; Page 637: Amy Cannon/Beyond Benign. **Chapter 19** Opener: Daxiao Productions/Shutterstock; Page 638 (top): m.pilot/Shutterstock; Page 638 (bottom): KPG_Payless/Shutterstock; Page 639: montian noowong/123RF; Figure 19.3: Eric Schrader/Pearson Education, Inc.; Figure 19.4: Rick Dalton - Ag/Alamy Stock Photo; Figure 19.6: Eric Schrader/Pearson Education, Inc.; Figure 19.7: Torychemistry/Shutterstock; Page 644: Sarah Cuttle/Dorling Kindersley Limited; Page 645: wrangel/123RF; Figure 19.9: Randy Olson/National Geographic/Getty Images; Figure 19.12: danymages/Shutterstock; Page 650: Jonathan Pearson/Alamy Stock Photo; Figure 19.13: USDA; Page 654: Yellowj/Shutterstock; Page 660: Paul Looyen/123RF; Page 666: Kate Anderson. **Chapter 20** Opener: Erin Lester/Getty Images; Page 667 (top): Patryk Kosmider/Shutterstock; Page 667 (bottom): YuriyZhuravov/Shutterstock; Figure 20.1: Richard Megna/Fundamental Photographs; Page 669: Classic Image/Alamy Stock Photo; Figure 20.5: Richard Megna/Fundamental Photographs; Figure 20.7a: Inc/Shutterstock; Figure 20.7b: Richard Megna/Fundamental Photographs; Figure 20.8a: Janice Haney Carr/CDC; Page 676: Eric Schrader/Pearson Education, Inc.; Figure 20.9: Alfred Pasieka/Science Source; Page 679: Eric Schrader/Pearson Education, Inc.; Figure 20.10: Richard Megna/Fundamental Photographs; Figure 20.11a: ChandraPhoto/Shutterstock; Figure 20.11b: Eric Schrader/Pearson Education, Inc.; Figure 20.11c: Prof. Genevieve Anderson; Page 689, 692: Eric Schrader/Pearson Education, Inc.; Page 694: Tanya Little/Shutterstock; Page 702 (left): David Smart/Shutterstock; Page 702 (top right): Phil Masturzo/Newscom; Page 702 (bottom right): Fernando Madeira/Fotolia. **Chapter 21** Opener: Studio Porto Sabbia/Fotolia; Page 703 (top): wolfman57/Shutterstock; Page 703 (bottom): hivaka/Fotolia; Figure 21.1: Gerard Sauze/Fotolia; Page 705 (top): Sufi/Shutterstock; Page 705 (bottom): Stu Smucker/age fotostock/Getty Images; Page 709: Kevin Wolf/Associated Press; Page 716 (left): Picsfive/Shutterstock; Page 716 (right): New York Public Library/Science Source; Page 720: US Food and Drug Administration FDA; Page 729: Kate Anderson.

Text Credits

Chapter 1 page 3 "Better Living Through Chemistry" Du Pont Ads. Darren Brouwer, BETTER LIVING THROUGH CHEMISTRY? Why chemists need to be humanists. Article. (A Cardus Project). page 30 1.10 Critical Thinking: Lett, James, "A Field Guide to Critical Thinking," *The Skeptical Inquirer*, Winter 1990, pp. 153–160. page 35 "Pure shampoo, with nothing artificial added." 20+ Best Organic Shampoos That Are Actually Non-Toxic. SKINCARE OX. page 37 "the face that launched a thousand ships" Christopher Marlowe. *The Tragical History of Doctor Faustus*. (J. M. Dent and Company, 1897). **Chapter 4** page 137 "Some of these hydrocarbons are very light, like methane gas—just a single carbon molecule attached to three hydrogen molecules." Lisa Marshall and Matthew Spiegelman: Fried Ice: Should we torch oil spills off Alaska with napalm? (Discover Magazine, November 08, 2003 page 02: http://discovermagazine.com/2003/nov/fried-ices) Kalmbach Media Co. page 138 "separates the positive and negative charges in the hydrocarbon fuel molecules, increasing their polarity and allowing them to react more readily with oxygen." Robert E. Hinchee and Robert F. Olfenbuttel: In Situ Bioreclamation: Applications and Investigations for Hydrocarbon and Contaminated Site Remediation. (Elsevier, 2013). page 138 "Fifty years ago, the hydrogen bond angle in water was 108? and you rarely heard of anyone with cancer. Today, it's only 104? and, as a result, cancer is an epidemic!" Orac: Your Friday Dose of Woo: Just what your water needs–more electrons! (August 25, 2006). **Chapter 5** page 168 "What goes in must come out." Seyi Hopewell: *Why Must I Marry This Bruvva?* (Xlibris Corporation, 2011). **Chapter 6** page 177 "Like dissolves like." Steven S. Zumdahl & Susan A. Zumdahl: Chemistry (Cengage Learning, 2013). **Chapter 8** page 229 "the charge an atom would have in a formula if all the bonds were entirely ionic." Kaplan Test Prep, MCAT Organic Chemistry Review 2018–2019: Online + Book (Simon and Schuster, 2017). **Chapter 10** page 327 Slogan: "Yes, this new product is greener!" London Court of International Arbitration (LCIA). page 333 "plastic bags are . . . an environmentally responsible choice." Olivia Legan: City Council poised to vote on plastic bag ban. (October 19, 2010) The Samohi. http://www.thesamohi.com/news/city-council-poised-to-vote-on-plastic-bag-ban. **Chapter 11** page 365 "May you live in interesting times" Tereze Glück: *May You Live in Interesting Times* (University of Iowa Press, 1995). **Chapter 12** page 391 Figure 12.9 Population timeline from 1800-2050 Source: United Nations, World Population Prospects: The 2012 Revision. New York: 2012. **Chapter 13** page 398 "I saw the blackness of space, and then the bright blue Earth. And then it looked as if someone had taken a royal blue crayon and traced along Earth's horizon. And then I realized that blue line, that really thin royal blue line, was Earth's atmosphere, and that was all there was of it. And it's so clear from that perspective how fragile our existence is." Sally Ride, Sher, Lynn. 2014. *Sally Ride: America's First Woman in Space*. New York: Simon & Schuster. page 415 "to use proven methods of controlling indoor air pollution. These methods include eliminating or controlling pollutant sources, increasing outdoor air ventilation, and using proven methods of air cleaning." U.S. Environmental Protection Agency, "Ozone Generators That are Sold as Air Cleaners." page 416 Figure 13.11 The ozone hole (purple) over Antarctica Ozone Hole, September 16, 2013: NASA; http://earthobservatory.nasa.gov/IOTD/view.php?id=82235. page 423 "We have met the enemy, and he is us." Walt Kelly: *Pogo* (Simon and Schuster, 1972). page 429 "There are more molecules in one breath of air than there are breaths in Earth's entire atmosphere." C. Donald Ahrens: *Meteorology Today: An Introduction to Weather, Climate, and the Environment* (Cengage Learning, 2012). **Chapter 14** page 437 Figure 14.2 Where Is Earth's Water? Data from Igor Shiklomanov's chapter "World fresh water resources" in *Water in Crisis: A Guide to the World's Fresh Water Resources*. page 438 Figure 14.3 The water (hydrological) cycle Source: USGS: http://water.usgs.gove/edu/watercyclessummary.html. page 457 "The water we drink in the present day may be some of the same water that flowed in the time of the dinosaurs." UNESCO: Free flow: Reaching Water Security through Cooperation (UNESCO, 2013). page 457 "oxygenated and structured water has smaller molecules that penetrate more quickly into the cells, and therefore hydrate your body faster and more efficiently." Denice D. Cook M.D.: *Choose Life: Optimizing Your Health and Functioning Toward 100 Years and Beyond* (Author House, 2010). **Chapter 15** page 467 "The forms of energy available for useful work are continually decreasing; energy spontaneously tends to distribute itself among the objects in the universe." Frank B. Salisbury: *"Units, Symbols, and Terminology for Plant Physiology: A Reference for Presentation of Research Results in the Plant Sciences."* (Oxford University Press, 1996). page 467 "Heat always flows from a hot object to a cooler one" George I. Sackheim: *Practical Physics for Nurses*. (Saunders, 1957). page 470 Table 15.3 Estimated U.S. and World Reserves of Economically Recoverable Fuels and Annual Consumption of Fossil Fuels Source: U.S. Energy Information Administration, April 2014 report; Data are as of January 1, 2013, for the year ending December 31, 2012. page 472 Table 15.4 Table: Major Components of Various Types of Coal (Minor Components Include Volatile Materials and Sulfur) Source: Data from Diessel, C. F. K. Coal-Bearing Depositional Systems. New York: Springer-Verlag, 1992. page 482 Table 15.6 Electricity Generation Using Nuclear Power Plants (Selected Countries) International Atomic Energy Agency, PRIS Database. Vienna: 2010. page 492 Table 15.7 Table: Total Primary Energy Consumption Per Capita, (quadrillion BTU's in 2011) Source: U.S. Energy Information Administration. **Chapter 17** page 561 "This statement has not been evaluated... treat, cure, or prevent any disease." Source: Dietary Supplement Health and Education Act of 1994. **Chapter 18** page 624 "leads to drug abuse" and "solves the teacher's problem, not the kid's." Quinn McGavin: Stimulant, depressant, and psychotropic drugs. Prezi, Inc. (July 2018). page 625 Table 18.5 Caffeine in Soft Drinks Source: National Soft Drink Association, FDA. page 625 "no one has ever become a cigarette smoker by smoking cigarettes without nicotine", "think of the cigarette as a dispenser for a dose unit of nicotine." Source: Philip Morris quoted in Internal Report for Philip Morris (1972), written by William L Dunn, Jr, cited in Douglas, op. cit., at 3. **Chapter 19** page 646 Table 19.3 Approximate Acute Toxicity of Insecticidal Preparations Administered Orally to Rats. Sources: O'Neill, Maryadele J., Patricia E. Heckelman, Cherie B. Koch, Kristin J. Roman, and Catherine M. Kenny, The Merck Index, 14th ed. (Whitehouse Station, NJ: Merck & Co., 2006), and the Extension Toxicology Network. page 654 "a weed is a plant whose virtues have not yet been discovered." Ralph Waldo Emerson, *The Later Lectures of Ralph Waldo Emerson*, 1843–1871, Volume 2 Edited by: Ronald A. Bosco, Joel Myerson. (University of Georgia Press, 2010). page 657 Figure 19.15 Energy use in modern agriculture and food production. Data from Center for Sustainable Systems, University of Michigan. 2013. "U.S. Food System Factsheet." Pub. No. CSS01-06. October 2013. page 665 "Try out tobacco 'sun tea' as a nontoxic pesticide." "This special tea can help you eliminate most garden pests without any unwanted toxicity." Kathy Van Mullekom: Homemade Tobacco 'Tea' Wards Off Pests (Daily Press. May 08, 2005). **Chapter 20** page 681 "antibacterial agents [including triclosan] to be used with water are not generally recognized as safe." U.S. Food and Drug Administration (September 02, 2016). page 685 "articles intended to be rubbed, poured, sprinkled or sprayed on, introduced into, or otherwise applied to the human body or any part thereof, for cleansing, beautifying, promoting attractiveness or altering the appearance." U.S. Federal Food, Drug, and Cosmetic Act of 1938. page 701 "reduces wrinkles and prevents or reverses damage caused by aging and sun exposure." Mayo Clinic Staff: Wrinkle creams: Your guide to younger looking skin. Mayo Foundation for Medical Education and Research (MFMER). page 701 "pH-balanced for a woman's tender skin" Lindsey Gremont: Easy Homemade Deodorant for Sensitive Skin. Homemade Mommy (January 14, 2014). **Chapter 21** page 703 "the dose makes the poison." Paracelsus Qouted in *Hugh D. Crone: The Man who Defied Medicine : His Real Contribution to Medicine and Science.* Albarello Press, 2004.

Index